GEOCHRONOLOGY, TIME SCALES AND GLOBAL STRATIGRAPHIC CORRELATION

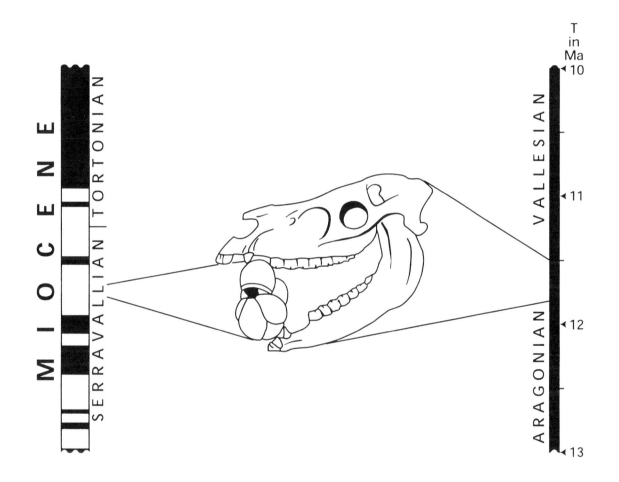

Edited by

William A. Berggren, *Woods Hole Oceanographic Institution, Woods Hole, MA*
Dennis V. Kent, *Lamont-Doherty Earth Observatory, Palisades, NY*
Marie-Pierre Aubry, *Institut des Sciences de l'Evolutión, Université de Montpellier, Montpellier, France*
and
Jan Hardenbol, *826 Plainwood Drive, Houston, TX*

Copyright 1995 by
SEPM (Society for Sedimentary Geology)

Peter A. Scholle, Editor of Special Publications
Special Publication No. 54

Tulsa, Oklahoma, U.S.A. *December, 1995*

A PUBLICATION OF

SEPM (SOCIETY FOR SEDIMENTARY GEOLOGY)

ISBN 1-56576-024-7

© 1995 by
SEPM (Society for Sedimentary Geology)
1731 East 71st Street
Tulsa, Oklahoma 74136-5108

Printed in the United States of America

CONTENTS

GEOCHRONOLOGY, TIME SCALES AND GLOBAL STRATIGRAPHIC CORRELATION: UNIFIED TEMPORAL FRAMEWORK FOR AN HISTORICAL GEOLOGY

INTRODUCTION

The last decade has witnessed significant advances in analytic techniques and methodologic approaches to understanding earth history. It was with this in mind that we convened a research symposium at the 67th annual meeting of SEPM in New Orleans (27 April 1993) entitled Geochronology, Time Scales and Stratigraphic Correlation: Framework for a Historical Geology. In a one day (27 April) symposium, 15 (of 18 scheduled) papers were presented on all aspects of the symposium topic. Oral presentations were complemented by a half day (28 April) poster session. The present volume represents a distillation of the symposium presentations with the inclusion of two of the three that were not presented owing to unavoidable circumstances, the addition of two new papers and the elimination of two of the papers presented at the symposium. The result is the 16 papers that constitute this volume and which represent a broad spectrum of approaches to understanding earth history and the passage of (geologic) time.

But let's back off a minute and ask what is time anyway? Time has been defined as a measured or measurable period, a continuum that lacks spatial dimensions. In the 5th century A.D., the Platonist St. Augustine wrestled with the inadequacy of words to define the concept. Time is of philosophical interest to us humans not the least because we regret the past and fear the (uncertainty and inevitability of) the future. Modern physics teaches us that time cannot, in fact, be treated in isolation from space. There is an interdependence shown in the case of measurement; the measurement of time assigned by a clock depends on the path and speed with which it is moved. The measurement of time involves establishing a precise system of reference for specifying when any event occurs-specifying the epoch and establishing a standard interval of time. The fundamental unit of time is the second. The reference scale could be based on any phenomenon that involves change with time such as the rotation of the earth about its axis. Until recently this has been the standard unit of time: mean solar time. Other independent systems have recently been developed: atomic time; coordinated universal time; dynamical time; ephemeris time; sideral time; Greenwich mean time; solar time; standard time; universal time; and radiometric time which brings us full cycle to the topic of our symposium.

With regard to the history of the earth, the rock record is the ultimate arbiter of time and its passage (no rock, no time). As earth scientists, our goal is the placement of events in the history of the earth in a temporal framework so that we can describe the evolution of the various components of the earth system in an integrated, unified manner. Call it the Grand Unified Theory of Earth Sciences if you will. Why not? Why should Particle Physics appropriate to itself such grandiose descriptors in their search for the grand theory of the unification of the space-time continuum? Our goal is no less noble! In order to understand the dynamic relationship which certainly exists between the evolution of ocean-continental geometries and concomitant changes in the climate and ocean circulation system and the evolution of life itself in the marine and terrestrial biosphere, we must situate the progression of events in this intricately related system in a precise temporal framework. That is the essence of Stratigraphy and its handmaiden Geochronology. One deals with spatial correlation, the other with temporal correlation.

Definition(s) of geochronology, or geologic time scale, have certainly varied over time (Schoch, 1989), but they all agree in the general concept that it represents a fusion, or synthesis, of material and non-material components of the phenomenologic world. For instance, Harland (1978) and Harland and others (1982, 1990) refer to the chronometric (units of duration, based on a standardized second, and year) and chronostratic (scale of rock sequences based on standardized reference points, GSSP, or "golden spikes") components of a geochronologic scale. Berggren and van Couvering (1978) and Van Couvering and Berggren (1977) refer to the essential iterative (stratigraphy) and ordinal progression (linking a series of events in a system of irreversibly varying properties) components of a valid system of earth history, or geochronology. Geologic time is perceived in the form of a progression in one or another ordinal series of events. Ergo, geochronology! Hedberg (1978, p. 33) chose a more picturesque metaphor in identifying a geochronologic scale as "a ladder with rungs indicating position in geologic time. It is a sequential arrangement of identifiable time horizons (chronohorizons) and units of geologic time. This time scale provides a standard reference system for expressing the age of a geologic event and its position with respect to earth history." Hedberg (1978) went on to indicate that an international geochronologic scale is one with worldwide acceptance and which provides a single universal reference standard for dating rock strata or events in earth history with respect to the passage of geologic time. He identified geologic time simply as "time used in the context of earth history, determined by geologic methods" and went on to enumerate the familiar methods of delineating/determining position/passage of time: fossils/evolution; isotopic decay of certain elements; growth rates (tree rings, coral rings, etc.; sedimentation rates, varves, etc.).

Geochronology is not simply a matter of measuring/applying numerical values to different parts of the stratigraphic record. That is chronometry (see Harland, 1978; Harland and others, 1982, 1990; Blow, 1979). Rather, the construction of a geologic time scale involves the synthesis/integration of four independent data sets: bio- and chronostratigraphy, radioisotopic dating, paleomagnetic stratigraphy and sea floor spreading anomaly patterns (Aubry and others, 1988). A geochronologic framework thus constructed provides a means for understanding and estimating rates of geologic processes, of correlating both contiguous and disjunct stratigraphies, and of placing discrete events in a temporal order, which together consitute the essence of historical geology. Perhaps the best examples that can be provided are the applications to marine geologic processes (as delineated in over 25 years of DSDP and ODP drilling and related coring activities) and the recent advances in correlations between Cenozoic marine and terrestrial stratigraphies, including several of the contributions to this volume.

There are many means at our disposal to achieve our glorious goals and we shall explore some of them in this volume. There are presentations from those dealing with the relativistic (iterative) methodologies such as paleomagnetism and chemo- or isotopic stratigraphy, as well as the non-iterative subdisciplines of isotopic radiochronology. And we hear from those practitioners of that most reliable and esoteric of geochronologic methodologies—biostratigraphers! Biostratigraphy is indeed the handmaiden, the link as it were, of both the above mentioned methodologies. The durability of biostratigraphic results is an important feature to bear in mind because in the perception of some geologists biostratigraphy is frequently seen as a rather routine procedure by comparison with the grander sciences of structural deduction or analytic empiricism. The biostratigrapher deals not so much with falsification of rival hypotheses, the definitive mode of scientific reasoning described by Karl Popper, but rather with the progressive refinement of what is already known. Indeed, as one of us (WAB) has indicated elsewhere, biostratigraphers are the High Priests of Time themselves, holding in their hands as they do the key to unlocking the veiled or hidden door of time itself. The responsibility is great but the rewards are intellectually and emotionally satisfying when progress can be demonstrated. The capacity of good biostratigraphic schemes to evade subsequent falsification is certainly one of the main achievements of this branch of geology.

We have prepared a volume that treats essentially the entire Phanerozoic record. We proceed in stratigraphic sequence, from oldest to younger and treat the Paleozoic (3 papers), Mesozoic (4 papers) and Cenozoic (9 papers) in successive order. The symposium on Geochronology was planned with the combined purpose of entertainment, enjoyment and intellectual stimulation. If the papers in this volume are able to convey in a small way some of the current intellectual fervor going on in the field of Geochronology we shall be fulfilled. Sit back, read and enjoy. Welcome to Deep Time!

William A. Berggren, Woods Hole, MA
Dennis V. Kent, Palisades, NY
Marie-Pierre Aubry, Montpellier, France
Jan Hardenbol, Houston, TX

April, 1995

REFERENCES

AUBRY, M.-P., BERGGREN, W. A., KENT, D. V., FLYNN, J. J., KLITGORD, K. D., OBRADOVICH, J. D., AND PROTHERO, D. R., 1988, Paleogene geochronology: an integrated approach, Paleoceanography, v. 6, p. 707–742.

BERGGREN, W. A. AND VAN COUVERING, J. A., 1978, Biochronology, in Cohee, G. V., Glaessner, M. F., and Hedberg, H. D., eds., Contributions to the Geologic Time Scale: Tulsa, American Association of Petroleum Geologists, Studies in Geology 6, p. 39–55.

BLOW, W. H., 1979, The Cenozoic Globigerinida: A Study of the Morphology, Taxonomy, Evolutionary Relationships and the Stratigraphical Distribution of Some Globigerinida (mainly Globigerinacea): Leiden, E.J. Brill (3 vols.), 1462 p.

HARLAND, W. B., 1978, Geochronologic scales, in Cohee, G. V., Glaessner, M. F., and Hedberg, H .D., eds., Contribiutions to the Geologic Time Scale: Tulsa, American Association of Petroleum Geologists, Studies in Geology 6, p. 9–32.

HARLAND, W. B., COX, A. V., LLEWELLYN, P. G., PICKTON, C. A. G., SMITH, A. G. and WALTERS, R., 1982, A Geologic Time Scale: Cambridge, Cambridge University Press, 131 p.

HARLAND, W. B., ARMSTRONG, R. L., COX, A. V., CRAIG, L. E., SMITH, A. G. and SMITH, D. G., 1990, A Geologic Time Scale 1989: Cambridge, Cambridge University Press, 263 p.

HEDBERG, H. D., 1978, Stratotypes and an International Geochronologic Scale, in Cohee, G. V., Glaessner, M. F., and Hedberg, H. D., eds., Contributions to the Geologic Time Scale: Tulsa, American Association of Petroleum Geologists Studies in Geology 6, p. 33–38.

SCHOCH, R. M., 1989. Stratigraphy: Principles and Methods: New York, Van Nostrand Reinhold, 375 p.

VAN COUVERING, J. A. AND BERGGREN, W. A., 1977, Biostratigraphical basis of the Neogene time scale, in Kauffman, E. G and Hazel, J. E., eds., Concepts and Methods of Biostratigraphy: Stroudsburg, PA Dowden, Hutchison and Ross, Inc., p. 283–306.

PART I
PALEOZOIC ERA

TWO CARBONIFEROUS AGES: A COMPARISON OF SHRIMP ZIRCON DATING WITH CONVENTIONAL ZIRCON AGES AND ^{40}Ar/^{39}Ar ANALYSIS

JONATHAN C. CLAOUÉ-LONG

Australian Geological Survey Organisation, GPO Box 378, Canberra, ACT 2601, Australia

WILLIAM COMPSTON

Research School of Earth Sciences, Australian National University, Canberra, ACT 0200, Australia

JOHN ROBERTS

Department of Applied Geology, University of New South Wales, Kensington, NSW 2033, Australia

AND

C. MARK FANNING

Research School of Earth Sciences, Australian National University, Canberra, ACT 0200, Australia

ABSTRACT: Using replicate measurements of a homogeneous reference zircon, the discrimination of Pb$^+$ and UO$^+$ ions relative to U$^+$ observed in zircon analysis with the SHRIMP ion microprobe has been established as a power law relationship. This relationship minimizes uncertainty in comparative measurement of ^{206}Pb/^{238}U ages in zircons. Ages thus obtained have been compared with isotope dilution thermal ionisation mass spectrometric (IDTIMS) analysis of zircons in the Paterson Volcanics (Carboniferous, Australia) and ^{40}Ar/^{39}Ar dating of sanidines in the Z1 tonstein (Carboniferous, Germany). No bias can be detected between the three dating methods, confirming that SHRIMP zircon ^{206}Pb/^{238}U ages are accurate and comparable with other well-measured Phanerozoic ages.

The age of the Westphalian B Stage of Europe is constrained by the Z1 tonstein; SHRIMP zircon analysis supports the ^{40}Ar/^{39}Ar age of 311 Ma and improves on its precision by a factor of two. The Paterson Volcanics mark both the Australian base of the Kiaman magnetic superchron and onset of the major, Late Palaeozoic period of Gondwana glaciation; SHRIMP and IDTIMS zircon dating reassign these processes to the Namurian from the late Westphalian assignment used until now for global correlations.

INTRODUCTION

For the first ten years of its application, zircon dating with SHRIMP, the Sensitive High Resolution Ion Micro Probe at the Australian National University, was directed primarily towards Archaean and Early Proterozoic timing questions in which ^{207}Pb/^{206}Pb ratios were the most sensitive indicator of age. In numerous studies, the advantage of SHRIMP over conventional means of zircon analysis was shown to be the ability to target zones within crystals at the 20–30 μm scale, and so address histories of inheritance, magmatism, metamorphism, and alteration at a scale approaching that at which crystals grow and alteration proceeds. The need to discriminate all these age components is equally true of younger terrains, and systematic SHRIMP study is now being made of Phanerozoic rocks. In such relatively young materials, ages are principally based on ^{206}Pb/^{238}U ratios and therefore have analytical controls which differ in important respects from ^{207}Pb/^{206}Pb ages; ion microprobe measurement of ^{207}Pb/^{206}Pb is essentially free of variable discrimination and accurate to within counting precision, whereas measurement of ^{206}Pb/^{238}U ratios in zircon sensitively depends on comparison with a zircon standard.

This contribution documents analytical practices that have been developed for SHRIMP measurement of ^{206}Pb/^{238}U ages. Their application is illustrated by comparison of SHRIMP zircon ages with ^{40}Ar/^{39}Ar dating and conventional zircon dating of two internationally important Carboniferous volcanic horizons: the Z1 tonstein from the Ruhr Coal Basin of Germany (which constrains the age of the uppermost Westphalian B Stage in Europe); and the Paterson Volcanics of Australia (whose age defines the onset of the Kiaman magnetic reversal and the major late Carboniferous glaciation of Gondwana). The Paterson Volcanics have been correlated with the uppermost Westphalian Stage on the basis of K-Ar dating, and the expectation at the outset was that a firm correlation would be established between these European and Australian markers.

ZIRCON ANALYTICAL PROCEDURES

The original SHRIMP I ion microprobe at the Research School of Earth Sciences, Australian National University, has been joined by an operating second-generation SHRIMP II; a third generation design is now being constructed. This discussion reports data produced with SHRIMPs I and II; similar procedures are applied with both instruments and indistinguishable age results are obtained. Procedures for SHRIMP zircon analysis were originally described by Compston and others (1984), and a subsequent change in the correction for variable discrimination in ^{206}Pb/^{238}U was outlined by Williams and Claesson (1987). A further refinement to the correction procedure is documented below with comments on analytical procedures for zircons of Phanerozoic age. All ages in this paper are calculated using the decay constants recommended by Steiger and Jäger (1977).

Sample Preparation

Grains of zircon are separated using the standard methods of crushing, heavy liquid separation, magnetic separation, and hand picking, with particular care to minimize particulate contamination of the small samples that are used. So that all varieties of zircon in the sample are available for probing, no selection of grains is made during separation; instead the total zircon population (or a representative fraction if the yield is large) is mounted onto the surface of an epoxy disc. A chip of standard zircon SL13 is mounted with the unknown zircons so that measurement conditions for both standard and unknowns are as similar as possible. Fine sandpaper is applied to the mount to expose and section the zircons on its surface, the analytical surface is then polished using 1 μm diamond paste. The grains are photographically mapped at high magnification to guide probing, the surface is cleaned, and a conductive coat of gold is applied to the mount surface to define the electrical potential of the mount as required for fixed and optimal extraction of secondary ions.

Geochronology Time Scales and Global Stratigraphic Correlation, SEPM Special Publication No. 54

SHRIMP Operational Practice

To sample the zircons, both SHRIMP I and SHRIMP II use a primary beam of O_2^- ions produced from a hollow cathode duoplasmatron. The beam is accelerated, focussed, strikes the target at an angle of $45°$ with an ion current of 2 to 4 nA, and sputters a crater 20 to 30 μm in diameter at a rate of approximately 2 μm per hour. A Wein filter ensures the isotopic purity of the primary beam, particularly by eliminating OH^- which would otherwise be a potential source of hydride interferences. Köhler illumination ensures even illumination of the sputtered area, so the crater has sharply defined edges and a flat bottom. A proportion of the material sputtered from the crater becomes ionized and is extracted normally by a 10 keV accelerating potential for transfer via matching lenses to the entrance slit of the mass spectrometer. This is a double focussing design comprising a cylindrical $85°$ electrostatic analyser with a turning radius of 1.27 m, a quadropole correcting lens, and a $72.5°$ magnet sector with a turning radius of 1 m. The beam is measured by ion counting with a single electron multiplier into which the masses of interest are directed by switching the field of the magnet. The wide magnet turning radius and consequent wide separation of masses at the collector allow the use of a wide source slit, thus combining high mass resolution with high sensitivity. A slit width of ~80 μm is employed for analysis of U-Th-Pb isotopes in zircon, giving a mass resolution of approximately 6000 at which all significant interferences with the masses of interest are removed (excepting Pb hydrides). At these settings all the peaks are flat-topped, indicating an image width less than the width of the collector slit, and the sensitivity of SHRIMP I for Pb isotopes is approximately 15 counts per second per ppm of Pb in the target. The sensitivity of SHRIMP II is significantly greater, approximately 60 counts per second per ppm of Pb in the target.

Before measuring U-Th-Pb isotopes, the sampling beam is first used to clean the surface of the target for approximately 5 minutes. Then the secondary beam of sputtered ions is collected while the magnet is cycled through field positions equivalent to nine masses of interest: $^{90}Zr_2^{16}O$, ^{204}Pb, ^{206}Pb, ^{207}Pb, ^{208}Pb, ^{238}U, $^{232}Th^{16}O$, $^{238}U^{16}O$, and a background reading 0.04 AMU heavier than ^{204}Pb. For Phanerozoic samples, usually five scans through these masses are combined as a single analysis over a period of 15 minutes. Ratios between the isotopes are not determined directly during the peak-stepping because several seconds intervene between collection of counts for one mass and the next; during this time, counts for some masses rise and others fall as a function of the increase in crater depth. The five scans are therefore used to establish a best fit track of the beam intensity for each mass over time. These tracks are then integrated over the time between the last mass of the first scan and the first mass of the last scan, and ratioed to produce a single average set of mass ratios that constitute the analysis.

Two stages of quality control are applied in measuring the beam intensities. First, during ion counting the total count period for each mass (from 2 s to 40 s depending on the particular mass) is divided into 10 time segments. The observed scatter of subcounts is compared with the Poisson counting uncertainty, and any significant outliers are rejected. This procedure monitors very short term beam instabilities; longer term variations, on the scale of minutes, are similarly monitored by assessing the fit of each scan to the track of beam intensities over time.

The Standard Zircon

The reference materials used to monitor biases and machine drift during zircon analysis are natural gem quality zircons from Sri Lanka. During most of the 1980s, zircon SL3 was used; detailed study in 1988 showed that some parts of this crystal contain heterogeneities, and it has been replaced by zircon SL13 as a routine standard. The reference composition of SL13 is based on conventional mass spectrometric isotope dilution measurement of 19 separate chips performed in three laboratories and using different spike solutions over a period of several years. Outcomes of these measurements are listed in Table 1. In Figure 1A the measured $^{206}Pb^*/^{238}U$ ratios are plotted as a series in which the data from different laboratories can be compared, and the high precision 1990's measurements can be set against data collected in the early 1980's. The internal consistency of all these data is evident; the 19 chips have the same composition within their assigned errors and combine to give a weighted mean $^{206}Pb^*/^{238}U$ ratio of 0.092821 ± .000054 (2σ), equivalent to an age of 572.2 ± 0.4 Ma (2σ). On this basis, the reference $^{206}Pb^*/^{238}U$ ratio of the SL13 crystal is taken as 0.0928 correct to four decimal places.

Figure 1B shows the conventionally measured $^{207}Pb^*/^{206}Pb^*$ ratios for the same 19 chips of SL13. Some complexity is evident in these data as measurements made in the early 1980s are dispersed beyond their assigned uncertainties. In contrast, the 10 measurements made in two laboratories in 1991 indicate the same composition within their assigned errors with a weighted mean ratio of 0.059250 ± .000018 (2σ). The dispersion of the earlier analyses could signify either real variation in composition or over-optimistic estimates of precision. Averaged without weightings, the mean of the whole data set is 0.059247 ± .000068 (2σ), and this is indistinguishable from the weighted

TABLE 1.—THERMAL IONISATION U-Pb ISOTOPIC DATA FOR CHIPS OF STANDARD ZIRCON SL13.

Sample	$^{206}Pb/^{238}U$	$^{207}Pb/^{235}U$	$^{207}Pb/^{206}Pb$
1	0.09269 ± 46	0.7513 ± 52	0.05886 ± 11
2	0.09302 ± 47	0.7600 ± 41	0.05932 ± 2
3	0.09298 ± 47	0.7656 ± 116	0.05978 ± 61
4	0.09300 ± 47	0.7600 ± 41	0.05934 ± 2
5	0.09249 ± 46	0.7539 ± 42	0.05919 ± 3
6	0.09251 ± 46	0.7563 ± 46	0.05936 ± 7
7	0.09238 ± 46	0.7528 ± 41	0.05917 ± 3
8	0.09286 ± 46	0.7583 ± 42	0.05929 ± 3
9	0.09309 ± 47	0.7584 ± 42	0.05916 ± 3
10	0.09286 ± 7	0.7586 ± 8	0.05925 ± 3
11	0.09287 ± 9	0.7584 ± 12	0.05923 ± 6
12	0.09282 ± 9	0.7568 ± 14	0.05913 ± 8
13	0.09282 ± 7	0.7583 ± 9	0.05925 ± 4
14	0.09286 ± 7	0.7583 ± 8	0.05923 ± 2
15	0.09278 ± 7	0.7578 ± 8	0.05924 ± 2
16	0.09278 ± 8	0.7584 ± 8	0.05928 ± 2
17	0.09286 ± 35	0.7573 ± 31	0.05915 ± 7
18	0.09292 ± 32	0.7584 ± 30	0.05919 ± 10
19	0.09277 ± 9	0.7581 ± 80	0.05927 ± 2

Analysts:

1–9 by R.D. Page, L.P. Black, I.S. Williams, J.J. Foster and W. Compston at the Australian National University;

10–16 by J.C. Roddick at the Geological Survey of Canada;

17–19 by C.M. Fanning at the Australian National University.

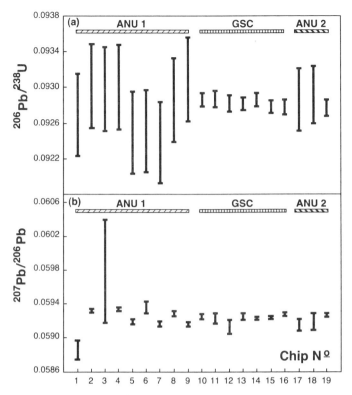

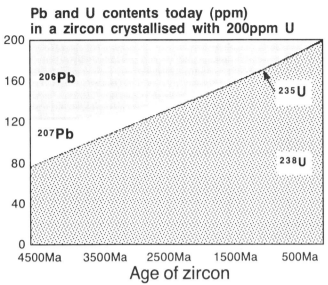

FIG. 2.—Chart of the present day concentrations of U and Pb in zircons of various ages. The present day concentrations are shown by the vertical extent of the shaded area for each isotope at the age of interest, and this shows the changes consequent on the rapid decay of ^{235}U and the slower decay of ^{238}U. Archaean zircons formed with a high proportion of ^{235}U, which has decayed rapidly to form readily measured abundances of ^{207}Pb and makes ages based on ^{207}Pb/^{206}Pb ratios very accurate. Very little ^{235}U is contained in Phanerozoic zircons; the tiny amount of ^{207}Pb formed is difficult to measure and ages based on ^{206}Pb/^{238}U ratios are more useful.

FIG. 1.—Comparison of (a) radiogenic ^{206}Pb*/^{238}U ratios, and (b) radiogenic ^{207}Pb*/^{206}Pb* ratios measured by thermal ionisation for 19 separate chips of standard zircon SL13. Error bars plotted are 1σ. ANU 1: data measured at the Australian National University by R. D. Page, L .P. Black, I. S. Williams, J. J. Foster, and W. Compston in the early 1980s; GSC: data measured at the Geological Survey of Canada by J. C. Roddick in 1991; ANU 2: data measured at the Australian National University by C. M. Fanning in 1991.

mean of the 1990s measurements alone. On this basis, the reference ^{207}Pb*/^{206}Pb* ratio of SL13 is taken as 0.05925 correct to five decimal places.

Measurement of ^{207}Pb/^{206}Pb Ratios

Ion microprobe ages based on ^{207}Pb/^{206}Pb ratios are rarely useful in Phanerozoic zircons. The reason for this is evident in Figure 2, which charts the present day abundances of U and Pb in zircons formed at different times. Decay of ^{238}U has completed one half-life since early Archaean time whereas ^{235}U has completed six, and very little ^{235}U is now available. In Archaean time, 5–20% of all uranium was ^{235}U and most of this has now decayed to ^{207}Pb; the rapid change with time of the proportions of the ^{207}Pb and ^{206}Pb isotopes can therefore be tracked with good measurement precision in Archaean zircons. In Carboniferous zircons, less than 2 ppm of a 200 ppm U zircon is ^{235}U; the amount of ^{207}Pb formed is correspondingly small and poses a measurement difficulty because the signal of radiogenic ^{207}Pb* is weak, sensitive to any small unresolved interferences, and depends on perfectly measured correction for common ^{207}Pb.

Calculation of ^{207}Pb/^{206}Pb ratios, and the correction for common Pb, follow procedures outlined by Compston and others (1984). Any hydride contribution as PbH to the ^{207}Pb and ^{208}Pb counts is not resolved at the operating mass resolution of approximately 6000 and would lead to overestimation of ^{207}Pb/

^{206}Pb ratios in young zircons with a low intrinsic content of ^{207}Pb. Hydrides are therefore minimized by lengthy overnight pumping before analysis. SHRIMP I has a liquid nitrogen cold trap in the vacuum located adjacent to the sample, and SHRIMP II employs a cryopump on the sample chamber; these effectively minimise OH in the vaccuum. A Wein filter in the path of the primary oxygen ion beam eliminates the contribution of OH impurity, so the beam sputtering the target is pure O$_2$. Any residual PbH can be detected from differences between the ratio of radiogenic ^{207}Pb*/^{206}Pb* measured for the standard zircon and its reference ratio of 0.05925. There was no detectable hydride interference during the analytical runs described in this paper.

The count rate of ^{204}Pb in SHRIMP analysis of low common Pb zircons is now close to zero as a result of reductions in the contribution of "laboratory blank" to the common Pb budget. The only possible source of contamination is the polished mount surface, and this is coated with 99.999% pure gold after cleaning. Before analysis, the surface of each probe site is cleaned by rastering the primary beam for 5 minutes, and Köhler focussing results in a flat-bottomed pit with well-defined edges, so there is no contribution of surface Pb from the edge of the probe site. The resulting weak signal of ^{204}Pb makes this mass sensitive to any small unresolved potential interferences, such as ^{186}W^{18}O, as a result of which over-correction for common Pb would be observed. Any such interference is monitored by reference to the standard zircon, whose Th and U decay systems are concordant and in which common Pb can be calculated independently of ^{204}Pb using the measured ^{208}Pb/^{206}Pb ratio. There was no detectable interference with ^{204}Pb while measuring the samples described in this paper.

Uncertainties in ^{207}Pb/^{206}Pb ratios measured by SHRIMP are determined principally by counting statistics on the various masses, plus the uncertainty introduced by correction for common Pb.

Measurement Of ^{206}Pb/^{238}U Ratios

In the sputtering process, most of the Pb in zircon is yielded as Pb$^+$ ions, whereas most of the uranium becomes UO$^+$ instead of U$^+$. The ^{206}Pb$^+$/^{238}U$^+$ ratio observed by SHRIMP is therefore much higher than the true ratio of ^{206}Pb/^{238}U in the target zircon. The bias is not constant, but at low sputtering rates in homogeneous zircons, a correlation is observed between the emission of ^{206}Pb$^+$/^{238}U$^+$ and ^{238}U^{16}O$^+$/^{238}U$^+$ (Compston and others, 1984). This forms the basis for comparative analysis which assumes only that the bias of observed ^{206}Pb$^+$/^{238}U$^+$ relative to the real ^{206}Pb/^{238}U in the target is the same for the standard zircon as for unknowns:

$$\frac{(^{206}\text{Pb}/^{238}\text{U})_{\text{unknown}}}{(^{206}\text{Pb}^+/^{238}\text{U}^+)_{\text{unknown}}} = \frac{(^{206}\text{Pb}/^{238}\text{U})_{\text{standard}}}{(^{206}\text{Pb}^+/^{238}\text{U}^+)_{\text{standard}}} \qquad (1)$$

Given knowledge that the ^{206}Pb/^{238}U ratio of standard SL13 is 0.0928, this equation allows the ^{206}Pb/^{238}U of an unknown zircon to be measured from the difference between its observed ^{206}Pb$^+$/^{238}U$^+$ and that measured for the standard at the same UO$^+$/U$^+$.

The reference ^{206}Pb/^{238}U ratio measurements of standard zircon SL13 are described above and the error of the mean composition is within 0.06% (2σ), a negligible uncertainty. Accuracy of the calibration, therefore, depends mainly on the procedure for estimating the Pb$^+$/U$^+$ ratio of the standard at UO$^+$/U$^+$ ratios appropriate to the unknown zircons. This requires close definition of the correlation relating Pb$^+$/U$^+$ and UO$^+$/U$^+$, which is observed to shift in $x - y$ space from day to day but has a constant form. Early work was based on a range in UO$^+$/U$^+$ restricted between 5.5–7, within which the relationship was approximated closely by a straight line (Compston and others, 1984). Over wider ranges, the relationship is observed to be curved, and a particular analytical run that exhibited extreme variation in UO$^+$/U$^+$ allowed a quadratic approximation to be defined by Williams and Claesson (1987). This curve, based on the observations of a single day, has been the basis of the SHRIMP ^{206}Pb/^{238}U calibration for several years.

Some discrepancies have been observed using large extrapolations of the quadratic curve, and recently, the need has arisen to define the curve as closely as possible for accurate ^{206}Pb/^{238}U age measurement in Phanerozoic zircons. Over the past three years, abundant ion microprobe measurements of the SL13 standard zircon have been accumulated from Phanerozoic dating work in which it is the practice to monitor the standard frequently for maximum control. Analytical runs which have a sufficient quantity of data, and a sufficient observed spread in UO$^+$/U$^+$ to define the correlation with Pb$^+$/U$^+$, have been collated; data meeting these criteria comprise 66 days of measurement by 9 independent SHRIMP operators and cover a range in UO$^+$/U$^+$ ratios from 5 to 9.

The 66 data sets have been assessed against the quadratic curve previously used, and against other candidate relationships including the exponential form ($y = a.e^{bx}$) and the power law form ($y = a.x^b$). Figure 3 shows that these curves differ prin-

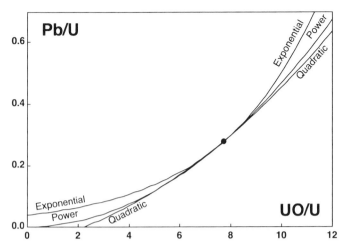

FIG. 3.—Comparison of candidate types of SHRIMP ^{206}Pb/^{238}U calibration curve. The three curves are all best fits of their type through the same data set whose centroid is indicated by the black dot. The quadratic curve is the previous SHRIMP calibration curve, and the power law and exponential fits impose progressively tighter degrees of curvature on the data. The power law solution agrees most closely with the data collected in this study.

cipally in the degree of curvature imposed. Over the normal range of operating conditions the difference between these curve types is small in relation to the natural scatter of measurement. However, study of residuals from attempts to match these curves to the wide coverage of new data has shown that, in general, quadratic fits are slightly too straight and that exponential fits impose more curvature than is suggested by the data. The intermediate degree of curvature offered by fitting a power law achieves the closest empirical solution to the whole range of measurements and is a refinement on the quadratic approximation previously used.

The exact form of power law that applies has been assessed by conversion to logarithms, which conveniently transform the power curve to a straight line. Regressing the individual data sets in ln(Pb/U) vs. ln(UO/U) space then allows determination of the exponent 'b' from the slopes, as illustrated in Figure 4.

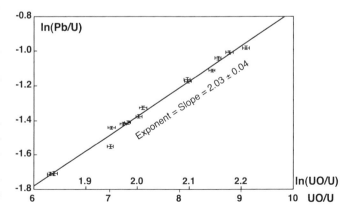

FIG. 4.—A single day's observations of the standard zircon plotted on a logarithmic calibration graph. The exponent of the power law correlation between ^{206}Pb/^{238}U and UO/U is given by the slope of the best fit regression through the data. Figure 5 summarise the results of 66 measurements of the exponent measured in this way.

Measurements of 'b' determined in this way from the 66 analytical runs are plotted against the regression uncertainties in Figure 5A, where it can be seen that the data are dispersed in a coherent fashion about a value of 2. The dispersion of data is a simple function of the uncertainty of the individual regressions, with the more precise determinations converging closely on the mean value. In Figure 5B the same data are related to the mean UO/U ratio of each analytical run. The constancy of 'b' across the wide variation in UO/U gives confidence that the power law form of curve is appropriate across the whole range of operating conditions represented by this large data set and that the exponent of the power relationship is a constant. The weighted mean value of the exponent 'b' is 1.979 ± .026 (2σ) with Mean Square of Weighted Deviates (MSWD) = 1.27, within the MSWD value expected at 95% confidence with 65 degrees of freedom. On this basis the value of the exponent is taken as 2.

These empirical observations lead us to conclude that the SHRIMP I variable discrimination relating the ionic ratio of Pb^+/U^+ to UO^+/U^+ follows the relationship:

$$\frac{Pb^+}{U^+} = a \cdot \left(\frac{UO^+}{U^+}\right)^2 \qquad (2)$$

within the normal operating range in UO^+/U^+ between 5 and 9.

The small uncertainty in the exponent is a negligible contribution to the error of extrapolating this curve. The uncertainty remaining in the calibration process lies in determination of the factor 'a', whose value varies from one analytical run to another and which is therefore measured empirically each day from the mean of several direct analyses of the standard zircon. The uncertainty of the daily mean value varies from one sample to another because it is controlled in part by the operator's choice of the number of measurements of the standard on which to base determination of 'a'. It is usual to base the calibration on 20 or more observations when dating Phanerozoic zircons; this equates to a daily net error of approximately 0.5% (σ_{mean}) or better.

The total uncertainty in $^{206}Pb/^{238}U$ ratios is determined by combining counting statistics on the various masses, uncertainty introduced by correction for common Pb, and uncertainty in the factor 'a' that describes the position of the calibration curve. As $^{206}Pb/^{238}U$ ratios are measured by difference from the determination of the position of the calibration curve, the uncertainty in calibration is the principal control on the accuracy obtained.

Measurement Of Elemental Abundances

Abundances of U, Th and Pb in zircons are calculated using an empirical approach similar to that applied to $^{206}Pb/^{238}U$ ratios (Compston and others, 1984). Uranium counts are normalized to the Zr content of the zircons which is assumed to be constant for all grains, and the approach assumes that the proportion of U/Zr to the observed $UO^+/Zr2O^+$ ionic ratio is the same for unknown zircons as for the standard. Previous work used an observed linear correlation between UO^+/Zr_2O^+ and UO^+/U^+ as a calibration line to relate UO^+/Zr_2O^+ observations of unknowns to measurements of the standard at the same UO^+/U^+, and this method held valid for U abundances over the range

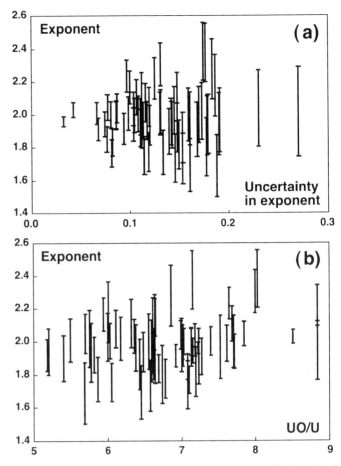

FIG. 5.—Measurements of the exponent of the power law calibration related to (a) the uncertainties of the regressions on which they are based, and (b) the mean UO/U of the individual data sets. The data group around a value of 2 and are dispersed in proportion to their individual measurement errors: those with small errors converge closely on the mean value. The constancy of the value over a wide range of UO/U gives confidence that the form of curve being applied is appropriate across the whole range of operating conditions included in this large data set.

200–3500 ppm for which there was data from standards. However, extrapolation of this relationship to low U abundances has produced some discrepancies.

Study of the same analyses of SL13 used to define the new calibration for $^{206}Pb/^{238}U$ ratios shows that the U abundance calibration can also be modeled by a power law curve. Observations are not as coherent as those for $^{206}Pb/^{238}U$ ratios because SL13 is less uniform in abundances than in ratios, but the available data are consistent with correlation of the form:

$$\frac{Zr_2O^+}{U^+} = a \cdot \left(\frac{UO^+}{U^+}\right)^{.66} \qquad (3)$$

where 'a' is a factor which varies from one analytical run to another and is determined from the mean of each day's measurements of the standard zircon. This calibration curve is now applied to calculation of U abundances. Th and Pb abundances then follow from the Th/U and Pb/U ratios. Uncertainty in the exact form of the abundance calibration curve is subordinate to the known variation of at least 15% in the U content of standard SL13, and this governs the uncertainty in abundance estimates.

THE PATERSON VOLCANICS, AUSTRALIA

The Paterson Volcanics are the Australian marker for the onset of the Kiaman magnetic superchron, the major reversed polarity interval of the Late Palaeozoic, which was defined on the basis of pioneering palaeomagnetic data collected from Carboniferous rocks of the Hunter Valley and Illawara Coast region of New South Wales in Australia (Irving and Parry, 1963; Irving, 1963, 1966). The Paterson Volcanics overlie fluvial sediments of the Mt. Johnstone Formation and have normal polarity, whereas the overlying glacigene sediments of the Seaham Formation are reversed, as are other late Carboniferous and Permian rocks up to the Illawara Reversal. McElhinny (1973) considered that the reversal above the Paterson Volcanics, termed the Paterson Reversal, could be correlated with similar reversals in the United States and Russia, and the Kiaman reversed magnetic superchron has been used widely for global correlation and dating. The section where the normal magnetic signature of the Paterson Volcanics was measured is a railway cutting near Paterson railway station.

The Paterson Volcanics are also the Australian marker for the onset of the major late Palaeozoic glaciation of Gondwana. The Seaham Formation, which immediately overlies the Paterson Volcanics, is the major glacigene interval of the late Carboniferous Period in Australia and is correlated with similar cold climate facies elsewhere in Gondwana (Roberts and others, 1991). It contains tillite and varved shales interlayered with tuffs, lithic sandstone, and conglomerate; the varved shales contain dropstones, and the conglomerates contain striated and faceted boulders.

The international significance of the Paterson Volcanics makes it important to determine the biostratigraphic position and numerical age for correlation with other regions and continents. Unfortunately, both underlying and overlying sediments are non-marine; they contain macro- and palyno-floras whose correlation is restricted mainly within Gondwanan terms. The underlying Mt. Johnstone Formation, which is fluvial in origin, contains the first occurrence of the *Nothorhacopteris* Flora and a palynoflora referred to the *Grandispora maculosa* Assemblage. The *Nothorhacopteris* Flora has always been interpreted as Late Carboniferous age; however, the zircon age for the Paterson Volcanics reported below indicates that it first appears in late Viséan time. Both Kemp and others (1977) and Jones and Truswell (1992) interpreted the age of the *G. maculosa* Assemblage to be late Viséan to earliest Namurian age, but the absence of monosaccate pollen (C. Foster, pers. commun., 1992) suggests that the Zone is unlikely to range into the Namurian (see also Jones and Truswell, 1992). Overlying the Paterson Volcanics is the Seaham Formation (as defined by Briggs and Archbold, 1990, an equivalent of the glacigene Grahamstown Lake Formation of Rattigan, 1967), which is glacial, fluvial and lacustrine in origin, and contains a palynoflora termed the *Potonieisporites* Assemblage (Helby, 1969). A re-examination of this Assemblage by C. Foster (pers. commun., 1992) suggests assignment to the *Asperispora reticulatisporites* Zone (Zone D) of Jones and Truswell (1992) in the Galilee Basin of Queensland or to the Late Carboniferous *Diatomozonotriletes birk-*

headensis Zone in the Bonaparte Basin of northern Australia. Although Jones and Truswell (1992) suggested assignment of Zone D to the latest Westphalian to Autunian, there is no independent evidence to support the younger age limit and Zone D is interpreted here as late Westphalian age. Examination by Briggs and Archbold (1990) of the section containing Stage 2 and 3 palynofloras and the marine fauna referred to the *Trigonotreta campbelli* Zone (= *Trigonotreta* n. sp. Zone of Archbold and Dickins, 1991), formerly referred to the Seaham Formation, has shown it to belong instead to the lower Permian Lochinvar Formation. In terms of previous usage (Kemp and others, 1977; Roberts and others, 1985), the Seaham Formation is therefore restricted to the Carboniferous Period (e.g., Jones and others, 1973).

These correlations of Formations bracketing the Paterson Volcanics are mainly restricted to Gondwana and leave wide ambiguity about the precise stratigraphic position within the Carboniferous Period. Intercontinental correlation of both the Kiaman magnetic superchron and Gondwanan glaciation has, therefore, depended on a much quoted age of 308 Ma for the Paterson Volcanics, which is based on a single hornblende K-Ar measurement of 308 ± 7 Ma of a Paterson Volcanics sample from a road cutting near Fal Brook. Supporting this result is a hornblende K-Ar age of 307 ± 4 Ma for a dacite at Raymond Terrace which lies stratigraphically above a porcellanite identified as an equivalent of the Paterson Volcanics. In fault blocks north of Raymond Terrace Quarry, this porcellanite forms the uppermost unit of Rattigan's (1967) Italia Road (= Mt. Johnstone) Formation, but the dacite is not represented within the overlying Grahamstown Lake Formation (= Seaham Formation). Samples adjacent to the palaeomagnetic type section at Paterson yielded much younger apparent K-Ar ages, these were interpreted as reflecting Ar loss and were not used in assigning the age of the Paterson Volcanics. Details of all these K-Ar measurements are given in Roberts and others (1991). On the basis of the 308 Ma age interpreted from K-Ar analysis, the Paterson Volcanics and the associated glaciation and magnetic reversal have been assigned a stratigraphic position close to the latest Carboniferous Westphalian C-D boundary.

In an effort to define the age of the Paterson Volcanics with greater confidence, zircons from two samples, Z765 and Z850, have been studied using the SHRIMP ion microprobe.

Sample Z765—SHRIMP I Analysis

Sample Z765 is from the section of Paterson Volcanics in the railway cutting at Paterson where their magnetic signature was measured by Irving (1966); samples from this location had previously yielded the young apparent K-Ar ages discussed above. Zircons in the sample have euhedral and elongate prismatic shapes with length:breadth ratios of more than 5:1. Most are internally structureless, but a few have the oscillatory growth zoning associated with crystallisation in a viscous felsic melt. They are characterized by a high abundance of inclusions of other phases, and the centres of some elongate crystals are occupied by amorphous material interpreted as devitrified glass. These features indicate, respectively, late crystallisation seeded on other phases, and skeletal crystal growth, both of which are consistent with rapid late crystallisation following eruption.

For this sample, a total of 15 measurements of standard zircon SL13 were collected together with 21 analyses of Paterson zircons (Table 2). Compositions measured for the standard adhere to the expected power law correlation between ^{206}Pb/^{238}U and UO/U (Fig. 6). The data were measured over a period of two days during which adjustments were made to the focussing and intensity of the primary beam. This influences the sputtering conditions and has the effect of slightly moving the position of the calibration curve. Thus the standards define three calibration batches slightly offset to one another with observed reproducibilities of, respectively, 1.82%, 2.18% and 1.65% (σ); these reproducibilities are included in the uncertainties in ^{206}Pb/^{238}U ratio of the Paterson zircons.

Also shown in Figure 6 are the data for the Z765 zircons. These adhere to a line of the same form as that through the standard analyses, but at lower ratios of ^{206}Pb/^{238}U appropriate to their younger age. Without any further processing of the analyses, the raw data on the calibration graph indicate that most of the Paterson zircons have coherent ^{206}Pb/^{238}U compositions with a reproducibility similar to that of the standard, with only one analysis having a ^{206}Pb/^{238}U ratio slightly lower than the main group. This homogeneity indicates that most of the zircons have the same age within measurement error.

Calibrated ^{206}Pb/^{238}U ratios for the Paterson zircons are measured on the calibration graph from the difference between their ^{206}Pb/^{238}U ratio and the mean position of the standard calibration curve at the same UO/U. The calibrated data are plotted without correction for common Pb on a ^{207}Pb/^{206}Pb-^{238}U/^{206}Pb evolution diagram in Figure 7. This version of Concordia diagram (Tera and Wasserburg, 1974) offers several advantages over the more commonly used Wetherill (1956) diagram when plotting Phanerozoic zircon compositions. It uses directly measured ratios, whereas the ^{207}Pb/^{235}U axis of the Wetherill diagram is derived from ^{206}Pb/^{238}U measurement, resulting in strongly correlated errors. The variation of ^{206}Pb/^{238}U ratios over Phanerozoic time is emphasised by the x-axis, while the ^{207}Pb/^{206}Pb ratio axis and corresponding ages highlight the difficulty of deriving Phanerozoic ages from ^{207}Pb/^{206}Pb measurement where small errors

TABLE 2.—SHRIMP I U-Pb ISOTOPIC DATA FOR THE PATERSON VOLCANICS, AUSTRALIA

Grain area	U (ppm)	Th (ppm)	Th/U	^{204}Pb* (ppb)	f^{206}Pb‡** (%)	Calibrated total Pb compositions			Radiogenic composition*	
						^{206}Pb/^{238}U ±1σ	^{207}Pb/^{235}U ±1σ	^{207}Pb/^{206}Pb ±1σ	^{206}Pb/^{238}U ±1σ	Apparent age# (Ma) ± 1σ
Mount Z765										
1.3	452	332	.73	3	.28	0.0530 ± 9	0.386 ± 9	0.0553 ± 7	0.0505 ± 9	318 ± 6
2.1	509	310	.61	9	.69	0.0506 ± 10	0.428 ± 10	0.0586 ± 7	0.0526 ± 10	330 ± 6
3.1	337	228	.68	3	.34	0.0522 ± 10	0.401 ± 10	0.0557 ± 8	0.0520 ± 10	327 ± 6
4.1	649	404	.62	15	.89	0.0537 ± 10	0.445 ± 10	0.0602 ± 6	0.0532 ± 10	334 ± 6
5.2	235	192	.82	4	.65	0.0551 ± 10	0.442 ± 12	0.0582 ± 10	0.0547 ± 10	344 ± 6
6.1	267	118	.44	2	.36	0.0525 ± 10	0.404 ± 11	0.0559 ± 9	0.0523 ± 10	329 ± 6
7.1	197	125	.63	3	.67	0.0522 ± 11	0.420 ± 12	0.0584 ± 11	0.0518 ± 10	326 ± 6
8.1	729	685	.94	8	.45	0.0546 ± 10	0.427 ± 9	0.0567 ± 5	0.0543 ± 10	341 ± 6
9.1	65	29	.45	3	1.95	0.0502 ± 10	0.476 ± 19	0.0687 ± 21	0.0493 ± 10	310 ± 6
10.1	357	135	.38	4	.47	0.0546 ± 10	0.428 ± 11	0.0568 ± 8	0.0544 ± 10	341 ± 6
11.1	458	176	.38	7	.63	0.0529 ± 10	0.424 ± 10	0.0581 ± 7	0.0526 ± 10	330 ± 6
12.1	456	174	.38	9	.82	0.0531 ± 12	0.437 ± 12	0.0596 ± 7	0.0538 ± 10	331 ± 7
13.1	541	277	.51	10	.73	0.0529 ± 12	0.430 ± 11	0.0589 ± 7	0.0536 ± 9	330 ± 7
14.1	530	368	.69	8	.59	0.0525 ± 12	0.419 ± 11	0.0578 ± 7	0.0533 ± 9	328 ± 7
15.1	301	197	.66	19	2.52	0.0546 ± 12	0.551 ± 16	0.0733 ± 11	0.0543 ± 10	334 ± 7
16.1	424	364	.86	8	.82	0.0513 ± 9	0.421 ± 9	0.0596 ± 7	0.0505 ± 9	320 ± 5
17.1	550	525	.95	2	.14	0.0518 ± 9	0.386 ± 8	0.0541 ± 6	0.0513 ± 9	325 ± 5
18.1	282	203	.72	6	.90	0.0479 ± 8	0.398 ± 10	0.0603 ± 10	0.0471 ± 8	299 ± 5
19.1	329	163	.50	5	.68	0.0509 ± 9	0.411 ± 10	0.0585 ± 8	0.0502 ± 9	318 ± 5
20.1	416	210	.51	2	.17	0.0529 ± 9	0.397 ± 9	0.0544 ± 7	0.0524 ± 9	332 ± 5
21.1	520	412	.79	30	2.33	0.0526 ± 9	0.520 ± 11	0.0717 ± 7	0.0510 ± 9	323 ± 5
22.1	453	368	.81	4	.34	0.0522 ± 9	0.401 ± 9	0.0557 ± 7	0.0516 ± 9	327 ± 5
23.1	278	87	.31	4	.54	0.0529 ± 9	0.418 ± 10	0.0574 ± 9	0.0522 ± 9	330 ± 5
Mount Z850										
1.1	547	544	.99	4	.32	0.0520 ± 11	0.418 ± 19	0.0582 ± 22	0.0517 ± 11	325 ± 7
2.1	560	310	.55	1	.06	0.0541 ± 12	0.400 ± 15	0.0537 ± 15	0.0540 ± 12	339 ± 7
3.1	154	72	.46	2	.54	0.0520 ± 11	0.354 ± 23	0.0494 ± 29	0.0522 ± 11	328 ± 7
5.1	545	246	.45	3	.26	0.0513 ± 11	0.375 ± 14	0.0530 ± 14	0.0513 ± 11	323 ± 7
6.1	336	284	.84	2	.21	0.0524 ± 11	0.406 ± 21	0.0562 ± 25	0.0522 ± 11	328 ± 7
7.1	232	97	.42	1	.25	0.0525 ± 11	0.396 ± 18	0.0547 ± 20	0.0524 ± 11	329 ± 7
8.1	110	60	54	2	.60	0.0531 ± 12	0.422 ± 28	0.0576 ± 34	0.0528 ± 12	332 ± 7
9.1	258	136	.53	−1	−.12	0.0522 ± 11	0.371 ± 18	0.0515 ± 22	0.0523 ± 11	329 ± 7
10.1	276	122	.44	3	.40	0.0522 ± 11	0.399 ± 18	0.0555 ± 20	0.0520 ± 11	327 ± 7
11.1	460	301	.66	2	.16	0.0532 ± 12	0.413 ± 17	0.0564 ± 18	0.0530 ± 11	333 ± 7
12.1	271	160	.59	2	.33	0.0547 ± 12	0.433 ± 19	0.0574 ± 21	0.0544 ± 12	341 ± 7
13.1	99	51	.52	1	.38	0.0517 ± 11	0.391 ± 29	0.0550 ± 37	0.0515 ± 11	324 ± 7
14.1	90	77	.85	2	.77	0.0509 ± 11	0.380 ± 38	0.0542 ± 51	0.0508 ± 11	320 ± 7
15.1	184	143	.78	2	.40	0.0524 ± 11	0.423 ± 26	0.0585 ± 32	0.0520 ± 11	327 ± 7
15.2	338	212	.63	−1	−.10	0.0543 ± 12	0.402 ± 18	0.0537 ± 20	0.0543 ± 12	341 ± 7
16.1	158	148	.93	0	.10	0.0528 ± 12	0.386 ± 31	0.0531 ± 40	0.0528 ± 11	332 ± 7
17.1	445	762	1.71	1	.08	0.0521 ± 11	0.400 ± 30	0.0557 ± 39	0.0519 ± 11	326 ± 7
18.1	377	310	.82	0	.03	0.0521 ± 11	0.400 ± 20	0.0558 ± 24	0.0519 ± 11	326 ± 7
19.1	288	193	.67	0	−.05	0.0515 ± 11	0.379 ± 20	0.0533 ± 25	0.0515 ± 11	324 ± 7
20.1	575	486	.84	4	.28	0.0522 ± 11	0.389 ± 18	0.0540 ± 20	0.0521 ± 11	328 ± 7

*^{207}Pb corrected data
†f^{206}Pb indicates the percentage of common ^{206}Pb in the total measured ^{206}Pb.
#Apparent age is the ^{206}Pb/^{238}U age.

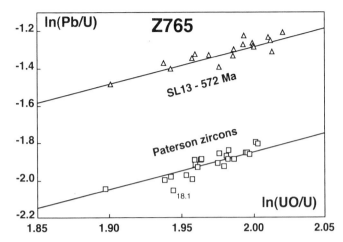

FIG. 6.—Data for standard SL13 and the Paterson zircons in Mount Z765 compared with the power law calibration relationship. Both sets of analyses adhere closely to the expected line. Calibrated $^{206}Pb/^{238}U$ ratios for the Paterson zircons are calculated from the difference between their observed $^{206}Pb/^{238}U$ ratios, and the $^{206}Pb/^{238}U$ ratio of the standard calibration line at the same UO/U.

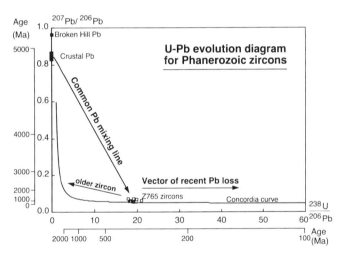

FIG. 7.—Geometry of influences on Phanerozoic zircon compositions in a $^{207}Pb/^{206}Pb$-$^{238}U/^{206}Pb$ Concordia diagram. Total Pb compositions of the Paterson zircons in mount Z765 are plotted, together with the range of candidate common Pb compositions. See text for discussion.

in ratio result in large uncertainties in age. Plotted on the $^{207}Pb/^{206}Pb$ ratio axis is the range of crustal Pb compositions for Phanerozoic time (Cumming and Richards, 1975), and Broken Hill Pb which is the composition of any "laboratory blank" Pb on the surface of ion probe mounts. These two compositions combine as the common Pb; measured zircon compositions, not corrected for common Pb, will lie on a mixing line between this narrow zone on the $^{207}Pb/^{206}Pb$ axis and totally radiogenic Pb on the Concordia curve. The closeness to Concordia of the Z765 total Pb compositions highlights the low common Pb content of these zircons, and any uncertainty over the composition of common Pb, within the permitted range, cannot materially affect the vector of common Pb correction. The common Pb mixing line does not cross Concordia at older apparent ages, so the vector of inheritance of any older zircon lies distinctly to

the left; conversely, any zero-age Pb loss would draw compositions to the right, parallel to the x-axis. The divergence of the vectors of inheritance and Pb loss from the common Pb mixing line is important in Phanerozoic zircons because of the dependence of their radiogenic $^{207}Pb*/^{206}Pb*$ ratios on perfect correction for common Pb, and the inherent difficulty of achieving it in zircons that contain little radiogenic Pb and barely measurable common Pb. In this context, it is useful to plot total Pb compositions not corrected for common Pb. Radiogenic compositions are then given by the high angle intercept of the common Pb mixing line with Concordia, and zircons of like age will lie within measurement error of the mixing line. The geometry of the diagram is such that inheritance and Pb loss result in deviations subparallel to the x-axis, and their recognition therefore depends on $^{206}Pb/^{238}U$ ratios which are the most reliably measured parameter in SHRIMP zircon analyses. This is the converse of the practice in conventional zircon laboratories, where the larger amounts of zircon analysed permit $^{207}Pb/^{206}Pb$ ratios to be determined with useful precision.

Figure 8 is an expanded Concordia view of the Z765 total Pb compositions. The data lie close to Concordia, indicating low contents of common Pb, and are dispersed within measurement error of the common Pb mixing line. Zircon 18 is the only significant departure from the main group. Its position to the right of the population is consistent with recent loss of Pb, as is its situation below the main group of data in the calibration graph. There is no indication of data tracking along an inheritance path, which would indicate admixture of zircons older than the principal population. The grouping of radiogenic $^{206}Pb*/^{238}U$ ratios, measured by projection of the total Pb compositions onto Concordia, is illustrated in Figure 9 which is a gaussian summation of the individual analyses and uncertainties. This highlights the significantly offset $^{206}Pb/^{238}U$ ratio of zircon 18, which is confidently rejected as an outlier that has lost Pb, and illustrates the close approach of the main popula-

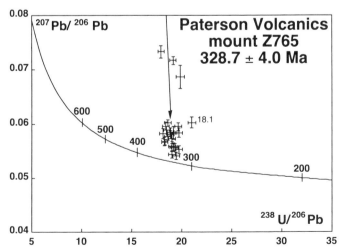

FIG. 8.—Expansion of the Concordia diagram in Figure 7, showing the dispersion of Z765 Paterson zircon total Pb compositions, without correction for common Pb. The arrow indicates the vector of common Pb correction; radiogenic $^{206}Pb*/^{238}U$ compositions are calculated by projection of the data onto Concordia along this vector. Dispersion to the right of this vector (beyond measurement uncertainty) indicates Pb loss from zircon 18, whereas the others group within error as an homogenous $^{206}Pb/^{238}U$ composition.

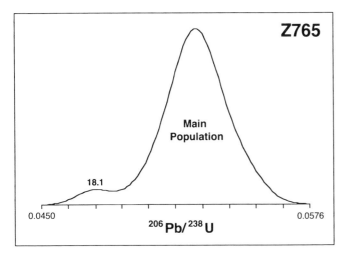

FIG. 9.—Histogram (gaussian summation) of $^{206}Pb*/^{238}U$ ages for the Paterson zircons in mount Z765, showing the close approach to a normal distribution of the main population of data and the significantly offset composition of zircon 18.

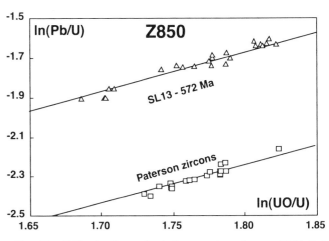

FIG. 10.—Calibration diagram for the Paterson zircons in mount Z850. Both unknowns and standard have similar reproducibility about the expected line, indicating that the unknowns have an homogeneous $^{206}Pb/^{238}U$ ratio within measurement uncertainty.

tion of radiogenic compositions to a normal distribution. The weighted mean radiogenic $^{206}Pb*/^{238}U$ ratio of the population is 0.05231 ± .00070 (2σ), equivalent to an age of 328.7 ± 4.0 Ma (2σ).

Sample Z850—SHRIMP I Analysis

The 328.7 ± 4.0 Ma (2σ) zircon age of sample Z765 is very different from the 308 Ma K-Ar age upon which previous intercontinental correlations of the Paterson Volcanics have been based. The K-Ar age is for a sample geographically distant from, but correlated to, the Paterson type locality. The question therefore arises of whether or not this correlation is sound. To resolve this ambiguity, SHRIMP zircon analysis has been made of a sample from the K-Ar dating locality, which is now obscured by a dam. A split of the original specimen used for K-Ar dating was made available from the collection of the Department of Applied Geology, University of New South Wales. Zircons in this sample (Z850) are morphologically identical to those in Z765 from Paterson.

Nineteen analyses of Z850 Paterson zircons were alternated with 23 measurements of SL13 during a single day. The pattern of compositions is simple (Table 2; Fig. 10). The analyses of the standard group as a single calibration run with reproducibility of 2.15% (σ); this uncertainty is included in the quoted errors of calibrated $^{206}Pb/^{238}U$ ratios. Uncertainty in the mean position of the calibration curve is 0.45% (σ), and this is added (in quadrature) to the uncertainty of the mean age of the unknowns. Data for the Paterson zircons likewise adhere closely to the power law relationship with a reproducibility comparable to that of the standard.

The total Pb compositions of the Z850 zircons exhibit a pattern very similar to that of Z765 when plotted on a Concordia diagram (Fig. 11). They are dispersed within measurement error of the common Pb mixing line and are substantially lower in common Pb than the Z765 analyses; Table 2 indicates the Z765 zircons to have between 0.5%–3% of common ^{206}Pb, whereas those in Z850 have less than 0.5% of common ^{206}Pb. This dif-

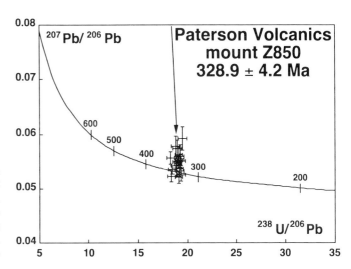

FIG. 11.—Concordia diagram for Paterson zircons in Mount Z850, showing the total Pb compositions without correction for common Pb. The data lie close to Concordia, indicating very low common Pb contents, along the vector of common Pb correction indicated by the arrow. There is no dispersion from this vector to indicate Pb loss or inheritance of older zircon.

ference corresponds to a change from coating sample mounts with carbon (which was applied to Z765) to using 99.999% pure gold (Z850). The contribution of "laboratory blank" Pb on the mount surface has been reduced substantially by this practice. With such low measured abundances of common Pb in the analyses, statistically negative readings for $f^{206}Pb$ (the fraction of common ^{206}Pb) are to be expected, and are indeed recorded for Z850 in Table 2; in Figure 11 these analyses lie slightly below, but within measurement error of, the Concordia curve.

The radiogenic $^{206}Pb*/^{238}U$ ratios for zircons in Z850, calculated by projecting the total Pb compositions onto the Concordia curve, form a simple homogeneous grouping without outliers. The weighted mean $^{206}Pb*/^{238}U$ ratio of the group is 0.05234 ± 0.00052 (2σ) with MSWD = 0.7, equivalent to an age of 328.9 ± 4.2 Ma (2σ).

Sample Z850—SHRIMP II Analysis

The SHRIMP II ion microprobe is a second-generation SHRIMP design. Sample Z850 has been analysed with the new instrument to establish whether comparable measurements would be obtained by applying SHRIMP I procedures to SHRIMP II Pb/U analysis. Specifically, this is a test of whether the Pb/U calibration relationship established above for SHRIMP I can be applied to SHRIMP II measurements.

For this exercise, the surface of mount Z850 was repolished with 0.25 μm diamond paste to remove the 1 μm deep sputtering craters created by the SHRIMP I analysis, and the surface was recoated with gold. Analytical procedures for SHRIMP II were the same as those applied with SHRIMP I: sampling craters approximately 25 μm in diameter were excavated with a 3 nA Köhler-focussed primary beam, and the secondary beam was collected by stepping the masses of interest into a single electron multiplier. Sensitivity achieved was 60 counts per second per ppm of Pb, which compares with 15 cps/ppm Pb obtained with SHRIMP I.

Over a two-day period, 34 analyses of Paterson zircons were collected together with 38 analyses of standard zircon SL13 (Table 3). The SHRIMP II data for the standard zircon are compared with the SHRIMP I-derived calibration relationship in Figure 12. The raw measured compositions adhere closely to the expected power law correlation between ^{206}Pb/^{238}U and UO/U over a range in observed UO/U between 5 and 6, indicating that the calibration relationship does apply to SHRIMP II measurements. The two days group as a single calibration run with an observed reproducibility of 1.89%, and this uncertainty is included in the uncertainties in ^{206}Pb/^{238}U ratio of the Paterson zircons. Also shown in Figure 12 are the SHRIMP II data for the Z850 zircons, which adhere to the power law relationship with a reproducibility comparable to that of the standard.

Plotted in a Concordia diagram (Fig. 13), the total Pb compositions measured with SHRIMP II are similar to those obtained with SHRIMP I (see Fig. 11). They have low contents of common Pb and lie within error of the common Pb mixing line. Projection of the total Pb compositions onto Concordia along this mixing line yields a simple grouping of radiogenic ^{206}Pb*/^{238}U ratios with a weighted mean age of 330.1 ± 3.0 Ma (2σ) and MSWD of 1.13. This SHRIMP II result is within the analytical uncertainty of 328.9 ± 4.2 Ma (2σ) as measured by SHRIMP I.

Conventional Zircon Analysis

The above SHRIMP ^{206}Pb/^{238}U ages are measured by difference from concurrent analyses of standard zircon SL13, whose

TABLE 3.—SHRIMP II U-Pb ISOTOPIC DATA FOR THE PATERSON VOLCANICS, AUSTRALIA

Grain area	U (ppm)	Th (ppm)	Th/U	^{204}Pb* (ppb)	f^{206}Pb*† (%)	Calibrated total Pb compositions			Radiogenic compositions	
						^{206}Pb/^{238}U ± 1σ	^{207}Pb/^{235}U ± 1σ	^{207}Pb/^{206}Pb ± 1σ	*^{206}Pb/^{238}U ± 1σ	Apparent age# (Ma) ± 1σ
Mount Z850										
1.1	620	257	.41	5	.33	0.0517 ± 10	0.396 ± 9	0.0556 ± 5	0.0515 ± 10	324 ± 6
1.2	508	346	.68	6	.46	0.0530 ± 10	0.415 ± 9	0.0567 ± 6	0.0528 ± 10	331 ± 6
4.1	308	195	.63	6	.79	0.0502 ± 10	0.411 ± 10	0.0594 ± 7	0.0498 ± 9	313 ± 6
6.1	497	201	.40	4	.37	0.0514 ± 10	0.397 ± 9	0.0560 ± 5	0.0512 ± 10	322 ± 6
6.2	134	63	.47	5	1.48	0.0548 ± 10	0.490 ± 13	0.0649 ± 11	0.0540 ± 10	339 ± 6
7.1	432	171	.39	5	.48	0.0518 ± 10	0.406 ± 9	0.0568 ± 6	0.0516 ± 10	324 ± 6
7.2	381	141	.37	4	.48	0.0502 ± 10	0.394 ± 9	0.0569 ± 6	0.0500 ± 10	314 ± 6
8.1	152	73	.48	3	.88	0.0527 ± 10	0.437 ± 12	0.0601 ± 10	0.0523 ± 10	328 ± 6
10.1	768	395	.51	2	.13	0.0523 ± 10	0.389 ± 8	0.0540 ± 4	0.0522 ± 10	328 ± 6
10.2	301	180	.60	5	.62	0.0519 ± 10	0.415 ± 10	0.0580 ± 7	0.0516 ± 10	324 ± 6
11.1	599	523	.87	4	.31	0.0520 ± 10	0.398 ± 9	0.0555 ± 5	0.0519 ± 10	326 ± 6
13.1	128	71	.55	5	1.42	0.0533 ± 10	0.473 ± 13	0.0644 ± 11	0.0525 ± 10	330 ± 6
13.2	230	137	.59	3	.58	0.0531 ± 10	0.422 ± 11	0.0576 ± 8	0.0528 ± 10	332 ± 6
14.1	79	71	.89	3	1.62	0.0525 ± 10	0.478 ± 15	0.0660 ± 14	0.0517 ± 10	325 ± 6
14.2	167	99	.59	4	1.08	0.0531 ± 10	0.452 ± 12	0.0617 ± 10	0.0526 ± 10	330 ± 6
15.1	82	84	1.03	4	2.05	0.0538 ± 10	0.516 ± 15	0.0695 ± 15	0.0527 ± 10	331 ± 6
16.1	228	226	.99	5	.84	0.0536 ± 10	0.441 ± 11	0.0598 ± 8	0.0531 ± 10	334 ± 6
16.2	201	142	.71	5	.90	0.0537 ± 10	0.446 ± 11	0.0602 ± 9	0.0532 ± 10	334 ± 6
18.1	151	54	.36	4	1.08	0.0554 ± 10	0.472 ± 12	0.0617 ± 10	0.0548 ± 10	344 ± 6
18.2	175	63	.36	4	.84	0.0518 ± 10	0.427 ± 11	0.0598 ± 10	0.0514 ± 10	323 ± 6
20.1	590	235	.40	3	.20	0.0526 ± 10	0.396 ± 9	0.0546 ± 5	0.0525 ± 10	330 ± 6
20.1	476	278	.58	4	.33	0.0530 ± 10	0.406 ± 9	0.0556 ± 5	0.0528 ± 10	332 ± 6
21.2	473	318	.67	8	.66	0.0526 ± 10	0.423 ± 9	0.0583 ± 6	0.0522 ± 10	328 ± 6
22.2	355	148	.42	7	.79	0.0541 ± 10	0.442 ± 10	0.0594 ± 7	0.0536 ± 10	337 ± 6
23.2	306	192	.63	3	.40	0.0540 ± 10	0.419 ± 10	0.0562 ± 6	0.0538 ± 10	338 ± 6
24.1	406	276	.68	4	.37	0.0555 ± 11	0.429 ± 10	0.0560 ± 6	0.0553 ± 11	347 ± 6
24.2	270	172	.63	6	.79	0.0560 ± 11	0.458 ± 11	0.0594 ± 7	0.0555 ± 11	348 ± 6
25.1	467	284	.61	4	.38	0.0527 ± 10	0.407 ± 9	0.0561 ± 6	0.0525 ± 10	330 ± 6
26.1	210	138	.66	5	.98	0.0532 ± 10	0.447 ± 11	0.0609 ± 8	0.0527 ± 10	331 ± 6
26.2	171	94	.55	3	.81	0.0526 ± 10	0.432 ± 11	0.0595 ± 10	0.0522 ± 10	328 ± 6
29.1	313	173	.55	4	.52	0.0515 ± 10	0.406 ± 10	0.0572 ± 7	0.0512 ± 10	322 ± 6
29.2	640	304	.47	4	.28	0.0522 ± 10	0.398 ± 9	0.0552 ± 5	0.0521 ± 10	327 ± 6
30.1	546	313	.57	3	.25	0.0529 ± 10	0.401 ± 9	0.0550 ± 5	0.0527 ± 10	331 ± 6
30.2	798	577	.72	6	.27	0.0543 ± 10	0.413 ± 9	0.0552 ± 4	0.0541 ± 10	340 ± 6
33.1	547	606	1.11	4	.29	0.0520 ± 10	0.396 ± 9	0.0553 ± 5	0.0518 ± 10	326 ± 6

*^{207}Pb correction
†f^{206}Pb indicates the percentage of common ^{206}Pb in the total measured ^{206}Pb
#Apparent age is the ^{206}Pb/^{238}U age

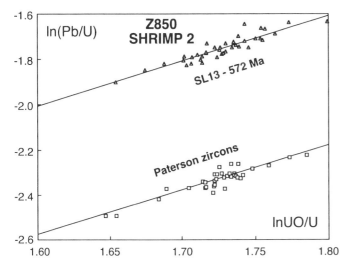

FIG. 12.—Calibration diagram for SHRIMP II using data for the Paterson zircons in mount Z850. The calibration line is that derived from SHRIMP I. Adherence of data to this relationship indicates that the SHRIMP I Pb/U calibration characteristics apply also to SHRIMP II.

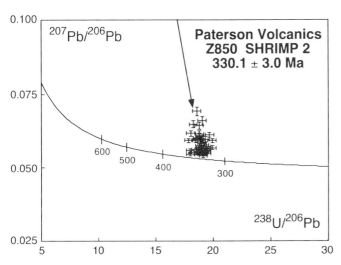

FIG. 13.—Concordia diagram for Paterson zircons in Mount Z850, showing the total Pb compositions without correction for common Pb. The compositions are essentially indistinguishable from the data collected using SHRIMP I (compare Fig. 11)

composition is known accurately from replicate conventional analyses. They are therefore free of bias provided there is no source of measurement bias between standard and unknown zircons within the ion microprobe; such bias is assumed to be absent because the standard and unknowns are probed at the same time under the same conditions. As a test of this assumption, isotope dilution thermal ionisation mass spectrometry (IDTIMS) measurement has been made of the Paterson zircons. Procedures used were similar to those described by Fanning and others (1988) for small sample analysis, using vapour transfer dissolution (Krogh, 1978); a mixed ^{205}Pb-^{235}U tracer was used to determine U-Pb ratios and concentrations. This is part of a wider study comparing SHRIMP and IDTIMS U-Pb zircon analyses through geological time (Fanning and others, 1991).

Two types of sample were prepared. First, six Paterson zircons previously probed by SHRIMP and selected on the basis of optical clarity and U content, were extracted from the Z765 probe mount and individually dissolved for IDTIMS measurement. These analyses represent an attempt to obtain for IDTIMS as closely as possible the same samples probed by SHRIMP. Second, six optically simple zircons from the Z765 mineral separation but not mounted for probing, were grouped as a multigrain dissolution for precise IDTIMS measurement. These zircons were abraded (Krogh, 1982) and dissolution was by vapour transfer as above; Pb and U were extracted following dissolution using HCl/H$_2$O with small-scale anion exchange columns. The IDTIMS analyses are listed in Table 4, where uncertainties and error-correlations are calculated at the 95% confidence limits following Ludwig (1980). Total procedural blanks measured in conjunction with these zircons were 14 pg of Pb and 0.1 pg of U for both unextracted and extracted analyses. The data have been corrected for the laboratory induced blank and for an assumed initial Pb isotope composition using the composition of the long term laboratory blank (17.97:15.55:37.71); this composition is applied because the total Pb ranged from about 30 to 180 pg, and, consequently, the small amounts of common Pb are considered to be dominated by the laboratory blank.

The single grain and multigrain IDTIMS analyses are overlain on the SHRIMP data for Z765 on a Concordia plot in Figure 14; the SHRIMP data are plotted as 1σ error boxes and the more precise IDTIMS data as 95% confidence boxes. The greater precision of ^{207}Pb/^{206}Pb measurement obtained for the larger IDTIMS samples is evident, particularly the multigrain analysis which combines the Pb and U of six zircons. The close comparison of the IDTIMS and SHRIMP data sets is also evident; the single grain data lie within the dispersion of ion probe analyses, and the precise multigrain analysis lies at the centroid of the data set.

All seven IDTIMS analyses have the same ^{206}Pb/^{238}U ratio within measurement uncertainty. The weighted mean ^{206}Pb/^{238}U ratio of the single-grain measurements is 0.05193 ± 40 (2σ) (MSWD 1.06), equivalent to an age of 326.4 ± 2.4 Ma (2σ). The single multigrain IDTIMS sample has a ^{206}Pb/^{238}U ratio of 0.05257 ± 58 (2σ), equivalent to an age of 330.3 ± 3.6 Ma (2σ). The weighted mean of all seven IDTIMS analyses is 0.05214 ± 34 (MSWD = 1.4), equivalent to an age of 327.7 ± 2.0 Ma (2σ). These IDTIMS ^{206}Pb/^{238}U results are indistinguishable from the SHRIMP ^{206}Pb/^{238}U ages.

The ^{207}Pb/^{206}Pb ratios measured by IDTIMS are more precise than their SHRIMP equivalents, but they contrast with the internal consistency of ^{206}Pb/^{238}U compositions by not forming a simple group. The very precise ^{207}Pb/^{206}Pb ratio obtained from the multigrain sample is 0.05329 ± 24 (2σ), equivalent to an age of 341.3 ± 10 (2σ). This is at the older end of the envelope of uncertainty on the ^{206}Pb/^{238}U age. The other IDTIMS ^{207}Pb/^{206}Pb measurements for single grains are dispersed well beyond their assigned uncertainties (MSWD = 7) with both normally and reversely discordant apparent ages ranging from 227 ± 49 Ma (grain 7) to 348 ± 22 Ma (grain 15); the weighted average ^{207}Pb/^{206}Pb composition is 0.05310 ± 30 (2σ), equivalent to an age of 333 ± 26 Ma (2σ); averaged without weightings, they have an age of 316 ± 36 Ma (2σ). Any similar dispersion of ^{207}Pb/^{206}Pb ratios that might be present

TABLE 4.—IDTIMS U-Pb ISOTOPIC DATA FOR THE PATERSON VOLCANICS

Zircon grains	Weight (μg)	Conc (ppm)		$^{206}Pb/^{204}Pb$ measured	Atomic Rations (radiogenic)			Model Ages (Ma)		
		U	Pb		$^{206}Pb/^{238}Pb$ measured	$^{207}Pb/^{235}U$	$^{207}Pb/^{206}Pb$	$^{206}Pb/^{238}U$	$^{207}Pb/^{235}U$	$^{207}Pb/^{206}Pb$
Single grains previously probed with SHRIMP (grain numbers refer to the analyses in Table 2)										
grain 4	2	565	31.6	819	0.05178 (±71)	0.3736 (±107)	0.05233 (±130)	325.4	322.3	299.9 (±57)
grain 5	2	415	23.5	273	0.05221 (±286)	0.3847 (±235)	0.05344 (±132)	328.1	330.5	347.5 (±56)
grain 6	2	467	27.5	685	0.05255 (±81)	0.3764 (±72)	0.05195 (±55)	330.1	324.4	283.2 (±24)
grain 7	2	269	14.8	561	0.05191 (±133)	0.3628 (±123)	0.05069 (±107)	326.3	314.3	226.8 (±49)
grain 9	5	162	8.95	643	0.05110 (±92)	0.3790 (±84)	0.05379 (±66)	321.3	326.3	362.3 (±28)
grain 15	8	254	14.8	567	0.05239 (±122)	0.3861 (±99)	0.05344 (±52)	329.2	331.5	347.7 (±22)
Analysis of abraded multigrain sample										
6 grains	12	263	14.8	888	0.05257 (±58)	0.3862 (±47)	0.05329 (±24)	330.3	331.7	341.3 (±10)

Note: 1. Single grains analysed unextracted; mutigrains: Pb and U extracted by small scale anion exchange.
Total procedural blank measured at 14 pg Pb and 0.1pg U for both procedures.
3. $^{206}Pb/^{204}Pb$ as measured, uncorrected for fractionation, blank and spike composition.
4. Uncertainties given in parentheses are at the 95% confidence level.

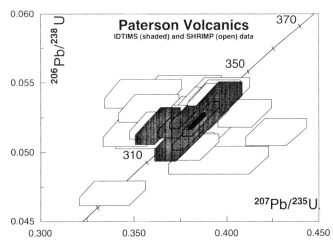

FIG. 14.—Multigrain (black) and single grain (shaded) IDTIMS analyses plotted as 2σ error boxes, compared with SHRIMP data (1σ open error boxes) for zircons in the Z765 Paterson Volcanics sample on a Wetherill (1956) Concordia diagram. Radiogenic compositions for the SHRIMP data have been calculated by the ^{208}Pb method of correcting for common Pb (Compston and others, 1984). There is no detectable difference between the ages and compositions measured by SHRIMP and IDTIMS analysis.

in the SHRIMP data is hidden by the larger uncertainties in $^{207}Pb/^{206}Pb$ measurement. This dispersion does not point to a unique age solution; instead it requires either real compositional variation among the samples or a source of uncertainty in $^{207}Pb/^{206}Pb$ measurement not accounted for within the assigned errors. The available data do not permit choice between these two possibilities, so the $^{207}Pb/^{206}Pb$ measurements do not contribute an independent measure of age in this sample. Instead, it is clear that the age is recorded in the IDTIMS analyses by the agreement of all the $^{206}Pb/^{238}U$ ratios at 327.7 ± 2.0 Ma (2σ).

THE Z1 TONSTEIN, GERMANY

In the absence of definitive biostratigraphy, stratigraphic assignment of the Paterson Volcanics depends on correlation of their numerical age to the international time scale for the Upper Carboniferous. A recent advance in this area of the Phanerozoic time scale has been the publication of precisely measured ^{40}Ar/

^{39}Ar ages for the European Upper Carboniferous Stages (Hess and Lippolt, 1986). These ages were measured for sanidines in tuffs and tonsteins having very tight biostratigraphic constraints; they exceed in biostratigraphic and analytical accuracy all the preexisting constraints on Upper Carboniferous time. The ages measured for Australian rocks are based on SHRIMP zircon U-Pb dating, so it is appropriate to explore the comparability of these two very different dating methods, which are based on different minerals and different decay constants.

Such a comparison is made here by SHRIMP zircon dating of Z1 coal tonstein from the Upper Carboniferous of the Ruhr in northwest Germany, one of the samples used to define the Upper Carboniferous time scale with ^{40}Ar/^{39}Ar dating (Hess and Lippolt, 1986). The expectation is that ages measured by ^{40}Ar/^{39}Ar in feldspars, and by U-Pb in zircons, should be equal in unaltered samples of rapidly-cooled volcanics. Owing to its importance in correlation of coal seams, the biostratigraphic position of this volcanic horizon is known precisely; it belongs to the uppermost Westphalian B Stage (Lippolt and others, 1984). Sanidine feldspar in this sample, given the sample label COT-Z by Hess and Lippolt (1986), yielded a simple ^{40}Ar/^{39}Ar stepheating pattern, indicating only minor secondary alteration of its K-Ar system, and a plateau age of 310.7 ± 7.8 Ma (2σ). Comparison of this age with the previously used 308 Ma K-Ar age for the Paterson Volcanics has been the basis of the uppermost Westphalian correlation for the base of the Kiaman magnetic reversal. A split of the original specimen dated by ^{40}Ar/^{39}Ar was made available by its collector, K. Burger, for zircon analysis.

Two morphological forms of zircon were found in the Z1 tonstein specimen and mounted for the ion microprobe as sample Z1029. One form is euhedral and prismatic with oscillatory growth zoning; some needle-like crystals include amorphous material at their centres, interpreted as devitrified glass. These features are consistent with rapid zircon crystallisation following eruption of the volcanic ash. This magmatic form is admixed with abundant rounded to subrounded grains having pitted surfaces, interpreted as representing sedimentary detritus. The ion probe was used to target forms interpreted as having grown during the eruption.

For the Z1 tonstein sample, 39 measurements of 37 zircons, together with 39 concurrent observations of the standard, were collected over a period of two days (Table 5). A large quantity

TABLE 5.—SHRIMP I U-Pb ISOTOPIC DATA FOR THE Z1 TONSTEIN, GERMANY

Grain area	U (ppm)	Th (ppm)	Th/U	204Pb* (ppb)	f206Pb*† (%)	Calibrated total Pb compositions			Radiogenic compositions*	
						206Pb/238U ±1σ	207Pb/235U ±1σ	207Pb/206Pb ±1σ	206Pb/238U ±1σ	Apparent age# (Ma) ±1σ
1.1	86	56	.65	0	.18	0.0508 ± 13	0.378 ± 21	0.0540 ± 24	0.0507 ± 13	319 ± 8
2.1	209	115	.55	1	.13	0.0506 ± 13	0.384 ± 16	0.0536 ± 16	0.0505 ± 13	318 ± 8
3.1	290	84	.29	1	.10	0.0522 ± 13	0.339 ± 14	0.0534 ± 13	0.0522 ± 13	328 ± 8
4.1	530	125	.24	− 1	− .12	0.0477 ± 12	0.356 ± 11	0.0516 ± 9	0.0478 ± 12	301 ± 7
5.1	305	197	.65	0	− .04	0.0494 ± 12	0.352 ± 13	0.0523 ± 12	0.0494 ± 13	311 ± 8
6.1	384	174	.45	0	.04	0.0482 ± 12	0.333 ± 12	0.0529 ± 12	0.0482 ± 12	304 ± 8
7.1	708	369	.52	11	.84	0.0408 ± 10	0.380 ± 10	0.0593 ± 9	0.0404 ± 10	255 ± 6
8.1	463	323	.70	1	.08	0.0517 ± 13	0.437 ± 12	0.0533 ± 9	0.0517 ± 13	325 ± 8
9.1	136	107	.79	3	.97	0.0526 ± 13	0.373 ± 18	0.0603 ± 18	0.0521 ± 13	327 ± 8
9.2	84	54	.65	1	.71	0.0465 ± 12	0.374 ± 21	0.0583 ± 27	0.0462 ± 12	291 ± 7
9.3	237	172	.73	2	.30	0.0507 ± 13	0.384 ± 16	0.0550 ± 16	0.0505 ± 13	318 ± 8
10.1	361	198	.55	4	.49	0.0479 ± 12	0.373 ± 14	0.0565 ± 14	0.0477 ± 12	300 ± 7
11.1	211	127	.60	2	.36	0.0500 ± 13	0.383 ± 17	0.0555 ± 18	0.0498 ± 13	313 ± 8
12.1	412	260	.63	4	.43	0.0488 ± 12	0.377 ± 14	0.0560 ± 14	0.0486 ± 12	306 ± 8
13.1	86	32	.37	1	.25	0.0505 ± 11	0.380 ± 20	0.0546 ± 24	0.0504 ± 11	317 ± 7
14.1	354	235	.66	3	.42	0.0489 ± 11	0.377 ± 12	0.0560 ± 12	0.0487 ± 11	306 ± 6
15.1	3601	3345	.93	82	2.86	0.0170 ± 4	0.177 ± 5	0.0755 ± 9	0.0165 ± 4	106 ± 2
16.1	275	123	.44	82	11.57	0.0546 ± 12	1.096 ± 32	0.1454 ± 25	0.0483 ± 11	304 ± 7
17.1	255	107	.42	− 1	− .11	0.0486 ± 11	0.347 ± 18	0.0517 ± 23	0.0487 ± 11	307 ± 7
18.1	126	80	.63	0	− .05	0.0476 ± 11	0.342 ± 25	0.0521 ± 34	0.0477 ± 11	300 ± 7
19.1	138	83	.60	3	.94	0.0488 ± 11	0.404 ± 26	0.0601 ± 34	0.0483 ± 11	304 ± 7
20.1	382	284	.74	0	− .01	0.0478 ± 11	0.347 ± 16	0.0525 ± 19	0.0478 ± 11	301 ± 7
21.1	128	77	.61	2	.56	0.0516 ± 12	0.405 ± 27	0.0569 ± 34	0.0513 ± 12	323 ± 7
22.1	1115	15	.01	5	.16	0.0542 ± 12	0.403 ± 11	0.0539 ± 6	0.0541 ± 12	340 ± 7
23.1	450	7	.01	3	.23	0.0535 ± 12	0.402 ± 12	0.0545 ± 9	0.0534 ± 12	335 ± 7
24.1	298	91	.31	1	.20	0.0480 ± 11	0.359 ± 12	0.0542 ± 12	0.0479 ± 11	301 ± 7
25.1	133	91	.68	1	.38	0.0499 ± 11	0.383 ± 15	0.0556 ± 17	0.0498 ± 11	313 + 7
26.1	653	344	.53	53	4.34	0.0403 ± 9	0.485 ± 13	0.0874 ± 10	0.0385 ± 9	244 ± 5
27.1	566	189	.33	0	.03	0.0510 ± 12	0.371 ± 11	0.0528 ± 8	0.0509 ± 12	320 ± 7
28.1	191	108	.57	1	.31	0.0478 ± 11	0.363 ± 13	0.0551 ± 14	0.0477 ± 11	300 ± 7
29.1	201	99	.49	0	.03	0.0505 ± 12	0.368 ± 13	0.0528 ± 13	0.0505 ± 12	318 ± 7
30.1	279	123	.44	1	.20	0.0502 ± 11	0.375 ± 12	0.0542 ± 12	0.0501 ± 11	315 ± 7
31.1	648	302	.47	10	.73	0.0456 ± 10	0.368 + 10	0.0584 ± 8	0.0453 ± 10	286 ± 6
32.1	40	19	.48	1	.76	0.0514 ± 12	0.416 ± 24	0.0587 ± 30	0.0510 ± 12	321 ± 7
33.1	341	237	.70	1	.12	0.0493 ± 11	0.364 ± 11	0.0535 ± 10	0.0492 ± 11	310 ± 7
34.1	64	28	.44	2	1.20	0.0496 ± 12	0.426 ± 21	0.0622 ± 25	0.0490 ± 12	309 ± 7
35.1	393	333	.85	5	.59	0.0418 ± 9	0.330 ± 10	0.0573 ± 11	0.0416 ± 9	263 ± 6
36.1	669	492	.74	− 1	− .05	0.0494 ± 11	0.356 ± 10	0.0522 ± 8	0.0494 ± 11	311 ± 7
37.1	238	158	.66	0	.09	0.0502 ± 11	0.368 ± 14	0.0533 ± 14	0.0501 ± 11	315 ± 7

*207Pb correction

†f206Pb indicates the percentage of common 206Pb in the total measured 206Pb

#Apparent age is the 206Pb/238U age

of data was collected because loss of Pb is evident in some zircons, and it is therefore important to define outlying compositions with good statistical confidence. The reference for a homogeneous composition is the 2.34% (σ) reproducibility of the measurements of the standard zircon, and this uncertainty is included in the uncertainties in $^{206}Pb/^{238}U$ ratio of the unknown zircons. The mean position of the calibration was determined with uncertainty of 0.37% (σ), and this error is included in the error of the mean age of the sample.

A visual comparison with $^{40}Ar/^{39}Ar$ step heating patterns is provided in Figure 15, where the Pb/U age measured for each tonstein zircon is plotted against the time of analysis. Data for the unknown zircons and the concurrently measured data for the standard are compared directly on this diagram, and the ages displayed have been calculated by projection of the raw measured ratios onto Concordia as described above for the Paterson zircons. The measurements of the standard zircon clearly define the 2.34% envelope of reproducibility of a homogeneous composition that was a constant over the 30 hours of analysis; this is the dominant component of the error bar plotted for each of the unknown zircons. Among the unknown zircons, 31 of the

39 analyses are within measurement uncertainty of each other and define a constant composition within the terms of the 2.34% reproducibility achieved during the analytical run. The homogeneity of this dominant population of analyses is analogous to the criteria that define acceptable plateaux in $^{40}Ar/^{39}Ar$ stepheating patterns. The weighted mean radiogenic $^{206}Pb*/^{238}U$ ratio of this group of zircons is 0.04943 ± .00048 (2σ) with MSWD = 1.29, within the MSWD value expected at 95% confidence with 30 degrees of freedom. This equates to an age of 311.0 ± 3.4 Ma (2σ).

Departing significantly from the main population of $^{206}Pb/^{238}U$ ages are 6 zircon compositions having variable and young (Permian and Cretaceous) apparent ages. With a large quantity of data in hand for the main population, these are treated with confidence as outlier compositions beyond the 95% uncertainty envelope. It is inferred that these zircons have lost variable proportions of their radiogenic Pb. Also present are two zircons with ages detectably (beyond 95% confidence) older than the main group. These grains were targeted as likely contributors to defining the crystallisation age of the sample but their older ages identify them as xenocrysts included in the volcanic ash

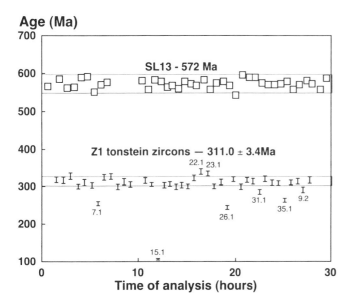

FIG. 15.—^{206}Pb/^{238}U ages measured for zircons in the Z1 tonstein plotted as a series against time and compared with concurrent measurements of the standard zircon. Data for the standard define a reproducibility of 2.34% (σ) for a constant composition, and the equivalent 95% confidence envelope is indicated. A similar 95% confidence envelope is overlain on the data for the tonstein zircons and shows 31 of the 39 analyses to be homogeneous within measurement uncertainty. Low ^{206}Pb/^{238}U ratos indicate Pb loss from six analysed grains, and two zircons are xenocrysts detectably older than the main group. The age of the sample is given by the weighted mean of the main group of compositions.

either from sedimentary processes or as grains inherited from the source rocks of the felsic magma.

DISCUSSION

The Age Of The Westphalian B Stage

Studies of stratigraphy and basin evolution require accurate numerical ages to derive rates of subsidence, sedimentation and sea-level change, to correlate sedimentary sequences that lack zone fossils, and to relate stratigraphy to igneous and structural events. The reference database for such calculations is the international time scale, a calibration of available radiometric ages with associated biostratigraphy of which several different versions are available (e.g., Odin, 1982; Snelling, 1985; Jones, 1991; Harland and others, 1990). Fundamental to use of the time scale is cross-correlation, the confidence with which ages input from one line of evidence (a fossil zonation; a particular radiometric dating method) can be compared with another type of measurement. The Upper Carboniferous time scale is essentially constrained by the ^{40}Ar/^{39}Ar sanidine ages measured by Hess and Lippolt (1986) in the Silesian coal basins of central Europe. These are based on precise analytical work and well-shaped argon release patterns and are for volcanic horizons with well-established biostratigraphic assignment. However, they are based on a single analytical method which opens the question of their comparability with ages measured by other techniques. SHRIMP zircon dating of one of the original ^{40}Ar/^{39}Ar dating samples offers a test of the degree to which the ^{40}Ar/^{39}Ar ages can withstand criticism on geochronological grounds, and is an

indication of how closely they can be compared with zircon ages being measured in the Carboniferous of Australia.

Details of the ^{40}Ar/^{39}Ar measurements on sanidines in the Z1 tonstein are given by Lippolt and others (1984), and their argon release spectrum is replotted as Figure 16 for comparison with the zircon data. The measurement consists of a relatively small number of argon release steps (five), and this hampers confident recognition of the plateau and significantly outlying ^{40}Ar/^{39}Ar ratios. As described by Lippolt and others (1984), combining all five steps gives a total gas age of 310.2 Ma, but the MSWD of 3.1 (compared with an expected value of 2.37 for 95% confidence with 4 degrees of freedom) indicates dispersion beyond measurement uncertainty. The excess dispersion is contributed by the first two release steps which have slightly low ^{40}Ar/^{39}Ar ratios, a familiar pattern ascribed to minor Ar diffusion losses at low temperatures. A "plateau" of homogeneous compositions is defined by the remaining three steps and these indicate a slightly higher age of 310.7 Ma with an acceptable MSWD; Lippolt and others (1984) interpret this plateau age as recording the crystallisation age of the sanidines.

The zircon ages are derived in a comparable way as the mean of a homogeneous group of isotopic ratios, but uncertainties on the zircon U-Pb age are not calculated in a comparable manner. Errors on ^{40}Ar/^{39}Ar ages are customarily quoted as standard deviations of the population, whereas SHRIMP U-Pb ages are quoted as standard deviations of the weighted mean. Fortunately, recalculating the error of the argon plateau as the standard deviation of the mean has a negligible effect on the net uncertainty of this ^{40}Ar/^{39}Ar age, because the error is dominated by uncertainty associated with the irradiation standards used. The reproducibility of the plateau composition is ±2.4 Ma ($2\sigma_{pop}$) or ±1.4 Ma ($2\sigma_{mean}$), but errors associated with the standards result in a wider net uncertainty of ±8 Ma (2σ). The ±8 Ma (2σ) uncertainty suggested by Hess and Lippolt (1986) is therefore an appropriate measure of the uncertainty on the ^{40}Ar/^{39}Ar age of the Z1 tonstein for the purpose of this comparison. In contrast, the age uncertainty of the zircon standard is a negligible contributor to the error on the SHRIMP U-Pb age because its composition is known within 0.06% (2σ) from mul-

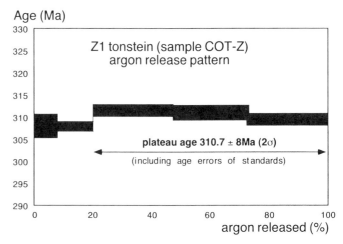

FIG. 16.—Argon release spectrum for sanidine in the Z1 tonstein, plotted from the data presented by Lippolt and others (1984). Compare with the zircon data in Figure 15.

tiple conventional U-Pb measurements. The error of the zircon age is dominated by the reproducibility of the $^{206}Pb/^{238}U$ ratio of the zircons and the concurrent measured reproducibility of the standard.

The age and net external uncertainty of $^{40}Ar/^{39}Ar$ measurement for sanidine in the Z1 tonstein is therefore 310.7 ± 8 Ma (2σ). The equivalent SHRIMP U-Pb measurement for zircons in the same sample is 311.0 ± 3.4 Ma (2σ). There is no detectable bias between these ages at the level of uncertainty of the measurements. The zircon age has a precision of about 1% (2σ), an improvement on the $^{40}Ar/^{39}Ar$ precision of about 2% (2σ), and this serves to confirm the viability of SHRIMP as tool for measuring precise Phanerozoic ages. It should be noted that the accuracy of these ages rests on the ability of both SHRIMP and the $^{40}Ar/^{39}Ar$ step heating method to unravel patterns of alteration and inheritance from the age of primary crystal formation that is being sought; the interpretive power offered by both techniques probably outweighs any gain in measurement precision *per se*. It can be concluded from this experiment that the sanidine $^{40}Ar/^{39}Ar$ and zircon $^{206}Pb/^{238}U$ ages for the Z1 tonstein are mutually supporting, and that SHRIMP zircon U-Pb ages for Australian Carboniferous can be compared directly with the $^{40}Ar/^{39}Ar$ ages measured by Lippolt and others (1984) for the European Stages. The two sets of measurements combine to indicate a mean age of 311.0 ± 3.1 Ma (2σ) for the Z1 tonstein and the uppermost Westphalian B Stage.

The Permian-Carboniferous Reversed Magnetic Superchron

The well-established, magnetically quiet period spanning the Late Carboniferous and Permian Periods is potentially of great significance for stratigraphic correlation because its length and consistency make it one of few magnetic zones likely to be recognized with confidence in the fragmentary pre-Jurassic magnetic record. This reversed period was first recognized by Irving and Parry (1963) and Irving (1966), who named it the Kiaman magnetic interval and defined its base in the Westphalian strata of New South Wales at the reversal between the normal Paterson Volcanics, then dated at 308 Ma, and the reversed Seaham Formation glacigene sediments. As more magnetic data became available from Europe, Russia and North America, frequent reversals were observed extending down into Namurian rocks. These observations resulted in confusion about the position of the Paterson Reversal in other continents; stars in Figure 17 indicate the variety of correlations which have been suggested. Irving and Pullaiah (1976) resolved this difficulty by redefining the Kiaman as the Permian-Carboniferous Reversed (PCR) Magnetic Superchron, and redefining its base at a Namurian reversal in North America. In this scheme, the Paterson Volcanics became one of several brief normal periods within the PCR and lost their significance as the international base of the superchron.

Correlation of the Paterson Volcanics is addressed in this paper with independent SHRIMP and IDTIMS U-Pb dating of zircons in a sample from the Paterson Reversal type locality, and a split of the same sample from which the previous 308 Ma K-Ar age and Westphalian correlation was derived. The four independent measurements group with an MSWD of 0.46, and the age of the Paterson Volcanics given by their mean is 328.5 ± 1.4 Ma (2σ).

Consequences of this revision of the age of the Paterson Volcanics are illustrated in Figure 17. This plots the $^{40}Ar/^{39}Ar$ ages for the European Stages and indicates the conformity of the SHRIMP zircon dating with sanidine $^{40}Ar/^{39}Ar$ dating of the Z1 tonstein. Against this scale is charted the magnetic stratigraphy proposed for America, Europe, and Russia, and the presently recognized base of the PCR in North America. The previous Australian scale correlated the Paterson Volcanics with a variety of Westphalian reversals within the PCR in other continents. The zircon dating moves the Paterson volcanics by 20 Ma from the Westphalian into the Viséan, significantly below the PCR as recognized elsewhere, so their normal polarity must now be seen as coinciding with one of the Early Carboniferous normal periods recorded in Europe and Russia.

The question then arises as to how the overlying glacigene sediments of the Seaham Formation, which are reversed (Irving, 1966), correlate with the magnetic signature of other continents. Regional stratigraphic data suggest a significant disconformity beneath the Seaham Formation in the area sampled by Irving (1966); the top of the Paterson Volcanics has ice scouring, and volcanic and sedimentary components of the lower Seaham Formation appear to be absent. This absence implies that the reversed polarities were measured only in mid-to-upper parts of the Seaham Formation, and that the polarity of lower parts of the formation is yet to be established. Internationally significant biozone correlations are lacking, but a significant time interval above the Viséan Paterson Volcanics is supported by correlation of the *Potonieisporites* Assemblage palynoflora of the upper Seaham Formation with the Westphalian Stage (C. Foster, pers. commun., 1992). Magnetically reversed samples from other formations studied by Irving (1966) also appear to be Late Carboniferous in age; his Currabubula Formation samples came from beneath the Taggarts Mountain Ignimbrite Member (McPhie, 1983) which has hornblende K-Ar ages of 293 ± 4 Ma and 280 ± 4 Ma (Roberts, unpubl. data); and his samples from Rocky Creek are from a pyroclastic unit immediately overlying the Rocky Creek Conglomerate and stratigraphically above tuffs having K-Ar ages of 319 ± 5 Ma and 313 ± 5 Ma (Roberts and others, 1985). All the 'Upper Kuttung' sections measured as magnetically reversed by Irving (1966) are therefore likely to be Westphalian or younger in age.

These observations lead to interpretation that the reversed Seaham Formation, Currabubula Formation and Rocky Creek Conglomerate are significantly younger than the normal Paterson Volcanics and lie within the PCR superchron as is suggested by their uniformly reversed magnetic signature. The Australian record in Figure 17 is constructed on this basis. By correlation to North America, the lower part of the PCR is recorded in Australia by the Yetholme Adamellite, a component pluton of the Bathurst Batholith for which Facer (1976) measured a reversed polarity, a K-Ar biotite age of 318 ± 8 Ma, and a Rb-Sr whole rock isochron age of 318 ± 17 Ma.

Gondwana Glaciation

Gondwanan continents are typified by successions containing Late Palaeozoic sediments of glacigene origin (Crowell and Frakes, 1975; Frakes, 1979; Dickins, 1985; Powell and Veevers, 1987), but there exists a variety of opinion on the nature of sediments ascribed a glacigene origin, the timing, regional ex-

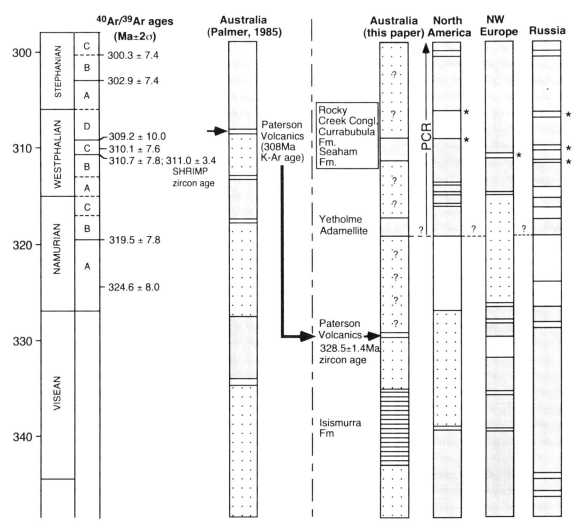

Fig. 17.—Tentative correlations of Carboniferous magnetic pole stratigraphy. Shown against the European stratigraphic scale are the ^{40}Ar/^{39}Ar ages of Hess and Lippolt (1986) and the correspondence of our SHRIMP zircon age with their sanidine age for the uppermost Westphalian B. North American, European, and Russian magnetostratigraphy follows Palmer and others (1985), whose Australian scale was based on a Paterson Volcanics age of 308 Ma; stars indicate correlations of the Paterson Reversal that have been proposed on the basis of this age. Reassignment to the Visean places the Paterson Volcanics significantly below the base of the PCR as recognised in North America. We infer a time gap before deposition of the magnetically reversed portion of the Seaham and Currububula Formations and Irving's (1966) samples at Rocky Creek. A suggested correlation for the base of the PCR is indicated by the dashed line. Grey = reversed polarity; white = normal; stripe = mixed; stipple = insufficient data. Magnetic data for Australia is from Irving (1966), Luck (1973), Facer (1976), Embleton (1977), and Palmer and others (1985).

tent, type, and period of greatest intensity of glaciation. The common view that glaciation was most intense during the latest Carboniferous and early Permian appears to be an oversimplification; Powell and Veevers (1984, 1987) argued for a Namurian onset of glaciation throughout Gondwana and the development of an Australian mid-Carboniferous ice cap, despite only fragmentary evidence for such an early glaciation; Dickins (1985) suggested that alpine glaciation prevailed throughout the late Carboniferous and that there was a more extensive early Permian glaciation; and Gonzalez (1990) provided evidence for a late Carboniferous interglacial period separating a late Visean-Namurian glaciation from a more widespread early Permian glaciation in South America.

Evidence for early commencement of glaciation (pre-Stephanian) exists in both Australia and South America. The Namurian Spion Kop Conglomerate in northwest New South Wales contains a single tillite amongst fluvial sediments (McKelvey and McPhie, 1985) and is tentatively dated by the overlying Ermelo Dacite Tuff with a hornblende K-Ar age of 319 ± 5 Ma (analysis by AMDEL for sample 154-8 of J. Roberts). Late Visean and Namurian glacials and interglacials in the El Paso and Hoyada Verde Formations in the Calingasta-Upsallata Basin of Argentina (Gonzalez, 1990) are correlated on the basis of brachiopods with latest Visean-Namurian zones in eastern Australia (Talboada, 1989). Other South American examples are given by Powell and Veevers (1987).

Correlation of the new zircon age of 328.5 ± 1.4 Ma (2σ) for the Paterson Volcanics to the European isotopic scale of Hess and Lippolt (1986) indicates that the horizon belongs in the Visean Stage, in contrast to the upper Westphalian assignment based on the superseded K-Ar dating. Together with latest Namurian to Westphalian ages for tuffs within the Seaham Formation (Roberts and others, pers. commun.), this correlation now suggests that deposition of glacigene sediments began in the early Namurian in southeastern Australia. The Seaham Formation, originally ascribed a glacigene origin by Sussmilch and David (1920), contains a wide variety of lithologies including volcanogenic lithic sandstone, conglomerate, tuff, varved shale and diamictite, some of which may be tillite. Although many of these sediments appear to be of fluvial or lacustrine origin, striated pavements on the underlying Paterson Volcanics (Osborne and Browne, 1921) and striated and faceted clasts within some of the conglomerates (Dickins, 1985) indicate ice movement and a glacigene origin, respectively. Doubt was cast by Dickins (1985) on the occurrence of terrestrial moraine (i.e., true tillite) and by Coombs (1958) on the origin of some of the varved shales, Coombs describing the latter as graded ash deposits. However, some of the varves contain dropped pebbles and can be used to indicate ice rafting and a cold climate, though not necessarily glaciation. Hence, although the evidence for glaciation is meagre, it is clear that the Seaham Formation was deposited under cold climatic conditions and derived some of its sediments from a glacial source.

The onset of cold conditions and possibly restricted glaciation in early Namurian time can now be seen to correspond with an abrupt drop in diversity of cosmopolitan marine invertebrate faunas at the end of the Viséan age and their replacement by the low diversity, cold water, Gondwanan *Levipustula levis* Zone (Roberts, 1981). The cold water fauna was present throughout the New England Orogen, extending as far north as Broad Sound between Rockhampton and Mackay, but continental glaciation appears to have been confined to the south until the latter part of the Carboniferous Period, the Galilee Basin in Queensland receiving its first glacial sediments in latest Westphalian and Stephanian time (Jones and Truswell, 1992). Although Namurian glaciation may have been more widespread throughout southeastern Australia than envisaged by many authors, the existence of a large mid-Carboniferous ice cap such as that proposed by Powell and Veevers (1984) remains to be confirmed.

Accuracy of SHRIMP $^{206}Pb/^{238}U$ Ages

This study is a direct test of SHRIMP zircon $^{206}Pb/^{238}U$ dating against other methods of dating Phanerozoic rocks. Descriptions above of independent efforts to date the Paterson Volcanics illustrate the different merits of SHRIMP and IDTIMS analysis of Phanerozoic zircons. One point of interest is that neither method has in this case been able to derive useful age information from $^{207}Pb/^{206}Pb$ ratios or test for concordance of the compositions in the usual manner of zircon data interpretation. In the SHRIMP data, this is because the very small probed volumes of zircon do not permit measurement of ^{207}Pb with the precision needed. Better precision is obtained from the larger sample volumes of IDTIMS analysis, but an analytical or compositional source of dispersion has frustrated interpretation of the $^{207}Pb/^{206}Pb$ ratios. This highlights the difficulty of obtaining reliable age information from Phanerozoic $^{207}Pb/^{206}Pb$ ratios, where very small deviations in measured composition result in large differences in apparent age. In contrast, the high angle interception of the common Pb mixing projection with the Tera-Wasserburg (1974) Concordia is insensitive to such uncertain-

ties and yields reliably measured $^{206}Pb/^{238}U$ ages. Testing for Pb loss then depends on detecting a dispersion of $^{206}Pb/^{238}U$ ratios: of all the grains analysed in the Paterson Volcanics, only a single zircon probed in mount Z765 detectably departs from the group of $^{207}Pb/^{238}U$ compositions, and the absence of Pb loss is indicated by the agreement of all other SHRIMP and IDTIMS $^{206}Pb/^{238}U$ measurements about a single composition. For the Paterson Volcanics, the IDTIMS age of 327.7 ± 2.0 Ma (2σ), which includes zircons previously probed by SHRIMP, cannot be distinguished from the SHRIMP I ages of 328.7 ± 4.0 Ma (2σ) for the same sample and 328.9 ± 4.2 Ma (2σ) for sample Z850 or from SHRIMP II measurement of 330.1 ± 3.0 Ma (2σ) for Z850. This revision of the age of the Paterson Volcanics serves as a reminder of the caution with which Palaeozoic K-Ar ages should be treated.

Benefits of within-grain ion probe analysis at the 20–30 μm scale include the ability to target inclusion-free areas of zircons and date individual growth zones separately. Equally important is the ability to acquire rapidly a large body of data from which patterns of Pb loss and/or inheritance can be construed with statistical confidence. The large quantity of data obtained with SHRIMP is then interpreted in a manner analogous to the study of argon-release spectra in $^{40}Ar/^{39}Ar$ dating: Pb loss can be recognized from deviations in composition from a 'plateau' of data defining the principal $^{206}Pb/^{238}U$ age population. The comparison of $^{40}Ar/^{39}Ar$ and SHRIMP U-Pb dating of the Z1 tonstein illustrates this similarity of interpretive method and detects no difference between the SHRIMP zircon U-Pb age and the sanidine $^{40}Ar/^{39}Ar$ age for the same sample.

The Carboniferous SHRIMP ages described here have precision comparable with that of the alternative analytical methods, and this confirms that SHRIMP can be used with confidence for dating and correlation in Phanerozoic sedimentary basins.

ADDENDUM

Two published SHRIMP zircon ages normalized to the superseded quadratic calibration are now markers in the international Phanerozoic time scale. Renormalisation to the new calibration refines these ages slightly.

The Age of the Devonian/Carboniferous Boundary

The Devonian/Carboniferous boundary age has been constrained by SHRIMP zircon dating of the Bed 79 bentonite in the Hasselbachtal auxiliary global stratotype for the boundary in Germany (Claoué-Long and others, 1992). The bentonite is a few centimetres above the biostratigraphic boundary and lies within the lower half of the *Siphonodella sulcata* Zone whose base defines the boundary. The published age of the Bed 79 bentonite, based on the quadratic calibration of $^{206}Pb/^{238}U$, is 353.2 ± 4.0 Ma (2σ); renormalisation to the power law calibration refines this age to 353.7 ± 4.2 Ma (2σ).

The Age of the Permian/Triassic Boundary

The Permian/Triassic boundary age has been constrained by SHRIMP zircon dating of the boundary clay bed in the Changxing stratotype in China, a candidate section for the global stratotype (Claoué-Long and others, 1991). The clay is a bentonite and is the bed marking the Permian/Triassic boundary in the section. The published age of the Changxing boundary clay, based on the quadratic calibration, is 251.2 ± 3.4 Ma (2σ); renormalisation to the power law calibration refines this age to 251.1 ± 3.6 Ma (2σ).

ACKNOWLEDGMENTS

We are indebted to E. Paproth and K. Burger for their assistance in obtaining European Carboniferous samples, and to J. Hess and H. Lippolt who made time to discuss the comparison of zircon and sanidine dating. S. Maxwell and D. Maidment assisted with collecting SHRIMP data, and we thank members of the SHRIMP group who contributed SL13 measurements to the study. I. S. Williams, R. D. Page, L. P. Black, J. J. Foster, and the late J. C. Roddick kindly gave permission for their unpublished thermal ionisation analyses of standard zircon SL13 to be discussed here. L. P. Black, P. J. Jones and four SEPM reviewers suggested improvements to an earlier draft of the manuscript. This work was supported by an Australian Research Council grant to J. R. and a Bureau of Mineral Resources Fellowship to J. C. L., who conducted the work as a Visiting Fellow of the Research School of Earth Sciences, Australian National University, and publishes with the permission of the Executive Director of the Australian Geological Survey Organisation.

REFERENCES

ARCHBOLD, N. W. AND DICKINS, J. M., 1991, Australian Phanerozoic time scales 6, Permian: Canberra, Australian Bureau of Mineral Resources Record 36, 17 p.

BRIGGS, D. AND ARCHBOLD, N., 1990, Late Carboniferous-Early Permian, Cranky Corner Outlier, northern Sydney Basin: Newsletter on Carboniferous Stratigraphy, v. 8, p. 17–18.

CLAOUE-LONG, J. C., ZHANG ZICHAO, MA GUOGAN, AND DU SHAOHUA, 1991, The age of the Permian-Triassic boundary: Earth and Planetary Science Letters, v. 105, p. 182–90.

CLAOUE-LONG, J. C., JONES, P. J., ROBERTS, J., AND MAXWELL, S., 1992, The numerical age of the Devonian-Carboniferous boundary: Geological Magazine, v. 129, p. 281–91.

COMPSTON, W., WILLIAMS, I. S., AND MEYER, C., 1984, U-Pb geochronology of zircons from lunar breccia 73217 using a sensitive high mass-resolution ion microprobe: Journal of Geophysical Research, v. 89, p. B525–534.

COOMBS, D. S., 1958,. Zeolitised tuffs from the Kuttung glacial beds near Seaham, New South Wales: Australian Journal of Earth Science, v. 21, p. 18–19.

CROWELL, J. C. AND FRAKES, L. A., 1975, The Late Palaeozoic glaciation, in Campbell, K. S. W., ed., Gondwana Geology: Canberra, Australian National University Press, p. 313–331.

CUMMING, G. L. AND RICHARDS, J. R., 1975, Ore lead isotope ratios in a continually changing Earth: Earth and Planetary Science Letters, v. 28, p. 155–71.

DICKINS, J. M., 1985, Late Palaeozoic glaciation: BMR Journal of Australian Geology and Geophysics, v. 9, p. 163–169.

EMBLETON, B. J. J., 1977, Discussion: Palaeomagnetism, radiometric age and geochemistry of an adamellite at Yetholme, N.S.W.: Journal of the Geological Society of Australia, v. 24, p. 121–123.

FACER, R. A., 1976, Palaeomagnetism, radiometric age and geochemistry of an adamellite at Yetholme, N.S.W.: Journal of the Geological Society of Australia, v. 23, p. 243–248.

FANNING, C. M., FLINT, R. B., PARKER, A. J., LUDWIG, K. R., AND BLISSETT, A. H., 1988, Refined Proterozoic tectonic evolution of the Gawler Craton, South Australia through U-Pb zircon geochronology: Precambrian Research, v. 40/41, p. 363–386.

FANNING, C. M., WILLIAMS, I. S., MCCULLOCH, M. T., AND COMPSTON, W., 1991, A comparison of U-Pb techniques through geologic time: is there a best method?: GAC-MAC Program with Abstracts, v. 16, p. A35.

FRAKES, L. A., 1979, Climates Throughout Geologic Time: Amsterdam, Elsevier, 310 p.

GONZALEZ, C. R., 1990, Development of the Late Paleozoic glaciations of the South American Gondwana in western Argentina: Palaeogeography, Palaeoclimatology, Palaeoecology, v. 79, p. 275–287.

HARLAND, W. B., ARMSTRONG, R. L., COX, A. V., CRAIG, L. E., SMITH, A. G., AND SMITH, D. G., 1990, A Geologic Time Scale 1989: Cambridge, Cambridge University Press, 263 p.

HELBY, R. J., 1969, Preliminary palynological study of Kuttung sediments in central eastern New South Wales: Geological Survey of New South Wales Records, v. 11, p. 5–14.

HESS, J. C. AND LIPPOLT, H. J., 1986, $^{40}Ar/^{39}Ar$ ages of tonstein and tuff sanidines: new calibration points for the improvement of the Upper Carboniferous time scale: Chemical Geology (Isotope Geoscience Section), v. 59, p. 143–54.

IRVING, E., 1963, Paleomagnetism of the Narrabeen Chocolate Shales and the Tasmanian Dolerite: Journal of Geophysical Research, v. 68, p. 2283–87.

IRVING, E., 1966, Paleomagnetism of some Carboniferous rocks from New South Wales and its relation to geological events: Journal of Geophysical Research, v. 71, p. 6025–6051.

IRVING, E. AND PARRY, L. G., 1963, The magnetism of some Permian rocks from New South Wales: Geophysics Journal of the Royal Astronomical Society, v. 7, p. 395–411.

IRVING, E. AND PULLAIAH, G., 1976, Reversals of the geomagnetic field, magnetostratigraphy, and relative magnitude of paleosecular variation in the Phanerozoic: Earth Science Reviews, v. 12, p. 35–64.

JONES, P. J., 1991, Australian Phanerozoic Time scales: 5. Carboniferous. Biostratigraphic charts and explanatory notes: Canberra, Australian Bureau of Mineral Resources Record 1989/35, 43 p.

JONES, P. J., CAMPBELL, K. S. W., AND ROBERTS, J., 1973, Correlation chart for the Carboniferous System of Australia: Australian Bureau of Mineral Resources, Bulletin 156A, 40 p.

JONES, M. J. AND TRUSWELL, E. M., 1992, Late Carboniferous and Early Permian palynostratigraphy of the Joe Joe Group, southern Galilee Basin, Queensland, and implications for Gondwanan stratigraphy: BMR Journal of Australian Geology and Geophysics, v. 13, p. 143–185.

KEMP, E. M., BALME, B. E., HELBY, R. J., KYLE, R. A., PLAYFORD, G., AND PRICE, P. L., 1977, Carboniferous and Permian palynostratigraphy in Australia and Antarctica: a review: BMR Journal of Australian Geology and Geophysics, v. 2, p. 177–208.

KROGH T. E., 1978, Vapour transfer for the dissolution of zircons in a multisample capsule at high pressure: Washington, D.C., United States Geological Survey Open File Report 78–701, p. 233–234.

KROGH T. E., 1982, Improved accuracy of U-Pb zircon ages by the creation of more concordant systems using an air abrasion technique: Geochimica et Cosmochimica Acta, v. 46, p. 637–649.

LIPPOLT, H. J., HESS, J. C. AND BURGER, K., 1984, Isotopische Alter von pyroklastischen Sanidinen aus Kaolin-Kohlentonsteinen als Korrelationsmarken fur das mitteleuropaische Oberkarbon: Fortschreifer Geologisches Rheinland und Westfalen, v. 32, p. 119–50.

LUCK, G. R., 1973, Palaeomagnetic results from Palaeozoic rocks of southeast Australia: Geophysical Journal of the Royal Australian Society v. 32, p. 35–52.

LUDWIG K. R., 1980, Calculation of uncertainties of U-Pb isotope data: Earth and Planetary Science Letters v. 46, p. 212–220.

MCELHINNY, M. W., 1973, Palaeomagnetism and Plate Tectonics. Cambridge University Press, Cambridge, 358 p.

MCKELVEY, B. C. AND MCPHIE, J., 1985, Tamworth Belt, in Diaz, C. M., ed., The Carboniferous of the World, II: Australia, Indian Subcontinent, South Africa, South America and North Africa: IUGS Publication 20, Madrid, Instituto Geologicol y Minero de Espana and Empressa Nacional adaro de Investigaciones Mineras SA, p. 15–23.

MCPHIE, J., 1983, Outflow ignimbrite sheets from Late Carboniferous calderas, Currabubula Formation, New South Wales, Australia: Geological Magazine, v. 120, p. 487–503.

ODIN, G. S., ed., 1982, Numerical Dating in Stratigraphy: Chichester, Wiley-Interscience (2 vols), 1040 p.

OSBORNE, G. D. AND BROWNE, W. R., 1921, Note on the glacially striated pavement in the Kuttung Series of the Maitland District: Proceedings of the Linean Society of New South Wales, v. 46, p. 259–262.

PALMER, J. A., PERRY, S. P. G., AND TARLING, D. H., 1985, Carboniferous magnetostratigraphy: Journal of the Geological Society of London, v. 142, p. 945–955.

POWELL, C. MCA. AND VEEVERS, J. J., 1984, Termination of the Uluru Regime: the mid-Carboniferous lacuna, *in* Veevers, J. J., ed., Phanerozoic Earth History of Australia: Oxford, Clarendon Press, p. 348–350.

POWELL, C. MCA. AND VEEVERS, J. J., 1987, Namurian uplift in Australia and South America triggered the main Gondwanan glaciation: Nature, v. 326, p. 177–179.

RATTIGAN, J. H., 1967, The Balickera section of the Carboniferous Kuttung facies, New south Wales: Journal and Proceedings of the Royal Society of New South Wales, v. 100, p. 75–84.

ROBERTS, J., 1981, Control mechanisms of Carboniferous brachiopod zones in eastern Australaia: Lethaia, v. 14, p. 123–134.

ROBERTS, J., CAMPBELL, K. S. W., CLARKE, M. J., ENGEL, B. A., JELL, J. S., JENKINS, T. B. H., MCKELVEY, B. C., MCPHIE, J., MARSDEN, M. A. H., MORRIS, L. N., NICOLL, R., OVERSBY, B. S., AND PLAYFORD, G., 1985, Australia, South Africa, South America and North Africa: Madrid, Instituto y Minero de Espana and Empressa Nacional adaro de Investigaciones Mineras SA, IUGS Publication 20, p. 9–145.

ROBERTS, J., ENGEL, B., AND CHAPMAN, J., 1991, Geology of the Camberwell, Dungog and Bulahdelah 1:100,000 Sheets 9133, 9233, 9333: Sydney, New South Wales Geological Survey, 382 p.

SNELLING, N. J., ed., 1985, The Chronology of the Geological Record: London, Geological Society of London Memoir 10, 343 p.

STEIGER, R. H. AND JAEGER, E., 1977, Subcommission of geochronology: convention on the use of decay constants in geo- and cosmochronology: Earth and Planetary Science Letters, v. 36, p. 359–62.

SUSSMILCH, C. A. AND DAVID, T. W. E., 1920, Sequence, glaciation and correlation of the Hunter River district, New South Wales: Journal and Proceedings of the Royal Society of New South Wales, v. 53, p. 310–322.

TALBOADA, A. C., 1989, La fauna de la Formacion El Paso, Carbonifero Inferior de la Precordillera Sanjuanina: Acta Geologica Lilloana, v. 17, p. 113–129.

TERA, F. AND WASSERBURG, G. J., 1974, U-Th-Pb systematics on lunar rocks and inferences about lunar evolution and the age of the moon: Proceedings of the Fifth Lunar Conference (Supplement 5, Geochemica et Cosmochimica Acta), v. 2, p. 1571–99.

WETHERILL, G. W., 1956, Discordant uranium-lead ages: Transactions of the American Geophysical Union, v. 37, p. 320–26.

WILLIAMS, I. S. AND CLAESSON, S., 1987, Isotopic evidence for the Precambrian provenance and Caledonian metamorphism of high grade paragneisses from the Seve Nappes, Scandinavian Caledonides: Contributions to Mineralogy and Petrology, v. 97, p. 205–17.

AUSTRALIAN EARLY CARBONIFEROUS TIME

JOHN ROBERTS

Department of Applied Geology, University of New South Wales, Kensington, NSW 2033, Australia.

AND

JONATHAN C. CLAOUÉ-LONG AND PETER J. JONES

Australian Geological Survey Organisation, GPO Box 378, Canberra, ACT 2601, Australia.

ABSTRACT: Zircon U-Pb ages measured with the SHRIMP ion microprobe have been used to date volcanic horizons associated with Early Carboniferous sediments of the Southern New England Orogen in eastern Australia. These results calibrate the numerical ages of eastern Australian biozones, including the four Early Carboniferous brachiopod zones between the *Schellwienella burlingtonensis* and *Rhipidomella fortimuscula* Zones, which include cosmopolitan faunas and so also constrain ages for European and international substages of the Carboniferous System. Refinements to the Early Carboniferous timescale include the ages of the Devonian-Carboniferous boundary, the Tournaisian-Viséan boundary, and correlation with the Holkerian Stage of Britain. New biostratigraphic correlations and ages constrain the associations of the *Granulatisporites frustulensis* Microflora and *Grandispora maculosa* Assemblage and indicate that the *Nothorhacopteris* flora, previously thought to be confined to Late Carboniferous units, ranges down into the upper Viséan Series (V3a). The zircon ages delimit the durations of the major volcanic events within the Southern New England Orogen and the time significance of depositional hiatuses within the stratigraphic succession, and revise correlations and the associated Viséan palaeogeography of eastern Australia.

INTRODUCTION

A major development in the study of Carboniferous processes was the publication by Hess and Lippolt (1986) of accurately measured $^{40}Ar/^{39}Ar$ ages for the Late Carboniferous Series of Europe. The isotopic and biostratigraphic accuracy of these new data effectively superseded the previous Carboniferous time scale assessments of Harland and others (1982), De Souza (1982), and Forster and Warrington (1985), which were based on less precise constraints. As a resource with which to calculate rates of Upper Carboniferous processes, the importance of this $^{40}Ar/^{39}Ar$ age control cannot be over-emphasised. Carboniferous time is less well constrained below the Namurian Series; except for zircon dating of the base of the System by Claoué-Long and others (1992), no new data have been published since the authoritative review by Forster and Warrington (1985) emphasised the dependence of the Dinantian time scale on K-Ar ages for intrusions and continental volcanics having uncertain correlation with the type marine stratigraphy.

In an effort to refine the Early Carboniferous time scale and to gain time control of local Australian stratigraphic processes, zircon U-Pb dating by SHRIMP ion microprobe has been applied to tuffs, ashes, and ignimbrite flows interlayered with Early Carboniferous sediments of the Southern New England Orogen (SNEO) in eastern Australia. The SNEO (Fig. 1) is part of an ancient fold mountain belt with a depositional and tectonic history extending from the Silurian to the Late Triassic Periods. Early Carboniferous processes within the orogen were controlled by a subduction zone that dipped westwards beneath the Australian craton (Aitchison and others, 1992) and generated a subduction complex, fore arc basin, and dacitic to andesitic volcanic arc. Stratigraphically coherent Early Carboniferous sequences within the fore arc basin, also termed the Tamworth Belt (Fig. 1), contain intertonguing marine and non-marine successions with numerous laterally extensive felsic volcanic eruptions (Roberts and Oversby, 1974; Roberts and others, 1991). The marine sediments contain warm water cosmopolitan faunas including conodonts, ammonoids, foraminifera, algae, brachiopods, and trilobites, which can be correlated with forms in the northern hemisphere. The international importance of these Australian biozones has significantly increased with the recent description of Viséan conodonts from eastern Australia by Jen-

kins and others (1993), compilation of detailed international correlation charts by Jones (1991), and adjustment of the correlation of biozones by Roberts and others (1993b) to accommodate changes to ammonoid zones in Europe. Felsic volcanics within this cosmopolitan faunal sequence are therefore well suited to constraining the ages of both local and international biostratigraphic zones.

The Late Carboniferous succession of the SNEO is dominated by cold climate conditions and endemic faunas, and can only be correlated faunally with other Gondwanan terrains. Zircon dates from this part of the Australian column enable Gondwanan stratigraphy to be placed in an international context by correlation with the European Upper Carboniferous $^{40}Ar/^{39}Ar$ ages of Hess and Lippolt (1986); these will be discussed elsewhere.

Zircon analytical results and associated biostratigraphy in the SNEO are discussed in full in this contribution, which supersedes an interim account of the work given at the 1991 International Congress on Carboniferous-Permian (Roberts and others, 1993a).

THE SOUTHERN NEW ENGLAND OROGEN

Detailed mapping in the Southern New England Orogen (SNEO) of Australia by Roberts and Engel (1987) and Roberts and others (1991) has demonstrated that the Early Carboniferous succession of the southern part of the Tamworth Belt reflects four main influences: gradual progradation of continental over marine environments; interruption of this progression by a series of three marine transgressions; structural control of deposition by active faults; and the asynchronous activity of three different volcanic centres. These volcanic centres provided the major source for sediments deposited in three regions bounded by syn-depositional faults within the fore arc basin: the Rouchel, Gresford and Myall Blocks whose outlines are mapped in Figure 1. The development of relationships over time between these blocks and between the marine and continental successions, is portrayed in Figure 2 that also highlights the volcanic units dated in this contribution. Each of the blocks has its own characteristic stratigraphic succession, but some ignimbrite eruptions and other units are laterally extensive and present across the blocks. Accumulation of erupted volcanic material

Geochronology Time Scales and Global Stratigraphic Correlation, SEPM Special Publication No. 54

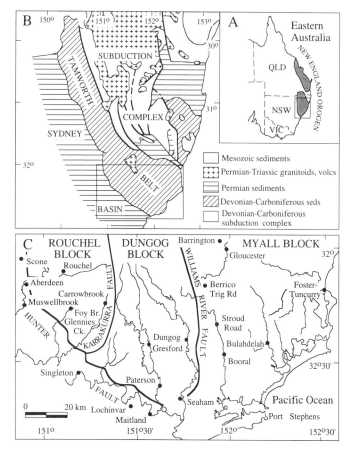

FIG. 1.—Maps showing locations referred to in this study. (A) location of the Southern New England Orogen (SNEO) in eastern Australia. (B) Major tectonic elements of the SNEO (adapted from Collins and others, 1993)—the study area lies in Devonian and Carboniferous sediments in the south of the Tamworth Belt. (C) Geography of the area outlined in (B) showing place names referred to in the text and the division of the area into major Blocks by fault systems.

and volcaniclastic detritus around the western and southern margins of the Tamworth Belt was responsible for the north-easterly progradation of regressive, continental sediments and volcanics, and a gradual displacement of both shallow and deeper marine shelf areas of deposition throughout Early Carboniferous time. Brief but extensive marine transgressions, the first in the early to late Tournaisian, the second in the early Viséan, and the third in the latest Viséan Epoch, are numbered 1–3 in Figure 2.

The Early Carboniferous biostratigraphic zonation of eastern Australia is presented in Figure 3. The principal control is provided by brachiopod zones with supporting information from conodonts and macro- and palynoflora (see also Fig. 12). Correlations of these zones to the standard European zonations, stages and series follow Roberts and others (1993b; Fig. 2) with slight modifications in the upper Viséan Series (V3) to accommodate both the chronostratigraphic implications of Riley's (1993) Dinantian classification and zonation of Britain, and the *Goniocyclus-Protocanites* Zone of Kullmann and others (1990) for the middle Tournaisian (Tn2) part of the ammonoid sequence of western Europe. The relationship to the stratigraphy

of volcanic horizons chosen for zircon dating is indicated by the stars in Figure 2. Samples were obtained from measured stratigraphic sections described in detail by Roberts and Oversby (1974), Roberts and others (1991), or as otherwise specified in the text. Biostratigraphic data constraining the location of each sample are outlined below, together with an appraisal of the quality of isotopic age information obtained for each sample.

SHRIMP ZIRCON DATING

Attempts have been made previously to date volcanics in the SNEO, using the K-Ar method of analysis, but with inconclusive results. For example, Claoué-Long and others (this volume) show that the K-Ar apparent age of the Paterson Volcanics is 20 Ma too young, a result attributable to unresolved loss of Ar. It is possible that $^{40}Ar/^{39}Ar$ analysis would yield more reliable age information by separating Ar loss components from compositions retaining the ages of samples, but rocks in the area have been subject to low grade metamorphism and deep surface weathering, and few samples preserve intact the phases (such as sanidine feldspar) suited to Ar stepheating study. The mineral zircon, on the other hand, is a ubiquitous component of volcanic rocks in the SNEO and has survived alteration processes; frequently it is the only igneous phase to have survived alteration of volcanic ashes to bentonite. The recent development of zircon U-Pb dating by ion microprobe, by the ability to probe and date separately the different altered, inherited, and magmatic zones within single crystals, has opened zircon dating to the same interpretation methods offered by $^{40}Ar/^{39}Ar$ stepheating analysis. SHRIMP zircon analysis has therefore been applied to the problem of dating volcanic horizons in the SNEO. Analytical and calibration procedures follow those documented in detail by Claoué-Long and others (this volume); the following section discusses the zircon age patterns obtained in order of the stratigraphic position of the samples.

Kingsfield Formation

The Kingsfield Formation is at the base of the Carboniferous succession in the Rouchel Block near Rouchel (Fig. 1) and contains felsic volcanics and volcaniclastics interlayered in shallow and deep water marine strata. Claoué-Long and others (1992) describe SHRIMP zircon dating of a tuff from this unit, giving an age of 355.8 ± 5.6 Ma. The tuff is bracketed by the conodont *Siphonodella sulcata* and lies 180 m stratigraphically below limestone containing the lower *crenulata* Zone and brachiopods tentatively referred to the *Spirifer sol* Zone, inferring a location in either the basal Carboniferous *S. sulcata* Zone or the Lower *S. duplicata* Zone. Both the biostratigraphic zonation and numerical age are indistinguishable from the position of bentonite Bed 79 in the auxiliary global stratotype through the Devonian/Carboniferous boundary at Hasselbachtal in Germany, which has been used to define the international reference age of the Devonian/Carboniferous boundary at 353.7 ± 4.2 Ma (Claoué-Long and others, 1992; renormalized by Claoué-Long and others, this volume).

Waverley Formation

The age of an andesite in the Waverley Formation at Foy Brook has appeared as a component of the IUGS Carboniferous

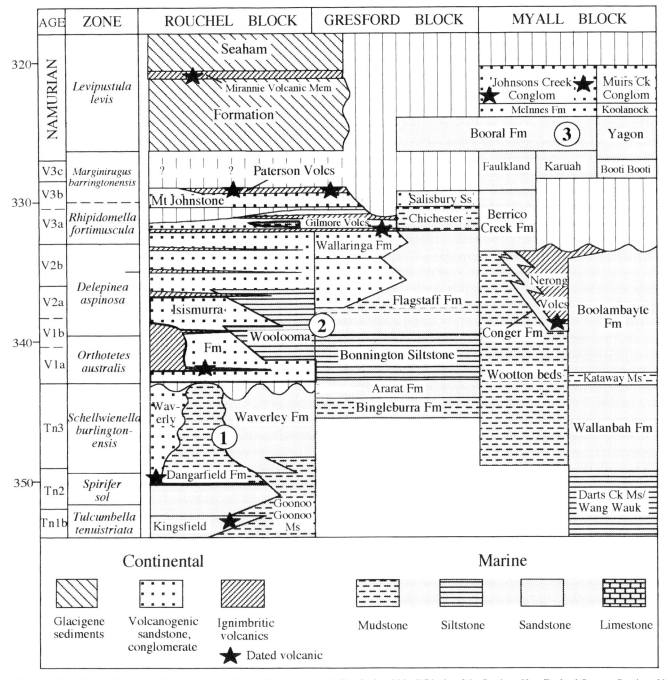

FIG. 2.—Early Carboniferous stratigraphic relationships within the Rouchel, Gresford and Myall Blocks of the Southern New England Orogen. Stratigraphic relationships are based on those of Roberts and others (1993a) adjusted for the biostratigraphic correlations revised by Roberts and others (1993b) and numerical ages presented in this paper. Numbers in circles refer to the three main marine transgressions (grey stippling, from right) into the terrestrial environment of fluvial and volcanogenic sediments (left, open stippling). Vertical lines indicate hiatus: broken, possible hiatus; solid, demonstrable hiatus. Tournaisian and Viséan marine stratigraphy is controlled mainly by brachiopod zonation. Onset of glacigene sedimentation in the Namurian (Seaham Formation) is matched in the marine sequence by low diversity cold water faunas (*Levipustula levis* Zone). Stars indicate volcanic horizons dated by SHRIMP zircon analysis.

timescale (Cowie and Bassett, 1989). This datum was based on a then-unpublished K-Ar plagioclase age of 341 ± 5 Ma and discussion of the stratigraphy by Jones (1988) suggesting a position for the andesite near the Tournaisian/Viséan boundary. The K-Ar data have subsequently been reported by Roberts and others (1991). We revisit both the age and biostratigraphy of

the Foy Brook andesite in this contribution and conclude that it does not merit inclusion as a constraint in the international Phanerozoic timescale because its biostratigraphic affinities are equivocal.

The unnamed andesite is within non-marine sediments of the Waverley Formation on the western flank of the Albano Syn-

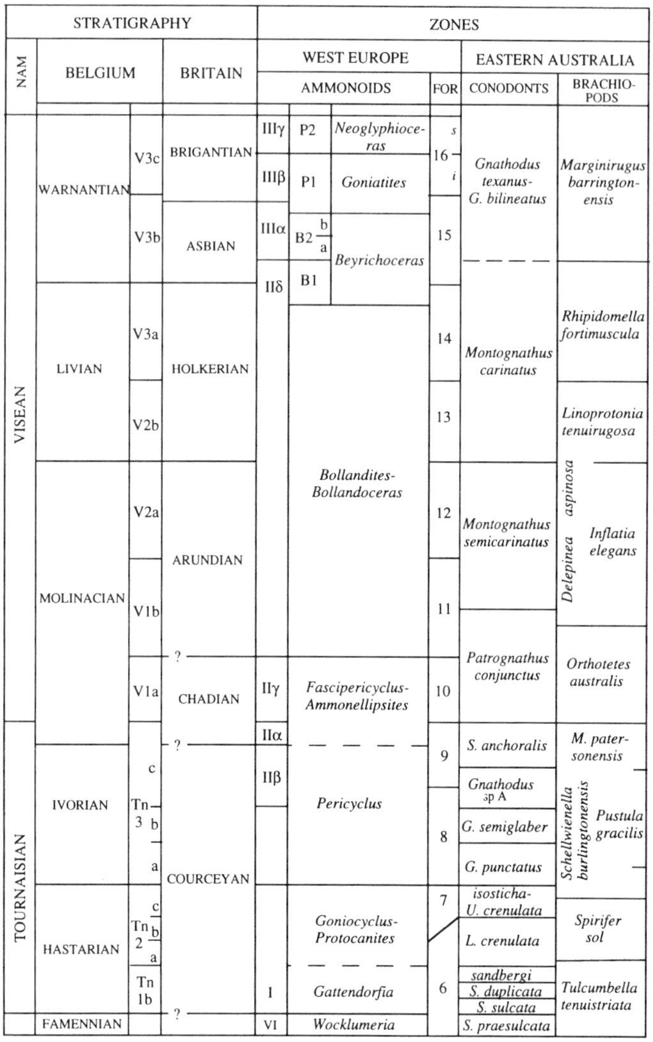

STRATIGRAPHY			ZONES					
NAM	BELGIUM	BRITAIN	WEST EUROPE				EASTERN AUSTRALIA	
			AMMONOIDS			FOR	CONODONTS	BRACHIOPODS
VISEAN / WARNANTIAN	V3c	BRIGANTIAN	IIIγ	P2	Neoglyphioceras	s / 16 / i	Gnathodus texanus-G. bilineatus	Marginirugus barringtonensis
			IIIβ	P1	Goniatites			
	V3b	ASBIAN	IIIα	B2 b/a	Beyrichoceras	15		
			IIδ	B1			Montognathus carinatus	Rhipidomella fortimuscula
VISEAN / LIVIAN	V3a	HOLKERIAN				14		
	V2b					13		Linoprotonia tenuirugosa
VISEAN / MOLINACIAN	V2a	ARUNDIAN			Bollandites-Bollandoceras	12	Montognathus semicarinatus	Inflatia elegans / Delepinea aspinosa
	V1b					11		
	V1a	CHADIAN	IIγ		Fascipericyclus-Ammonellipsites	10	Patrognathus conjunctus	Orthotetes australis
TOURNAISIAN		?	IIα			9	S. anchoralis	M. patersonensis
			IIβ		Pericyclus		Gnathodus sp A	
	IVORIAN Tn3 b	COURCEYAN				8	G. semiglaber	Pustula gracilis / Schellwienella burlingtonensis
	a						G. punctatus	
	HASTARIAN Tn2 c/b/a				Goniocyclus-Protocanites	7	isosticha-U. crenulata / L. crenulata	Spirifer sol
	Tn1b	?	I		Gattendorfia	6	sandbergi S. duplicata S. sulcata	Tulcumbella tenuistriata
FAMENNIAN			VI		Wocklumeria		S. praesulcata	

FIG. 3.—Correlation of Early Carboniferous biostratigraphic units in the Southern New England Orogen with those of Western Europe. Adapted from Roberts and others (1993).

cline northwest of Foy Brook (Roberts and others, 1991), where the sequence is overlain, probably disconformably, by the Native Dog Ignimbrite Member of the Isismurra Formation (Roberts and Oversby, 1974). Fossils are lacking in adjacent sediments, and this frustrates direct biostratigraphic assignment. Twelve kilometres to the northwest, the Native Dog Ignimbrite overlies similar Waverley Formation non-marine sediments, which here interfinger with marine sediments (Dangarfield Formation) containing brachiopods of the *Schellwienella burlingtonensis* Zone and conodonts of the *Gnathodus punctatus* and *G. semiglaber* Zones of Jenkins (1974). The andesite is not present in the northern exposures of the Waverley Formation and there is a variable amount of erosion of the top of the Waverley Formation below the Native Dog Ignimbrite and Isismurra Formation, so correlation of these zonations to the andesite is uncertain. Biostratigraphic constraints on the andesite are therefore also uncertain.

Zircon data from the andesite at Foy Brook are listed in Table 1 and plotted without correction for common Pb in a Tera and Wasserburg (1974) Concordia diagram in Figure 4. The geometry of Phanerozoic zircon compositions in this type of diagram is discussed by Claoué-Long and others (this volume). The zircon compositions group close to Concordia, indicating low contents of common Pb, with a single exception having a higher common Pb content within error of the common Pb mixing vector. Without further processing, it can be seen from these raw data that a homogeneous age population is indicated within analytical error.

This homogeneity is tested in Figure 5, where the ^{206}Pb/^{238}U ages of the zircons are plotted in a probability diagram. The x-axis of this diagram is scaled as probability, which has the effect of normalizing data to the statistically ideal Normal Distribution that appears on the diagram as a straight line of slope proportional to the standard deviation of the distribution. The slope plotted is that of the reproducibility of measurements of the standard zircon obtained concurrently with the Foy Brook zircons; the line, therefore, forms the reference for the normal statistical dispersion expected in replicate measurements of a homogenous composition. The age spectrum of the Foy Brook andesite zircons is indistinguishable from the Normal Distribution. The mathematical counterpart to this graphical matching of data to the Normal Distribution is a Mean Square of Weighted Deviates (MSWD) of 1.06, well within the figure of 1.44 that indicates acceptable homogeneity at 95% confidence with 30 degrees of freedom. The data therefore indicate a single age population, which is the main defining feature of zircons crystallised together following cooling of an erupted magma. The weighted mean radiogenic ^{206}Pb/^{238}U ratio of the population is 0.05595 ± .00064 (2σ), equivalent to an age of 350.9 ± 4.0 Ma (2σ), and this is interpreted with confidence as the crystallization age of the andesite.

Curra Keith Tongue

The Curra Keith Tongue is a stratigraphically extensive ignimbrite unit forming a projection from the lower part of the Native Dog Ignimbrite Member of the Isismurra Formation; it was sampled for zircon dating in the Rouchel Block (Roberts and Oversby, 1974). In the vicinity of the collecting site, the Curra Keith Tongue lies above brachiopods of the upper *S. burlingtonensis* Zone in the Waverley Formation and below brachiopods of the *Inflatia elegans* Subzone, foraminifera and algae of the Mamet and Skipp (1970) Zones 11 and 12, and conodonts of the *Montognathus semicarinatus* Zone of Jenkins and others (1993), which are in the overlying Woolooma Formation (section 6 of Roberts and Oversby, 1974). At its easternmost extent at Malumla, the Curra Keith Tongue lies above the Waverley Formation with upper *S. burlingtonensis* Zone brachiopods and below brachiopods of the *Orthotetes australis* Zone in the Woolooma Formation. The Curra Keith Tongue itself does not extend into the Gresford Block, but its stratigraphic correlatives are located in the upper part of the Ararat Formation above the upper *S. burlingtonensis* and *Scaliognathus anchoralis* Zones and below the *Orthotetes australis* Zone and conodont zone *Patrognathus conjunctus* of Jenkins and others (1993) in the basal Bonnington Siltstone. Roberts and others (1993b) demonstrate that this stratigraphic position is equivalent to the boundary between the Tournaisian and Viséan Series of Europe.

Compositions measured for zircons in the Curra Keith Tongue are listed in Table 2, and the ages are plotted in a prob-

TABLE 1.— SHRIMP U-Pb ISOTOPIC DATA FOR THE FOY BROOK ANDESITE

Grain area	U (ppm)	Th (ppm)	Th/U	^{204}Pb* (ppb)	f^{206}Pb*† (%)	Calibrated total Pb compositions			Radiogenic compositions*	
						^{206}Pb/^{238}U ± 1σ	^{207}Pb/^{235}U ± 1σ	^{207}Pb/^{206}Pb ± 1σ	^{206}Pb/^{238}U ± 1σ	Apparent age# (Ma) ± 1σ
1.1	153	145	.95	7	1.78	0.0571 ± 11	0.532 ± 17	0.0676 ± 15	0.0561 ± 11	352 ± 7
2.1	562	705	1.25	6	.39	0.0561 ± 11	0.436 ± 11	0.0564 ± 7	0.0559 ± 11	350 ± 7
2.2	204	141	.69	6	1.04	0.0559 ± 11	0.475 ± 16	0.0616 ± 16	0.0553 ± 11	347 ± 7
3.1	248	259	1.04	5	.76	0.0549 ± 11	0.450 ± 13	0.0594 ± 11	0.0545 ± 11	342 ± 7
3.2	201	199	.99	5	.88	0.0574 ± 11	0.477 ± 16	0.0603 ± 15	0.0569 ± 11	356 ± 7
3.3	144	134	.92	5	1.21	0.0584 ± 12	0.508 ± 19	0.0630 ± 18	0.0577 ± 12	362 ± 7
4.1	145	84	.58	4	1.13	0.0564 ± 11	0.484 ± 16	0.0623 ± 15	0.0557 ± 11	350 ± 7
5.1	270	326	1.21	5	.73	0.0553 ± 11	0.450 ± 13	0.0591 ± 11	0.0548 ± 11	344 ± 7
5.2	161	139	.86	4	.90	0.0548 ± 11	0.457 ± 17	0.0605 ± 17	0.0543 ± 11	341 ± 7
6.1	202	198	.98	6	1.14	0.0572 ± 11	0.493 ± 15	0.0625 ± 14	0.0565 ± 11	355 ± 7
7.1	143	101	.71	4	.94	0.0562 ± 11	0.472 ± 16	0.0608 ± 15	0.0557 ± 11	349 ± 7
8.1	252	195	.77	4	.60	0.0577 ± 11	0.462 ± 13	0.0581 ± 11	0.0573 ± 11	359 ± 7
9.1	110	77	.70	5	1.74	0.0582 ± 12	0.540 ± 19	0.0673 ± 18	0.0571 ± 12	358 ± 7
10.1	162	118	.73	4	1.02	0.0554 ± 11	0.469 ± 15	0.0615 ± 14	0.0548 ± 11	344 ± 7
11.1	175	101	.58	7	1.62	0.0557 ± 11	0.509 ± 16	0.0662 ± 14	0.0548 ± 11	344 ± 7
12.1	337	267	.79	5	.54	0.0570 ± 11	0.452 ± 12	0.0576 ± 9	0.0567 ± 11	355 ± 7
13.1	238	131	.55	3	.43	0.0560 ± 11	0.438 ± 13	0.0567 ± 11	0.0557 ± 11	350 ± 7
14.1	327	231	.71	4	.41	0.0572 ± 11	0.446 ± 12	0.0566 ± 9	0.0570 ± 11	357 ± 7
15.1	177	133	.75	27	5.67	0.0578 ± 11	0.788 ± 23	0.0988 ± 19	0.0546 ± 11	343 ± 7
16.1	321	262	.82	9	1.02	0.0566 ± 11	0.480 ± 13	0.0615 ± 10	0.0560 ± 11	351 ± 7
16.2	351	309	.88	8	.81	0.0585 ± 11	0.482 ± 14	0.0598 ± 12	0.0580 ± 11	363 ± 7
17.1	261	289	1.11	4	.57	0.0566 ± 11	0.451 ± 14	0.0578 ± 13	0.0562 ± 11	353 ± 7
18.1	102	66	.65	5	1.74	0.0547 ± 11	0.508 ± 21	0.0673 ± 22	0.0538 ± 11	338 ± 7
19.2	281	193	.69	2	.32	0.0560 ± 11	0.432 ± 13	0.0559 ± 12	0.0559 ± 11	350 ± 7
20.1	219	195	.89	5	.85	0.0586 ± 12	0.486 ± 15	0.0601 ± 14	0.0581 ± 11	364 ± 7
21.1	773	952	1.23	6	.27	0.0574 ± 11	0.439 ± 11	0.0554 ± 7	0.0573 ± 11	359 ± 7
21.2	441	503	1.14	7	.63	0.0551 ± 11	0.443 ± 12	0.0583 ± 10	0.0547 ± 11	344 ± 7
21.3	696	1145	1.64	10	.53	0.0554 ± 11	0.440 ± 11	0.0575 ± 8	0.0552 ± 11	346 ± 7
23.1	241	232	.96	4	.68	0.0556 ± 11	0.450 ± 15	0.0587 ± 14	0.0552 ± 11	347 ± 7
24.1	172	121	.71	7	1.42	0.0574 ± 11	0.512 ± 17	0.0647 ± 16	0.0566 ± 11	355 ± 7
25.1	185	166	.89	6	1.23	0.0579 ± 11	0.504 ± 17	0.0631 ± 15	0.0572 ± 11	359 ± 7

*^{207}Pb correction

†f^{206}Pb indicates the percentage of common ^{206}Pb in the total measured ^{206}Pb

#Apparent age is the ^{206}Pb/^{238}U age

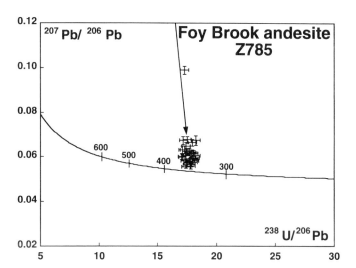

FIG. 4.—Compositions of zircons in the Foy Brook andesite plotted without correction for common Pb on a Concordia diagram. Arrow indicates the vector of common Pb correction. Radiogenic ^{206}Pb/^{238}U compositions and corresponding ages are calculated by projection of the data along this vector on to Concordia. Absence of Pb loss and inheritance is indicated by the adherence of the data, within error, to this mixing vector.

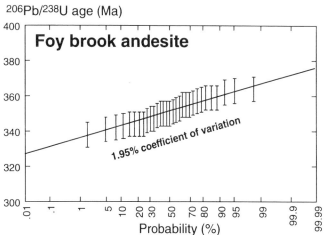

FIG. 5.—Probability plot of the Foy Brook zircon ages. ^{206}Pb/^{238}U ages have been calculated by projection of the raw measured data onto Concordia as described in Figure 4. This plot normalizes the age population to the Normal Probability Distribution indicated by the diagonal straight line, which is the locus of reproducibility of measurements of the standard zircon obtained during this analytical run (coefficient of variation 1.95%), and indicates the statistical dispersion expected of measurements of a homogeneous composition. The Foy Brook zircon compositions are indistinguishable from the Normal Distribution, indicating that they all have the same age within error. The mathematical counterpart to this visual matching of data to the Normal Distribution is a Mean Square of Weighted Deviates (MSWD) less than 1.06, the figure appropriate to 95% confidence at 30 degrees of freedom.

JOHN ROBERTS, JONATHAN C. CLAOUÉ-LONG, AND PETER J. JONES

TABLE 2.—SHRIMP U-Pb ISOTOPIC DATA FOR THE CURRA KEITH TONGUE

Grain area	U (ppm)	Th (ppm)	Th/U	^{204}Pb* (ppb)	f^{206}Pb*† (%)	Calibrated total Pb compositions			Radiogenic compositions*	
						^{206}Pb/^{238}U ± 1σ	^{207}Pb/^{235}U ± 1σ	^{207}Pb/^{206}Pb ± 1σ	^{206}Pb/^{238}U ± 1σ	Apparent age# (Ma) ± 1σ
1.1	77	55	.71	7	3.38	0.0566 ± 8	0.628 ± 21	0.0804 ± 23	0.0547 ± 8	343 ± 5
1.2	142	102	.72	30	7.54	0.0597 ± 9	0.937 ± 29	0.1138 ± 30	0.0552 ± 8	346 ± 5
2.1	212	101	.48	10	1.81	0.0546 ± 8	0.511 ± 13	0.0678 ± 12	0.0536 ± 8	337 ± 5
3.1	155	99	.64	6	1.53	0.0546 ± 8	0.494 ± 14	0.0655 ± 14	0.0538 ± 8	338 ± 5
4.1	172	93	.54	4	1.01	0.0543 ± 8	0.459 ± 13	0.0614 ± 13	0.0537 ± 8	337 ± 5
5.1	250	170	.68	7	1.07	0.0542 ± 8	0.462 ± 11	0.0619 ± 11	0.0536 ± 8	337 ± 5
5.2	215	127	.59	31	5.34	0.0568 ± 10	0.753 ± 23	0.0962 ± 22	0.0538 ± 9	338 ± 6
5.3	173	115	.67	43	8.95	0.0591 ± 10	1.021 ± 31	0.1252 ± 29	0.0538 ± 10	338 ± 6
5.4	363	329	.90	24	2.37	0.0604 ± 11	0.602 ± 17	0.0723 ± 15	0.0590 ± 10	369 ± 6
6.1	294	233	.79	4	0.49	0.0534 ± 8	0.422 ± 10	0.0572 ± 9	0.0532 ± 8	334 ± 5
6.2	140	87	.62	14	3.81	0.0546 ± 10	0.632 ± 22	0.0839 ± 23	0.0525 ± 10	330 ± 6
6.3	180	131	.73	29	5.88	0.0590 ± 11	0.818 ± 26	0.1005 ± 23	0.0555 ± 10	349 ± 6
6.4	248	232	.94	20	3.41	0.0506 ± 9	0.562 ± 18	0.0806 ± 19	0.0488 ± 9	307 ± 6
7.1	309	234	.76	9	1.13	0.0544 ± 8	0.468 ± 10	0.0623 ± 9	0.0538 ± 8	338 ± 5
7.2	178	90	.50	36	7.19	0.0591 ± 11	0.905 ± 26	0.1110 ± 23	0.0548 ± 10	344 ± 6
7.3	199	133	.66	37	7.11	0.0554 ± 10	0.843 ± 25	0.1104 ± 23	0.0515 ± 10	323 ± 6
7.4	225	135	.60	53	7.48	0.0667 ± 12	1.043 ± 30	0.1134 ± 23	0.0617 ± 11	386 ± 7
8.1	122	68	.55	5	1.71	0.0549 ± 8	0.507 ± 15	0.0670 ± 16	0.0539 ± 8	339 ± 5
9.1	342	269	.79	24	2.62	0.0565 ± 8	0.579 ± 12	0.0743 ± 10	0.0550 ± 8	345 ± 5
9.2	133	72	.54	11	3.09	0.0557 ± 9	0.599 ± 20	0.0781 ± 21	0.0539 ± 9	339 ± 5
9.3	232	162	.70	13	1.88	0.0613 ± 11	0.579 ± 17	0.0684 ± 14	0.0602 ± 11	377 ± 7
9.4	141	87	.62	15	4.06	0.0560 ± 10	0.664 ± 22	0.0859 ± 22	0.0538 ± 10	338 ± 6
9.5	128	78	.61	12	3.51	0.0577 ± 11	0.648 ± 22	0.0815 ± 22	0.0557 ± 11	349 ± 6
10.1	139	85	.61	7	1.97	0.0575 ± 8	0.548 ± 15	0.0691 ± 15	0.0564 ± 8	353 ± 5
11.1	159	80	.50	7	1.66	0.0555 ± 8	0.510 ± 13	0.0666 ± 14	0.0546 ± 8	343 ± 5
11.2	167	89	.53	6	1.49	0.0552 ± 8	0.497 ± 16	0.0653 ± 18	0.0544 ± 8	341 ± 5
12.1	218	132	.60	5	0.86	0.0548 ± 8	0.454 ± 11	0.0601 ± 11	0.0543 ± 8	341 ± 5
13.1	137	66	.48	5	1.37	0.0577 ± 8	0.511 ± 14	0.0642 ± 14	0.0569 ± 8	357 ± 5
13.2	148	76	.52	42	9.42	0.0635 ± 11	1.130 ± 35	0.1290 ± 30	0.0575 ± 11	361 ± 6
14.1	338	286	.85	4	0.43	0.0571 ± 8	0.447 ± 10	0.0567 ± 8	0.0569 ± 8	357 ± 5
14.2	423	362	.86	13	1.23	0.0549 ± 9	0.478 ± 11	0.0632 ± 10	0.0542 ± 9	340 ± 5
14.3	125	79	.63	14	4.31	0.0547 ± 9	0.663 ± 22	0.0879 ± 23	0.0524 ± 9	329 ± 5
14.4	285	196	.69	13	1.67	0.0567 ± 9	0.395 ± 24	0.0514 ± 28	0.0558 ± 9	350 ± 5
14.5	238	185	.78	12	1.98	0.0523 ± 12	0.522 ± 14	0.0667 ± 13	0.0524 ± 10	329 ± 6
15.1	234	132	.56	5	0.85	0.0574 ± 8	0.476 ± 11	0.0601 ± 10	0.0569 ± 8	357 ± 5
15.2	251	127	.51	27	4.05	0.0565 ± 10	0.669 ± 19	0.0858 ± 17	0.0542 ± 10	340 ± 6
15.3	100	47	.47	23	8.31	0.0581 ± 11	0.962 ± 34	0.1200 ± 33	0.0533 ± 10	335 ± 6
15.4	183	118	.64	22	4.18	0.0605 ± 11	0.725 ± 23	0.0869 ± 20	0.0580 ± 11	363 ± 7
15.5	114	66	.58	14	4.63	0.0563 ± 11	0.703 ± 27	0.0905 ± 28	0.0537 ± 10	337 ± 6
16.1	78	52	.66	15	6.81	0.0610 ± 15	0.908 ± 40	0.1080 ± 36	0.0568 ± 15	356 ± 9
16.2	48	36	.75	39	24.93	0.0690 ± 10	2.415 ± 90	0.2537 ± 82	0.0518 ± 11	326 ± 6
16.3	49	33	.69	4	2.57	0.0607 ± 9	0.619 ± 31	0.0739 ± 34	0.0591 ± 9	370 ± 5
16.4	74	48	.65	8	3.87	0.0574 ± 8	0.668 ± 26	0.0843 ± 29	0.0552 ± 8	346 ± 5
17.1	183	209	1.14	9	1.83	0.0552 ± 14	0.518 ± 20	0.0680 ± 18	0.0542 ± 14	340 ± 8
17.2	189	176	.93	31	5.84	0.0600 ± 8	0.829 ± 21	0.1002 ± 20	0.0565 ± 8	355 ± 5
18.1	99	69	.70	50	18.47	0.0577 ± 14	1.604 ± 63	0.2018 ± 55	0.0470 ± 12	296 ± 8
18.2	71	43	.60	17	8.46	0.0608 ± 9	1.017 ± 37	0.1213 ± 38	0.0557 ± 9	349 ± 5
19.1	265	265	1.00	6	0.82	0.0559 ± 14	0.461 ± 18	0.0599 ± 16	0.0554 ± 14	348 ± 8
19.2	237	204	.86	6	1.00	0.0527 ± 7	0.445 ± 14	0.0613 ± 16	0.0521 ± 7	328 ± 4
20.1	235	150	.64	7	1.04	0.0573 ± 14	0.487 ± 18	0.0616 ± 15	0.0567 ± 14	356 ± 9
21.1	422	358	.85	23	1.99	0.0579 ± 14	0.552 ± 18	0.0692 ± 12	0.0567 ± 14	356 ± 8
22.1	275	235	.85	4	0.57	0.0547 ± 8	0.437 ± 12	0.0578 ± 13	0.0544 ± 8	342 ± 5
23.1	371	210	.57	6	0.64	0.0558 ± 8	0.449 ± 11	0.0584 ± 11	0.0554 ± 8	348 ± 5
24.1	168	83	.49	8	2.45	0.0422 ± 6	0.424 ± 19	0.0729 ± 29	0.0411 ± 6	260 ± 4
24.2	212	145	.69	4	0.82	0.0528 ± 7	0.435 ± 14	0.0599 ± 16	0.0523 ± 7	329 ± 5
25.1	204	202	.99	8	1.71	0.0469 ± 7	0.434 ± 14	0.0670 ± 18	0.0461 ± 7	291 ± 4
26.1	145	102	.70	57	12.78	0.0651 ± 9	1.401 ± 39	0.1560 ± 34	0.0568 ± 9	356 ± 5

*^{207}Pb correction #Apparent age is the ^{206}Pb/^{238}U age
†f^{206}Pb indicates the percentage of common ^{206}Pb in the total measured ^{206}Pb

ability diagram in Figure 6 where the data are overlain on the diagonal straight line of the expected Normal Distribution derived from the reproducibility of concurrent measurements of the homogeneous reference zircon. In marked contrast to the simple age spectrum found in the Foy Brook andesite zircons, considerable heterogeneity is evident in the data, and there is real age variation in the sample beyond analytical error. In the middle of the spectrum, there is a large body of data within error of the Normal Distribution line. These grains all have the same age within error and comprise the dominant population of zircons in the sample. Falling significantly below the line are five zircons with low apparent ages; evidently these grains lost Pb subsequent to crystallization, and their ^{206}Pb/^{238}U ages no longer record the eruption age of the ignimbrite. Six grains have ages significantly above the Normal Distribution line, and their older ages identify them as zircon xenocrysts inherited by the ignimbrite; these could derive from inclusions of wall rock mechanically included in the explosive eruption. They could be zircons that grew in the magma before eruption, or they could be restite zircon remnants of the crustal rocks that were partially

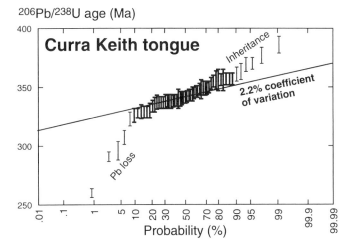

206Pb/238U age (Ma)

FIG. 6.—Probability plot of zircon ages in the Curra Keith tongue. Normal Distribution is indicated by the diagonal straight line plotted from the concurrent reproducibility measured for the standard zircon. Five zircons have young apparent ages significantly below the Normal Distribution line and are interpreted to have lost Pb subsequent to crystallization. A few grains have older ages significantly above the Normal Distribution line and are identified as xenocrysts. Data marked bold are not distinguishable beyond error from Normally distributed measurements of a homogeneous age and form the principal age population from which the age of the sample is calculated.

melted to form the felsic magma. The presence in the sample of evidence for both inheritance and Pb loss complicates interpretation of the crystallization age of the ignimbrite, which must be defined separately from these older and younger age components. The data define a continuum of ages, with some of the inherited and altered grains having ages only slightly beyond the normal statistical dispersion of the main group. A large quantity of data has therefore been collected so that statistical confidence can be brought to bear on the problem of separating the three age components in the rock. The principal feature expected of zircons crystallized in an eruptive event is a dominant group of grains all having the same age within error; the guide to defining the magmatic population is therefore the reference Normal Distribution plotted from concurrent analyses of the homogeneous reference zircon; this line defines the maximum dispersion that can be attributed to analytical error in zircons of the same age. The analyses that cannot be proved to depart from this expected distribution are marked in bold in Figure 6. The MSWD of this group is 1.28 which is within the value of 1.37 that indicates homogeneity at 95% confidence, and this provides a definition of the principal population. A supporting definition is given by the breaks in slope at the extremities of the age spectrum of the main group; those compositions that depart significantly from the Normal Distribution also define markedly steeper slopes than that of the main spectrum of data, which indicates that they belong to wider standard deviations. This is the clearest and most defensible interpretation of the complex age spectrum presented by the Curra Keith zircons, but the interpretation is not unique; it remains possible (if unlikely) that too young or too old an age is being interpreted by inclusion of one or more altered or inherited grains in the group defined as magmatic. It will be noticed that the construction applied here to recognizing the magmatic zircon age is the

same as that used in defining plateau ages from Ar release patterns in $^{40}Ar/^{39}Ar$ dating. Thus defined, the weighted mean $^{206}Pb/^{238}U$ ratio of the main group of zircons is 0.05451 with a standard deviation of the mean of ± 0.00052 (2σ); this is equivalent to an age of 342.1 ± 3.2 Ma (2σ) and is interpreted as the crystallization age of the Curra Keith Tongue.

Nerong Volcanics

The type section of the Nerong Volcanics is on the eastern limb of the Girvan Anticline west of Bulahdelah (Roberts and others, 1991); from this locality, a hornblende dacite 520–555 m above the base of the section was sampled for zircon dating. The volcanics overlie marine to continental sediments of the Conger Formation (Engel, 1962; Roberts and others, 1991) which contains brachiopods referable to the *Delepinea aspinosa* Zone; the assemblage from the *D. aspinosa* Zone is not sufficiently diverse to be assigned to a subdivision of that zone. In turn, they are disconformably overlain by Karuah Formation that contains the *Marginirugus barringtonensis* Zone. *Linoprotonia tenuirugosa* Subzone faunas are present in the Conger Formation beneath the Buggs Creek Volcanic Member in the basal Berrico Creek Formation, which is interpreted as a distal extremity of the Nerong Volcanics on the northwestern limb of the Stroud-Gloucester Syncline. Here, the volcanics are conformably overlain by a marine succession containing the *Rhipidomella fortimuscula* and *M. barringtonensis* Zones. Possible *L. tenuirugosa* Subzone faunas are also present beneath Nerong Volcanics in the Boolambayte Formation of the eastern Myall Block. In the type area, the Nerong Volcanics are therefore located within the interval of the *D. aspinosa* Zone, but possible distal extremities are younger and appear to represent a final, geographically widespread phase of volcanism.

Zircons in the Nerong Volcanics have a simple distribution of ages (Table 3). The spectrum is plotted in a probability diagram in Figure 7 and overlain on the reference Normal Distribution defined from concurrent measurements of the standard zircon. The compositions of two zircon grains are off scale and have ages of 455 ± 16 Ma (2σ) and 603 ± 22 Ma (2σ) (Table 3); these are clearly xenocrysts incorporated in the dacite, whose eruption age is recorded by the dominant group of zircons. There is no further indication of excess dispersion, and all the remaining zircons have the same age within error. The weighted mean $^{206}Pb/^{238}U$ ratio of this group of zircons is 0.05392 ± .00031 (2σ) (MSWD = 1.53, equal to the reference value of 1.53), equivalent to an age of 338.6 ± 3.8 Ma (2σ), and this is interpreted as the crystallization age of the dacite.

Martins Creek Ignimbrite Member

The Martins Creek Ignimbrite is a geographically widespread marker in the Carboniferous of eastern Australia. It has been used widely as a reference datum within the Carboniferous of NSW and in the international Phanerozoic timescale (Harland and others, 1964; Harland and others, 1990). The ignimbrite constitutes the lowest widespread member of the Gilmore Volcanic Group (Hamilton and others, 1974) and is the second highest volcanic member of the Isismurra Formation (Roberts and Engel, 1987). In biostratigraphic terms, the ignimbrite lies above the *L. tenuirugosa* Subzone and either below or within the lower part of the *R. fortimuscula* Zone (see Roberts and

TABLE 3.—SHRIMP U-Pb ISOTOPIC DATA FOR THE NERONG VOLCANICS

Grain area	U (ppm)	Th (ppm)	Th/U	$^{204}Pb^*$ (ppb)	$f^{206}Pb^*$† (%)	Calibrated total Pb compositions			Radiogenic compositions*	
						$^{206}Pb/^{238}U$ ± 1σ	$^{207}Pb/^{235}U$ ± 1σ	$^{207}Pb/^{206}Pb$ ± 1σ	$^{206}Pb/^{238}U$ ± 1σ	Apparent age# (Ma) ± 1σ
1.1	59	41	.69	6	4.24	0.0545 ± 13	0.657 ± 29	0.0875 ± 31	0.0522 ± 13	328 ± 8
2.1	75	24	.31	6	3.01	0.0569 ± 14	0.608 ± 25	0.0776 ± 24	0.0552 ± 13	346 ± 8
3.1	411	159	.39	11	1.06	0.0552 ± 13	0.472 ± 14	0.0619 ± 9	0.0546 ± 13	343 ± 8
4.1	254	144	.57	10	1.49	0.0569 ± 14	0.512 ± 17	0.0654 ± 13	0.0560 ± 13	351 ± 8
5.1	163	104	.64	8	1.92	0.0541 ± 16	0.513 ± 20	0.0688 ± 15	0.0530 ± 16	333 ± 10
6.1	388	452	1.16	7	.72	0.0526 ± 16	0.429 ± 15	0.0591 ± 9	0.0522 ± 16	328 ± 10
7.1	71	44	.63	6	3.19	0.0531 ± 16	0.578 ± 26	0.0791 ± 24	0.0514 ± 16	323 ± 10
8.1	68	61	.89	5	2.56	0.0566 ± 17	0.577 ± 26	0.0739 ± 23	0.0552 ± 17	346 ± 10
9.1	450	262	.58	5	.42	0.0578 ± 17	0.452 ± 16	0.0568 ± 7	0.0576 ± 17	361 ± 11
10.1	313	161	.51	5	.47	0.0735 ± 13	0.579 ± 14	0.0572 ± 9	0.0731 ± 13	455 ± 8
11.1	96	80	.84	5	1.82	0.0551 ± 9	0.516 ± 18	0.0680 ± 19	0.0541 ± 9	340 ± 6
12.1	40	40	.98	3	2.34	0.0572 ± 12	0.569 ± 25	0.0722 ± 28	0.0558 ± 10	350 ± 6
13.1	220	211	.96	2	.40	0.0546 ± 9	0.426 ± 12	0.0566 ± 11	0.0543 ± 9	341 ± 6
14.1	143	74	.52	4	1.16	0.0541 ± 9	0.467 ± 15	0.0627 ± 16	0.0535 ± 9	336 ± 6
15.1	97	63	.65	4	1.63	0.0529 ± 9	0.484 ± 17	0.0665 ± 18	0.0520 ± 9	327 ± 6
16.1	249	101	.40	5	.74	0.0526 ± 9	0.430 ± 11	0.0593 ± 11	0.0522 ± 9	328 ± 6
17.1	398	206	.52	4	.43	0.0531 ± 9	0.417 ± 10	0.0569 ± 8	0.0529 ± 9	332 ± 6
18.1	207	103	.50	3	.51	0.0543 ± 9	0.430 ± 12	0.0574 ± 11	0.0540 ± 9	339 ± 6
19.1	292	191	.65	4	.55	0.0548 ± 9	0.437 ± 11	0.0578 ± 9	0.0545 ± 9	342 ± 6
20.1	363	216	.59	11	1.26	0.0531 ± 10	0.465 ± 13	0.0635 ± 12	0.0524 ± 9	329 ± 6
21.1	114	64	.56	11	3.52	0.0566 ± 10	0.637 ± 22	0.0817 ± 23	0.0546 ± 10	343 ± 6
22.1	238	147	.62	10	1.55	0.0559 ± 10	0.507 ± 15	0.0658 ± 14	0.0550 ± 10	345 ± 6
23.1	113	227	2.01	24	4.48	0.1027 ± 19	1.267 ± 36	0.0894 ± 17	0.0981 ± 18	603 ± 11
24.1	420	627	1.49	5	.50	0.0547 ± 10	0.433 ± 11	0.0574 ± 9	0.0544 ± 10	342 ± 6
25.1	295	137	.47	2	.20	0.0555 ± 10	0.421 ± 12	0.0550 ± 11	0.0554 ± 10	347 ± 6

*^{207}Pb correction
†$f^{206}Pb$ indicates the percentage of common ^{206}Pb in the total measured ^{206}Pb
#Apparent age is the $^{206}Pb/^{238}U$ age

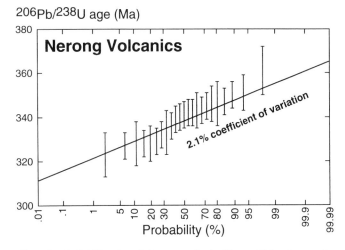

FIG. 7.—Probability plot of zircon ages in the Nerong Volcanics sample. These zircons form a simple homogeneous population relative to the Normal Distribution defined by the reproducibility of concurrent measurements of the standard zircon. Two xenocryst ages are off scale.

others, 1991, p. 93). On the western limb of the Wallarobba Syncline, the ignimbrite overlies non-marine sediments of the Wallaringa Formation, which rest disconformably on the Flagstaff Formation containing the *L. tenuirugosa* Subzone (Roberts and others, 1991). In the Black Jack Range near Carrowbrook, it is overlain by the marine Chichester Formation containing the *R. fortimuscula* Zone, which in turn is overlain by the Lambs Valley Ignimbrite Member near the top of the Isismurra Formation.

Six K-Ar ages for the Martins Creek Ignimbrite (Evernden and Richards, 1962; Roberts and Oversby, 1974; Roberts and others, 1991) have been variously averaged at 328 or 331 Ma. Evernden and Richards' (1962) ages for the Martins Creek Ignimbrite and for a hornblende andesite at Hudsons Peak, now identified as the Martins Creek Ignimbrite, were used by Harland and others (1990) to indicate an age for the Asbian of the European Carboniferous Stages. Revision of the biostratigraphic relationships by Roberts and others (1993b) now correlates it to the Holkerian.

The Martins Creek Ignimbrite was sampled at Martins Creek Quarry, 5 km north of Paterson. This is a large, deep, active quarry, and the sample was chosen in the hope of avoiding the Pb loss evident in other surface samples from the district. Some analytical attributes of the data (Table 4) differ from other samples described in this paper. First, there are higher concentrations of common Pb in some of the analyses; these are attributed, not to indigenous initial Pb in the zircons, but to carbon coating of the ion probe mount surface; other samples described in this paper were coated with 99.999% pure gold which has significantly reduced the laboratory blank Pb in SHRIMP analyses. Calculated $^{206}Pb/^{238}U$ ages are insensitive to the correction for common Pb (Claoué-Long and others, this volume), so the higher contents of common Pb do not materially affect interpretation of the age of the sample. Second, the data were collected when reproducibility of the SHRIMP Pb/U calibration was stable at better than 1.4% (a figure of 2% or more applies to other analytical runs described in this paper); the tighter reproducibility results in more precise individual measurements and more confident recognition of outlier compositions.

Data measured for zircons in the Martins Creek Ignimbrite are plotted without correction for common Pb in a Concordia

TABLE 4.—SHRIMP U-Pb ISOTOPIC DATA FOR THE MARTINS CREEK IGNIMBRITE

Grain area	U (ppm)	Th (ppm)	Th/U	^{204}Pb* (ppb)	f^{206}Pb*† (%)	Calibrated total Pb compositions			Radiogenic compositions	
						^{206}Pb/^{238}U ±1σ	^{207}Pb/^{235}U ±1σ	^{207}Pb/^{206}Pb ±1σ	^{206}Pb/^{238}U ±1σ	Apprent age# (Ma) ±1σ
2.1	202	157	.78	14	2.73	0.0553 ± 6	0.571 ± 10	0.0749 ± 9	0.0538 ± 6	337 ± 4
3.1	189	106	.56	10	2.21	0.0526 ± 6	0.514 ± 9	0.0708 ± 9	0.0515 ± 6	324 ± 3
4.1	456	284	.62	12	1.12	0.0512 ± 5	0.437 ± 6	0.0620 ± 5	0.0506 ± 5	318 ± 3
5.1	221	179	.81	12	2.08	0.0538 ± 8	0.517 ± 10	0.0698 ± 9	0.0527 ± 8	331 ± 5
6.1	240	123	.51	12	1.99	0.0534 ± 8	0.508 ± 10	0.0690 ± 8	0.0523 ± 8	329 ± 5
7.1	403	219	.54	8	0.85	0.0520 ± 7	0.429 ± 8	0.0599 ± 6	0.0516 ± 8	324 ± 5
8.1	322	233	.72	9	1.14	0.0541 ± 8	0.464 ± 9	0.0622 ± 7	0.0535 ± 8	336 ± 5
8.2	190	109	.57	54	10.15	0.0592 ± 9	1.100 ± 19	0.1347 ± 11	0.0532 ± 8	334 ± 5
8.3	220	119	.54	53	8.70	0.0594 ± 9	1.007 ± 18	0.1231 ± 10	0.0542 ± 8	340 ± 5
9.1	273	203	.74	9	1.36	0.0533 ± 8	0.470 ± 9	0.0640 ± 7	0.0526 ± 8	330 ± 5
10.1	171	104	.61	10	2.31	0.0538 ± 8	0.531 ± 11	0.0716 ± 9	0.0526 ± 8	330 ± 5
11.1	256	213	.83	15	2.23	0.0544 ± 6	0.532 ± 8	0.0709 ± 7	0.0532 ± 6	334 ± 3
11.2	241	168	.70	60	9.06	0.0586 ± 8	1.018 ± 17	0.1260 ± 9	0.0533 ± 8	335 ± 5
11.3	237	145	.61	52	7.99	0.0585 ± 8	0.947 ± 16	0.1174 ± 9	0.0538 ± 8	338 ± 5
11.4	182	123	.67	42	8.35	0.0583 ± 9	0.967 ± 17	0.1202 ± 11	0.0535 ± 8	336 ± 5
12.1	131	74	.56	6	1.78	0.0546 ± 6	0.507 ± 10	0.0673 ± 10	0.0536 ± 6	337 ± 4
12.3	214	134	.63	16	2.96	0.0543 ± 8	0.576 ± 11	0.0769 ± 8	0.0527 ± 8	331 ± 5
12.4	193	105	.54	78	14.28	0.0604 ± 9	1.400 ± 24	0.1679 ± 12	0.0518 ± 8	326 ± 5
12.5	203	106	.52	63	11.43	0.0580 ± 8	1.160 ± 20	0.1450 ± 12	0.0514 ± 8	323 ± 5
12.6	242	179	.74	73	10.72	0.0596 ± 9	1.144 ± 20	0.1393 ± 11	0.0532 ± 8	334 ± 5
13.1	167	85	.51	7	1.86	0.0511 ± 5	0.479 ± 9	0.0679 ± 9	0.0501 ± 6	315 ± 3
14.1	220	112	.51	16	3.01	0.0514 ± 8	0.547 ± 11	0.0772 ± 9	0.0499 ± 8	314 ± 5
15.1	121	56	.46	6	2.24	0.0454 ± 7	0.445 ± 11	0.0710 ± 12	0.0444 ± 7	280 ± 4
16.1	180	112	.62	25	5.28	0.0551 ± 9	0.725 ± 15	0.0954 ± 11	0.0522 ± 8	328 ± 5
17.1	394	294	.75	12	1.28	0.0528 ± 8	0.461 ± 9	0.0633 ± 6	0.0521 ± 8	328 ± 5
18.1	371	175	.47	11	1.22	0.0531 ± 8	0.460 ± 9	0.0629 ± 6	0.0525 ± 8	330 ± 5
19.1	364	69	.19	14	1.19	0.0670 ± 10	0.596 ± 11	0.0646 ± 6	0.0662 ± 10	413 ± 6
19.2	93	82	.88	15	1.60	0.1916 ± 32	2.380 ± 47	0.0901 ± 8	0.0743 ± 20⁺	1050 ± 54^
20.1	137	88	.65	11	3.13	0.0539 ± 9	0.581 ± 13	0.0782 ± 11	0.0522 ± 8	328 ± 5
21.1	207	111	.54	9	1.83	0.0530 ± 8	0.495 ± 11	0.0677 ± 9	0.0520 ± 8	327 ± 5
22.1	394	324	.82	13	1.35	0.0522 ± 8	0.460 ± 9	0.0638 ± 6	0.0515 ± 8	324 ± 5
23.1	155	62	.40	10	2.55	0.0551 ± 9	0.559 ± 12	0.0735 ± 10	0.0537 ± 9	337 ± 5
23.2	325	108	.33	35	4.04	0.0573 ± 8	0.677 ± 11	0.0856 ± 7	0.0550 ± 8	345 ± 5
23.3	351	115	.33	55	5.75	0.0575 ± 8	0.788 ± 13	0.0994 ± 7	0.0542 ± 8	340 ± 5
23.4	183	82	.45	5	1.03	0.0545 ± 8	0.462 ± 10	0.0614 ± 8	0.0539 ± 8	339 ± 5
23.5	452	264	.59	40	0.73	0.2056 ± 29	4.323 ± 64	0.1525 ± 4	0.1775 ± 57⁺	2630 ± 53^
24.1	384	298	.78	10	1.08	0.0533 ± 8	0.453 ± 9	0.0617 ± 6	0.0527 ± 8	331 ± 5
25.1	659	688	1.04	10	0.61	0.0524 ± 8	0.418 ± 7	0.0579 ± 4	0.0521 ± 8	327 ± 5
26.1	274	143	.52	27	3.87	0.0541 ± 8	0.627 ± 12	0.0841 ± 9	0.0520 ± 8	327 ± 5
27.1	259	159	.62	17	2.69	0.0522 ± 6	0.537 ± 9	0.0747 ± 9	0.0508 ± 6	319 ± 3
28.1	159	83	.52	22	5.29	0.0553 ± 6	0.728 ± 13	0.0956 ± 12	0.0524 ± 6	329 ± 4
29.1	253	129	.51	14	2.44	0.0482 ± 5	0.482 ± 8	0.0726 ± 9	0.0470 ± 5	296 ± 3
29.2	138	76	.55	15	4.29	0.0528 ± 6	0.636 ± 12	0.0875 ± 12	0.0505 ± 6	318 ± 4
30.1	186	124	.67	20	4.23	0.0543 ± 6	0.652 ± 11	0.0870 ± 11	0.0521 ± 6	327 ± 4
31.1	276	232	.84	22	3.06	0.0556 ± 6	0.595 ± 9	0.0776 ± 8	0.0539 ± 6	338 ± 4
31.3	242	196	.81	20	3.37	0.0526 ± 7	0.582 ± 11	0.0803 ± 8	0.0508 ± 7	320 ± 4
32.1	202	105	.52	13	2.55	0.0517 ± 6	0.525 ± 10	0.7350 ± 10	0.0504 ± 6	317 ± 3

*^{207}Pb correction

†f^{206}Pb indicates the percentage of common ^{206}Pb in the total measured ^{206}Pb

#Apparent age is the ^{206}Pb/^{238}U age

⁺^{208}Pb corrected ^{207}Pb/^{206}Pb ratio

^^{208}Pb corrected ^{207}Pb/^{206}Pb age

diagram in Figure 8, where the main group of compositions adheres closely to the vector of common Pb correction. Three analyses have significantly older ages and lie to the left of the main group in this diagram. The core of grain 19 has a Proterozoic age, and its rim is Silurian (Table 4); although targeted as optically similar to other zircons in the sample, this is evidently a xenocryst incorporated in the ignimbrite eruption. An Archaean age was measured in the core of grain 23; four analyses of a thick overgrowth on this core all have ages within the main Carboniferous population (Table 4), which suggests that the Archaean core formed a seed upon which new zircon crystallized following eruption and cooling of the magma. The main population of ages is plotted in a probability diagram in Figure 9, where the youngest zircon xenocryst is clearly resolved from the main group of Carboniferous ages. Despite efforts to obtain

a sample that had not been susceptible to Pb loss, there is a dispersion of young apparent ages departing significantly from the expected Normal Distribution. This minor departure from normality can be defined with confidence because a large quantity of precisely measured ^{206}Pb/^{238}U ratios is available; the excess dispersion is contributed by nine of the thirty-two zircons that depart from the main group by more than 95% confidence. The distribution of the remaining data is not provably different from the Normal Distribution, and this defines the population of magmatic zircons that have not lost Pb. The weighted mean ^{206}Pb/^{238}U ratio of this group is 0.05290 ± .00018 (2σ), equivalent to an age of 332.3 ± 2.2 Ma (2σ) (MSWD = 1.41, within the reference value of 1.43), and this is interpreted as the crystallization age of the Martins Creek Ignimbrite.

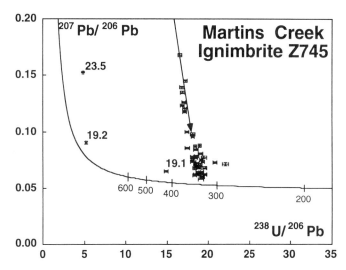

FIG. 8.—Zircon data from the Martins Creek Ignimbrite plotted without correction for common Pb on a Concordia diagram. Adherence of compositions to the common Pb mixing vector (arrow) is evident. Grain 19 is a Proterozoic xenocryst with a Silurian rim; grain 23 has a Carboniferous overgrowth on an Archaean core; others group as the magmatic population of zircons with minor overprinting by Pb loss.

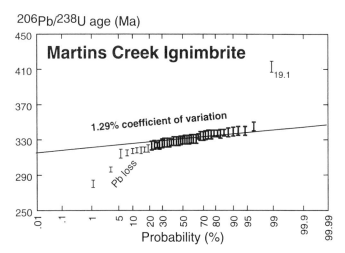

FIG. 9.—Probability plot of the main group of zircon ages in the Martins Creek Ignimbrite. Xenocryst grain 19 is readily resolved from the main group and the two other xenocryst ages are off scale. Most data are within error of the Normal Distribution defined by the concurrent reproducibility of the standard zircon. Nine zircons whose compositions lie significantly below the line are interpreted to have lost Pb subsequent to crystallization; the age of the sample has been calculated from the mean of the homogeneous population marked in bold.

Lambs Valley Ignimbrite Member

The Lambs Valley Ignimbrite Member is confined to southern parts of the Rouchel and Gresford Blocks; it constitutes the uppermost volcanic unit of the Isismurra Formation in the northwest, and of the Gilmore Volcanic Group in the east and southeast (Roberts and others, 1991; Hamilton and others, 1974; Browne, 1927). Erosion removed the Lambs Valley Ignimbrite from many areas before deposition of the disconform-

ably overlying Mt. Johnstone Formation. Beneath the Lambs Valley Ignimbrite, an intercalation of the marine Chichester Formation near Carrowbrook contains faunas of the *R. fortimuscula* Zone; it is likely that this marginal intercalation does not reflect the entire range of this zone (Roberts and others, 1993b), but this is the closest direct biostratigraphic constraint on the Lambs Valley Ignimbrite. The only constraint above the ignimbrite is the incoming of the *Nothorhacopteris* flora in the Mt. Johnstone Formation; more generally, the ignimbrite lies within the range of the *Grandispora maculosa* palynoflora.

Zircons in the Lambs Valley Ignimbrite were dated in the hope of measuring the age of the uppermost Isismurra Formation and placing a constraint on the timing of the overlying unconformity. Instead, the range of zircon ages found in the sample serves to illustrate the difficulty that can be experienced in dating ignimbritic eruptions. The data are listed in Table 5 and plotted in a Concordia diagram in Figure 10, where the compositions appear superficially to indicate homogeneity similar to that found in the Foy Brook andesite. Most of the analyses cluster close to Concordia, a reflection of low common Pb contents, and projection of their compositions on to Concordia would seem to produce a simple cluster of $^{206}Pb/^{238}U$ ages. This apparent simplicity is belied in Figure 11 by a close comparison of measured ages with the expectation of a Normal Distribution. If the zircons all had the same age, within error, the data would adhere in this diagram to the Normal Distribution, which is represented by straight lines of appropriate slope plotted from the reproducibility of concurrent measurements of the standard zircon. The data are dispersed well beyond this reference distribution, and it is not possible to discern a main population of homogeneous analyses that would define a magmatic crystallization age. This dispersion was reproduced on three analytical days separated by periods of several weeks and so is not an analytical artifact: it must therefore be real age variation within the sample. Multiple analyses within individual grains (Table 5) agree at precise mean ages for individual crystals that are significantly different one from another; for example, four analyses of zircon #1 agree at an age of 319 ± 8 Ma, significantly different from grain #9 which has a mean age of 347 ± 7 Ma. These data do not permit a unique solution to the age of the sample to be interpreted. Instead, they are a reminder that ignimbrite eruptions produce mixed rocks containing not only crystallization products of the magma but also mechanically included wall rock and some inheritance from the partial melting process and magma chamber crystallization; zircon ages representing all these components may be present and may also be overprinted by Pb loss. Other ignimbrites studied in this paper contain this sort of complexity but demonstrably have a predominance of zircons grown in the eruption from which the crystallization age of the unit can be drawn. Magmatic zircons might be present in the Lambs Valley Ignimbrite sample, but they are not sufficiently predominant to stand out from grains that are inherited or have lost Pb; the age of this unit is therefore not determined.

Paterson Volcanics

The Paterson Volcanics form an important rhyodacitic marker horizon between the Mt. Johnstone and Seaham For-

TABLE 5.—SHRIMP U-Pb ISOTOPIC DATA FOR THE LAMBS VALLEY IGNIMBRITE

Grain area	U (ppm)	Th (ppm)	Th/U	^{204}Pb* (ppb)	f^{206}Pb*† (%)	Calibrated total Pb compositions			Radiogenic compositions*	
						^{206}Pb/^{238}U ± 1σ	^{207}Pb/^{235}U ± 1σ	^{207}Pb/^{206}Pb ± 1σ	^{206}Pb/^{238}U ± 1σ	Apparent age# (Ma) ± 1σ
1.1	205	341	1.67	12	2.34	0.0513 ± 12	0.508 ± 15	0.0718 ± 12	0.0501 ± 11	315 ± 7
1.2	190	339	1.78	8	1.62	0.0524 ± 15	0.477 ± 19	0.0661 ± 16	0.0515 ± 14	324 ± 9
1.6	128	200	1.56	5	1.60	0.0516 ± 5	0.468 ± 12	0.0659 ± 16	0.0508 ± 5	319 ± 3
2.1	113	72	.64	8	2.94	0.0521 ± 12	0.550 ± 18	0.0767 ± 16	0.0505 ± 12	318 ± 7
2.2	117	80	.68	7	2.41	0.0518 ± 14	0.517 ± 23	0.0724 ± 23	0.0505 ± 14	318 ± 9
2.5	106	76	.72	2	.83	0.0519 ± 10	0.427 ± 15	0.0597 ± 16	0.0515 ± 10	323 ± 6
3.1	115	50	.43	7	2.32	0.0549 ± 12	0.542 ± 18	0.0717 ± 15	0.0536 ± 12	337 ± 7
3.2	193	97	.50	9	1.79	0.0562 ± 16	0.522 ± 21	0.0674 ± 18	0.0552 ± 15	346 ± 9
3.5	128	68	.53	4	1.21	0.0538 ± 5	0.466 ± 13	0.0628 ± 15	0.0532 ± 5	334 ± 3
4.1	126	78	.62	5	1.49	0.0552 ± 13	0.494 ± 16	0.0650 ± 13	0.0543 ± 12	341 ± 8
4.2	187	195	1.04	8	1.61	0.0542 ± 15	0.493 ± 20	0.0660 ± 18	0.0533 ± 15	335 ± 9
4.3	170	186	1.09	5	1.29	0.0522 ± 10	0.456 ± 14	0.0634 ± 14	0.0516 ± 10	324 ± 6
5.1	141	86	.61	6	1.81	0.0526 ± 12	0.490 ± 15	0.0675 ± 13	0.0517 ± 12	325 ± 7
6.1	167	75	.45	5	1.11	0.0535 ± 12	0.457 ± 14	0.0620 ± 11	0.0529 ± 12	332 ± 7
6.2	240	276	1.15	9	1.37	0.0567 ± 16	0.500 ± 19	0.0640 ± 15	0.0559 ± 16	351 ± 10
7.1	94	67	.72	4	1.84	0.0553 ± 13	0.517 ± 18	0.0678 ± 16	0.0543 ± 12	341 ± 8
7.2	111	103	.93	7	2.66	0.0525 ± 15	0.538 ± 24	0.0744 ± 24	0.0511 ± 14	321 ± 9
7.3	137	80	.58	10	2.85	0.0530 ± 15	0.555 ± 24	0.0759 ± 23	0.0515 ± 14	324 ± 9
7.5	126	102	.81	2	.60	0.0516 ± 5	0.411 ± 12	0.0578 ± 15	0.0513 ± 5	322 ± 3
7.6	121	108	.89	1	.39	0.0497 ± 9	0.385 ± 13	0.0562 ± 14	0.0495 ± 9	311 ± 6
8.1	122	74	.60	5	1.47	0.0579 ± 13	0.517 ± 16	0.0649 ± 13	0.0570 ± 13	357 ± 8
8.2	134	80	.60	7	2.08	0.0573 ± 16	0.551 ± 24	0.0697 ± 21	0.0561 ± 16	352 ± 10
8.4	171	103	.60	2	.49	0.0538 ± 10	0.423 ± 13	0.0570 ± 12	0.0536 ± 10	336 ± 6
9.1	455	228	.50	7	.61	0.0559 ± 13	0.447 ± 12	0.0579 ± 6	0.0556 ± 13	349 ± 8
9.2	171	101	.59	9	2.00	0.0561 ± 16	0.535 ± 21	0.0691 ± 17	0.0550 ± 15	345 ± 9
9.4	430	232	.54	4	.37	0.0554 ± 5	0.428 ± 8	0.0560 ± 8	0.0552 ± 5	347 ± 3
9.5	201	134	.67	3	.55	0.0553 ± 10	0.438 ± 13	0.0574 ± 11	0.0550 ± 10	345 ± 6
10.1	67	62	.93	4	2.67	0.0539 ± 14	0.553 ± 22	0.0745 ± 20	0.0524 ± 14	329 ± 8
10.2	102	76	.75	9	3.62	0.0538 ± 15	0.609 ± 27	0.0821 ± 26	0.0519 ± 15	326 ± 9
10.5	92	79	.86	1	.51	0.0521 ± 5	0.411 ± 13	0.0572 ± 17	0.0519 ± 5	326 ± 3
10.6	62	47	.76	2	1.38	0.0509 ± 10	0.450 ± 19	0.0641 ± 22	0.0502 ± 10	316 ± 6
11.1	218	137	.63	10	1.78	0.0535 ± 14	0.497 ± 16	0.0673 ± 10	0.0526 ± 14	330 ± 8
12.1	340	360	1.06	6	.67	0.0556 ± 14	0.448 ± 14	0.0584 ± 7	0.0553 ± 14	347 ± 9
12.2	186	108	.58	31	5.97	0.0596 ± 17	0.829 ± 32	0.1010 ± 23	0.0560 ± 16	351 ± 10
12.3	297	250	.84	8	1.03	0.0557 ± 11	0.471 ± 12	0.0613 ± 10	0.0551 ± 10	346 ± 6
13.1	376	455	1.21	4	.46	0.0529 ± 14	0.413 ± 12	0.0567 ± 7	0.0526 ± 14	331 ± 8
13.2	276	328	1.19	9	1.25	0.0540 ± 15	0.470 ± 18	0.0631 ± 15	0.0533 ± 15	335 ± 9
14.1	144	104	.72	3	.84	0.0575 ± 15	0.474 ± 16	0.0598 ± 11	0.0570 ± 15	357 ± 9
14.2	118	69	.59	7	2.30	0.0588 ± 16	0.579 ± 27	0.0715 ± 25	0.0574 ± 16	360 ± 10
15.1	73	65	.89	4	2.13	0.0519 ± 13	0.502 ± 20	0.0701 ± 19	0.0508 ± 13	319 ± 8
15.2	110	118	1.07	6	2.01	0.0546 ± 16	0.521 ± 26	0.0692 ± 26	0.0535 ± 15	336 ± 9
15.4	55	77	1.41	2	1.85	0.0519 ± 5	0.486 ± 18	0.0679 ± 24	0.0510 ± 5	321 ± 3
16.1	162	106	.66	3	.62	0.0532 ± 14	0.426 ± 14	0.0580 ± 11	0.0529 ± 14	332 ± 8
17.1	124	62	.50	4	1.21	0.0532 ± 14	0.460 ± 16	0.0627 ± 13	0.0526 ± 14	330 ± 8
18.1	151	254	1.68	3	.68	0.0533 ± 14	0.430 ± 15	0.0585 ± 11	0.0530 ± 14	333 ± 8
19.1	503	508	1.01	5	.38	0.0551 ± 14	0.425 ± 12	0.0561 ± 6	0.0549 ± 14	344 ± 9
19.2	84	109	1.29	8	4.01	0.0531 ± 15	0.624 ± 30	0.0852 ± 31	0.0510 ± 14	320 ± 9
19.4	223	408	1.83	3	.54	0.0533 ± 5	0.422 ± 9	0.0574 ± 11	0.0530 ± 5	333 ± 3
20.1	189	123	.65	3	.60	0.0539 ± 14	0.430 ± 14	0.0579 ± 10	0.0536 ± 14	336 ± 8
20.2	95	60	.63	7	2.88	0.0522 ± 15	0.548 ± 26	0.0762 ± 27	0.0507 ± 14	319 ± 9
20.5	106	74	.70	2	.84	0.0528 ± 5	0.435 ± 13	0.0598 ± 16	0.0523 ± 5	329 ± 3
20.6	179	133	.74	2	.51	0.0515 ± 10	0.405 ± 13	0.0571 ± 13	0.0513 ± 10	322 ± 6
21.1	105	55	.52	13	5.29	0.0511 ± 14	0.673 ± 29	0.0955 ± 28	0.0484 ± 14	304 ± 8
22.1	525	490	.93	10	.72	0.0568 ± 16	0.461 ± 15	0.0588 ± 9	0.0564 ± 16	354 ± 10
23.1	158	132	.84	6	1.57	0.0533 ± 15	0.482 ± 21	0.0656 ± 20	0.0525 ± 15	330 ± 9
24.1	92	48	.52	8	3.66	0.0529 ± 15	0.601 ± 30	0.0824 ± 31	0.0509 ± 14	320 ± 9

*^{207}Pb correction
†f^{206}Pb indicates the percentage of common ^{206}Pb in the total measured ^{206}Pb
#Apparent age is the ^{206}Pb/^{238}U age

mations throughout most of the Gresford Block. A biostratigraphic constraint below the Paterson Volcanics is provided by *R. fortimuscula* Zone faunas in the Chichester Formation 290 m beneath the volcanics and palynofloras of the *Grandispora maculosa* Assemblage in the underlying Mt. Johnstone Formation (Playford and Helby, 1968; Helby, 1969). The "enriched *Nothorhacopteris, Sphenopteridium* macroflora" of Morris (1985) below the Paterson Volcanics is now known to range from the upper Viséan to lower Namurian Series (Roberts and

others, 1995). Palynofloras identified as *Potonieisporites* Assemblage (Helby, 1969) are present at Seaham (locality 656 = University of Newcastle locality 1914) within slumped, laminated siltstone of the Seaham Formation 180 m above the Paterson Volcanics and in a similar stratigraphic position 2.5 km southeast of Paterson (locality 648 = University of Newcastle localities 1915 and 2464). Re-examination of Helby's (1969) material by C. Foster (in Roberts and others, 1995) suggests that this assemblage belongs to palynofloral Zone D of the Gal-

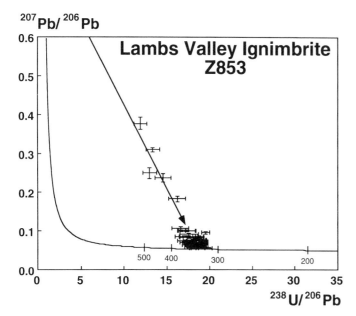

FIG. 10.—Concordia diagram of zircon compositions in the Lambs Valley Ignimbrite showing the apparent simplicity of the data set.

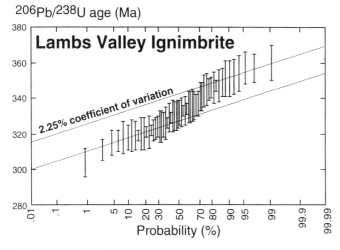

FIG. 11.—Probability plot of the ages of the same data as Figure 10, showing that the data are dispersed beyond the analytical error, which is represented by diagonal lines whose slope is calculated from the concurrent reproducibility of the standard zircon. Zircons in this sample have a range of ages with individual grains measurably distinct in age from one another. There is no grouping of ages from which the magmatic population might be recognized.

ilee Basin in Queensland (Jones and Truswell, 1992) or to the upper part of the *Diatomozonotriletes birkheadensis* Zone in the Bonaparte Basin of northwestern Australia. Jones (in Roberts and others, 1995) considers these units to be late Westphalian and mid-late Westphalian, respectively.

Claoué-Long and others (this volume) show that the age of the Paterson Volcanics is 328.5 ± 1.4 Ma (2σ) and discuss the magnetostratigraphic and biostratigraphic significance in greater detail.

DISCUSSION

Regional Correlations Within the SNEO

The new zircon ages for volcanic markers require significant alteration to some correlations within the Southern New England Orogen and consequent changes to stratigraphic and palaeogeographic reconstructions of Early Carboniferous eastern Australia.

Two major hiatuses at the sites of regional disconformities are reduced in duration, and an additional hiatus is inferred beneath all or part of the glacigene Seaham Formation. The first of these unconformities is the one below the Isismurra Formation in Figure 2. The hiatus represented by this unconformity was considered to span most of middle Viséan time by Roberts (1985), partly on the basis that the disconformity is geographically widespread beneath the Isismurra Formation and within southwestern parts of the Ararat Formation and supported by a late Viséan correlation for the *O. australis* Zone in the succeeding Bonnington Siltstone and Woolooma Formation (Roberts 1975). New biostratigraphic data from conodonts and the age of 342.1 ± 3.2 Ma (2σ) measured for lower Isismurra Formation (Curra Keith Tongue, located close to the Tournaisian/Viséan boundary), now permit little time for this hiatus. Conodonts of the Tournaisian *S. anchoralis* Zone have been recognized in the lower Ararat Formation beneath the disconformity, and early Viséan conodonts have been reported from limestone containing the *O. australis* Zone in the basal Bonnington Siltstone above the disconformity (Roberts and orthers, 1991; Jenkins and others, 1993). It is probable that this disconformity is a reflection of local uplift associated with the onset of widespread magmatic activity rather than a result of falling sea level (Roberts, 1985).

The second hiatus beneath the Mt. Johnstone Formation (Fig. 2) has been inferred from a mappable erosion surface on the upper part of the Isismurra Formation and Gilmore Volcanic Group and a Late Carboniferous age suggested for the Mt. Johnstone Formation. The Late Carboniferous age was inferred by the *Nothorhacopteris* and "enriched" *Nothorhacopteris* flora within the upper Mt. Johnstone Formation and apparent conformity between that formation and the overlying Paterson Volcanics, the latter then being considered Westphalian in age on the basis of a K-Ar age of 308 Ma. Demonstration by Claoué-Long and others (this volume) that the Paterson Volcanics are 328.5 ± 1.4 Ma (2σ) old (= Viséan) requires that the underlying Mt. Johnstone Formation and its contained flora are confined to the Viséan Series. An age for the Lambs Valley Ignimbrite would constrain the interval of the unconformity closely, but this unit remains undated. Nevertheless, the brief time measured between Martins Creek Ignimbrite (332.3 ± 2.2 Ma (2σ)) and the Paterson Volcanics requires rapid deposition of the Mt. Johnstone Formation and only a short interval for development of the unconformity. The suggestion that the uplift and erosion were caused by accretion of the SNEO on to the Australian craton at the beginning of the Namurian (Roberts, 1987) has been questioned by Flood and Aitchison (1992) who presented limited evidence for Late Devonian accretion.

A new and lengthy depositional hiatus is required by the reassignment of the Paterson Volcanics from Westphalian to Viséan age by Claoué-Long and others (this volume). This important hiatus is inferred beneath the Seaham Formation and most marine formations in the Gresford and Myall Blocks containing *M. barringtonensis* Zone faunas, and affects all areas of the shelf except the Dungog Embayment (Roberts and others, 1991), a marine re-entrant in the north. The duration of the hiatus beneath the Seaham Formation cannot be determined precisely. The lower part of the overlying Seaham Formation is inferred to be early Namurian by unpublished zircon dating showing that the Mirannie Ignimbrite 220 m above its base is about 320 Ma old (Claoué-Long, unpublished analyses, 1992), and correlation to the Viséan/Namurian boundary at about 327 Ma in Europe (see Fig. 12) following work by Roberts and others (1995). The base of the glacigene Seaham Formation is younger elsewhere. East of the Cranky Corner Basin, the formation appears to consist of younger sediments and volcanics, and in the Rosebrook Range mapping by Benson (pers. commun., 1976) has demonstrated the removal of units down to and including the upper part of the Gilmore Volcanic Group.

The revised Viséan age for the Mt. Johnstone Formation provides a much more understandable palaeogeographic picture for the Hunter-Myall region of the SNEO in the middle and late Viséan. In the southern Rouchel and Gresford Blocks, the conglomeratic basal portion of the Mt. Johnstone Formation and equivalent Balickera Conglomerate (Rattigan, 1967) represents a wedge deposited adjacent to an uplifted volcanic arc. The conglomerates pass into cyclical fluvial facies in the upper Mt. Johnstone and Italia Road Formations, a facies also present within the Salisbury Sandstone in the northern Gresford Block. Contemporaneous marine sediments in the Dungog Embayment are represented by lower parts of the Berrico Creek and Copeland Road Formations at Rawdon Vale and Barrington, respectively.

The suggestion that some parts of the lower Seaham Formation are early Namurian in age implies contemporaneity of this glacigene unit with marine sediments containing the *Levipustula levis* Zone. This contemporaneity is supported by correlated indicators of a deteriorating climate, including an abrupt lowering of diversity in marine invertebrate faunas in the *M. barringtonensis* Zone and particularly the *L. levis* Zone (Roberts, 1981), and in the striated pavements on the upper surface of the Paterson Volcanics and continental nature of glacigene sediments in the Seaham Formation. There is also a close similarity with the timing of glaciations in Argentina (Gonzalez, 1990) which is further elaborated by Roberts and others (1995).

The zircon ages also constrain the timing and duration of major Early Carboniferous eruptive episodes in the SNEO. The Isismurra Formation, which is the major volcanogenic unit within the Rouchel Block, spans an interval of about 10 Ma, from the base of the Viséan (Curra Keith Tongue, 342.1 ± 3.2 Ma (2σ)) into the upper part of the Viséan Series above the Martins Creek Ignimbrite (332.3 ± 2.2 Ma (2σ)). Lower parts of the Isismurra Formation were derived from the Muswellbrook Volcanic Centre, but the uppermost two ignimbrites, which are better represented in the Gilmore Volcanic Group in the Gresford Block, were derived from the Maitland Volcanic Centre (Buck, 1986; Roberts and others, 1991). The new timing information suggests that the Muswellbrook Volcanic Centre was active over a period of about 9 Ma. The Maitland Volcanic Centre, which produced the Gilmore Volcanic Group, appears to have had a much shorter history limited to the brief span between the Martins Creek Ignimbrite (which represents the

base of this volcanic centre in most sections) and the undated Lambs Valley Ignimbrite which represents its top. This range is a minimum because some additional volcanic horizons lie below the Martins Creek Ignimbrite in sections of the Rosebrook Range (Browne, 1927), and the source of the Paterson Volcanics is unknown.

The Nerong Volcanics in the eastern Gresford and Myall Blocks are less well known than the other two units and have previously been equated with the lower part of the Gilmore Volcanic Group (Roberts and Engel, 1987). The zircon age of 338.6 ± 3.8 Ma for the lower part of the Nerong Volcanics indicates that volcanism in the type area commenced well before that responsible for the Gilmore Volcanic Group. The Nerong Volcanics are centred on Port Stephens and have a substantial thickness in the type area west of Bulahdelah. Thin extensions project northwards almost to Gloucester and eastwards towards Forster; erosion in the east, indicated by the absence of *R. fortimuscula* Zone faunas, may have removed substantial quantities of volcanics before deposition of the overlying Booti Booti Sandstone and Karuah Formation. Poorly preserved brachiopods from immediately beneath the type section of the Nerong Volcanics in the Conger Formation belong to the *D. aspinosa* Zone (undifferentiated), which is compatible with the zircon age. At Boolambayte, however, a distal extremity of the volcanics overlies sediments with probable *L. tenuirugosa* Subzone fossils, and in the north near Gloucester another possible tongue (Buggs Creek Volcanic Member) overlies units with *L. tenuirugosa* Subzone faunas and is overlain by sediments containing the *R. fortimuscula* Zone. These distal extremities are therefore considerably younger than basal parts of the volcanic pile in the south, and a diachronous base to the Nerong Volcanics is inferred (Fig. 2). The zircon age from the base and palaeontological evidence from the top suggest that the Nerong Volcanics accumulated during an interval of about 6 Ma, probably originating from the Port Stephens Volcanic Centre (Buck, 1986). Minor volcanic units exist within the base of the Chichester Formation in the northern Gresford Block between the *L. tenuirugosa* Subzone and the *R. fortimuscula* Zone (Roberts and others, 1991). These have not been dated, but they correlate with the Buggs Creek Member and the base of the Gilmore Volcanic Group. The source of these volcanics appears to be local.

Ages of Australian Palyno- and Macrofloras

Zircon ages from non-marine parts of the SNEO succession profoundly affect the interpretation of Carboniferous floras and palynofloras. Kemp and others (1977) recognized two Carboniferous microfloras throughout Australia, the *Granulatisporities frustulentus* Microflora (Tournaisian-Viséan) containing the *Grandispora spiculifera* and *Anapiculatisporites largus* Assemblages, and the *Secarisporites* Microflora (late Viséan-Late Carboniferous) with the *Grandispora maculosa*, *Anabaculites yberti* and *Potonieisporites* Assemblages. Jones and Truswell (1992) have modified the Namurian and younger parts of this zonation from work in the Galilee Basin, Queensland. SHRIMP results from the Early Carboniferous part of the succession significantly affect the age of the *G. maculosa* Assemblage, which is present in the Wallaringa, Italia Road, and upper Mt. Johnstone Formations (Playford and Helby, 1968; Helby, 1969; Rob-

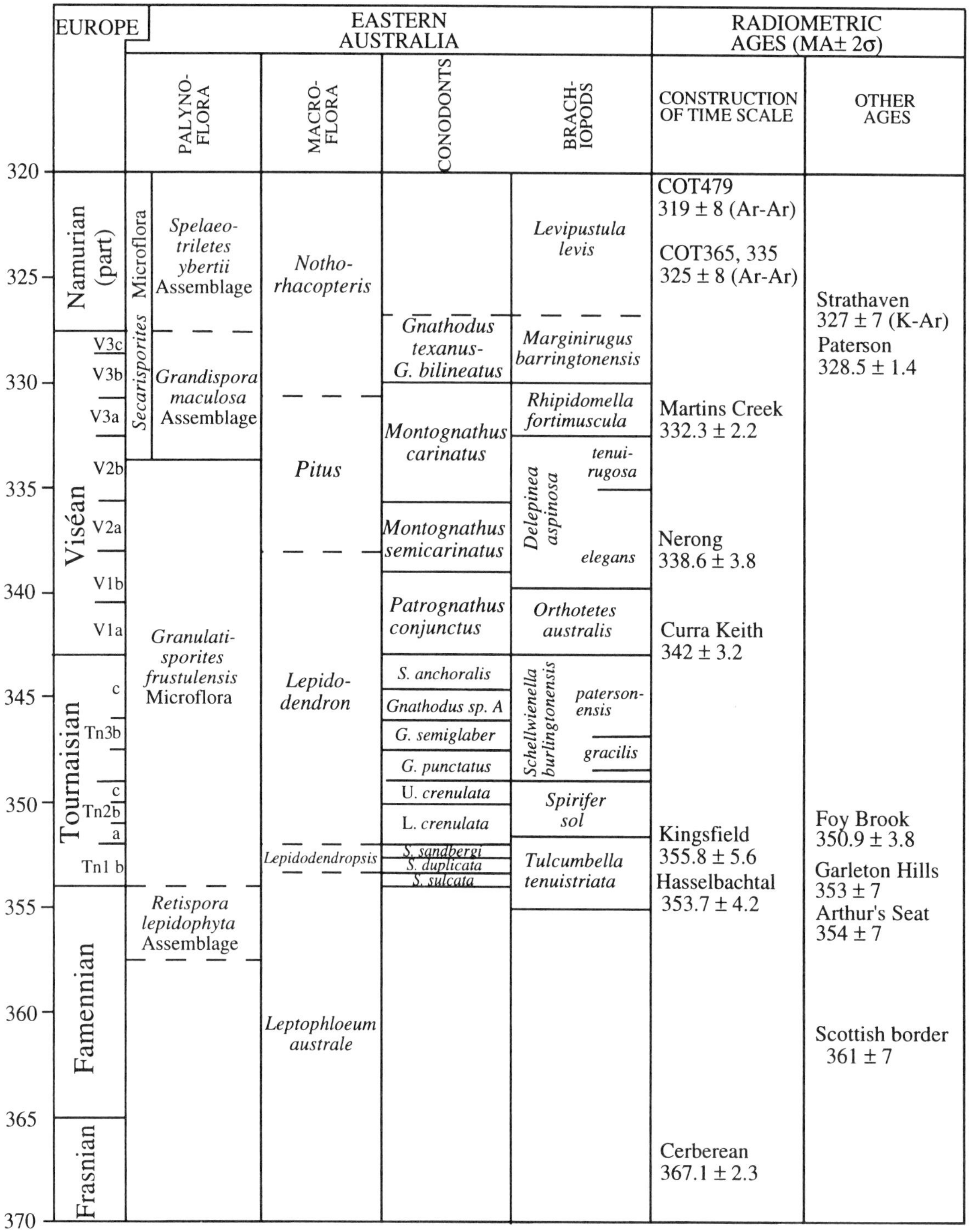

AGE (Ma)	EUROPE	EASTERN AUSTRALIA PALYNO-FLORA	MACRO-FLORA	CONODONTS	BRACH-IOPODS	RADIOMETRIC AGES (MA± 2σ) CONSTRUCTION OF TIME SCALE	OTHER AGES
320	Namurian (part)	Microflora — Spelaeo-triletes ybertii Assemblage	Notho-rhacopteris		Levipustula levis	COT479 319 ± 8 (Ar-Ar) COT365, 335 325 ± 8 (Ar-Ar)	
325				Gnathodus texanus- G. bilineatus	Marginirugus barringtonensis		Strathaven 327 ± 7 (K-Ar) Paterson 328.5 ± 1.4
330	Viséan V3c V3b V3a	Secarisporites — Grandispora maculosa Assemblage	Pitus	Montognathus carinatus	Rhipidomella fortimuscula / tenui-rugosa	Martins Creek 332.3 ± 2.2	
335	V2b V2a			Montognathus semicarinatus	Delepinea aspinosa / elegans	Nerong 338.6 ± 3.8	
340	V1b V1a	Granulati-sporites frustulensis Microflora		Patrognathus conjunctus	Orthotetes australis	Curra Keith 342 ± 3.2	
345	Tournaisian c Tn3b		Lepido-dendron	S. anchoralis / Gnathodus sp. A / G. semiglaber / G. punctatus	Schellwienella burlingtonensis / paterson-ensis / gracilis		
350	c Tn2b a Tn1 b			U. crenulata / L. crenulata / S. sandbergi / S. duplicata / S. sulcata	Spirifer sol / Tulcumbella tenuistriata	Kingsfield 355.8 ± 5.6 Hasselbachtal 353.7 ± 4.2	Foy Brook 350.9 ± 3.8 Garleton Hills 353 ± 7 Arthur's Seat 354 ± 7
355	Famennian	Retispora lepidophyta Assemblage	Lepidodendropsis				
360			Leptophloeum australe				Scottish border 361 ± 7
365	Frasnian					Cerberean 367.1 ± 2.3	
370							

Fig. 12.—Early Carboniferous time scale constructed from the zircon U-Pb ages described in this paper which are linked with eastern Australian biozones and the standard Western European Series. Biozone correlations after Jones (1991) and Roberts and others (1993). The right hand column compares earlier constraints on the Carboniferous timescale; see text for discussion. Except where marked otherwise, the numerical ages are zircon dates.

erts and others, 1991). The *G. maculosa* Assemblage extends from slightly below the Martins Creek Ignimbrite Member of the Gilmore Volcanic Group to the Paterson Volcanics. The assemblage therefore appears to range from the late middle to the late Viséan and has an age from about 334 to 328 Ma. A suggestion by Helby (in Roberts and others, 1991) that some upper Mt. Johnstone Formation assemblages could be post-*G. maculosa* and pre-*Potonieisporites* Assemblage is not sustained following reprocessing of material (Foster, pers. commun., 1993). The older *G. frustulentus* Microflora is tentatively identified within the Flagstaff Formation at Greenhills (GR731006 Paterson 1:100,000 sheet), 9 km north-northeast of Paterson (R.

J. Helby, pers. commun., 1993) from beneath the lowest occurrence of the *G. maculosa* Assemblage. The microflora, which is identified from the restricted *Dibolisporites distinctus/acritachus plexus,* accompanies brachiopods of the lower *D. aspinosa* Zone and ammonoids (*Irinoceras tuba* and *Merocanites* [*Erdbachites*] sp. B. of Campbell and others, 1983), indicative of an early to possibly middle Viséan age (Roberts and others, 1993b). The SHRIMP zircon ages indicate that the *G. frustulentus* Microflora ranges between about 354 and 334 Ma.

Eastern Australian Carboniferous macrofloras were subdivided by Morris (1985) into the *Lepidodendropsis, Lepidodendron, Pitus,* and *Nothorhacopteris* floras, the latter containing an "enriched" upper portion. Changed correlations in both marine and non-marine parts of the succession have an impact on the interpretation of three of the floras, particularly the *Nothorhacopteris* flora. The *Lepidodendron* flora is confined beneath the Nerong Volcanics and hence is older than about 339 Ma. The Pitus or Viséan petrifaction flora is present in the Wallaringa Formation, Nerong Volcanics, and the Caroda Formation (*L. tenuirugosa* Subzone) in northwestern New South Wales (Morris 1985). Zircon dates indicate that the age of the Pitus flora is between about 339 and 332 Ma. The *Nothorhacopteris* flora was previously interpreted as Late Carboniferous in age because it occurs in the Mt. Johnstone Formation, disconformably above the Gilmore Volcanic Group, and in the Seaham Formation from above the Paterson Volcanics (previously assigned as Westphalian from K-Ar dating); it was also present in the Late Carboniferous Booral, McInnes and Johnsons Creek Formations and Koolanock Sandstone of the SNEO, and Majors Creek and Mingaletta Formations in the Hastings Terrane. Zircon dates now indicate that the *Nothorhacopteris* flora first appeared at around 330 Ma within the Mt. Johnstone Formation and equivalents and before the eruption of the Paterson Volcanics (328.5 ± 1.4 Ma (2σ)). The flora extends into the Late Carboniferous succession, making its final appearance in the Seaham Formation. A diverse assemblage termed the "enriched *Nothorhacopteris, Sphenopteridium* flora" (Morris, 1985), recognized in the Joe Joe Group of Queensland, Currabubula Formation in northwestern NSW, and upper Mt. Johnstone, Italia Road and upper McInnes Formations, and the Johnsons Creek Conglomerate in the SNEO, was considered to be in the upper part of the range of *Nothorhacopteris.* (The localities for such records in the Joe Joe Group and Currabubula Formation require confirmation). The revised late Viséan age of the Mt. Johnstone Formation beneath the Paterson Volcanics and the occurrence of the "enriched" *Nothorhacopteris* flora above the *Levipustula levis* Zone (in the McInnes Formation and Johnsons Creek Conglomerate) indicates that the "enriched" *Nothorhacopteris* flora has a longer stratigraphic range than originally envisaged. Roberts and others (1995) have utilised zircon ages to indicate a range of Late Viséan to early Namurian within the SNEO.

The International Carboniferous Time Scale

Existing time scales for the Early Carboniferous (de Souza, 1982; Forster and Warrington, 1985; Harland and others, 1990) are based almost exclusively on K-Ar dating of continental basaltic volcanism in Britain. Most of these basalts have equivocal correlation with the marine faunas that are the basis of the bio-

stratigraphic scale, and it is not surprising that some revision of the time scale is required by dating of volcanics more directly associated with marine fossils. The ages discussed here date only three positions in the Early Carboniferous: its base, the Tournaisian/Viséan boundary, and a position within the Holkerian of the British Stages; much detail of the Early Carboniferous scale remains to be constrained. A suggested construction of the time scale is given in Figure 12.

The reference datum for the base of the Carboniferous is now the zircon age of 353.7 ± 4.2 Ma (2 s) for bentonite Bed 79 in the Hasselbachtal auxiliary stratotype section in Germany, 35 cm above the boundary and within the lower part of the *S. sulcata* conodont Zone (Claoué-Long and others, 1992; renormalized by Claoué-Long and others, this volume). In eastern Australia, the age for a volcanic in the Kingsfield Formation constrained by the same conodonts is within measurement uncertainty of the reference Hasselbachtal age. Figure 12 shows how previously applied ages for British basalts compare with this age for the base of the Carboniferous. Of these data, the 361 ± 7 Ma K-Ar age for the Scottish Border lavas has most influenced reviewers; these lavas are within unfossiliferous strata and could occupy a stratigraphic position anywhere between the entire Famennian and late Tournaisian (Odin, 1982; Claoué-Long and others, 1992), but the boundary has long been interpreted to predate them on the basis of a preferred interpretation that they are Carboniferous (Forster and Warrington, 1985). Comparison with the Hasselbachtal boundary age indicates that the Scottish Border lavas are in fact Late Devonian. The Cerberean Volcanics in Victoria, Australia, have also been used as a constraint on the base of the Carboniferous on the basis of previous correlation of associated fish faunas to the late Famennian; revision of this correlation means that their age is now a constraint on the Frasnian (see discussion by Claoué-Long and others, 1993). The difference in age between the Hasselbachtal bed 79 bentonite and the Cerberean Volcanics allows a duration of about 10 Ma for the Late Devonian Famennian Stage.

The Curra Keith Tongue (342.1 ± 3.2 Ma (2σ)) is a constraint on the age of the Tournaisian/Viséan boundary. The age of this boundary merits further investigation, however, in view of the interpretation-dependent age of the sample, which contained both inheritance and Pb loss. The Curra Keith Tongue is located between the top of the *S. burlingtonensis* brachiopod Zone and base of the *O. australis* brachiopod Zone at Rouchel (Roberts and Oversby, 1974), a level equivalent to sediments immediately above the top of the *S. anchoralis* Zone in the Gresford Block (Roberts and others, 1993b). At Gresford, a marine sequence containing the *S. anchoralis* Zone within the lower part of the Ararat Formation is disconformably overlain by non-marine sediments of the upper Ararat Formation that revert to marginal marine sandstone followed by a limestone at the base of the overlying Bonnington Siltstone (Roberts and others, 1991); the limestone contains the first occurrence of the *O. australis* Zone and early Viséan cosmopolitan conodonts of *Patrognathus conjunctus* Zone (Jenkins and others, 1993). This position closely coincides with the Tournaisian/Viséan boundary (Roberts and others, 1993b). The unnamed andesite from the Waverley Formation at Foy Brook was used by Cowie and Bassett (1989) to provide an age for the same boundary on the latest IUGS Global Stratigraphic Chart. The equivocal biostratigraphic position of this andesite is discussed above; its zircon age shows it to belong to the middle Tournaisian Series (Fig. 12). In apparent conflict with the revised age for the Tournaisian/Viséan boundary are the K-Ar ages for the Arthur's Seat and Garleton Hills lavas in Britain. These lavas have been inferred to be Viséan on miospore evidence, but their K-Ar ages correspond to the base of the Tournaisian, close to the age of the Hasselbachtal section. Jones (1988) has questioned the validity of these K-Ar age determinations, and Patterson and Hall (1986) have queried aspects of their palynological correlation. The duration of the Tournaisian Epoch is measured at approximately 12 Ma from the difference in age between the Hasselbachtal Bed 79 (353.7 ± 4.2 Ma (2σ)) and the Curra Keith Tongue (342.1 ± 3.2 Ma (2σ)).

The Martins Creek Ignimbrite (332.3 ± 2.2 Ma (2σ)) dates a position between the *D. aspinosa* and *R. fortimuscula* Zones, equivalent to a position within the Holkerian Stage of Britain (Roberts and others, 1993b). The Chichester Formation at Carrow Brook, which is bracketed by the volcanics, appears to contain only a lower part of the *R. fortimuscula* Zone, but younger parts of the zone are represented in the Berrico Creek Formation and in Queensland, where ammonoids suggest extension into the upper *Beyrichoceras* Zone (Roberts and others, 1993b). However, the first occurrence of the conodont *Gnathodus bilineatus* within the succeeding *M. barringtonensis* Zone in Queensland (Jenkins and others, 1993) may indicate an older age for the top of the *R. fortimuscula* Zone. Alone among the Scottish K-Ar ages, the 327 ± 7 Ma age of the Strathaven lavas, attributed to the upper Viséan, appears to coincide with the zircon age constraints (Fig. 12); however, the correspondence of these lavas to biostratigraphy is based on lithologic correlation (Forster and Warrington, 1985).

The age of the top of the Viséan is not constrained by data in this paper, and a range is permitted by dates in the literature. The most useful constraints are in the Upper Carboniferous of Europe, from where Hess and Lippolt (1986) have reported well-shaped [40]Ar/[39]Ar plateau ages that supersede preexisting data for the Upper Carboniferous in both measurement precision and biostratigraphic constraints. These are directly comparable with zircon U-Pb dating (Claoué-Long and others, this volume). Two possible ages within the Namurian A are proposed by Hess and Lippolt (1986): one at 319 ± 8 Ma and another at 325 ± 8 Ma; they construct their Upper Carboniferous timescale on the basis that the older of these constrains the base of the Namurian. We take the Viséan/Namurian boundary at 327 Ma following work by Roberts and others (1995).

Ages of Upper Carboniferous Series are constrained by the [40]Ar/[39]Ar dating of Hess and Lippolt (1986). When these are combined with the new zircon dating of the Early Carboniferous reported in this paper, a radically adjusted Carboniferous timescale is produced. Figure 13 compares the revised Carboniferous timescale with the recent compilations of De Souza (1982), Forster and Warrington (1985), and Harland and others (1990).

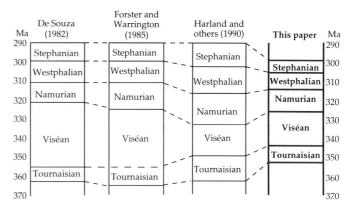

FIG. 13—A timescale for the Carboniferous Period, calculated from zircon dating in this paper (Early Carboniferous) and [40]Ar/[39]Ar dating of the European Upper Carboniferous (Hess and Lippolt, 1986) and compared with previous scales compiled prior to the new dating work. The Carboniferous Period overall is reduced by 20 Ma, but the relative proportions of the Series remain comparable to those inferred by De Souza (1982) and Forster and Warrington (1985).

Ma and younger than about 350 Ma. The base of the *O. australis* Zone is younger than about 342 Ma. The top of the *D. aspinosa* and basal part of the *R. fortimuscula* Zone (at Carrowbrook) is about 332 Ma old.

2. Palynofloras of the *Granulatisporities frustulentus* Microflora are older than about 335 Ma and confined to the Tournaisian to middle Viséan interval. Those of the *Grandispora maculosa* Assemblage span an interval of about 334–328 Ma in the middle to late Viséan.

3. The *Nothorhacopteris* flora first appeared in the late Viséan at about 332 Ma and extends to the top of the Seaham Formation (Westphalian; Roberts and others, in prep.) The "enriched *Nothorhacopteris, Sphenopteridium* flora" of Morris (1985) is present in units ranging in age from late Viséan to early Namurian (Roberts and others, in prep.).

4. The duration of Early Carboniferous volcanic centres in the SNEO of Australia ranges from ⩾1 Ma in the Maitland Volcanic Centre (which produced the Gilmore Volcanic Group), to about 6 Ma in the Port Stephens Centre (Nerong Volcanics), and 9 Ma in the Muswellbrook Centre (feeding the Isismurra Formation volcanics).

5. The non-marine Mt. Johnstone Formation of the SNEO is confined to the upper Viséan Series, and may be a proximal equivalent of the Salisbury Sandstone; it is also correlative with the mainly marine Berrico Creek and Copeland Road Formations.

6. A major hiatus during the middle to late Viséan Epoch affected all areas outside the Dungog Embayment in the SNEO. Shorter breaks at disconformities are inferred beneath the Mt. Johnstone Formation (early late Viséan), and beneath the Isismurra Formation and within the Ararat Formation (late Tournaisian).

7. The Tournaisian/Viséan boundary is placed close to 343 Ma on the basis that the Curra Keith Tongue of the Isismurra Formation can be interpreted as 342.0 ± 3.6 Ma old and is located above the upper *S. burlingtonensis* and *S. anchoralis* Zones and below the *O. australis* Zone and the *Patrognathus conjunctus* conodont Zone (Jenkins and others, 1993; Rob-

SUMMARY AND CONCLUSIONS

1. The numerical ages of Early Carboniferous brachiopod zones in eastern Australia are constrained by zircon U-Pb dating. The *S. burlingtonensis* Zone is older than about 342

erts and others, 1993b). The pyroxene andesite from Foy Brook used by Cowie and Bassett (1989) to indicate an older age for the Tournaisian/Viséan boundary on the IUGS Global Stratigraphic Chart is not reliably constrained biostratigraphically, its zircon age placing it within the middle Tournaisian.

8. The age of the Holkerian Stage of Britain can also be constrained from the Australian numerical time scale by correlation of the age of the Martins Creek Ignimbrite (332.3 ± 2.2 Ma), which lies between the *L. tenuirugosa* Subzone and the basal *R. fortimuscula* Zone.

9. The difference between the Hasselbachtal Bed 79 in Germany (and Kingsfield Formation in the SNEO) and the Curra Keith Tongue measures the duration of the Tournaisian Epoch of the Carboniferous as about 12 Ma. The Famennian Age of the Upper Devonian has a similar duration.

ACKNOWLEDGMENTS

This study was initiated by an Australian Research Council grant to J. R. and a Bureau of Mineral Resources Fellowship to J. C. L. who conducted the work as a Visiting Fellow at the Research School of Earth Sciences, Australian National University. L. P. Black and G. C. Young are thanked for reviewing early drafts, and the expert assistance of S. Maxwell and D. Maidment is gratefully acknowledged. The contribution of J. C. L. and P. J. J. is published with the permission of the Executive Director of the Australian Geological Survey Organisation.

REFERENCES

AITCHISON, J. C., FLOOD, P. G., AND SPILLER, F. C. P., 1992, Tectonic setting and palaeoenvironment of terranes in the southern New England orogen, eastern Australia, as constrained by radiolarian biostratigraphy: Palaeogeography, Palaeoclimatology, Palaeoecology, v. 94, p. 31–54.

BROWNE, W. R., 1927, The geology of the Gosforth district, NSW: Journal and Proceedings of the Royal Society of New South Wales, v. 60, p. 213–277.

BUCK, M. D., 1986, Early Carboniferous volcanic centres in eastern Australia—revealed by measurements of flow direction indicators in ignimbrites: Auckland-Hamilton-Rotorua International Volcanological Congress Abstracts, 33 p.

CAMPBELL, K. S. W., BROWN, D. A., AND COLEMAN, A. R., 1983, Ammonoids and the correlation of the Lower Carboniferous rocks of eastern Australia: Alcheringa, v. 7, p. 75–123.

CLAOUÉ-LONG, J. C., JONES, P. J., ROBERTS, J., AND MAXWELL, S., 1992, The numerical age of the Devonian-Carboniferous boundary: Geological Magazine, v. 129, p. 281–291.

CLAOUÉ-LONG, J. C., JONES, P. J., AND ROBERTS, J., 1993, The age of the Devonian-Carboniferous boundary: Annales de la Société géologique de Belgique, T. 115-fasc. 2, p. 531–549.

COLLINS, W. J., OFFLER, R., FARRELL, T. R. AND LANDENBERGER, B., 1993, A revised Late Palaeozoic-Early Mesozoic tectonic history for the southern New England Fold Belt, in Flood, P. G. and Aitchison, J. C., eds., New England Orogen, Eastern Australia: Armidale, Department of Geology and Geophysics, University of New England, NSW, Australia, p. 69–84.

COWIE, J. W. AND BASSETT, M. G., 1989. IUGS 1989, Global Stratigraphic Chart, with geochronometric and magnetostratigraphic calibration. Supplement to Episodes v. 12 (2).

DE SOUZA, H. A. F., 1982, Age data from Scotland for the Carboniferous time scale, in Odin, G. S., ed., Numerical Dating in Stratigraphy, Part 1: Chichester, Wiley, p. 455–460.

ENGEL, B. A., 1962, Geology of the Bulahdelah-Port Stephens district, NSW: Journal and Proceedings of the Royal Society of New South Wales, v. 95, p. 197–215.

EVERNDEN, J. F. AND RICHARDS, J. R., 1962, Potassium-argon ages in eastern Australia: Journal of the Geological Society of Australia, v. 9, p. 1–49.

FLOOD, P. G. AND AITCHISON, J. C., 1992, Late Devonian accretion of the Gamilaroi Terrane to eastern Gondwana: provenance linkage suggested by the first appearance of Lachlan Fold Belt-derived quartzarenite: Australian Journal of Earth Sciences, v. 39, p. 539–544.

FORSTER, S. C. AND WARRINGTON, G., 1985, Geochronology of the Carboniferous, Permian and Triassic, in Snelling, N. J., ed., The Chronology of the Geological Record: London, Geological Society of London Memoir 10, p. 99–113.

GONZALEZ, C. R., 1990, Development of the Late Palaeozoic glaciations of the South American Gondwana in western Argentina: Palaeogeography, Palaeoclimatology, Palaeoecology, v. 79, p. 275–285.

HAMILTON, G., HALL, G. C., AND ROBERTS, J., 1974, The Carboniferous nonmarine stratigraphy of the Paterson-Gresford district, New South Wales: Journal and Proceedings of the Royal Society of New South Wales, v. 107, p. 76–86.

HARLAND, W. B., SMITH, A. G., AND WILCOCK. B., eds, 1964, The Phanerozoic time scale: Supplement to Quarterly Journal of the Geological Society of London, v. 120, p. 1–458.

HARLAND, W. B., COX, A. V., LLEWELLYN, P. G., PICKTON, C. A. G., SMITH, A. G., AND WALTERS, R., 1982, A Geologic Time Scale: Cambridge, Cambridge University Press, 131 p.

HARLAND, W. B., ARMSTRONG, R. L., COX, A. V., CRAIG, L. E., SMITH, A. G., AND SMITH, D. G., 1990, A Geologic Time Scale 1989: Cambridge, Cambridge University Press, 263 p.

HELBY, R. J., 1969, Preliminary palynological study of Kuttung sediments in central eastern New South Wales: Geological Survey of New South Wales Records, v. 11, p. 5–14.

HESS, J. C. AND LIPPOLT, H. J., 1986, [40]Ar/[39]Ar ages of tonstein and tuff sanidines: new calibration points for the improvement of the Upper Carboniferous time scale: Chemical Geology (Isotope Geoscience Section), v. 59, p. 143–154.

JENKINS, T. B. H., 1974, Lower Carboniferous conodont biostratigraphy: Palaeontology, v. 17, p. 909–924.

JENKINS, T. B. H., CRANE, D. T., AND MORY, A. J., 1993, Conodont biostratigraphy of the Viséan Series in eastern Australia: Alcheringa, v. 17, p. 211–283.

JONES, M. J. AND TRUSWELL, E. M., 1992, Late Carboniferous and Early Permian palynostratigraphy of the Joe Joe Group, southern Galilee Basin, Queensland, and implications for Gondwanan stratigraphy: Bureau of Mineral Resources Journal of Australian Geology and Geophysics, v. 13, p. 143–185.

JONES, P. J., 1988, Comments on some Australian, British and German isotopic aga data for the Carboniferous System: Newsletter on Carboniferous Stratigraphy, v. 6, p. 26–29.

JONES, P. J., 1991, Australian Phanerozoic Timescales 5. Carboniferous biostratigraphic charts and explanatory notes: Canberra, Bureau of Mineral Resources, Geology and Geophysics, Record 1989/3, 43 p.

KEMP, E. M., BALME, B. E., HELBY, R. J., KYLE, R. A., PLAYFORD, G., AND PRICE, P. L., 1977, Carboniferous and Permian palynostratigraphy in Australia and Antarctica: a review: Bureau of Mineral Resources Journal of Australian Geology and Geophysics, v. 2, p. 177–208.

KULLMANN, J., KORN, D., AND WEYER, D. 1990, Ammonoid zonation of the lower Carboniferous subsystem, in Brenckle, P. L. and Manger, W. L. eds., International correlation and division of the Carboniferous system: Courier Forschungsinstitut Senckenberg, v. 130, p. 127–131.

MAMET, B. AND SKIPP, B., 1970, Lower Carboniferous calcareous foraminifera; preliminary zonation and stratigraphic implications for the Mississippian of North America: Sheffield, 6iéme Congres International de Stratigraphie et de Géologie du Carbonifére, Compte Rendu, 1967, p. 1129–1146.

MORRIS, L. N., 1985, The floral succession in eastern Australia, in Diaz, C. M., ed., The Carboniferous of the World: Madrid, IUGS Publication 16, Instituto Geologico y Mineras de Espana and Empressa Nacional Adaro de Investigaciones Mineras SA, p. 118–123.

ODIN, G. S., ed., 1982, ed., Numerical Dating in Stratigraphy: London, Wiley, 1094 p.

PATTERSON, I. B. AND HALL, I. H. S., 1986, Lithostratigraphy of the Devonian and early Carboniferous rocks of the Midland Valley of Scotland: British Geological Survey Report, v. 18, p. 1–19.

PLAYFORD, G. AND HELBY, R., 1968, Spores from a Carboniferous section in the Hunter Valley, New South Wales: Journal of the Geological Society of Australia, v. 15, p. 103–119.

RATTIGAN, J., 1967, The Balickera Section of the Carboniferous Kuttung Facies, New South Wales: Journal and Proceedings of the Royal Society of New South Wales, v. 100, p. 75–84.

RILEY, N. J. 1993, Dinantian (lower Carboniferous) biostratigraphy and chronostratigraphy in the British Isles: Journal of the Geological Society, London, v. 150, p. 427–446.

ROBERTS, J., 1975, Early Carboniferous brachiopod zones of eastern Australia: Journal of the Geological Society of Australia, v 22, p. 1–32.

ROBERTS, J., 1981, Control mechanisms of Carboniferous brachiopod zones in eastern Australia: Lethaia, v. 14, p. 123–134.

ROBERTS, J., 1985, Carboniferous sea level changes derived from depositional patterns in Australia: 10iéme Congrés International de Stratigraphie et de Géologie du Carbonifére, Compte Rendu Madrid 1983, v. 4, p. 43–64.

ROBERTS, J., 1987, Carboniferous faunas: their role in the recognition of tectonostratigraphic terranes in the Tasman Belt, eastern Australia, *in* Leitch, E. C. and Scheibner, E., eds., Terrane Accretion and Orogenic Belts, Geodynamics Series 19: Washington, American Geophysical Union Geological Society of America, p. 93–102.

ROBERTS, J., CLAOUÉ-LONG, J. C., AND JONES, P. J., 1993a, SHRIMP zircon dating and Australian Carboniferous time. 11iéme Congres International de Stratigraphie et de Géologie du Carbonifére, Buenos Aires 1991, Compte Rendu, v. 2, p. 319–338.

ROBERTS, J., CLAOUÉ-LONG, J. C., JONES, P. J., AND FOSTER, C. B., 1995, SHRIMP zircon age control of Gondwana sequences in late Carboniferous and Early Permian Australia, *in* Dunay, R. L. and Hailwood, E. A., eds., Non-biostratigraphical Methods of Dating and Correlation: Geological Society Special Publication 89, p. 145–174.

ROBERTS, J. AND ENGEL, B. A., 1987, Depositional and tectonic history of the southern New England Orogen: Australian Journal of Earth Sciences, v. 34, p. 1–20.

ROBERTS, J., ENGEL, B. A., AND CHAPMAN, J., 1991, Geology of the Camberwell 9133, Dungog 9233, and Bulahdelah 9333 1:100,000 sheets (Hunter-Myall Region) New South Wales: Sydney, Geological Survey of New South Wales, 382 p.

ROBERTS, J., JONES, P. J., AND JENKINS, T. B. H., 1993b, Revised correlations for Carboniferous marine invertebrate zones of eastern Australia: Alcheringa, v. 17, p. 353–376.

ROBERTS, J. AND OVERSBY, B. S., 1974, The Lower Carboniferous geology of the Rouchel district, New South Wales: Bureau of Mineral Resources, Geology and Geophysics Bulletin, v. 147, p. 1–93.

TERA, F. AND WASSERBURG, G. J. 1974, U-Th-Pb systematics on lunar rocks and inferences about lunar evolution and the age of the moon: Proceedings of the Fifth Lunar Conference (Supplement 5, Geochemica et Cosmochimica Acta), v. 2, p. 1571–1599.

MAGNETOSTRATIGRAPHY OF PERMO-CARBONIFEROUS TIME

NEIL D. OPDYKE

Department of Geology, University of Florida, Gainesville, FL 32611

ABSTRACT: New data has recently become available on the length and duration of the Permo-Carboniferous reversed superchron (PCRS, Kiaman). This data set is reviewed, and the upper boundary is found to occur at approximately 260 Ma and the lower boundary at about 315 Ma. The frequency of reversal before and after the onset of the PCRS is similar to that observed before and after the Cretaceous quiet zone (KQZ). The earth's magnetic field seems to have two states, one with reversal frequencies exceeding 1 per million years and the other with no or few reversals for 10's of millions of years. The KQZ occurred at a time of continental dispersal and the PCRS at a time of continental amalgamation.

PERMIAN PERIOD

Irving and Parry (1964) were the first to recognize the long period of reversed polarity spanning the late Carboniferous and Permian time. They named it the Kiaman Magnetic Interval, because it was from near the town of that name in New South Wales, Australia where a rock showing reversed polarity was first reported (Mercanton, 1926) for the Kiaman region of eastern Australia. The name Kiaman has been retained by Russian workers but was abandoned by Irving and Pulliah (1976) and replaced by the Permo-Carboniferous Reversed Superchron (PCRS). The PCRS is known to terminate within late Permian time. Irving and Parry (1963) originally placed the termination of the Kiaman Magnetic Interval at the beginning of the Narabeen Chocolate shales of the Sidney Basin (Australia) which are late Permian and Early Triassic age. They called this polarity change the Illawara reversal. A restudy of this formation was carried out by Embleton and McDonnell (1981) who confirmed a reversed to normal polarity change in red claystones near the base of the Triassic units. The older sediments in this sequence are unstably magnetized so that a sequence of magnetozones could not be established.

The later Permian and Early Triassic sequences in North America, Europe and in many of the Gondwanaland continents often contain large unconformities. Marine sediments are present, however, in upper Permian sediments along the northern margin of Gondwanaland (Pakistan and India) as well as China. The best magnetic stratigraphy for later Permian time comes from the Wargal and Chidru Formations of the Salt Range in Northern Pakistan (Haag and Heller, 1991; Fig. 1, column 13). These formations are dominantly carbonate and yield an excellent fauna aiding correlation to other sections. The magnetic record in these formations, which span late Kazanian and Tartarian time, shows as many as six normal and five reversed magnetozones; however, the top of the PCRS is not determined since the lower part of Wargal Formation is normally magnetized. Unfortunately, it has become apparent that there is missing time at the top of the Permian System in this section, and sediments equivalent to Changxingian time in the Chinese sequence are missing (Wignall and Hallam, 1993). The biostratigraphic and magnetostratigraphic correlation of the Salt Range sequence to China (Steiner and others, 1989; Heller and others, 1988) is therefore tenuous, although the correlation of the long normal magnetozone at the top of the sequence in both China and Pakistan is reasonable and is indicated on Figure 1.

The correlation of the sequence from Pakistan to non-marine red beds which dominate the rock record in the Soviet Union, western Europe, and North America and South America is not clear. A recent study on the Ochoan (Tartarian) Dewey Lake Formation in northern Texas shows at least two normal magnetozones in late Permian time; however, the top of the PCRS was not determined as in the Pakistan and Chinese sequences (Fig. 1, column 4). Khramov (1974) shows at least 5 normal magnetozones in sediments from the Upper Permian type section of the Ural Mountains (Fig. 1, column 3). The data quality appears to be high, and the results internally consistent. Unfortunately, these sections have unconformities across the Permo-Triassic boundary, and it is clear that some amount of time is missing from the rock sequences. This situation is similar to that in the western United States.

A considerable amount of work has been done on the magnetostratigraphy of the Permian units from eastern Germany by Menning and others (1988) who have systematically investigated the sediments of the Rotliegende and Zechstein Formations for reversal stratigraphy in both outcrop and bore holes (Fig. 1, column 9). Normally magnetized sediments have been found in the upper Rotliegende and throughout the Zechstein Formation. Menning and others (1988) show at least seven normal magnetozones in the late Permian Period. From what we know from other areas, this number of magnetozones probably is close to being correct since at least five normal magnetozones are known from the USSR and six from Pakistan. In the composite column for the Permian Period given in Figure 1, eight normal magnetozones are given for the late Permian strata derived principally from the Chinese, Pakistan and central European records.

The position of the top of the PCRS is constrained in the rock record to be older than the Tartarian units of the USSR, the Zechstein and upper Rotliegende Formations of central Europe, and the base of the Wargal Formation of the Salt Range in Pakistan which is Murgabian in age; correlative to the Guadalupian or Wordian sequence of North America; and to the base of the Kazanian stage (Haag and Heller, #1991; Catalano and others, 1991). The PCRS North American boundary lies beneath Ochoan units of west Texas which has an associated radiometric date of 251 ± 4 Ma on a volcanic ash (Molina-Garza and others, 1990). It should be noted that the Permian/Triassic boundary in China has been dated at 251 MA by Claoué-Long and others (1991) using the SHRIMP ion probe at the ANU. Taken at face value, the date from Texas would place the Ochoan magnetic stratigraphy very close to the Permian/Triassic boundary. Harland and others (1990) have subdivided the Permian into the Rotliegende and Zechstein Epochs in a bipartite division. The base of the Zechstein Epoch is given a numerical age of 256 Ma by Harland and others (1990) and a chronogram age of 259 Ma. Valencio and others (1977) in Argentina have reported a Normal magnetozone lying above an

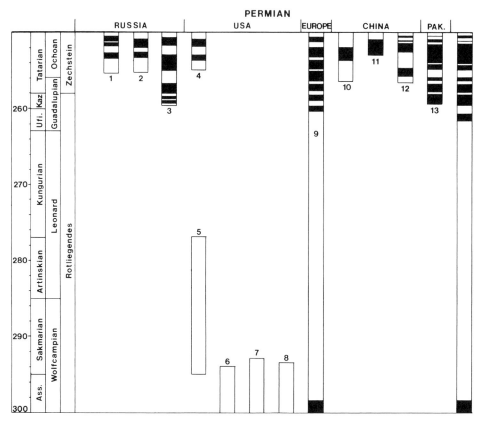

Fig. 1.—Compilation of Permain magnetostratigraphic data. Black represents normal field; white, reversed field. 1. Molostovsky (1983), 2. Gurevitch and Stautsitays (1985), 3. Khramov (1974), 4. Molina-Garza and others (1989), 5. Gose and Helsley (1972) 6. Miller and Opdyke (1985), 7. Diehl and Shive (1981), 8. Diehl and Shive (1979), 9. Menning and others (1988), 10. McFadden and others (1988), 11. Heller and others (1988), 12. Steiner and others (1988), 13. Haag and Heller (1991), and 14. Composite stratigraphic column.

intrusion dated at 259 ± 7 Ma which they called the Q. Del Pimiento event. Creer and others (1971) also noted normally magnetized lavas at 258 ± 5 Ma. It is possible that these normal magnetozones may mark the end of the PCRS and not a subchron within the PCRS. Unfortunately, because of a large unsampled interval above this normal zone, a pattern cannot be discerned since 55% of the section was not sampled. This Argentine section may offer an opportunity to date the end of the Kiaman. Reliable radiometric dates directly associated with sediments of late Permian age are rare, and the late Permian stage boundaries are largely unconstrained in time. This fact is reflected in the chronogram ages given in Harland and others (1990). The precise date of the end of the Kiaman Magnetic Interval is therefore unknown; Haag and Heller (1992) believe it to be older than 261 Ma and Menning (pers. commun., 1992) gives an ages of 265 Ma. In Figure 1, an age of 262 Ma is chosen for this boundary; however, it should be understood that nowhere is it directly dated.

Over the years, reports of short normal subchrons within the PCRS have been published both in Soviet and western literature. Helsley (1963) published initial results of normally magnetized sediments from the Dunkard Formation of West Virginia which is believed to be early Permian age. The initial study was done with alternating field demagnetization which is notoriously ineffective in removal of secondary overprints from hematite-bearing samples. However, samples from the normal

horizon were subjected to thermal demagnetization by Gose and Helsley (1972) as an adjunct to the study of Permian sediments from New Mexico and were found to retain their normal polarity. Recently isolated, normally magnetized sites have been obtained from lower Permian Pictou age sediments from Prince Edward Island by Symonds (1990). The normal polarity in the Pictou beds may well be correlative with that within the Dunkard Series of West Virginia. Menning (1988) in his study of the Rotliegende of eastern Germany found normally magnetized sediments in the lower Permian part of this sequence which he places at the Permian/Pennsylvanian boundary, implying a correlation with the Dunkard series of West Virginia. It is not clear whether these results from different areas are correlative or not. It does seem certain that one or more normal subchrons occur within the lower Permian strata even if they are difficult to detect in magnetostratigraphic studies designed for this purpose. Several sections have been sampled across the Permo-Carboniferous boundary in the western United States (Diehl and Shive, 1979, 1981; Miller and Opdyke, 1985), and no normally magnetized rocks were detected. Sinito and others (1979) give evidence for a further short subchron within the PCRS, positioned 100 m below a lava dated at 288 ± 5 Ma. This may be equivalent to the normal subchron seen in the Dunkard series, although it was originally thought to be the base of the PCRS. Normally magnetized sites of lower Permian age have been reported in several pole position studies (Halverson and

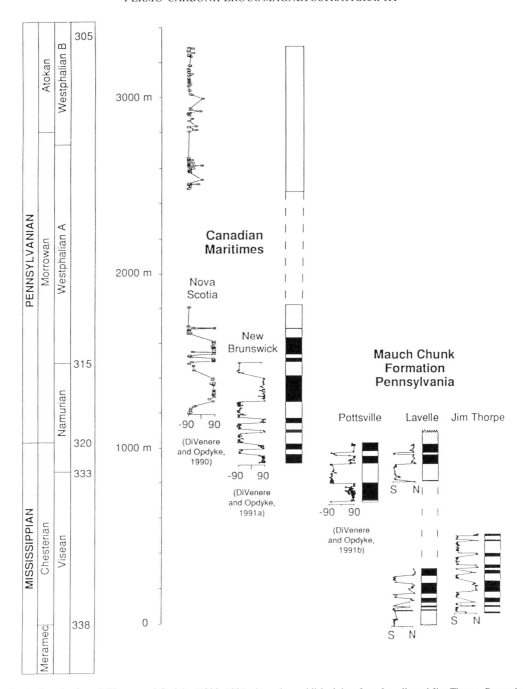

FIG. 2.—Magnetic stratigraphy from DiVenere and Opdyke (1990, 1991a, b) and unpublished data from Lavelle and Jim Thorpe, Pennsylvania and Opdyke and DiVenere (1994). VGP plotted against stratigraphic position and time.

others, 1989; Wynne and others, 1983; Irving and Monger, 1987) and may be correlative with the Dunkard subchron. Although normal subchrons occur within the PCRS, recent studies by Diehl and Shive (1979, 1981), Miller and Opdyke (1985), and Magnus and Opdyke (1991) have shown that the PCRS is dominantly of reversed polarity as claimed by Irving and Pulliah (1976; Fig. 1, columns 6–9).

CARBONIFEROUS PERIOD

The position of the base of the PCRS has not been fully resolved. In North America, McMahon and Strangway (1968)

reported normal polarity magnetozones in sediments of the Late Demoinesian Minturn Formation from Colorado. The section was restudied by Miller and Opdyke (1985) who did not find normal polarity in this section. The study was extended throughout the Minturn formation by Magnus and Opdyke (1991) who determined that reversed polarity extended to the base of the formation which is earliest Demoinesian or late Atokan age.

Work in the Maritime provinces of Canada by Roy and Morris (1983) has shown that sediments with dual polarity were present in rocks as young as Westphalian A age. DiVenere and

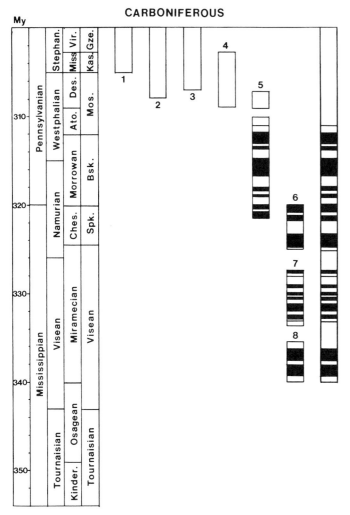

FIG. 3.—Time scale modified from Harland and others (1990), Hess and Lippolt (1986), Roberts and others (this volume). Black represents normal field; white reversed field. 1. Steiner, (1988), 2. Diehl and Shive (1981), 3. Miller and Opdyke (1985), 4. Magnus and Opdyke (1991), 5. DiVenere and Opdyke (1990) 6. DiVenere and Opdyke (1991a, b), 7. Opdyke and DiVenere (1994), and 8. Torsvik and others (1989).

Opdyke (1990, 1991) restudied their sections and revealed that these sediments contained an internally coherent set of magnetozones in the Maringouin, Shepody, and Claremont Formations of Namurian and Westphalian A age. At least seven normal magnetozones are present which form a pattern yielding a reversal rate of about one field reversal per million years (Fig. 2). The age of the magnetic stratigraphy is constrained to be late Carboniferous age by a positive fold test. An attempt was made to extend the magnetic stratigraphy to younger formations exposed along the shore of the Bay of Fundy. All samples were reversely magnetized; however, a data gap of 700 m occurs in the sequence limiting the precision of the study. These data yield the best and most complete reversal sequence now available for early Upper Carboniferous time. The dual polarity for rocks of early Pennsylvanian age is supported by a study on red Morrowan paleosols from Arizona (Nick and others, 1991). A coherent pattern of reversals is hard to recognize in the Arizona

study because of data gaps and the fact that both polarities occur at the same horizon.

The age of the beginning of the PCRS was originally determined to be 310 Ma based on the age of the Paterson Volcanics in Australia. These volcanics were shown by Irving (1966) to be normally magnetized and defined as the base of the PCRS (Kiaman). Sites taken from the Seaham Formation, directly overlying the Paterson Volcanics, were reversely magnetized and were thought to be late Carboniferous age. Late Carboniferous sites from the Currabubula and the Lark Hill Formations were also reversely magnetized and passed the fold test. The base of the Kiaman Magnetic Interval was, therefore, thought to be firmly determined. However, a recent study of the Paterson Volcanics using single crystal zircon dating by Claoue-Long and others (this volume) has yielded a date of 328 Ma. This date places these rocks within Visean time, much older than the normally magnetized sequences from the Canadian Maritimes. Therefore, the Paterson Volcanics cannot be used to fix the beginning of the PCRS.

The other significant data set that bears on this problem is the magnetostratigraphy of Middle Carboniferous strata of the Donetz basin in the Ukraine. Khramov (1973) has presented a magnetic stratigraphy for the upper Carboniferous Period which shows normal magnetozones as young as the Upper Moscovian time. Opdyke and others (1993) have done a preliminary restudy of these sediments, and although it was possible to recover a prefolding characteristic direction of magnetization which is probably Carboniferous age, it was not possible to confirm Khramov's (1973) original magnetic stratigraphy. At this time, the only unimpeachable data that bears on the age of the base of the MPTS is that from Maritime Canada. The highest normal magnetozone in this sequence occurs within sediments of Westphalian A time at about 312 Ma and yields a duration for the PCRS of approximately 50 my (from 262–312 Ma).

In North America, the reversal pattern determined in the lower Pennsylvanian strata has been extended into the Upper Mississippian (Visean) sediments of the Mauch Chunk Formation of Pennsylvania (DiVenere and Opdyke, 1991a; Fig. 2). Opdyke and DiVenere (1994) have reported a minimum of nine normal magnetozones within the Mauch Chunk Formation with a reversal frequency of 1.5/my, similar to that of early Pennsylvanian time. The earth's field would appear to be about 50% normal and 50% reversed in Middle Carboniferous time (Fig. 3).

Little is known of the pattern of reversal for the rest of lower Carboniferous time; however, both polarities are present in sediments of Visean and Tournaisian age (Irving and Strong, 1985). Recently, the beginnings of a magnetic stratigraphy has been obtained from lower Carboniferous lavas of the Midland Valley of Scotland by Torsvik and others (1989). In Figures 3 column 8, Palmer and others (1985), have previously reported magnetic stratigraphy from Carboniferous limestones from Great Britain. A recent study by McCabe and Channell (1994) has shown that some of these limestones are remagnetized. Therefore, the stratigraphy shown by Palmer and others (1985) is in serious doubt.

DISCUSSION

As a result of the improved state of knowledge of the behavior of the earth's magnetic field in the Late Paleozoic Era, it is

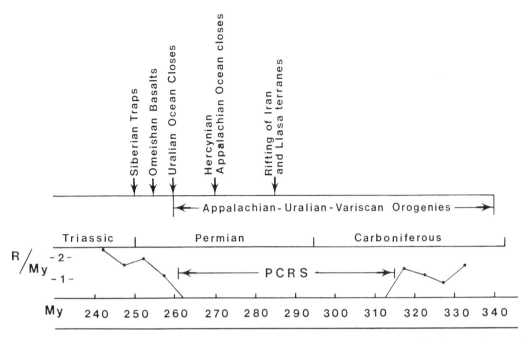

FIG. 4.—Reversal frequency in the Carboniferous, Permian and early Triassic Periods with timing of major tectonic and magmatic events for the interval.

possible to attempt a comparison of the Kiaman Magnetic Interval with the Cretaceous Quiet Zone or KQZ. The reversal record is reasonably complete for Middle Carboniferous and younger times.

The two intervals of quiet field behavior are similar in length, 40 my for the KQZ and 50 my for the PCRS. The reversal frequencies preceding and following the quiet intervals are similar, dropping to about 1 reversal/my prior to the base of the quiet intervals and increasing to a rate of 2/my within 20 my of the recommencement of field reversals. These quiet intervals represent about 25% of the elapsed time from Middle Carboniferous to Recent time. The record would seem to indicate that these are two states for the earth's magnetic field, one which is operative for 75% of the time where the field reverses at a ratio of 2/my, or more, and the other in which field reversals are rare or absent.

It has often been suggested that the tectonic and volcanic effects observed in earth's history are perhaps correlative with the behavior of the earth's magnetic field. How events observed at the surface of the earth are related to the core of the earth where the earth's geomagnetic field is generated is not clear. In recent years, several authors have considered the importance of the D″ layer at the base of the mantle as the source for plumes which are recorded at the earth's surface as large igneous provinces such as the Siberian Traps and the Ontong-Java Plateau. Two recent studies (Courtillet and Besse, 1987; Larson and Olson, 1991) have suggested that plumes leaving the D″ region would carry heat away from the core-mantle boundary thus affecting the generation of the magnetic field. Courtillot and Besse (1987) suggested that the superplume which gave rise to the Deccan Traps had caused the Earth's field to begin reversing polarity. Larson and Olsen (1991) point out that the most voluminous plume heads are manifested as large oceanic plateaus such as the Ontong-Java Plateau which arrived at the surface

early in the KQZ. They, therefore, argue that the removal of material from the D″ layer effectively turns off the reversal process. Both hypotheses cannot be correct. The tectonic situation during the KQZ is essentially the continuing breakup of Pangea.

We may now consider the situation during the PCRS (Fig. 4). Unlike the KQZ, the Kiaman Magnetic Interval is on the other side of the Wilson cycle and is coincident with the closing of ocean basins and not their opening. It is the time of closure of the Proto-Atlantic and the Uralian Ocean and the final episode in the formation of Pangea. There are at least two episodes of plume generation. The most important of the two is the emplacement of the Siberian Traps close in time to the Permian/Triassic boundary and a possible analogue to the Deccan Traps. The second is the emplacement of the Omeishan basalts of southwest China that were emplaced in the late Permian time. Oceanic plateaus generated during late Paleozoic time would have been subducted and would have left no trace. Larson (1991) has suggested that other data may be used as proxies for the generation of oceanic plateau's by superplumes. He points out that the Permo-Pennsylvanian flooding of Pangaea as represented by the Absaroka sedimentary sequence of Sloss (1963) is analogous to the Cretaceous Zuni sequence correlating in each case to non reversal of the field. He also suggests a link between the creation of oil and gas in the Cretaceous Period and increasing CO_2. This was then equated to the formation of Permo-Pennsylvanian coals. General warming in the Permian Period is, however, hard to support because of the extensive southern hemisphere glaciation. In the future, it might be possible to deconvolve the signal from the $^{87}Sr/^{86}Sr$ values from marine sediments to identify the portion that is derived from spreading ridges and superplumes. At present, however, the evidence is too uncertain to lead to a conclusion. A better understanding of the middle and lower Paleozoic reversal sequence is needed in order to solve this problem. If sedimentary sequences are a predictor of superplume generation, then the next older superchron can be expected in late Ordovician and early Silurian time.

REFERENCES

CATALANO, R., STEFANO, P., AND KOSUR, H, 1991, Permain circumpacific deep water faunas from the western Tethy's, (Sicily, Italy)- new evidence for the position of the Permian Tethy's: Palaeogeography, Palaeoclimatology, Palaeoecology, v. 87, p. 75–108.

CLAOUÉ-LONG, J. C., ZICHAO, Z., GUOGAN, M., AND SCHAOHUA, D., 1991, The age of the Permian—Triassic boundary: Earth and Planetary Science Letters, v. 105, p. 182–190.

COURTILLOT, V. AND BESSE, J., 1987, Magnetic field reversals, polar wander, and core mantle coupling: Science, v. 237, p. 1140–1147.

CREER, K. M., MITCHELL, J. G., AND VALENCIO, D. A., 1971, Evidence for a normal geomagnetic field polarity event at 263 ± 5 My B.P. within the late Paleozoic reversal interval: Nature, v. 233, p. 87–89.

DIEHL, J. F. AND SHIVE, P. N., 1979, Paleomagnetic studies of the early Permian Ingelside formation of northern Colorado: Geophysical Journal of the Royal Astronomical Society, v. 56, p. 278–282.

DIEHL, J. F. AND SHIVE, P. N., 1981, Paleomagnetic results from the Late Carboniferous/Early Permian Casper Formation: Implications for Northern Appalachian Tectonics: Earth and Planetary Science Letters, v. 54, p. 281–291.

DIVENERE, V. J. AND OPDYKE, N. D., 1990, Paleomagnetism of the Maringouin and Shepody formations, New Brunswick: A Namurian Magnetic Stratigraphy: Canadian Journal of Earth Science, v. 27, p. 803–810.

DIVENERE, V. J. AND OPDYKE, N. D., 1991a, Magnetic polarity stratigraphy in the uppermost Mississippian Mauch Chunk formation, Pottsville, Pennsylvania: Geology, v. 19, p. 127–130.

DIVENERE, V. J. AND OPDYKE, N. D., 1991b, Magnetic polarity stratigraphy and carboniferous paleopole positions form the Joggins Section, Cumberland Basin, Nova Scotia: Journal of Geophysical Research, v. 96, p. 4051–4064.

EMBLETON, B. J. J. AND MCDONNELL, K. L., 1981, Magnetostratigraphy in the Sydney Basin, Southeastern Australia, in McElhinny and others, eds., Global Reconstructions and the Geomagnetic Field During the Paleozoic: Tokyo, Center for Academic Publishing, p. 1–10.

GOSE, W. A. AND HELSLEY, C .E., 1972, Paleomagnetism and rock magnetism of the Permian Cutler and Elephant Canyon Formations in Utah: Journal of Geophysical Research, v. 77, p. 1534–1548.

GUREVICH, Y. L. AND SLAUTSITAYS, I. P., 1985, A paleomagnetic section in the Upper Permian and Triassic deposits on Novaya Zemlya: International Geological Review, v. 27, p. 168–177.

HAAG, M. AND HELLER, F., 1991, Late Permian to Early Triassic magnetostratigraphy: Earth and Planetary Science Letters, v. 107, p. 42–54.

HALVERSON, E., LOWANDOWSKI, M., AND JELENSKA, M., 1989, Paleomagnetism of the upper Caroniferous: Physics of Earth and Planetary Interiors, v. 55, p. 54–64.

HARLAND, W. B., ARMSTRONG, R. L., COX, A. V., CRAIG, L. E., SMITH, A. G., AND SMITH, D. G., 1990, A Geologic Time Scale, 1989: Cambridge, Cambridge University Press, 263 p.

HELLER, F., LOWRIE, W., LI, H., AND WANG, J., 1988, Magnetostratigraphy of the Permio-Triassic boundary section at Shangsi: Earth and Planetary Science Letters, v. 88, p. 348–356.

HELSLEY, C. E., 1965, Paleomagnetic results from the lower Permain Dunkard series of West Virginia: Journal of Geophysical Research, v. 70, p. 413–424.

HESS, J. C. AND LIPPOLT, H. J., 1986, 40Ar/39Ar ages of Tonstein and Tuff sanidines: New calibration points for the improvment of the upper Carboniferous time scale: Chemical Geology (Isotope Geoscience Section), v. 59, p. 143–154.

IRVING, E., 1966, Paleomagetism of some Carboniferous rocks from New South Wales and its relation to geological events: Journal of Geophysical Research, v. 71, p. 6025–6051.

IRVING, E. AND MONGER, J. W. H., 1987, Preliminary paleomagnetic results from the Permian Asitka group, British Columbia: Canadian Journal of Earth Science, v. 24, p. 1490–1497.

IRVING, E. AND PARRY, L. G., 1963, The magnetism of some Permian rocks from New South Wales: Geophysical Journal of the Royal Astronomical Society, v. 7, p. 395–411.

IRVING, E. AND PULLIAH, G., 1976, Reversals of the geomagnetic field, magnetostratigraphy, and the relative magnitude of Paleosecular variation in the Phanerozoic: Earth Science Review, v. 12, p. 35–64.

IRVING, E. AND STRONG, D. F., 1985, Paleomagnetism of rock from Burin Peninsula Newfoundland: hypothesis of late paleozoic displacement of Acadia criticized: Journal of Geophysical Research, v. 90, p. 1949–1962.

KHRAMOV, A. N., 1974, Paleomagnetism of the Paleozoic: Moscow, NEDRA, 236 p. (in Russian).

KHRAMOV, A. N. AND RODIONOV, R., 1981, The geomagnetic field during Paleozoic time, in McElhinny and others, eds., Global Reconstructions and the Geomagnetic Field During the Paleozoic: Tokyo, Center for Academic Publishing, p. 99–116.

LARSON, R. L., 1991a, Latest pulse of Earth: evidence for a mid-Cretaceous super plume: Geology, v. 19, p. 547–550.

LARSON, R. L., 1991a, Geological consequences of super plumes: Geology, v. 19, p. 963–969.

LARSON, R. L. AND OLSON, P., 1991, Mantle plumes control magnetic reversal frequency: Earth and Planetary Science Letters, v. 107, p. 437–447.

MAGNUS, G. AND OPDYKE, N. D., 1991, A paleomagnetic investigation of the Minturn Formation, Colorado: A study in establishing the timing of remanence acquisition: Tectonophysics, v. 187, p. 181–189.

MCCABE, C. AND CHANNELL, J. E. T., 1994, Late paleozoic remagnetization in limestones of the Craven Basin (northern England) and the rock magnetic fingerprint of remagnetized sedimentary carbonates: Journal of Geophysical Research, v. 99, p. 4603–4612.

MCFADDEN, P. L., MA, M.W., MCELHINNY, M. W., AND ZHANG, Z. K., 1988, Permo-Triassic magnetostratigraphy in China: Northern Tarim: Earth and Planetary Science Letters, v. 87, p. 152–160.

MCMAHON, B. E. AND STRANGWAY, D. W., 1968, Stratigraphic implications of paleomagneitc data from upper Paleozoic — lower Triassic redbeds of Colorado: Geological Society of America Bulletin, v. 79, p. 417–428.

MENNING, M., KATZUNG, G., AND LUTZNER, H., 1988, Magnetostratigraphic investigation in the Rotliegendes (300–252 Ma) of Central Europe: Zeitschrieff Geologia Wissenschaften Berlin, v. 16, p. 1045–1063.

MERCANTON, P. L., 1926, Inversion de l'inclination aux âges geologique: Terrestrial Magnetism and Atmospheric Electricity, v. 31, p. 187–190.

MILLER, J. D. AND OPDYKE, N. D., 1985, The magnetostratigraphy of the red sandstone creek section, Vail, Colorado: Geophysical Research Letters, v. 12, p. 133–136.

MOLINA-GARZA, R. S., VAN DER VOO, R., AND GEISSMAN, J. W., 1989, Paleomagnetism of the Dewey Lake Formation Northwest Texas: End of the Kiaman superchron in North America: Journal of Geophysical Research, v. 94, p. 17881–17888.

MOLOSTOVSKY, E. A., 1983, Paleomagnetic stratigraphy of the upper Permian and Triassic of the Eastern European part of the USSR: Unpublished Thesis, University of Saratov, Saratov, 162 p.

NICK, K., XIA, K., AND ELMORE, R. D., 1991, Paleomagnetic and petrographic evidence for early magnetizations in successive Terra Rosa Paleosols, Lower Pennsylvanian Black Price limestone Arizona: Journal of Geophysical Research, v. 96, p. 9873–9885.

OPDYKE, N. D. AND DIVENERE, V. J., 1994, Magnetostratigraphy of the Mississippian Mauch Chunk formation of Pennsylvania: EOS, p. 130.

OPDYKE, N. D., KHRAMOV, A. N., GUREVITCH, E., IOSIFIDI, A. G., AND MAKAROV, I. A., 1993, A paleomagnetic study of the middle Carboniferous of the Donetz Basin, Ukraine: EOS, p. 118.

PALMER, J. A., PERRY, S. P. G., AND TARLING, D. H., 1985, Carboniferous Magnetostratigraphy: Journal of the Geological Society of London, v. 142, p. 945–955.

ROY, J. L. AND MORRIS, W. A., 1983, A review of paleomagnetic results from the Carboniferous of North America: the concept of Carboniferous geomagnetic field horizon markers: Earth and Planetary Science Letters, v. 65, p. 167–181.

SINITO, A.M., VALENCIO, D. A., AND VILAS, J. F., 1979, Paleomagnetism of a sequence of upper Paleozoic — lower Mesozoic red beds from Argentina: Geophysical Journal of the Royal Astronomical Society, v. 58, p. 237–247.

SLOSS, L. L., 1963, Sequences in the cratonic interior of North America, Geological Society of America Bulletin, v. 74, p. 93–114.

STEINER, M., 1988, Paleomagnetism of the late Pennsylvanian and Permian: Journal of Geophysical Research, v. 93, p. 2201–2215.

STEINER, M., OGG, J., ZHANG, Z., AND SUN, S., 1989, The late Permian/early Triassic magnetic polarity timescale and plate motions of South China: Journal of Geophysical Research, v. 84, p. 7343–7363.

SYMONDS, D. T. A., 1990, Early Permian pole evidence from the Pictou red beds, Prince Edward Island, Canada: Geology, v. 18, p. 234–237.

TORSVIK, T.H., LYSE, O., ATTARAS, G., AND BLUCK, B. J., 1989, Paleozoic paleomagnetic results from Scotland and their bearing on the British apparent polar wandering path: Physics of Earth and Planetary Interiors, v. 55, p. 93–105.

TURNER, P. AND TARLING, D. H., 1975, Implications of new paleomagnetic results from the Carboniferous system of Britain: Journal of the Geological Society of London, v. 131, p. 469–488.

VALENCIO, D. A., VILAS, J. F., AND MENDIA, J. E., 1977, Paleomagnetism of a sequence of red beds of the Middle and Upper sections of Paganzo group (Argentina) and the correlation of Upper Paleozoic-Lower Mesozoic rocks: Geophysical Journal of the Royal Astronomical Society, v. 51, p. 59–74.

WIGNALL, P. B. AND HALLAM, A., 1993, Greisbachian (Earliest Triassic) palaeoenvironmental changes in the Salt Range, Pakistan and southeast China and their bearing on the Permo-Triassic mass extinction Tethy's: Palaeogeography, Palaeoclimatology, Palaeoecology, v. 102, p. 215–237.

WYNNE, P. J., IRVING, E., AND OSADETZ, K., 1983, Paleomagentism of the Esayoo Formation (Permian) of northern Ellesmere Island: possible clue for the solution of the Nares Strait dilema: Tectonophysics, v. 100, p. 241–256.

PART II
MESOZOIC ERA

LATE JURASSIC-EARLY CRETACEOUS TIME SCALES AND OCEANIC MAGNETIC ANOMALY BLOCK MODELS

JAMES E. T. CHANNELL
Department of Geology, University of Florida, Gainesville, FL 32611
ELISABETTA ERBA
Dipartimento di Scienze della Terra, Università di Milano, 20133 Milano, Italy
MASAO NAKANISHI
Ocean Research Institute, University of Tokyo, Tokyo 164, Japan and Scripps Institution of Oceanography, La Jolla, CA 92093
AND
KENSAKU TAMAKI
Ocean Research Institute, University of Tokyo, Tokyo 164, Japan

ABSTRACT: Comparison of oceanic anomaly block models in the M0-M29 interval from the Japanese, Phoenix, Hawaiian and Keathley lineations indicates that the Hawaiian block model represents the closest approximation to a constant spreading rate record. The new Hawaiian block model differs slightly from that of Larson and Hilde (1975). Currently popular numerical age estimates for polarity chrons, base CM0 (121 Ma), CM16-CM15 (137 Ma) and top CM25 (154 Ma), are consistent with constant spreading rate in the new Hawaiian block model but inconsistent with constant spreading in the Larson and Hilde (1975) block model. A new time scale (CENT94) is based on the above ages and constant spreading in the new Hawaiian block model.

Land section magnetostratigraphy, mainly from Italy and Spain, has provided direct correlations of polarity chrons to stage boundaries through ammonite biozones, and indirect correlation through nannofossil and calpionellid biozonations: Barremian-Aptian (base of CM0), Hauterivian-Barremian (upper part of CM4), Valanginian-Hauterivian (base of CM11n), Berriasian-Valanginian (CM15n), Tithonian-Berriasian (base of CM18), Kimmeridgian-Tithonian (CM22A) and Oxfordian-Kimmeridgian (top CM25). These correlations yield the following stage boundary ages using CENT94: Barremian-Aptian (121 Ma), Hauterivian-Barremian (126 Ma), Valanginian-Hauterivian (131.5 Ma), Berriasian-Valanginian (135.8 Ma), Tithonian-Berriasian (141.6 Ma), Kimmeridgian-Tithonian (150 Ma), and Oxfordian-Kimmeridgian (154 Ma).

INTRODUCTION

The Late Jurassic-Early Cretaceous time scales of Kent and Gradstein (1985, KG85) and Harland and others (1990, GTS89) imply constant spreading rate in the M0-M25 Hawaiian oceanic anomaly block model of Larson and Hilde (1975, LH75). The KG85 time scale uses the constant spreading rate assumption to interpolate between 119 Ma for the Barremian-Aptian boundary (base CM0*) and 156 Ma for the Oxfordian-Kimmeridgian boundary (CM25n). The GTS89 time scale is based on chronogram ages for the Late Jurassic-Early Cretaceous stage boundaries. These ages were used to make a linear recalibration of the KG85 time scale, thereby inheriting the constant spreading rate assumption. The chronograms of GTS89 for this interval are poorly constrained and therefore do little to validate the constant spreading rate assumption.

Gradstein and others (1994, this volume) have presented integrated Triassic to Cretaceous time scales. GRAD94 supersedes an earlier version (GRAD93) presented at the AAPG/SEPM special symposium in New Orleans in April, 1993. In both time scales, radiometric ages were used to construct maximum likelihood chronograms for stage boundary ages. The maximum likelihood age estimates were then combined (weighted average) with stage boundary age estimates from magnetochronology. The magnetochronological estimates utilize magnetostratigraphic correlations of polarity chrons to stage boundaries, and the LH75 constant spreading rate block model to interpolate between polarity chron ages. For GRAD93, the magnetochronological interpolation is among five polarity chron age estimates: CM0 base (122 Ma), CM10N base (130 Ma), CM16n (137 Ma), CM21n base (145 Ma) and CM25n top (154 Ma). The GRAD93 time scale ages differ significantly

*polarity chron nomenclature after Harland and others (1982, p. 80–82) with the prefix "C" to distinguish polarity chrons from oceanic magnetic anomalies

from these magnetochronological tie-point ages, due to the weighting of the radiometric chronogram estimates which incorporate both high and low temperature radiometric ages. For GRAD94, the magnetochronological LH75 interpolation is between two ages: CM1 top (at 123.5 Ma) and CM26 (at 155.3 Ma), and radiometric control is restricted to high temperature ages.

Obradovich (1993) presented a Cretaceous time scale (OBRAD93) in which the Early Cretaceous portion is based on constant, but different, spreading rates for two intervals of LH75 defined by three age tie points: CM0 at 121 Ma, CM16/CM16n at 137 Ma and CM25n at 156 Ma. The resulting time scale implies a spreading rate change in LH75 at CM16/CM16n.

In this paper, we present block models derived from the Japanese, Hawaiian and Phoenix magnetic anomaly lineations, and compare these models with published data from the Hawaiian and Keathley lineations (Larson and Hilde, 1975; Klitgord and Schouten, 1986). The availability of block models from different seafloor spreading centers allows apparent spreading rate changes to be assessed and provides an independent means of evaluating time scales.

OCEANIC ANOMALY BLOCK MODELS

The Japanese, Hawaiian and Phoenix block models are based on selected magnetic anomaly profiles (Table 1, Fig. 1) from the cruise data used by Nakanishi and others (1989, 1992). For each lineation set (Fig. 2), the linkage between individual profiles was based on matching overlapping profile segments. Each selected profile was projected perpendicular to the local strike of the magnetic anomaly lineation using Generic Mapping Tools (GMT, Wessel and Smith, 1991). Inverse earth and phase filters (Schouten and McCamy, 1972) were applied to each profile, and profiles were filtered to remove the short and long wavelength components. Reversal boundaries between the prin-

TABLE 1.—CRUISE PROFILES USED TO CONSTRUCT OCEANIC MAGNETIC ANOMALY BLOCK MODELS (SEE NAKANISHI AND OTHERS, 1989, 1992, FOR COMPLETE LIST OF CRUISE PROFILES FROM WHICH CHOICE WAS MADE).

Japanese	Phoenix	Hawaiian
ZTES03AR (M1–M3)	7TOW3BWT (M1–M4)	POL7201 (M0–M29)
V2006 (M4–M10)	GH7801 (M5–M11)	
V3212 (M10N–M16)	C1205 (M12–M17)	
V3214 (M17–M21)	V3214 (M18–M29)	
JPYN04BD (M22–M27)		
KH87–3 (M28–M29)		

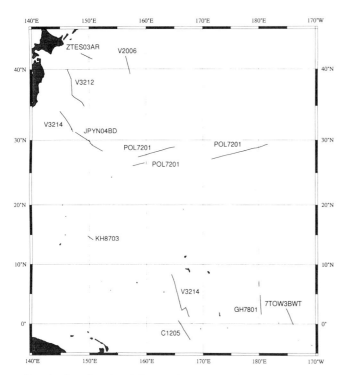

FIG. 1.—Location of track lines used to construct the block models for this study.

cipal magnetic anomalies were determined on the basis of local zero crossings (see example in Fig. 3). In a few cases, the procedure did not produce a zero crossing for a major anomaly, and reversal boundaries had to be estimated by eye. The method is more objective than picking anomaly boundaries by eye, but has the disadvantage that short wavelength minor anomalies are obscured by the filtering procedure. We therefore consider only the major anomaly boundaries (Fig. 4, Table 2). No attempt was made to place reversals associated with small amplitude anomalies in the block models. The distance data for the Keathley record (Table 2) are based on spreading rates from Klitgord and Schouten (1986).

The distances between reversals in the Japanese, Hawaiian and Phoenix block models can now be compared with one another, and with LH75 and the Keathley block model. The Phoenix block model reveals the most variable spreading rate. The similarity in shape of the plot of the Japanese, Hawaiian and Keathley block model distances against the Phoenix block model distances (Fig. 5) indicates that the spreading rate changes are predominantly in the Phoenix record. The plot of

the Japanese and Hawaiian block models against the Keathley record (Fig. 6) indicates a spreading rate change at M21 which is most likely to be a change in the Keathley spreading rate rather than in the Japanese, Hawaiian and LH75 block models (see also Sundvik and Larson, 1988). The Hawaiian block model differs slightly from LH75 as illustrated by the gradual inflexion in the plot of these two block models at about M16 time (Fig. 7). This slight inflexion at about M16 time is also seen in the plot of the LH75 record against the Japanese and Keathley records (Fig. 7), and therefore, is probably an artifact of LH75.

The POL7201 tracks used as the basis for our Hawaiian block model (Figs. 1, 2; Table 1) are part of the NOAA1-4 tracks used as the basis for LH75. We attribute the differences between the new Hawaiian block model and LH75 to different emphasis given to the various tracks in constructing the block models, and to different methods used in picking anomaly boundaries and in splicing of profiles across the Waghenaer (Mendocino) fracture zones.

SELECTION OF ABSOLUTE AGES FOR A REVISED TIME SCALE

Mahoney and others (1993) obtained a ^{40}Ar/^{39}Ar age of 122.3 ± 1 Ma for the basaltic basement at ODP Site 807 on the Ontong Java Plateau. Tarduno and others (1991) have argued that Ontong Java volcanism is constrained to a short Early Aptian interval. This interpretation was based on the observation that sediments overlying basaltic basement, and volcanoclastic sediments at distal sites, belong to the *Globigerinelloides blowi* planktonic foraminiferal zone, which was considered to be confined to the Early Aptian interval (Sliter, 1992). Recent Lower Cretaceous planktonic foraminiferal biostratigraphy in sections dated with ammonites and magnetostratigraphy (Coccioni and others, 1992) has indicated that the *G. blowi* Zone extends into the Barremian stage (see Fig. 8). The first occurrence (FO) of *G. blowi* has now been correlated to the upper part of CM3 (middle Barremian) and the presence of the *G. blowi* Biozone does not, therefore, restrict Ontong Java volcanism to the Early Aptian interval. In the standard low-latitude nannofossil biozonation of Thierstein (1971, 1973, 1976), two zones were identified in the Aptian (Fig. 8). The base of the *Chiastozygus litterarius* Zone was identified by the FO of *C. litterarius*, the FO of *Rucinolithus irregularis* and the last occurrence (LO) of *Nannoconus colomii*. Recent biostratigraphic studies (Coccioni and others, 1992; Erba, in prep.) have shown that of these three events only the FO of *R. irregularis* is a reliable marker for the base of this zone (Fig. 8). In Thierstein's zonation, the base of the *Rhagodiscus angustus* Zone was defined by the FO of *R. angustus* and the FO of *Eprolithus floralis*. Recent biostratigraphic studies have confirmed that the FO of *E. floralis* is reliable, being consistently correlative to the base of the *L. cabri* foraminiferal Zone, whereas *R. angustus* often occurs in the underlying *C. litterarius* Zone and *G. blowi* Zone (Larson and others, 1993; Fig. 8). At Site 807 on the Ontong Java Plateau, the oldest sediments above basement can be attributed to the *C. litterarius* Zone based on the FO of *R. irregularis*. Moreover, they correlate to the upper part of this biozone, above the "nannoconid crisis" which is a distinct nannofossil event occurring post CM0 (Erba, in prep.; Fig. 8). The ^{40}Ar/^{39}Ar age of 122.3 ± 1 Ma for the basaltic basement at this site (Mahoney and others, 1993) obviously predates these sediments.

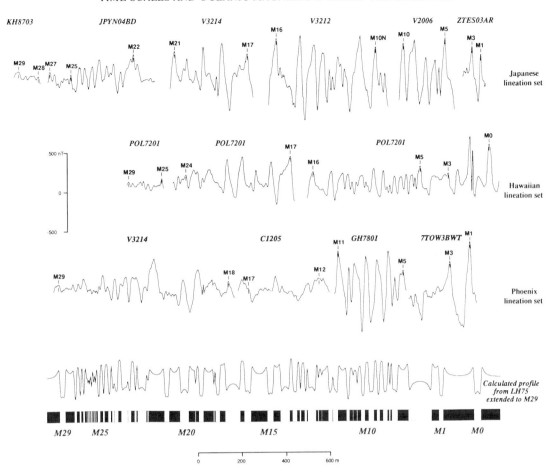

Fɪɢ. 2.—Profiles used to construct the block models. The calculated profile is the simulation from the Larson and Hilde (1975) block model extended to M29.

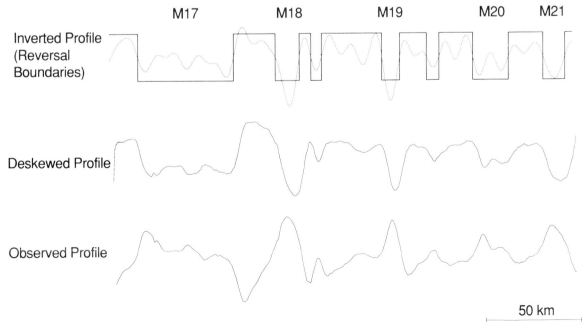

Fɪɢ. 3.—An example of the analytical procedure, illustrated using part of the profile from V3214 (Japanese lineations).

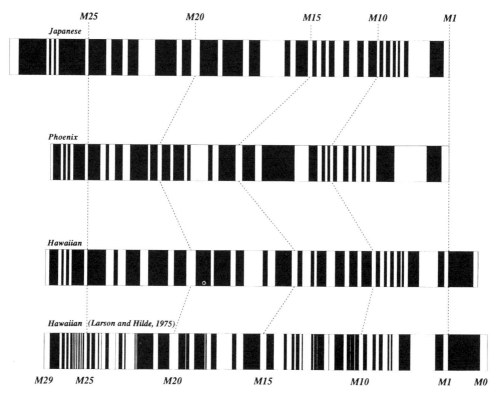

FIG. 4.—Comparison of Japanese, Phoenix, new Hawaiian and Larson and Hilde (1975) (Hawaiian) block models.

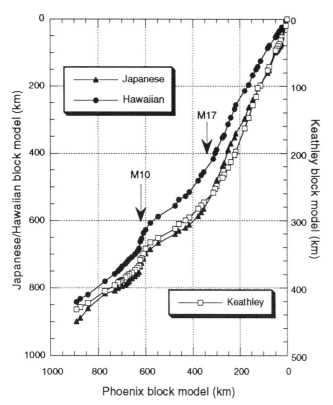

FIG. 5.—Block model distances for Japanese/Hawaiian/Keathley lineations plotted against Phoenix block model distances.

At Site 878 on MIT Guyot, the oldest dated sediments (at 25m above basement) are from the lower part of the *C. litterarius* nannofossil Zone, based on the FO of *R. irregularis* and their location below the "nannoconid crisis" (Erba and others, in prep.). Pringle and others (1993) obtained a $^{40}Ar/^{39}Ar$ age of 121.8 ± 0.5 Ma from these basalts. The standard used by Pringle and others (1993) (MMhb-1 513.9 Ma) is different from that used by Mahoney and others (1993; MMhb-1 520.4 Ma), therefore the $^{40}Ar/^{39}Ar$ ages at the two sites cannot be directly compared. Applying the conversion factor (*1.01417) to the age obtained by Pringle and others (1993) from the MIT Guyot, we obtain an age of 123.5 ± 0.5 Ma (Pringle, pers. commun., 1993). The remanent magnetization in the basaltic basement indicates a reversed polarity zone overlying a normal polarity zone (Nakanishi and Gee, in prep.; Fig. 8). The reversed polarity zone may be correlative to CM0 or CM1, however, in view of the earliest Aptian age of sediments located 25 m above basement, we correlate the reversed polarity zone to CM1.

Coleman and Bralower (1993) obtained a U-Pb zircon age of 122 ± 0.3 Ma from a bentonite in the Great Valley Group (Northern California). This level has been correlated to the *C. litterarius* Zone based on the FO of *C. litterarius*. Unfortunately, *R. irregularis* has not been found in the section (Bralower, pers. commun., 1993). As discussed above, *C. litterarius* has been reported from several Barremian sections and cannot, therefore, be used to define the Barremian/Aptian boundary, and the 122 ± 0.3 Ma bentonite layer from the Great Valley Group may be Barremian.

TABLE 2.—BLOCK MODEL DISTANCES FOR JAPANESE, PHOENIX, HAWAIIAN AND KEATHLEY BLOCK MODELS. KEATHLEY DATA AFTER KLITGORD AND SCHOUTEN (1986).

Anomaly	Japanese (km)	Phoenix (km)	Hawaiian (km)	Keathley (km)
M0			904.125	461.361
			894.317	455.971
M1	899.962	891.679	840.710	432.024
	889.484	874.779	831.878	428.636
M3	860.998	842.187	819.704	422.630
	816.742	770.319	780.132	404.689
M5	808.158	730.382	758.022	396.142
	800.642	716.484	749.639	391.676
M6		704.612	744.905	390.444
		709.693	741.760	389.443
M7	795.222	698.741	735.535	388.057
	790.992	685.407	728.085	384.592
M8	785.415	677.625	721.222	381.820
	780.724	669.519	714.643	379.741
M9	772.382	657.244	707.912	377.354
	765.783	644.258	700.255	373.350
M10	759.656	635.594	692.277	370.347
	753.641	628.089	683.139	367.498
M10N	736.367	622.042	662.104	358.489
	726.183	616.665	653.439	356.256
M11	714.193	609.815	637.234	349.480
	697.508	600.153	627.371	342.011
M12	685.752	580.878	606.520	332.386
	667.295	547.587	588.169	326.149
M13	656.820		569.502	317.833
	649.101		563.041	314.291
M14	640.983	476.678	556.149	312.058
	631.128	460.983	537.733	305.051
M15	622.379	432.058	528.053	299.584
	612.653	416.286	515.153	295.195
M16	587.276	378.327	481.845	282.182
	575.722	364.615	465.852	277.716
M17	565.080	354.385	456.384	273.516
	515.423	314.054	416.824	258.616
M18	493.424	307.036	399.402	252.916
	480.920	300.691	389.178	248.716
M19	437.869	275.552	354.056	236.416
	428.366	269.072	346.919	231.816
M20	389.600	251.352	314.145	221.516
	371.087	240.451	294.031	212.916
M21	353.017	223.436	268.698	202.016
	341.516	218.444	256.219	197.316
M22	298.568	178.237	213.902	163.296
	263.921	162.243	196.841	147.420
M23	241.840	145.420	167.381	128.709
	230.237	130.744	150.026	116.802
M24	207.765	124.212	142.248	103.005
	196.626	111.833	124.954	99.225
M25	160.481	82.707	86.363	76.356
	152.703	76.295	78.936	71.442
M26	99.523	50.143	54.196	46.872
	94.767	43.932	48.441	43.848
M27	89.763	38.609	42.383	40.068
	84.781	33.861	36.349	37.044
M28	79.566	28.003	32.711	31.563
	74.040	22.995	24.874	27.594
M29	17.972	5.606	6.480	10.017
	0.000	0.000	0.000	0.000

The $^{40}Ar/^{39}Ar$ age of 124 ± 1 Ma obtained from reversely magnetized granitic plutons from Québec (Foland and others, 1986) may be correlative to CM3, or to the shorter duration CM0 or CM1. Ten major igneous complexes with $^{40}Ar/^{39}Ar$ ages tightly bunched around 124 Ma record only reversed polarity (Foster and Symons, 1979), and it seems likely that the reversed polarity chron is CM3 which has more than three times the duration of any reversed polarity chron in the younger part of the M-sequence. The duration of the interval between CM0 and CM3 is about 3.5 my, leading to an age for the base of CM0 of about 120.5 Ma.

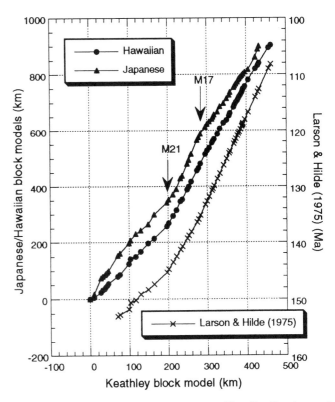

FIG. 6.—Block model distances from Japanese/Hawaiian lineations and Larson and Hilde (1975) block model plotted against Keathley block model distances.

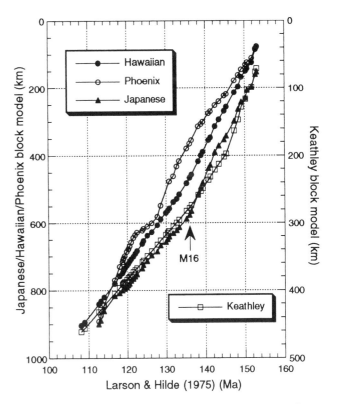

FIG. 7.—Japanese, Hawaiian, Phoenix and Keathley block model distances plotted against the Hawaiian block model of Larson and Hilde (1975).

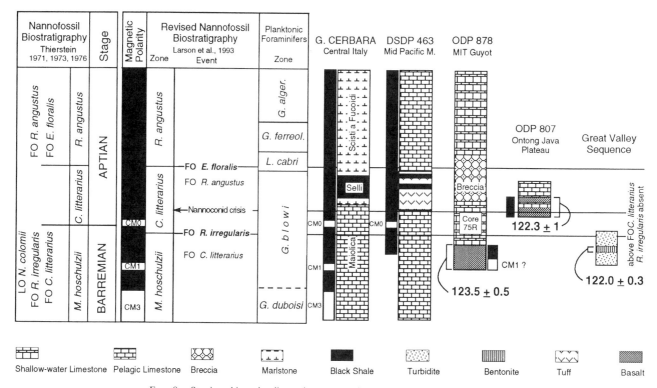

FIG. 8.—Stratigraphic and radiometric age control at the Barremian-Aptian boundary.

Duration estimates for the Aptian, Albian and Cenomanian stages have been obtained from lithologic cyclostratigraphy in Italian pelagic limestones (Herbert and others, this volume). These authors utilized the $^{40}Ar/^{39}Ar$ age estimate for the Cenomanian/Turonian boundary (93.5 ± 0.2 Ma) (Obradovich, 1993) and deduced an age of 122 Ma for the Barremian/Aptian boundary (base of CM0). Using similar reasoning based on cyclostratigraphic data (cited in Herbert and others, this volume), Obradovich (1993) arrived at an estimate of 121 Ma for this stage boundary.

A U-Pb zircon age estimate of 137.1 (+1.6/-0.6) Ma from the Great Valley Sequence of Northern California has been correlated to CM16 or CM16n (Bralower and others, 1990). This is an indirect correlation to the polarity time scale using nannofossil stratigraphy. The interval containing the two dated tuff layers was attributed to the Upper Berriasian *Assipetra infracretacea* subzone based on three nannofossil events: (1) the first occurrence (FO) of *Cretarhabdus angustiforatus* which was used to define the base of the *A. infracretacea* subzone, (2) the occurrence of *Rhagodiscus nebulosus* and (3) the absence of *Percivalia fenestrata,* the FO of which defines the top of the assigned subzone. Close inspection of the ranges of the three marker species suggests that the correlation of the two volcanic layers to CM16 or CM16n is too restrictive. Although *C. angustiforatus* was first observed at the base of Horizon A (Bralower and others, 1990), its FO cannot be placed at that level with certainty. Calcareous nannofossils are documented in only two samples in more than 40 m of section below Horizon A (Bralower and others, 1990) and *C. angustiforatus* generally has rare and scattered occurrence in the lower part of its range. For these reasons, we consider that the presence of *C. angustiforatus* indicates that this level is correlative to polarity chron CM16 or younger (Fig. 9) but does not provide further resolution. Bralower and others (1990) considered *R. nebulosus* to be restricted to the Upper Berriasian/Lower Valanginian interval. The FO of *R. nebulosus* ranges from CM17n to CM15 in sections where it has been correlated to polarity chrons (Fig. 9), but this species has been observed in very few sections. It has been observed in the lowermost Berriasian at Broyon, where it occurs below the *B. grandis* ammonite Zone and within the B calpionellid Zone (Bralower and others, 1989). The FO of *P. fenestrata* ranges from CM16n to CM14 (Fig. 9) and this species is generally rare and discontinuous in distribution. The correlation of this event to polarity CM14 has been observed at Fonte Giordano (Bralower and others, 1989); however, nannofossil perservation is very poor at this location. At Capriolo, the FO of *P. fenestrata* was observed within CM15n (Channell and others, 1987). An additional biostratigraphic constraint on the 137.1 Ma volcanic horizon is given by the absence of *Calcicalathina oblongata* (see Fig. 2 in Bralower and others, 1990); the FO has been correlated to CM14 and occasionally to CM13 (Fig. 9). If the absence of this species is considered as age diagnostic, then the 137.1 Ma datum is correlative to the CM16-CM15 interval (Fig. 9).

The age of 154 Ma for the young end of CM25 is based on the correlation of this polarity chron boundary to the Oxfordian/Kimmeridgian boundary (Ogg and others, 1984) and to age estimates for this stage boundary of about 154–156 Ma from California (Schweickert and others, 1984) and Oregon (Pessagno and Blome, 1990). In the Californian Sierra Nevada, the ammonite-bearing Mariposa Formation is a synorogenic flysch supposed to contain the Oxfordian/Kimmeridgian boundary

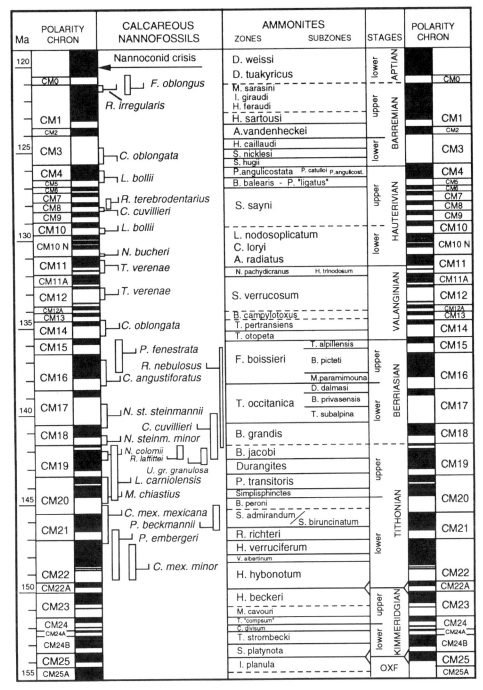

FIG. 9.—Stratigraphic framework for the Oxfordian to Aptian interval. Open bars indicate the range of the nannofossil events with respect to polarity chrons.

(Imlay, 1961) and is affected by Nevadan orogeny. K-Ar horn-blende ages and U-Pb zircon ages on dykes and plutons appear to constrain the age of the Mariposa Formation and the Nevadan orogeny to the 154–159 Ma interval (see Schweickert and others, 1984) but many of the age determinations are not high quality by modern standards. More recent ⁴⁰Ar/³⁹Ar and con-cordant U/Pb zircon ages from the Klamath Mountains appar-ently constrain the Oxfordian/Kimmeridgian boundary to the 150–157 Ma interval (see Pessagno and Blome, 1990). The critical age determination is a concordant U-Pb age of 157 ±

1.5 Ma (Saleeby, 1984; Harper and others, 1994) underlying a macrofossil assemblage correlated to the middle and upper Ox-fordian interval (Imlay, 1980). This age determination corre-lates to the overlapping occurrence of radiolaria *Mirifusus* and *Xiphostylus*, considered to mark the middle-upper Oxfordian interval in the Klamath Mountains (Pessagno and Blome, 1990), however, this is controversial as *Mirifusus* and *Xiphos-tylus* are both present from Aalenian time in Mediterranean sec-tions (see Baumgartner, 1987). A 154 Ma age for M25 is con-sistent with a K-Ar age of 155 ± 3.4 Ma from a celadonite vein

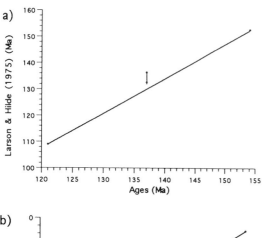

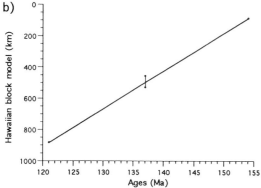

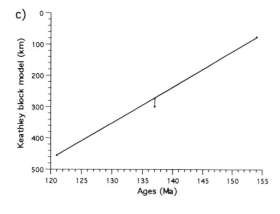

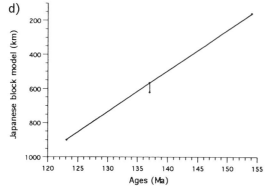

FIG. 10.—The selected radiometric age determinations for M0 (121 Ma), M16-M15 (137 Ma) and M25 (154 Ma) plotted against (a) the Larson and Hilde (1975) block model, (b) Hawaiian block model, (c) Keathley block model, (d) Japanese block model. M0 is not present in the Japanese lineations (Table 2) and for (d) the young end of M1 is assigned an age of 123.2 Ma.

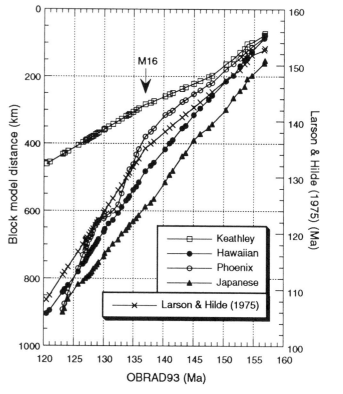

FIG. 11.—OBRAD93 time scale (Obradovich, 1993) compared to oceanic magnetic anomaly block models.

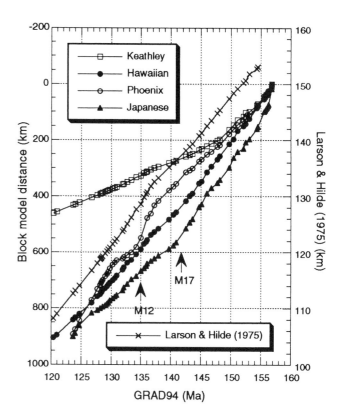

FIG. 12.—GRAD94 time scale (Gradstein and others, 1994, this volume) compared to oceanic magnetic anomaly block models.

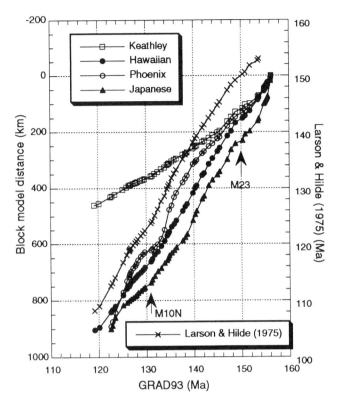

FIG. 13.—GRAD93 time scale (Gradstein and others, this volume) compared to oceanic magnetic anomaly block models.

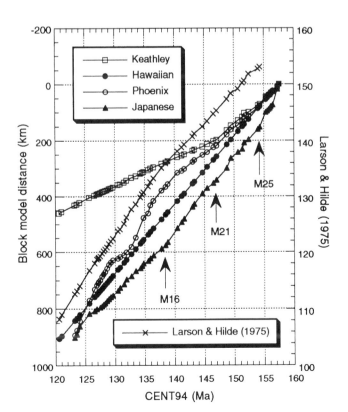

FIG. 14.—CENT94 time scale (this paper) compared to oceanic magnetic anomaly block models.

TABLE 3.—TIME SCALES FOR THE M0-M25 AND M0-M29 INTERVALS. LH75 (LARSON AND HILDE, 1975), KG85 (KENT AND GRADSTEIN, 1985), GTS89 (HARLAND AND OTHERS, 1990), GRAD93 AND GRAD94 (GRADSTEIN AND OTHERS, 1994, THIS VOLUME), CENT94 (THIS PAPER). POLARITY CHRON NOMENCLATURE AFTER HARLAND AND OTHERS (1982).

Reversed Chron	LH75 (Ma)	KG85 (Ma)	GTS89 (Ma)	GRAD93 (Ma)	GRAD94 (Ma)	CENT94 (Ma)
M0 (top)	108.19	118.00	124.32	119.15	120.38	120.60
(base)	109.01	118.70	124.88	120.10	120.98	121.00
M1	112.62	121.81	127.35	122.56	123.67	123.19
	113.14	122.25	127.70	122.88	124.04	123.55
M3	114.05	123.03	128.32	123.45	124.72	124.05
	116.75	125.36	130.17	125.15	126.73	125.67
M5	118.03	126.46	131.05	125.96	127.68	126.57
	118.72	127.05	131.51	126.39	128.19	126.91
M6	118.91	127.21	131.64	126.50	128.33	127.11
	119.06	127.34	131.74	126.60	128.44	127.23
M7	119.27	127.52	131.89	126.80	128.59	127.49
	119.79	127.97	132.25	127.25	128.98	127.79
M8	120.21	128.33	132.53	127.68	129.29	128.07
	120.52	128.60	132.75	127.98	129.53	128.34
M9	120.88	128.91	132.99	128.32	129.79	128.62
	121.49	129.43	133.41	128.88	130.24	128.93
M10	121.94	129.82	133.72	129.31	130.58	129.25
	122.37	130.19	134.01	129.71	130.90	129.63
M10Nn-1	122.82	130.57	134.31	130.13	131.23	129.91
	122.88	130.63	134.36	130.20	131.28	129.95
M10Nn-2	123.31	131.00	134.65	130.60	131.60	130.22
	123.34	131.02	134.67	130.62	131.62	130.24
M10N	123.73	131.36	134.94	130.99	131.91	130.49
	124.07	131.65	135.17	131.26	132.10	130.84
M11	125.10	132.53	135.87	131.81	132.70	131.50
	125.68	133.03	136.27	132.12	133.05	131.71
M11	125.74	133.08	136.31	132.15	133.08	131.73
	126.22	133.50	136.64	132.42	133.37	131.91
M11An-1				132.66	133.64	132.35
				132.68	133.66	132.40
M11A	127.16	134.01	137.30	132.92	133.93	132.47
	127.29	134.42	137.37	132.99	134.00	132.55
M12.1	127.68	134.75	137.63	133.19	134.23	132.76
	128.62	135.56	138.28	133.70	134.79	133.51
M12.2	128.74	135.66	138.36	133.75	134.85	133.58
	128.99	135.88	138.53	133.90	135.01	133.73
M12A	129.41	136.24	138.82	134.12	135.25	133.99
	129.56	136.37	138.92	134.20	135.34	134.08
M13	129.87	136.64	139.14	134.37	135.53	134.27
	130.41	137.10	139.50	134.66	135.84	134.53
M14	130.75	137.39	139.73	134.84	136.04	134.81
	131.81	138.30	140.46	135.41	136.67	135.57
M15	132.63	139.01	141.02	135.90	137.21	135.96
	133.30	139.58	141.47	136.38	137.89	136.49
M16	135.18	141.20	142.76	137.72	139.68	137.85
	135.94	141.85	143.28	138.26	140.40	138.50
M17	136.42	142.27	143.61	138.51	140.86	138.89
	138.16	143.76	144.80	139.86	142.51	140.51
M18	138.82	144.33	145.25	140.33	143.14	141.22
	139.31	144.75	145.58	140.68	143.60	141.63
M19n-1	139.46	144.88	145.69	140.80	143.72	141.78
	139.55	144.96	145.75	140.88	143.79	141.88
M19	140.74	145.98	146.56	141.82	144.68	143.07
	141.28	146.44	146.93	142.25	145.08	143.36
M20n-1	141.64	146.75	147.17	142.54	145.35	143.77
	141.71	146.81	147.22	142.60	145.41	143.84
M20	142.47	147.47	147.75	143.21	145.99	144.70
	143.47	148.33	148.43	144.01	146.74	145.52
M21	144.74	149.42	149.30	145.02	147.69	146.56
	145.29	149.89	149.67	145.46	148.11	147.06
M22n-1	147.11	151.46	150.92	146.92	149.48	148.57
	147.17	151.51	150.96	146.95	149.52	148.62
M22n-2	147.23	151.56	151.00	147.01	149.57	148.67
	147.29	151.61	151.04	147.06	149.61	148.72
M22	147.38	151.69	151.10	147.13	149.68	148.79
	148.36	152.53	151.77	147.91	150.42	149.49
M22A	148.51	152.66	151.87	148.03	150.53	149.72
	148.72	152.84	152.01	148.20	150.69	150.04
M23	149.15	153.21	152.31	148.80	151.09	150.69
	149.48	153.49	152.53	149.24	151.39	150.91
M23	149.51	153.52	152.56	149.30	151.42	150.93
	150.24	154.15	153.06	150.31	152.10	151.40
M24	150.63	154.48	153.32	150.85	152.46	151.72
	151.06	154.85	153.61	151.44	152.86	151.98

TABLE 3.—*Continued*

Reversed Chron	LH75 (Ma)	KG85 (Ma)	GTS89 (Ma)	GRAD93 (Ma)	GRAD94 (Ma)	CENT94 (Ma)
M24	151.09	154.88	153.64	151.49	152.89	152.00
	151.33	155.08	153.80	151.82	153.11	152.15
M24A	151.48	155.21	153.90	152.02	153.25	152.24
	151.79	155.48	154.11	152.46	153.54	152.43
M24B	152.21	155.84	154.40	153.04	153.93	153.13
	152.39	156.00	154.53	153.30	154.10	153.43
M25	152.73	156.29	154.76	153.52	154.31	154.00
	153.03	156.55	154.96	153.72	154.49	154.31
M26					155.51	155.32
					155.69	155.55
M27					155.83	155.80
					156.00	156.05
M28					156.14	156.19
					156.29	156.51
M29					156.77	157.27
					156.85	157.53

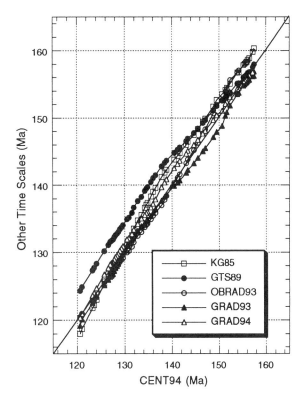

FIG. 15.—Plot of various time scales against CENT94. LH75 (Larson and Hilde, 1975), KG85 (Kent and Gradstein, 1985), GTS89 (Harland and others, 1990), GRAD93 and GRAD94 (Gradstein and others, 1994, this volume), CENT94 (this paper).

(Ludden, 1992) in basaltic crust from the Argo Abyssal Plain (ODP Site 765) which lies between oceanic magnetic anomalies interpreted as M26 and M25A (Sager and others, 1992).

A REVISED TIME SCALE BASED ON THE NEW HAWAIIAN BLOCK MODEL

The three age estimates discussed above for the base of CM0 (121 Ma), CM16-CM15 (137 Ma) and the top CM25 (154 Ma) are similar to the polarity chron age estimates adopted in other recent time scales (e.g., Obradovich, 1993). These age estimates

are not consistent with constant spreading rate in LH75 (Fig. 10A); they are consistent with constant spreading in the new Hawaiian block model (Fig. 10B) and marginally consistent with constant spreading in the Keathley and Japanese block models (Figs. 10C, D).

In OBRAD93 (Obradovich, 1993), the Early Cretaceous time scale is constructed by interpolation using LH75 between three tie points: base CM0 at 121 Ma, CM16/CM16n chron boundary at 137 Ma, and CM25n at 156 Ma. This implies an abrupt change in spreading rate at 137 Ma in LH75 and other block models (Fig. 11). GRAD94 implies constant spreading rate for most of LH75, with a spreading rate change in the M12-M17 interval (Fig. 12). For GRAD94, base CM0 is at 120.98 Ma, 137 Ma lies in CM15n, and top CM25 is at 154.3 Ma, consistent with the popular age estimates cited above. For GRAD93 (Fig. 13), inflexions in the distance-time plots at about 130 Ma (M10N) at about 150 Ma (M23) for the Hawaiian, Japanese and Keathley records appear to be an artifact of the GRAD93.

As mentioned above, the method used to construct the new Hawaiian block model, although optimal for determining the bounds of the principal anomalies, does not allow the minor anomalies to be incorporated in the block model (Fig. 4, Table 2). For the minor polarity chrons, we adopt the relative position within the individual major polarity chron from LH75. In our time scale based on the new Hawaiian block model (CENT94), the only modification to the number of polarity chrons in LH75 is the inclusion of an additional reversed polarity chron (CM11An-1, Table 3). This results in two reversed polarity chrons between CM11 and CM12, as proposed by Tamaki and Larson (1988). The two reversed polarity chrons between CM11 and CM12 were recognized in the Capriolo land section (Channell and others, 1987) at 24–34% and 42–48% (from the base) of the normal polarity zone between CM11 and CM12. We adopt this spacing of the two reversed polarity chrons.

The resulting time scale (CENT94) implies a spreading rate decrease for M16-M25 in the LH75 block model, and an increase in spreading rate for M21-M29 time (148–157 Ma) in the Keathley block model (Fig. 14). CENT94 implies a spreading rate increase for M16-M29 (139–157 Ma) in the Japanese block model and variable spreading rates in the Phoenix block model (Fig. 14). Figure 15 illustrates the relationship between CENT94 and other time scales.

Cyclostratigraphy permits polarity chron duration to be estimated where Milankovitch-type cycles can be documented and provides a potential means of testing time scales such as GTS89, GRAD94 and the new time scale based on the new Hawaiian block model (CENT94). For two sections in the underlying Maiolica Limestones (Gorgo a Cerbara and Cismon) which record polarity zones correlative to CM0-CM3, bedding rhythms appear to record both the eccentricity and precession periodicities (Herbert, 1992). The durations of CM0-CM3 in CENT94 are generally lower than predicted from the cyclostratigraphy (Fig. 16).

CORRELATION OF STAGE BOUNDARIES TO THE POLARITY TIME SCALE

In the past 15 years, Cretaceous polarity chrons have been correlated with calpionellid, nannofossil and foraminiferal events in numerous pelagic limestone sections from Southern Alps, Central Italy, Atlantic and Pacific oceans, although poor

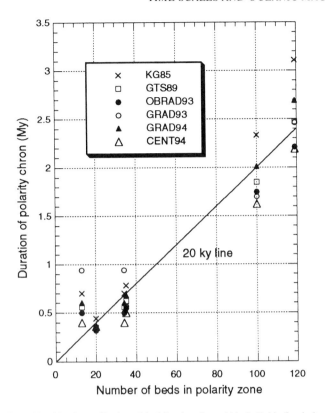

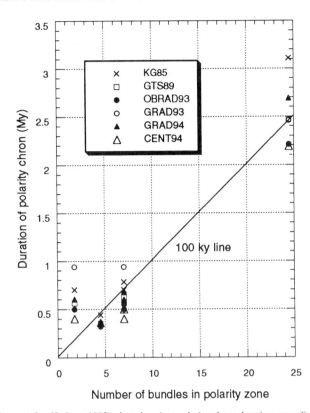

FIG. 16.—Numbers of beds and bedding bundles within individual polarity chrons (after Herbert, 1992) plotted against polarity chron duration according to various time scales. KG85 (Kent and Gradstein, 1985), GTS89 (Harland and others, 1990), GRAD93 and GRAD94 (Gradstein and others, 1994, this volume), OBRAD93 (Obradovich, 1993), CENT94 (this paper).

TABLE 4.—LATE JURASSIC-EARLY CRETACEOUS STAGE BOUNDARY AGES. GTS82 (HARLAND AND OTHERS, 1982), KG85 (KENT AND GRADSTEIN, 1985), GTS89 (HARLAND AND OTHERS, 1990), OBRAD93 (OBRADOVICH, 1993), GRAD93 AND GRAD94 (GRADSTEIN AND OTHERS, 1994, THIS VOLUME), CENT94 (THIS PAPER).

Stage Boundary	GTS82 (Ma)	KG85 (Ma)	GTS89 (Ma)	OBRAD93 (Ma)	GRAD93 (Ma)	GRAD94 (Ma)	CENT94 (Ma)
Barr.-Apt.	119	119	124.5	121	119.0	121.0 ± 1.4	121.0
Haut. Barr.	125	124	131.8	127	127.3	127.0 ± 1.6	126.0
Val.-Haut.	131	131	135.0	130	131.6	132.0 ± 1.9	131.5
Berr.-Val.	138	138	140.7	135	138.8	137.0 ± 2.2	135.8
Tith.-Berr.	144	144	145.6	142	144.8	144.2 ± 2.6	141.6
Kimm.-Tith.	150	152	152.1		152.8	150.7 ± 3.0	150.0
Oxf.-Kimm.	156	156	154.7			154.1 ± 3.2	154.0

to moderate preservation of nannoflora accounts for some variability in their correlation (Fig. 9). Late Jurassic and Cretaceous chronostratigraphy is based on stage stratotypes defined by ammonite zones. Due to absence or uneven distribution of ammonite faunas in land sections and oceanic cores, correlation of Cretaceous stage boundaries to polarity chrons has been based on the supposed correlation of calcareous microplankton events to ammonites (e.g., Thierstein, 1973). In the last few years, ammonite biozones have been directly correlated to magnetostratigraphy in parts of the Oxfordian-lowermost Valanginian (see Ogg and others, 1991) and Valanginian-Barremian intervals (Cecca and others, 1994; Channell and others, 1994).

The Oxfordian/Kimmeridgian boundary is correlative to the base of the *Sutneria platynota* ammonite Zone which has been correlated to CM25 in southern Spain (Ogg and others, 1984). The Kimmeridgian/Tithonian boundary is correlative to the base of the *Hybonoticeras hybonotum* ammonite Zone which

lies close to CM22A also in southern Spain (Ogg and others, 1984). The Tithonian/Berriasian (Jurassic/Cretaceous) boundary does not have a universally accepted definition. Many of the candidate markers have been correlated to the polarity chrons (Ogg and Lowrie, 1986; Channell and Grandesso, 1987; Bralower and others, 1989; Ogg and others, 1991). The base of CM18 (top of *Berriasella jacobi* ammonite Zone) has become the generally accepted correlation to the Tithonian-Berriasian boundary. The Berriasian/Valanginian boundary is defined by the base of the *T. otopeta* ammonite Zone, and falls within CM15n (Ogg and others, 1988) and between the FO of *Cretarhabdus angustiforatus* and the FO of *Calcicalathina oblongata*. The Valanginian/Hauterivian boundary coincides with the base of the *A. radiatus* ammonite Zone and correlates to the base of CM11n (Channell and others, 1994). The base of the *Spitidiscus hugii* ammonite Zone defines the Hauterivian/Barremian boundary, which falls between the LO of *Lihtraphidites bollii* and the LO of *Calcicalathina oblongata,* and in the upper part of CM4 (Cecca and others, 1994). The Barremian/Aptian boundary was formally defined at the first occurrence of *Deshayesites.* None of the nannofossil events proposed by Thierstein (1973) to define this boundary have proved to be reliable. The FO of *Rucinolithus irregularis* is correlated to the upper part of the *M. sarasini* ammonite Zone and is therefore slightly older than the Barremian/Aptian boundary (Channell and Erba, 1992; Coccioni and others, 1992). The base of CM0 coincides closely to this boundary. Direct correlation of this polarity chron with ammonite biozones has not been well documented, although a single specimen of *Prodeshayesites* sp. occurs near the top of CM0 at Gorgo a Cerbara indicating that at least part of this polarity chron is Early Aptian in age (Channell and others, 1994). The correlation of polarity chrons to ammonite zones in the Early Cretaceous (Fig. 9) indicates substantial variation in duration of ammonite zones.

CONCLUSIONS

Although considerable progress has been made in the correlation of paleontological events and biozones to the M-sequence polarity chrons, the absolute age control on the Late Jurassic-Early Cretaceous time scale remains poor. We consider that the available absolute age control is insufficient to justify abandoning the constant oceanic spreading rate assumption for this part of the time scale. Comparison of block models from Japanese, Hawaiian, Phoenix and Keathley lineations indicates that the Hawaiian record is closest to a constant spreading rate record. The new Hawaiian block model differs slightly from that published by Larson and Hilde (1975). The absolute ages for CM0 (121 Ma), CM16-M15 (137 Ma) and CM25 (154 Ma) are considered to be the only reliable estimates for M-sequence polarity chrons, and these ages are consistent with constant spreading in the new Hawaiian block model.

The resulting time scale (CENT94) based on the new Hawaiian block model yields the following stage boundary ages: Barremian/Aptian (121 Ma), Hauterivian/Barremian (126 Ma), Valanginian/Hauterivian (131.5 Ma), Berriasian/Valanginian (135.8 Ma), Tithonian/Berriasian (141.6 Ma), Kimmeridgian/Tithonian (150 Ma), Oxfordian/Kimmeridgian (154 Ma). See Figure 15, Table 3 and Table 4 for comparisons among published time scales.

ACKNOWLEDGMENTS

We are grateful to Tim Bralower, Tim Herbert, Roger Larson and Neil Opdyke for reviews of an early draft of this manuscript, and to Malcolm Pringle for discussions. J. C. acknowledges financial support from the United States National Science Foundation (OCE 8915697).

REFERENCES

BAUMGARTNER, P. O., 1987, Age and genesis of Tethyan Jurassic radiolarites: Eclogae Geologicae Helvetiae, v. 80, p. 831–879.

BRALOWER, T. J., MONECHI, S., AND THIERSTEIN, H. R., 1989, Calcareous nannofossil zonation of the Jurassic-Cretaceous boundary interval and correlation with the geomagnetic polarity timescale: Marine Micropaleontology, v. 14, p. 153–235.

BRALOWER, T. J., LUDWIG, K. R., OBRADOVICH, J. D., AND JONES, D. L., 1990, Berriasian (Early Cretaceous) radiometric ages from the Grindstone Creek Section, Sacramento Valley, California: Earth and Planetary Science Letters, v. 98, p. 62–73.

CECCA, F., PALLINI, G., ERKA, E., PREMOLI-SILVA, I., AND COCCIONI, R., 1994, Hauterivian-Barremian chronostratigraphy based on ammonites, nannofossils, planktonic foraminifera and magnetic chrons from the Mediterranean domain: Cretaceous Research, v. 15, p. 457–467.

CHANNELL, J. E. T., BRALOWER, T. J., AND GRANDESSO, P., 1987, Biostratigraphic correlation of Mesozoic polarity chrons CM1 to CM23 at Capriolo and Xausa (Southern Alps, Italy): Earth and Planetary Science Letters, v. 85, p. 203–221.

CHANNELL, J. E. T., CECCA, F., AND ERBA, E., 1994, Correlations of Hauterivian and Barremian (Early Cretaceous) stage boundaries to polarity chrons: Transactions American Geophysical Union (EOS), v. 75, p. 202.

CHANNELL, J. E. T. AND ERBA, E., 1992, Early Cretaceous polarity chrons CM0 to CM11 recorded in Northern Italian land sections near Brescia (Northern Italy): Earth and Planetary Science Letters, v. 108, p. 161–179.

CHANNELL, J. E. T. AND GRANDESSO, P., 1987, A revised correlation of Mesozoic polarity chrons and calpionellid zones: Earth and Planetary Science Letters, v. 85, p. 222–240.

COCCIONI, R., ERBA, E., AND PREMOLI SILVA, I., 1992, Barremian-Aptian calcareous plankton biostratigraphy from the Gorgo Cerbara section (Marche, central Italy) and implications for plankton evolution: Cretaceous Research, v. 13, p. 517–537.

COLEMAN, D. S. AND BRALOWER, T. J., 1993, New U-Pb zircon age constraints on the Early Cretaceous time scale: Transactions American Geophysical Union (EOS), v. 73, p. 556.

FOLAND, K. A., GILBERT, L. A., SEBRING, C. A., AND JIANG-FENG, C., 1986, ^{40}Ar/^{39}Ar ages for plutons of the Monteregian Hills, Quebec: evidence for a single episode of Cretaceous magmatism: Geological Society America Bulletin, v. 97, p. 966–974.

FOSTER, J. AND SYMONS, D. T. A., 1979, Defining a paleomagnetic polarity pattern in the Monteregian intrusives: Canadian Journal of Earth Sciences, v. 16, p. 1716–1725.

GRADSTEIN, F. M., AGTERBERG, F. P., OGG, J. G., HARDENBOL, J., VAN VEEN, P., THIERRY, J., AND HUANG, Z., 1994, A Mesozoic time scale: Journal of Geophysical Research, v. 99, p. 24,051–24,074.

HARLAND, W. B., COX, A. V., LLEWELLYN, P. G., PICKTON, C. A. G., SMITH, A. G., AND WALTERS, R., 1982, A Geologic Time Scale: Cambridge, Cambridge University Press, 131 p.

HARLAND, W. B., ARMSTRONG, R. L., COX, A. V., CRAIG, L. E., SMITH, A. G. AND SMITH, D. G., 1990, A Geologic Time Scale: Cambridge, Cambridge University Press, 263 p.

HARPER, G. D., SALEEBY, J. B., AND HEIZLER, M., 1994, Formation and emplacement of the Josephine ophiolite and the Nevadan orogeny in the Klamath Mountains, California-Oregon: U/Pb zircon and 40Ar/39Ar geochronology: Journal of Geophysical Research, v. 99, p. 4293–4321.

HERBERT, T. D., 1992, Paleomagnetic calibration of Milankovitch cyclicity in Lower Cretaceous sediments: Earth and Planetary Science Letters, v. 112, p. 15–28.

IMLAY, R. W., 1961, Late Jurassic ammonites from the western Sierra Nevada, California: Washington, D.C., United States Geological Survey Professional Paper 374D, p. D1-D30.

IMLAY, R. W., 1980, Jurassic paleobiogeography of the western conterminous United states in its continental setting: Washington, D.C., United States Geological Survey Professional Paper 1062, 134 p.

KENT, D. V. AND GRADSTEIN, F. M., 1985, A Cretaceous and Jurassic geochronology: Geological Society of America Bulletin, v. 96, p. 1419–1427.

KLITGORD, K. D. AND SCHOUTEN, H., 1986, Plate kinematic of the central Atlantic, in Vogt, P. R. and Tucholke, B. E., eds., The Geology of North America, The Western North Atlantic Region, v. M: Boulder, Geological Society of America, p. 351–378.

LARSON, R. L., FISCHER, A. G., ERBA, E., AND PREMOLI SILVA, I., eds., 1993, APTICORE-ALBICORE: A Workshop Report on Global Events and Rhythms of the mid-Cretaceous, 4–9 October 1992, Perugia, Italy: Washington, D.C., Joint Oceanographic Institutions Inc., 56 p.

LARSON, R. L. AND HILDE, T. W. C., 1975, A revised time scale of magnetic reversals for the Early Cretaceous and Late Jurassic: Journal of Geophysical Research, v. 80, p. 2586–2594.

LUDDEN, J. N., 1992, Radiometric age determinations for basement from sites 765 and 766, Argo Abyssal Plain and northwestern Australian margin, in Gradstein, F. M., Ludden, J. N., and others, eds., Proceedings of Ocean Drilling Program, Scientific Results 123: College Station, Ocean Drilling Program, p. 557–559.

MAHONEY, J. J., STOREY, M., DUNCAN, R., SPENCER, K. J., AND PRINGLE, M., 1993, Geochemistry and age of the Ontong-Java Plateau, *in* Pringle, M. S., Sager, W. W., Sliter W. V., and Stein, S., eds., Monograph on the Mesozoic Pacific: Washington, D.C., American Geophysical Union, p. 233–261.

NAKANISHI, M., TAMAKI, K., AND KOBAYASHI, K., 1989, Mesozoic magnetic anomaly lineations and seafloor spreading history of the northwestern Pacific: Journal of Geophysical Research, v. 94, p. 15437–15462.

NAKANISHI, M., TAMAKI, K., AND KOBAYASHI, K., 1992, Magnetic anomaly lineations from Late Jurassic to Early Cretaceous in the west-central Pacific Ocean: Geophysical Journal International, v. 109, p. 701–719.

OBRADOVICH, J. D., 1993, A Cretaceous time scale, *in* Caldwell, W. G. E., ed., Evolution of the Western Interior Basin: St. Johns, Geological Association of Canada Special Paper 39, p. 379–396.

OGG, J. G., COMPANY, M., STEINER, M. B., AND TAVERA, J. M., 1988, Magnetostratigraphy across the Berriasian-Valanginian boundary (Early Cretaceous) at Cehegin (Murcia Province, southern Spain): Earth and Planetary Science Letters, v. 87, p. 205–215.

OGG, J. G., HASENYAGER, R. W., WIMBLEDON, W. A., CHANNELL, J. E. T., AND BRALOWER, T. J., 1991, Magnetostratigraphy of the Jurassic-Cretaceous boundary interval—Tethyan and English faunal realms: Cretaceous Research, v. 12, p. 455–482.

OGG, J. G. AND LOWRIE, W., 1986, Magnetostratigraphy of the Jurassic/Cretaceous boundary: Geology, v. 14, p. 547–550.

OGG, J. G., STEINER, M. B., OLORIZ, F., AND TAVERA, J. M., 1984, Jurassic magnetostratigraphy: 1. Kimmeridgian-Tithonian of Sierra Gorda and Carcabuey, southern Spain: Earth and Planetary Science Letters, v. 71, p. 147–162.

PESSAGNO, E. A. AND BLOME, C. D., 1990, Implications of new Jurassic stratigraphic, geochrononometric and paleolatitudinal data from the western Klamath terrane (Smith River and Rogue Valley subterranes): Geology, v. 18, p. 665–668.

PRINGLE, M. S., STAUDIGEL, H., DUNCAN, R. A., CHRISTIE, D. M., ODP LEG 143 AND 144 SCIENTIFIC STAFFS, 1993, 40Ar/39Ar ages of basement lavas at Resolution, MIT, and Wodejebato Guyots compared with magneto- and bio-stratigraphic results from ODP Legs 143/144: Transactions American Geophysical Union (EOS), v. 73, p. 353.

SAGER, W. W., FULLERTON, L. G., BUFFLER, R. T., AND HANDSCHUMACHER, D. W., 1992, Argo Abyssal Plain magnetic lineations revisited: implications for the onset of seafloor spreading and tectonic evolution of the eastern Indian ocean, *in* Gradstein, F. M., Ludden, J. N., and others, eds., Proceedings of Ocean Drilling Program, Scientific Results 123: College Station, Ocean Drilling Program, p. 659–669.

SALEEBY, J. B., 1984, Pb/U zircon ages from the Rogue River area, western Jurassic belt, Klamath Mountains, Oregon: Geological Society America Abstracts with Programs, v. 16, p. 331.

SCHOUTEN, H. AND MCCAMY, K., 1972, Filtering marine magnetic anomalies: Journal of Geophysical Research, v. 77, p. 7089–7099.

SCHWEICKERT, R. A., BOGEN, A., GIRTY, N. L., HANSON, R. E., AND MERGUERIAN, C., 1984, Timing and structural expression of the Nevadan orogeny, Sierra Nevada, California: Geological Society of America Bulletin, v. 95, p. 967–979.

SLITER, W. V., 1992, Biostratigraphic zonation for Cretaceous planktonic foraminifers examined in thin section: Journal of Foraminiferal Research, v. 19, p. 1–19.

SUNDVIK, M. T. AND LARSON, R. L., 1988, Seafloor spreading history of the western North Atlantic Basin derived from the Keathley sequence and computer graphics: Tectonophysics, v. 155, p. 49–71.

TAMAKI, K. AND LARSON, R. L., 1988, The Mesozoic tectonic history of the Magellan microplate in the western central Pacific: Journal of Geophysical Research, v. 93, p. 2857–2874.

TARDUNO, J. A., SLITER, W. V., KROENKE, L., LECKIE, M., MAYER, H., MAHONEY, J. J., MUSGRAVE, R., STOREY, M., AND WINTERER, E. L., 1991, Rapid formation of the Ontong Java Plateau by Aptian mantle plume volcanism: Science, v. 254, p. 399–403.

THIERSTEIN, H. R., 1971, Tentative Lower Cretaceous calcareous nannoplankton zonation: Eclogae Geologicae Helvetiae, v. 64, p. 459–488.

THIERSTEIN, H. R., 1973, Lower Cretaceous calcareous nannoplankton biostratigraphy: Abhandlungen der Geologischen Bundesanstalt, v. A 29, p. 1–52.

THIERSTEIN, H. R., 1976, Mesozoic calcareous nannoplankton biostratigraphy of marine sediments: Marine Micropaleontology, v. 1, p. 325–362.

WESSEL, P. AND SMITH, W. H., 1991, Free software helps map and display data: Transactions American Geophysical Union (EOS), v. 72, p. 441–446.

AN INTEGRATED CRETACEOUS MICROFOSSIL BIOSTRATIGRAPHY

TIMOTHY J. BRALOWER
Department of Geology, University of North Carolina, Chapel Hill, NC 27599-3315
R. MARK LECKIE
Department of Geology and Geography, University of Massachusetts, Amherst, MA 01003
WILLIAM V. SLITER
Division of Paleontology and Stratigraphy, United States Geological Survey, Menlo Park, CA 94025
AND
HANS R. THIERSTEIN
Geological Institute, ETH-Zentrum, CH-8092 Zürich, SWITZERLAND

ABSTRACT: We have constructed an integrated calcareous nannoplankton, calpionellid, and planktonic foraminifer biostratigraphy for the Cretaceous Period. This biostratigraphy, which consists of 73 informal zones, is based upon a literature survey of numerous DSDP/ODP and land sections as well as our own investigation of several of these sequences. Although the sections included are from low, mid, and high latitudes, from all of the major ocean basins and from epicontinental seaways, the integrated scheme is most applicable in mid- and low-latitude sequences. Current nannofossil, calpionellid, and planktonic foraminiferal zonations offer limited resolution (2–6 my/zone) in the Cretaceous Period. The integrated zonation scheme proposed significantly increases potential biostratigraphic resolution to between 0.5 and 1.5 my/zone because these fossil groups are often worked on collectively, and because the correlation between the groups is reasonably well known in most intervals. This zonation holds great promise for improving the chronostratigraphic framework and biostratigraphic correlations needed in paleoenvironmental and paleoceanographic investigations.

INTRODUCTION

Planktonic foraminifers, nannofossils, and calpionellids are the primary fossil groups used in the biostratigraphy of Mesozoic calcareous deep-sea sediments. Cenozoic zonation schemes of the former two groups possess sufficient resolution for most detailed applications (e.g., Moore and Romine, 1981; Berggren and others, 1985, this volume). Mesozoic microfossil zonations, on the other hand, offer diminished resolution and thus biostratigraphy becomes a weak link in many geological studies.

Original zonations for Cretaceous planktonic foraminifers, calcareous nannofossils and calpionellids were largely developed in Tethyan and other low-latitude sections over 20 years ago (e.g., Bolli, 1957, 1959, 1966; Moullade, 1966; Pessagno, 1967; Sigal, 1967, 1977; Allemann and others, 1971; Thierstein, 1971, 1973; Van Hinte, 1972, 1976; Longoria, 1974; Sissingh, 1977). These zonations were modified in part by the addition of subzones (e.g., Manivit and others, 1977); however, the resolution for all groups remains low at an average of 2–6 my/zone and 1–3 my/subzone. At the same time, there have been advances in techniques and taxonomy. For planktonic foraminifers and nannofossils, processing techniques have evolved allowing the study of indurated sedimentary rocks such as limestones (e.g., Postuma, 1971; Premoli Silva, 1977; Monechi and Thierstein, 1985; Sliter, 1989; Premoli Silva and Sliter, unpubl. data) which had previously only been datable using calpionellids within the Tithonian (Upper Jurassic) to Valanginian strata. These new techniques have led to the study of numerous additional land sections (e.g., Erba and Quadrio, 1987; Tornaghi and others, 1989). Taxonomic advances which include the clarification of existing and the discovery of new taxa, also have lead to increased potential resolution as there are more species available for biostratigraphy.

The Deep Sea Drilling Project and Ocean Drilling Program have recovered many new Cretaceous sections in the last twenty years. These include sequences from high latitudes and from all of the major ocean basins. Many of these sections are more complete than those in which original zonations were developed. Therefore, through the addition of zones and subzones, new zonal schemes (e.g., Bralower and others, 1993) offer higher resolution than those previously proposed. Several biostratigraphic schemes (e.g., Bralower and others, 1989) have proposed a series of non-zonal biohorizons. These are events which are, largely for preservational and biogeographic reasons, undefinable in a majority of sequences, but they offer the ability to obtain higher resolution than traditional zonations. However, because of the limited number of sections in which they can be currently defined, the relative order of subsidiary biohorizons is still not well established in most Cretaceous intervals. Since the correlation between different microfossil biostratigraphies is becoming better established in most intervals, higher biostratigraphic resolution with maximum precision is best obtained by integrating different schemes.

In this paper, we propose an informal integrated Cretaceous calcareous nannofossil, calpionellid and planktonic foraminifer biostratigraphic scheme consisting of 73 zones. This zonation is informal as we have not formally defined each unit according to the criteria recommended by the International Subcommission on Stratigraphic Classification (Hedberg, 1976). Advantages of integration include: (1) ability to refine the stratigraphy of sequences with poor preservation by combining events based on the most resistant taxa of two microfossil groups; (2) correlation of biostratigraphies where one fossil group is better preserved than the other as is often the case in Cretaceous sections (e.g., Tarduno and others, 1992); (3) checking results of one fossil group with the other permitting increased biostratigraphic precision; and (4) planktonic foraminifers generally provide higher biostratigraphic resolution in certain intervals (e.g., Aptian, lower to mid Cenomanian) than calcareous nannofossils, while the opposite is true for the Barremian, Albian and upper Cenomanian to lower Turonian. In this zonation, we utilize calcareous nannofossils which range throughout the Cretaceous Period; calpionellids, which range from the Tithonian (uppermost Jurassic) to the middle part of the Valanginian; and planktonic foraminifers, which originate in the Jurassic, but do not become biostratigraphically useful until the Barremian. Because the ranges of calpionellids and planktonic foraminifers do not overlap, the integrated scheme is only based on two microfossil groups in any one interval.

CRETACEOUS MICROFOSSIL BIOSTRATIGRAPHY

This investigation is based upon compilation of previous planktonic foraminifer and calcareous nannofossil biostratigraphies as well as selective reinvestigation of parts of particular sections. Range charts for these reinvestigated sections are published elsewhere (Bralower and others, 1994). Correlations between calcareous nannofossils and calpionellids are taken entirely from the literature. In this section, we briefly discuss the current Cretaceous zonations of the three microfossil groups and the correlations between them, followed by the procedure with which the integrated zonation is constructed.

Calcareous Nannofossil Zonation

Cretaceous calcareous nannofossil zonations have developed over the last twenty years, yet the early schemes of Thierstein (1971, 1973; both Lower Cretaceous), Sissingh (1977) and Roth (1978; both cover the entire Cretaceous) are still widely applied. A comprehensive summary of these zonations is given in Perch-Nielsen (1985). High latitude zonal schemes have been proposed by Wise and Wind (1977) and Wise (1983; Aptian through Maastrichtian) and Watkins (1992; Coniacian through Maastrichtian). These schemes as well as those of Manivit and others (1977; Aptian through Coniacian) and Applegate and Bergen (1988; Valanginian through Cenomanian) have included subzonal units in order to improve biostratigraphic resolution. Detailed evaluation of zonations, including the choice of markers, and the availability of alternative events is given in Bralower and others (1989) for the Berriasian and Valanginian stages; Bralower (1987), Channell and others (1987) and Erba and Quadrio (1987) for the Valanginian through Aptian; Erba (1988) for the Aptian and Albian; Bralower and others (1993) for the Aptian and Albian, Bralower and others (1994) for the upper Barremian to lower Aptian, Bralower (1988) for the Cenomanian/Turonian boundary interval; Bralower and Siesser (1992) for the Aptian through Maastrichtian; and Mutterlose (1992) for the entire Early Cretaceous Period. Several of these studies have proposed alternative zonations for particular intervals. For example, Bralower and others (1989) proposed a series of new subzones for the Berriasian and lower Valanginian stage; Bralower (1987) defined several subzones in the upper Hauterivian and Barremian intervals; Erba (1988) defined a series of new zones for the Aptian and Albian intervals; Bralower and others (1993) proposed several informal subzones for the Aptian and Albian intervals which were tied to the zonation of Roth (1978); and Bralower (1988) defined new zones and subzones in the Cenomanian/Turonian boundary interval.

Several of these studies also led to the addition of non-zonal nannofossil events or biohorizons to biostratigraphic schemes (e.g., Bralower and others, 1989; Bralower and others, 1993). These events can usually be determined in expanded DSDP/ODP sites with suitable preservation but are more difficult to detect in land sequences. Because of the limited number of sequences in which these events have currently been established, their relative order is still not entirely certain. Recovery of more suitable sequences in the future should allow these

events to become more standard components of nannofossil biostratigraphies.

Even though certain schemes can be widely applied in particular time intervals, there is no zonal scheme which is globally applicable. This situation is especially apparent in high latitude sequences. The geographic range over which zonations are applicable depends on relative temperature gradients in the oceans (e.g., Mutterlose, 1989). These gradients changed throughout the Cretaceous Period the result being that at certain times cosmopolitan schemes are widely applicable, but at others, the correlation between high and low latitude schemes is still uncertain. For Lower Cretaceous strata, well defined zonations for Boreal sections exposed in Northern Europe (e.g., Crux, 1989) are still poorly correlated to low latitude zonal schemes due to the absence of common marker taxa. Aptian and Albian high- and low-latitude zonal schemes have been correlated based on the co-occurrence of markers at temperate ODP Site 763 on the Exmouth Plateau, Indian Ocean (Bralower, 1992). Similarly, Watkins (1992) showed that cosmopolitan schemes were applicable from the Coniacian to the lower Campanian on the Kerguelen Plateau (Indian Ocean), but that Maastrichtian schemes were not.

In general, Lower Cretaceous zonations for low latitude and temperate areas are rather uniform, but Upper Cretaceous zonations are not. The two most commonly applied schemes, those of Sissingh (1977) and Roth (1978), have a limited number of common events in Upper Cretaceous units and neither is entirely satisfactory for this interval. This has been clearly demonstrated in ODP Sites 762 and 763 on the Exmouth Plateau (Bralower and Siesser, 1992).

Correlations between nannofossil zones, biohorizons, and stage boundaries are moderately well determined for Lower Cretaceous units where stratotype and parastratotype sequences are, for the most part, apparently complete and fossiliferous. More uncertainties exist in Upper Cretaceous units where some boundaries are poorly exposed, sparsely fossiliferous, and apparently incomplete (e.g., Burnett and others, 1992). Correlation between Upper Cretaceous nannofossil zones and stage boundaries, shown in Figure 1, may, therefore, undergo future revision.

The calcareous nannofossil scheme utilized here is that of Roth (1978; Fig. 1). This scheme has been selected because: (1) a scheme is needed to cover the entire Cretaceous Period, and (2) the correlation between nannofossil and planktonic foraminiferal biostratigraphies is largely based on results from DSDP/ODP sequences on which the Roth (1978) scheme was also constructed. The Sissingh (1977) scheme was based on results from land sequences (largely in Europe) where planktonic foraminiferal data are often unavailable or where the markers used here are unreported.

In order to improve resolution in Upper Cretaceous intervals where there are few widely applicable, reliable microfossil events, we have proposed four new informal subzonal divisions of Roth's (1978) zones (Fig. 1). Campanian zone NC19 is divided into two subzones, NC19A (from the base of *Ceratolithoides aculeus* to the base of *Tetralithus gothicus*) and NC19B (from the base of *T. gothicus* to the base of *T. trifidus*). Maastrichtian Zone NC21 (from the top of *T. trifidus* to the base of *Lithraphidites quadratus*) is divided into two subzones, NC21A and NC21B, by the last occurrence of *Reinhardtites levis*. The

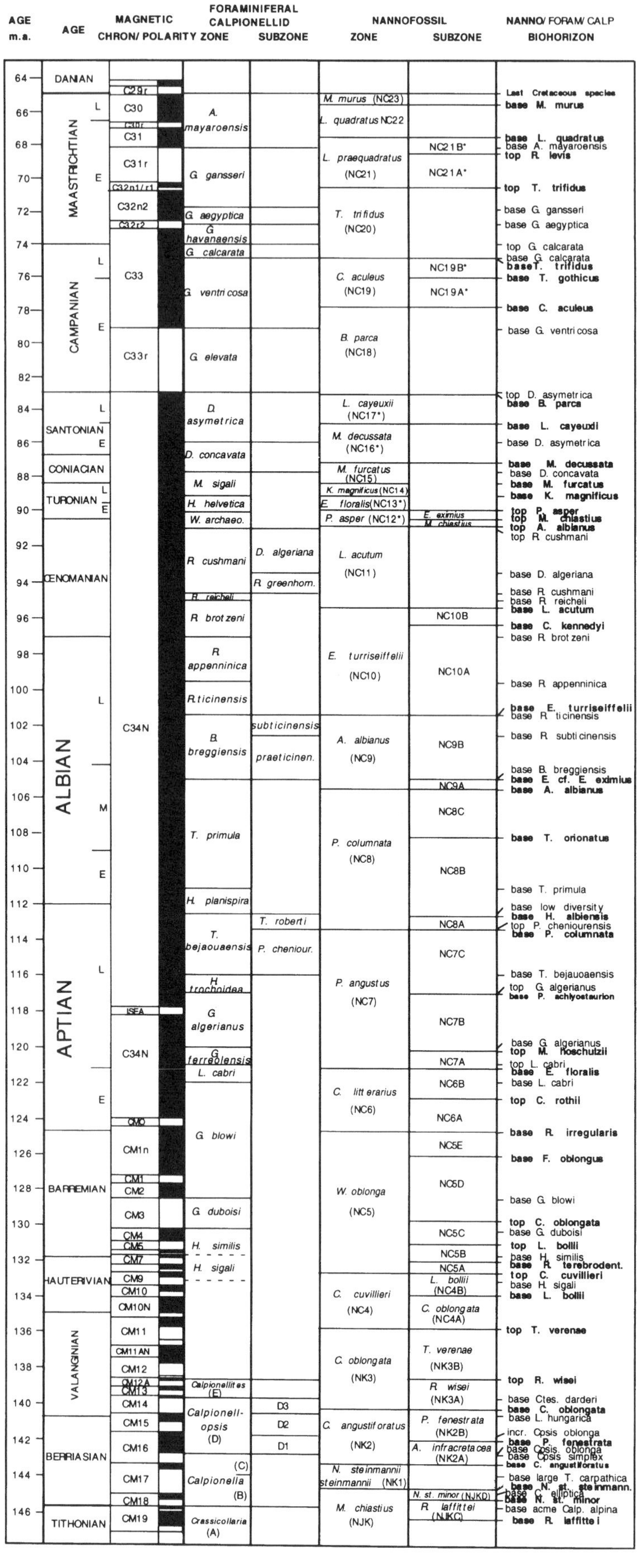

AGE m.a.	AGE	MAGNETIC CHRON/POLARITY ZONE	FORAMINIFERAL CALPIONELLID ZONE	SUBZONE	NANNOFOSSIL ZONE	SUBZONE	NANNO/FORAM/CALP BIOHORIZON
64	DANIAN	C29r	A. mayaroensis		M. murus (NC23)		Last Cretaceous species — base M. murus
66	MAASTRICHTIAN L	C30, C30r, C31	A. mayaroensis		L. quadratus NC22		base L. quadratus — base A. mayaroensis — top R. levis
68		C31			L. praequadratus (NC21)	NC21B* / NC21A*	
70	MAASTRICHTIAN E	C31r, C32n1/r1	G. gansseri				top T. trifidus
72		C32n2, C32r2	G. aegyptica		T. trifidus (NC20)		base G. gansseri — base G. aegyptica
74			G. havanaensis / G. calcarata				top G. calcarata — base G. calcarata — base T. trifidus
76	CAMPANIAN L	C33	G. ventricosa		C. aculeus (NC19)	NC19B* / NC19A*	base T. gothicus — base C. aculeus
78							base G. ventricosa
80	CAMPANIAN E	C33r	G. elevata		B. parca (NC18)		
82							
84	SANTONIAN L	C34N	D. asymetrica		L. cayeuxii (NC17*)		top D. asymetrica — base B. parca — base L. cayeuxii
86	SANTONIAN E		D. concavata		M. decussata (NC16*)		base D. asymetrica — base M. decussata
88	CONIACIAN		M. sigali		M. furcatus (NC15)		base D. concavata — base M. furcatus
90	TURONIAN L / E		H. helvetica / W. archaeo.		K. magnificus (NC14) / E. floralis (NC13*) / P. asper (NC12*)	E. eximius / M. chiastius	base K. magnificus — top P. asper — top M. chiastius — top A. albianus
92			R. cushmani	D. algeriana	L. acutum (NC11)		top R. cushmani
94	CENOMANIAN			R. greenhorn.			base D. algeriana — base R. cushmani — base R. reicheli — base L. acutum
96			R. reicheli / R. brotzeni			NC10B	base C. kennedyi — base R. brotzeni
98			R. appenninica		E. turriseiffelii (NC10)	NC10A	
100	ALBIAN L		R. ticinensis				base R. appenninica
102			B. breggiensis	subticinensis	A. albianus (NC9)	NC9B	base E. turriseiffelii — base R. ticinensis — base R. subticinensis
104				praeticinen.		NC9A	base B. breggiensis — base E. cf. E. eximius — base A. albianus
106	ALBIAN M		T. primula			NC8C	
108					P. columnata (NC8)	NC8B	base T. orionatus
110	ALBIAN E						base T. primula
112			H. planispira				base low diversity — base H. albiensis
114			T. bejaouaensis	T. roberti		NC8A	top P. cheniourensis — base P. columnata
116	APTIAN L		H. trochoidea	P. cheniour.	P. angustus (NC7)	NC7C	base T. bejauoaensis — top G. algerianus — base P. achlyostaurion
118		ISEA	G. algerianus			NC7B	
120		C34N	G. ferreolensis			NC7A	base G. algerianus — top M. floschutzii — top L. cabri — base E. floralis
122	APTIAN E		L. cabri		C. litterarius (NC6)	NC6B	base L. cabri — top C. rothii
124		CM0	G. blowi			NC6A	base R. irregularis
126		CM1n				NC5E	base F. oblongus
128	BARREMIAN	CM1, CM2			W. oblonga (NC5)	NC5D	base G. blowi
130		CM3	G. duboisi			NC5C	top C. oblongata — base G. duboisi
132		CM4, CM5, CM7	H. similis			NC5B / NC5A	top L. bolii — base H. similis — base R. terebrodent.
134	HAUTERIVIAN	CM9, CM10, CM10N	H. sigali		C. cuvillieri (NC4)	L. bolii (NC4B) / C. oblongata (NC4A)	top C. cuvillieri — base H. sigali — base L. bolii
136		CM11					top T. verenae
138	VALANGINIAN	CM11AN, CM12			G. oblongata (NK3)	T. verenae (NK3B) / R. wisei (NK3A)	top R. wisei
140		CM12A, CM13, CM14	Calpionellites (E)	D3			base Ctes. darderi — base L. hungarica — base C. oblongata
142	BERRIASIAN	CM15, CM16	Calpionellopsis (D)	D2 / D1	C. angustiforatus (NK2)	P. fenestrata (NK2B) / A. infracretacea (NK2A)	incr. Cpsis oblonga — base P. fenestrata — base Cpsis oblonga — base Cpsis simplex — base C. angustiforatus
144		CM17, CM18	Calpionella (C) / Calpionella (B)		N. steinmannii steinmannii (NK1)	N. st. minor (NJKD)	base large T. carpathica — base N. st. steinmann. — base N. st. elliptica — base N. st. minor
146	TITHONIAN	CM19	Crassicollaria (A)		M. chiastius (NJK)	R. laffittei (NJKC)	base acme Calp. alpina — base R. laffittei

Fig. 1.—Integrated Cretaceous biochronology and magnetochronology. Chronostratigraphy and geomagnetic polarity timescales are after Harland and others (1989). Tithonian through Valanginian calpionellid zones/subzones and Hauterivian through Maastrichtian planktonic foraminiferal zones/subzones are shown in the same column. New nannofossil subzones are shown with asterices. Zonal markers are shown on the right. Nannofossil events are in bold type; planktonic foraminiferal/calpionellid events are in light type. See text for details on zonation schemes and correlations between biostratigraphies and the geomagnetic polarity timescale.

markers for these subzones are solution resistant and observed over a broad geographic area.

Calpionellid Zonation

The calpionellid biostratigraphy utilized here is the standard zonation of Allemann and others (1971) which has been widely applied in sequences across the Tethyan Realm from southern Europe to Mexico. This scheme (Fig. 1) has been augmented by the addition of several subzones (e.g., Remane, 1971). Problems related to the statistical basis of several events in these zonations, which are not based on first occurrences but on changes in relative abundances, have been discussed by Remane (1985) and Channell and Grandesso (1987). This calpionellid zonation has been accurately correlated with ammonite zonations (e.g., Allemann and Remane, 1979) and with stratotype stage boundaries.

Planktonic Foraminiferal Zonation

Cretaceous planktonic foraminiferal biostratigraphy has developed over several decades (e.g., Bolli, 1966; Moullade, 1966; Barr, 1972; Premoli Silva and Bolli, 1973; Longoria, 1974, 1984; Moullade, 1974; van Hinte, 1976; Masters, 1977; Sigal, 1977; Robaszynski, Caron and others, 1979, 1984; Wonders, 1980; Leckie, 1984; Caron, 1985; Sliter, 1989; Huber, 1992). A wide variety of different zonal schemes is still utilized. Evolution of the different schemes showing their correlation is illustrated in Figure 2. Three major problems exist in the application of planktonic foraminiferal zonation: (1) the correlation between high- and low-latitude zonations is not well established (e.g., Huber, 1992), (2) a number of events (e.g., the first occurrence of *Abathomphalus mayaroensis*) have been shown to be distinctly diachronous, and (3) the biostratigraphy of the Hauterivian through Aptian interval of earliest evolution of planktonic foraminifers is in a state of flux.

The planktonic foraminiferal biostratigraphic scheme utilized here (Fig. 1) is that of Sliter (1989, 1992) which is based largely on the summary of Caron (1985). The Caron (1985) scheme is a compilation of zonations for various intervals of the Cretaceous Period. This scheme can be applied to washed residues as well as to thin sections of indurated sediments (e.g., Sliter and Leckie, 1993). The Barremian and lower Aptian intervals are in a state of flux and need a re-examination of the tiny, early species of planktonic foraminifera, particularly of the genus *Globigerinelloides*, from deep sea and land sections (see below). With the exception of the Hauterivian/Barremian boundary, which is likely to undergo future revision, the status of correlation between planktonic foraminifer zones and stage boundaries (Fig. 1) is similar to that of the calcareous nannofossils. Uncertainties exist for particular Upper Cretaceous boundaries through lack of suitable exposure, presence of condensed horizons, and absence of zonal marker taxa. For example, no foraminiferal events lie close to the Santonian/Campanian boundary, and the upper and lower limits of the Coniacian are hard to detect using planktonic foraminifera.

Problems with the Barremian-Lower Aptian Interval

The first occurrence (FO) of the genus *Globigerinelloides* (*G. duboisi, G. gottisi, G. blowi*) has previously been used as an important biostratigraphic datum within basal Aptian strata (Moullade, 1966; Van Hinte, 1976; Sigal, 1977; Caron, 1985). However, more recent work has shown that various "species" of *Globigerinelloides* occur within Barremian strata (e.g., Sigal, 1979; Moullade and others, 1988; Leckie and Bralower, 1991; Coccioni and others, 1992). The most reliable datum for the Barremian/Aptian boundary is the FO of the calcareous nannofossil *Rucinolithus irregularis* (Thierstein, 1973; Bralower, 1987; Coccioni and others, 1992; Bralower and others, 1993). Chron M0, the uppermost magnetic anomaly in the Early Cretaceous M-sequence, occurs just above the boundary (Channell and others, 1979; Tarduno and others, 1989; Channell and Erba, 1992; Coccioni and others, 1992). A revised but preliminary planktonic foraminiferal zonal scheme for the lower Barremian through lower Aptian is needed in view of the recent and ongoing work on this interval.

Two different zonal schemes have emerged. Coccioni and others (1992), in a study of limestone thin sections from the Umbrian Apennines of central Italy (Gorgo a Cerbara), have documented the FO of *G. duboisi* within the *Micrantholithus hoschulzii* nannofossil Zone, just below polarity zone Chron M1. This correlates with the uppermost ammonite zone of the lower Barremian in southern Spain. The FO of *G. blowi* is in the upper part of the *M. hoschulzii* Zone (upper Barremian) between Chrons M1 and M0. The FO of *Hedbergella similis* predates both of these *Globigerinelloides* datums according to Coccioni and others (1992). Their lower Barremian-lower Aptian planktonic foraminiferal biostratigraphy is as follows: lower Barremian *H. similis* Zone, mid-Barremian *G. duboisi* Zone, upper Barremian-lower Aptian *G. blowi* Zone, and lower Aptian *Leupoldina cabri* Zone.

In a study of well preserved specimens from ODP Site 641 in the eastern North Atlantic and a reassessment of planktonic foraminiferal distributions in nearby DSDP Site 398 (Sigal, 1979), Leckie has established a different zonal scheme for the mid-Barremian-lower Aptian interval (Leckie and Bralower, 1991; Bralower and others, 1994; Leckie, unpubl. data). Following Moullade (1966; *Hastigerina* aff. *H. simplex* Zone) and Sigal (1977), the *"Clavihedbergella" eocretacea* Zone represents a total range zone in the upper lower or mid-Barremian interval (*Watznaueria oblonga* nannofossil Zone; Bralower and others, 1994). The FO of *Globigerinelloides* aff. *G. duboisi-gottisi* group is in the lower part of the *"C." eocretacea* Zone. The FO of large, well-developed *Hedbergella similis* s.s. and the FO of *G.* aff. *G. blowi* are near the last occurrence (LO) of *"C." eocretacea* in the upper Barremian interval of Site 641. A similar sequence of datums was recorded by Sigal (1979) in nearby Site 398. The FO of *H. similis* above the FO of the *G.* aff. *G. duboisi-gottisi* group is at apparent odds with the findings of Coccioni and others (1992). The LO of *H. similis* at Site 641 is just above Chron M0, but other studies have shown this taxon to range into the mid-Aptian (e.g., Longoria, 1974; Sigal, 1977, 1979; Sliter, 1992). The discrepancy in the stratigraphic range data for *H. similis* may result from taxonomic differences between workers, particularly in the differentiation of ancestral and descendant forms (Leckie, unpubl. data). Given the present taxonomic and stratigraphic uncertainties, not only in *H. similis* but also in the tiny early species of *Globigerinelloides*, the following alternate informal zonal scheme is proposed for the lower Barremian through lower Aptian interval: lower and mid-

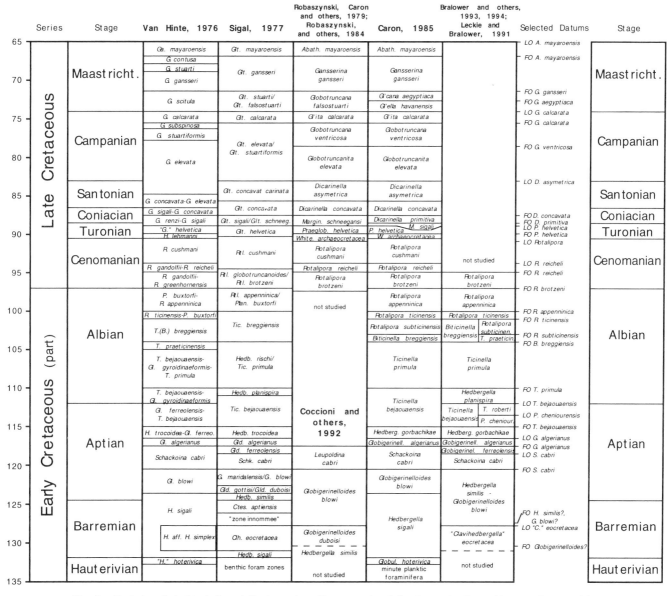

Fig. 2.—Evolution of planktonic foraminiferal zonations. Datums used to define the zonal units used here are shown at right.

Barremian *"C." eocretacea* Zone (total range zone), upper Barremian and lower Aptian *H. similis-G. blowi* Zone, and lower Aptian *Schackoina cabri* Zone (Fig. 2).

CORRELATION BETWEEN MICROFOSSIL BIOSTRATIGRAPHIES

Detailed correlation between calcareous nannofossils and planktonic foraminiferal zonations can best be achieved in expanded sections with good microfossil preservation. Such sections are mainly found in the deep sea. Land sections contain poorer microfossil preservation, especially of calcareous nannofossils, and numerous condensed intervals. Exceptions to this are the sections in the *Fosse Vocontienne* (Moullade, 1966; Thierstein, 1973) and some Upper Cretaceous intervals in the Western Interior Basin (e.g., Leckie, 1985; Bralower 1988) which contain moderate microfossil preservation, are relatively

expanded, and provide critical macrofossil biostratigraphic, and thus chronostratigraphic, correlations. In addition, detailed studies of Cretaceous sections in the Umbrian Apennines of Italy (e.g., Premoli Silva, 1977; Monechi and Thierstein, 1985; Coccioni and others, 1992; Premoli Silva and Sliter, unpubl. data) provide important information. Our correlation between nannofossil and planktonic foraminiferal biostratigraphies is obtained from the *Fosse Vocontienne* sections studied by Moullade (1966) and Thierstein (1973), from the Piobbico Core drilled in the Umbrian Apennines of Italy investigated by Erba (1988) and Tornaghi and others (1989), from DSDP and ODP sites in the Atlantic (Sites 545 and 547 in the Moroccan basin and Site 641 on the Galicia Margin; Leckie, 1984; Wiegand, 1984; Bralower, 1992; Bralower and others, 1993; Bralower and others, 1994) and Pacific (Site 167 on the Magellan Rise and Site 463 in the Mid Pacific Mountains; Tarduno and others,

LOCATION	MACROFOSSIL ZONE	STAGE	40Ar/39Ar AGE (Ma)	NANNOFOSSIL MARKERS	ZONE R78	S77	STAGE
Montrose, CO Mancos Shale (O&C, Loc. 5)	*E. jenneyi*	U. Camp.	74.76±0.45	*A. cymbiformis* *B. parca* *E. eximius* *R. anthophorus*	NC20	CC22	Top Camp.
Cedar Creek, SD Pierre Shale (O&C, Loc. 7)	*D. nebrascense*	U. Camp.	75.89±0.72	*A. cymbiformis* *B. parca* *E. eximius* *R. anthophorus*	NC20	CC22	Top Camp.
Winnett, MT Telegr. Ck. Fm. (O&C, Loc. 11)	*D. bassleri*	Top Sant.	83.83±0.43	*P. regularis* *L. grillii* *R. anthophorus* *M. staurophora* *B. "preparca"* No *B. parca s.s.*	NC17	CC17	Top Sant.
Marias River, MT Colorado Shale (O&C, Loc. 13)	*S. preventric.*	m. Con.	88.34±0.60	*E. floralis* *M. furcatus* *M. staurophora* *R. anthophorus*	NC15	CC14-CC15	U. Con-L. Sant
Cone Hill, MT Marias River Sh. (O&C, Loc. 14)	*N. judii*	Top Cen.	93.55±0.47	*E. eximius* *Q. gartneri* No *A. albianus* No *M. chiastius* No *L. maleformis*	NC12	CC11	L. Tur.
Carbon County, WY (O&C, Loc. 15)	*D. pondi*	U. Cen.	94.63±0.61	*A. albianus* *C. kennedyi* *M. chiastius* *G. segmentatum* *V. octoradiata*	NC11-NC12	CC10	U. Cen.
Niobrara, WY Greenhorn Lst. (O&C, Loc. 16)	*A. amphibolum*	m. Cen.	94.93±0.53	*A. albianus* *L. acutum* *C. kennedyi* No *G. segmentatum* No *V. octoradiata*	NC11	CC9	m. Cen.

O&C---Obradovich and Cobban (1975), zonal, stage designations and Ar/Ar ages from Obradovich (1993). R78-Roth (1978); S77-Sissingh (1977)

FIG. 2.—Continued.

1989), and from numerous sections in the Western Interior Basin of North America (Leckie, 1985; Bralower, 1988) and Northern Europe (Hart and Bigg, 1981; Bralower, 1988; Jarvis and others, 1988). Correlation between Upper Cretaceous zonations are based on the Gubbio sections (Premoli Silva, 1977; Monechi and Thierstein, 1985; Premoli Silva and Sliter, unpubl. data) and DSDP sites in the South Atlantic Ocean (Poore and others, 1983; Boersma, 1984; Manivit, 1984).

Numerous other deep-sea sections which serve as a basis for our high-resolution, calcareous nannofossil biostratigraphy are in the process of detailed planktonic foraminifer biostratigraphic investigation. These include: North Atlantic DSDP Site 398 and ODP Site 641 and Indian Ocean ODP Sites 761, 762

and 763. South Atlantic DSDP Site 511 has a detailed planktonic foraminifer biostratigraphy (Krashenninikov and Basov, 1983) but lacks traditional marker taxa which are useful in the other sections. Somewhat different correlations between calcareous nannofossil and planktonic foraminiferal zonations exist in high-latitude sections (e.g., Huber and Watkins, 1992) where different zonal markers have proven to be useful (e.g., Huber, 1992; Watkins and others, 1992).

Correlation between calcareous nannofossil and calpionellid biostratigraphy is based on various sequences in the *Fosse Vocontienne* (e.g., Le Hēgarat and Remane, 1968; Thierstein, 1973; Thierstein, 1975; Bralower and others, 1989), the Umbrian Apennines (e.g., Cirilli and others, 1984; Lowrie and

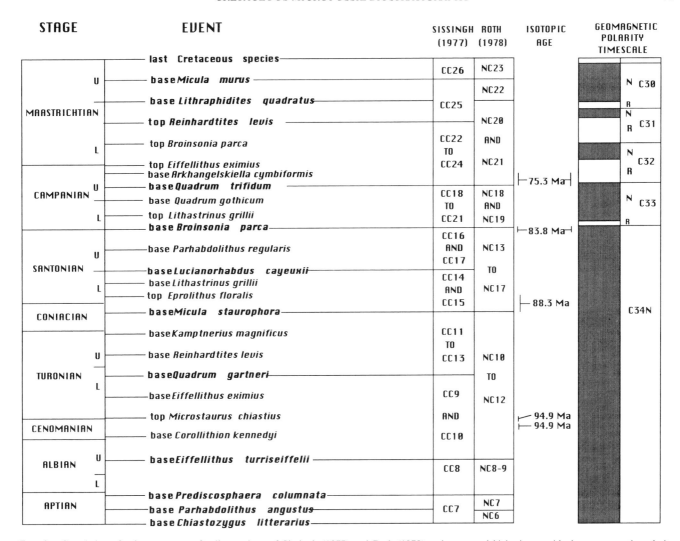

FIG. 3.—Correlation of calcareous nannofossil zonations of Sissingh (1977) and Roth (1978) and non-zonal biohorizons with the geomagnetic polarity timescale. Only those zones relevant to the discussion are divided. The correlation shown is the scheme discussed in detail by Bralower and Siesser (1992). This correlation allows indirect ties between Western Interior bentonite isotopic ages and the geomagnetic polarity timescale (see text for complete discussion).

Channell, 1984; Bralower and others, 1989), and Southern Alps (Ogg, 1981; Channell and Grandesso, 1987; Channell and others, 1987) of Italy and DSDP Site 534 in the western North Atlantic (Roth, 1983; Remane, 1983; Bralower and others, 1989). The numerous sequences in which these correlations have been established indicate few problems with the ties shown (Fig. 1).

BIOMAGNETOSTRATIGRAPHIC CORRELATION

Microfossil biostratigraphic correlation with the geomagnetic polarity timescale has been the subject of intense recent investigation. Correlation of M-sequence chrons with nannofossils and calpionellids is based on studies of sequences in the Umbrian Apennines (Lowrie and others, 1980; Lowrie and Alvarez, 1984; Cirilli and others, 1984; Bralower, 1987; Bralower and others, 1989) and Southern Alps (Channell and others, 1979; Ogg, 1981; Bralower, 1987; Channell and Grandesso, 1987; Channell and others, 1987; Bralower and others, 1989; Channell and Erba, 1992) of Italy, DSDP Site 534 in the western

North Atlantic (Wind, 1978; Roth, 1978; Ogg, 1983; Roth, 1983; Bralower and others, 1989) and Pacific DSDP Sites 167 and 463 (Tarduno and others, 1989). Planktonic foraminifer zones have been correlated to the upper part of the M-sequence only in one section, at Gorgo a Cerbara in the Umbrian Apennines (Coccioni and others, 1992; Premoli Silva, pers. commun., 1992). Correlation of stage boundaries and the geomagnetic polarity timescale has been discussed in most of these previously cited references. We have modified the correlation of the Hauterivian/Barremian boundary to Chron CM4 as reported by Cecca and others (1994) based on ammonite biostratigraphy of the Gorgo a Cerbara section. Reversed magnetic polarity zones within the long normal polarity interval (C34N) have been reported by Vandenberg and others (1978), Tarduno (1990) and Tarduno and others (1992). High-frequency reversals in the middle Albian interval (Tarduno and others, 1992) have only been reported from one sequence and are not included in our chronology although the biostratigraphic correlation of the late Aptian ISEA reversal is well established (Tarduno, 1990; Fig. 1). Correlation between Upper Cretaceous magnetic

TABLE 1.—INTEGRATED ZONES

ZONE	BASE	TOP
IC73	base *Micula murus*	Last **Cretaceous species**
IC72	base *Lithraphidites quadratus*	base *Micula murus*
IC71	base *Abathomphalus mayaroensis*	base *Lithraphidites quadratus*
IC70	top *Reinhardtites levis*	base *Abathomphalus mayaroensis*
IC69	top *Tetralithus trifidus*	top *Reinhardtites levis*
IC68	base *Gansserina gansseri*	top *Tetralithus trifidus*
IC67	base *Globotruncana aegyptica*	base *Gansserina gansseri*
IC66	top *Globotruncanita calcarata*	base *Globotruncana aegyptica*
IC65	base *Globotruncanita calcarata*	top *Globotruncanita calcarata*
IC64	base *Tetralithus gothicus*	base *Globotruncanita calcarata*
IC63	base *Ceratolithus aculeus*	base *Tetralithus gothicus*
IC62	base *Globotruncana ventricosa*	base *Ceratolithus aculeus*
IC61	top *Lithraphidites grillii*	base *Globotruncana ventricosa*
IC60	base *Broinsonia parca*	top *Lithraphidites grillii*
IC59	base *Lucianorhabdus cayeuxii*	base *Broinsonia parca*
IC58	base *Dicarinella asymetrica*	base *Lucianorhabdus cayeuxii*
IC57	base *Micula decussata*	base *Dicarinella asymetrica*
IC56	base *Dicarinella concavata*	base *Micula decussata*
IC55	base *Marthasterites furcatus*	base *Dicarinella concavata*
IC54	base *Kamptnerius magnificus*	base *Marthasterites furcatus*
IC53	top *Parhabdolithus asper*	base *Kamptnerius magnificus*
IC52	top *Microstaurus chiastius*	top *Parhabdolithus asper*
IC51	top *Axopodorhabdus albianus*	top *Microstaurus chiastius*
IC50	top *Rotalipora cushmani*	top *Axopodorhabdus albianus*
IC49	top *Corollithion kennedyi*	top *Rotalipora cushmani*
IC48	base *Vagalapilla octoradiata*	top *Corollithion kennedyi*
IC47	base *Dicarinella algeriana*	base *Vagalapilla octoradiata*
IC46	base *Rotalipora cushmani*	base *Dicarinella algeriana*
IC45	base *Rotalipora reicheli*	base *Rotalipora cushmani*
IC44	base *Lithraphidites acutum*	base *Rotalipora reicheli*
IC43	base *Corollithion kennedyi*	base *Lithraphidites acutum*
IC42	base *Rotalipora brotzeni*	base *Corollithion kennedyi*
IC41	base *Gartnerago nanum*	base *Rotalipora brotzeni*
IC40	base *Rotalipora appenninica*	base *Gartnerago nanum*
IC39	base *Eiffellithus turriseiffelii*	base *Rotalipora appenninica*
IC38	base *Rotalipora subticinensis*	base *Eiffellithus turriseiffelii*
IC37	base *Biticinella breggiensis*	base *Rotalipora subticinensis*
IC36	base *Axopodorhabdus albianus*	base *Biticinella breggiensis*
IC35	base *Tranolithus orionatus*	base *Axopodorhabdus albianus*
IC34	base *Corollithion signum*	base *Tranolithus orionatus*
IC33	base *Ticinella primula*	base *Corollithion signum*
IC32	base low diversity forams	base *Ticinella primula*
IC31	base *Hayesites albiensis*	base low diversity forams
IC30	base *Prediscosphaera columnata*	base *Hayesites albiensis*
IC29	base *Ticinella bejaouaensis*	base *Prediscosphaera columnata*
IC28	top *Globigerinelloides algerianus*	base *Ticinella bejaouaensis*
IC27	base *Globigerinelloides algerianus*	top *Globigerinelloides algerianus*
IC26	top *Leupoldina cabri*	base *Globigerinelloides algerianus*
IC25	base *Eprolithus floralis*	top *Leupoldina cabri*
IC24	base *Leupoldina cabri*	base *Eprolithus floralis*
IC23	top *Conusphaera rothii*	base *Leupoldina cabri*
IC22	base *Rucinolithus irregularis*	top *Conusphaera rothii*
IC21	base *Flabellites oblongus*	base *Rucinolithus irregularis*
IC20	base *G. blowi, H. similis*	base *Flabellites oblongus*
IC19	top *Calcicalathina oblongata*	base *G. blowi, H. similis*
IC18	base *Globigerinelloides?*	top *Calcicalathina oblongata*
IC17	top *Lithraphidites bollii*	base *Globigerinelloides?*
IC16	base *Rucinolithus terebrodentarius*	top *Lithraphidites bollii*
IC15	top *Cruciellipsis cuvillieri*	base *Rucinolithus terebrodentarius*
IC14	base *Lithraphidites bollii*	top *Cruciellipsis cuvillieri*
IC13	top *Tubodiscus verenae*	base *Lithraphidites bollii*
IC12	top *Rucinolithus wisei*	top *Tubodiscus verenae*
IC11	base <u>Calpionellites darderi</u>	top *Rucinolithus wisei*
IC10	base *Calcicalathina oblongata*	base <u>Calpionellites darderi</u>
IC9	base <u>Lorenziella hungarica</u>	base *Calcicalathina oblongata*
IC8	incr <u>Calpionellopsis oblonga</u>	base <u>Lorenziella hungarica</u>
IC7	base *Percivalia fenestrata*	incr <u>Calpionellopsis oblonga</u>
IC6	base <u>Calpionellopsis simplex</u>	base *Percivalia fenestrata*
IC5	base *Cretarhabdus angustiforatus*	base <u>Calpionellopsis simplex</u>
IC4	base large <u>Tintinnopsella carpathica</u>	base *Cretarhabdus angustiforatus*
IC3	base *Nannoconus st. steinmannii*	base large <u>Tintinnopsella carpathica</u>
IC2	base *Nannoconus st. minor*	base *Nannoconus st. steinmannii*
IC1	base acme <u>Calpionella alpina</u>	base *Nannoconus st. minor*

Taxa in bold: calcareous nannofossils, light type: planktonic foraminifers, underlined: calpionellids.

chrons and calcareous nannofossil and planktonic foraminiferal zones is based on investigations of the Gubbio sequences (Alvarez and others, 1977; Monechi and Thierstein, 1985; Premoli Silva and Sliter, unpubl. data), sections in the South Atlantic (Poore and others, 1983; Tauxe and others, 1983; Boersma, 1984; Manivit, 1984; Chave, 1984; Stradner and Steinmetz, 1984; Huber, 1991), Southern Ocean (Huber, 1990; Pospichal and Wise, 1990); Pacific (Monechi and others, 1985) and the eastern Indian Ocean (Galbrun, 1992; Bralower and Siesser, 1992). Minor modifications of the current scheme (Fig. 1) are

to be expected in the future as new, expanded sequences with suitable microfossil preservation are recovered. Once again, the correlations illustrated are most consistent for low- and mid-latitude sections. Paleobiogeographic factors causing the diachroneity of particular planktonic foraminiferal and nannofossil events with respect to magnetostratigraphy at high-latitude sites are discussed by Huber and Watkins (1992).

DIRECT CALIBRATION OF RADIOMETRIC AGES AND
NANNOFOSSIL BIOSTRATIGRAPHY

Inherent in the construction of any time scale is the correlation between biostratigraphy and isotopic ages measured on a variety of different igneous and sedimentary materials. Cretaceous time scales differ significantly according to their selectivity of radiometric data. Some studies, for example, Kent and Gradstein (1985) and Obradovich (1993) use only those isotopic dates with the lowest geochemical uncertainties and exclusively volcanic and pyroclastic horizons. Others, such as Harland and others (1989) and Gradstein and others (1993), utilize isotopic data less selectively and apply statistical techniques to arrive at the "best" age estimate for individual boundaries. No matter which approach is chosen, however, the precision of time scales depends partly on the precision with which an isotopic age is correlated with biostratigraphic elements.

Early Cretaceous Ages

Two recently published Early Cretaceous isotopic ages have precise biostratigraphic ties. Zircon fractions from closely spaced bentonites in the middle Berriasian of the Great Valley Group, northern California have been dated at 137 + 1.6/-0.6 Ma using U/Pb. These horizons have been correlated to the *Cretarhabdus angustiforatus* (NK-2) nannofossil Zone, *Assipetra infracretacea* Subzone (Bralower and others, 1990) and indirectly to magnetic chrons CM16 and CM16n. Basalt samples recovered from the Ontong Java Plateau at ODP Site 807 have been dated using $^{40}Ar/^{39}Ar$ providing a mean of 122 Ma (Mahoney and others, 1993). Directly-overlying sediments lie in the lower Aptian *Chiastozygus litterarius* (NK-6) nannofossil Zone (E. Erba, pers. commun., 1993) and *G. blowi* foraminifer Zone (Sliter and Leckie, 1993). The normal magnetic polarity of this basement indicates that it correlates to the base of the Cretaceous long normal polarity interval (Tarduno and others, 1991).

Late Cretaceous Ages

Common elements of all Cretaceous time scales have been the K/Ar ages of sanidines and biotites from Upper Cretaceous bentonites deposited in macrofossiliferous sections of the Western Interior Basin (Obradovich and Cobban, 1975). These bentonite ages have been correlated with macrofossil zones, mainly those of ammonites, allowing indirect ties to European stage stratotype and parastratotype sequences. Recently, Obradovich (1993) remeasured many of these same units, as well as newly collected layers, with the extremely precise $^{40}Ar/^{39}Ar$ radiometric system. These dates have led to a precise Upper Cretaceous time scale with tie points to stages at over 30 different levels. This part of the Cretaceous time scale is now well established and should change very little in the future.

One possible problem with the established Upper Cretaceous time scale is that several of the correlations between Western

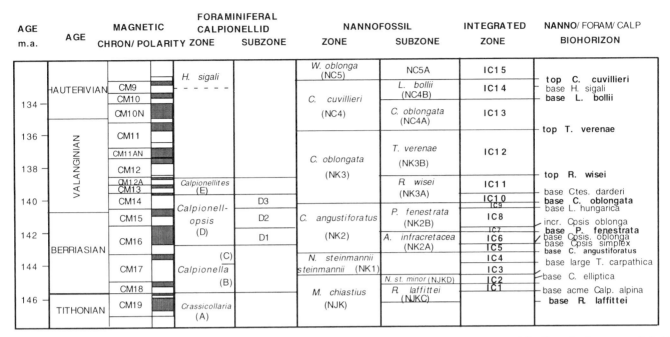

FIG. 4.—Integrated microfossil zonation for the Berriasian and Valanginian interval. Zonal markers are shown on the right. Nannofossil events are in bold type; planktonic foraminiferal/calpionellid events are in light type. See text for details on zonation schemes and correlations between biostratigraphies and the geomagnetic polarity timescale.

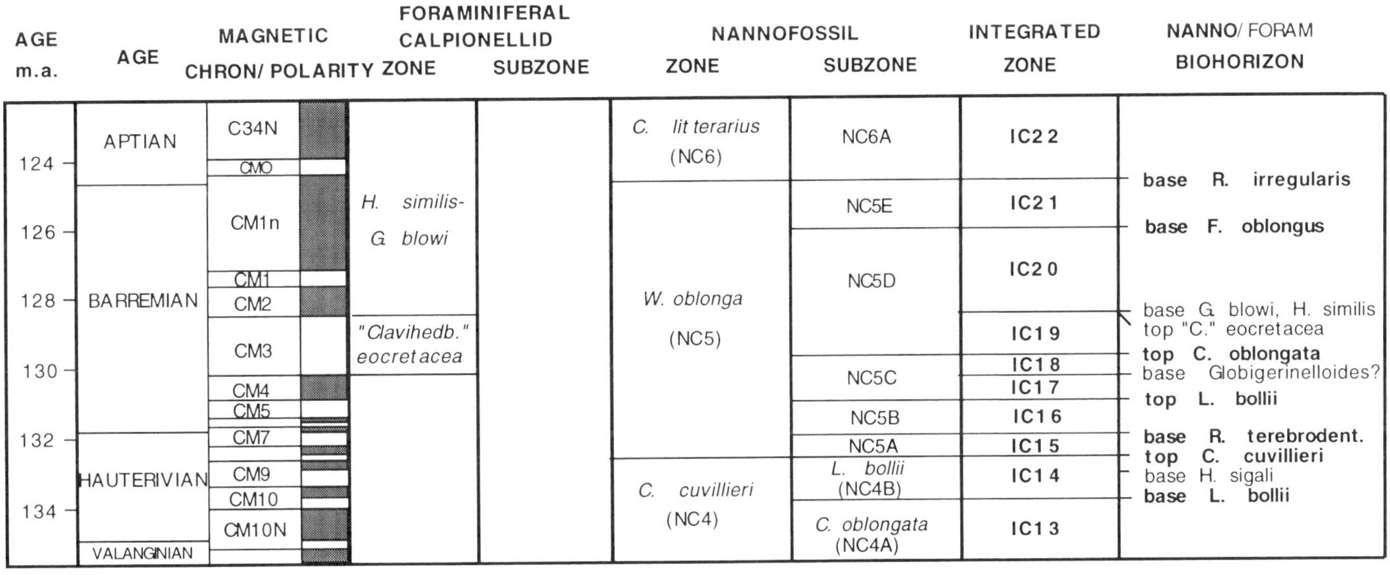

FIG. 5.—Integrated microfossil zonation for the Hauterivian through Aptian interval. Zonal markers are shown on the right. Nannofossil events are in bold type; planktonic foraminiferal events are in light type. See text for details on zonation schemes and correlations between biostratigraphies and the geomagnetic polarity timescale.

Interior macrofossil zones and European stage boundaries are indirect (e.g., see discussion in Hancock and others, 1993). This results from the rather different assemblages found in these two distant regions. In order to address this problem, we collected detailed samples from fourteen of the seventeen original Obradovich and Cobban (1975) localities and processed them for standard calcareous nannofossil biostratigraphic analysis. In all, some 470 samples were processed. Of these, less than ten percent (40) were nannofossiliferous. The results are compiled in Table 1. Seven localities were datable. Locations 2, 3, 4, 8, 9, 12, and 17 were sampled but found to be almost entirely barren of nannofossils and therefore are not discussed further. Fossiliferous sequences have been correlated with the zonation scheme of Roth (1978) and Sissingh (1977) and thereby more directly with the European stage stratotypes (Fig. 3). Key markers are listed in Table 1. The correlations with zonations are

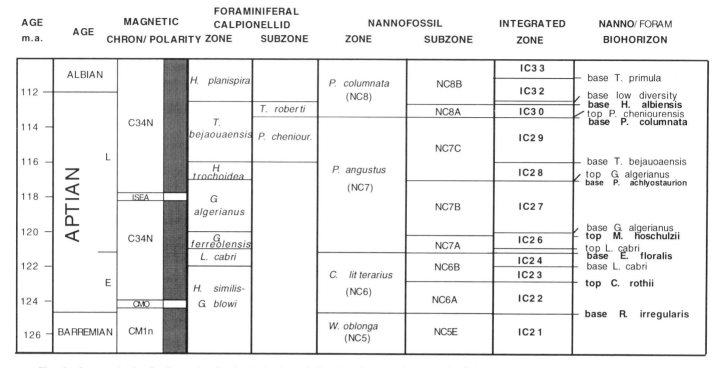

Fig. 6.—Integrated microfossil zonation for the Aptian interval. Zonal markers are shown on the right. Nannofossil events are in bold type; planktonic foraminiferal events are in light type. See text for details on zonation schemes and correlations between biostratigraphies and the geomagnetic polarity timescale.

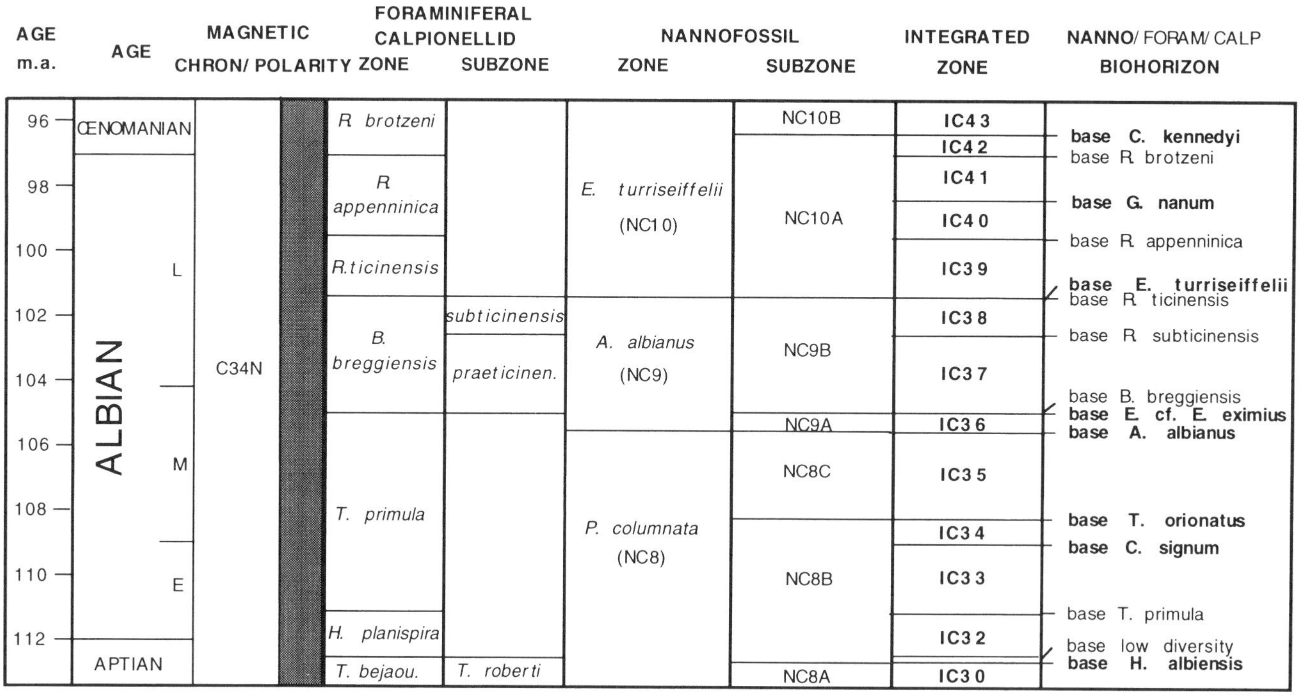

Fig. 7.—Integrated microfossil zonation for the Albian interval. Zonal markers are shown on the right. Nannofossil events are in bold type; planktonic foraminiferal events are in light type. See text for details on zonation schemes and correlations between biostratigraphies.

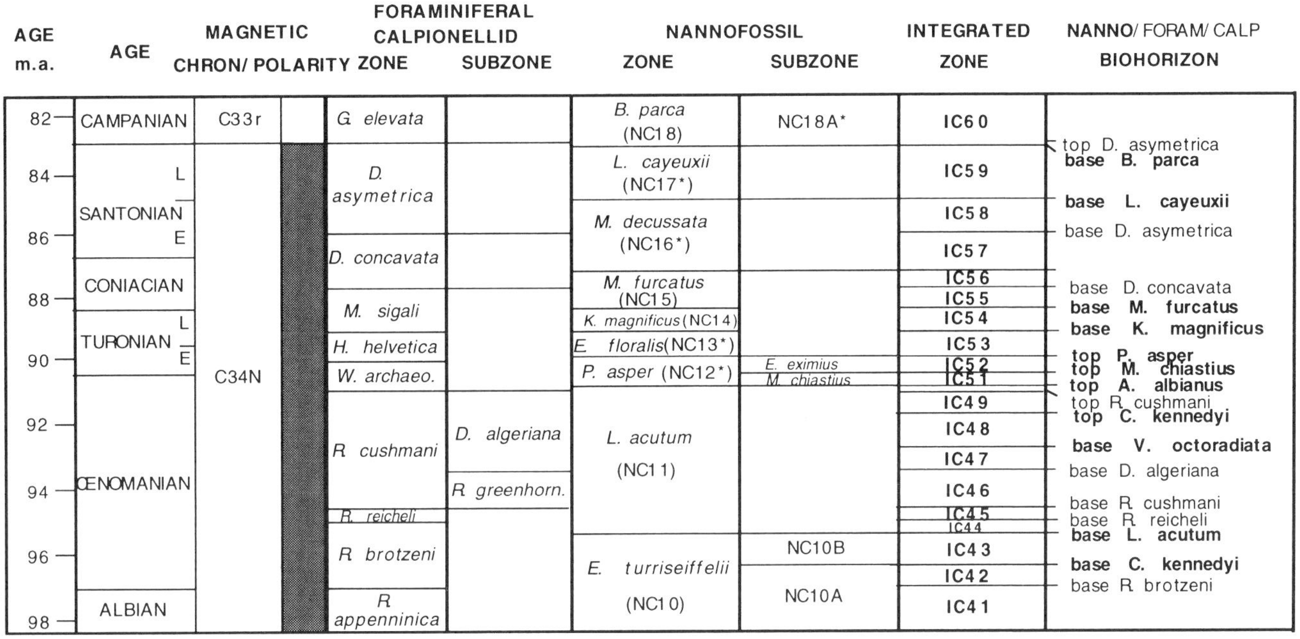

AGE m.a.	AGE	MAGNETIC CHRON/POLARITY	FORAMINIFERAL CALPIONELLID ZONE	SUBZONE	NANNOFOSSIL ZONE	SUBZONE	INTEGRATED ZONE	NANNO/FORAM/CALP BIOHORIZON
82	CAMPANIAN	C33r	G. elevata		B. parca (NC18)	NC18A*	IC60	top D. asymetrica / base B. parca
84	L SANTONIAN E		D. asymetrica		L. cayeuxii (NC17*)		IC59	base L. cayeuxii
86					M. decussata (NC16*)		IC58	base D. asymetrica
	CONIACIAN		D. concavata				IC57	base D. concavata
88	TURONIAN L E	C34N	M. sigali		M. furcatus (NC15)		IC56 IC55	base M. furcatus
					K. magnificus (NC14)		IC54	base K. magnificus
90			H. helvetica		E. floralis (NC13*)		IC53	top P. asper
			W. archaeo.		P. asper (NC12*)	E. eximius	IC52	top M. chiastius
						M. chiastius	IC51	top A. albianus
							IC49	top R. cushmani
92			R. cushmani	D. algeriana	L. acutum (NC11)		IC48	top C. kennedyi
							IC47	base V. octoradiata
94	CENOMANIAN			R. greenhorn.			IC46	base D. algeriana
			R. reicheli				IC45	base R. cushmani
							IC44	base R. reicheli / base L. acutum
96			R. brotzeni		E. turriseiffelii (NC10)	NC10B	IC43	base C. kennedyi
							IC42	base R. brotzeni
98	ALBIAN		R. appenninica			NC10A	IC41	

Fig. 8.—Integrated microfossil zonation for the Cenomanian through Santonian interval. Zonal markers are shown on the right. Nannofossil events are in bold type; planktonic foraminiferal events are in light type. See text for details on zonation schemes and correlations between biostratigraphies and the geomagnetic polarity timescale.

AGE m.a.	AGE	MAGNETIC CHRON/POLARITY	FORAMINIFERAL CALPIONELLID ZONE	SUBZONE	NANNOFOSSIL ZONE	SUBZONE	INTEGRATED ZONE	NANNO/FORAM BIOHORIZON
64	DANIAN	C29r			M. murus (NC23)		IC73	Last Cretaceous species / base M. murus
66	L	C30	A. mayaroensis		L. quadratus (NC22)		IC72	
68		C30r C31				NC21B*	IC71 IC70	base L. quadratus / base A. mayaroensis / top R. levis
70	E	C31r	G. gansseri		L. praequadratus (NC21)	NC21A*	IC69	top T. trifidus
72		C32n1/r1 C32n2	G. aegyptica		T. trifidus (NC20)		IC68	base G. gansseri
		C32r2	G. havanaensis				IC67	base G. aegyptica
74			G. calcarata				IC66	top G. calcarata
	L	C33			C. aculeus (NC19)	NC19B*	IC65	base G. calcarata / base T. trifidus
76			G. ventricosa			NC19A*	IC64	base T. gothicus
78	E						IC63	base C. aculeus
					B. parca (NC18)	NC18B*	IC62	base G. ventricosa
80		C33r	G. elevata				IC61	top L. grillii
82						NC18A*	IC60	top D. asymetrica
84	SANTONIAN		D. asymetrica		L. cayeuxii (NC17*)		IC59	base B. parca

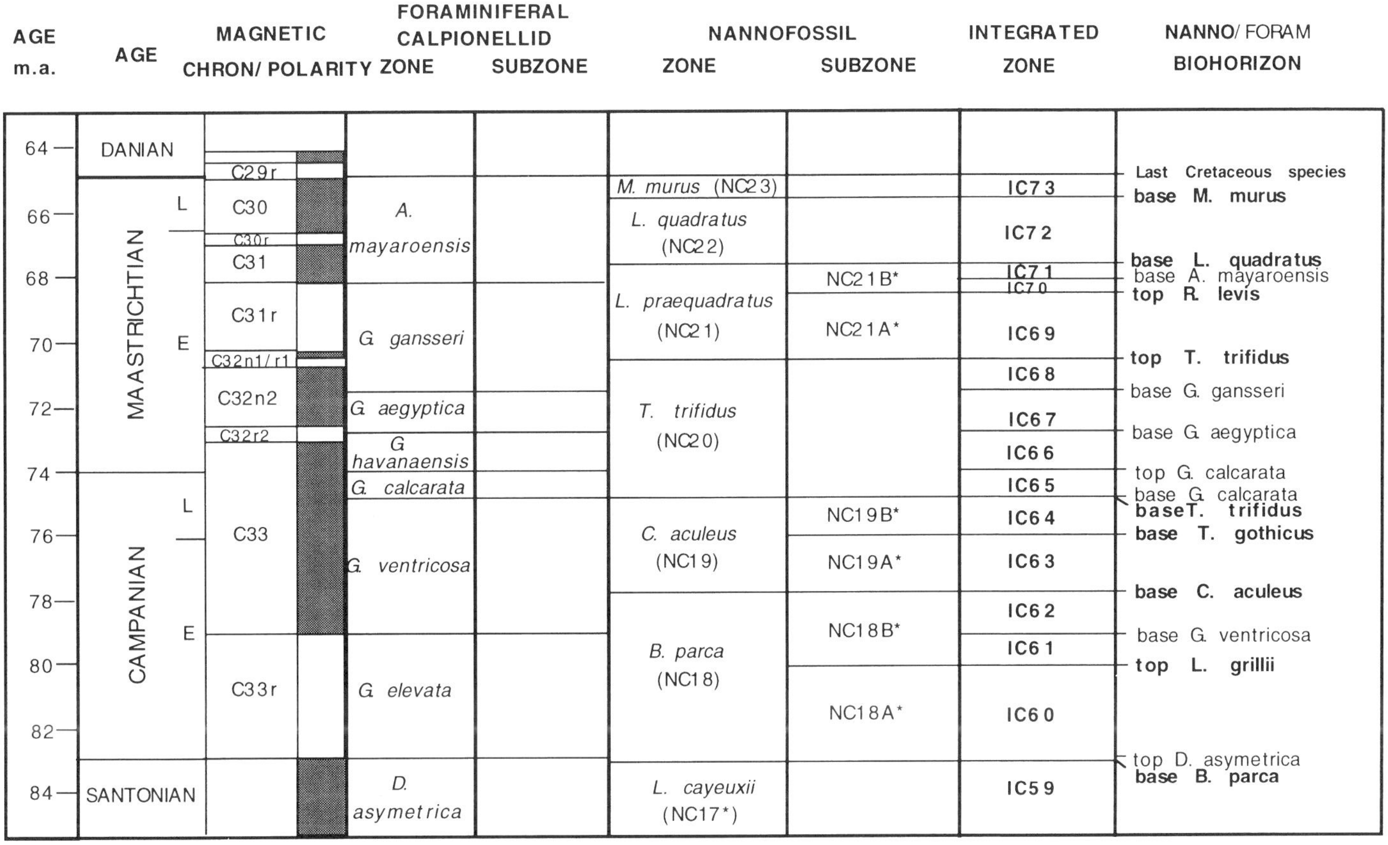

Fig. 9.—Integrated microfossil zonation for the Campanian and Maastrichtian interval. Zonal markers are shown on the right. Nannofossil events are in bold type; planktonic foraminiferal events are in light type. See text for details on zonation schemes and correlations between biostratigraphies and the geomagnetic polarity timescale.

straight-forward with one exception: the informal taxon *Broinsonia "preparca"* is a form which is smaller (5–6 µm long) than true *B. parca* and has a minuscule central area. This form occurs just below the first *B. parca* in two ODP sequences (Holes 762C and 763B) from the Exmouth Plateau (Bralower and Siesser, 1992). Perfect agreement with the published stage assignments occurs in three cases (Locations 11, 15, and 16; Table 1). In two cases (Locations 13 and 14), minor differences between the stage assignments predicted by nannofossils and by macrofossils (Obradovich and Cobban, 1975; Obradovich, 1993) may arise from uncertainty in the correlation of our sample set to the dated bentonites; in these localities the exact pyroclastic horizon dated was uncertain to us. Two upper Campanian levels (Locations 5 and 7) may have been placed slightly too low with respect to the Campanian/Maastrichtian boundary. Samples from around these levels contain an assemblage with *B. parca, Eiffellithus eximius, Reinhardtites anthophorus,* and *Arkhangelskiella cymbiformis.* We have not differentiated species of *Reinhardtites* (*R. anthophorus* and *R. levis*). This problem is discussed by Burnett and others (1992), but we have been unable to separate with the light microscope differences among these taxa which are primary from those which are diagenetic. Therefore, we cannot utilize these taxa in dating Campanian-Maastrichtian sediments with the exception of the last occurrence of the genus (e.g., Bralower and Siesser, 1992; see above). No specimens of *Quadrum trifidum* have been observed in these locations. The taxonomy and biostratigraphy of the nannofossil genus *Arkhangelskiella* is fraught with uncertainties. The specimens of *A. cymbiformis* observed here are well preserved with a fairly reduced central area size and few (15–20) pores in the central area, and are clearly differentiable from *A. speciallata* (which first occurs in the lower Campanian; Perch-Nielsen, 1985) and *B. enormis.* The first occurrence of such clear specimens of *A. cymbiformis* has been observed only in Maastrichtian units in Europe (Sissingh, 1977; Perch-Nielsen, 1985) and other locations in the Western Interior Basin (D. Watkins, pers. commun., 1993). However, similar specimens have been identified co-occuring with *Q. trifidum, B. parca,* and *E. eximius* in sediments of the upper Campanian of the Exmouth Plateau (Bralower and Siesser, 1992). Thus the co-occurrence of *A. cymbiformis* with *B. parca* (last occurrence in lower Maastrichtian) and *E. eximius* (last occurrence close to Campanian/Maastrichtian boundary; Thierstein, 1976) at Locations 5 and 7 suggests rather indirectly that these two levels are in the uppermost Campanian. A conclusive solution of this question must await an in-depth study of the range of *A. cymbiformis* in northern European sections. It is clear from the range of this taxon described by Burnett and others (1992) that these authors have much wider taxonomic concepts than those utilized here. The macrofossil definition of the Campanian/Maastrichtian boundary has been the topic of much previous discussion (e.g., Kennedy and others, 1992) and is currently under close scrutiny (Premoli Silva and Sliter, unpubl. data). Thus, even though the zonal correlations should remain firm, the implications of our findings on the age of the Campanian/Maastrichtian boundary may change.

Besides more direct correlation with the European stage stratotypes, calcareous nannofossil biostratigraphy provides direct ties between the bentonite ages and the Geomagnetic Polarity Timescale (GPT). Correlation of nannofossil datums, the Sis-

singh (1977) and Roth (1978) zonation schemes, and the GPT as discussed in detail by Bralower and Siesser (1992) is shown in Figure 3. These correlations are not without problems, and different solutions have been described (e.g., Premoli Silva and Sliter, unpubl. data). However, our results allow two direct tie points, 75.3 Ma (the average age between locations 5 and 7) and the top of Chron C32R, and 83.8 Ma and the top of the long normal interval, C34N. Clearly much further work is required to refine many of these correlations, and for this reason we have not applied the correlations of nannofossil zones and isotopic ages and their implications to the GPT in the integrated biostratigraphy proposed here.

AN INTEGRATED CRETACEOUS MICROFOSSIL BIOSTRATIGRAPHY

Construction of the integrated microfossil scheme follows the correlation between the three microfossil biostratigraphies illustrated in Figure 1. The chronology of Harland and others (1989) is used for the ages of magnetic chrons and stage boundaries (with the exception of the Hauterivian/Barremian boundary, which has been moved to Chron CM4 according to the results of Cecca and others (1994)) and thus is assigned an age of 130.6 Ma instead of 131.8 Ma in Harland and others (1989). Microfossil-magnetostratigraphic correlations are used to place events within the chronology in the Lower Cretaceous M-sequence and the Upper Cretaceous reversed polarity interval. The ages of events within the Cretaceous long normal polarity interval are determined by relative sedimentary thicknesses in European land sequences (Tornaghi and others, 1989; Premoli Silva and Sliter, unpubl. data). The integrated scheme utilizes all potential markers except where two markers lie close to one another. In these cases the event which is more widely applicable is utilized. Zones are numbered from the base of the Berriasian stage upwards and given a prefix of IC (Integrated Cretaceous). Seventy three zones have been defined in this way (Table 2; Figs. 4–9). We stress that this is an informal scheme (ie, we have not defined zonal units formally as recommended by the International Subcommission on Stratigraphic Classification (Hedberg, 1976). In addition, since uncertainties remain with the correlations between the microfossil groups, we stress that this zonation can be entirely replaced by updated schemes in the future.

CONCLUSIONS

An informal, integrated, calcareous nannoplankton, calpionellid, and planktonic foraminifer biostratigraphy is proposed for the Cretaceous Period. This biostratigraphy consists of 73 zones and is based upon a literature survey of numerous DSDP/ODP and land sections as well as our own investigation of several of these sequences. The integrated scheme is most applicable in mid- and low-latitude sites and significantly increases potential biostratigraphic resolution to between 0.5 and 1.5 my/zone.

ACKNOWLEDGMENTS

The authors are grateful to J. Channell, W. Cobban, E. Erba, J. Mutterlose, J. Obradovich, I. Premoli Silva, J. Tarduno, and D. Watkins for stimulating discussions. We thank B. Huber, K. McDougall, J. Pospichal, and P. Quinterno for reviewing an early version of the manuscript. Research supported by NSF (EAR 8313213; EAR-8721350) (to TJB and HRT) and ACS-PRF grants (to TJB and RML).

REFERENCES

ALLEMANN, F. AND REMANE, J., 1979, Les faunes de Calpionelles du Berriasien Supérieur/Valanginien, in Busnardo, R., Thieuloy, J.-P., and Moullade, M., eds.,Stratotypes Français, 6. Hypostratotype mésogén de l'étage Valanginien (Sud-Est de la France): Paris, CNRS, pp. 99–109.

ALLEMANN, F., CATALANO, R., FARES, F., AND REMANE, J., 1971, Standard calpionellid zonation (Upper Tithonian-Valanginian) of the western Mediterranean Province: Rome, Proceedings Second Planktonic Conference, v. 2, p. 1337–1340.

ALVAREZ, W., ARTHUR, M. A., FISCHER, A. G., LOWRIE, W., NAPOLEONE, G., PREMOLI SILVA, I., AND ROGGENTHEN, W. M., 1977, Upper Cretaceous-Paleocene magnetic stratigraphy at Gubbio, Italy. V. Type section for the Late Cretaceous-Paleocene geomagnetic polarity timescale: Bulletin Geological Society of America, v. 88, p. 367–389.

APPLEGATE, J. L. AND BERGEN, J. A., 1988, Cretaceous calcareous nannofossil biostratigraphy of sediments recovered from the Galicia Margin, ODP Leg 103: Scientific Results Ocean Drilling Program, v. 103, p. 293–319.

BARR, F. T., 1972, Cretaceous biostratigraphy and planktonic foraminifera from Libya: Micropaleontology, v. 18, p. 1–46.

BERGGREN, W. A., KENT, D. V., AND FLYNN, J. J., 1985, Jurassic to Paleogene: Part 2. Paleogene geochronology and chronostratigraphy, in Snelling, N. J., ed., The Chronology of the Geologic Record: London, Geological Society of London Memoir 10, p. 141–195.

BOERSMA, A., 1984, Cretaceous-Tertiary planktonic foraminifers from the Southeastern Atlantic, Walvis Ridge area, DSDP Leg 74: Initial Reports Deep Sea Drilling Project, v. 74, p. 501–523.

BOLLI, H. M., 1957, The genera Praeglobotruncana, Rotalipora, Globotruncana and Abathomphalus in the Upper Cretaceous of Trinidad, B.W.I.: Bulletin of the United States. National Museum, v. 215, p. 51–60.

BOLLI, H. M., 1959, Cretaceous planktonic foraminifera from the Cretaceous of Trinidad, B.W.I.: Bulletin of American Paleontologist, v. 39, p. 257–277.

BOLLI, H. M., 1966, Zonation of Cretaceous to Pliocene marine sediments based on planktonic foraminifera: Boletin Informativo Asociacion Venezolana de Geologia, Mineria y Petroleo, v. 9, p. 3–32.

BRALOWER, T. J., 1987, Valanginian to Aptian calcareous nannofossil stratigraphy and correlation with the upper M-sequence magnetic anomalies: Marine Micropaleontology, v. 17, p. 293–310.

BRALOWER, T. J., 1988, Calcareous nannofossil biostratigraphy and assemblages of the Cenomanian-Turonian boundary interval: implications for the origin and timing of oceanic anoxia: Paleoceanography, v. 3, p. 275–316.

BRALOWER, T. J., 1992, Aptian-Albian calcareous nannofossil biostratigraphy of ODP Site 763 and the correlation between high- and low-latitude zonations, in Duncan, R. A., Rea, D. K., Kidd, R. B., von Rad, U., and Weissel, J. K., eds., Synthesis of Results from Scientific Drilling in the Indian Ocean: Washington, D. C., American Geophysical Union Monograph 70, p. 245–252.

BRALOWER, T. J., ARTHUR, M. A., LECKIE, R. M., SLITER, W. V., ALLARD, D. J., AND SCHLANGER, S. O., 1994, Timing and paleoceanography of oceanic anoxia/ dysoxia in the late Barremian-early Aptian: Palaios, v. 9, p. 335–369.

BRALOWER, T. J., LUDWIG, K. R., OBRADOVICH, J. D., AND JONES, D. L., 1990, Berriasian (Early Cretaceous) radiometric ages from the Grindstone Creek Section, Sacramento Valley, California: Earth Planetary Science Letters, v. 98, p. 62–73.

BRALOWER, T. J., MONECHI, S., AND THIERSTEIN, H. R., 1989, Calcareous nannofossil zonation of the Jurassic-Cretaceous boundary interval and correlation with the geomagnetic polarity timescale: Marine Micropaleontology, v. 11, p. 153–235.

BRALOWER, T. J. AND SIESSER, W. G., 1992, Cretaceous calcareous nannofossil biostratigraphy of ODP Leg 122 Sites 761, 762 and 763, Exmouth and Wombat Plateaus, N.W. Australia: Proceedings Ocean Drilling Program Scientific Results, v. 122, p. 529–556.

BRALOWER, T. J., SLITER, W. V., ARTHUR, M. A., LECKIE, R. M., ALLARD, D. J., AND SCHLANGER, S. O., 1993, Dysoxic/anoxic episodes in the Aptian-Albian (Early Cretaceous): Washington, D. C., Schlanger Memorial Volume, American Geophysical Union Monograph 73, p. 5–37.

BURNETT, J. A., HANCOCK, J. M., KENNEDY, W. J., AND LORD, A. R., 1992, Macrofossil, planktonic foraminiferal and nannofossil zonation at the Campanian/ Maastrichtian boundary: Newsletters in Stratigraphy, v. 27, p. 157–172.

CARON, M., 1985, Cretaceous planktonic foraminifera, in Bolli, H. M., Saunders, J. B., and Perch-Nielsen, K., eds., Plankton Stratigraphy: Cambridge, Cambridge University Press, p. 17–86.

CECCA, F., PALLINI, G., ERBA, E., PREMOLI SILVA, I., AND COCCIONI, R., 1994, Hauterivian-Barremian chronostratigraphy based on ammonites, nannofossils, planktonic foraminifera and magnetic chrons from Mediterranean Domain: Cretaceous Research, v. 15, p. 457–468.

CHANNELL, J. E. T. AND GRANDESSO, P., 1987, A revised correlation of Mesozoic polarity chrons and calpionellid zones: Earth Planetary Science Letters, v. 85, p. 222–240.

CHANNELL, J. E. T. AND ERBA, E., 1992, Early Cretaceous polarity Chrons CM0 to CM11 recorded in northern Italian land sections near Brescia: Earth Planetary Science Letters, v. 108, p. 161–179.

CHANNELL, J. E. T., LOWRIE, W., AND MEDIZZA, F., 1979, Middle and Early Cretaceous magnetic stratigraphy from the Cismon section, Northern Italy: Earth Planetary ScienceLetters, v. 42, p. 153–166.

CHANNELL, J. E. T., BRALOWER, T. J., AND GRANDESSO, P., 1987, Biostratigraphic correlation of Mesozoic polarity chrons CM1 to CM23 at Capriolo and Xausa (S. Alps, Italy): Earth Planetary Science Letters, v. 85, p. 203–221.

CHAVE, A. D., 1984, Lower Paleocene-Upper Cretaceous magnetic stratigraphy from Sites 525,527,528 and 529, DSDP leg 74: Initial Reports of the Deep Sea Drilling Project, v. 74, p. 525–531.

CIRILLI, S., MARTON, P., AND VIGLI, L., 1984, Implications of a combined biostratigraphic and paleomagnetic study of the Umbrian Maiolica Formation: Earth Planetary Science Letters, v. 69, p. 203–214.

COCCIONI, R., ERBA, E., AND PREMOLI SILVA, I., 1992, Barremian-Aptian calcareous plankton biostratigraphy from the Gorgo a Cerbara section (Marche, central Italy) and implications for plankton evolution: Cretaceous Research, v. 13, p. 517–537.

CRUX, J. A., 1989, Biostratigraphy and paleogeographical applications of Lower Cretaceous nannofossils from north-western Europe, in Crux, J. A. and van Heck, S. E., eds., Nannofossils and Their Applications: Chichester, Ellis Horwood, p. 143–211.

ERBA, E., 1988, Aptian-Albian calcareous nannofossil biostratigraphy of the Scisti a Fucoidi cored at Piobbico (Central Italy): Rivista Italiana Paleontogia Stratigrafia, v. 94, p. 249–284.

ERBA, E. AND QUADRIO, B., 1987, Biostratigrafia a Nannofossili calcarei, Calpionellidi e Foraminiferi planctonici della Maiolica (Titoniano superiore-Aptiano) nelle Prealpi Bresciane (Italia settentrionale): Rivista Italiana Paleontogia Stratigrafia, v. 93, p. 3–108.

GALBRUN, B., 1992, Magnetostratigraphy of Upper Cretaceous and Lower Tertiary sediments, Sites 761 and 762, Exmouth Plateau, Northwest Australia: Proceedings Ocean Drilling Program Scientific Results, v. 122, p. 699–716.

GRADSTEIN, F. M., HUANG, Z., AGTERBERG, F. P., OGG, J. G., AND HARDENBOL, J., 1993, A Mesozoic time scale: American Association Petroleum Geologists Abstracts, p. 65.

HANCOCK, J. M., KENNEDY, W. J., AND COBBAN, W. A., 1993, A correlation of upper Albian to basal Coniacian sequences of northwest Europe, Texas and the United States Western Interior, in Caldwell, W. G. E. and Kauffman, E. G., eds., Evolution of the Western Interior Basin: St. Johns, Geological Association of Canada Special Paper 39, p. 453–476.

HARLAND, W. B., ARMSTRONG, R. L., COX, A. V., CRAIG, L. E., SMITH, A. G., AND SMITH, D. G., 1989, A Geologic Time Scale, 1989: Cambridge, Cambridge University Press, 263 p.

HART, M. B. AND BIGG, P. J., 1981, Anoxic events in the late Cretaceous chalk seas of North-West Europe, in Neale, J. W. and Brasier, M. D., eds., Microfossils From Recent and Fossil Shelf Seas: Chichester, Horwood, p. 177–185.

HEDBERG, H. D., 1976, International Stratigraphic Guide. A Guide to Stratigraphic Classification, Terminology, and Procedure: New York, J. Wiley and Sons, 200 p.

HUBER, B. T., 1990, Maestrichtian planktonic foraminifer biostratigraphy of the Maud Rise (Weddell Sea, Antarctica): ODP Leg 113 holes 689B and 690C: Proceedings Ocean Drilling Program Scientific Results, v. 113, p. 489–513.

HUBER, B. T., 1991, Planktnic foraminifer biostratigraphy of Campanian-Maestrichtian sediments from ODP Leg 114, sites 698 and 700, southern Atlantic Ocean: Proceedings Ocean Drilling Program Scientific Results, v. 114, p. 281–297.

HUBER, B. T., 1992, Upper Cretaceous planktonic foraminiferal biozonation for the Austral Realm: Marine Micropaleontology, v. 20, p. 107–128.

HUBER, B. T. AND WATKINS, D. K., 1992, Biogeography of Campanian-Maastrichtian calcareous plankton in the region of the Southern Ocean: paleogeographic and paleoclimatic implications: The Antarctic Paleoenvironment: a perspective on global change: Antarctic Research Series, v. 56, p. 31–60.

JARVIS, I., CARSON, G. A., COOPER, M. K. E., HART, M. B., LEARY, P. N., TOCHER, B. A., HORNE, D., AND ROSENFELD, A., 1988, Microfossil assemblages and the Cenomanian-Turonian (late Cretaceous) Oceanic Anoxic Event: Cretaceous Research, v. 9, p. 3–103.

KENNEDY, W. J., COOBAN, W. A., AND SCOTT, G. R., 1992, Ammonite correlation of the uppermost Campanian of Western Europe, the U.S. Gulf Coast, Atlantic Seaboard and Western Interior, and the numerical age of the base of the Maastrichtian: Geological Magazine, v. 129, p. 497–500.

KENT, D. V. AND GRADSTEIN, F. M., 1985, A Cretaceous and Jurassic geochronology: Bulletin Geological Society of America, v. 96, p. 1419–1427.

KRASHENINNIKOV, V. A. AND BASOV, I. A., 1983, Stratigraphy of Cretaceous sediments of the Falkland Plateau based on planktonic foraminifers, Deep Sea Drilling Project, Leg 71: Initial Reports Deep Sea Drilling Project, v. 71, p. 789–820.

LECKIE, R. M., 1984, Mid-Cretaceous planktonic foraminiferal biostratigraphy off Central Morocco, Deep Sea Drilling Project Leg 79, Sites 545 and 547: Initial Reports Deep Sea Drilling Project, v. 79, p. 579–620.

LECKIE, R. M., 1985, Foraminifera of the Cenomanian-Turonian boundary interval, Greenhorn Formation, Rock Canyon anticline, Pueblo, Colorado, in Pratt, L. M., Kauffman, E. G., and Zelt, F. B., eds., Fine-Grained Deposits and Biofacies of the Cretaceous Western Interior Seaway: Evidence of Cyclic Sedimentary Processes: Tulsa, Society of Economic Paleontologists and Mineralogists, Field Trip Guidebook 4, p. 139–150.

LECKIE, R. M. AND BRALOWER, T. J., 1991, New ages for planktonic foraminiferal datums in the Barremian-Lower Aptian: Geological Society of America, Abstracts with Programs, v. 23, p. 167.

LE HÉGARAT, G. AND REMANE, J., 1968, Tithonique superieur et Berriasian de l'Ardéche et de l'Hérault: correlation des Ammonites et des Calpionelles: Geobios Faculté Science Lyon, v. 1, p. 7–70.

LONGORIA, J. F., 1974, Stratigraphic, morphologic and taxonomic studies of Aptian planktonic foraminifera: Revista Espaola Paleontologia Num. Extra., 107 p.

LONGORIA, J. F., 1984, Cretaceous biochronology from the Gulf of Mexico region based on planktonic microfossils: Micropaleontology, v. 30, p. 225–242.

LOWRIE, W. AND ALVAREZ, W., 1984, Lower Cretaceous magnetic stratigraphy in Umbrian pelagic limestone sections: Earth Planetary Science Letters, v. 71, p. 315–328.

LOWRIE, W. AND CHANNELL, J. E. T., 1984, Magnetostratigraphy of the Jurassic-Cretaceous boundary in the Maiolica limestone (Umbria, Italy): Geology, v. 12, p. 44–47.

LOWRIE, W., ALVAREZ, W., PREMOI SILVA, I., AND MONECHI, S, 1980, Lower Cretaceous magnetic stratigraphy in Umbrian carbonate rocks: Geophysical Journal Royal Astronomical Society, v. 60, p. 263–281.

MAHONEY, J. J., STOREY, M., DUNCAN, R. A., SPENCER, K. J., AND PRINGLE, M., 1993, Geochemistry and geochronology of Leg 130 basement lavas: nature and origin of the Ontong Java Plateau: Proceedings Ocean Drilling Program, Scientific Results, v. 130, p. 3–22.

MANIVIT, H., 1984, Paleogene and Upper Cretaceous calcareous nannofossils from DSDP Leg 74: Initial Reports of the Deep Sea Drilling Project, v. 74, p. 475–501.

MANIVIT, H., PERCH-NIELSEN, K., PRINS, B., AND VERBEEK, J., 1977, Mid Cretaceous calcareous nannofossil biostratigraphy: Koninklijke Nederlandse Akademie van Wetenschappen, v. 80, p. 169–181.

MASTERS, B. A., 1977, Mesozoic planktonic foraminifera. A world-wide review and analysis, in Ramsay, A. T. S., ed., Oceanic Micropaleontology: London, Academic Press, p. 301–731.

MONECHI, S., BLEIL, U., AND BACKMAN, J., 1985, Magnetobiochronology of Late Cretaceous-Paleocene and late Cenozoic pelagic sedimentary sequences from the Northwest Pacific (Deep Sea Drilling Project, Leg 86, Site 577): Initial Reports Deep Sea Drilling Project, v. 86, p. 787–798.

MONECHI, S. AND THIERSTEIN, H. R., 1985, Late Cretaceous-Eocene nannofossil and magnetostratigraphic correlations near Gubbio, Italy: Marine Micropaleontology, v. 9, p. 419–440.

MOORE, T. C., JR. AND ROMINE, K., 1981, In search of biostratigraphic resolution, in Warme, J. E., Douglas, R. G., and Winterer, E. L., eds., The Deep Sea Drilling Project: a Decade of Progress: Tulsa, Society of Economic Paleontologists and Mineralogists Special Publication 32, p. 317–334.

MOULLADE, M., 1966, Etude stratigraphique et micropaleontologique du Crétacé inferieur de le Fosse Vocontien: Documents des Laboratoires de Géologie de la Faculte des Sciences de Lyon, v. 15, p. 369.

MOULLADE, M., 1974, Zones de foraminiféres du Crétacéinférieur mésogéen: Comptes rendus des séances de l'Académie des Sciences (Paris), Série D, v. 278, p. 1813–1816.

MOULLADE, M., APPLEGATE, J. L., BERGEN, J. A., THUROW, J., DOYLE, P. S., DRUGG, W. S., HABIB, D., MASURE, E., OGG, J., AND TAUGOURDEAU-LANTZ, J., 1988, Ocean Drilling Program Leg 103 biostratigraphic synthesis: Proceedings of the Ocean Drilling Program, Scientific Results, v. 103, p. 685–695.

MUTTERLOSE, J., 1989, Temperature-controlled migration of calcareous nannofloras in the north-west European Aptian, in Crux, J.A. and van Heck, S. E., eds., Nannofossils and Their Applications: Chichester, Ellis Horwood, p. 122–142.

MUTTERLOSE, J., 1992, Biostratigraphy and paleobiogeography of Early Cretaceous calcareous nannofossils: Cretaceous Research, v. 13, p. 167–189.

OBRADOVICH, J. D., 1993, A Cretaceous time-scale, in Caldwell, W. G. E. and Kauffman, E. G., eds., Evolution of the Western Interior Basin: St. Johns, Geological Association of Canada Special Paper 39, p. 379–396.

OBRADOVICH, J. D. AND COBBAN, W. A., 1975, A time scale for the Late Cretaceous Western Interior of North America, in Caldwell, W. G. E., ed., Cretaceous System in the Western Interior of North America: Toronto, Geological Association Canada Special Paper 13, p. 31–54.

OGG, J. G., 1981, Sedimentology and paleomagnetism of Jurassic pelagic limestones: Unpublished Ph.D. Dissertation, University of California, San Diego, 203 p.

OGG, J. G., 1983, Magnetostratigraphy of Upper Jurassic and lowest Cretaceous sediments, DSDP Site 534A, Western North Atlantic: Initial Reports of the Deep Sea Drilling Project, v. 76, p. 685–697.

PERCH-NIELSEN, K., 1985, Mesozoic calcareous nannofossils, in Bolli, H. M., Saunders, J. B., and Perch-Nielsen, K., eds., Plankton Stratigraphy: Cambridge, Cambridge University Press, p. 329–426.

PESSAGNO, E. A., 1967, Upper Cretaceous planktonic foraminifera from the Western Gulf Coastal Plain: Palaeontographica Americana, v. 5, p. 259–441.

POORE, R. Z., TAUXE, L., PERCIVAL, S. F., Jr., LABRECQUE, J. L., WRIGHT, R., PETERSEN, N. P., SMITH, C. S., TUCKER, P., AND HSU, K. J., 1983, Late Cretaceous-Cenozoic magnetostratigraphic and biostratigraphic correlations of the South Atlantic Ocean: DSDP Leg 73: Palaeogeography, Palaeoclimatology, Palaeoecology, v. 42, p. 127–149.

POSPICHAL, J. J. AND WISE, S. W., JR., 1990, Maestrichtian calcareous nannofossil biostratigraphy of Maud Rise ODP Leg 113 sites 689 and 690, Weddell Sea: Proceedings of the Ocean Drilling Program, Scientific Results, v. 113, p. 465–487.

POSTUMA, J., 1971, Manual of Planktonic Foraminifera: Amsterdam, Elsevier, 420 p.

PREMOLI SILVA, I., 1977, Upper Cretaceous-Paleocene magnetic stratigraphy at Gubbio, Italy. II Biostratigraphy: Geological Society of America Bulletin, v. 88, p. 371–374.

PREMOLI SILVA, I. AND BOLLI, H. M., 1973, Late Cretaceous to Paleogene planktonic foraminifera and stratigraphy of Leg 15 sites in the Caribbean Sea: Initial Reports of the Deep Sea Drilling Project, v. 15, p. 499–547.

REMANE, J., 1971, Les calpionelles, protozaires planktonique des mers mésogéenees de l'époque secondaire: Annales Guebhard, v. 47, p. 1–25.

REMANE, J., 1983, Calpionellids and the Jurassic/Cretaceous boundary at Deep Sea Drilling Project Site 534, Western North Atlantic Ocean: Initial Reports of the Deep Sea Drilling Project, v. 76, p. 561–568.

REMANE, J., 1985, Calpionellids, in Bolli, H. M., Saunders, J. B., and Perch-Nielsen, K., eds., Plankton Stratigraphy: Cambridge, Cambridge University Press, p. 555–572.

ROBASZYNSKI, F., CARON, M., AND OTHERS, 1979, Atlas de Foraminiféres planctoniques du Crétacé moyen, parts 1–2: Cahiers Micropaleontologie, 1–2, 1–185 and 1–181.

ROBASZYNSKI, F., CARON, M., AND OTHERS, 1984, Atlas of Late Cretaceous planktonic foraminifera: Revue de Micropaleontologie, 26, v. 3–4, p. 145–305.

ROTH, P. H., 1978, Cretaceous nannoplankton biostratigraphy and oceanography of the northwestern Atlantic Ocean: Initial Reports of the Deep Sea Drilling Project, v. 44, p. 731–759.

ROTH, P. H., 1983, Jurassic and Lower Cretaceous calcareous nannofossils in the western North Atlantic (Site 534): biostratigraphy, preservation and some observations on biogeography and paleoceanography: Initial Reports of the Deep Sea Drilling Project, v. 76, p. 587–621.

SIGAL, J., 1967, Essai sur l'état actuel d'une zonation stratigraphique á l'aide des principales espéces de Rosalines (Foraminiféres): Comptes rendus Sommaire des sances de la Sociétogéologique de France, v. 2, 48 p.

SIGAL, J., 1977, Essai de zonation du Crétacé méditerranéen á l'aide des foraminiféres planctoniques: Géologie Méditerranéenne, v. 4, p. 99–108.

SIGAL, J., 1979, Chronostratigraphy and ecostratigraphy of Cretaceous formations recovered on DSDP Leg 47B, Site 398: Initial Reports of the Deep Sea Drilling Project, v. 47B, p. 287–327.

SISSINGH, W., 1977, Biostratigraphy of Cretaceous calcareous nannoplankton: Geologie en Mijnbouw, v. 56, p. 37–65.

SLITER, W. V., 1989, Biostratigraphic zonation for Cretaceous planktonic foraminifers examined in thin section: Journal of Foraminiferal Research, v. 19, p. 1–19.

SLITER, W. V., 1992, Cretaceous planktonic foraminiferal biostratigraphy and paleoceanographic events in the Pacific Ocean with emphasis on indurated sediment, *in* Ishizaki, K. and Saito, T., eds., Century of Japanese Micropaleontology: Tokyo, Terra Scientific Publishing Co., p. 281–299.

SLITER, W. V. AND LECKIE, R. M., 1993. Cretaceous planktonic foraminifers and depositional environments from the Ongtong Java Plateau with emphasis on Sites 803 and 807: Proceedings Ocean Drilling Program, Scientific Results, v. 130, p. 63–84.

STRADNER, H. AND STEINMETZ, J., 1984, Cretaceous calcareous nannofossils from the Angola Basin, DSDP Site 530, Initial Reports of the Deep Sea Drilling Project, v. 75, p. 565–650.

TARDUNO, J. A., 1990, Brief reversed polarity interval during the Cretaceous Normal Polarity Superchron: Geology, v. 18, p. 683–686.

TARDUNO, J. A., SLITER, W. V., BRALOWER, T. J., McWILLIAMS, M., PREMOLI SILVA, I., AND OGG, J. G., 1989, M-sequence reversals recorded in DSDP Sediment Cores from the Western Mid-Pacific Mountains and Magellan Rise: Geological Society America Bulletin, v. 101, p. 1306–1316.

TARDUNO, J. A., SLITER, W. V., KROENKE, L., LECKIE, R. M., MAYER, H., MAHONEY, J. J., MUSGRAVE, R., STOREY, M., AND WINTERER, E. L., 1991, Rapid formation of Ontong Java Plateau by Aptian mantle volcanism: Science, v. 254, p. 399–403.

TARDUNO, J. A., LOWRIE, W., SLITER, W. V., BRALOWER, T. J ., AND HELLER, F., 1992, Albian geomagnetic reversals from the Valle della Contessa (Umbrian Apennines, Italy): implications for a Mid-Cretaceous mixed polarity interval: Journal of Geophysical Research, v. 97, p. 241–271.

TAUXE, L., TUCKER, P., PETERSEN, N. P., AND LaBRECQUE, J. L., 1983, The magnetostratigraphy of leg 73 sediments: Palaeogeography, Palaeoclimatology, Palaeoecology, v. 42, p. 65–90.

THIERSTEIN, H. R., 1971, Tentative lower Cretaceous calcareous nannoplankton zonation: Eclogae geologicae Helvetiae, v. 64, p. 459–488.

THIERSTEIN, H. R., 1973, Lower Cretaceous calcareous nannoplankton biostratigraphy: Abhandlungen der Geologischen Bundesanstalt, v. 29, p. 3–53.

THIERSTEIN, H. R., 1975, Calcareous nannoplankton biostratigraphy at the Jurassic-Cretaceous boundary, *in* Colloque sur la limite Jurassique-Crétacé: Lyon, Mémoires du Bureau de Recherches Géologiques et Minières, v. 86, p. 84–94.

THIERSTEIN, H. R., 1976, Mesozoic calcareous nannoplankton biostratigraphy of marine sediments: Marine Micropaleontology, v. 1, p. 325–362.

TORNAGHI, M. E., PREMOLI SILVA, I., AND RIPEPE, M., 1989, Lithostratigraphy and planktonic foraminiferal biostratigraphy of the Aptian-Albian "Scisti a Fucoid" in the Piobbico core, Marche, Italy: background for cyclostratigraphy: Rivista Italiana Paleontogia Stratigrafia, v. 95, p. 223–264.

VANDENBERG, J., KLOOTWIJK, C. T., AND WONDERS, A. A. H., 1978, The Late Mesozoic and Cenozoic movements of the Umbrian Peninsula: further paleomagnetic data from the Umbrian sequence: Geological Society of America Bulletin, v. 89, p. 133–150.

VAN HINTE, J. E., 1972, The Cretaceous time scale and planktonic foraminiferal zones: Koninklijke Nederlandse Akademie van Wettenschappen, Proceedings, Series B, no. 1, p. 1–8.

VAN HINTE, J. E., 1976, A Cretaceous time scale: Bulletin American Association of Petroleum Geologists, v. 60, p. 498–516.

WATKINS, D. K., 1992, Upper Cretaceous nannofossils from Leg 120, Kerguelen Plateau, Southern Ocean: Proceedings Ocean Drilling Program, Scientific Results, v. 120, p. 343–370.

WATKINS, D. K., WISE, S. W., JR., POSPICHAL, J. J., AND CRUX, J. A., 1992, Upper Cretaceous calcareous nannofossil biostratigraphy of the Southern Ocean: Marine Micropaleontology, v. 19, p. 123–145.

WIEGAND, G. E., 1984, Cretaceous nannofossils from the Northwest African margin, Deep Sea Drilling Project Leg 79: Initial Reports of the Deep Sea Drilling Project, v. 79, p. 563–578.

WIND, F. H., 1978, Western North Atlantic Upper Jurassic-Cretaceous calcareous nannofossil biostratigraphy: Initial Reports of the Deep Sea Drilling Project, v. 44, p. 761–773.

WISE, S. W., JR., 1983, Mesozoic and Cenozoic calcareous nannofossils recovered by Deep Sea Drilling Project Leg 71 in the Falkland Plateau Region, Southwest Atlantic Ocean: Initial Reports of the Deep Sea Drilling Project, v. 71, p. 481–550.

WISE, S. W. JR. AND WIND, F. H., 1977, Mesozoic and Cenozoic calcareous nannofossils recovered by DSDP Leg 36 drilling on the Falkland Plateau, Southwest Atlantic sector of the Southern Ocean: Initial Reports of the Deep Sea Drilling Project, v. 36, p. 269–492.

WONDERS, A. A. H., 1980, Middle and Late Cretaceous planktonic foraminifera of the Western Mediterranean area: Utrecht Micropaleontological Bulletins, v. 24, p. 1–158.

ORBITAL CHRONOLOGY OF CRETACEOUS-PALEOCENE MARINE SEDIMENTS

TIMOTHY D. HERBERT
Geological Research Division, Scripps Institution of Oceanography, La Jolla, CA 92093-0215
ISABELLA PREMOLI SILVA AND ELISABETTA ERBA
Dipartimento Scienze della Terra, Via Mangiagalli 34, Universita degli Studi di Milano, 20133 Milano, Italy
AND
ALFRED G. FISCHER
Department of Geological Sciences, University of Southern California, Los Angeles, CA 90089

ABSTRACT: Changes made to the Pliocene and Pleistocene portions of the Geomagnetic Polarity Time Scale (GPTS) by matching paleoclimatic oscillations in marine sediments to variations in the earth's orbital elements have been validated by recent $^{40}Ar/^{39}Ar$ dating. We explore here the potential for orbital chronology to improve and refine the Cretaceous and early Paleocene portion of the GPTS. Because orbital cycles mark off time in geologically short increments, their sedimentary imprint may be used to measure elapsed time between events such as magnetic reversals or biostratigraphic datums very precisely. A large number of Deep Sea Drilling Project drill sites of late Cretaceous to early Paleocene age display carbonate cycles whose mean period, estimated by paleomagnetically determined sedimentation rates, is close to 20 ky, the expected mean period of the precessional cycle. Because the cycles can be detected before and after the Cretaceous-Tertiary boundary, they offer a new tool to date the position of the extinction level within magnetochron C29R and to measure the rates of environmental and evolutionary changes across the boundary. Cycle-by-cycle correlations appear possible between South Atlantic sites; recent information from sections in Spain suggests that interhemispheric correlations can be made. We also present an astronomical time scale for the durations of the Cenomanian, Albian, and Aptian stages as defined by marine microfossil datums and the top of reversed polarity chron M-0. The durations of these stages are estimated to be 6.0 ± 0.5 my, 11.9 ± 0.2 my, and 10.6 ± 0.2 my, respectively. A cumulative cyclo-chronology from the well-dated Cenomanian-Turonian boundary places the top of polarity chron M-0, a tie point for the calibration of the M-series anomaly sequence, at 121.6 Ma.

INTRODUCTION

Cyclical variations in the composition of pelagic and hemipelagic sediments ripple in an almost unbroken wavetrain from the Pleistocene Ice Age world into the warm Cretaceous Period (Moore and others, 1982; Imbrie and others, 1984; Hilgen, 1991; Mead and others, 1986; Schwarzacher and Fischer, 1982). Variations in insolation as a function of latitude and season, caused by quasi-cyclical changes in the earth's orbit, appear to be the driving forces that induce these patterns. These so-called "Milankovitch cycles" can be partitioned into the precessional index (modern mean period circa 21 ky), obliquity, (modern mean period circa 41 ky), and eccentricity (modern mean periods of 95, 123, and 413 ky). In order to link sedimentary cycles to orbital forcing, geologists have always relied upon standard geochronometric methods for estimating sedimentation rate. For example, the landmark paper of Hays and others (1976) used a combination of carbon-14 and uranium series dating, and the accepted date of the Brunhes-Matuyama boundary (then placed at 730 ka) to construct a late Pleistocene time scale for spectral analyses of $\delta^{18}O$ and other proxy climate data. The good match between the estimated durations of the Pleistocene climatic cycles and the periods of the orbital parameters provided convincing evidence for the orbital theory of the ice ages. Paleomagnetic and/or biochronological time scales have been called on to make the case that older sedimentary cycles likewise represent orbital pacing (e.g., Schwarzacher, 1987; Herbert and D'Hondt, 1990; Huang and others, 1993).

Stratigraphers may now be in a position to invert the usual argument; the "Milankovitch" hypothesis may be used *in appropriate settings* to improve standard time scales (House, 1985). Just as the tick marks of a yardstick measure distance, sedimentary cycles of orbital origin count time at high precision as compared with most chronological methods. Furthermore, because cyclic marine sediments generally contain stratigraphically useful microfossils and often retain polarity reversal information, the "Milankovitch" clock could be widely applicable to time scale problems. When anchored accurately in a magnetochronologic and biochronologic framework (which must ultimately be calibrated by radiometric methods), orbital cycles in sediments provide a chronometer that functions at one to two orders of magnitude better resolution than standard techniques.

The concept of orbital chronometry, first put forward by G. K. Gilbert in 1895 (Fischer, 1980), has recently won some important victories in the Plio-Pleistocene portion of the GPTS. Shackleton and co-workers (1990) used a high-resolution $\delta^{18}O$ curve to show that the earliest Brunhes portion of Pleistocene time contains one more obliquity cycle than suggested by the SPECMAP study (Imbrie and others, 1984), confirming an earlier suggestion of Johnson (1982). Not only did this model push back the Brunhes/Matuyama boundary by some 50 ka, but Shackleton and others (1990) found discrepancies continuing into the Matuyama reversed Chron that suggest a systematic bias toward overly young ages in the conventional GPTS. Hilgen (1991) used lithological cycles exposed in southern Italy to extend the orbital calibration of the polarity time scale into latest Miocene time. Again, the astronomical model suggested that the conventional K-Ar-based GPTS underestimated Plio-Pleistocene time by about 8%. These predictions have recently been tested by investigators using the $^{40}Ar/^{39}Ar$ method to redate polarity transitions. The new dates are in very close agreement with ages predicted by Shackleton and Hilgen (Baksi and others, 1992; Tauxe and others, 1992).

Although extending the orbital clocks from the last few million years into the Mesozoic remains an enormous project, in this article we suggest that orbital chronometry of sediments will play a useful role in calibrating the GPTS at least into middle Cretaceous time. Our information comes from cores recovered by the Deep Sea Drilling Project (DSDP), Ocean Drilling Program (ODP), and from outcrop and core studies of pelagic sediments preserved in Europe. The potential improvement in chronostratigraphy possible with orbital methods will be important for studying climate dynamics, sedimentary fluxes, and the rates of tectonic processes in the late Mesozoic with Pleistocene-like resolution.

Geochronology Time Scales and Global Stratigraphic Correlation, SEPM Special Publication No. 54

CHARACTERISTICS OF THE ORBITAL CLOCKS

The orbital rhythms imprinted on sediments do not constitute perfect chronometers. Repeat times are only quasi-periodic, due to the complicated modulation patterns of the basic cycles (Fig. 1). Where the orbital parameters may be calculated with high precision, as Berger and Loutre (1988) have done over the past 5 Ma, cycles in sediments may conceivable be correlated to the astronomical template with an error of only a few thousand years, or the error inherent in assuming a fixed phase lag between orbital forcing and the sedimentary response. Orbital cycles can therefore be used for numerical dating of the Pliocene, Pleistocene, and latest Miocene time scales (Shackleton and others, 1990; Hilgen, 1991). For earlier periods of time, the accuracy of current orbital solutions degrades such that we can only be confident of the mean periods and general behavior of the astronomical cycles (Berger and others, 1992). Every indication from celestial mechanical calculations is that the main eccentricity periods should stay constant over time. Such is not the case for the mean periods of obliquity and precessional index, which are expected to increase gradually over the Earth's history due to tidal friction. The obliquity and precessional periods have not been measured accurately enough in Mesozoic sediments to derive an empirical rate of change. Berger and others (1992) estimate that at 100 Ma the precessional and

obliquity repeat times should diminish from their modern values by 2.8 and 5.7%, respectively. This small change indicates that even if the tidal friction estimates are incorrect by a factor of 2, the error in using assumed Cretaceous orbital periods to measure *elapsed* time between horizons 10^5 to 10^7 yr apart will be comparable to or smaller than radiometry.

Orbital cycles can therefore still be useful in measuring time between datum points in early Neogene and older age sediments, if not for numerical dating in the usual sense. One can choose to analyze sedimentary oscillations in the depth domain by visual inspection, in the frequency domain by one of several techniques of frequency analysis or both. Working in the frequency domain has the advantage of objectively partitioning the energy of often complicated waveforms into different frequencies, and following certain assumptions, offers some methods of quantifying uncertainties in frequency estimates. Furthermore, it may be possible to "lock on" to the correct time-depth scaling in a sedimentary sequence by obtaining a diagnostic match of the fine-scale frequency structure of a geological record to that of the orbital forcing spectrum (for example, a precise match to the circa 23- and 19-ky lines of precession or the 95- and 123-ky lines of the short cycle of eccentricity).

Working in the frequency domain also has several disadvantages. In most cases, relatively long (order of 0.4–1 my) records are needed to produce reliable spectra with sufficient frequency resolution to see features such as the twin peaks of precession. It is unfortunately also true that most frequency spectra of geological records are substantially degraded by changes in sedimentation rate and by drifts in the response of the sedimentary system to orbital forcing. "Tuning" of the time-depth relationship can remove much of this distortion (Schiffelbein and Dorman, 1986; Martinson and others, 1987; Park and Herbert, 1987; Kominz and Bond, 1990) but requires choosing an appropriate set of assumptions.

It will therefore generally be necessary to work in the depth as well as the frequency domain to produce orbital chronologies. Provided that maxima and minima of orbital cycles are correctly identified in the stratigraphic series, many sediments will contain an internal clock; on average, each orbital cycle ticks off elapsed time according to its mean period. Berger and Loutre's (1988) 5-my calculation of the precessional index, obliquity, and eccentricity can be studied to determine the relative error of this metronomic assumption. The mean repeat times of the orbital cycles over this record are 20.8 ky for precession, 40.4 ky for obliquity, and 94.9 ky for the short cycle of eccentricity (note that in the frequency domain, precession is represented by twin peaks at about 23 and 19 ky, and the short cycle of eccentricity is seen as split peaks at about 95 and 123 ky). However, there must be a tradeoff between resolution and accuracy when one assumes that sedimentary proxies of precessional, obliquity, and eccentricity cycles measure time in even increments. As the metronomic assumption is driven to its limits to measure time with fewer and fewer cycles, the relative error in estimating elapsed time increases (Fig. 1). The relative error in calculating durations by cyclostratigraphy can be reduced to only a few percent but at a cost of averaging over more cycles and more time. Nevertheless, the potential gain in resolution given by orbital cycles is still impressive when one realizes that they are subdividing time in increments of a few tens of thousands of years.

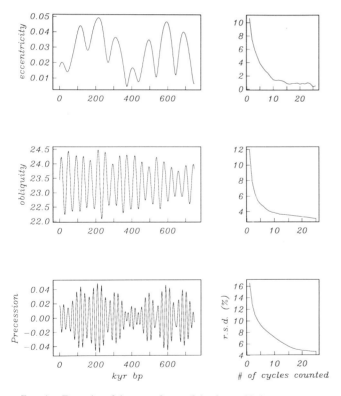

Fig. 1.—Examples of the wave forms of the three orbital parameters over the past 750 ka, taken from Berger and Loutre (1988) and the uncertainty in using the cycles to measure time in sediments if the cycles are assumed to mark off uniform time increments. The panel to the right shows the error relative to the mean repeat time respectively of the short cycle of eccentricity, the cycle of obliquity, and the precessional index, as a function of the number of cycles averaged. For example, the relative error in assuming that any one precessional cycle marks off exactly 20.8 ky is about 16%; the numerical error is 3.4 ky. The relative error drops as more cycles or longer times, are integrated.

The question naturally arises as to how orbital cycles can be detected confidently in pre-Pliocene strata. Like most techniques in stratigraphy, the process should be an iterative one, in which a considerable amount of cross-checking with independent time reference points is combined with statistical tests for the consistency of sedimentary data with the orbital model. Sections should be chosen that appear to meet the criterion of continuous deposition. All available bio- and magnetostratigraphic data should be incorporated into an initial time model before the orbital method is applied. Sections should be sampled with attention to principles of time series analysis. Spectral analyses may then be a powerful tool for verifying cyclic patterns in sedimentation, with the caution that some harmonic distortion inevitably intervenes between orbital forcing and sedimentary patterns.

To create a convincing orbital time scale, the time series should show some of the characteristic "fingerprints" of the orbital wave forms. Some examples include the diagnostic 100- and 400-ky modulations of precessional amplitude, and the beating pattern of the circa 100-ky cycle of eccentricity, which has a prominent envelope at the 413-ky wavelength. The presence of multiple strong bands of energy in the frequency spectrum can act like a tuning fork, because they help to narrow the match of periodicities in a geological series to the orbital template. For example, periodicities with a nearly 1:5 frequency ratio are diagnostic of a combination of 100-ky eccentricity and 21-ky precessional cycles; a frequency spectrum with the 1:2.5:5 ratios of the short cycle of eccentricity, obliquity, and precessional index would be even more compelling evidence for a unique match of sedimentary rhythmicity to orbital forcing. Examples will follow where "fingerprinting" of sedimentary cycles in middle and latest Cretaceous strata does indeed seem possible.

LATE CRETACEOUS/EARLY PALEOCENE ORBITAL CHRONOLOGY

New approaches are needed to improve the late Cretaceous and early Cenozoic time scale. It is well recognized that the South Atlantic, which provides the sea-floor magnetic anomaly template for early Cenozoic and late Cretaceous time, underwent pronounced changes in spreading rate in latest Cretaceous and early Paleocene time (see Cande and Kent, 1992). Because the timing of spreading rate changes is not well-constrained by radiometric dates, there is considerable play in the relative durations of Chrons C26 through C30. The result is that, while the aggregate time scale errors are probably no worse than 5%, there may be significant uncertainties in the durations of individual polarity zones. These uncertainties frustrate attempts to measure rates of geological processes during a particularly interesting time, the terminal Cretaceous extinctions (c.f. Alvarez and others, 1980; Keller, 1989; Smit and Romein, 1985).

Limestone-marl and redox oscillations of apparent regularity have been described from pelagic and hemipelagic sections of Campanian-Paleocene age from the Indian Ocean (Huang and others, 1992), from the South Atlantic Ocean (Herbert and D'Hondt, 1990, Park and others, 1993) and from Tethyan sections in Spain (Ten Kate and Sprenger, 1992). Many of the sections studied have coherent magnetostratigraphies, which allow estimates to be made of sedimentation rates at roughly 0.5- to 1-my intervals. Nannofossil and foraminiferal biostratigraphies, calibrated to the polarity sequence at a number of sites, provide further time constraints to test the hypothesis of the orbital origin of the sedimentary variations. The mean periods of the cycles can be determined to an accuracy of some 5–15%, an uncertainty that depends on the combined errors of radiometric dates of the Santonian/Campanian and Paleocene/Eocene boundary, correlations of stage boundaries to polarity zones, and choice of the correct sea-floor template for the reversal time scale between these tie-points. Our original objective was simply to test the astronomical hypothesis in reasonably well-dated upper Cretaceous sections, before using cyclicity in sediments of Long Normal age to refine the GPTS. However, it became apparent that orbital cyclicity, because of its superior time resolution as compared to more established chronostratigraphic methods, could contribute to the chronology of events across the Cretaceous-Tertiary (K/T) boundary (see Fig. 2 for a standard magnetochronology). We now believe that precessional cyclicity of pelagic sediments can situate the position of the K/T boundary and associated radiometric ages

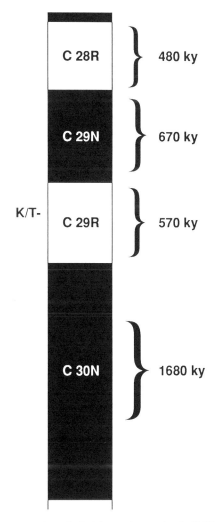

FIG. 2.—Magnetochronological time scale spanning the K/T boundary, after Berggren and others (1985). While the numerical boundaries must be revised, the durations of the magnetochrons do not change much with new radiometric information.

(c.f. Izett and others, 1991; Swisher, this volume) in polarity interval C29R to within 10–20 ky.

Upper Cretaceous and lower Tertiary sediments often show a striking pattern of color oscillations (Fig. 3). Color and carbonate content generally correspond inversely; darkly colored (whether green or red) beds are clay-rich and lightly colored intervals are highly calcareous. That the oscillations are primary is clearly shown by the bioturbational mixing of one lithology into its neighbor; such burrow mottling could only occur if the contrast originated at the time of deposition. Huang and others (1992) created a semi-quantitative color depth series by assigning a visual color index to oscillations of upper Campanian to lower Maastrichtian age at ODP Site 762. Walsh spectral analysis was then applied to the step-like series in order to test for periodic structure. Two components stood out prominently at a frequency ratio of about 2:1. Biostratigraphically estimated sedimentation rates gave numerical periods of about 21 and 45 ky to the cycles; a reasonable match to the expected precessional and obliquity repeat times (Huang and others, 1992).

Maastrichtian through Paleocene oscillations have also been described in outcrop sections. Ten Kate and Sprenger (1992) have produced careful measurements of calcium carbonate content and magnetic susceptibility at three Tethyan sections from Spain: Zumaya, Agost, and Relleu. Paleomagnetic and microfossil stratigraphies indicate that the sections span the interval from upper Maastrichtian (later portion of magnetochron C30N) through early Paleocene (magnetochron 28R) time. Ten Kate and Sprenger (1992) suggest that the carbonate cycles, whose mean recurrence is about 20 ky, are correlative between the sections.

The longest available record of paleomagnetically calibrated orbital cycles of upper Cretaceous and lower Tertiary age comes from a suite of DSDP holes drilled in the South Atlantic. Carbonate and redox cycles are present at DSDP Sites 356, 357, 516F, 525, 527, 528, and 529, located on the Rio Grande Rise and the Walvis Ridge. At paleodepths less than about 1.2 km, the sediments alternate between white and green/drab colors on a scale of 10–30 cm, depending on the time interval. Below this depth, the stratification is comprised of white and red colors. Carbonate content varies from highs of 80–85% in white beds to lows of 50–60% in the darker zones.

Digital time series of the South Atlantic sites have been prepared by digitizing photographs (Herbert and D'Hondt, 1990), by recording visible light reflectance spectra on core surfaces (this paper) and by measuring detailed magnetic susceptibility profiles (Park and others, 1993; D'Hondt and others, unpubl. data). The site with the longest record of cyclicity is DSDP Site 516F, where carbonate cycles run from near the base of C33R into C28N, a time span of nearly 20 my. Representative time series and frequency spectra of the South Atlantic cores are shown in Figure 4. In all cases, a concentration of spectral energy coincides with the lithological spacing evident to the eye. The mean period of the carbonate rhythms (*excluding* C29R, which does not figure into the calibration) is 23.5 ± 4.4 ky (Berggren and others, 1985, time scale). Not only do the repeat times of the cycles match the precessional period, but the char-

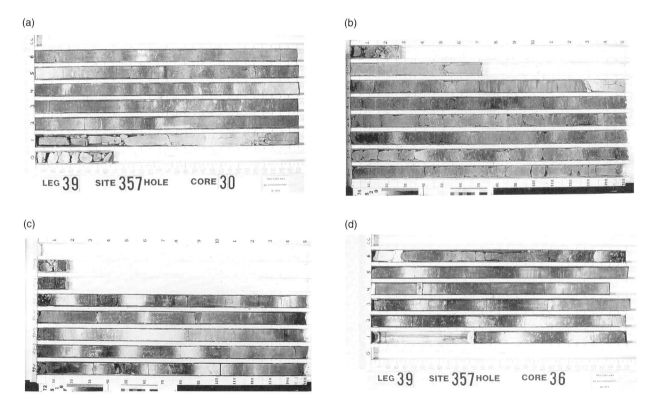

FIG. 3.—Cyclic DSDP cores of late Cretaceous age from the South Atlantic. (A) DSDP 357, core 30 is of early Paleocene age. (B) DSDP Site 529, core 41, contains the K/T boundary at 110 cm of section 6. (C) DSDP 516F, core 93 is of Maastrichtian age (middle of polarity zone C30N). (D) DSDP 357, core 36 is of Campanian age. The color banding reflects cycles in calcium carbonate deposition. The spacing of the cycles, divided by the mean repeat time of the precessional index, can be used to calculate sedimentation rate at high resolution (from Herbert and D'Hondt (1990)

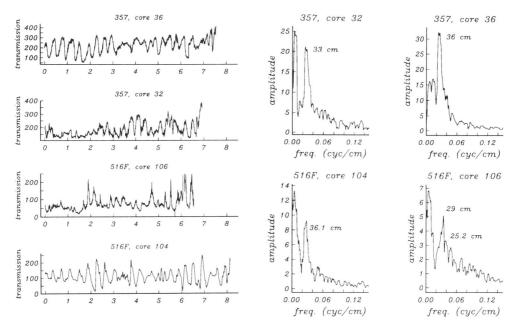

FIG. 4.—Digitized curves of the variations in sediment color in the South Atlantic drill sites correspond to carbonate cycles. The curves do not have long-wavelength lighting biases removed. Right-hand panel shows spectral analyses of time series, identifying precessional (20 ky) wavelength (from Herbert and D'Hondt, 1990).

acter often shows the modulation (see especially Site 357, cores 32 and 36) of precession imparted by the eccentricity cycle. In the frequency domain, this modulation is represented by split or composite spectral peaks strongly resembling the modern 23- and 19-ky components of precession (Herbert and D'Hondt, 1990; Park and others, 1993).

If the South Atlantic carbonate cycles are indeed a nearly metronomic response to precessional forcing, then they may be used to fine-tune the GPTS and to establish a quantitative time framework for events at time scales below those resolvable by polarity stratigraphy. Because DSDP cores did not fully recover the drilled intervals, it is difficult to propose a definitive precessional tuning of the entire late Cretaceous GPTS through anomaly 33. However, enough of the cored sites recovered the K/T and C29R/C30N boundaries to permit a high-resolution precessionally-based chronology to be constructed for events across the K/T transition (for original shipboard magnetostratigraphy, see Hamilton and Suzyumov, 1983 and Chave, 1984). Numerical ages are assigned according to the Berggren and others (1985) time scale, because the Cande and Kent (1992) time scale has to be readjusted to correct for an erroneous numerical tie point in C29R (S. Cande, pers. commun., 1993). Because the carbonate cycles tick off time in approximately 20 ky increments, they may be used to place events quite precisely within the magnetic polarity skeleton. Orbital cycles can place extinction and speciation events across the K/T boundary in a time framework that does not a priori assume the synchroneity of evolutionary events at all study locations. At the high resolution required of K/T boundary studies, such questions will always bedevil purely biostratigraphic approaches (c.f. Smit and Romein, 1985; Keller, 1989; MacLeod and Keller, 1991). Additionally, the carbonate cyclicity can be used to measure instantaneous sedimentation rates, which can be calculated by dividing the cycle spacing by the 20.4-ky mean period of the latest Cretaceous precessional cycle (subject to the error limits shown in Fig. 1).

To arrive at a high-resolution stratigraphy across the K/T boundary, we anchored cycle counts on magnetic reversal boundaries and attempted to correlate cycles between sites. Six locations were used (DSDP 356, 516, 525, 527, 528, and 529). Using a number of sites helps to compensate for the imperfect recovery of sediments at each drill hole. Since the publication of Herbert and D'Hondt (1990) we have improved the resolution of polarity boundaries at DSDP 516F, 527, and 528, by conducting AF and thermal demagnetization of samples in the vicinity of the C29R/C30N and C28R/C29N transitions. One important improvement has been to locate the C29R/C30N boundary at DSDP Site 516F in core 90, section 3, between 83 and 94 cm. The equivalent boundary at DSDP Site 528 is located in core 32, section 5, between 13 and 33 cm.

The improved polarity stratigraphy allows us to establish cycle-by-cycle correlations of some South Atlantic sections (Fig. 5) and to propose a cycle nomenclature within the magnetostratigraphic framework (29M1 refers to the uppermost Maastrichtian carbonate maximum within C29R, 29M10 the tenth cycle, etc.). We rely on the characteristic modulating pattern of the carbonate cycles to propose a composite stratigraphy between Sites 516F and 528. For example, maximal precessional amplitudes occur at cycles 29M1, 29M5, 29M9 and 29M10, 29M14, and 30N2. Note that this correlation leads us to infer a gap at Site 528 between cores 31 and 32 that omits 2 of the uppermost Maastrichtian carbonate cycles. This inferred gap, whose existence must be tested by further work, may be best explained by the incomplete recovery of the top of core 32, which is missing about 1.5 m of section. Likewise, we attribute the absence of cycle 29M8 at Site 516F to an approximately

30-cm break in core recovery at the top of core 90, and insert a 70-cm break between sections 3 and 4 of core 40 at Site 525, which show signs of drilling damage.

The position of the C29R/C30N boundary is well defined at Sites 516F, 525A, 527, and 528. In all cases, it occurs at the transition between the underlying marl and an overlying calcareous bed, at the base of a node of four to five low-amplitude cycles. This low-amplitude interval represents a minimum in the modulation of precession by the 413 ky component of eccentricity. Our precessional chronology recognizes 18.5 couplets (each of estimated 20.4-ky mean duration) in the Maastrichtian portion of C29R. This chronology is consistent with the recognition of about 3 and one-half 100-ky eccentricity cycles noted above. It is slightly longer (by 1.5 precessional cycles) than we had proposed prior to the new paleomagnetic sampling (Herbert and D'Hondt, 1990).

Unfortunately, the earliest Paleocene sediments are sufficiently condensed at nearly all South Atlantic sites that it is difficult to recognize carbonate cycles to complete the C29R chronology. DSDP Sites 356 and 529 have an expanded and relatively clear record of cyclicity immediately after the K/T boundary, but only Site 529 has a reliable magnetostratigraphy. Herbert and D'Hondt (1990) recognized 11 cycles in the Danian portion of C29R at Site 529 or about 224-ky duration. As discussed below, new data produced by Ten Kate and Sprenger (1993) suggest that the correct number of carbonate cycles in the earliest Paleocene interval of C29R may be as high as 14.5 or 296 ky.

The orbital timekeeper also allows us to test whether the pelagic system showed perturbations before the K/T boundary, and how it responded following the terminal Cretaceous planktonic extinctions. Consider the case of a gradual decline in sedimentation rate up section, as might have occcured if precursor events affected planktonic production of carbonate tests prior to the main extinction. The sequential thickness of orbital cycles would trace this decrease, as the cycles would gradually condense in spacing proportionally to the reduction in sedimentation rate. An increase in sedimentation rate up section would of course generate the opposite pattern. An abrupt perturbation of sedimentation rate would be recorded by a step-like change in the spacing of orbital cycles in the sediment. This use of the spacing of orbital cycles was suggested to us after compiling low-resolution sedimentation rate estimates on a core-by-core (9.5-m length) spacing (Herbert and D'Hondt, 1990).

A high-resolution, stacked sedimentation rate across the K/T boundary (Fig. 6) was compiled by determining the sequential changes in 20-ky carbonate cycle spacing at the study sites. We continue to assign numerical ages according to the Berggren and others (1985) time scale, because the Cande and Kent (1992) time scale has to be readjusted to correct for an erroneous numerical tie point in C29R. In order to produce a composite record between sites with differing sedimentation rates, we normalized the sequence of cycle spacings at each site by their Maastrichtian mean thickness. The normalized cycle thickness variations, a proxy for sedimentation rate, were combined by hanging the records on reversal boundaries and the K/T horizon. As before, our confidence in the Danian estimate is lower than for the uppermost Maastrichtian, because fewer sites yield reliable oscillations of earliest Tertiary age. Nevertheless, the pattern strikingly resembles the step-function drop in sedimentation rates that might be expected from the impact

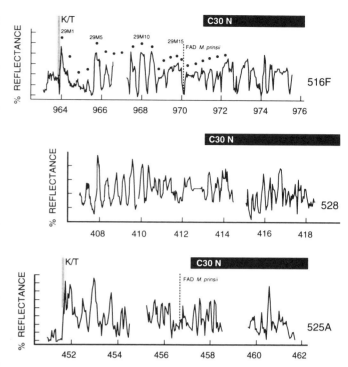

FIG. 5.—Correlation of latest Maastrichtian South Atlantic precessional stratigraphy between DSDP Sites 516F (Rio Grande Rise) and 525A and 528 (Walvis Ridge) using percent reflectance values measured by a Minolta visible light reflectance spectrometer. Because percent reflectance is highly correlated to carbonate content, the spectral data serve as a carbonate proxy. Normal magnetozone C30N, which extends below the interval illustrated, is indicated by the dark pattern. Below the K/T boundary, carbonate maxima record individual precessional cycles (referred to in text). Note that the carbonate cycles are bundled into groups of 4 to 5 by a probable eccentricity cycle. The position of the M. prinsii first appearance datum has been carefully measured by Henrikkson (1993) at Sites 516F and 525A.

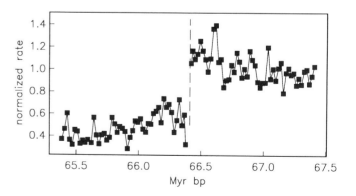

FIG. 6.—High resolution sedimentation rates made on a composite of South Atlantic DSDP sites from bed-by-bed estimates of precessional cycle spacing (from Herbert and D'Hondt, 1990). Dashed line represents position of K/T boundary. Since each precessional carbonate cycle represents approximately the same amount of time, the thickness of the cycle is proportional to the sedimentation rate. Sequential thickness data from various sites were tied to magnetic reversal boundaries and to the K/T level, normalized to the mean late Maastrichtian cycle length, and stacked to produce a smoother, more reliable record.

extinction mechanism first envisaged by Alvarez and others (1980). As best as can be measured by the existing cyclostratigraphic data, the microfossil extinctions and the drop in sedimentation rate happened within one precessional period or within the resolution of one "tick" of our dating method.

While the South Atlantic precessional chronology seems consistent with independent evidence of abrupt environmental perturbations (Alvarez and others, 1980; Zachos and others, 1989) and with the expected duration of chron C29R (570 ky according to Berggren and others, 1985, 580 ky according to Harland and others, 1989), it is clearly desirable to reproduce the results in another location. The recent study of Ten Kate and Sprenger (1993) offers an unusual chance to compare the pattern of South Atlantic carbonate cycles across the K/T boundary with sediments a hemisphere away. Figure 7 presents their results from one of three Spanish sections studied. Maastrichtian sedimentation rates at the two southern sections, Agost and Relleu, are 1.5–1.8 cm/ky. The Zumaya section from northern Spain represents an expanded K/T boundary sequence with high-amplitude carbonate fluctuations.

Fine-scale carbonate oscillations in the Spanish sections seem likely to represent circa 20-ky precessional cycles (Ten Kate and Sprenger, 1993). In addition to chron C29R, the one complete polarity zone recorded in the Spanish sections is chron C29N at Zumaya. Here 32 ± 2 carbonate and magnetic susceptibility cycles are present, yielding an average duration of 22.6 ky (Berggren and others, 1985, time scale). While the Zumaya record does not extend quite to the C30N/C30R boundary, the number of cycles recorded by Ten Kate and Sprenger (1993) is consistent with a 20-ky repeat time in the Maastrichtian as well. Furthermore, the decimetric carbonate variations seem to be correlatable between sections, as would be expected for orbitally imprinted sedimentation.

In Table 1 we propose a precessionally based subdivision of chron C29R that incorporates information from the Spanish and South Atlantic sections. Our examination of Ten Kate and Sprenger's (1993) precessional chronology of the Maastrichtian portion of C29R finds 16 carbonate cycles at Relleu, 19 carbonate cycles at Agost, and 18 carbonate cycles at Zumaya, in

close agreement with the 18.5 cycles (373 ky) recorded in the South Atlantic. Our cyclostratigraphic correlation links the node of low precessional amplitude between cycles 29M13 through 29M16 in the South Atlantic to the node of low-amplitude carbonate variations at the Spanish sections near the position of the C29R/C30 boundary. It appears that our previous estimate for the duration of the Danian portion of C29R (11 cycles or 224 ky), based solely on color variations at DSDP Site 529, may be too short. New analysis of magnetic susceptibility data from Site 529 indicate 13 precessional cycles of earliest Danian age in C29R (D'Hondt and others, unpubl. data). A cycle count of the most condensed Spanish section, Relleu, suggests 12 ± 2 Paleogene C29R cycles, although the low amplitude of the carbonate variations makes this estimate less than secure. At Agost, we count 14 ± 1 carbonate cycles of this age. Counting high-amplitude carbonate variations in the Paleogene portion of C29R at the Zumaya section, in which cycle identification is the most unambiguous, yields 14.5 ± 1 cycles or a 296-ky duration. We view these results as strong confirmation of the South Atlantic chronology.

The entire duration of chron C29R, is then estimated by counting precessional cycles to have occupied 673 ky. Cyclochronology partitions this time into 18.5 precessional couplets before the boundary and 14.5 couplets after the extinction level. Using the Zumaya record as the best template for chron C29N, we estimate that the ratio of the durations of chrons C29N and C29R is 32:33 precessional cycles or 0.97. For comparison, the Cande and Kent (1992) seafloor magnetic anomaly template gives a ratio of 1.02 for the two anomaly widths.

It may be possible to establish an essentially one-to-one correlation of carbonate cycles between the South Atlantic and the Spanish sections in the K/T boundary interval. In this comparison, we are guided by reversal boundaries, the extinction level, and by the characteristic enveloping pattern of the carbonate cycles, which, while not perfectly expressed, is similar to the 100-ky modulation of precessional amplitude. The three more slowly deposited sites (Agost, Relleu, and DSDP 529) show almost 3 circa 100-ky bundles in the Paleogene portion of C29R. At all sites the C29R/C30N boundary is located in a node of low-amplitude cyclicity. Note also that carbonate couplets 29M5–29M11 form a set of seven relatively high-amplitude cycles at most study sites.

The good agreement between chronologies at the two study regions suggests that the precessional clock can be a reliable timekeeper, and that none of the sections sampled is significantly incomplete. The position of the K/T boundary within C29R according to the orbital chronology differs from standard estimates (c.f. Berggren and others, 1985), which place the extinction level at about 2/3 above the base of the reversed polarity zone. The latter estimate uses the reasonable assumption that the average stratigraphic position of the boundary relative to C29R is equal to its chronological position. Our data indicate, however, that the K/T boundary occurred *in time* very near the middle of the reversed polarity interval; extremely slow deposition rates in the Danian portion of the stratigraphic column make it appear that the extinction horizon occurs later in chron C29R than its true position.

ORBITAL CALIBRATION OF MID AND LOWER CRETACEOUS TIME SCALES

The absence of polarity reversals for Aptian through Santonian time (a nearly 34-my span) deprives stratigraphers of useful

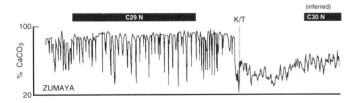

Fig. 7.—Carbonate cycles at Zumaya, Spain (from Ten Kate and Sprenger, 1993) can be used to extend the South Atlantic precessional chronology. The position of the C29R/C30N boundary was not determined at Zumaya but has been inferred by correlation to the Sopelana section (see Ten Kate and Sprenger, 1993).

TABLE 1.—K/T CHRONOLOGY

	C29N	C29R (Paleocene)	C29R (Maastrichtian)
Prec. cycles	32 ± 2	14.5 ± 2	18.5 ± 1
Duration (ky)	653 ± 41	296 ± 41	377 ± 20

isochronous horizons for much of middle Cretaceous time. The sea-floor anomaly template of chrons M-0 and older becomes progressively less reliable, as smaller and smaller areas of sea-floor remain. And finally, there are few high-quality radiometric dates in stratigraphically useful sequences of mid and lower Cretaceous age. Orbital cycles, securely anchored to a biostratigraphic, and where possible, magnetostratigraphic, framework, have great potential to refine chronologies in an interval of time now only poorly resolved by conventional stratigraphic methods. Such improvement is important if we are to understand the behavior of the earth during the peak "greenhouse" climates of the mid-Cretaceous (c.f. Savin, 1977; Barron, 1983).

A number of field studies point to the influence of orbital forcing in the mid-Cretaceous (the significant work of Gale, 1989, in the Cenomanian of England and France; of the ROCC group, 1986, in the Cenomanian-Turonian of the U.S. Western Interior, of Cotillon and Rio, 1984, and Huang and others, 1993, on the Barremian-Hauterivian of the Atlantic and Vocontian basin of France; and of Ten Kate and Sprenger, 1992 in the Berriasian of Spain and France). Of these the studies of de Boer (1982), Park and Herbert (1987), Gale (1989), Herbert (1992), Schwarzacher (1993), and Huang and others (1993) are most relevant to orbital calibration of the middle Cretaceous time scale.

The most thoroughly documented evidence of orbitally-forced sedimentation of mid-Cretaceous age comes from studies of pelagic sediments of Barremian to Cenomanian age exposed in northern and central Italy (de Boer, 1982; Schwarzacher and Fischer, 1982; Weissert and others, 1985; Herbert and Fischer, 1986). Sedimentation here was exceptionally continuous. Nannofossil (Bralower, 1987; Channell and others, 1987; Erba, 1988) and foraminiferal biostratigraphies (Premoli Silva and Paggi, 1976, Tornaghi and others, 1989) consistent with the standard Tethyan biozonations (c.f. Thierstein, 1973; Sigal, 1979) exist for many of the studied sections that display lithological rhythmicity. Furthermore, paleomagnetic studies have identified the late Jurassic through middle Cretaceous M-series magnetic anomaly patterns in a number of outcrops (Channell and others, 1979; Channell and Erba, 1992). As measured by paleomagnetic and biostratigraphic dates, sedimentation rates in the Italian pelagic sequences do not vary by more than a factor of three from Barremian through Cenomanian times.

A consistent "bundling" of carbonate couplets runs from Barremian through most of Cenomanian time or nearly 30 my of sedimentation in the Italian sections. This bundling is defined by an approximately 5:1 modulation of couplet thickness and, where carbonate content has been sampled in detail, by a prominent cycle in carbonate content (Herbert and Fischer, 1986). In some cases, the carbonate variations are in phase with redox cycles, including "black shale" deposition in parts of the upper Barremian, lowermost Aptian-Albian, and lower Cenomanian sediments. Organic carbon-enriched strata nearly always coincide with the carbonate-poor member of the couplet. In other examples, as for the dominantly reddish middle Aptian strata of Umbria, the carbonate cycles persist through a rather uniform sediment redox state. A number of criteria establish the primary origin of the carbonate cycles, including the burrow mixing of lithologies, trace fossil variations, and several lines of geochemical evidence.

De Boer (1982) and Schwarzacher and Fischer (1982) proposed that the carbonate rhythms represented precessional forcing, with the bundling cycle an expression of the 100-ky eccentricity envelope. Their time control was poor; sedimentation rates were estimated from the highly uncertain (at the time) duration of the Cenomanian stage. In addition, the reliability of the field measurements depended on the skill and judgement of the observer. In part to solve the problem of creating an "objective" data base for studies of Cretaceous cyclicity, and in part to solve the problem of poor outcrop exposure in the shaly and organic-rich Aptian and Albian part of the Italian pelagic sequence, an 84-m core was drilled near the town of Piobbico (Marche), Italy as a joint project of the U.S. National Science Foundation and Italian Consiglio Nazionale della Richerca. The resulting integration of geochemical, trace fossil, and biostratigraphic information provided an unusual opportunity to develop the cyclostratigraphy of the Aptian and Albian stages (Herbert and Fischer, 1986; Erba, 1988; Tornaghi and other, 1989).

Figure 8 summarizes the foraminiferal and nannofossil biostratigraphies of the Piobbico core (*C. litterarius* through *E. turriseiffeli* nannofossil zones, *G. blowi* through *R. ticinensis* planktonic foraminiferal zones), outrop studies and thin-section analyses at Monte Petrano and Contessa Quarry (*P. buxtorfi* through *R. brotzeni* foraminiferal zones), and outcrop studies at

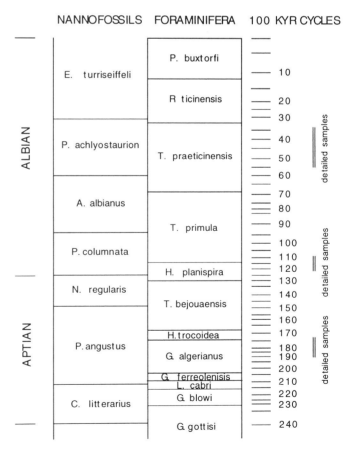

FIG. 8.—Aptian/Albian nannofossil and planktonic foraminiferal biostratigraphies of the Italian pelagic series, correlated to the right to a "bundle" (95 ky) chronology based on outcrop and core studies. Biostratigraphic data are reported in Erba (1988) and Tornaghi and others (1989). The vertical bars show intervals of the Piobbico core sampled in detail (see Fig. 9).

the Gorgo a Cerbara and Cismon sections (early Aptian through early Barremian age). Vertical bars to the right indicate the three intervals of the core, totaling 18 meters, sampled intensively for the time series analyses reported here. Although reversed polarity chron M-0 was not intersected at the base of the core, core-to-outcrop correlation with the Gorga a Cerbara reversal sequence established by Lowrie and Alvarez (1984) indicates that the core terminated 1 to 2 meters above this horizon. Additional field measurements, combined with thin-section identification of planktonic foraminiferal datums, were made at Monte Petrano and Gubbio to extend the cyclostratigraphy across the Albian/Cenomanian boundary. The measured sections were keyed to a distinctive set of black shale markers at about meter level 4.5 of the Piobbico core, and sequential bedding thicknesses recorded for about 35 m of section. The agreement between biostratigraphy (three foraminiferal datum events) and cyclostratigraphy was to within one bundle (estimated 100 ky) between the two studied sections.

Closely-spaced sampling of representative intervals of the Piobbico core clearly confirms the basic validity of the cyclicities recognized previously in outcrop (Fig. 9). The metronomic alternation of carbonate maxima and minima at a roughly 10-cm scale can be followed throughout the Piobbico core. These variations are grouped into 4–6 couplets by the "bundling" rhythm at a spacing of 40–60 cm. In Figure 8, 21 bundle cycles can be recognized in the upper Albian depth series, 10 for the Aptian-Albian section, and 13 for the middle Aptian interval. A longer-term enveloping pattern groups 3–5 bundles into nodes of higher and lower amplitude. Silica of biogenic origin follows carbonate content in lockstep, suggesting that the basic cycle is one of both carbonate and siliceous surface productivity. Frequency analysis (Fig. 10) confirms the statistical significance of periodicities in the time series, especially the prominence of the bundling cycle, which has a spatial period of about 55 cm in the upper Albian time series.

One constraint on the duration of these cycles is based on assuming constant sedimentation rates for the Aptian and Albian stages in the Piobbico core, and dividing the stage thicknesses by their durations, estimated from various time scales. As admittedly imperfect as this calibration is, the estimated periods fall in the range of 20 ky for the repeat time between carbonate maxima, 100 ky for the bundling cycle, and 400 ky for long wavelength modulation of the bundling amplitude.

The resemblance of the Piobbico time series to the orbital metronomes goes beyond this, however. Figure 11 shows the Albian *T. breggiensis* time series smoothed to remove the short, circa 20 ky rhythms, compared to a 1.6-my record of eccentricity taken from Berger and Loutre (1988). We do not intend a literal comparison but rather wish to highlight the qualitative similarity of the time series. What is diagnostic of the circa 100-

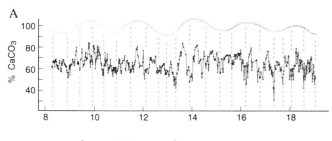

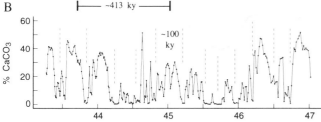

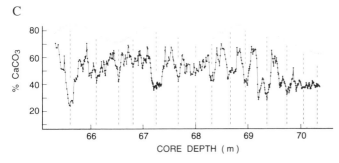

FIG. 9.—Calcium carbonate stratigraphic series from the upper Albian T. breggiensis foraminiferal zone (A), from the lowermost Albian P. columnata zone (B), and from the middle Aptian P. angustus nannofossil zone (C). The finest scale of variation recurs on a scale of 6–12 cm and represents precessional forcing. 4–6 couplets are grouped into the "bundle" cycle, representing the 95 ky repeat time of the short cycle of eccentricity. Dashed lines drawn through the centers of carbonate minima to define these cycles. The 413 ky eccentricity is expressed as a long wavelength enveloping pattern of the bundle cycle.

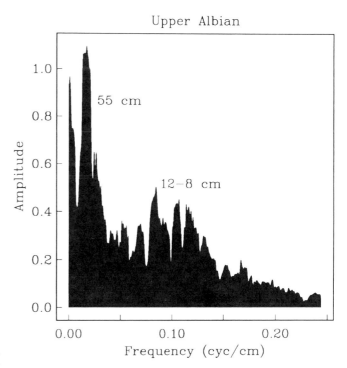

FIG. 10.—Frequency spectrum of the upper Albian time series shown in Figure 9A. The large broad peak centered at 1/55 cm⁻¹ is the bundling (short eccentricity) cycle; split peaks at 1/12, 1/10, and 1/8 cm⁻¹ represent components of precession. Spectra of the other time series shown in Figure 9 also confirm the picks of the bundle cycles shown in that figure.

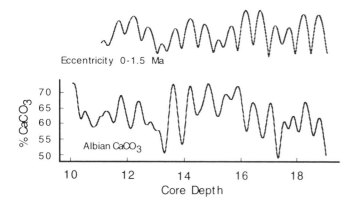

TABLE 2.—PERIODS OF BARREMIAN CARBONATE CYCLES, ITALY

Chron	Sed. rate (m/My)	# bundles	# couplets	Bundle period (ky)	Couplet period (ky)
Cerbara section					
M0	1.86 [2.32]	2	13	350 [281]	53.8 [43.2]
M1n	7.0 [8.9]	24.5	119	127 [101]	26.1 [20.8]
M1	10.0 [12.6]	4.5	20	97 [78]	21.0 [17.5]
M2	13.8 [17.4]	7	35	111 [89]	22.4 [17.7]
M3	11.5 [14.5]	10.5 in 14.3 m	100	118 [94]	23.3 [18.5]
Cismon section					
M0	5.4 [6.8]	7	34	100 [80]	20.6 [16.5]

Durations estimated using Kent and Gradstein (1985) Jurassic-Cretaceous time scale (no brackets) and Harland and others (1989) time scale (brackets).

FIG. 11.—Smoothed version of the Upper Albian series, compared with eccentricity record 0–1.522 Ma. The comparison is qualitative, since the eccentricity series never exactly repeats itself. The most important point of similarity is the amplitude modulation of the short cycle of eccentricity with its node of low energy at 413-ky increments, which is mimicked in the Cretaceous paleoclimatic series.

ky short eccentricy cycle is its nodal pattern, which follows the longer 413 ky cycle of eccentricity. The amplitude modulation of the short eccentricity cycle breaks the Fourier series representation into components centered around 95 and 123 ky. Similar structure may be embedded within the "bundling" spectral peak, which actually spans a range of 1/45 to 1/60 cm^{-1} (Fig. 10). The spectral resolution of the time series, which must be degraded to some degree by changes in sedimentation rate, can be sharpened by removing the first-order variations in the spacing of the bundling cycle (Park and Herbert, 1987). Statistical tests on the resulting "tuned" record using a multiple-taper spectral method give twin peaks near 100 ky with a frequency ratio of 1.3 or within a few percent of the ratio of the two dominant frequency modes of the short cycle of eccentricity. Park and Herbert (1987) proposed that the twin circa 100-ky peaks represent the 95- and 123-ky numerical periods and can be used as a stratigraphic tuning fork to derive sedimentation rate to a very high precision in the Piobbico sequence.

Since that analysis was published, the correspondence between the bundled carbonate cycles and "Milankovitch" periods has been strengthened by outcrop studies of the upper Barremian strata of Italy, which permit the durations of the sedimentary cycles to be estimated by magnetostratigraphy (Herbert, 1992). The M-series reversals improve the numerical resolution of the sedimentary time series, so that the constancy of sedimentary periods can be better tested than in the Long Normal superchron. Furthermore, it may be possible in the future to demonstrate that the cycles correlate precisely across true isochrons defined by polarity reversals. Herbert (1992) produced bedding thickness time series at two Italian sections with reversal stratigraphies: Cismon (northern Italy), previously zoned biostratigraphically and paleomagnetically by Channell and others (1979) and the Gorga a Cerbara section of Lowrie and Alvarez (1984). The periods of the carbonate couplets and their five-fold spacing modulation were estimated over chrons M-0 through M-3 or a time span of about 5 my. The numerical periods (Table 2) are in close agreement with the periods expected of precession and eccentricity, with the lone discrepancy of chron M-0 in the Gorgo a Cerbara section. An optimal match

of cycle periods to orbital forcing can be achieved if the correct M-series numerical calibration lies between those proposed by Kent and Gradstein (1985) and Harland and others (1989) (see also Channell and others, this volume).

With some confidence that the combination of numerical constraints and internal characteristics of the Italian mid-Cretaceous carbonate rhythms are diagnostic of circa 20- and 100-ky cycles, we now offer an estimate of the durations of the Aptian and Albian Stages, as defined by planktonic foraminiferal and calcareous nannofossil datums. The position of each carbonate maximum and minimum in the Piobbico core was logged visually above, below, and in between the data series shown in Figure 9. Outcrop studies at four locations have verified the completeness of the Piobbico sequence down to the Aptian/Albian boundary, with the exception of a fault that removes 4 bundle cycles (400 ky) at the core depth of 29.5 m. Likewise, correlation of the core sequence to outcrops indicates that there is no identifiable discontinuity in the lower 15–20 m. Poor outcrop exposure of the shaly upper Aptian strata currently prevents us from documenting the continuity of about 15 m of the uppermost Aptian interval of the core. The bundle chronology can be extended down to reversed chron M-0 from field work reported in Herbert (1992) and updated with new biostratigraphic information from Erba (1994).

To simplify the data, we have placed the microplankton biostratigraphy next to a count of the bundle cycles, beginnning with the latest Albian P. buxtorfi planktonic foraminiferal Zone (Figure 8). The 4 bundles removed by faulting at meter 29 of the core have been restored to the chronology. Following the analysis of Park and Herbert (1987), we match the time domain expression of the bundle cycle to the 95-ky component of eccentricity. Elapsed time can then simply be scaled to bundle count. We must caution that the correlation of microfossil zones to the Tethyan type sections that define stage boundaries is always subject to revision. For example, the correlation of foraminiferal and nannofossil zones to the Aptian/Albian stratotype is in flux; we place the Aptian/Albian boundary at the base of the P. columnata nannofossil Zone. The Albian/Cenomanian boundary located in outcrop is placed at the first occurrence of the planktonic foraminifer R. brotzeni.

Our estimate for the duration of the Albian stage, combining cycle counts from the Piobbico core and from outcrop studies, is that it encompasses 125 bundle cycles or 11.9 my, and that the interval from the Aptian/Albian boundary to the top of chron M-0 occupies 106 bundles or 10.1 my. If the Barremian/Aptian boundary coincides with the base of chron M-0, then its total duration is calculated to be 10.6 my. The duration of the Albian given by cyclochronometry is 2–3 my shorter than that given by Kent and Gradstein (1985) or Harland and others (1989), while the Aptian duration is longer than most previous time scales suggest but shorter than the 12.5-my duration proposed by Harland and others (1989). Of particular importance is the orbital estimate of 1.3 my for the duration of the *G. blowi* foraminiferal Zone, the biostratigraphic link to a recently published [40]Ar/ [39]Ar date of 122.4 ± 0.8 Ma (Mahoney and others, 1993).

Integrating these durations into the numerical time scale requires correlating Tethyan biostratigraphic datums to high-quality radiometric ages. We feel that the Cenomanian/Turonian boundary is the most precisely correlatable horizon at present and accept J. Obradovich's (1993) [40]Ar/[39]Ar date of 93.5 ± 0.2 Ma from the western interior of the United States. Our estimate for the duration of the Cenomanian, as defined by Tethyan planktonic foraminifera, comes from bedding studies by de Boer (1982), Schwarzacher (1993), and Herbert (unpubl. data), and from geochemical arguments for the duration of the uppermost Cenomanian Bonarelli Horizon (Arthur and Premoli Silva, 1982). We prefer a duration of 6 ± 0.5 my (the error estimate indicates that the middle portion of the Cenomanian of central Italy is less clearly cyclic than the upper and lower portions, and therefore harder to date with the orbital method), significantly longer than the orbital chronology reported by Gale (1989) of 4.4 my (our estimate is slightly shorter than the 7.2 my estimated by de Boer, 1982, because we use the base of the *R. brotzeni* rather than the base of the *R. appeninica* foraminiferal Zone to define the Albian/Cenomanian boundary). It is not clear if the discrepancy between Gale's determinations and ours reflects problems in biostratigraphic correlation, inadequacies of either the Paris Basin or Umbrian sections or misidentification at one or both sections of the orbital periodicities. We note that while Gale (1989) assumed that the Chalk carbonate variations were arranged in 20- and 100-ky units, Hart (1987) identified the rhythms as 41-ky obliquity cycles and that neither study offered a statistical analysis of cyclicity. For now, we adopt the longer estimate from Umbria, and place the Albian/Cenomanian boundary at 99.5 Ma. The Aptian/Albian boundary, as defined here, would then lie at 111.4 Ma, and the Barremian/Aptian boundary (base of M-0) would occur at 122.2 Ma, with the Long Normal/M-0 boundary at 121.6 Ma. As the age of M-0 is an important tie-point for the dating of the M-series anomaly sequence, our cumulative chronology should significantly improve the accuracy of the lower Cretaceous GPTS (see Channell and others, this volume).

Two further studies that propose orbital calibration of Cretaceous stage durations are those of Cottle (1989) for the Turonian (orbital estimate of 3.5 my) and of Huang and others (1993) for the Hauterivian and Valanginian sequences exposed in the Angles-Vergons region of France. The latter study proposes a chronology based on obliquity cycles that gives a duration of 5.3 my for the Hauterivian and 5.9 my for the Valanginian stage. These durations are shorter than given by most time scales (e.g., Kent and Gradstein, 1985; Harland and others, 1989) but are plausible. Unfortunately, Huang and others (1993) study cannot yet be tied to microplankton datum events or magnetic polarity zones. Future work will need to identify cyclical sections with a good M-series reversal stratigraphy.

CONCLUSIONS

The data base for Cretaceous and early Tertiary orbital chronometry is still small, and it is unlikely that our proposed chronologies are as accurate as they are precise. However, at least two regions of the world, the South Atlantic for the upper Cretaceous and the Italian sections for the middle Cretaceous, are now established in which cyclic sedimentation, defined objectively by spectral analysis, runs for at least 20 and 30 my, respectively. The probability that the periods and characters of these carbonate cycles could coincidentally match precessional and eccentricity orbital cycles seems small. Where a number of sections can be compared, as across the K/T boundary, counting orbital cycles in sediments gives surprisingly consistent estimates of time. We have elsewhere (Herbert, 1992) taken a first step toward a similar test with early Aptian strata. Cycle counts through the early Aptian "Selli Horizon" (Coccioni and others, 1989) interval of organic carbon-enriched sediments gave durations consistent to 10% between northern and central Italian sections (in this case, the isochrons used were the top of polarity interval M-0 and the first appearance of the nannofossil *P. angustus*). One of the future challenges of Cretaceous cyclostratigraphy will be to determine whether cycle correlations can be made over much longer distances, so that mid-Cretaceous sequences can be tied globally at very high resolution.

The advantages of using the orbital approach, in conjunction with other stratigraphic techniques, are potentially very great. Unlike magnetostratigraphical resolution, set by the vagaries of the earth's field reversal behavior, or radiometric time scale calibrations, limited by the rarity of datable materials in biostratigraphically useful sequences, orbital rhythms run continuously through many marine sections. Most sequences studied to date contain the microplankton datum events most useful for routine relative age assignment of sediments. It is seems likely that the durations of these biozones can be estimated quite confidently by orbital methods. It is also conceivable that a composite record of orbital variations will eventually stretch from Pleistocene through Mesozoic time and that this record can be used to fine-tune the polarity time scale, particularly during intervals such as the early Paleogene, where a change in seafloor spreading rate of the South Atlantic template may complicate assigning durations to magnetic polarity zones (Cande and Kent, 1992). We emphasize, however, that much careful work is needed before the potentially continuous orbital timekeeper can be integrated into the GPTS. More studies will need to produce Cretaceous time series that can be evaluated statistically, and more sequences must be continuously cored to permit high-resolution stratigraphic work.

ACKNOWLEDGMENTS

We thank J. Obradovich and M. Kominz for helpful reviews, S. DiDonna, L. Tauxe, and J. Gee for help in paleomagnetic analyses, and S. D'Hondt for sharing results of magnetic susceptibility analyses ahead of publication. This research was funded by the U.S. National Science Foundation, the Italian Consiglio Nazionale della Richerca, and by the donors of the Petroleum Research Fund of the American Chemical Society. We thank the N.S.F. for curation of the DSDP cores used in our research.

REFERENCES

ALVAREZ, L. W., ALVAREZ, W., ASARO, F., AND MICHEL, H. V., 1980, Extraterrestrial cause for the Cretaceous-Tertiary extinction: Science, v. 208, p. 1095–1108.

ARTHUR, M. A. AND PREMOLI SILVA, I., 1982, Development of widespread organic-rich strata in the Mediterranean Tethys, in Schlanger, S. O. and Cita, M. B., eds., Nature and Origin of Cretaceous Carbon-Rich Facies: New York, Academic Press, p. 7–54.

BAKSI, A. K., HSU, V., McWILLIAMS, M. O., AND FARRAR, E., 1992, ⁴⁰Ar/³⁹Ar dating of the Brunhes-Matuyama geomagnetic field reversal: Science, v. 256, p. 356–357.

BARRON, E. J., 1983, A warm, equable Cretaceous: the nature of the problem: Earth Science Reviews, v. 19, p. 305–338.

BERGER, A. L. AND LOUTRE, M. F., 1988, New insolation values for the climate of the last 10 my: Institut d'Astronomie et de Geophysique G. Lemaitre, University Catholique Louvain-la-Neuve, Science Report 88–13.

BERGER, A. L., LOUTRE, M. F., AND LASKAR, J., 1992, Stability of the astronomical frequencies of the earth's orbit for paleoclimate studies: Science, v 255, p. 560–566.

BERGGREN, W. A., KENT, D. V., FLYNN, J. J., AND VANCOUVERING, J. A., 1985, Cenozoic geochronology: Geological Society of America Bulletin, v. 96, p. 1407–1418.

BRALOWER, T. J., 1987, Valanginian to Aptian calcareous nannofossil stratigraphy and correlation with the upper M-sequence magnetic anomalies: Marine Micropaleontology, v. 11, p. 293–310.

CANDE, S. C. AND KENT, D. V., 1992, A new geomagnetic polarity time scale for the late Cretaceous and Cenozoic: Journal of Geophysical Research, v. 97, p. 13917–13951.

CHANNELL, J. E. T., BRALOWER, T. J., AND GRANDESSO, P., 1987, Biostratigraphic correlation of Mesozoic polarity chrons CM1-CM23 at Capriolo and Xausa (S. Alps, Italy): Earth and Planetary Science Letters, v. 85, p. 203–221.

CHANNELL, J. E. T. AND ERBA, E., 1992, Early Cretaceous polarity chrons CM0 to CM11 recorded in northern Italian land sections near Brescia: Earth and Planetary Science Letters, v. 108, p. 161–180.

CHANNELL, J. E. T., LOWRIE, W., AND MEDIZZA, F., 1979, Middle and early Cretaceous magnetic stratigraphy from the Cismon section, northern Italy: Earth and Planetary Science Letters, v. 42, p. 153–166.

CHAVE, A. D., 1984, Lower Paleocene-Upper Cretaceous magnetic stratigraphy from Sites 525, 527, 528, and 529, DSDP Leg 74: Initial Reports of the Deep Sea Drilling Project, v. 74, p. 525–531.

COCCIONI, R., FRANCHI, R., NESCI, O., WEZEL, F.-C., BATTISTINI, F., AND PALLECHI, P., 1989, Stratigraphy and mineralogy of the Selli Level (Early Aptian) at the base of the Marne a Fucoidi in the Umbro-Marchean Apennines (Italy), in Wiedmann, J., ed., Cretaceous of the Western Tethys: Tubingen, Proceedings 3rd International Cretaceous Symposium, p. 563–584.

COTILLON, P. AND RIO, M., 1984, Cyclic sedimentation in the Cretaceous of Deep Sea Drilling Project Sites 535 and 540 (Gulf of Mexico), 534 (central Atlantic) and in the Vocontian Basin (France): Initial Reports of the Deep Sea Drilling Project, v. 77, p. 339–376.

COTTLE, R. A., 1989, Orbitally mediated cycles from the Turonian of southern England: their potential for high-resolution stratigraphic correlation: Terra Nova, v. 1, p. 426–431.

DE BOER, P. L., 1982, Cyclicity and storage of organic matter in middle Cretaceous pelagic sediments, in Einsele, G. and Seilacher, A., eds., Cyclic and Event Stratification: Berlin, Springer-Verlag, p. 456–475.

ERBA, E., 1988, Aptian-Albian calcareous nannofossil biostratigraphy of the Scisti a Fucoidi cored at Piobbico (central Italy): Rivista Italiana Paleontologia e Stratigrafia, v. 94, p. 294–284.

ERBA, E., 1994, Nannofossils and superplumes: The early Aptian "nannoconid crisis": Paleoceanography, v. 9, p. 483–501.

FISCHER, A. G., 1980, Gilbert-bedding rhythms and geochronlogy: Boulder, Geological Society of America Special Publication 183, p. 93–104.

GALE, A. S., 1989, A Milankovitch scale for Cenomanian time: Terra Nova, v. 1, p. 420–425.

HAMILTON, N. AND SUZYUMOV, A. E., 1983, Late Cretaceous magnetostratigraphy of Site 516, Rio Grande Rise, Southwestern Atlantic Ocean, Deep Sea Drilling Project, Leg 72: Initial Reports of the Deep Sea Drilling Project, v. 72, p. 723–730.

HARLAND, W. B., ARMSTRONG, R. L., COX, A. V., CRAIG, L. E., SMITH, A. G., AND SMITH, D. G., 1989, A Geologic Time Scale: Cambridge, Cambridge University Press, 263 p.

HART, M., 1987, Orbitally induced cycles in the chalk facies of the United Kingdom: Cretaceous Research, v. 8, p. 335–348.

HAYS, J. D., IMBRIE, J., AND SHACKLETON, N. J., 1976, Variations in the earth's orbit: pacemaker of the ice ages: Science, v. 194, p. 1121–1132.

HENRIKSSON, A. S., 1993, Biochronology of the terminal Cretaceous calcareous nannofossil zone of Micula prinsii: Cretaceous Research, v. 14, p. 59–68.

HERBERT, T. D., 1992, Paleomagnetic calibration of Milankovitch cyclicity in Lower Cretaceous sediments: Earth and Planetary Science Letters, v. 112, p. 15–28.

HERBERT, T. D. AND FISCHER, A. G., 1986, Milankovitch climate origin of mid-Cretaceous black shale rhythms in central Italy: Nature, v. 321, p. 739–743.

HERBERT, T. D. AND D'HONDT, S. L., 1990, Precessional climate cyclicity in Late Cretaceous-Early Tertiary marine sediments: a high resolution chronometer of Cretaceous-Tertiary boundary events: Earth and Planetary Science Letters, v. 99, p. 263–275.

HILGEN, F. J., 1991, Astronomical calibration of Gauss to Matuyama sapropels in the Mediterranean and implication for the Geomagnetic Polarity Time Scale: Earth and Planetary Science Letters, v. 104: 226–244.

HOUSE, M. R., 1985, A new approach to an absolute time scale from measurements of orbital cycles and sedimentary microrhythms: Nature, v. 315, p. 721–725.

HUANG, Z., BOYD, R., AND O'CONNELL, S., 1992, Upper Cretaceous cyclic sediments from ODP Hole 122-762C—Exmouth Plateau, N.W. Australia: Scientific Results of the Ocean Drilling Program, v. 122, p. 259–277.

HUANG, Z., OGG, J. G., AND GRADSTEIN, F. M., 1993, A quantitative study of Lower Cretaceous cyclic sequences from the Atlantic Ocean and Vocontian Basin (SE France): Paleoceanography, v. 8, p. 275–291.

IMBRIE, J., SHACKLETON, N. J., PISIAS, N. G., MORLEY, J. J., PRELL, W. L., MARTINSON, D. G., HAYS, J. D., McINTYRE, A., AND MIX, A. C., 1984, The orbital theory of Pleistocene climate: support from a revised chronology of the marine δ¹⁸O record, in Berger,. A. and others, eds., Milankovitch and Climate: Hingham, D. Reidel, p. 269–306.

IZETT, G. A., DALRYMPLE, G. B., AND SNEE, G. B., 1991, ⁴⁰Ar/³⁹Ar age of the Cretaceous-Tertiary boundary tektites from Haiti: Science, v. 252, p. 1539–1542.

JOHNSON, R. G., 1982, Brunhes-Matuyama magnetic reversal dated at 790,000 yr B.P. by marine-astronomical correlations: Quaternary Research, v. 17, p. 135–147.

KELLER, G., 1989, Extended Cretaceous/Tertiary boundary extinctions and delayed population changes in planktonic foraminifera from Brazos River, Texas: Paleoceanography, v. 4, p. 287–332.

KENT, D. V. AND GRADSTEIN, F. M., 1985, A Cretaceous and Jurassic geochronology: Geological Society of America Bulletin, v. 96, p. 1419–1427.

KOMINZ, M. AND BOND, G., 1990, A new method for testing periodicity in cyclic sediments:application to the Newark Supergroup: Earth and Planetary Science Letters, v. 98, p. 23–244.

LOWRIE, W. AND ALVAREZ, W., 1984, Lower Cretaceous magnetic stratigraphy in Umbrian pelagic limestone sections: Earth and Planetary Science Letters, v. 71, p. 315–328.

MACLEOD, N. AND KELLER, G., 1991, Hiatus distribution and mass extinctions at the Cretaceous/Tertiary boundary: Geology, v. 19, p. 497–501.

MAHONEY, J. J., STOREY, M., DUNCAN, R. A., SPENCER, K. J., AND PRINGLE, M. S., 1993, Geochemistry and Age of the Ontong Java Plateau, in Pringle, M. S., Sager, W. V., Sliter, W. V., and Stein, S., eds., The Mesozoic Pacific: Geology, Tectonics, and Volcanism: Washington, American Geophysical Union Monograph 77, p. 233–261.

MARTINSON, D. G., PISIAS, N. G., HAYS, J. D., IMBRIE, J., MOORE, T. C. JR., AND SHACKLETON, N. J., 1987, Age dating and the orbital theory of the Ice Ages: Development of a high-resolution 0 to 300,000 year chronostratigraphy: Quaternary Research, v. 27, p. 1–29.

MOORE, T. C. JR., PISIAS, N. G., AND DUNN, D. A., 1982, Carbonate time series of Quaternary and Miocene sediments in the Pacific Ocean: a spectral comparison: Marine Geology, v. 46, p. 217–233.

MEAD, G. A., TAUXE, L., AND LABREQUE, J. L., 1986, Oligocene paleoceanography of the South Atlantic: paleoclimatic implications of sediment accumulation rates and magnetic susceptibility measurements: Paleoceanography, v. 1, p. 273–284.

OBRADOVICH, J. D., 1993, A Cretaceous time scale, in Caldwell, W. G. and Kauffman, E. G., eds., Evolution of the Western Interior Basin: St. Johns, Geological Association of Canada, Special Paper 39, p. 379–396.

PARK, J. AND HERBERT, T. D., 1987, Hunting for paleoclimatic periodicities in a sedimentary series with uncertain time scale: Journal of Geophysical Research, v. 92, p. 14027–14040.

PARK, J., D'HONDT, S. L., KING, J. W., AND GIBSON, C., 1993, Late Cretaceous precessional cycles in double time: a warm-earth Milankovitch response: Science, v. 261, p. 1431–1434.

PREMOLI SILVA, I. AND PAGGI, L., 1976, Cretaceous through Paleocene biostratigraphy of the pelagic sequence at Gubbio, Italy: Memorie della Societa Geologica Italiana, v. 15, p. 21–32.

R.O.C.C. GROUP (ARTHUR, M. S., BOTTJER, D. J., DEAN, W. E., FISCHER, A. G., HATTIN, D. E., KAUFFMAN, E. G., PRATT, L. M., AND SCHOLLE, P. A.), 1986, Rhythmic bedding in Upper Cretaceous pelagic carbonate sequences: Varying sedimentary response to climatic forcing: Geology, v. 14, p. 153–156.

SAVIN, S. M., 1977, The history of the earth's surface temperature during the past 100 million years: Annual Review of Earth and Planetary Sciences, v. 5, p. 319–355.

SCHIFFELBEIN, P. AND DORMAN, L., 1986, Spectral effects of time-depth nonlinearities in deep sea sediment records: a demodulation technique for realigning time and depth scales: Journal of Geophysical Research, v. 19, p. 3821–3835.

SCHWARZACHER, W., 1987, Astronomically controlled cycles in the lower Tertiary of Gubbio (Italy): Earth and Planetary Science Letters, v. 84, p. 22–26.

SCHWARZACHER, W., 1993, Cyclostratigraphy and the Milankovitch Theory, Elsevier, Amsterdam, 225 p.

SCHWARZACHER, W. AND FISCHER, A. G., 1982, Limestone-shale bedding and perturbations in the earth's orbit, *in* Einsele, G. and Seilacher, A., eds., Cyclic and Event Stratification: Berlin, Springer-Verlag, p. 72–95.

SHACKLETON, N. J., BERGER, A., AND PELTIER, W. R., 1990, An alternative astronomical calibration of the lower Pleistocene time scale based on ODP Site 677: Transactions of the Royal Society of Edinburgh: Earth Science, v. 81, p. 251–261.

SIGAL, J., 1977, Essai de zonation du Cretace mediterraneen a l'aide des foraminiferes planctoniques: Geologie Mediterraneean, v. 4, p. 99–108.

SMIT, J. AND ROMEIN, A. J. T., 1985, A sequence of events across the Cretaceous-Tertiary boundary: Earth and Planetary Science Letters, v. 74, p. 155–170.

TAUXE, L., DEINO, A. D., BEHRENSMEYER, A. K., AND POTTS, R., 1992, Pinning down the Brunhes-Matuyama and upper Jaramillo boundaries: a reconciliation of orbital and isotopic time scales: Earth and Planetary Science Leters, v. 109, p. 561–572.

TEN KATE, W. G. H. AND SPRENGER, A., 1993, Orbital cyclicities above and below the Cretaceous/Paleogene boundary at Zumaya (N. Spain), Agost and Relleu (SE Spain): Sedimentary Geology, v. 87, p. 69–101.

TEN KATE, W. G. H. AND SPRENGER, A., 1992, Rhythmicity in deep-water sediments, documentation and interpretation by pattern and spectral analysis: Ph.D. Dissertation, Free University Utrecht, 244 p.

THIERSTEIN, H. R., 1973, Mesozoic calcareous nannoplankton biostratigraphy of marine sediments: Marine Micropaleontology, v. 1, p. 325–362.

TORNAGHI, M. E., PREMOLI SILVA, I., AND RIPEPE, M., 1989, Lithostratigraphy and planktonic foraminiferal biostratigraphy of the Aptian-Albian "Scisti a Fucoidi" in the Piobbico core, Marche, Italy: background for cyclostratigraphy: Rivista Italiana Paleontologia e Stratigrafia, v. 95, p. 223–264.

WEISSERT, H., MCKENZIE, J. A., AND CHANNELL, J. E. T., 1985, Natural variations in the carbon cycle during the early Cretaceous, *in* Sundquist, E. T. and Broecker, W. S., eds., The Carbon Cycle and Atmospheric CO_2: Natural Variations Archean to Present: Washington, American Geophysical Union Monograph 32, p. 531–545.

ZACHOS, J. C., ARTHUR, M. A., AND DEAN, W. E., 1989, Geochemical evidence for the suppression of pelagic marine productivity at the Cretaceous/Tertiary boundary: Nature, v. 337, p. 61–64.

A TRIASSIC, JURASSIC AND CRETACEOUS TIME SCALE

FELIX M. GRADSTEIN
Saga Petroleum a.s., Kjorboveien 16, Postboks 490, N-1301 Sandvika, Norway
FRITS P. AGTERBERG
Geological Survey of Canada, Ottawa, Ontario, K1A OE8, Canada
JAMES G. OGG
Department of Earth and Atmospheric Sciences, Purdue University, West Lafayette, IN 47907
JAN HARDENBOL
826 Plainwood Drive, Houston, TX 77079-4227
PAUL VAN VEEN
Norsk Hydro Research Centre, N-5020 Bergen, Norway
JACQUES THIERRY
6 Université de Bourgogne, Centre des Sciences de la Terre, 6 Bd. Gabriel, 2100 Dijon, France
AND
ZEHUI HUANG
Atlantic Geoscience Centre, P.O. Box 1006, Dartmouth, Nova Scotia, B2Y 4A2, Canada

ABSTRACT: We present an integrated geomagnetic polarity and stratigraphic time scale for the Triassic, Jurassic, and Cretaceous Periods of the Mesozoic Era, with age estimates and uncertainty limits for stage boundaries. The time scale uses a suite of 324 radiometric dates, including high-resolution $^{40}Ar/^{39}Ar$ age estimates. This framework involves the observed ties between (1) radiometric dates, biozones, and stage boundaries and (2) between biozones and magnetic reversals on the seafloor and in sediments. Detailed attention is given to chronostratigraphic calibration of stage boundaries using tethyan and boreal biozonations. Interpolation techniques to arrive at a geochronology include maximum likelihood estimation, smoothing cubic spline fitting, and magnetochronology. The age estimates for the 31 stage boundaries (Ma with uncertainty in my to 2 standard deviations), and the duration of the preceding stages (in my) are:

Maastrichtian/Danian (Cretaceous/Cenozoic)	65.0 ± 0.1 Ma	6.3 my
Campanian/Maastrichtian	71.3 ± 0.5	12.2
Santonian/Campanian	83.5 ± 0.5	2.3
Coniacian/Santonian	85.8 ± 0.5	3.2
Turonian/Coniacian	89.0 ± 0.5	4.5
Cenomanian/Turonian	93.5 ± 0.2	5.4
Albian/Cenomanian	98.9 ± 0.6	13.3
Aptian/Albian	112.2 ± 1.1	8.8
Barremian/Aptian	121.0 ± 1.4	6.0
Hauterivian/Barremian	127.0 ± 1.6	5.0
Valanginian/Hauterivian	132.0 ± 1.9	5.0
Berriasian/Valanginian	137.0 ± 2.2	7.2
Tithonian/Berriasian (Jurassic/Cretaceous)	144.2 ± 2.6	6.5
Kimmeridgian/Tithonian	150.7 ± 3.0	3.4
Oxfordian/Kimmeridgian	154.1 ± 3.2	5.3
Callovian/Oxfordian	159.4 ± 3.6	5.0
Bathonian/Callovian	164.4 ± 3.8	4.8
Bajocian/Bathonian	169.2 ± 4.0	7.3
Aalenian/Bajocian	176.5 ± 4.0	3.6
Toarcian/Aalenian	180.1 ± 4.0	9.5
Pliensbachian/Toarcian	189.6 ± 4.0	5.7
Sinemurian/Pliensbachian	195.3 ± 3.9	6.6
Hettangian/Sinemurian	201.9 ± 3.9	3.8
Rhaetian/Hettangian (Triassic/Jurassic)	205.7 ± 4.0	3.9
Norian/Rhaetian	209.6 ± 4.1	11.1
Carnian/Norian	220.7 ± 4.4	6.7
Ladinian/Carnian	227.4 ± 4.5	6.9
Anisian/Ladinian	234.3 ± 4.6	7.4
Olenekian/Anisian	241.7 ± 4.7	3.1
Induan/Olenekian	244.8 ± 4.8	3.4
Tatarian/Induan (Permian/Triassic)	248.2 ± 4.8	

The uncertainty in the relative duration of each individual stage is much less than the uncertainties on the ages of the stage boundaries.

INTRODUCTION

The geological time scale for the Triassic, Jurassic, and Cretaceous Periods proposed in this study is composed of standard stratigraphic units (stages) calibrated in millions of years, with an uncertainty estimate at each level. Stratigraphically critical data underpinning this large framework of isotopic age dates involve the observed ties between (1) radiometric dates, biozones, and stage boundaries and (2) biozones and magnetic reversals on the seafloor and in sediments.

The Cenozoic time scale is primarily calibrated from biostratigraphic correlations to magnetic polarity chrons, which in turn are scaled according to magnetic anomaly profiles from the

Geochronology Time Scales and Global Stratigraphic Correlation, SEPM Special Publication No. 54

South Atlantic pinned to a selected set of radiometric ages (e.g., Berggren and others, 1992; Cande and Kent, 1992; Berggren et al., this volume). In contrast, the Mesozoic time scale lacks a single unifying interpolation concept, because magnetic anomaly profiles only extend back to the Callovian stage, and much of the middle Cretaceous Period lacks a magnetic anomaly signature. In addition, the radiometric database has inadequate precision to contrain the age assignment of most stage boundaries. Therefore, whereas portions of the Mesozoic time scale can be exactly determined by a combination of radiometric ages on biostratigraphically controlled sections, the majority of the stage boundaries have been assigned through geological and mathematical interpolation methods.

Differences in selection criteria for radiometric ages and variations in interpolation methods have led to a bewildering array of Mesozoic time scales within the past two decades (Figs. 1–3).

It is possible to calculate a Mesozoic time scale with estimates of uncertainty for the age of each stage boundary. The derivation of such a numerical scale depends on sufficient and stratigraphically meaningful ages from isotopic measurements in magmatic intrusions, bentonites, glauconites and other rocks. In this regard, there have been approximately 65 well-calibrated Mesozoic age dates published since the previous array of time

scales (e.g., Odin, 1982; Snelling, 1985; Kent and Gradstein, 1985, 1986; Haq and others, 1988; Harland and others, 1990). Also, improvements in calibrations of stages and standard zones and development of better interpolation techniques allow significant updates to previously proposed age estimates for the stage boundaries in the Triassic, Jurassic, and Cretaceous Periods.

Frequently cited geologic time scale studies are abbreviated as follows: Harland and others (1982) is PTS82; Odin (1982) is NDS82; Kent and Gradstein (1985) is KG85, also referred to as part of the Decade of North America Geology (DNAG) scale (Palmer, 1983); Haq and others (1988) is EX88; Harland and others (1990) is PTS89; Cande and Kent (1992) is CK92; and Obradovich (1993) is OB93.

In this study we present the data and methods selected to arrive at the age estimates for the 31 Mesozoic stage boundaries. All age dates and stage boundary estimates are accompanied by two standard deviation ($\pm 2\sigma$) error bars. Portions of the methodology and the Mesozoic time scale were reported earlier (Gradstein and others, 1994), and the numerical ages for the standard stages are incorporated in a series of Mesozoic-Cenozoic chronostratigraphic charts (Hardenbol and others, *in* de Graciansky and others, Mesozoic-Cenozoic Sequence Stratigraphy of European Basins, SEPM Special Publication, in prep.).

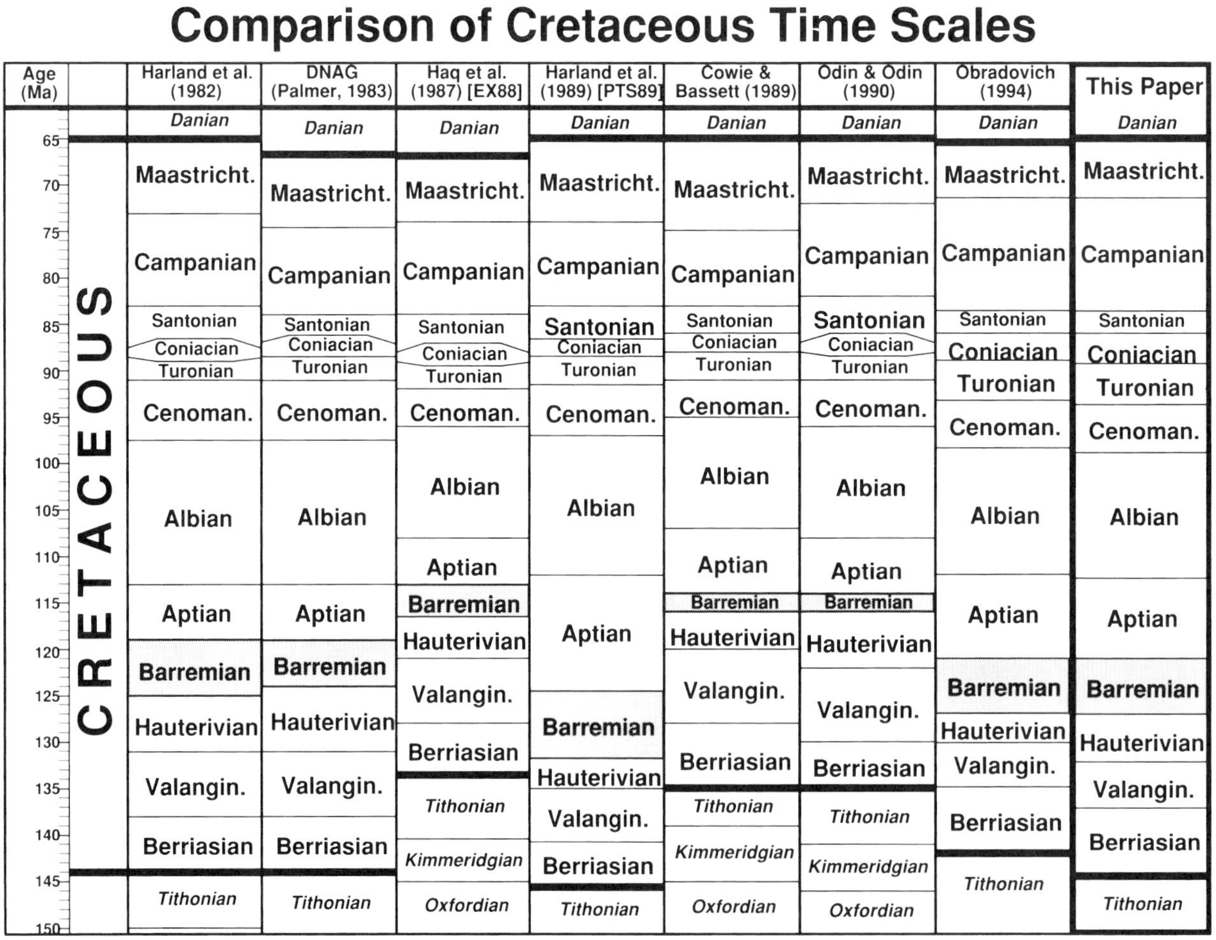

Comparison of Cretaceous Time Scales

Age (Ma)	Harland et al. (1982)	DNAG (Palmer, 1983)	Haq et al. (1987) [EX88]	Harland et al. (1989) [PTS89]	Cowie & Bassett (1989)	Odin & Odin (1990)	Obradovich (1994)	This Paper
65	Danian	Danian	Danian	Danian	Danian	Danian	Danian	Danian
70	Maastricht.	Maastricht.	Maastricht.	Maastricht.	Maastricht.	Maastricht.	Maastricht.	Maastricht.
75	Campanian	Campanian	Campanian	Campanian	Campanian	Campanian	Campanian	Campanian
80								
85	Santonian	Santonian	Santonian	Santonian	Santonian	Santonian	Santonian	Santonian
	Coniacian	Coniacian	Coniacian	Coniacian	Coniacian	Coniacian	Coniacian	Coniacian
90	Turonian	Turonian	Turonian	Turonian	Turonian	Turonian	Turonian	Turonian
95	Cenoman.	Cenoman.	Cenoman.	Cenoman.	Cenoman.	Cenoman.	Cenoman.	Cenoman.
100			Albian		Albian	Albian		Albian
105	Albian	Albian		Albian			Albian	
110			Aptian		Aptian	Aptian		
115	Aptian	Aptian	Barremian	Aptian	Barremian	Barremian	Aptian	Aptian
120			Hauterivian		Hauterivian	Hauterivian		
125	Barremian	Barremian	Valangin.		Valangin.	Valangin.	Barremian	Barremian
130	Hauterivian	Hauterivian	Berriasian	Barremian	Berriasian	Berriasian	Hauterivian	Hauterivian
135	Valangin.	Valangin.	Tithonian	Hauterivian	Tithonian	Tithonian	Valangin.	Valangin.
140	Berriasian	Berriasian	Kimmeridgian	Valangin.	Kimmeridgian	Kimmeridgian	Berriasian	Berriasian
145				Berriasian			Tithonian	
150	Tithonian	Tithonian	Oxfordian	Tithonian	Oxfordian	Oxfordian		Tithonian

FIG. 1.—Comparison of Cretaceous time scales.

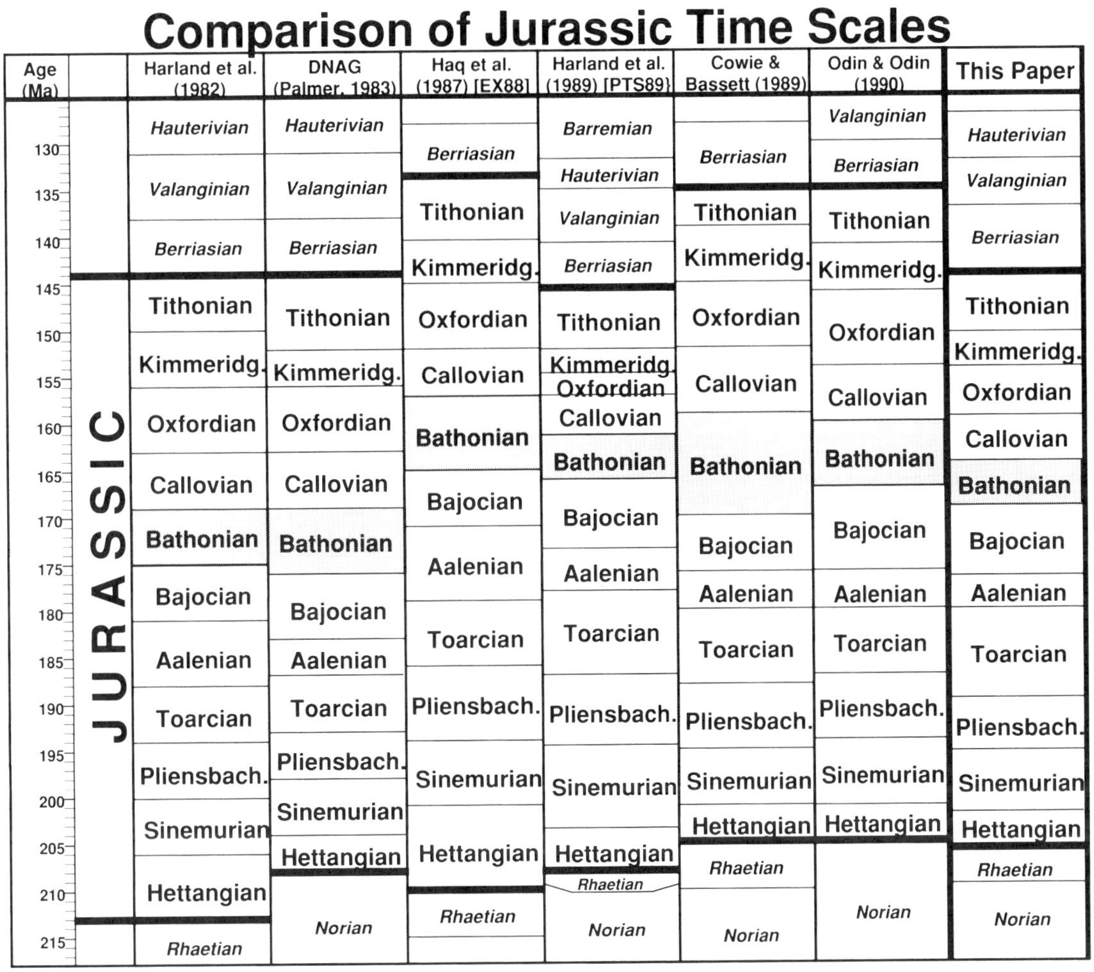

Comparison of Jurassic Time Scales

Age (Ma)	Harland et al. (1982)	DNAG (Palmer, 1983)	Haq et al. (1987) [EX88]	Harland et al. (1989) [PTS89]	Cowie & Bassett (1989)	Odin & Odin (1990)	This Paper
130	Hauterivian	Hauterivian	Berriasian	Barremian	Berriasian	Valanginian	Hauterivian
135	Valanginian	Valanginian		Hauterivian		Berriasian	Valanginian
			Tithonian	Valanginian	Tithonian	Tithonian	
140	Berriasian	Berriasian		Berriasian			Berriasian
			Kimmeridg.		Kimmeridg.	Kimmeridg.	
145	Tithonian	Tithonian	Oxfordian	Tithonian	Oxfordian	Oxfordian	Tithonian
150							
	Kimmeridg.	Kimmeridg.	Callovian	Kimmeridg.	Callovian	Callovian	Kimmeridg.
155				Oxfordian			Oxfordian
	Oxfordian	Oxfordian	Bathonian	Callovian		Bathonian	
160					Bathonian		Callovian
165	Callovian	Callovian	Bajocian	Bathonian		Bajocian	Bathonian
170	Bathonian	Bathonian		Bajocian	Bajocian		Bajocian
175	Bajocian	Bajocian	Aalenian	Aalenian	Aalenian	Aalenian	Aalenian
180	Aalenian	Aalenian	Toarcian	Toarcian	Toarcian	Toarcian	Toarcian
185							
190	Toarcian	Toarcian	Pliensbach.	Pliensbach.	Pliensbach.	Pliensbach.	Pliensbach.
195	Pliensbach.	Pliensbach.	Sinemurian	Sinemurian	Sinemurian	Sinemurian	Sinemurian
200	Sinemurian	Sinemurian			Hettangian	Hettangian	Hettangian
205		Hettangian	Hettangian	Hettangian	Rhaetian		Rhaetian
210	Hettangian	Norian	Rhaetian	Rhaetian	Norian	Norian	Norian
215	Rhaetian			Norian			

Fig. 2.—Comparison of Jurassic time scales.

PREVIOUS MESOZOIC TIME SCALES

During the last two decades, the Mesozoic geologic time scale has undergone various improvements. The stratigraphic calibration of the Hawaiian lineation spreading profile (Larson and Hilde, 1975) was adapted for scaling of geologic stages by Larson and Hilde (1975), DNAG, and OB93. Databases of radiometric ages have been statistically analyzed with various best-fit methods to estimate ages of stage boundaries (PTS82 and PTS89). Comparisons in duration of stages and age of stage boundaries for selected Mesozoic geologic time scales are documented in Figures 1, 2 and 3.

Among widely used scales, PTS89 stands out because of its authoritative radiometric database, developed and expanded since Armstrong (1978) and PTS82, its elegant use of the chronogram method for unbiased age interpolation, and its efforts to provide biochronologic underpinning. Unfortunately, no error bars on stage boundary age estimates were calculated, but its balanced approach in dealing with glauconite versus high-temperature age dates has provided a measure of stability for age estimates on Early Cretaceous and Jurassic stage boundaries. PTS89 takes glauconite-based age estimates as minimum ages for the age calibration of Mesozoic stratigraphic boundaries older than 115 Ma. Indeed, many Jurassic-Cretaceous glauco-nite ages are significantly younger than their high-temperature counterparts (e.g., Gradstein and others, 1988). PTS89 relied heavily on chronograms (see below) and smoothing of its results in a plot of chrons versus ages, where chrons are a substitute for biozones per stage unit. For example, the Jurassic/Cretaceous boundary is between 13 and 17 my older using high-temperature minerals rather than glauconite dates (PTS89, and Table 1).

If we take a brief look at changes in method philosophy among the three key scales available to date for the whole Mesozoic Era (i.e., PTS89, Odin and Odin, 1990, and EX88), we see a greater reliance on direct linkage of low- and high-temperature radiometric dates with stratigraphic levels by Odin and Odin (1990) and more reliance on interpolation methods with the other two scales. Ideally, precise dating at distinct stratigraphic boundaries is preferred; however, such precision is scarce below the Upper Cretaceous Period. EX88 largely used glauconites, which explains why the Jurasssic/Cretaceous boundary in EX88 is at 131 Ma and in PTS89 at 145 Ma. EX88 employed a composite of several spreading profiles in different oceans to scale the Late Jurassic through Early Cretaceous time scales and also used the zones versus ages plots for further smoothing of results. It is not clear if the composite spreading profile in fact increases reliability of the interpolation.

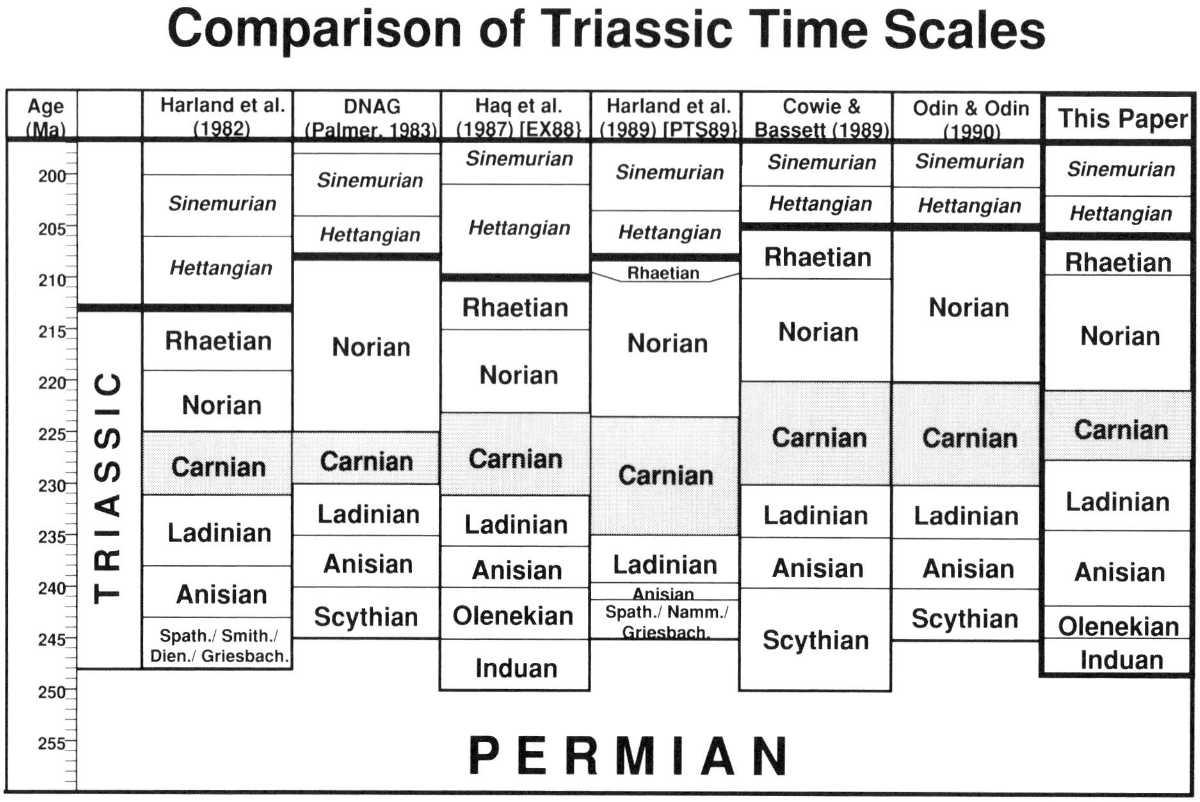

Comparison of Triassic Time Scales

Age (Ma)		Harland et al. (1982)	DNAG (Palmer, 1983)	Haq et al. (1987) [EX88]	Harland et al. (1989) [PTS89]	Cowie & Bassett (1989)	Odin & Odin (1990)	This Paper
200			Sinemurian	Sinemurian	Sinemurian	Sinemurian	Sinemurian	Sinemurian
205		Sinemurian	Hettangian	Hettangian	Hettangian	Hettangian	Hettangian	Hettangian
210		Hettangian			Rhaetian	Rhaetian		Rhaetian
215		Rhaetian	Norian	Rhaetian	Norian	Norian	Norian	Norian
220		Norian		Norian				
225		Carnian	Carnian	Carnian	Carnian	Carnian	Carnian	Carnian
230								
235		Ladinian	Ladinian	Ladinian		Ladinian	Ladinian	Ladinian
240		Anisian	Anisian	Anisian	Ladinian	Anisian	Anisian	Anisian
245		Spath./ Smith./ Dien./ Griesbach.	Scythian	Olenekian	Anisian Spath./ Namm./ Griesbach.	Scythian	Scythian	Olenekian
250				Induan				Induan
255				**PERMIAN**				

Fig. 3.—Comparison of Triassic time scales.

EX88, KG85, NDS82, and van Hinte (1976) relied on biochronology to interpolate the duration of Jurassic stages. The theory assumes that the numerous, short-lived ammonite zones for the Jurassic Period have approximately equal duration, the result of regular evolutionary change. Assuming age tiepoints at the top and bottom of Jurassic Period, the average duration of each zone is estimated (approximately 1 my). From the number of zones per stage, the duration of that stage was estimated. KG85 also took into account some intra-Jurassic control points. The method is not necessarily accurate, but it is detailed, because the number of standard ammonite zones is relatively large. To minimize taxonomic bias, Westermann (1988) compiled independent sets of Jurassic zones and subzones from the different faunal provinces. Subzones were calibrated as 75% of the undivided length of zonal units and on average lasted 0.45 my.

The advent of $^{40}Ar/^{39}Ar$ radiometric dates for a significant part of the U.S. Western Interior Cretaceous (OB93), coupled to a detailed ammonite biostratigraphy (Cobban and others, 1994), is another significant improvement of late Mesozoic geochronology.

A Global Stratigraphic Chart was compiled under the auspices of the International Union of Geological Sciences (IUGS) by Cowie and Bassett (1989) incorporating scales from the stratigraphic working groups of IUGS. The chart reflects much of current stratigraphic usage. The Bureau de Recherches Géologiques et Minières and the Société Géologique de France published a stratigraphic scale and time scale compiled by Odin and Odin (1990). Of the more than 110 defined boundaries, 20 lack radiometric data.

A major difference among time scales, for example, is the duration of the Aalenian Stage, using either KG85, EX88 or PTS89. Admittedly, there is a paucity of reliable radiometric dates in the middle Jurassic (Appendices 1, 2), but Jurassic standard ammonite zonations (e.g., Westermann, 1984, 1988; PTS89) indicate that the adjacent Toarcian and Bajocian Stages have double the number of ammonite subzones over the Aalenian, which casts doubt on the (unusually) long Aalenian Stage in EX88. As pointed out by Westermann (1984, 1988), the relative brevity of the Aalenian Stage in Europe has caused many debates. Despite the fact that the uncertainties on the ages of the Aalenian Stage boundaries exceed the estimate of the stage duration (see below), we predict with confidence that the Aalenian Stage is significantly shorter than the underlying Toarcian and the overlying Bajocian stage.

DATA AND STEPS IN TIME SCALE ANALYSIS

As explained earlier, no single unifying concept is available for Mesozoic time scale analysis and interpolation. In order to achieve a measure of objectivity and standardization, we combined several geological and statistical methods, while trying to assess and preserve estimates of analytical and stratigraphic uncertainty in the data. Statistics played a more major role for the parts of the Mesozoic time scale where age control is limited.

TABLE 1.—COMPARISON OF MAXIMUM LIKELIHOOD ESTIMATES

Stage	Parabola Peak Estimates[a]	Local Maximum Obtained by Scoring Method			Probability[b]
		High-Temperature Dates	Low-Temperature Dates	Combined High and Low Temperature	
Maa	66.06 (0.41)	65.98 (0.69)	66.18 (0.51)	6.11 (0.40)	0.145
Cam	71.58 (0.42)	71.99 (1.38)	71.96 (0.63)	71.96 (0.58)	0.508
San	82.54 (0.21)	82.32 (1.08)	83.12 (1.30)	82.61 (0.92)	0.320
Con	86.04 (0.26)	86.51 (0.59)	85.02 (0.73)	85.40 (0.58)	0.940
Tur	89.11 (0.27)	89.33 (1.26)	87.73 (0.44)	88.31 (0.28)	0.884
Cen	93.26 (0.20)	93.40 (0.16)	90.24 (0.41)	92.73 (0.18)	1.000
Alb	100.48 (0.34)	98.94 (0.30)[c]	96.62 (0.46)	98.41 (0.15)	0.992
Apt	108.33 (0.69)	112.18 (0.87)[c]	109.43 (0.62)	110.15 (0.41)	0.995
Bar	120.26 (0.74)[c]	121.43 (20.48)[c]	118.52 (0.98)	118.53 (0.95)	0.991
Hau	124.96 (0.53)	127.11 (0.88)[c]	118.52 (0.97)	125.01 (1.39)	1.000
Val	133.90 (0.93)	135.07 (2.59)	126.05 (1.64)	128.71 (1.59)	0.998
Ber	133.73 (0.95)	136.43 (0.96)[c]	129.40 (1.87)	132.37 (1.16)	1.000
Tit	147.75 (3.66)	148.77 (3.10)	130.75 (1.60)	136.36 (0.38)	1.000
Kim	145.72 (1.71)	149.81 (3.83)	139.32 (0.97)	139.99 (0.89)	0.996
Oxf	156.83 (0.69)	155.09 (1.80)[c]	147.26 (1.31)	153.42 (0.25)	1.000
Cal	161.85 (1.01)	159.05 (2.32)[c]	no solution	159.28 (2.25)	
Bat	163.27 (2.07)	164.28 (2.60)[c]	no solution	164.32 (2.56)	
Baj	165.22 (1.90)	166.31 (2.37)[c]	no solution	166.32 (2.40)	
Aal	178.10 (2.69)	177.32 (4.26)	no solution	178.18 (3.64)	
Toa	177.96 (2.81)	177.32 (4.10)	no solution	178.63 (3.38)	
Pli	181.05 (6.61)	180.15 (10.89)	187.82 (20.47)	182.24 (7.57)	0.370
Sin	196.30 (1.01)	196.47 (2.23)[c]	no solution	196.35 (2.29)	
Het	197.92 (0.70)	200.57 (0.67)[c]	no solution	200.55 (0.67)	
Rha	209.06 (1.65)	208.66 (3.32)[c]	no solution	208.69 (3.16)	
Nor	211.49 (2.61)	209.76 (4.06)[c]	no solution	210.76 (3.74)	
Car	223.67 (3.02)	223.44 (6.13)[c]	no solution	223.48 (5.68)	
Lad	233.89 (2.22)	233.69 (2.61)[c]	no solution	233.69 (2.59)	
Ani	236.55 (1.73)	236.51 (1.96)[c]	239.51 (8.63)	236.77 (1.77)	0.367
Ole	242.38 (2.81)	242.27 (3.54)[c]	239.51 (9.05)	241.99 (3.32)	0.612
Tat	243.63 (2.22)	244.75 (4.06)[c]	no solution	244.75 (4.06)	

Standard deviations in parentheses. For top of Barremian, standard deviation selected is parabola peak estimate instead of scoring result because of exceptionally flat local maximum.

[a] Based on wider-range, high-temperature dates.

[b] Probability that low-temperature estimate for low-temperature dates is younger than the high-temperature estimate of local maximum obtained by scoring method high-temperature dates.

[c] Results used in Table 2.

Distribution of Radiometric Dates

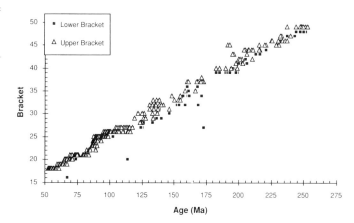

FIG. 4.—Distribution of radiometric dates. The bracket refers to the relative age of sediments immediately above and below the dated level.

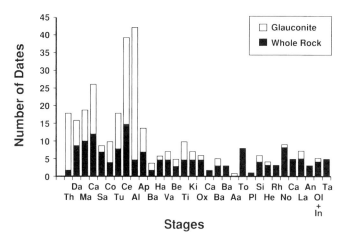

FIG. 5.—Radiometric dates by type and by stage.

The first step to build the Mesozoic time scale was to assess the age of key stratigraphic boundaries. Next, a database was built of selected radiometric estimates throughout the Mesozoic that are stratigraphically closely bracketed, analytically standardized and reasonably documented (Figs. 4, 5; Appendix 1). The majority of items were taken from the detailed list in PTS89 (their Table 4.2) that screened dates on excessive analytical uncertainty and anomalous stratigraphic position. The selection of Mesozoic dates in Appendix B of EX88, which form the basis of that time scale, was also reassessed, and the weighted means and standard errors recalculated (Appendix 1). Over 65 new and stratigraphically significant Triassic, Jurassic, and Cretaceous dates published since PTS89 and EX88 were added to the data file. The ^{40}Ar/^{39}Ar ages were recomputed to be in accordance with a 520.4 Ma age for the MMhb-1 laboratory standard.

The time scale proposed in this study uses only high-temperature radiometric age assignments. Low-temperature dates are significantly younger for a number of stage boundaries where there are many low- and high-temperature dates (Table 1); this is a good reason for not using low-temperature dates, of which there are few in pre-Callovian rocks (Fig. 5).

Next, we interpolated the Maastrichtian through Cenomanian analytical data from OB93, in combination with the new esti-

mate for the K/T boundary to arrive at dates with narrow uncertainty limits for the Late Cretaceous stage boundaries. Aptian and Albian boundary ages were calculated with maximum likelihood estimation. The Barremian through Oxfordian interval was age calibrated with the Hawaiian seafloor spreading profile and then averaged with the best estimates from maximum likelihood age estimation. The ages of pre-Oxfordian Stage boundaries come from maximum likelihood estimation; poor estimates for the Pliensbachian and Anisian/Ladinian Stage boundaries were constrained with single, well-calibrated direct age estimates from the literature.

The resulting age estimates for Triassic and Early-Middle Jurassic stage boundaries were smoothed with a spline to achieve statistical improvements for age estimates of stage boundaries with larger uncertainty. This spline scaling incorporated a suite of detailed ammonite zonations for the Tethyan and Boreal faunal realms.

Each difference between ages of two successive stage boundaries provides an estimate of the duration of the enclosed stage. It is inversely proportional to average sedimentation rate for this stage. The uncertainty in relative duration of each individual stage is much less than the uncertainties on the ages of the stage boundaries. It is not possible to estimate standard deviations for the durations because of lack of information on serial correlation of the stage boundary age estimates. As shown in the last column of Table 1, the ratio of stage duration estimates for this Mesozoic time scale and PTS89 fluctuates between 0.2 and 1.7. Our duration estimates are relatively precise. Nevertheless, the large differences between results based on successive time scales suggest that the Mesozoic time scale needs further improvement, especially if it is to be used for basin modelling involving sedimentation rates.

Three statistical methods were used repeatedly during the construction of the Mesozoic time scale: (1) maximum likelihood estimation to combine stratigraphic with geochronological uncertainty for individual stage boundaries; (2) averaging of statistically independent estimates for the same stage boundary; and (3) spline-curve fitting for adjustment of individual stage boundary estimates using variable precision estimates for adjacent stages. These procedures will be discussed in the following sections.

AGE ESTIMATES FOR SELECTED STRATIGRAPHIC BOUNDARIES

Below, we outline differing age assignments for key Mesozoic stratigraphic boundaries and their age estimates entered in calculation of the Mesozoic time scale.

Permian/Triassic Boundary

Estimates for the age of the Permian/Triassic boundary vary between 248 Ma in EX88 and 245 Ma in PTS89. As pointed out by these authors, the available data were such that the age could be placed in a window between 255/260 and 240 Ma. Recently, new evidence has become available to better constrain the age of this boundary between Paleozoic and Mesozoic strata.

Claoue-Long and others (1991) used ion microprobe dating of zircons on a bentonite in the Chinese stratotype section through the Permian-Triassic boundary (Item 319 in Appendix 1). The bentonite is 5 cm thick and occurs at the boundary, between the uppermost Permian Changxing Formation and the overlying Mixed Fauna Bed 1 of the Chinglung Formation with (?) *Otoceras* ammonites, and below the Mixed Fauna Bed 2 with *Hindeodus parvus* conodonts of definite earliest Triassic age. The Pb/Th/U radiometric age for the boundary bed is 251.2 ± 3.4 Ma (2σ). Since the *Otoceras* identifications are uncertain, the age may be a maximum one for the boundary (E. T. Tozer, as cited by Claoue-Long and others (1991)). The mixed fauna beds indicate slight reworking, which may also pertain to the zircon in the bentonite. Nakazawa (1992) suggests a latest Permian age for the boundary clay; on the basis of the occurrence of Permian conodonts, Yin (1993) supports a latest Permian age of this horizon at the Meishan locality. Hence, the boundary bed may actually be slightly older than the Permian/Triassic boundary in less condensed sections.

New data have become available using sanidine from the same rock sample in the bentonite bed (Zhang Zichao, as cited by Odin, 1992b; Item 320 in Appendix 1). Ar/Ar dates average at 255.9 ± 5.6 Ma (2σ). The Rb/Sr isochron age on the same

sanidine is 250 ± 6 Ma (2σ). The Ar/Ar date uses a standard different from MMhb-1 at 520.4 Ma.

Further radiometric evidence for the age of the Permian/Triassic boundary comes from Ar/Ar and K/Ar dates on flood basalts in Siberia at the Permian-Triassic boundary (Renne and Basu, 1991). Detailed sampling and analysis resolve that virtually all flood basalt erupted between 248.3 ± 0.3 and 247.5 ± 0.7 Ma (Items 317 and 318 in Appendix 1). Following Nalivkin (1973), the Siberian traps straddle the Permian/Triassic boundary with Tatarian ostracod clays interbedded in its lower tuffaceous part and Early Triassic Estheria molluscs in shales of the upper tuffs.

If we consider the Chinese dates to be maximum ones, which overlap at the 95% confidence level with the Siberian flood basalt dates (which admittedly are biostratigraphically less well constrained), a 248 Ma age for the Permian/Triassic boundary is reasonable. This age is a best estimate for the onset of the major volcanic activity that may have associated the major chemical and biotic changes that separate the Paleozoic from the Mesozoic Era (e.g., Holser and Magaritz, 1987; Holser and others, 1991).

Ladinian/Anisian Boundary

An important anchor point in the Triassic Period is dating of the Ladinian/Anisian boundary in the Grenzbitumen horizon in Switzerland (Hellmann and Lippolt, 1981; Forster and Warrington, 1985). Both K/Ar and Ar/Ar dates on alkali feldspars were done from tuff layers in the basal part of the lowermost Ladinian *Nevadites* (tethyan) ammonite Zone (Brack and Rieber, 1993). The best dates are derived from homogenous and clear, high-sanidine feldspars (type G), which average at 232 ± 9 Ma (2σ) for K/Ar and 233 ± 7 Ma (2σ) for Ar/Ar, from which Hellmann and Lippolt (1981) estimate the boundary to be 232 ± 9 Ma (2σ) (Items 309 and 310 in Appendix 1). Odin (1982) suggests 232.7 ± 4.5 Ma (2σ) for K/Ar and 232 Ma plateau age for Ar/Ar in the waterclear sanidine feldspars (NDS196).

Triassic/Jurassic Boundary

The current working definition of the Triassic/Jurassic boundary is the migration influx of the ammonite *Planorbis* observed in the Blue Lias of the United Kingdom. Age estimates for the Triassic/Jurassic boundary range from 205 Ma to 213 Ma (Fig. 3).

EX88 quote an age of 206.0 ± 12.0 Ma for the Hettangian/Sinemurian boundary, using data referred to NDS82 listing 202. NDS82 quote 190–200 Ma as a most likely age bracket for these dates from the West Rock and Mt. Carmel sills, Hartford Basin, eastern United States. PTS89 uses an age of 198.50 ± 5.78 Ma referred to the Hettangian-Sinemurian Stages, for the same item NDS202. Forster and Warrington (1985) give a detailed account of all the dates and errors and use 196 ± 4.5 Ma for the Palisades Sills and 195 ± 4.2 Ma for the Mt. Carmel sills (in agreement with NDS202, p. 878). The relative age using palynofloral evidence, is quoted as early Liassic, which unfortunately is broad. We follow Forster and Warrington's (1985) use of the data and assign the 195 and 196 Ma estimates (Appendix 1, items 278 and 279) an Hettangian-Sinemurian age.

The 196 Ma age for Hettangian-Sinemurian strata above the Triassic/Jurassic boundary agrees with the average U/Pb age of 194.12 ± 0.62 Ma (1σ) on zircons in the Cold Fish volcanics, British Columbia (Thorkelson and others, 1991, 1994; J. Mortensen, pers. commun., 1993; item 276, Appendix 1). Ammonites in sedimentary beds alternating with the volcanics include *Miltoceras* sp., *Tropidoceras* sp., *Metaderoceras* sp. aff. *talkeetnaense, Acanthopleuroceras* sp. aff *A. stahli*, and *Metaderoceras evolutum* and assign the Cold Fish volcanics an early to middle Carixian (Early Pliensbachian) age (Thomson and Smith, 1992).

The post Triassic/Jurassic boundary ages of 196 and 194 Ma are consistent with 201–202 Ma as an approximate estimate for the Triassic/Jurassic boundary in eastern North America. The latter estimate derives from U/Pb zircon dating of the North Mountain basalt, Fundy Basin at 202 ± 1 Ma (1σ) and U/Pb and Ar/Ar ages of 201 ± 1 Ma (1σ) for correlative intrusives in the Newark Basin (Hodych and Dunning, 1992; Fowell and Olsen, 1993; Appendix 1, items 285 and 286).

However, interpretation of paleontologic criteria for correlation above and below the Triassic/Jurassic boundary to the basalt levels is far from straightforward. Post-Sinemurian *Callialasporites* is recorded above the lower basalts. Palynological assemblages at 30 m below the Jacksonwald Basalt in the Exeter Village section (Newark basin) contain predominantly vesicate pollen (*Vallasporites ignacii, Patinasporites densus, Enzonalasporites* spp.), referred to as latest Triassic age (Cornet and Olsen, 1985, Fowell and Olsen, 1993). The criterium for the Triassic/Jurassic boundary in the Newark basins is the preponderance of the pollen *Corollina meyeriana*, occurring 10 m below the lower Newark Basin basalts (Fowell and Olsen, 1993). Palynofloras in the basal Liassic Planorbis beds in southern England are also dominated by *Corollina* (e.g., Orbell, 1983). On the basis of Milankovitch cyclicity criteria, Fowell and Olsen (1993) suggest (a) that the disappearance of vesicate pollen predates the *Corollina meyeriana* dominated assemblages by a maximum of 21,000 years, and (b) that the palynological placement of the Triassic/Jurassic boundary predates the Jacksonwald Basalt by an additional 21,000 years. Yet, in Alpine Europe and in contemporaneous sections in Western Europe, vesicate pollen (e.g., *Vallasporites ignacii, Enzonalasporites* sp. div.) disappear at the base of the *suessi* ammonoid zone (Warrington, 1974; Mostler and others, 1978; Morbey, 1978; Schuurman, 1979), probably below the base of the Rhaetian, approximately 3–4 my before the Triassic/Jurassic boundary. This suggests that Milankovitch cyclicity may not be applicable to approximate the time span between the disappearance of vesicate pollen, the palynologically assigned Triassic/Jurassic boundary and the onset of basalt deposition in the Newark Basin.

The 201–202 Ma dates from the lower basalts of the Newark basins are assigned a Hettangian age (Items 285 and 286 in Appendix 1).

Oxfordian/Kimmeridgian Boundary

Dates near the Oxfordian/Kimmerdigian boundary anchor the beginning of the Hawaiian sequence of marine magnetic anomalies in magnetochronology.

U/Pb dating of zircons from igneous rocks pre- and post-Galice Formation (see discussion in Harper, 1984) of northern California and southwestern Oregon bracket the *Buchia concentrica* mollusc Assemblage between 150.5 ± 2.0 and 157 ± 2.0 Ma, mid-middle Oxfordian to mid-late Kimmeridgian age (Appendix 1, items 244 and 253). This also constrains the Oxfordian/Kimmeridgian boundary to fall somewhere in this age bracket, with an average estimate near 154 Ma.

Pessagno and Blome (1990) combined new radiolaria and ammonite data in the lower level of the (basal) Galice Formation, Oregon, USA with *Buchia concentrica, Dichotosphinctes*, and *Miritusus*. This assemblage is assigned to the mid-Middle Oxfordian and only slightly younger than a level in the uppermost Rogue Formation with dacitic tuff breccia dated with U/Pb on zircons as 157.00 ± 2.00 Ma (Saleeby, 1984; item 253 in Appendix 1). By extrapolation, assuming a duration of 3 my for the younger Oxfordian, the Oxfordian/Kimmeridgian boundary is estimated to be 154 Ma.

Magnetic polarity chron M25 is at the Oxfordian/Kimmeridgian boundary in ammonite-zoned sections in Spain (Ogg and others, 1984). Ocean crust of marine magnetic anomaly M26r (Late Oxfordian) at Ocean Drilling Program Site 765, Argo Abyssal Plain, has an Ar/Ar age of 155.3 ± 3.4 Ma (Ludden, 1992; Item 251 in Appendix 1).

Jurassic/Cretaceous Boundary and Berriasian/Valanginian Boundary

The Tithonian/Berriasian Stage boundary or Jurassic/Cretaceous boundary does not have an internationally accepted definition, although several possible biostratigraphic markers have been correlated to magnetostratigraphy (Ogg and others, 1991a). A placement of the Jurassic/Cretaceous boundary at the top of the *Berriasella jacobi* ammonite Zone corresponds approximately to the base of magnetic polarity chron M18r (Ogg and Lowrie, 1986).

An alternate placement of the Jurassic/Cretaceous boundary at the base of this *B. jacobi* Zone (top of *Durangites* ammonite Zone) is used in our compilation in accordance with recent proposed boundary definitions (J. Remane, pers. commun., 1993).

Previous time scales have generally not specified the definition of the Jurassic/Cretaceous boundary and have proposed boundary ages ranging from 133 Ma to 146 Ma (Fig. 2). A contributing factor is the younger definition of the Jurassic/Cretaceous boundary used within the Boreal realm. The top of the regional Portland Stage of "uppermost Jurassic" strata is approximately equivalent to the top of the regional Volgian Stage, and these levels correlate approximately to the middle of the Berriasian Stage of the Tethyan realm (Mesezhnikov, 1988; Ogg and others, 1991a; Ph. Hoedemaeker, pers. commun., 1994). These regional Boreal stages are terminated by a significant unconformity, and the base of the overlying regional Ryazinian Stage of "lowest Cretaceous" may locally be equivalent to the upper Berriasian in the Tethyan realm.

The age of the Berriasian Stage is partially constrained by a U/Pb date of 137.1 ± 0.6 Ma from volcanic horizons in the Grindstone Creek section in northern California (Bralower and others, 1990; Appendix 1, item 232). Bralower and others (1990) assigned this level to late Middle Berriasian Age according to nannofossil assemblage ranges, correlated the level to polarity chron M16 and projected the age of the Jurassic/

Cretaceous boundary to be 141 Ma. This radiometric date and assigned stratigraphic age was a key control in the Early Cretaceous portion of the OB93 time scale (Obradovitch, 1993; Fig. 1). However, the age assignment from the mollusc and nannofossil assemblages may be less precise than published (Ph. Hoedemaeker and E. Erba, pers. communs., 1993), and a stratigraphic age as young as earliest Valanginian Stage is possible. Nevertheless, we maintained the Berriasian Age assignment for this 137.1 Ma date. Therefore, its high-precision error limit constrains the Berriasian/Valanginian boundary to be not older than 137 Ma. The Berriasian/Valanginian boundary has been correlated to polarity chron M15n.4 (Ogg and others, 1988). The relative stratigraphic timing of an event within the polarity zone is denoted by using the appropriate fractional subdivision within the zone (modified from Hallam and others, 1985; Ogg and others, 1991a). Therefore, an event at "C33n.85" implies that 85% of normal-polarity chron C33n precedes the event. However, CK92 used this notation in opposite sense; therefore, C33n.15 in their notation denotes that 15% of polarity chron C33n followed the event.

Barremian/Aptian Boundary

Dates near the Barremian/Aptian boundary anchor the youngest part of the Hawaiian sequence of marine magnetic anomalies in magnetochronology. Magnetic polarity chron M0r occurs in the Early Aptian *Chiastozyus litterarius* calcareous nannofossil Zone. The base of polarity zone M0r may coincide with the Barremian/Aptian Stage boundary (E. Erba, pers. commun., 1993).

Pringle (1995) obtains an Ar/Ar date at the top of polarity zone M1r in seamount basalts of MIT Guyot of 123.51 ± 0.5 Ma (1σ), which implies an age of approximately 121 Ma for the base of polarity chron M0r and the Barremian/Aptian boundary (Appendix 1, item 213).

Campanian/Maastrichtian Boundary

The former calibration of microfossil datums and magnetic polarity chrons to the macrofossil placement of the Campanian/ Maastrichtian boundary has been revised upward.

The Campanian/Maastrichtian boundary was correlated to the top of the *Globotruncana calcarata* foraminiferal Zone in the Gubbio Section, Italy, and this datum occurs in the upper part of polarity zone C33n (C33n.85) (Alvarez and others, 1977).

This foraminiferal datum may be correlated to bentonites in the U.S. Western Interior, yielding K/Ar and Ar/Ar dates of 74.5 Ma, thereby providing a tie point for the Late Cretaceous magnetic polarity scale (Cande and Kent, 1992).

The base of the Maastrichtian is close to the appearance of *Belemnella lanceolata*, a well-defined belemnite datum in the boreal realm (Birkelund and others, 1984). In Kronsmoor, Germany, a suitable boundary stratotype for the Campanian/Maastrichtian boundary, *Belemnella lanceolata*, appears 3.5–5 m below *Hoploscaphites constrictus*, an ammonite datum also used for recognition of the Campanian/Maastrichtian boundary in the boreal realm.

Kennedy and others (1992) and McArthur and others (1993) have established detailed macrofossil and strontium isotope correlations, respectively, between the Kronsmoor section, the

Campanian/Maastrichtian boundary in the English Chalk and the Upper Campanian/Lower Maastrichtian of the U.S. Western Interior. Kennedy and others (1992) found that the top of the *Globotruncana calcarata* foraminiferal Zone occurs significantly older than the macrofossil assignment of the Campanian/ Maastrichtian boundary. These studies also correlate the Campanian/Maastrichtian boundary in Kronsmoor to a level within the *Baculites jenseni* ammonite Zone or possibly the overlying *Baculites eliasi* Zone in the United States Western Interior. The *Baculites jenseni* Zone is in a series of eight *Baculites* zones spanning the Upper Campanian-Lower Maastrichtian of the Western Interior (Obradovich, 1988). The isotope correlations suggest a moderate uniformity of zonal duration from the *B. compressus* zone to the *Baculites grandis* Zone. Obradovich (1988) gives K/Ar ages for bentonites in the *B. grandis* Zone of 70.1 ± 0.7 Ma and for the *B. compressus* Zone of 73.2 ± 0.7 Ma. A linear interpolation results in an estimate for the Campanian/Maastrichtian boundary of 71.6 ± 0.7 Ma. This estimate closely matches that of PTS89 based on chronogram interpolations of 40 Maastrichtian and Campanian radiometric age dates and is the preferred one in our analysis (items 52, 53 and 56 in Appendix 1).

Therefore, the CK92 age of 74.5 Ma for a level in the upper part of polarity zone C33n is satisfactory for magnetochronology but is not satisfactory for dating the Campanian/Maastrichtian boundary. Hicks (1993) correlated magnetic polarity zones C29n through C33r to ammonite zones in the United States Western Interior and interpolated ages for the magnetic reversals from the dated bentonite horizons (Obradovich, 1988, 1993). We combined these ages on the magnetic polarity chrons and associated error limits derived by Hicks (1993) with the spacing of magnetic anomalies in CK92 to obtain a best-fit, smoothed seafloor spreading curve and associated ages of the magnetic polarity chron boundaries.

Cretaceous/Paleogene Boundary

The Cretaceous/Paleogene (Cretaceous/Tertiary) or Maastrichtian/Danian boundary occurs in the upper portion of polarity chron C29r (approximately C29r.7), according to an average of 5 magnetostratigraphic sections (CK92). PTS89 obtained a chronogram age of 66 Ma for the K/T boundary, using 20 age dates for the Maastrichtian and 15 for the Danian. Obradovich (1988) arrived at the same 66 Ma value, based on a magnetostratigraphic section in Red Desert Valley, Alberta (item 31, Appendix 1). The 66 Ma age estimate was also supported by laser fusion ^{40}Ar/^{39}Ar dates (unpubl.) on single crystals of sanidine extracted from the iridium-bearing lower Z coal in Montana, which yielded a mean age of 66.1 Ma (C. Swisher, as cited by Berggren and others, 1992). However, these dates were later rejected as being analytically suspect (C. Swisher, pers. commun., 1993).

Our age of the Maastrichtian/Danian (Cretaceous/Paleogene) boundary combines the standardized Ar/Ar dates of 64.98 ± 0.05 Ma and 65.2 ± 0.4 Ma at the Chicxulub impact crater, with age estimates for the Arroyo el Mimbral (Mexico) and Beloc (Haiti) tektites of 65.07 ± 0.10 Ma and 65.01 ± 0.08 Ma (Swisher and others, 1992; Sharpton and others, 1992) to yield 65.00 ± 0.04 Ma (item 34 in Appendix 1).

MAXIMUM LIKELIHOOD ESTIMATION

The maximum likelihood method is suitable for estimation of the age of stage boundaries from a radiometric database if most rock samples used for age determination are subject to significant stratigraphic uncertainty. The method is particularly applicable where a group of radiometric dates with differing precision is available near, but not at, a stage boundary. It provides a way of combining the measurement errors of the dates with their stratigraphic uncertainty.

In the method of maximum likelihood, one or more parameters of a statistical population are estimated by maximizing the likelihood that a sample of observed values was drawn at random from the population. Each parameter to be estimated is considered as a variable in the equation of the frequency distribution for the population which has to be known beforehand. The frequency distribution of a maximum likelihood estimator itself converges to normal (Gaussian) form with increasing sample size.

A statistical model for the construction of chronograms was originally proposed by Cox and Dalrymple (1967). It is based on the following two assumptions: (1) the true ages of the rock samples subjected to radiometric age determination are uniformly distributed over time in the vicinity of the chronostratigraphic boundary of which the age is to be estimated (i.e., the number of dates per time interval remains constant on the average), and (2) each date is derived from a normal (Gaussian) distribution centered around the true age of the rock sample with the reported measurement error as its standard deviation. This model allows for a date to be inconsistent in two ways: a sample known to be stratigraphically younger than a given boundary may have a measured date that is older than the age of the boundary, and a stratigraphically older sample may produce a measured date that is younger.

PTS82 and PTS89 constructed chronograms in the following way. For a number of equally spaced test ages in the vicinity of a stage boundary, the sum of squares is computed for standardized differences between each test age and the dates that are inconsistent for it. Each difference is standardized by dividing it by the standard deviation of the date. The number of inconsistent dates for a test age increases when the test age becomes either significantly younger or older than the true age of the boundary that is to be estimated. The graph representing the relation between the sum of squares plotted in the vertical direction and the test age (horizontal axis) is a chronogram. Typically, a chronogram of this type resembles a basket with a relatively flat bottom around its minimum value which is selected as the best estimate of the age of the stage boundary. Updated chronograms were constructed for PTS89.

The preceding method can be improved by applying the method of maximum likelihood, using the consistent dates in addition to the inconsistent dates for each test date (Agterberg, 1988, 1990). If the two basic assumptions (uniform distribution of samples through time with superimposed Gaussian errors) hold true, the probability P_{it} that a date x_i with standard deviation σ_i differs from a given test age μ_t satisfies $P_{it} = 1 - \Phi(z_{it})$, where $z_{it} = (x_i - \mu_t)H_i/\sigma_i$ is the standardized date and Φ represents cumulative frequency of the normal (Gaussian) distribution in standard form. In the equation for z_{it}, H_i denotes stratigraphic relation between sample and boundary: $H_i = 1$ for

older samples and $H_i = -1$ for younger samples. The product of the probabilities, which may be written as πP_{it}, reaches its maximum for the test age which is closest to the true age.

The product of the probabilities also known as the likelihood function $L_t = \Pi P_{it}$ tends to become normal (Gaussian) as the number of independent observations increases. In general, the log-likelihood function $\log_e L_t = \Sigma \log_e P_{it}$, where Σ denotes summation, assumes a bell-shaped curve. This alternative form is more suitable for further calculations and is used for graphical representation in the log-likelihood graphs (see Fig. 6). A procedure commonly used is to locate the maximum of the log-likelihood function by means of the method of scoring (e.g., Rao, 1973) which involves a local, iterative search.

Inconsistent dates have negative values of z_{it} and result in values of P_{it} which are less than 0.5. For example, if an inconsistent date is three standard deviations away from a test age (i.e., $z_{it} = -3.0$ or $h_t - x_i = 3\sigma_i$), then $P_{it} = 0.00135$. A small value like this leads to a significant decrease in ΠP_{it}. On the other hand, consistent dates have positive values of z_{it} resulting in values of P_{it} which are greater than 0.5 but less than 1.0. For $z_{it} = 3.0$, $P_{it} = 0.99865$ which is nearly equal to 1.0. This implies that consistent dates more than three standard deviations away from a test age have almost no influence on the product ΠP_{it}. The standard deviations si are the measurement errors of the dates.

When the number of dates is large, the maximum likelihood method gives approximately the same results as the method used in PTS89. However, when there are relatively few dates, as in the pre-Aptian time interval, use of the consistent dates yields significantly better results. Moreover, the method provides a standard deviation for the best estimate. Locally, the end product of maximum likelihood resembles a Gaussian curve of which the standard deviation can be determined and used as the standard deviation of the best estimate.

The maximum likelihood method was used to estimate the ages of 30 Mesozoic stage boundaries from the Permian/Triassic to the Cretaceous/Paleogene boundary (Table 1). Triassic through Albian estimates obtained by scoring applied to high-temperature dates (highlighted in Table 1) were later combined with other types of estimates (Table 2) for estimation of the 31 stage boundaries of the Mesozoic Era.

Local Versus Wider-Range Log-Likelihood Maxima

Initially, a trial age was selected for each boundary in order to define the subset of dates from the Mesozoic radiometric database as well as the sequence of test ages to be used for that boundary. The ranges for the dates and trial ages for the stage boundaries were set as follows. In the Late Cretaceous Period, all dates within 10 my from a prior estimate of the age of the stage boundary were used, and the corresponding range of the test ages was set half as wide (trial age $\pm$ 5 my). For the older stages, these two windows were set at trial age $\pm$ 20 my and $\pm$ 10 my, respectively. The trial ages, listed in the database of Appendix 1, are our best subjective estimates for the age of Mesozoic stage boundaries.

Log-likelihood graphs were constructed for all 30 stage boundaries estimated (see Fig. 6). The method of scoring was applied to locate the maxima of the log-likelihood functions. These estimates (m) are shown in Figure 6 with their 95% con-

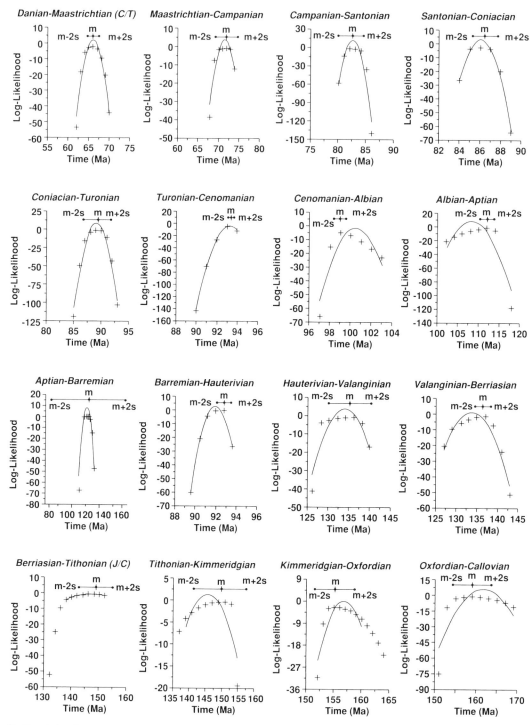

Fig. 6.—Log-likelihood plots for 30 Mesozoic stage boundaries (high-temperature dates). Log-likelihood is plotted against test ages in geological time. Lack of data prevented using maximum likelihood estimation for the Induan-Olenekian boundary. Largest log-likelihood value (m) and corresponding standard deviation (σ) found by method of scoring were used for approximate 95% confidence intervals (m ± 2σ). Parabolas were fitted to log-likelihood values for wider neighborhoods around prior estimates (trial ages). Each parabola provides other estimates: m_p (parabola peak) and σ_p (parabola standard deviation). Asymmetry of the log-likelihood function results in differences between m and mp; and $\sigma > \sigma_p$, if the likelihood function has a flatter peak than the parabola; in Table 1, σ is replaced by σ_p for the Barremian-Aptian boundary which is exceptionally flat. These types of discrepancies are primarily due to the influence of relatively few recent dates that are much more precise than the earlier dates.

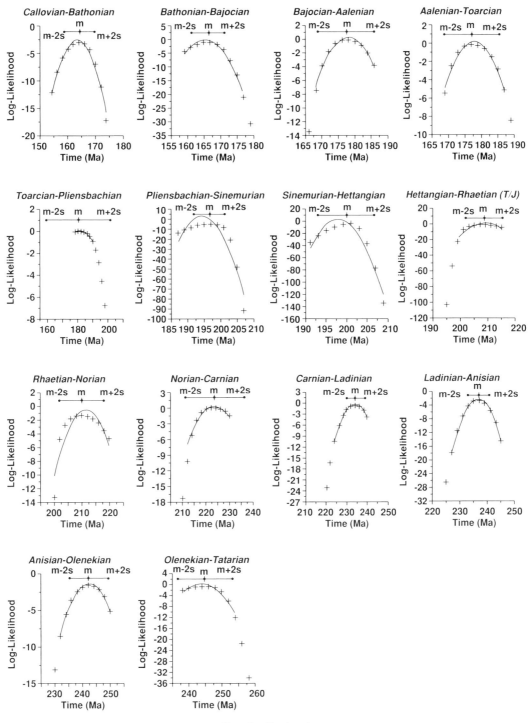

FIG. 6.—Continued.

fidence interval (m ± 2σ). Scoring is equivalent to constructing a parabola through three points over a very narrow range at the peak of the log-likelihood function. Additionally, parabolas were fitted to the likelihood values from wider ranges defined by the test ages by using the method of least squares. An advantage of this replacement of the calculated log-likelihood values by a continuous curve may be to reduce the effect of local (random) fluctuations. The peaks of the parabolas also can be used to estimate the stage boundary ages. In all cases scoring and parabola-fitting methods gave approximately the same age, but the corresponding standard deviations are only approximately equal if the parabola provides a good fit to the log-likelihood function.

In most graphs, the log-likelihood function has a flatter top than the best-fitting parabola and asymmetry occurs in some graphs (e.g., Hauterivian/Barremian, Valanginian/Hauterivian,

TABLE 2.—STEPWISE CALCULATION OF MESOZOIC TIME SCALE

Stage	Individual Age Estimates		Spline Computations		
	Initial[a]	Modified[b]	Ages[c]	Deviations[d]	Final Ages[e]
Maa	65.00 ± 0.04				65.0 ± 0.1
Cam	71.30 ± 0.25	71.29 ± 0.25			71.3 ± 0.5
San	83.50 ± 0.25	83.46 ± 0.24			83.5 ± 0.5
Con	86.30 ± 0.25	85.79 ± 0.23			85.8 ± 0.5
Tur	88.70 ± 0.25	88.96 ± 0.25			89.0 ± 0.5
Cen	93.30 ± 0.10	93.49 ± 0.08			93.5 ± 0.2
Alb				(0.30 0.31)	98.9 ± 0.6
Apt				(0.87 0.53)	112.2 ± 1.1
Bar	120.77 ± 0.50	120.98 ± 0.41		(0.41 0.69)	121.0 ± 1.4
Hau	126.93 ± 1.00	127.03 ± 0.66		(0.66 0.81)	127.0 ± 1.6
Val	132.03 ± 1.00			(1.00 0.94)	132.0 ± 1.9
Ber	138.36 ± 1.50	136.99 ± 0.81		(0.81 1.08)	137.0 ± 2.2
Tit	144.19 ± 1.50			(1.50 1.31)	144.2 ± 2.6
Kim	150.69 ± 1.50			(1.50 1.52)	150.7 ± 3.0
Oxf	153.46 ± 2.00	154.36 ± 1.34	154.10	(1.34 1.62)	154.1 ± 3.2
Cal	159.99 ± 2.00	159.59 ± 1.51	159.36	(1.51 1.78)	159.4 ± 3.6
Bat			164.43	(2.60 1.90)	164.4 ± 3.8
Baj			169.20	(2.37 1.98)	169.2 ± 4.0
Aal			176.54	(2.02)	176.5 ± 4.0
Toa			180.08	(2.02)	180.1 ± 4.0
Pli			189.63	(1.98)	189.6 ± 4.0
Sin			195.27	(2.23 1.96)	195.3 ± 3.9
Het			201.89	(0.67 1.97)	201.9 ± 3.9
Rha			205.66	(1.65 2.00)	205.7 ± 4.0
Nor			209.59	(2.61 2.06)	209.6 ± 4.1
Car			220.74	(3.02 2.21)	220.7 ± 4.4
Lad			227.40	(2.22 2.27)	227.4 ± 4.5
Ani	232.62 ± 2.76	235.21 ± 1.60	234.29	(1.73 2.32)	234.3 ± 4.6
Ole			241.66	(2.81 2.36)	241.7 ± 4.7
Ind			244.82	(2.38)	244.8 ± 4.8
Tat	248.18 ± 0.28	248.16 ± 0.28	248.20	(2.22 2.40)	248.2 ± 4.8

Standard deviations in parentheses except in last column for final age with estimated 2σ precision. Selected maximum likelihood results in Table 1 (see footnote) were combined with other estimates.

[a] Initial estimates for Cretaceous-Cenozoic boundary (Maa), OB93 (Cen-Cam), Hawaiian magnetochronology (Cal-Bar), and direct estimates for Tat and Ani.

[b] Final estimates for Middle-Late Cretaceous. Other modified estimates (Tat-Bar) are averages of direct estimates of column 2 combined with maximum likelihood estimates of Table 1.

[c] Smoothing and interpolation results obtained by spline-curve fitting.

[d] Input and output for separate spline-curve fitted to standard deviations.

[e] Final ages with 2σ-precision.

and Berriasian/Valanginian boundaries in Fig. 6). These discrepancies are due to local scarcity of dates and sensitivity of the method to relatively few, more recent dates that have a standard deviation which is much smaller than the majority of the dates used; also, the frequency distribution of the ages of the dated samples is not everywhere approximately uniform, and this results in asymmetric log-likelihood functions. The validity of these statements is illustrated by using computer simulation experiments (Agterberg, in press). A comparison between estimates for the wider-range (parabola) and local (scoring) maxima is given in Table 1 for high-temperature dates only.

Estimates based on situations where the parabola provides a good fit to the log-likelihood function are the best. In the other cases, the parabolas fitted to the chronograms provide answers that are fairly good approximations. In all cases, the position of the local (scoring) maximum was taken as the best estimate of the mean. Several of the estimates, especially in the Jurassic, have relatively large standard deviations due to scarcity of dates and were not used (e.g., Valanginian/Hauterivian, Aalenian/Bajocian, and Toarcian/Aalenian boundaries in Fig. 6).

For high-temperature dates, the largest deviations (those >2 my) in location of the peak of the best-fitting parabola occur at the upper stratigraphic boundary of Kimmeridgian (−4.1 my),

Aptian (−3.9 my), Callovian (2.8 my), Berriasian (−2.7 my), Hettangian (−2.7 my), and Hauterivian (−2.2 my). Most of these differences are negative with the parabola peak older than the local maximum. In part, this reflects asymmetry of the log-likelihood function caused by decreasing frequency of dates toward older stages. The local maximum provides a better estimate than the parabola peak in all cases. Most standard deviations for local maxima as derived by the scoring method are greater than those based on the parabolas. This is because approximate normality of the likelihood function has not been reached because of insufficient data or order of magnitude differences in precision of individual age estimates, and most local maxima remain flatter than the peaks of the wider-range parabolas. Consequently, most standard deviations obtained by scoring are slightly too large.

High-Temperature Versus Low-Temperature Dates

Three maximum likelihood functions were developed for each boundary (Table 1). Unless there were insufficient data, high- and low-temperature dates were analyzed separately as well as combined. Dates determined by methods and materials with closure temperature less than 250°C (cf. Harland and others, 1990, Table 4.1) were classified as low-temperature dates. These are reset more readily than high-temperature dates. The low-temperature dates are almost exclusively K/Ar dates on glauconites with closure temperature of about 200°C. Probabilities that the low-temperature mean is younger than the high-temperature mean were computed as follows. For each boundary, the low-temperature estimate was subtracted from the high-temperature estimate. The resulting number was divided by the combined standard deviation for the two estimates. The resulting standardized value was converted into the probability that the low-temperature estimate is younger than the high-temperature estimate (one-tailed significance test). As shown in Table 1, ten successive maximum likelihood estimates based on low-temperature dates from top Cenomanian downward are significantly younger with probability greater than 99%.

The systematic discrepancy between high-temperature and low-temperature estimates fluctuates between 3 my and 18 my. The preceding, statistical significance test merely indicates that, on the average, either the low-temperature dates are too young, the high-temperature dates are too old, or both these test hypotheses could be true. We have assumed that only the high-temperature dates are unbiased. From top Callovian downward, there are very few low-temperature dates, and the maximum likelihood estimates are not changed significantly whether or not these low-temperature dates are included (Table 1, column 5). For all boundaries, only the high-temperature estimates were used.

Within the Mesozoic Era, the systematic difference between high- and low-temperature estimates is restricted to the time interval between the Oxfordian and the Santonian Stages. This bias was not found in the younger stages where low-temperature dates are abundant nor in the older stages where they are scarce. This indicates that the problem of bias is not related to choice of decay constants. The low-temperature dates are primarily K/Ar dates for glauconites and may provide minimum ages instead of unbiased estimates. This topic is discussed in more detail in other publications (e.g., Gradstein and others, 1988; Harland and others, 1990).

CALLOVIAN THROUGH BARREMIAN MAGNETOCHRONOLOGY

The magnetic anomaly record of the oceanic crust coupled with magneto-biostratigraphic correlations provides a powerful means for relative scaling of the durations of the associated geological stages (e.g., Cande and Kent, 1992). Estimates of the ages of biostratigraphic datums and stage boundaries are possible once the magnetic anomaly scale has been pinned to several radiometric tie points.

This scaling procedure is not possible for the middle portion of the Cretaceous because a long interval of normal polarity, the "Cretaceous Long Normal-Polarity Chron" or polarity chron C34n, extends from the early Aptian to the Santonian/Campanian boundary. Within the late Aptian through late Albian portion of the Cretaceous Long Normal-Polarity Chron, brief subchrons of reversed-polarity have been observed in pelagic sediment sections although there are no well-defined marine magnetic anomalies recognized in ocean crust of these ages (reviewed by Ogg, 1995).

The spacing of the Hawaiian sequence of marine magnetic anomaly lineations M0 through M25 is used here to interpolate the Oxfordian-Aptian time interval and arrive at ages for the intervening stage boundaries. The same procedure was utilized by the DNAG scale (Palmer, 1983; KG85). This "M-sequence" of polarity chrons M0r through M25r is from the Larson and Hilde (1975) model with nomenclature after A. V. Cox (in PTS82) and with a brief reversed-polarity subchron inserted in M11n (Tamaki and Larson, 1988).

Cande and others (1978) identified a series of Jurassic magnetic anomalies in the western Pacific, which they called M26 through M29. Later, Handschumacher and others (1988) used more detailed surveys to propose a revised M26 through M38 pattern, which retained in modified form the M26–M28 portion from Cande and others (1978) and extended the series by approximately 5 my. Handschumacher and others (1988) assigned ages to their M26–M38 pattern by assuming a constant spreading rate from M21–M38 in the Japanese lineations and using the M21–M25 ages of Larson and Hilde (1975) scale.

The anomaly numbering of Handschumacher and others (1988) was retained in this new scale, but their polarity chron nomenclature was revised to be consistent with the younger portion of the M-sequence (nomenclature system of Cox). Therefore, subchrons are labeled according to whether they are subdivisions of a major normal-polarity chron or of the underlying reversed-polarity chron. For example, the normal-polarity chron "35a" of Handschumacher and others (1988) is labeled as subchron "M35r.1n" occurring within reversed-polarity chron M35r. There may be a one-chron duplication within the marine magnetic anomaly pattern of Handschumacher and others (1988) due to a spreading ridge jump (R. Larson, pers. commun., 1991).

Correlation of the M-sequence polarity chrons to ammonite, calcareous nannofossil, dinoflagellate and calpionellid zones and datums are compiled by Ogg (1988, 1995), Bralower and others (1989) and Ogg and others (1991a). Additional magnetostratigraphy studies within Lower Cretaceous strata (e.g., Channell and Erba, 1992) have indicated that further refinement of stage boundary positions relative to the magnetic polarity pattern is constrained by the use of micropaleontological markers rather than ammonite zones to recognize stage boundaries

and by the variable preservation and reworking of those micropaleontological markers.

The Barremian/Aptian boundary is correlated to the base of polarity chron M0r. This magnetostratigraphic correlation, which is under consideration by the Subcommission on Cretaceous Stratigraphy (E. Erba, pers. commun., 1993), is only slightly older than the magnetochronologic placement in previous scales (e.g., KG83).

The Hauterivian/Barremian stage boundary occurs approximately two-thirds from the base of polarity chron M5n (M5n.66; Cecca and others, 1994). Previous magnetic polarity time scales had placed this boundary near polarity chron M7 based on microfossil assignments, but these microfossil datums and associated polarity chron ages have now been recalibrated to ammonite zones.

There have not been any precise ammonite or nannofossil markers for the Valanginian/Hauterivian boundary in magnetostratigraphic sections, and the observed variability in the dinoflagellate marker (last appearance datum of *Scriniodinium dictyotum*) brackets polarity zone M10Nr. Therefore, we provisionally assign this boundary to the middle of polarity chron M10Nr.

The base of the *Thurmanniceras pertransiens* ammonite Zone in southern Spain occurs just below the middle of polarity chron M15n, and the corresponding Berriasian/Valanginian Stage boundary is placed at M15n.4.

The Tithonian/Berriasian stage boundary at the base of the *Berriasella jacobi* Zone is in the upper portion of polarity chron M19n (approximately M19n.2n.5; Ogg and others, 1991a).

The base of the *Hybonoticeras hybonotum* ammonite Zone is the Kimmeridgian/Tithonian boundary. This horizon occurs in southern Spain within polarity chrons M23n to M22An (Ogg and others, 1984). In this study, the boundary is assigned to the top of polarity chron M23n.

The base of the *Sutneria platynota* ammonite Zone appears to occur within polarity chron M25r or M24Br within southern Spain (Ogg and others, 1984). Therefore, pending detailed magnetostratigraphic studies, a working definition for the Oxfordian/Kimmeridgian boundary is the top of polarity chron M25n.

Close-spaced magnetic anomalies M26 through M39 are observed in Pacific crust of presumed Early Callovian-Oxfordian Age (Handschumacher and others, 1988). Magnetostratigraphy of Callovian-Oxfordian sections have also indicated rapidly changing polarity (Steiner and others, 1986; Channell and others, 1990; Ogg and others, 1991b). However, the correlation of the Callovian/Oxfordian boundary to the magnetic anomaly sequence remains ambiguous. Pre-M39 oceanic crust in the Pacific is called the "Jurassic Quiet Zone," and the indistinct nature of the oceanic crust magnetization of this region may indicate the blurring effect caused by a high frequency of magnetic reversals. Ocean Drilling Program (ODP) Site 801 was drilled in oceanic crust older than magnetic anomaly M39, and the basement basalts are dated by radiolarian assemblages as latest Bathonian or earliest Callovian (Larson, and others, 1992).

The ages of these stage boundaries can be estimated by assuming a constant rate of spreading of the Hawaiian magnetic anomaly lineations during the Late Jurassic and Early Cretaceous Periods. There are only two direct radiometric datings of magnetic anomalies: (1) the top of magnetic polarity zone M1r as 123.5 ± 0.5 Ma at ODP Site 878 on MIT guyot basalts (M. Pringle, pers. commun., 1993, 1995) and (2) magnetic anomaly M26r as 155.3 ± 3.4 Ma at ODP Site 765 in the Argo Abyssal Plain (Ludden, 1992). These tie points were used to project the ages of the magnetic polarity chrons associated with the stage boundaries:

Barremian/Aptian	120.77 Ma
Hauterivian/Barremian	126.93
Valanginian/Hauterivian	132.03
Berriasian/Valanginian	138.36
Tithonian/Berriasian	144.19
Kimmeridgian/Tithonian	150.69
Oxfordian/Kimmeridgian	153.46

Magnetochronology does not easily allow an estimate of uncertainty since no error is available for the seafloor lineations distances, or for the age estimates of the tiepoints (e.g., Cande and Kent, 1992). We have assigned arbitrary uncertainties on these constant-spreading estimates based upon the standard deviations of the ages on the magnetic polarity chron M1r and M26r tie points (Appendix 1).

These ages derived from a constant spreading rate assumption must be reconciled with the radiometric databases and associated maximum likelihood age estimates. For this purpose, we performed a weighted average of the boundary age estimates obtained independently from the constant spreading rate model and the maximum likelihood. For the Barremian/Aptian, Hauterivian/Barremian, and Oxfordian/Kimmeridgian boundaries, the maximum likelihood ages are essentially identical as the ages from the constant spreading rate model, indicating that the choice of the two age tie points for the Hawaiian lineation scaling are consistent with the radiometric database as a whole. For the Kimmeridgian/Tithonian/Berriasian boundary suite and the Valanginian/Hauterivian boundary, where there are so few high temperature radiometric ages that the maximum likelihood age estimates for the adjacent boundaries overlap (Table 1), the spreading rate estimates and uncertainties were used exclusively for the boundary age interval for the following spline fit.

However, for the Berriasian/Valanginian boundary, the maximum-likelihood age estimate is significantly older than the estimate based on the constant spreading-rate model (Table 1). This suggests that the spreading rate for the Berriasian may slower than during the Valanginian stage. As explained in the next section, further scaling using spline-fitting of the age sets, supports the interpretation that the average spreading rate for the latest Jurassic and earliest Cretaceous Periods was slower than during the Valanginian-Hauterivian-Barremian stages.

COMBINING MAXIMUM LIKELIHOOD AND
MAGNETOCHRONOLOGY RESULTS

The next step in time scale calculation was to combine the Triassic-Early Cretaceous maximum likelihood age estimates with direct ages estimates of stage boundaries based on Oxfordian-Aptian magnetochronology and with direct estimates for the Olenekian/Anisian and Permian/Triassic boundaries. This approach of weighted averaging combines stratigraphic and mathematical interpolations on the age of those stage boundaries for which no detailed, high-resolution geochronologic estimate is available similar to that for Late Cretaceous time.

A basic assumption underlying the maximum likelihood method is that the dates are randomly distributed over time in the vicinity of each stage boundary. This uniform distribution is assumed to be independent of the age of the stage boundary. On the other hand, direct estimates of stage boundary ages are available in a number of situations. These direct estimates are either dates for samples that were almost exactly taken at a given stage boundary or those derived from magnetostratigraphic correlation of a stage boundary to a calibrated seafloor spreading model. Provided that the direct estimates are independent and were not used for constructing the maximum likelihood graph, they may be combined with one another and the maximum likelihood estimate by assuming that a single overall mean age for the stage boundary is being estimated.

Suppose that n estimates X_j ($j = 1, \ldots, n$) with variances σ_j^2 are to be combined into a single estimate $X = \Sigma w_j X_j$ where the coefficients w_j are relative weights. It can be shown (e.g., Deming, 1948) that the optimum choice of weights is $w_j = \sigma_j^{-2}/\Sigma \sigma_j^{-2}$. The variance of X then satisfies $\sigma^2(X) = \Sigma w_j^2 \sigma_j^2$ (also see Taylor, 1982). This procedure was followed to combine the maximum likelihood estimates (Table 1, column 3) with those from Oxfordian-Aptian magnetochronology, the Olenekian/Anisian boundary, and the Permian/Triassic boundary (Table 2, column 2). The results obtained this way are shown in Table 2, column 3.

Due to a general lack of radiometric dates for Early Triassic and Early and Middle Jurassic time, the corresponding absolute time scale is poorly constrained by the maximum likelihood method. For these intervals, a secondary scaling using relative numbers of biostratigraphic subdivisions provided improved age estimates of stage boundaries (Fig. 7). This secondary scaling is based upon high-resolution ammonite zonal subdivisions.

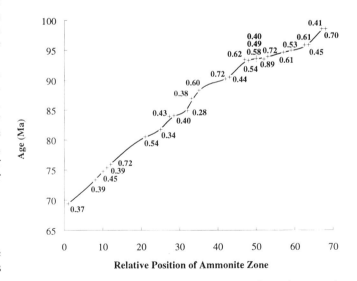

FIG. 7.—Cubic smoothing spline fitted to 27 $^{40}Ar/^{39}Ar$ dates (shown as +'s with 2s-values) from Obradovich (1993) for bentonites in 25 biozones, United States Western Interior. These biozones belong to a continuous sequence of 65 biozones extending from the base of the mid-Cenomanian to the middle of the early Maastrichtian time. The bentonites were spaced assuming equal duration of these 65 biozones. This interpolation gave precise results because the standard deviations of the 27 dates are relatively small, ranging from 0.14 to 0.45 my, and the observations nearly fall on a continuous curve.

MESOZOIC AMMONITE ZONATION CALIBRATION

Mesozoic stages are traditionally recognized by ammonite zonal boundaries. Considerable consensus exists between specialists in different geographic areas of Europe about the placement of ammonite zones in particular stages although some ambiguity remains, especially in the lower Cretaceous Period. Only a few Mesozoic stage boundaries have been formally defined by a boundary stratotype.

The ammonite zonation calibrated here with the time scale (Fig. 8) is from "The Mesozoic-Cenozoic Chronostratigraphic Framework" (Hardenbol and others, in de Graciansky and others, Mesozoic-Cenozoic Sequence Stratigraphy of European Basins, SEPM Special Publication, in prep.) and represents a highest-resolution composite made up of the greatest number of ammonite zones or subzones described from any Mesozoic stage in Boreal or Tethyan areas of Europe. In addition, some upper Cretaceous high-resolution data from the Western Interior of the United States and Triassic data from British Columbia in Canada and of Siberia are included to supplement the western European data. The number of ammonite zonal units for each stage is listed in Appendix 1.

Since few ammonite zones are directly dated by radiometric methods, all zonal units within a stage have been arbitrarily assigned equal duration unless data were available to demonstrate otherwise. For example, the relative thicknesses of ammonite zones in outcrop may reflect either relative duration or rates of deposition. Exceptions to this equal-duration scaling are in portions of the Late Cretaceous where several Western Interior ammonite zones are well dated (Obradovich, 1993) and in the Kimmeridgian to Valanginian interval where calibrations with the magnetic polarity history are available (e.g., Ogg and others, 1991a).

Triassic Period

The Triassic ammonoid zonation has seen several important changes since the publication of Zapfe (1983), reflecting the improved knowledge of the Tethyan (Alpine) as well as Boreal Triassic sequences. For the Tethyan area, the subdivision of Krystyn (as cited by Zapfe, 1983) has been applied. Krystyn's subdivision has remained relatively unchanged as far as it concerns the Lower Triassic Period, but it is considerably modified in the Anisian, Ladinian and lowermost Carnian Stages following the zonation proposed by Mietto and Manfrin (cited in De Zanche and others, 1993). De Zanche and others (1993) used this zonation as a tool for their sequence stratigraphic subdivision of the Triassic in the Italian Dolomites. In the Alpine Triassic sequence, the lower part of the Anisian Stage contains no ammonoid assemblages. A detailed subdivision for the lower Anisian Stage is in Bucher (1989), who reported on the ammonoid succession from beds straddling the Spathian/Anisian boundary in Nevada, but the implications for correlations of these results are not fully understood (compare, e.g., Dagys, 1988).

The placement of the Anisian/Ladinian boundary in the Alpine Triassic strata is subject to a thorough re-evaluation (cf. Gaetani, 1993a, b). The boundary betwen the Anisian and Ladinian stages is placed at the base of the *Nevadites* Zone, following Krystyn (as cited by Zapfe, 1983). Brack and Rieber (1993), however, propose a redefinition of the Anisian/Ladinian bound-

ary at the base of the *Eoprotrachyceras curioni* Zone, a similar level as in the Boreal Triassic (cf. E. Tozer, cited in Zapfe, 1983). Mietto and Manfrin (cited in De Zanche and others, 1993) add the *Daxatina cf. canadensis* Subzone to their *Trachyceras* Zone and suggest an earliest Carnian Age for this subzone. Pending an internationally accepted decision on the Ladinian/Carnian boundary, the *Trachyceras aon* Zone is taken as the base of the Carnian Stage (cf. L. Krystyn, cited in Zapfe, 1983). This implies that the *D. canadensis* Subzone is provisionally retained in the Late Ladinian Stage. The rest of the Carnian subdivision follows L. Krystyn (as cited in Zapfe, 1983). The *Tropites dilleri* Zone is assigned a similar status as most of the Carnian, Norian and Rhaetian zones, each comprising two or more zonal units. The same reasoning is followed for the Norian, where the assemblages recognized by Tozer (as cited by Zapfe, 1983) are applied for the unit subdivsion. The lower boundary of the Rhaetian Stage is provisionally taken at the base of the *Sagenites reticulatus* Zone (Gaetani, 1992). We consider the three ammonoid zones that consitute the Rhaetian, as being of equal status to those in the Carnian and Norian Stages, and hence, comprise each of two units.

Considerable progess has also been made in refinement of the biostratigraphic subdivision of the Boreal Triassic strata. Dagys and Weitschat (1993) provide an up-to-date discussion concerning the ammonoid and pelecypod zones of the Boreal Triassic strata, comprising Siberia, Svalbard (Spitsbergen), the Sverdrup Basin, and British Columbia. The Lower Triassic sequence of Siberia provides most of the units for the time scale. Tozer (as cited by Zapfe, 1983) subdivided the *Wasatchites tardus* Zone in two assemblages, each constituting one zonal unit.

We follow Dagys and Weitschat (1993) for the subdivision of the Spathian into four zones but refrain from applying the very detailed subdivision, which would imply assigning ten units for the Spathian alone.

For Middle Triassic units, the Siberian succession provides optimum resolution (Weitschat and Dagys, 1989), with a subdivision of the *Grambergia talmyrensis* Zone into four subzones. One of the subzones, *Karangites evolutus*, has been recognized on Spitsbergen (Weitschat and Dagys, 1989), supporting its correlative value. Dagys and Weitschat (1989) propose an improved correlation between Anisian ammonoid zones of the Boreal succession and British Columbia, particularly affecting correlation of the *deleeni* and *chischa* Zones. In the Boreal Triassic sequences, the Anisian/Ladinian boundary is taken at the base of the *Eoprotrachyceras subasperum* Zone (Tozer, 1967; Tozer in Zapfe, 1983). This level corresponds to the base of the *E. curionii* Zone in the Alpine Triassic strata. Units in the Upper Ladinian are derived from the nathorstid ammonoid zonation decribed from Siberia (Dagys and Konstantinov, 1992). Dagys and others (1993) reported new data concerning the Ladinian/Carnian boundary in the Boreal Triassic strata, proposing the *Nathorstites lindstromi* Zone as the first Ladinian ammonoid zone, including the *Daxatina canadensis* Zone on Bear Island (Svalbard). Dagys and others (1993) subdivide the lowermost Carnian *Stolleyites tenuis* Zone into two subzones, in Siberia as well as in Svalbard. Dagys and Weitschat (1993) propose an improved correlation of the Carnian and Norian ammonoid and bivalve zones beween Siberia and British Columbia. As mentioned earlier, Upper Norian and Rhaetian ammonoid zones are given the same status as those in

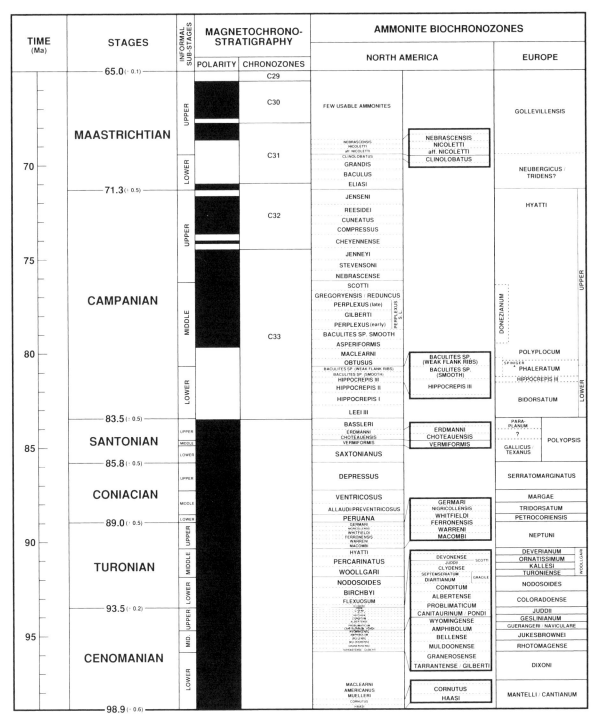

FIG. 8.—Highest-resolution Mesozoic ammonite subdivisions and magnetic polarity time scale. (A) Upper Cretaceous magnetic polarity time scale calibration with North American and European highest resolution ammonite subdivisions. Direct calibration between ammonites and polarity exists for the middle Campanian through lower Maastrichtian stages (Hicks, 1993). North American ammonite calibration from Cobban and others (1994) and Obradovich (1993), European ammonites from J. M. Hancock (pers. commun., 1994) and F. Amédro (pers. commun., 1994).

the Norian and Carnian Stages, although no subzones have been recognized in this interval. Support for this practice is found in Kazakov and Kurushin (1992), who, based upon new finds from Siberia, subdivided the traditional Norian bivalve zones, such as the *Monotis ochotica* Zone, into several subzones.

Caution should be used when correlating the ammonoid zones between the Boreal and Tethyan Triassic strata. Due to endemism in each faunal realm, a direct comparison through correlation of different assemblages often is difficult. Only through integration of all zonal disciplines (e.g., zonations

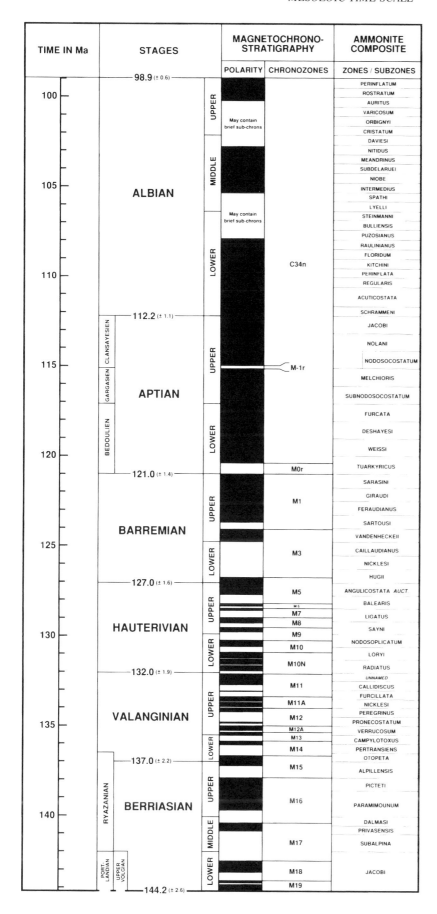

FIG. 8.—(B) Lower Cretaceous magnetic polarity time scale comparison with European highest-resolution ammonite subdivisions (Hoedemaker and others, 1993). Ammonites are calibrated with standard stages and the magnetic polarity history in the Berriasian and lower Valanginian. For the upper Valanginian through Albian, ammonites zones are calibrated with the stages only and are assumed to have equal duration with each stage.

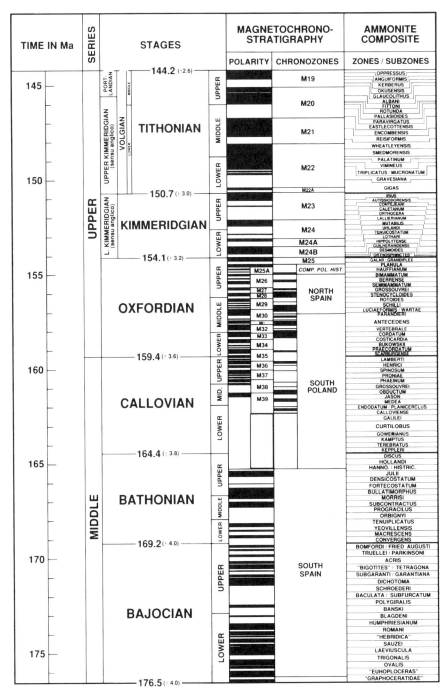

FIG. 8.—(C) Upper and upper Middle Jurassic magnetic polarity time scale comparison with European highest-resolution ammonite subdivisions from Boreal and Tethyan regions. Ammonites are calibrated with standard stages and the magnetic polarity history in the Kimmeridgian and Tithonian stages. For the Bajocian through Oxfordian stages, ammonite zones are calibrated with the stages only and are assumed to have equal duration within each stage.

based on conodonts and palynomorphs) and magnetic reversals will it be possible to understand the mutual relationship. Attempts to apply sequence stratigraphic correlations between widely different areas should bear in mind this limitation.

Direct calibration of magnetostratigraphy to ammonite zones has been accomplished in portions of the Early Triassic (Ogg and Steiner, 1991), Middle Triassic (Muttoni and others, 1994), and Late Triassic strata (Gallet and others, 1992, 1993; Marcoux, 1993).

Jurassic Period

Following the Tethyan convention the Jurassic/Cretaceous boundary is placed between the Tithonian and Berriasian stages. The boreal Portland and Volgian regional stages overlap a portion of the Berriasian stage (Fig. 8B–C). The Jurassic System is subdivided into a high-resolution "composite" of 180 units or subzones. The "composite" is made up of zones and subzones from Boreal as well as Tethyan areas to achieve maximal

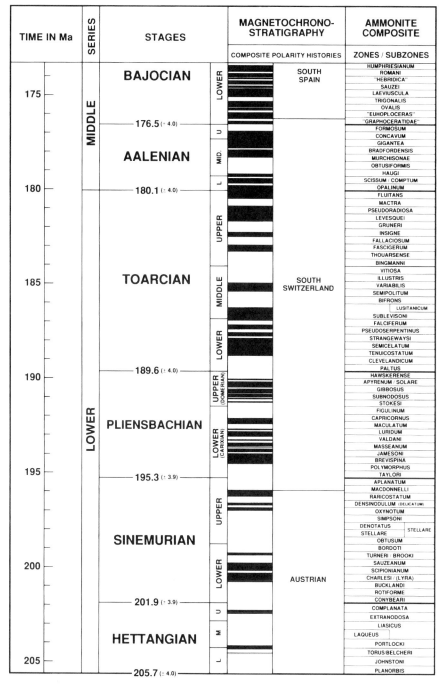

FIG. 8.—(D) Lower and lower Middle Jurassic magnetic polarity time scale comparison with European highest-resolution ammonite subdivisions from Boreal and Tethyan regions. Ammonite zones are calibrated with the stages only and are assumed to have equal duration within each stage.

resolution since no individual basin or region appears to provide high resolution throughout the Jurassic Period. The principal references for the composite zonation are Cope and others (1980a, b) and the Groupe Français d'Etudes du Jurassique (1991). A more comprehensive historic review of the Jurassic ammonite subdivisions and a calibration of Boreal and Tethyan zonal schemes is given in Hardenbol and others (in prep.). For the Hettangian through Oxfordian Stages, all ammonite zones/subzones in the composite zonation have been assigned equal

duration since no unambiguous information is available to decide otherwise. For the Kimmeridgian through Tithonian Stages, paleomagnetic calibration provided independent information for ammonite zonal durations (Ogg and others, 1984; Bralower and others, 1989; Ogg and others, 1991a).

The magnetic polarity pattern for several ammonite zones prior to the Kimmeridgian Stage is partially known. We have used the composite scale compiled by Ogg (1995), but with the Toarcian magnetic polarity pattern derived from a revised bio-

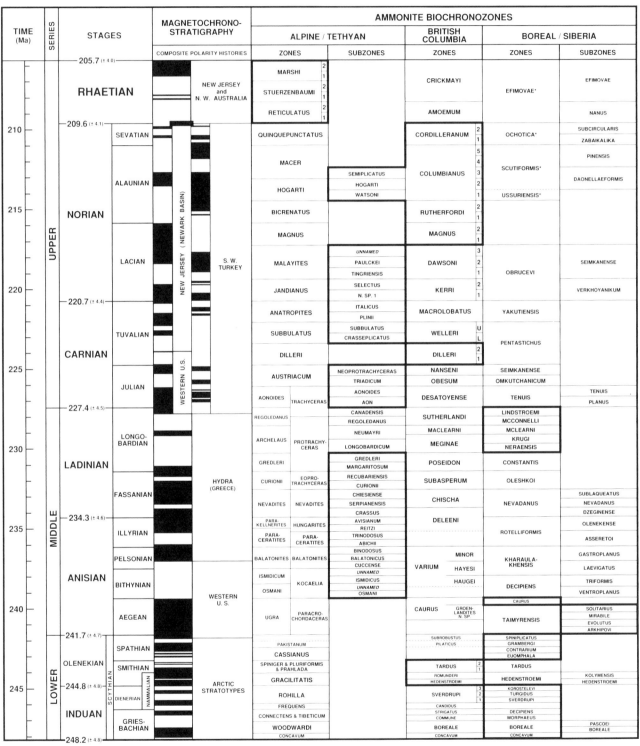

FIG. 8.—(E) Triassic magnetic polarity time scale comparison with highest-resolution ammonite subdivisions. The composite ammonite subdivision (heavy outline) is based on highest resolution imformation from Tethyan areas (De Zanche and others, 1993), British Columbia (L. Krystyn, as cited by Zapfe, 1993) and Boreal regions including Siberia. Ammonites are calibrated with standard stages and are assumed to have equal duration within each stage. The magnetic polarity history is calibrated with ammonite zones.

stratigraphic correlation of three independent studies (Horner, 1983; Galbrun and others, 1988; Galbrun and others, 1990).

Cretaceous Period

Cretaceous ammonite biostratigraphy in Europe is primarily based on Robaszynski and others (1980), Owen (1984, 1988), Hoedemaeker and Bulot (1990), Hancock (1991), Amédro (1992), Bulot and others (1993), Bulot and Thieuloy (1993), and Hoedemaeker and others (1993).

In the upper Cretaceous strata of Europe, high-resolution ammonite zonations exist for the Cenomanian through Santonian Stages, but the Campanian and Maastrichtian Stages are characterized by a much lower resolution ammonite biostratigraphy. In northern Europe, belemnites provide an alternative high-resolution subdivision for the Campanian through Maastrichtian Stages although prospects for correlation with this group appear limited. In North America, however, a high-resolution ammonite zonation (Cobban and others, 1994) is available for the Cenomanian through lower Maastrichtian interval. A significant number of ammonite zones in the Western Interior are radiometrically well dated (Obradovich, 1993), while the remainder of the zones are interpolated by assigning equal time to each zone (Fig. 8A). The Campanian/Maastrichtian boundary in Europe is between the *hyatti* and *neubergicus/tridens* ammonite Zones corresponds to the base of the *Belemnella lanceolata* Zone (Birkelund and others, 1984). This boundary is correlated to the base of the *Baculites eliasi* ammonite Zone (Kennedy and others, 1992) or slightly below (McArthur and others, 1993) in the Western Interior.

The lower Cretaceous time is subdivided into 66 ammonite zones and subzones (Fig. 8B). From the upper Valanginian to the Albian interval, zones/subzones were assigned equal duration in the absence of data. From the Berriasian through the lower Valanginian interval, ammonite zones were calibrated to some extent with paleomagnetic polarity information (Galbrun, 1984; Ogg and others, 1988; Ogg and others, 1991a).

TRIASSIC THROUGH CRETACEOUS SPLINES

Triassic and Early-Middle Jurassic Ages

In this section we discuss the final scaling of the Mesozoic ages. A secondary scaling is useful because, due to a general lack of radiometric dates, a drawback of the time scale constructed so far is that it is poorly defined for Early Triassic and Early and Middle Jurassic time.

PTS82 proposed the following method for dealing with segments of the geologic time scale where dates are scarce. Linear interpolation can be used to estimate the ages of stage boundaries between tie-points located near the beginning and the end of such a segment using the hypothesis of equal duration of stages between the tie-points. The tie-points themselves are stage boundaries with estimated ages that are relatively good. This method has the advantage of simplicity. However, it has a number of drawbacks: (1) certainty (zero standard deviation) is assigned to the tie-point estimates, and the estimates of uncertainty for all other stage boundaries are disregarded; (2) the length of duration of the stages, which is constant between tie-points, changes abruptly at the tie-points; and (3) it is more reasonable to assume equal duration of zones within stages.

PTS89 used equal duration of chrons in Mesozoic stages to further smooth chronogram age estimates for stage boundaries where chrons are units of equal duration within stages as derived from zones. The units were interpolated with "dog-legged" linear segments, showing abrupt changes at tiepoints, probably the result from hand fitting.

In this study, the equal duration of ammonite zones is used as a secondary scaling method to smooth Triassic and Jurassic stage boundary ages. These ages were scaled initially with maximum likelihood estimates. The approach, outlined below, uses a smoothing cubic spline where the uniform spacing of (sub)zones is the independent variable, and the irregularly spaced age estimates for stage boundaries from maximum likelihood (partly combined with direct estimates; Table 2, column 3) are the dependent variables. The approach is a statistical improvement on the maximum likelihood estimates and reduces uncertainty for stages with relatively large errors. For example, insufficient radiometric dates are available for the Induan, Olenekian, and Pliensbachian stages (Appendix 1), and errors are relatively large for Early and Middle Jurassic time. The effect of this scaling tool is directly proportional to the magnitude of the uncertainty, as outlined below. The tool has minimal effect on the duration of stages with small error bars.

The use of smoothing splines for interpolation over segments where dates are scarce is discussed by Agterberg (1988, 1990). A cubic spline function consists of a chain of cubic polynomials bounded by consecutive points (knots). The function represents the relation between a dependent variable (estimated ages of stage boundaries) and an independent variable (e.g., unequally spaced stage boundaries). The spline curve is smooth because it has continuous first and second derivatives (i.e., abrupt changes in slope and rate of change of slope are not permitted). It can be fitted by means of the least squares method. We used the algorithm originally developed by Reinsch (1967). This implies that the knots were selected at the observation points, the second derivative was set equal to zero at the first and last point, and the sum of squares of standardized residuals was set equal to the number of observations.

Two other properties of smoothing splines are as follows. Each observation is weighted according to the inverse of the square of its standard deviation. For the maximum likelihood estimates, the ratio of largest and smallest standard deviation exceeds 20. This means that the boundary with the smallest standard deviation (Permian/Triassic boundary with $\sigma = 0.28$, Table 2, column 3) will receive more than $400\times$ as much weight as the boundary with the largest standard deviation (Carnian/Norian with $\sigma = 6.13$, Table 1, column 3). Another property is that, if all standard deviations are relatively large, the smoothing spline becomes a best-fitting straight line.

In total, 14 knots were used for the spline fitting. These are the 13 pre-Kimmeridgian stage boundaries with maximum likelihood estimates highlighted in Table 1 (see Table 2, column 3 for four revisions) plus the estimate for the Lower Pliensbachian Coldfish volcanics in British Columbia (Appendix 1, item 276). These knots were spaced such that distances between stratigraphic boundaries are proportional to the number of (sub)zones per stage; equal duration of ammonite zones was assumed within the Triassic and Jurassic separately. The knot for the Pliensbachian stage (Appendix 1, item 276) was spaced at one third of the total duration above the base of this stage.

In total, 17 ages were estimated by smoothing and interpolation with the cubic spline curve (Table 2, column 4). The final value is close to input if its standard deviation is small. The final value is primarily determined by the number of ammonite (sub)zones if its standard deviation is large or not used because of scarcity of dates.

Late Jurassic and Early Cretaceous Ages

Initially in this study, the Late Jurassic through Early Cretaceous stage boundaries were estimated by applying a variation of the method used to compute the earlier stage boundaries. The array of maximum likelihood estimates for these stage boundaries were combined with the independent estimates derived from the magnetochronology of a fixed spreading-rate for Hawaiian lineations. This merger of the two independent age estimates incorporated a weighted average based upon the relative standard deviations on each age. A cubic smoothing spline was passed through the array of merged stage boundary estimates using the uniform spacing of subzones as the independent variable. This statistical optimization of the radiometric, magnetic and biostratigraphic data yielded the following suite of age estimates:

Aptian/Albian	112.4 ± 0.9 (Ma ± 2s)
Barremian/Aptian	120.5 ± 1.6
Hauterivian/Barremian	126.6 ± 1.8
Valanginian/Hauterivian	131.2 ± 1.7
Berriasian/Valanginian	135.7 ± 1.4
Tithonian/Berriasian	141.3 ± 1.9
Kimmeridgian/Tithonian	148.2 ± 1.9
Oxfordian/Kimmeridgian	153.3 ± 1.5
Callovian/Oxfordian	159.1 ± 1.6

The Kimmeridgian through Barremian Stage boundaries are systemmatically younger by approximately 0.5–1 my than the estimates derived entirely from the constant-spreading rate model for the Hawaiian magnetic lineation set. The main causes of this offset are (1) some relatively young ages reported for some of these stages and (2) the large number of ammonite subzones within the Kimmeridgian and Tithonian Stages within the Boreal realm as compared to Early Cretaceous stages (Appendix 1). The computed relative duration of stages, which is influenced by the relative number of ammonite subzones, would imply relatively faster Pacific spreading rates during the Oxfordian, Valanginian, and Barremian intervals, and relatively slower spreading rates during the Kimmeridgian-Berriasian and Hauterivian intervals. In contrast to the variable rates in the Pacific, this scale would imply that the Keathley sequence of the Central Atlantic was formed by a relatively smooth spreading rate. To a first approximation, these changes in Pacific spreading rates were suggested to be a factor in long-term changes in eustatic sea level.

However, Channell (this volume) further analyzed the implications of this initial scale ("GRAD93") and other time scales for spreading rates in different ocean basins and accumulation rates of pelagic sediments, and he concluded that the constant spreading-rate model for the Hawaiian magnetic lineation set is a preferable scale that does not cause artifacts of irregular spreading and accumulation rates. The constant-rate Hawaiian

scale is within the error limits of the maximum likelihood estimates from radiometric ages. Therefore, we have used a constant spreading rate for the Hawaiian lineations as the main control on derivation of stage boundaries within this interval and have not used the number of ammonite subzones for a secondary scaling factor. An implication is that the durations of ammonite subzones within the Boreal realm during the Kimmeridgian-Tithonian intervals were very brief (Fig. 8C).

Late Cretaceous Ages

As pointed out in the section on "Age Estimates for Selected Stratigraphic Boundaries," our best estimate for the age of the Maastrichtian-Danian (Cretaceou/Paleogene) boundary is 65.00 ± 0.04 Ma. This new estimate is significantly different from the maximum likelihood value of 66.06 ± 0.41 Ma (Table 1, column 2) which was rejected for that reason.

Obradovich (1993) directly interpolated the age of Late Cretaceous stage boundaries from over 30 closely spaced, $^{40}Ar/^{39}Ar$ radiometric estimates in United States Western Interior bentonites dated with ammonites (Table 3). All Ar/Ar age estimates were calibrated with 520 MMhb-1.

According to OB93, the age estimates for the stage boundaries are as follows: Campanian/Maastrichtian is 71.30 ± 0.25 Ma, Santonian/Campanian is 83.50 ± 0.25 Ma, Coniacian/Santonian is 86.30 ± 0.25 Ma, Turonian/Coniacian is 88.70 ± 0.25 Ma, Cenomanian/Turonian is 93.30 ± 0.10, and Albian/Cenomanian is 98.50 ± 0.25 Ma.

We re-evaluated the local ammonite calibration in terms of European stage assignments and accepted the chronostratigraphy of OB93 accompanying the radiometric age estimates. Stratigraphic details on the relatively young Campanian/Maastrichtian boundary were discussed earlier. In this study, all $^{40}Ar/$ ^{39}Ar age estimates were entered in Appendix 1.

TABLE 3.—U. S. WESTERN INTERIOR AMMONITE ZONES AND Ar/Ar DATES

Zone	40Ar/39Ar Ages for Bentonites, Ma	Geologic Age
Baculites clinolobatus	69.42 ± 0.37	early Maastrichtian
Baculites compressus	73.35 ± 0.39	late Campanian
Exiteloceras jenneyi	74.76 ± 0.45	late Campanian
Didymoceras stevensoni	75.37 ± 0.39	late Campanian
Didymoceras nebrascense	75.89 ± 0.72	late Campanian
Baculites obtusus	80.54 ± 0.55	mid Campanian
Scaphites hippocrepis II	81.71 ± 0.34	early Campanian
Desmoscaphites bassleri	83.91 ± 0.43	late Santonian
Just below *Boehmoceras assemblage*	84.09 ± 0.40	late Santonian
Top of *Inoceramus undulatoplicatus* Zone	84.88 ± 0.28	early Santonian
Protexanites bourgeoisianus	86.92 ± 0.39	late Coniacian
Scaphites preventricosus	88.34 ± 0.60	mid Coniacian
Prionocyclus macombi	90.21 ± 0.72	mid Turonian
Prionocyclus hyatti	90.51 ± 0.45	mid Turonian
Vascoceras birchbyi	93.40 ± 0.63	early Turonian
Pseudaspidoceras flexuosum	93.25 ± 0.55	early Turonian
Neocardioceras judii	93.30 ± 0.40	late Cenomanian
N. judii	93.78 ± 0.49	late Cenomanian
N. judii	93.59 ± 0.58	late Cenomanian
Euomphaloceras septemseriatum	93.49 ± 0.89	late Cenomanian
Vascoceras diartianum	93.90 ± 0.72	late Cenomanian
Dunveganoceras pondi	94.63 ± 0.61	late Cenomanian
Acanthoceras amphibolum	94.93 ± 0.53	mid Cenomanian
Conlinoceras gilberti	95.78 ± 0.61	mid Cenomanian
27 m below *C. gilberti*	95.86 ± 0.45	mid Cenomanian
Neogastroplites cornutus	98.52 ± 0.41	early Cenomanian
N. haasi	98.54 ± 0.70	early Cenomanian

A cubic spline curve was fitted to the first 25 Ar/Ar dates of Table 3 for 6 stages (Cenomanian through Maastrichtian; Fig. 7). The relative positions of the 23 biozones associated with these bentonites were used for location in the curve-fitting exercise. These 23 biozones belong to a continuous sequence of 65 biozones extending from the base of the mid-Cenomanian to the middle of the early Maastrichtian. This interpolation gave precise results because the standard deviations of the 25 dates are relatively small, ranging from 0.14 to 0.45 Ma, and the observations fall nearly on a continuous curve. This interpolation is more accurate than the dog-legged (?hand-fitted) interpolation lines through the age estimated used in OB93. The following values were obtained by a method similar to the one explained earlier when discussing Triassic through Middle Jurassic spline smoothing:

Campanian/Maastrichtian	71.29 ± 0.25 Ma
Santonian/Campanian	83.46 ± 0.24 Ma
Coniacian/Santonian	85.79 ± 0.23 Ma
Turonian/Coniacian	88.96 ± 0.25 Ma
Cenomanian/Turonian	93.49 ± 0.08 Ma

Pre-Cenomanian Standard Deviations

Final standard deviations for the boundary ages older than Cenomanian were determined by fitting a separate spline-curve with input and output for the dependent variable as shown in Table 2, column 4. The final ages were used for the independent variable, and the degree of smoothing was determined by assigning equal weights to the input values, followed by cross-validation (Agterberg, 1990). Input values for the standard deviations of the Triassic stage boundaries were based on the parabola peaks (Table 1, column 1), because Triassic likelihood functions have relatively flat tops. Uncertainty of the final ages is expressed by means of 2σ precision (Table 2, column 5). It is not possible to estimate standard deviations for the durations because of lack of information on serial correlation of the stage boundary age estimates.

THE MESOZOIC TIME SCALE

The age estimates for the 30 stage boundaries (in Ma units ±2s) are tabulated in the last column of Table 1 and in Table 4. Final age estimates were rounded to one decimal position and standard deviations are expressed at 2σ precision, which approximates the 95% confidence level. The uncertainty in the relative duration of each individual stage is much less than indicated by the uncertainties on the ages of the boundaries.

The ages of period boundaries are (in Ma ± 2σ):

Permian/Triassic	248.2 ± 4.8 Ma
Triassic/Jurassic	205.7 ± 4.0
Jurassic/Cretaceous	144.2 ± 2.6
Cretaceous/Paleogene	65.0 ± 0.1 Ma

The boundary of the Induan and Olenekian stages of the Early Triassic is assigned solely according to the relative numbers of ammonite zonal units, 9 vs. 8 respectively. Radiometric age estimates are needed to corroborate and refine this age. The five longest stages are Norian at 11.1 my, Toarcian at 9.5 my, Aptian at 8.8 my, Albian at 13.3 my, and Campanian at 12.2

TABLE 4.—MESOZOIC TIME SCALE

Stage Boundary	Age, Ma	Duration of Preceding Stage, my
Maastrichtian/Danian (Cretaceous/Tertiary)	65.0 ± 0.1	6.3
Campanian/Maastrichtian	71.3 ± 0.5	12.2
Santonian/Campania	83.5 ± 0.5	2.3
Coniacian/Santonian	85.8 ± 0.5	3.2
Turonian/Coniacian	89.0 ± 0.5	4.5
Cenomanian/Turonian	93.5 ± 0.2	5.4
Albian/Cenomanian	98.9 ± 0.61	3.3
Aptian/Albian	112.2 ± 1.1	8.8
Barremian/Aptian	121.0 ± 1.4	6.0
Hauterivian/Barremian	127.0 ± 1.6	5.0
Valanginian/Hauterivian	132.0 ± 1.9	5.0
Berriasian/Valanginian	137.0 ± 2.2	7.2
Tithonian/Berriasian (Jurassic/Cretaceous)	144.2 ± 2.6	6.5
Kimmeridgian/Tithonian	150.7 ± 3.0	3.4
Oxfordian/Kimmeridgian	154.1 ± 3.3	5.3
Callovian/Oxfordian	159.4 ± 3.6	5.0
Bathonian/Callovian	164.4 ± 3.8	4.8
Bajocian/Bathonian	169.2 ± 4.0	7.3
Aalenian/Bajocian	176.5 ± 4.0	3.6
Toarcian/Aalenian	180.1 ± 4.0	9.5
Pliensbachian/Toarcian	189.6 ± 4.0	5.7
Sinemurian/Pliensbachian	195.3 ± 3.9	6.6
Hettangian/Sinemurian	201.9 ± 3.9	3.8
Rhaetian/Hettangian (Triassic/Jurassic)	205.7 ± 4.0	3.9
Norian/Rhaetian	209.6 ± 4.11	1.1
Carnian/Norian	220.7 ± 4.4	6.7
Ladinian/Carnian	227.4 ± 4.5	6.9
Anisian/Ladinian	234.3 ± 4.6	7.4
Olenekian/Anisian	241.7 ± 4.7	3.1
Induan/Olenekian	244.8 ± 4.8	3.4
Tatarian/Induan (Permian/Triassic)	248.2 ± 4.8	

Uncertainty on ages of stage boundaries is given to 2 standard deviations.

my. Most stages are 5–6 my in duration. The Santonian is the shortest stage at 2.3 my.

The Mesozoic magnetic polarity reversals compiled by Ogg (1995) are displayed along the geologic time scale in Figures 8A–E. There are no magnetic anomalies preserved prior to the Callovian stage, and the pre-Oxfordian polarity time scale is tentative pending further verification in multiple fossiliferous sedimentary section.

When using the estimates of uncertainty, no simple solution exists for combining stratigraphic uncertainty of relative age data with analytical uncertainty of radiometric age estimates in a single statistical number. In this study, two approaches are combined in time scale calculation:

1. Stratigraphic interpolation of analytically precise age estimates close to key stage boundaries, and
2. Stratigraphic/statistical analysis with maximum likelihood estimation and spline smoothing of numerous, stratigraphically meaningful radiometric age estimates.

The second approach allows estimation of error bars. In order to further reduce uncertainty levels, more radiometric dates are needed in pre-Aptian marine strata, with particular emphasis on the Jurassic Period.

The advent of radiometric dating techniques with less than 1% analytical error, as demonstrated in the new Cretaceous Ar/Ar dates, furnishes a major challenge to biostratigraphy. It used to be that biostratigraphic resolution in the Mesozoic Era largely surpassed that based on radiometric dates, but the gap is closing. Even the most detailed biostratigraphic scheme probably has no biozonal units of less than 0.5–1.0 my duration, not to speak of the actual precision in dating a particular "piercing point" for which an Ar/Ar age estimate would be available with an analytical uncertainty of 0.1 to 0.5 my. The combination of such dates with high-resolution biostratigraphy, magnetostratigraphy, or Milankovitch cyclicity is a major challenge for Mesozoic geochronologic studies.

SUMMARY AND CONCLUSIONS

In this study, we derive an absolute geologic time scale, integrated with ammonite zones and magnetic polarity chrons for the Triassic, Jurassic and Cretaceous Periods, together comprising the Mesozoic Era. The geologic time scale is composed of standard stratigraphic units (stages), calibrated in millions of years, with an uncertainty estimate at each level (Fig. 9). Stratigraphically critical data, underpinning this isotopic age dates framework, incorporate the observed ties between (1) radiometric dates, biozones and stage boundaries, and (2) biozones and magnetic reversals on the seafloor and in sediments.

One reason to propose the new time scale for the Mesozoic is that over 65 new Triassic, Jurassic and Cretaceous age dates have become available since 1989 (an increase of 25% since PTS89). Also, improvements in calibrations of stages, standard zones, and magnetic polarity reversals, and development of better interpolation techniques where age control is sparse, enable updating previously proposed age estimates for the stage boundaries in the Triassic, Jurassic and Cretaceous Periods. We have preserved estimates of uncertainty for the age of each stage boundary. The uncertainty in relative duration of each individual stage is much less than the uncertainties on the ages of the stage boundaries. It is not possible to estimate standard deviations for the durations because of lack of information on serial correlation of the stage boundary age estimates. As shown in the last column of Table 1, the ratio of stage duration estimates for the Mesozoic time scale fluctuates between 0.2 and 1.7. The stage duration estimates are relatively precise. Nevertheless, the large differences between results based on successive time scales, (Figs. 1–3) suggest that the Mesozoic time scale needs further improvement, especially the Jurassic and Lower Cretaceous parts.

The ages of period boundaries are (in Ma ± 2s): Permian/Triassic = 248.2 ± 4.8 Ma, Triassic/Jurassic = 205.7 ± 4.0 Ma, Jurassic/Cretaceous = 144.2 ± 2.6 Ma, and Cretaceous/Paleogene = 65.0 ± 0.1 Ma.

Note.—In June 1995, Brack et al. reported single-grain zircon U/Pb age dates from tuffaceous layers associated with the Anisian/Ladinian boundary interval in sections near Bagolino in northern Italy. This region is a proposed candidate for the Global Stratotype Section and Point of the base of the Ladinian stage, although disagreements on the precise level have not yet been resolved. A suite of seven zircons from a thin crystal tuff in the lower part of the Secedensis [Nevadites] ammonite zone yielded a weighted Pb/U mean age of 241.0 ± 0.5 Ma. This same bed can be traced to equivalent tuffs in the Grenzbitumenzone at Monte San Giorgio in southern Switzerland (Brack and Rieber, 1993). As summarized in our text, sanidine feldspars from these tuffs in the Grenzbitumenzon had previously yielded a mean K/Ar age of 233 ± 4.5 Ma (2s) and a plateau age of approximately 232 ± 4.5 Ma (2s) (Hellmann and Lippolt, 1981; discussed as NDS196 in Odin, 1982; incorporated as items 309 and 310 in Appendix), which were key age controls on the final computed age of 234.3 ± 4.6 Ma for this stage boundary. However, Brack et al. (1995) report that preliminary zircon ages from this same Grenzbitumenzone are consistent with their age from Bagolino.

Incorporation of the new radiometric date for this boundary interval and other zircon ages (Brack et al., 1995) for overlying middle Ladinian tuffs (238.8 ± 0.4 Ma in the middle of Gredleri Zone; 237.7 ± 0.5 Ma in the middle of Archelaus Zone) into the maximum-likelihood database and new spline computations imply a longer time span for the Ladinian Stage (8.3 instead of 6.9 my), and a greatly reduced extent of the Anisian Stage (3.1 instead of 7.4 my). However, Brack et al. (1995) recommend placement of the Nevadites Zone into the Anisian Stage, rather than its traditional placement in the Ladinian, which would result in a younger Anisian/Ladinian boundary.

ACKNOWLEDGMENTS

The following colleagues provided valuable criticism and advice on data and methods adopted in this study or supplied new stratigraphic information: Sven Backstrom, William Berggren, Timothy Bralower, James Channell, Elisabetta Erba, Martin Farley, Jason Hicks, Ph. Hoedemaker, Thierry Jacquin, Dennis

Ma	Era	Period	Epoch	Stage	Age Ma	2s error in myr
65	Czn.	Paleogene	Paleocene	Danian	65.0 ± 0.1	
70				Maastrichtian	71.3 ± 0.5	
80			Late	Campanian		
				Santonian	83.5 ± 0.5	
90				Coniacian	85.8 ± 0.5	
				Turonian	89.0 ± 0.5	
				Cenomanian	93.5 ± 0.2	
100		Cretaceous			98.9 ± 0.6	
110				Albian		
120				Aptian	112.2 ± 1.1	
			Early	Barremian	121.0 ± 1.4	
130	Mesozoic			Hauterivian	127.0 ± 1.6	
				Valanginian	132.0 ± 1.9	
140				Berriasian	137.0 ± 2.2	
150				Tithonian	144.2 ± 2.6	
			Late	Kimmeridgian	150.7 ± 3.0	
160				Oxfordian	154.1 ± 3.3	
				Callovian	159.4 ± 3.6	
170			Middle	Bathonian	164.4 ± 3.8	
		Jurassic		Bajocian	169.2 ± 4.0	
180				Aalenian	176.5 ± 4.0	
				Toarcian	180.1 ± 4.0	
190			Early	Pliensbachian	189.6 ± 4.0	
200				Sinemurian	195.3 ± 3.9	
				Hettangian	201.9 ± 3.9	
210				Rhaetian	205.7 ± 4.0	
			Late	Norian	209.6 ± 4.1	
220				Carnian	220.7 ± 4.4	
230		Triassic	Middle	Ladinian	227.4 ± 4.5	
				Anisian	234.3 ± 4.6	
240			Early	Olenekian	241.7 ± 4.7	
				Induan	244.8 ± 4.8	
250	Pzc.	Permian		Tatarian	248.2 ± 4.8	

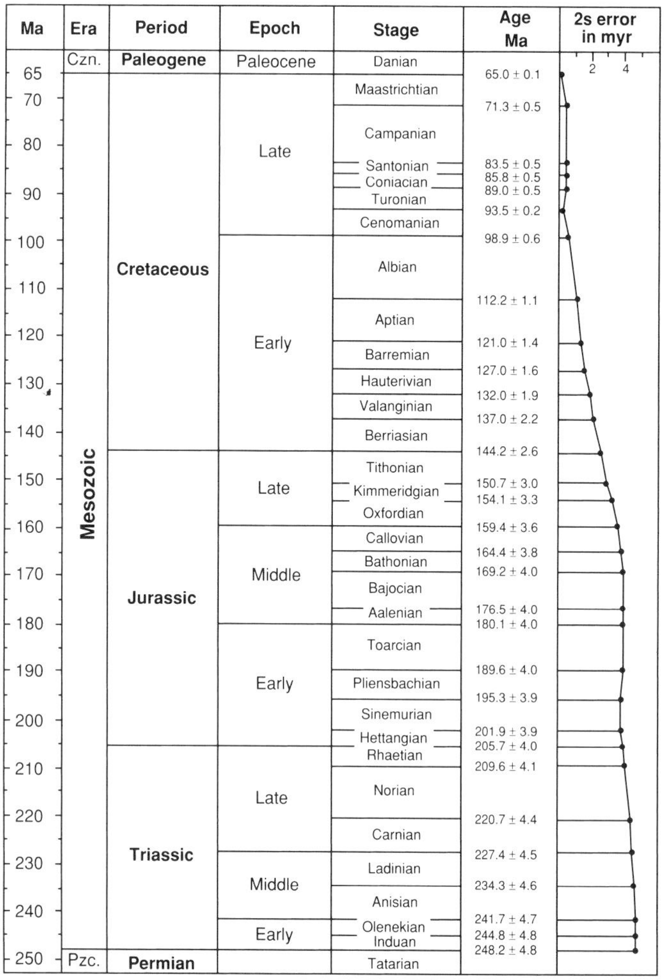

FIG. 9.—The Mesozoic geologic time scale.

Kent, Han Leerveld, Ian Lerche, J.M. McArthur, Peter Mc-Laughlin, J. Mortensen, Giovanni Muttoni, John Obradovich, Neil Opdyke, Serge Odin, Katherina Perch-Nielsen, Malcolm Pringle, Will Sager, P. Smith, Henk Visscher, and Gerd Westermann. We thank Exxon Production Research Company for supporting this project. Portions of the text and figures from "A Mesozoic time scale" by Gradstein and others (1994) are used by permission of the American Geophysical Union.

REFERENCES

AGTERBERG, F. P., 1988, Quality of time scales: a statistical appraisal, *in* Merriam, D. F., ed., Current Trends in Geomathematics: New York, Plenum, p. 57–103.

AGTERBERG, F. P., 1990, Automated Stratigraphic Correlation: New York, Elsevier, 424 p.

ALVAREZ, W., ARTHUR, M. A., FISCHER, A. G., LOWRIE, W., NAPOLEONE, G., PREMOLI-SILVA, I., AND ROGGENTHEN, W. M., 1977, Upper Cretaceous-Paleocene magnetic stratigraphy at Gubbio, Italy. V. Type section for the Late Cretaceous-Paleocene geomagnetic reversal time scale: Bulletin of the Geological Society of America, v. 88, p. 383–389.

AMÉDRO, F., 1992, L'Albian du Bassin Anglo-Parisiens: Ammonites, zonation phyletic, sequences: Bulletin des Centres de Recherches Exploration-Production Elf-Aquitaine, v. 16, p. 187–233.

ARMSTRONG, R. L., 1978, Pre-Cenozoic Phanerozoic time scale: Computer file of critical dates and consequences of new and in-progress decay-constant revisions, *in* Cohee, G. V., Glaessner, M. F., and Hedberg, H. D., eds., The Geologic Time Scale: Tulsa, American Association of Petroleum Geologists, Studies in Geology 6, p. 73–91.

BERGGREN, W. A., KENT, D. V., OBRADOVICH, J. D., AND SWISHER, III, C. C., 1992, Towards a revised Paleogene geochronology, *in* Prothero, D. R. and Berggren, W. A., eds., Eocene-Oligocene Climatic and Biotic Evolution: Princeton, Princeton University Press, p. 29–45.

BIRKELUND, T., HANCOCK, J. M., HART, M. B., RAWSON, P. F., REMANE, J., ROBASZYNSKI, F., SCHMID, F., AND SURLYK, F., 1984, Cretaceous stage boundaries: proposals: Bulletin of the Geological Society of Denmark, v. 33, p. 3–20.

BOLES, J. R. AND LANDIS, C. A., 1984, Jurassic sedimentary melange and associated facies, Baja California, Mexico: Geological Society of American Bulletin, v. 95, p. 513–521.

BRACK, P. AND RIEBER, H., 1993, Towards a better definition of the Anisian/Ladinian boundary: New biostratigraphic data and correlations of boundary sections from the Southern Alps: Eclogae Geologicae Helvetiae, v. 86, p. 415–427.

BRACK, P., RIEBER, H., AND MUNDI, R., 1995. The Anisian/Ladinian boundary interval at Bagolino (southern Alps, Italy): I. Summary and new results on ammonoid horizons and radiometric age dating: Albertiana 15: 45–56.

BRALOWER, T. J., LUDWIG, K. R., OBRADOVICH, J. D., AND JONES, D. L., 1990, Berriasian (Early Cretaceous) radiometric ages from the Grindstone Creek Section, Sacramento Valley, California: Earth and Planetary Science Letters, v. 98, p. 62–73.

BRALOWER, T. J., THIERSTEIN, H. R., AND MONECHI, S., 1989, Calcareous nannofossil zonation of the Jurassic-Cretaceous boundary interval and correlation with the geomagnetic polarity time scale: Marine Micropaleontology, v. 14, p. 153–235.

BUCHER, H., 1989, Lower Anisian ammoinoids from the northern Humbolt Range (northwest Nevada, USA) and their bearing upon the Lower-Middle Triassic boundary: Eclogae Geologicae Helvetiae, v. 82, p. 945–1002.

BULOT, L., BLANC, E., THIEULOY, J. P., AND REMANE, J., 1993, La limite Berriasien-Valanginien dans le Sud-Est de la France: donnees biostratigraphiques nouvelles: Comptes Rendus de l'Academie des Sciences, Série 2, Mecanique, Physique, Chimie, Sciences de l'Univers, Sciences de la Terre, v. 316, p. 1771–1778.

BULOT, L. AND THIEULOY, J. P., 1993, Implications chronostratigraphiques de la revision de l'échelle biostratigraphique du Valanginien supérieur et de l'Hauterivien du Sud-Est de la France: Comptes Rendus de l'Academie des Sciences, Série 2, Mecanique, Physique, Chimie, Sciences de l'Univers, Sciences de la Terre, v. 317, p. 387–394.

CANDE, S. C. AND KENT, D. V., 1992, A new geomagnetic polarity time scale for the Late Cretaceous and Cenozoic: Journal of Geophysical Research, v. 97, p. 13917–13951.

CANDE, S. C., LARSON, R. L., AND LaBRECQUE, J. L., 1978, Magnetic lineations in the Pacific Jurassic Quiet Zone: Earth and Planetary Science Letters, v. 41, p. 434–440.

CECCA, F., PALINI, G., ERBA, E., PREMOLI-SILVA, I., AND COCCIONI, R., 1994, Hauterivian-Barremian chronostratigraphy based on ammonites, nannofossils, planktonic foraminifera, and magnetic chrons from the Mediterranean domain: Cretaceous Research, v. 15, p. 457–467.

CHANNELL, J. E. T. AND ERBA, E., 1992, Early Cretaceous polarity chrons CM0 to CM11 recorded in northern Italian land sections near Brescia: Earth and Planetary Science Letters, v. 108, p. 161–179.

CHANNELL, J. E. T., MASSARI, F., BENETTI, A., AND PEZZONI, N., 1990, Magneto-stratigraphy and biostratigraphy of Callovian-Oxfordian limestones from the Trento Plateau (Monti Lessini, northern Italy): Palaeogeography, Palaeoclimatology, Palaeoecology, v. 79, p. 289–303.

CLAOUE-LONG, J. C., ZICHAO, Z., GUOGAN, M., AND SHAOHUA, D., 1991, The age of the Permian-Triassic boundary: Earth and Planetary Science Letters, v. 105, p. 182–190.

COBBAN, W. A., MEREWETHER, E. A., FOUCH, T. D., AND OBRADOVICH, J. D., 1994, Some Cretaceous shorelines in the Western Interior of the United States, *in* Caputo, M. V., Peterson, J. A., and Franczyk, K. J., eds., Mesozoic Systems of the Rocky Mountain Region, USA: p. 393–425.

COPE, J. C. W., GETTY, T. A., HOWARTH, M. K., MORTON, N., AND TORRENS, T. S., 1980a, A correlation of Jurassic rocks in the British Isles. Part One: Introduction and Lower Jurassic: Oxford, Geological Society of London Special Report 14, 73 p.

COPE, J. C. W., DUFF, K. L., PARSONS, C. F., TORRENS, T. S., WIMBLEDON, W. A., AND WRIGHT, J. K., 1980b, A correlation of Jurassic rocks in the British Isles. Part Two: Middle and Upper Jurassic: Oxford, Geological Society of London Special Report 15, 109 p.

CORNET, B. AND OLSEN, P. E., 1985, A summary of the biostratigraphy of the Newark Supergroup of Eastern North America, *in* Weber, R., ed., III Congreso Latinamericano de Paleontologia, Mexico, Simposia Floras del Triasico Tardio, su Fitogeografia y Paleoecologia, Memoria: Mexico City, Universidad National Autonoma de Mexico, Instituto de Geologia, p. 67–81.

COWIE, J. W. AND BASSETT, M. G., 1989, 1989 Global Stratigraphic Chart: Ottawa, International Union of Geological Sciences.

COX, A. V. AND DALRYMPLE, G. B., 1967, Statistical analysis of geomagnetic reversal data and the precision of potassium-argon dating: Journal of Geophysical Research, v. 72, p. 2603–2614.

DAGYS, A. S., 1988, Boundary of the Lower and Middle Triassic in Boreal and Tethyan regions and correlation of Anisian deposits: Geologiya i Geofizika, v. 29, p. 3–9.

DAGYS, A. S. AND KONSTANTINOV, A. G., 1992, A new zonal scheme of the boreal Ladinian: Albertiana, v. 10, p. 17–21.

DAGYS, A. S. AND WEITSCHAT, W., 1993, Correlation of the Boreal Triassic: Mitteilungen aus dem Geologisch Paläontologischen Institut der Universität Hamburg, v. 75, p. 249–256.

DAGYS, A. S., WEITSCHAT, W., KOSTANTINOV, A., AND SOBOLEV, E., 1993, Evolution of the boreal marine biota and biostratigraphy at the Middle/Upper Triassic boundary: Mitteilungen aus dem Geologisch Paläontologischen Institut der Universität Hamburg, v. 75, p. 193–209.

DE ZANCHE, V., GIANOLLA, P., MIETTO, P., SIORPAES, C., AND VAIL, P., 1993, Triassic sequence stratigraphy in the Dolomites (Italy): Memorie di Scienze Geologiche, v. 45, p. 1–27.

DEMING, W. E., 1948, Statistical Adjustment of Data: New York, Wiley, 261 p.

FOLAND, K. A., GILBERT, L. A., SEBRING, C. A., AND JIANG-FENG, C., 1986, $^{40}Ar/^{39}Ar$ ages for plutons of the Monteregian Hills, Quebec: Evidence for a single episode of Cretaceous magmatism: Bulletin of the Geological Society of America, v. 97, p. 966–974.

FORSTER, S. C. AND WARRINGTON, G., 1985, Geochronology of the Carboniferous, Permian and Triassic, *in* Snelling, N. J., ed., The Chronology of the Geological Record: Oxford, Blackwell Scientific Publishers, Geological Society of London Memoir 10, p. 99–113.

FOWELL, S. J. AND OLSEN, P. E., 1993, Time calibration of Triassic/Jurassic microfloral turnover, eastern North America: Tectonophysics, v. 222, p. 361–369.

GAETANI, M., 1992, Report on the Symposium on Triassic Stratigraphy (Lausanne, October 20–23, 1991): Albertiana, v. 10, p. 6–9.

GAETANI, M. (ed.), 1993a, I.U.G.S. Subcommission on Triassic Stratigraphy, Anisian/Ladinian boundary field workshop, Southern Alps-Balaton Highlands, 27 June–4 July 1993: Milano, Dipartimento di Scienze della Terra, Universitá degli Studi, 117 p.

GAETANI, M., 1993b, Anisian/Ladinian Boundary Field Workshop, Report: Albertiana, v. 12, p. 5–9.

GALBRUN, B., 1984, Magnétostratigraphie de la limité Jurassique-Crétacé. Proposition d'une échelle de polarité á partir du stratotype du Berriasien (Berrias, Ardéche, France) et la Sierra de Lugar (Province de Murcie, Espagne): Paris, Mémoires des Sciences della Terre, Univ. Pierre et Marie Curie, 95 p.

GALBRUN, B., BAUDIN, F., FOURCADE, E., AND RIVAS, P., 1990, Magnetostratigraphy of the Toarcian Ammonitico Rosso limestone at Iznalloz, Spain: Geophysical Research Letters, v. 17, p. 2441–2444.

GALBRUN, B., GABILLY, J., AND RASPLUS, L., 1988, Magnetostratigraphy of the Toarcian stratotype sections at Thouars and Airvault (Deux-Sevres, France): Earth and Planetary Science Letters, v. 87, p. 453–462.

GALLET, Y., BESSE, J., KRYSTYN, L., MARCOUX, J., AND THÉVENIAUT, H., 1992, Magnetostratigraphy of the Late Triassic Bolücektasi Tepe section (southwestern Turkey): implications for changes in magnetic reversal frequency: Physics of the Earth and Planetary Interiors, v. 73, p. 85–108.

GALLET, Y., BESSE, J., KRYSTYN, L., THÉVENIAUT, H., AND MARCOUX, J., 1993, Magnetostratigraphy of the Kavur Tepe section (southwestern Turkey): A magnetic polarity time scale for the Norian: Earth and Planetary Science Letters, v. 117, p. 443–456.

GRADSTEIN, F. M., AGTERBERG, F. P., AUBRY, M.-P., BERGGREN, W. A., FLYNN, J. J., HEWITT, R., KENT, D. V., KLITGORD, K. D., MILLER, K. G., OBRADOVICH, J. D., OGG, J. G., PROTHERO, D. R., AND WESTERMANN, G. E. G., 1988, Sea level history: Science, v. 214, p. 599–601.

GRADSTEIN, F. M., AGTERBERG, F. P., OGG, J. G., HARDENBOL, J., VAN VEEN, P., THIERRY, J., AND HUANG, Z., 1994, A Mesozoic time scale: Journal of Geophysical Research, v. 99, p. 24051–24074.

GROUPE FRANÇAIS D'ETUDES DU JURASSIQUE, 1991, Réactualisation des zones d'ammonites, in 3rd International Symposium on Jurassic Stratigraphy: Poitiers, International Subcommission of Jurassic Stratigraphy, Lyon, France, abstract volume, p. 124–134.

HALLAM, A. J. M., HANCOCK, J. M., LaBREQUE, J. L., LOWRIE, W., AND CHANNELL, J. E. T., 1985, Jurassic and Cretaceous geochronology and Jurassic to Paleogene magnetostratigraphy, in Snelling, N. J., ed., The Chronology of the Geological Record: Oxford, Blackwell Scientific Publishers, Geological Society of London Memoir 10, p. 118–140.

HANCOCK, J. M., 1991, Ammonites scales for the Cretaceous System: Cretaceous Research, v. 12, p. 259–291.

HANDSCHUMACHER, D. W., SAGER, W. W., HILDE, T. W. C., AND BRACEY, D. R., 1988, Pre-Cretaceous evolution of the Pacific plate and extension of the geomagnetic polarity reversal time scale with implications for the origin of the Jurassic "Quiet Zone": Tectonophysics, v. 155, p. 365–380.

HAQ, B. U., HARDENBOL, J., AND VAIL, P. R., 1988, Mesozoic and Cenozoic chronostratigraphy and cycles of sea-level change, in Wilgus, C. K., Hastings, B. S., Ross, C. A., Posamentier, H. W., Van Wagoner, J., and Kendall, G. St. C., eds., Sea-Level Changes: An Integrated Approach: Tulsa, Society of Economic Paleontologists and Mineralogists Special Publication 42, p. 71–108.

HARLAND, W. B., COX, A. V., LLEWELLYN, P. G., PICKTON, C. A. G., SMITH, A. G., AND WALTERS, R., 1982, A Geologic Time Scale: Cambridge, Cambridge University Press, 131 p.

HARLAND, W. B., ARMSTRONG, R. L., COX, A. V., CRAIG, L. E., SMITH, A. G., AND SMITH, D. G., 1990, A Geologic Time Scale 1989: Cambridge, Cambridge University Press, 265 p.

HARPER, G. D., 1984, The Josephine ophiolite, northwestern California: Bulletin of the Geological Society of America, v. 95, p. 1009–1026.

HELLMANN, K. N. AND LIPPOLT, H. J., 1981, Calibration of the Middle Triassic time scale by conventional K-Ar and Ar-Ar dating of alkali feldspars: Journal of Geophysics, v. 50, p. 73–86.

HERBERT, T. D., 1992, Paleomagnetic calibration of Milankovitch cyclicity in Lower Cretaceous sediments: Earth and Planetary Science Letters, v. 112, p. 15–28.

HICKS, J. F., 1993, Chronostratigraphic analysis of the foreland basin sediments of the latest Cretaceous, Wyoming, U.S.A.: Unpublished Ph.D. Dissertation, Yale University, New Haven, 250 p.

HODYCH, J. P. AND DUNNING, G. R., 1992, Did the Manicouagan impact trigger end-of-Triassic mass extinction?: Geology, v. 20, p. 51–54.

HORNER, F., 1993, Palaeomagnetismus von Karbonatsedimenten der Südlichen Tethys: Implikationen für die Polarität des Erdmagnetfeldes im untern Jura und für die tektonik der Ionischen zone Griechenlands: Unpublished Ph.D. Dissertation, ETH Zürich, 139 p.

HOEDEMAEKER, PH. J. AND BULOT, L., 1990, Preliminary ammonite zonation of the Lower Cretaceous of the Mediterranean region: Géologie Alpine, v. 66, p. 123–127.

HOEDEMAEKER, PH. J., COMPANY, M., AGUIRRE-URRETTA, B., AVRAME, BOGDANOVA, T. N., BULOT, L., CECCA, F., DELANNOY, G., ETTACHFINI, M., MEMMIL, OWEN, H. G., RAWSON, A. F., SANDOVAL, J., TAVERA, J. M., THIEULOY, J. P., TOVBINA, S. Z., AND VASICEK, Z., 1993, Ammonite zonation for the Lower Cretaceous of the Mediterranean region, basis for stratigraphic correlations within IGCP Project 262: Revista Española Paleontologia, v. 8, p. 117–120.

HOLSER, W. T. AND MAGARITZ, M., 1987, Events near the Permian-Triassic boundary: Modern Geology, v. 11, p. 155–180.

HOLSER, W. T., SCHÖNLAUB, H.-P., BOECKELMANN, K., AND MAGARITZ, M., 1991, The Permian-Triassic of the Gartnerfokel-1 Core (Carnic Alps, Austria): synthesis and conclusions: Abhandlungen der Geologischen Bundesanstalt, v. 45, p. 213–232.

KAZAKOV, A. M. AND KURUSHIN, N. I., 1992, Stratigraphy of Norian and Rhaetian deposits in the northern middle Siberia: Russian Geology and Geophysics, v. 33, p. 1–8.

KENNEDY, W. J., COBBAN, W. A., AND SCOTT, G. R., 1992, Ammonite correlation of the uppermost Campanian of Western Europe, the U.S. Gulf Coast, Atlantic Seaboard and Western Interior, and the numerical age of the base of the Maastrichtian: Geological Magazine, v. 129, p. 497–500.

KENT, D. V. AND GRADSTEIN, F. M., 1985, A Cretaceous and Jurassic chronology: Bulletin of the Geological Society of America, v. 96, p. 1419–1427.

KENT, D. V. AND GRADSTEIN, F. M., 1986, A Jurassic to Recent chronology, in Vogt, P. R. and Tucholke, B. E., eds., The Western North Atlantic Region: Boulder, Geological Society of America, The Geology of North America, Volume M, p. 45–50.

LARSON, R. L. AND HILDE, T. W. C., 1975, A revised time scale of magnetic reversals for the Early Cretaceous and Late Jurassic: Journal of Geophysical Research, v. 80, p. 2586–2594.

LARSON, R .L., LANCELOT, Y., AND OTHERS, 1992, Proceedings of the Ocean Drilling Program, Scientific Results, v. 129: College Station, Ocean Drilling Program, 745 p.

LUDDEN, J., 1992, Radiometric age determinations for basement from Sites 765 and 766, Argo Abyssal Plain and Northwestern Australia: Proceedings of the Ocean Drilling Program, Scientific Results, v. 123, p. 557–559.

MARCOUX, J., 1993, Comparison between the magnetostratigraphic sequences obtained from Bolücektasi Tepe and the Kavur Tepe sections (southwestern Turkey): Albertiana, v. 12, p. 67.

MARCOUX, J., GIRARDEAU, J., FOURCADE, E., BASSOULLET, J. P., PHILIP, J., JAFFREZO, M., XUCHANG, X., AND CHENGFA, C., 1987, Geology and biostratigraphy of the Jurassic and lower Cretaceous series to the north of the Lhasa Block (Tibet, China): Geodinamica Acta, v. 1, p. 313–325.

McARTHUR, J. M., THIRWALL, M. F., CHEN, M., GALE, A. S., AND KENNEDY, W. J., 1993, Strontium isotope stratigraphy in the Late Cretaceous: Numerical calibration of the Sr isotope curve and intercontinental correlation for the Campanian: Paleoceanography, v. 8, p. 859–873.

MESEZHNIKOV, M. S., 1988, Tithonian (Volgian), in Krymholts, G. Ya., Mesezhnikov, M. S., and Westermann, G. E. G., eds., The Jurassic Ammonite Zones of the Soviet Union: Boulder, Geological Society of America Special Paper 223, p. 50–62.

MORBEY, S. J., 1978, Late Triassic and Early Jurassic subsurface palynostratigraphy in northwestern Europe, in Cramer, F. H., Diez, M. D. C. R., and Gutierrez, M., eds., First International Palynological Conference, Palinologia, extraordinario 1, p. 355–366.

MOSTLER, H., SCHEURING, B., AND URLICHS, M., 1978, Zur Mega-, Mikrofauna und Mikroflora der Koessener Schichten (alpine Obertrias) vom Weissloferbach in Tirol unter besonderer Berücksichtigung der in der suessi- und marshi-Zone auftretenden Conodonten: Schriftenreihe der Erdwissenschaftlichen Kommissionen, Österreichische Akademie der Wissenschaften, v. 4, p. 141–171.

MUTTONI, G., CHANNELL, J. E. T., NICORA, A., AND RETTORI, R., 1994, Magnetostratigraphy and biostratigraphy of an Anisian-Ladinian (Middle Triassic) boundary section from Hydra (Greece): Palaeogeography, Palaeoclimatology, Palaeoecology, v. 111, p. 249–262.

NAKAZAWA, K., 1992, The Permian-Triassic boundary: Albertiana, v. 10, p. 23–30.

NALIVKIN, D. V. (N. Rast, translator), 1973, in Geology of the USSR: Edinburgh, Oliver and Boyd, 855 p.

OBRADOVICH, J. D., 1988, A different perspective on glauconite as a chronometer for geologic time scale studies: Paleoceanography, v. 3, p. 757–770.

OBRADOVICH, J. D., 1993, A Cretaceous time scale, *in* Caldwell, W. G. E., ed., Evolution of the Western Interior Basin: Ottawa, Geological Association of Canada Special Paper 39, p. 379–396.

ODIN, G. S. (ed.), 1982, Numerical Dating in Stratigraphy, Pt. 1 and 2: Chichester, J. Wiley and Sons, 1040 p.

ODIN, G. S. (ed.), 1992a, Phanerozoic Time Scale: Paris, Bulletin of Liason and Informations, IUGS Subcommission on Geochronolology, v. 9.

ODIN, G. S. (ed.), 1992b, Phanerozoic Time Scale: Paris, Bulletin of Liason and Informations, IUGS Subcommission on Geochronolology, v. 10, p. 1–56.

ODIN, G. S. AND ODIN, C., 1990, Echelle Numerique des Temps Geologiques: Geochronologie, v. 35, p. 12–20.

OGG, J. G., 1988, Early Cretaceous and Tithonian magnetostratigraphy of the Galicia margin (Ocean Drilling Program Leg 103): College Station, Proceedings of the Ocean Drilling Program, Scientific Results, v. 103, p. 659–681.

OGG, J. G., 1995, Magnetic polarity time scale of the Phanerozoic, *in* Ahrens, T., ed., Global Earth Physics: A Handbook of Physical Constants: Washington, American Geophysical Union Reference Shelf, v. 1, p. 240–270.

OGG, J. G., COMPANY, M., STEINER, M. B., AND TAVERA, J. M., 1988, Magnetostratigraphy across the Berriasian-Valanginian stage boundary (Early Cretaceous) at Cehegin (Murcia Province, southern Spain): Earth and Planetary Science Letters, v. 87, p. 205–215.

OGG, J. G., HASENYAGER, R. W., WIMBLEDON, W. A., CHANNELL, J. E. T., AND BRALOWER, T. J., 1991a, Magnetostratigraphy of the Jurassic-Cretaceous boundary interval, Tethyan and English faunal realms: Cretaceous Research, v. 12, p. 455–482.

OGG, J. G. AND LOWRIE, W., 1986, Magnetostratigraphy of the Jurassic/Cretaceous boundary: Geology, v. 14, p. 547–550.

OGG, J. G. AND STEINER, M. B., 1991, Early Triassic magnetic polarity time scale—Integration of magnetostratigraphy, ammonite zonation and sequence stratigraphy from stratotype sections (Canadian Arctic Archipelago): Earth and Planetary Science Letters, v. 107, p. 69–89.

OGG, J. G., STEINER, M. B., OLORIZ, F., AND TAVERA, J. M., 1984, Jurassic magnetostratigraphy: 1. Kimmeridgian-Tithonian of Sierra Gorda and Carcabuey, southern Spain: Earth and Planetary Science Letters, v. 71, p. 147–162.

OGG, J. G., WIECZOREK, J., HOFFMANN, M., AND STEINER, M. B., 1991b, Jurassic magnetostratigraphy: 4. Early Callovian through Middle Oxfordian of Krakow Uplands (Poland): Earth and Planetary Science Letters, v. 104, p. 289–303.

ORBELL, G., 1983, Palynology of the British Rhaeto-Liassic: Bulletin of the Geological Survey of Great Britian, v. 44, p. 1–44.

OTTONE, E. G. AND GARCIA, G. B., 1991, A Lower Triassic microspore assemblage from the Puesto Viejo Formation, Argentina: Reviews of Palaeobotany and Palynology, v. 68, p. 217–232.

OWEN, H. G., 1984, Albian stage and substage boundaries: Bulletin of the Geological Society of Denmark, v. 33, p. 183–189.

OWEN, H. G., 1988, The ammonite zonal sequence and ammonite taxonomy in the *Douvilleiceras mammillatum* Superzone (Lower Albian) in Europe: Bulletin of the British Museum of Natural History (Geology), v. 44, p. 177–231.

PALMER, A. R., compiler, 1983, Decade of North American Geology (DNAG). Geologic Time Scale: Geology, v. 11, p. 503–504.

PESSAGNO, E. A. AND BLOME, C. D., 1990, Implications of new Jurassic stratigraphic, geochronometric and paleolatitudinal data from the western Klamath terrane (Smith River and Rogue Valley subsititute): Geology, v. 18, p. 665–668.

PRINGLE, M. S., 1992, Radiometric ages of basaltic basement recovered at sites 800, 801 and 802, Leg 129, western Pacific Ocean: College Station, Proceedings of the Ocean Drilling Program, Scientific Results, v. 129, p. 389–404.

PRINGLE, M. S., 1995, Radiometric dating of seamount edifices of guyots, western Pacific: Proceedings of the Ocean Drilling Program, Scientific Results, v. 144, in press.

PRINGLE, M. S., OBRADOVICH, J. D., AND DUNCAN, R. A., 1992, Estimated age for magnetic anomaly M0 and Interval ISEA, and a minimum estimate for the duration of the Aptian: Eos, v. 73, p. 633.

RAO, C. R., 1973, Linear Statistical Inference and Its Applications: New York, Wiley, 625 p.

REINSCH, C. H., 1967, Smoothing by spline functions: Numerische Mathematik, v. 16, p. 451–454.

RENNE, P. R. AND BASU, A. R., 1991, Rapid eruption of the Siberian Traps flood basalts at the Permo-Triassic boundary: Science, v. 253, p. 176–179.

ROBASZYNSKI, F., AMÉDRO, F., (coord.), FOUCHER, J.-C., GASPARD, D., MAGNIEZ-JANNIN, F., MANIVIT, H., AND SORNAY, J., 1980, Synthése biostratigraphique de l'Aptien au Santonien du Boulonnais. A partir de sept groupes paléontologique: Foraminiféres, nannoplancton, dinoflagellés et macrofaunes. Zonations micropaléontologiques intégrées dans le cadre du Crétacé boréal nord-européen: Revue de Micropaléontologie, v. 22, p. 195–321.

SALEEBY, J. B., 1984, Pb/U zircon ages from the Rogue River area, western Klamath belt, Klamath Mountains, Oregon: Geological Society of America, Abstracts with Programs, v. 16, p. 331.

SCHUURMAN, W. M. N., 1979, Aspects of Late Triassic palynology. 3 Palynology of the latest Triassic and earliest Jurassic deposits of the Northern Limestone Alps in Austria and southern Germany, with special reference to a palynologial characterization of the Rhaetian Stage in Europe: Reviews of Palaeobotany and Palynology, v. 27, p. 53–75.

SHARPTON, V. L., DALRYMPLE, G. B., MARIN, L. E., RYDER, G., SCHURAYTZ, B. C., AND URUTTIA-FUCUGAUCHI, J., 1992, New links between the Chicxulub impact structure and the Cretaceous/Tertiary boundary: Nature, v. 359, p. 819–821.

SNELLING, N. J., ed., 1985, The Chronology of the Geological Record: Oxford, Blackwell Scientific Publishers, Geological Society of London Memoir 10, 343 p.

STEINER, M. B., OGG, J. G., MELENDEZ, G., AND SEQUIEROS, L., 1986, Jurassic magnetostratigraphy: 2. Middle-Late Oxfordian of Aguilon, Iberian Cordillera, northern Spain: Earth and Planetary Science Letters, v. 76, p. 151–166.

SWISHER, C. C., GRAJALES-NISHIMURA, J. M., MONTANARI, A., MARGOLIS, S. V., CLAEYES, P., ALVAREZ, W., RENNE, P., CEDILLO-PARDO, E, MUARASSEE, F., CURTIS, G. H., SMIT, J., AND MCWILLIAMS, M. O., 1992, Coeval ^{40}Ar/^{39}Ar ages of 65.0 Million years ago from Chicxulub Crater melt rock and Cretaceous-Tertiary boundary tektites: Science, v. 257, p. 954–958.

TAMAKI, K. AND LARSON, R. L., 1988, The Mesozoic tectonic history of the Magellan Microplate in the western Central Pacific: Journal of Geophysical Research, v. 93, p. 2857–2874.

TAYLOR, J. R., 1982, An Introduction to Error Analysis: Mill Valley, University Science Books, 253 p.

THORKELSON, D. J., MARSDEN, H., AND MORTENSEN J., 1991, Early Jurassic volcanism north of the Bowser Basin and its role in paired subduction beneath Stikinia: Geological Society of America Abstracts and Programs, v. 23, 5, p. A191.

THORKELSON, D. J., MORTENSEN, J. K., MARSDEN, H., AND TAYLOR, R. P., 1994, Age and tectonic setting of Early Jurassic episodic volcanism along the northeastern margin of the Hazelton Trough, northern British Columbia: Boulder, Geological Society of America, Special Paper, in press.

THOMSON, R. C. AND SMITH, P. L., 1992, Pliensbachian (Lower Jurassic) biostratigraphy and ammonite fauna of the Spatsizi area, North-Central British Columbia: Bulletin of the Geological Survey of Canada , v. 437, p. 1–73.

TOZER, E. T., 1967, A standard for Triassic time: Bulletin of the Geological Survey of Canada, v. 156, 103 p.

VAN HINTE, J., 1976, A Jurassic time scale: American Association of Petroleum Geologists Bulletin, v. 60, p. 489–497.

WARRINGTON, G., 1974, Studies in the palynological biostratigraphy of the British Trias. 1. Reference sections in West Lancashire and North Somerset: Reviews of Palaeobotany and Palynology, v. 17, p. 133–147.

WEITSCHAT, W. AND DAGYS, A. S., 1989, Triassic biostratigraphy of Svalbard and a comparison with NE-Siberia: Mitteilungen aus dem Geologisch Paläontologischen Institut der Universitét Hamburg, v. 68, p. 179–213.

WESTERMANN, G. E. G., 1984, Gauging the duration of stages: A new approach for the Jurassic: Episodes, v. 7, p. 26–28.

WESTERMANN, G. E. G., 1988, Duration of Jurassic stages based on averaged and scaled subzones, *in* Agterberg, F. P. and Rao, C. N., eds., Recent Advances in Quantitative Stratigraphic Correlation: Dehli, Hindustan Publishing Company, p. 90–100.

YIN H., 1993, A proposal for the global stratotype section and point (GSSP) of the Permian-Triassic boundary: Albertiana, v. 11, p. 4–30.

ZAPFE, H., 1983, Das Forschungsprojekt "Trias of the Tethys Realm" (I.G.C.P. Project No. 4). Abschlussbericht, *in* Zapfe, H., ed., Neue Beitraege zur Biostratigraphie der Tethys Trias: Schriftenreihe der Erdwissenschaftlichen Kommissionen, Österreichisch Akademie Wissenschaft, v. 5, p. 7–16.

APPENDIX 1.—RADIOMETRIC DATA BASE FOR MESOZOIC STRATA

List of entire dataset and selected radiometric dates employed in the maximum likelihood estimates for Mesozoic stage boundaries. For further explanation see text.

Stages: The stage units heading groups of dates are approximate classification units, and bear no influence on the use of the maximum likelihood method used to calculate the age of stage boundaries. For the latter the age brackets (see above) are important. The stage bracket numbering largely corresponds to the coding used in PTS89.

Trial age: Initial estimate of stage boundary age for maximum likelihood calculations.

Number of zones: The number of (sub)zones is based on the highest-resolution ammonite zonal path.

Listing: Unique reference number for date; most items listed as new or revised are discussed in Appendix 2. Total number of dates = 323.

Date: Age date in Ma units.

Standard deviation (s): 1 standard deviation. Items listed as "revised" have had the standard deviation recalculated according to weighted averaging (Taylor, 1982) with the assumptions that the dates involved are from populations of radiometric dates with different standard deviations in the same or closely comparable samples, as listed in the cited literature. A "*" indicates an arbitrary standard deviation of 1 my (1 σ). See explanations under "Brackets" and in text.

Brackets: This is the principal standardized tool to assign chronostratigraphic uncertainty to the relative age of sediments with a specific radiometric age date. The bracket refers to the relative age of sediments immediately above and below the dated level. The A (above) and B (below) coding of geologic stages follows the scheme of PTS89. A number "1" in column C marks a directly estimated age for a stage boundary (see text); these items include numerically interpolated ages for late Cretaceous stage boundaries, and magnetochronological interpolated ages for late Jurassic to early Cretaceous stage boundaries using the Hawaiian marine magnetic anomaly sequence, and radiometric age estimates in Pliensbachian, Hettangian, and at the Anisian/Ladinian and Permian/Triassic boundaries.

Method: "1" = high-temperature; "2" = low-temperature minerals; KAr WR or KAr Bi indicate ages from bentonites coded as high-temperature.

Comments: Reference or source of age data and brief annotation. The majority of items are taken from Table 4.2 in PTS89, cited according to this table, and from the smaller Appendix B in EX88, also cited accordingly. "Revised" means that the data quoted by these authors has been re-interpreted using standard averaging procedures of analytical data, as documented in Appendix 2. "New" means that the item was not listed previously in Table 4.2 in PTS89, Appendix B in EX88, or NDS82.

Stage (Bracket); Number of Zones

Listing	Date	Standard Deviation	A	B	C	Method	Comments
Thanetian (bracket 18)							
1	52.60	1.20	18	18		2	NDS27
2	53.10	1.65	18	18		2	NDS17
3	53.50	1.90	18	18		1	B/MH ArAr WR
4	54.10	1.00	18	18		2	NDS28
5	54.40	0.97	18	18		2	NDS38
6	54.80	1.75	18	18		2	NDS16a
7	55.10	0.20	18	18		1	new; CK92
8	56.00	0.95	18	18		2	NDS39
9	56.60	3.40	18	18		2	NDS22

Stage (Bracket); Number of Zones continued

Listing	Date	Standard Deviation	A	B	C	Method	Comments
10	57.50	0.67	18	18		2	NDS113
11	57.50	1.50	18	18		2	NDS16b
12	58.25	0.31	18	18		2	FIT77
13	58.87	2.01	18	18		2	PTS105
14	59.90	2.53	18	18		2	PTS113
15	60.92	3.00	18	18		2	PTS106
16	61.02	0.45	18	18		2	FHMB1
17	59.44	0.70	18	19		2	NDS55
18	60.50	0.35	18	19		2	NDS247
Danian (bracket 19)							
19	60.10	1.82	1	19		1	PTS306 KAr WR
20	61.25	1.08	19	19		2	NDS92
21	62.06	1.50	19	19		1	PTS362 KAr BiSd
22	62.10	1.55	19	19		2	NDS114
23	63.00	0.32	19	19		1	NDS126a KAr Sd
24	63.03	0.63	19	19		2	NDS120
25	63.70	0.32	19	19		1	NDS127a
26	63.90	0.35	19	19		1	NDS127b UPb Zr
27	65.40	1.10	19	19		2	BLM
28	65.50	1.00	19	19		1	OSK ArAr Sd
29	65.80	0.70	19	19		2	NDS103
30	65.80	0.80	19	19		2	OB2
31	66.00	0.90	19	20		1	OB1 ArAr Sd
32	66.34	2.41	19	19		2	PTS307 KAr Fl
33	67.14	3.02	19	19		1	PTS329 KAr WR
34	65.00	0.04	19	20		1	new; see text
Maastrichtian (bracket 20); trial age 65.0 Ma number of zones = 15 (approx)							
35	67.17	5.00	16	20		2	PTS198a
36	65.41	1.00	20	20		2	PTS364 KAr Fl
37	66.70	1.00	20	20		2	NDS36
38	66.80	2.00	20	20		1	S3 KAr WR
39	66.80	0.55	20	20		1	new; ArAr Sd; OB93
40	67.00	0.70	20	20		1	PTS365 KAr BiSd
41	67.05	3.02	20	20		2	A449
42	69.00	0.70	20	20		2	NDS104
43	69.80	2.10	20	20		1	S8
44	69.80	2.10	20	20		1	S7 KAr WR
45	69.42	0.18	20	20		1	new; ArAr Sd; OB93
46	70.40	2.11	20	20		2	S6
47	70.63	1.93	20	20		2	PTS54
48	71.50	0.74	20	20		2	NDS139
49	69.50	2.10	20	21		1	S9
50	70.30	2.11	20	21		2	S4
51	74.18	0.77	20	21		2	PTS365 KAr BiSd
52	71.60	0.70	20	21		1	new; McArthur and others (1993)
53	71.30	0.25	20	21	1	1	new; OB93
Campanian (bracket 21); trial age 71.3 Ma; number of zones = 23							
54	71.60	1.35	21	21		2	NDS116
55	72.98	0.70	21	21		2	NDS105 KAr Fl
56	73.19	0.70	21	21		2	NDS105; McArthur and others (1993)
57	73.40	1.00	21	21		1	OSK ArAr Sd
58	73.35	0.17	21	21		1	new; ArAr Sd; OB93
59	73.90	0.70	21	21		2	NDS105
60	74.00	2.22	21	21		1	S5 KAr WR
61	74.31	0.70	21	21		2	NDS105
62	74.52	0.70	21	21		2	NDS105
63	74.60	1.35	21	21		2	NDS116
64	75.33	0.66	21	21		2	NDS140
65	75.37	0.18	21	21		1	new; ArAr Sd; OB93
66	75.89	0.36	21	21		1	new; ArAr Sd; OB93
67	75.50	0.60	21	21		1	AB ArAr Sd
68	75.20	0.50	21	21		1	new; CK92
69	79.12	0.80	21	21		2	NDS106
70	79.30	0.67	21	21		2	NDS117
71	79.73	0.80	21	21		2	NDS106
72	79.83	2.00	21	21		1	A410
73	80.04	0.80	21	21		2	NDS106

Stage (Bracket); Number of Zones continued

Listing	Date	Standard Deviation	A	B	C	Method	Comments
74	80.54	0.55	21	21		1	new; Ar/Ar Sd; OB93
75	81.50	1.50	21	21		1	NDS163 KAr WR
76	81.71	0.17	21	21		1	new; ArAr Sd; OB93
77	81.88	3.00	21	21		1	A508 KAr SdBi
78	82.90	2.84	21	21		2	PTS62
79	84.02	5.00	21	21		2	A417
80	83.50	0.25	21	22	1	1	new; OB93

Santonian (bracket 22); trial age 83.5 Ma; number of zones = 5

Listing	Date	Standard Deviation	A	B	C	Method	Comments
81	82.38	1.50	22	22		1	HMFT RbSr WRBi
82	83.91	0.21	22	22		1	new; ArAr Sd; OB93
83	84.09	0.20	22	22		1	new; ArAr Sd; OB93
84	84.90	0.15	22	22		1	new; ArAr Sd; OB93
85	84.40	1.00	22	22		1	OSK ArAr Sd
86	84.43	0.80	22	22		2	NDS107
87	84.94	2.99	22	22		2	PTS229
88	89.03	3.00	22	22		1	A509 KAr Sd
89	86.30	0.25	22	23	1	1	new; OB93

Coniacian (bracket 23); trial age 86.3 Ma; number of zones = 4

Listing	Date	Standard Deviation	A	B	C	Method	Comments
90	85.87	0.89	23	23		2	NDS83a
91	86.00	4.00	23	23		1	MB ArAr WR
92	86.80	1.65	23	23		2	NDS86
93	86.92	0.20	23	23		1	new; ArAr Sd; OB93
94	87.02	0.65	23	23		2	NDS83b
95	88.34	0.30	23	23		1	new; ArAr Sd; OB93
96	88.67	3.07	23	23		2	PTS57
97	90.50	1.05	23	23		2	NDS60
98	88.82	0.90	23	24		2	NDS108
99	88.70	0.25	23	24	1	1	new; OB93

Turonian (bracket 24); trial age 88.7 Ma; number of zones = 13

Listing	Date	Standard Deviation	A	B	C	Method	Comments
100	92.20	2.00	1	24		1	LAN1 ArAr Hb
101	87.30	2.05	24	24		2	NDS94b
102	87.60	1.30	24	24		2	NDS82b
103	88.10	0.75	24	24		2	NDS227
104	88.17	1.50	24	24		2	NDS82a
105	88.70	1.10	24	24		2	NDS164
106	89.50	0.75	24	24		2	NDS226a
107	90.10	3.60	24	24		1	SM1
108	90.21	0.36	24	24		1	new; ArAr Sd; OB93
109	90.51	0.22	24	24		1	new; ArAr Sd; OB93
110	90.97	0.90	24	24		2	NDS109
111	93.69	2.40	24	24		2	SM2
112	88.75	1.26	24	25		2	NDS95
113	92.09	3.00	24	25		1	A510 KAr Sd
114	92.76	0.43	24	25		2	NDS118
115	93.25	0.28	24	24		1	new; ArAr Sd; OB93
116	93.40	0.31	24	24		1	new; ArAr Sd; OB93
117	93.30	0.10	24	25	1	1	new; OB93

Cenomanian (bracket 25); trial age 93.3 Ma; number of zones = 23 (approx)

Listing	Date	Standard Deviation	A	B	C	Method	Comments
118	91.83	3.67	1	25		1	PTS335 KAr WR
119	87.60	1.60	25	25		2	NDS69
120	87.93	4.00	25	25		2	A418
121	89.50	1.65	25	25		2	NDS81
122	89.80	1.80	25	25		2	NDS59
123	90.50	3.00	25	25		2	NDS62
124	90.60	0.75	25	25		2	NDS226b
125	90.86	3.63	25	25		2	PTS209
126	91.20	0.73	25	25		2	NDS81
127	92.20	1.55	25	25		2	NDS85
128	92.40	0.82	25	25		2	NDS119a
129	93.30	0.78	25	25		2	NDS80a
130	93.42	0.90	25	25		2	NDS110
131	93.51	0.14	25	25		1	new; ArAr Sd; OB93
132	93.90	0.36	25	25		1	new; ArAr Sd; OB93
133	93.49	0.45	25	25		1	new; ArAr Sd; OB93
134	94.20	0.35	25	25		2	NDS211
135	94.24	0.90	25	25		2	NDS110
136	94.63	0.30	25	25		1	new; ArAr Sd; OB93
137	94.70	0.55	25	25		2	NDS62
138	94.93	0.27	25	25		1	new; ArAr Sd; OB93

Stage (Bracket); Number of Zones continued

Listing	Date	Standard Deviation	A	B	C	Method	Comments
139	95.00	1.00	25	25		2	NDS64
140	95.00	1.00	25	25		2	NDS96
141	95.32	0.66	25	25		2	NDS80b
142	95.40	0.75	25	25		2	NDS67
143	95.78	0.30	25	25		1	new; ArAr Sd; OB93
144	95.86	0.22	25	25		1	new; ArAr Sd; OB93
145	96.53	1.07	25	25		2	NDS119b
146	97.40	1.25	25	25		2	NDS68
147	97.69	3.91	25	25		2	PTS211
148	97.17	0.34	25	25		1	new; ArAr Sd; OB93
149	98.22	3.04	25	25		2	PTS226
150	98.52	0.20	25	25		1	new; ArAr Sd; OB93
151	98.54	0.35	25	25		1	new; ArAr Sd; OB93
152	98.74	0.30	25	25		1	new; ArAr Sd; OB93
153	96.50	0.60	25	128		1	TWH1 UPb Zr
154	102.31	3.00	25	26		1	PTS202 KAr BiSd
155	95.00	1.00	25	26		2	revised; NDS96
156	98.50	0.25	25	26	1	1	new; OB93

Albian (bracket 26); trial age 98.5 Ma; number of zones = 25

Listing	Date	Standard Deviation	A	B	C	Method	Comments
157	113.85	2.00	20	26		1	PTS217 UPb Zr
158	96.18	3.11	26	26		2	PTS230
159	96.18	3.14	26	26		2	PTS51
160	96.50	1.35	26	26		2	NDS145
161	97.50	1.00	26	26		2	NDS111
162	97.60	0.48	26	26		2	NDS144b
163	97.60	1.00	26	26		1	NDS111 KAr Sd
164	98.22	2.00	26	26		1	PTS204 KAr BiSd
165	98.22	3.22	26	26		2	PTS56
166	98.35	1.16	26	26		2	NDS65
167	98.70	2.50	26	26		2	NDS61
168	98.90	1.23	26	26		2	NDS97a
169	99.00	1.12	26	26		2	NDS63
170	99.24	3.38	26	26		2	PTS228
171	99.25	1.39	26	26		2	NDS66
172	99.50	1.55	26	26		2	new; KAr; Odin and Hunziker (cited in OB93)
173	98.40	1.60	26	26		2	new; KAr; Odin and Hunziker (cited in OB93)
174	99.40	0.65	26	26		2	NDS157a
175	99.60	2.50	26	26		2	NDS67
176	99.70	1.10	26	26		2	NDS144d
177	99.72	0.76	26	26		2	NDS79b
178	99.77	0.98	26	26		2	NDS78a
179	100.00	0.80	26	26		2	NDS144a
180	100.27	3.00	26	26		2	PTS242
181	100.60	0.50	26	26		2	NDS144c
182	100.60	2.50	26	26		2	NDS97
183	100.62	4.02	26	26		2	PTS212
184	100.62	4.00	26	26		2	PTS220
185	102.57	4.10	26	26		1	PTS336 KAr WR
186	103.10	0.95	26	26		2	NDS144e
187	103.55	4.00	26	26		2	A428
188	103.58	0.72	26	26		2	NDS70
189	104.40	1.05	26	26		2	NDS157b
190	105.36	0.91	26	26		2	NDS78b
191	106.00	0.50	26	26		2	NDS143
191	107.45	5.00	26	26		2	A429
192	107.10	0.15	26	26		2	new; OB93
193	110.48	3.87	26	26		2	PTS219
194	114.76	4.01	26	26		2	PTS223
195	116.05	1.24	26	26		2	NDS98a
196	104.53	3.89	26	27		2	PTS60
197	111.00	2.00	26	27		1	new; Marcoux and others (1987)
198	112.00	0.50	26	27	1	1	new; OB93

Aptian (bracket 27); trial age 112 Ma; number of zones = 9

Listing	Date	Standard Deviation	A	B	C	Method	Comments
199	105.80	1.31	27	27		2	NDS77a
200	107.30	1.95	27	27		2	NDS71
201	108.20	0.95	27	27		2	NDS146
202	109.30	3.00	27	27		2	NDS188 KAr Fl

Stage (Bracket); Number of Zones continued

Listing	Date	Standard Deviation	A	B	C	Method	Comments
203	112.05	1.17	27	27		2	NDS98b
204	113.00	1.00	27	27		1	JE1 ArAr Sd
205	113.00	0.70	27	27		1	OSK ArAr Sd
206	114.00	0.65	27	27		1	new; ArAr Sd; OB93
207	115.00	0.03	27	27		1	new; Pringle and others (1992)
208	115.47	1.26	27	27		2	NDS77b; revised PTS89
209	117.62	3.88	27	27		2	PTS50
210	125.26	6.00	27	28		1	SMYH1
211	124.00	1.00	27	28		1	new; Ar/Ar; Herbert (1992)
212	121.00	0.50*	27	28		1	1 new; OB93 and this study

Barremian (bracket 28); trial age 122 Ma; number of zones = 9

Listing	Date	Standard Deviation	A	B	C	Method	Comments
213	123.50	0.50	28	28		1	new; Pringle (pers. commun., 1993, see text)
214	132.83	5.31	28	30		2	PTS215
215	126.00	1.50	28	29		2	NDS162b
216	126.90	1.00*	28	29	1	1	new; this study

Hauterivian (bracket 29); trial age 127 Ma; number of zones = 12

Listing	Date	Standard Deviation	A	B	C	Method	Comments
217	133.50	2.84	1	29		1	LBM1
218	119.30	0.45	29	29		2	NDS162a
219	133.60	2.00	29	33		1	PTS75R KAr BiHb
220	127.60	0.20	29	29		1	new; Pringle (pers. commun., 1993, see text)
221	136.00	2.00	29	33		1	LJ3
222	132.00	1.00*	29	30	1	1	new; this study

Valanginian (bracket 30); trial age 130 Ma; number of zones = 10

Listing	Date	Standard Deviation	A	B	C	Method	Comments
223	126.81	3.80	1	30		1	A406
224	122.09	4.88	30	30		2	PTS322
225	125.02	5.00	30	30		2	A430
226	136.50	2.70	30	30		1	LJ4 KAr Bi
227	146.00	4.26	30	31		1	LBM2
228	138.40	1.00*	30	31	1	1	new; this study

Berriasian (bracket 31); trial age 135 Ma; number of zones = 7

Listing	Date	Standard Deviation	A	B	C	Method	Comments
229	130.88	5.24	1	31		1	PTS238 KAr Bi
230	133.95	4.00	31	31		2	PTS177
231	138.00	4.00	31	32		1	W misc
232	137.10	0.60	31	31		1	new; Bralower and others (1990)
233	144.20	1.00*	31	32	1	1	new; this study

Tithonian (bracket 32); trial age 142 Ma; number of zones = 20

Listing	Date	Standard Deviation	A	B	C	Method	Comments
234	131.10	1.50	32	32		2	revised EX88; NDS76
235	132.83	5.31	32	32		2	PTS322
236	134.97	4.00	32	32		2	PTS178
237	135.53	1.20	32	32		2	revised EX88, PTS89; NDS99
238	141.40	1.75	32	32		2	NDS228
239	152.50	2.00	32	128		1	HMP2 UPb Zr
240	154.00	2.00	32	128		1	HMP3 UPb Zr
241	158.42	3.57	32	128		1	H445 KAr Hb
242	150.70	1.50*	32	33	1	1	new; this study
243	139.66	4.00	29	32		1	PTS76

Kimmeridgian (bracket 33); trial age 147 ma; number of zones = 15

Listing	Date	Standard Deviation	A	B	C	Method	Comments
244	150.50	2.00	1	33		1	HS1 UPb Zr
245	155.00	3.00	33	34		1	revised; SCH1; UPb Zr
246	153.32	4.60	1	33		2	A480
247	155.00	3.00	1	33		1	SBG UPb Zr
248	155.28	4.66	1	33		1	A481
249	138.45	1.06	33	33		2	revised EX88; NDS142
250	153.50	2.00*	33	34	1	1	new; this study

Oxfordian (bracket 34); trial age 154 Ma; number of zones = 20

Listing	Date	Standard Deviation	A	B	C	Method	Comments
251	155.30	3.40	34	34		1	new; Ludden (1992)
252	148.00	0.87	34	34		2	revised EX88; NDS141 (KAr)
253	157.00	2.00	34	128		1	HS2 UPb Zr
254	161.00	2.00	34	128		1	HMP1 UPb Zr
255	162.00	2.00	34	128		1	HMP4 UPb Zr

Stage (Bracket); Number of Zones continued

Listing	Date	Standard Deviation	A	B	C	Method	Comments
256	160.00	2.00*	34	34		1	new; this study

Callovian (bracket 35); trial age 159 Ma; number of zones = 18

Listing	Date	Standard Deviation	A	B	C	Method	Comments
257	159.00	5.00	32	35		1	revised PTS89; BL1 KAr Hb
258	166.80	4.50	35	35		1	new; Pringle (1992)

Bathonian (bracket 36); trial age 164 Ma; number of zones = 15

Listing	Date	Standard Deviation	A	B	C	Method	Comments
259	160.14	6.00	36	37		1	A419 KAr WR
260	168.50	4.00	36	37		1	BFSB6R KAr WR
261	173.00	4.00	27	37		2	WEST2 KAr Fl
262	168.92	2.00	32	37		2	PTS90
263	161.30	3.20	35	37		1	new; Odin (1992, Bull 9)

Bajocian (bracket 37); trial age 169 Ma; number of zones = 20

Listing	Date	Standard Deviation	A	B	C	Method	Comments
264	168.00	2.00	37	37		1	new; KAr, WHR; Gillot in Odin (1992b)
265	172.00	2.00	37	37		1	new; KAr, WHR; Gillot in Odin (1992b)
266	171.00	3.50	37	37		1	NDS182 KAr Hb

Aalenian (bracket 38); trial age 176 Ma; number of zones = 9

Listing	Date	Standard Deviation	A	B	C	Method	Comments
267	171.67	5.15	34	38		2	A479

Toarcian (bracket 39); trial age 179 Ma; number of zones = 24

Listing	Date	Standard Deviation	A	B	C	Method	Comments
268	182.50	2.84	39	40		1	NDS183a KAr Hb
269	185.00	3.00	39	40		1	HLB1 ArAr BiFl
270	190.00	7.00	39	40		1	NDS183b RbSr Hb
271	198.86	2.65	39	40		1	NDS184R KAr BiHb
272	197.60	2.50	39	41		1	NDS184R ArAr Bi
273	191.25	3.38	39	45		1	FW40 KAr WRFl
274	193.00	1.50	39	45		1	FW42 KAr WR
275	193.00	1.50	39	45		1	FW41 RbSr WR

Pliensbachian (bracket 40); trial age 188 Ma; number of zones = 15

Listing	Date	Standard Deviation	A	B	C	Method	Comments
276	194.12	0.62	40	40	1	1	new; Thorkelsen and others (1994)

Sinemurian (bracket 41); trial age 197 Ma; number of zones = 17

Listing	Date	Standard Deviation	A	B	C	Method	Comments
277	203.20	1.95	1	41		1	revised; NDS181 KAr Hb
278	196.00	4.50	40	43		1	revised; Forster and Warrington (1985)
279	195.00	4.20	40	43		1	revised; Forster and Warrington (1985)
280	200.00	5.00	41	42		1	NDS203 KAr WR
281	202.40	2.13	41	42		2	NDS248
282	206.00	6.00	41	43		2	WEST1 KAr Fl

Hettangian (bracket 42); trial age 201 Ma; Number of zones = 8

Listing	Date	Standard Deviation	A	B	C	Method	Comments
283	210.36	6.00	1	42		1	A478 KAr Hb
284	205.00	2.50	42	44		2	NDS177
285	201.00	0.70	42	42	1	1	new; Hodych and Dunning (1992)
286	201.10	0.71	42	42	1	1	" "

Rhaetian (bracket 43); trial age 205 Ma; number of zones = 6

Listing	Date	Standard Deviation	A	B	C	Method	Comments
287	216.00	2.00*	43	44		1	NDS137a (T13 in EX88) KAr BiHb
288	216.00	3.00	43	44		1	NDS137b RbSr WRBi
289	213.00	3.50	43	128		1	NDS199 KAr WR

Norian (bracket 44); trial age 210 Ma; number of zones = 16

Listing	Date	Standard Deviation	A	B	C	Method	Comments
290	204.00	4.50	1	44		1	NDS174 KAr WR
291	207.00	1.03	1	44		1	FW36 RbSr WR
292	207.00	4.00	1	44		1	NDS178b
293	208.00	2.00	1	44		1	NDS179 KAr Hb
294	210.00	1.05	1	44		2	NDS204
295	210.00	4.00	1	44		1	NDS178a KAr HbBi
296	215.70	2.70	1	44		1	NDS178R ArAr HbBi
297	211.00	4.22	44	44		1	KM1 UPb Zr
298	221.00	6.00	44	45		1	NDS170 KAr Hb

Carnian (bracket 45); trial age 220 Ma; number of zones = 10

Listing	Date	Standard Deviation	A	B	C	Method	Comments
299	215.82	5.00	1	45		1	PTS160 UPb Ur
300	220.00	5.00	1	45		1	NDS171 KAr HbBi
301	231.00	2.48	1	45		1	NDS201a KAr Ph
302	229.00	2.50	45	45		1	NDS187 KAr WR
303	230.50	2.48	45	46		1	NDS193 KAr WR

Ladinian (bracket 46); trial age 230 Ma; number of zones = 12

Listing	Date	Standard Deviation	A	B	C	Method	Comments
304	225.00	2.00	1	46		2	FW33

Stage (Bracket); Number of Zones continued

Listing	Date	Standard Deviation	Brackets A	B	C	Method	Comments
305	225.68	8.00	1	46		1	PTS358b
306	233.50	2.48	1	46		2	NDS194
307	235.78	7.07	1	46		1	PTS361
308	237.06	3.00	1	46		1	PTS361
309	233.00	4.50	46	47	1	1	NDS196
310	232.00	4.50	46	47	1	1	"

Anisian (bracket 47); trial age 235 Ma; number of zones = 16

311	237.00	2.20	47	49		1	NDS186; revised
312	240.00	4.80	47	47		1	KM2
313	243.00	2.12	47	128		1	NDS194

Olenekian and Induan (bracket 48); trial age 240 Ma; number of zones: Olenekian-8, Induan-9

314	248.00	7.00	48	49		1	NDS158
315	243.92	7.32	48	49		1	PTS357
316	250.00	4.94	48	84		2	PTS346
317	248.30	0.30	48	49	1	1	new; ArAr, WHR; Renne and Basu (1991)
318	247.50	0.70	48	49	1	1	new; ArAr, WHR; Renne and Basu (1991)

Tatarian (bracket 49); trial age 248 Ma

319	251.20	1.70	48	49		1	new; ArAr, Zr; Claoue-Long and others (1992)
320	250.00	3.00	48	49		1	new; RbSr, Sd; Odin (1992b)
321	248.85	8.71	49	49		1	A434; KAr, WR
322	250.00	1.79	49	49		1	NDS205; KAr, WR
323	253.00	2.50	49	49		1	FW22; KAr, WR

APPENDIX 2.—COMMENTS ON AGES IN DATABASE

Comments on selected radiometric and chronostratigraphic items, listed in Appendix 1 as "new" or "revised."

Items 39, 45, 58, 65, 66, 68, 76, 82, 83, 84, 93, 95, 108, 109, 115, 116, 131, 132, 133, 136, 138, 143, 144, 148, 150, 151, 152, 172, 173, 192, 206, 211, 232. These items are Ar/Ar dates on sanidine in bentonites as listed in OB93 and have been normalized to an age of 520.4 Ma for the lab calibration standard MMhb-1.

Item 155: NDS96 (95.00 ± 1.00 Ma). PTS89 assigns the relative age as intra Albian. The original designation in NDS82 is late Albian to early Cenomanian Age; hence, we recode 26–26 as 25–26.

Item 203: NDS98b (112.05 ± 1.17 Ma). EX88 (item C26) quote an age of 112.0 ± 3.3 Ma (2σ). The actual dates are 111.7 ± 1.75 Ma (1σ) and 112.4 ± 1.75 Ma (1σ), which averages as 112.05 ± 1.17 Ma. The latter date is also used in PTS89.

Item 208: NDS77 (115.47 ± 1.26 Ma). PTS89 cite an age of 115.80 ± 1.24 Ma for the average of two Rb/Sr ages on Late Aptian glauconites. Our averaging procedure yields 115.47 ± 1.26 Ma.

Item 211: (124.00 ± 1.00 Ma). Herbert (1992) quotes an average Ar/Ar age of 124.00 ± 1.00 Ma on reversed magnetized plutons from Canada (Foland and others, 1986), which may fit at the top of the M-reversal sequence. According to Foland and others (1986) paleo-inclination studies and the constant reversed polarity of the extensive pluton complex are indicative of emplacement within 2 my and certainly less than 5 my. Hence, the plutons may belong in one of the longer reversal sequences at the top of the M-sequence, and M3r is a good

candidate. This suggests the relative age for the plutons may be Barremian stage.

Item 213: (123.5 ± 0.5 Ma). Pringle (pers. commun., 1993) reports 123.5 ± 0.5 Ma (1σ) for polarity chron M1r on MIT seamount (Ar/Ar on oceanic basement lavas). The relative age is latest Barremian.

Items 215, 218: NDS162a,b. PTS89 twice use the same age of 126.0 ± 1.5 Ma, listed as NDS162b, once for Hauterivian strata and once for Barremian strata. The glauconites are from a level at about the Hauterivian-Barremian boundary, hence we assign a Hauterivian-Barremian age. EX88 does not utilize this age, but only NDS162a of 119.3 ± 0.45 Ma for latest Hauterivian glauconites. NDS82 mentions that this second and younger age might be unreliable due to upwarping of the adjacent salt diaper. However, NDS162b also maybe unreliable due to mineral "inclusions."

Item 220: (127.6 ± 0.2 Ma). Using Ar/Ar dating of basement basalts of Resolution Guyot (ODP Site 866A), Pringle (pers. commun., 1993) dates polarity chron M5r of ?late Hauterivian age, as 127.6 ± 0.2 Ma (1σ).

Item 243: (139.66 ± 4.0 Ma). PTS89 reference PTS #76 is cited as Kimmeridgian age, at location Loomis, Sierra Nevada, California. The original age is given as 143 Ma but based on a different noralization. These quartz-diorite or gabbro intrusions are associated with the "Nevadan orogeny" and intrude the Mariposa Formation, which is dated by ammonites as Late Oxfordian- Early Kimmeridgian Age. The upper age is less certain, but is definitely pre-Barremian, because related batholiths (e.g., Shasta Bally) are overlain by sediments of this age. There is inadequate control on the age of the intrusions, with a possible Kimmeridgian through Hauterivian range.

Item 251: (155.30 ± 3.40 Ma). M26r (Late Oxfordian) age ocean crust at ODP Site 765, Argo Abyssal Plain, is dated at 155.3 ± 3.4 Ma, using the Ar/Ar method (Ludden, 1992).

Item 257: Callovian-Tithonian interval. PTS89 quotes a K/Ar age of 159 Ma for intra Callovian volcanics, dated by ammonites in the Coloradito Formation of California (Boles and Landis, 1984). This age is revised to involve volcanics underlain by Callovian strata and overlain by Tithonian radiolarites, hence the bracket changes from 35–35 to 32–35.

Item 258: Callovian Stage. Pringle (1992) dates oceanic crust at ODP Site 801 at 166.8 ± 4.5 Ma; sediment in the upper crust contain radiolarians of early Callovian or latest Bathonian Age. The crust slightly predates M38.

Item 263: brackets 35/37; "Bathonian" age. High-temperature radiometrics of 161.3 ± 3.2 Ma, on lavas associated with sediments younger than Bajocian Age, and older than Callovian sediments in the Caucasus (Odin, 1992b).

Items 264 and 265: Bajocian Stage. Gillot in Odin (1992b) dated volcanic hornblende with K/Ar as 168–172 ± 2.0 Ma. The rocks are identified as of Bajocian Age.

Item 276: Early Pliensbachian Stage. Thorkelsen and others (1994), and J. Mortensen (pers. commun., 1993) report three U/Pb ages of 194.8 ± 1.0 Ma, 193.7 ± 1.4 Ma, and 193.7 ± 0.95 Ma (1s) on zircons in the Cold Fish volcanics, British Columbia. Ammonites in sedimentary beds alternating with the volcanics, include *Miltoceras* sp., *Tropidoceras* sp., *Metaderoceras* sp. aff. *talkeetnaense*, *Acanthopleuroceras* sp. aff *A. stahli*, and *Metaderoceras evolutum*, and assign the Cold Fish volcanics an early to middle Carixian (Early Pliensbachian) Age (Thomson and Smith, 1992).

Item 277: NDS181; earliest Jurassic Period. EX88 item J2 quote 204 ± 5.0 Ma for item NDS181 of Sinemurian-Early Pliensbachian relative age. The same item is quoted in PTS89 as 203 ± 1.95 Ma between Sinemurian and Recent. In fact, NDS181 quotes two ages: one at 204 ± 5 Ma and one 202 ± 6 Ma (2σ). PTS89 pooled these two dates and divided the combined standard deviation by the square root of the number of analysis. However, this is only correct if the standard deviation for both analysis are equal. The correctly interpolated age is 203.2 ± 1.95 Ma (1σ).

Item 280: NDS203. EX88 item J3 is NDS203. EX88 picks one value (202 ± 6 Ma) from NDS203 where in fact there are 2 age dates: 202 ± 6 Ma and 206 ± 6 Ma. The correct (pooled) age is 204 ± 4.24 Ma, used in this study. PTS89 gives an age for item NDS203 of 200 ± 5 Ma.

Item 311: NDS186. Brackets Anis/Induan interval: 237 ± 2.2 Ma (1σ) in tuffs; K/Ar high-temperature volcanic/sedimentary Puesto Viejo Formation in Argentina (NDS 186; item 28 in Forster and Warrington, 1985). The basal part of the formation has a palynological date, suggesting an Early Triassic Age (Induan, ?Olenekian; Ottone and Garcia, 1991).

PART III
CENOZOIC ERA

A REVISED CENOZOIC GEOCHRONOLOGY AND CHRONOSTRATIGRAPHY

WILLIAM A. BERGGREN
Department of Geology and Geophysics, Woods Hole Oceanographic Institution, Woods Hole, MA 02543
DENNIS V. KENT
Lamont-Doherty Earth Observatory of Columbia University, Palisades, NY 10964
CARL C. SWISHER, III
Berkeley Geochronology Center, Berkeley, CA 94709
AND
MARIE-PIERRE AUBRY
Institute des Sciences de l'Evolution, Université Montpellier II, 34095 Montpellier, Cedex 05, France

ABSTRACT: Since the publication of our previous time scale (Berggren and others, 1985c = BKFV85) a large amount of new magneto- and biostratigraphic data and radioisotopic ages have become available. An evaluation of some of the key magnetobiostratigraphic calibration points used in BKFV85, as suggested by high precision ^{40}Ar/^{39}Ar dating (e.g., Montanari and others, 1988; Swisher and Prothero, 1990; Prothero and Swisher, 1992; Prothero, 1994), has served as a catalyst for us in developing a revised Cenozoic time scale. For the Neogene Period, astrochronologic data (Shackleton and others, 1990; Hilgen, 1991) required re-evaluation of the calibration of the Pliocene and Pleistocene Epochs. The significantly older ages for the Pliocene-Pleistocene Epochs predicted by astronomical calibrations were soon corroborated by high precision ^{40}Ar/^{39}Ar dating (e.g., Baksi and others, 1992; McDougall and others, 1992; Tauxe and others, 1992; Walter and others, 1991; Renne and others, 1993). At the same time, a new and improved definition of the Late Cretaceous and Cenozoic polarity sequence was achieved based on a comprehensive evaluation of global sea-floor magnetic anomaly profiles (Cande and Kent, 1992). This, in turn, led to a revised Cenozoic geomagnetic polarity time scale (GPTS) based on standardization to a model of South Atlantic spreading history (Cande and Kent, 1992/1995 = CK92/95).

This paper presents a revised (integrated magnetobiochronologic) Cenozoic time scale (IMBTS) based on an assessment and integration of data from several sources. Biostratigraphic events are correlated to the recently revised global polarity time scale (CK95). The construction of the new GPTS is outlined with emphasis on methodology and newly developed polarity history nomenclature. The radioisotopic calibration points (as well as other relevant data) used to constrain the GPTS are reviewed in their (bio)stratigraphic context.

An updated magnetobiostratigraphic (re)assessment of about 150 pre-Pliocene planktonic foraminiferal datum events (including recently available high southern (austral) latitude data) and a new/modified zonal biostratigraphy provides an essentially global biostratigraphic correlation framework. This is complemented by a (re)assessment of nearly 100 calcareous nannofossil datum events. Unrecognized unconformities in the stratigraphic record (and to a lesser extent differences in taxonomic concepts), rather than latitudinal diachrony, is shown to account for discrepancies in magnetobiostratigraphic correlations in many instances, particularly in the Paleogene Period. Claims of diachrony of low amplitude (<2 my) are poorly substantiated, at least in the Paleocene and Eocene Epochs.

Finally, we (re)assess the current status of Cenozoic chronostratigraphy and present estimates of the chronology of lower (stage) and higher (system) level units. Although the numerical values of chronostratigraphic units (and their boundaries) have changed in the decade since the previous version of the Cenozoic time scale, the relative duration of these units has remained essentially the same. This is particularly true of the Paleogene Period, where the Paleocene/Eocene and Eocene/Oligocene boundaries have been shifted ~2 my younger and the Cretaceous/Paleogene boundary ~1 my younger. Changes in the Neogene time scale are relatively minor and reflect primarily improved magnetobiostratigraphic calibrations, better understanding of chronostratigraphic and magnetobiostratigraphic relationships, and the introduction of a congruent astronomical/paleomagnetic chronology for the past 6 my (and concomitant adjustments to magnetochron age estimates).

INTRODUCTION

The historical and methodologic background to the construction of the Paleogene (Berggren and others, 1985b = BKF85; Aubry and others, 1988) and Neogene (Berggren and others, 1985a = BKV85) components of an integrated Cenozoic time scale have been explained in detail. In the decade since the publication of our previous Cenozoic time scale (BKFV85), major advances in radioisotopic dating, magnetostratigraphy and (calcareous) plankton biostratigraphy, as well as the introduction of new approaches to geochronology (astrochronology; e.g., Shackleton and others, 1990; Hilgen, 1991) have forced a reevaluation and eventual updating of that study.

Cande and Kent (1992) presented a revised Cenozoic magnetochronology based on an evaluation of sea-floor magnetic anomalies. The geomagnetic polarity sequence was based primarily on data from the South Atlantic (with fine-scale data inserted from faster spreading regions of the Pacific and Indian Ocean) using a combination of finite rotation and averages of anomaly spacings from stacked profiles projected onto a synthetic sea-floor spreading flow line. The time scale was then generated by fitting a spline function to a set of nine age calibration points plus the zero-age ridge axis to the composite polarity sequence.

The present version of the global polarity time scale (GPTS) differs from that presented in Cande and Kent (1992) in the following respects: (1) the astronomical time scale values of polarity boundaries from Chron C1n to Subchron C3n.4n$_{(0)}$ (Thvera) (0–5.23 Ma) are accepted as standard which results in coherent and congruent magnetostratigraphic and astronomic chronologies back to 5.23 Ma, and (2) an age of 65.0 Ma is used for the Cretaceous/Paleocene (K/P) boundary (vs. 66.0 Ma in CK92). These changes, and the resulting modification to the GPTS, were revised in CK95.

We assess over 250 pre-Pliocene first-order (and to a lesser extent, second) correlations between calcareous plankton datum events and the GPTS, resulting in a significant improvement in biochronologic resolution and accuracy over a wide range of biogeographies and biostratigraphies compared to that proposed in 1985.

In recent years, we have witnessed a dramatic improvement in both geochronologic precision and accuracy (e.g., ^{40}Ar/^{39}Ar dating; astronomical chronology). The puzzling discordance between astrochronology and magnetochronology beyond ~3.5 Ma has been recently resolved and we can now look forward to the next generation of time scales, improvements to which we believe will come in the following areas: (1) extension of the astronomical time scale into the Miocene Epoch (e.g., Krijgsman and others, 1994; Shackleton and others, 1995), (2) fine tuning of the chronology of the astronomical time scale (e.g., Langereis and others, 1994), and (3) application of Milankovitch climatic cyclicity for high-resolution relative chronology in older parts of the stratigraphic record (e.g., Herbert and d'Hondt, 1990).

In the discussion below and elsewhere in the text, repeated reference to various time scale studies has led us to use the following abbreviations for specific references (in addition to those listed above): HB78 for Hardenbol and Berggren (1978); GTS82 for Harland and others (1982) and GTS89 for Harland and others (1990).

GEOMAGNETIC POLARITY TIME SCALE

The first extended GPTS for the Cenozoic (and Late Cretaceous Period) era was based largely on a single magnetic anomaly profile from the South Atlantic (Heirtzler and others, 1968). The near-ridge axis polarity sequence was correlated to the independently developed radiometric polarity time scale (e.g., Cox and others, 1964), and the ages of older reversals were inferred by extrapolation from the older end of Anomaly 2A (Gilbert/Gauss boundary at ~3.4 Ma) assuming a constant rate of sea-floor spreading. Segments of the reversal sequence in the Neogene and Late Cretaceous Periods have been subsequently revised, for example, from Anomalies 1 to 3A (Klitgord and others, 1975), Anomalies 2 to 4A (Talwani and others, 1971), Anomalies 4A to 6 (Blakely, 1974), and Anomalies 30 to 34 (Cande and Kristofferson, 1977) after the addition of Anomalies 33 and 34 (Larson and Pitman, 1972). These changes were incorporated in the revised time scale of LaBrecque and others (1977). The only modification made to the Paleogene sequence (~Anomalies 6C to 29) was the deletion of Anomaly 14 as an artifact.

Age calibration has been the focus of changes to the GPTS since the compilation of LaBrecque and others (1977). As additional age control became available based on developing magnetobiostratigraphic ties, the implicit assumption of constant spreading in the South Atlantic was relaxed to smaller time intervals (e.g., Ness and others, 1980; Lowrie and Alvarez, 1981; GTS82, GTS89; BKFV85; Haq and others, 1987, 1988).

For the first time since the magnetic anomaly time scale of Heirtzler and others (1968), the relative widths of the entire Late Cretaceous and Cenozoic polarity intervals were systematically determined from magnetic anomaly profiles (CK92). The geomagnetic polarity sequence was based primarily on data from the South Atlantic using a combination of finite rotation poles (Cande and others, 1988) and averages of anomaly spacings from stacked profiles projected onto a synthetic sea-floor spreading flowline. The South Atlantic has a long, continuous history of spreading that is well documented on both sides of the ridge axis, making it possible to compensate for asymmetric spreading, ridge jumps and similar processes that could distort the magnetic anomaly sequence. Finer scale information where necessary and possible was derived from magnetic profiles on faster spreading ridges in the Pacific and Indian Oceans and inserted into the framework of South Atlantic spreading. The composite reference sequence was scaled to distance from the ridge axis on a model flow line in the South Atlantic. The conversion of the anomaly sequence into a geomagnetic polarity

time scale is thus reduced to the determination of the seafloor spreading history in the South Atlantic.

In comparison to the frequency of geomagnetic reversals, there have been relatively few reliable, analytically precise and stratigraphically well-controlled radioisotopic ages available to establish a chronology: oceanic basalts, the source of the magnetic anomalies, are notoriously difficult to date isotopically, and datable horizons in magnetostratigraphic sections have been infrequently reported. Hence, time scales have relied on interpolation between selected calibration datums that have been correlated to the characteristic reversal pattern.

Cande and Kent (1992) assumed that spreading on the South Atlantic ridge system was smoothly varying but not necessarily constant over the Late Cretaceous and Cenozoic. A time scale of reversals was generated by interpolating the age of polarity intervals using a cubic spline function fitted to 9 age calibration-anomaly distance tie-points (plus the zero-axis ridge axis). This GPTS is referred to as CK92. A total of 92 normal polarity and a like number of reversed polarity intervals are recognized between the end of the Cretaceous Long Normal at 83.0 Ma and Chron C1n (Bruhnes). The available magnetic anomaly data indicated that virtually all polarity intervals longer than ~30 ky are likely to have been identified. However, very short wavelength magnetic anomalies of ambiguous origin and with apparent durations less than 30 ky were not uniformly resolved over the entire anomaly sequence. These shortest features, numbering 54 in the present data, are referred to as cryptochrons and have been excluded from the GPTS.

To complement the comprehensive revision to the relative spacing of the polarity intervals, a revised polarity chron nomenclature was introduced. This scheme provides all polarity intervals with unique designations that is consistent with prior usage and that is amenable to accomodate further revision. This nomenclature, adapted from that used by Tauxe and others (1983), LaBrecque and others (1983) and Harland and others (1982, 1990), is described in detail in Cande and Kent (1992: Appendix) and only salient features are summarized here.

Because marine magnetic anomalies are effectively the standard global reference for geomagnetic reversals extending back to the Jurassic Period, the most useful polarity chron nomenclature is based on the numbered anomaly identifications (see Hailwood (1989) for a discussion of alternate schemes). For Late Cretaceous and Cenozoic time, prominent positive anomalies have been numbered (Pitman and others, 1968) from 1 (Central Anomaly) to 34 (younger end of the Cretaceous Quiet Zone). They are associated with oceanic crust magnetized more or less along the present field direction and correspond to time intervals (chrons) of normal geomagnetic polarity. Each of these intervals of predominantly normal polarity is designated by the anomaly number followed by the suffix n, whereas the preceding interval of predominantly reversed polarity is designated by the suffix r. For example, Chron C4n corresponds to Anomaly 4, and Chron 4r to the reversed interval between Anomalies 4 and 4A. When chrons are subdivided into shorter polarity intervals or subchrons, they are designated by a sequence number (from youngest to oldest within the chron) that is appended after a decimal point, with the suffix n for normal polarity and r for reversed polarity. For example, Chron C4n is subdivided into Subchrons C4n.1n, C4n.1r and C4n.2n, whereas Chron C4r is comprised of Subchrons C4r.1r, C4r.1n and C4r.2r. There will aways be an odd number of polarity intervals within each chron so that the earliest subchron, if present, will not have a formally designated complementary interval of opposite polarity that is of comparable subchron rank. For example, it would be inappropriate to refer to Chron C4r as Subchron C4n.2r which is not formally recognized.

The young or old end of a chron or subchron can be conveniently designated by appending (y) or (o), respectively, as in Chron C4n(y) for the younger end of Chron C4n or the Chron C4n/Chron C3Br boundary. For more precise correlation, the relative or proportional position within a chron or subchron from its younger end can be specified by appending the equivalent decimal value within the parenthesis, as in Chron C4n(0.25) for one-quarter of the duration of Chron C4n from its young end.

An earlier system for identifying polarity chrons has traditionally been used for the radioisotopically dated part of the reversal time scale, i.e., the last ~5 my (Cox and others, 1964). Names of geomagneticians (Brunhes, Matuyama, Gauss and Gilbert) are used for the chrons (previously referred to as epochs), and names of localities (e.g., Jaramillo, Olduvai, etc.) for subchrons (previously referred to as events). This nomenclature is widely used and understood for magnetostratigraphic correlations in the Pliocene-Pleistocene Epochs and presents no conflict with the alternative magnetic anomaly system with which there is clear correlation. Efforts have been made to extend the magnetostratigraphic nomenclature below the Gilbert, by numbering intervals of alternating predominant polarity identified in deep-sea cores to about Chron 23, encompassing most of the Neogene Period (Theyer and Hammond, 1974; Opdyke and others, 1974). The magnetostratigraphic chrons, however, become increasingly difficult to correlate to the magnetic anomaly sequence below the Gilbert (implied as Chron 4) and are thus problematical in their utility for global correlation. To differentiate the anomaly designations for chrons from the superceded magnetostratigraphic numbering scheme that has been used in the literature (e.g., Ryan and others, 1974), the magnetic anomaly chrons have the prefix 'C' (LaBrecque and others, 1983), as in Chron C5n (a prominent normal polarity interval in the early late Miocene, which has variously been correlated to either Chron 9 or Chron 11) compared to Chron 5 (a predominantly normal polarity interval just prior to the Gilbert in the late Miocene and which is correlated to Chron C3An).

Since the publication of CK92, it has become apparent that the calibration tie-points at the Cretaceous/Paleocene boundary and in the Pliocene-Pleistocene Epochs should be modified. An age of 66 Ma for the Cretaceous/Paleocene (K/P) boundary was used in CK92, based on a chronogram estimate in GTS89 that was apparently supported by a series of high precision laser fusion Ar/Ar sanidine single crystal dates related to the iridium-bearing lower Z coal in Montana (see discussion in Berggren and others, 1992). The 66 Ma age for the K/P boundary, however, has become problematic because the 66 Ma dates from Montana are now believed to be systematically too old due to peculiarities of sample processing (Swisher and others, 1992). $^{40}Ar/^{39}Ar$ redating of the K/P boundary in Montana has yielded an age of ~65.0 Ma (Swisher and others, 1993). New, highly consistent $^{40}Ar/^{39}Ar$ dates on K/P boundary tektite glass from marine sections in Haiti and Mexico strongly suggest an age close to 65 Ma for the K/P boundary (Izett and others, 1991;

Swisher and others, 1992; Dalrymple and others, 1993; see also review in Item 1 in section on Radioisotopic Chronology below). A 65 Ma age is consequently now widely accepted for the K/P boundary and is used, for example, as a key tie-point in revisions to the Mesozoic geologic time scale (e.g., Gradstein and others, 1994; this volume).

In the Pliocene-Pleistocene Epochs, a new approach to geochronology has been to date high-resolution climate records by assuming that their characteristic variability is related to the well-known variations in Earth's orbital parameters, resulting in Milankovitch cyclicity. Unlike the discreteness of radioisotopic dates which requires interpolation using sedimentation or spreading rates to extend their range of usefulness, the astrochronologic method can effectively provide a continuum of independent age control, with a precision that is essentially a function of the shortest Milankovitch orbital variation (~20-ky precession cycle) and an accuracy that ultimately depends on the reliability of the extrapolation of the modern orbital parameters back in time. On the basis of climatic records from the equatorial Pacific and the Mediterranean, astrochronologic control for geomagnetic reversals is now available to the Thvera (Subchron C3n.4n) in the early part of the Gilbert (Shackleton and others, 1990; Hilgen, 1991).

The astrochronologic estimates of Shackleton and others (1990) and Hilgen (1991) have been confirmed to ~3 Ma using high precision $^{40}Ar/^{39}Ar$ dating (e.g., Baksi and others, 1992; Tauxe and others, 1992; Walter and others, 1991; Renne and others, 1993; see also review in Item 7 in section on Radioisotopic Chronology, below). A comparison of the astrochronology with CK92 shows good agreement, typically to within 30 ky, to the earliest Gauss (Chron C2An) (Berggren and others, 1995). This is not unexpected because the direct calibration point at the Gauss/Matuyama boundary (Chron C2An(y)) in CK92 was based on this astrochronology. An appreciable discrepancy, however, appears in Chron C3n, where the astronomical time scale of Hilgen (1991) gives ages for the constituent subchrons (Subchrons C3n.1n, C3n.2n, C3n.3n, and C3n.4n, or Cochiti, Nunivak, Sidufjall, and Thvera, respectively) that are systematically 150 to 180 ky older than the interpolated ages in CK92. High precision radiometric dating that is relevant to this problem is presently lacking. However, it is now apparent that the magnetic anomaly sequence for this interval used by CK92 is the likely source of the problem. This was demonstrated by Wilson (1993) who showed that the astrochronology gives a more consistent spreading history when applied to a set of revised spacings of anomalies on the Cocos-Nazca and other Pacific spreading ridges.

The success of the continuous chronology provided by the Pliocene-Pleistocene astronomical time scale means that it is not necessary to interpolate between a few discrete dated levels to construct a geomagnetic polarity time scale, as has been done heretofore. Instead, the ages of reversals in the interval where astrochronology has been well developed simply become equivalent to the astrochronological values, thereby avoiding the promulgation of separate time scales. Astrochronological estimates are currently published for geomagnetic reversals back to the earliest Thvera Subchron (Subchron C3n.4n) (Shackleton and others, 1990; Hilgen, 1991), and these are considered to provide the best geomagnetic polarity time scale for the Pliocene-Pleistocene Epochs (Berggren and others, 1995). Older than this, interpolation using marine magnetic anomalies is still necessary pending extension of the astrochronology.

A revised geomagnetic polarity time scale was generated using a radioisotopic age of 65 Ma rather than 66 Ma for the K/P boundary, and an astronomical age of 5.23 Ma (Hilgen, 1991) for the older end of Subchron C3n.4n (Thvera Subchron) rather than the age of 2.69 Ma for the younger boundary of Chron C2A (Gauss/Matuyama boundary) used in CK92, for the cubic spline interpolation. Calibration data (Table 1) are otherwise the same as by Cande and Kent (1992). The ages of Pliocene and Pleistocene polarity intevals, corresponding to Subchron C3n.4n and younger subchrons, are then inserted from the astrochronology of Shackleton and others (1990) and Hilgen (1991) with a refined astronomical age recently suggested for the Gauss/Matuyama boundary by Langereis and others (1994). With the revised K/P calibration, all of the 9 calibration tie points have direct marine biostratigraphic constraints for correlation. The revised geomagnetic polarity time scale (CK92/95) is listed for convenience with the age range and chron nomenclature of both normal polarity intervals (Table 2) as well as reversed polarity intervals (Table 3).

RADIOISOTOPIC CHRONOLOGY

In this section we discuss recent (mostly post-1990) radioisotopic data that have a bearing upon the revised integrated Cenozoic time scale presented here. Additional data, essentially presented between 1985 and 1990, were reviewed in Berggren and others (1992). Since the publication of BKF85, the field of geochronology has undergone a dramatic transformation culminating in the virtual replacement of conventional $^{40}K/^{40}Ar$ dating with $^{40}Ar/^{39}Ar$ dating as the primary method of choice in the calibration of the geologic time scale. This transformation has been a direct consequence of the development of low-blank high-resolution mass spectrometers. Coupled with micro-volume extraction lines, laser- and/or low blank furnace-heating, the $^{40}Ar/^{39}Ar$ dating method is capable of yielding highly reproducible ages (0.1% standard error on high potassium-bearing minerals) for much of the Cenozoic era. Automation of these dating systems further permits replicate ages to be easily obtained (a single age determination takes approximately 20 minutes), allowing recognition of multiple age components due to the presence of detrital crystals and/or alteration.

TABLE 1.—REVISED AGE CALIBRATIONS FOR GEOMAGNETIC POLARITY TIME SCALE, SOUTH ATLANTIC

Chron	Distance (km)	Age (Ma)
C3n.4n(o)	84.68	5.23*
C5Bn(y)	290.17	14.8
C6Cn.2r(y)	501.55	23.8
C13r(.14)	759.49	33.7
C21n(.33)	1071.62	46.8
C24r(.66)	1221.20	55.0
C29r(.3)	1364.37	65.0**
C33n(.15)	1575.56	74.5
C34n(y)	1862.32	83.0

*Ages for reversals for Subchron C3n.4n(o) and younger are made equivalent to astronomical time scale of Shackleton and others (1990) and Hilgen (1991), with refinement of Langereis and others (1994).

**Revised K/P boundary age, see text. Other age calibration data from Cande and Kent (1992).

TABLE 2.—NORMAL POLARITY INTERVALS.

Interval (Ma)	Chron
0.000–0.780	C1n (BRUNHES)
0.990–1.070	C1r.1n (Jaramillo)
1.201–1.211	C1r.2r-1n (Cobb Mountain)
1.770–1.950	C2n (Olduvai)
2.140–2.150	C2r.1n (Reunion)
2.581–3.040	C2An.1n (GAUSS)
3.110–3.220	C2An.2n (GAUSS)
3.330–3.580	C2An.3n (GAUSS)
4.180–4.290	C3n.1n (Cochiti)
4.480–4.620	C3n.2n (Nunivak)
4.800–4.890	C3n.3n (Sidufjall)
4.980–5.230	C3n.4n (Thvera)
5.894–6.137	C3An.1n
6.269–6.567	C3An.2n
6.935–7.091	C3Bn
7.135–7.170	C3Br.1n
7.341–7.375	C3Br.2n
7.432–7.562	C4n.1n
7.650–8.072	C4n.2n
8.225–8.257	C4r.1n
8.699–9.025	C4An
9.230–9.308	C4Ar.1n
9.580–9.642	C4Ar.2n
9.740–9.880	C5n.1n
9.920–10.949	C5n.2n
11.052–11.099	C5r.1n
11.476–11.531	C5r.2n
11.935–12.078	C5An.1n
12.184–12.401	C5An.2n
12.678–12.708	C5Ar.1n
12.775–12.819	C5Ar.2n
12.991–13.139	C5AAn
13.302–13.510	C5ABn
13.703–14.076	C5ACn
14.178–14.612	C5ADn
14.800–14.888	C5Bn.1n
15.034–15.155	C5Bn.2n
16.014–16.293	C5Cn.1n
16.327–16.488	C5Cn.2n
16.556–16.726	C5Cn.3n
17.277–17.615	C5Dn
18.281–18.781	C5En
19.048–20.131	C6n
20.518–20.725	C6An.1n
20.996–21.320	C6An.2n
21.768–21.859	C6AAn
22.151–22.248	C6AAr.1n
22.459–22.493	C6AAr.2n
22.588–22.750	C6Bn.1n
22.804–23.069	C6Bn.2n
23.353–23.535	C6Cn.1n
23.677–23.800	C6Cn.2n
23.999–24.118	C6Cn.3n
24.730–24.781	C7n.1n
24.835–25.183	C7n.2n
25.496–25.648	C7An
25.823–25.951	C8n.1n
25.992–26.554	C8n.2n
27.027–27.972	C9n
28.283–28.512	C10n.1n
28.578–28.745	C10n.2n
29.401–29.662	C11n.1n
29.765–30.098	C11n.2n
30.479–30.939	C12n
33.058–33.545	C13n
34.655–34.940	C15n
35.343–35.526	C16n.1n
35.685–36.341	C16n.2n
36.618–37.473	C17n.1n
37.604–37.848	C17n.2n
37.920–38.113	C17n.3n
38.426–39.552	C18n.1n
39.631–40.130	C18n.2n
41.257–41.521	C19n
42.536–43.789	C20n
46.264–47.906	C21n
49.037–49.714	C22n
50.778–50.946	C23n.1n

TABLE 2.—*Continued.*

Interval (Ma)	Chron
51.047–51.743	C23n.2n
52.364–52.663	C24n.1n
52.757–52.801	C24n.2n
52.903–53.347	C24n.3n
55.904–56.391	C25n
57.554–57.911	C26n
60.920–61.276	C27n
62.499–63.634	C28n
63.976–64.745	C29n
65.578–67.610	C30n
67.735–68.737	C31n
71.071–71.338	C32n.1n
71.587–73.004	C32n.2n
73.291–73.374	C32r.1n
73.619–79.075	C33n
83.000–118.000	C34n

TABLE 3.—REVERSE POLARITY INTERVALS.

Polarity Interval (Ma)	Chron
0.780–0.990	C1r.1r (MATUYAMA)
1.070–1.201	C1r.2r.1r (MATUYAMA)
1.211–1.770	C1r.2r.2r (MATUYAMA)
1.950–2.140	C2r.1r (MATUYAMA)
2.150–2.581	C2r.2r (MATUYAMA)
3.040–3.110	C2An.1r (Kaena)
3.220–3.330	C2An.2r (Mammoth)
3.580–4.180	C2Ar (GILBERT)
4.290–4.480	C3n.1r (GILBERT)
4.620–4.800	C3n.2r (GILBERT)
4.890–4.980	C3n.3r (GILBERT)
5.230–5.894	C3r (GILBERT)
6.137–6.269	C3An.1r
6.567–6.935	C3Ar
7.091–7.135	C3Br.1r
7.170–7.341	C3Br.2r
7.375–7.432	C3Br.3r
7.562–7.650	C4n.1r
8.072–8.225	C4r.1r
8.257–8.699	C4r.2r
9.025–9.230	C4An.1r
9.308–9.580	C4Ar.1r
9.642–9.740	C4Ar.2r
9.880–9.920	C4Ar.3r
10.949–11.052	C5n.1r
11.099–11.476	C5r.1r
11.531–11.935	C5r.2r
12.078–12.184	C5r.3r
12.401–12.678	C5An.1r
12.708–12.775	C5Ar.1r
12.819–12.991	C5Ar.2r
13.139–13.302	C5Ar.3r
13.510–13.703	C5AAr
14.076–14.178	C5ABr
14.612–14.800	C5ACr
14.888–15.034	C5ADr
15.155–16.014	C5Bn.1r
16.293–16.327	C5Br
16.488–16.556	C5Cn.1r
16.726–17.277	C5Cn.2r
17.615–18.281	C5Cr
18.781–19.048	C5Dr
20.131–20518	C5Er
20.725–20.996	C6r
21.320–21.768	C6An.1r
21.859–22.151	C6Ar
22.248–22.459	C6AAr.1r
22.493–22.588	C6AAr.2r
22.750–22.804	C6AAr.3r
23.069–23.353	C6Bn.1r
23.535–23.677	C6Br
23.800–23.999	C6Cn.1r
24.118–24.730	C6Cn.2r
24.781–24.835	C6Cr
25.183–25.496	C7n.1r
25.648–25.823	C7r
25.951–25.992	C7Ar
26.554–27.027	C8n.1r
27.972–28.283	C8r
28.512–28.578	C9r
28.745–29.401	C10n.1r
29.662–29.765	C10r
30.098–30.479	C11n.1r
30.939–33.058	C11r
33.545–34.655	C12r
34.940–35.343	C13r
35.526–35.685	C15r
36.341–36.618	C16n.1r
37.473–37.604	C16r
37.848–37.920	C17n.1r
38.113–38.426	C17n.2r
39.552–39.631	C17r
40.130–41.257	C18n.1r
41.521–42.536	C18r
43.789–46.264	C19r
47.906–49.037	C20r
49.714–50.778	C21r
50.946–51.047	C22r
	C23n.1r

The ^{40}Ar/^{39}Ar dating method differs from conventional ^{40}K/^{40}Ar dating in that samples to be dated are first irradiated in a nuclear reactor to convert a portion of ^{39}K to ^{39}Ar by nuclear bombardment of fast neutrons. This conversion permits the measurement of the potassium (now ^{39}Ar) content of the sample as well as the daughter radiogenic argon (^{40}Ar) to be made on the same sample aliquot, by the same method (mass spectrometry), at the same time. The assumption is that the ^{39}K/^{40}K ratio in nature has been essentially constant, the amount produced during irradiation being dependent upon the total amount of potassium present in the sample and the duration of neutron irradiation. Because the actual amount of ^{39}K converted to ^{39}Ar during irradiation is not known, the samples are irradiated with a fluence monitor mineral or standard of known age permitting the calculation of an irradiation coefficient, *J*, which is then applied to the unknowns.

The calibration of the fluence monitor mineral thus directly affects the accuracy of the ^{40}Ar/^{39}Ar ages calculated for the unknowns. The developments outlined above now permit ^{40}Ar/^{39}Ar ages with a precision that far exceeds the accuracy of the age of any currently available monitor mineral. For intercalibration purposes, most laboratories now report ^{40}Ar/^{39}Ar ages with reference to the age of international monitor minerals such as McClure Mountain Hornblende (MMhb-I) or Fish Canyon Tuff Sanidine (FCTS). At present, however, there appears to be no consensus among geochronologists as to the "correct" age of these monitor minerals. The reason for this is primarily historical. The ages of the ^{40}Ar/^{39}Ar monitor minerals are derived from the age of first principal ^{40}K/^{40}Ar standards calibrated directly against atmospheric air. Most first principal ^{40}K/^{40}Ar standards, however, are primarily Ar standards, used for the calibration of a ^{38}Ar spike or tracer. Thus while the radiogenic ^{40}Ar moles/g of the ^{40}K/^{40}Ar standard is well known, the ^{40}K content and consequently, the age of the standard, is not. The application of ^{40}K/^{40}Ar standards as ^{40}Ar/^{39}Ar monitor minerals in the calculation of the irradiation parameter *J*, requires that the age and, consequently, the ^{40}K be accurately known. The age of the ^{40}Ar/^{39}Ar monitor mineral directly affects the age of the unknown sample being dated. Currently, different laboratories report conflicting ages for international monitor minerals such as MMhb-I or FCTS, thus adding an unknown uncertainty to the accuracy of ^{40}Ar/^{39}Ar ages.

TABLE 3.—Continued.

Polarity Interval (Ma)	Chron
51.743–52.364	C23r
52.663–52.757	C24n.1r
52.801–52.903	C24n.2r
53.347–55.904	C24r
56.391–57.554	C25r
57.911–60.920	C26r
61.276–62.499	C27r
63.634–63.976	C28r
64.745–65.578	C29r
67.610–67.735	C30r
68.737–71.071	C31r
71.338–71.587	C32n.1r
73.004–73.291	C32r.1r
73.374–73.619	C32r.2r
79.075–83.000	C33r

Clearly, the "absolute" age of the $^{40}Ar/^{39}Ar$ monitor minerals is in serious need of review. In the interim, however, we stress the importance of internal consistency in the calibration of geological time scale tie-points. When it is deemed necessary to modify this calibration following refinements in the age of the standards, the time scale can be adjusted accordingly. Consequently, we would caution that any $^{40}Ar/^{39}Ar$ age used in comparison with this time scale must be adjusted to conform with the ages adopted here for the following international standards or monitor minerals: MMhb-I at 520.4 ± 1.7 Ma (Samson and Alexander, 1987), FCTS at 27.84 Ma (modified slightly from Cebula and others, 1986). The age used for one or both of these standards currently accompanies most $^{40}Ar/^{39}Ar$ dating studies which can be used for comparison with this time scale. Other frequently used ages for these standards seen in the literature that bear directly upon time scale calibration are those adopted by the Menlo Park U.S.G.S. laboratory at 513.9 ± 2.3 Ma and 27.55 ± 0.10 Ma, respectively, and the Australian National University at 524.2 Ma for MMhb-I.

Work in progress suggests that the older age reported for FCTS may be closer to its true geologic age. New $^{40}K/^{40}Ar$ ages for two biotite standards (the University of California, Berkeley GHC305 and the Australian National University GA1550) that are intercalibrated with FCTS by $^{40}Ar/^{39}Ar$, indicate the age of FCTS should be 28.05 ± 0.02 Ma (Swisher and others, 1994).

Independent support for this age comes from the age of FCTS predicted from the astronomically-calibrated geomagnetic polarity time scale (APTS) developed by Shackleton and others, (1990) and Hilgen (1991). The use of seven dated volcanic layers that can be directly tied to the APTS via magnetochronology, and solving of the $^{40}Ar/^{39}Ar$ age equation for the age of FCTS required to produce ages for the volcanic layers coincident with those predicted by the APTS, yields mutually indistinguishable estimates for each of these seven reversals ranging from 27.78 to 28.09 Ma, with an inverse variance weighted mean of 27.95 ± 0.09 Ma (Renne and others, 1994). While we do not recommend the use of the APTS to calibrate the K/Ar system, the APTS can serve as an independent means of evaluating the accuracy in the calibration of radioisotopic decay systems. Additional intercalibration studies are needed to confirm this age.

We expect that there will be additional refinements in the age of $^{40}Ar/^{39}Ar$ fluence monitors over the next few years. The con-

version of the $^{40}Ar/^{39}Ar$ ages used in this time scale to a different monitor age can be achieved following Dalrymple and others, (1993):

$$t_2 = 1/l \, \log_e[e^{lt_{m2}} - 1/e^{lt_{m1}} - 1(e^{lt_1} - 1) + 1]$$

where t_1 = the published age, t_{m1} = the age of the monitor used to obtain t_1, t_{m2} = the new age for the monitor mineral, and t_2 = the converted age; l = the decay constant 5.543 × 10^{-10} yr^{-1}.

The intercalibration of other geochronometers used in time scale calibration such as Rb-Sr, U-Pb and F-T faces similar problems and we caution the reader when comparing these different types of data. The accuracy and intercalibration of these geochronometers is beyond the scope of this paper, but will surely be the focus of studies in the near future.

Discussion of the Cenozoic radioisotopic data base is presented under seven items (older to younger) which correspond to seven of the nine calibration points (the other two calibration points are in the upper Cretaceous and beyong the scope of this paper) used in CK92, with modifications in CK92/95, and retained in this work.

Cretaceous/Paleogene (K/P) Boundary (Chron C29 to C28)

Much of the variation in currently available ages for the K/P boundary results primarily from interlaboratory differences in methodology and ages adopted for $^{40}Ar/^{39}Ar$ monitor minerals (Obradovich, 1984; Hall and others, 1991; Izett and others, 1991; Gillot and others, 1991; Swisher and others, 1992; McWilliams and others, 1992; Sharpton and others, 1992; Swisher and others, 1993; Dalrymple and others, 1993). However, $^{40}Ar/^{39}Ar$ dating has been shown to yield intralaboratory ages with reproducibility as low as 0.1% for K/P samples. Consequently, by minimizing variation due to calibration, it is possible for $^{40}Ar/^{39}Ar$ ages to provide precise tests of correlation between geographically distant sites and among various environments of deposition.

Compilations of the radioisotopic ages of the K/P boundary were recently made by Dalrymple and others (1993) and Swisher and others (1993, 1995); (Table 4). The extrapolated age for the K/P boundary at Agost, Spain, is 64.95 ± 0.07 Ma. The weighted mean of the $^{40}Ar/^{39}Ar$ ages compiled in Table 4 (omitting the age of the Chicxulub melt-rock because of lack of biostratigraphic control) is 65.06 ± 0.02 Ma. This age for the K/P boundary is in agreement with $^{40}Ar/^{39}Ar$ data for Haiti tektites from the U.S. Geological Survey, initially reported in Izett and others (1991) and subsequently thoroughly reviewed and summarized by Dalrymple and others (1993). For direct comparison with the above ages, the $^{40}Ar/^{39}Ar$ ages reported in Dalrymple and others (1993) are converted here using the above discussed algorithm for the monitor mineral Fish Canyon Tuff Sanidine at 27.55 Ma to 27.84 Ma (Table 5).

$^{40}Ar/^{39}Ar$ ages for the K/P boundary in the terrestrial record of eastern Montana are also summarized in Dalrymple and others (1993). The weighted mean of these analyses is 65.44 ± 0.07 Ma, slightly older than those obtained for the Haiti tektites. The differences are attributed by Dalrymple and others (1993) to be a result of imprecision in determining the neutron-efficiency factor, J, for the different irradiations and different sample positions and are not considered geologically significant.

TABLE 4.—SUMMARY OF ^{40}Ar/^{39}Ar LASER TOTAL FUSION AND INCREMENTAL HEATING AGES ASSOCIATED WITH IMPACT GENERATED SIGNATURES AT AGOST, SPAIN; EASTERN MONTANA, HAITI/MIMBRAL AND CHICXULUB (FROM SWISHER AND OTHERS, 1993, 1995).

Sample Location	Material Analyzed	Analysis Type	Total Number Of Analyses	Weighted Mean Age (Ma ± SE[1])
Spain, Agost	Biotite	TF	8	64.81 ± 0.13[2]
Spain, Agost	Biotite	IH	1	64.86 ± 0.09[2]
Montana, Ir.Z	Sanidine	TF	11	65.16 ± 0.04[2]
Montana, Z-Hell Creek	Sanidine	TF	13	65.00 ± 0.05[2]
Montana, Z-McGuire Creek	Sanidine	TF	20	65.03 ± .04[2]
Haiti	Tektite glass	IH	7	65.01 ± 0.08[3]
Haiti/Nimbral	Tektite glass	TF	5	65.07 ± 0.10[3]
Mexico, Chicxulub	Melt-rock	IH	3	64.98 ± 0.05[3]

TF = Laser Total Fusion ^{40}Ar/^{39}Ar analysis
IH = Laser Incremental Heating ^{40}Ar/^{39}Ar analysis
[1]IH analyses are plateau ages
[2]Data from Swisher and others, 1995
[3]Data from Swisher and others, 1992

TABLE 5.—K/P BOUNDARY AGES FROM HAITI (DALRYMPLE AND OTHERS, 1993).

Sample Location	Material Anlyzed	Analysis Type	Total Number Of Analyses	Weighted Mean Age (Ma ± SE[1])
Haiti	Tektite glass	TF	52	65.09 ± 0.06
Haiti	Tektite glass	IH	4	65.05 ± 0.18
Haiti	Tektite glass	IH bulk	2	65.16 ± 0.10

For example, if the analyses from one of the irradiations that yielded older ages are omitted from the calculation of the mean, the revised weighted mean, 65.20 ± 0.05 Ma, is within analytical error of the U.S. Geological Survey ages for the Haiti tektites.

These ages are approximately 1.0% younger than preliminary results reported in BKF85, Berggren and others (1992) and CK92. Some of the 66 Ma ages reported in Berggren and others (1992) and used in CK92 were derived from preliminary dating by Swisher (in Berggren and others, 1992). They are now known to be a result of an unorthodox method used in preparing samples from eastern Montana and are considered too old (Swisher and others, 1993). Although the K/P boundary itself was not used as a tie-point in BKF85, its broadly interpolated age of 66.4 Ma was supported by ^{40}Ar/^{39}Ar dates from K/P boundary sections in Montana and Colorado obtained by Obradovich (1984).

Gillot and others (1991) reported two ^{40}K/^{40}Ar ages on bulk samples of the Haiti tektites with a mean value of 64.0 ± 0.35 Ma. Other ^{40}Ar/^{39}Ar ages for the K/P boundary, reported in abstracts only, have yielded similar ages to those reported above. However, they are not further discussed as the data are unavailable for evaluation; these are the Haiti tektites (Hall and others, 1991), and the sanidines from Montana, Alberta and Saskatchewan (McWilliams and others, 1991a, b, 1992). The latter were from splits of K-Ar samples reported by Baadsgaard and others (1988).

^{40}Ar/^{39}Ar dating of sanidine from units overlying the K/P boundary in the terrestrial record of eastern Montana in sections with magnetostratigraphy helps to constrain the age of the K/P boundary and provides additional calibration for Chrons C29 and C28. The weighted mean ^{40}Ar/^{39}Ar ages given in Table 6 are based on multiple analyses of single crystals of sanidine separated from bentonites interbedded in coal beds of the Tullock Formation, Garfield and McCone counties, eastern Montana (Swisher and others, 1993, 1995). According to these data, an age of ~64.6 Ma is suggested for Chron C29n(o), 64.1 Ma for Chron C29n(y), and 63.8 Ma for Chron C28n(o). These estimates are seen to be close to those estimated in CK92/95 based on a revised age of 65 Ma for the K/P boundary (Chron C29(o): 64.745 Ma; C29n(y): 63.976 Ma; Chron C28n(y): 63.634 Ma.

Paleocene/Eocene Boundary (Chron C24r)

Berggren and others (1992) reported a revised age estimate of 55 Ma for the Paleocene/Eocene boundary based primarily on the age of the Mo Clay, or −17 Ash in Denmark. A bulk incremental-heating ^{40}Ar/^{39}Ar plateau age on sanidine of 55.07 ± 0.16 Ma was obtained by J. D. Obradovich (U.S.G.S., Denver; see Wing and others, 1991; Berggren and others, 1992). Additional laser total-fusion ^{40}Ar/^{39}Ar ages (Swisher and Knox, 1991; unpubl. data) on sanidine for the −17 Ash as well as its correlative N° 70 Ash in SE England gave concordant weighted mean ages of 54.51 ± 0.05 Ma and 54.56 ± 0.14 Ma, respectively. Also bearing on the age of the P/E boundary are laser total-fusion ^{40}Ar/^{39}Ar ages on sanidine from the Danish +19 Ash and its correlative N° 60 Ash in SE England that gave weighted mean ages of 54.0 ± 0.53 Ma and 54.04 ± 0.33 Ma, respectively. The −17 Ash coincides closely with the base of the London Clay Formation in SE England, and in DSDP Hole 550 it lies 7 m above the (unconformable) NP9/NP10 calcareous nannofossil zonal boundary and approximately 2/3 of the way down Chron C24r (see below for further discussion). An age of 55 Ma was estimated by Swisher and Knox (1991) for the NP9/NP10 zonal boundary and was utilized by CK92 as a calibration point and proxy for the Paleocene/Eocene boundary in the formulation of the GPTS.

The −17 Ash and the +19 Ash bracket the planktonic foraminiferal P5/P6a zonal boundary in DSDP Hole 550. The

TABLE 6.—AGES ON CHRONS C28 AND C29 IN THE TERRESTRIAL RECORD OF EASTERN MONTANA (SWISHER AND OTHERS, 1993, 1995).

Unit Name	Chron	Lab No	Total Number of Analyses	Age (Ma ± SE)
U-Coal	C28r	6587	10	63.90 ± 0.04
W-Coal	C29n	6584	15	64.13 ± 0.04
		6753	20	64.09 ± 0.03
		overall weighted mean	(35 analyses)	64.11 ± 0.02
HFZ-Coal	C29r	6581	10	64.77 ± 0.06
		6581 (plagioclase)	9	64.76 ± 0.38

Paleocene/Eocene boundary which awaits determination of a global stratigraphic section and point (GSSP) is bracketed by the P5/P6a zonal boundary (LAD of *Morozovella velascoensis*) and the $\delta^{13}C$ spike (and associated events) in mid-Zone NP9. The "boundary interval" encompasses the NP9/NP10 calcareous nannofossil zonal boundary at ~55 Ma and the base of the London Clay Formation (see Berggren and Aubry, 1995, and Aubry and others, 1995, for further discussion). These ages aid in constraining the age of the Paleocene/Eocene boundary provisionally taken at the base of the London Clay Formation, which now differs significantly (~54.8 Ma, based on sediment rate interpolation in DSDP Hole 550; see Berggren and Aubry, 1995, for discussion of this estimate) from the estimates of 57.8 Ma in BKF85 and 57 Ma in Aubry and others (1988).

Early Middle Eocene Epoch (Chron C21n)

K/Ar dating of a biotite recovered from DSDP Hole 516F-76-4, 107–115 cm, gave dates of 45.8 ± 0.5 Ma and 46.8 ± 0.5 Ma (Bryan and Duncan, 1983). The biotite was recorded as occurring within an interval identified as Zone P10 and lower part of Zone NP15 (Barker and others, 1983; Wei and Wise, 1989). The interval falls within a normal polarity magnetozone correlated with the mid to upper part of Chronozone C21n (Berggren and others, 1983b; Berggren and others, 1985b; Berggren and others, 1992; Aubry, this volume). In constructing the GPTS, CK92 used 46.8 Ma as a calibration for the mid-younger part of Chron C21n (see Table 1). A small amount of the rock from this interval was made available to us by A. Montanari for $^{40}Ar/^{39}Ar$ dating. $^{40}Ar/^{39}Ar$ laser total-fusion analyses of sanidine and biotite from this same interval yielded dates concordant with the younger biotite date reported by Bryan and Duncan (1983). The weighted mean ages of the analyses are 45.66 ± 0.05 Ma and 45.60 ± 0.20 Ma, respectively (Swisher and Montanari, in prep).

The $^{40}Ar/^{39}Ar$ ages for the lower part of planktonic foraminiferal Zone P10, lower part of calcareous nannofossil Zone NP15, and mid to upper Magnetic Polarity Chronozone C21n are supported by a number of K/Ar dates on biotite and glauconite from the Gulf and Atlantic Coastal Plain. They are, however, significantly older than Rb/Sr ages from the same samples. Harris and Fullagar (1989, 1991) in their review of the Gulf and Atlantic Coastal Plain Rb/Sr and K/Ar data on stratigraphic levels correlative with Zone NP15, show clearly that $^{40}K/^{40}Ar$ dating has yielded dates in the range of 46–47 Ma, whereas Rb-Sr dates are centered on 42–45 Ma. Thus, while there is a slight degree of overlap in the two data sets, it is clear that, for the most part, the K/Ar dates are, on average, 1 to 2 my older than the Rb/Sr dates. One possible explanation for this difference could reside in the difference in calibration of the K/Ar and Rb/Sr systems.

Since BKF85, Harris and Fullagar (1991) reported a Rb/Sr isochron date of 41.6 ± 1.5 Ma on the Warley Hill Formation (= Zone NP15), which is substantially younger than the age span of ~44–47.5 Ma that conversion of the chronology of Zone NP15 (in BKF85) to this time scale suggests. Harris and Fullagar (1991) observed that the two K/Ar dates of 43.5 ± 1.7 Ma and 46.6 ± 1.8 Ma are in analytic agreement with each other and that the younger of these two dates is in essential agreement with a previously published (Harris and Fullagar, 1989) Rb/Sr date of 42.0 ± 0.5 Ma on glauconites at Wilson's Landing which were correlated to the type locality of the Warley Hill Formation, Calhoun County, South Carolina. The older conventional K/Ar date (46.6 ± 1.8 Ma) does not agree analytically with the Rb/Sr date(s) but falls within the limits estimated here for the lower part of Zone NP15 as well as with those in HB78, BKF85 and Haq and others (1987, 1988). The older conventional K/Ar date is, however, consistent with K/Ar ages reported by Harris and Fullagar (1989) for Zone NP15 from the Castle Hayne Limestone of the Atlantic Coastal Plain. K/Ar ages on biotite from a bentonite and on glauconite below it yielded similar ages of 46.2 ± 1.8 Ma and 46.7 ± 1.8 Ma. Rb/Sr isochron ages on the same biotite and glauconite yielded consistently younger ages of 45.7 ± 0.7 Ma and 45.3 ± 0.3 Ma, respectively. It also yielded Rb/Sr and K/Ar ages on a glauconite overlying the bentonite of 43.1 ± 1.2 Ma and 44.5 ± 1.7 Ma. These findings are consistent with the anomalously younger Rb-Sr isochron age of 41.6 ± 1.5 Ma for the Warley Hill Formation. However, Harris and Fullagar (1991) noted that previously published Rb/Sr and K/Ar dates on younger stratigraphic units containing (or stratigraphically correlative with) *Cubitostrea sellaeformis* (in calcareous nannofossil Zone NP16) are essentially in agreement over the interval of ~39–42 Ma. In summary, the K/Ar dates reported on Zone NP15 by Harris and Fullagar (1989, 1991) are in agreement with the $^{40}Ar/^{39}Ar$ ages discussed above. Also, they are more reliable and agree more closely with the revised estimated age span of Zone NP15 than do the Rb/Sr dates.

Volcanic activity built up a shallow marine edifice at ODP Site 713 that yielded a weighted mean whole rock $^{40}Ar/^{39}Ar$ age of 49.3 ± 0.6 Ma (Duncan and Hargraves, 1990). Sediments interbedded with these basalts yield calcareous nannofossil assemblages indicative of Subzone CP13b (Subzone NP15b), and the reversed magnetized basalts are correlated with Chron C20r in BKF85. Given the current age calibration for Chron C20 presented here, in Berggren and others (1992), and used in CK92/95, the age of the Site 713 basalts appears too old.

Eocene/Oligocene Boundary (Chron C13r)

The calibration of the Eocene/Oligocene boundary at 33.7 Ma used in CK92 and retained here is discussed thoroughly in

Berggren and others (1992). Indeed, the history of radioisotopic studies for the early middle Eocene Epoch (Chron C21n) and bracketing the Eocene/Oligocene boundary (Chron C13r) which led to the revised GPTS of CK92/95, and, ultimately, this paper, has been reviewed in detail by Prothero (1994). Additional age data published since on bentonites from the upper Eocene Yazoo Formation, west-central Mississippi, assigned to planktonic foraminifera Zone P17 and to Chron C13r, were reported in Obradovich and others (1993). A single incremental-heating experiment (J.D.O.) on sanidine (Sample 1 of Satartia) yielded a plateau age of 34.32 $\pm$ 0.05 Ma. Laser total-fusion analyses (C.C.S.) of the sanidine from the same unit yielded a weighted mean age of 34.30 $\pm$ 0.06 Ma, while biotite from the nearby Society Ridge Core, level 80.0–81.0 feet, yielded a concordant age of 34.29 $\pm$ 0.11 Ma. The overall weighted mean of the analyses of 34.31 $\pm$ 0.04 Ma provides additional calibration of Zone P17 and Chron C13r.

Oligocene/Miocene Boundary (Chron C6Cn)

CK92 used the Chron C6Cn.2n/r boundary as a mid-Cenozoic calibration point for the Cenozoic time scale. CK92 cited BKF85 for an estimate of 23.7 Ma and the chronogram estimate of 23.8 Ma from GTS89 as the age of the Oligocene/Miocene boundary. A GSSP for the Oligocene/Miocene boundary is currently being proposed at the Chron C6Cn.2n/r boundary in the Carrioso-Lemme Section, NE Italy (Steininger, 1994) and this proposal is accepted here.

Miocene Epoch (Chron C6Cn to Chron C3r)

The CK92/95 age for the Oligocene/Miocene boundary is consistent with dating of the Livello Raffaello Tuff in the Contessa CQ-CT section, Italy, by Montanari and others (1991). This tuff yields a mean $^{40}Ar/^{39}Ar$ age of 21.17 $\pm$ 0.23 Ma. It occurs in sediments with reversed polarity correlated in conjunction with biostratigraphic constraints with Chron C6Ar which has age limits of 21.320 to 21.768 Ma in our time scale.

CK92 incorporated an age of 14.8 Ma for the younger end of Chron C5Bn. This age is based on age constraints on the correlative N9/N10 foraminiferal zonal boundary (Miller and others, 1985; Berggren and others, 1985a) as estimated in Japan at 14.6 $\pm$ 0.4 Ma (Tsuchi and others, 1981; Saito, 1984) and in Martinique at 15.0 $\pm$ 0.3 Ma (Andreieff and others, 1976; pers. commun. in BKF85).

An alternative calibration for Chron C5Bn has been recently proposed by Baksi (1993, and refs. therein) largely on the basis of geomagnetic polarity and $^{40}Ar/^{39}Ar$ age data from the Columbia River Basalts (CRB), primarily from that of the Imnaha, Grande Ronde and Steen Basalts. Ages derived from $^{40}Ar/^{39}Ar$ incremental-heating of whole rock basalts led Baksi (1993) to suggest that Chron C5Bn is approximately 7% (or approximately 1 my) older than that indicated by CK92.

A key calibration point of the revised time scale proposed by Baksi (1993) is the CRB N_1-R_1 transition dated at 16.32 $\pm$ 0.1 Ma and thought to correspond to Chron C5Bn(o) and ~1.2 my older than Chron C5Bn(o) in CK92 or CK92/95. This age is approximately 0.8 my older than the $^{40}K/^{40}Ar$ age (15.5 $\pm$ 0.3 Ma) reported for this transition (Baksi and others, 1967) and ~0.2 my older than the previously reported $^{40}Ar/^{39}Ar$ age (Baksi and Farrar, 1990). As pointed out by Baksi and Farrar (1990),

however, the accuracy of the CRB calibration of the GPST requires that the magnetic polarities of the CRB are stratigraphically unique, that no magnetic polarity unit greater than 0.1 my is missing, that the correlation of the CRB polarity transitions to the GPTS are correct, and that the newly revised $^{40}Ar/^{39}Ar$ ages are accurate.

Against the older age of the proposed calibration point for Chron C5Bn in Baksi (1993) and in addition to the above mentioned ages from Japan and Martinique, we cite the $^{40}K/^{40}Ar$ and $^{40}Ar/^{39}Ar$ data from the continental Barstow Formation of southern California (MacFadden and others , 1990). These authors reported ages of 15.9 $\pm$ 0.06 Ma on a tuff located within Chron C5Br, 14.8 $\pm$ 0.06 Ma on a tuff located within Chron C5ADr, and 14.0 $\pm$ 0.09 Ma on a tuff located within Chron C5ACn. For the interval between 16 to 14 Ma, the Barstow data agree quite well with the predictions of CK92/95. However, if we attempt to correlate the Barstow tuffs with the time scale proposed by Baksi (1993), each of the Barstow tuffs is predicted to occur either in intervals of opposite polarity from that observed, or at reversal boundaries.

Another calibration point used by Baksi (1993) is a reversed-normal transition recorded in the Akaroa volcanic field, New Zealand, which Baksi and others (1993a) correlate with the top of Chron C5n. Three basalts spanning this transition yielded a mean age of 9.67 $\pm$ 0.11 Ma. We note that this suggested younger limit of Chron C5n is consistent with that of 9.74 Ma in this work and of 9.6 Ma by McDougall and others (1984) from Icelandic basalts. However, we note that the reversed-normal transition in the Akaroa volcanic field can be correlated either with the Chron C5n.1r/C5n.2n or the Chron C4Ar.2 r/n boundary.

We conclude that the time scale that Baksi (1993) proposed for the ~10–17 Ma interval suffers from ambiguous correlation of isolated magnetic reversals with the GPTS. For example, assuming that the new $^{40}Ar/^{39}Ar$ ages of Baksi (1993) are accurate, an alternative correlation for the dated CRB units would be that the N_1-R_1 transition dated at 16.32 $\pm$ 0.1 Ma correlates with one of the short reversed intervals within Chron 5Cn and that the CRB R_1-N_0 transition, dated at 17.5 $\pm$ 0.2 Ma, correlates with Chron C5Cr/C5Dn. This revised interpretation would be consistent with the CK92/95 ages in this interval. However, this would imply that the erosional unconformity between the Grande Ronde and Wanapum basalts represents more time than previously recognized or that the poorly calibrated Wanapum basalts are in need of $^{40}Ar/^{39}Ar$ dating.

We also cite ten $^{40}Ar/^{39}Ar$ dates on the magnetostratigraphically measured sanidine-bearing tuffs of the Ngorora Formation, Kenya (Deino and others, 1990). These range between 13.0 Ma at the base of Chron C5AA and 11.7 Ma at the top of Chron C5A. Correlation of the magnetostratigraphy of the Ngorora Formation with the time-scale proposed by Baksi (1993) results in opposite polarity assignments for the $^{40}Ar/^{39}Ar$ ages obtained on interbedded tuffs.

An age of 7.26 $\pm$ 0.1 Ma for the Tortonian/Messinian boundary in the northern Apennines of Romagna was recently suggested by Vai and others (1993) based on a K/Ar (biotite) and a $^{40}Ar/^{39}Ar$ (plagioclase) date of 7.33 Ma on volcanogenic horizons a few meters below the FAD of *Globorotalia conomiozea* and *Gt. mediterranea* and a K/Ar (biotite) date of 7.72 $\pm$ 0.15 Ma on the stratigraphically lower FAD of *Globorotalia suterae* (which agrees closely with the magnetochronologic estimate for this datum event proposed here). The radioisotopic age estimate of 7.26 Ma for the Tortonian/Messinian boundary is seen to agree closely with the magnetochonologic age estimate of 7.12 Ma in this paper.

Pliocene-Pleistocene Epochs (Chron C1 to Chron C3 = Brunhes to Gilbert)

The youngest calibration point used in virtually all previous time scales is the Gilbert/Gauss boundary. Based on conventional ^{40}K/^{40}Ar dates, Mankinen and Dalrymple (1979) estimated the age of the Gilbert/Gauss (Chrons C2An/C2Ar) boundary to be 3.40 Ma. However, age estimates for this boundary based on astronomical predictions suggest that the age of many of the Plio-Pleistocene geomagnetic reversal boundaries as calibrated by conventional ^{40}K/^{40}Ar dates are too young. The astronomical predictions are derived by assuming that the variability recorded in high-resolution climatic records was forced by the well-known variations in the Earth's orbital parameters. Forcing the climatic records from the equatorial Pacific (Shackleton and others, 1990) and the Mediterranean region (Hilgen, 1991) to fit with the Earth's orbital parameter variations, suggests that the conventional K/Ar ages reported by Mankinen and Dalrymple (1979) for the estimation of the age of Plio-Pleistocene magnetic reversal boundaries were indeed too young by ~5 to 7%. As a consequence of these studies, the CK92 time-scale departed from the conventional use of only radioisotopic data and incorporated the astronomical data for the calibration of the Pliocene-Pleistocene Epochs (Cande and Kent, 1992). CK92 used the astronomically derived age of the Gauss/Matayama boundary (= C2An(y)) as determined independently by both Shackleton and others (1990) and Hilgen (1991) at 2.60 Ma. The astronomically derived age is approximately 5% older than the 2.48 Ma estimate for the Gauss/Matayama boundary based on previously reported K/Ar data (Mankinen and Dalrymple, 1979). More recently, Langereis and others (1994) provided a refined astronomically derived age of 2.581 Ma for the Gauss/Matuyama boundary.

Recent ^{40}Ar/^{39}Ar dating of stratigraphic levels representative of the last 4 my of the Neogene Period now confirms the astronomical age predictions based on high-resolution climatic records for the Pliocene-Pleistocene Epochs. In a series of recent papers, conventional ^{40}K/^{40}Ar ages on anorthoclase (McDougall and others, 1992) as well as both ^{40}Ar/^{39}Ar incremental-heating of bulk separates and replicate laser total-fusion of single volcanic crystals (Baksi and others, 1992; Baksi, and others, 1993b; Izett and Obradovich, 1991; Renne and others, 1993; Spell and McDougall, 1992; Spell and Harrison, 1993; Tauxe and others, 1992; Turrin and others, 1994; Izett and Obradovich, 1994) now indicate congruence between the astronomically and radioisotopically derived ages for the Plio-Pleistocene interval. Accordingly, we have adopted the astronomical time scale of Shackleton and others (1990) and Hilgen (1991) as standard for the interval Chron C1-Chron C3 (0 to 5.23 Ma). Table 7 summarizes these data (see also Berggren and others, 1995).

INTEGRATED CENOZOIC MAGNETOBIOCHRONOLOGY

At this point, we introduce the revised, integrated Cenozoic magnetobiochronologic scale which has resulted from our com-

TABLE 7.— ISOTOPIC VS. ASTRONOMICAL AGES FOR EARLY PLIOCENE-PLEISTOCENE MAGNETOCHRONS.

Chron	Radioisotopic Age Estimates	Astronomically Estimated Ages
C1n/1r (= Brunhes/Matuyama)	>0.746[1] >0.780[2] >0.783[3] ~0.780[4]	0.780[S]
C1r.1n (Jaramillo)	0.992–?[1]	0.990–1.070[S]
C2n (Olduvai)	1.78–1.96[4]	1.79–1.95[H] 1.95–1.77[S]
C2An.1n	2.60–3.02[4]	2.60–3.04[H]
C2An.2n	3.09–3.21[4,5]	3.11–3.22[H]
C2An.3n	3.29–3.57[4,5]	3.33–3.58[H]

[1]Tauxe and others, 1992
[2]Izett and Obradovich, 1991
[3]Baksi and others, 1992
[4]Spell and McDougall, 1992, and Spell and Harrison, 1993
[5]McDougall and others, 1992
[6]Renne and others, 1993
[S]Shackleton and others, 1990
[H]Hilgen, 1991

bined efforts (Figs. 1–6). In this way, the time scale figures can serve as a link between the geomagnetic polarity/sea-floor anomaly and radioisotopic data base which have been discussed above and the revised correlation and calibration of planktonic foraminiferal and calcareous nannofossil biostratigraphic datum events to the GPTS, which are discussed below. The planktonic foraminiferal and calcareous nannofossil datum events which have served to construct the magnetobiochronologic framework we present are discussed in Tables 8 to 13 and 14 to 17, respectively.

Planktonic Foraminiferal Biostratigraphy

Paleogene Zonation.—

The Paleogene planktonic foraminiferal zonation of Berggren and Miller (1988) is adopted here for essentially tropical-subtropical stratigraphies (with modifications noted below). At high austral latitudes a Paleogene scheme has been developed by Stott and Kennett (1990) and modified by Huber (1991) or augmented by Berggren (1992). Cross correlation of these zonal stratigraphies is afforded, in some instances, by means of magnetostratigraphy but in other cases problems in taxonomy (particularly among Paleocene and Eocene acarininids) or lack of precise magnetobiostratigraphic calibration hinder true interhemispherical correlation and calibration (see also relevant comments by Bolli and Krasheninnikov, 1978 from an earlier period of studies). Zonal magnetobiochronology has been based, as in the case of BKF85, on a compilation and evaluation of first-order correlations between biostratigraphic datums and magnetostratigraphy in DSDP and ODP boreholes as well as landsections. Our own studies on planktonic foraminiferal biostratigraphy (WAB) have been incorporated in this study where relevant. For example, Paleocene planktonic foraminiferal stratigraphy of DSDP/ODP Holes 384 (NW Atlantic), 577 and 465 (western Pacific), 213, 758 and 761 (Indian Ocean) and 750A (Kerguelen Plateau) have been studied in detail to supplement existing data. A paleomagnetic record spanning virtually the entire Paleocene in DSDP Hole 384 (Berggren and others, 1994) has provided a much needed improvement to first-order

PALEOCENE TIME SCALE

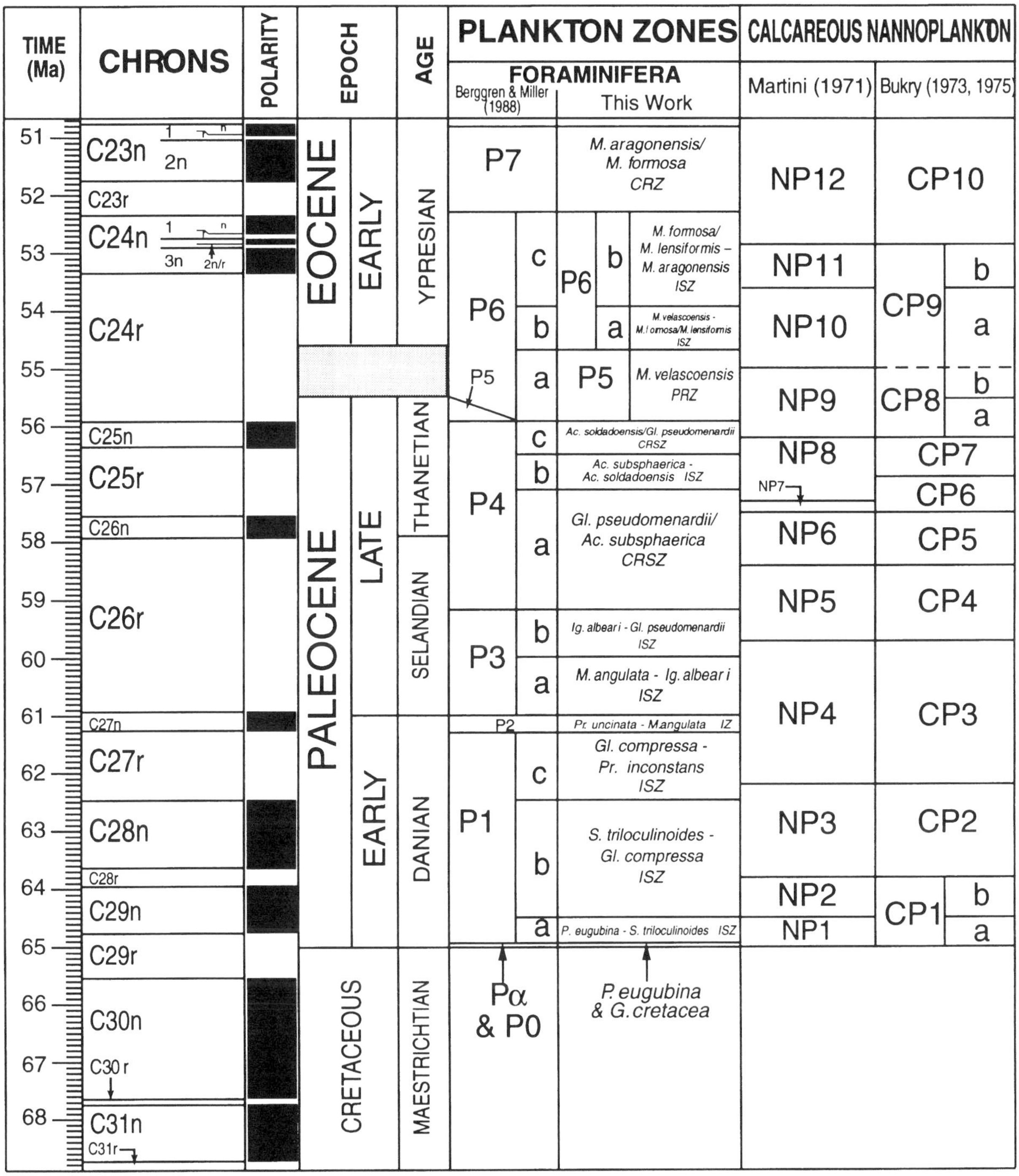

TIME (Ma)	CHRONS	POLARITY	EPOCH	AGE	PLANKTON ZONES FORAMINIFERA					CALCAREOUS NANNOPLANKTON		
					Berggren & Miller (1988)	This Work				Martini (1971)	Bukry (1973, 1975)	
51 52	C23n 1/2n n		EOCENE	YPRESIAN	P7	M. aragonensis/ M. formosa CRZ				NP12	CP10	
53	C23r C24n 1/3n n 2n/r					c	P6	b	M. formosa/ M. lensiformis – M. aragonensis ISZ	NP11	CP9	b
54	C24r				P6	b		a	M. velascoensis - M.f omosa/M.lensiformis ISZ	NP10		a
55					P5	a	P5		M. velascoensis PRZ	NP9	CP8	b
56	C25n		PALEOCENE	THANETIAN		c			Ac. soldadoensis/Gl. pseudomenardii CRSZ	NP8	CP7	a
57	C25r					b			Ac. subsphaerica - Ac. soldadoensis ISZ	NP7	CP6	
58	C26n				P4	a			Gl. pseudomenardii/ Ac. subsphaerica CRSZ	NP6	CP5	
59	C26r			SELANDIAN						NP5	CP4	
60					P3	b			Ig. albeari - Gl. pseudomenardii ISZ			
						a			M. angulata - Ig. albeari ISZ	NP4	CP3	
61	C27n			DANIAN	P2				Pr. uncinata - M.angulata IZ			
62	C27r					c			Gl. compressa - Pr. inconstans ISZ			
63	C28n				P1					NP3	CP2	
64	C28r C29n					b			S. triloculinoides - Gl. compressa ISZ	NP2	CP1	b
65	C29r					a			P. eugubina - S. triloculinoides ISZ	NP1		a
66	C30n		CRETACEOUS	MAESTRICHTIAN	Pα & P0	P. eugubina & G.cretacea						
67	C30 r											
68	C31n C31r											

FIG. 1.—The chronology of the Cenozoic Era (and its chronostratigraphic subdivisions) presented in this paper has been developed by fitting a spline fit to a set of 9 calibration points (described in greater detail in the section on the geomagnetic polarity time scale) and represents a modification of the chronology of CK92 to that of CK92/95. The various datum events used to define the Paleogene and Neogene planktonic foraminiferal biostratigraphic zones have then been correlated to the resulting magnetochronology resulting in a magnetobiochronology. Most of the correlations are by direct, first-order correlation (see Tables 8 to 17); others represent indirect, second-order correlations. The relatively broad gray band spanning the interval between ~54.6 to 55.5 Ma reflects current opinion on the position of the Paleocene/Eocene boundary. Taken at the base of the type Ypresian in the Belgian basin or the base of the London Clay in the London-Hampshire Basin its position would be at ~54.6 Ma to 54.8 Ma, respectively. Planktonic foraminiferal specialists commonly use the P5/P6 boundary (~54.7 Ma) and calcareous nannoplankton specialists the NP9/NP10 zonal boundary (~55.0 Ma) as denotative of the boundary for purpose of correlation. More recently the major $\delta^{13}C$ spike at ~55.5 Ma has been suggested as providing a useful means of correlation between marine and terrestrial stratigraphies. Lacking a clearly defined Global Stratotype Section and Point (GSSP), the Paleocene/Eocene boundary awaits an unequivocal definition (IGCP Project 308) in 1996. See text for further discussion of the intricate (bio)stratigraphic problems associated with the Paleocene/Eocene boundary. The other epoch/series boundaries of the Cenozoic Era have GSSP's so that physical position and corresponding age estimates are relatively well established.

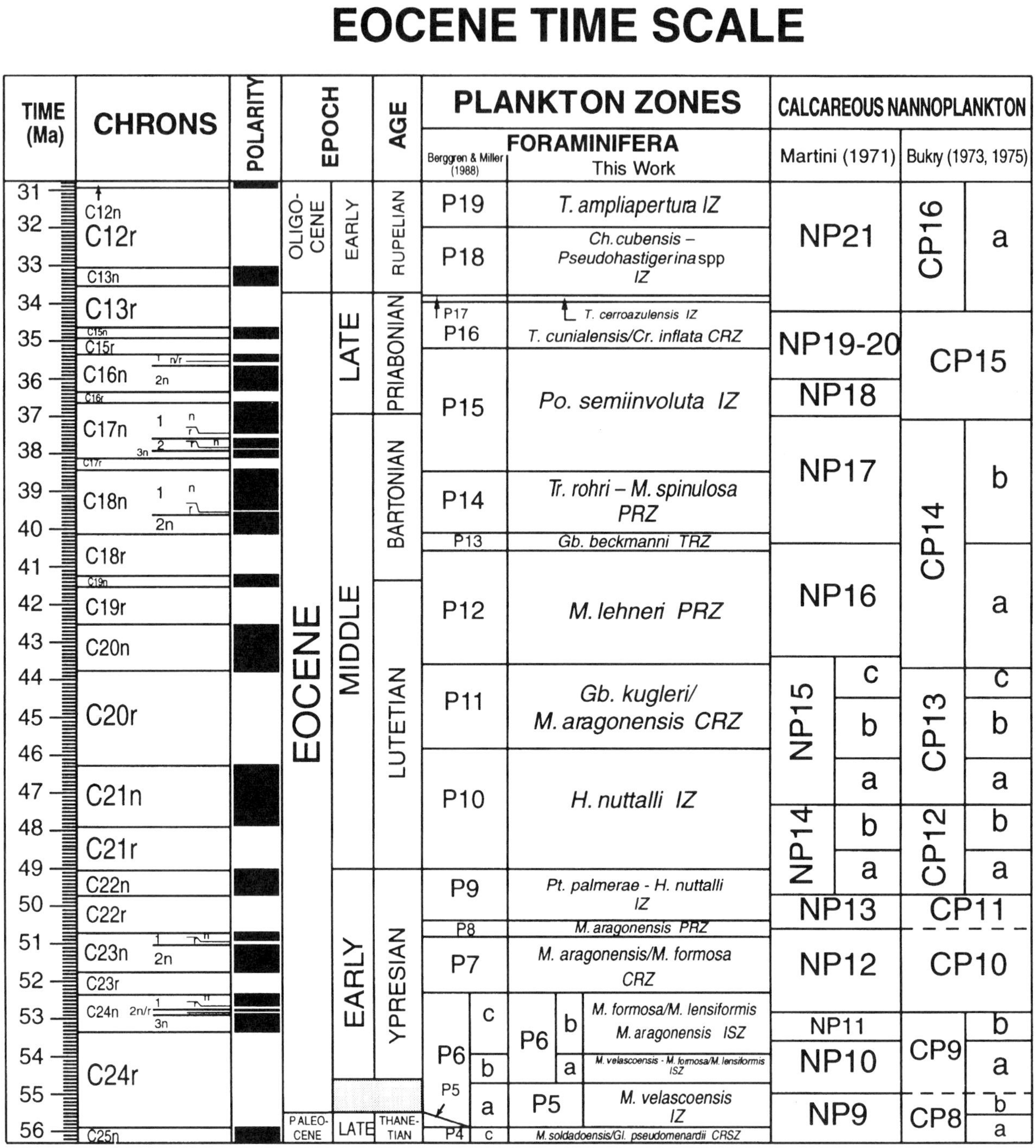

EOCENE TIME SCALE

TIME (Ma)	CHRONS	POLARITY	EPOCH	AGE	FORAMINIFERA Berggren & Miller (1988)	FORAMINIFERA This Work	CALCAREOUS NANNOPLANKTON Martini (1971)	CALCAREOUS NANNOPLANKTON Bukry (1973, 1975)
31–33	C12n, C12r, C13n		OLIGO-CENE EARLY	RUPELIAN	P19	T. ampliapertura IZ	NP21	CP16 a
					P18	Ch. cubensis – Pseudohastigerina spp IZ		
34–35	C13r, C15n, C15r		LATE	PRIABONIAN	P17	T. cerroazulensis IZ	NP19-20	CP15
					P16	T. cunialensis/Cr. inflata CRZ		
36–37	C16n (1 n/r, 2n), C16r, C17n (1 n, 2, 3n)				P15	Po. semiinvoluta IZ	NP18	
38–40	C17r, C18n (1 n, 2n)			BARTONIAN	P14	Tr. rohri – M. spinulosa PRZ	NP17	CP14 b
40–41	C18r, C19n		MIDDLE		P13	Gb. beckmanni TRZ		
41–43	C19r, C20n		EOCENE	LUTETIAN	P12	M. lehneri PRZ	NP16	CP14 a
44–46	C20r				P11	Gb. kugleri/ M. aragonensis CRZ	NP15 c / b / a	CP13 c / b / a
46–48	C21n, C21r				P10	H. nuttalli IZ	NP14 b / a	CP12 b / a
49–50	C22n, C22r		EARLY	YPRESIAN	P9	Pt. palmerae - H. nuttalli IZ	NP13	CP11
50–51					P8	M. aragonensis PRZ		
51–52	C23n (1, 2n), C23r				P7	M. aragonensis/M. formosa CRZ	NP12	CP10
53	C24n (2n/r 1, 3n)				P6 c / P6 b	M. formosa/M. lensiformis M. aragonensis ISZ	NP11	CP9 b
54	C24r				P6 b / P6 a	M. velascoensis - M. formosa/M. lensiformis ISZ	NP10	CP9 a
55			PALEO-CENE LATE	THANE-TIAN	P5 a / P5	M. velascoensis IZ	NP9	CP8 b
56	C25n				P4 c	M. soldadoensis/Gl. pseudomenardii CRSZ		CP8 a

FIG. 2.—See Figure 1 for further explanation.

magnetobiostratigraphic correlations spanning Chrons C29-C25. A similar improvement to correlations in Chron C24r has been incorporated from studies in DSDP Hole 550 (Berggren and Aubry, 1995). Second order, or inferred, (magnetobiostratigraphic) correlations are used where considered relevant and discussed at the appropriate place(s) in Tables 8 to 13. Additional background information on the historical development of Paleogene planktonic foraminiferal zonations can be found in Berggren and Miller (1988).

Paleocene Zonation.—The notation "P" used by Berggren and Miller (1988) to denote a series of subtropical-tropical zones applicable on a global (exclusive of high southern and northern latitudes) is maintained here. However, several modifications to the existing zonal scheme are made based on studies made since 1988 and with a view toward bringing the zonal terminology in line with suggested modifications being made in the revision to the International Guide to Stratigraphic Principles (Salvador, 1994). A comparison between the Paleocene zonation

OLIGOCENE TIME SCALE

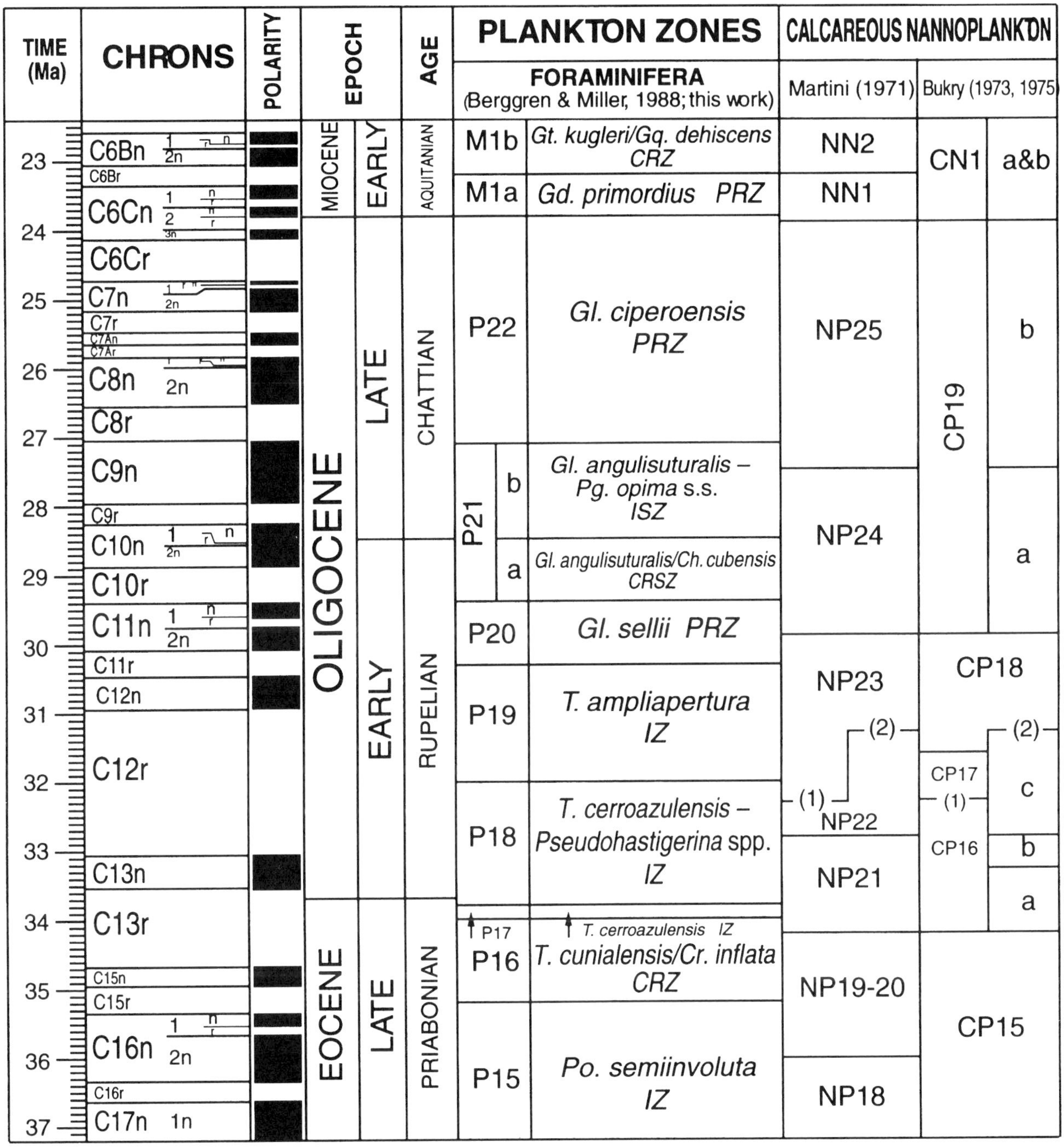

TIME (Ma)	CHRONS	POLARITY	EPOCH	AGE	PLANKTON ZONES FORAMINIFERA (Berggren & Miller, 1988; this work)		CALCAREOUS NANNOPLANKTON Martini (1971)	Bukry (1973, 1975)	
23	C6Bn 1 2n / C6Br		MIOCENE	EARLY / AQUITANIAN	M1b	*Gt. kugleri/Gq. dehiscens* CRZ	NN2	CN1	a&b
	C6Cn 1 2 3n				M1a	*Gd. primordius* PRZ	NN1		
24	C6Cr								
25	C7n 1 2n / C7r / C7An / C7Ar			LATE / CHATTIAN	P22	*Gl. ciperoensis* PRZ	NP25	CP19	b
26	C8n 2n								
27	C8r		OLIGOCENE						
28	C9n / C9r				P21 b	*Gl. angulisuturalis – Pg. opima* s.s. ISZ	NP24		a
	C10n 1 2n				P21 a	*Gl. angulisuturalis/Ch. cubensis* CRSZ			
29	C10r			EARLY / RUPELIAN					
30	C11n 1 2n / C11r / C12n				P20	*Gl. sellii* PRZ	NP23	CP18	
31					P19	*T. ampliapertura* IZ	(2) NP22	CP18 / CP17 (2) (1)	c
32	C12r				P18	*T. cerroazulensis – Pseudohastigerina* spp. IZ	(1) NP21	CP16	b
33	C13n								a
34	C13r		EOCENE	LATE / PRIABONIAN	↑P17 / P16	↑ *T. cerroazulensis* IZ / *T. cunialensis/Cr. inflata* CRZ	NP19-20	CP15	
35	C15n / C15r								
36	C16n 1 2n / C16r				P15	*Po. semiinvoluta* IZ	NP18		
37	C17n 1n								

FIG. 3.—See Figure 1 for further explanation.

(re)defined in this paper and several earlier schemes is shown in Figure 7. The chronology of Paleocene planktonic foraminiferal datum events/zones is shown in Figure 8.

P0. *Guembelitria cretacea* Partial Range Zone (P0; Keller, 1988, emend. of Smit, 1982)

Definition: Biostratigraphic interval characterized by the partial range of the nominate taxon between the Last Appearance Datum (LAD) of Cretaceous taxa (*Globotruncana, Rugoglobigerina, Globigerinelloides,* among others) at the K/P boundary as delineated by the essentially global iridium spike and the First Appearance Datum (FAD) of *Parvularugoglobigerina eugubina*.

Magnetochronologic calibration: Chron C29r (late part)

Estimated age: 65.0–64.97 Ma; earliest Paleocene (Danian)

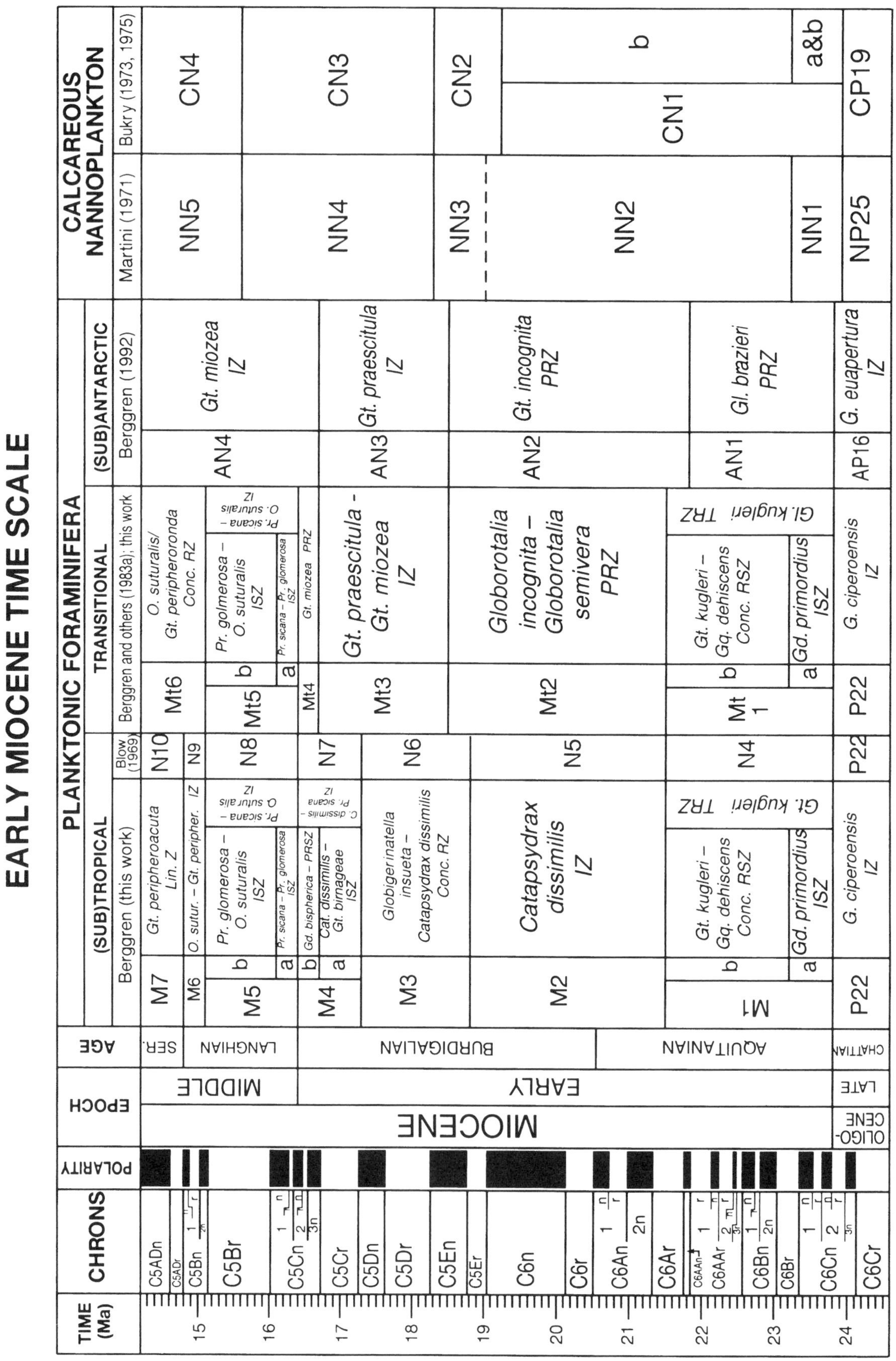

EARLY MIOCENE TIME SCALE

CALCAREOUS NANNOPLANKTON

Bukry (1973, 1975): CN4, CN3, CN2, CN1 (b, a&b), CP19

Martini (1971): NN5, NN4, NN3, NN2, NN1, NP25

PLANKTONIC FORAMINIFERA

(SUB)ANTARCTIC — Berggren (1992):
- Gt. miozea IZ — AN4
- Gt. praescitula IZ — AN3
- Gt. incognita PRZ — AN2
- Gl. brazieri PRZ — AN1
- G. euapertura IZ — AP16

TRANSITIONAL — Berggren and others (1983a); this work:
- O. suturalis/Gt. peripheroronda Conc. RZ — Mt6
- Pr. sicana – O. suturalis IZ — b
- Pr. golmerosa – O. suturalis ISZ — Mt5
- Pr. sicana – Pr. glomerosa ISZ — a
- Gt. miozea PRZ — Mt4
- Gt. praescitula – Gt. miozea IZ — Mt3
- Globorotalia incognita – Globorotalia semivera PRZ — Mt2
- Gt. kugleri TRZ
- Gt. kugleri – Gq. dehiscens Conc. RSZ — b — Mt1
- Gd. primordius ISZ — a
- G. ciperoensis IZ — P22

(SUB)TROPICAL — Berggren (this work):
- Gt. peripheroacuta Lin. Z — M7
- O. sutur. – Gt. peripher. IZ — M6
- Pr. golmerosa – O. suturalis ISZ — b — M5
- Pr. sicana – Pr. glomerosa ISZ — a
- Gd. bispherica – PRSZ — b — M4
- Cat. dissimilis – Gt. birnageae ISZ — a
- Globigerinatella insueta – Catapsydrax dissimilis Conc. RZ — M3
- Catapsydrax dissimilis IZ — M2
- Gt. kugleri TRZ
- Gt. kugleri – Gq. dehiscens Conc. RSZ — b — M1
- Gd. primordius ISZ — a
- G. ciperoensis IZ — P22

Blow (1969): N10, N9, N8, N7, N6, N5, N4, P22

Subtropical: M7, M6, M5, M4, M3, M2, M1, P22

AGE: SER. LANGHIAN, BURDIGALIAN, AQUITANIAN, CHATTIAN

EPOCH: MIDDLE, EARLY, LATE — MIOCENE / OLIGOCENE

POLARITY

CHRONS: C5ADn, C5ADr, C5Bn, C5Br, C5Cn, C5Cr, C5Dn, C5Dr, C5En, C5Er, C6n, C6r, C6An, C6Ar, C6AAn, C6AAr, C6Bn, C6Br, C6Cn, C6Cr

TIME (Ma): 15, 16, 17, 18, 19, 20, 21, 22, 23, 24

FIG. 4.—See Figure 1 for further explanation.

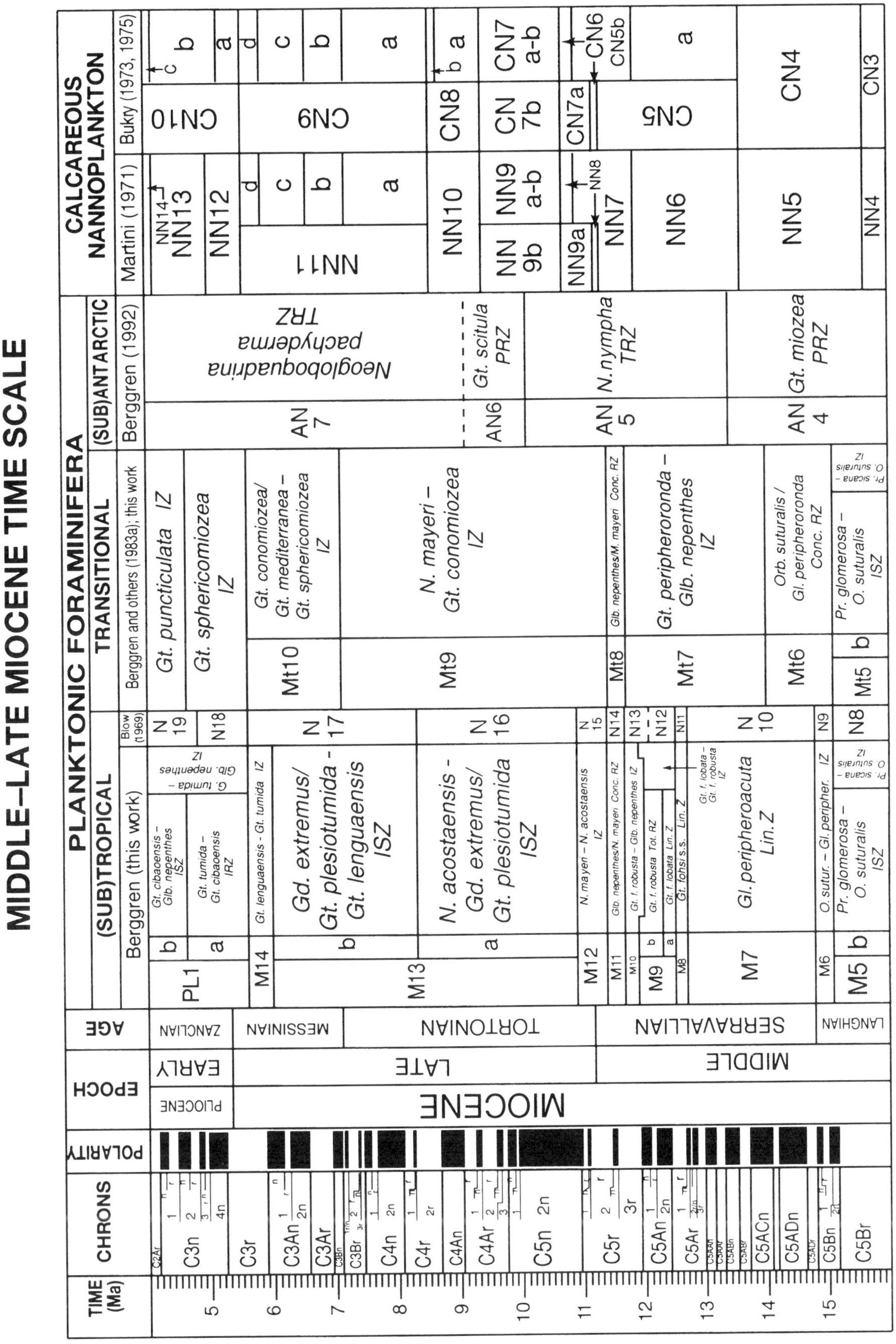

FIG. 5.—See Figure 1 for further explanation.

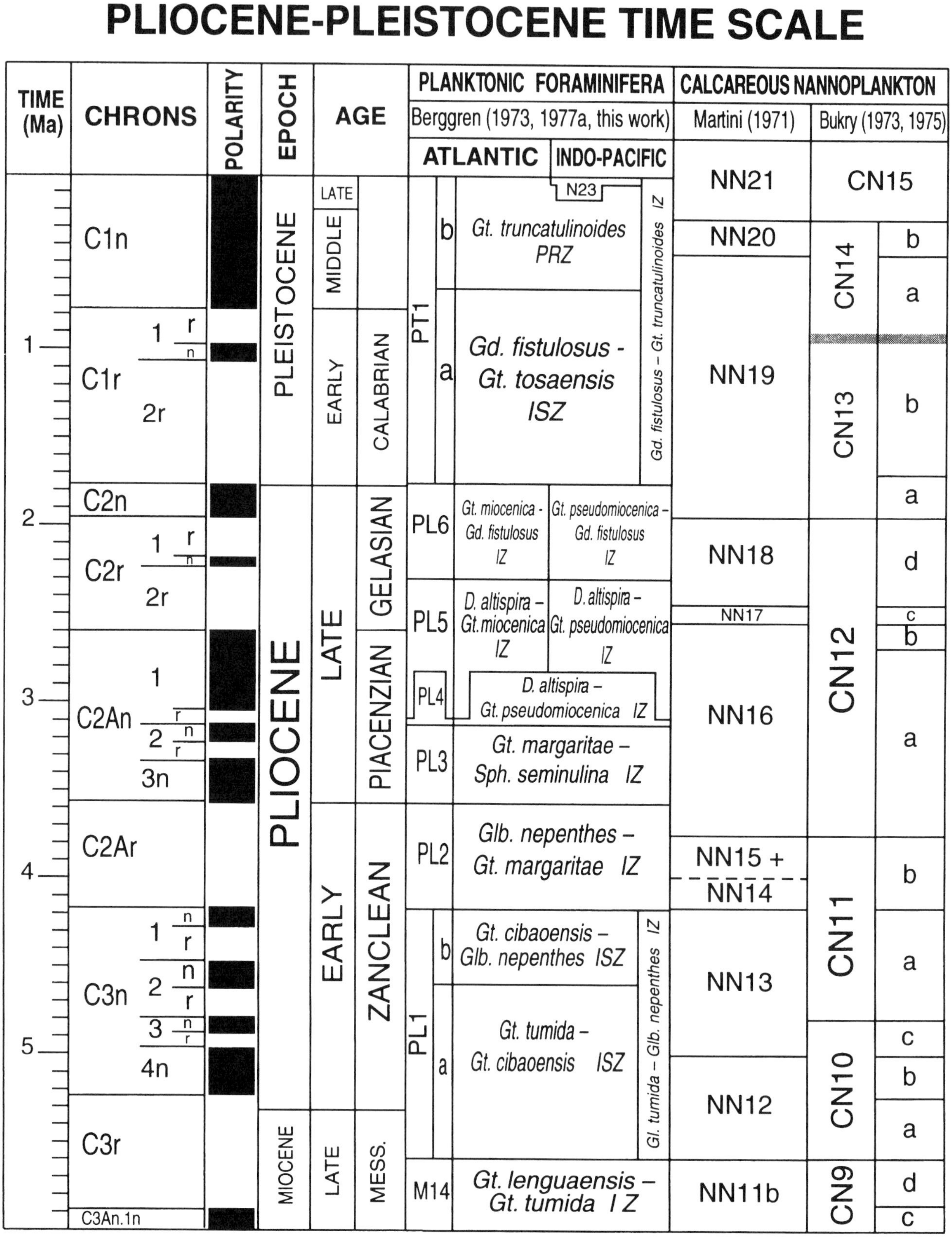

PLIOCENE-PLEISTOCENE TIME SCALE

FIG. 6.—See Figure 1 for further explanation.

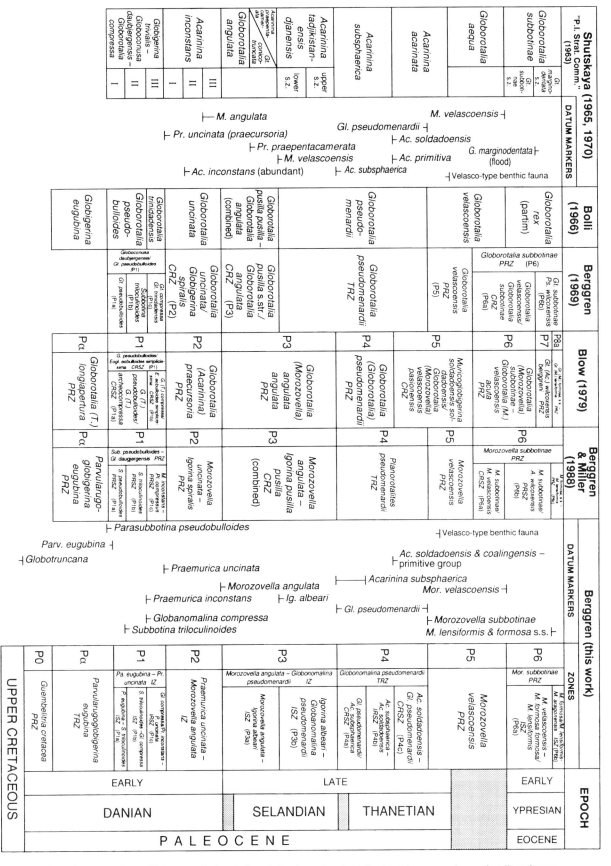

FIG. 7.—Comparison of Paleocene planktonic foraminiferal zonation (re)defined in this paper and several earlier schemes.

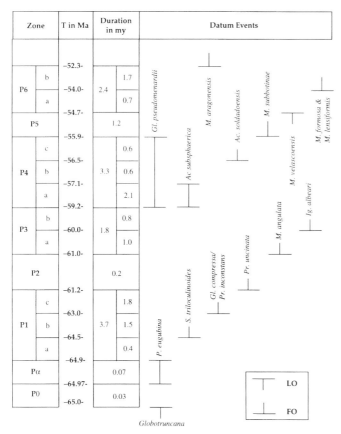

FIG. 8.—Chronology of Paleocene (sub)tropical planktonic foraminiferal zones.

Remarks: See below under *Parvularugoglobigerina eugubina* Total Range Zone.

Pa. *Parvularugoglobigerina eugubina* Total Range Zone (Liu, 1993, emend. of Pa of Blow, 1979; Luterbacher and Premoli Silva, 1964)

Definition: Biostratigraphic interval characterized by the total range of the nominate taxon

Magnetochronologic calibration: Chron C29r (later part)

Estimated age: 64.97–64.9 Ma; earliest Paleocene (Danian)

Remarks: The Cretaceous/Paleogene (K/P) boundary is located about two-thirds of the way up within Chron C29r with an assigned age of 65.0 Ma (CK92/95) vs. 66 Ma (CK92) and corresponds to the mass extinction of Cretaceous species of planktonic foraminifera at the recommended (and approved but currently contested) boundary stratotype at El Kef, Tunisia (Liu, 1993) and correlative levels elsewhere. The earliest Paleogene (Danian) planktonic foraminiferal Zones P0 and Pa (defined by the partial range of *Guembelitria cretacea* following the extinction of Cretaceous taxa at the K/P boundary and the initial appearance of *P. eugubina*, respectively; Keller, 1988, emend., Smit, 1982), now appear to be well recognized, although Zone P0 appears to be restricted to nearshore, rather than open ocean, environments and to be of extremely short duration. In view of the less equivocal taxonomic problems associated with the recognition of *P. eugubina* vs. the earliest representatives of *Parasubbotina pseudobulloides*, we follow Liu (1993) in recogniz-

ing the terminal occurrence of the former rather than the initial appearance of the latter in defining the boundary between Zones Pa and P1.

Estimation of the age and duration of Zones P0 and Pa remains fraught with difficulty. Liu (1993) has pointed to the hazard of assuming constant rates of sedimentation across the K/P boundary in estimating biochronological datums owing to fluctuating (predominantly declining) productivity across (and particularly above) the boundary interval. He estimated a duration for the K/P boundary interval (i.e., the stratigraphic interval between the extinction of the Cretaceous taxa and the initial appearance of Paleogene eoglobigerinids in the assumedly continuously deposited section at Miller's Ferry, Alabama) of ~3.53 ky and 3.34 ky based on data in BKF85 (South Atlantic) and Groot and others (1989; Agost, Spain), respectively. He did this by using the estimated ages for the LAD of *P. eugubina* and the FAD of *Pa. pseudobulloides* in these two studies as calibration points. The FAD of *P. eugubina* was estimated to have occurred ~50 ky and 20 ky after the K/P boundary by BKF85 and Groot and others (1989), respectively. However, it should be borne in mind that the duration of Chron C29r is significantly different in BKF85 (570 ky) vs CK92 (869 ky) or CK92/95 (833 ky), and that the duration of the stratigraphic interval from the K/P boundary to the Chron C29n/r boundary is correspondingly lengthened and then shortened from 230 ky (BKF85) to 268 ky (CK92) to 255 ky (CK92/95). This difference simply has the effect of extending the duration of Zone P0 by about 30–35%. The duration of Zone Pa is also uncertain in view of conflicting data which suggest: a) biostratigraphic exclusion of the ranges of *P. eugubina* and *Pa. pseudobulloides* in the later part of Chron C29r at Agost (Groot and others, 1989) and b) an overlap (37 cm) in the stratigraphic range of these two taxa at Miller's Ferry (Liu, 1993). Despite the taxonomic problems associated with the recognition of the earliest morphotypes of *Pa. pseudobulloides* (Olsson and others, 1992) mentioned above, it would appear that most specialists agree that the FAD of *Pa. pseudobulloides* and the LAD of *P. eugubina* are closely associated (Poore and others, 1984; Liu, 1993; see also Blow, 1979), and that these two events occur within the later part of Chron C29r, somewhat older than that estimated in BKF85. We take a compromise view here between the data in Poore and others (1984), Groot and others (1989) and Liu (1993) in viewing the FAD of *P. eugubina* in the later part of Chron C29r (Poore and others, 1984; Groot and others, 1989) with an estimated calibration of 66.43 Ma (by Groot and others, 1989) as definitive; a revised age of 64.97 Ma is suggested here based on calibration to CK92/95. With regard to the upper limit of *eugubina*, we view the overlap in the ranges of *eugubina* and *pseudobulloides* as real (Blow, 1979; Poore and others, 1984; Liu, 1993; pers. observ., WAB) and place the Pa/P1 zonal boundary just below the Chron C29n/r boundary, at (admittedly somewhat arbitrarily) 64.8 Ma (CK92/95). The P0 and Pa Zones would then have estimated durations of 30 ky and 170 ky, respectively. The difference between Liu's (1990) estimate of ~3-ky duration for Zone P0 vs our estimate of 30 ky illustrate the difficulty and uncertainty in estimating durations of the earliest Paleocene Biozones.

P1. *Parvularugoglobigerina eugubina-Praemurica uncinata* Interval Zone (P1; defined herein; emend. of Berggren and Miller, 1988).

Definition: Biostratigraphic interval between the LAD of *Parvularugoglobigerina eugubina* and the FAD of *Praemurica uncinata*.

Magnetochronologic calibration: Chron C29r (later part)-Chron C27n(0)

Estimated age: 64.9–61.2 Ma; early Paleocene (Danian)

Remarks: Zone P1 (the biostratigraphic interval between the LAD of *P. eugubina* and the FAD of *Pr. uncinata*) has been subdivided into three subzones based on the sequential appearances of *Subbotina triloculinoides* and *Globanomalina compressa/Pr. inconstans* (Berggren and Miller, 1988). The FAD of *S. triloculinoides* (= P1a/b) occurs in the early to mid-part of Chron C29n at DSDP Hole 384 with an estimated age of 64.5 Ma and that of *Gl. compressa* (= P1b/c) in mid-Chron C28n in Hole 384 (63.0 Ma, Table 8). The zone as redefined here is essentially the same as Zone P1 in Berggren and Miller (1988), except that the LAD of *P. eugubina*, rather than the FAD of *Pa. pseudobulloides*, has been adopted to characterize the base of the zone (following Liu, 1993). Definition as an interval zone allows its recognition in a relatively unequivocal manner.

P1a. *Parvularugoglobigerina eugubina-Subbotina triloculinoides* Interval Subzone (P1a; herein defined; emendation of *Pa. pseudobulloides* Subzone (P1a) in Berggren and Miller, 1988)

Definition: Biostratigraphic interval between the LAD of *Parvularugoglobigerina eugubina* and the FAD of *Subbotina triloculinoides*

Magnetochronologic calibration: Chron C29r (later part)-Chron C29n (mid part)

Estimated age: 64.9–64.5 Ma; early Paleocene (early Danian)

Remarks: The biochronologic denotation of this subzone is virtually the same as that of Subzone P1a in Berggren and Miller (1988), but the biostratigraphic connotation has been emended to the extent that the LAD of *P. eugubina*, rather than the FAD of *Pa. pseudobulloides*, has been used as the criterion for the definition of its base following Liu (1993). Characteristic elements of this subzone include: spinose eoglobigerinids (*Eoglobigerina eobulloides*, *E. edita*), parasubbotinids (*Pa. pseudobulloides*), non spinose praemorozovellids (*Pr. pseudoinconstans*, *Pr. taurica*) and globanomalinids (*Gl. archaeocompressa*, *Gl. planocompressa*) and *Globoconusa daubjergensis*.

P1b. *Subbotina triloculinoides-Globanomalina compressa/ Praemurica inconstans* Interval Subzone (P1b; herein defined; emendation of, but equivalent to, Subzone P1b in Berggren and Miller, 1988)

Definition: Biostratigraphic interval between the FAD of *Subbotina triloculinoides* and the FADs of *Globanomalina compressa* and/or *Praemurica inconstans*

Magnetochronologic calibration: Chron C29n (mid part)-Chron C28n (mid part)

Estimated age: 64.5–63.0 Ma

Remarks: This subzone has the same biochronologic denotation as Subzone P1b in Berggren and Miller (1988) but its biostratigraphic connotation is emended to the extent that it is defined here as an interval subzone rather than as a partial range subzone of the taxon *S. triloculinoides*. *Pa. varianta* has its FAD within this biostratigraphic interval, although it does not become a significant and morphologically distinct element in Paleocene faunas until Zone P3.

P1c. *Globanomalina compressa/Praemurica inconstans-Praemurica uncinata* Interval Subzone (P1c; herein defined; emendation of, but equivalent to, Subzone P1c in Berggren and Miller, 1988).

Definition: Biostratigraphic interval between the FAD of *Globanomalina compressa* and/or *Praemurica inconstans* and the FAD of *Praemurica uncinata*

Magnetochronologic calibration: Chron C28n (mid)-Chron C27n(o)

Estimated age: 63.0–61.2 Ma

Remarks: Our recent studies (Berggren and Norris, unpubl. data) do not support Blow's (1979, p. 944) suggestion of the synonymy of *Pr. uncinata* (Bolli) and *Pr. praecursoria* (Morozova) (cf., Berggren and Miller, 1988, p. 368), but we would agree with the latter as to the synonymy between *Pr. inconstans* (Subbotina) and *Pr. trinidadensis* (Bolli). We regard *praecursoria* as an advanced morphotype of *inconstans*.

It should be noted that the FAD of *Pr. trinidadensis* was used as denotatative of the P1b/c boundary in BKF85. The FADs of *inconstans (trinidadensis)* and *compressa* are virtually simultaneous in Chron C28n and are retained here as alternate definitive elements in recognizing this subzone.

P2. *Praemurica uncinata-Morozovella angulata* Interval Zone (P2; herein defined; emend. of, but biostratigraphically equivalent to, Zone P2 in Berggren and Miller, 1988)

Definition: Biostratigraphic interval between the FAD of *Praemurica uncinata* and the FAD of *Morozovella angulata*

Magnetochronologic calibration: Chron C27n(o)-Chron C27n(y)

Estimated age: 61.2–61.0 Ma; late early Paleocene (late Danian)

Remarks: Zone P2, as used here, has the same biostratigraphic extent but somewhat different taxonomic connotation; work by WAB and Richard Norris (WHOI) suggests that the taxa *praecursoria* and *uncinata* are not synonymous (cf. Berggren and Miller, 1988, p. 368) as suggested by Blow (1979, p. 944) but that *trinidadensis* and *inconstans* (as well as *praecursoria*) are conspecific (cf. Blow, 1979) and the taxon *uncinata* is retained here as the nominative taxon of Zone P2. In DSDP Hole 384, we have recently found (Berggren and others, 1994) that the range of *Pr. uncinata* effectively brackets Chron C27n (Table 8). At Hole 577, the transition from *Pr. inconstans-trinidadensis* to *Pr. uncinata* occurs in the upper part of Core 12 and in a reversed interval immediately above Chron C28n which, by correlation, is close to Chron C27n in the adjacent Hole 577A; a brief hiatus is suspected to occur at this level in Hole 577 (see also discussion, below, of the FAD of *Ellipsolithus macellus* in section on Calcareous Nannofossil Magnetobiochronology: Paleogene). At any rate, the FAD of this taxon here is seen to be consistent with the earlier (BKF85) calibration to Chron C27n and correlation with Zone NP4 (Table 8, Item 24). However, that was based primarily on calibration in the northern Apennines based on thin-section analysis. The more precise determination provided here is based on well preserved material in deep-sea cores. In DSDP Hole 384, *Pr. uncinata* has its FAD (and can be seen to evolve from *Pr. inconstans*) at the base of Chron C27n at the same level as the FAD of (the muricate but non-carinate) *M. praeangulata* in Zone NP4 which is believed to lie at the base of the morozovellid radiation. It has its FAD at a comparable level in ODP Hole 758A (Indian

TABLE 8.—RELATIONSHIP OF PALEOCENE PLANKTONIC FORAMINIFERAL DATUM LEVELS TO OBSERVED (AND INTERPRETED) MAGNETIC POLARITY STRATIGRAPHY. AGE ESTIMATES (MA) DERIVED FROM THE GPTS OF CANDE AND KENT (1992/1995) AND ADOPTED IN THIS PAPER. INFERRED ESTIMATES (I.E., WHERE DIRECT MAGNETOBIOCHRONOLOGIC CALIBRATION IS LACKING) ARE PRESENTED IN PARENTHESIS. THESE DATA HAVE PROVIDED THE BASIC MAGNETOBIOCHRONOLOGIC FRAMEWORK FOR ESTIMATING THE CHRONOLOGY OF STANDARD TIME-STRATIGRAPHIC UNITS AND STAGE STRATOTYPES.

PALEOCENE

Datum	FAD	LAD	Paleomagnetic Chron	Age in Ma	Reference	Remarks
1. *Morozovella acuta*		X	C24r	54.7	1	Occurs midway between −17 (54.5 Ma) and + 19 (54.0 Ma) Ash Beds in Hole 550 and taken as proxy for the P5/P6 a zonal boundary there.
2. *Morozovella velascoensis*		X	C24r	54.7	2	P6a/b boundary in Berggren and Miller (1988); Zone P5/P6 boundary in this paper; generally used to approximate the Paleocene/Eocene boundary.
3. *Globanomalina australiformis*	X		C24r	55.5	3	Holes 689B, 690B; used to denote the P/E boundary in high southern latitudes (ref. 3).
4. *Morozovella subbotinae*	X		C25n$_{(y)}$	55.9	4,5	Hole 577.
5. *Globanomalina pseudomenardii*		X	C25n$_{(y)}$	55.9	1,4–6	P4/P5 boundary. Recorded in base C24r (ref. 4) and top C25n (ref. 5) in Hole 577 and top C25n in Hole 752 (ref 6). Juxtaposition/overlap of *Gl. pseudomenardii* LAD and *M. subbotinae* FAD observed in Holes 465 and 758 (ref. 7).
6. *Acarinina acarinata*		X	C25r	56.3	1	Hole 384; at comparable levels (without paleomagnetic control) in Holes 465 and 758A (ref. 1).
7. *Acarinina mckannai*		X	C25r	56.3	1	Hole 384; at comparable levels (without paleomagnetic control) in Holes 465 and 758A (ref. 1).
8. *Morozovella aequa*	X		C25r	56.5	1	Holes 384 (N. Atlantic), 465 (Ontong–Java Plateau, equatorial Pacific), 758A (Indian Ocean) in association with FAD of *Acarinina coalingensis*-triplex group and *Ac. soldadoensis* (see below).
9. *Acarinina coalingensis— triplex*	X		C25r	56.5	1	Hole 384; at comparable levels (without paleomagnetic control) in Holes 465 and 758A (ref. 1).
10. *Acarinina soldadoensis*	X		C25r	56.5	1	P4 b/c boundary (this paper). Hole 384; at comparable levels (without paleomagnetic control) in Holes 465 and 758A (ref. 1).
11. *Acarinina subsphaerica*		X	C26r	57.1	1	P4 a/b boundary (this paper). Hole 384; at comparable levels (without paleomagnetic control) in Holes 465 and 758A (ref. 1). The LAD of this taxon occurs approximately mid-way in Zone P4.
12. *Acarinina mckannai*	X		C26r	59.1	1	Hole 384; at comparable levels (without paleomagnetic control) in Holes 465 and 758A (ref. 1).
13. *Acarinina subsphaerica*	X		C26r	59.2	1	Hole 384; at comparable levels (without paleomagnetic control) in Holes 465 and 758A (ref. 1).
14. *Acarinina acarinata*	X		C26r	59.2	6	Hole 384; at comparable levels (without paleomagnetic control) in Holes 465 and 758A (ref. 1) The FAD of three acarininids (i.e., *acarinata, subsphaerica, mckannai*) has been observed to coincide closely with the FAD of *Gl. pseudomenardii* (this paper).
15. *Globanomalina pseudomenardii*	X		C26r	59.2	2,5	Hole 384; at comparable levels (without paleomagnetic control) in Holes 465 and 758A (ref. 1) Younger part of C26r in Hole 577 (ref. 4) within Zones CP5-CP6 (undiff.); FAD recorded in C25r in Zone CP5 (= NP6) in Hole 605 (ref. 8) but in C26n in CP5 (= N P5) in Hole 605 in refs. 8, 9; see also ref. 2.
16. *Parasubbotina varianta*		X	C26r	59.2	1	See under 17 below.
17. *Parasubbotina variolaria*		X	C26r	59.2	1	*P. variolaria* and *P. varianta* have simultaneous LADs in Holes 465 and 384.
18. *Morozovella velascoensis*	X		C26r	60.0	1,2	Hole 384; at comparable levels (without paleomagnetic control) in Holes 465 and 758A (ref. 1).
19. *Muricella albeari (= M. laevigata)*	X		C26r	60.0	1	P3a/b subzonal boundary of this paper (not P3a/b subzonal boundary in Berggren and Miller, 1988); occurs approximately mid-way in Zone P3 in Holes 463 and 758A and in the mid-part of C26r in Hole 384.
20. *Praemurica strabocella*	X		C26r	60.5	1	Occurs in Holes 384 and 758A midway between FAD *M. angulata* and FAD *M. albeari* and in Zone NP5; recorded in Hole 690 in lower Zone CP5 (= NP6 , ref. 2); in Hole 689B at level in Zone CP8 (= NP9) but equated with C26n (ref. 2); recorded in Hole 738 at CP4 /CP5 (= NP5/N P6) zonal boundary (ref. 10).
21. *Morozovella conicotruncata*	X		C26r (early part)	60.9	1	Occurs at virtually the same level as FAD *M. angulata* in Holes 384, 465, 758A (ref. 1) and 577 (refs 6,7). Definite position in Hole 577 difficult to determine because of absence of C27n (Bleil, 1985).
22. *Muricella pusilla*	X		C26r	61.0	1,5	Occurs in early C26r and top Zone CP3 (= NP3) in Hole 577 (refs. 4, 5) but precise position difficult to determine because of absence of C27n and possible unconformity just above C28n at 102.95 m (see Bleil, 1985, p. 449–450). Occurs at some level as FAD of *M. angulata* and *M. conicotruncata* in Holes 465 and 758A (ref. 1).
23. *Morozovella angulata*	X		C27n$_{(y)}$	61.0	1,2	In Hole 577 occurs in early C26r (ref. 1, 5), (but precise position difficult to determine because of absence of C27n (Bleil, 1985); in Hole 384 it occurs <2-m above FAD of *P. uncinata* in NP4 and C27n$_{(y)}$, 2-m above FAD *P. uncinata* in Holes 465 and 758A (ref. 1).
24. *Praemurica uncinata*	X		C27n$_{(o)}$	61.2	2	The FAD of *P. uncinata* occurs in Hole 577 (refs. 1, 5) just above a N–R polarity reversal (C28n); C27n is missing in Hole 577 (Bleil, 1985, p. 446, Fig. 5, p. 449, Fig 7B) and there may be a short hiatus at/near the level if FAD of this taxon in C27n (ref. 2) is correct; alternatively C27n may be missing in the core break between Cores 11 and 12 (~101.8 m) in which case the FAD of this taxon would be near the C27r/C28n boundary (see ref. 2); in Hole 384 it occurs at the same level as FAD *M. praeangulata* in Zone NP4 (ref. 1) at the base of C27n, as well as in Holes 465 and 758A (ref. 1).
25. *Morozovella praeangulata*	X		C27n$_{(o)}$	61.2	1	Hole 384; at comparable levels (without paleomagnetic control) in Holes 465 and 758A (ref. 1); see remarks above (item 23).
26. *Globanomalina imitata*	X		(?C2 7n$_{(o)}$)	(61.3)	3	Holes 689B, 690B (ref. 3).
27. *Praemurica inconstans*	X		C28n(mi d)	63.0	3	Holes 689B, 690C (ref. 3) and 384 (ref. 1)

TABLE 8.—*Continued.*
PALEOCENE

Datum	FAD	LAD	Paleomagnetic Chron	Age in Ma	Reference	Remarks
28. *Globanomalina compressa*	X		C28n(mid)	63.0	1,2	Hole 384 (ref. 1); also occurs at comparable level (by correlation) in Hole 750A in interval with no polarity data just above incomplete C28n in 750A/14/2:44 to 750A/14/2:40 (ref. 6; see also ref. 1).
29. *Parasubbotina varianta*	X		C28n(mid)	63.0	1	Hole 384, mid-Zone NP2 (ref. 1) and at comparable level by correlation in Hole 750A (ref. 1).
30. *Parvularugoglobigerina eugubina*	X		C29r	64.7	1,3,10,11	Hole 738 (ref. 10), Hole 384 (ref. 1) and Agost, Spain (ref. 11); first common occurrence (FC O) noted at C29n$_{(o)}$ in Holes 689, 690 (ref. 2).
31. *Subbotina triloculinoides*	X		C29n	64.3	1	Hole 384 (ref. 1).
32. *Parvularugoglobigerina eugubina*	X		C29r	64.97	11	Agost, Spain (ref. 11; see discussion in text).
33. *Globotruncana*		X	C29r	65.00	12	

[1]this work
[2]Berggren and others, 1985b
[3]Stott and Kennett, 1990
[4]Corfield, 1987
[5]Liu and Olsson (pers. commun., 1993)
[6]Pospichal and others, 1991

[7]Nederbragt and van Hinte, 1987
[8]Saint Marc, 1987
[9]Moullade, 1987
[10]Huber, 1991
[11]Groot and others, 1989
[12]multiple sources

Ocean). In ODP Holes 738 and 750A the FAD of *M. praeangulata* occurs at apparently the same stratigraphic level (within Zone NP4) and in Hole 750A this occurs near the base of Core 11 within an interval of no magnetostratigraphy slightly above a short (and incomplete) normal event identified here (and in Schlich and others, 1989, p. 308) as Chron C27n. However, reference to the Initial Reports of Site 750 (Schlich and others, 1989, p. 308, 576) reveals that there is an approximately 7-m coring gap (no recovery) between the lowest occurrence of *M. praeangulata* (base Core 11 at ~310 m) and the top of Core 12 at ~317 m (where the (incomplete) top of Chronozone C27n occurs). Thus, the incomplete record in the southern Indian Ocean supports the FAD of *M. praeangulata* (= FAD of *Pr. uncinata*) at least as early as the younger part of Chron C27n as recorded earlier in BKF85 (p. 189). Further, but inconclusive, data come from Hole 577 where the FAD of *Pr. uncinata* occurs just above Chron C28n, but Chron C27n is absent at this location because of an unconformity at the break between Cores 11 and 12 at ~101.87 m (see Bleil, 1985, and section on Calcareous Nannofossil Magnetobiochronology: Paleogene, Table 14) thus making it difficult to determine whether the reversed interval above Chron C28n is Chron C27r or C26r. The relatively thin stratigraphic interval of Zone P2 (~2–3 m at several deep-sea sites) supports the relatively short temporal interval estimated for this zone. For example in DSDP Hole 384, the FAD of *Pr. uncinata* (150.75 m), *M. angulata* (148.70 m), and *M. conicotruncata* (146.3 m) over a 3-m interval attests to the short temporal span of Zone P2 (see BKF85) and the very rapid branching of the morozovellid lineage between about 61–60 Ma.

P3. *Morozovella angulata-Globanomalina pseudomenardii* Interval Zone (P3, herein defined; emendation of Zone P3 in Berggren and Miller, 1988).

Definition: Biostratigraphic interval between the FAD of *Praemurica angulata* and the FAD of *Globanomalina pseudomenardii*

Magnetochronologic calibration: Chron C27n(y)-Chron C26r (mid)

Estimated age: 61.0–59.2 Ma; late Paleocene (Selandian)

Remarks: Zone P3 is the biostratigraphic interval between the FAD of *M. angulata* and that of *Globanomalina pseudomenardii* (Berggren and Miller, 1988, p. 368). It has been subdivided into a lower (a) and an upper (b) subzone based on the (presumed) FAD of *Igorina pusilla* in the lower third of the biostratigraphic interval. However our recent studies on DSDP Holes 384, 465 and ODP Hole 758A have shown clearly that the FAD of *pusilla* coincides with that of *M. angulata* and *M. conicotruncata* at the base of Zone P3 (see also Blow, 1979: 1109) but that the FAD of its descendant form *Ig. albeari* (= *Ig. pusilla laevigata*) occurs about midway within Zone P3 and this occurrence has been used to define a new subdivision of Zone P3 (Berggren and Norris, 1993).

P3a. *Morozovella angulata-Igorina albeari* Interval Subzone (P3a; herein defined)

Definition: Biostratigraphic interval between FAD of *Morozovella angulata* and FAD of *Igorina albeari*

Magnetochronologic calibration: Chron C27n(y)-Chron C26r (early)

Estimated age: 61.0–60.0 Ma; early late Paleocene (Selandian)

Remarks: We have chosen to use *M. angulata* as the denominative form for the base of this subzone because of the rarity or sporadic occurrence (in some cases) of *Ig. pusilla* in the lower part of this subzone. Indeed, the FAD of *Ig. pusilla* has been clearly observed to occur together with the FAD of *M. angulata* and *M. conicotruncata* (at the base of Zone P3) in DSDP/ODP Holes 465 and 758A.

P3b. *Igorina albeari-Globanomalina pseudomenardii* Interval Subzone (P3b; herein defined)

Definition: Biostratigraphic interval between FAD of *Igorina albeari* and the FAD of *Globanomalina pseudomenardii*

Magnetochronologic calibration: Chron C26r (early)-Chron C26r (mid)

Estimated age: 60.0–59.2 Ma; late Paleocene (Selandian)

Remarks: The FAD of the weakly but densely muricate, keeled *Igorina albeari* is a distinct biostratigraphic datum event that is clearly seen about midway in the biostratigraphic interval between the FAD of *M. angulata* and the FAD of *Gl. pseudo-*

menardii. It occurs a short distance just below (brief time interval before) the FAD of *M. velascoensis* from its ancestor *M. conicotruncata*. A distinctive feature of the upper part of this (sub)zone is the common occurrence of large, robust representatives of *Pa. varianta* and (in some areas) of the large, robust, and uniquely umbilically-toothed *Pa. variolaria*; both these taxa have their LAD near the P3/P4 zonal boundary.

P4. *Globanomalina pseudomenardii* Total Range Zone (P4; Bolli, 1957a)

Definition: Biostratigraphic interval of the total range of the nominate taxon, *Globanomalina pseudomenardii*

Magnetochronologic calibration: Chron C26r (mid)-Chron C25n(y)

Estimated age: 59.2–55.9 Ma; middle part of late Paleocene (late Selandian-Thanetian)

Remarks: Zone P4 is the biostratigraphic interval characterized by the total range of the nominate taxon, *Gl. pseudomenardii* (Berggren and Miller, 1988, p. 370). The FAD of this taxon has been recorded in Chron C26n and in Zone CP4 (=NP5) by Moullade (1987) but in Chron C25r and in Zone CP5 (=NP6) by Nederbragt and van Hinte (1987) in DSDP Hole 605. However, the calcareous nannoplankton magneto-biostratigraphic correlations of Lang and Wise (1987) in Hole 605 differ significantly (by about 1/2 a chron and >1 my) from those indicated in BKF85. The anomalously young "FAD" of *Gl. peudomenardii* and nannoplankton "datum events" at Site 605 may be due to generally poor preservation at this location. In DSDP Hole 577, the FAD of this taxon occurs in the younger part of Chron C26r in Zone CP5–6 (undifferentiated; =NP5–6; Table 8). In DSDP Hole 384, we have recently found that the FAD of *Gl. pseudomenardii* ocurs in mid-Chron C26r with an estimated age of 59.2 Ma (Berggren and others, 1994).

The upper limit of Zone P4 has been the subject of considerable debate (see discussion in Berggren and Miller, 1988, p. 370). The succeeding Zone P5 (of Berggren and Miller, 1988, p. 370) has been defined as the partial range of *M. velascoensis* between the LAD of *Gl. pseudomenardii* and the FAD of *M. subbotinae*. In BKF85 this interval was shown to be of relatively short duration (~0.5 my). However, recent studies at DSDP Site 577 (Corfield, 1987; Liu and Olsson, pers. commun. 1992) show a juxtaposition of these two biostratigraphic events in the early part of Chron C24r in the case of the former and at the top of Chron C25n in the case of the latter. A similar juxtaposition/overlap has been observed in DSDP/ODP Holes 465 and 758A (WAB). In DSDP Hole 550 (NE Atlantic) the lowest occurrence of *M. subbotinae* is near the top of a dissolution interval which is itself ~8–13 m above the top of Chronozone C25n; thus its FAD, there, is a minimum/delayed occurrence, although it has not been observed in the dissolution free interval just above Chron C25n. In the stratigraphic sections Ermua and Trabakua of the deep-water Basque Basin (Western Pyrenees, Spain) Orue-Etxebarria and others (1992) record the simultaneous LAD of *Gl. pseudomenardii* and FAD of *M. subbotinae*, respectively. At Zumaya (Spain), Canudo and Molina (1992) recorded a short stratigraphic interval between these two biostratigraphic datums and indicate the presence of Zone P5, although they substitute a redefined *M. aequa* Zone for the biostratigraphic interval between the LAD of *Gl. pseudomenardii* and the FAD of *Pseudohastigerina wilcoxensis*, in the general

absence of *M. velascoensis* in the relatively shallow water Pyrenees sections.

Thus, on the basis of available data, it would appear that Zone P6 (as defined) succeeds directly Zone P4 under ideal conditions, but that Zone P5 (Berggren and Miller, 1988) exists under conditions of delayed entry of *M. subbotinae*. To avoid confusion caused by the contiguous FAD of *subbotinae* and LAD of *pseudomenardii*, Zone P5 is redefined here to conform with earlier usage (see below).

We find it useful to subdivide Zone P4 into three subzones based on the appearance and/or ranges of specific muricate acarininids which make their first appearance within this zone. This subdivision materially improves the biostratigraphic/biochronologic resolution within this relatively long (~2 my) zone.

P4a. *Globanomalina pseudomenardii/Acarinina subsphaerica* Concurrent Range Subzone (P4a; herein defined)

Definition: Biostratigraphic interval characterized by the concurrent range of the two nominate taxa between the FAD of *Globanomalina pseudomenardii* and the LAD of *Acarinina subsphaerica*

Magnetochronologic calibration: Chron C26r (mid)-Chron C25r (early)

Estimated age: 59.2–57.1 Ma; late Paleocene (latest Selandian-early Thanetian)

Remarks: The small, tightly coiled and high spired muricate form *Ac. subsphaerica* is a distinct component of lower P4 faunas and has its FAD essentially coincident with that of *Ac. acarinata* and *Gl. pseudomenardii*, nominate taxon of Zone P4, in DSDP/ODP Holes 384, 465 and 758A (Berggren and Norris, unpubl.; Table 8). Its range is relatively short and it has its LAD approximately midway through Zone P4 . As such, it provides a useful means of differentiating the lower and upper parts of Zone P4. In its present denotation, this subzone corresponds closely to the *Ac. subsphaerica* Zone as used in the Kuban River section (northern Caucasus) which has served as the classic reference section for pre-Oligocene Paleogene planktonic foraminiferal zonation of the SW part of the former Soviet Union (as defined in the Permanent Interdepartmental Stratigraphic Commission for the Paleogene of the U.S.S.R., 1963). It was also used in a more extended sense in the northern Caucasus and Crimea (Shutskaya, 1953, 1956, 1960a, b, 1962; Alimarina, 1962, 1963; Leonov and Alimarina, 1961, 1964; and Morozova, 1960) as the biostratigraphic interval between the LAD of *Globorotalia angulata* and the FAD of *Globorotalia subbotinae* (which was considered to coincide approximately with that of *Nummulites planulatus*; see Yanshin, 1960: Table 1) and it should be borne in mind that this taxon was sometimes identified as *Acarinina spiralis* by some Russian authors (e.g., Leonov and Alimarina, 1961). These data pertain to an earlier era when detailed biostratigraphic data were unavailable owing to the broad spacing of samples. The significant fact is that we are returning here, and in the case of Subzone P4b (below) to a zonal system which was correct and appropriate in its essentials (if not details) over 30 years ago.

Characteristic elements of this subzone include *M. velascoensis*, *M. angulata-conicotruncata* complex (which has its LAD in this interval), *M. apanthesma*, *Ac. mackanni* (which has its FAD in this interval), *Ig. albeari*, and *S. triloculinoides*, *S. triangularis* and *S. velascoensis*.

P4b. *Acarinina subsphaerica-Acarinina soldadoensis* Interval Subzone (P4b: herein defined)

Definition: Biostratigraphic interval from the LAD of *Acarinina subsphaerica* to the FAD of *Acarinina soladoensis*

Magnetochronologic calibration: Chron C25r (early)-Chron C25r (late)

Estimated age: 57.1–56.5 Ma; late Paleocene (Thanetian)

Remarks: As defined here this subzone corresponds essentially to the *Ac. acarinata* Zone of the northern Caucasus as codified by the "Soviet Stratigraphic Commission" in 1963 apparently for the biostratigraphic interval between the LAD of *Ac. subsphaerica* and the FAD of *Gr. aequa*. Our studies have shown that the FAD of *M. aequa* essentially coincides with that of *Ac. soldadoensis* and the *Ac. coalingensis-triplex* complex so that biostratigraphic usage similar, if not identical, to that employed over 30 years ago in the former Soviet Union is implied. In DSDP Hole 384, the FAD of *Ac. soldadoensis* (and the *Ac. coalingensis-triplex* plexus) occurs in late Chron C25r (Berggren and others, 1994; Table 8). Characteristic elements of this subzone include, i.al., *Ac. acarinata, Ac. mckannai, Gl. pseudomenardii, M. apanthesma, M. velascoensis, S. velascoensis* and *S. triangularis.*

P4c. *Acarinina soldadoensis-Globanomalina pseudomenardii* Concurrent range Subzone (P4c; herein defined)

Definition: Biostratigraphic interval containing the concurrent range of the nominate taxa from the FAD of *Acarinina soldadoensis* to the LAD of *Globanomalina pseudomenardii*

Magnetochronologic calibration: Chron C25r (late)-Chron C25n(y)

Estimated age: 56.5–55.9 Ma; late Paleocene (late Thanetian)

Remarks: Blow (1979, p. 267–269) used the concurrent range of *Ac. soldadoensis* (FAD at base) and *M. velascoensis pasionensis* (LAD at top) to denominate his Zone P5. Our studies do not corroborate Blow's (1979) upper limit of *pasionensis* prior to that of *velascoensis*, but his recognition of the LAD of *Ac. acarinata* and *Ac. mckannai* at the top of his Zone P5, essentially coincident with the FAD of *M. subbotinae*, agrees well with our own observations and suggests that Subzone P4c (herein defined) coincides closely, if not precisely, with Zone P5 of Blow (1979). On the other hand we have not observed indigenous occurrences of *Gl. pseudomenardii* together with *M. subbotinae*, other than their juxtaposition/brief overlap at the base of Zone P5 (= Subzone P6a of Berggren and Miller, 1988). For example in DSDP Site 213 (Indian Ocean), a typical Zone P5 fauna (with *M. velascoensis* and *M. subbotinae*) occurs for at least 8.56 m above basalt basement with no *Gl. pseudomenardii*. We believe that records of (sporadic) occurrences of *Gl. pseudomenardii* in Zone P5 are due to reworking or may include specimens referable to taxa other than *pseudomenardii* (*Gl. chapmani, Gl. australiformis, Gl. planoconica*, etc.).

P5. *Morozovella velascoensis* Interval Zone (P5; Bolli, 1957a; P5 and P6a of Berggren and Miller, 1988)

Definition: Biostratigraphic interval between the LAD of *Globanomalina pseudomenardii* and the LAD of *Morozovella velascoensis*

Magnetochronologic calibration: Chron C25n(y)-Chron C24r (mid)

Estimated age: 55.9–54.7 Ma; latest Paleocene-earliest Eocene (latest Thanetian-earliest Ypresian)

Remarks: Zone P5, with a different denotation (partial range of the nominate taxon between the LAD of *Gl. pseudomenardii* and the FAD of *M. subbotinae*) was defined (Berggren and Miller, 1988) before the evidence for the juxtaposition/overlap of these two datum events which were used to define the top of Zone P4 and the base of Zone P6a, respectively, became available (cf. Blow, 1979, p. 265–267), although we would disagree with Blow (1979) on the upper limit of *pseudomenardii* in his Zone P7 (= P6b of Berggren and Miller (1988); ≃ P6a herein). However, in some instances the initial appearance of *M. subbotinae* was delayed by the widespread occurrence of one or more distinct dissolution events that span the lower part of magnetozone C24r and Zone P5. The chronology of such a zone is obviously very tenuous and approximate, but this estimate is based on an assessment of conditions observed in DSDP Hole 550 and other deep-sea sites as well as various outcrop sections in the North Caucasus and the Mediterranean region. For this reason we have decided to revert to the previously, relatively unequivocal, usage of Bolli (1957a) in which the sequential LADs of *pseudomenardii* and *velascoensis* are used to bracket the partial, terminal range of *velascoensis* and which characterize a distinct biostratigraphic interval which spans the Paleocene/Eocene boundary as currently recognized by at least some (bio)stratigraphers. As a result of this modification to Zone P5, Zone P6 of Berggren and Miller (1988) is also modified. Zone P5 essentially contains the concurrent range of *M. subbotinae* (FAD) and *M. velascoensis* (LAD), but it is defined as a partial range zone because the definition of the top of Zone P4 is defined as the LAD of *Gl. pseudomenardii*.

Characteristic features of this zone include a relatively distinct turnover in the planktonic foraminifera including the appearance of *M. marginodentata, M. formosa gracilis, Ig. broedermanni, Ac. wilcoxensis, Turborotalia pseudoimitata* and the relatively common occurrence of strongly muricate large acarininids (*soldadoensis, coalingensis-triplex* group).

The Paleocene/Eocene boundary is usually correlated with the P5/P6 (= P6a/b of Berggren and Miller, 1988) zonal boundary by planktonic foraminiferal specialists and is estimated at 54.7 Ma here (see discussion under Radiosotopic Chronology and in Berggren and Aubry, 1995). Calcareous nannoplankton specialists usually consider the NP9/NP10 zonal boundary, estimated here at ~55 Ma, as denotative of this boundary. The base of the London Clay Formation (Oldhaven Bed = Hales Clay = base Eocene = Ypresian/Thanetian boundary) has been cross correlated to the − 17 Ash in DSDP Hole 550 and dated in NW Europe at 54.5 Ma (see discussion under Radioisotopic Chronology). However, we choose to use an age estimate of 54.8 Ma for the base of the London Clay Formation based on sedimentation rates in DSDP Hole 550, rather than the − 17 Ash date. The problems associated with the identification and delineation of events suitable for the determination of an appropriate Paleocene/Eocene boundary Global Stratotype Section and Point (GSSP) are currently being examined by IGCP Project 308 (Paleocene/Eocene Boundary Events in Time and Space) and have been discussed in greater detail elswhere (Berggren and Aubry, 1995; Aubry and others, 1995).

Eocene Zonation.—The chronology of Eocene (sub)tropical planktonic foraminiferal datum events/zones is shown in Figure 9.

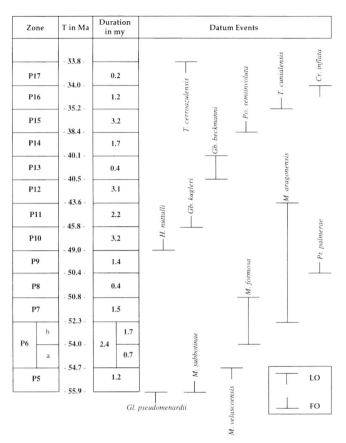

Zone	T in Ma	Duration in my	Datum Events
P17	- 33.8 - - 34.0 -	0.2	
P16	- 35.2 -	1.2	
P15	- 38.4 -	3.2	
P14	- 40.1 -	1.7	
P13	- 40.5 -	0.4	
P12	- 43.6 -	3.1	
P11	- 45.8 -	2.2	
P10	- 49.0 -	3.2	
P9	- 50.4 -	1.4	
P8	- 50.8 -	0.4	
P7	- 52.3 -	1.5	
P6 b	- 54.0 -	2.4	1.7
P6 a	- 54.7 -		0.7
P5	- 55.9 -	1.2	

FIG. 9.—Chronology of Paleocene (sub)tropical planktonic foraminiferal zones.

P6. *Morozovella subbotinae* Partial Range Zone (P6 of Berggren and Miller, 1988, partim; herein emended and redefined; = *Morozovella subbotinae* Zone of Luterbacher and Premoli Silva in Caro and others, 1975, partim)

Definition: Biostratigraphic interval characterized by the partial range of the nominate taxon between the LAD of *Morozovella velascoensis* and the FAD of *Morozovella aragonensis*

Magnetochronologic calibration: Chron C24r (mid)-Chron C23r

Estimated age: 54.7–52.3 Ma; early Eocene (early Ypresian)

Remarks: Berggren and Miller (1988, p. 370) defined the *M. subbotinae* Partial Range Zone (P6) as the partial range of the nominate taxon between its supposed FAD and the FAD of *M. aragonensis*. Investigations on a number of deep-sea cores and outcrop sections have now shown that the FAD of *M. subbotinae* coincides essentially with the LAD of *Gl. pseudomenardii* and that the supposed gap between the LAD of *pseudomenardii* and the FAD of *subbotinae* is illusory, although the delayed entry of *subbotinae* is often caused by strong dissolution in some sections within Zone NP9 (cf., Molina and others, 1992). Thus Zones P5 and P6a, as defined by Berggren and Miller (1988), are essentially equivalent. As emended here, Zone P6 coincides essentially with the *M. subbotinae* Zone of Premoli Silva and Bolli (1973) and Luterbacher and Premoli Silva in Caro and others, (1975), except for the (apparently brief) temporal interval between the LAD of *M. velascensis* and the

LAD of the small, enigmatic taxon *M. edgari* (Premoli Silva and Bolli, 1973) (? = *M. finchi* Blow, 1979) which was shown by Toumarkine and Luterbacher (1985, Figure 5, p. 100) to occur only slighly below the simultaneous FADs of *M. formosa formosa* and *M. lensiformis* (cf. Blow, 1979, Figs. 48, 50, in which the LAD of *M. finchi* is shown to occur at essentially the same level). The temporal span between the LAD of *M. velascoensis* and the FAD of *M. formosa formosa* is estimated here at ~0.8 my.

In order to maintain numerical and biostratigraphic continuity with the zonation of Berggren and Miller (1988), we redefine Zone P6 in such a manner that its two subzones, a and b, now correspond to Subzones P6b and P6c, respectively, of Berggren and Miller (1988, p. 371).

P6a. *Morozovella velascoensis-Morozovella formosa formosa* and/or *Morozovella lensiformis* Interval Zone (P6a; = P6b of Berggren and Miller, 1988; numerical notation herein emended; = *Gr. edgari* Zone of Premoli Silva and Bolli, 1973, partim)

Definition: Biostratigraphic interval between the LAD of *Morozovella velascoensis* and the simultaneous FAD of *Morozovella formosa formosa* and/or *Morozovella lensiformis*

Magnetochronologic calibration: Chron C24r (mid)-Chron C24r (late)

Estimated age: 54.7–54.0 Ma; earliest Eocene (earliest Ypresian)

Remarks: In normal open ocean stratigraphic successions, the sequence of FAD of *M. subbotinae*, LAD of *M. velascoensis/ acuta*, FADs of *M. formosa formosa/M. lensiformis* and the FAD of *M. aragonensis* serve as a means of providing a discrete biostratigraphic subdivision of the uppermost Paleocene-lowermost Eocene Series (Berggren and Miller, 1988). The subdivision made there has been followed, with minor modifications, in this paper.

This subzone is redefined here as an interval subzone so as to avoid conceptual confusion/overlap with the use of *M. subbotinae* as a nominate form of both Zone P6 and Subzone P6a in its original definition (Berggren and Miller, 1988, p. 370, 371) and for Zones P6 and Subzone P6a (emended here). At the same time we remove *Ph. wilcoxensis* as one of the nominate forms of Subzone P6a (P6b of Berggren and Miller, 1988); this form appears under ideal conditions at the P5/P6 (= P6a/b boundary in Berggren and Miller, 1988) in fully tropical assemblages but has a demonstrably delayed entry in mid-high latitude regions within the P6b-P7 (= P6c-P7 of Berggren and Miller, 1988) biostratigraphic interval.

The LAD of *S. velascoensis* occurs within this zone and the common and widespread early Eocene taxon *S. patagonica* assumes its characteristic morphologic features in this subzone. Additional comments on this subzone (as P6b) are to be found in Berggren and Miller (1988, p. 371).

P6b. *Morozovella formosa formosa/Morozovella lensiformis-Morozovella aragonensis* Interval Zone (P6b; = P6c of Berggren and Miller, 1988; numerical notation herein emended; Blow, 1979, as Subzone P8a)

Definition: Biostratigraphic interval between the virtually simultaneous FADs of *Morozovella formosa formosa* and/or *Morozovella lensiformis* and the FAD of *Morozovella aragonensis*

Magnetochronologic calibration: Chron C24r (late)-Chron C23r

Estimated age: 54.0–52.3 Ma; early Eocene (early Ypresian)

Remarks: This is a biostratigraphically distinct interval (see also Premoli Silva and Bolli, 1973; Blow, 1979) characterized by the essentially simultaneous FADs of the *M. formosa formosa* and *M. lensiformis* and the LADs of *M. subbotinae* and *M. aequa.*

P7. *Morozovella aragonensis/M. formosa formosa* Concurrent Range Zone (P7; Blow, 1979, as Subzone P8b; see also Berggren and Miller, 1988, p. 371, as Zone P7; Berggren, 1969 as *G. formosa* Zone; Bolli, 1966, as *Globorotalia formosa* and *G. aragonensis* Zones (combined)

Definition: Concurrent range of the nominate taxa between the FAD of *Morozovella aragonensis* and the LAD of *Morozovella formosa formosa*

Magnetochronologic calibration: Chron C23r-Chron C23n(y)

Estimated age: 52.3–50.8 Ma; middle early Eocene (mid-Ypresian)

Remarks: See Berggren and Miller (1988: 371).

P8. *Morozovella aragonensis* Partial Range Zone (P8; Berggren, 1969; see also Berggren and Miller, 1988, p. 371)

Definition: Partial range of the nominate taxon between the LAD of *Morozovella formosa formosa* and the FAD of *Planorotalites palmerae*

Magnetochronologic calibration: Chon C23n(y)-Chron C22r (estimated)

Estimated age: 50.8–50.4 Ma; late early Eocene (late Ypresian)

Remarks: See Berggren and Miller (1988, p. 371).

P9. *Planorotalites palmerae-Hantkenina nuttalli* Interval Zone (P9 of Berggren and Miller, 1988; herein emended)

Definition: Biostratigraphic interval between the FAD of *Planorotalites palmerae* and the FAD of *Hantkenina nuttalli*

Magnetochronologic calibration: Chron C22r (estimated)-Chron C22n(y)

Estimated age: 50.4–49.0 Ma; late early Eocene (latest Ypresian)

Remarks: This zone was originally denoted as the partial range of *S. inaequispira* with the same boundary criteria as employed here. By redefining this as an interval range zone we have simply lifted the onerous burden placed upon *inaequispira* as the name bearer of this extremely complex and controversial biostratigraphic interval (see Berggren and Miller, 1988, p. 372, 373 for further discussion). Characteristic elements of this zone include *S. inaequispira, S. frontosa, Ac. bullbrooki, Ac. pentacamerata, Ac. aspensis, Planorotalites palmerae* (lower part), *M. aragonensis, M. caucasica, Ph. wilcoxensis, Ph. sharkriverensis, Globigerinatheka senni* and, in the upper part, the stellate, lobulate clavigerinellids.

P10. *Hantkenina nuttalli* Interval Zone (P10 of Berggren and Miller, 1988, p. 373; = *Hantkenina aragonensis* Zone of Bolli, 1957c, emended by Stainforth and others, 1975, and renamed by Toumarkine, 1981)

Definition: Biostratigraphic interval characterized by the presence/partial range of the nominate taxon between its FAD and the FAD of *Globigerapsis kugleri*

Magnetochronologic calibration: Chron C22n(y)-Chron C20r (estimated)

Estimated age: 49.0–45.8 Ma; early middle Eocene (early Lutetian)

Remarks: Characteristic elements of this zone include, in addition to the nominate form and *H. mexicana* s.l., clavigerinellids, *S. griffinae, S. inaequispira,* and (the virtually planispiral, inflated) *S. bolivariana, M. spinulosa, M. aragonensis, Ac. aspensis, Ac. mathewsae, Ac. bullbrooki,* and the distinctive, multi-apertured *Globigerinoides*-homeomorph *Guembelitrioides higginsi*

The muricate multiapertured *Truncorotaloides collactea* and *Tr. rohri* make their initial, tentative appearance within this zone as well.

P11. *Globigerapsis kugleri/Morozovella aragonensis* Concurrent Range Zone (herein defined; = Bolli, 1957c, 1966, (upper part only); Berggren, 1969; P11 of Berggren and Miller, 1988)

Definition: Biostratigraphic interval characterized by the concurrent range of the nominate taxa between the FAD of *Globigerapsis kugleri* and the LAD of *Morozovella aragonensis*

Magnetochronologic calibration: Chron C20r (estimated)-Chron C20n(o)

Estimated age: 45.8–43.6 Ma; middle Eocene (Lutetian)

Remarks: At the time of its introduction (Berggren, 1969) the *Globigerapsis kugleri* Zone (P11) was based on the partial range of *Gb. kugleri* between its FAD and the LAD of *M. aragonensis.* As thus defined it was an interval zone, then corresponding to Bolli's (1957c) zone of the same name. The latter zone was emended by Proto Decima and Bolli (1970) and renamed by Bolli (1972) the *Globigerinatheka subconglobata subconglobata* Zone based on interim taxonomic studies of the group of early globigerinathekids. Its identification as a concurrent range zone (as the biostratigraphic interval with the nominate taxon from the FAD of *Globigerinatheka mexicana mexicana* to the LAD of *M. aragonensis*; see redefinition by Stainforth and others, 1975; Toumarkine and Luterbacher, 1985) is puzzling inasmuch as this definition is that of an interval zone, not a concurrent range zone. Berggren and Miller (1988, p. 373) denominated this zone as the *Gb. kugleri-S. frontosa* Partial Range Zone (based on the concurrent partial ranges of the nominate taxa between the FAD of *Gb. kugleri* and the LAD of *M. aragonensis*). As thus defined it was a hybrid of an interval and a concurrent range zone. Blow (1979, p. 285, 286) used the concurrent ranges of *Gb. kugleri* and *S. frontosa boweri* from the FAD of the former to the LAD of the latter to define his Zone P11. He considered the LAD of *frontosa boweri* to coincide essentially with the LAD of *M. aragonensis,* but Berggren and Miller (1988, p. 373) pointed out that *frontosa* has been reported to range as high as levels correlative with Zone P14 in Chronozone C18 (BKF85, p. 190). In following Blow's (1979) intent in retaining the concept, and delineating the limits, of Zone P11, we redefine it here insofar as we use the LAD of *M. aragonensis* (rather than the LAD of *S. frontosa*) to define its upper limit (with Zone P12).

A significant turnover in planktonic faunas occurs at, or near, the P11/P12 zonal boundary including the LADs of *M. aragonensis, S. griffinae, Turborotalia praecentralis, Ac. mathewsae,* the *Ig. broedermanni* group and *Clavigerinella akersi.*

At/near the P10/P11 zonal boundary the FADs of *M. lehneri*, *Gb. kugleri*, *T. pomeroli* (= *T. centralis* s.l.) and the LAD of "*Subbotina*" *bolivariana* are characteristic features. See Berggren and Miller (1988, p. 373) for additional discussion of this zone.

P12. *Morozovella lehneri* Partial Range Zone (Bolli, 1957c, 1966, emended; Berggren, 1969; P11 of Berggren and Miller, 1988)

Definition: Biostratigraphic interval characterized by the presence/partial range of nominate taxon between the LAD of *Morozovella aragonensis* and the FAD of *Globigerapsis beckmanni*

Magnetochronologic calibration: Chron C20n(o)-Chron C18r

Estimated age: 43.6–40.5 Ma; middle Eocene (Lutetian-early Bartonian)

Remarks: This zone is characterized by the relatively common occurrence of the nominate form together with *M. spinulosa*, *M. coronata*, *Tr. rohri*, *Tr. topilensis* and the transition from the primitive (*mexicana*) to more advanced (*alabamensis*) hantkeninids. As originally defined by Bolli (1957; see also Toumarkine and Luterbacher, 1985) this is a partial range (not an interval) zone.

P13. *Globigerapsis beckmanni* Total Range Zone (Bolli, 1957c, 1966; Blow, 1979, as P13; Berggren, 1969; Berggren and Miller, 1988, as P13)

Definition: Total range of the nominate taxon

Magnetochronologic calibration: Chron C18r-Chron C18n(o)

Estimated age: 40.5–40.1 Ma; late middle Eocene (Bartonian)

Remarks: Low conical morozovellids (*coronata*, *lehneri*) exhibit a rapid reduction within, and eventual extinction near the top of, this zone and only *spinulosa* continues to the overlying Zone P14 where it has its LAD.

P14. *Truncorotaloides rohri-Morozovella spinulosa* Partial Range Zone (P14 of Berggren and Miller, 1988; see also Blow, 1979)

Definition: Concurrent partial ranges of the nominate taxa between the LAD of *Globigerapsis beckmanni* and the FAD of *Porticulasphaera semiinvoluta*

Magnetochronologic calibration: Chron C18n(o)-Chron C18n(y)

Estimated age: 40.1–38.4 Ma; late middle Eocene (late Bartonian)

Remarks: The convoluted history of this zone is explained in more detail in Blow (1979, p. 291) and Berggren and Miller (1988, p. 373).

P15. *Porticulasphaera semiinvoluta* Interval Zone (P15, herein defined; similar, to Zone P15 of Blow, 1979; P15 of Berggren and Miller, 1988 but with different denotation at top)

Definition: Biostratigraphic interval between the FAD of *Porticulopshaera semiinvoluta* and the FAD of *Turborotalia cunialensis*

Magnetochronologic calibration: Chron C18n(y)-Chron C15r(o)

Estimated age: 38.4–35.2 Ma; late middle to early late Eocene (late Bartonian to early Priabonian)

Remarks: We have chosen to modify the denotation of this zone from its usual sense by redefining the top of the zone using the FAD of the relatively common and morphologically distinct (carinate periphery) and GPTS calibrated *T. cunialensis*, rather than that of *Cribrohantkenina inflata*, a taxon which is less common (relative to *cunialensis*) in some (sub)tropical stratigraphies and whose FAD remains poorly calibrated to the GPTS (Table 9, item 8). Both events occur close together in time but appear to be separated by a short/brief stratigraphic/temporal interval originally defined (with a somewhat different denotation) as an interval zone by Bolli (1957c), Blow (1979, p. 292) emended it so that the base coincided with the FAD of the nominate taxon, rendering it an interval, not a partial range, zone. See Berggren and Miller (1988, p. 374) for further discussion of the history of this zone (with a slightly different denotation).

Dentoglobigerinids (*galavisi, pseudovenezuelana, tripartita*) subbotinids (*angiporoides, linaperta, hagni, eocaena*), catapsydracids (*unicavus, dissimilis, cryptomphala, pera*), globigerapsids (*index, tropicalis*) and (cribro)hantkeninids characterize this zonal interval. The LAD of the nominate taxon occurs only slightly older (~0.3 my) than the base of this zone.

P16. *Turborotalia cunialensis/Cribrohantkenina inflata* Concurrent-Range Zone (P16, herein defined; *non* Blow, 1969, 1979; *non* P16 of Berggren, 1969, and Berggren and Miller, 1988)

Definition: Biostratigraphic interval characterized by the concurrent range of the nominate taxa between the FAD of *Turborotalia cunialensis* and the LAD of *Cribrohantkenina inflata*

Magnetochronologic calibration: Chron C15r(o)-Chron C13r(mid)

Estimated age: 35.2–34.0 Ma; late late Eocene (late late Priabonian)

Remarks: In modifying the earlier definition(s) of Zone P16 we have chosen to stress the concurrent ranges of two distinct latest Eocene taxa. The LAD of *Cr. inflata* occurs in mid-Chron C13r, approximately midway temporally between the LADs of *Globigerapsis index* (earlier) and the hantkeninids and the *Turborotalia cerroazulensis* group (later) in the younger part of Chron C13r at the top of the Eocene. Characteristic features of this zonal interval include, among others, development of the acutely peripheried *Turborotalia cunialensis* (at the base), the FAD of "*Turborotalia*" *ampliapertura* and *T. pseudoampliapertura*, and the LAD of *Globigerapsis index*. Additional details on the calibration of faunal events in this zone can be found in Berggren and Miller (1988, p. 374) and Berggren and others (1992).

P17. *Turborotalia cerroazulensis* Interval Zone (P17 of Berggren and Miller, 1988)

Definition: Biostratigraphic interval between the LAD of *Cribrohantkenina inflata* and the extinction of *Turborotalia cerroazulensis/Turborotalia cumialensis*

Magnetochronologic calibration: Chron C13r (mid)-Chron C13r (late)

Estimated age: 34.0–33.8 Ma; latest Eocene (latest Priabonian)

Remarks: We retain the LAD of the *cerroazulensis/cunialensis* group to denominate the boundary between Zones P17 and

TABLE 9.—EOCENE PLANKTONIC FORAMINIFERAL MAGNETOBIOCHRONOLOGY.
EOCENE

Datum	FAD	LAD	Paleomagnetic Chron	Age in Ma	Reference	Remarks
1. *Hantkenina* spp.		X	C13r	33.7	1–4	Located just above LAD of *T. cerroazulensis cunialensis* and youngest of 3 normal "events" in C13r in Contessa Highway section (refs. 2, 3) and above normal "event" = C13n.2n at Massignano (ref. 4).
2. *Turborotalia cerroazulensis*		X	C13r	33.8[s]	1–4	Associated with youngest of 3 normal "events" in C13r in Contessa Section (ref. 2, 3) and above normal "event" = C13n.2n at Massignano (ref. 4). It is not absolutely clear whether the LAD of *Hantkenina* and *T. cerroazulensis* group are synchronous or sequential. Berggren and Miller (1988) used the HO of *T. cerroazulensis* (rather than that of *Hantkenina*) to denote the P17/P18 zonal boundary because of its greater potential for preservation.
3. *Cribrohantkenina inflata*		X	C13r	34.0	1–4	= P16/P17 boundary of Blow (1979) and Berggren and Miller (1988). Located between LAD of *Gl. index* and *T. cunialensis* and *Hantkenina* in mid-part of C13r at Massignano (ref. 4) and just below youngest of 3 normal "events" in C13r in Contessa Highway section (ref. 2).
4. *Globigerapsis index*		X	C13r	34.3	1–5	Essentially coincident with the LAD of *Discoaster saipanensis* and *Discoaster barbadoensis* and with lower of 3 normal "events" in C13r in Contessa Highway section (refs. 2, 3). Occurs in younger part of C13r on Kerguelen Plateau (Site 748) (ref. 5) and then appears to be globally reliable datum. Recorded at a level in Hole 689B (Weddell Sea) identified as C16n (ref. 6), but an alternative interpretation of the paleomagnetic stratigraphy (ref. 5) suggests this may be C15n, in which case the (apparent) latitudinal diachrony of the HO of *Gl. index* would be significantly reduced, if not eliminated , inasmuch as dissolution is strong across the Eocene/Oligocene boundary interval at high latitudes.
5. *Turborotalia cunialensis*	X		C15r	35.2	2–4	Just above (i.e. , younger than) LAD of *P. semiinvoluta* (refs. 2, 3).
6. *Turborotalia pomeroli*		X	C15r	35.3	1–3	HO of *P. semiinvoluta* in Massignano section (ref. 4).
7. *Porticulasphaera semiinvoluta*		X	C15r(o)	35.3	1–3	Located in C16n.2n in Massignano section (ref. 4).
8. *Cribrohantkenina inflata*	X		(C16n or oldest part of C15r)	(35.5)	3,4	= P15/P16 boundary of Blow (1979) and Berggren and Miller (1988). Occurs between FAD of *Isthmolithus recurvus* and LAD of *P. semiinvoluta* and *T. pomeroli* in Spain (Molina, 1986; Monechi, 1986; Molina and others, 1986). FAD of *I. recurvus* is in C16n.2n (refs. 1–4) and LADs of *T. pomeroli* and *P. semiinvoluta* occur in C15r in Contessa Highway (refs . 1–3) or C16n.2n, in Massignano section (ref. 4) which suggests that FAD of *C. inflata* corresponds approximately with C16n.2n. FAD not considered unequivocally determined (ref. 3). Brief overlap of *C inflata* and *P. semiinvoluta* shown in the upper part of *G. semiinvoluta* Zone = lower Zone P16 of Blow (1979) in Toumarkine and Luterbacher (1985).
9. *Acarinina collactea*		X	C17n	37.7	1	Recorded in interval of uninterpretable paleomagnetic data just above level identified as C17n in Hole 748B (ref. 5) and at level interpreted as C18n in Hole 690B (Weddell Sea) (ref. 6).
10. *Subbotina linaperta*		X	C17n.1n	37.7	5,6	Recorded in mid-C17n in Hole 689B (ref. 6) and in C17n in Hole 748B (ref. 6).
11. *Morozovella spinulosa*		X	C17n(o)	38.1	1	
12. *Porticulasphaera semiinvoluta*	X		C18n(y)	38.4	1,2	= P14/P15 zonal boundary of Blow (1979) and Berggren and Miller (1988). Note brief overlap of *Truncorotaloides* and *P. semiinvoluta* (Blow, 1979) but brief but distinct, separation of *Acarinina* and *P. semiinvoluta* in ref. 2, Fig. 1.
13. *Planorotalites*		X	C18n(y)	38.5	2	
14. *Acarinina primitiva*		X	C18n	39.0	1,6	Recorded in mid-part of C18n in Hole 689 (ref. 6) and in interval of uninterpretable paleomagnetic data just above probable C18n in Hole 748B (ref. 5).
15. *Acarinina* spp.		X	C17n-C18n	37.5–38.5	1,2	Mid-C17n (ref. 1) or top C18n (refs. 2, 3) in the Apennines.
16. *Subbotina frontosa*		X	C18n (mid)	39.3	1	
17. *Globigerapsis beckmanni*		X	C18n(o)	40.1	1	P13/P14 boundary in Berggren and Miller (1988).
18. *Globigerapsis beckmanni*	X		C18r	40.5	1	P12/P13 boundary in Berggren and Miller (1988).
19. *Acarinina bullbrooki*		X	C18r	40.5	1	
20. *Turborotalia pomeroli*	X		C19r	42.4	1	
21. *Globigerapsis index*		X	C20n	42.9	1	Recorded near C20n/C20r boundary in Hole 689B (ref. 6).
22. *Morozovella lehneri*	X		C20n	43.5	1	
23. *Morozovella aragonensis*		X	C20n	43.6	1	= P11/P12 boundary in Berggren and Miller (1988).
24. *Turborotalia possagnoensis*	X		C20r	46.0		
25. *Hantkenina nuttalli*	X		C22n(y)	49.0	1	= P9/P10 zonal boundary in Berggren and Miller (1988).
26. *Planorotalites palmerae*	X		C22r	50.4	1	= P8/P9 zonal boundary in Berggren and Miller (1988).
27. *Acarinina pentacamerata*	X		C23n.1n(y)	50.8	6	Recorded at top of C23n in Hole 689 (ref. 6).
28. *Morozovella aragonensis*	X		C23n.2r	52.3	7–10	= P6c/P7 zonal boundary of Berggren and Miller (1988). Recorded in C23r in Holes 550 (refs. 7, 8), 577 (refs. 9, 10) and 527 (ref. 9).
29. *Morozovella marginodentata*		X	C24n.1r	52.5	7,8	Hole 550 (refs. 7, 8).
30. *Morozovella lensiformis*		X	C24n.1r	52.7	7–9	Holes 550 (refs. 7, 8) and 577 (ref. 9).
31. *Subbotina velascoensis*		X	C24r	53.5	7,8	Hole 550 (refs. 7, 8).
32. *Morozovella aequa*		X	C24r	53.6	7–9	Hole 550 (refs. 7,8); C24n.1n in Hole 577 (ref. 9).
33. *Morozovella formosa*	X		C24r	54.0	9	= P6b/c subzonal boundary of Berggren and Miller (1988). Hole 577 (ref. 9).
34. *Morozovella lensiformis*	X		C24r	54.0	7–9	Holes 550 (refs. 7, 8) and 577 (ref. 9).

TABLE 9.—*Continued.*
EOCENE

Datum	FAD	LAD	Paleomagnetic Chron	Age in Ma	Reference	Remarks
35. *Morozovella velascoensis*		X	C24r	54.7	1	Recorded in younger part of C24r in Hole 577 (ref. 9) and in mid-part C24r in Hole 577 (at 81.98 m) (ref. 11). However, paleomagnetic data (Bleil, 1985, p. 449, Fig. 7b) suggest that a hiatus occurs between ~82 and 83 m in Hole 577, which would have the effect of depressing (rendering older) the HO of *M. velascoensis*. The HO of *M. velascoensis* in Hole 577 is in CP8 (= NP9) (Monechi, 1985, p. 310, Table 5) which supports interpretation of hiatus in Hole 577, because the LAD of *M. velascoensis* is in Zone NP10. Furthermore Monechi (1985, p. 307) indicates absence of Subzone CP9a at ~81-m in Hole 577 and/or an interval of extremely condensed sedimentation across the Paleocene/Eocene boundary interval.
36. *Morozovella acuta*		X	C24r (mid)	54.7	8	Hole 550; located midway between −17 and + 19 (54.0 Ma) Ashes.
37. *Morozovella formosa gracilis*	X		C24r (mid)	54.7	7–9	Hole 550 (refs. 7, 8).
38. *Muricella broedermanni*	X		C24r (mid)	54.7	7–9	Hole 550 (refs. 7, 8).
39. *Morozovella marginodentata*	X		C24r (lower)	54.8	7–9	Hole 550 (refs. 7, 8); Hole 577, top C25n (ref. 9).
40. *Globanomalina australiformis*	X		C24r	55.5	6	= AP4/AP5 zonal boundary of Stott and Kennett (1990); Holes 689B, 690B (ref. 6), used to denote P/E boundary in high austral latitudes. Recorded also at Sites 738 (ref. 12) 747 and 748 (ref. 5) but with no paleomagnetic control.

[1]Berggren and others, 1985b
[2]Nocchi and others, 1986
[3]Premoli-Silva and others, 1988a
[4]Coccioni and others, 1988
[5]Berggren, 1992
[6]Stott and Kennett, 1990

[7]Snyder and Waters, 1985
[8]Berggren (this work)
[9]Corfield, 1987
[10]Liu and Olsson (pers. commun., 1993)
[11]Pak and Miller, 1993
[12]Huber, 1991

P18 rather than the LAD of *Hantkenina* spp. for the reasons cited in Berggren and Miller (1988, p. 375), namely the greater robustness and frequency of the former in latest Eocene assemblages, while at the same time recognizing that the LAD of both these groups may ultimately be shown to be simultaneous. As defined and applied here the P17/P18 zonal boundary would be about 0.1 my older than the Eocene/Oligocene boundary as denoted (not defined) by the LAD of *Hantkenina* spp.

Oligocene zonation.—The chronology of Oligocene (sub)tropical planktonic foraminiferal datum events/zones is shown in Figure 10.

P18. *Turborotalia cerroazulensis-Pseudohastigerina* spp. Interval Zone (herein defined; emendation of Zone P18 of Berggren and Miller, 1988)

Definition: Biostratigraphic interval between the LAD of *Turborotalia cerroazulensis* and the LAD of *Pseudohastigerina* spp.

Magnetochronologic calibration: Chron C13r (late)-Chron C12r

Estimated age: 33.8–32.0 Ma; early Oligocene (early Rupelian)

Remarks: Berggren and Miller (1988, p. 375) substituted the partial range of *Chiloguembelina cubensis* for *Cassigerinella chipolensis* in modifying Bolli and Saunders' (1985) lowermost Oligocene zone because of the irregular occurrence of *chipolensis* in lower Oligocene sections. Zone P18 is exactly correlative with the *C. chipolensis/Ph. micra* Concurrent Range Zone of Bolli and Saunders (1985) and Zones P18 and P19 of Blow (1969; see further discussion in Berggren and Miller, 1988, p. 375). The zonal denotation has been changed from a partial range to an interval zone here to provide easier recognition of this zone on a regional basis (i.e., the erratic occurrence of *Ch. cubensis* in shallower (marginal) deposits renders it less useful in these areas).

P19. *"Turborotalia" ampliapertura* Interval Zone (P19 of Berggren and Miller, 1988)

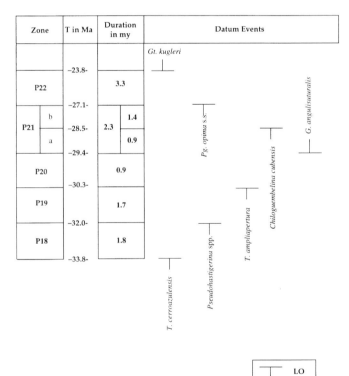

FIG. 10.—Chronology of Oligocene (sub)tropical planktonic foraminiferal zones.

Definition: Biostratigraphic interval between the LAD of *Pseudohastigerina* spp. and the LAD of *"Turborotalia" ampliapertura*

Magnetochronologic calibration: Chron C12r-Chron C11r

Estimated age: 32.0–30.3 Ma; late early Oligocene (late Rupelian)

Remarks: The historical vicissitudes associated with this zone are discussed in detail by Berggren and Miller (1988, p. 375).

P20. *Globigerina sellii* Partial Range Zone (P20 of Berggren and Miller, 1988)
Definition: Biostratigraphic interval characterized by the partial range of the nominate taxon between the LAD of "*Turborotalia*" *ampliapertura* and the FAD of *G. angulisuturalis*
Magnetochronologic calibration: Chron C11r-Chron C11n(y)
Estimated age: 30.3–29.4 Ma; late early Oligocene (late Rupelian)
Remarks: Spinose paragloborotaliids (*opima* s.s.), small (*ciperoensis, praebulloides, anguliofficinalis*) and large (*tapuriensis, sellii*) globigerinids and subbotinids (*angiporoides*, which has its LAD within the mid-part of this zone), and microperforate tenuitellids (*gemma, munda*) characterize Zone P20. Further discussion may be found in Berggren and Miller (1988, p. 376).

P21. *Globigerina angulisuturalis/Paragloborotalia opima opima* Concurrent Range Zone (P21 of Blow, 1969, 1979; P21 of Berggren, 1969; Berggren and Miller, 1988)
Definition: Biostratigraphic interval characterized by the concurrent range of the nominate taxa between the FAD of *Globigerina angulisuturalis* and the LAD of *Paragloborotalia opima opima*
Magnetochronologic calibration: Chron C11n(y)-Chron C9n(y)
Estimated age: 29.4–27.1 Ma; latest early-early late Oligocene (latest Rupelian-early Chattian)
Remarks: This zone is subdivided into two subzones based on the virtual extinction of chiloguembelinids in the midpart of the zone. Further discussion is given in Blow (1969) and Berggren and Miller (1988, p. 376).

P21a. *Globigerina angulisuturalis/Chiloguembelina cubensis* Concurrent Range Subzone (P21a of Berggren and Miller, 1988)
Definition: Biostratigraphic interval characterized by the concurrent range of the nominate taxa between the FAD of *Globigerina angulisuturalis* and the LAD of *Chiloguembelina cubensis*
Magnetochronologic calibration: Chron C11n(y)-Subchron C10n.1n
Estimated age: 29.4–28.5 Ma; latest early Oligocene (latest Rupelian)
Remarks: See Berggren and Miller (1988, p. 376) for additional information.

P21b. *Globigerina angulisuturalis-Paragloborotalia opima opima* Interval Subzone (P21b of Berggren and Miller, 1988; see also Jenkins and Orr, 1972)
Definition: Biostratigraphic interval between the (virtual) LAD of *Chiloguembilina cubensis* and the LAD of *Paragloborotalia opima opima*
Magnetochronologic calibration: Chron C10n.1n-Chron C9n(y)
Estimated age: 28.5–27.1 Ma; early late Oligocene (early Chattian)

Remarks: The use of the LAD of *Ch. cubensis* to subdivide the P21 Zone of Blow (1969) was initially suggested by Jenkins and Orr (1972) in the eastern Equatorial Pacific. It has since found wide application in both (sub)tropical (BKF85; Berggren and Miller, 1988) as well as in high southern (Berggren, 1992) latitude biostratigraphies.

P22. *Globigerina ciperoensis* Partial Range Zone (Cushman and Stainforth, 1945; emended by Bolli, 1957b; P22 of Berggren and Miller, 1988)
Definition: Biostratigraphic interval characterized by the partial range of the nominate taxon between the LAD of *Paragloborotalia opima opima* and the FAD of *Globorotalia kugleri* sensu stricto
Magnetochronologic calibration: Chron C9n(y)-Chron C6Cn.2n(o)
Estimated age: 27.1–23.8 Ma; late Oligocene (Chattian)
Remarks: The historical vicissitudes of this zone are discussed in greater detail in Berggren and Miller (1988, p. 376).

Neogene zonation.—

The use of planktonic foraminifera to biostratigraphically subdivide (sub)tropical marine Neogene stratigraphies began essentially in the 1940s with petroleum exploration in the Caribbean region (Cushman and Stainforth, 1945). The zonal schemes of Bolli (1957b, 1966), Banner and Blow (1965), Blow (1969, 1979), developed primarily in connection with petroleum exploration in the Caribbean region in the post-World War II years, have been reviewed by Stainforth and others (1975). Modifications to Blow's (1969, 1979) N-zone scheme have been made by Srinivasan and Kennett (1981a, b) and alternate/modified zonal stratigraphies developed by Thunell (1981) and Berggren and others (1985a). More recently Saunders and Bolli (1983) and Bolli and Saunders (1985) have provided an overview of the history of (sub)tropical planktonic foraminiferal biostratigraphic zonal schemes. Various zonal schemes have also been adopted for use in the Mediterranean region, the most commonly used one for the Pliocene Series being that of Cita (1973, 1975).

Mid-(temperate/transitional) and high-latitude Neogene biostratigraphies and evolutionary trends have been developed/reviewed by Jenkins (1967, 1971, 1975, 1992, 1993a, b), Srinivasan and Kennett (1981a, b; 1983), Kennett and Srinivasan (1983, 1984), Berggren (1973, 1977a, b, 1992) and Berggren and others (1983a), predominantly for Southern Hemisphere stratigraphies, whereas Northern Hemisphere zonal stratigraphies have been developed/discussed by, among others, Berggren (1972), Poore and Berggren (1975) and Weaver and Clement (1986; North Atlantic) and Keller (1979a-c, 1980a, b, 1981; Pacific).

As part of our attempt to provide a revised calibration of Neogene planktonic foraminiferal biostratigraphy to the new GPTS of CK92/95, we have reviewed the current status of existing planktonic zonal schemes and attempted to formulate a unified one that reflects our current understanding of planktonic foraminiferal biogeography. The problem we have faced is that while various aspects of both the Bolli (1957b, 1966) and Blow (1969, 1979) (sub)tropical scheme are fully acceptable for a zonal stratigraphy, other parts of the zonal schemes are less applicable (see Berggren, 1973, 1977a, 1993). Berggren and others (1983a) and Bronnimann and Resig (1971) reviewed some of the shortcomings of the zonal definitions and applications of Zones N17-N21 of Blow (1969) shortly after their formulation. Recall that Blow's ideas on Neogene planktonic foraminiferal zonation were already formed in the late 1960s; he died in 1972 but revisions to his Neogene zonal scheme were only published posthumously in 1979 and the need for a set of multiple zones reflecting regional biogeographies has been stressed by Jenkins and Orr (1972), Kennett (1973), Srinivasan and Kennett (1983), among others. It is with this in mind that we have attempted a modification/unification of these schemes in the past and herein. In applying a numerical notation scheme to what are, in some instances, the same zonal concepts as previous authors, we wish to emphasize that we do not seek, nor accept, credit for creating a "new" zonal stratigraphy (other than for those (sub)zones defined as "new" here or in previous work(s)). Unification of existing schemes with a view to providing improved biochronologic subdivisions reflecting regional biogeographies is the main goal of this exercise. For instance, in a recent study of some of Blow's (1969) Jamaican (para)type localities for Zones N16-N21, Berggren (1993) has demonstrated the inadequacies of Zones N18-N21 (at least) and suggested the use of alternate schemes. In the sections below, we shall review the state of Neogene planktonic foraminiferal biostratigraphy and suggest zonal biostratigraphies (converted to biochronologies) appropriate to regional biogeographies. In the case of the (sub)tropical regions, some workers may prefer simply to refer to, or utilize, the Bolli or Blow zonal stratigraphy in their work. The M-zonal scheme developed here is simply meant to provide continuity with the Pliocene Pl-zonal scheme developed earlier (and modified here) for (sub)tropical regions by combining the best (in our judgment) components of previous zonal schemes into a single, unified system.

Miocene Zonation of (Sub)Tropical Regions.—The chronology of early and middle-late Miocene (sub)tropical planktonic foraminiferal datum events/zones is shown in Figures 11 and 12, respectively.

The notation "M" used by Berggren (in Berggren and others, 1983a) to denote a series of Miocene subtropical-transitional zones for (predominantly) South Atlantic subtropical-transitional stratigraphies is preempted here to denote a zonal scheme applicable to tropical-subtropical regions. The notation "Mt" is used to denote a zonal scheme applicable to (predominantly) transitional Miocene faunas (see below).

M1. *Globorotalia kugleri* Total Range Zone (M1; Berggren and others, 1983a)
Definition: Total range of the nominate taxon
Magnetochronologic calibration: Subchron C6Cn. 2n(o)-Chron C6Ar
Estimated age: 23.8–21.5 Ma; early Miocene (Aquitanian)
Remarks: See comments by Blow (1969), Srinivasan and Kennett (1981b, 1983) and Berggren and others (1983a) regarding the history of this, and closely related zones.

M1a. *Globigerinoides primordius* Interval Subzone (M1a; Berggren and others, 1983a)
Definition: The biostratigraphic interval characterized by the nominate taxon between the FAD of *Globorotalia kugleri* and the FAD of *Globoquadrina dehiscens*

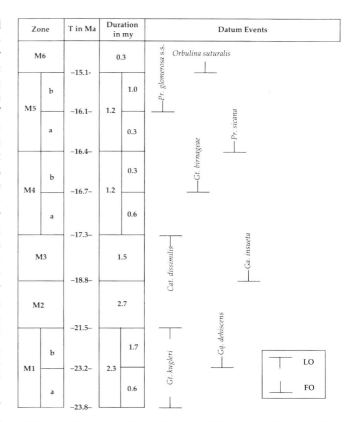

FIG. 11.—Chronology of early Miocene (sub)tropical planktonic foraminiferal zones.

Magnetochronologic calibration: Subchron C6Cn.2n(o)-Chron C6Br
Estimated age: 23.8–23.2 Ma; early Miocene (Aquitanian)
Remarks: Srinivasan and Kennett (1981b) emended Blow's (1969) *Globoquadrina quadrilobatus primordius/Globorotalia (Turborotalia) kugleri* Concurrent Range Zone (N4) and subdivided it into a lower *Gl.primordius/Gt. kugleri* "Concurrent Range" Subzone using the FAD of *Gd. primordius* and the FAD of *Gq. dehiscens* as lower and upper boundary criteria, respectively. This definition, in effect, renders this an interval zone; it would represent a concurrent range zone if the LAD of *Gt. kugleri* were used to denote its upper boundary. Berggren in Berggren and others (1983a) returned to the original concept of a *Gt. kugleri* Total Range and subdivided it into two parts based on the basis of the FAD of *Gq. dehiscens* within the range of *Gt. kugleri*. A characteristic element of upper Oligocene and lowermost Miocene faunas is the development of early forms of *Globigerinoides* (*Gd. primordius*) and this has been found to be useful to denote this subzone.

M1b. *Globorotalia kugleri/Globoquadrina dehiscens* Concurrent Range Subzone (M1b; Berggren and others, 1983a = Zone N4b of Srinivasan and Kennett, 1981b)
Definition: The biostratigraphic interval characterized by the concurrent range of the nominate taxa between the FAD of *Globoquadrina dehiscens* and the LAD of *Globorotalia kugleri*
Magnetochronologic calibration: Chron C6Br-Chron C6Ar
Estimated age: 23.2–21.5 Ma

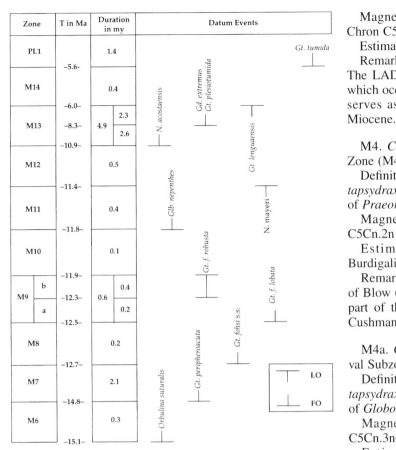

Fig. 12.—Chronology of middle-late Miocene (sub)tropical planktonic foraminiferal zones.

Remarks: This subzone corresponds to the upper part of the *Globoquadrina quadrilobatus primordius/Globorotalia (Turborotalia) kugleri* Zone of Blow (1969, 1979).

M2. *Catapsydrax dissimilis* Partial Range Zone (M2; Cushman and Renz, 1947, emend. Bolli, 1957b; = Zone N5 of Blow, 1969, 1979)

Definition: Biostratigraphic interval characterized by the partial range of the nominate taxon between the LAD of *Globorotalia kugleri* and the FAD of *Globigerinatella insueta*

Magnetochronologic calibration: Chron C6Ar-Chron C5En (inferred)

Estimated age: 21.5–18.8 Ma; early Miocene (Burdigalian)

Remarks: This zone (as emended by Bolli, 1957b) and Zone N5 (Blow, 1969, 1979) have the same boundary criteria. The common occurrence of *Cat. dissimilis* in this interval prompts us to utilize this zone in the sense of Bolli (1957b). The magnetochronologic calibration of the M2/M3 zonal boundary is inferred inasmuch as there is no direct calibration for the FAD of *Ga. insueta* at present.

M3. *Globigerinatella insueta/Catapsydrax dissimilis* Concurrent Range Zone (M3; Blow, 1969, 1979; =N6)

Definition: The concurrent range of the nominate taxa between the FAD of *Globorotalia insueta* and the LAD of *Catapsydrax dissimilis* and/or *Catapsydrax unicavus*

Magnetochronologic calibration: Chron C5En (inferred)-Chron C5Dn(y)

Estimated age: 18.8–17.3 Ma; early Miocene (Burdigalian)

Remarks: We use this zone as defined by Blow (1969, 1979). The LAD of *Catapsydrax* spp. represents a distinct bioevent which occurs in transitional and high-latitude areas as well and serves as a point of regional correlation in the late early Miocene.

M4. *Catapsydrax dissimilis-Praeorbulina sicana* Interval Zone (M4; here defined; = N7 of Blow, 1969, 1979)

Definition: Biostratigraphic interval between the LAD of *Catapsydrax dissimilis* and/or *Catapsydrax unicavus* and the FAD of *Praeorbulina sicana*

Magnetochronologic calibration: Chron C5Dn(y)-Chron C5Cn.2n

Estimated age: 17.3–16.4 Ma; early Miocene (late Burdigalian)

Remarks: As defined here this zone corresponds to Zone N7 of Blow (1969, 1979) in extent, if not intent, and to the lower part of the *Globigerinatella insueta* Interval Range Zone of Cushman and Stainforth (1945; as emended by Bolli, 1957b).

M4a. *Catapsydrax dissimilis-Globorotalia birnageae* Interval Subzone (M4a; here defined)

Definition: Biostratigraphic interval between the LAD of *Catapsydrax dissimilis* and/or *Catapsydrax unicavus* and the FAD of *Globorotalia birnageae*

Magnetochronologic calibration: Chron C5Dn(y)-Chron C5Cn.3n(o)

Estimated age: 17.3–16.7 Ma; early Miocene (late Burdigalian)

Remarks: Zone M4 is a relatively short zone (<1 my) within which biostratigraphic differentiation has been found possible under certain circumstances. The FAD of *Gt. birnageae* occurs within the biostratigraphic interval denoted by the bounding criteria of Zone M4 and has been found useful in various tropical-subtropical stratigraphies (e.g., Gulf of Mexico, Caribbean, Mediterranean).

M4b. *Globigerinoides bisphericus* Partial Range Zone (M4b; here defined)

Definition: Biostratigraphic interval characterized by the partial range of the nominate taxon between the FAD of *Globorotalia birnageae* and the FAD of *Praeorbulina sicana*

Magnetochronologic calibration: Subchron C5Cn.3n(o)-Subchron C5Cn.2n

Estimated age: 16.7–16.4 Ma; late early Miocene (late Burdigalian)

Remarks: We follow the differentiation made by Jenkins and others (1981) between *Pr. sicana* and *Gd. bisphericus* (see discussion below under Zone M5). Precise determination of the FAD of *Gd. bisphericus* is difficult because of its intergradation with *Gl. trilobus* sensu stricto, and we have recently found (Zhang and others, 1993) morphotypes referable to this taxon in Gulf of Mexico Eureka cores at levels referable to Zone M3 (= N6 of Blow, 1979). However, the taxon is well developed in its typical morphology in the younger part of the early Miocene (latest Burdigalian) and for that reason we use it as a partial range subzone to characterize/subdivide Zone M4.

M5. *Praeorbulina sicana-Orbulina suturalis* Interval Zone (M5; here defined; = N8 of Blow, 1969, 1979)

Definition: Biostratigraphic interval between the FAD of *Praeorbulina sicana* and the FAD of *Orbulina suturalis*

Magnetochronologic calibration: Subchron C5Cn.2n-Subchron C5Bn.2n

Estimated age: 16.4–15.1 Ma; middle Miocene (Langhian)

Remarks: This zone corresponds in extent to the upper part of the *Globigerinatella insueta* Zone of Cushman and Stainforth (1945; as emended by Bolli, 1957b) and to Zone N8 of Blow (1969, 1979). The evolutionary transition from *Praeorbulina* to *Orbulina* occurs within this zone and while the relatively rapid morphologic transitions between "taxa" of the *glomerosa* group in Zone M5 are often difficult to discern owing to preservational bias; the sequential development of the component morphotypes allows a high-resolution discrimination under ideal preservational circumstances. Under these, a twofold subdivision has been found practical (see below). This zone corresponds fully to Zone Mt5 in the temperate-transitional regions where the praeorbulinid-orbulinid bioseries is generally found owing to the significant late early and early middle Miocene warming which brought these forms to higher latitudes.

An examination of the type material led Jenkins and others (1981) to differentiate the taxa *bisphericus* and *sicana* (formerly considered synonymous by many workers) on the basis of the greater degree of enrollment of the test by the final chambers, reduction in size of, and concomitant increase in number of, supplementary apertures, absence of discernible umbilicus and greater test sphericity in *sicana*. They noted that the base of Blow's (1969, 1979) Zone N8 was defined using the concept of *sicana* (non *bisphericus*), redefined the N7/N8 zonal boundary to coincide with the subsequent FAD of *Pr. glomerosa curva* and observed that there is no significant temporal reduction in Zone N8 as previously defined. Indeed, our magnetostratigraphic tabulations suggest that the FADs of the two taxa are separated by only 0.1 my. We have chosen to use the FAD of *Pr. glomerosa glomerosa*, rather than that of *Pr. glomerosa curva*, to subdivide Zone M5, because of the somewhat larger temporal and morphologic separation between the FAD of these two taxa which we believe allows more reliable and consistent recognition and differentiation of these two biostratigraphic intervals.

The FAD of *Pr. sicana* is used to denote the base of the Langhian Stage and the lower/middle Miocene boundary (Cita and Blow, 1969) although precise correlation has not been demonstrated between the two. If the FAD of this taxon occurs somewhat lower (in the Cortemilia Formation) than the base of the stratotype Langhian Stage, the Burdigalian/Langhian (lower/middle Miocene) boundary may be more closely associated with the FAD of *Pr. glomerosa glomerosa* (Chron C5Cn.1n; 16.1 Ma).

M5a. *Praeorbulina sicana-Praeorbulina glomerosa* sensu stricto Interval Subzone (M5a; here defined)

Definition: Biotratigraphic interval between the LAD of *Praeorbulina sicana* and the FAD of *Praeorbulina glomerosa* senu stricto

Magnetochronologic calibration: Subchron C5Cn.2n-Subchron C5Cn.1n

Estimated age: 16.4–16.1 Ma; earliest middle Miocene (earliest Langhian)

Remarks: This subzone corresponds to the upper part of the *Globigerinatella insueta* Zone of Cushman and Stainforth (1945; as emended in Bolli, 1957b) and to the lower part of Zone N8 of Blow (1969, 1979).

M5b. *Praeorbulina glomerosa* sensu stricto-*Orbulina suturalis* Interval Subzone (M5b; here defined)

Definition: Biostratigraphic interval between the FAD of *Praeorbulina glomerosa* sensu stricto and the FAD of *Orbulina suturalis*

Magnetochronologic calibration: Subchron C5Cn.1n-Subchron C5Bn.2n

Estimated age: 16.1–15.1 Ma; early Miocene (Langhian)

Remarks: This subzone corresponds closely, but not precisely, to the *Praeorbulina glomerosa* Zone of Jenkins (1967) and Srinivasan and Kennett (1981b, 1983) in which the FAD of *Pr. glomerosa curva* was used to denote the base of the zone. It also corresponds to the upper part of Zone N8 of Blow (1969, 1979).

M6. *Globorotalia peripheroronda* Partial Range Zone (M6; = Zone N9 of Blow (1969, 1979)

Definition: Biostratigraphic interval characterized by the partial range of the nominate taxon between the FAD of *Orbulina suturalis* and the FAD of *Globorotalia peripheroacuta*

Magnetochronologic calibration: Subchron C5Bn.2n-Chron C5Bn(y)

Estimated age: 15.1–14.8 Ma; middle Miocene (Langhian)

Remarks: This zone corresponds fully to Blow's (1969, 1979) Zone N9, and we would agree with Blow's (1969, p. 231) assessment that the FAD of *O. suturalis* provides a more widespread and easily recognizable datum event than the LAD of *Ga. insueta* used by Bolli (1957b, 1966, and retained by Bolli and Saunders, 1985) for denoting the base of the *Gt. peripheroronda* "Interval Range Zone" (actually a partial range zone as defined by Bolli, 1957b).

M7. *Globorotalia peripheroacuta* Lineage Zone (M7; here defined; = Zone N10 of Blow, 1969; 1979)

Definition: Biostratigraphic interval between the initial evolutionary appearance of the nominate taxon and the initial evolutionary appearance of *Globorotalia fohsi* sensu stricto

Magnetochronologic calibration: Chron C5Bn(y)-Subchron C5Ar.1n

Estimated age: 14.8–12.7 Ma; middle Miocene (Serravallian)

Remarks: Bolli and Saunders (1985, p. 215) have pointed out the advanced biocharacters of the holotype of *Gt. praefohsi* Blow and Banner (1966) and suggested that it is a form intermediate between *Gt. fohsi* sensu stricto and *Gt. fohsi lobata*, a view supported by the fact that the holotype of *praefohsi* is from the *Globorotalia fohsi lobata* Zone in Venezuela. The paratype of *praefohsi* appears to be intermediate between *peripheroacuta* and *fohsi* sensu stricto, but inasmuch as the species concept is based upon the holotype, this observation is irrelevant to the definition of *Gt. praefohsi*. Thus, we prefer to use *Gt. fohsi* sensu stricto as the nominate form of Zone M8. In all likelihood, Blow (1969, 1979), in using both the holotype and paratype of *praefohsi* as concept of the species, would have placed the N10/N11 zonal boundary at the same level as we have placed the Zone M7/M8 boundary, inasmuch as the FAD of *Gt. fohsi* sensu stricto and *Gt. praefohsi* (sensu paratype) are at about the same level according to our investigations.

M8. *Globorotalia fohsi* sensu stricto Lineage Zone (M8; = *Globorotalia fohsi fohsi* Lineage Zone of Cushman and Stainforth, 1945; emended by Bolli, 1957b)

Definition: Biostratigraphic interval between the initial evolutionary appearance of the nomninate taxon and the initial evolutionary appearance of *Globorotalia fohsi lobata*

Magnetochronologic calibration: Subchron C5Ar.1n-Subchron C5Ar.1r

Estimated age: 12.7–12.5 Ma; middle Miocene (Serravallian)

Remarks: The revised isotopically based interpretation of the magnetostratigraphy of DSDP Hole 563 (Wright and Miller, 1992) indicates that the FAD of *Gt. fohsi lobata* occurs in Chron C5Ar.1r rather than in Chron C5AAr (Miller and others, 1985).

M9. *Globorotalia fohsi lobata-Globorotalia fohsi robusta* Interval Zone (M9; = combined *Gt. fohsi lobata* Interval Zone of Bolli, 1957b and *Gt. fohsi robusta* Taxon Range Zone of Bolli, 1957b)

Definition: Biostratigraphic interval between the initial (evolutionary) appearance of *Globorotalia fohsi lobata* to the LAD of its descendant *Globorotalia fohsi robusta*

Magnetochronologic calibration: Subchron C5Ar.1r-Subchron C5An.1n

Estimated age: 12.5–11.9 Ma; middle Miocene (Serravallian)

Remarks: This zone is combined in the sense of Stainforth and others (1975, p. 82) because of the difficulty in some instances/areas in discerning the component parts of the evolutionary sequence from *lobata* to *robusta*. Zone M9 is subdivided into two subzones based on the sequential FADs of these two "subspecies" for use in those circumstances where this sequence is recognizable. The magnetochronologic calibration of the nominate taxa of the boundaries of Subzones M9a and M9b is based on a reinterpretation of the original magnetostratigraphy in DSDP Hole 563 (Wright and Miller, 1992).

M9a. *Globorotalia fohsi lobata* Lineage Zone (M9a; = *Globorotalia fohsi lobata* Lineage Zone of Bolli, 1957b)

Definition: Biostratigraphic interval between initial evolutionary appearance of nominate taxon and that of its descendant *Globorotalia fohsi robusta*

Magnetochronologic calibration: Subchron C5Ar.1r-Subchron C5An.2n

Estimated age: 12.5–12.3 Ma

Remarks: See remarks on Zone M9 above.

M9b. *Globorotalia fohsi robusta* Total Range Zone (M9b; = *Globorotalia fohsi robusta* Total Range Zone of Bolli, 1957b)

Definition: Biostratigraphic interval characterized by the total range of the nominate taxon.

Magnetochronologic calibration: Subchron C5An.2n-Subchron C5An.1n

Estimated age: 12.3–11.9 Ma; middle Miocene (Serravallian)

Remarks: See remarks under Zone M9 above.

M10. *Globorotalia robusta-Globoturborotalita nepenthes* Interval Zone (M10; here defined)

Definition: Biostratigraphic interval between the LAD of *Globorotalia fohsi robusta/lobata* and the FAD of *Globoturborotalita nepenthes*

Magnetochronologic calibration: Subchron C5An.1n-Subchron C5r.3r

Estimated age: 11.9–11.80 Ma; middle Miocene (Serravallian)

Remarks: In Jamaica we (Berggren, 1993) have observed the same sequence of bioevents as denoted by Bolli (1966; see also Bolli and Saunders, 1985, p.167) as well as in the expanded mid-Miocene section in the Bodjonegoro-1 well of Indonesia, namely LAD of *Gt. fohsi robusta*, LAD of *Gl. subquadratus*, LAD of *N. mayeri* (with the FAD of *Glb. nepenthes* occurring approximately midway within the 25-m interval separating the latter two datums in Jamaica). Using the magnetochronologic calibration of the LAD of *Gt. fohsi robusta* of Chron C5An.1n (DSDP Hole 563: Miller and others, 1985), the FAD of *Glb. nepenthes* of Chron C5r.3r (DSDP Hole 563; Miller and others, 1985) and the LAD of *N. mayeri* of Chron C5r.2r (Buff Bay section; Miller and others, 1994), we have calculated that the LAD of *Gd. subquadratus* is ~0.05 my younger than the LAD of *Gt. fohsi robusta*. In view of the fact that we have estimated the duration of Zone M10 at only 0.10 my, it seems unnecessary to subdivide this biostratigraphic interval formally into two subzones. However, in areas where this biostratigraphic interval is expanded an informal subdivision into a lower Subzone a (*Globorotalia lobata/robusta-Globigerinoides subquadratus* Interval Range Subzone-Bolli's (1966) *Globigerinoides ruber* Zone) and upper Subzone b (*Globigerinoides subquadratus-Globoturborotalita nepenthes* Interval Zone) may be appropriate.

M11. *Globoturborotalita nepenthes/Neogloboquadrina mayeri* Concurrent Range Zone (M11; = Zone N14 of Blow, 1969, 1979)

Definition: Biostratigraphic interval with the concurrent range of the two nominate taxa from the FAD of *Globoturborotalita nepenthes* (base) to the LAD of *Neogloboquadrina mayeri* (top)

Magnetochronologic calibration: Subchron C5r.3r-Subchron C5r.2r

Estimated age: 11.80–11.4 Ma; middle Miocene (Serravallian)

Remarks: The FAD of *Glb. nepenthes* has been calibrated to early Chron C5r in DSDP Holes 563 (Miller and others, 1985) and 608 (Miller and others, 1991) and occurs at approximately the same level in the Buff Bay section of Jamaica (Berggren, 1993) with an age estimate of 11.8 Ma, only ~0.1 my younger than the LAD of the *Gt. fohsi lobata/robusta* group. It would appear that this relatively brief temporal juxtaposition between the LAD of the *lobata/robusta* group and the FAD of *Glb. nepenthes* may explain the various records of overlap in the stratigraphic range of these two taxon groups in (relatively condensed) oceanic cores rather than diachrony. Determination of the FAD of *Glb. nepenthes* can also be a difficult taxonomic decision and requires high quality preservation. In Jamaica, the highest ocurrence of *Gt. lobata/robusta* and the lowest occurrence of *Glb. nepenthes* are separated by a 22-m interval within which the LAD of *Gd. subquadratus* occurs approximately midway between these two events (Berggren, 1993).

M12. *Neogloboquadrina mayeri-Neogloboquadrina acostaensis* Interval Zone (M12; here defined; = N15 of Blow, 1969, 1979, but with different denotation)

Definition: Biostratigraphic interval between the LAD of *Neogloboquadrina mayeri* and the FAD of *Neogloboquadrina acostaensis*

Magnetochronologic calibration: Subchron C5r.2r-Subchron C5n.2n (early)

Estimated age: 11.4–10.9 Ma; middle-late Miocene (Serravallian-Tortonian)

Remarks: Bolli (1966) redefined earlier versions of this biostratigraphic interval as the *Globorotalia menardii* Interval Range Zone with the same boundary criteria used here and by Blow (1969, 1979) for his Zone N15. However in denoting it the *Globorotalia menardii* Zone becomes a partial, rather than an interval, range zone. Although *Gt. menardii* "A" (of Bolli) is often a common component of faunas of this biostratigraphic interval, we believe it more appropriate to denominate this an interval range zone to free the zone from the burden of containing the nominate taxon (and because of continuing problems of mid-Miocene menardine taxonomy) in order that it be recognized (cf. Berggren, 1993, in which the name *Gt. menardii* "A" was used informally for this interval in a biostratigraphic study of Jamaica). Another taxon commonly occurring with *Gt. menardii* over this biostratigraphic interval is the distinctive *Gt. lenguaensis*, which has its FAD near the FAD of *Glb. nepenthes* in Gulf Coast and Caribbean stratigraphic sections. At the same time, we have found the use of *Neogloboquadrina continuosa* as the nominate form of this biostratigraphic interval (Blow, 1969, 1979) unsatisfactory, inasmuch as that taxon is frequently absent or, at best, sporadic in many (sub)tropical late middle-early late Miocene stratigraphies including Jamaica (contra Blow, 1969, p. 248).

The LAD of *N. mayeri* in Jamaica is in an interval interpreted as Chron C5r.2r (= 11.4 Ma) and the FAD of *N. acostaensis* (interpreted as occurring in the *Globorotalia plesiotumida* Zone (= N17 of Blow, 1969, above an unconformity) is at least 45 m higher (Berggren, 1993). The duration of Zone M12 is estimated at ~0.5 my, and the zone is hereby reinstated in middle Miocene zonal hagiography. The juxtaposition of these two datum events in North Atlantic DSDP Holes 558, 563 (Miller and others, 1985) and 608 (Miller and others, 1991) suggests the presence of a relatively brief early late Miocene hiatus in this region. The alternate interpretation of a later (diachronous) LAD of *N. mayeri* in mid-high latitudes is less appealing/parsimonious in view of the essentially (sub)tropical distribution of this taxon (see Miller and others, 1991, 1994, for additional discussion, and an alternate interpretation, based on integrated stable isotope-biostratigraphic correlations).

M13. *Neogloboquadrina acostaensis/Globorotalia lenguaensis* Concurrent Range Zone (M13; herein defined)

Definition: Biostratigraphic interval with the concurrent range of the nominate taxa between the FAD of *Neogloboquadrina acostaensis* and the LAD of *Globorotalia lenguaensis*

Magnetochronologic calibration: Subchron C5n.2n (early)-Subchron C3An.1n (inferred)

Estimated Age: 10.9–6.0 Ma; late Miocene (Tortonian-Messinian)

Remarks: We have found that both the initial evolutionary appearance of *Gt. plesiotumida* from its (putative) ancestor *Globorotalia merotumida* (= Zone N16/N17 boundary) (Blow, 1969, p. 52) and the initial appearance of *Neogloboquadrina*

humerosa (Bolli and Bermudez, 1965; Bolli and Saunders, 1985, p. 170) are difficult to apply as zonal criteria in the upper Miocene Series. The differing records of the FAD of *Gt. plesiotumida* in the literature attest to the difficulties in recognizing this taxon in both Atlantic and Indo-Pacific stratigraphies. Berggren (1993) drew attention to the difficulties encountered in distinguishing typical representatives of the predominantly Indo-Pacific *Gt. plesiotumida* group in late Miocene assemblages of the Atlantic-Caribbean region. On the other hand, we believe that the relatively small but morphogically distinct and geographically widespread taxon *Gt. lenguaensis* is an appropriate form for late Miocene biostratigraphy. Unfortunately, until recently, the only direct magnetochronologic calibration of the LAD of *Gt. lenguaensis* was that by Poore and others (1983, 1984) at the younger end of Chron C4An in DSDP Hole 519 (South Atantic). However, Chaproniere and others (1994, Fig. 9) have calibrated this datum to the early part of Chronozone C3An.1n (6.0 Ma, this paper) in ODP Hole 840 (Tongan Platform). At this site, it occurs ~10 m below the FAD of *Pulleniatina primalis* and ~40 m above the FAD of *Globorotalia margaritae* which occurs within the upper third of Chronozone C3An.2n. Chaisson and Leckie (1993, Tables 1, 3) demonstrated that *Gt. lenguaensis* stratigraphically overlaps the lower part of the range of *Pu. primalis* in the uppermost Miocene of ODP Hole 806B (Ontong-Java Plateau) and that its LAD was ~5.5 Ma (using BKV85; = ~5.9 Ma in CK92 and 6.0 Ma in CK92/95 and this paper). This range is consistent with that delineated by Berggren (1977a) on the Rio Grande Rise (South Atlantic), although the uppermost part of the range of the taxon there may have been truncated by a regional unconformity (Hodell and Kennett, 1986; Aubry, pers. commun.). In Gulf of Mexico Eureka coreholes E66–73 and E68–136, the LAD of *Gt. lenguaensis* occurs virtually, and anomalously, at the same level as the FAD of *Gt. plesiotumida* and within Zone NN10 (i.e., ~8 Ma; Zhang and others, 1993; Aubry, 1993b).

We have placed the LAD of *Gt. lenguaensis* stratigraphically equivalent to the middle part of Chron C3An.1n (with an estimated age of 6.0 Ma) following Chaisson and Leckie (1993). Srinivasan and Kennett (1981b) subdivided this biostratigraphic interval (= Zone N17 of Blow, 1969, 1979) in the Indo-Pacific based on the FAD of the (geographically restricted) *Pulleniatina primalis*; Chaisson and Leckie (1993, Table 2) estimated this event to have occurred at 5.8 Ma (using BKV85; = 6.2 Ma in CK92; = 6.4 Ma in CK92/95 and this paper) in Hole 806B.

M13a. *Neogloboquadrina acostaensis-Globigerinoides extremus/Globorotalia plesiotumida* Interval Subzone (here defined; M13a)

Definition: Biostratratigraphic interval between the FAD of *N. acostaensis* and the FAD of *Globigerinoides extremus* and/or *Globorotalia plesiotumida*

Magnetochronologic calibration: Subchron C5n.2n (early)-Subchron C4r.1n (estimated)

Estimate age: 10.9–8.3 Ma; late Miocene (Tortonian)

Remarks: The FAD of *Gt. plesiotumida* is difficult to determine under some circumstances owing to problems with general late Miocene menardine morphology and geographic distribution (predominantly an Indo-Pacific taxon; see discussion above). A common and geographically widely distributed late Neogene taxon (often associated with *Gt. plesiotumida*) is *Glo-*

bigerinoides extremus with its distinctly flattened (asymmetric) terminal chamber. This form is particularly well developed in the Mediterranean region where it appears together with another distinct form which is (apparently) restricted to the late Miocene (Messinian) *Turborotalita multiloba*.

Data relevant to determining the relationship of the FAD of *Gt. plesiotumida* and *Gl. extremus* include the following:

1. In Jamaica, the type level of the *Globorotalia plesiotumida* Zone is in Zone NN10 (Berggren, 1993) not in Zone NN11 (cf. Berggren and others 1983a; Aubry, 1993a) as previously thought (BKV85).

2. Backman and others (1990, p. 275) have provided new calibrations for the NN9/NN10 (C4Ar = 9.20 Ma in CK92 and 9.30 Ma in CK92/95) and NN10/NN11 (C4r.1r = 8.10 Ma in CK92 and 8.3 Ma in CK92/95 and this paper) zonal boundaries which means that Zone NN10 is ~1 my long.

3. Hodell and Kennett (1986, p. 294) provided a magnetochronologic calibration of Chron 7N2 (= Chron C4n.2n) for the FAD of *Gt. plesiotumida* which corresponds to a level within Zone NN11, that appears too young based on evidence cited in points 1 and 2 above.

4. Chaisson and Leckie (1993) denote the FAD of *Gt. plesiotumida* and *Gd. extremus* within lower Zone NN11 and uppermost Zone NN10, respectively (Takayama, 1993, Table 12). The age of 7.1 Ma for the FAD of *Gt. plesiotumida* (BKF85) translates into 7.7 Ma (CK92/95 and this paper) and that of *Gd. extremus* to ~8.2 Ma (CK92/95 and this paper). At the same time, Chaisson and Leckie (1993, Table 1) extend questionable occurrences of *Gt. plesiotumida* as low as the basal part of 806B/31H (i.e., to about the same level as the FAD of *Gd. extremus* and within uppermost Zone NN10). However, the age estimates of the FAD of *Gd. extremus* and several other Neogene datums were incorrectly estimated by Chaisson and Leckie (1993, Table 3; see Table 13 of this paper for further explanation).

5. In Gulf of Mexico Eureka coreholes E66–73 and E68–136 as well as the Mt. Giammoia section (Sicily), the FAD of *Gd. extremus* is within Zone NN10 (WAB, unpubl. data), suggesting an age of 8.3 Ma (or older) for this datum.

We cite these data here by way of demonstrating a general consistency (at least in some instances) of the estimated FAD of *Gt. plesiotumida*, its close relationship with the FAD of *Gd. extremus*, and their possible use separately or together as subzonal indicators in upper Miocene (sub)tropical biostratigraphies under certain conditions. We have (somewhat arbitrarily) estimated the FAD of both *extremus* and *plesiotumida* at ~8.3 Ma (i.e., near or within Chron C4r.1n) in order to bring the FAD of *Gt. plesiotumida* in line with the general evidence reviewed above. Clearly additional verification of the FAD of these two Subzone M13a index taxa is needed.

While the study by Chaisson and Leckie (1993) suggests a short stratigraphic interval between the FADs of *Gd. extremus* and *Gt. plesiotumida*, their close juxtaposition in time suggests that either or both of these forms are appropriate as markers for subdivision of Zone M13 on a regional basis.

M13b. Globigerinoides extremus/Globorotalia plesiotumida-Globorotalia lenguaensis Interval Subzone (M13b; here defined)

Definition: Biostratigraphic interval between the FAD of *Globigerionoides extremus* and/or *Globorotalia plesiotumida* and the LAD of *Globorotalia lenguaensis*

Magnetochronologic calibration: Subchron C4r.1n (inferred)

Estimate age: 8.3–6.0 Ma

Remarks: See discussion above

M14. Globorotalia lenguaensis-Globorotalia tumida Interval Zone (M14; here defined)

Definition: Biostratigraphic interval between the LAD of *Globorotalia lenguaensis* and the FAD of *Globorotalia tumida*

Magnetochronologic calibration: Subchron C3An.1n (inferred)-Chron C3r (early Gilbert)

Estimated age: 6.0–5.6 Ma; latest Miocene (Messinian)

Remarks: The FAD of *Gt. tumida* has been substituted for the FAD of *Gt. margaritae* for the base of Zone Pl1 (see remarks under Zone Pl1 below) with the result that the M14/Pl1 zonal boundary corresponds closely to (but is somewhat older than) the base of the commonly accepted Miocene/Pliocene (Messinian/Zanclean) boundary with a revised age estimate of 5.32 Ma (Hilgen, 1991).

Pliocene zonation of (Sub)Tropical Regions.—The Pliocene Pl1-Pl6 zonal system of Berggren (1973, 1977a, b) was intended to serve as a zonation for Pliocene sediments of (predominantly) (sub)tropical-transitional areas of the Atlantic Ocean. Recent reviews by Weaver and Clement (1986, 1987) suggest that this zonal system is applicable with reasonable consistency to 41°–42° north latitude and our experience suggests comparable application in the South Atlantic. Farther to the north and south an alternative zonal scheme is necessary to reflect the provincial development of temperate taxa; this is discussed below. In fact, this zonation is equally adaptable to Indo-Pacific and Mediterranean stratigraphies with several shortcomings discussed below at appropriate places. In particular there are problems associated with the application of Zone Pl1 (as defined), which were recognized by the author shortly after its formulation. It was the critical examination of the biostratigraphic events of the upper Miocene-lower Pliocene interval by Hodell and Kennett (1986) that clarified the source of these problems. The reexamination conducted here of Neogene zonal schemes in connection with a revision of the Neogene time scale has afforded an opportunity of rectifying shortcomings in this, and other, zonal stratigraphies.

The Pl-zonal scheme is discussed below and the zones redefined to conform with accepted and applied biostratigraphic concepts and terminology (see review by Schoch, 1989). Zone Pl6 is renamed and redefined for reasons given below.

Pl1. Globorotalia tumida/Globoturborotalita nepenthes Concurrent Range Zone (Pl1; herein redefined; emendation of Berggren, 1973, 1977b)

Definition: The biostratigraphic interval denoted by the concurrent range of the nominate taxa between the FAD of *Globorotalia tumida* and the LAD of *Globoturborotalita nepenthes*

Magnetochronologic calibration: Chron C3r (early Gilbert)-Subchron C3n.1n(y) (top Cochiti Subchron)

Estimated age: 5.6–4.18 Ma; early Pliocene (early Zanclean)

Remarks: Originally defined as the "concurrent range" of *Gl. margaritae* and *G. nepenthes* between the LAD of *Globoquadrina dehiscens* and the LAD of *G. nepenthes*, this zone corresponded to a two-taxon partial range zone (see discussion in

Berggren and Miller, 1988, p. 364–367). Its redefinition here as a concurrent range zone with substitution of the FAD of *Gt. tumida* as the defining element of the base of the zone represents a return to the original intent/denotation of the zone (Berggren, 1973) and is consistent with appropriate and accepted stratigraphic usage. It also reflects recent data (Weaver and Clement, 1986, 1987; Hodell and Kennett, 1986) which suggest that *Gt. margaritae* has an erratic open ocean (non-Mediterranean) FAD spanning over 1 my from Chron C3An to basal Chron C3r (early Gilbert) which is lower (older) in some (sub)tropical stratigraphies of the western Pacific (DSDP Site 317, Manihiki Plateau; DSDP Site 588), Indian Ocean (DSDP Site 214, Ninety East Ridge) and the south equatorial Atlantic (DSDP Site 366, Sierra Leone Rise) than the LAD of *Gt. tumida* (Thunell, 1981; Hodell and Kennett, 1986). A brief overlap between the ranges of *Globoquadrina dehiscens* and *Gt. tumida* was observed only at DSDP Site 317 in the eastern equatorial Pacific (Thunell, 1981); indeed Hodell and Kennett (1986) have shown convincingly that the LAD of *Gl. dehiscens* had a diachronous span of ~1 my in the late Miocene Epoch, ranging from Chron C3Bn (~6.8 Ma; near the $\delta^{13}C$ shift) in subtropical regions to basal Chron C3r (early Gilbert; 5.5 Ma) in the tropics. At ODP Hole 806B, in which there is an expanded Miocene-Pliocene carbonate section, Chaisson and Leckie (1993) have observed a short (<3 m) overlap between *dehiscens* and *margaritae* at a level approximately coincident with the FAD of *Sphaeroidinella dehiscens* s.l. and ~5 m above the FAD of *Gt. tumida*. In view of the rather erratic FAD of *Gt. margaritae* and the fact that at SW Pacific DSDP Site 588 the FAD of *Gt. tumida* and *Gt. sphericomiozea* (used in SW Pacific biostratigraphies to denote the Miocene/Pliocene boundary) can be linked via strontium isotopes to a level near the base of the stratotype Zanclean in Sicily (Hodell and Kennett, 1986) at ~5.3 Ma (in the chronology of CK92; = 5.6 Ma in CK92/95 and this paper), the use of the FAD of *Gt. tumida* to denote the base of Zone Pl1 provides a means of regional correlation at a level close to, but slightly older than, the base of the Pliocene Epoch (as defined by the base of the Zanclean).

In the Mediterranean, the entry of *Gt. margaritae* was delayed by the reestablishment of normal marine conditions following the "terminal Miocene" desiccation phase. Although apparently present in rare numbers at the base of the Zanclean transgression a short distance/brief time below/before the Thvera (C3n.4n) Subchron, its lowest common occurrence (LCO) is normally associated with the Thvera (C3n.4n) Subchron (Channell and others, 1990; Langereis and Hilgen, 1991) with an estimated age of 5.2 Ma.

Thus it should be borne in mind that the co-occurrence of *Glb. nepenthes* and *Gt. margaritae* may serve as an approximation of Zone Pl1 equivalence (in predominantly subtropical stratigraphies) but may actually be somewhat older in some tropical areas. Under favorable circumstances, the FAD of *Sphaeroidinella dehiscens* sensu lato may serve to denote a level approximately equivalent to (but slightly younger than) the base of Zone Pl1.

Pl1a. *Globorotalia tumida/Globorotalia cibaoensis* Concurrent Range Subzone (Pl1a; Berggren, 1977b; herein redefined)
Definition: Biostratigraphic interval between the FAD of *Globorotalia tumida* and the LAD of *Globorotalia cibaoensis*

Magnetochronologic calibration: Chron C3r (early Gilbert)-Subchron C3n.2n (Nunivak Subchron)
Estimated age: 5.6–4.6 Ma; latest Miocene to early Pliocene (latest Messinian to early Zanclean)
Remarks: This subzone was originally denominated the *Globorotalia cibaoensis* Partial Range Zone, using as bracketing criteria the LAD of *Gq. dehiscens* (base) and FAD of *Globorotalia puncticulata* (top). The fact that as defined Subzone Pl1a did not correspond strictly to a partial range (sub)zone and that *Gt. puncticulata* does not (consistently) occur in lower Pliocene (sub)tropical stratigraphies lends a degree of urgency to the need for redefinition of this subzone. The essential ubiquity of *Gt. cibaoensis* in low latitude assemblages (Thunell, 1981) and its LAD concomitant with the FAD of *Gt. puncticulata* in transitional waters provides a means of correlation between low and mid-latitude biostratigraphies during the latest Miocene and earliest Pliocene Epochs. It should be noted that in its original definition, and as redefined here, Subzone Pl1a has a different (and more extensive) denotation than the *Gt. cibaoensis* Partial Range Zone of Thunell (1981, p. 82). The latter zone was defined as the partial range of the nominate taxon between the LAD of *Globoquadrina dehiscens* and the FAD of *Sphaeroidinella dehiscens* which is essentially correlative with Zone N18 of Blow (1969; but see discussion in Berggren, 1993, of problems with using Zone N18).

Pl1b. *Globorotalia cibaoensis-Globoturborotalita nepenthes* Interval Subzone (Pl1b + Pl1c combined; Berggren, 1977b; herein redefined)
Definition: Biostratigraphic interval between the LAD of *Globorotalia cibaoensis* and the LAD of *Globoturborotalita nepenthes*
Magnetochronologic calibration: Subchron 3n.2n (Ninuvak Subchron)-Subchron C3n.1n (late Cochiti Subchron)
Estimated age: 4.6–4.18 Ma; early Pliocene (early Zanclian)
Remarks: *Globorotalia puncticulata* and/or *Gt. crassaformis* are relatively common components of early Pliocene temperate areas (Berggren, 1977b; Keller, 1978, 1979a-c) but are absent, rare or delayed in most low-latitude areas (e.g., Gulf of Mexico, Keigwin, 1982; southern equatorial Atlantic and Indian Ocean, Thunell, 1981; Srinivasan and Chaturvedi, 1992) to levels close to, or slighly younger than, the LAD of *Glb. nepenthes* which has been correlated with the Cochiti Subchron. On the other hand, the FAD of *Gt. crassaformis* at cool subtropical South Atlantic Site 519 has been observed to occur in the lower part of Chron C3r (early Gilbert) prior to that of *Gt. puncticulata* which suggests that *crassaformis* may have evolved in the South Atlantic and migrated to lower latitudes in the early-mid Pliocene. In the Mediterranean, the inital appearance of *Gt. crassaformis* is associated with the Gilbert/Gauss boundary at ~3.55 Ma, over 1 my later than its earliest occurrence in temperate areas.

In its original definition, this biostratigraphic interval was subdivided into two subzones: a lower *Globorotalia puncticulata* Consecutive Range (=Lineage or Phylozone) Subzone (Pl1b) and an upper *Globorotalia crassaformis* sensu lato Consecutive Range Subzone (Pl1c) to denote the (interpreted) phylogenetic linkage between *Gt. cibaoensis* and *Gt. puncticulata*, and between *Gt. puncticulata* and *Gt. crassaformis*, respectively. In view of the continuing controversy over the ancestry of *Gt. puncticulata* (Berggren, 1977b; Arnold, 1983 vs Wei, 1987; Scott, 1982) and *Gt. crassaformis* (Cifelli and Scott, 1986, p. 50), their essentially simultaneous FADs in some temperate stratigraphies (Keller, 1978, 1979a-c; Weaver and Clement, 1986, 1987; Scott and others, 1990) and the general absence or extreme rarity of *Gt. puncticulata* and/or *Gt. crassaformis* in lower Pliocene (sub)tropical assemblages it would seem more appropriate to simply use the LAD of the relatively ubiquitous (sub)tropical forms *Gt. cibaoensis* and *Glb. nepenthes* to denote the boundary between Subzones Pl1a/b and Pl2, respectively, and to designate this biostratigraphic interval as an interval zone, thus eliminating Subzone Pl1c from the (sub)tropical zonal scheme.

Pl2. *Globoturborotalita nepenthes-Globorotalia margaritae* Interval Zone (Pl2; herein redefined; emendation of Berggren, 1973; 1977b)

Definition: Biostratigraphic interval between the LAD of *Globoturborotalita nepenthes* and the LAD of *Globorotalia margaritae*

Magnetochronologic calibration: Subchron C3n.1n (late Cochiti Subchron)-Chron C2Ar(y) (Gauss/Gilbert boundary)

Estimate age: 4.18–3.58 Ma; early Pliocene (late Zanclean)

Remarks: Subzone Pl2 was originally designated the *Sphaeroidinellopsis subdehiscens-Globorotalia margaritae* Concurrent Range Zone (Berggren, 1973, 1977b) bracketed by the LAD of *Glb. nepenthes* (base) and the LAD of *Gt. margaritae* (top). As defined in this manner it corresponds, in fact, to a two-taxon Partial Range Zone (see discussion in Berggren and Miller, 1988). The zone is redefined here as interval range zone delimited by the LAD of two of the major and relatively ubiquitous (sub)tropical mid-Pliocene planktonic foraminiferal taxa.

Pl3. *Globorotalia margaritae-Sphaeroidinellopsis seminulina* Interval Zone (Pl3; herein redefined; emendation of Berggren, 1973, 1977b)

Definition: Biostratigraphic interval between the LAD of *Globorotalia margaritae* and the LAD of *Sphaerodinellopsis seminulina*

Magnetochronologic calibration: Chron C2Ar(y) (Gauss/Gilbert boundary)-Subchron C2An.1r (earliest Kaena Subchron)

Estimated age: 3.58–3.12 Ma; late Pliocene (Piacenzian)

Remarks: Zone Pl3 was originally designated the *Globoquadrina altispira-Sphaeroidinellopsis subdehiscens* Concurrent Range Zone (Berggren, 1973, 1977b) between the bracketing events of the LAD of *Gt. margaritae* (base) and the LAD of *Sphaerodinellopsis subdehiscens* (top). As such it corresponded to a two-taxon partial range zone (see remarks above). It is here redefined as an interval range zone using two distinct and relatively ubiquitous (sub)tropical mid-Pliocene biostratigraphic events to denote its boundaries. The LAD of the *Globorotalia limbata* ecophenotypic variant *Globorotalia multicamerata* and the development of *S. dehiscens* sensu stricto (with everted test—encircling girdle) occurred in the upper part of this zone in tropical-subtropical regions. The change in *Sphaeroidinellopsis* taxonomy follows recent studies in that group (see discussion in Berggren, 1993).

Pl4. *Sphaeroidinellopsis seminulina-Dentoglobigerina altispira* Interval Zone (Pl4; herein redefined; emendation of Berggren, 1973, 1977b)

Definition: Biostratigraphic interval from the LAD of *Sphaeroidinellopsis seminulina* to the LAD of *Dentoglobigerina altispira*

Magnetochronologic calibration: Subchron C2An.1r (earliest Kaena Subchron)-Subchron C2An.1r (latest Kaena Subchron)

Estimated age: 3.12–3.09 Ma; late Pliocene (Piacenzian)

Remarks: Originally designated the *Gt. multicamerata-Globoquadrina altispira* Concurrent Range Zone between the LAD of *Sphaeroidinellopsis subdehiscens* (base) and the virtually simultaneous LADs of the two nominate taxa (top), this zone as thus defined actually corresponds to a two-taxon partial range zone (see discussion in Berggren and Miller, 1988). As redefined here this extremely short (30 ky) zone characterizes a biostratigraphically distinct mid-Pliocene interval. In tropical-subtropical sequences, the FAD of *Globigerinoides fistulosus* is associated with the Kaena (C2An.1r) Subchron also.

Pl5. *Dentoglobigerina altispira-Globorotalia miocenica* Interval Zone (Pl5; herein redefined; emendation of Berggren, 1973, 1977b)

Definition: Biostratigraphic interval between the LAD of *Dentoglobigerina altispira* and the LAD of *Globorotalia miocenica*

Magnetochronologic calibration: Subchron C2An.1r (latest Kaena Subchron)-Subchron C2r.2r (early Matuyama; just prior to Reunion Subchron)

Estimated age: 3.09–2.30 Ma; late Pliocene (late Piacenzian)

Remarks: This zone was originally designated the *Globorotalia miocenica-Globorotalia exilis* Concurrent Zone delimited by the LAD of *Gq. altispira/Gt. multicamerata* (base) and the LAD of *Gt. miocenica* (top). As defined, it corresponded to a two taxon partial range zone and is redefined here as an interval range zone to conform to commonly accepted and applied biostratigraphic usage. A number of significant events characterize the stratigraphic interval delimited by the zonal boundary events: the LAD of *Globorotalia pertenuis* occurred in the latest part of the Gauss (2.63 Ma), the LAD of *Gt. puncticulata* occurred in temperate-subtropical regions of the Atlantic and Mediterranean in the early Matuyama (2.45 Ma), and *Pulleniatina* reappeared in the Atlantic Ocean (from the Indo-Pacific region) at the top of the zone after an approximately 1-my absence. This zone is applicable to Atlantic Ocean stratigraphy only inasmuch as *Gt. miocenica* (and *Globorotalia exilis*) are restricted to tropical-temperate regions of this ocean (Berggren, 1973, 1977a, b; Weaver and Clement, 1986, 1987). Alternative biostratigraphic zonal criteria for this interval suitable for Indo-Pacific stratigraphies are discussed below.

Pl6. *Globorotalia miocenica-Globigerinoides fistulosus* Interval Zone (Pl6; herein defined; modified from Berggren, 1973, 1977b)

Definition: Biostratigraphic interval from the LAD of *Globorotalia miocenica* to the LAD of *Globigerinoides fistulosus*

Magnetochronologic calibration: Subchron C2r.2r (early Matuyama; just below Reunion Subchron)-Chron C2n (latest Olduvai Subchron)

Estimated age: 2.30–1.77 Ma; latest Pliocene (latest Piacenzian)

Remarks: The original definition of this zone, as the *Globigerinoides obliquus* Partial Range Zone between the LAD of *Gt. miocenica* (base) and the initial evolutionary appearance of

Globorotalia truncatulinoides (top), is redefined here to avoid the problems familiar to biostratigraphers of the ephemeral and quixotic FAD of *Gt. truncatulinoides* (Dowsett, 1988, 1990). The FAD of *Gt. truncatulinoides* has been recorded as early as the Subchron C2An.1n (Gauss) in the S. Equatorial Pacific (Dowsett, 1988, 1990), although in most subtropical regions of the Atlantic and Indo-Pacific Oceans, the FAD of *Gt. truncatulinoides* occurs generally just below the base of the Olduvai Subchron. It can also be (considerably) delayed in temperate-subantarctic regions. On the other hand, the LAD of *Globigerinoides extremus* has been reported at various levels ranging from just below the LAD of *Gt. miocenica* (early Matuyama, slightly older than the Reunion Subchron) (see discussion in Bolli and Saunders, 1985, relating to morphologic-taxonomic problems associated with recognition of *Gd. extremus*) to earliest to latest Olduvai Subchron. The latter level is taken as the true LAD of *Gd. extremus* (under ideal conditions). Srinivasan and Sinha (1992) have found that the LAD of *Gd. obliquus* sensu lato (= *Gd. extremus*) was diachronous between the SW Pacific (2.88 Ma; BKV85) and the Indian Ocean (1.8 Ma; BKV85). While this diachrony may be real, we suspect that this (apparent) diachrony is due, at least in part, to the inconsistent taxonomic usage referred to above. On the other hand, the LAD of *Gd. fistulosus* appears to be synchronous in the SW Pacific and Indian Oceans at the end of the Olduvai Subchron (Srinivasan and Sinha, 1992), consistent with its disappearance in the Atlantic (BKV85). For this reason, its LAD is used as the defining criterion for the top of Zone Pl6. The LAD of *Gd. extremus* and the FAD of *Gt. truncatulinoides* in (some) low latitude stratigraphies may serve to denote the uppermost part of Zone Pl6.

Late Pliocene zonation of the Indo-Pacific Region.—Following the closure of the Isthmus of Panama at ~3 Ma (Keigwin, 1982), planktonic foraminiferal faunas in the Atlantic and Indo-Pacific regions underwent a biogeographic provincialization that left its imprint on late Pliocene-Pleistocene faunas. Among the more characteristic features of the late Pliocene was the absence of typical Atlantic taxa *Gt. miocenica* and *Gt. exilis* in the Indo-Pacific region precluding their use in upper Pliocene (Zones Pl5 and Pl6) zonal stratigraphy there. However, alternate taxa (and zonal definitions) are suggested here following a discussion of some of the taxonomic problems that first require resolution.

Taxa that may serve as late Pliocene boundary criteria include *Globigerinoides fistulosus* and *Gt. pseudomiocenica*.

a. Jenkins and Orr (1972) proposed the *Globigerinoides fistulosus* Total Range Zone for late Pliocene tropical-subtropical stratigraphies of the eastern equatorial Pacific. This zone did not appear in subsequent (Jenkins and Srinivasan, 1986) late Neogene (sub)tropical zonal schemes nor did it figure in the tropical zonal scheme(s) adopted by Kennett and Srinivasan (1983). The magnetostratigraphically calibrated range of *Gd. fistulosus* is from the Mammoth (Subchron C2An.2r, or just below) to the top of the Olduvai Subchron (Chron C2n, or slightly above, i.e., ~3.33–1.56 Ma, using BKV85; = 3.58–1.70 Ma in CK92/95 and this paper). Srinivasan and Sinha (1992) have shown that the LAD of *Gd. extremus* is synchronous at the top of the Olduvai Subchron (C2n) in the Indian and SW Pacific Oceans. By contrast, Bolli and Premoli Silva (1973; see review in Bolli and Saunders, 1985) proposed a Pli-

ocene-Pleistocene zonal scheme for the Caribbean which included a "mid-Pliocene" *Globigerinoides trilobus fistulosus* Interval Range Subzone (of the *Gt. miocenica* Interval Range Zone) bracketed by the LAD of *Gt. margaritae evoluta* and the LAD of the nominate taxon. The LAD of *Gd. fistulosus* was shown to occur prior to that of *Gt. miocenica* and *Gt. exilis* coincident with the LAD of *Dentoglobigerina altispira* at ~2.8 Ma. A *Globorotalia exilis* Interval Subzone and *Globorotalia tosaensis* sensu stricto Interval Range Zone were inserted between the LAD of *Gd. fistulosus* and the FAD of *Gt. truncatulinoides* in this Caribbean zonation. On the other hand, Keigwin (1982) showed that *Gd. fistulosus* overlapped with *Gt. miocenica* and *Gt. exilis* at DSDP Site 502 (Colombia Basin) and overlapped with *Gt. truncatulinoides* at eastern tropical Pacific Site 503. This apparent discrepancy has its explanation in the fact that Bolli and Premoli Silva (1973) restricted their concept of the taxon *fistulosus* sensu stricto to forms on which the bizarre arborescent, staghorn and cockscomb ornament is developed on the chambers of the last whorl, while reserving the appelation *Globigerinoides trilobus* cf. *fistulosus* for presumably latest Pliocene forms with reduced chamber ornament generally restricted to the last chamber. However, most workers would include all fistulose forms in *fistulosus* sensu stricto, and indeed, Kennett and Srinivasan (1983, Plate 14, Figs. 7–9) illustrate fully fistulose forms from the early Pleistocene *Gl. tosaensis-Gt. truncatulinoides* "overlap" (= Interval) Range Zone. A similar overlap has been demonstrated in west equatorial Pacific ODP Site 806 (Chaisson and Leckie, 1993) just above the LAD of *Discoaster brouweri* (Takayama, 1993).

b. The taxon *Gt. pseudomiocenica*, the ancestor of *Gt. miocenica* in the Atlantic Ocean, occurs in Pliocene faunas of the Indo-Pacific region. It was recorded by Keigwin (1982) as part of the *Gt. menardii* sensu lato group in the Gulf of Mexico and eastern equatorial Pacific and by Jenkins and Orr (1972, Plate 23, Figs. 1–3, 7–9; cf. Bolli and Saunders, 1985, Fig. 35, 1–3 of *Gt. pseudomiocenica*) as *Gt. exilis* in the eastern equatorial Pacific. In the Red Sea, Fleisher (1974, Plate 21, Figs. 1–4) recorded *pseudomiocenica* as *Globorotalia* sp. cf. *praemiocenica*. Together with other records from DSDP and ODP cruises in which *Gt. pseudomiocenica* is couched in other menardine taxonomies, these records indicate that *pseudomiocenica* had an extensive geographic distribution in the Indo-Pacific region during the Pliocene Epoch. Fleisher (1974) recorded *Gt. praemiocenica* (= *pseudomiocenica*) from the late Pliocene *Globigerinoides ruber* Assemblage Zone of the Arabian Sea. Jenkins and Orr (1972, p. 1098) indicated that "*exilis*" (= *pseudomiocenica*) was predominantly dextrally coiled and that its extinction denoted a "relatively consistent datum in the *G. fistulosus* Zone". These observations are consistent with a study by Berggren (unpubl. data) of the stratigraphic range of the *Gt. miocenica-pseudomiocenica* group at DSDP Sites 502 and 503. While the evolutionary transition from *Gt. pseudomiocenica* to *Gt. miocenica* can be observed from Chron C2An.3n to C2An.2n (early to middle Gauss) at DSDP Site 502 (Colombia Basin), *Gt. pseudomiocenica* continued at DSDP Site 503 (eastern equatorial Pacific) until Chron C2r.2r (early Matuyama) and had its LAD at essentially the same stratigraphic level as *Gt. miocenica* in the Atlantic-Caribbean region (i.e., 2.30 Ma). It is dextrally coiled from the late Gilbert (just below the Cochiti Subchron; Chron C3n.1n) to its extinction in the early Matuyama, similar to its coiling pattern at Site 502 except that at the latter site the dextral coiling pattern began during the Thvera Subchron (Subchron C3n.4n).

This review indicates that the taxa *fistulosus* and *pseudomiocenica* were extensively distributed in the Indo-Pacific during the late Pliocene and suitable for substitution for boundary definition and/or characterization of upper Pliocene Zones Pl5 and Pl6 in these regions with the caveat that *fistulosus* is restricted to tropical-subtropical regions and less applicable to temperate regions.

Pl5. *Dentoglobigerina altispira-Globorotalia pseudomiocenica* Interval Zone (ZonePl5; Indo-Pacific; here defined)

Definition: Biostratigraphic interval from the LAD of *Dentoglobigerina altispira* to the LAD of *Globorotalia pseudomiocenica*

Magnetochronologic calibration: Subchron C2An.1r (Kaena Subchron)-Subchron C2r.2r (early Matuyama)

Estimated age: 3.09–2.30 Ma; late Pliocene (late Piacenzian)

Remarks: As defined here Zone Pl5 in the Indo-Pacific has the same temporal extent as Zone Pl5 of the Atlantic-Caribbean region; see further discussion above.

Pl6. *Globorotalia pseudomiocenica-Globigerinoides fistulosus* Interval Zone (Pl6; Indo-Pacific; here defined)

Definition: Biostratigraphic interval from LAD of *Globorotalia pseudomiocenica* to the LAD of *Globigerinoides fistulosus*

Magnetochronologic calibration: Subchron C2r.2r (early Matuyama)-Chron C2n (latest Olduvai)

Estimated age: 2.30–1.77 Ma; latest Pliocene (latest Piacenzian)

Remarks: As defined here this zone has the same temporal duration as Zone Pl6 of the Atlantic-Caribbean region. The range of *Globigerinoides fistulosus* is essentially concomitant with the combined interval of Zones Pl5 and Pl6 and may serve as an adjunct/substitute for recognizing part/all of this biostratigraphic interval in the absence of one or the other of the boundary denominating taxa. For further discussion see above.

Pleistocene Zonation of (Sub)Tropical Regions.—The inadequacy of Blow's (1969, 1979) Pleistocene Zone N23 was already pointed out by Bronnimann and Resig (1971, p. 1248) based on the general absence of *Sphaerodinella dehiscens excavata* in bottom samples in the vicinity of DSDP Leg 7 sites in the equatorial Pacific and on the subjectivity involved in distinguishing clearly between *Globigerina (vel Bolliella) calida calida* and *B. calida praecalida* (see also Bolli and Saunders, 1985, p. 176, who noted its (?local) late Pleistocene FAD in the Caribbean-Atlantic region while referring to its apparently sporadic and restricted occurrence in the early Pliocene Epoch of the Indo-Pacific). Chaproniere and others (1994) have recently correlated this event to the late Brunhes (Chron C1n) oxygen isotope stage 7 (~0.22 Ma). Bronnimann and Resig (1971) suggested substituting the FAD of *Hastigerinella adamsi* as the denominative form for a formally defined Zone N23 but were unable to provide an estimated age for this datum event, and it remains uncalibrated to either stable isotope stratigraphic zones or paleomagnetic stratigraphy to this date. The recent discovery that the FAD of *Gt. truncatulinoides* in the southeast Pacific is at the late Pliocene Gauss/Matuyama (Chron C2r/C2An) boundary with an estimated age of 2.581 Ma (Jenkins and Gamson, 1993; Langereis and others, 1994; this paper) renders Zone

N22 as defined by Blow (1969) difficult to apply globally, because in the southern hemisphere it overlaps with the upper part of Zone Pl5 and basal Pl6. Its use as a partial range or interval zone (with a different definition for its base) could be justified, and we have used the latter course in (re)defining the N22 and N23 zonal interval. We recognize but a single zone in the Pleistocene (with two subzones; Fig. 6) and designate it as Pt1 (with the Pt standing for Pleistocene in the same manner that Pl refers to the Pliocene; see above). More detailed biostratigraphic/biochronologic resolution in the Pleistocene is best achieved by using repetitive/cyclical/periodic changes such as coiling changes, (local to regional) climatically controlled taxon entries/disappearance and other local zonations.

The chronology of Pleistocene (sub)tropical planktonic foraminiferal datum events/zones is shown in Figure 13.

Pt1. *Globigerinoides fistulosus-Globorotalia truncatulinoides* Interval Zone (Pt 1; herein defined)

Definition: Biostratigraphic interval characterized by *Globorotalia truncatulinoides* between the LAD of *Globigerinoides fistulosus* and the present day

Magnetochronologic calibration: Chron C2n (Olduvai Subchron)-present day

Estimated age: 1.77 Ma to present time; Pleistocene (including Holocene)

Remarks: The LAD of *Globigerinoides fistulosus* has been found to occur at the top of the Olduvai (C2n) Subchron in both Atlantic and Pacific (sub)tropical sites at ~1.77 Ma. *Gt. trun-*

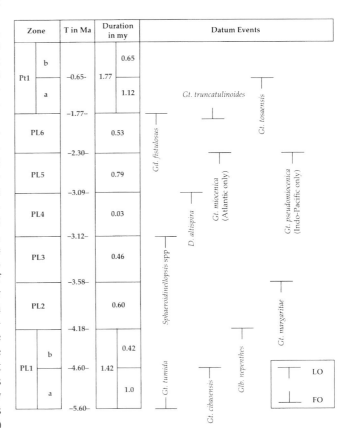

FIG. 13.—Chronology of Pliocene-Pleistocene (sub)tropical planktonic foraminiferal zones.

catulinoides characteristically appeared in northern hemisphere (sub)tropical and transitional geographies near the Chron C2n/C2r boundary (~2 Ma) and is a widespread element in Pleistocene faunas.

Pt1a. *Globigerinoides fistulosus-Globorotalia tosaensis* Interval Subzone (Pt1a; herein defined)

Definition: Biostratigraphic interval between the LAD of *Globigerinoides fistulosus* and the LAD of *Globorotalia tosaensis*

Magnetostratigraphically calibrated range: Chron C2n (Olduvai Subchron)-Chron C1n (early Brunhes)

Estimated age: 1.77–0.65 Ma; early to earliest middle Pleistocene

Remarks: The early/middle Pleistocene LAD of *Gt. tosaensis* is used as the denotative criterion for this subzone. The association of *Gt. tosaensis* and its descendant *Gt. truncatulinoides* following the FAD of the latter was used to denote a (predominantly) lower Pleistocene cool subtropical-temperate zone by Srinivasan and Kennett (1981b).

Pt1b. *Globorotalia truncatulinoides* Partial Range Subzone (Pt1b; Srinivasan and Kennett, 1981b)

Definition: Biostratigraphic interval characterized by the partial range of the nominate taxon following the LAD of *Globorotalia tosaensis*

Magnetostratigraphically calibrated range: Chron C1n to present day (early-late Brunhes)

Estimated age: 0.65 Ma to present time; early Pleistocene-Holocene

Remarks: We use this subzone in the same sense as that defined by Srinivasan and Kennett (1981b). Subdivision of the middle and late Pleistocene Period may be achieved by using several "datum events" recently calibrated to the magnetochronologic scale as a result of ODP drilling in the Atlantic and Indo-Pacific oceans, although the use of some of these events may require further verification in different biogeographies.

Miocene Zonation of Temperate-Transitional Regions.—Miocene temperate-transitional planktonic foraminiferal zonal schemes have been developed for the Indo-Pacific region by Jenkins (1967, 1975), Srinivasan and Kennett (1981a, b), Kennett and Srinivasan (1983); by Berggren (1973, 1977a, b, 1992) and Berggren and others (1983a) predominantly for Southern Hemisphere (predominantly Atlantic) stratigraphies, whereas Northern Hemisphere zonal stratigraphies have been developed/discussed by, among others, Berggren (1972) and Poore and Berggren (1975) for the North Atlantic, and Keller (1979a-c, 1980a, b, 1981) for the North Pacific. In this section, we have attempted to combine elements of these biostratigraphic schemes into a unified zonation which is broadly applicable over most of these areas. The notation Mt is used to denote application to temperate-transitional regions; the notation M was used earlier by Berggren (1977b) for this zonal scheme.

Mt1. *Globorotalia kugleri* Total Range Zone (Bolli, 1957b; Mt1; Berggren in Berggren and others, 1983a; = N4 of Berggren and Miller, 1988)

Definition: Total range of the nominate taxon

Magnetochronologic calibration: Subchron C6Cn.2n(o)-Chron C6Ar (mid)

Estimated Age: 23.8–21.5 Ma; early Aquitanian

Remarks: This zone (and its subzonal subdivision) is the same as the *Gt. kugleri* Total Range Zone (M1) described above under the (sub)tropical region(s) inasmuch as the nominate taxon(a) of the (sub)zones have been found to occur in temperate-transitional areas as well (see also Srinivasan and Kennett, 1981a, b; Kennett and Srinivasan, 1983, 1984). Further details on this zone are provided under the description of this zone in the section on (sub)tropical zones described above.

Mt2. *Globorotalia incognita/Globorotalia semivera* Partial Range Zone (Mt2; Berggren, in Berggren and others, 1983a)

Definition: The partial range of the nominate taxa between the LAD of *Globorotalia kugleri* and the FAD of *Globorotalia praescitula*

Magnetochronologic calibration: Chron C6Ar (mid)-Chron C5En (mid)

Estimated age: 21.5–18.5 Ma; early Miocene (Burdigalian)

Remarks: The FAD of *Gt. praescitula* has been calibrated to mid-Chron C5En at North Atlantic (42°N) DSDP Site 608 (Miller and others, 1991) and Southern Indian Ocean-Subantarctic (58°S) ODP Sites 748 and 751 (Berggren, 1992) providing a reliable, regional magnetobiostratigraphic correlation datum.

Mt3. *Globorotalia praescitula-Globorotalia miozea* Interval Zone (Mt3; herein redefined from Berggren, 1992, as informal zone NK2 of Kerguelen Plateau)

Definition: Biostratigraphic interval between the FAD of the nominate taxon to the FAD of its descendant *Globorotalia miozea*

Magnetochronologic calibration: Chron C5En (mid)-Chron C5Cn.3n (early)

Estimated age: 18.5–16.7 Ma; early Miocene (Burdigalian)

Remarks: This zone was originally defined (informally) in the subantarctic region of the southern Indian Ocean (Berggren, 1992). The widespread occurrence of both of the nominate taxa in transitional and subantarctic biostratigraphies make them suitable as zonal boundary markers and as defined here the zone corresponds exactly with (informal) zone AN3 (see below) of the Subantarctic region. This zone corresponds in part to the *Globorotalia zealandica-Globorotalia pseudomiozea* (M3) Zone of the South Atlantic (Berggren, 1977a) whose base was correlated with a level within Chron C5Dn. However, continuing problems with determining the precise relationships between, and the sequential appearance of *Gt. zealandica* and *Gt. praescitula* (see Scott and others, 1990) have led us to abandon that zone in favor of a zone based on the less equivocal FAD of *G. praescitula*.

Mt4. *Globorotalia miozea* Partial Range Zone (Mt4; = Zone M4 of Berggren, 1977b)

Definition: Biostratigraphic interval between the FAD of *Globorotalia miozea* and the FAD of *Praeorbulina sicana*

Magnetochronologic calibration: Subchron C5Cn.3n (early)-Subchron C5Cn.2n

Estimated age: 16.7–16.4 Ma; latest early Miocene (latest Burdigalian)

Remarks: This zone is retained as originally defined (Berggren, 1977b) as the partial range of *miozea* between its FAD and that of *sicana*. This zone corresponds to temperate/transitional Zone M4 of Berggren (1977b) and to the *Globigerinoides bisphericus* Partial Range Subzone (M4b) of (sub)tropical regions (= upper part of Zone N7 of Blow, 1969, 1979).

Mt5. *Praeorbulina sicana-Orbulina suturalis* Interval Zone (Mt5; = combined Zones M5 and M6 of Berggren, 1977b)

Definition: Biostratigraphic interval between the FAD of *Praeorbulina sicana* and the FAD of *Orbulina suturalis*

Magnetochronologic calibration: Subchron C5Cn.2n-Subchron C5Bn.2n

Estimated age: 16.4–15.1 Ma; middle Miocene (Langhian)

Remarks: This zone corresponds to Zones M5 and M6 of Berggren (1977b) of the temperate/transitional regions and to Zone M5 of (sub)tropical regions (= Zone N8 of Blow, 1969, 1979). Considering the relatively brief interval between the estimated ages of the FAD of *Pr. sicana* (16.4 Ma) and the FAD of *Pr. glomerosa* (16.1 Ma), recognition of these discrete biostratigraphic events may not be uniformly possible. For those instances where they are, we propose a two-fold subdivision of this biostratigraphical interval identical to the (sub)tropical regions.

Mt5a. *Praeorbulina sicana-Praeorbulina glomerosa glomerosa* Interval Subzone (Mt5a; = Zone M5 of Berggren, 1977b, emended)

Definition: Biostratigraphic interval between the FAD of *Praeorbulina sicana* and the FAD of *Praeorbulina glomerosa glomerosa*

Magnetochronologic calibration: Subchron C5Cn.2n-Subchron C5Cn.1n

Estimated age: 16.4–16.1 Ma; earliest middle Miocene (earliest Langhian)

Remarks: As originally defined (Berggren, 1977b), this (sub)zone represented the partial range of two taxa: *Pr. sicana* and *Gt. miozea* between the FAD of *sicana* and the FAD of *glomerosa* sensu stricto. We restrict this here simply to the interval between the same criteria.

This subzone is correlative with and identical to Subzone M5a (*Pr. sicana—Pr. glomerosa glomerosa* Interval Range Subzone) of the (sub)tropical region (= upper part of Zone N8 of Blow, 1969, 1979) (see above).

Mt5b. *Praeorbulina glomerosa glomerosa-Orbulina suturalis* Interval Subzone (Mt5b; = emendation of Zone M6 of Berggren, 1977b)

Definition: Biostratigraphic interval between the FAD of *Praeorbulina glomerosa glomerosa* and the FAD of *Orbulina suturalis*

Magnetochronologic calibration: Subchron C5Cn.1n-Subchron C5Bn.2n

Estimated age: 16.1–15.1 Ma; middle Miocene (Langhian)

Remarks: As originally defined (Berggren, 1977b) with the same bounding criteria this was a partial range zone of two taxa: *Pr. sicana* and *Pr. glomerosa glomerosa*. It is emended here to an (more appropriate) interval range subzone. This zone corresponds precisely to (sub)tropical Subzone M5b (see above) and is correlative with Zone M6 of Berggren (1977b) and the upper part of Zone N8 (Blow, 1969, 1979). The occurrence of the praeorbulinid-orbulinid bioseries at relatively high latitudes reflects the global warming trend (and expansion of low latitude warm water masses) during the latest early and earliest middle Miocene.

Mt6. *Orbulina suturalis/Globorotalia peripheroronda* Concurrent Range Zone (Mt6; = Zone M7 of Berggren, 1977b)

Definition: Concurrent range of the nominate taxa between the FAD of *Orbulina suturalis* and the LAD of *Globorotalia peripheroronda*

Magnetochronologic calibration: Subchron C5Bn.2n-Chron C5ACn (estimated)

Estimated age: 15.1–14.0 Ma; middle Miocene (latest Langhian-early Serravallian)

Remarks: The LAD of *Gt. peripheroronda* is estimated here to correspond to Chron C5ACn (in the absence of a direct magnetostratigraphic correlation) based on the observed overlap in the stratigraphic ranges of *Gt. peripheroronda* and *Gt. peripheroacuta* (Bolli and Saunders, 1985; pers. observ., WAB) and the fact that the FAD of the latter taxon is calibrated to Chron C5Bn(y). The Mt6/Mt7 zonal boundary is thus somewhat stratigraphically higher (younger) than the (sub)tropical M5/M6 zonal boundary.

Mt7. *Globorotalia peripheroronda-Globoturborotalita nepenthes* Interval Zone (Mt7; here defined)

Definition: Biostratigraphic interval from LAD of *Globorotalia peripheroronda* to the FAD of *Globoturborotalita nepenthes*

Magnetochronologic calibration: Chron C5ACn (estimated)-Subchron C5r.3r

Estimated age: 14.0–11.8 Ma; middle Miocene (Serravallian)

Remarks: This biostratigraphic interval corresponds to the *Globigerina druryi* (M8) Partial Range Zone of Berggren (1977b) and is characterized by the (sporadic) presence of that taxon, *Globoturborotalita decoraperta*, *Globigerina woodi*, *Neogloboquadrina mayeri* and the presence of, and transition between, *Globorotalia miozea* and *Globorotalia conoidea*, among others. In view of the sporadic occurrence of *G. duryi* in this interval, it is thought more appropriate to designate this an interval range zone.

Mt8. *Globoturborotalita nepenthes/Neogloboquadrina mayeri* Concurrent Range Zone (Mt 8; = Zone N14 of Blow, 1969, 1979)

Definition: Concurrent range of the nominate taxa between the FAD of *Globoturboralita nepenthes* and the LAD of *Neogloboquadina mayeri*

Magnetochronologic calibration: Subchron C5r.3r-Subchron C5r.2r

Estimated age: 11.8–11.4 Ma; middle Miocene (Serravallian)

Remarks: This zone is generally applicable to (sub)tropical and transitional biostratigraphies, although the lowest (earliest) occurrence of *Glb. nepenthes* may be delayed in some localities (e.g., Srinivasan and Kennett, 1981b). Miller and others (1994) argue that the LAD of *N. mayeri* is latitudinally diachronous, persisting to the earliest part of Chron C5n at mid-latitudes (DSDP Site 563, among others), in the North Atlantic, although an unconformity at this (and other localities) could equally well have juxtaposed the LAD of *N. mayeri* and FAD of *N. acostaensis* (Aubry, 1993a; Berggren, 1993).

Mt9. *Neogloboquadrina mayeri-Globorotalia conomiozea* Interval Zone herein defined; Mt9)

Definition: Biostratigraphic interval between the LAD of *Neogloboquadrina mayeri* and the FAD of *Globorotalia conomiozea*

Magnetochronologic calibration: Subchron C5r.1r-Subchron C3Br.1r

Estimated age: 11.4–7.12 Ma; latest middle to earliest late Miocene (latest Serravallian-earliest Messinian)

Remarks: The rarity or absence of *N. acostaensis* in transitional biostratigraphies precludes its application outside of (sub)tropical regions. There have been numerous attempts at zonation of the upper Miocene Series of temperate-cool subtropical regions. Kennett (1973) used the sequential partial ranges of *Neogloboquadrina continuosa*, *Glb. nepenthes* and *Gt. conomiozea* following the extinction of *N. mayeri* in the Southwest Pacific, while Jenkins (1967, 1971, 1975, 1993a) and Berggren (1977a; Berggren and others, 1983a) and Poore and Berggren (1975) utilized the presence of members of the *Gt. miotumida* (including *Gt. conoidea*)-*Gt. conomiozea* complex in developing an upper Miocene zonation applicable to the Southwest Pacific, South Atlantic and North Atlantic, respectively. We retain these forms as definitive and denotative of two upper Miocene zones which would appear to find application in both northern and southern ocean late Miocene biogeographies. Characteristic elements of this biostratigraphic interval include, among others, *Glb. nepenthes*, *Gl. woodi*, *Gt. conoidea*, *Gt. lenguaensis*, *Gt. menardii* sensu lato, *Gt. scitula* and *Neogloboquadrina nympha* (a transitional-subantarctic isomorph of *N. acostaensis* having a considerably earlier FAD than *N. acostaensis*). This zone corresponds to the *Globorotalia miotumida* Zone of Jenkins (1967, et seq.) and to the *Globorotalia paralenguaensis-Neogloboquadrina continuosa* (M10) Partial Range Zone and the succeeding *Globorotalia miozea-Globorotalia conoidea* (M11) Partial Range Zone (Berggren, 1977a, b) which used the FAD of *N. acostaensis* as the defining criterion separating the two at DSDP Site 516. The Rio Grande Rise is located in a region which was alternately influenced by subtropical and transitional waters which accounts for the alternating presence of (subtropical) elements as the *fohsi* group and the *miozea-conoidea-conomiozea* group (Berggren and others, 1983a). *Globorotalia paralenguaensis* is now included in the concept of *Gt. lenguaensis* and the sporadic distribution (and apparently premature LAD) of this taxon and of *N. acostaensis* in transitional biostratigraphies suggests that a more appropriate zonation be found for this interval.

Mt10. *Globorotalia conomiozea/Globorotalia mediterranea-Globorotalia sphericomiozea* Interval Zone (Mt10; here defined)

Definition: Biostratigraphic interval between the FAD of *Globorotalia conomiozea* and/or *Globorotalia mediterranea* and the FAD of *Globorotalia sphericomiozea*

Magnetochronologic calibration: Subchron C3Br.1r-Chron C3r (mid)

Estimated age: 7.12–5.6 Ma; late Miocene (Messinian)

Remarks: The review by Hodell and Kennett (1986) suggests that the FAD of *G. conomiozea* (as morphometrically distinguished from its sister taxon/ancestor *Gt. conoidea* by Malmgren and Kennett, 1982; cf. Scott and others, 1990) occurred in Subchron 5n.1r (BKV85; = Subchron C3An.1r of CK92/95) at South Atlantic DSDP Site 588 which appears at first glance to have been consistent with the record in Subchron C3An.1r in Crete by Langereis and others, 1984; cf. BKVC85) which is above the Chron 6 (= Chron C3Ar) carbon isotope shift of Haq

and others (1980). On the other hand, a somewhat earlier initial occurrence of *Gt. conomiozea* was recorded by Kennett and Watkins (1974) in the Blind River section of New Zealand within Subchron 6.1r (= Chron C3Ar) coincident with the carbon shift (see also Loutit and Kennett, 1979; Kennett and Srinivasan, 1984; cf. recently revised magnetobiostratigraphic interpretation by Roberts and others, 1994, of the Blind River section which suggests a local FAD of *conomiozea* in Subchron C3An.1r, at ~6.2 Ma in the chronology of CK92/95 and this paper). In view of the fact that the earliest occurrence of this taxon has been used to denote the base of the Messinian Stage in the Mediterranean, this point assumes a significance beyond mere provincial correlation. The recent modification to the GPTS by Cande and Kent (1992; 1995 and this paper) shows that the earlier Chron 6 (in BKV85; = Chron C3Ar to Chron C3Br of this paper) contains two additional normal subchrons which now allows an unambiguous correlation of the FAD of *conomiozea* in the Cretan sections to the new GPTS in Subchron C3Br.1r (inadvertently attributed to Chron C3Bn.1r in Krijgsman and others, 1994) with an estimated age of 6.92 Ma in CK92 and 7.12 Ma in the chronology adopted here. The FAD of *Gt. conomiozea*, nominate taxon for the boundary between Zones Mt10 and Mt11, is correlated with Subchron C3Br.1r at 7.12 Ma and equated with the Tortonian/Messinian boundary.

The initial appearance of *Globorotalia sphericomiozea* at transitional South Atlantic DSDP Site 588 has been found to coincide with that of *Gt. tumida* (Hodell and Kennett, 1986) and to be only nominally younger than its FAD at more southerly DSDP Sites 590 and 284. We have also observed this form (i.e., *sphericomiozea*) in high southern latitude ODP Site 747 at about the same level (Berggren, 1992). This zone corresponds closely to the *Globorotalia conomiozea-Globorotalia mediterranea* (M12) Partial Range Zone (Berggren, 1977a, b) which used the FAD of *Gt. cibaoensis* as its upper boundary. However, it would appear that the FAD of *Gt. cibaoensis* is somewhat older than that suggested by Berggren (1977a, b) on the Rio Grande Rise, thus necessitating a revised definition of this zone. This zone corresponds (in greater part but not in its entirety) to the *Gt. conomiozea* Zone of Jenkins (1971, et seq.) and Kennett (1973).

The chronology of early and middle-late Miocene transitional planktonic foraminiferal datum events/zones is shown in Figures 14 and 15, respectively.

Pliocene-Pleistocene Zonation of North Atlantic Temperate Region.—Weaver and Clement (1986, 1987) have reviewed the historical development of temperate-subarctic late Neogene planktonic foraminiferal zonal schemes from Berggren (1972) through the studies of Poore and Berggren (1974, 1975) and assessed these zones in terms of paleomagnetic calibrations made possible by North Atlantic DSDP Leg 94. Elements of the Pl zonation of Berggren (1973, 1977b) were retained for the lower Pliocene Series, but for the mid-upper Pliocene Series, other taxa were used to reflect the biogeographic differentiation which occurred as a result of significant high-latitude cooling. This zonal scheme is adopted here for temperate-subantarctic stratigraphies of the North Atlantic. It should be pointed out in passing that the zones of Weaver and Clement (1986) are interval (rather than partial) range zones as defined by the ISSC (Hedberg, 1976; recently revised in Salvador, 1994).

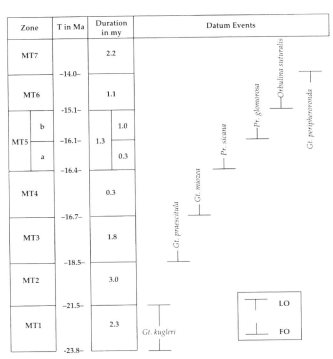

FIG. 14.—Chronology of early Miocene transitional planktonic foraminiferal zones.

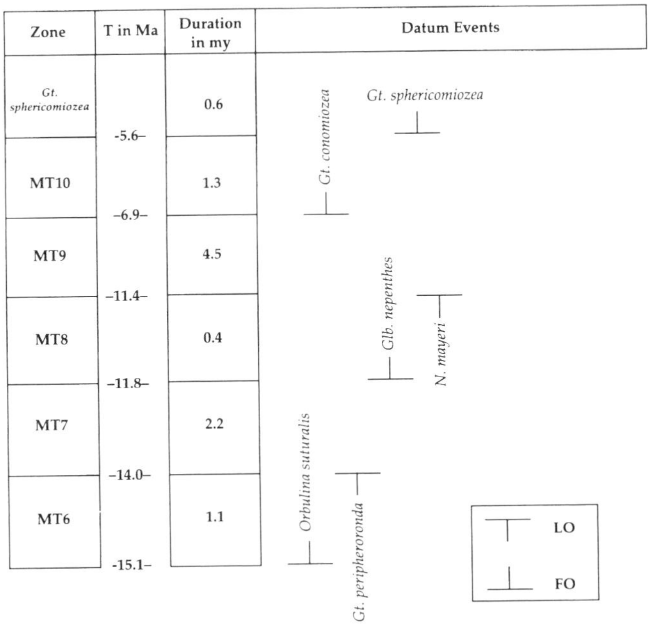

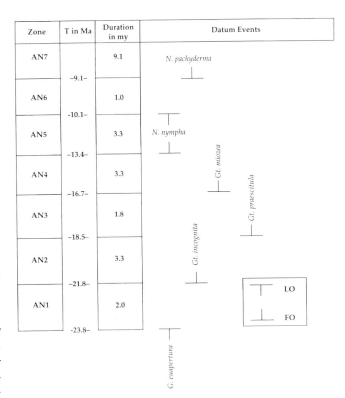

FIG. 15.—Chronology of middle and late Miocene transitional planktonic foraminiferal zones.

Neogene zonation of Austral-(Sub)Antarctic Regions.—A seven-fold Neogene biostratigraphic zonation was developed for the austral-subantarctic region of the Kerguelen Plateau (Berggren, 1992; see also Jenkins, 1993a). Correlation of Zones An1–4 (lower to lower middle Miocene) with contemporary temperate-transitional biostratigraphies is relatively straightforward owing to comparable (if fewer) elements in common between the two regions and direct, independent, magnetostratigraphic correlation in both cases. With increased middle Miocene high-latitude cooling, faunal diversity was reduced, endemism enhanced and biostratigraphic resolution decreased. Indeed, the introduction of *Neogloboquadrina pachyderma* to (sub)antarctic region(s) at ~9 Ma was the last major, innovative event in planktonic foraminiferal biostratigraphy and the circumantarctic region has been dominated by virtually monospecific faunas of this taxon since then (with minor contributions from *Globigerina bulloides* and several small sized taxa). The chronology of austral/(sub)antarctic planktonic foraminiferal datum events/zones is shown in Figure 16.

Planktonic Foraminiferal Magnetobiochronology Paleogene Period.—

Paleocene Epoch.—Over 30 datum events (Table 8) have been recognized in the Paleocene Epoch including two (LAD of *M. velascoensis* and FAD of *Gl. australiformis*) which serve to denote the approximate position of the Paleocene/Eocene boundary at low and high austral latitudes, respectively. These two datum events are, however, not correlative; the FAD of *Globanomalina australiformis* has been linked in high southern latitude stratigraphies directly with the major $\delta^{13}C$ spike that

Pliocene-Pleistocene Zonation of Transitional Austral (Southern Hemisphere) Regions.—The zonal scheme of Kennett (1973) and Srinivasan and Kennett (1981a, b) is used for late Neogene (Pliocene-Pleistocene) transitional biostratigraphies. Jenkins (1993a) has recently presented a review of mid- and high-latitude planktonic foraminiferal biostratigraphy of this region.

FIG. 16.—Chronology of subantarctic Neogene planktonic foraminiferal zones.

occurs in mid Zone NP9 and in early Chron C24r (Stott and Kennett, 1990), whereas the LAD of *Morozovella acuta* and *M. velascoensis* have been shown to occur within Zone NP10 and the mid part of Chron C24r (Table 8). In most instances where direct magnetobiostratigraphic correlation of a datum event has not been achieved, the relative position within a zone in which other elements are directly calibrated by magnetobiostratigraphy can be determined so that a reasonable estimate of position in a magnetic polarity chron and a corresponding age estimate can be made. The revised magnetochronology used here (based on CK92/95) has the effect, essentially, of shifting the estimated ages of Paleocene datum levels in BKF85 to younger ages by ~1.5–2.7 my.

Eocene Epoch.—Nearly 40 datum levels (excluding the two which serve to delineate the approximate position of the Paleocene/Eocene boundary) have been recognized in the Eocene Epoch (Table 9), many of which serve to denote the 12 zonal (and 2 subzonal) boundaries of this epoch. Of particular interest is a series of new and revised calibrations of early Eocene datums afforded by a virtually complete stratigraphic sequence from Chron C23n to mid-Chron C24r in NE Atlantic DSDP Hole 550 (Berggren and Aubry, 1995). One problem associated with magnetobiostratigraphic calibrations across the Paleocene/Eocene boundary interval is the presence of hiatuses of brief duration which tend to concatenate datum levels; the problem is discussed elsewhere in more detail (Aubry and others, 1995; Berggren and Aubry, 1995).

Middle Eocene datum levels remain essentially those delineated in BKF85, the lack of new information attesting to the difficulty of obtaining good magnetobiostratigraphic records in the (predominantly) low latitude chert bedeviled sections, although some useful datums have been delineated at high austral latitudes for acarininid extinctions and that of *Subbotina linaperta.*

A relatively high-resolution, although not completely unequivocal, succession of datum events has been delineated for the late Eocene Epoch based on systematic, integrated studies in the Apennines and tabulated originally by Berggren and others (1992; see also Table 9). It should be noted that the LAD of *Turborotalia cerroazulensis* sensu lato is used in preference to that of *Hantkenina* spp. to denote the P17/P18 zonal boundary (Berggren and Miller, 1988) owing to the greater preservational potential of the former. Whether there is a real sequential extinction of these two taxa remains a moot point; recent studies by Leckie and others(1993) suggest that they disappear simultaneously.

The revised magnetochronology of CK92/95 (and that adopted in this paper) has the effect of shifting Eocene datum levels to younger ages by ~2.5–3 my from those suggested in BKF85. The relative age spacing is similar; only the numerical values have been altered.

Oligocene Epoch.—Eighteen datum levels have been identified in the Oligocene Epoch (Table 10), their relative scarcity reflecting the significant paleoceanographic/paleoclimatic changes which occurred during this epoch. Most of the datum levels have been recognized previously (BKF85), but the calibrations are refined and/or modified here owing to new data available since 1985. However, several significant datum levels have been identified in high austral latitudes which enables relatively precise interhemispherical correlation between the

"standard" low (Berggren and Miller, 1988) and high austral (Stott and Kennett, 1990) latitude biostratigraphies.

The magnetochronology of CK92/95 has the effect of shifting the Oligocene datum levels recognized in BKF85 from ~3 my (early Oligocene) to 1.5 my (mid-early late Oligocene) younger; at the Oligocene/Miocene boundary, values are essentially the same because a comparable value was used to anchor the time scale at the older end of Subchron C6C.2n (CK92/95).

Neogene Period.—

Miocene Epoch.—About 70 planktonic foraminiferal datum levels have been identified in the Miocene Epoch (Tables 11 to 13), many of which were already recognized in BKV85. We have (re)calibrated these datum levels to the revised magnetochronology of CK92/95 and added several more, generally from high austral latitudes. In some instances, revisions/reinterpretations of magnetostratigraphy has resulted in revised age estimates for several datum levels as well. In general, we have found our current analysis and/or evaluation of new (post 1985) data to be quite consistent with earlier interpretations and/or calibrations. The major difference between this work and BKVC85 is the firmer documentation of the regional correlation of the biogeographically overlapping zonal schemes adopted here within a more precise magnetochronologic framework. The relatively high degree of stratigraphic/chronologic resolution achievable in (sub)tropical and transitional regions should be contrasted with the significantly lower resolution in high-latitude (sub)polar latitudes. This latter fact reflects the increased, and accelerated, late Neogene cooling at high latitudes with concomitant replacement of calcareous microfaunas and floras by biosiliceous (diatom, radiolarians) components.

We have recognized some 22 datum events spanning the 7.4-my interval of the early Miocene (23.8–16.4 Ma; Table 11), or an average of about 2.9 events /1 my. Zones M1 (2.24 my), M2 (2.75 my) and M3 (1.50 my) can be contrasted with the much shorter Zones M4 and M5 (and indeed middle Miocene Zones M6, M8–10 as well), reflecting the late early to early middle Miocene global warming trend and concomitant flurry of speciation events in (sub)tropical environments. This allows fine-scaled biostratigraphic subdivision of the upper lower and middle Miocene Series. Of some consternation is the lack of a direct magnetobiostratigraphic correlation of the FAD of *Globigerinatella insueta* (which defines the base of Zone M3).

Some 25 datum eventslevels have been identified in the middle Miocene Series (16.4–11.2 Ma; Table 12) interval of ~5 my or an average of 4.8 events/1 my which provides the highest degree of biostratigraphic resolution for the Cenozoic Era, except for the Pliocene Epoch where nearly 45 different types of biostratigraphic events spread over the ~3-my extent of the Pliocene Epoch provides some 15 datum events/1 my (Berggren and others, 1995). The early middle Miocene warming trend followed by a middle middle Miocene cooling trend was probably responsible for the high degree of middle Miocene biostratigraphic resolution. This may be contrasted and compared with the accelerated (and punctuated) cooling trends of the Pliocene Epoch responsible for the relatively high number of LADs of numerous (predominantly Miocene) taxa, the various biogeographic immigration/disappearance events, and FADs of several taxa.

TABLE 10.—OLIGOCENE PLANKTONIC FORAMINIFERAL MAGNETOBIOCHRONOLOGY.
OLIGOCENE

Datum	FO	LO	Paleomagnetic Chron	Age in Ma	Reference	Remarks
1. *Globorotalia kugleri*	X		C6Cn.2n	23.8	1	= Oligocene/Miocene boundary; Holes 516F, 558, 563 (ref. 1)
2. *Globigerina euapertura*		X	C6Cn.2n	23.8	2	Recorded near top C6Cn in Hole 748B and near base C6Cn in Holes 747A AND 747B (ref. 2)
3. *Globorotalia mendacis*	X		C6Cn.2n	23.8	1	Hole 516F (ref. 1)
4. *Tenuitella gemma*		X	C6Cr	24.3	5	Hole 747A (ref. 5)
5. *Globigerinoides primordius* (common occurrence)	X		C6Cr	24.3	1	Hole 516F (ref. 1)
6. *Globorotalia pseudokugleri*	X		C8n	25.9	3	Recorded in C8n in Hole 803D (ref. 3) AND in C7n in Hole 628A (ref. 3)
7. *Globigerinoides primordius*	X		C8r	26.7	3	Hole 628A (ref. 3); recorded in C7n (ref. 1).
8. *Paragloborotalia opima*		X	C9n$_{(y)}$	27.1	1	= P21/P22 boundary in Berggren AND Miller (1988). Recorded between C9n AND C8n in Hole 516F (ref. 1); individuals become significantly smaller above C9n AND are not typical of *P. opima* sensu stricto in Hole 558 (ref. 1; see also discussion in Berggren AND Miller, 1988, p. 377).
9. *Globigerina labiacrassata*		X	C9n$_{(y)}$	27.1	2, 4	= AP15/AP16 boundary in Berggren (1992); recorded at top C9n in Holes 747A and 748B (ref. 2), 689B and 690B (ref. 4) and in C8n$_{(o)}$ in Hole 747A (ref. 5).
10. *Globigerinita boweni*	X		C10n (mid)	28.5	5	Hole 747A (ref. 5)
11. *Chiloguembelina cubensis* (common occurrence)		X	C10n (mid)	28.5	1,2,5	= P21a/b boundary in Berggren and Miller (1988). Recorded in mid-part of C10n in Holes 747A (ref. 1, 5), 749 (ref. 2) and in mid-part of concatenated C9n AND C10n in Hole 748B (ref. 2) and associated approximately with C10n in Holes 689B and 690B (ref. 4). Whether this datum refers to extinction or strong reduction in numbers remains a moot point; records of discontinuous presence in reduced numbers of this taxon into basal Miocene levels continue (ref. 3).
12. *Globigerina angulisuturalis*	X		C11n$_{(y)}$	29.4	1,3	Berggren and Miller (1988); Holes 516 and 558 (ref. 1), 628A and 803C (ref. 3).
13. *Tenuitellinata juvenilis*	X		(C11n)	(29.7)	5	Interval of no polarity data between normal polarity intervals interpreted as parts of C11 in Hole 747A (ref. 5).
14. *Subbotina angiporoides*		X	C11n$_{(o)}$	30.0	1,2,4	= P19/P20 boundary in Berggren and Miller (1988) and AP13/AP14a zonal boundary in Stott and Kennett (1990). Recorded in early part of C11n in Holes 516F (ref. 1), 748B (ref. 2), and 689B and 690B (ref. 4).
15. *"Turborotalia" ampliapertura*		X	C11r	30.3	1,6	Recorded in C11r in Holes 516F and 558 (ref. 1) and St. Stephen's Quarry, Alabama borehole (ref. 4).
16. *Paragloborotalia opima opima*	X		C12n (mid)	30.6	1	Hole 558 and elsewhere (ref. 1).
17. *Pseudohastigerina* spp.		X	C11r	32.0	1,6	Holes 516F, 558, 563 (ref. 1) and St. Stephen's Quarry, Alabama borehole (ref. 6).
18. *Cassigerinella chipolensis*	X		C13r	33.65	6	St. Stephen's Quarry, Alabama borehole (ref. 6).

[1] Berggren and others, 1985b
[2] Berggren, 1992
[3] Leckie and others, 1993
[4] Stott and Kennett, 1990
[5] Li and others, 1992
[6] Miller and others, 1993

Some 23 datum events have been recognized in the 5.9-my interval of the late Miocene (ll.2–5.3 Ma; Table 13), or an average of ~3.9 events/1 my. Of particular significance is the replacement of the *Gt. merotumida-plesiotumida* group by *Gt. lenguaensis* in subdividing the upper Miocene Series (see discussion above) of (sub)tropical and transitional regions which provides more confident calibration to the GPTS. The joint occurrence of the FAD of *Gt. sphericomiozea* and *Gt. tumida* at DSDP Hole 519 has provided a means of regional correlation between (sub)tropical and transitional regions at the Miocene/Pliocene boundary (5.2 Ma in the chronology of CK92; 5.3 Ma in CK92/95 and herein; see discussion below under Chronostratigraphy).

Pliocene-Pleistocene Epochs.—A comprehensive compilation has recently been made of some 45 Pliocene and 10 Pleistocene planktonic foraminiferal datum events in connection with a larger review of the current status of late Neogene (Pliocene-Pleistocene) astro-and magnetobiochronology (Berggren and others, 1995) and the interested reader is referred to that source for additional information.

Calcareous Nannofossil Magnetobiochronology

Paleogene Period.—

Progress in Paleogene calcareous nannofossil magnetochronology has been uneven since the publication of Berggren and others (1985b). No synthesis of post 1985 magnetobiostratigraphic correlation has been published for the Cenozoic. This includes the magnetobiochronology of Wei and Peleo-Alampay (1993) which is mostly a conversion of the biochronology in Berggren and others (1985b) to the (obsolete) magnetochronology of Cande and Kent (1992). Little attention has been paid to Paleocene biochronology, except for the Southern Ocean (Wei, 1992), whereas efforts have been concentrated on late middle Eocene to Oligocene biochronology (Premoli Silva and others, 1988a; Wei, 1991; Wei and Wise, 1992). The compila-

TABLE 11.—EARLY MIOCENE PLANKTONIC FORAMINIFERAL MAGNETOBIOCHRONOLOGY.
EARLY MIOCENE

Datum	FAD	LAD	Paleomagnetic Chron	Age in Ma	Reference	Remarks
1. *Praeorbulina sicana*	X		C5Cn.2n (mid)	16.4	1	= M4b/c boundary.
2. *Globorotalia incognita*		X	C5Cn.2n (mid)	16.4	2	
3. *Globorotalia miozea*	X		C5Cn.3n$_{(o)}$	16.7	2	Hole 751, Kerguelen Plateau.
4. *Globorotalia birnageae*	X		C5Cn.3n$_{(o)}$	16.7	1	= M4a/b boundary; Hole 516, South Atlantic (ref. 1).
5. *Globorotalia zealandica*		X	C5Dn$_{(y)}$	17.3	3	Hole 747A, (ref. 3); observed in C5Dn$_{(o)}$ in Hole 751 (ref. 2).
6. *Globorotalia semivera*		X	C5Dn$_{(y)}$	17.3	1	
7. *Catapsydrax dissimilis*		X	C5Dn$_{(y)}$	17.3	1	= M3/M4 boundary; Holes 558, 563 North Atlantic (ref. 1); observed at top of C5Dr in Hole 608 in interval of strong dissolution (refs. 2, 3).
8. *Globoquadrina dehiscens forma spinosa*		X	C5Dr	17.9	1	Hole 516, South Atlantic (ref. 1).
9. *Globorotalia praescitula*	X		C5En	18.5	2,4	Holes 747A, 748B, 751 (ref. 2), and 608 (ref. 4); recorded in earliest part of C5En in Hole 516F (ref. 1).
10. *Globigerinatella insueta*	X		(C5En)	(18.8)	1	= M2/M3 boundary; polarity chron and age estimate inferred here (see text for discussion).
11. *Globigerinoides altiaperturus*	X		C6An$_{(y)}$	20.5	1,5	Hole 516F, South Atlantic (ref. 1); recorded in older part of C6r in Contessa Highway Section (ref. 5).
12. *Tenuitella munda*		X	C6Ar	21.4	3	Hole 747A, Kerguelen Plateau (ref. 3); magnetostratigraphy reinterpreted in ref. 6.
13. *Globorotalia kugleri*		X	C6Ar	21.5	1,5	= M1/M2 boundary; Hole 516 F, South Atlantic (ref. 1) and at base C6An in Contessa Highway Section (ref. 5).
14. *Globorotalia incognita*	X		C6Ar	21.6	2	Hole 747A (ref. 2); magnetostratigraphy reinterpreted in ref. 6.
15. *Globoturborotalita angulisuturalis*		X	C6Ar	21.6	1	Hole 516F (ref. 1); magnetic stratigraphy reinterpreted in ref. 6.
16. *Globorotalia pseudokugleri*		X	C6Ar	21.6	1	Hole 516F (ref. 1); magnetic stratigraphy reinterpreted in ref. 6.
17. *Globoquadrina dehiscens forma spinosa*	X		C6AAr.1n	22.2	1	Hole 515F (ref. 1).
18. *Globoquadrina globularis*		X	C6Bn.1r	22.8	1	Hole 516F, South Atlantic (ref. 1).
19. *Globoquadrina dehiscens*	X	X	C6Br	23.2	1	= M1a/b boundary; Holes 516F, 558, 563 (ref. 1); South Atlantic (refs. 7, 8) and Contessa Highway Section.
20. *Globigerina euapertura*		X	C6Cn.2n	23.8	2	Holes 747A, 747B (ref. 2); reported in C6Cn.1n in Hole 747A in ref. 3.
21. *Globorotalia kugleri*	X		C6Cn.2n	23.8	1	= P22/M1 boundary; Holes 558, 563 (North Atlantic) and 516F (South Atlantic) (ref. 1).

[1]Berggren and others, 1985a
[2]Berggren, 1992
[3]Li and others, 1992
[4]Miller and others, 1994

[5]Montanari and others, 1991
[6]Wright and Miller, 1992
[7]Poore and others, 1983
[8]Poore and others, 1984

tion given below (Tables 14–17) shows dramatically that the number of sections with a reliable magnetostratigraphy is extremely small, and that there are intervals such as the middle Eocene Series (Zones NP14 to NP16 in particular) where biostratigraphic events are still poorly tied to magnetochronology. It is clear that the advances in magnetobiochronology around the Eocene/Oligocene boundary (Premoli Silva and others, 1988a; Berggren and others, 1992) and more recently around the Paleocene/Eocene boundary (Aubry and others, 1995; Berggren and Aubry, 1995) have been fostered by the activities of Working Groups (in these cases IGCP Projects 174 (Terminal Eocene Events) and 308 (Paleocene/Eocene Boundary Events in Time and Space)). We expect that with the recent creation of Working Groups on Paleogene stages, additional improvements to magnetobiostratigraphic correlations in the least well documented stratigraphic intervals will be made in the next few years.

The major question when establishing a magnetobiochronologic framework concerns the temporal reliability of the paleontologic events. The view that many FADs and LADs of Cenozoic calcareous nannofossil species are unreliable owing to preservational problems or to latitudinal diachrony has been amply discussed (Backman, 1987; Dowsett, 1988; Miller and others, 1994; Monechi and Thierstein, 1985; Spencer-Cervato and others, 1994; Wei 1991, 1992; Wei and Wise, 1989, 1990). This view is based on the assumption that discrepancies in mag-

netobiostratigraphic correlations between sections reflect diachrony. We suggest instead that undeciphered unconformities in the stratigraphic record (and to a lesser extent differences in taxonomic concepts) are more likely to account for such discrepancies (see Aubry, this volume; Aubry and others, 1995), although we recognize that diachrony occurs (Aubry, 1992a, b). It is beyond the scope of this paper to discuss in detail the reliability of every Cenozoic calcareous nannofossil datum. Comments are made in Tables 14 to 17 to this effect. They reveal that claims of diachrony of small amplitude (<2 my) are poorly substantiated, at least in the Paleocene and Eocene Epochs (Tables 14, 15). Two examples taken from the Paleocene Epoch—the FAD of *Ellipsolithus macellus* and the FAD of *Heliolithus kleinpelli*—are however discussed in detail below to illustrate that stratigraphic sections must be submitted to a comprehensive examination before magnetobiostratigraphic correlations are established.

The lowest occurrence (LO) of *Ellipsolithus macellus* defines the NP3/NP4 and the CP2/CP4 zonal boundaries in the zonal schemes of Martini (1971) and Bukry (1973), respectively. This datum is generally regarded as temporally unreliable owing to the assumed susceptibilty of the coccolith towards dissolution or to environmental restriction of the species (e.g., Monechi and Thierstein, 1985; Wei and Wise, 1989). Monechi and Thierstein (1985) observed that in the Contessa and the Bottacione sections, Italy, the LO of *E. macellus* is in Chron C28r whereas in

TABLE 12.—MIDDLE MIOCENE PLANKTONIC FORAMINIFERAL MAGNETOBIOCHRONOLOGY.
MIDDLE MIOCENE

Datum	FAD	LAD	Paleomagnetic Chron	Age in Ma	Reference	Remarks
1. *Neogloboquadrina mayeri*		X	C5r.2r	11.4	1–4	See discussion in text on Late Miocene and refs. 1–4.
2. *Globoturborotalita nepenthes*	X		C5r.3r	11.8	1–4	M10/M11 boundary; occurs in North Atlantic Sites 563 (ref. 1) and 608 (ref. 3) and basal Buff Bay Formation, Jamaica (refs. 2, 4).
3. *Globorotalia panda*		X	C5r.3r	11.8	2	Hole 747A (Kerguelen Plateau, S. Indian Area); magnetic stratigraphy reinterpreted in ref. 5.
4. *Globorotalia praescitula*		X	C5An.1n	11.9	6	Hole 747A; see remarks in item 3 above.
5. *Globorotalia fohsi robusta*		X	C5An.1n	11.9	1	M9/M10 boundary; Hole 563 (ref. 1).
6. *Globorotalia fohsi lobata*		X	C5An.1r	12.1	1,5	= M9a/b boundary; based on reinterpretation of paleomagnetics in Hole 563, based in turn on stable isotope studies at Kerguelen Plateau Site 747 and North Atlantic Hole 608 (ref. 5).
7. *Globorotalia fohsi robusta*	X		C5An.2n	12.3	1,6	See revised interpretation of magnetostratigraphy of 563 based on stable isotope studies in ref. 5.
8. *Tenuitella clemenciae*		X	C4An.2N	12.3	7	Hole 747A.
9. *Tenuitella minutissima*		X	C5An.2n	12.3	7	Hole 747A.
10. *Tenuitella pseudoedita*		X	C5An.2n	12.3	7	Hole 747A.
11. *Tenuitella selleyi*		X	C5An.2n	12.3	7	Hole 747A.
12. *Globorotalia fohsi lobata*	X		C5Ar.1r	12.5		= M8/M9 boundary; revised interpretation of magnetostratigraphy of Hole 563 (ref. 5) indicates FAD of *Gt. fohsi lobata* in C5Ar.1r (ref. 5) rather than C5Ar (ref. 1).
13. *Globorotalia fohsi* s.str.	X		C5Ar.1n–2n (undiff.)	12.7	1,5	= M7/M8 boundary; reinterpretation of magnetostratigraphy at Hole 563 based on stable isotope studies at 747A and 608 (see ref. 5).
14. *Globorotalia praefohsi*	X		C5Ar.1n–2n (undiff.)	12.7	1,5	See remarks, item 13 above.
15. *Neogloboquadrina nympha*	X		C5ABn	13.4	5,6	Holes 747A, 751, Kerguelen Plateau (ref. 6). In Hole 608 occurs in sample assigned to C5AAr, just above C5ABn (ref. 5).
16. *Globorotalia peripheroronda*		X	C5ADn$_{(o)}$	14.6	1	
17. *Globorotalia peripheroacuta*	X		C5Bn.1n	14.8	1	= M6/M7 boundary.
18. *Praeorbulina sicana*		X	C5Bn.1n	14.8	1	
19. *Praeorbulina glomerosa* s. str.		X	C5Bn.1n	14.8	1	
20. *Orbulina suturalis*	X		C5Bn.2n	15.1	1	= N5/N6 boundary.
21. *Globorotalia miozea*		X	C5Br$_{(o)}$	15.9	6	Hole 747A; recorded in C5B.1n in ref. 7.
22. *Praeorbulina circularis*	X		C5Cn.1n$_{(y)}$	16.0	1	
23. *Praeorbulina glomerosa* s.str.	X		C5Cn.1n$_{(y)}$	16.10	1	= M4/M5 boundary.
24. *Globigerinoides diminutus*	X		(C5Cn.2n)	(16.1)	1	See remarks in ref. 1.
25. *Praeorbulina curva*	X		C5Cn (undiff., mid part)	16.3	1	
26. *Praeorbulina sicana*	X		C5Cn.2n (mid)	16.4	1	= Subzone M4b/c boundary.
27. *Globorotalia incognita*		X	C5Cn.2n (mid)	16.4	6	Hole 747C, Kerguelen Plateau.

[1]Berggren and others, 1985a
[2]Berggren, 1993
[3]Miller and others, 1991
[4]Miller and others, 1994
[5]Wright and Miller, 1992
[6]Berggren, 1992
[7]Li and others, 1992

DSDP Holes 577 and 577A on the Shatsky Rise (NW equatorial Pacific), and in DSDP Hole 527 on the Walvis Ridge (South Atlantic), it lies in Chron C26r. Subsequently, Wei and Wise (1989) determined that *E. macellus* occurred 1 my earlier at DSDP Site 516 in the western South Atlantic (Rio Grande Rise) than at eastern South Atlantic sites (Walvis Ridge) as determined by Shackleton and others (1984). Referring to Monechi and Thierstein (1985), they remarked on the unreliablitity of the datum.

The age derived by Shackleton and others (1984) for the FAD of *E. macellus* was based in particular on DSDP Site 527 where Manivit (1984) recorded the LO of the species between samples 527–30–2, 150 cm [?50 cm] and 527–30–2, 70 cm, that is, slightly above the magnetic polarity reversal interpreted as the Chron C26r/C27n boundary in the hole (Chave, 1984; Shackleton and others, 1984). Our work (MPA) shows that the LO of *E. macellus* in Hole 527 lies between Core 31–1, 59–61 cm and 31–2, 82–83 cm, that is, in an interval of undetermined polarity

well below (~8 m) the Chron C26r/C27n boundary and slightly above (~1 m) Chron C28n (the Chron C27r/C28n boundary is not located in Hole 527). As all other Paleocene species, *E. macellus* shows a rapid increase in size in its lower range. Its earliest morphotypes are small and the central area of the coccolith may be more prone to dissolution than in later morphotypes. For these reasons, they are easily overlooked although well characterized by the extinction pattern typical of the taxon. In both Holes 527 and 516F, the LO of *E. macellus* is in Chron C27r. The precise position of this event with respect to the Chron C27r/C28n boundary cannot be determined however because of poor recovery in Hole 516F and because the chron boundary is not located in Hole 527.

The apparent location of the LO of *E. macellus* in Chron C26r in Hole 528 by Shackleton and others (1984) results from a misinterpretation of the interval with normal polarity in Core 30, sections 3 and 4, possibly based on the belief of the stratigraphic relation between Chron C27n and the LO of *E. macel-*

TABLE 13.—LATE MIOCENE PLANKTONIC FORAMINIFERAL MAGNETOBIOCHRONOLOGY.
LATE MIOCENE

Datum	FAD	LAD	Paleomagnetic Chron	Age in Ma	Reference	Remarks
1. *Globorotalia sphericomiozea*	X		C3r	5.6	1	Hole 588 (SW Pacific).
2. *Globorotalia pliozea*	X		C3r	5.6	1	Hole 588.
3. *Globorotalia tumida*	X		C3r	5.6	1–4	M14/PL1 boundary; Hole 588; synchronous in the SW Pacific (586B, 587, 588, 590) and Indian (114, 219, 237, 238) Oceans (ref. 3). Identified at comparable position without paleomagnetic control at ODP Hole 806B (Ontong-Java Plateau equatorial Pacific; ref. 4).
4. *Globoquadrina dehiscens*		X	C3r	5.8	1–4	See remarks under 3 above.
5. *Globorotalia lenguaensis*		X	C3An.1n	6.0	4	M13/M14 boundary. Directly calibrated to C3An.1n in ODP Hole 840 (Tonga Platform; ref. 4) and directly to this level in Hole 806B (Ontong-Java Plateau equatorial west Pacific; ref. 5); LAD occurs just below FAD of *P. primalis* at ODP 840 and just above it at ODP 806B (see below). In Hole 519 (South Atlantic LAD of *Gt. lenguaensis* has been calibrated to C4An(y) (~8.7 Ma; CK 95) (Poore and others, 1983, 1984). =M13/M14 boundary; Hole 519, South Atlantic. The LAD of *G. lenguaensis* in E68–136 (Gulf of Mexico) occurs about 35′ above the FAD of *G. plesiotumida* (=N16/N17) and within the range of *Minylitha convallis* and above the LAD of *Discoaster bollii*.. In E66–73 (Gulf of Mexico) the LAD of *G. lenguaensis* occurs at same level as FAD of *G. plesiotumida* (=N16/N17) and below the FAD of *M. convallis* and the LAD of *D. bollii* (Zhang and others, 1993, Aubry, 1993b).
6. *Globorotalia margaritae*	X		C3An (mid)	6.0	1–7	North Atlantic; refs. 2, 6, 7; late Miocene at Site 588 (ref. 1) and considered asynchronous between SW Pacific and Indian Ocean (ref. 3); recorded above FAD *G. tumida* in Hole 806B (equatorial Pacific) (ref. 4). FAD calibrated to C3An.2n in ODP Hole 840 (Tonga Platform; ref. 4, below FAD of *G. tumida*. Initial occurrence in Mediterranean recorded immediately above base Zanclean in mid-Thvera at 4.93 Ma (Langereis and Higen (1991) and First Common Occurrence (FCO) only shortly thereafter at 4.89 Ma (Langereis and Hilgen 1991; ref. 7).
7. *Pulleniatina primalis*	X		C3An.2n	6.4	2–3,5	LAD simultaneous in SW Pacific (DSDP Sites 587, 588, 590) and Indian Ocean (DSDP Sites 214, 219, 238; ref. 3). Also recorded at comparable level in ODP Hole 806B (ref. 4) without paleomagnetic calibration. Recorded slightly above FAD *G. lenguaensis* at ODP Hole 840 (ref. 4).
8. *Neogloboquadrina acostaensis* (S to D)	X		C3An.2n	6.2	3	
9. Dextral menardine globorotaliids (=*Globorotalia menardii* form 5)	X		C3An.2r	6.4	3	
10. *Neogloboquadrina acostaensis* (D to S)	X		C3An.2r	6.6	3	
11. *Neogloboquadrina atlantica* (D to S)	X		C3Ar	6.8	7,8	North Atlantic DSDP Hole 609, 611 (age inferred; ref. 7); 642 (paleomag calibration, ref. 8).
12. *Globorotalia conomiozea*	X		C3Br.1r	7.12	9	=Mt 8/Mt9 boundary = Tortonian/Messinian boundary; Cretan sections (ref. 8); recorded in mid-C3An in Mediterranean in ref. 7; see also ref. 1, 2. Paleomagnetic age estimated (this paper) agrees closely with an estimated age of 7.26 ± 5.1 Ma for Tortonian/Messinian boundary in NE Apennines based on ^{40}Ar/^{39}Ar age of 7.33 Ma just below FAD of *Gt. conomiozea* (Vai and others, 1993).
13. *Globorotalia menardii* form 5	X		C3Br.2r	7.2	9	Cretan sections (ref. 9).
14. Sinistral menardine globorotaliids (*G. menardii* form 4)		X	C3Br.3r	7.4	9	Cretan sections (ref. 9); see discussion in refs. 2, 9, 10.
15. *Globorotalia suterae*	X		C4n.2n	7.8	10	Hole 654, Mediterranean. Paleomagnetic age estimate (this paper) agrees perfectly with ^{40}Ar/^{39}Ar age of 7.72 ± 0.15 Ma on this datum in the northern Apennines (Vai and others, 1993).
16. *Globorotalia cibaoensis*	X		(C4n.2n)	(7.8)	5	ODP Hole 806B (Ontong-Java Plateau). Age estimate in ref. 5, Table 3 of 6.7 Ma calibrated to BKVC85 is incorrect. Note that FAD of *Gt. cibaoensis* is at virtually same level as FAD *Gt. plesiotumida* (~266-m) but that this level is incorrectly estimated (296.75-m) in Table 2 and plotted Figure 6. The FADs of *Gt. cibaoensis* (6.7 Ma), *Gt. juanai* (6.8 Ma), *Candeina nitida* (6.8 Ma) and *Globigerinoides extremus* (6.9 Ma) are all shown to be younger than that of *Gt. plesiotumida* (7.1 Ma = BKVC 85) although their FADs range from ~3 to 23-m below the FAD of *Gt. plesiotumida*. The incorrect ages were derived by misplotting the FAD of *Gt. plesiotumida* (7.1 Ma) at 297-m on Figure 6 and then calculating ages of the other datums by interpolating upward to their correct depths. Correct age estimates have been made here based on the stratigraphic position of these datums in Hole 806B.

TABLE 13.—*Continued.*
LATE MIOCENE

Datum	FAD	LAD	Paleomagnetic Chron	Age in Ma	Reference	Remarks
16. *Globorotalia juanai*	X		(C4r.1r)	(8.1)	5	ODP Hole 806B; see remarks under 14 above.
17. *Candeina nitida*	X		(C4r.1r)	(8.1)	5	ODP Hole 806 B; see remarks under 14 above.
18. *Globigerinoides extremus*	X		(C4r.2r)	(8.3)	5	ODP Hole 806 B; See remarks above.
19. *Globorotalia plesiotumida*	X		(C4r.2r)	(8.3)	5	ODP Hole 806B. The FAD of this form is shown to occur in Core 29H (with the FAD of *Gt. cibaoensis*) in Zone NN11, but questionable occurrences are recorded as low as Core 31H at the same level as the FAD of *Gl. extremus* (in low NN10). FAD recorded in upper part of Zone NN10, overlapping upper part of range of *Discoaster bollii* for almost 3 m and ~4 m below FAD of *Minylitha convallis* at E68–136 (Gulf of Mexico) (Zhang and others, 1993; Aubry, 1993b). FAD occurs ~20 m below *M. convallis* in E66–73 (Zhang and others, 1993; Aubry, 1993b). At Buff Bay, Jamaica, the type level of *G. plesiotumida* (Blow, 1969, 1979) is in Zone NN10 (Berggren, 1993; Aubry, 1993a)
20. *Neogloboquadrina humerosa*	X		C4r.2r	8.5	2	
21. *Neogloboquadrina pachyderma*	X		C4Ar.1n	>9.2	11	Holes 748, 751 (Kerguelen Plateau, South Indian Ocean).
22. *Neogloboquadrina nympha*		X	C5n.2n	10.1	11	Holes 748, 751 (Kerguelen Plateau, South Indian Ocean).
23. *Neogloboquadrina acostaensis*	X		C5n.2n	10.9	2,12	Holes 558 and 563 (ref. 2) and 608 (ref. 12) in North Atlantic.
24. *Neogloboquadrina mayeri*		X	C5r.2r	11.4	2,12,13	M12/M13 boundary; lower part of Buff Bay Formation at Buff Bay, Jamaica, 3 m above FAD of *Discoaster hamatus* (Berggren, 1993; Aubry, 1993a). In Bodjonegoro 1 borehole (Indonesia) these two datum levels occur together (pers. observ., WAB, MPA). In North Atlantic Sites 558, 563 (ref. 2) and 608 (ref. 12) LAD of *N. mayeri* was observed in close juxtaposition with FAD of *N. acostaensis*, essentially eliminating Zone N15 (=M12). This has been variously interpreted as due to a hiatus in the North Atlantic (Aubry, 1993a) or diachrony (ref. 12, 13).

[1]Hodell and Kennett, 1986
[2]Berggren and others, 1985a
[3]Srinivasan and Sinha, 1992
[4]Chaproniere and others, 1994
[5]Chaisson and Leckie, 1993
[6]Weaver and Clement, 1986

[7]Weaver and Clement, 1987
[8]Spiegler and Jansen, 1989
[9]Krijgsman and others, 1994
[10]Glaçon and others, 1990
[11]Berggren, 1992
[12]Miller and others, 1991
[13]Miller and others, 1994

lus in Hole 527. Planktonic foraminiferal stratigraphy which indicates that Core 30–2 belongs to Zone P1b (Boersma, 1984) supports our revised identification of the normal polarity interval in Core 30 as Chron C28n rather than C27n.

Monechi and Thierstein (1985) recorded the lowest occurrence of *E. macellus* in Chron C27r in both the Bottacione and Contessa Highway sections (Monechi and Thierstein, 1985, Figs. 2, 6), not in Chron C28r as claimed by Monechi and others (1985). The discrepancy between the Italian and the South Atlantic sections is thus considerably reduced if not annuled. Monechi and others (1985) correlated the LO of *E. macellus* with a level low in Chron C26r in Holes 577 and 577A. Although no interval with a normal polarity representing Chron C27n was recorded in Hole 577, they assumed that this chron corresponded to a gap in the sample record correlative with a normal polarity interval between 100.95 and 101.95 mbsf in Hole 577A and identified by them as Chron C27n (Monechi and others, 1985, Fig. 2). In Hole 577, the LO of *E. macellus* (at 100.20 mbsf) is slightly above the LO of *Sphenolithus primus* and just below the LO of *Fasciculithus pileatus*. In Hole 577A, it is just above the LO of *F. pileatus* and immediately below the LO of *S. primus*. In both holes, it is (~2 m) below the LO of *Fasciculithus tympaniformis* (at 98 mbsf in Hole 577; at 97.50 mbsf in Hole 577A) and in an interval with reversed polarity which Monechi and others (1985) identified as Chron C26r. Magnetobiostratigraphic correlations in other Paleocene sections (Sites 384, 527, 772C, Bottacione) support their chron identification.

However, in these sections, the LOs of *S. primus, F. pileatus* and *F. tympaniformis* lie in mid Chron C26r, not immediately above the Chron C26r/C27n boundary. In addition, the close association of the LO of *E. macellus, S. primus* and *F. pileatus* indicates that only the upper part of Zone NP4 occurs in Holes 577 and 577A. Both the biostratigraphic evidence and the magnetobiostratigraphic correlations imply that the range of *E. macellus* is truncated in the two holes. An unconformity occurs at ~100.9 mbsf in Hole 577A and ~101.8 mbsf in Hole 577. The stratigraphic gap encompasses the lower part of Zone NP4 and an undetermined part of Zone NP3 and the hiatus comprises the early part of Chron C26r, Chron C27n, and most of Chron C27r (represented between 101.8 and 102.95 mbsf) in Hole 577 and all of Chron C27r in Hole 577A. We deduce from this that the normal polarity interval between 100.95 and 101.95 mbsf in Hole 577A does not represent Chron C27n but is part of Chron C28n.

We conclude that diachrony of the FAD of *E. macellus* between the Atlantic Ocean and the eastern equatorial Pacific is not demonstrated. Instead sections with a good magnetostratigraphic record show that the FAD of *E. macellus* lies in the early part of Chron C27r.

Heliolithus kleinpelli is another species whose lowest occurrence has been suggested to be diachronous. Wei and Wise (1989) indicated that it occurs at various levels in Chron C26r (DSDP Sites 516, 527, 577), Chron C25n (Bottacione and Contessa Highway sections) and Chron C25r (DSDP Hole 528)

TABLE 14.—RELATIONSHIP OF PALEOCENE CALCAREOUS NANNOFOSSIL DATUM LEVELS TO OBSERVED (AND INTERPRETED) MAGNETIC POLARITY STRATIGRAPHY, AND CALCAREOUS NANNOFOSSIL PALEOCENE CHRONOLOGY. AGE ESTIMATES (MA) DERIVED FROM THE GPTS OF CANDE AND KENT (1992/1995) AND ADOPTED IN THIS PAPER.
PALEOCENE

Datum	FAD	LAD	Paleomagnetic Chron	Age in Ma	Reference	Remarks
1. *Tribrachiatus bramlettei*	X		Chron C24r	55	2,3	See text for discussion
2. *Rhomboaster cuspis*	X		Chron C24r	55.16	2,3	See text for discussion
3. *Fasciculithus tympaniformis*		X	Chron C24r	55.33	2,3	See text for discussion
4. *Cruciplacolithus eodelus*	X		Chron C24r	55.5	7	LO of this species precedes closely the carbon isotope excursion in many DSDP sites, in particular at Site 577 (ref. 7).
5. *Discoaster multiradiatus*	X		Chron C25n	56.2	9	Mid Chron C25n in Holes 577 (ref. 9), 752A (ref. 12) and in Contessa Highway section (ref. 8). Lower part of magnetozone assigned to Chron C25n in Hole 384 but exact location indeterminate because of poor recovery and unconformity (ref. 1). Just above magnetozone interpreted as Chron C25n in Hole 689 (ref. 15) and Hole 690 (ref. 2, but see remark on extreme scarcity of discoasters at this site).
6. *Discoaster okadai*	X		Chron C25r	56.8	1	Slightly above FAD of *D nobilis* in Hole 384 (ref. 1); coincident with FAD of *D nobilis* in Hole 577A and slightly below FAD of *D nobilis* in Hole 577 (ref. 7).
7. *Discoaster nobilis*	X		Chron C25r	56.9	1	Mid part of magnetozone interpreted as Chron C25r in Hole 384 (ref. 1). Uppermost part of magnetozone interpreted as Chron C25r in the Contessa Highway section (ref. 8) and in Hole 577 (ref. 9). Basal part of magnetozone interpreted as Chron C25n in Hole 752A (ref. 12). Lowermost part of magnetozone assigned to Chron C25r in Hole 577A (ref. 9); however identification of the normal polarity interval between 88.91 and 89.81 mbsf in Hole 577A (Bleil, 1985) as Chron C26n (ref. 9) is incorrect. Similarly, identification of the normal polarity interval between 91.47 and 92.30 mbsf in Hole 577 (Bleil, 1985) as Chron C26n (ref. 9) is improper.
8. *Heliolithus riedelii*	X		Chron C25r	57.3	1	Lower part of magnetozone interpreted as Chron C25r in Hole 384 (ref. 1). Occurs in magnetozone interpreted as Chron C25r in the Thanet Beds.
9. *Discoaster mohleri*	X		Chron C25r	57.5	1	Lowermost part of magnetozone interpreted as Chron C25r in Holes 384 (ref. 1) and 762 C (cf. refs. 6 and 14); poorly constrained in magnetozone assigned to Chron C25r in Holes 527 (ref. 1) and 752A (ref. 12) because of poor recovery. But mid part of upper normal polarity interval assigned to Chron C26n (Alvarez and others, 1977; Napoleone and others, 1983) in the Bottacione section and in upper part of middle normal polarity interval assigned to Chron C26n (Lowrie and others, 1982) in the Contessa Highway section (ref. 8).
10. *Hornibrookina teuriensis*		X	Chron C26r	58.3	11	HO recorded just above LO of *H. kleinpelli* in Hole 690B (ref. 11) in an interval without magnetostratigraphy but likely below Chron C26n following Spiess's data (1990).
11. *Sphenolithus anarrhopus*	X		Chron C26r	58.4	1	Essentially coincident with the FAD of *H. kleinpelli* in Holes 384 and 527 (ref. 1). Slightly below the FAD of *H. kleinpelli* in Hole 577 (ref. 7). Slightly above the FAD of *H. kleinpelli* in Hole 577 and coincident with the FAD of *D mohleri* in Hole 577A (ref. 7); these relationships are interpreted here as indicative of an unconformity which implies that no magnetozone corresponding to Chron C26n was recovered in Hole 577A contrary to Bleil (1985) and ref. 9 (see text).
12. *Heliolithus kleinpelli*	X		Chron C26r	58.4	1	Upper fourth of magnetozone interpreted as Chron C26r in Hole 527 (cf. refs. 1 and 13), in Hole 762C (cf. refs. 6 and 14 following reinterpretation of Galbrun's identification (ref. 6) of polarity intervals in Core 31) and in Hole 752A (ref. 12). But lower part of upper normal polarity interval assigned to Chron C26n (Alvarez and others, 1977, Napoleone and others, 1983) in the Bottacione section and mid part of lower normal polarity interval assigned to Chron C26n (Lowrie and others, 1982) in the Contessa Highway section (ref. 8). The different correlations between magnetozones and the LO's of *H. kleinpelli* and *D. mohleri* in the Italian sections compared to other sections lead to question the identification of Chron C26n in the Bottacione and the Contessa Highway sections.
13. *Chiasmolithus consuetus*	X		Chron C26r	59.7	1	Mid part of magnetozone assigned to Chron C26r in Holes 384 and 527 (ref. 1); upper part of magnetozone assigned to Chron C26r in Hole 762C (cf. refs. 6 and 14).
14. *Fasciculithus tympaniformis*	X		Chron C26r	59.7	1	Mid part of magnetozone assigned to Chron C26r in Hole 384 (ref. 1) and in the Bottacione section (ref. 8); lower part of magnetozone interpreted as Chron C26r in Holes 752A (ref. 12), 762C (cf. refs. 6 and 14) and in the Contessa Highway section (ref. 8).
15. *Fasciculithus ulii*	X		Chron C26r	59.9	1	Lower third of magnetozone interpreted as Chron C26r in Holes 384 (ref. 1) and 762C (cf. refs. 6 and 14). LO poorly constrained in Hole 527 because of poor recovery (ref. 1).
16. *Sphenolithus primus*	X		Chron C26r	60.6	1	1/5th above the magnetic polarity reversal interpreted as the Chron C26r/C27n boundary in Holes 527 and 384 (ref. 1).

TABLE 14.—*Continued.*
PALEOCENE

Datum	FAD	LAD	Paleomagnetic Chron	Age in Ma	Reference	Remarks
17. *Chiasmolithus bidens*	X		Chron C26r	60.7		Lowermost part of magnetozone interpreted as Chron C26r in Hole 384 (ref. 1). Lower part of magnetozone interpreted as Chron C26r in Hole 762C (cf. refs. 6 and 14); upper part of magnetozone interpreted as Chron C26r in Hole 577A, but lower part of magnetozone interpreted as Chron C27r in Hole 577 (cf. refs. 7 and 9; see text for discussion). Mid part of magnetotozone interpreted as chron C26r in Hole 527 (ref. 1) and (through indirect correlation) in Hole 738C (ref. 16).
18. *Ellipsolithus macellus*	X		Chron C27r	62.2	1	Associated with lower half of magnetozone assigned to Chron C27r in Hole 384 (ref. 1); with the lower part of an interval of unknown polarity between Chron C27n and C28n in Hole 527 (ref. 1); with an interval of unknown polarity immediately above the magnetozone interpreted as Chron C28n in Hole 516F (cf. refs. 2 and 17); with mid part of magnetozone interpreted as Chron C27r in the Contessa Highway section (ref. 8); also clearly associated with magnetozone representing Chron C27r in the Bottacione section (ref. 8), but reported from lowermost part of magnetozone interpreted as Chron C26r in Hole 524 (ref. 10), Hole 527 (ref. 13) and from lower part of magnetozone assigned to Chron C26r in Holes 577 and 577A (ref. 9, but see text for discussion).
19. *Chiasmolithus danicus*	X		Chron C28r	63.8	1	Associated with mid part of magnetozone interpreted as Chron C28r in Holes 527 (cf. refs. 1 and 13) and 384 (ref. 1), but with lower part of magnetozone interpreted as Chron C28n in Hole 577 and 577A (ref. 9) and in the Bottacione section (ref. 8). Associated with mid part of magnetozone representing Chron C28n in Hole 762C following Galbrun's interpretation (ref. 6) but with lower part of magnetozone representing Chron C28n following reinterpretation herein (see text).
20. *Cruciplacolithus tenuis*	X		Chron C29n	64.5	1	Associated with the mid to lower third of magnetozone interpreted as Chron C29n in Holes 527 (refs. 1, 13), 762C (cf. refs. 6 and 14), 577 (ref. 9); with basal magnetozone representing Chron C29n in the Contessa Highway Section (ref. 8); but reported from magnetozone interpreted as Chron C29r in Hole 738 (ref. 16).
21. *Cruciplacolithus primus*	X		Chron C29r	64.8		Slightly below the magnetic polarity reversal interpreted as the Chron C29n/C29r boundary in Holes 762C (cf. refs. 5 and 13), 527 (refs. 1, 13) and 577 (ref. 9).
22. *Hornibrookina teuriensis*	X		Chron C29r	64.9		Restricted to southern high latitudes. Slightly below the magnetic polarity reversal representing the Chron C29n/C29r boundary in Holes 738 (ref. 16) and 690C (ref. 11). Also present in lowermost Paleocene (Zone NP1b) in Hole 750 (ref. 5). LO used to define regional high latitude Zone NA2 (ref. 16).
23. *Biantholithus sparsus*	X		Chron C29r	65		First exclusively Cenozoic taxon. Occurs immediately above the K/P boundary in sections considered to be essentially continuous across the boundary such as at Sites 527 (ref. 1) and 762C (cf. refs. 6 and 13).
24. *Micula* spp.		X	Chron C29r	65	16	

[1]this work
[2]Aubry and others, 1995
[3]Berggren and Aubry, 1995
[4]Berggren and others, 1983b
[5]Ehrendorfer and Aubry, 1992
[6]Galbrun, 1992
[7]Monechi, 1985
[8]Monechi and Thierstein, 1985
[9]Monechi and others, 1985

[10]Poore and others, 1983
[11]Pospichal and Wise, 1990
[12]Pospichal and others, 1991
[13]Shackleton and Members of the Shipboard Scientific Party, 1984
[14]Siesser and Bralower, 1992
[15]Thomas and others, 1990
[16]Wei and Pospichal, 1991
[17]Wei and Wise, 1989
[18]multiple references

based on data from Monechi and Thierstein (1985), Monechi and others (1985), Shackleton and others (1984) and their own data (for Site 516). Correlation between calcareous nannofossil stratigraphy (Siesser and Bralower, 1992) and the magnetostratigraphic record interpreted by Galbrun (1992) could also be used to indicate that the LO of *H. kleinpelli* in ODP Hole 762C (Exmouth Plateau, eastern Indian Ocean) is located in Chron C25r. We do not comment here on the location of the LO of *H. kleinpelli* in Chron C26n in the Italian sections, but we address the other reports.

Regarding the inconsistent location of the LO of *H. kleinpelli* within Chron C26r, we question the reliability of the ages de-

termined by Wei and Wise (1989). Of the three sites they cite, only Site 527 allows determination of the age of the LO of *H. kleinpelli* through magnetochronologic interpolation. In Hole 516F, Chron C27n was questionably identified by Berggren and others (1983b), and no normal polarity interval corresponding to Chron C26n was recovered. In addition, there are several problems (e.g., the normal polarity interval identified as Chron C25n extends from Zone NP8 to Zone NP10; thick intervals without magnetic data in Cores 81 to 85) associated with the identification of Chron C25n by Berggren and others (1983b) and with the interpretation of the NP9-NP10 zonal interval in the hole. As for the determination of the age of the LO of *H.*

kleinpelli at Site 577, we conclude from the preceding discussion that it cannot be achieved through magnetochronology; Chron C27n being missing owing to a stratigraphic gap (we note in passing that the normal polarity intervals between 91.47 and 92.30 mbsf in Hole 577 and between 88.91 and 89.81 mbsf in Hole 577A (Bleil, 1985), assigned to Chron C26n by Bleil (1985) and Monechi and others (1985) are not identifiable).

Wei and Wise (1989) indicate a 1.8-my diachrony for the lowest occurrence of *H. kleinpelli* between Sites 527 and 528, a surprising fact considering that the two sites are separated by ~100 km. This important diachrony contradicts the results of Shackleton and others (1984) who, based on an inappropriate assumption/methodology, derived an age of 61.6 Ma for the LO of *H. kleinpelli* in Hole 528, equal to the age they determined for this event in Hole 527. Hole 528 has a very poor magnetostratigraphic record and poor recovery. Zone NP6 was not determined in the hole. It may or may not occur in the interval of no recovery between 343 and 350 mbsf (Core 25–2 to Core 26–1). Chron C25r is represented as a thin sliver of sediments in Core 24–3, but the sediments below yield no magnetostratigraphic information down to Core 30 (Chron C28n). It can be seen that it is an incorrect procedure to determine the age of a datum based on an assumed stratigraphic position in a section through extrapolation from a magnetic reversal (Chron C25n/C25r in Core 24) to a level where there is no magnetic signal. The claim that the LO of *H. kleinpelli* is associated with Chron C25r in Hole 528 is thus invalid.

Integration of magneto- and biostratigraphy in Hole 762C (Galbrun, 1992) indicates that the NP5/NP6 zonal boundary (LO of *H. kleinpelli*) occurs in Chron C26r. However, the lowest reported occurrence of *H. kleinpelli* in the hole (Siesser and Bralower, 1992) lies in an interval of unknown polarity slightly below a reversed polarity interval assigned by Galbrun to Chron C25r. There is a 7-m uncertainty as to the exact location of the LO of *H. kleinpelli* (between Core 32-CC and Core 33–1) due to poor recovery, and possibly for the same reason, a normal polarity interval corresponding to Chron C26n was not recovered. Galbrun (1992) assigned the thick interval of predominantly reversed polarity in Core 29–3 to Core 32–2 to Chron C25r. Four thin normal polarity events occur in this interval, and we note that the three upper ones that cluster around 443 mbsf (Core 31, sections 1–3) are immediately below the LO of *Discoaster mohleri* and correlate with the upper part of Zone NP6. Although we recognize that these three thin normal events have weak magnetization intensity and are represented by single points, we suggest that they represent Chron C26n based on their close association with the NP6/NP7 zonal boundary. Alternatively, Chron C26n was not identified because of the recovery gap in Core 30. This nevertheless establishes that the LO of *H. kleinpelli* in Hole 762C is in Chron C26r. Except for the reports on the two Italian sections, the LO of *H. kleinpelli* most probably lies in Chron C26r at all sites considered.

The two examples discussed above and discussions on the upper Paleocene and lower Eocene Series (Aubry and others, 1995), the lower and middle Eocene Series (Aubry, this volume), and the Neogene System (Aubry, 1991, 1993a, b) indicate that stratigraphic sections must be submitted to great scrutiny before the relationships between (paleontologic, magnetic, isotopic) events can be established and used for the purpose of magnetobiochronology. We emphasize, however, that although

we believe that unconformities are more widespread in the stratigraphic record than once thought, we do not deny that diachrony of several millions of years occurs. But diachrony of low amplitude (<2 my) can only be demonstrated once the completeness of the sections that are compared has been established, and from the "remarks" in Tables 14 to 17 it can be deduced that most claims for small latidudinal diachrony are not warranted.

Discrepancies in magnetobiostratigraphic correlations between sections are also sometimes the result of differences in taxonomic concepts. This can be seen in particular for the Eocene taxon *Reticulofenestra umbilicus* in which large changes in size occur. Although size is not a good specific criteria for calcareous nannoplankton species, size is often used by specialists to differentiate taxa. It may also be that a confusion of 5-rayed morphotypes of *Discoaster lodoensis* with *D. sublodoensis* when overgrowth occurs, accounts for the discrepancies in the correlation between the lowest occurrence of *D. sublodoensis* and the magnetic reversal pattern (see Table 15, Item 18). Although the role of differences in taxonomic concepts is probably limited for Paleogene markers, it may be substantial for other taxa (e.g., *Ericsonia subpertusa* whose LO is given as low as NP1 and as high as NP5), thus reducing the number of taxa useful to construct a magnetobiochronologic framework.

The limiting factor in this compilation remains the lack of quality of the magnetobiostratigraphic correlations in many sections, due to poor recovery in some sections, or/and to the ambiguity or insufficient quality of the magnetic polarity signal in others. For example, poor recovery in Holes 527, 528, 577 limits considerably the interpretation of the Paleogene sections (see Aubry, this volume, for Sites 527 and 528), and prevents tying a datum event to a precise level in a reversal, while ambiguous magnetic patterns in Hole 516F (see Table 15 and Aubry, this volume) prevents a confident tie between middle Eocene datums and magnetic reversals. Ambiguous magnetic records in many Oligocene sections have led to unwarranted descriptions of latitudinal diachrony (see below). Thus there is an uncertainty of variable importance (<50,000 yr to >200,000 yr) in the age estimates proposed here. Also, these impediments reduce the use of other valuable markers in constructing a biochronologic framework (e.g., *Sphenolithus furcatolithoides, S. obtusus, Cruciplacolithus delus*), a situation well reflected by the low number of datums for the middle Eocene Series which contrasts with the fact that this was the time of greatest diversity in the Cenozoic (Aubry, 1992a). We point out that we use LO and HO in Tables 14 to 17, as strictly stratigraphic terms meaning respectively "lowest occurrence" (not "last occurrence") and "highest occurrence", in contrast with FAD and LAD, terms with temporal significance.

Paleocene Epoch.—Twenty four Paleocene datums are established. Few Paleocene sections with reliable magnetobiostratigraphic correlations are currently available, either due to poor recovery (e.g., Hole 762C) or to the presence of unconformities (e.g., Holes 577, 577A). Although it may eventually yield presently undeciphered unconformites, DSDP Site 384 appears to offer the greatest completeness between Chron C25n and C29n. DSDP Site 527 is another quite complete section but precise magnetobiostratigraphic correlations are difficult to establish because of poor recovery.

TABLE 15.—EOCENE CALCAREOUS NANNOFOSSIL MAGNETOCHRONOLOGY. SEE TABLE 14 FOR EXPLANATION.
EOCENE

Datum	FAD	LAD	Paleomagnetic Chron	Age in Ma	Reference	Remarks
1. *Discoaster saipanensis*		X	Chron C13r in low-mid latitudes	34.2	1,12	This datum is clearly diachronous between low-mid latitudes and southern high latitudes:
			Chron C16n in southern high latitudes	35.4	33	*Low-mid latitudes*: Lower part of magnetozone corresponding to Chron C13r in Massignano (ref. 12) and Contessa Highway (cf. refs. 19 and 25) sections; Lower part of magnetozone representing Chron C13r (ref. 1) but upper part of it (ref. 22) in Hole 522 LAD claimed to be younger in Hole 516F (30° S) than in Contessa Highway section (43°N) (ref. 34) but magnetostratigraphic record in Hole 516F for the Chron C13n to C16r interval is very poor, magnetozones representing Chrons C15n and C16n being very thin (ref. 8) and unconformities are suggested. LAD claimed to be younger at Site 528 than at other localities (ref. 34). HO of *D. saipanensis* immediately below the magnetozone interpreted as Chron C13n in Hole 528, but sediments are assigned to planktonic foraminiferal Zone P15 to P16 immediately underlying Zone P18 (ref. 11) which implies a stratigraphic gap. *Southern high latitudes*: HO in lowermost magnetozone interpreted as Subchron C16n.1n in Hole 744A (ref. 33) and within magnetozone assigned to Chron C16n in Hole 748B (ref. 3).
2. *Discoaster barbadiensis*		X	Chron C13r in low-mid latitudes	34.3	2,21	This datum is clearly diachronous between low-mid latitudes and southern high latitudes:
			Chron C18n in southern high latitudes	~39	3	*Low-mid latitudes*: HO at the same level as HO of *D. saipanensis* (lower third of magneto zone representing Chron C13r) in Massignano (ref. 12) and Contessa Highway (refs. 19, 25) sections, but slightly below HO of *D. saipanensis* in expanded oceanic sections such as Hole 522 (ref. 22) and Hole 612 (ref. 21). Same comments as for *D. saipanensis* regarding diachrony of the LAD of this taxon between Sites 527, 528, 516 and Contess a Highway section (ref. 34). *Southern high latitudes*: HO within magnetozone representing Chron C18n (through indirect correlation) in Hole 749 (ref. 3).
3. *Reticulofenestra reticulata*		X	Chron C15r in mid-low latitudes	35	6,21,25	This datum is clearly diachronous between low-mid latitudes and southern high latitudes:
			Subchron C16n.2n at southern high latitudes	36.1	33	*Low-mid latitudes*: HO in mid magnetozone representing Subchron C16n.1n in the Massignano section (ref. 12) but near the magnetic polarity reversal corresponding to the Chron C15n/C15r boundary in the Contessa Highway section (ref. 25) and in Hole 522 (ref. 6; not in C15n as shown in refs. 34 and 35). Has been tied to Chron C15n in Hole 516F (ref. 34), but the paleomagnetic data (ref. 8) are insufficient to warrant this correlation. See also discussion in ref. 21.
		X				*Southern high latitudes*: HO in magnetozone interpreted as Chron C16n in Holes 689 and 690B (ref. 35), 744A (ref. 33), 748B (ref. 3) and 703A (ref. 32; we agree with Wei (ref. 32) that the normal polarity interval between 130.70 and 136.55 mbsf in Hole 703A represents Chron C16n rather than Chron C15n in Hailwood and Clement (1991b). Diachrony occurs, but is not as large as suggested in reference 2.
4. *Reticulofenestra oamaruensis*	X		Subchron C16n.1n	35.4	32	Just below (between 130.80 and 132.30 mbsf) the magnetic polarity reversal interpreted as the Chron C15r/C16n boundary (at 130.70 mbsf) in Hole 703A (ref. 32; see additional comment regarding identification of Chron C16n in Item 3). This species is restricted to southern high latitudes.
5. *Isthmolithus recurvus*	X		Subchron C16n.2n	36	25,33	Mid part of magnetozone representing Chron C15n (ref. 14) but upper part of magnetozone assigned to Chron C15r (ref. 25) in the Contessa Quarry section. Uppermost part of magnetozone representing Chron C16n (ref. 12) but at the magnetic polarity reversal corresponding to the Chron C15r/C16n boundary (ref. 15) and upper part of magnetozone assigned to Subchron C16n.2n (ref. 25) in the Contessa Highway section. Basal part of magnetozone corresponding to Chron C15r in the Bottacione section (ref. 15) but mid part of magnetozone interpreted as Subchron C16n.2n in the Massignano section (ref. 25). Mid upper part of magnetozone assigned to Chron C15r (ref. 22) but at least as low as upper part of magnetozone interpreted as Chron C16n (ref. 6) in Hole 522A. Not delineated be cause of a recovery gap (ref. 23) but as low as magnetozone interpreted as Chron C17n (ref. 6) in Hole 523. Tied to latest Chron C17n in Hole 516F (ref. 34) but uncertainty in the magnetostratigraphic record in the hole (ref. 8) does not warrant this claim. Mid part of magneto zone interpreted as Subchron C16n.2n in Hole 744A (ref. 33); Mid part of magnetozone assigned to Chron C16n in Holes 703A (ref. 32), 748 (ref. 3), 689 and 690 (ref. 35); Polarity chronozone C16n is truncated by an unconformity in Hole 689 (refs. 28, 30) so that it is misleading to suggest that the LO of *I. recurvus* is older at Site 689 (64° S) than at Site 690 (65° S) as done in ref. 35. See also discussion in ref. 21.

TABLE 15.—Continued.
EOCENE

Datum	FAD	LAD	Paleomagnetic Chron	Age in Ma	Reference	Remarks
6. *Chiasmolithus oamaruensis*	X		Subchron C17n.1n	37	23	Located in magnetozone interpreted as Subchron C17n.1n in Hole 523 (ref. 23; the species was recorded from a single level which likely corresponds to its FAD since this level is immediately above the HO of *C. grandis*, a relationship seen in many middle-upper Eocene sections). Located in magnetozone assigned to Chron C17n in Holes 689 (ref. 35) and 748 (ref. 3); indeterminate position in Hole 516F (ref. 34; large uncertainty in the magnetostratigraphic record), and in Hole 690 (ref. 35; Polarity Chronozone C17n is truncated by an unconformity which also truncates the lower range of *C. oamaruensis* in the hole (refs. 28, 30, 35). Based on these remarks, it is clear that the claim that the FO of *C. oamaruensis* is diachronous with latitude (refs. 34, 35) is unsubstantiated.
7. *Chiasmolithus grandis*		X	Subchron C17n.1n	37.1	23	Located in magnetozone assigned to Subchron C17n.1n in Hole 523 (refs. 6, 23). Tied to Chron C17n in Hole 516F (ref. 34) but location imprecise because of large uncertainty in the magnetostratigraphic record in the hole with neither the Chron C17n/C17r nor C16r/C17n boundaries delineated (see ref. 8). Mid part of magnetozone assigned to Chron C18n in the Bottacione section but lowermost part of magnetozone in terpreted as Chron C18n in the Contessa Highway section (cf. refs. 15 and 19) and upper part of magnetozone assigned to Chron C18n in the Contessa Quarry section (ref. 14). However no magnetozone representing Chron C17n was recorded from the Bottacione section (ref. 18). We note the same relation ship between the HO of *C. grandis* and the LO of *G. semiinvoluta* in this section and in Hole 516F (cf. refs. 8 and 26). In this latter the 2 events are closely associated with Polarity Chronozone C17n. *Chiasmolithus grandis* does not occur in southern high latitudes.
8. *Reticulofenestra bisecta*	X		Subchron C17n.3n	38	20,23	Tied to Polarity Chronozone C17n.3n in Hole 523 (cf. refs. 20 and 23). Lower part of magnetozone interpreted as Chron C18n in the Contessa Highway section and mid part of magnetozone assigned to Chron C18r in the Bottacione section (ref. 15). The relationship between the LO of *R. bisecta* and Polarity Chronozone C17r in Hole 523 supports the sugges (Item 7) of a miscorrelation between normal polarity intervals and seafloor anomalies in the Italian sections around the Chrons C19n–C16n interval.
9. *Chiasmolithus solitus*		X	Chron C18r	40.4	23	Tied to upper Polarity Chronozone C18r in Hole 523 (ref. 23). Tentatively with in Polarity Chronozone C18n in Hole 748 (ref. 3). Because of discontinuous occurrence, HO in Hole 516F is difficult to determine. Given in Polarity Chronozone C18n (ref. 34) but may be better located in Polarity Chronozone C18r (using ref. 8). Possibly Polarity Chronozone C17r in Hole 689 (ref. 35). Cannot be located in Hole 690B (contrary to ref. 35) owing to a stratigraphic gap which truncates the upper range of *C. solitus* (refs. 28, 30).
10. *Reticulofenestra reticulata*			Chron C19r	42	24,28,30	Mid Polarity Chronozone C19r in Hole 689 (based on data in refs 24 and 28, and contrary to ref. 35 which places this LO in mid Chronozone C18n). Given in mid Chronozone C18n in Hole 690B (ref. 35) but falls in a reversed interval (ref. 28) assigned to Chron C19r (ref. 30) in this hole. Falls at a magnetic polarity reversal interpreted as the Chron C19n/C18r boundary in Hole 516F (refs. 34, 35) but magnetostratigraphic succession in this section (ref. 8) is unclear and polarity chronozone assignments were mostly tentative between Polarity Chronozones C20n and C18n.
11. *Nannotetrina fulgens*		X	Chron C20r	43.1	9,23	Mid Polarity Chronozone C20n in Hole 523 (ref. 23). Given in Polarity Chronozone C19r in Hole 516F (ref. 34) but see comments regarding item 13. In the epicontinental sediments of the Hampshire Basin, the HO of *N. fulgens* is in lower Zone NP16 (above the HO of *B. gladius*, ref. 2)
12. *Blackites gladius*		X	Chron C20n	43.4	2,4,34	Tied to Polarity Chronozone C19n in Hole 516F (ref. 34). However magnetostratigraphic pattern in Hole 516F is unclear between Polarity chronozone C20n and C18n (ref. 8). Data from ref. 3 suggest that the normal polarity interval interpreted as Chron C19n in ref. 8 may best be identified as Chron C20n. If that is correct, the LO of *B. gladius* is tied to early Chron C20n, in agreement with magnetobiostratigraphic correlations in the Bracklesham beds of the Hampshire Basin (ref. 2 and Aubry and others, 1986).

TABLE 15.—*Continued.*
EOCENE

Datum	FAD	LAD	Paleomagnetic Chron	Age in Ma	Reference	Remarks
13. *Reticulofenestra umbilicus*	X		Chron C20n	43.7	9,15,19	Located at the magnetic polarity reversal interpreted as the Chron C20n/C20r boundary (refs. 15, 19) but in the mid part of the magnetozone interpreted as Chron C20r (ref. 14) in the Contessa Highway section. Polarity Chronozone C19r in Hole 528 (ref. 23). Mid part of magnetozone interpreted as Chron C19r in Hole 523 (ref. 6, based on >14μ specimens as established in ref. 7). Upper part of magnetozone interpreted as Chron C20n in Hole 516F (ref. 34, using the >14μ size limit). See further comments in ref. 9.
14. *Chiasmolithus gigas*		X	Chron C20r	44.5	4,9,34	Mid part of magnetozone intepreted as Chron C19r in Hole 516F (ref. 34) but in the upper third of magnetozone interpreted as Chron C20r in this hole based on reinterpretation of the magnetostratigraphy following data in ref. 4 (see also Item 13); upper part of magnetozone assigned to Chron C20r in Hole 523 (ref. 23)
15. *Chiasmolithus gigas*	X		Chron C20r	46.1	4,9,34	Slightly above magnetozone assigned to Chron C21n in Hole 516F (ref. 34). Not delineated (base of hole) in lower Polarity Chronozone C20r or below in Hole 523 (ref. 23).
16. *Nannotetrina fulgens*	X		Chron C21n	47.3	9,15	Mid Polarity Chron ozone C21n in the Bottacione and Contessa Highway sections (ref. 15). See also comments in ref. 9 and interpretations of various middle Eocene sections from the Atlantic Ocean (ref. 4).
17. *Blackites inflatus*	X		Chron C21r	48.5	9,15	Upper Polarity Chronozone C21r in the Bottacione section (ref. 15).
18. *Discoaster sublodoensis*	X		Chron C22n	49.7	9,15	Lower Polarity Chronozone C22n in the Contessa Highway section, but Polarity Chronozone C22r in the Bottacione section (ref. 15). Mid Polarity Chronozone C22n in Hole 530A (cf. refs. 13 and 29, see also ref. 4); upper Polarity Chronozone C22n in Hole 577 (ref. 16)
19. *Tribrachiatus orthostylus*		X	Chron C22r	50.6	4,9	Lower Polarity Chronozone C22r in the Bottacione section (ref. 15). Mid to lower Polarity Chronozone C22r in Holes 550 and 549 (cf. refs. 17 and 31; see also ref. 4), Holes 528 and 527 (ref. 27; see also ref. 4).
20. *Discoaster lodoensis*	X		Subchron C24n.2r	52.85	4,9	Mid Polarity Chronozone C24n in the Bottacione section (ref. 15). Polarity Subchronozone C24n.2r in Holes 549, 550 and in Hole 530 (cf. respectively refs. 17 and 31 and refs. 13 and 29; see also ref. 3 for discussion).
21. *Tribrachiatus contortus* Morphotype B		X	Chron C24r	53.61	5,10	see text
22. *Tribrachiatus orthostylus*	X		Chron C24r	53.64	5,10	see text
23. *Tribrachiatus bramlettei*		X	Chron C24r	53.89	5,10	see text
24. *Tribrachiatus contortus* Morphotype B	X		Chron C24r	53.93	5,10	see text
25. *Tribrachiatus contortus* Morphotype A		X	Chron C24r	54.17	5,10	see text
26. *Tribrachiatus contortus* Morphotype A	X		Chron C24r	54.37	5,10	see text

[1]This work
[2]Aubry, 1983
[3]Aubry, 1992
[4]Aubry, this volume
[5]Aubry and others, 1995
[6]Backman, 1987
[7]Backman and Hermelin, 1986
[8]Berggren and others, 1983b
[9]Berggren and others, 1985b
[10]Berggren and Aubry, 1995
[11]Boersma, 1984
[12]Coccioni and others, 1988
[13]Keating and Herrera-Bervera, 1984
[14]Lowrie and others, 1982
[15]Monechi and Thierstein, 1985
[16]Monechi and others, 1985
[17]Müller, 1985
[18]Napoleone and others, 1983
[19]Nocci and others, 1986
[20]Percival, 1984
[21]Poag and Aubry, 1995
[22]Poore and other, 1982
[23]Poore and others, 1983
[24]Pospichal and Wise, 1990
[25]Premoli Silva and others, 1988a
[26]Pujol, 1983
[27]Shackleton and others, 1984
[28]Spiess, 1990
[29]Steinmetz and Stradner, 1984
[30]Thomas and others, 1990
[31]Townsend, 1985
[32]Wei, 1991
[33]Wei and Thierstein, 1991
[34]Wei and Wise, 1989
[35]Wei and Wise, 1990

Latest Paleocene-earliest Eocene Epochs.—The complexity of the stratigraphic relationships between paleontologic events and magnetic events around the Paleocene/Eocene boundary have been unravelled, implying the extensive development of unconformities in deep-sea and shallow water sections around the Paleocene/Eocene boundary (Aubry and others, 1995). The intricacies involved in tying a magnetochronologic and a chronostratigraphic framework (Cande and Kent, 1992, 1995) and the lack of (magnetic) resolution in the long Chron C24r interval have resulted in an unsatisfactory magnetobiochronologic framework for the latest Paleocene-earliest Eocene Epochs. This is discussed at length in Aubry and others (1995) and Berggren and Aubry (1995). We point to the fact that there is

as yet no firm magnetochronologic calibration for *Discoaster diastypus.*

Eocene Epoch.—Twenty five datums are established. Some are tentative because of insufficient documentation of their relationship to magnetic stratigraphy. The poorest documented datums are the FAD of *Blackites inflatus* (largely due to widespread unconformities around the lower/middle Eocene boundary (Aubry, this volume), the LAD of *Blackites gladius* (largely due its restriction to epicontinental areas), the LAD and the FAD of *Chiasmolithus gigas* (largely due to ambiguous interpretation of the magnetic record in sections where the species occurs). Their age estimates are somewhat arbitrary.

Although Hole 762C represents an expanded Eocene section, it has not been used in this compilation because the magnetic reversals identified by Galbrun (1992) on the basis of numerical succession need to be reinterpreted. Regardless of Galbrun's interpretation, the magnetobiostratigraphic correlations in the site appear to be rather ambiguous.

Oligocene Epoch.—Magnetobiochronology in the Oligocene Epoch is difficult to achieve because of the complexity of the reversal pattern, particularly after Chron C9, and because of biogeographic provincialism, many Oligocene marker species being excluded from the southern high latitudes as a result of progressive middle Eocene cooling and subsequent early Oligocene individualization of a southern water mass (see review in Aubry, 1992a). At these latitudes, identification of the upper Oligocene reversal pattern is mostly dependant upon pattern matching without the benefit of the firm biostratigraphic control established at mid-low latitudes. This results in weak high-latitude magnetobiostratigraphic correlations which do not sustain the patterns of diachrony of paleontologic events described by some (e.g., Wei, 1991, 1992; Wei and Wise, 1992).

A case in point concerns the HO of *Chiasmolithus altus* reported by Wei (1992, p. 162) to "have a fairly consistent age of ~26.7 Ma in the Southern Ocean except in the extreme high latitudes where the datum appears to be substantially younger." Wei (1992, Fig. 8) shows that the HO of *C. altus* is very near the Chron C7Ar/C8n boundary at DSDP Site 516 (30°S), in Chron C7Ar at ODP Sites 699 (52°S) and 748 (58°S), in Chron C8n at Site 744 (62°S), at least as young as Chron C7n at Site 689 (65°S) but in Chron C10n at Site 690 (65°S).

The HO of *C. altus* in Hole 516F is located between Cores 13–1 and 13–2 (Wei and Wise, 1989; however, we note the discontinuous occurrence of the species above Core 17–3). This falls in an interval with poor paleomagnetic control between upper Core 12 and lower Core 15, bracketed by two thin normal intervals. Berggren and others (1983b) identified the upper normal polarity interval as Chon C7An, assigned the lower one to Chron C8 and located the Chron C7A/C8 boundary at the lower boundary of the upper normal. Thus, if anything, the HO of *C. altus* in Hole 516F is located in Chron C8n or C8r, not in Chron C7Ar or at the Chron C7A/C8n boundary (unrecovered in the hole). Considering the insufficient magnetic control and the ambiguous correlation between planktonic foraminiferal and calcareous nannofossil stratigraphy in Hole 516F (Berggren and others, 1983b), it is only safe to conclude that the HO of *C. altus* at Site 516 is likely associated with Chron C8.

Wei (1991) located the HO of *C. altus* in mid Chron C7Ar (between Cores 15H-5 and 16H-1) at Site 699. Firstly, we note that direct correlation between magneto- and biostratigraphy in Hole 699A (Hailwood and Clement, 1991a; Wei, 1991) does not allow such precise location. Due to incomplete recovery in Core 15, the Chron C7An/C7Ar boundary was not delineated in the hole. Secondly, we question the interpretation of the upper Oligocene magnetic reversal stratigraphy. As noted by Hailwood and Clement (1991a), there are few biostratigraphic constraints to correctly identify the magnetic reversals in Hole 699A, and these authors relied heavily on the FAD of *Rocella gelida* to this end. Using an age estimate of 26 Ma for the FAD of this diatom species, derived from Fenner (1984) through implied and indirect correlation to the then as yet unpublished chronology of BKF85, they assigned the reversed polarity in-terval in the lower part of Core 14 and the upper part of Core 15 to Chron C7r, and consequently determined the normal polarity interval in lower Core 15 to Chron C7An, the reversed polarity interval in upper Core 16 to Chron C7Ar, and the underlying normal polarity interval in Core 16 to Chron C8n. Different extrapolated age estimates have been given for the FAD of *R. gelida* (Fenner, 1984; Barron, 1985), and as indicated above, the age estimate by Fenner (1984) has served for an inferred correlation between the FAD of *R. gelida* and Chron C7r (Cieselski and others, 1988). However, the first available direct magnetobiostratigraphic correlation in Hole 744A shows that the FAD of *R. gelida* falls in lower Chron C8n (Baldauf and Barron, 1991), indicating that the normal polarity interval in Core 15H (Hole 699A) represents Chron C8n rather than Chron C7An. Hence, the HO of *C. altus* in Hole 699A is associated with Chron C8n, just as it is in Hole 744A (Wei and Thierstein, 1991).

The HO of *C. altus* at Site 748 is located by Wei (1991) in Chron C7Ar (and apparently slightly younger than at Site 699). There are several problems associated with this correlation. First, it is difficult to determine the highest occurrence of *C. altus* in Hole 748B because there is an abundance change above Core 9H-6 (Aubry, 1992b, Wei and others, 1992) and there is no *a priori* reason to suspect that *C. altus* is reworked above this level. In any case, the HO of *C. altus* in Hole 748B (and of all other associated species in Core 9, including *R. bisecta*, see, Table 16, Item 2) has little stratigraphic significance since no *in situ* Paleogene calcareous nannofossil assemblages are preserved in this hole above Core 9 (Aubry, 1992b). Second, if a straightforward correlation is established between calcareous nannofossil (Aubry, 1992b) and magnetostratigraphy following the interpretation by Inokuchi and Heider (1992), the HO of *C. altus* in Hole 748B falls in Chron C6r (following Aubry, 1992b) not in Chron C7Ar (following Wei and others, 1992). Second, the magnetostratigraphic record in Hole 748B is ambiguous and every reversal was interpreted as corresponding to a different chron by Inokuchi and Heider (1992). The lowest occurrence of *R. gelida* lies in Core 10 in Hole 748B (Schlich and others, 1989) indicating that the normal polarity interval between ~77 and ~85 mbsf in the hole corresponds to Chron C8n. This is in agreement with Harwood and Maruyama (1992, Fig. 16) who inferred from diatom stratigraphy the presence of an unconformity in the lower part of Core 9H with a hiatus encompassing the younger part of Chron C8n to mid Chron C7n. The change in abundance of *C. altus* indicated above occurs on both sides of the unconformity (from abundant to common above, see Aubry, 1992b, Table 3, and from abundant to rare, see Wei and others, 1992, Table 3). We thus interpret the presence of *C. altus* above it as reflecting intensive reworking (reworking of *C. altus* and other Paleogene taxa occur throughout the Neogene section in Hole 748B) and conclude that the range of *C. altus* is truncated in Hole 748B, and that there is poor evidence that the species ranges higher than Chron C8n.

Because the upper range of *C. altus* is truncated in Holes 689B and 690B, Wei (1991) concluded that the HO of *C. altus* is younger than "25.6 Ma and 26.6 Ma at these two sites respectively" (1991, p. 159). This corresponds to a correlation with Chron C7n or younger at Site 689 and with Chron C7Ar or younger at Site 690, although Wei (1991, Fig. 8) shows a correlation with the Chron C10n/C10r boundary. We note that

TABLE 16.—OLIGOCENE CALCAREOUS NANNOFOSSIL MAGNETOCHRONOLOGY. SEE TABLE 14 FOR EXPLANATION.
OLIGOCENE

Datum	FAD	LAD	Paleomagnetic Chron	Age in Ma	Reference	Remarks
1. *Reticulofenestra bisecta*		X	Subchron C6Cn.2r	23.9	21	Located in mid part of magnetozone interpreted as Chron C6Cn in Hole 522 (refs. 1, 16); in magnetozone interpreted as Subchron C6Cn.2r at Site 703 (47° S; ref. 21); just below the magnetic polarity reversal interpreted as the Chron C6Cn/C6r boundary in Holes 563 and 558 (ref. 12) but Chron C6Cn is represented by only two normal polarity intervals in these holes. Given in mid Polarity Chronozone C6Cn in Hole 516F (and slightly younger than at Sites 522, 558 and 563; refs 21, 23) although we do not understand this determination, Wei and Wise (ref. 23, p. 134) indicating the presence of a hiatus encompassing most of Chrons C6Cn and C6Br. Shown as being younger (in Subchron C6Cn.2r) at Site 748 (58° S) than at any other southern latitude sites (Ref. 21, 25). This is however not supported by diatom stratigraphy (ref. 9) which indicates that the upper range of *R. bisecta* is truncated by an unconformity in Hole 748B (see Item 6 for discussion). Shown to be slightly older (in Suchron C6Cn.3n) at Site 699 (52° S) than at Site 516 (ref. 21) but magnetostratigraphic record in Hole 699A is ambiguous and does not allow this claim. HO of *R. bisecta* in Holes 744A (ref. 22), 689A and 690B (ref. 24) not recorded due to unconformities. Thus the claim that the LAD of *R. bisecta* is not reliable at extreme southern high latitudes (ref. 21) is inappropriate.
2. *Sphenolithus delphix*	X		Chron C6Cr	24.3	12	Upper Polarity Chronozone C6Cr in Hole 563 (ref. 12)
3. *Zygrhablithus bijugatus*		X	Chron C6Cr	24.5	12	Located in lower magnetozone assigned to Chron C6Cr in Holes 558 and 563 (ref. 12); indeterminate position in Polarity Chronozone C6Cr in Hole 516F because of insufficiently detailed magnetic record (cf. refs. 5 and 23)
4. *Sphenolithus ciperoensis*		X	Chron C6Cr/C7n	24.75	19	Tied to latest Chron C7n in Holes 522 (refs. 1, 16; however, the polarity reversal representing the Chron C6Cr/C7n boundary is located at a core break), 558 and 563 (ref. 12), and 608 (ref. 6; but the Polarity Chronozone C6Cr/C7n boundary is not delineated in the hole, ref. 7). Probably associated with Polarity Chronozone C7n in Hole 516F (cf. refs. 5 and 23) but insufficient magnetostratigraphic control does not allow greater detail. Located in lowest part of magnetozone representing Chron C6Cr in Hole 528 (ref. 19, but Chron C7n is represented by only one (thin ?) normal polarity interval in this hole, see ref. 1, Fig. 3) and in the Contessa Highway section (ref. 11). The quality of the magnetostratigraphic records in Holes 558, 563, 522, 528 and 516F does not allow a precise comparison between geographic locations of the HO of *S. ciperoensis* with respect to Chron C7n as done by Wei and Wise (ref. 23, Fig. 21).
5. *Chiasmolithus altus*		X	Chron C8n	26.1	22	Located in upper Polarity Chronozone C8n in Hole 744A (ref. 22). Located in basal Polarity Chronozone C8r in Hole 522 (cf. refs. 15 and 16). This record provides the only indication that the LAD of *C. altus* may be diachronous between low-mid and high southern latitudes. See text for other reported correlations to the magnetic record.
6. *Sphenolithus umbrellus*	X		Chron C9n	~27.5	12	Indeterminate position in Polarity Chronozone C9n in Hole 563 (ref. 12).
7. *Sphenolithus distentus*		X	Chron C9n	27.5	1,16	Located in magnetozone interpreted as Chron C9n (ref. 16) but as Chron C7n (ref. 23, Fig. 20) in Hole 522. *Sphenolithus distentus* is sporadic in Hole 522 but it is clearly reworked above Core 20 (i.e., in the Polarity Chronozone C7n-C8r interval in the hole) as indicated by its discontinuous occurrence (ref. 1). Located in a reversed polarity interval above Polarity Chronozone C9n in Hole 528 (ref. 23) if the magnetic reversal at 182.06 mbsf in the hole represents the top of Polarity Chronozone C9n as interpreted by Shackleton and others (ref. 19). However, confident identification of the reversed polarity interval is not possible: following Shackleton and others's interpretation the stratigraphic interval representing Chron C7n to Chron C11n(y) is only 19.40 m, which corresponds to an unlikely sedimentation rate of 0.3 cm/10³ yr. Instead we suggest that the interval between 170.96 and 190.36 mbsf in Hole 528 comprises one or more unconformities. Located in an interval of unknown magnetic polarity bounded by normal polarity intervals in Hole 516F (ref. 23) assigned to polarity Chronozone C9n (ref. 5). However, this interpretation of the magnetostratigraphic record in Hole 516F is questionable (see Item 11). Tied to earliest Chron C10n in Hole 558 (ref. 12), but see comments at Item 9. Precise position of the HO of *S. distentus* with respect to magnetochronology at different geographic locations as proposed by Wei and Wise (ref. 23) is not acceptable considering the uncertainty in the magnetic stratigraphy at these locations .

TABLE 16.—*Continued.*
OLIGOCENE

Datum	FAD	LAD	Paleomagnetic Chron	Age in Ma	Reference	Remarks
8. *Sphenolithus predistentus*		X	Chron C9n	27.5	1	Closely associated with the HO of *S. distentus* in many sections but the stratigraphic succession of the 2 events is variable. Located in Polarity Chronozone C9n in Hole 522 (slightly below the HO of *S. distentus*, refs. 1, 16). HO of *S. predistentus* below that of *S. distentus* in Hole 516F where it is located in an interval of unknown magnetic polarity bounded by 2 normal polarity intervals (ref. 23) interpreted as Chron C9n (ref. 5). However, this interpretation of the magnetic reversal stratigraphy in Hole 516F is questionable (see Item 11). Located in Polarity Chronozone C10n, (slightly above the HO of *S. distentus*) in Hole 558 (ref. 12), but the magnetostratigraphic succession in the hole is unclear between Polarity Chronozones C7n and C11n and it may be that either the interval assigned to Chron C10n by Miller and others (ref. 12) represents Chron C9n, or that an unconformity occurs just above it. The magnetobiostratigraphic record in this hole cannot serve to evaluate possible diachrony of the LAD of *S. distentus* and *S. predistentus.*
9. *Sphenolithus pseudoradians*		X	Chron C10r	29.1	16	Mid Polarity Chronozone C10r in Holes 522 (ref. 16) and 558 (ref. 12)
10. *Sphenolithus ciperoensis*	X	X	Subchron C11n.2n	29.9	1	Located in Polarity Subchronozone C11n.2n (refs. 1, 15) but in Subchronozone C10n.2n (ref. 16) in Hole 522; in Polarity Subchronozone C11n.2n in Hole 558 (ref. 12). Located in Polarity Chronozone C11n in Hole 528 if the normal polarity interval between 190.36 and 201.36 mbsf in the hole is identified as Chronozone C11n rather than as Chronozones C11n and C12n concatenated as proposed by Shackleton and others (ref. 19). LO not paleomagnetically constrained in Hole 563 (ref. 12) because of lack of magnetic record contrary to Wei and Wise (ref. 23). Falls in an interval of unknown polarity (in Core 19) bounded by normal polarity intervals in Cores 19 and 20 in Hole 516F (ref. 23), assigned to Chron C9n (ref. 5). However, because of insufficient magnetic control with almost no reversed polarity intervals delineated in the Polarity Chronozone C7A-C11n interval in the hole, this correlation may be erroneous. Also, the correlation between planktonic foraminifera and calcareous nannoplankton biozones in Hole 516F (ref. 5) is problematic, and there is no overlap of the ranges of *S. distentus* and *S. predistentus* with that of *S. ciperoensis* in Hole 516F unlike in other sections. This may indicate the absence of Zone NP24 (unconformity) as an alternative explanation to the diachrony of the LAD of *S. ciperoensis* suggested by Wei and Wise (ref. 23).
11. *Sphenolithus distentus*	X		Chron C12r	31.5 to 33.1	17,23	Located in upper Polarity Chronozone C12r in the Contessa Highway section (refs. 14, 18) and in Hole 523 (ref. 17); in mid Polarity Chronozone C12r in Hole 522 (refs. 1, 16); in lower Polarity Chronozone C12r in Hole 563 (slightly lower than the HO of *E. formosa*) and in Hole 558 (at a level between the HO of *E. formosa* and that of *R. umbilicus*; ref. 12); in lower Polarity Chronozone C13n in Hole 516F (ref. 23). This is a very inconsistent datum which may occur as low as Zone NP21 (e.g., Sites 516, 563, Oceanic Formation, Barbados), in Zone NP22 (e.g., Site 558) or as high as in Zone NP23 (e.g., Contessa Highway section and many DSDP/ODP sites).
12. *Reticulofenestra umbilicus/ R. hillae*		X	early Chron C12r in low-mid latitudes	32.3	1,12	This datum is clearly diachronous between low-mid latitudes and southern high latitudes:
			late Chron C12r in southern high latitudes	31.3	22	*Low-mid latitudes*: Located in lower Polarity Chronozone C12r (ref. 18) or slightly above Polarity Chronozone C13n (ref. 14) in the Contessa Highway section; in lower third of Polarity Chronozone C12r in Holes 558 (ref. 12) and 523 (ref. 17); in mid part of magnetozone interpreted as Chron C12r in Hole 522 (refs. 1, 3, 16). Indeterminate position in Polarity Chronozone C12r in Hole 563 (ref. 12); located slightly above magnetic polarity reversal corresponding to the Chron C13n/C12r boundary in Hole 516F (ref. 23) but this boundary is poorly delineated in the hole (ref. 5). Given in Polarity Chronozone C12r in Hole 528 (ref. 23 using ref. 19). However, we suggest that the normal polarity interval between 190.36 and 201.36 mbsf in Hole 528 corresponds to Chron C11n (see Item 11) rather than to the concatenation of Polarity Chronozones C11n and C12n as interpreted by Shackleton and others (ref. 19). Also, Chron C13n is poorly characterized in Hole 528 and sampling in the interval between 201.36 mbsf (polarity reversal N to R) and 221.83 mbsf (polarity reversal R to N) is insufficient to confidently characterize Chron C12r (see ref. 13, Fig. 3). *Southern high latitudes*: Located in upper Polarity Chronozone C12r in Holes 690B (ref. 24), 744A (ref. 22), 748 (refs. 2 (but misprinted in Fig. 3, cf. refs 2 and 10), 23), and 703A (ref. 21); in uppermost Polarity Chronozone C12r in Hole 699A (ref. 21) where Chron C12r is poorly characterized due to insufficient sampling (ref. 9) and in Hole 689B (refs. 21, 24) where the magnetic record is also ambiguous, with Chronozone C11r considerably thinner than Chronozone C12n following Spiess's interpretation (ref. 20) which would result in sedimentation rates of 0.56 cm/10^3 yr for the mixed calcareous and siliceous sediments (ref. 4) representing Chron C12n.

TABLE 16.—*Continued.*
OLIGOCENE

Datum	FAD	LAD	Paleomagnetic Chron	Age in Ma	Reference	Remarks
13. *Isthmolithus recurvus*		X	Chron C12r	31.8 to 33.1	16,22	One of the most inconsistent datums. Falls in upper Polarity Chronozone C13n in Hole 522 (ref. 16); in lowermost Polarity Chronozone C12r in the Contessa Highway section (ref. 18); in lower Chronozone C12r, between the HO's of *R. umbilicus* and *E. formosa* in Hole 523 (ref. 17); in lower Chronozone C12r at the same level as the HO of *R. umbilicus* in Hole 558 (ref. 12); in Chronozone C12r at the same level as the HO of *R. umbilicus* in Hole 563 (ref. 12); slightly above mid Chronozone C12r in Hole 744A (ref. 22), in upper Chronozone C12r (ref. 2 but misprinted in Fig. 3, cf. refs. 2 and 10) but in mid Chronozone C12r (ref. 25) in Hole 748B. Given in lowermost Chronozone C12r in Hole 516F (refs. 21, 23) but the Chronozone C12r/C13n boundary is poorly characterized in this hole (ref. 5); given in mid Chronozone C12r (ref. 21) in Hole 699A, but falls in an ~15 m thick interval without magnetic data (ref. 8) and the Chronozone C12r/C13n boundary is not delineated in the hole; given in mid Chronozone C12r in Hole 689B (ref. 21) but identification of Chronozone C12n in this hole is questionable; given in mid Chronozone C12r in Hole 690B (ref. 21) but precise location cannot be determined as the Chronozone C12r/C13n boundary was not recovered due to an unconformity (ref. 20).
14. *Ericsonia formosa*		X	Chron C12r in mid-low latitudes	32.8	1,3	This datum is clearly diachronous between low-mid and high southern latitudes:
			Chron C18 in southern high latitudes	~39.7	2,20,24	*Low-mid latitudes*: Located in uppermost Chronozone C13n (ref. 14) but immediately above Chronozone C13n (ref. 18) in the Contessa Highway section. Located in lowermost Chronozone C13n in Hole 558 (ref. 12), in lower Chronozone C12r in Holes 522 (refs. 1, 3, 16), 523 (ref. 17), 528 (ref. 19, but see remarks regarding Item 13) and Hole 747C (ref. 2). Given in upper Chronozone C13n in Hole 516F (ref. 23) but the Chronozone C12r/C13n boundary is poorly characterized in this hole (ref. 5).
						High southern latitudes: HO associated with polarity Chronozone C18 (through indirect correlation) in Holes 748C and 749 (ref. 2) and possibly in Hole 689B (cf. refs. 20 and 24). There is no magnetic stratigraphy in Holes 748C and 749, and the magnetostratigraphic record is insufficiently documented in Hole 689B to allow a more precise location in Chon C18 of this event.
15. *Clausicoccus subdistichus*	Acme		Chron C13n	33.3	18	Located (= *Ericsonia obruta*) in lower Chronozone C13n in the Massignano section (ref. 18).
16. *Reticulofenestra oamaruensis*		X	Chron C13r	33.7	22	Located in uppermost Chronozone C13r in Holes 744A (ref. 22), 748B (ref. 25), 689B (ref. 24), and possibly at the Chronozone C13n/C13r boundary in Hole 699A (ref. 21, Fig. 4). This species is restricted to southern high latitudes.

[1]This work
[2]Aubry, 1992b
[3]Backman, 1987
[4]Barker and others, 1988
[5]Berggren and others, 1983b
[6]Clement and Robinson, 1987
[7]Gartner, 1992
[8]Hailwood and Clement, 1991a
[9]Harwood and Maruyama, 1992
[10]Inokuchi and Heider, 1992
[11]Lowrie and others, 1982
[12]Miller and others, 1985
[13]Moore, and others, 1984

[14]Nocci and others, 1986
[15]Percival, 1984
[16]Poore and others, 1982
[17]Poore and others, 1983
[18]Premoli Silva and others, 1988a
[19]Shackleton and others, 1984
[20]Spiess, 1990
[21]Wei, 1991
[22]Wei and Thierstein, 1991
[23]Wei and Wise, 1989
[24]Wei and Wise, 1990
[25]Wei and others, 1992

these correlations are strictly dependant upon the interpretation of the magnetic reversal succession in the holes by Spiess (1990) who noted in particular the lack of biostratigraphic constraints below the unconformable Oligocene/Miocene contact at 67 mbsf in Hole 689B. Spiess (1990) derived a chron assignment in the upper Oligocene section in Hole 689 through upwards extrapolation from Chron C18 to C10 to the unconformity based on a depth/age curve. Spiess (1990) remarked that this curve indicates constant sedimentation rates and does not provide evidence of hiatuses in the interval considered. However, he failed to notice the extraordinarily low sedimentation rates (~0.1 cm/10³ yr) resulting from his interpretation,

a rate incompatible with the lithology of these upper Oligocene sediments (diatom-rich calcareous nannofossil oozes, Kennett, Barker and others, 1988). The lowest occurrence of *R. gelida* is located in lower Chron C8n in Hole 744A (Baldauf and Barron, 1991). In Hole 689B, it is located between Cores 8H-4, 74–76 cm and 8H-5, 74–76 cm (Kennett and others, 1988a) or in Core 8H-6 (Gersonde and Burckle, 1990; i.e., in the upper part of the normal polarity interval identified by Spiess, 1990, as Chron C8n). On this basis, we suggest that this interval corresponds to Subchron C8n.2n, that the interval of primarily reversed polarity that Spiess (1990) assigned to Chron C7r and C7A may represent Subchron C8n.1r, and that the normal polarity interval immediately below the unconformity at 67 mbsf (Chron C7n in Spiess, 1990) represents part of Subchron C8n.1n.

The situation is similar in Hole 690B. Spiess (1990) assigned the polarity intervals below the unconformable Oligocene/Miocene contact at 51 mbsf in Hole 690B to Chron C7Ar, Chron C8 and Chron C9. Whereas Spiess (1990, p. 292) noted that "Subchron C11n between 72 and 74 mbsf is extraordinarily short in comparison to C11r" (and interpreted by him as indicative of a stratigraphic gap encompassing the upper part of Chron C11n), Spiess did not remark on the similarly remarkably thin interval representing Chron C8r that results from his interpretation. We suggest that the magnetic reversal succession was not correctly interpreted by Spiess (1990). The LO of *R. gelida* is located in Core 7H-3 (Gersonde and Burckle, 1990) which indicates that the reversed polarity interval in Core 7H-1, 100–102 cm to Core 7H-2, 124–126 cm likely corresponds to Subchron C8n.1r. If these reinterpretations are correct, Holes 689 and 690 do not provide evidence that at the extreme southern high latitudes, the HO of *C. altus* is younger than Chron C8n.

This review shows that there is no firm evidence that the HO of *C. altus* is diachronous across latitudes with diachrony increasing with increasing latitude. However, there may be slight (~1 my) diachrony between low-mid and high southern latitudes, as suggested by the different magnetobiostratigraphic correlations at southern high-latitude sites (in Chron C8n) compared to Site 522 (in Chron C8r; cf. Percival, 1984; Poore and others, 1982).

The review of the reliability of other datums in Table 16 shows that the situation is similar for many other taxa. In particular, we point to the fact that the magnetobiostratigraphic records in many sections (discussed in Table 16) do not allow the precise chronologic ties between paleontologic datums and magnetic reversals shown by some (Wei, 1991, 1992; Wei and Wise, 1989, 1990, 1992; Wei and others, 1992). These records are not clean and detailed enough to support comparisons between sections at the resolution of 10,000 to 100,000 yr claimed by these authors. Discrimination of 200,000 yr for any datum between sections should already be regarded questionable. Thus the claim of progressively increasing diachrony with increasing latitude is considered unwarranted for most datums at this time (see Table 16, Items 9 to 11 in particular). There are, however, datums that are diachronous between low-mid and southern high latitudes. This is the case of *Reticulofenestra umbilicus/R. hillae* (two close morphotypes that are treated as one species herein for convenience) and *Ericsonia formosa*. The former disappeared earlier from the low-mid latitudes than from the southern latitudes, whereas the latter disappeared very early from the southern latitudes. The diachrony involved is ~1 my for *R. umbilicus* and perhaps as much as 7 my for *E. formosa* and is likely directly linked to water mass evolution rather than to a latitudinal (temperature) effect.

A few datums with chronologic potential that needs to be explored are tentatively given (e.g., FAD and LAD of *Sphenolithus umbrellus*; LAD of *S. pseudoradians*). Altogether 16 Oligocene datums are discussed.

Neogene Period.—

Miocene Epoch.—Although many Miocene sections have been drilled by DSDP/ODP, few sections have yielded a magneto-

biostratigraphic record of sufficient quality from which to derive a reliable calcareous nannofossil biochronology. The Miocene calcareous nannofossil chronology in BKVC85 relied heavily on the magnetobiostratigraphic correlations at DSDP Site 516 (for the early and late Miocene; Berggren and others, 1983b), 558 and 563 (for the middle Miocene; Miller and others, 1985), and to a lesser extent on DSDP Sites 519 and 521 (for the middle and late Miocene; Poore and others, 1983). The best sections subsequently recovered are from DSDP/ODP Sites 608 (Olafsson, 1991; Gartner, 1992), 710 (Backman and others, 1990; Rio and others, 1990), 711 (Rio and others, 1990) and more recently from ODP leg 138 Sites 844, 845, 848, 852 and 853 (Raffi and Flores, 1995; Raffi and others, 1995). The magnetostratigraphic record at some of the ODP Leg 138 sites is of excellent quality (Mayer and others, 1992; Schneider, 1995), thus allowing exceptionally precise ties between biostratigraphic and magnetostratigraphic events (Raffi and Flores, 1995; Raffi and others, 1995). There remain, however, inconsistencies between the correlations established from these sites and those in other sections. The synchrony of biostratigraphic events is currently being challenged by most specialists, and despite the fact that excellent synchrony (with a resolution better than 50,000 yr) has been demonstrated for some Pliocene markers (Backman and Shackleton, 1983), the consensus remains that diachrony is widespread among calcareous nannofossils accounting for different magnetobiostratigraphic correlations at different locations. As for the Paleogene Period we do not share this view and would argue that undeciphered unconformities of regional extent are a greater impediment than diachrony in constructing a biochronologic framework.

There are several problems specific to Miocene magnetobiochronology. Firstly, the polarity reversal pattern is far more complex than for any interval in the Paleogene System, with a high reversal frequency. While this increases the potential for greater chronologic resolution, it also makes the interpretation of the magnetostratigraphic record difficult unless the magnetic reversal stratigraphy is of excellent quality. Comparison between the two interpretations given of the upper middle Miocene magnetostratigraphic record in Hole 563 by Miller and others (1985) and by Miller and others (1994) illustrates well the seriousness of the problem. Secondly, whereas the integration of calcareous nannofossil and planktonic foraminiferal stratigraphy with magnetostratigraphy is particularly critical for interpreting Miocene sections, many DSDP/ODP sections suffer from the lack of correlations between the two microfossil groups. This is the case in particular for the Indian Ocean and the equatorial Pacific sites. Thirdly, some of the magnetobiostratigraphic correlations proposed here are based on a single record or on magnetic records that are less than straightforward, and there is a certain amount of circular reasoning involved in revising polarity interval idenfication based on newer but ambiguous magnetobiostratigraphic correlations (as for instance in Hole 563, Miller and others, 1994, see below). For these reasons, we believe that Miocene biochronology will still undergo major changes. Progress in isotope stratigraphy will help in improving the temporal interpretation of stratigraphic sections (e.g., Wright and Miller, 1992; Miller and others, 1994), but even so, pattern matching between isotopic peaks can be deceiving as exemplified in Aubry and others (1995). A particularly difficult interval is the upper middle Miocene Series

(Zones NN8-NN10; Zones N15-N16). The problem is very complex and involves the stratigraphic relationships between several calcareous nannofossil (see Table 17, Items 9 to 15) and planktonic foraminiferal species. In particular, the relationships among the LOs of *Discoaster hamatus, Catinaster calyculus, Neogloboquadrina acostaensis* and *N. mayeri,* and between these and the magnetic polarity record vary between sections (see discussion in BKVC85, Aubry (1993a), Berggren (1993) and Tables 13, 17). This has been differently interpreted as reflecting diachrony (Miller and others, 1994) or unconformities (Aubry, 1993a, b), but there is no definitive evidence at this time to support one or the other explanation. The difficulty is exacerbated by the fact that, contrary to Miller and others's (1994) claim of a dichotomy between tropical and subtropical late middle and early late Miocene markers, no clear pattern in the inconsistencies emerges which would support diachrony (see Table 17). We thus provide, when necessary, two alternative sets of correlation between calcareous nannofossil events and magnetic reversals (Table 17).

Another difficult relationship to establish is that between the LAD of *S. heteromorphus* and the magnetic reversal polarity. This LAD was tied to Chron C5ADn and given an estimated age of 14.4 Ma in BKF85, based on the polarity record in Hole 563 (Miller and others, 1985). A revised age estimate of 13.6 Ma was subsequently proposed by Backman and others (1990) based on Site 710 where the datum occurs in an interval without interpretable magnetic polarity record due to slumping. The validity of this inexplicit revision (in an interval which displays numerous slumps) was emphasized by Backman and others (1990), because of the apparent agreement between the magnetobiostratigraphic correlation (with Chron C5ABr) inferred from this age and that directly established in Hole 608. We caution that despite the apparently straightforward magnetobiostratigraphic record in Hole 608 (Gartner, 1992), the magnetic polarity pattern is no less ambiguous in this hole than in Holes 563 and 558 (Miller and others, 1985). The normal polarity intervals interpreted as Chrons C5ADn and C5ACn (Clement and Robinson, 1987) are anomalously thin (and thinner than Chrons C5AAn and C5ABn) in Hole 608, and Chron C5ACr was not recovered possibly due to no recovery in lower Core 33. However, identification of the reversal pattern between Cores 34 and 31 is strictly based on numerical succession, assuming stratigraphic completeness and accepting tremendous variations in rates of sedimentation at this location. Miller and others (1994) followed the magnetic interpretation in Hole 608 and reinterpreted the magnetic reversal pattern in Hole 563. We agree with their assignment to Chron C5AAn of the normal polarity interval originally ascribed to Chron C5ACn (Miller and others, 1985), but do not support their reassignment to Chron C5AB-C5AD of the normal polarity interval initially assigned to Chron C5AD-C5B (Miller and others, 1985). If the LAD of *S. heteromorphus* is located in Chron C5ABr following Gartner (1992), the upper part of this latter normal polarity interval can only represent Chron C5ACn or older. This indicates the absence of Chron C5ABn in Hole 563. Considering the ambiguous magnetic polarity record in Hole 608 around the Chrons C5AB-C5AD interval, it seems reasonable to await confirmation from another magnetobiostratigraphic record to firmly establish the age of the LAD of *S. heteromorphus.* Recent magnetobiostratigraphic correlations in the central Paratethys Basin

(Steininger and others, 1995) show that the Badenian/Sarmatian (sensu Suess) boundary lies within calcareous nannofossil Zone NN6 and correlates with Chron C5ABr at 13.6 Ma which supports an older age of the LAD of *S. heteromorphus.*

Following Raffi and Flores (1995), we incorporate the recently defined subzonal interval which corresponds to the total range of *Amaurolithus amplificus* in the reference biostratigraphic frameworks of Okada and Bukry (1980) and Martini (1971). We also tentatively include a few datums that are very useful for lower Miocene stratigraphy but are poorly tied to the magnetic polarity record. We prefer to refrain from citing datums which are stratigraphically very useful (e.g., HO of *Discoaster neohamatus,* LO of *D. pentaradiatus, D. surculus,* LO of *Helicosphaera ampliaperta,* see Gartner and others, 1983; Aubry, 1993a, b) but which may not be satisfactorily tied to the magnetic scale. We do not believe that the LAD of *Triquetrorhabdulus carinatus,* one of Martini's (1971) zonal markers, has yet been satisfactorily tied to the magnetic polarity reversal. Thirty Miocene datums, most of which were considered by Gartner (1992), Raffi and Flores (1995) and Raffi and others (1995), are further discussed here (Table 17).

Pliocene and Pleistocene Epochs.—For calcareous nannofossil chronology in the Pliocene and Pleistocene Epochs, we refer the reader to the work of Berggren and others (1995).

The biozonal criteria and the duration of the Paleocene to upper/late Miocene biozones/biochrons based on Martini's (1971), Bukry's (1973, 1975) and Okada and Bukry's (1980) zonal schemes are summarized in Figures 17 to 22.

CHRONOSTRATIGRAPHY

Paleogene Period

A revised geochronology for the Paleogene Period based on the integration of calcareous plankton biostratigraphic data to the revised magnetochronology of CK92/95 has been presented above in Figures 1 (Paleocene), 2 (Eocene) and 3 (Oligocene) with corresponding chronology of the (sub)tropical planktonic foraminiferal datum events used to delineate zonal boundaries shown in Figures 8, 9 and 10, respectively, and the chronology of calcareous nannofossil zonal boundaries shown in Figures 17 to 19. In this section, we discuss (predominantly post-1985) biostratigraphic data relevant to identification and correlation of the limits and extent of Paleogene chronostratigraphic units.

The terms Paleogene and Neogene are used here as period/system subdivisions of the Cenozoic Era/Erathem. The term Tertiary is considered, like the other antiquarian subdivisions of earth history (Primary, Secondary and Quaternary), as inappropriate, if not "obsolete" (GTS89; p. 3). A thorough discussion of Paleogene chronostratigraphic terminology and recommended usage was presented in BKF85 (see also Berggren, 1971). Discussion here is limited to aspects of Paleogene chronostratigraphy requiring elaboration or modification in the context of the revised geochronology adopted here.

The Cretaceous/Paleogene boundary is located within the younger part of Chron C29r with an estimated age (CK92) of 66.0 Ma and 65 Ma (CK92/95; adopted here). The age estimate in CK92 was based upon a "compromise" between problems current at that time in obtaining consistent radioisotopic isotopic ages for the K/P boundary (see discussion by Swisher in

TABLE 17.—MIOCENE CALCAREOUS NANNOFOSSIL MAGNETOCHRONOLOGY. SEE TABLE 14 FOR EXPLANATION.
MIOCENE

Datum	FAD	LAD	Paleomagnetic Chron	Age in Ma	Reference	Remarks
1. *Discoaster quinqueramus*		X	Mid Chron C3r	5.6	16,17,19	HO tied to mid Chron C3r in Holes 710, 844, and to late Chron C3r in Holes 845 and 854 (refs. 16, 17).
2. *Amaurolithus amplificus*		X	Chron C3r/C3 An.1n boundary	5.9	16,17	HO tied to the Chron C3r/Subchron C3An.1n boundary in Holes 844 and 853, and to mid Subchron C3An.1n in Hole 845 (refs. 16, 17).
3. *Amaurolithus amplificus*	X		Subchron C3An.2n/ Chron C3 Ar boundary	6.6	16,17	LO tied to the Subchron C3An.2n/Chron C3Ar boundary in Holes 844 and 853 (refs., 16, 17).
4. *Amaurolithus primus*	X		Subchron 3Br.2r	7.2	16,17	LO tied to mid Subchron C3Br.2r in Holes 710, 844, 845, 848 (although only the early part of Chron C3Br is represented in the hole) and 853 (refs. 16, 17).
5. *Discoaster loeblichii*		X	Chron C4n/C3Br	7.4	16,17	Tied to latest Chron C4n in Hole 744 but to mid Chron C4n in Hole 853 (refs. 16, 17); to early Chron C3A in Hole 608 (ref. 8), but the Subchron C3An.2n/C3Ar boundary is not delineated in the hole.
6. *Minylitha convallis*		X	Subchron C4n.2n	7.8	16,17	HO slightly below the LO of *A. primus* in many sections including cores from the Gulf of Mexico (refs. 2, 9), Indian Ocean Sites 709 and 710 (ref. 18) and equatorial Pacific Sites 844 and 845 (refs. 16, 17). HO tied to late Subchron C4n.2n in Holes 710, 844, 854 (refs. 16, 17); HO tied to Chron C4n in Holes 698 (ref. 10) and 848 (refs. 16, 17).
7. *Discoaster berggrenii*	X		Subchron C4r.2r	8.6	16,17	We agree with Raffi and Flores (ref. 16) that *D. berggreni* and *D. quinqueramus* are distinct morphotypes of the same taxon, with slightly different ranges. LO tied to earliest Chron C4r.2r in Holes 710, 844, 845 and 848 (refs. 16, 17); to late Chron C4n in Hole 608 (ref. 8) but we note the anomalously close proximity of this LO and the LO of *A. primus* in this hole.
8. *Discoaster loeblichii*	X		Chron 4r/C4An boundary	8.7	16,17	LO tied to the Chron C4r/C4An boundary in Hole 844 (refs. 16, 17) and to late Chron C4An in Hole 608 (ref. 8).
9. *Discoaster bollii*		X	Atlantic-Caribbean realm: Subchron C4Ar.1r	9.1	8,10	Given in Chronozone C4Ar.2r in Holes 844 and 710 at a position similar to that in Holes 563 and 608 (ref. 17). However, in Holes 844 and 710, the HO of *D. bollii* is immediately above the LO of *D. hamatus* (ref. 17) whereas
			Indo-Pacific realm: Subchron C4Ar.2r	9.4	16,17	in Holes 563 (ref. 10) and 608 (ref. 8), in several cores from the Gulf of Mexico (ref. 2) and in the Buff Bay section, Jamaica, (ref. 1), the two events are well separated. The magnetic polarity record is insufficient to locate precisely in Chronozone C4Ar the HO of *D. bollii* in Holes 563 and 608, but it is more appropriately located in the upper part than in the lower part of the chrono zone. We tentatively tie the LAD of *D. bollii* to Subchron C4Ar.1r in the Atlantic-Caribbean realm where this event is younger than the LAD of *D. hamatus* and the FAD of *M. convallis*. (see item 10).
10. *Minylitha convallis*	X		Equatorial Pacific: Subchron C4Ar.1n	9.3	16,17	LO juxtaposed with the HO of *D. hamatus* in Chronozone C4Ar.2r in Hole 710 (refs. 16, 17). LO considered restricted in the equatorial Pacific Ocean compared to the tropical
			Atlantic and Indian Ocean regions: Subchron C4Ar.2r	9.5	17	Indian Ocean (ref. 17): At equatorial Pacific Sites 844 and 848, its LO, well into Zone NN10, is well above the HO of *D. hamatus* and located in Subchronozone C4Ar.1n (refs. 16, 17). We point out that in several cores from the Gulf of Mexico, the LO of *M. convallis* is consistently found well above the HO of *D. hamatus* (ref. 2), as in the equatorial Pacific cores, and this corresponds well with the range given by Bukry (ref. 6) for this species. Also, in the Gulf of Mexico cores and in the Buff Bay section, Jamaica, there is a substantial overlap of the ranges of *D. bollii* and *M. convallis* (refs. 1, 2) whereas in Holes 710 and 844, the LO of the latter is immediately above the HO of the former (refs. 17). Until further studies establish whether these discrepancies reflect diachrony, we tentatively propose two distinct ages for the FAD of *M. convallis*, for the equatorial Pacific Ocean, and for the Atlantic-Indian ocean regions.
11. *Discoaster hamatus*		X	Subchron C4Ar.2r	9.4	10,16,17	HO located in mid Chron C4Ar.2r in Holes 844 (but the HO is poorly defined in this hole because it is located just above a dissolution interval) and 848 (refs. 16, 17); in Chron C4Ar in Holes 608 (ref. 8) and 710 (refs. 3, 17, 18); in upper Chron C5n in Hole 563 (ref. 10). See Items 9, 10 for comments regarding the relationships between the HO of *D. hamatus* and the HO/LO of other species in different regions.

TABLE 17.—*Continued.*
MIOCENE

Datum	FAD	LAD	Paleomagnetic Chron	Age in Ma	Reference	Remarks
12. *Catinaster calyculus*	X		Subchron C5n.2n	10.7	10,16,17	The relationship between this datum and the FAD 's of *C. coalitus* and *D. hamatus* are still poorly understood. In Hole 608 (ref. 8), in the Buff bay section, Jamaica (ref. 1), and in several Eureka cores from the Gulf of Mexico (ref. 2), the LO of *C. calyculus* falls in the upper range of *D. hamatus* (well justifying Bukry's subdivision of Zone CN7, ref. 7). In Holes 563 (ref. 10) and in Hole 710 (refs. 16, 17), the LO of *C. calyculus* is at the same level as the LO of *D. hamatus*. In Hole 714 the LO of *C. calyculus* precedes that of *D. hamatus* (ref. 17). In Hole 521A where *D. hamatus* is absent, the LO of *C. calyculus* is above the LO of *C. coalitus* and the two species co-occur for part of their ranges (ref. 20), where as in Hole 519 where *D. hamatus* is also absent, the LO of *C. calyculus* immediately follows the HO of *C. coalitus* (ref. 15). In Hole 563, the LO of *C. calyculus* is located in early Chron C5n.2n (ref. 10); in Hole 608, it is located in Chron C4Ar (ref. 8; but this may not be reliable, the species being very rare); in Hole 710, it is located below Polarity Subchronozone C5n.2n (ref. 18) although the reliability of this correlation is questioned by Raffi and others (ref. 17) owing to slumping below the normal polarity interval assigned to Chron C5n.2n. Finally, in Holes 844 and 845, the LO of *C. calyculus* is located in lower Subchron C5n.2n, slightly below the LO of *D. hamatus* (in a position similar to that in Hole 563).
13. *Discoaster hamatus*	X		Some locations: Subchron C5n.2n	10.7	10,11,16,17	Controversial datum. Its relationships to other datums (planktonic foraminifera as well as calcareous nannofossils) and to the magnetic polarity is highly inconsistent between sites, *regardless of latitude*, and is difficult to interpret because the
			Other locations: Subchron C5r.2r	11.2	11	species is absent from critical sites (e.g., Sites 519 and 521), or the magnetic polarity pattern is ambiguous (e.g., Buff Bay section, Jamaica). LO located in mid Subchronozone C5n.2n in Hole 845 and 848 (refs. 16, 17) and 608 (ref. 8) but in lower Chronozone C5n.2n in Hole 844 (refs. 16, 17). LO located in lower Subchronozone C5n.2n in Hole 563 (ref. 10), in a reversed polarity interval below Subchronozone C5n.2n in Hole 710 (ref. 18) but the validity of this correlation is questioned owing to slumping below the normal polarity interval (ref. 17). Located in an interval of primarily reversed polarity below Chronozone C5n (interpreted as Chronozone C5r.2r) in the Buff Bay section, Jamaica (ref. 11).
14. *Coccolithus miopelagicus*		X	Equatorial Pacific: Subchron C5n.2n	10.8	16,17	HO located in the younger part of Zone NN8 (ref. 1, 2, 6, 17). Located in lower Subchronozone C5r.1r in Hole 608 (ref. 8), but in lower Subchronozone C5n.2n in Holes 844 and 845
			North Atlantic Ocean: Subchron C5r.1r	11.0	8	(where it is clearly below the LO of *D. hamatus*) and in mid Subchronozone C5n.2n in Hole 848 (where it is juxtaposed with the LO of *D. hamatus*; refs. 16, 17).
15. *Catinaster coalitus*	X		Equatorial Pacific: Subchron C5n.2n	10.9	16,17	One of the most controversial Miocene datums because of inconsistent correlations to the magnetic polarity record. There are currently two sets of magnetobiostratigraphic correlations:
			Atlantic and (?) Indian Oceans: Subchron C5r.2r	11.3	10	1) FAD located in early Chron C5n.2n as determined from Hole 608 (ref. 12; but if so, LO anomalously above the HO of *C. miopelagicus* in the hole, see ref. 8 and Item 14), 519 (ref. 15), 521A (ref. 20), and 845 (ref. 16, 17). The position of the FAD is not constant in the chron, but varies from early Chron C5n.2n as in Hole 519 (ref. 15) to earliest Chron C5n.2n as in Hole 845 (ref. 17).

2) the FAD is located in Chron C5r at a moment that remains uncertain because of insufficient magnetic records. This is seen in Hole 563 (ref. 10), 710 (ref. 17) and in the Buff Bay section, Jamaica, which yields a magnetic polarity record difficult to interpret but where the LO of *C. coalitus* is located in an interval of predominantly reversed polarity assignable to Chron C5r (ref. 11).

This second correlation was followed in Berggren and others (ref. 5). Despite the excellent resolution of the magnetobiostratigraphy in the equatorial Pacific sites (refs. 16, 17), we do not believe that the conundrum has been resolved because it involves not only *C. coalitus* but several other species among calcareous nannofossils (e.g., *C. miopelagicus, D. hamatus, C. calyculus*; see Items 12 to 14) and among the planktonic foraminifera (e.g., *N. acostaensis, N. mayeri*; see Tables 12, 13). Until the reason for the discrepancy is firmly established, we provisionally propose two ages for this datum.

TABLE 17.—*Continued.*
MIOCENE

Datum	FAD	LAD	Paleomagnetic Chron	Age in Ma	Reference	Remarks
16. *Discoaster kugleri*		X	Subchron C5r.2n	11.5	8	Located in Polarity Subchronozone C5r.2n in Holes 608 (ref. 8) and 845 (as a common occurrence; ref. 16, 17). Raffi and Flores (ref. 16) indicate that the LAD of *D. kugleri* is in fact in Subchron C5n.2n as determined from Hole 845. Because of the discontinuous occurrence of the species in the upper part of its range in Hole 845 we question the validity of this determination. See also Item 17.
17. *Discoaster kugleri*	X		Subchron C5r.3r	11.8	8	LO in lower Subchronozone C5r.3r in Hole 845 (as common occurrence, refs. 16, 17) but inupper Subchronozone C5r.3r in Hole 608 (ref. 8). LO given in Chronozone C5AA in Hole 563 (ref. 10) based on a broad concept of the species (see ref. 1). Revised LO located in the lower reversed magnetic polarity interval in Chronozone C5 in this hole, in good agreement with Site 608. This species is usually accorded a very short range but Raffi and Flores and Raffi and others (refs. 16, 17) indicate that this species has in fact along range which extends from Subchron C5An.1n to C5n.2n as determined from Hole 845 (ref. 16). The species shows however discontinuous occurrence in the hole which is unfamiliar to us and the determination of Raffi and Flores (ref. 16) requires confirmation.
18. *Triquetrorhabdus rugosus*	X		Chron C5AAn	13.2	8	Located in mid Chronozone C5AAn in Hole 608 (ref. 8) and at base of normal polarity interval reinterpreted as Chron C5AAn in Hole 563 (ref. 11), but in Chronozone C5Ar.1n in Hole 845 (ref. 16, 17).
19. *Sphenolithus heteromorphus*		X	Chron C5ABr	13.6	8	Located in lower Chronozone C5ABr in Hole 608 (ref. 8). Located in normal polarity interval questionably assigned to Chron C5AD in Hole 558 (ref. 10). Located in normal polarity interval assigned to Chron C5AD (ref. 10) but reinterpreted as Chron C5ABn (ref. 11) in Hole 563. This latter revised interpretation is not satisfactory however, the LAD of *S. heteromorphus* being tied to Chron C5ABr in Hole 608. At best this normal represents Chron C5ACn, and Chron C5ABn is not represented in Hole 563 (an inferred unconformity is well supported by the juxtaposition of the HO of *S. heteromorphus* and the LO of *T. rugosus* in the hole (ref. 10)).
20. *Helicosphaera ampliaperta*		X	Chron C5Br	15.6	8	Located in the lower part of Chronozone C5Br in Hole 608 (ref. 8) and in lowest Chronozone C5Br in Hole 558 (ref. 10). Associated with a thin normal polarity interval questionably assigned to Chron C5C in Hole 563 (ref. 10; however, the three normal polarity intervals in Hole 563 are very thin and poorly representative of Chron C5C).
21. *Sphenolithus heteromorphus*	X		Chron C5Dr	18.2	8	LO follows closely the HO of *S. belemnos* in many sections (e.g., Site 667, ref. 13). Located in lower Chronozone C5D in Hole 608 (ref. 8) and in mid Chronozone C5Dr in Hole 710 (refs. 3, 18). Given just older than Chronozone C5Cn in Hole 516 (ref. 4) but cross correlation with planktonic for aminifer a between Holes 516 and 563 justifies a revised assignment to Chronozone C5Dn.
22. *Sphenolithus belemnos*		X	Chron C5En	18.3	8	Located in uppermost Chronozone C5En in Hole 608 (ref. 8). Also located in Chronozone C5En in Hole 710 (ref. 3) but this chronozone constitute a very thin interval in the hole. Given in a reversed polarity interval just younger than Chron C5D in Hole 516 (ref. 4) and not recovered in Holes 558 and 563 where a hiatus encompasses early Chron C5Dr to late Chron C6n (ref. 10). Cross correlations between Holes 516, 558 and 563 based on planktonic for aminifera and calcareous nannofossil do not support the magnetic polarity interpretation in ref. 4.
23. *Triquetrorhabdulus carinatus*		X	?	?	—	Difficult datum which may not be applicable in some sections where it ranges well above the HO of *S. belemnos* (e.g., Holes 516F (ref. 4) and 710 (ref. 3). Not recovered in Holes 558 and 563 because of an unconformity (ref. 10). HO located in lower Chronozone C6Bn in Hole 608 (ref. 8), well below the LO of *S. belemnos*, and therefore does not correspond to the LAD of the species.
24. *Sphenolithus belemnos*	X		Chron C6n	19.2	8,10,18	Tied to Chron C6n in Holes 563 and 558 (ref. 10). Tied to late Chron C6n in Hole 710 (ref. 18) but Chron C5E is poorly represented in the hole. Tied to late Chron C6n in Hole 608 (ref. 8), but the HO of *S. belemnos* is located in the younger part of a normal polarity interval assigned to Chrons C6 and C6A and the magnetic reversal pattern is obscure between Chron C5E and C6B in the hole. Located in a normal polarity interval identified as Chron C5ADn in Hole 516F (ref. 4) but see comments regarding Item 22

TABLE 17.—*Continued.*
MIOCENE

Datum	FAD	LAD	Paleomagnetic Chron	Age in Ma	Reference	Remarks
25. *Sphenoltithus sp. aff. S. belemnos*	X		? Chron C6Cr	—	—	Characteristic form regarded as an early morphotype of *S. belemnos* with a short spine (see ref. 2). Has its LO in mid to upper Zone NN2. LO located slightly below the normal polarity interval tentatively assigned to Chron C6An in Hole 563 (ref. 10) but interpreted here as part of Chron C6C. This datum needs to be firmly tied to the magnetic polarity pattern.
26. *Triquetrorhadulus serratus*		X	Chron C6Br	23.2	4	LO located at the same level as the LO of *D. druggii* in Hole 516F (ref. 4).
27. *Discoaster druggii*	X		Chron C6Br	23.2	4	Located in Chronozone C6Br in Holes 516F (ref. 4; but the chron C6Br/C6Cn boundary is not delineated in the hole), 558, 563 (ref. 10) and 608 (cf. ref. 8, Figs. 1, 2). FAD not in Chron C6A as in Young and others (ref. 21) based on a mis-interpretation of the magnetic polarity stratigraphy in Hole 522 (ref. 15).
28. *Sphenolithus umbrellus*		X	Subchron C6Cn.1r	23.6	10	Tied to younger normal polarity interval of Chron C6Cn in Hole 563 but Chron C6Cn is represented by only 2 normal polarity intervals in the hole (ref. 12). Tentatively correlated with mid Subchron C6Cn.1n which makes it an earliest Miocene datum.
29. *Sphenolithus capricornutus*		X	Subchron C6Cn.2n	23.7	10	Given in Subchron C6cn.2n (ref. 8) but see comments for item 29
30. *Sphenolithus delphix*		X	Subchron C6Cn.2n.	23.8	8,10	LAD given in Chron C6Cn.2n (ref. 8). Chron C6Cn is poorly characterized in Hole 608 and the three subchrons C6Cn.1n, 2n and 3n are undifferentiated, which precludes a precise location in Chron C6Cn.2n. In Hole 608, the HO of *S. delphix* is at the same level as that of *S. capricornutus* (ref. 8) contrary to Hole 558 (ref. 10). In Hole 563, the HO of *S. delphix* is at the same level as the HO's of *R. bisecta* and *S. capricornutus*, at the Subchronozone C6Cn/C6Cr boundary (ref. 10). However, in this hole, Chronozone C6Cn comprises only 2 normal events and the simultaneous HO's of the three species suggest the presence of an unconformity at the polarity reversal.

[1]Aubry, 1993a
[2]Aubry, 1993b
[3]Backman and others, 1990
[4]Berggren and others, 1983a
[5]Berggren and others, 1985a
[6]Bukry, 1973
[7]Bukry, 1975
[8]Gartner, 1992
[9]Gartner and others, 1983
[10]Miller and others, 1985
[11]Miller and others, 1994

[12]Olafsson, 1989
[13]Olafsson, 1991
[14]Poore and others, 1982
[15]Poore and others, 1983
[16]Raffi and Flores, 1995
[17]Raffi and others, 1995
[18]Rio and others, 1990
[19]Shackleton and others, 1995
[20]Van Salis, 1984
[21]Young and others, 1994

Berggren and others, 1992, and discussion above in this paper); these ages currently range from <65 to >66 Ma.

Paleocene Series.—

The Paleocene Series consists of three stages: the Danian, Selandian and Thanetian. The Danian Stage was discussed at length by Berggren (1971) and BKF85. In BKF85, the Thanetian was considered an upper substage of the Selandian Stage with the lower substage (=lower Selandian) remaining unnamed. This is recognized as a completely unsatisfactory situation and the three units are used here in sequential order to represent the Paleocene Series. The base of (exposed) Thanetian sediments in SE England is within Chron C26n and (probably) correlative with a level in Zone NP6 (Knox and others, 1994) and corresponds to the TA2 (2.1/2.2) (56 Ma) 3rd-order cycle of Haq and others (1988). The age of the base of the Thanetian Stage is estimated at 58 Ma here. Between the top of the Danian (61.0 Ma) and the base of the Thanetian (58 Ma) are ~3 my unrepresented by an appropriate chronostratigraphic term. The Danian Stage could be extended upward to coincide with the base of the next younger (Thanetian) stage or the Thanetian could be lowered to coincide with the top of the Danian Stage.

An alternative, and more appropriate, procedure would be to insert the Selandian Stage for the intervening stratigraphic interval between the top and base of the Danian and Thanetian Stages, respectively (and this has been now accepted by the Paleogene Subcommission on Stratigraphy). The Selandian Stage described by Rosenkrantz (1924) consists of lower (Lellinge Greensand) and middle (Kerteminde Clay) fossilferous units and an upper ("gray unfossiliferous clay") and Holmehus unit. It lies unconformably upon the Danian Chalk Formation and contains clasts of this unit in its basal part. It is overlain, in turn, by the well known "Ash series" (=Mo Clay) which characterizes the upper part of the Holmehus and Fur/Ølst Formations. The major lithologic change from (Danian) carbonates (below) to (Selandian) clastics (above) reflects a major change in the geotectonic evolution of the NE Atlantic (i.e., marginal uplift and basinal subsidence prior to the initiation of late Paleocene sea-floor spreading in the Norwegian-Greenland Sea during Chrons C25-C26).

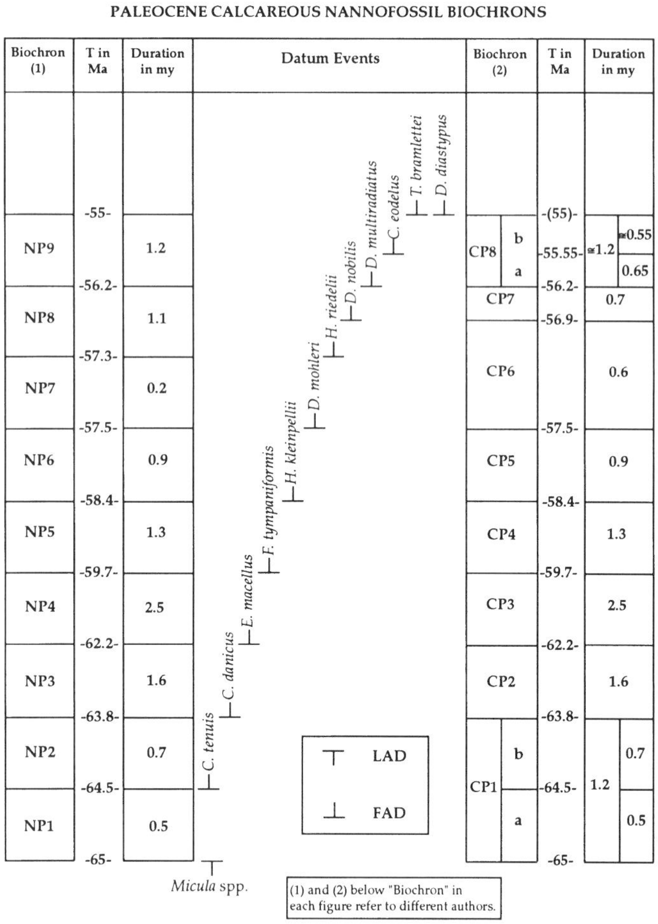

PALEOCENE CALCAREOUS NANNOFOSSIL BIOCHRONS

EOCENE CALCAREOUS NANNOFOSSIL BIOCHRONS

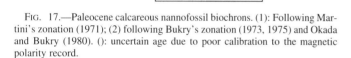

FIG. 17.—Paleocene calcareous nannofossil biochrons. (1): Following Martini's zonation (1971); (2) following Bukry's zonation (1973, 1975) and Okada and Bukry (1980). (): uncertain age due to poor calibration to the magnetic polarity record.

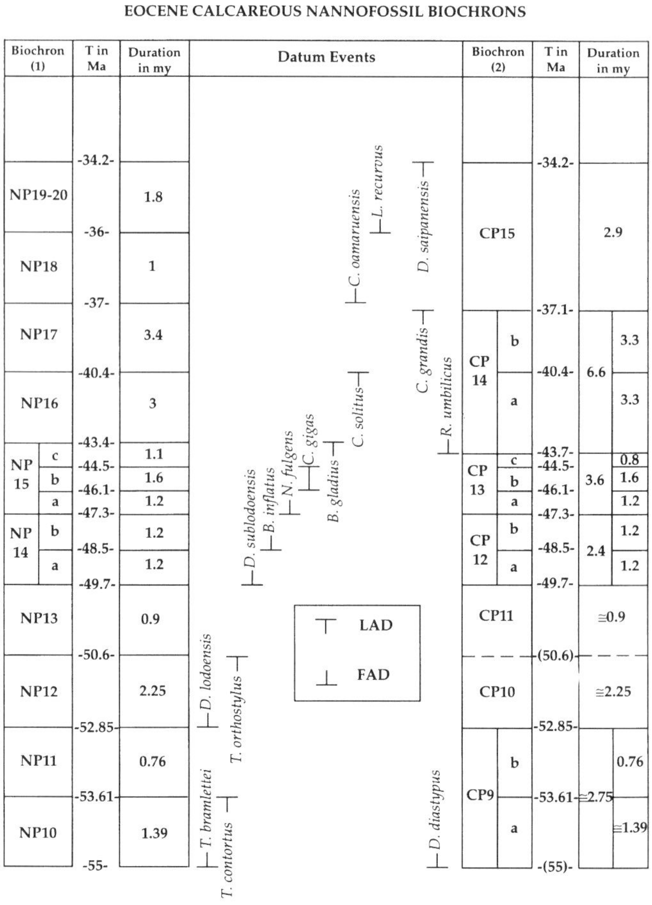

FIG. 18.—Eocene calcareous nannofossil biochrons. See Figure 17 for explanation.

Integration of bio- and magnetostratigraphic studies, particularly those recently completed at DSDP Hole 384, suggests that:

1. The top of the Danian sensu stricto correlates approximately with Chron C27n(o) (estimated age: 61.2 Ma), and that of the Danian sensu lato (= Montian Stage sensu stricto) with Chron C27n(y) (estimated age 61.0 Ma; cf. earlier correlations in BKF85 and Berggren, 1994, based on suggested correlations of these levels with C27n(y) and C26r (early part)).

2. The base of the Lellinge Greensand corresponds to a level in the lower part of Chronozone C26r and to the TA1/TA2 relative coastal onlap cycle boundary of Haq and others (1988) with an age estimated here of ~60.5 Ma, whereas the Danian/Selandian boundary is (arbitrarily) placed at the Zone P2/P3 boundary (FAD of *M. angulata* and *Ig. pusilla*) with an estimated age of 61.0 Ma (Chron C27n(y); cf. Berggren, 1994, in which this level was estimated to have an age of 60.4 Ma, based on the belief that the P2/P3 zonal boundary lay in early Chron C26r). Recent data from DSDP Hole 384 have shown that the P2/P3 zonal boundary is in Chron C27n(y).

3. The lower and middle parts of the Selandian Stage are stratigraphically correlative with Zones P3 and NP4-NP5 (undifferentiated) and the *Deflandrea* (vel *Cerodinium*) *speciosa* and *Palaeoperidinium pyrophorum* (dinoflagellate) Zones. The base of the latter zone is defined by the FAD of *Alisocysta margarita* which appears to be approximately correlative with the NP5/NP6 zonal boundary. The overlying *Alisocysta margarita* Zone extends to (at least) the middle part of the Holmehus Formation (Denmark), although dinocysts are generally rare in this unit, and the overlying (calcareous barren) "Gray Clay" unit belongs to the (lower) *Apectodinium hyperacanthum* Zone characterized by the earliest representatives of the wetzeliellids which become prominent biostratigraphic markers in the lower Eocene Series.

4. The base of the Thanetian Stage lies in the *Alisocysta margarita* Zone (defined by the FAD of *Deflandrea denticulata*), within Zone NP6 and is correlated with the basal part of Chronozone C26n (Knox and others, 1994) with an age estimate of 57.9 Ma here. The *A. margarita* Zone extends into mid-Thanetian levels which are referable to Zone NP8 (i.e., equivalent to the Reculver Silts but older than the Lambeth Group, formally the Woolwich-Reading series).

5. It would appear that the Selandian and base Thanetian stages overlap within the *Paleoperidinium pyrophorum* Zone (i.e.,

OLIGOCENE CALCAREOUS NANNOFOSSIL BIOCHRONS (LOW-MID LATITUDES)

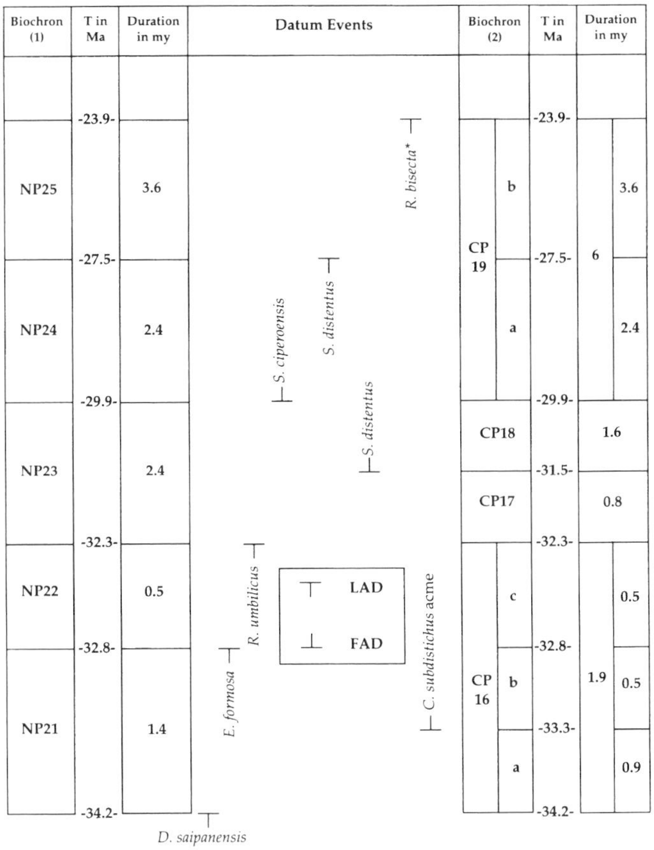

Biochron (1)	T in Ma	Duration in my	Datum Events	Biochron (2)	T in Ma	Duration in my
	-23.9-				-23.9-	
NP25		3.6		CP19 b		3.6
	-27.5-		R. bisecta*		-27.5-	
NP24		2.4	S. ciperoensis / S. distentus	a		2.4
	-29.9-				-29.9-	
NP23		2.4	S. distentus	CP18		1.6
					-31.5-	
	-32.3-			CP17		0.8
NP22		0.5	R. umbilicus		-32.3-	
	-32.8-			c		0.5
			E. formosa		-32.8-	
NP21		1.4	C. subdistichus acme	CP16 b	1.9	0.5
					-33.3-	
				a		0.9
	-34.2-		D. saipanensis		-34.2-	

Legend: ⊤ LAD ⊥ FAD

FIG. 19.—Oligocene calcareous nannofossil biochrons (for the low and mid latitudes; for ages of datum events and duration of chrons at southern high latitudes, see Table 16). See Figure 17 for explanation. *: *R. bisecta* is substituted for *H. recta*, the marker of the NP15/NN1 boundary in Martini's (1971) zonal scheme; +: end of acme.

EARLY AND MIDDLE MIOCENE CALCAREOUS NANNOFOSSIL BIOCHRONS

Biochron (1)	T in Ma	Duration in my	Datum Events	Biochron (2)	T in Ma	Duration in my
	-11.3/ 10.9-		C. coalitus		-11.3/ 10.9-	
NN7		0.5		CN5 b		0.5
	-11.8-		D. kugleri		-11.8-	2.3
NN6		1.8		a		1.8
	-13.6-				-13.6-	
NN5		2	S. heteromorphus	CN4		2
	-15.6-				-15.6-	
NN4		2.7	H. ampliaperta	CN3		2.7
	-18.3-				-18.3-	
NN3		≡0.7	S. belemnos	CN2		0.9
	-(19)-				-19.2-	
NN2		≡4.2	D. druggii / T. carinatus / S. belemnos	CN1 c		4
	-23.2-			a + b	-23.2-	4.7 / 0.7
NN1		0.7				
	-23.9-		R. bisecta*		-23.9-	

Legend: ⊤ LAD ⊥ FAD

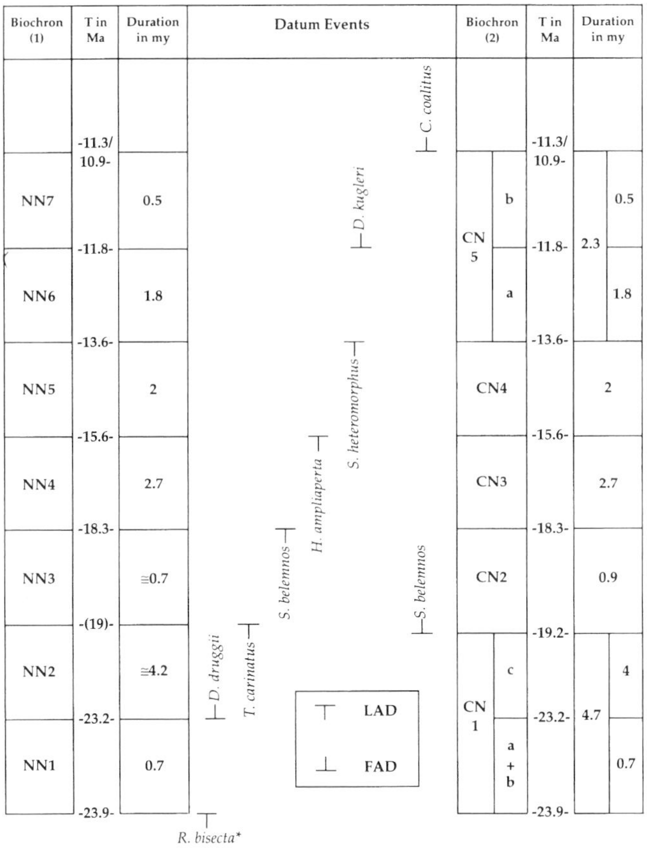

FIG. 20.—Early and middle Miocene calcareous nannofossil biochrons in the (applicable only to mid and low latitudes, the calcareous nannofossil assemblages being extremely poorly diversified at southern high latitudes). See Figure 17 for explanation. ( ): uncertain age due to poor calibration to the magnetic polarity record.

within Zone NP6) or *Alisocysta margarita* Zone (basal Zone NP7) and at a level approximately correlative with the lower part of Zone P4 (estimated here to be only slightly older than Chron C26n(o)). When it is possible to clearly constrain the limits of these two units, it would be appropriate to formally "decapitate" the upper (predominantly un-, or sparsely, fossiliferous) part of the Selandian Stage and use the term Thanetian Stage for the remainder of the (upper) Paleocene Series.

6. The base of the Selandian Stage is ~3 my older than the base of the Thanetian Stage.

7. The hiatus between the Danian sensu stricto and the base of the Lellinge Greensand (=base Selandian Stage s.str.) is about/less than 1 my and between the Danian sensu lato and that between the base of the Lellinge Greensand is ~0.5 my.

The Selandian Stage is seen to span the entire post-Danian Paleocene Series and corresponds in its middle and upper part with the Thanetian Stage of England and to the Woolwich-Reading Beds (=Sparnacian of France). The Selandian Stage (restricted sense) can be conveniently accomodated in the Paleocene chronostratigraphic hagiography between the top of the Danian (P2=NP4) and the base of the Thanetian Stages

(=NP6, basal P4a). This approach would have the advantage of having the stratotype area (and concomitant type sections) of two successive time-stratigraphic units, the Danian and Selandian, lying in spatial continuity (i.e., in Denmark), although a distinct unconformity representing ~1-my duration separates the lithostratigraphic units of the upper Danian and lower Selandian Stages. It also results in a more appropriate three-fold chronostratigraphic subdivision of the Paleocene Series in which the temporal gap between the top of the Danian Stage sensu stricto/sensu lato and the base of the Thanetian Stage is represented (i.e., stratotypified in the rock record). A number of distinct biostratigraphic events (particularly within the dinoflagellates and perhaps within the calcareous nannofossils) can serve as biostratigraphic guideposts to delimiting the boundaries of these units in NW Europe. Hopefully the magnetostratigraphic record will provide additional, critically needed information on the position of these units within the framework of the GPTS.

Paleocene/Eocene boundary.—

The Paleocene/Eocene boundary was a focal point of controversy and discussion in our original (BKF85) and subsequent modification (Aubry and others, 1988) of the Paleogene time

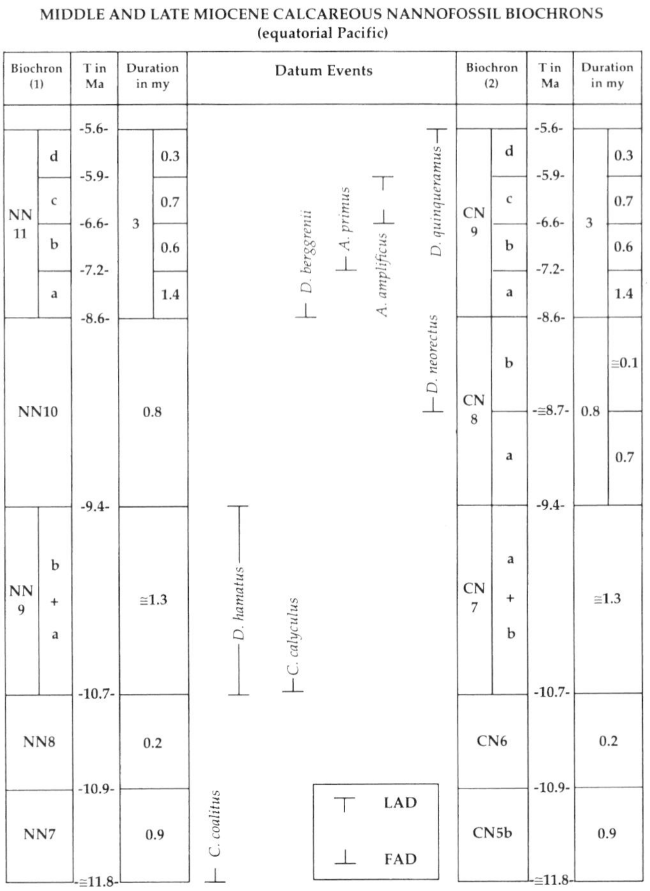

MIDDLE AND LATE MIOCENE CALCAREOUS NANNOFOSSIL BIOCHRONS
(equatorial Pacific)

Biochron (1)	T in Ma	Duration in my	Datum Events	Biochron (2)	T in Ma	Duration in my
NN11 d	-5.6- / -5.9-	0.3		CN9 d	-5.6- / -5.9-	0.3
NN11 c	-6.6-	0.7		CN9 c	-6.6-	0.7
NN11 b	-7.2-	0.6 (3)		CN9 b	-7.2-	0.6 (3)
NN11 a	-8.6-	1.4		CN9 a	-8.6-	1.4
NN10		0.8		CN8 b		≅0.1
	-9.4-			CN8 a	-≅8.7-	0.8 / 0.7
NN9 b + a	-10.7-	≅1.3		CN7 a + b	-9.4- / -10.7-	≅1.3
NN8	-10.9-	0.2		CN6	-10.9-	0.2
NN7	-≅11.8-	0.9		CN5b	-≅11.8-	0.9

Datum Events: D. berggrenii, A. primus, A. amplificus, D. quinqueramus, D. neorectus, D. hamatus, C. calyculus, C. coalitus

LAD / FAD

FIG. 21.—Middle and late Miocene calcareous nannofossil biochrons, Atlantic Ocean and (?) Indian Ocean. See Figures 17 and 20 for explanation.

MIDDLE AND LATE MIOCENE CALCAREOUS NANNOFOSSIL BIOCHRONS
(Atlantic and (?) Indian Ocean)

Biochron (1)	T in Ma	Duration in my	Datum Events	Biochron (2)	T in Ma	Duration in my
NN11 d	-5.6- / -5.9-	0.3		CN9 d	-5.6- / -5.9-	0.3
NN11 c	-6.6-	0.7 (3)		CN9 c	-6.6-	0.7 (3)
NN11 b	-7.2-	0.6		CN9 b	-7.2-	0.6
NN11 a	-8.6-	1.4		CN9 a	-8.6-	1.4
NN10		0.8		CN8 b		0.1
	-9.4-			CN8 a	-8.7-	0.8 / 0.7
NN9 b	-10.7-	1.3		CN7 a	-9.4- / -10.7-	1.3
NN9 a	-11.2-	0.5		CN7 b		0.5
NN8	-11.3-	0.1		CN6	-11.2- / -11.3-	0.1
NN7	-11.8-	0.5		CN5b	-11.8-	0.5

Datum Events: D. berggrenii, A. primus, A. amplificus, D. quinqueramus, D. neorectus, D. hamatus, C. calyculus, C. coalitus, D. kugleri

LAD / FAD

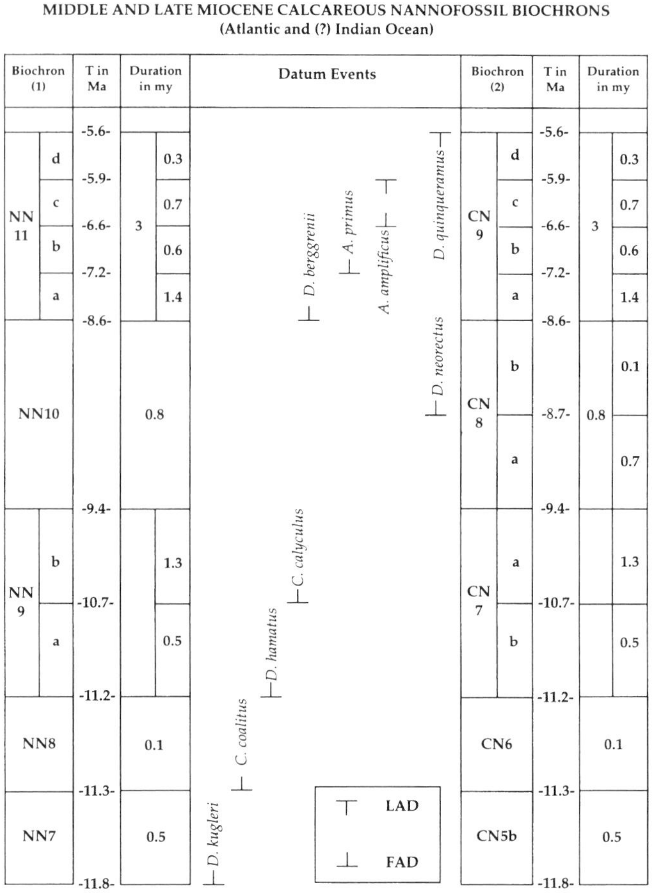

FIG. 22.—Middle and late Miocene calcareous nannofossil biochrons, equatorial Pacific Ocean. See Figures 17 and 20 for explanation.

scale, and it remains so at the present time. Indeed, in the past few years, the jointly sponsored IUGS-UNESCO International Geological Correlation Project (IGCP) 308 "Paleocene/Eocene Boundary Events in Time and Space" has been focusing attention on the various events in the stratigraphic record that may serve to delineate (and correlate) the boundary between these two units at an as yet to be determined boundary stratotype. The subject has been discussed in considerable detail in a recent overview by Berggren and Aubry (1995) and thus only a brief summary will be presented here in the context of the time scale presented in this paper.

The Paleocene/Eocene boundary interval is located within Chron C24r, a chron with a duration of at least 2.55 my (CK92/95). Thus magnetostratigraphy is limited in this instance for more precise recognition/correlation, and other methods must be used to delineate the Paleocene/Eocene boundary (wherever it is ultimately placed) within Chron C24r. Cyclostratigraphy may be expected to aid in identifying and correlating the position of the "Golden Spike" once it has been "nailed down."

An intricate, integrated correlation network links the base of the London Clay Formation (Ypresian Stage) with the -17 Ash Bed of Denmark and the North Sea and DSDP Hole 550 in the NE Atlantic, now dated at 55.0 Ma by Obradovich (in Berggren and others, 1992) in Denmark and at 54.5 Ma by Swisher and Knox (1991) in Denmark and offshore England to a level within

the lower part of Zone NP10 = lower CP9a and near the P5/P6 zonal boundary and to the 2.2/2.3 (TA2) 3rd-order cycle (54.5 Ma) boundary of Haq and others (1988). The estimated age of the base of the London Clay (= -17 Ash Bed) is 54.8 Ma and that of the NP9/NP10 zonal boundary 55 Ma in our chronology (see CK92/95 and discussion in Berggren and Aubry, 1995, on the methodology used in deriving these estimates). A distinct unconformity (with an estimated hiatus of ~0.5 my; Berggren and Aubry, 1995) separates the base of the London Clay Formation from the underlying Reading-Woolwich beds in SE England. While biostratigraphers often use the NP9/NP10 (calcareous nannoplankton) or the P6a/b (= P5/P6a zonal boundary of this paper; planktonic foraminifera) zonal boundary to *define* the Paleocene/Eocene boundary, this approach is obviously inappropriate and incorrect. A chronostratigraphic boundary is defined by a lithostratigraphic Global Stratotype Section and Point (GSSP), the so called "golden spike", which serves to integrate and unify the correspondence between rock and time at a unique stratigraphic level. Magneto-, bio-, chemostratigraphic datum events are then used to correlate this unique level in the stratigraphic record away from the boundary stratotype area. Insofar as the Paleocene/Eocene boundary is concerned, at the present time the best we can say is that the boundary, as typified by the oldest rocks attributed to the London Clay Formation, is within Zone NP10 and near the P5/P6 zonal boundary, whereas if we refer to the Ieper Clay Formation (= Ypresian Stage of Belgium), the boundary is only slightly higher/younger. There are a large number of bio- (extinction of deep sea benthic foraminifera), chemo- ($\delta^{13}C$ isotope spike), and climato- (strong reduction in windblown aerosols) stratigraphic events associated with a level in mid-upper NP9 and P5 Zones. If the Paleocene/Eocene boundary GSSP were eventually to be placed at this level, its age would be ~55.5 Ma in our chronology. Suffice to say that the Paleocene/Eocene boundary lies within the stratigraphic interval bracketed by the P5/P6 zonal boundary (at ~54.7 Ma) and the chemo- and biostratigraphic events mentioned above which appear to occur at a level closely associated with, or slightly older than, the NP9/NP10 zonal boundary (55 Ma) and encompassing the base of the London Clay Formation (= Thanetian/Ypresian boundary) at ~54.8 Ma, and the NP9/NP10 zonal boundary at 55 Ma. As we have mentioned above, we have chosen not to use the radioisotopic ages of 54.5 Ma or 55.0 Ma to denote the base of the London Clay Formation, preferring to use instead estimated rates of sedimentation within the ash series in DSDP Hole 550; the reasons for this and the methods used are described in detail in Berggren and Aubry (1995).

Eocene Series.—

The Eocene Series is seen to span about 21 my in our revised chronology, essentially the same duration as that estimated in the revised chronology of Aubry and others (1988). The main difference in the two scales is, of course, the younger ages assigned to the lower and upper boundaries of the Eocene Series. The lower Eocene Series of SE England and Belgium (stratotype of the Ypresian Stage) have now been precisely correlated based on integrated litho- and magnetostratigraphic studies (Ali and others, 1993), and the base of the London Clay and Ieper Formations are now known to lie within the lower part of Chronozone C24r with age estimates of ~54.8 and 54.7 Ma, respectively.

*Lower/middle Eocene boundary.—*The lower/middle Eocene (Ypresian/Lutetian) boundary, calibrated to Chron C22n(y), is now estimated at ~49 Ma. Haq and others (1988) interpreted the lower Lutetian deposits of the Lutetian stratotype (at St. Leu d' Esserent, near Paris) as representing their third-order cycle TA3.1, with the Cuisian/Lutetian boundary corresponding to the 49.5 Ma sea-level fall (chronology of Haq and others, 1988). We do not agree with this interpretation inasmuch as the Cuisian/Lutetian unconformable contact reflects a stratigraphic gap which encompasses the upper part of calcareous nannofossil Zone NP12, Zone NP13, and Subzone NP14a (Aubry, 1983, 1986). The upper surface of the unconformity (= the base of the type Lutetian) lies in Subzone NP14b. Thus as discussed in Aubry (1991), the lower Lutetian Stage, would more likely represent third-order cycle TA3.2, whereas the Cuisian/Lutetian boundary reflects several offlap events, including the 49.5 Ma and the 48.5 Ma events (chronology of Haq and others, 1988).

The Lutetian/Bartonian boundary is correlated to a level within Zone NP16 and closely associated with the LAD of *Reticulofenestra reticulata* which is calibrated to Chron C19n(y) (Aubry, 1986, p. 322, Fig. 23; see also BKF85 and Haq and others, 1988, for similar placement of this boundary) with an estimate here of 41.25 Ma.

*Middle/upper Eocene boundary.—*The problem with the placement of the middle/upper Eocene boundary is intimately linked with the difficulties in biostratigraphic correlation between the geographically disjunct location of the upper middle Eocene Bartonian (NW Europe) and upper Eocene Priabonian (NE Italy) Stage stratotypes (see discussion in BKF85). The Bartonian Stage is assigned to Zones NP16 and NP17 (and questionably to a part of Zone NP18; Aubry, 1986), whereas the Priabonian has been assigned to Zones NP19-NP21 (Verhallen and Romein, 1983) or to NP17 (*partim*) to NP21 (*partim*; (Barbin, 1986). Brinkhuis (1992, p. 85–87, Figs. 4.2, 4.3; 1994, p. 135–137, Figs. 6, 9) indicated that the lower part of the Priabona section belongs to the *Melitasphaeridium pseudorecurvatum* (Mps) Zone (with an undefined and undelimited lower boundary) which is correlative with Chrons C15 and C16 and Zones NP19–20 (= CP15b) and P16 at Massignano. This zone extends lower in the Massignano section into Zones NP18 (= CP15a) and P15 and lower Chronozone C16n, which suggested to Brinkhuis (1992) that the lowermost part of the Priabonian Stage sensu lato may not be present at Priabona if the commonly accepted correlation of the base of the Priabonian with the NP17/NP18 boundary is followed. The latter interpretation, however, is an arbitrary one in the absence of a more definitive GSSP for the base of the Priabonian (BKF85) on the basis of which a definitive biostratigraphic criterion(a) could be chosen for the purpose of regional/global correlation. Precise numerical age constraints on the base of the Priabonian are also lacking. Odin (1982, p. 786) listed an average whole rock age of 39.9 ± 1.1 Ma on pre-Priabonian lavas near the village of Priabona, suggesting a maximum age for the base of the Priabonian, and more recently Odin and Montanari (1989) have suggested an age of 37 ± 2 Ma for a stratigraphic level about 5-m above the base of the Priabona section. We follow earlier placement of the Bartonian/Priabonian boundary at the NP17/NP18 zonal boundary (BKF85; Aubry, 1986, p. 322, Fig. 16; Haq and others, 1988, p. 92, Fig. 14) which remains correlated with younger part of Chron C17n with an estimated age of 36.9 Ma (40 Ma in BKF 85 and 39.4 Ma in Haq and others, 1988).

Eocene/Oligocene boundary.—

The position and age of the Eocene/Oligocene boundary are intimately associated with problems pertaining to the litho- and biostratigraphic characteristics and limits of the upper Eocene Priabonian and lower Oligocene "standard" stages. Since 1985, the Priabonian Stage has been the focus of considerable, and renewed, interest, and three studies, in particular, are significant in the context of this paper: Barbin (1986), Brinkhuis (1992) and Brinkhuis and Biffi (1993). At the same time, a series of studies devoted to determining a GSSP for the Eocene/Oligocene boundary (Pomerol and Premoli Silva, 1986; Premoli Silva and others, 1988b) have been published. A GSSP for the Eocene/Oligocene boundary was determined in the Massignano section of the Apennines at a level denoted by the LAD of *Hantkenina* spp. which is located in the younger part of Chron C13r (= Chron C13r (0.14); Berggren and others, 1992) with an age estimated at 33.7 Ma. More recently, however, Brinkhuis (1992) and Brinkhuis and Biffi (1993) have indicated that at Massignano the *Hantkenina* LAD occurs just below the boundary between the *Achomosphaera alcicornu* (Aal) and *Glaphyrocysta semitecta* (Gse) Zones which in the section at Priabona falls near the boundary between the Nodular Limestone and the *Asterodiscus* Beds, that is within the mid-part of the lithostratigraphic sequence assigned to the Priabonian. The top of the Priabonian Stage at Priabona (= the "Bryozoan Limestone") is assignable to the *Areosphaeridium diktyoplocus* (Adi) Zone which is correlated with the TA4.3/4.4 third-order cycle boundary of Haq and others (1988) at a level equivalent to Chron C13n. At the nearby Bressana section, this level is equivalent to the boundary between the *Areosphaeridium diktyoplocus* (Adi) and *Reticulatosphaera actinocoronata* (Rac) Zones and to the LAD of the Discocyclinidae.

Brinkhuis (1988, p. 90) makes the case for (re)defining the Eocene/Oligocene boundary in the Priabonian section(s) to correspond to the lithostratigraphic top of the Priabonian Stage which would be correlative with the TA4.3/4.4 cycle boundary and the $\delta^{18}O$ event believed to indicate a major expansion (if not initial growth) of a continental ice sheet on Antarctica. The temporal difference between these two placements of the Eocene/Oligocene boundary is on the order of 0.5 my or less. However, in stratotypifying the Eocene/Oligocene boundary in the Massignano section, the precept that "base defines unit" and that the boundaries of lower level units in the chronostratigraphic hierarchy automatically encompass the boundaries in the next higher level (on the principle of coterminous boundaries) was ignored. Instead, it will be seen that stratotypification of the Eocene/Oligocene boundary in the Massignano section automatically delimits (by decapitation) the upper level(s) of the Priabonian Stage and the lower limit of the Rupelian Stage (by temporal extension) so that they become coterminous. Realism, and a respect for the history of debate on chronostratigraphic boundaries would suggest, however, that we have not heard the last word on the subject of the Eocene/Oligocene boundary.

Lower/upper Oligocene boundary.—

We retain an informal two-fold chronostratigraphic subdivision of the Oligocene Series: a lower Rupelian and upper Chattian Stage. We correlate (as before, BKF85) the boundary between these two stages to the P21a/b boundary (= LAD of (common) chiloguembelinids) in (mid) Chron C10n with an estimated age of 28.5 Ma (= 30 Ma in BKF85 and Haq and others, 1988).

Oligocene/Miocene boundary.—

The Paleogene/Neogene Working Group of the IUGS Neogene Subcommission (under the leadership of Professor Fritz Steininger) has been focusing on the definition of a GSSP for the Paleogene/Neogene boundary for over a decade. It has now decided (Steininger, 1994) to recommend that the GSSP be located at the 35-m level of the Rigorosa Formation in the Carrosio-Lemme section of NW Italy corresponding to the base of Chron C6Cn.2n and the FAD of *Globorotalia kugleri* with an age estimate of 23.8 Ma (CK92/95 this paper; for a somewhat different point of view see Srinivasan and Kennett, 1983).

Chron C6Cn has a duration of about 0.76 my (CK92/95). By placing the boundary within the chron (rather than at either extreme), global correlation is assured (when only part of this normal subchron is present) with a relatively high degree of precision (< 0.5 my) whereas if the boundary were to have been placed at either extreme the degree of precision would essen-

tially be delimited by the duration of the entire subchron (under the most unfavorable circumstance, which is frequently the case in hiatus ridden sections on both land and in the deep sea).

Neogene Period

Miocene Series.—

A three-fold subdivision of the Miocene Series is generally accepted by most stratigraphers (Berggren, 1971; BKV85). The relationship between calcareous plankton biostratigraphy and standard chronostratigraphic units was discussed at length in BKVC85 and the reader is referred to that source for background information. We review here only several (minor) adjustments and/or studies that have been made since 1985.

The Aquitanian/Burdigalian boundary in the Contessa section (Gubbio, Italy) has been correlated with the FAD of *Globigerinoides altiaperturus* (Iaccarino, 1985), and the Burdigalian/Langhian boundary with the FAD of *Pr. glomerosa* (Cita and Premoli Silva, 1968). The base of the Langhian Stage was subsequently (Cita and Blow, 1969) said to correspond to the FAD of *Pr. bisphericus* (vel *sicana*) which is stratigraphically slightly lower/older than the FAD of *Pr. glomerosa glomerosa* (Blow, 1969; Jenkins and others, 1981). We use the FAD of *Pr. sicana* (non *bisphericus*) to denote the base of the Langhian here. The Langhian/Serravallian boundary remains somewhat controversial (see discussion in BKV85 and Iaccarino, 1985). Originally defined so as to coincide with the LAD of *Pr. glomerosa* (Cita and Premoli Silva, 1968) which occurs within the interval of Zone N9 (Blow, 1969), it was subsequently defined to coincide with the N10/N11 zonal boundary of Blow (1969) in Cita and Blow (1969). Iaccarino (1985) drew attention to the fact that the upper limit of the type Langhian coincides with the top of the Cessole Formation as originally defined by Cita and Premoli Silva (1960), coincident in turn, with the top of their *Orbulina suturalis* Zone, a level which is within the upper part of Zone N9 of Blow (1969). Iaccarino (1985) accordingly equates the Langhian/Serravallian boundary with the LAD of *Pr. glomerosa* which she indicates is approximately correlative with the upper part of the *Orbulina suturalis* Zone which is correlative, in turn, with the upper part of Zone N9. Our data (see Table 11) indicates that these events occur essentially simultaneously and coincident with the FAD of *Globorotalia peripheroacuta* (= Zone N9/N10 boundary of Blow, 1969, 1979, and M6/M7 zonal boundary of this paper) and we have thus drawn the Langhian/Serravallian boundary coincident with the N9/N10 (= M6/M7) zonal boundary (Figs. 4, 5). This level would appear to be consistent with, and equivalent to, the base of the Serravallian type section as (re)defined by Vervloet (1966) and correlated with the base of his *Globorotalia menardii* (including *Globorotalia praemenardii*) Zone. The FAD of *Gt. praemenardii* coincides essentially with the base of Zone N10 (Bolli and Saunders, 1985).

The upper limit of the Serravallian Stage, as redefined by Boni (1967) and Mosna and Micheletti (1968) in the Gavi section is older than the FAD of *Gt. lenguaensis*, whereas the Arguello-Lequio parastratotype section of Cita and Premoli Silva (1968) included a *Gt. mayeri—Gt. lenguaensis* Zone in its upper part, indicating that the upper stratigraphic limit of the Serravallian Stage is within Zone N14 (Blow, 1969; = M11 of this paper). There is thus a short gap between the top of the Serravallian and the base of the Tortonian Stage which lies within (the lower part of) Zone N15 (=M12) and perhaps upper N14 (=M11).

The lower boundary of the Messinian Stage has undergone several modifications over the past 40 years (compare Gianotti, 1953; Selli, 1960; d'Onofrio and others, 1975; Colalongo and others, 1979) with the latter study suggesting that the GSSP of the Tortonian/Messinian boundary stratotype be linked with the FAD of *Gt. conomiozea* in the Falconara section in Sicily. We follow this convention in this study.

Pliocene Series.—

The Miocene/Pliocene boundary is equated here with the base of the Zanclean Stage as stratotypified at Capo Rosello in Sicily (Cita, 1975; Hilgen, 1991; Hilgen and Langereis, 1993; Langereis and Hilgen, 1991) which appears to be bracketed by the FADs of *Gt. tumida* and *Gt. sphericomiozea* (below) and the FADs of *Ceratolithus acutus* and *Gt. puncticulata* and the LAD of *Discocaster quinqueramus* (above). An alternative point of view suggests that in view of the perimediterranean lithologic unconformity between the nonmarine Messinian (below) and marine Zanclean (above) sediments and the difficulty of biostratigraphically extending the lithostratigraphic base of the Zanclean Stage away from its stratotype area, a boundary stratotype section should be sought in a continuous marine section outside the Mediterranean (for further discussion see Berggren and others, 1995; Benson and Hodell, 1994). For the purpose of this paper, we follow the commonly acccepted usage of the base of the Pliocene equals the base of the Zanclean Stage (*sensu* Cita, 1975; see also Hilgen and Langereis, 1993).

A comprehensive discussion on Pliocene and Pleistocene chronostratigraphy and chronology is presented in Berggren and others (1995).

CHRONOLOGY OF CHRONOSTRATIGRAPHY

Paleogene Period

The relationship of Paleogene epoch/series and age/stage boundaries to the revised magnetochronology of CK92/95 and estimated duration (in my) of these units are shown in Figure 23. The age estimates of the older and younger boundaries of the Paleogene Period are similar to those in BKF85 (65.0 vs. 66.4 Ma and 23.8 vs 23.7 Ma, respectively; Fig. 23). The main changes relative to BKF85 involve the younger age estimates of the Eocene/Oligocene boundary (33.7 Ma) and for the Paleocene/Eocene boundary (54.8 Ma) based on new (post-1985) data, resulting in a somewhat longer Paleocene (11.6 my vs. 8.6 my in BKF85) and shorter Oligocene Epoch (9.9 vs. 12.6 my in BKF85) with the Eocene Epoch remaining essentially of the same duration. The basic reasons requiring a change in the Eocene/Oligocene boundary estimate have been reviewed by Swisher and Prothero (1990), Prothero and Swisher (1992) and Berggren and others (1992) and need not be repeated here. With regards to the Paleocene/Eocene boundary, the new ^{40}Ar-^{39}Ar dates of 54.5 and 55 Ma on the -17 Ash Bed in NW Europe which lies virtually at the Paleocene/Eocene boundary results in a boundary estimate some 2 my younger than that estimated in BKF85 (Swisher and Knox, 1991).

Paleogene ages vary in duration by a factor of two with the Selandian (3.0 my) and the Lutetian Age (7.65 my) as the short-

EPOCH/SERIES Ages in Ma	AGE/STAGE Duration in my	PALEOMAGNETIC CALIBRATION
MIOCENE	Aquitanian	
———23.8———		——— C6Cn.2n$_{(o)}$
	Chattian (4.7)	
	———28.5———	——— C10n
OLIGOCENE 9.9	Rupelian (5.2)	
———33.7———		——— C13r(.14)
	Priabonian (3.3)	
	———37———	——— C17n$_{(y)}$
	Bartonian (4.3)	
EOCENE 21.0 [21.3]	———41.3———	——— C19n$_{(y)}$
	Lutetian (7.7)	
	———49———	——— C22n$_{(y)}$
	Ypresian (5.5) [5.8]	
——54.5 [54.8]——		——— C24r
	Thanetian (3.4) [3.1]	
	———57.9———	——— C26n
PALEOCENE 10.5 [10.2]	Selandian (3.0)	
	———60.90———	——— C27n$_{(y)}$
	Danian (4.1)	
———65.0———		——— C29r
CRETACEOUS	Maestrichtian	

FIG. 23.—Chronology of Paleogene chronostratigraphy. Different ages and duration of Paleocene and Eocene and Thanetian and Ypresian ages due to estimated age of Paleocene/Eocene boundary on base of London Clay Formation at 54.5 Ma (−17 Ash date of Swisher and Knox, 1991) or 54.8 Ma (based on sedimentation rate estimates in DSDP Hole 550; Berggren and Aubry, 1995; Aubry and others, 1995). The latter age is preferred here.

est and longest, respectively (Table 5). The late Paleocene Selandian (3.0 my) and Thanetian (3.4/3.1 my) Ages are seen (Fig. 23) to have individual duration comparable to the early Paleocene Danian Age (4.1 my). The 3.4/3.1-my duration of the Thanetian Age compares with 2.8 my in BKF85, while the 7.65-my duration of the Lutetian compares with the 8.35-my duration in BKF85. The shorter Oligocene results in a shorter duration for the Rupelian (5.2 vs. 6.3 my in BKF85) and Chattian (5.2 vs. 6.3 my) Ages.

The revised Paleogene time scale presented here is seen to have elements in common with earlier, and other, versions. There appears to be general agreement among most time scale proponents of the older and younger limits of the Paleogene, although the current controversy among specialists as to the precise age of the Cretaceous/Paleogene boundary (see Obradovich, 1993, for example) remains unresolved at this time. The boundary age estimate of 65.0 Ma used here has been adopted by Gradstein and others (1994; this volume) for their newly proposed Mesozoic time scale providing continuity with our time scale. At the same time, it must be remembered that proponents of different time scales have chosen different (bio)stratigraphic levels for their boundary estimates in some instances, and these must be considered in interpreting the different numerical values used in different scales. For instance, the level at which the Oligocene/Miocene boundary is placed in GTS89 is slightly different than that used in BKF85 and significantly different than that in Haq and others (1988), while the level at which GTS89 place the Rupelian/Chattian boundary differs significantly from that used in BKF85 and this paper; the chronologic difference is almost 2 my! Within the Paleogene Period, we have adopted the values proposed by Montanari and others (1988; see also McIntosh and others, 1992) for the Eocene/Oligocene boundary. Other and earlier versions appear to have underestimated (by 1 to 2 my) or overestimated (by 1 to 5 my; see Berggren and others, 1992, Fig. 13) the age of this boundary, and there is now agreement on this value in the present scale and that recently published by Odin and Odin (1990).

The Paleocene/Eocene boundary is now seen to have an age of about 54.5–54.8 Ma. Earlier and other versions have hovered tantalizingly close to these values, although deviations of as much as 4 my younger (Curry and Odin, 1982) and nearly 3 my older (BKF85) have occurred in the past decade.

Age estimates of the various Paleogene stage boundaries in the various chronologies published over the past twenty years vary as a function of the ages assigned to the major epoch/series boundaries (Fig. 24) and the methodologies used to derive the time scale itself (see discussion of this problem in BKF85; Berggren, 1986; Berggren and others, 1992; Odin and Luterbacher, 1992). In general, the numerical estimates derived here are similar to those in Haq and others (1988) for the Paleocene-middle Eocene Epochs. Upper Eocene and Oligocene age estimates are unique to this time scale because of the revised age of the Eocene/Oligocene boundary, but bear similarities to those proposed in Curry and Odin (1982) and Odin and Luterbacher (1992). The main differences remaining between the geochronology presented here and that suggested by Odin and Luterbacher (1992) are at the Paleocene/Eocene boundary (~54.5–55 Ma, this paper, vs. 53 Ma in Odin and Luterbacher, 1992) and the Ypresian/Lutetian boundary (49 Ma, this paper, vs. 46 Ma in Odin and Luterbacher, 1992). It would appear that we have reached, after two decades of concentrated research on Paleogene chronology, a point where a general unity of opinion is possible on the numerical ages of the major (epoch/series) and minor (age/stage) boundaries. We believe that future changes to the time scale presented here will involve relatively minor adjustments based on (possible) modifications to the age of the Cretaceous/Paleogene boundary, modifications to the present biostratigraphic correlation of age/stage boundaries based on boundary stratotype studies currently underway, and refinements based on the application of Milankovitch identified cyclicity to the stratigraphic record.

Neogene Period

The relationship of Neogene chronostratigraphic boundaries to the revised GPTS of this paper is shown in Figure 25. The Oligocene/Miocene boundary, as calibrated to Chron C6C.2n has an estimated age of 23.8 Ma and corresponds to the FAD of *Gt. kugleri* (BKV85; CK92/95 and this paper). The early Miocene Epoch is biostratigraphically bracketed by the FAD of *Gt. kugleri* and the FAD of *Pr. sicana* (Chron C5Cn.2n; 16.4 Ma) giving the early Miocene Epoch a duration of 7.4 my. The Aquitanian/Burdigalian boundary has been correlated with the LAD of *Gt. kugleri* (= N4/N5 boundary of Blow, 1969, 1979; M1/M2 zonal boundary of Berggren, this paper), but it has also been correlated with the FAD of *Gd. altiaperturus* in Mediterranean and Aquitaine Basin stratigraphies (see discussion in Montanari and others, 1991). The LAD of *Gt. kugleri* has been observed in Chron C6Ar (with an estimated age of 21.5 Ma) in Hole 516F, whereas it has been recorded at the base of Chron

	Berggren (1972)	Hardenbol and Berggren (1978)	Curry and Odin (1982)	Harland and others (1982)	Berggren and others (1985b)	Haq and others (1987,1988)	Harland and others (1990)	Odin and Odin (1990)	This Paper
OLIGOCENE	22.5	24	23	24.6	23.7	25.2 (24.2)	23.3	23.5	23.8
EOCENE	37.5	37	33±2	38.0	36.3	36(36.3)	35.9	34	33.7
	53.5	53.5	51±1.5	54.9	57.8	54	56.5	53	54.8
PALEOCENE	65	65	65±1.5	65.0	66.4	66.5	65	65	65

FIG. 24.—Age estimates of Paleogene epoch/series boundaries. Ages in parentheses are recalibrated to same magnetostratigraphic level as in Berggren and others (1985b).

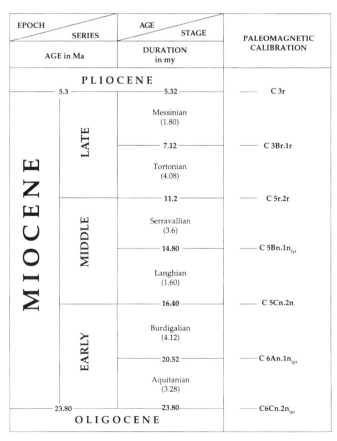

FIG. 25.—Chronology of Neogene chronostratigraphy.

C6An (with an estimated age of 21.32 Ma) in the Contessa Highway section (Montanari and others, 1990, 1991). The FAD of *Gd. altiaperturus* has been recorded at the top of Chronozone C6An (with an estimated age of 20.52 Ma (CK95) in DSDP Hole 516F, whereas it has been recorded in the older part of Chron C6r (with an estimated age that is indistinguishable from that of 20.5 Ma of the record in Hole 516F, in view of the fact that the entire Chron C6r is only 0.387 my long in the chronology of CK92/95; this paper) in the Contessa Highway section (Montanari and others, 1991). Accordingly, we would recommend correlation of the Aquitanian/Burdigalian boundary with the top of Chron C6An (with an estimated age of 20.52 Ma; CK92/95 and this paper); this procedure would facilitate regional/global correlation and would correspond with generally accepted practice in Mediterranean stratigraphies where Neogene chronostratigraphy is rooted (van Couvering and Berggren, 1977). We note, in passing, the close temporal correspondence between the estimated ages of Chron C6An in the chronologies of CK92/95 and this paper, and Montanari and others (1991). The latter have obtained an isochron age of 21.17 ± 0.23 Ma based on $^{40}Ar/^{39}Ar$ dating of plagioclase on an ash termed the Livello Rafaello, about a half meter below the LAD of *Gt. kugleri* and the base of Chron C6An. Chron C6An has an estimated age of 20.518–21.320 Ma in the magnetochronology of CK92/95. The Burdigalian/Langhian boundary (= FAD of *Pr. sicana*) is calibrated to Chron C5Cn.2n (16.4 Ma; CK92/95 and this paper).

Middle Miocene time is biostratigraphically bracketed by the FAD of *Pr. sicana* and a level within Zone M12 (i.e., between the LAD of *N. mayeri* and the FAD of *N. acostaensis* (Cita and others, 1965; cf. BKVC85 in which the middle/upper Miocene boundary was pragmatically, but incorrectly, correlated with the FAD of *N. acostaensis*). The restoration of Zone M12 (= N15; Berggren, 1993) with an estimated duration of ~0.5 my between 11.4 and 10.9 Ma (this paper) suggests that the middle/late (Serravallian/Tortonian) boundary is in Chron C5r.2r at ~11.2 Ma. The Langhian/Serravallian boundary is correlated here with the FAD of *Gt. peripheroacuta* at the top of Chron C5Bn (14.8 Ma; CK 92/95 and this paper). The middle Miocene Epoch has a time span of ~5.2 my (16.4 — 11.2 Ma).

The upper Miocene (Tortonian and Messinian Stages) series is bracketed biostratigraphically by a level within Zone M12 (= N15, ~11.2 Ma) to a level slightly higher/younger than the FAD of *Gt. tumida* (M14/Pl1 boundary) and/or *Gt. sphericomiozea* which lies within Chron C3r with an estimated age of 5.6 Ma (this paper). The Tortonian/Messinian boundary is correlated here with the FAD of *Gt. conomiozea* which has been magnetobiostratigraphically correlated in Crete with Chron C3Br.1r with an estimated (astrochronologic) age of 7.1 Ma (Krijgsman and others, 1994). Calibration to CK92 yields an age estimate of 6.92 Ma and, in the chronology subsequently derived by CK92/95, a magnetochronologic age estimate of 7.12 Ma, essentially identical to the age estimate of Krijgsman and others (1994). A slightly younger position for the FAD of *conomiozea* in Chron C3Bn has been reported by Benson and Rakic-El Bied (1991) in the Vera Basin of Spain which would place this event at ~7.0 Ma in the chronology adopted here. Thus, the astrochronologic and magnetochronologic scales are seen to be coherent and concordant back to Chron C3Br at ~7 Ma. At the same time, we note that the late Miocene (Chron 6)

carbon shift has been observed to start near the base of Chron 6 in Hole 588 (Hodell and Kennett, 1986) equal Chron C3Br.2r (7.20 Ma in CK92; 7.4 Ma in CK92/95 and this paper). An age of 7.26 ± 0.1 Ma for the Tortonian/Messinian boundary in the northern Apennines of Romagna was recently suggested by Vai and others (1993) based on a K-Ar (biotite) and a $^{40}Ar/^{39}Ar$ (plagioclase) date of 7.33 Ma on volcanogenic horizons a few meters below the FAD of *Gt. conomiozea* and *Gt. mediterranea* and a K-Ar (biotite) date of 7.72 ± 0.15 Ma on the stratigraphically lower FAD of *Globorotalia suterae* (which agrees closely with the magnetochronologic estimate of 7.12 Ma for this datum event proposed here; Table 11; cf. the estimate of 5.6 Ma and correlation to Chron C3An for the Tortonian/Messinian boundary by Langereis and Dekkers, 1992). The late Miocene Epoch thus has a span of ~6 my (11.2 — 5.3 Ma). In a recent integrated magnetobiostratigraphic study of the Sorbas (Andalusia, Spain) and Caltanissetta (Sicily, Italy) Basins, the evaporitic phase ("salinity crisis") of the Mediterranean has been shown to be restricted to Chron C3r (Gilbert reversed) and to have had a duration of <0.5 my (from ~5.8–5.32 Ma in the chronology of this paper; Gaultier and others, 1994). The implication of these results is that the pre-evaporitic Messinian (~7.12 to ~5.8 Ma (~1.32 my) represents about 2/3 of the duration of the Messinian Age itself (7.12–5.32 Ma, or ~1.80 my), whereas the "late" (evaporitic) Messinian represents but 1/3 the duration of the Messinian Age (~5.8–5.32 Ma, or ~0.48 my). Comparable studies on a Moroccan drill hole have yielded a high-resolution integrated magnetobiostratigraphic and stable isotope stratigraphy of correlative upper Miocene non-evaporitic (Atlantic facies) strata (Hodell and others, 1994).

There is ongoing debate regarding the adoption of an astronomically calibrated (Hilgen and Langereis, 1989) versus anomaly interpolated (Cande and Kent, 1992) based time scale for the Neogene Period, but this is beyond the scope of this paper. At the present time, the Cande and Kent (1995) modification adopts directly the astronomical time scale to the Thvera Event (Subchron C3n.4n) which results in good concordance of the time scales back to about 7 Ma (see above). The situation is discussed at length in Berggren and others (1995) to which the reader is referred.

Postscript.—Following the completion and submission of this paper, our attention has been drawn to the paper by Wei (1995) which questions the validity of several of the calibration points of the geomagnetic polarity time scale of Cande and Kent (1992, 1995). Space does not allow a detailed discussion of this work but we comment on several points of concern relative to the time scale published here.

1. For the early middle Miocene calibration, Wei (1995, p. 957) substitutes an age of 16.3 Ma for Chron C5Br(y)based on the "clear correlation of the magnetic reversal sequence of the Columbia River Basalt with seafloor anomaly pattern (Baksi and Farrar, 1990)." The interpretation of this magnetic reversal sequence is anything but unambiguous, as we have discussed above, especially when contradictory evidence from the Barstow Formation (Wyoming) and the Ngorora Formation (Kenya) are considered. In the construction of our time scale, we have assiduously avoided the use of nonmarine calibration(s) because of lack of independent (i.e., marine biostratigraphic) constraints (cf. BKVF85).

Thus we used an estimate of 14.8 Ma for the N9/N10 zonal boundary based on an assessment of radioisotopic dates near the *Orbulina* Datum in Japan (Tsuchi and others, 1981; Shibata and others, 1984; Tsuchi, 1984) and, particularly, in Martinique (Andreieff, 1985; Andreieff, person. commun. to WAB, 1985) which were correlated albeit indirectly to the geomagnetic polarity time scale as the best available calibration for early middle Miocene time. We take this opportunity of clarifying several points regarding our choice of the 14.8 Ma calibration point at the N9/N10 zonal boundary

a. At Le François, Martinique, sediments with a *Globorotalia peripheroronda* (N9) Zone fauna are overlain by basalts of the Vert-Pré volcanic episode dated at 15.1 ± 0.3 Ma (Pacquemar Quarry).

b. At Bassignac, Martinique, the "Tufs de Bassignac" belonging to the same Zone N9 are directly underlain by andesites of the François-Robert volcanic episode dated at 15.9 ± 1 Ma (Bois-soldat Quarry) and 15.69 ± 1 Ma (Ilet à Bau, Le Robert Bay).

c. At Le Marin, Martinique, sediments assigned to the early *Globorotalia fohsi* (N10) Biochron are underlain by the Vert-Pré volcanics (15.1 ± 0.3 Ma, cf. supra).

Andreieff subsequently suggested in a written communication to WAB (1985) that the best estimate for the N9/N10 zonal boundary was 14.8 Ma, the value we adopted in BKFV85 and which was accepted in CK92, 95. The work by Andreieff (1985) was an internal report prepared for the Bureau de Recherches Géologiques et Minières as part of the "Explanatory Notes" to accompany the new geological map of Martinique (scale 1:50,000) which was eventually published in 1989. The information reviewed above was published in more detail as Andreieff and others (1988).

The use of a 16.3 Ma calibration for Chron C5Br(y) by Wei results in age estimates ~1 my older than those in CK95 but are contradicted by a large body of integrated magnetobiostratigraphic and radioisotopic data in the early-middle Miocene, not the least of which are several dates of ~15 Ma near the *Orbulina* datum (Chron C5Bn.2n) and ~16 Ma near the *Praeorbulina glomerosa* Datum (early Biochron N8 = Chron C5Cn.1n; Tsuchi and others, 1981; Shibata and others, 1984; Andreieff, 1985; Andreieff and others, 1988), ~1 my younger than inferred by Wei (1995).

2. As pointed out by Wei (1995), Bryan and Duncan (1983) reported dates of 45.8 Ma and 46.8 Ma on duplicate samples of biotite from a clastic horizon in DSDP Hole 516F considered to lie in the younger part of C21n (Berggren and others, 1983). In a review of Paleogene geochronology by Berggren and others (1992), the 45.8 Ma date was inadvertently omitted in a tabulation of Eocene-Oligocene radioisotopic ages and magnetobiostratigraphic controls. This omissions was propagated in CK92/95 who used the 46.8 Ma date as a calibration point for the middle Eocene (Chron C21n(0.33)). The mean in the Hole 516F biotite ages, given as 46.3 ± 0.7 Ma by Bryan and Duncan (1983), in conjunction with the dates of 45.7 Ma and 46.2 Ma on bentonite from the lower Castle Hayne Formation (Harris and Fullager, 1989) which is biostratigraphically constrained to the older part of Chron C20r (Berggren and others, 1992) would

nevertheless suggest an age of ~46 Ma for the Chron C20r/C21n boundary, not very different from 46.264 Ma derived in CK95 and used here. ^{40}Ar/^{39}Ar laser total-fusion analyses of sanidine and biotite from the clastic horizon in Hole 516F (Swisher and Montanari, in prep.) now appear to provide dates of ~45.6 Ma that are concordant with the younger biotite date (45.8 Ma) reported by Bryan and Duncan (1983). Thus, Wei's remark (1995) that the isotopic age on the biotite from Hole 516F may have been unsuitable as a calibration point due to the large age uncertainty is well taken. The isotopic dates by Swisher and Montanari (in prep.) however, became available after CK95 was in press and at a late stage in the preparation of this time scale. On the other hand, Wei (1995) did not use a middle Eocene calibration point, which results in a rather long (17.6 my) gap between calibration points in the early and late Eocene.

Wei (1995) also casts doubt on the reliability of the 46.8 Ma calibration point in CK92/95 based on the fact that the biotite bearing clastic levels in Hole 516F are turbidite horizons and that their position in Chron C21n is uncertain because of the presence of a slump block below the normal polarity interval assigned to Chron C21n in the hole. The turbiditic nature of the biotic-bearing horizons is of minor concern in as much as (1) they are considered "locally derived from a mixed volcanic/plutonic terrain . . . , alkaline volcanoes active on the Rio Grand Rise in the middle Eocene" being "the most probable source for this ash" (= "the [dated] biotite separated from Sample 516F-76-4, 107–115 cm") (Bryan and Ducan, 1983, p. 475); and (2) they occur in sediments assigned to Zone NP15 by Wei and Wise (1989). From the range of the calcareous nannofossil species recorded by these authors, it is unquestionable that the normal polarity interval in Core 516F-76-5 to 75-1, correlative with the lower part of Zone NP15, represents Chron C21n. However, of greater concern to us than the presence of a Maestrichtian slump in Core 516F-78 and 79 (*partim*) is the fact that the normal polarity inverval interpreted as Chron C21n in Berggren and others (1983) corresponds in fact to the concatenation of two normal magnetozones, one of which corresponds to the late part of Chron C21n in the other to Chron C22n (Aubry, this volume). Yet, using the high sedimentation rates calculated by Aubry, it is serendipitously likely that the dated biotite in Hole 516F lies at a young level in Chron C21n (0.33 or less). In view of the rapidly changing dating scenario of Chron C21n, we chose to retain the calibration value of 46.8 Ma in CK95 for this paper rather than generate yet another permutation in the Cenozoic time scale until a stable solution is achieved.

3. Wei (1995) substitutes an age calibration of 52.8 Ma at Chron C24n.1r(y) from the nonmarine Willwood Formation in Wyoming (Tauxe and others, 1994; Wing and others, 1991) for the 55 Ma age estimate of CK95 at the NP9/NP10 zonal boundary in Chron C24r. There are several points here:

a. Wei (1995) states that the 55 Ma at the NP9/NP10 zonal boundary was based on "the ^{40}Ar/^{39}Ar date of 54.5 Ma for the -17 Ash by Swisher and Knox (1991), who estimated an age of 55 Ma for the nannofossil NP9/NP10 boundary (~Paleocene/Eocene boundary) based on the short distance from the -17 Ash to the NP9/NP10 boundary in DSDP Hole 550". Actually, Swisher and Knox

(1991) based their age of 55 Ma on the NP9/NP10 zonal boundary on sedimentation-rate extrapolation over the 7-m interval separating the -17 Ash and the NP9/NP10 zonal boundary based on Swisher's ages of 54.0 Ma and 54.5 Ma on the +19 and -17 Ash beds, respectively, themselves separated by 7 m. Parenthetically, there are two sets of ages on these two ashes, one by Carl Swisher and the other by John Obradovich (abbreviated here to CCS and JDO, respectively). The ages are:

+ 19 Ash: 54.0 Ma (CCS) and 54.5 Ma (JDO)
- 17 Ash: 54.5 Ma (CCS) and 55.0 Ma (JDO)

The ages on the +19 Ash have been disavowed by both workers (pers. communs. to WAB, 1994) owing to inhomogeneities in the mineral populations. Nonetheless, radioisotopic dating of the -17 Ash indicates an age of 54.5 to 55 Ma for the Paleocene/Eocene boundary, as denoted at the base of the Ieper Clay in Belgium and/or the base of the London Clay Formation (Oldhaven/Hales Clay Beds) in the London—Hampshire Basin.

b. Wei (1995) states that the NP9/NP10 zonal boundary lies at an unconformity at 408 m in DSDP Hole 550, resulting in an uncertain position of CK95's tie point within Chron C24r. Although we agree with the presence of an unconformity, we view the evidence that he cites as supportive (juxtaposed disappearance of *Fasciculithus tympaniformis* and appearances of *Discoaster diastypus* and *Coccolithus formosus* at a lithologic boundary) inadequate for this assertion. We also point out that the presence of an unconformity at 408 m in Hole 550 does not necessarily jeopardize the validity of the NP9/NP10 tie point.

c. Subsequent to CK92, a hiatus at 408 m in DSDP Hole 550 has been determined based on integrated magneto-, bio-, and isotope stratigraphy (Aubry and others, 1995) and its implication for geochronology has been discussed in detail (Aubry and others, 1995; Berggren and Aubry, 1995). It has been shown that the hiatus is substantive (~0.3 to 0.5 my) and encompasses the later part of Biochron NP9 and the earlier part of Biochron NP10. It is now clear that the position of the NP9/NP10 biochronal boundary in Chron C24r is younger than accepted in CK95, but the ages of Subchron C24n estimated in CK95 are nevertheless internally consistent and acceptable. Indeed, the calibration age of 52.8 Ma at the Chron C24n.1r(y) used by Wei (1995) differs insignificantly from the value (52.663 Ma) for this reversal in CK95.

4. With regard to the other calibrations proposed by Wei (1995), 9.67 Ma for Chron C5n(y) differs by only 0.07 my from the estimate of 9.740 Ma derived in CK95, 28.1 Ma for Chron C9r(y) differs by <0.15 my from the estimate of 27.972 Ma, and 35.2 Ma for Chron C15r(y) differs by <0.3 my from the estimate of 34.940 Ma. Indeed as Wei (1995) observes, with the exception of the early Miocene where the differences are up to 1.3 my due to what we regard as the ambiguous Columbia River Basalt tie point, the differences over the Cenozoic are generally less than 0.3 my. The revised calibration of Wei (1995) scarcely offers a significant improvement to the GPTS of CK95 and that presented here and indeed, supports its robustness.

With the increasing generation of high precision radioisotopic dates, we anticipate that the next significant advance in age calibrations will include more precise methods (e.g., astronomical climate cycles) to relate the dated level(s) in stratigraphic sections to the isochronous boundaries (e.g., magnetochrons) that are used for global correlations.

ACKNOWLEDGMENTS

This paper is the outgrowth of nearly a decade of work because almost as soon as BKFV85 was published, we (WAB, DVK) realized that revisions would be required. Rather than rush to print with (periodic and piecemeal) modifications and updates, we have chosen to wait until a sufficient amount of data are in hand which warrant a wholescale revision to the version published 10 years ago. During this time, we have had the pleasure of interacting with many of our colleagues (too numerous to mention) who have given unselfishly of their data and ideas as we began to compile and integrate the various data sets into a harmonious and unified time scale. We would like to particularly acknowledge the continued and stimulating input of, discussions with, and critical and helpful reviews of various incarnations of the nascent paper by, among others, J. Ali, S. C. Cande, J. Hardenbol, E. A. Hailwood, R. O'B. Knox, C. Liu, K. G. Miller, R. K. Norris, J. D. Obradovich, R. K. Olsson, D. R. Prothero, F. Steininger, J. A. van Couvering, M. O. Woodburne. This research was supported in part by NSF grants OCE 80 (WAB), 0CE91/n-/04447 (DVK) and by a consortium of oil companies (AMOCO, BP, CHEVRON, EXXON, MARATHON, UNOCAL). This is Woods Hole Oeanographic Institution Contribution No. 8853, Lamont-Doherty Earth Observatory of Columbia University Contribution No. 5389 and Institut des Sciences de l'Evolution de Montpellier No. 95047.

REFERENCES

ALI, J. R., KING, C., AND HAILWOOD, E., 1993, Magnetostratigraphic calibration of early Eocene depositional sequences in the southern North Sea Basin, *in* Hailwood, E. A. and Kidd, R. B., eds., High Resolution Stratigraphy: London, Geological Society Special Publication 70, p. 99–125.

ALIMARINA, V. P., 1962, Observations on the evolution of planktonic foraminifera in the early Paleogene of the northern Caucasus: Biulletin Moskovsk Obschchii ispytatelni prirody, otdel geologia, v. 37, p. 128–129 (in Russian).

ALIMARINA, V. P., 1963, Some peculiarities in the development of planktonic foraminifera and zonation of the Lower Paleogene of the northern Caucasus: Voprosy Mikropaleontologii, v. 7, p. 158–195 (in Russian).

ALVAREZ, W., ARTHUR, M. A., FISCHER, A. G., LOWRIE, W., NAPOLEONE, G., PREMOLI SILVA, I., AND ROGGENTHEN, W. M., 1977, Upper Cretaceous-Paleocene magnetic stratigraphy at Gubbio, Italy. V. Type section for the Late Cretaceous-Paleocene geomagnetic reversal time scale: Geological Society of America Bulletin, v. 88, p. 383–389.

ANDREIEFF, P., 1985, Biostratigraphie, chronométrie et géochronologie des formations sédimentaires anté—Quaternaires de la Martinique: Bureau de Recherches Géologiques et Minières, 18 p. (unpublished internal report).

ANDREIEFF, P., BAUDRON, J. C., AND WESTERCAMP, D., 1988, Histoire géologique de la Martinique (Petites Antilles): Biostratigraphie (foraminifères). radiochronologie (potassium-argon), évolution volcano-structurale: Géologie de la France, no. 2–3, p. 39–70.

ANDREIEFF, P., BELLON, H., AND WESTERKAMP, D., 1976, Chronométrie et stratigraphie comparée des édifices volcaniques et formations sédimentaires de la Martinique (Antilles françaises): Bulletin du Bureau de Recherche Géologiques et Minières, 2è série, section 4, n° 4, p. 335–346.

ARNOLD, A. J., 1983, Phyletic evolution in the *Globorotalia crassaformis* (Galloway and Wissler) lineage: a preliminary report: Paleobiology, v. 9, p. 390–397.

AUBRY, M.-P., 1983, Biostratigraphie du Paléogène épicontinental de l'Europe du Nord-Ouest. Etude fondée sur les nannofossiles calcaires: Documents des Laboratoires de Géologie de Lyon, v. 89, 317 p.

AUBRY, M.-P., 1986, Paleogene calcareous nannoplankton biostratigraphy of northwestern Europe: Palaeogeography, Paleoclimatology, Palaeoecology, v. 55, p. 267–334.

AUBRY, M.-P., 1991, Sequence stratigraphy: Eustasy or tectonic imprint?: Journal of Geophysical Research, v. 96, p. 6641–6679.

AUBRY, M.-P., 1992a, Late Paleogene calcareous nannoplankton evolution: A tale of climatic deterioration, in Prothero, D. R. and Berggren, W. A., eds., Eocene-Oligocene Climatic and Biotic Evolution: Princeton, Princeton University Press, p. 272–309.

AUBRY, M.-P., 1992b, Paleogene calcareous nannofossils from the Kerguelen Plateau, Leg 120: College Station, Proceedings of the Ocean Drilling Program, Scientific Results, v. 120, p. 471–491.

AUBRY, M.-P., 1993a, Calcareous nannofossil stratigraphy of the Neogene formations of eastern Jamaica, in Robinson, E. R. and Wright, R. M., eds, Biostratigraphy of Jamaica: Boulder, Geological Society of America Memoir 182, p. 131–178.

AUBRY, M.-P., 1993b, Neogene allostratigraphy and depositional history of the DeSoto Canyon area, northern Gulf of Mexico: Micropaleontology, v. 39, p. 327–366.

AUBRY, M.-P., BERGGREN, W. A., KENT, D. V., FLYNN, J. J., KLITGORD, K., OBRADOVICH, J. D., AND PROTHERO, D. R., 1988, Paleogene geochronology: an integrated approach: Paleoceanography, v. 3, p. 707–742.

AUBRY, M.-P., BERGGREN, W. A., STOTT, L. D., AND SINHA, A., 1995, The upper Paleocene-lower Eocene stratigraphic record and the Paleocene/Eocene boundary isotope excursion: implications for geoochronology, in Knox, R. O'B, Corfield, R. M., and Dunnay, R. E., eds., Correlation of the Early Paleogene in Northwestern Europe: London, Special Publication of the Geological Society (in press).

AUBRY, M.-P., HAILWOOD, E., AND TOWNSEND, H., 1986, Magneto- and calcareous nannofossil stratigraphy of the lower Paleogene of the Hampshire-London Basin: Journal of the Geological Society of London, v. 143, p. 729–735.

BACKMAN, J., 1987, Quantitative calcareous nannofossil biochronology of middle Eocene through early Oligocene sediment from DSDP Sites 522 and 523: Abhandlungen der Geologische Bundesanstalt, v. 39, p. 21–31.

BACKMAN, J. AND HERMELIN, J. O. R., 1986, Morphometry of the Eocene nannofossil Reticulofenestra umbilicus lineage and its biochronological consequenses: Palaeogeography, Palaeoclimatology, Palaeoecology, v. 57, p. 103–106.

BACKMAN, J., SCHNEIDER, D. A., RIO, D., AND OKADA, H., 1990, Neogene low-latitude magnetostratigraphy from Site 710 and revised age estimates of Miocene nannofossil datum events: College Station, Proceedings of Ocean Drilling Program, Scientific Results, v. 115, p. 271–276.

BACKMAN, J., AND SHACKLETON, N., 1983, Quantitative biochronology of Pliocene and early Pleistocene calcareous nannofossils from the Atlantic, Indian and Pacific Oceans: Marine Micropaleontology, v. 8, p. 141–170.

BAKSI, A. K. 1993, A Geomagnetic Polarity Time Scale for the period 0–17 Ma, based on ⁴⁰Ar/³⁹Ar plateau ages for selected field reversals: Geophysical Research Letters, v. 20, p. 1607–1610.

BAKSI, A. K. AND FARRAR, E., 1990, Evidence for errors in the Geomagnetic Polarity Time Scale at 17 to 15 Ma: ⁴⁰Ar/³⁹Ar dating of basalts from the Pacific Northwest, USA: Geophysical Research Letters, v. 17, p. 1117–1120.

BAKSI, A. K., HOFFMAN, K. A., AND FARRAR, E., 1993a, A new calibration point for the late Miocene section of the Geomagnetic Polarity Time Scale: ⁴⁰Ar/³⁹Ar dating of lava flows from Akaroa Volcano, New Zealand: Geophysical Research Letters, v. 20, p. 667–670.

BAKSI, A. K., HOFFMAN, K. A., AND McWILLIAMS, M., 1993b, Testing the accuracy of the geomagnetic polarity time-scale (GPTS) at 2.5 Ma, utilizing ⁴⁰Ar/³⁹Ar incremental heating data on whole-rock basalts: Earth and Planetary Science Letters, v. 118, p. 135–144.

BAKSI, A. K., HSU, V., McWILLIAMS, M. O., AND FARRAR, E., 1992, ⁴⁰Ar/³⁹Ar dating of the Brunhes-Matuyama geomagnetic field reversal: Science, v. 256, p. 356–357.

BAKSI, A. K., YORK, D., AND WATKINS, N. D., 1967, The age of the Steens Mountain geomagnetic polarity transition: Journal of Geophysical Research, v. 22, p. 6299–6308.

BALDAUF, J. G. AND BARRON, J. A., 1991, Diatom biostratigraphy: Kerguelen Plateau and Prydz Bay regions of the southern ocean: College Station, Proceedings of the Ocean Drilling Program, Scientific Results, v. 119, p. 547–598.

BANNER, F. T. AND BLOW, W. H., 1965, Progress in the planktonic foraminiferal biostratigraphy of the Neogene: Nature, v. 208, p. 1164–1166.

BARBIN, V., 1986, Le Priabonian dans sa région-type (Vicentin, Italie du Nord). Stratigraphie; Micropaléontologie: Essai d'intégration dans l'échelle chrono-

stratigraphique: Unpublished Ph.D. Dissertation, University Pierre et Marie Curie, Paris, 227 p.

BARKER, P. F., JOHNSON, D., AND OTHERS, 1983, Site 516: Initial Reports of the Deep Sea Drilling Project, v. 72, p. 155–338.

BARKER, P. F., KENNETT, J. P., AND OTHERS, 1988, Site 689: Proceedings of the Ocean Drilling Program, Initial Reports, v. 113, p. 89–181.

BARRON, J. A., 1985, Late Eocene to Holocene diatom biostratigraphy of the equatorial Pacific Ocean, Deep Sea Drilling Project Leg 85: College Station, Initial Reports of the Deep Sea Drilling Project, v. 85, p. 413–456.

BENSON, R. H., AND HODELL, D.A., 1994, Comment on "A critical re-evaluation of the Miocene/Pliocene boundary as defined in the Mediterranean" by F. G. Hilgen and C. C. Langereis: Earth and Planetary Science Letters, v. 124, p. 245–250.

BENSON, R., AND RAKIC EL BIED, K., 1991, The Messinian parastratotype at Cuevas del Almanzora, Vera Basin, SE Spain: refutation of the deep-basin, shallow-water hypothesis?: Micropaleontology, v. 37, p. 289–402.

BERGGREN, W. A., 1969, Rates of evolution of some Cenozoic planktonic foraminifera: Micropaleontology, v. 15, p. 351–365.

BERGGREN, W. A., 1971, Tertiary boundaries, in Funnell, B. F. and Riedel, W. R., eds., The Micropalaeontology of the Oceans: Cambridge, Cambridge University Press, p. 693–808.

BERGGREN, W. A., 1972, A Cenozoic time scale—some implications for regional geology and paleobiogeography: Lethaia, v. 5, p. 195–215.

BERGGREN, W. A., 1973, The Pliocene time scale: calibration of planktonic foraminifera and calcareous nannoplankton zones: Nature, v. 243, p. 391–397.

BERGGREN, W. A., 1977a, Late Neogene planktonic foraminiferal biostratigraphy of the Rio Grande Rise (South Atlantic): Marine Micropaleontology, v. 2, p. 265–313.

BERGGREN, W. A., 1977b, Late Neogene planktonic foraminiferal biostratigraphy of DSDP Site 357 (Rio Grande Rise): Washington, D. C., Initial Reports of the Deep Sea Drilling Project, v. 29, p. 591–614.

BERGGREN, W. A., 1986, Geochronology of the Eocene/Oligocene, in Pomerol, C. and Premoli Silva, I., Terminal Eocene Events: Amsterdam, Elsevier, Developments in Stratigraphy 9, p. 349–356.

BERGGREN, W. A., 1992, Neogene planktonic foraminifer magnetobiostratigraphy of the southern Kerguelen Plateau (Sites 747, 748 and 751): College Station, Proceedings of the Ocean Drilling Program, Scientific Results, v. 120, p. 631–647.

BERGGREN, W. A., 1993, Neogene planktonic foraminiferal biostratigraphy of eastern Jamaica, in Wright, R. M. and Robinson, eds., Biostratigraphy of Jamaica: Boulder, Geological Society of America Memoir 182, p. 179–217.

BERGGREN, W. A., 1994, In defense of the Selandian Age/Stage: Geologiska Foreningen Forhandlingen (Stockholm), v. 116, p. 6–8.

BERGGREN, W. A., AND AUBRY, M.-P., 1995, A late Paleocene-early Eocene NW European and North Sea magnetobiochronologic correlation network: A sequence stratigraphic network, in Knox, R. O'B., Corfield, R. M., and Dunnay, R. E., eds., Correlation of the Early Paleogene in Northwestern Europe: London, Special Publication of the Geological Society (in press).

BERGGREN, W. A., AUBRY, M.-P., AND HAMILTON, N., 1983a, Neogene magnetobiostratigraphy of DSDP Site 516 (Rio Grande Rise, South Atlantic): Washington, D.C., Initial Reports of the Deep Sea Drilling Project, v. 72, p. 675–706.

BERGGREN, W. A., HAMILTON, N., JOHNSON, D. A., PUJOL, C., WEISS, W., CEPEK, P., AND GOMBOS, JR., A. M., 1983b, Magnetobiostratigraphy of Deep Sea Drilling Leg 72, Sites 515–518, Rio Grande Rise (South Atlantic): Washington, D.C., Initial Reports of the Deep Sea Drilling Project, v. 72, p. 939–948.

BERGGREN, W. A., HILGEN, F. J., LANGEREIS, C. G., KENT, D. V., OBRADOVICH, J. D., RAFFI, I., RAYMO, M., AND SHACKLETON, N. J., 1995, Late Neogene (Pliocene-Pleistocene) chronology: New perspectives in high resolution stratigraphy: Geological Society of America Bulletin, v. 107, p. 1272–1287.

BERGGREN, W. A., KENT, D. V., AND COUVERING, J. A., VAN, 1985a, Neogene geochronology and chronostratigraphy, in Snelling, N. J., ed., The Chronology of the Geological Record: London, Geological Society of London Memoir 10, p. 211–260.

BERGGREN, W. A., KENT, D. V. AND FLYNN, J. J., 1985b, Paleogene geochronology and chronostratigraphy, in Snelling, N. J., ed., The Chronology of the Geological Record: London, Geological Society of London Memoir 10, p. 141–195.

BERGGREN, W. A., KENT, D. V., FLYNN, J. J., AND COUVERING, J. A., van, 1985c, Cenozoic geochronology: Geological Society of America Bulletin, v. 96, p. 1407–1418.

BERGGREN, W. A., KENT, D. V., OBRADOVICH, J. D., AND SWISHER, C. C., III, 1992, Towards a revised Paleogene geochronology, *in* Prothero, D. R. and Berggren, W. A., eds., Eocene-Oligocene Climatic and Biotic Evolution: Princeton, Princeton University Press, p. 29–45.

BERGGREN, W. A. AND MILLER, K. G., 1988, Paleogene tropical planktonic foraminiferal biostratigraphy and magnetobiochronology: Micropaleontology, v. 34, p. 362–380.

BERGGREN, W. A. AND NORRIS, R. D., 1993, Origin of the genus *Acarinina* and revision to Paleocene biostratigraphy: Geological Society of America, Annual Meeting, Abstract with Programs, v. 25, p. A359.

BERGGREN, W. A., NORRIS, R. D., AUBRY, M.-P. AND VAN FOSSEN, M., 1994, A Paleocene magnetic, biochronologic and isotopic reference section at DSDP Site 384: EOS, Transactions, American Geophysical Union, Spring Meeting, v. 75, no. 16, p. 52.

BLAKELY, R. J., 1974, Geomagnetic reversals and crustal spreading rates during the Miocene: Journal of Geophysical Research, v. 79, p. 2979–2985.

BLEIL, U., 1985, The magnetostratigraphy of northwest Pacific sediments, Deep Sea Drilling Project Leg 86: Washington, D.C., Initial Results of the Deep Sea Drilling Project, v. 86, p. 441–458.

BLOW, W. H., 1969, Late Middle Eocene to Recent planktonic foraminiferal biostratigraphy, *in* Bronnimann, P. and Renz, H. H., eds., Proceedings of the First International Conference on Planktonic Microfossils (Geneva, 1967): Leiden, E. J. Brill, v. 1, p. 199–421.

BLOW, W. H., 1979, The Cenozoic Globigerinida: A Study of the Morphology, Taxonomy, Evolutionary Relationships and the Stratigraphical Distribution of some Globigerinida (mainly Globigerinacea), 3 vols: Leiden, E. J. Brill, 1413 p.

BLOW, W. H. AND BANNER, F., 1966, The morphology, taxonomy and biostratigraphy of *Globorotalia barisanensis* LeRoy, *Globorotalia fohsi* Cushman and Ellisor, and related taxa: Micropaleontology, v. 12, p. 286–303.

BOERSMA, A., 1984, Cretaceous-Tertiary planktonic foraminifers from the southeastern Atlantic, Walvis Ridge Area, Deep Sea Drilling Project Leg 74: Washington, D.C., Initial Reports of the Deep Sea Drilling Project, v. 74, p. 501–523.

BOLLI, H. M., 1957a, The genera *Globigerina* and *Globorotalia* in the Paleocene-lower Eocene Lizard Springs Formation of Trinidad, B.W.I.: United States National Museum Bulletin 215, p. 51–81.

BOLLI, H. M., 1957b, Planktonic foraminifera from the Oligocene-Miocene Cipero and Lengua Formations of Trinidad, B.W.I.: United States National Museum Bulletin 215, p. 97–123.

BOLLI, H. M., 1957c, Planktonic foraminifera from the Eocene Navet and San Fernando Formtions of Trinidad, B.W.I.: United States National Museum Bulletin 215, p. 155–172.

BOLLI, H. M., 1966, Zonation of Cretaceous to Pliocene marine sediments based on planktonic foraminifera: Boletin Informativo, Asociacion Venezolano de Geologia, Mineraria y Petroleo, v. 9, p. 3–32.

BOLLI, H. M., 1970, The foraminifera of Sites 23–31, Leg 4: Washington D.C., Initial Results of the Deep Sea Drilling Project, v. 4, p. 577–643.

BOLLI, H. M., 1972, The genus *Globigerinatheka* Bronnimann: Journal of Foraminiferal Resarch, v. 2, p. 109–136.

BOLLI, H. M. AND BERMUDEZ, P. J., 1965, Zonation based on planktonic foraminifera of Middle Miocene to Pliocene warm-water sediments: Bolletin Informativo, Asociacion Venezolana di geologia, Mineria y Petroleo, v. 8, p. 119–149.

BOLLI, H. M . AND KRASHENINNIKOV, V. A., 1978, Problems in Paleogene and Neogene correlations based on planktonic foraminifera: Micropaleontology, v. 23, p. 436–452.

BOLLI, H. M. AND PREMOLI SILVA, I., 1973, Oligocene to Recent plantkonic foraminifera and stratigraphy of the Leg 15 Sites in the Caribbean Sea: Washington, D.C., Initial Reports of the Deep Sea Drilling Project, v. 15, p. 475–497.

BOLLI, H. M. AND SAUNDERS, J. B., 1985, Oligocene to Holocene low latitude planktic foraminifera, *in* Bolli, H. M., Saunders, J. B., and Perch-Nilsen, K., eds., Plankton Stratigraphy: Cambridge, Cambridge University Press, p.155–262.

BONI, A., 1967, Notizie sul Serravalliano tipo, *in* Selli, R., ed., Guida alle escursioni del IV Congresso: Committee on Mediterranean Neogene Stratigraphy, International Union of Geological Sciences, 4th International Congress, p. 47–63.

BRINKHUIS, H., 1988, Late Eocene to early Oligocene dinoflagellate cysts from the Priabonian type-area (northeast Italy): biostratigraphy and paleoenvironmental interpretation: Palaeogeography, Palaeoclimatology, Palaeoecology, v. 107, p. 121–163.

BRINKHUIS, H., 1992, Late Eocene to Early Oligocene Dinoflagellate Cysts from Central and Northeast Italy: Doctoral Dissertation, University of Utrecht, 169 p. (Published privately by the author).

BRINKHUIS, H. AND BIFFI, U., 1993, Dinoflagellate cyst stratigraphy of the Eocene/Oligocene transition in central Italy: Marine Micropaleontology, v. 22, p. 131–183.

BRONNIMANN, P. AND RESIG, J., 1971, A Neogene globigerinacean biochronologic time-scale of the southwestern Pacific: Washington, D.C., Initial Reports of the Deep Sea Drilling Project, v. 7, p. 1235–1469.

BRYAN, W. B. AND DUNCAN, R. A., 1983, Age and provenance of clastic horizons from Hole 516F: Washington, D.C., Initial Reports of the Deep Sea Drilling Project, v. 72, p. 475–477.

BUKRY, D., 1973, Low-latitude coccolith biostratigraphic zonation: Washington, D.C., Initial Reports of the Deep Sea Drilling Project, v. 15, p. 127–149.

BUKRY, D., 1975, Coccolith and silicoflagellate stratigraphy, northwestern Pacific Ocean, Deep Sea Drilling Project Leg 32: Washington D.C., Initial reports of the Deep Sea Drilling Project, v. 32, p. 677–701.

CANDE, S. C. AND KENT, D. V., 1992, A new geomagnetic polarity time scale for the Late Cretaceous and Cenozoic: Journal of Geophysical Research, v. 97, p. 13,917–13,951.

CANDE, S. C. AND KENT, D. V., 1995, Revised calibration of the geomagnetic polarity time scale for the Late Cretaceous and Cenozoic: Journal of Geophysical Research, v. 100, p. 6093–6095.

CANDE, S. C. AND KRISTOFFERSEN, Y., 1977, Late Cretaceous magnetic anomalies in the North Atlantic: Earth and Planetary Science Letters, v. 35, p. 215–224.

CANDE, S., LABRECQUE, J. L., AND HAXBY, W. F., 1988, Plate kinematics of the South Atlantic: Chron C34 to present: Journal of Geophysical Research, v. 93, p. 13,479–13,492.

CANUDO, J. I. AND MOLINA, E., 1992, Planktic foraminiferal faunal turnover and bio-chronostratigraphy of the Paleocene-Eocene boundary at Zumaya, northern Spain: Revista Societa Geologia Espana, v. 5, p. 145–157.

CARO, Y., LUTERBACHER, H. P., PERCH-NIELSEN, K., PREMOLI SILVA, I., RIEDEL, W. R., AND SANFILIPPO, A., 1975, Zonations à l'aide de microfossiles pélagiques du Paléocène supérieur et de l' Eocène inférieur: Bulletin de la Société géologique de France, v. 17, p. 125–147.

CEBULA, G. T., KUNK, M. J., MEHNERT, H. H., NAESER, C. W., OBRADOVICH, J. O., AND SUTTER J. F., 1986, The Fish Canyon Tuff, a potential standard for the $^{40}Ar/^{39}Ar$ and Fission-Track methods: Terra Cognita, Abstract with Programs, v. 6, p. 139–140.

CHAISSON, W. P. AND LECKIE, R. M.,1993, High-resolution Neogene planktonic foraminiferal biostratigraphy of ODP Site 806, Ontong Java Plateau (Western Equatorial Pacific): College Station, Proceedings of the Ocean Drilling Program, Scientific Results, v. 130, p. 137–178.

CHANNELL, J. E. T., RIO, D., SPROVIERI, R., AND GLAÇON, G., 1990, Biomagnetostratigraphic correlations from Leg 107 in the Tyrrhenian Sea: College Station, Proceedings of the Ocean Drilling Program, Scientific Results, v. 107, p. 669–682.

CHAPRONIERE, G. C. H., STYZEN, M. J., SAGER, W. W., NISHI, H., QUINTERNO, P. J., AND ABRAHAMSEN, N., 1994, Late Neogene biostratigraphic and magnetostratigraphic synthesis, Leg 135: College Station, Proceedings of the Ocean Drilling Program, Scientific Results, v. 135, p. 857–877.

CHAVE, A. D., 1984, Lower Paleocene-upper Cretaceous magnetostratigraphy, Sites 525, 527, 528, and 529, Deep Sea Drilling Project Leg 74: Washington, D.C., Initial Reports of the Deep Sea Drilling Project, v. 74, p. 525–531.

CIESELSKI, P. F., KRISTOFFERSEN, Y., AND OTHERS, 1988, Site 699: College Station, Proceedings of the Ocean Drilling Program, Initial Reports, v. 114, p. 151–254.

CIFELLI, R., AND SCOTT, G., 1986, Stratigraphic Record of the Neogene Globorotaliid Radiation (Panktonic Foraminifera): Smithsonian Contributions to Paleobiology 58, 101 p.

CITA, M. B., 1973, Pliocene biostratigraphy and chronostratigraphy: Washington, D.C., Initial Reports of the Deep Sea Drilling Project, v. 13, p. 1343–1379.

CITA, M. B., 1975, The Miocene/Pliocene boundary: history and definition, *in* Saito, T. and Burckle, L. H., eds., Late Neogene Epoch Boundaries: Micropaleontology, Special Publication, v. 1, p. 1–30.

CITA, M. B. AND BLOW, W. H., 1969, The biostratigraphy of the Langhian, Serravallian and Tortonian stages in the type-sections in Italy: Rivista Italiana Paleontologia, v. 75, p. 549–603.

CITA, M. B. AND PREMOLI SILVA, I., 1960, Pelagic foraminifera from the type Langhian: Proceedings of the International Paleontological Union Norden 1960, part XXII, p. 39–50.

CITA, M. B. AND PREMOLI SILVA, I., 1968, Evolution of the planktonic foraminiferal assemblages in the stratigraphical interval between the type-Langhian and the type-Tortonian and biozonation of the Miocene of the Piedmont: Giornale Geologia, v. 35, p. 1–27.

CITA, M. B., PREMOLI SILVA, I., AND ROSSI, R., 1965, Foraminifero planktonici del Tortoniano-tipo: Rivista Italiana Paleontologia, v. 71, p. 217–308.

CLEMENT, B. M. AND ROBINSON, F., 1987, The magnetostratigraphy of Leg 94 sediments: Washington D. C., Initial Reports of the Deep Sea Drilling Project, v. 94, p. 635–650.

COCCIONI, R., MONACO, P., MONECHI, S., NOCCHI, M., AND PARISI, G., 1988, Biostratigraphy of the Eocene-Oligocene boundary at Massignano (Ancona, Italy), in Premoli Silva, I., Coccioni, R., and Montanari, A., eds., The Eocene-Oligocene Boundary in the Marche-Umbria Basin (Italy): Ancona, International Subcommission on Paleogene Stratigraphy, E/O Meeting, Special Publication II, p. 59–74.

COLALONGO, M. L., DIGRANDE, A., D'ONOFRIO, S., GIANELLI, L., IACCARINO, S., MAZZEI, R., ROMEO, M., AND SALVATORINI, G., 1979, Stratigraphy of Late Miocene Italian sections straddling the Tortonian/Messinian boundary: Bolletino Societa Paleontologia Italiana, v. 18, p. 258–302.

CORFIELD, R. M., 1987, Patterns of evolution in Palaeocene and Eocene planktonic Foraminifera, in Hart, M., ed., The Micropalaeontology of Carbonate Environments: Chichester, Ellis Horwood, p. 93–110.

COX, A., DOELL, R. R., AND DALRYMPLE, G. B., 1964, Geomagnetic polarity epochs: Science, v. 143, p. 351–352.

CURRY, D. AND ODIN, G. S., 1982, Dating of the Palaeogene, in Odin, G. S., ed., Numerical Dating in Stratigraphy: New York, John Wiley and Sons, v. 1, p. 607–630.

CUSHMAN, J. A. AND RENZ, H. H., 1947, The foraminiferal fauna of the Oligocene, Ste Croix Formation of Trinidad: Bulletin of the Western Indies, Special Publication of the Cushman Laboratory, v. 22, p. 1–46.

CUSHMAN, J. A. AND STAINFORTH, R. M., 1945, The Foraminifera of the Cipero Marl Formation of Trinidad, British West Indies: Washington, D.C., Cushman Laboratory for Foraminiferal Research Special Publication 14, 75 p.

DALRYMPLE, G. B., IZETT, G. A., SNEE, L. W., AND OBRADOVICH, J. D., 1993, ^{40}Ar/^{39}Ar age spectra and total-fusion ages of tektites from Cretaceous-Tertiary Boundary sedimentary rocks in the Beloc Formation, Haiti: United States Geological Bulletin, v. 2065, p. 1–20.

DEINO, A., TAUXE, L., MONOGHAN, M., AND DRAKE, R., 1990, ^{40}Ar/^{39}Ar age calibration of the litho- and paleomagnetic stratigraphies of the Ngorora Formation, Kenya: Journal of Geology, v. 98, p. 567–587.

DOWSETT, H., 1988, Diachrony of late Neogene microfossils in the southwest Pacific ocean: application of the graphic correlation method: Paleoceanography, v. 3, p. 209–222.

DOWSETT, H., 1989, Application of the graphic correlation method to Pliocene marine sequences: Marine Micropalentology, v. 14, p. 3–32.

DUNCAN, R. A. AND HARGRAVES, R. B., 1990, ^{40}Ar/^{39}Ar geochronology of basement rocks from the Mascarene Plateau, the Chagos Bank, and Maldives Ridge: College Station, Proceedings of the Ocean Drilling Program, Scientific Results, v. 115, p. 43–51.

EHRENDORFER, T. AND AUBRY, M.-P., 1992, Calcareous nannoplankton changes across the Cretaceous/Paleocene boundary in the southern Indian Ocean (Site 750): College Station, Proceedings of the Ocean Drilling Program, Scientific Results, v. 120, p. 451–470.

FENNER, J., 1984, Eocene-Oligocene planktic diatom stratigraphy in the low latitudes and the high southern latitudes: Micropaleontology, v. 30, p. 319–342.

FLEISHER, R. L., 1974, Cenozoic planktonic foraminifera and biostratigraphy, Arabian Sea Deep Sea Drilling Project, Leg 23A: Washington, D.C., Initial Reports of the Deep Sea Drilling Project, v. 23, p. 1001–1072.

GALBRUN, B., 1992, Magnetostratigraphy of upper Cretaceous and lower Tertiary sediments, Sites 761 and 762, Exmouth Plateau, northwest Australia: College Station, Proceedings of the Ocean Drilling Program, Scientific Results, v. 122, p. 699–716.

GARTNER, S., 1992, Miocene nannofossil chronology in the North Atlantic, DSDP Site 608: Marine Micropaleontology, v. 18, p. 307–331.

GARTNER, S., CHEN, M. P., AND STANTON, R. J., 1983, Late Neogene nannofossil biostratigraphy and paleoceanography of the Northeastern Gulf of Mexico and adjacent areas: Marine Micropaleontology, v. 8, p. 17–50.

GAULTIER, F., CLAUZON, G., SUC, J.-P., CRAVATTE, J., AND VIOLANTI, D., 1994, Age et durée de la crise de salinité messinienne: Comptes Rendus de l' Académie des Sciences de Paris, t. 318, séries II, p. 1103–1109.

GERSONDE, R. AND BURCKLE, L. H., 1990, Neogene diatom biostratigraphy of ODP Leg 113, Weddell Sea (Antarctic Ocean): College Station, Proceedings of the Ocean Drilling Program, Initial Reports, v. 113, 761–789.

GIANOTTI, A., 1953, Microfaune della serie tortoniano del Rio Mazzapiedi-Castellania (Tortona-Allesandria): Rivista Italiana Paleontologia, Memoir, v. 6, p. 167–308.

GILLOT, P. Y., JEHANNO, C., ROCCHIA, R., AND H. SIGURDSSON, H., 1991, Datation par la méthode potassium-argon de la limite Crétacé-Paléogène en milieu marin: age des verres de Béloc (Haiti): Comptes Rendus de l' Académie des Sciences de Paris, t. 313, séries II, p. 193–199.

GLAÇON, G., VERGNAUD GRAZZINI, C., IACCARINO, S., REHAULT, J.-P., RANDRIANASOLO, A., SIERRO, J. V., WEAVER, P., CHANNELL, J., TORII, M., AND HAWTHORNE, T., 1990, Planktonic foraminiferal events and stable isotope records in the upper Miocene, Site 654: College Station, Proceedings of the Ocean Drilling Program, Scientific Results, v. 107, p. 415–427.

GRADSTEIN, F. M., AGTERBERG, F. P., OGG, J. G., HARDENBOL, J., Veen P. V., THIERRY, J., AND HUANG, Z., 1994, A Mesozoic time scale: Journal of Geophysical Research, v. 99, p. 24.051–24,074.

GROOT, J. J., JONGE, R. B. G., de, LANGEREIS, C. G., KATE, W. G. H. Z. TEN, AND SMIT, J., 1989, Magnetostratigraphy of the Cretaceous-Tertiary boundary at Agost (Spain): Earth and Planetary Science Letters, v. 94, p. 385–397.

HAILWOOD, E. A., 1989, The role of magnetostratigraphy in the development of geological time scales: Paleoceanography, v. 4, p. 1–18.

HAILWOOD, E. A. AND CLEMENT, B. M., 1991a, Magnetostratigraphy of Sites 699 and 700, East Georgia Basin: College Station, Proceedings of the Ocean Drilling Program, Scientific Results, v. 114, p. 337–357.

HAILWOOD, E. A. AND CLEMENT, B. M., 1991b, Magnetostratigraphy of Sites 703 and 704, Meteor Rise, southeastern South Atlantic: College Station, Proceedings of the Ocean Drilling Program, Scientific Results, v. 114, p. 367–385.

HALL, C. M., YORK, D., AND H. SIGURDSSON, H., 1991, Laser ^{40}Ar/^{39}Ar stepheating ages from single Cretaceous-Tertiary boundary glass spherules (abs.): EOS, Transactions, American Geophysical Union, v. 72, p. 531.

HAQ, B. U., HARDENBOL, J., AND VAIL, P. R., 1987, The chronology of fluctuating sea level since the Triassic: Science, v. 235, p. 1156–1167.

HAQ, B. U., HARDENBOL, J., AND VAIL, P. R., 1988, Mesozoic and Cenozoic chronostratigraphy and cycles of sea-level change, in Wilgus, C. K., Kendall, C. G. St. C., Posamentier, H. W., Ross, C. A., and Van Wagoner, J. C., eds., Sea Level Changes- An Integrated Approach: Tulsa, Society of Economic Paleontologists and Mineralogists Special Publication 42, p. 71–108.

HAQ, B. U., WORSLEY, T. R., BURCKLE, L. H., DOUGLAS, R. G., KEIGWIN, L. D., Jr., OPDYKE, N. D., SAVIN, S. M., SOMMER, M. A., III, VINCENT, E., AND WOODRUFF, F., 1980, Late Miocene marine carbon-isotope shift and synchroneity of some phytoplanktonic biostratigraphic events: Geology, v. 8, p. 427–431.

HARDENBOL, J. AND BERGGREN, W. A., 1978, A new Paleogene numerical time scale, in Cohee, G. V., Giaessner, M. F., and Hedberg, H. D., eds., The Geologic Time Scale: Tulsa, American Association of Petroleum Geology, Studies in Geology 6, p. 213–234.

HARLAND, W. B., ARMSTRONG, R. L., COX, A. V., CRAIG, L. E., SMITH, A. G., AND SMITH, D. G., 1990, A Geologic Time Scale 1989, revised edition: Cambridge, Cambridge University Press, 263 p.

HARLAND, W. B., COX, A. V., LLEWELLYN, P. G., PICKTON, C. A. G., SMITH, A. G., AND WALTERS, R., 1982, A Geologic Time Scale: Cambridge, Cambridge University Press, 131 p.

HARRIS, W. B. AND FULLAGAR, P. D., 1989, Comparison of Rb-Sr and K-Ar dates of middle Eocene bentonite and glauconite, southeastern Atlantic Coastal Plain: Geological Society of America Bulletin, v. 101, p. 573–577.

HARRIS, W. B. AND FULLAGAR, P. D., 1991, Middle Eocene and Late Oligocene isotopic dates of glauconitic mica from the Santee River area, South Carolina: Southeastern Geology, v. 32, p. 1–19.

HARWOOD, D. M. AND MARUYAMA, T., 1992, Middle Eocene to Pleistocene diatom biostratigraphy of southern Ocean sediment from the Kerguelen Plateau: College Station, Proceedings of the Ocean Drilling Program, Scientific Results, v. 120, p. 683–733.

HEDBERG, H. D., ed., 1976, International Stratigraphic Guide: A Guide to Stratigraphic Classification, Terminology and Procedure: New York, John Wiley and Sons, 200 p.

HEIRTZLER, J. R., DICKSON, G. O., HERRON, E. M., PITMAN, W. C., AND LE PICHON, X., 1968, Marine magnetic anomalies, geomagnetic field reversals, and motions of the ocean floor and continents: Journal of Geophysical Research, v. 73, p. 2119–2136.

HERBERT, T. D. AND D'HONDT, S. L., 1990, Precessional climate cyclicity in Late Cretaceous-early Tertiary marine sediments: a high resolution chronometer of Cretaceous-Tertiary boundary events: Earth and Planetary Science Letters, v. 99, p. 263–275.

HILGEN, F., 1991, Extension of the astronomically calibrated (polarity) time scale to the Miocene/Pliocene boundary: Earth and Planetary Science Letters, v. 107, p. 349–368.

HILGEN, F. AND LANGEREIS, C. G., 1989, Periodicities of CaCO₃ cycles in the Mediterranean Pliocene: discrepancies with the quasi-periods of the Earth's orbital cycles?: Terra Nova, v. 1, p. 409–415.

HILGEN, F. AND LANGEREIS, C. G., 1993, A critical (re)evaluation of the Miocene/Pliocene boundary as defined in the Mediterranean: Earth and Planetary Science Letters, v. 118, p. 167–179.

HODELL, D. A. AND KENNETT, J. P., 1986, Late Miocene-early Pliocene stratigraphy and paleoceanography of the South Atlantic and southwest Pacific Oceans: a synthesis: Paleoceanography, v. 1, p. 285–311.

HODELL, D., BENSON, R. H., KENT, D. V., BOERSMA, A., AND RAKIC-EL BIED, K., 1994, Magnetostratigraphic, biostratigraphic, and stable isotope stratigraphy of an Upper Miocene drill core from the Sale Briqueterie (northwestern Morocco): a high-resolution chronology for the Messinian stage: Paleoceanography, v. 9, p. 835–855.

HUBER, B. T., 1991, Paleogene and early Neogene planktonic foraminifer biostratigraphy of Sites 738 and 744, Kerguelen Plateau (southern Indian Ocean): College Station, Proceedings of the Ocean Drilling Program, Scientific Results, v. 119, p. 427–449.

IACCARINO, S., 1985, Mediterranean Miocene and Pliocene planktic foraminifera, *in* Bolli, H. M., Saunders, J. B., and Perch-Nielsen, K., Plankton Stratigraphy: Cambridge, Cambridge University Press, p. 283–314.

INOKUCHI, H. AND HEIDER, F., 1992, Magnetostratigraphy of sediments from Sites 748 and 750, Leg 120: College Station, Proceedings of the Ocean Drilling Program, Scientific Results, v. 120, p. 247–252.

IZETT, G. A., DALRYMPLE, G. B., AND SNEE, L. W. 1991, ⁴⁰Ar/³⁹Ar age of the Cretaceous -Tertiary boundary tektites from Haiti: Science, v. 252, p. 1539–1542.

IZETT, G. A. AND OBRADOVICH, J. D., 1991, Dating of the Matayama-Brunhes boundary based on ⁴⁰Ar/³⁹Ar ages of the Bishop Tuff and Cerro San Luis Rhyolite (abs.): Geological Society of America Abstracts with Programs, v. 23, p. A105.

IZETT, G. A. AND OBRADOVICH, J. D., 1994, ⁴⁰Ar/³⁹Ar age constraints for the Jaramillo normal Subchron and the Matuyama-Brunhes geomagnetic boundary. Journal of Geophysical Research, v. 90, p. 2925–2934.

JENKINS, D. G., 1967, Planktonic foraminiferal zones and new taxa from the lower Miocene to the Pleistocene of New Zeland: New Zeland Journal of Geology and Geophysics, v. 10, p. 1064–1078.

JENKINS, D. G., 1971, New Zeland Cenozoic Planktonic Foraminifera: Palaeontological Bulletin of the New Zealand Geological Survey, v. 42, p. 1–278.

JENKINS, D. G., 1975, Cenozoic planktonic foraminiferal biostratigraphy of the southwestern Pacific and Tasman Sea: Washington, D.C., Initial Reports of the Deep Sea Drilling Project, v. 29, p. 449–467.

JENKINS, D. G., 1992, The paleogeography, evolution and extinction of Late Miocene-Pleistocene planktonic foraminifera from the Southwest Pacific, *in* Ishizaki, K., and Saito, T., eds., Centenary of Japanese Micropaleontology, Contributed papers in Honor of Professor Yokichi Takayanagi: Tokyo, Terra Scientific Publishing Co., p. 27–35.

JENKINS, D. G., 1993a, Cenozoic southern middle and high latitude biostratigraphy and chronostratigraphy based on planktonic foraminifera, *in* Kennett, J. P. and Warnke, D. A., eds., The Antarctic Paleoenvironment: A Perspective on Global Change: Washington, D.C., American Geophysical Union, Antarctic Research Series, vol. 60, p. 125–144.

JENKINS, D. G., 1993b, The evolution of the Cenozoic southern high- and mid-latitude planktonic foraminiferal faunas, *in* Kennett, J. P. and Warnke, D. A., eds., The Antarctic Paleoenvironment: A Perspective on Global Change: Washington, D.C., American Geophysical Union, Antarctic Research Series, vol. 60, p. 175–194.

JENKINS, D. G. AND GAMSON, P., 1993, The late Cenozoic *Globorotalia truncatulinoides* datum- plane in the Atlantic, Pacific and Indian Oceans, *in* Hailwood, E. A. and Kidd, R. B., eds., High Resolution Stratigraphy: London, Geological Society of London Special Publication 70, p. 127–130.

JENKINS, D. G. AND ORR, W., 1972, Planktonic foraminiferal biostratigraphy of the Eastern Equatorial Pacific: Washington, D.C., Initial Reports of the Deep Sea Drilling Project, v. 9, p. 1059–1193.

JENKINS, D. G., SAUNDERS, J. B., AND CIFELLI, R., 1981, The relationship of *Globigerinoides bisphericus* Todd 1954 to *Praeorbulina sicana* (De Stephani) 1952: Journal of Foraminiferal Research, v. 11, p. 262–267.

JENKINS, D. G. AND SRINIVASAN, M. S., 1986, Cenozoic planktonic foraminifers from the Equator to the Sub-Antarctic of the southwest Pacific: Washington, D.C., Initial Reports of the Deep Sea Drilling Project, v. 90, p. 795–834.

KEATING, B. H. AND HERRERO-BERVERA, E., 1984, Magnetostratigraphy of the Cretaceous and early Cenozoic sediments of Deep Sea Drilling Project Site 530, Angola: Washington, D.C., Initial Reports of the Deep Sea Drilling Project, v. 75, p. 1211–1226.

KEIGWIN, L. D., JR., 1982, Neogene planktonic foraminifers from Deep Sea Drilling Project Sites 502 and 503: Washington, D.C., Initial Reports of the Deep Sea Drilling Project, v. 68, p. 269–288 (see also Appendix, p. 493–495).

KELLER, G., 1978, Late Neogene biostratigraphy and paleoceanography of DSDP Site 310 Central North Pacific and correlation with the southwest Pacific: Marine Micropaleontology, v. 3, p. 97–119.

KELLER, G., 1979a, Late Neogene planktonic foraminiferal biostratigraphy and paleoceanography of the northwest Pacific DSDP Site 296: Palaeogeography, Palaeoclimatology, Palaeoecology, v. 27, p. 129–154.

KELLER, G., 1979b, Late Neogene paleoceanography of the North Pacific DSDP Sites 173, 310, and 296: Marine Mivcropaleontology, v. 4, p. 159–172.

KELLER, G., 1979c, Early PLiocene to Pleistocene planktonic foraminiferal datum levels in the North Pacific: DSDP Sites 173, 310, 296: Marine Micropaleontology, v. 4 , p. 281–294.

KELLER, G., 1980a, Early to middle Miocene planktonic foraminiferal datum levels of the equatorial and subtropical Pacific: Micropaleontology, v. 6, p. 372–391.

KELLER, G., 1980b, Middle to late Miocene planktonic foraminiferal datum levels and paleoceanography of the North and southeastern Pacific Ocean: Marine Micropaleontology, v. 5, p. 249–281.

KELLER, G., 1981, Miocene biochronology and paleoceanography of the North Pacific: Marine Micropaleontology, v. 6, p. 535–551.

KELLER, G., 1988, Extinction, survivorship and evolution of planktic foraminifers across the Cretaceous/Tertiary boundary in El Kef, Tunisia: Marine Micropaleontology. v. 13, p. 239–263.

KENNETT, J. P., 1973, Middle and late Cenozoic planktonic foraminiferal biostratigraphy of the southwest Pacific- DSDP Leg 21: Washington, D.C., Initial Reports of the Deep Sea Drilling Project, v. 21, p. 575–640.

KENNETT, J. P. AND SRINIVASAN, M., 1983, Neogene Planktonic Foraminifera: A Phylogenetic Atlas: Stroudsburg, Hutchinson Ross Publishing Co., 265 p.

KENNETT, J. P. AND SRINIVASAN, M. S., 1984, Neogene planktonic foraminiferal datum planes of the South Pacific: mid to equatorial latitudes, *in* Tsuchi, R. and Ikebe, N., eds., Pacific Neogene Datum Planes — Conribution to Biostratigraphy and Chronology: Tokyo, University of Tokyo Press, p. 11–25.

KENNETT, J. P. AND WATKINS, N. D., 1974, Late Miocene-early Pliocene paleomagnetic stratigraphy, paleoclimatology and biostratigraphy in New Zeland: Geological Society America Bulletin, v. 85, p. 1385–1398.

KLITGORD, K. D., HEUSTIS, S. P., MUDIE, J. D., AND PARKER, R. L., 1975, An analysis of near-bottom magnetic anomalies: sea floor spreading and the magnetized layer: Geophysical Journal of the Royal Astronomical Society, v. 43, p. 387–424.

KNOX, R. W. O'B., HINE, N. M., AND ALI, J., 1994, New information on the age and sequence stratigraphy of the type Thanetian of southeast England: Newsletters in Stratigraphy, v. 30, p. 45–60.

KRIJGSMAN, W., HILGEN, F. J., LANGEREIS, C. G., AND ZACHARIASSE, W. J, 1994, The age of the Tortonian/Messinian boundary: Earth and Planetary Sciences Letters, v. 121, p. 533–547.

LABRECQUE, J. L., HSU, K. J., CARMAN, M. F. J., KARPOFF, A. M., MCKENZIE, J. A., PERCIVAL, S. F. P., PETERSEN, N. P., PISCIOTTO, K. A., SCHREIBER, E., TAUXE, L., TUCKER, P., WEISSERT, H. J., AND WRIGHT, R., 1983, DSDP Leg 73: Contributions to Paleogene stratigraphy in nomenclature, chronology and sedimentation rates: Palaeogeography, Palaeoclimatology, Palaeoecology, v. 42, p. 91–125.

LABRECQUE, J. L., KENT, D. V., AND CANDE, S. C., 1977, Revised magnetic polarity time scale for the Late Cretaceous and Cenozoic time: Geology, v. 5, p. 330–335.

LANG, T. H. AND WISE, S. W., JR., 1987, Neogene and Paleocene-Maestrichtian calcareous nannofossil stratigraphy, Deep Sea Drilling Project Sites 604 and 605, upper continentsal rise of New Jersey: sedimentation rates, hiatuses, and correlations with seismic stratigraphy: Washington, D.C., Initial Reports of the Deep Sea Drilling Project, v. 93, p. 661–683.

LANGEREIS, C. G. AND DEKKERS, M. J., 1992, Paleomagnetism and rock magnetism of the Tortonian-Messinian boundary stratotype at Falconara, Sicily: Physics of the Earth and Planetary Interiors, v. 71, p. 100–111.

LANGEREIS, C. G. AND HILGEN, F. G., 1991, The Rosello composite: a Mediterranean and global reference section for the Early to early Late Pliocene: Earth and Planetary Science Letters, v. 104, p. 211–225.

LANGEREIS, C. G., VAN HOOF, A. A. M., AND HILGEN, F. J., 1994, Steadying the rates: Nature, v. 369, p. 615.

LANGEREIS, C. G., ZACHARIASSE, W. J., AND ZIJDERVELD, J. D. A., 1984, Late Miocene magnetobiostratigraphy of Crete: Marine Micropaleontology, v. 8, p. 261–281.

LARSON, R. L. AND PITMAN, W. C., 1972, Worldwide correlation of Mesozoic magnetic anomalies, and its implications: Geological Society of America Bulletin, v. 83, p. 3645–3662.

LECKIE, M., FARNHAMN, C., AND SCHMIDT, M. G., 1993, Oligocene planktonic foraminiferal biostratigraphy of ODP Hole 130–803D (Ontong Java Plateau) and ODP Hole 101–628A (Little Bahama Bank) and comparison with the southern high latitudes: College Station, Proceedings of the Ocean Drilling Program, Scientific Results, v. 130, p. 113–136.

LEONOV, G. P. AND ALIMARINA, V. P., 1961, Stratigraphy and planktonic foraminifera of the Cretaceous-Paleogene "Transition" beds of the central part of the North Caucasus: Collected Papers Geological Faculty, University of Moscow, to XXI International Geological Congress, p. 29–60 (in Russian with an extensive English abstract).

LEONOV, G. P. AND ALIMARINA, V. P., 1964, Stratigraphical problems of the Paleogene deposits of the northwestern Caucasus: Moscow, Moscow University Press, 204 p.

LI, Q., RADFORD, S. S., BANNER, F. T., 1992, Distribution of microperforate tenuitellid planktonic foraminifers in Holes 747A and 749B, Kerguelen Plateau: College Station, Proceedings of the Ocean Drilling Program, Scientific Results, v. 120, p. 569–602.

LIU, C., 1993, Uppermost Cretaceous-Lower Paleocene stratigraphy and turnover of planktonic foraminifera across the Cretaceous/Paleogene boundary: Unpublished Ph.D. Dissertation, Rutgers University, New Brunswick, 181 p.

LOUTIT, T. AND KENNETT, J. P., 1979, Application of carbon isotope stratigraphy to late Miocene shallow marine sediments, New Zeland: Science, v. 204, p. 1196–1199.

LOWRIE, W. AND ALVAREZ, W., 1981, One hundred million years of geomagnetic polarity history: Geology, v. 9, p. 392–397.

LOWRIE, W., ALVAREZ, W., NAPOLEONE, G., PERCH-NIELSEN, K., PREMOLI SILVA, I., AND TOUMARKINE, M., 1982, Paleogene magnetic stratigraphy in Umbrian pelagic carbonate rocks: The Contessa sections, Gubbio: Geological Society of America Bulletin, v. 93, p. 414–432.

LUTERBACHER, H.-P. AND PREMOLI SILVA, I., 1964, Biostratigrafia del limite Cretaceo-Terziaro nell'Appennino centrale: Rivista Italiana Paleontologia Stratigraphia, v. 70, p. 67–128.

MACFADDEN, B. J., SWISHER, C. C., OPDYKE, N. D., AND WOODBURNE, M. O., 1990, Paleomagnetism, geochronology, and possible tectonic rotation of the middle Miocene Barstow Formation, Mojave Desert, southern California: Geological Society of America Bulletin, v. 102, p. 478–493.

MALMGREN, B. AND KENNETT, J. P., 1982, The potential of morphometrically based phylozonation: application of a Late Cenozoic planktonic foraminiferal lineage: Marine Micropaleontology, v. 7, p. 285–296.

MANIVIT, H., 1984, Paleogene and upper Cretaceous calcareous nannofossils from Deep Sea Drilling Project Leg 74: Washington, D.C., Initial Reports of the Deep Sea Drilling Project, v. 74, p. 475–499.

MANKINEN, E. A., AND DALRYMPLE, G. B., 1979, Revised geomagnetic polarity time scale for the interval 0–5 my B.P.: Journal of Geophysical Research. v. 84, p. 615–626.

MARTINI, E., 1971, Standard Tertiary and Quaternary calcareous nannoplankton zonation, in Farinacci, A., ed., Proceedings of the Second Planktonic Conference, Roma 1970: Roma, Tecnoscienza, p. 739–785.

MAYER, L. C., PISIAS, N., AND JANACEK, T., AND OTHERS, 1992, Leg 138: College Station, Proceedings of the Ocean Drilling Porgram, Initial Reports, v. 138, 1462 p.

MCDOUGALL, I., BROWN, F. H., CERLING, T. E., AND HILLHOUSE, J. W., 1992, A reappraisal of the geomagnetic polarity time scale to 4 Ma using data from the Turkana Basin, East Africa: Geophysical Research Letters, v. 19, p. 2349–2352.

MCDOUGALL, I., KRISTJANSSON, L., AND SAEMUNDSSON, K., 1984, Magnetostratigraphy and geochronology of northwest Iceland: Journal of Geophysical Research, v. 89, p. 7029–7060.

MCINTOSH, W. C., GEISSMAN, J. W., CHAPIN, C. E., KUNK, M. J., AND HENRY, C. D., 1992, Calibration of the latest Eocene-Oligocene geomagnetic polarity timew scale using $^{40}Ar/^{39}Ar$ ignimbrites: Geology, v. 20, p. 459–463.

MCWILLIAMS, M. O., BAKSI, A. K., AND BAADSGAARD, H., 1991a, High resolution $^{40}Ar/^{39}Ar$ ages from Cretaceous-Tertiary boundary bentonites in Western North America (abs.): IGCP Projects 216, 293, 303, Event Markers in Earth History, p. 53.

MCWILLIAMS, M. O., BAKSI, A. K., AND BAADSGAARD, H., 1991b, New $^{40}Ar/$ ^{39}Ar ages from K-T boundary bentonites in Montana and Saskatchewan (abs.): EOS, Transactions, American Geophysical Union, v. 72, p. 301.

MCWILLIAMS, M. O., BAKSI, A. K., BOHOR, B. F., IZETT, G. A., AND MURALI, A. V., 1992, High precision relative ages of K/T boundary events in North America and Deccan Trap volcanism in India (Abs): EOS, Transactions, American Geophysical Union, v. 73, p. 363.

MILLER, K. G., AUBRY, M.-P., KHAN, M. J., MELILLO, A., KENT, D. V., AND BERGGREN, W. A., 1985, Oligocene-Miocene biostratigraphy, magnetostratigraphy, and isotopic stratigraphy of the western North Atlantic: Geology, v. 13, p. 257–261.

MILLER, K. G., FEIGENSON, M. D., WRIGHT, J. D., AND CLEMENT, B. M., 1991, Miocene isotope reference section, Deep Sea Drilling Project Site 608: an evaluation of isotope and biostratigraphic resolution: Paleoceanography, v. 6, p. 33–52.

MILLER, K. G., THOMPSON, P. R., AND KENT, D. V., 1993, Integrated late Eocene-Oligocene stratigraphy of the Alabama coastal Plain: Correlation of hiatuses and stratal surfaces to glacioeustatic lowerings: Paleoceanography, v. 8, p. 313–331.

MILLER, K. G., WRIGHT, J. D., FOSSEN, M. C., VAN, AND KENT, D. V., 1994, Miocene stable isotopic stratigraphy and magnetostratigraphy of Buff Bay, Jamaica: Geological Society of America Bulletin, v. 106, p. 1605–1620.

MOLINA, E., 1986, Description and biostratigraphy of the main reference section of the Eocene/Oligocene boundary in Spain: Fuente Caldera section, in Pomerol, C. and Premoli Silva, I., eds., Terminal Eocene Events: Amsterdam, Elsevier, Developments in Stratigraphy 9, p. 53–63.

MOLINA, E., CANUDO, J. I., GUERNET, C., MCDOUGALL, K., OTIZ, N., PASCUAL, J. O., PARES, J. M., SAMSO, J. M., SERRA-KIEL, J, AND TOSQUELLA, J., 1992, The stratotypic Ilerdian revisited: integrated stratigraphy across the Paleocene/Eocene boundary: Revue de Micropaléontologie, v. 35, p. 143–156.

MOLINA, E., MONACO, P., NOCCHI, M., AND PARISI, G., 1986, Biostratigraphic correlation between the Central Subbetic (Spain) and Umbro-Marchean (Italy) pelagic sequences at the Eocene/Oligocene boundary using foraminifera, in Pomerol, C. and Premoli Silva, I., eds., Terminal Eocene Events, Developments in Stratigraphy, v. 9: Amsterdam, Elsevier Science Publishers, p.75–85.

MONECHI, S., 1985, Campanian to Pleistocene calcareous nannofossil stratigraphy from the northwest Pacific Ocean, Deep Sea Drilling Project Leg 86: Washington, D.C., Initial Reports of the Deep Sea Drilling Project, v. 86, p. 301–36.

MONECHI, S., 1986, Calcareous nannofossil events around the Eocene-Oligocene boundary in the Umbrian sections (Italy): Palaeogeography, Palaeoclimatology, Palaeoecology, v. 57, p. 61–69.

MONECHI, S., BLEIL, U., AND BACKMAN, J., 1985, Magnetobiochronology of late Cretaceous—Paleogene and late Cenozoic pelagic sedimentary sequences from the northwest Pacific (Deep Sea Drilling Project, Leg 86, Site 577): Washington, D.C., Initial Reports of the Deep Sea Drilling Project, v. 86, p. 787–797.

MONECHI, S. AND THIERSTEIN, H., 1985, Late Cretaceous- Eocene nannofossil and magnetostratigraphic correlations near Gubbio, Italy: Marine Micropaleontology, v. 9, p. 419–440.

MONTANARI, A., DEINO, A., COCCIONI, R., LANGENHEIM, V. E., CAPO, R., AND MONECHI, S., 1991, Geochronology, Sr isotope analysis, magnetostratigraphy, and planktonioc stratigraphy across the Oligocene-Miocene boundary in the Contessa section (Gubbio, Italy): Newsletters in Stratigraphy, v. 23, p. 151–180.

MONTANARI, A., DEINO, A., DRAKE, R., TURRIN, B., DEPAOLO, D., ODIN, G., CURTIS, G. H., ALVAREZ, W., AND BICE, D. M., 1988, Radioisotopic dating of the Eocene/Oligocene boundary in the pelagic sequence of the northern Appennines, in Premoli Silva, I., Coccioni, R., and Montanari, A., eds., International Union of Geological Sciences Commission on Stratigraphy: Ancona, International Subcommission on Paleogene Stratigraphy report, p. 195–208.

MONTANARI, A., DEINO, A., LANGENHEIM, V. E., AND COCCIONI, R., 1990, $^{40}Ar/^{39}Ar$ laser-fusion dating of magnetrc polarity reversals and planktonic foraminiferal events across the Aquitanian-Burdigalian boundary at Gubbio, Italy (abs.): EOS, Transactions, American Geophysical Union, v. 71, p. 1295.

MOORE, T. C., Jr., RABINOWITZ, P. D., AND OTHERS, 1984, Site 528: Washington D. C., Initial Reports of the Deep Sea Drilling Project, v. 74, p. 307–405.

MOROZOVA, V. G., 1960, Zonal stratigraphy of Danian-Montian beds of the U.S.S.R. and the boundary between the Cretaceous and the Paleogene: International Geological Congress, XXI Session, Doklady Soviet Geologists, v. 5, p. 83–100 (in Russian, extended English abstract).

MOSNA, S. AND MICHELETTI, A.,1968, Microfaune del "Serravalliano", Committee on Mediterranean Neogene Stratigraphy, Proceedings of IV Session: Giornale di geologia, v. 35, p. 183–189.

MOULLADE, M., 1987, Deep Sea Drilling Project Leg 93: biostratigraphic synthesis: Washington, D.C., Initial Reports of the Deep Sea Drilling Project, v. 93, p. 1271–1282.

MÜLLER, C., 1985, Biostratigraphic and paleoenvironmental interpretation of the Goban Spur region based on a study of calcareous nannoplankton: Washington D. C., Initial Reports of the Deep Sea Drilling Project, v. 80, p. 573–599.

NAPOLEONE, G., PREMOLI SILVA, I., HELLER, F., CHELI, P., COREZZI, S., FISCHER, A. G., 1983, Eocene magnetic stratigraphy at Gubbio, Italy, and its implications for Paleogene geochronology: Geological Society of America Bulletin, v. 94, p. 181–191.

NEDERBRAGT, A. J. AND VAN HINTE, J. E., 1987, Biometric analysis of *Planorotalites pseudomenardii* (upper Paleocene) at Deep Sea Drilling Project Site 605, northwestern Atlantic: Washington, D.C. Initial Reports of the Deep Sea Drilling Project, v. 93, p. 577–591.

NESS, G., LEVI, S., AND COUCH, R., 1980, Marine magnetic anomaly timescales for the Cenozoic and Late Cretaceous: A precis, critique, and synthesis: Reviews of Geophysics and Space Physics, v. 18, p. 753–770.

NOCCHI, M., PARISI, G., MONACO, P., MONECHI, S., MANDILE, M., NAPOLEONE, G., RIPEPE, M., ORLANDO, M., PREMOLI SILVA, I., AND BICE, D. M., 1986, The Eocene-Oligocene boundary in the Cambrian pelagic regression, *in* Pomerol, C. and Premoli Silva, I., eds., Terminal Eocene Events: Amsterdam, Elsevier, Developments in Stratigraphy 9, p. 25–40.

OBRADOVICH, J. D., 1984, An overview of the measurement of geologic time and the paradox of geologic time scales: Stratigraphy, v. 1, p. 11–30.

OBRADOVICH, J. D., 1993, A Cretaceous time scale, *in* Caldwell, W. G. E. and Kauffman, E. G., eds., The Evolution of the Western Interior Basin: Ottawa, Geological Society of Canada Special Paper 39, p. 379–396.

OBRADOVICH, J. O., DOCKERY, D. D., AND SWISHER, C. C., 1993, $^{40}Ar/^{39}Ar$ ages of bentonites in the upper part of the Yazoo Formation (upper Eocene), west-central Mississippi: Mississippi Geology, v. 14, p. 1–9.

ODIN, G. S., ed., 1982, Numerical Dating in Stratigraphy, Part II: New York, John Wiley and Sons, p. 63–1040.

ODIN, G. S. AND LUTERBACHER, H., 1992, The age of the Paleogene stage boundaries, *in* Luterbacher, H., ed., Paleogene Stages and Their Boundaries: Neues Jahrbuch für Geologie und Paläontologie, v. 186, p. 21–48.

ODIN, G. S. AND MONTANARI, A., 1989, Age radiométrique et stratotype de la limite Eocène-Oligocène: Compts Rendus de l' Académie des Sciences de Paris, t. 309, series II: 1939–1945.

ODIN, G. S. AND ODIN, C., 1990, Echelle numérique des temps géologiques: Géochronique, v. 35, p. 12–21.

OKADA, H. AND BUKRY, D., 1980, Supplementary modification and introduction of code numbers to the low-latitude coccolith biostratigraphic zonation: Marine Micropaleontology, v. 5, p. 321–325.

OLAFSSON, G., 1989, Quantitative calcareous nannofossil biostratigraphy of upper Oligocene to middle Miocene sediment from ODP Hole 667A and middle Miocene sediment from DSDP Site 574: College Station, Proceedings of the Ocean Drilling Program, Scientific Results, v. 108, p. 9–22.

OLAFSSON, G., 1991, Quantitative calcareou nannofossil biostratigraphy and biochronology of early through late Miocene sediment from DSDP Hole 608: Meddelanden från Stockholms Universitets Institution for Geologi och Geokemi, v. 203, 23 pp.

OLSSON, R. K., HEMLEBEN, C., BERGGREN, W. A., AND LIU, C., 1992, Wall texture classification of planktonic foraminifera genera in the lower Danian: Journal of Foraminiferal Research, v. 22, p. 195–213.

ONOFRIO, d' S., GIANELLI, L., IACCARINO, S., MORLOTTI, E., ROMEO, M., SALVATORINI, G., SAMPO, M., AND SPROVIERI, R.,1975, Planktonic foraminifera from some Italian sections and the problem of the lower boundary of the Messinian: Bolletino Societa Paleontologia Italiana, v.14, p. 177–196.

OPDYKE, N. D., BURCKLE, L. H., AND TODD, A., 1974, The extension of the magnetic time scale in sediments of the Central Pacific Ocean: Earth and Planetary Science Letters, v. 22, p. 300–306.

ORUE-ETXEBARRIA, X., BACETA, J. I., PUJALTE, V., NUNEZ-BETELU, K., SERRA-KIEL, J., APELLANIZ, E., AND PAYROS, A., 1992, Evaluating prospective Paleocene-Eocene boundary parastratotypes in the deep water Basque Basin, western Pyrenees: Ermua and Trabakua Pass sections, Biscay, Basque Basin (abs.): Zaragoza, IGCP Project 308: Paleocene/Eocene Boundary Events, International Meeting and Field Conference, Abstract Volume, p. 9 and 10 page addendum.

PAK, D. AND MILLER, K. G., 1993, Late Paleocene to early Eocene benthic foraminiferal stable isotopes and assemblages: implications for deep-water circulation: Paleoceanography, v. 7, p. 405–422.

PERCIVAL, S. F., 1984, Late Cretaceous to Pleistocene calcareous nannofossils from the South Atlantic, Deep Sea Drilling Project 73: Washington, D.C., Initial Reports of the Deep Sea Drilling Project, v. 73, p. 391–424.

PERMANENT INDERDEPARTMENTAL STRATIGRAPHIC COMMISSION, 1963, Decision of the Permanent Interdepartmental Stratigraphic Commission on the Paleogene of the USSR: Sovietskaya Geologiya, v. 6, p. 145–154.

PITMAN, W. C., HERRON, E. M., AND HEIRTZLER, J. R., 1968, Magnetic anomalies in the Pacific and sea floor spreading: Journal of Geophysical Research, v. 73, p. 2069–2085.

POAG, W. C. AND AUBRY, M.-P., 1995, Upper Eocene impactites of the U.S. East Coast: Depositional origins, biostratigraphic framework, and correlation: Palaios, v. 10, p. 16–43.

POMEROL, C. AND PREMOLI SILVA, I., eds., 1986, Terminal Eocene Events: Amsterdam, Elsevier, Developments in Paleontology and Stratigraphy 9, 420 p.

POORE, R. Z. AND BERGGREN, W. A., 1974, Pliocene biostratigraphy of the Labrador Sea: calcareous plankton: Journal of Foraminiferal Research, v. 4, p. 91–108.

POORE, R. Z. AND BERGGREN, W. A., 1975, Late Cenozoic planktonic foraminiferal biostratigraphy and paleoecology of Hatton-Rockall Basin, DSDP Site 116: Journal of Foraminiferal Research, v. 5, p. 270–293.

POORE, R. Z., TAUXE, L., PERCIVAL, JR., S. F., LaBRECQUE, J. L., 1982, Late Eocene-Oligocene magnetostratigraphy and biostratigraphy at South Atlantic DSDP Site 522: Geology, v. 10, p. 508–511.

POORE, R. Z., TAUXE, L., PERCIVAL, Jr., S. F., LaBRECQUE, J. L., WRIGHT, R., PETERSON, N. P., SMITH, C. C., TUCKER, P., AND HSU, K. J., 1983, Late Cretaceous- Cenozoic magnetostratigraphy and biostratigraphy of the South Atantic Ocean: DSDP Leg 73: Palaeogeography, Palaeoclimatology, Palaeoecology, v. 42, p. 127–149.

POORE, R. Z., TAUXE, L., PERCIVAL, Jr., S. F., LaBRECQUE, J. L., WRIGHT, R., PETERSEN, N. P., SMITH, C. C., TUCKER, P., AND HSU, K. J., 1984, Late Cretaceous-Cenozoic magnetostratigraphy and biostratigraphic correlations of the South Atlantic Ocean: Washington, D.C., Initial Reports of the Deep Sea Drilling Project, v. 73, p. 645–656.

POSPICHAL, J. J., DEHN, J., DRISCOLL, N. W., EIJDEN, VAN, A. J. M., FARRELL, J. W., FOURTANIER, E., GAMSON, P., GEE, J., JANACEK, T. R., JENKINS, D. G., KLOOTWIJK, L. C., NOMURA, R., OWEN, R. M., REA, D., RESIWATI, P., SMIT, J., AND SMITH, G., 1991, Cretaceous-Paleogene biomagnetostratigraphy of Sites 752–755, Broken Ridge: A synthesis: College Station, Proceedings of the Ocean Drilling Program, Scientific Results, v. 121, p. 721–741.

POSPICHAL, J. J. AND WISE, S. W., 1990, Paleocene to middle Eocene calcareous nannofossils of ODP Sites 689 and 690, Maud Rise, Weddell Sea: College Station, Proceedings of the Ocean Drilling Program, Scientific Results, v. 113, p. 613–638.

PREMOLI SILVA, I. AND BOLLI, H. M., 1973, Late Cretaceous to Eocene planktonic foraminifera and stratigraphy of Leg 15 sites in the Caribbean Sea: Washington, D.C., Initial Reports of the Deep Sea Drilling Project, v. 15, p. 449–547.

PREMOLI SILVA, I., COCCIONI, R., AND MONTANARI, A., eds., 1988b, The Eocene-Oligocene Boundary in the Marche-Umbria Basin (Italy): Ancona, International Union of Geological Sciences Commission on Stratigraphy, International Subcommission on Paleogene Stratigraphy Report, 268 p.

PREMOLI SILVA, I., ORLANDO, M., MONECHI, S., MADILE, M., NAPOLEONE, G., AND RIPEPE, M., 1988a, Calcareous plankton biostratigraphy and magnetostratigraphy at the Eocene/Oligocene transition in the Gubbio area, *in* Premoli Silva, I., Coccioni, R., and Montanari, A., eds., The Eocene-Oligocene Boundary in the Marche-Umbria Basin (Italy): Ancona, International Union of Geological Sciences Commission on Stratigraphy, International Subcommission on Paleogene Stratigraphy report, p. 137–161.

PROTHERO, D. R., 1994, The Eocene-Oligocene Transition: Paradise Lost: New York, Columbia University Press, 291 p.

PROTHERO, D. R. AND SWISHER, C. C., III, 1992, Magnetostratigraphy and geochronology of the terrestrial Eocene-Oligocene transition in North America, *in* Prothero, D. R. and Berggren, W. A., eds., Eocene-Oligocene Climatic and Biotic Evolution: Princeton, Princeton University Press, p. 46–73.

PROTO DECIMA, F. AND BOLLI, H. M., 1970, Evolution and variability of *Orbulinoides beckmanni* (Saito): Eclogae geologiae Helvetiae, v. 63, p. 883–905.

PUJOL, C., 1983, Cenozoic planktonic foraminiferal biostratigraphy of the southwestern Atlantic (Rio Grande Rise): Deep Sea Drilling Project Leg 72:

Washington, D.C., Initial Reports of the Deep Sea Drilling Project, v. 72, p. 623–673.

RAFFI, I. AND FLORES, J.-A, 1995, Pleistocene through Miocene calcareous nannofossils from eastern equatorial Pacific Ocean (ODP leg 138): College Station, Proceedings of the Ocean Drilling Program, Scientific Results, v. 138, (in press).

RAFFI, I., RIO, D., d'AFRI, A., FORNACIARI, E., AND ROCHETTI, S., 1995, Quantitative distribution patterns and biomagnetostratigraphy of middle and late Miocene calcareous nannofossils from equatorial Indian and Pacific Oceans (legs 115, 130, and 138): College Station, Proceedings of the Ocean Drilling Program, Scientific Results, v. 138, (in press).

RENNE, P. R., DEINO, A. L., WALTER, R. C., TURRIN, B. D., SWISHER, C. C., BECKER, T. A., CURTIS, G. H., SHARP, W. D., AND JAOUNI, A. R., 1994, Intercalibration of astronomical and radioisotopic time: Geology, v. 22, p. 783–786.

RENNE, P. R., WALTER, R., VEROSUB, K., SWEITZER, M., AND ARONSON, J.,1993, New Data from Hadar (Ethiopia) support orbitally tuned timescale to 3.3 Ma: Geophysical Research Letters, v. 20, p. 1067–1070.

RIO, D., FORNACIARI, E., AND RAFFI, I., 1990, Late Oligocene through early Pleistocene calcareous nannofossils from western equatorial Indian Ocean (Leg 115): College Station, Proceedings of Ocean Drilling Program, Scientific Results, v. 115, p. 175–235.

ROBERTS, A. P., TURNER, G. M., AND VELLA, P. P., 1994, Magnetostratigraphic chronology of late Miocene to early Pliocene biostratigraphic and oceanographic events in New Zealand: Geological Society of America Bulletin, v. 106, p. 665–683.

ROSENKRANTZ, A., 1924, De københavnske Grønsandslag og deres Placering i den danske Lagraekke (med et Skema over det danske Palaeocaen): Meddelelser fra Dansk Geologisk Forening, v. 6, p. 22–39.

RYAN, W. B. F., CITA, M. B., RAWSON, M. D., BURCKLE, L. H., AND SAITO, T., 1974, A paleomagnetic assignment of Neogene stage boundaries and the development of isochronous datum planes between the Mediterranean, the Pacific and Indian Oceans in order to investigate the response of the World Ocean to the Mediterranean "Salinity Crisis": Rivista Italiana Paleontologia, v. 80, p. 631–688.

SAINT-MARC, P., 1987, Biostratigraphic and paleoenvironmental study of Paleocene benthic and planktonic foraminifers, Site 605, Deep Sea Drilling Project Leg. 93: Washington, D.C., Initial Reports of the Deep Sea Drilling Project, v. 93, p. 539–547.

SAITO, T., 1984, Planktonic foraminiferal datum planes for biostratigrapic correlation of Pacific Neogene sequences — 1982 status report, in Ikebe, N. and Tsuchi, R., eds, Pacific Neogene Datum Planes: Tokyo, University of Tokyo Press, p. 3–10.

SALVADOR, A., ed., 1994. International Stratigraphic Guide: Boulder, International Union of Geological Sciences and the Geological Society of America, 214 p.

SAMSON, S. D. AND ALEXANDER, JR., E. C., 1987, Calibration of the interlaboratory ⁴⁰Ar/³⁹Ar standard, MMhb-I: Chemical Geology, Isotope Geoscience Section, v. 66, p. 27–34.

SAUNDERS, J. B. AND BOLLI, H. M., 1983, Trinidad's contribution to world biostratigraphy: Trinidad, Transactions of the Fourth Latin American Congress, 1979.

SCHLICH, R., WISE, S. W., JR., AND OTHERS, Site 750: College Station, Proceedings of the Ocean Drilling Program, Initial Reports, v. 120, p. 277–337, 569–619.

SCHNEIDER, D., 1995. Paleomagnetism of some ODP Leg 138 sediments: Miocene magnetostratigraphic miscellany: College Station, Proceedings of the Ocean Drilling Program, Scientific Results, v. 138, (in press).

SCHOCH, H. M., 1989, Stratigraphy: Principles and Methods: New York, Van Nostrand and Reinhold, 375 p.

SCOTT, G. H. ,1982, Tempo and stratigraphic record of speciation in Globorotalia puncitculata: Journal of Foraminiferal Research, v. 12, p. 1–12.

SCOTT, G. H., BISHOP, S. AND BURT, B. J., 1990, Guide to Some Neogene Globorotaliids (Foraminiferida) from New Zeland: New Zeland Geological Survey Paleontological Bulletin, v. 61, 135 p.

SELLI, R., 1960, The Mayer-Eymar Messinian, 1867, proposal for a neostratotype: International Geological Congress, Report 21st Session, Norden, p. 311–333.

SHACKLETON, N. J., BERGER, A., AND PELTIER, W. R., 1990, An alternative astronomical calibration of the lower Pleistocene timescale based on ODP Site 677: Transactions of the Royal Society of Edinburgh, Earth Sciences, v. 81, p. 251–261.

SHACKLETON, N. J., CROWHURST, S., HAGELBERG, T., PISIAS, N. G., AND SCHNEIDER, D. A., 1995, A new Late Neogene time scale: Application to

Leg 138 Sites: College Station, Proceedings of the Ocean Drilling Program, Scientific Results, v. 138, p. 73–101.

SHACKLETON, N. J. AND MEMBERS OF THE SHIPBOARD SCIENTIFIC PARTY, 1984, Accumulation rates in leg 74 sediments: Washington, D.C., Initial Reports of the Deep Sea Drilling Project, v. 74, p. 621–644.

SHARPTON, V. L., DALRYMPLE, G. B., MARTIN, L. E., RYDER, SCHURAYTZ, B. C., AND FUCUGAUCHI, J. URRUTIA, 1992, New links between the Chicxulub impact structure and the Cretaceous/Tertiary boundary: Nature, v. 359, p. 819–821.

SHIBATA, K., NISHIMURA, S. AND CHINZEI, K. , 1984, Radiometric dating related to Pacific Neogene planktonic datum planes, in Ikebe, N. and Tsuchi, R., eds., Pacific Neogene Datum Planes: Contributions to Biostratigraphy and Chronology: Tokyo, Tokyo University Press, p. 85–89.

SHUTSKAYA, E. K., 1953, The subdivision of the Kuban and Elburgan Horizons of the northern Caucasus by means of Globigerinas: Biulletin Moskovski Obshchii isputatelni prirody, otdel geologia, v. 28, p. 71–79 (in Russian).

SHUTSKAYA, E. K., 1956, Stratigraphy of the lower horizons of the Paleogene in the central Precaucasus according to the foraminifera: Institut Geologicheskii Nauk Akademiia Nauk SSR, Trudy, v. 164, geol. ser., no. 70, p. 3–114 (in Russian).

SHUTSKAYA, E. K., 1960a, Stratigraphy of the Early Paleogene of Crimea and the northern Precaucasus, in Yanshin, A. L., Vyalov, O. S., Dolgopolov, N. N., and Menner, V. V., eds., Paleogene Deposits of the Southern European USSR: Moscow, Izdatel'stvo Akademiia Nauk, p. 207–229 (in Russian).

SHUTSKAYA, E. K., 1960b, Stratigraphy and Facies of the Lower Paleogene of the Precaucasus: Gostoptekhizdat, 104 p. (in Russian).

SHUTSAKAYA, E. K., 1962, Foraminifera of the Danian and Paleocene in pelagic facies of the Crimea, Precaucasus and Transcapsian regions: Biulletin Moskovski Obshchii ispytatelni prirody, otdel geologia, v. 37, p. 126–127 (in Russian).

SHUTSAKAYA, E. K., 1965, Nizhnii paleogen Kryma, Predkavkaz'ya i zapadnoi chasti Srednei Azii: Geologicheskaya Institut Nauk, Akademiia Nauk SSSR, Gosgeolkoma SSSR, Moskva.

SHUTSAKAYA, E. K., 1970, Stratigrafiya, foraminifery i paleogeografiya nizhnego paleogena Kryma, predkavkaz'ya i zapadnoi chasti srednei Azii, Vsesoyuznyi nauchno-issledovadetel'skii geologo-razvedochniyi neftyanoi institut (VNIGRI): Trudy, v. 70, 256 p.

SIESSER, W. G. AND BRALOWER, T. J., 1992, Cenozoic calcareous nannofossil biostratigraphy on the Exmouth Plateau, eastern Indian Ocean: College Station, Proceedings of the Ocean Drilling Program, Scientific Results, v. 122, p. 601–631.

SMIT, J., 1982, Extinction and evolution of planktonic foraminifera at the Cretaceous/Tertiary boundary after a major impact, in Silver, L. T. and Schultz, P. H., eds., Geological Implications of Impacts of Large Asteroids and Comets on the Earth: Boulder, Geological Society of America Special Paper 190, p. 329–352.

SNYDER, S. W. AND WATERS, V. J., 1985, Cenozoic planktonic foraminiferal biostratigraphy of the Goban Spur region, Deep Sea Drilling Project 80: Washington, D.C. Initial Reports of the Deep Sea Drilling Project, v. 80, p. 439–472.

SPELL, T. L. AND HARRISON, T. M., 1993, ⁴⁰Ar/³⁹Ar geochronology of the post-Valles Caldera rhyolites, Jemez volcanic field, New Mexico: Journal of Geophysical Research, v. 98, p. 8031–8051.

SPELL, T. L. AND McDOUGALL, I., 1992, Revisions to the age of the Brunhes-Matuyama boundary and the Pleistocene geomagnetic polarity time scale: Journal of Geophysical Research, v. 94, p. 10370–10396.

SPENCER-CERVATO, C., THIERSTEIN, H. R., LAZARUS, D. B., AND BECKMAN, J.-P., 1994, How synchronous are Neogene marine plankton events?: Paleoceanography, v. 9, p. 739–763.

SPIEGLER, D. AND JANSEN, E., 1989, Planktonic foraminifera biostratigraphy of Norwegian Sea sediments; ODP Leg 104: College Station, Proceedings of the Ocean Drilling Program, Scientific Results, v. 104, p. 681–696.

SPIESS, V., 1990, Cenozoic magnetostratigraphy of Leg 113 drill sites, Maud Rise, Weddell Sea, Antarctica: College Station, Proceedings of the Ocean Drilling Program, Scientific Results, v. 113, p. 261–315.

SRINIVASAN, M. S. AND CHATURVEDI, S. N., 1992, Neogene planktonic foraminiferal biochronology of the DSDP Sites along the Ninetyeast Ridge, northern Indian Ocean, in Ishizaki, K. and Saito, T., eds., Centenary of Japanese Micropaleontology: Tokyo, Terra Scientific Publishing Company, p. 175–188.

SRINIVASAN, M. S. AND KENNETT, J. P., 1981a, A review of Neogene planktonic foraminiferal biostratigraphy; applications in the Equatorial and South Pacific, in Warme, J. E., Douglas, R. G., and Winterer, E. L., eds., The Deep

Sea Drilling Project: A Decade of Progress: Tulsa, Society of Economic Paleontologists and Mineralogists Special Publication 32, p. 395–432.

SRINIVASAN, M. S. AND KENNETT, J. P., 1981b, Neogene planktonic foraminiferal biostratigraphy and evolution; equatorial to Subantarctic South Pacific: Marine Micropaleontology, v. 6, p. 499–533.

SRINIVASAN, M. S. AND KENNETT, J. P., 1983, The Oligocene-Miocene boundary in the South Pacific: Geological Society of America Bulletin, v. 94, p. 798–812.

SRINIVASAN, M. S. AND SINHA, D. K., 1992, Late Neogene planktonic foraminiferal events of the southwest Pacific and Indian Ocean: a comparison, in Tsuchi, R. and Ingle, Jr., J. C., eds., Pacific Neogene, Environment, Evolution and Events: Tokyo, University of Tokyo Press, p. 203–220.

STAINFORTH, R. M., LAMB, J. L., LUTERBACHER, H., BEARD, J. H., AND JEFFORDS, R. M., 1975, Cenozoic planktonic foraminifera zonation and characteristics of index forms: Lawrence, University of Kansas Paleontological Contributions 62, p. 1–425 (in two parts).

STEININGER, F. F., 1994, Proposal for the Global Stratotype Section and Point (GSSP) for the Base of the Neogene (the Palaeogene/Neogene boundary), International Commission on Stratigraphy, Subcommission on Neogene Stratigraphy; Working Group on the Palaeogene/Neogene Boundary: Vienna, Institute for Paleontology, University of Vienna, 41 p.

STEININGER, F. F., BERGGREN, W. A., KENT, D. V., BERNOR, R. C., SEN, S., AND AGUSTI, J., 1995, Circum Mediterranean (Miocene and Pliocene) marine-continental chronologic correlations of European mammal units and zones, in Bernor, R. L., Fahlbusch, V., and Rietschel, S., eds., Later Neogene European Biotic and Stratigraphic Correlation: New York, Columbia University Press (in press).

STEINMETZ, J. C. AND STRADNER, H., 1984, Cenozoic calcareous nannofossils from Deep Sea Drilling Project leg 75, southeast Atlantic Ocean: Washington, D.C., Initial Reports of the Deep Sea Drilling Project, v. 75, p. 671–753.

STOTT, L. D. AND KENNETT, J. P., 1990, Antarctic Paleogene planktonic foraminifer biostratigraphy: ODP Leg 113, Sites 689 and 690: College Station, Proceedings of the Ocean Drilling Program, Scientific Results, v. 113, p. 549–569.

SWISHER, C. C., III, DePAOLO, D., AND OWNES, T., 1994, Age of the Fish Canyon Tuff Sanidine (FCTS): A single crystal ^{40}Ar/^{39}Ar dating standard (abs.): Berkeley, Eighth International Conference on Geochronology, Cosmochronology and Isotope Geology, United States Geological Survey Circular 1107, p. 312.

SWISHER, C. C., DINGUS, L., AND BUTLER, R. F., 1993, ^{40}Ar/^{39}Ar dating and magnetostratigraphic correlation of the terrestrial Cretaceous-Paleogene boundary and Puercan Mammal Age, Hell Creek-Tullock formations, eastern Montana: Canadian Journal of Earth Sciences, v. 30, p. 1981–1996.

SWISHER, C. C., DINGUS, L., MONTANARI, A., AND SMIT, J., 1995, Terminal Cretaceous events in classic marine and terrestrial sections synchronous with Chicxulub impact: Geophyscal Research Letters, in press.

SWISHER, C. C., III, GRAJALES-NISHIMURA, J. M., MONTANARI, A., CEDILLO-PARDO, E., MARGOLIS, S. V., CLAEYS, P., ALVAREZ, W., SMIT, J., RENNE, P., MAURRASSE, F. J.-M. R., AND CURTIS, G. H., 1992, Chicxulub crater melt-rock and K-T boundary tektites from Mexico and Haiti yield coeval ^{40}Ar/^{39}Ar ages of 65 Ma: Science, v. 257, p. 954–958.

SWISHER, C. C., III, AND KNOX, R. O'B., 1991, The age of the Paleocene/Eocene boundary: ^{40}Ar/^{39}Ar dating of the lower part of NP10, North Sea Basin and Denmark (abs): IGCP 308 (Paleocene/Eocene boundary events), International Annual Meeting and Field Conference, 2–6 December 1991, Brussels, Abstracts with Program, p. 16

SWISHER, C. C., III, AND PROTHERO, D. R., 1990, Single-crystal ^{40}Ar/^{39}Ar dating of the Eocene-Oligocene transition in North America: Science, v. 249, p. 760–762.

TAKAYAMA, T., 1993, Notes on Neogene calcareous nannofossil biostratigraphy of the Ontong-Java Plateau and size variations of Reticulofenestra coccoliths: College Station, Proceedings of the Ocean Drilling Program, Scientific Results, v. 130, p. 179–229.

TALWANI, M., WINDISCH, C. C., AND LANGSETH, M. G., 1971, Rekjanes Ridge crest: A detailed geophysical study: Journal of Geophysical Research, v. 76, p. 473–517.

TAUXE, L., DEINO, A., BEHRENSMEYER, A., AND POTTS, R., 1992, Pinning down the Brunhes/Matayama and upper Jaramillo boundaries: a reconciliation of orbital and isotopic time scales: Earth and Planetary Science Letters, v. 109, p. 561–572.

TAUXE, L., GEE, Y., PICK, T., AND BOWN, T., 1994, Magnetostratigraphy of the Willwood Formation, Bighorn Basin, Wyoming: New constraints on the location of the Paleocene/Eocene boundary: Earth and Planetary Science Letters, v. 125, p. 159–172.

TAUXE, L., TUCKER, P., PETERSEN, N. P., AND LaABRECQUE, J. L., 1983, The magnetostratigraphy of Leg 73 sediments: Palaeogeography, Palaeoclimatology, Palaeoecology, v. 42, p. 65–90.

THEYER, F. AND HAMMOND, S. R., 1974, Cenozoic magnetic time scale in deep-sea cores: completion of the Neogene: Geology, v. 2, p. 487–512.

THOMAS, E., BARRERA, E., HAMILTON, N., HUBER, B. T., KENNETT, J. P., O'CONNELL, S. B., POSPICHAL, J. J., SPIESS, V., STOTT, L. D., WEI, W., AND WISE, S. W., Jr., 1990, Upper Cretaceous-Paleogene stratigraphy of Sites 689 and 690, Maud Rise (Antarctica): College Station, Proceedings of the Ocean Drilling Program, Scientific Results, v. 113, p. 901–914.

THUNELL, R., 1981, Late Miocene-early Pliocene planktonic foraminiferal biostratigraphy and paleoceanography of low-latitude marine sequences: Marine Micropaleontology, v. 6, p. 71–90.

TOUMARKINE, M., 1981, Discusssion de la validité de l'espèce Hantkenina aragonensis Nuttall, 1930. Description de Hantkenina nuttalli n.sp.: Cahiers de Micropaléontologie, Livre Jubilaire en l'honneur de Madame Y. Le-Calvez, fasc. 4, p. 109–119.

TOUMARKINE, M., AND LUTERBACHER, H., 1985, Paleocene and Eocene planktonic foraminifera, in Bolli, H. M., Saunders, J. B., and Perch-Nielsen, K., Plankton Stratigraphy: Cambridge, Cambridge University Press, p. 87–154.

TOWNSEND, H. A., 1985, The paleomagnetism of sediments acquired from the Goban Spur on Deep Sea Drilling Project 80: Washington, D.C., Initial Reports of the Deep Sea Drilling Project, v. 80, p. 389–421.

TSUCHI, R., 1984, Neogene biostratigraphy and chronology of Japan, in Ikebe, N. and Tsuchi, R., eds., Pacific Neogene Datum Planes: Contributions to Biostratigraphy and Chronology: Tokyo, Universioty of Tokyo Press, p. 223–233.

TSUCHI, R., TAKAYANAGI, Y., AND SHIBATA, K., 1981, Neogene bio-events in the Japanese islands, in Tsuchi, R., ed., Neogene of Japan — Its Biostratigraphy and Chronology: Shitzuoka, Faculty of Science, IGCP-114 National Working Group of Japan, p. 15–32.

TURRIN, B. D., DONNELLY-NOLAW, J. M., AND HEARN, B. C., 1994, ^{40}Ar/^{39}Ar ages from the rhyolite of Alder Creek, California: Age of the Cobb Mountain normal polarity Subchron revised: Geology, v. 22, p. 251–254.

VAI, G. B., VILLA, I. M., AND COLALONGO, M. L., 1993, First direct radiometric dating of the Tortonian/Messinian boundary: Comptes Rendus de l' Académie des Science de Paris, t. 316, sér. II, p. 1407–1414.

VAN COUVERING, J. A. AND BERGGREN, W. A., 1977, Biostratigraphical basis of the Neogene time scale, in Kauffman, E. G. and Hazel, J. E., eds., Concepts and Methods in Biostratigraphy: Stroudsburg, Dowden, Hutchinson and Ross, p. 283–306.

VERHALLEN, P. J. J. M. AND ROMEIN, A. J. T., 1983, Calcareous nannofossils from the Priabonian stratotype and correlation with the parastratotypes, in Setiawan J. R., ed., Foraminifera and microfacies of the type-Priabonian: Utrecht Micropaleontological Bulletin, v. 29, p. 163–173.

VERVLOET, C. C., 1966, Stratigraphical and Micropaleontological Data on the Tertiary of Southern Piemont (northern Italy): Utrecht, Scotanus and Jens, 88 p.

VON ALIS, K., 1984, Miocene calcareous nannofossil biostratigraphy of Deep Sea Drilling Project Hole 521A (Southeast Atlantic): Washington, D.C., Initial Reports of the Deep Sea Drilling Project, v. 73, p. 425–427.

WALTER, R. C., MANEGA, P. C., HAY, R. L., DRAKE, R. E., AND CURTIS, G. H., 1991, Laser-fusion ^{40}AR/^{39}Ar dating of Bed 1, Olduvai Gorge, Tanzania: Nature, v. 354, p. 145–149.

WEAVER, P. P. E. AND CLEMENT, B. M., 1986, Synchroneity of Pliocene planktonic foraminiferal datums in the North Atlantic: Marine Micropaleontology, v. 10, p. 295–307.

WEAVER, P. P. E. AND CLEMENT, B. M., 1987, Magnetobiostratigraphy of planktonic foraminiferal datums: Deep Sea Drilling Project Leg 94, North Atlantic: Washington, D.C., Initial Reports of the Deep Sea Drilling Project, v. 94, p. 815–829.

WEI, K. Y., 1987, Tempo and mode of evolution in Neogene planktonic foraminifera: taxonomic and morphometric evidence: Unpublished Ph.D. Dissertation, University of Rhode Island, Kingston, 397 p.

WEI, W., 1991, Middle Eocene-lower Miocene calcareous nannofossil magnetobiochronology of ODP Holes 699A and 703A in the Subantarctic South Atlantic: Marine Micropaleontology, v. 18, p. 143–165.

WEI, W., 1992, Paleogene chronology of Southern Ocean drill holes: An update, in Kennet, J. P. and Warnke, D. A., eds., The Antarctic Paleoeonvironment: A Perspective on Global Change: Washington, D.C., American Geophysical Union, Antarctic Research Series, v. 56, p. 75–96.

WEI, W. 1995, Revised age calibration points for the geomagnetic polarity time scale: Geophysical Research Letters, v. 22, p. 159–172.

WEI, W. AND PELEO-ALAMPAY, A., 1993, Updated Cenozoic nannofossil magnetobiochronology: INA Newsletter, v. 15, p. 15–17.

WEI, W. AND POSPICHAL, J. J., 1991, Danian calcareous nannofossil succession at Site 738 in the southern Indian Ocean: College Station, Proceedings of the Ocean Drilling Program, Scientific Results, v. 119, p. 495–512.

WEI, W. AND THIERSTEIN, H. P., 1991, Upper Cretaceous and Cenozoic calcareous nannofossils of the Kerguelen Plateau (southern Indian Ocean) and Prydz Bay (East Antarctica): College Station, Proceedings of the Ocean Drilling Program, Scientific Results, v. 119, p. 467–493.

WEI, W., VILLA, G., WISE, S. W., JR., 1992, Paleoceanographic implications of Eocene-Oligocene calcareous nannofossils from Sites 711–748 in the Indian Ocean: College Station, Proceedings of the Ocean Drilling Program, Scientific Results, v. 120, p. 979–999.

WEI, W. AND WISE, S. W., 1989, Paleogene calcareous nannofossil magnetobiochronology: Results from South Atlantic DSDP Site 516: Marine Micropaleontology, v. 14, p. 119–152.

WEI, W. AND WISE, S. W., 1990, Middle Eocene to Pleistocene calcareous nannofossils recovered by Ocean Drilling Program Leg 113 in the Weddell Sea: College Station, Proceedings of the Ocean Drilling Program, Initial Reports, v. 113, p. 639–666.

WEI, W. AND WISE, S. W., 1992, Eocene-Oligocene calcareous nannofossil magnetobiochronology of the southern Ocean: Newsletter in Stratigraphy, v. 26, p. 119–132.

WESTERCAMP, D. AND ANDREIEFF, p., 1983, Saint-Barthélémy et ses ilets, Antilles françaises: stratigraphie et évolution magmato-structurale: Bulletin Société Géologique France, v. 25, p. 873–883.

WILSON, D. S., 1993, Confirmation of the astronomical calibration of the magnetic polarity timescale from sea-floor spreading rates: Nature, v. 364, p. 788–790.

WING, S. L., BROWN, T. M., AND OBRADOVICH, J. D., 1991, Early Eocene biotic and climatic change in interior western North America: Geology, v. 19, p. 1189–1192.

WRIGHT, J. D. AND MILLER, K. G., 1992, Miocene stable isotope stratigraphy, Site 747, Kerguelen Plateau: College Station, Proceedings of the Ocean Drilling Program, Scientific Results, v. 120, p.855–866.

YANSHIN, L., ed., 1960, Paleogenovye otlozhenyia yuga evropeiskoi yasti SSSR: Moscow, Izdatel'stvo Akademiya Nauk SSSR, 312 p.

YOUNG, J. R., FLORES, J.-A., WEI, W., AND CONTRIBUTORS, 1994. A summary chart of Neogene nannofossil magnetobiostratigraphy: Journal of Nannoplankton Research, v. 16, p. 21–27.

ZHANG, J., MILLER, K. G., AND BERGGREN, W. A., 1993, Neogene planktonic foraminiferal biostratigraphy of the northeastern Gulf of Mexico: Micropaleontology, v. 39, p. 299–326.

FROM CHRONOLOGY TO STRATIGRAPHY: INTERPRETING THE LOWER AND MIDDLE EOCENE STRATIGRAPHIC RECORD IN THE ATLANTIC OCEAN

MARIE-PIERRE AUBRY

Laboratoire de Géologie du Quaternaire, CNRS-Luminy, 13288 Marseille, France

ABSTRACT: Whereas it is well established that on a broad scale the deep-sea stratigraphic record is punctuated by unconformities involving hiatuses as long as an epoch or more, it is generally believed that on a fine scale the deep-sea sedimentary sequences bounded by these large-scale gaps are essentially continuous. I have examined stratigraphic sections which span the lower/middle Eocene boundary at over 100 deep-sea sites in the North and South Atlantic Ocean. Integrated magnetobiochronology provides a framework for deciphering the presence of unconformities and for estimating hiatus duration. Based on nine temporal maps representing the lower and middle Eocene sedimentary record in different sectors of the Atlantic Ocean, I show that, even at a fine scale, the deep-sea record is punctuated by multiple unconformities. Continuous sequences representing 5- to 10-my intervals of geological time alternate with hiatuses of 1 to several million years. The distribution of the hiatuses presents an uneven pattern such that, in many regions — but not all— the early Eocene Epoch is well represented whereas the latest early Eocene and the early middle Eocene Epochs are not. In addition, hiatuses in the deep sea overlap with hiatuses on the shelves in contrast with reports by some authors for the Oligocene and the Miocene Epochs. Whereas a discussion of the mechanism(s) responsible for this sedimentary architecture is beyond the scope of the paper, these findings suggest that the overall sedimentary architecture is controlled by (a) fundamental mechanism(s) other than those (glacio-eustasy) applicable to the Neogene, and in particular to the Pleistocene Epoch. This study further suggests that accumulation rates of pelagic sediments remained relatively steady (rather than highly fluctuating) over intervals of several hundred thousand years to several million years. The discontinuities observed in the deep sea suggest that its architecture may be most appropriately described in terms of allostratigraphic units.

INTRODUCTION

The last decade has witnessed the development of two concepts which have had tremendous impact in stratigraphy. One is sequence stratigraphy, which represents the first step towards interpreting the stratigraphic record as a response to a unique set of globally forcing mechanisms (Vail and others, 1977). The impact that sequence stratigraphy has had in sedimentary basin analysis is well known and does not need elaboration here. The other is the integrated magnetobiochronologic scale (IMBS; Berggren and others, 1985) derived from the geomagnetic polarity time scale (GPTS). The IMBS consists in a magnetochronology derived from the sea floor lineation pattern and calibrated by isotopic dating (the GPTS), which provides the support for a biochronology through magnetobiostratigraphic correlations. The IMBS is a powerful chronologic framework whose accuracy and precision are constantly being improved as Milankovitch-based astrochronology contributes to the calibration of magnetochronology while chemochronologic schemes are being developed (Miller and others, 1991). The IMBS satisfies the increasing demand for an ever greater chronologic resolution, a prerequisite for deciphering distant contingent relationships at a time when the awareness of earth scientists of the complexity of the earth system is increasing.

The IMBS is generally applied under the assumption that the stratigraphic record is essentially continuous and that a one to one relationship exists between stratigraphic sections and the temporal interval they cover. This is particularly the case with regard to the deep-sea record (sediments deposited below 200 m, including those now exposed on land). It is clear that the so-called 'high-resolution stratigraphy', which has been developed in response to an increasing necessity to understand the modalities and to project the consequences of global changes, is currently regarded by most earth scientists as synonymous with high-resolution chronology (see for instance Hailwood and Kidd, 1993). Tied to a time scale, high resolution stratigraphy is thought to provide a direct numerical dating of stratigraphic horizons where age-calibrated events occur and to constitute the first and straightforward step towards quantifying rates and duration.

Yet, evidence has grown that the stratigraphic record may not be the faithful recorder of geological time that we would like to believe, as pointed out in fact in the late seventies and early eighties (see for instance Van Andel, 1981). It is now well documented (and accepted) that the Cenozoic stratigraphic record on epicontinental margins is punctuated by numerous stratigraphic gaps. Also, there are regional and even near-global unconformities in the deep sea (e.g., Rona, 1973; Davies and others, 1975; Pimm and Sclater, 1975; Keller and Barron, 1983; Tucholke and Mountain, 1979, 1986; Aubry, 1991; Poag, 1994). Still, despite this, the prevailing view is that "open ocean pelagic sequences deposited in this environment offer the greatest potential for stratigraphic completeness available in the stratigraphic record" (Kidd and Hailwood, 1993).

The stratigraphic record is the only means by which we grasp the elusive concept of geological time, this fourth dimension without which there is no geological history. As the stratigraphic record yields the documentation of the physical, chemical and biological changes that have affected the planet through time, while, concurrently, providing the chronology which allows us, through a complex network of stratigraphic correlations, to establish the causes of these changes, there is a great danger that circular reasoning affects our geological reconstructions. In assuming a one to one relationship between the stratigraphic record and geological time, there is a risk that we may merely demote this latter to the third spatial dimension. Witness the simple relationship often established between thickness of sequences and sedimentation rates.

In this study, I evaluate the current assumptions regarding the completeness of the deep-sea record and the relationship between the stratigraphic record and geological time. To this end and for reasons given below, I have chosen to examine the lower and middle Eocene sedimentary record of the Atlantic Ocean. Prior to discussing this record, I explain the methodology used and show that interpreting the stratigraphic record is not the easy exercise commonly thought.

INTERPRETING THE STRATIGRAPHIC RECORD

Stratigraphic Record and Geological Time

The stratigraphic record and geological time are closely linked because the latter can only be extracted from the former.

Geochronology Time Scales and Global Stratigraphic Correlation, SEPM Special Publication No. 54

By stratigraphic record, we usually mean the vertical succession of rock sequences. This is the record that serves as support for chronostratigraphy, and for the construction of the classical time scales (hereafter referred to as CTS) derived from the isotopic dating of bodies of rocks chosen to characterize certain temporal intervals (e.g., Odin, 1982; the pre-Cenozoic time scales in Harland and others, 1990). There is however a stratigraphic record of a different kind, produced at the crest of oceanic ridges. It consists of the subhorizontal succession of rocks that yield alternatively normal and reversed polarity and which forms the sea floor lineation pattern. This is the record that constitutes the backbone of the GPTS. Magnetochronology, which is central to the GPTS and in turn to the IMBS, results from the calibration of a reference profile based on a linear distance-age relationship using isotopic dating on selected magnetic reversals (see Aubry and others, 1988; Hailwood, 1989; Cande and Kent, 1992).

Geological time is thus embodied in two independent stratigraphic records (Fig. 1). In the CTS, chronology is derived solely from the stratigraphic record. For the CTS, it can be correctly argued that geochronology and chronostratigraphy are no more than "different aspects of a single procedure" (Harland and others, 1990, p. 3). But the IMBS is radically different from the CTS in that it uses the sea floor lineation pattern as an independant chronometer, in which geochronology and chronostratigraphy appear as two fundamentally distinct disciplines. It is for this very reason that the IMBS constitutes an unprecedented tool for stratigraphic analysis, although I would contend that its power has not yet been appropriately perceived.

The Integrated Magnetobiostratigraphic Scale

The IMBS is comprised of three constituents: a magnetochronology, a numerical scale and a magnetobiochronologic framework (Berggren and others, 1985; Berggren and others, this volume). Magnetochronology is the fundament of the GPTS and the IMBS. Just as stratigraphic sections are patched together to produce a reference chronostratigraphic sequence, overlapping sea-floor anomaly profiles are selected to construct a reference sequence of relative anomaly spacing (see Berggren and others, 1985; Aubry and others, 1988, Hailwood, 1989; Cande and Kent, 1992). It is this sequence and the selected isotopic dates on particular horizons that are essential in the IMBS and from which numerical ages for magnetic reversals and paleontologic events (and now pre-Pliocene isotopic events, Miller and others, 1991; Hodell and Woodruff, 1994) can be derived. The numerical scale is probably the most valued constituent of the IMBS, because it allows quantification of geo-

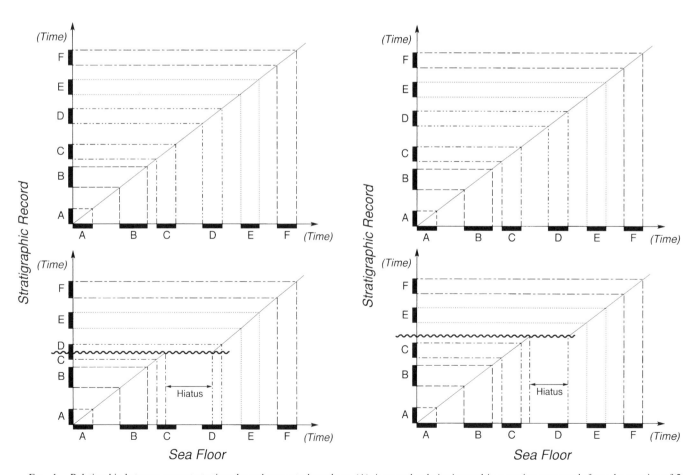

FIG. 1.—Relationship between magnetostratigraphy and magnetochronology. (A) A normal polarity interval in a section may result from the suturing of 2 partial normal polarity intervals across an unconformity. The intervals A–F are normal polarity chrons in time and normally polarized intervals in the stratigraphic record. (B) Similarly, a reversed polarity interval may result from the concatenation of two intervals with reversed polarity.

logical processes. However, I would argue that the magneto-biochronologic framework in the IMBS is equally, if not more, important and is more stable. While the magnetic reversals constitute the link between the stratigraphic record and the sea floor (Fig. 1), the magnetobiostratigraphic correlations on land sections and in the deep sea are turned into a magnetobiochronologic framework using a relative chronology (relative anomaly spacing) that is independent of the stratigraphic record. Because of this, the GPTS is an important but poorly recognized means to assess the continuity of sedimentary sections.

Reading Time in the Stratigraphic Record

Outside of radioisotopic dating, which is restricted to levels with datable minerals, there are only indirect ways to extract numerical ages from the stratigraphic record. The most ubiquitous is biostratigraphy, complemented for appropriate stratigraphic intervals by magnetostratigraphy, chemostratigraphy ($\delta^{18}O$, $\delta^{13}C$, $^{87}Sr/^{86}Sr$) and cyclostratigraphy. Although numerical dating has also been derived from seismic and sequence stratigraphies (and from logging), these methodologies are insufficiently documented to warrant the common use of such practices.

Because it is non-iterative, biostratigraphy remains an essential component in the science of stratigraphy. In most instances, magnetozones and stable isotopic events can only be confidently identified with the support of biostratigraphy, fortunately positioned radioisotopic dates helping only on rare occasions (e.g., East African and Paratethys Miocene). Two normal magnetozones may be concatenated into one by an unconformity (Fig. 1A). A normal magnetozone may be lost in a stratigraphic gap where an apparent single reversed magnetozone straddles an unconformity (Fig. 1B). In turn, magnetobiostratigraphic correlations considerably increase stratigraphic resolution in allowing characterization of parts of a biozone. In isotope stratigraphy, similar patterns can be deceiving, in particular when excursions (short abrupt changes in isotopic composition) are involved (see Aubry and others, 1995).

Diachrony of paleontologic events is a well recognized impediment to biostratigraphic resolution and a potential source of misdating sediments (see Hedberg, 1976). A biozone is a stratigraphic unit defined by two or more paleontologic events that are lowest (LO) and highest (HO) occurrences. In Figure 2, Biozone B is a total range zone between the LO and HO of species X. The temporal interval represented by Biozone B varies regionally as diachrony occurs (from times T1-T2 between latitude 0° and 30° to times T1'-T2' at latitude 60°). The maximum temporal interval it covers corresponds to the life span of species X. This maximum temporal interval is referred to as the biochron, bounded by the first (temporal; evolutionary) appearance datum (FAD) and the last (temporal; extinction) appearance datum (LAD) of the species. In the geographic province where species X has its maximum stratigraphic range, its LO corresponds to its FAD, its HO to its LAD. Outside of this province, the HO and LO of species X differ from its LAD and FAD, respectively, and the time interval that the zone represents is less than the biochron. Diachronous LO's and HO's have been documented for all microfossils groups (e.g., Johnson and Nigrini, 1985; Dowsett, 1988, 1989; Wei and Wise, 1990, 1992; Aubry, 1992).

In planktonic microfossils, that diachrony can involve time differences of several million years is unquestionable, as for instance the HO of *Ericsonia formosa* (calcareous nannoplankton) which is in the lower Oligocene Series (=LAD) at low latitudes and in the upper middle Eocene Series in the high latitudes (Aubry, 1992). In many instances, however, diachrony on the order of 1 my or less may be more apparent than real. Establishing diachrony of low amplitude requires demonstration that the sections that are compared are temporally continuous over the interval considered. While truncated ranges of taxa are commonly explained by diachrony, there is an alternative explanation. Truncation of ranges may simply result from stratigraphic gaps (Fig. 2). If an unrecognized unconformity truncates the upper range of species X (as in Fig. 3, section b), there will be apparent diachrony of the HO of the species. If unconformities truncate both the upper and the lower stratigraphic range of species X (Fig. 3, section c), the situation will be comparable to that produced by diachrony of the HO and LO of the species. In other words, a shortened stratigraphic range representing a temporal interval between times T'1 and T'2 (Fig. 2) may result equally from true diachrony or from a discontinuous stratigraphic record. While the stratigraphic patterns produced for species X are the same in both cases, they reflect disparate causes, and the implications are radically different. More troubling perhaps is the last situation described in Figure 3 (section d or e) where a stratigraphic gap occurs within the range of species X. In this case, as the LO and the HO of species X are correctly identified as the species's FAD and LAD, respectively, the thinness of Biozone B in the section may be interpreted as reflecting deposition at low accumulation rates. This, in turn, bears on the problem of condensed sections examined below.

Delineating Unconformities

Stratigraphy is concerned with correlations before being concerned with dating. It is only through correlation networks that the relative chronology of events can be established, a step which is preliminary to their numerical dating. The recognition of this fundamental property of stratigraphy has led to the development of quantitative stratigraphic methods to help in determining the most likely model of the full succession of events in particular stratigraphic intervals (Edwards, 1982; Gradstein and others, 1985; Kovach, 1993). Of these, graphic correlation methods (Shaw, 1964) and probabilistic stratigraphy (Hay, 1972) have been the most successful. Both methods have been described in great detail, their advantages and limitations have been discussed (Dowsett, 1988, 1989; Edwards, 1984; Miller, 1977; Southam and others, 1975; Gradstein and others, 1985), and graphic correlation methods have been extended from biostratigraphy to other stratigraphic means (e.g., Prell and others, 1986).

This is not the place to discuss quantitative methods in stratigraphy, but it is necessary to point out that while they provide the most probable succession of stratigraphic events, they do not yield inherent temporal information. As stated by Dowsett (1989, p. 7) the CU (composite units) are "units of accumulation expressed in terms of depth scale of the section originally chosen as the standard." Thus, although the CSRS (composite standard reference section) is often regarded as time significant,

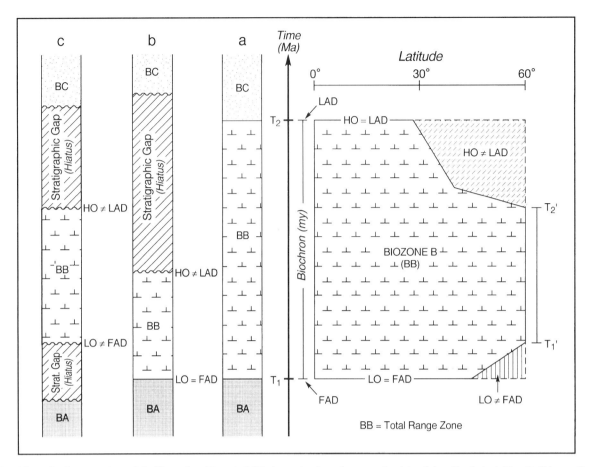

FIG. 2.—Alternative interpretations of the biostratigraphic record. Diachrony (such as shown on the right of time line in both Figs. 2, 3) is usually regarded as the primary impediment to reliable stratigraphic interpretation. However, the apparent range of a taxon may be reduced as the result of unconformities (left of time line in both figures). The effect of unconformities would be obvious in sections d and e (Fig. 3), but might easily be overlooked in sections b and c.

it is essentially an idealized stratigraphic section. It acquires a temporal significance only when calibrated to a numerical framework (as in Hazel and others, 1984; Dowsett, 1988, 1989) in which case the relationship CSRS-geological time is comparable to the relation of chronostratigraphy and geochronology in the CTS as discussed above. Another problem with the method of graphic correlation is that the LOC (line of correlation) is a function of sedimentation rates, not of time. Terraces in the LOC are justifiably interpreted as representing unconformities. However, not all unconformities may be apparent. The frequent changes in the slope of a LOC, particularly if not well constrained by data, may reflect additional unconformities which remain unrecognized, perhaps in part because of subjectivity in the construction of the LOC. As I have shown for Neogene units of the Gulf of Mexico and Jamaica (Aubry, 1993a, b) and as I will show below, some temporal intervals are well represented in the stratigraphic record, others correspond mostly to widespread unconformities. Considering the procedure used in the construction of the CSRS, there is little chance that the graphic correlation method allows characterization of these widespread unconformities unless a huge data base involving the comparison of numerous sections is employed in the construction of the CSRS.

The IMBS can be applied directly to assess the continuity of stratigraphic sections, and hence, to determine which strati-

graphic events in a given section have a truly temporal significance. The principle is simple: if a section is continuous, the thicknesses of the magnetozones and biozones it contains are proportional to their respective durations as magnetochrons and biochrons. The sedimentation rate curve is then a straight line. If the section is continuous, a change in the sedimentation rate will be marked by an inflection of the curve at the stratigraphic level where the change occurs. Even if the change is towards slower rates, all events which happened while the rates were low will be recorded in a sequential fashion in the section (Fig. 4). If instead a section is discontinuous (Fig. 5), events which are temporally distant will be juxtaposed at the same stratigraphic level (a common case for paleontologic events) or will not be represented (a common case for magnetic reversals). Of great importance is the fact that, as shown in Figure 5 which is based upon a real case study, a normal biozonal succession is no warranty of full stratigraphic continuity. It is only through the integration of independant controls such as magnetobiostratigraphic correlations and chemobiostratigraphic correlations (see Aubry and others, 1995) that we can gain a more accurate understanding of the relationship between the stratigraphic record and geological time.

The thickness versus time plot is particularly well suited for marine carbonate facies where microfossil zonations are applicable. It may be compared to some extent to a graphic corre-

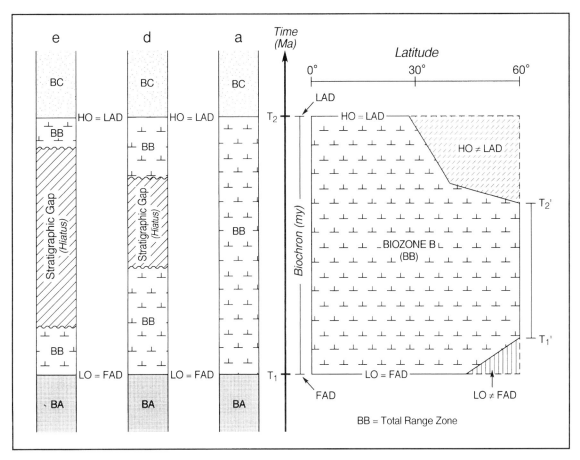

F_{IG}. 3.—Alternative interpretations of the biostratigraphic record. See Figure 2 for explanation.

lation technique in which the CSRS (1) results from the interregional integration of a large dataset and (2) has been calibrated to a numerical chronology. However, the thickness versus time plot is a more direct procedure which requires only selected events of global temporal significance.

Condensation and Truncation

Because we tend to view the stratigraphic record as a continuous representation of geological time, differences in thickness of a given stratigraphic interval in compared sections are usually considered to relate to differences in sedimentation rates. Yet, truncation of a given stratigraphic interval in different sections deposited at comparable sedimentation rates by an unconformity with an associated hiatus of variable duration between sections will also result in thickness differences (Fig. 3). Condensation—the fact that a stratigraphic interval in a given section is anomalously thin compared to other sections—is generally thought to result from slow sedimentation rates. Figure 4 illustrates condensation in a 12-m thick interval deposited over 5 my at an (arbitrary) sedimentation rate of 0.3 cm/10^3 yr compared to rates of 2 cm/10^3 yr and 1.5 cm/10^3 yr for the underlying and overlying sections, respectively. Figure 5 illustrates the same 12-m thick interval as being riddled with unconformities. In this case, although each unconformity-bounded segment of the 12-m thick interval was deposited at the same

rate as the underlying and overlying segments, the apparent sedimentation rate is 0.3 cm/10^3 yr, as for the condensed section in Figure 4, because the greater part of the temporal interval in this 12-m thick interval is represented by unconformities. As explained above (and shown in Figs. 4, 5), a condensed section (sensu lato) and a truncated section leave different signatures in the stratigraphic record. It is crucial to distinguish between them because they reflect fundamentally different genetic histories. In the example given in Figure 6 (left), condensation was initiated at 53 Ma and lasted until 48 Ma when normal sedimentation resumed. A truncated section which contains several unconformities cannot, however, result from a single event. In Figure 6 (right), condensation results from at least 4 events which may or may not have been related, and which produced a discontinuous stratigraphic record. The first event occurred between 53 Ma and 52.8 Ma; the second between 52.6 Ma and 50.2 Ma; the third between 49.6 Ma and 48.9 Ma; the fourth between 48.1 Ma and 48 Ma. It is worth pointing out that, considering the duration of the second hiatus, the associated stratigraphic gap may in fact reflect more than a single event.

Condensed sections hold a special place in sequence stratigraphy. Those which are interpreted as developing at times of maximum regional transgression are used as a correlation tool (Loutit and others, 1988; Armentrout, 1991). Defined as " . . . thin marine stratigraphic units consisting of pelagic to hemipelagic sediments characterized by very low sedimentation

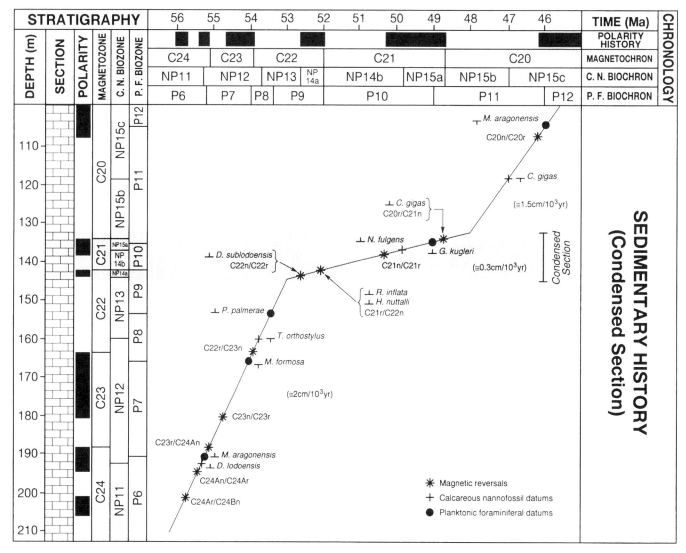

FIG. 4.—Sedimentary history of a section which contains a condensed interval, inferred from the sedimentation rate curve. The condensed section yields the sequential occurrences of paleontologic and magnetic reversals which characterize the temporal interval during which it was deposited.

rates . . . ," they are also said to be ". . . associated commonly with apparent marine hiatuses and often occur either as thin but continuous zones of burrowed, slightly lithified beds (omission surfaces) or marine hardgrounds . . ." (Loutit and others, 1988, p. 186). In this definition, a condensed section in a stratigraphic sequence appears to represent a hybrid between condensation and truncation. The published literature provides no real case illustration of the relationship between a condensed section and time (in a similar manner to the cases discussed in Fig. 6). It is now a challenge to establish such a relationship in order to improve our understanding of the relation between condensed sections and maximum regional transgression.

The Temporal Significance of Unconformities

Although it has long been recognized (see Blackwelder, 1909) that unconformities are part of the stratigraphic record, their temporal significance is currently often ignored. This may be in part because, using the CTS, it is difficult to determine

the temporal interval (herein strictly referred to as hiatus) associated with unconformities, although it may be more likely related to the basic assumption that most of the stratigraphic record is continuous.

An unconformity is not merely a physical surface that separates two sequences. It is the expression of a stratigraphic gap, the fact that a certain interval of geological time, possibly substantial, is not represented in the stratigraphic record. A stratigraphic gap in a section or in a basin can only be characterized by comparison with another section or basin where it can be shown to correspond to an x-m thick interval (in another section) or to a particular stratigraphic unit or succession of units (in another basin). A stratigraphic gap in a section cannot be associated with the thickness of sediment missing because the gap may result from erosion, non-deposition, or from a combination of both. It is only by inference that, in some instances, it will be possible to determine which mechanism(s) produced the gap.

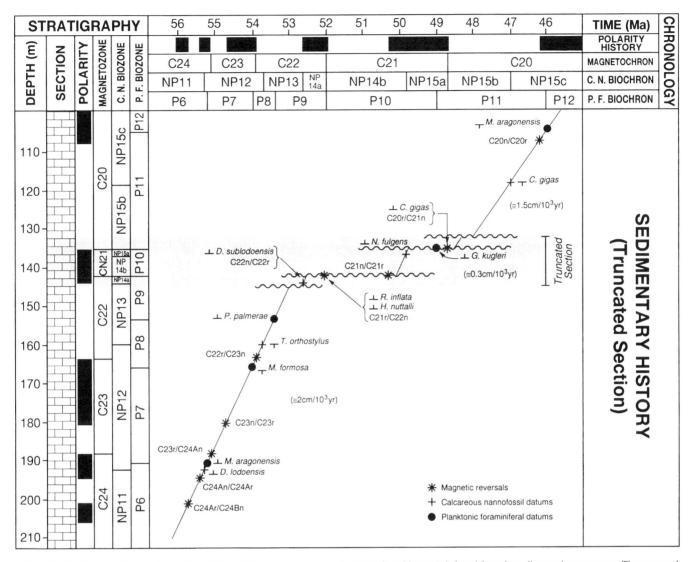

FIG. 5.—Sedimentary history of a section which contains successive truncations (stratigraphic gaps), inferred from the sedimentation rate curve. The truncated section is characterized by the anomalous stratigraphic juxtaposition of paleontologic events and magnetic reversals which are temporally well separated.

Whereas we cannot determine how much sediment is missing in a given unconformity, we can determine the hiatus, the amount of time that an unconformity represents. This requires the dating of the two genetic surfaces which compose the unconformity (Aubry, 1991), and it is one of the most valuable applications of the GPTS. I will discuss below the difficulties and uncertainties associated with the dating of unconformities. The two surfaces which compose an unconformity are of unequal genetic significance. The upper surface is little more than a bounding surface linked to the resumption of sedimentation. In contrast the lower surface yields the complete, but mostly undecipherable history of the stratigraphic gap (non deposition-without erosion, erosion, or single or multiple cycles of deposition and erosion). If the lower surface can be considered the unconformity proper, it must be acknowledged that the upper surface is of equal importance for the dating of the (last) event which has fashioned the unconformity, and for correlating between unconformities (see Aubry, 1991, for further explanation).

The dating of the surfaces of an unconformity requires three steps.

1. Delineation of the unconformity when no erosional contact occurs (following the principle shown in Fig. 5). Inferring the presence of an unconformity from magnetobiostratigraphic correlations is relatively easy (the longer the hiatus, the easier its delineation), but positioning it precisely may be difficult even with detailed stratigraphic analysis. Arbitrary choices between several alternatives may be necessary (see Aubry, 1991, Fig. 18; and below). However, many—but not all—unconformities that are inferred from magnetobiostratigraphic correlations correspond to contacts between lithologic units/subunits (see Aubry and others, 1995, and below).

2. Estimation of the hiatus. The estimate is based on the model duration of the biozones and magnetozones (or their parts) that are not represented. The uncertainty on the estimate is highly variable between cases, from less than a hundred

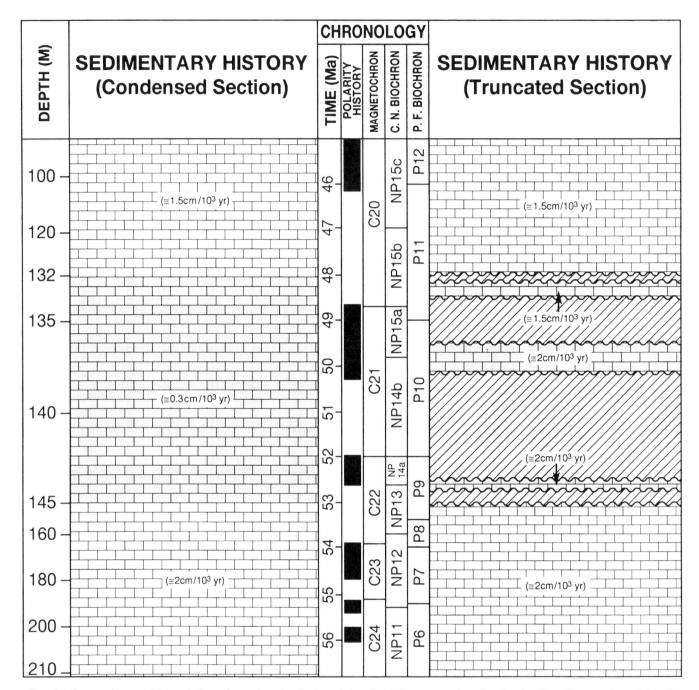

FIG. 6.—Compared temporal interpretations of a condensed and a truncated section. Whereas a condensed section is still continuous, a truncated section constitutes a discontinuous record of the temporal interval during which it was deposited. Whereas each condensed section is usually the reflection of a single event, a truncated section may contain the signature of several (4 in this figure), possibly unrelated events.

thousand to several hundred thousand (up to a few million) years (see discussion below).

3. Dating of the surfaces. This is a difficult exercise whose precision depends upon the precision reached in the two preceding steps. If an unconformity is marked by a lithologic contact, the age of the surfaces may be determined through extrapolation of the sedimentation rates below and above, providing that the underlying and overlying sections are suf-

ficiently extended (in Figs. 5, 6, the age of the surfaces at 145 m and at 132 m can be determined in this manner; see also the dating of the lower surface of the lower middle Eocene unconformity at DSDP Sites 261 and 613 in Aubry, 1991). If multiple truncation occurs (as in Figs. 5, 6, between 144 m and 132.5 m), this procedure is not applicable, and the age of the surfaces can only be approximated (with an uncertainty which varies with that of the hiatus).

Preliminary Conclusions

It is clear from the discussion above that there is not a single, straight, unequivocal interpretation of stratigraphic sections, but alternative interpretations. The discussion which follows on the lower and middle Eocene Atlantic record (1) illustrates the difficulties associated with interpreting the stratigraphic record and (2) demonstrates the significance of the temporal interpretation of stratigraphic sections. We can no longer interpret the geological history of the planet without evaluating our assumptions on the stratigraphic record and re-examining its relationship with geological time.

THE UPPER LOWER AND LOWER MIDDLE EOCENE STRATIGRAPHIC RECORD DURING THE INTERVAL BETWEEN ~CHRON C24N.1–2N AND CHRON C20N IN THE ATLANTIC OCEAN: DESCRIPTION

Objectives

Two categories of reasons of equal importance have guided my choice of the upper lower to lower middle Eocene stratigraphic record to assess the continuity of sedimentary sequences in the deep sea. In the first category, the reasons are thematic. Unconformities in broadly separated regions occur around the lower/middle Eocene boundary. It has been proposed that they are stratigraphically correlative and are the expression of a major sea-level fall in the latest early Eocene Epoch (see review in Aubry, 1991). Previous examination of several lower and middle Eocene sections (Aubry, 1991) suggested that a contrast occurs between the well preserved continuous lower Eocene Series and the poorly preserved, unconformity riddled, uppermost lower and lower middle Eocene Series, not only on the shelves (<200 m) but also in deeper parts of the continental margin (1000 to 2000 m). This work is thus intended to decipher the extent of unconformities around the lower/middle Eocene boundary in the deep sea as a logical continuation of my previous study, and to confirm its findings. The relationships between unconformities on the shelf and in the deep sea may bring pertinent clues in deciphering the mechanisms involved in eustatic changes (see Cathles and Hallam, 1992).

In the second category, my reasons relate to methodological advantages. The late early and early middle Eocene magnetochrons C23r to C20n are long and undivided (with the exception of Chron C23n), and they are unambiguously identifiable in the stratigraphic record through their association with zonal boundaries. In addition, early and middle Eocene calcareous and siliceous biochrons are of relatively short duration (0.6 to 2 my) so that the magnetobiochronologic framework currently available (Berggren and others, 1985; Aubry and others, 1988) provides a resolution that is unmatched for any other interval in the Paleogene Period (Table 1).

Chronologic Framework

This work was begun in 1990, at about the same time as preliminary modifications were being made (Berggren and others, 1992) to the Paleogene part of the Cenozoic time scale (Berggren and others, 1985). However, a revised IMBS has not been available during the course of preparation of this paper. Thus, the magnetochronologic framework in Berggren and others (1985) is followed in this work. There are, however, no reasons for substantial revisions to the lower and middle Eocene magnetobiostratigraphic correlations proposed in Berggren and others (1985) and Aubry and others (1988). Uncertainty of a couple hundred thousand years on FAD's and LAD's is negligible compared to the hiatuses (>1 my) delineated in this study and is thus irrelevant to the discussion that follows. The numerical ages are those in Berggren and others (1985) even though the early Paleogene magnetochronology has been revised (Cande and Kent, 1992, 1995). This is because early Paleogene numerical chronology is still unstable (see Aubry and others, 1995; Berggren and others, this volume). Fortunately, the fact that the magnetic reversals from Chron C20r to C24n are ~2 my younger in Cande and Kent (1995) compared to Berggren and others (1985) has little bearing on the results presented here because the differences in chron *duration* between the two magnetochronologies are negligible (< 20 x 10^3 yr to <250 x 10^3 yr; Table 2) compared to the hiatuses discussed. The only potential difficulties concern the shorter duration of Chron C21r (475 x 10^3 yr) and the longer duration of Chron C23n (378 x 10^3 yr) in Cande and Kent (1995). The latter implies that sediments belonging to Zone NP12 were deposited at slightly lower sedimentation rates than inferred here. The former implies that the hiatuses which encompass Chron C21r are somewhat shorter than shown. However, the longer durations of Chrons C21n and C22n (which together represent a difference of + 100 x 10^3 yr) compensate partly for the shorter duration of Chron C21r given in Berggren and others (this volume).

Geographic Setting

All DSDP and ODP sites drilled until 1988 in the North and South Atlantic Ocean, including the subantarctic South Atlantic, have been considered in this work. Of the 104 sites that penetrated through lower Eocene series, only 51 had sufficient recovery and stratigraphic control to allow a comprehensive interpretation of their lower and middle Eocene sections (Fig. 7). At 21 sites, coring gaps prevented the recovery of lower to middle Eocene sections if present (Appendix A). In 16 sites, a stratigraphic gap encompasses the lower to middle Eocene Series (Appendix B). Inappropriate lithology, probably combined with large stratigraphic gaps in some cases, eliminated another 15 sites (Appendix C). Among these are DSDP Sites 355, 367, 368 and 672 from which thick lower and middle Eocene sequences were recovered. The biostratigraphy of these sites, however, is based solely on radiolarians, and radiolarian zonal schemes are presently insufficiently well tied to the magnetochronologic framework of Berggren and others (1985) to allow comprehensive stratigraphic/temporal interpretations.

To ease their description, the sites have been regrouped regionally. The Atlantic Ocean has been subdivided more or less arbitrarily into 8 areas: eastern North Atlantic (Hatton drift-Rockall Plateau and Goban Spur-Biscay), central eastern Atlantic, Labrador Sea, western North Atlantic, Gulf of Mexico-Caribbean area, eastern South Atlantic, western South Atlantic and Subantarctic South Atlantic. Description of the sites according to their geographic distribution may result at first in apparently equivocal stratigraphic interpretations. The precepts

TABLE 1.—MAGNETOBIOCHRONOLOGIC FRAMEWORK USED IN THIS STUDY. AGES ARE FROM BERGGREN AND OTHERS (1985), AUBRY AND OTHERS (1988) AND BERGGREN AND MILLER (1988)

Magnetic reversals	Calcareous Nannofossil Datums	Planktonic foraminiferal datums	Radiolarian zonal boundaries	Ages (Ma)
		LAD *M. aragonensis*		46.0
	LAD *N. fulgens*			45.4
Chron C20N/C20R				46.17
	LAD *C. gigas*		Zone R20/R21	47.0
Chron C20R/C21N				48.75
		FAD *D. kugleri*		48.0
	FAD *C. gigas*			48.8
	FAD *N. fulgens*		Zone R21/R22	49.8
Chron C21N/C21R				50.34
			Zone R22/R23	51.4
Chron C21R/C22N				51.95
	FAD *R. inflata*	FAD *H. nuttalli*	Zone R23/R24	52.0
	FAD *D. sublodoensis*			52.6
Chron C22N/C22R				52.62
		FAD *P. palmerae*	Zone R24/R25	53.4
	LAD *T. orthostylus*			53.7
Chron C22R/C23AN				53.88
		LAD *M. formosa*		54.0
Chron C23AN/C23AR				54.03
Chron C23AR/C23BN				54.09
Chron C23BN/C23BR				54.70
Chron C23BR/C24AN				55.14
		FAD *M. aragonensis*	Zone R25/R26	55.2
	FAD *D. lodoensis*			55.3
Chron C24AN/C24AR				55.37
Chron C24AR/C24BN				55.66
		FAD *M. formosa*		56.1
Chron C24BN/C24BR				56.14

TABLE 2.—COMPARISON BETWEEN THE MAGNETOCHRONOLOGIC FRAMEWORKS OF BERGGREN AND OTHERS (1985) AND BERGGREN AND OTHERS (THIS VOLUME). THE DIFFERENCES IN CHRON DURATIONS REFLECT THE RE-EVALUATION BY CANDE AND KENT (1992) OF THE RELATIVE SEA FLOOR ANOMALY SPACING.

Chron	Berggren and others, 1985	Chron duration	Berggren and others, in press	Chron duration	Difference in duration
Chron C20n	44.66–46.17	1.51 my	42.536–43.789	1.253 my	−0.257 my
Chron C20r	46,17–48.75	2.58 my	43.789–46.264	2.475 my	−0.105 my
Chron C21n	48.75–50.34	1.55 my	46.264–47.906	1.642 my	+0.092 my
Chron C21r	50.34–51.95	1.61 my	47.906–49.037	1.131 my	−0.475 my
Chron C22n	51.95–52.62	0.67 my	49.037–49.714	0.677 my	+0.007 my
Chron C22r	52.62–53.88	1.28 my	49.714–50.778	1.064 my	−.216 my
Chron C23n.1n	53.88–54.03	0.15 my	50.778–50.946	0.168 my	+0.018 my
Chron C23n.1r	54.03–54.09	0.06 my	50.946–51.047	0.100 my	+0.040 my
Chron C23n.2n	54.09–54.70	0.61 my	51.047–51.743	0.690 my	+0.080 my
Chron C23r	54.70–55.14	0.44 my	51.743–52.364	0.620 my	+0.378 my
Chron C24n.1n			52.364–52.663	0.299 my	
Chron C24n.1r	55.14–55.37	0.23 my	52.663–52.757	0.094 my	+0.207 my
Chron C24n.2n			52.757–52.801	0.044 my	
Chron C24n.2r	55.37–55.66	0.29 my	52.801–52.903	0.102 my	−.188 my
Chron C24n.3n	55.66–56.14	0.48 my	52.903–53.347	0.444 my	−.038 my

which guided my interpretations do not transpire from the first sites discussed below. Thus, the reader is referred to the preliminary remarks which introduce the discussion below. The late early and early middle Eocene paleodepth of each site is given in Table 3.

Stratigraphic Interpretation of DSDP and ODP Lower to Middle Eocene Sections Recovered from the North Atlantic Ocean

Eastern North Atlantic: Hatton Drift-Rockall Plateau.—

DSDP Site 553 (56° 05.32'N, 23° 20.61'W).—DSDP Site 553 was drilled on the Hatton drift of the west margin of Rockall Plateau, in 2329-m water depth (Roberts and others, 1984).

A 25-meter thick section of upper lower and middle Eocene volcanic tuffs and foraminiferal nannofossil chalks, unconformable with upper Oligocene chalks, was recovered from Hole 553A (Fig. 8). Calcareous nannofossils (Backman, 1984) provide the major stratigraphic control for this Eocene interval supplemented by magnetostratigraphy (Krumsieck and Roberts, 1984). Planktonic foraminifera (Huddleston, 1984), diatoms (Baldauf, 1984), and radiolarians (Westberg-Smith and Riedel, 1984) yield no critical zonal information.

In addition to the intra-Eocene unconformity at ~240.70 m which results in the direct superposition of Zones NP14 and NP16 and in the concatenation of Chrons C22n and C20n (or C19n or C18Cn), an unconformity between 245 and 248 m is inferred from integrated magnetobiostratigraphy. The lower part

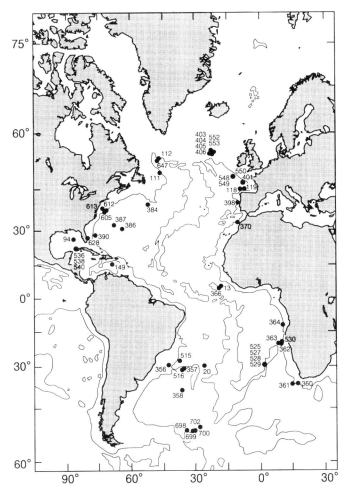

FIG. 7.—DSDP and ODP sites discussed in this study from which lower and middle Eocene sections were recovered.

TABLE 3.—LATE EARLY TO EARLY MIDDLE EOCENE PALEODEPTHS OF THE DSDP/ODP SITES CONSIDERED IN THIS STUDY.

DSDP/ODP Site	Paleodepth (Backtracking/'backstripping')	Paleodepth (benthic foraminiferal evidence)
20	~3200 m[19]	
94	2000–2100 m[19]	
111	~1000 m[11]	~1000 m[3]
		2000 to 3000 m[16]
112		~2500m [6]
118	~4190 m[9]	
119	~3500 m[11]	
356	~2500m[17]	
	~1700 m[19]	
357	~1200 m[1]	
	~4500 m[19]	
358	~3200 m[2]	
360	~3300 m[19]	
361	~4000 m[17]	
362	~1800 m[4]	
363	2000 to 2200 m[19,4]	
364	~1200 m[4]	
366	~1880 m[9]	
370	~4445 m[9]	
384	~3600 m[20]	
386	~5400 m[20]	
387	~5500 m[20]	
390	~2700 m[19]	
398	~3300 m[15]	
400		~4000 m[14]
401		~1800 m[14]
		m. Eocene: 250 m[12]
403		e. Eocene: <100 m
		m. Eocene: 200 m[12]
404		e. Eocene: <75 m
405		~1000 m[12]
406		~1000 m[12]
515	~3600 m[2]	
516	~800 m[1]	
525	~1700 m[11]	500–1500 m[18]
527	~3500 m[11]	
528	~3000 m[11]	
529	~2300 m[11]	
530	~4500 m[4]	
536	~2495 m[9]	
538	~2576 m[9]	
540	~2747 m[9]	
548	~1200 m[8]	
549	~2300 m[8]	
550	~4200 m[8]	
		m. Eocene: >700 m[13]
552		e Eocene: 75–200 m
		m. Eocene: >700 m [13]
553		e Eocene: 75–200 m
605	2000 to 2300 m[5]	
612		~1000 m[10]
613		~2000 m[10]
628		2000 to 3000 m[16]
647		~2500 m[6]
698	~900 m[7]	1000–2000 m[7]
699	~2900 m[7]	~2500 m[7]
700	~2000 m[7]	2500–3000 m[7]
702	~1800 m [7]	1000–2000 m[7]

The geophysically derived values were obtained using different backtracking/backstripping methods and may not be directly comparable. At some of the sites, paleodepth changed notably during early and middle Eocene Epochs. The values given here are paleodepth estimates for latest early Eocene Epoch.

(1): Barker (1983), (2): Barker and others (1983b), (3): Berggren and Aubert (1976), (4): Dean and others (1984), (5): Hulsbos (1987), (6): Kaminski and others (1989), (7): Katz and Miller (1991), (8): Masson and others (1985), (9): Miller, K. G. (pers. commun., 1992), (10): Miller and Katz (1987), (11): Moore and others (1984b), (12): Murray (1979), (13): Murray (1979), (14): Schnitker (1979), (15): Sibuet and Ryan (1979), (16): Srivastava and others (1987), (17): Thiede and Van Andel (1977), (18): Tjalsma (1983), (19): Tjalsma and Lohmann (1983), (20): Tucholke and Vogt (1979).

of the interval with normal polarity in Core 10 immediately below the unconformity at ~240.70 m correlates with the lower part of Zone NP14 (Subzone NP14a) as indicated by the co-occurrence of *D. sublodoensis, D. lodoensis* and the absence of *Rhabdosphaera inflata* (see Backman, 1984, Table 4). I assign it to Chron C22n. The underlying normal polarity interval in Core 11 correlates with Zone NP12 and represents Chron C23n while the next interval with normal polarity in Cores 11–6 and 12 represents (in part or entirely) Chron C24An. The sedimentation rate curve based on this interpretation of the magnetostratigraphic record in Hole 553A suggests stratigraphic gaps in Zone NP12 and near the Chrons C22r/C23n and/or C22n/C22r boundaries (Fig. 8). There are no data to constrain further the position of the unconformities. The younger unconformity may lie within Zone NP13, or straddle the NP13/NP14 or the NP12/NP13 zonal boundary. The Chron C22r/C23n boundary is not determined precisely enough to help resolve this question, but the apparent slow sedimentation rate (0.5 cm/10³ yr) for the interval assigned to Zone NP12 may reflect truncation of the zone. On this ground, the unconformity is tentatively drawn at 248 m (Figs. 8, 9). This is in reasonable agreement with Backman and others (1984a) who delineated an unconformity at ~250 m in DSDP Hole 553A.

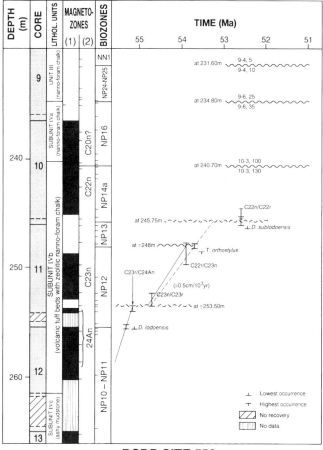

DSDP SITE 553

Fig. 8.—Stratigraphic interpretation and sedimentation rate curve(s) of the lower and middle Eocene interval recovered from DSDP Hole 553A. Unconformities at 234.80 m and at ~240.70 m (as inferred from calcareous nannofossil stratigraphy, Backman, 1984) or 240.32 m (based on sediment lithology, Roberts and others, 1984) are associated with intense burrowing. Unconformity between 252 and 251 m inferred from the sedimentation rate curve. Lithologic units are from Robert and others (1984). Datums used to constrain the curve are given in Appendix D, Table 1. (1) from Krumsieck and Roberts (1984) and (2) proposed interpretation. Biozones from Backman (1984), slightly modified. Note that the boundaries between magnetozones and calcareous nannofossil zones are drawn at half distance between analyzed samples and were not adjusted to fit the correlations suggested in Berggren and others (1985). Tick marks in columns (2) and biozones indicate levels sampled for magnetostratigraphy (Krumsieck and Roberts, 1984) and calcareous nannofossil stratigraphy (Backman, 1984), respectively.

DSDP Site 552 (56° 02.56′N, 23° 18.88′W).—Two holes were drilled at DSDP Site 552 also located on the Hatton Drift in 2103-m water depth (Roberts and others, 1984).

Extremely poor recovery and large sampling interval hampers a comprehensive stratigraphic interpretation of the expanded 90-m thick, upper lower to middle Eocene section recovered from Hole 552 (Fig. 10). An apparently continuous zonal succession NP12-NP14 occurs, but identification of zonal boundaries is uncertain, and correlations between calcareous nannofossil datums (Backman, 1984) and polarity intervals (Krumsieck and Roberts, 1984) differ from those expected from the framework of correlations of Berggren and others (1985). For instance, the HO of *Tribrachiatus orthostylus* may occur at

this site as low as between Core 16–1, 31 cm and 15–1, 10 cm, or as high as between Cores 14–2, 56–58 cm and 14–2, 26–27 cm (see discussion in Backman, 1984). In the first case, it may correlate with a reversed interval (in Core 16, identifiable as Chron C22r), so that the normal polarity interval in Core 17–1 would correspond to Chron C23n. In the second case, it would lie within a normal interval (in Core 14), while elsewhere (e.g., at nearby DSDP Hole 553A, cf. Fig. 8) it lies within Chron C22r. Backman and others (1984b) preferred to draw the NP12/NP13 zonal boundary in Core 14, and Backman and others (1984a) identified the normal polarity interval in Cores 13 and 14 as Chron C23n. There are good reasons, however, why the NP12/NP13 zonal boundary should be placed between Cores 15–1 and 16–1, in particular the lowest occurrence of *Reticulofenestra dictyoda* in Core 14–4 (see Backamn, 1984), the scarcity of *T. orthostylus* in Core 14, evidence of reworking (e.g., *Discoaster lodoensis* and *D. kuepperi* in upper Zone NP14, see below), and correlation to magnetostratigraphy. A (single?) prominent normal polarity interval occurs in Cores 14 to 12. If this interval is correlative with Zone NP13, it has only local significance. However, this normal polarity interval may represent Chron C22n. The interval with reversed polarity in Core 9 correlates with the upper part of Biozone NP14 (Backman, 1984) and best represents Chron C21r. The reversed polarity interval in Core 12 may as well represent Chron C21r. As Backman (1984) observes, discontinuous occurrence of taxa is common in Hole 552, and it can be speculated that *D. sublodoensis* occurs lower than Core 10-cc but was not recorded because of scarce occurrence. From this interpretation which attempts at reconciling at best magneto- and biostratigraphy, it results that Biozone NP13 is very thin. This suggests a stratigraphic gap corresponding to most of Zone NP13 between Cores 14 and 15, at about 220 m. Backman and others (1984a, Fig. 2) also concluded to the presence of a stratigraphic gap corresponding to most of Zone NP13 but located it at ~200 m following their interpretation of the normal polarity interval in Cores 14 and 13 as Chron C22n. Thus, although there are alternative interpretations of the stratigraphy of the lower and middle interval recovered from DSDP Hole 552, there is strong support for an unconformity near the lower/middle Eocene boundary. The age of its lower and upper surfaces would, however, differ depending upon the interpretation. As I undertand it, the stratigraphic record around the lower/middle Eocene boundary at DSDP Hole 552 is very similar to that at DSDP Hole 553A (Fig. 9).

Hole 552A was terminated in middle Eocene zeolitic mudstones, unconformably underlying a lower Oligocene chalk (Zone NP22). Calcareous nannofossil stratigraphy (Backman, 1984) and magnetostratigraphy (Krumsieck and Roberts, 1984) complement each other in the interpretation of the stratigraphy of the 9-m thick middle Eocene interval (Fig. 9) which comprises a thin veneer of Subzone NP15b (Core 37–1, 137–138) unconformable with Subzone NP14b (Core 38-cc to 37–2; Backman, 1984). While the age of the upper surface of the intramiddle Eocene unconformity cannot be determined precisely, that of the lower surface is approximated by correlation of the upper part of the interval assigned to Subzone NP14b with a normal polarity interval which likely represents Chron C21n. *DSDP Site 406 (55° 15.50′N, 22° 05.41′W).*—DSDP Site 406 is situated in 2907-m water depth at the foot of the east-west

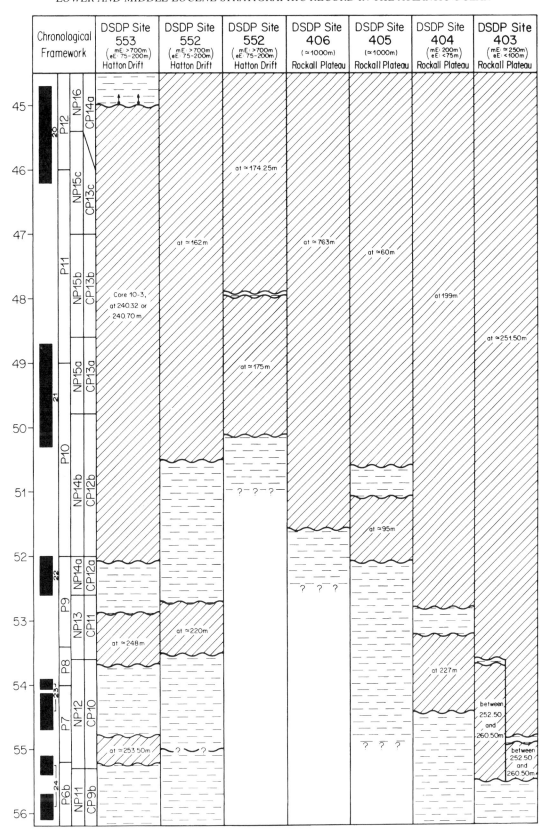

FIG. 9.—Temporal interpretation of 7 DSDP lower and middle Eocene sections recovered from the Hatton Drift-Rockall Plateau, eastern North Atlantic Ocean. Dashed pattern represents sediment; hatched pattern represents hiatuses. The chronologic framework is that of Berggren and others (1985). The planktonic foraminiferal chronology refers to the zonal scheme of Berggren (1969) and Blow (1979). The calcareous nannofossil chronology refers to the zonal schemes of Martini (1971; NP), and Bukry (1973b, 1975) codified in Okada and Bukry (1980, CP). The late early and early middle Eocene paleodepth is given for each site.

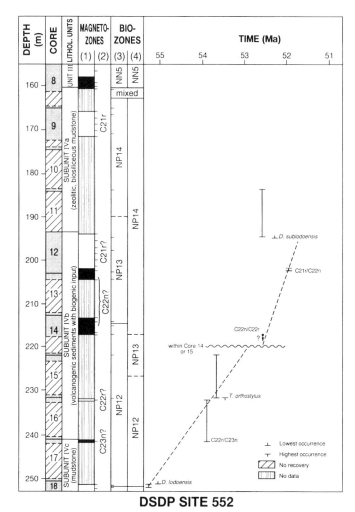

DSDP SITE 552

FIG. 10.—Stratigraphic interpretation of the lower and middle Eocene interval recovered from DSDP Site 552. Lithologic units are from Robert and others (1984). Datums used to constrain the curve are given in Appendix D, Table 2. (1) from Krumsieck and Roberts (1984), (2) proposed tentative interpretation, (3) from Backman (1984) and Backman and others (1984a) and (4) revised interpretation. Note that the boundaries between magnetozones and calcareous nannofossil zones are drawn at half distance between analyzed samples, and were not adjusted to fit the correlations suggested in Berggren and others (1985). Tick marks in (2) and (3) indicate levels sampled for magnetostratigraphy (Krumsieck and Roberts, 1984) and calcareous nannofossil stratigraphy (Backman, 1984), respectively.

trending transform fault which defines part of the southwest margin of Rockall Plateau (Montadert and Roberts, 1979a).

Sixty six meters of middle Eocene marly chalk unconformably underlie an upper Eocene (Biozone NP18) marly limestone characterized by numerous lower and middle Eocene slump beds (Montadert and Roberts, 1979a; Müller, 1979). Recovery was very poor, and stratigraphic interpretation of the middle Eocene interval relies primarily on calcareous nannofossils (Müller, 1979). The co-occurrence of *Discoaster sublodoensis* and *D. lodoensis* in Core 52 to 47 suggests Subzone NP14a. *Rhabdosphaera inflata* was not found in Core 46 but planktonic foraminiferal stratigraphy suggests assignment to Biozone P10 (Krasheninnikov, 1979), correlative with Subzone NP14b. The

interpretation of the middle Eocene interval recovered from DSDP Site 406 in Figure 9 should be regarded as tentative. *DSDP Site 405 (55° 20.18'N, 22° 03.49'W).*—A 350-m thick lower to middle Eocene interval of predominantly siliceous and calcareous mudstones, unconformable with upper Miocene foram nanno oozes, was recovered from DSDP Site 405 drilled 8 km south of DSDP Site 406 in 2958-m water depth (Montadert and Roberts, 1979a). Extremely poor recovery, magnetic instability in the lower part of the section (Hailwood, 1979), low-diversity planktonic foraminiferal assemblages (Krasheninnikov, 1979), and ambiguous identification of the magnetozones through correlation with calcareous nannofossil stratigraphy (Müller, 1979) result in a provisional stratigraphic interpretation of the Eocene interval recovered at this site (Figs. 9, 11).

Hailwood and others (1979) assigned the intervals with predominantly normal polarity in Cores 38 to 32 and in Cores 19 to 11, respectively, to Chron C24An and C23n and suggest that the thin normal polarity interval in Core 11–1 represents Chron C22n. While this latter assignment is supported by calcareous nannofossil and planktonic foraminiferal stratigraphy (Müller,

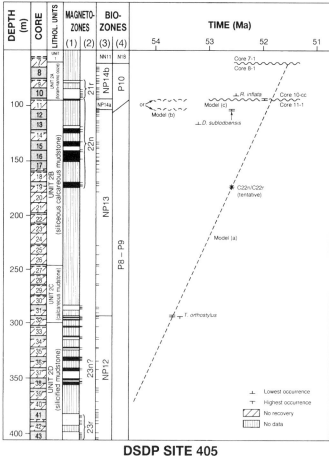

DSDP SITE 405

FIG. 11.—Stratigraphic interpretation of the lower and middle Eocene interval recovered from DSDP Site 405. Lithologic units are from Montadert and Roberts (1979a). Datums used to constrain the curve are given in Appendix D, Table 3. (1) from Hailwood (1979), (2) proposed interpretation, (3) from Müller (1979) slightly modified, and (4) from Krasheninnikov (1979).

1979; Krasheninnikov, 1979), correlation of the lower normal polarity interval (if reflecting primary magnetisation) with Biozone NP12 suggests assignment to Chron C23n rather than Chron C24An. Unless the normal intervals in Cores 19 to 14 have only local significance, Chron C22n may thus extend from Core 11–1 to Core 19, although the NP13/NP14 zonal boundary was drawn between Cores 12 and 11 (Müller, 1979). A similar situation was described in DSDP Hole 553A (see above). There are several indications that it may be the case. Radiolarians indicate a middle Eocene age for Cores 14 to 12, and an early Eocene age for Core 15 (Sanfilippo and Riedel, 1979). The record of *Nannotetrina cristata* down to Core 14 (Müller, 1979) suggests a middle Eocene rather than an early Eocene age. With the data presently available, three alternative stratigraphic interpretations can be given of the Eocene section at Site 405. Following the revised magnetostratigraphic assignment proposed here, the lower to middle Eocene interval would appear to be essentially continuous (model (a) in Fig. 11). Based on calcareous nannofossil stratigraphy alone, an unconformity would occur at the NP13/NP14 or NP14a/NP14b subzonal boundary (model (b) in Fig. 11). A mixed interpretation is preferred, in which the section is continuous up to the Chron C21r/C22n boundary (i.e., continuous from Core 43 to 11), but interrupted by an unconformity at this magnetic reversal boundary so that the upper part of Subzone NP14a and the lower part of Subzone NP14b are missing (model (c) in Fig. 11). The presence of an unconformity around 95 m is well-supported by the lithologic change that takes place at this level, from siliceous calcareous mudstone below to foraminiferal nannofossil ooze above.

DSDP Site 404 (56° 03.13′N, 23° 14.95′W).—DSDP Site 404 was drilled on the southwest margin of Rockall Plateau in 2306-m water depth (Montadert and Roberts, 1979b). Lower Eocene mudstones, over 100-m thick, were recovered underlying unconformably upper Miocene nannofossil oozes (Fig. 12). Despite poor recovery, integrated magneto- (Hailwood, 1979) and calcareous nannofossil (Müller, 1979) stratigraphy allows a comprehensive interpretation of this lower Eocene interval.

In agreement with Hailwood (1979), the interval with normal polarity in Cores 10 and 11 represents Chron C23n, but, based on its association with the lowest occurrence of *D. lodoensis*, the interval with normal polarity in Cores 15 and 16 is interpreted as Chron C24An rather than Chron C24Bn. Accordingly, the normal polarity intervals in Cores 7 and 8, Cores 12 and 14, and Cores 17 and 18 correspond to Chron C22r, C23r and C24Ar. The thinness of Chron C23n and the association of the highest occurrence of *T. orthostylus* with Chron C23n rather than with Chron C22r suggest that Zone NP12 is truncated as confirmed by the sedimentation rate curve (Fig. 12). It results from this that Biozone NP13 is likely truncated as well (at the base), and since it is truncated at the top, it is not possible to determine the duration of the hiatus. Location in Figure 9 of the NP13 zonal interval is thus arbitrary.

DSDP Site 403 (56° 08.31′N, 23° 17.64′W).—An expanded lower Eocene section was recovered from DSDP Site 403, also drilled on the southwest margin of the Rockall Plateau, in 2301-m water depth (Montadert and Roberts, 1979b). However, most of the section is earliest Eocene, and only less than a meter of chalk of Biochron NP12 age was recovered, unconformable with middle Eocene (Zone NP16) chalk (Müller, 1979). Re-

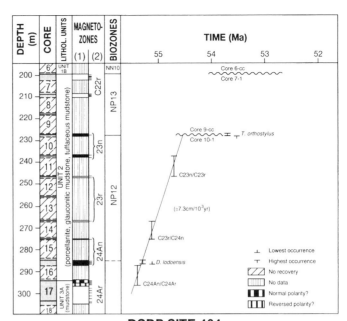

DSDP SITE 404

FIG. 12.—Stratigraphic interpretation of the lower Eocene interval recovered from DSDP Site 404. Lithologic units are from Montadert and Roberts (1979a). Datums used to constrain the curve are given in Appendix D, Table 4. (1) from Hailwood (1979) and (2) proposed interpretation. Biozones from Müller (1979).

covery was very poor which hampers magnetostratigraphic interpretation (Hailwood, 1979). As no normal polarity intervals were identified, it is likely that the NP11/NP12 zonal contact in Core 28 does not represent a normal succession. The interval of reversed polarity with which Biozone NP12 is associated (Core 28–1) may represent Chron C23r or C22r. Both interpretations are given in Figure 9.

Eastern North Atlantic-Goban Spur-Biscay.—

DSDP Hole 400A (47° 23.4′N, 09° 13.3′W).—Located in a half graben in a succession of tilted and rotated fault blocks near the continent/ocean boundary, Hole 400A was drilled at the foot of the Meriadzek escarpment of North Biscay in 4399-m water depth (Montadert and Roberts, 1979c).

A comprehensive interpretation of the stratigraphy of the 100-m thick lower to middle Eocene section recovered can be derived from integration of magneto- and biostratigraphies (Fig. 13). Calcareous nannofossils (Müller, 1979) provide the main biostratigraphic framework; the planktonic foraminifera being generally poorly preserved (Krasheninnikov, 1979). Despite a somewhat discontinuous record, the magnetostratigraphy (Hailwood, 1979) provides key elements for delineating unconformities.

The magnetostratigraphic interpretation proposed here differs strongly from that offered by Hailwood (1979). While the assignment of the normal polarity interval at 607.99 m to Chron C24An remains to be firmly established, it appears that the normal polarity interval at 589.45 m can be confidently assigned to Chron C23n. The interval of greater importance in this discussion is between 550 and 570 m. Two normal polarity intervals separated by a short reversed interval around 559.08 m

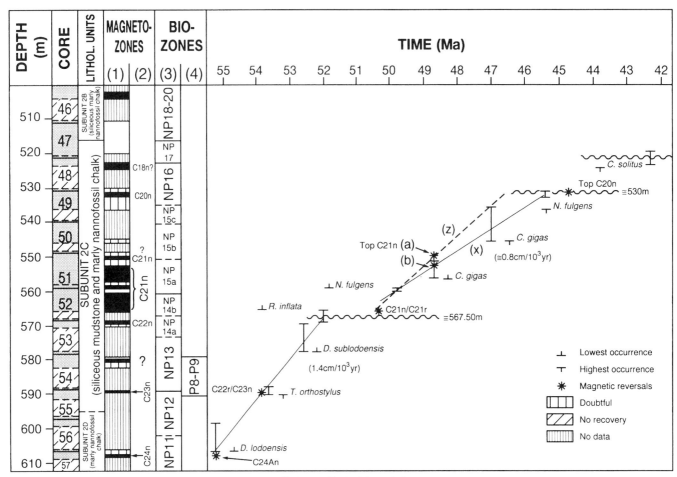

DSDP SITE 400

FIG. 13.—Stratigraphic interpretation of the lower and middle Eocene section recovered from DSDP Hole 400A and sedimentation rate curve. Three unconformities are inferred from calcareous nannofossil stratigraphy. The younger, at ~515 m, is at the contact between Subunits 2B and 2C. The next, at ~528 m, is reflected by the close juxtaposition of the last occurrences of *N. fulgens* and *C. solitus*. The oldest occurs at ~567.50 m as reflected by an inflexion of the sedimentation rate curve and the juxtaposition of the LO of *R. inflata* and the base of a normal polarity interval assignable to Chron C21n. Lithologic units are from Montadert and Roberts (1979c). Biostratigraphic datums and magnetic reversals used to constrain the sedimentation rate curve given in Appendix D, Table 5. (1) from Hailwood (1979, Table 1), (2) proposed re-interpretation, (3) from Müller (1979) and (4) from Krasheninnikov (1979). Equivalence to "P" Zones based on Berggren and Miller (1988, Text Fig. 1). (a) and (b) are different locations of the Chron C20r/C21n boundary depending of the interpretation of the magnetic reversal pattern.

occur between 553.77 and 565.94 m. No magnetic data are available between 565.94 and 568.94 m because of no recovery in Core 52, section 6. Normal polarity characterizes level 568.94 m followed by questionable reversed polarity at 569.95 m. No magnetic data are available for the underlying 10 m. These three intervals with normal polarity were regarded as a single magnetozone and interpreted as Chron C22n (Hailwood, 1979; in fact the thin normal polarity interval at 550.50 m was also included in this chron). Correlation with calcareous nannofossil stratigraphy indicates that this interval corresponds to 2 different chrons. The interval of essentially normal polarity between 553.77 and 565.94 m straddles the NP14b/NP15a subzonal boundary and thus represents Chron C21n. The normal polarity interval at 568.94 m on another hand correlates with Subzone NP14a and thus represents Chron C22n. The normal polarity interval at 550.50 m may be part of Chron C21n. The

two normal polarity intervals at 523.94 and 531.85 m may represent respectively Chron C18n and C20n.

Based on this reinterpretation of the magnetostratigraphy, an unconformity is delineated at ~567.50 m within Zone NP14 (Fig. 13). Its upper surface is within Chron C21n or close to the Chron C21n/C21r reversal boundary. Its lower surface is within Chron C22n or very close to the Chron C21r/C22n reversal boundary. The lower Eocene section below the unconformity appears to be continuous as well as the middle Eocene section above it. However, the middle Eocene section is difficult to interpret because of the poor magnetostratigraphic record and the poorly located nannofossil datums. As a result, the sedimentation rate curve for the middle Eocene interval is poorly constrained and at least 2 models are possible (Fig. 13). Model (x) suggests low sedimentation rates which in turn may reflect undeciphered unconformities. This model suggests the presence of an unconformity at ~530 m. Model (z) suggests higher sedimentation rates and possibly an unconformity within Zone NP15. In the absence of definitive evidence, model (x) is followed (Fig. 14) and an unconformity is delineated in the upper part of Subzone NP15c at a level such that Chron C20n is truncated.

DSDP Site 401 (47° 25.65′ N; 08°.62 W).—An 80-m thick lower to middle Eocene section was recovered from DSDP Site 401 (Fig. 15) which lies on the edge of a tilt fault block which underlies the southern edge of the Meriadzek terrace on the North Biscay margin, in 2495-m water depth (Montadert and Roberts, 1979d).

Planktonic foraminifera (Krasheninnikov, 1979) and calcareous nannofossils (Müller, 1979) provide a strong stratigraphic control for the lower Eocene section, while a few additional tie-points are gained from an otherwise uninterpretable magnetic polarity record (Hailwood, 1979). The magnetostratigraphy is partly reinterpreted based on correlation with calcareous nannofossil stratigraphy. In agreement with Hailwood (1979), the Chron C23n/C22n reversal boundary occurs between 181.27 and 180.47 m. However, the normal polarity interval at 171.72 m, at a level where *R. inflata* and *D. sublodoensis* co-occur (Core 11–2, 22–23 cm, see Müller, 1979), should be assigned to Chron C21n rather than to Chron C22n.

Good agreement is observed between early Eocene chronologic markers suggesting stratigraphic continuity of the lower Eocene section. It is obviously truncated by an unconformity at ~172 m as indicated by the simultaneous lowest occurrences of *D. sublodoensis* and *R. inflata* and their proximity to the lowest occurrence of *N. fulgens*. The middle Eocene section is characterized by a thick Subzone NP15b while Subzone NP15a is extremely thin. A lithologic boundary occurs at 171.50 m (Montadert and Roberts, 1979d), above the Zone NP13/Subzone NP14b unconformity. I suggest that this lithologic boundary reflects an unconformity between Zones NP14 and NP15. Two closely spaced unconformities thus occur around the lower/middle Eocene boundary at Site 401. The upper and the lower surfaces of, respectively, the upper and the lower unconformities are easy to locate. The upper surface of the Subzones NP14b/NP15a unconformity lies very close to the NP15a/NP15b subzonal boundary while the lower surface of the Zone NP13/Subzone NP14b unconformity lies close to the top of Zone NP13 (as determined through extrapolation of the early Eocene sedimentation rate curve). It is more difficult to locate the lower and the upper surfaces of, respectively, the upper and lower unconformities. Both are within Chron C21n.

A third unconformity is present in the section as indicated (weakly) by the anomalously low highest occurrence of *N. fulgens*, and (strongly) by the superficial alignment of the P11/P12 and NP15c/NP16 zonal boundaries (see Berggren and others, 1985). The hiatus is at least 0.5 my long and may be as long as 1.6 my. While the age of the lower surface can be established by extrapolation, the age of the upper surface cannot be determined without a reliable magnetic polarity record. This is the only surface arbitrarily located in Figure 14.

DSDP Sites 550, 549 and 548.—DSDP Sites 550 (Graciansky, Poag and others, 1985a), 549 (Graciansky and others, 1985b) and 548 (Graciansky and others, 1985c) are part of the Goban Spur transect (Graciansky and others, 1985). While Sites 548 and 549 were drilled on the seaward edge of a tilted block of

Hercynian basement, Site 550 lies on the Porcupine abyssal plain, 10 km southwest of the seaward edge of the Goban Spur. *DSDP Site 550 (48° 30.91′ N, 13° 26.37′ W).*—Lower to middle Eocene, ~50-m thick marly nannofossil chalks, unconformable with an Oligocene chalk, were recovered from DSDP Hole 550 drilled in 4420-m water depth. The unconformity at 310.24 m (Core 24–2, 34 cm) is marked by a series of manganese-rich black crusts in Core 24–2, 45–75 cm (Graciansky and others, 1985a).

Integration of magnetostratigraphy and calcareous nannofossil stratigraphy allows a comprehensive interpretation of the stratigraphy of the lower Eocene part of the section (Fig. 16). A remarkable agreement between magnetic reversal boundaries (Townsend, 1985) and calcareous nannofossil datums (Müller, 1985) in Cores 29 to 25 suggests continuous sedimentation at a constant rate of ~1.1 cm/10^3 yr during Biochrons NP11 and NP12.

Less straightforward is the stratigraphic interpretation of the upper part of the section. Townsend (1985) identified the interval with normal polarity which extends from 315.51 to 318.69 m as Chron C22n. This is in agreement with its correlation with Zone NP14 (Aubry, pers. obs., in Miller and others, 1985). The low stratigraphic position of the Chron C22n/C22r reversal boundary (in Core 25–1, between 69–71 cm and 128–130 cm) suggests that a stratigraphic gap occurs between Chrons C23n and C22n, within Zone NP13 and between 319 and 324 m. While the hiatus may be ~0.7 my long, no available data allows the position of the unconformity within this depth interval to be constrained. It is arbitrarily located at 320 m in Figure 16. Because of extremely poor preservation, biozonal subdivision in Core 24 is difficult. *Discoaster sublodoensis* occurs in sample 24–2, 86–87 cm which is almost barren as a result of secondary silicification. It is common in sample 24–2, 144–149 cm to sample 24–4, 54–59 cm, and rare in sample 24–4, 101–106 to 25–1, 12–15. From sample 24–4, 54–59 to 25–1, 12–15, it is associated with abundant *D. lodoensis* (with its 5-rayed morphotype common), which indicates Subzone NP14a. There is a sharp contrast in assemblage composition between samples 24–4, 54–59 cm (313.60 m) and 24–3, 107–112 cm (312.60 m). At this latter level (312.60 m), *D. sublodoensis* is associated with *D. barbadiensis* (abundant), *D. lodoensis* (rare), *D. cruciformis* (few), *T. orthostylus* (rare), *Chiphragmalithus calathus* (few), and *Nannotetrina cristata* (rare). In sample 24–2, 144–149 cm (311.50 m), a similar assemblage occurs but with *C. calathus* rare, *N. cristata* common and *N.* cf. *N. fulgens* rare. The co-occurrence of *C. calathus, N. cristata, T. orthostylus, D. lodoensis* and *D. sublodoensis* suggests massive reworking. The presence of *N. cristata, N.* cf. *N. fulgens* and *Reticulofenestra samodurovi* at 311.5 m suggests an NP15c subzonal assignment, while the co-occurrence of *D. sublodoensis* and *N. cristata* at 312.60 m suggests Subzone NP14b or lower NP15a. It is remarkable that the interval assigned to Subzone NP14b or NP15a yields a normal polarity, which suggests Chron C21n, while the interval assigned to Subzone NP15c is of reversed polarity and may represent Chron C20r. This zonal succession suggests two successive unconformities below the unconformity at 310.24 m. These are part of a series of unconformities described for Core 24, section 1 and upper part of section 2 (see Müller, 1985). There are no data to constrain the hiatuses

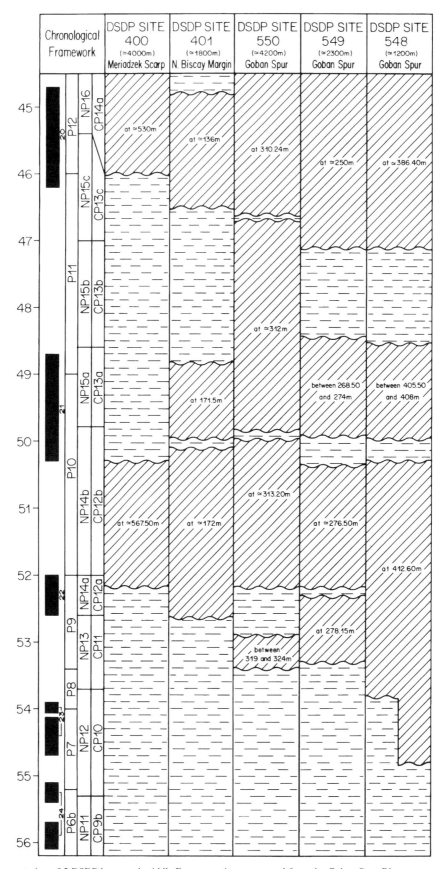

Fig. 14.—Temporal interpretation of 5 DSDP lower and middle Eocene sections recovered from the Goban Spur-Bicay area, eastern North Atlantic Ocean. See Figure 9 for description.

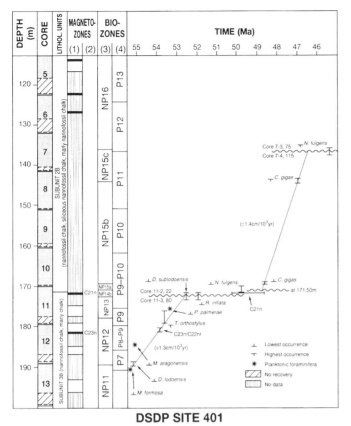

DSDP SITE 401

FIG. 15.—Stratigraphic interpretation of the lower and middle Eocene interval recovered from DSDP Site 401 and sedimentation rate curve(s). Two inferred unconformities occur at ~172.50 m and ~134 m. Lithologic units are from Montadert and Roberts (1979d). Biostratigraphic datums and magnetic reversals used to constraint the curve(s) are given in Appendix D, Table 6. (1) from Hailwood (1979, Table 1), (2) proposed re-interpretation, (3) from Müller (1979) and (4) from Krasheninnikov (1979). Equivalence to "P" Zones based on Berggren and Miller (1988, Text, Fig. 1).

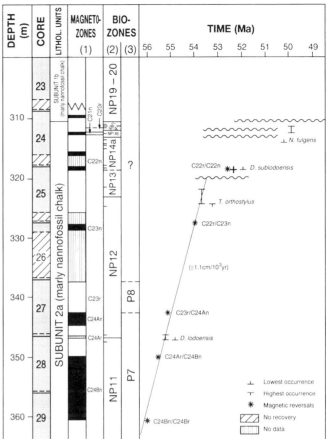

DSDP SITE 550

FIG. 16.—Stratigraphic interpretation of the lower and middle Eocene interval recovered from DSDP Site 550 and sedimentation rate curve. An unconformity is visible at 310.50 m at the boundary between lithologic Subunits 1b and 2a. An inferred unconformity occurs in Core 25. Lithologic units are from Graciansky and others (1985a). Datums used to constrain the curve are given in Appendix D, Table 7. (1) from Townsend (1985), (2) from Müller (1985) and (3) from Snyder and Waters (1985).

associated with these inconformities, and the interpretation in Figure 14 is arbitrary.

DSDP Site 549 (49° 05.28′N, 13° 05.88′W).—Eighty meters of lower and middle Eocene nannofossil chalks were recovered from DSDP Site 549 drilled in 2515-m water depth.

I have discussed the stratigraphic interpretation of this Eocene section (Aubry, 1991), but a few remarks are necessary following the replot of the magneto- and biostratigraphic data (Fig. 17). Calcareous nannofossils provide the main stratigraphic control while an ambiguous magnetostratigraphic record provides a few additional tie-points. While the normal polarity intervals which occur between 241 and 275 m are not interpretable, identification of the normal polarity intervals between 290 and 317 m is somewhat uncertain. The normal polarity interval which extends from 289 to 294.59 m can be confidently identified as Chron C23n (however, only the level where the reversal boundary occurs between Chrons C23n and C23r is known; the position in the core of the Chrons C23n/ C22r reversal boundary is not known). The normal polarity interval in Core 12, sections 3 and 4 may be identified as Chron C24An based on planktonic foraminiferal stratigraphy, but remains of questionable identification based on calcareous nan-

nofossil stratigraphy. The thin interval with normal polarity around 276 m correlates with Zone NP14 and was identified as Chron C22n (Townsend, 1985). It is reassigned here to Chron C21n based on its correlation with Subzone NP14b. The thickness of this interval is not known.

At least four unconformities occur in the middle Eocene section as indicated by the close stratigraphic juxtaposition of datums which are chronologically well separated, by the absence of Subzones NP15a and NP15c and by that of Chrons C21r and C22n.

The older unconformity occurs between Zone NP13 and Subzone NP14a and is marked by the lithologic contrast at 278.15 m, between lithologic Units 2 and 3 (although the contact appeared conformable, Graciansky and others, 1985b). An inferred unconformity occurs within Zone NP14. Its lower surface lies in Subzone NP14a; its upper surface in the lower part of Chron C21n or in the upper part of Chron C21r (because of large sampling intervals, the position of the NP14a/NP14b subzonal boundary cannot be determined accurately). Even though Subzone NP15a could be represented by a thin interval not

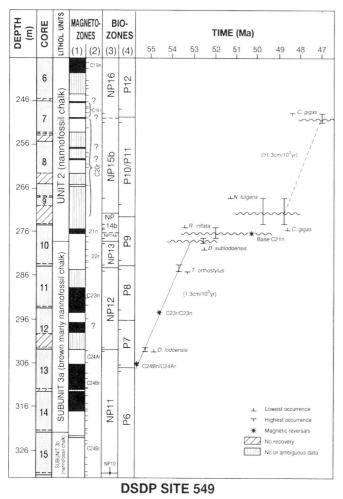

DSDP SITE 549

FIG. 17.—Stratigraphic interpretation of the lower and middle Eocene interval recovered from DSDP Site 549 and sedimentation rate curve(s). All unconformities are inferred. Lithologic units are from Graciansky and others (1985b). Datums used to constrain the curve are given in Appendix D, Table 8. (1) from Townsend (1985), (2) revised interpretation, (3) from Müller (1985), partly re-interpreted and (4) from Snyders and Waters (1985).

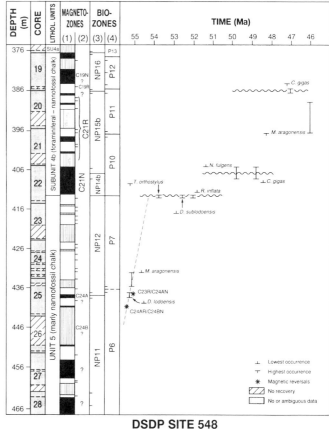

DSDP SITE 548

FIG. 18.—Stratigraphic interpretation of the lower and middle Eocene interval recovered from DSDP Hole 548A and sedimentation rate curve(s). An unconformity is visible at 406.60 m between lithologic Units 4 and 5. Unconformities at ~412.50 m and ~386.50 m are inferred. Lithologic units are from Graciansky and others (1985c). Datums used to constrain the curve are given in Appendix D, Table 9. (1) from Townsend (1985), (2) revised interpretation, (3) from Müller (1985) and (4) from Snyders and Waters (1985). Note that for DSDP Sites 550, 549 and 548, the magnetic reversals are drawn at the levels where normal polarity was identified (consequently, intervals with reversed polarity are probably thinner and intervals with normal polarity thicker than shown).

recovered in Core 9, sections 4 to 6, an unconformity between 268.50 and 274 m is inferred from the close juxtaposition of the lower occurrences of *N. fulgens* and *C. gigas*. The absence of Chron C20n and the anomalous alignment of the P11/P12 and NP15/NP16 zonal boundaries suggest the presence of an unconformity around 250 m. There are little data to constrain the ages of the upper and lower surfaces of these unconformities, so that their position in Figure 14 is somewhat arbitrary.
DSDP Hole 548A (48° 54.95'N, 12° 09.84'W).—DSDP Site 548 is the shallowest site (1248-m water depth) of the Goban Spur transect (Graciansky and others, 1985).

I have discussed and re-interpreted the stratigraphy of DSDP Hole 548A in an earlier paper (Aubry, 1991). A replot of the data (Fig. 18) shows clearly the position of the inferred middle Eocene unconformities, between 405.50 and 408 m and between 386 and 386.80 m. The stratigraphic gaps are so important as to encompass several biozones. It should be noted that several intervals with normal polarity (e.g., around 387 m, 398 m, 465 m) have no equivalent in the magnetic polarity history,

and that identification of others is obscured by long intervals of uncertainty (due to weak magnetic signal and/or poor recovery). It is possible that Chron C23n, not identified, occurs in the interval of no recovery of Core 23. Since the sedimentation rate curve is constrained by only few data (all occurring over a thin stratigraphic interval), it is not possible to determine whether Chron C23n may be present or not. The lower surface of the unconformity at 410.24 m may lie in Chron C23r or in Chron C22r. Both interpretations are given in Figure 14.

Contrary to Townsend (1985) and Aubry (1991), Chron C20n, which straddles the NP15/NP16 zonal boundary, does not occur in DSDP Hole 548A. Subzone NP15c is missing and the alignment of the biozonal boundaries P11/P12 and NP15b/NP16 indicates a stratigraphic gap. The lack of agreement between calcareous nannofossil datums and magnetic reversals in the middle Eocene section from Hole 548A as shown in Aubry (1991, Fig. 12) results from the misidentification of the magnetic reversals in Cores 20 and 21. Both surfaces of the inferred unconformity between 405.50 and 408 m lie within Subzone NP14b and Chron C21n (Fig. 14). The lower surface of the unconformity at ~386.25 m lies in Chron C21r (Fig. 14) while its upper surface lies close to the Chron C19n/C19r reversal boundary (Fig. 18).

Eastern central Atlantic.—

DSDP Site 366 (05° 40.7'N, 19° 51.1'W).—A 140-m thick section of middle lower to middle middle Eocene limestone, nannofossil chalk and porcellanite was recovered from DSDP Site 366 drilled on the Sierra Leone Rise in 2853-m water depth (Lancelot and others, 1978a). Despite the array of correlation tools available, a comprehensive stratigraphic interpretation (Fig. 19) cannot be fully achieved due to several limiting factors. Recovery was poor in critical intervals, in particular in Cores 29 and 30. Magnetostratigraphy is available for only part of the section (Core 27 to 44) and a number of intervals with normal polarity are questionable due to very low intensities of

NMR (Hailwood, 1978). Planktonic foraminiferal assemblages are well-diversified and yield index species in the lower part of the section (Cores 40 to 38 and 31 to 29) but impoverished, poorly preserved microfaunas occur from Cores 37 to 32 and 28 to 23, and zonal assignments in the middle Eocene are questionable (Krasheninnikov and Pflaumann, 1978). Radiolarians are restricted to the upper part of the section (above Core 32) and preservation decreases with increasing depth from Core 23 to 31, so that there is some uncertainty on zonal boundaries (Johnson, 1978). Calcareous nannofossils are rare to common, thickly overgrown (middle Eocene) or partly dissolved and partly overgrown (lower Eocene) so that not all zonal boundaries could be precisely delineated. However, and although no range chart is given, high confidence is placed in this group because of the excellent agreement between the zonal subdivision established onboard ship (Lancelot and others, 1978a) and the onshore study by Bukry (1978a).

While there is a general agreement between the biostratigraphic subdivisions based on different paleontologic groups, there are also obvious discrepancies, mainly centered around Cores 29 and 30 (Fig. 19). Correlation between radiolarian Zone 22, planktonic foraminiferal Zone P9 and calcareous nannoplankton Subzone NP15a in Core 29 (cf. Berggren and others, 1985) suggests mixing through reworking or intensive burrowing.

The magnetic reversal pattern in Core 27 through 34 is not straightforward and differs from what would be expected from magnetobiostratigraphic correlations as proposed in Berggren and others (1985). Normal polarity is dominant, and no thick reversal which would correlate with Zones NP14, P9 and R22 or 23 is observed, suggesting a stratigraphic gap. Hailwood (1978) identified the interval with predominantly normal polarity in Core 27 to 29 as Chron C21n. This is in agreement with calcareous nannofossil stratigraphy. Correlation with Subzone NP15a indicates that this interval represents the younger part of Chron C21n. Hailwood (1978) identified the interval with predominantly normal polarity in Core 32 to 34 as Chron C22n. This is also in agreement with calcareous nannofossil stratigraphy, indicating in turn the lower part of Zone NP14 (Subzone NP14a) as supported by the co-occurrence of *D. sublodoensis* and *D. lodoensis* in Core 33 (Bukry, 1978a). The normal polarity intervals in Cores 31 and 30 have inclination values which do not significantly differ from zero and may be questionable (Hailwood, 1978). Yet, the normal interval in Core 31 is thick, well-delineated (four samples were analyzed), and its correlation with Zones NP14, P9 and R23 suggests that it may also be part of Chron C22n. In the absence of decisive data which would support an upwards extent of Chron C22n, the Chron C21r/C22n boundary is drawn for the purpose of this study between sections 1 and 2 of Core 31. The normal interval in Core 30 is thin and may not have significance. No recovery in Core 29, sections 5 and 6 and in Core 30 precludes a definitive interpretation of the reversed intervals in Core 29–4, 145–147 cm and 31–1, 82–84 cm as part of Chron C21r, and the biostratigraphic data remain inconclusive either because insufficient (calcareous nannofossils) or contradictory (planktonic foraminifera and radiolarians). Based on the weak support provided by radiolarian stratigraphy, the reversal in Core 29–4, 145–147 cm is regarded here as part of Chron C21r.

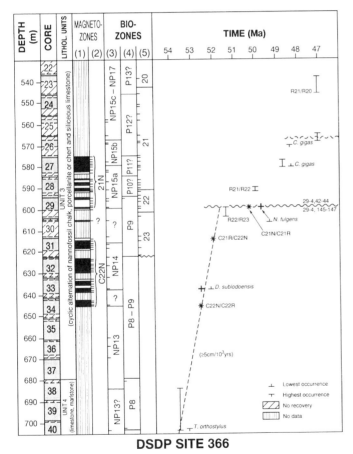

DSDP SITE 366

Fig. 19.—Stratigraphic interpretation of the lower and middle Eocene interval recovered from DSDP Site 366. Lithologic units are from Lancelot and others (1978a). Datums used to constrain the curve are given in Appendix D, Table 10. (1) from Hailwood (1978), (2) revised interpretation, (3) combined from Bukry (1978a) and Lancelot and others (1978a), converted to Martini's (1971) zonal scheme, (4) from Krasheninnikov and Pflaumann (1978) and (5) from Johnson (1978). Tick marks in (2), (3), (5) indicate levels analyzed. The unconformity at 599.50 m is inferred from magneto- and calcareous nannofossil stratigraphy and at ~560.70 m is inferred from calcareous nannofossil stratigraphy alone.

Based on the integration of this revised magnetostratigraphy with biostratigraphy, it appears that sedimentation was continuous through the late early Eocene Epoch (Core 40 to 31), although the sedimentation rate curve is poorly anchored, the highest occurrence of *Tribrachiatus orthostylus* in the section being poorly constrained, and the Chron C22n/C22r boundary being indeterminate. An unconformity at ~599.50 m is inferred from the stratigraphic juxtaposition of three events which are temporally well separated. It should be noted that while the position of this unconformity can be debated and depends upon the magnetostratigraphic interpretation chosen, the thinness of Chron C21r (minimal in the interpretation chosen here) in itself testifies to the presence of an unconformity, as mentionned above. According to the interpretation given in Figure 19, both Chron C21r (at the top) and Chron C21n (at the base) are truncated. The upper part of the section is extremely difficult to interpret because of lack of data and of agreement between the biozonal assignments derived from the three paleontologic groups. The thinness of Subzone NP15b is suggestive of an unconformity either between 568 and 580 m or at 580 m. This latter interpretation is followed in Figure 20.

DSDP Site 370 (32° 50.2′ N; 10° 46.6′ W).—DSDP Site 370 lies in the deep basin off the Atlantic coast of Morocco, very close to the base of the Moroccan continental slope in 4214-m water depth (Lancelot and others, 1978b).

About 50 m of lower to middle Eocene calcareous silty claystone were recovered from this site, overlying unconformably upper Paleocene nannofossil marls and clays. The interpretation of this stratigraphic interval is extremely limited because of large coring gaps, poor recovery, lack of magnetostratigraphy, and poor biostratigraphic control (Fig. 21). Radiolarians are not identifiable below Core 12 (Johnson, 1978); planktonic foraminiferal assemblages are poor (Krasheninnikov and Pflaumann, 1978); and calcareous nannofossils are rare to common and moderately to poorly strongly etched and fragmented (Lancelot and others, 1978b; Bukry, 1978a). In addition, there are enormous discrepancies between zonal assignments derived from different paleontologic groups, which probably reflect the turbiditic nature of the sediments and the presence of numerous slumps (Lancelot and others, 1978b). Whereas the interval from Core 18 to 13 is assigned to the undifferentiated planktonic foraminiferal Zones P8-P9 of the lower Eocene (Krasheninnikov and Pflaumann, 1978), Core 15 is assigned to the middle Eocene Zone NP15 (Lancelot and others, 1978b; Bukry, 1978a) and Core 12 is regarded of earliest middle Eocene age (Biochron R22) based on radiolarian occurrences (Johnson, 1978). Subzone NP14a constitutes a thick interval at DSDP Site 370, which contrasts with the thinness of Zone NP15 and the apparent absence of Subzone NP14b. Considering that there is no lithologic change in this turbiditic lower and middle Eocene interval, it is likely that an unconformity occurs between Cores 15 and 16, as indicated by the superposition of Zone NP15 and Subzone NP14a. As it is not known which Subzone of Zone NP15 is present, there may be only one stratigraphic gap in the section (encompassing Subzones NP14b, NP15a, NP15b and perhaps part of Subzone NP15c) or two successive gaps (the older encompassing Subzone NP14b and perhaps NP15a, the younger Subzone NP15c). Considering the poor record for this site, the simplest interpretation (a single unconformity) is given in Figure 20.

DSDP Site 398 (40° 57.6′N, 10° 43.1′W).—DSDP Site 398 is situated on the southern apron of Vigo seamount off the west coast of the Iberic peninsula in 3910-m water depth (Sibuet and others, 1979). An 80-m thick lower to middle Eocene section was recovered from Hole 398D, which comprises rythmically bedded siliceous marly nannofossil chalk and mudstone with thin intercalated sands and silts and with frequent slumps (Fig. 22). Because of its turbiditic nature, the stratigraphic interpretation of this lower to middle Eocene interval based on biostratigraphy alone is open to discussion. However, the magnetostratigraphic data available (Morgan, 1979), although scarce, contribute greatly to resolve the discrepancies between planktonic foraminiferal (Iaccarino and Premoli Silva, 1979) and calcareous nannofossil (Blechschmidt, 1979) zonal assignments.

Stratigraphic interpretation of the lower part of the section, between Cores 35 and 32, is straightforward, and the sedimentation rate curve based on integrated planktonic foraminiferal, calcareous nannofossils, and magnetostratigraphy suggests stratigraphic continuity. The occurrence of an unconformity at ~708 m (?708.07 m; HO of *T. orthostylus* in Core 32–1, 7–8 cm and LO of *Globigerinatheka* spp. in Core 32–1, 5–6 cm) is well delineated by the juxtaposed occurrence of the highest occurrence of *T. orthostylus* and lowest occurrences of *D. sublodoensis* and *Globigerinatheka* spp. Iaccarino and Premoli Silva (1979) recognized the presence of a stratigraphic gap in this section but draw the unconformity between Cores 33 and 32, estimating the gap to correspond to Zones P9 and P10. No planktonic foraminiferal zonal assignment was provided for Core 32 below sample 32–1, 5–6 cm (attributed to Zone P11), but a middle Eocene age was suggested. Since no firm data from the planktonic foraminifera indicate a definitive middle Eocene age for Core 32, except for its uppermost centimeters, and since there is a reasonably good correlation between calcareous nannofossil and magnetostratigraphy from Core 35 to 32, placement of the unconformity at the top of Core 32 rather than of Core 33 is preferred.

The sediments immediately above the unconformity are assigned to Zone NP14 (Blechschmidt, 1979) which is incompatible with a P11 zonal assignment (Iaccarino and Premoli Silva, 1979). Planktonic foraminiferal stratigraphy of the middle Eocene interval is hampered by very poor preservation at most levels, absence of index species and reworking (Iaccarino and Premoli Silva, 1979). The P11/P12 zonal boundary was not recognized and the base of Zone P13 was only approximated. Likewise, recognition of Zone P11 in Core 32 was not based on the occurrence of the marker species (*G. kugleri*) but approximated by the presence of *Globigerinatheka* spp. This is insufficient evidence for Zone P11, the genus *Globigerinatheka* being already represented in Zone P10 and possibly P9 (Toumarkine and Luterbacher, 1985). Reassignment of Cores 31 to 28 to Zone P10 is agreeable with assignment to Zone NP14 and with association with a thick interval of reversed polarity confidently assigned to Chron C21r.

The NP14/NP15 zonal boundary falls within Chron C21n (cf. Berggren and others, 1985). Core 26 and 27 are assigned to Zone NP15 and yield a predominantly normal polarity. Although magnetostratigraphic data from Cores 28 and 29 are scarce, it seems that the interval immediately below the NP15/NP14 zonal boundary may be of reversed rather than of normal polarity, which suggests that Zones NP15 and NP14 may be

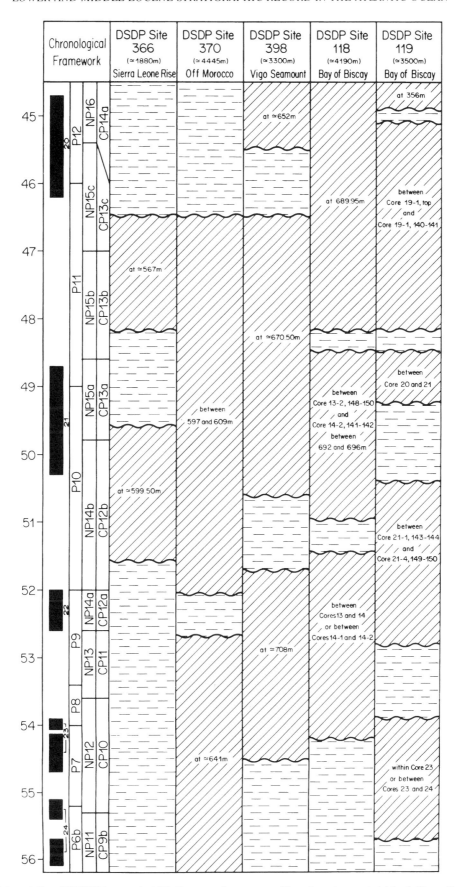

FIG. 20.—Temporal interpretation of 5 DSDP lower and middle Eocene sections recovered from the eastern central Atlantic Ocean. See Figure 9 for description.

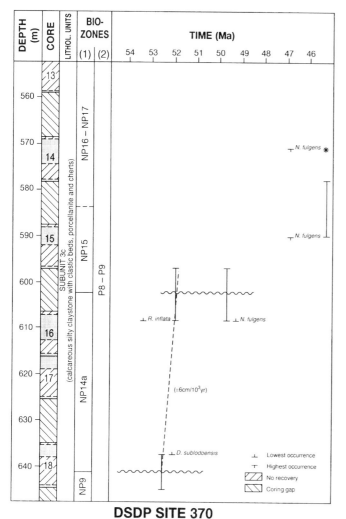

DSDP SITE 370

FIG. 21.—Stratigraphic interpretation of the lower and middle Eocene interval recovered from DSDP Site 370. See text for further explanation. Lithologic units are from Lancelot and others (1978b). Datums used to constrain the
curve are given in Appendix D, Table 11. (1) from Bukry (1978a) and Lancelot
and others (1978b), converted to Martini's (1971) zonal scheme, and (2) from
Krasheninnikov and Pflaumann (1978).

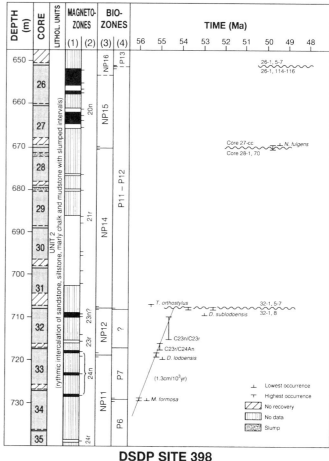

DSDP SITE 398

FIG. 22.—Stratigraphic interpretation of the lower and middle Eocene interval recovered from DSDP Hole 398D. Lithologic units are from Sibuet and
others (1979). Datums used to constrain the curve are given in Appendix D,
Table 12. (1) from Morgan (1979), (2) proposed interpretation, (3) from Blechschmidt (1979) and (4) from Iaccarino and Premoli Silva (1979). Tick marks in
columns (2), (3) and (4) indicate positions of samples analyzed.

unconformable. The presence of a stratigraphic gap at this level
is strongly supported by the low sedimentation rate of less than
2 cm/10^3 yr calculated for Zone NP14 (Cores 31 to 28), a value
anomalously low considering the intense reworking and the
presence of slumps in the section. Determining the extent of
the stratigraphic gap is, however, extremely difficult. *Chiasmolithus gigas* was not recorded in the interval assigned to Zone
NP15, which suggests that either Subzone NP15a or Subzone
NP15c is represented in Cores 26 and 27. Both subzones are
associated with normal polarity, Subzone NP15a with Chron
C21n (upper part), Subzone NP15c with Chron C20n (lower
part). The interval with predominantly normal polarity in Cores
26 and 27 thus represents either Chron C21n or Chron C20n.
Based on the high sedimentation rate expected for this turbiditic
section, this latter interpretation is chosen in Figure 20. Finally,
the proximity of the base of Zone P13 to the NP15/NP16 zonal

boundary (and to Chron C20n) suggests a third unconformity
at the top of Core 26.
DSDP Site 118 (45° 02.65'N, 9° 00.63'W).—DSDP Site 118
lies in the western part of the abyssal plain of Biscay in 4901-
m water depth (Laughton and others, 1972a). About 20 m of
upper lower and middle Eocene clays unconformable with
lower Miocene clays were recovered. Poor recovery, extensive
coring gaps, and evidence of reworking among the calcareous
nannofossils hamper a satisfactory interpretation of the stratigraphy of Cores 12 to 14 based on planktonic foraminifera
(Berggren, 1972a; Iaccarino and Premoli Silva, 1979) and calcareous nannofossils (Laughton and others, 1972a, Perch-Nielsen, 1972; Bukry, 1972).
A lower/middle Eocene unconformity between Cores 13 and
14 (Berggren, 1972a) or between sections 1 and 2 of Core 14
(Iaccarino and Premoli Silva, 1979) results in Zone P10 overlying Zone P7. The zonal succession NP14-NP15b over a 5-m
thick interval suggests at least one additional unconformity between Zones NP14 and NP15 (continuous sedimentation would
have occurred at an unlikely rate of 0.2 cm/10^3 yr). In the ab-

sence of further constraints on their ages, location of the unconformities in Figure 20 is arbitrary.

DSDP Site 119 (45° 01.90′N, 7° 58.49′W).—DSDP Site 119 was drilled on Cantabria Seamount in the Bay of Biscay in 4447-m water depth. About 40 m of lower and middle Eocene clays unconformable with lower Oligocene silty clays were recovered (Laughton and others, 1972b). Interpretation of the stratigraphy of Cores 19 to 24 is hampered by poor recovery and conflicting zonal assignments. Whereas Berggren (1972a) indicates that the planktonic foraminifera were too rare and poorly preserved to allow biozonal subdivision, Iaccarino and Premoli Silva (1979) propose that the section is continuous and extends from Zone P6 to P12. Whereas Laughton and others (1972b) assigned Core 22 to 19 to Zone NP15, Bukry (1972) recognized Zones NP13, Subzone NP14b and Subzone NP15b respectively in Core 21–4, 149–150 cm , Core 21–1, 143–144 cm and Core 19–1, 140–141 cm. Whereas Bukry (1972) and Laughton and others (1972b) agree on a zonal assignment of Core 23–1 to Zone NP12, Iaccarino and Premoli Silva (1979) indicate Zone P9. Despite these discrepancies, it is obvious that this section is riddled with unconformities. The lowest occurs in Core 23 where Zone NP12 or P9 overlaps Zone P6. A second unconformity occurs in Core 21 as indicated by the absence of Subzone NP14a. A third unconformity occurs between sections 1 and 2 of Core 19 as indicated by the zonal successions NP15b–NP16 and P11–P12. The thinness of the interval assigned to Subzone NP14b–Zone NP15(a?) may also suggest an unconformity in Core 19 or 20. Since there are no sufficient paleontologic data to determine precisely the extent of these stratigraphic gaps, the interpretation given in Figure 20 is arbitrary (the gap between Zones NP14 and15 being speculative).

Labrador Sea.—

ODP Site 647 (53° 19.9′N, 45° 15.7′W).—ODP Site 647 was drilled on the south flank of the Gloria Drift in the southern Labrador Sea in 3861.8-m water depth (Srivastava and others, 1987). One hundred and thirty meters of lower to middle Eocene nannofossil-bearing claystone were recovered from Hole 547A. Despite poor recovery, coring gaps and a 30-m interval barren of calcareous microfossils, calcareous nannofossil zonal subdivision (Firth, 1989) and dinocyst stratigraphy (Head and Norris, 1989) contribute to a comprehensive interpretation of the stratigraphy of this lower to middle Eocene interval (Fig. 23). Calcareous nannofossil Zones NP11, NP12 and NP13 were recognized (Core 71 to 66) separated from Subzone NP15b (Core 63–1, 36–38 cm) by a 30-m thick barren interval. Because of poor or no recovery in Cores 60 to 57, biozonal subdivision of this interval is uncertain and is not discussed here.

Good agreement between calcareous nannofossil zonal assignments and dinocyst stratigraphy in the lower Eocene record of this hole as discussed in Head and Norris (1989) suggests that stratigraphic continuity for this interval may be a reasonable assumption (Fig. 23). This, in turn, suggests that at least one unconformity occurs between Cores 65 and 63–1, 38 cm, in the interval barren of calcareous nannofossils. The alternative would be a decrease in the sedimentation rates (to ~4 mm/10³ yr) in this interval as questionably suggested by Srivastava and others (1987). Dinoflagellate cysts, well to moderately preserved in Core 65 to 63–1, 140 cm confirm that this interval comprises a stratigraphic gap (Head and Norris, 1989). These

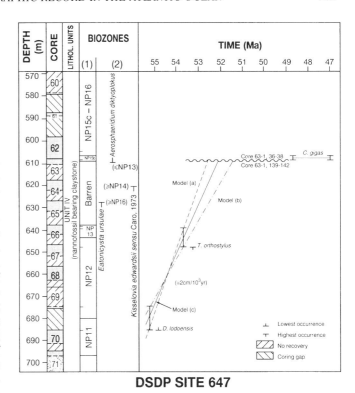

DSDP SITE 647

Fig. 23.—Stratigraphic interpretation of the lower and middle Eocene interval recovered from ODP Hole 647A. Lithologic units are from Srivastava and others (1987). Datums used to constrain the curve are given in Appendix D, Table 13. (1) from Firth (1989) and (2) from Head and Norris (1989).

authors suggest two possible locations for a gap, either between Core 66R-1, 66–69 cm and 65R-1, 117–121 cm, based on a sudden influx of species in Core 65R-1, 117–121 cm, or between Core 64R-3, 88–91 cm and 63R-1, 139–142 cm, based on the apparent truncated ranges of *Aerosphaeridium diktyoplokus* and *Kisselovia edwardsii sensu* Caro, 1973. While it seems well-established that the lowest occurrence of *A. diktyoplokus* is in Zone NP13, a stratigraphic range for *K. edwardsii sensu* Caro, 1973, corresponding to calcareous nannofossil Zones NP12 to NP14, seems more provisional (see discussion and Table 2 in Head and Norris, 1989). The lack of stratigraphic overlap between the two species in Hole 647A is difficult to explain if the former appears in Zone NP13 while the latter disappears in Zone NP14. Since the upper range of *K. edwardsii sensu* Caro, 1973 is poorly established, the lowest occurrence of *A. diktyoplokus* may be the most significant stratigraphic event in this section, which would suggest that Core 65R to 63R-1, 140 cm belongs to Zone NP13 equivalent. This would imply that an unconformity occurs in Core 63–1 between 36 cm and 142 cm, resulting in the juxtaposition of Zone NP13 and Subzone NP15b. This is compatible with the stratigraphic extent of Zone NP13 as deduced through extrapolation from the sedimentation rate curve if this latter is drawn using the extreme limits of uncertainty (model (a) in Fig. 23). The unconformity would occur at the very top of Zone NP13. However, because of poor recovery, the coring gap between cores 69 and 70, and large sampling intervals, there is a large uncertainty in the depth at which the calcareous nannofossil datums which constrain the curve occur. Using the known limits of occurrences of *D. lo-doensis* and *T. orthostylus* to constrain the curve (model (b) in Fig. 23) results in assigning the barren interval to Zone NP 13 and Subzone NP14b (partim) equivalent. While using model (a) the stratigraphic gap comprises all of Zone NP14 (= Subzones NP14a and NP14b), Subzone NP15a and part of Subzone NP15b, based on model (b) it includes part of Subzone NP14b, Subzone NP15a and part of Subzone NP15b.

The interpretation given in Figure 24 is based on a compro-mise between models (a) and (b), in which the nannofossil-barren interval is equivalent to Zone NP13 and Subzone NP14a (model (c) in Fig. 23). *Chiasmolithus gigas* occurs in only one sample (63R-1, 36–38 cm). Because other low to mid latitude species (e.g., *Nannotetrina cristata*) are sporadic in Hole 647A, it is uncertain whether this occurrence constitutes a reliable cri-terion to date the younger surface of the unconformity. In the absence of other chronologic criteria, this is arbitrarily followed in Figure 24. Finally, it must be emphasized that the interpre-tation made for the stratigraphy of Cores 65 to 63 involves the simplest sedimentary history, but that complex scenarios cannot be ruled out.

DSDP Site 112 (54° 01.00′ N, 46° 36.24′ W).—Firth (1989) sug-gested that the interval barren of calcareous nannofossils in Core 65 to 63 of ODP Hole 647A correlates with an Eocene interval recovered from DSDP Site 112, also drilled in the southern Labrador Sea in 3657-m water depth (Laughton and others, 1972c). Discontinuous coring at Site 112 resulted in the recovery of a single core (Core 15) of lower to middle Eocene nannofossil clay. Only 2.20 m of sediments were recovered (1 m in Core 15–1, 70 cm in Core 15–6, and 50 cm in Core 15-cc). Core 15 was dated as latest early Eocene based on plank-tonic foraminifera, and Core 15-cc was assigned to Zone P9 (Berggren, 1972a). On the basis of calcareous nannofossils, Core 15–1, 47–48 cm, was assigned to Zone NP 14 by Bukry (1972), and the entire core was assigned to the same zone by Perch-Nielsen (1972). Co-occurrence of *D. sublodoensis* and *R. inflata* (Bukry, 1972; Perch-Nielsen, 1972) and correlation with planktonic foraminifera indicates that Core 15 belongs to the basal part of Subzone NP14b. Firth (1989) revised the stratig-raphy of Core 15 and suggested that it yields the NP14b/NP15 zonal boundary (but see contradictory interpretations in text, Fig. 2 and Table 5). Based on the combined reports by Bukry (1972) and Firth (1989), this boundary occurs between Core 15–1, 5–6 cm and Core 15–1, 47–48 cm. Since the interval from Core 15-cc to Core 15–1, 47–48 cm is assigned to the lower Eocene Series (Berggren and others, 1972c), assignment of the upper few centimeters of Core 15 to the middle Eocene Zone NP15 would imply a stratigraphic gap corresponding to the lower middle Eocene Series. Reassignment of the upper part of Core 15 to Zone NP15 (Firth, 1989, Fig. 5), based on negative evidence, is however unwarranted. The marker for the base of Zone NP15 was not found, nor any other species which could be indicative of this zone. *Rhabdosphaera inflata* is rare in Core 15–6, 110–114 cm, and its absence in Core 15–1, 6–10 cm (see Firth, 1989, Table 5) may reflect its scarcity at high latitude. In Core 62 of ODP Hole 647A, we note also the scarcity and discontinuous occurrence of temperate and low latitude taxa (see Firth, 1989, Table 4). Simply applying Martini's (1971) zonal definition the presence of *Discoaster sublodoensis* and

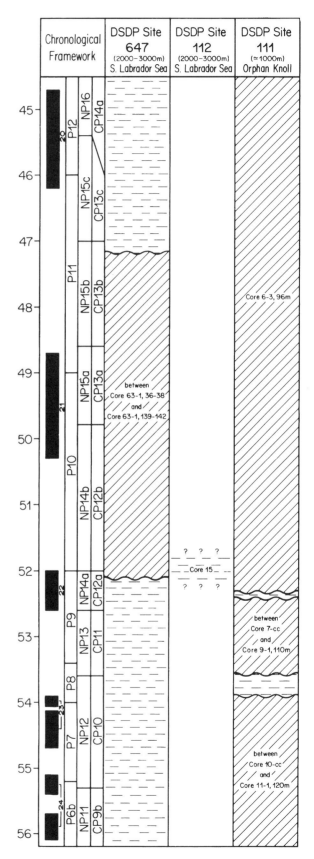

FIG. 24.—Temporal interpretation of 3 DSDP and ODP lower and middle Eocene sections recovered from the Labrador Sea. See Figure 9 for description.

the absence of *Nannotetrina fulgens* in Core 112–15–1 characterize Zone NP14.

The uppermost lower Eocene sediments recovered in Core 15 at DSDP Site 112 may or may not correlate with the interval barren of calcareous nannofossils recovered in Cores 65 to 63 in ODP Hole 647A (Fig. 24). If they do, they would correlate only with the upper part of this interval unless an unconformity occurs at the base of the nannofossil barren interval. Core 112–15 yields an NP14b subzonal assignment, while the interval from Cores 63 to 65 in Hole 647A likely yields a stratigraphic position equivalent to Zone NP13 and Subzone NP14a.

DSDP Site 111 (50° 25.57'N, 46° 22.05'W).—DSDP Site 111 was drilled on Orphan Knoll at the edge of the continental rise northeast of Newfoundland in 1797-m water depth (Laughton and others, 1972d). About 35 m of Eocene marls and clays (Core 6–3 to 10-cc) sandwiched between upper Miocene oozes and Maestrichtian chalks were recovered from Hole 111A (Fig. 25). The upper part of this Eocene section is highly discontinuous and at least 2 unconformities occur in Core 6, section 3, one separating upper Eocene (Zone NP19–20) from middle Eocene (Zone NP16) Series, the other separating middle Eocene from uppermost lower Eocene (Zone NP14) Series (see Laughton and others, 1972d). It is clear from Figure 25 that unconformities also occur in the lower part of this Eocene section, as indicated by the thinness of the interval assigned to Zone NP13. One or more unconformities may occur in Cores 8 and 9. Planktonic foraminiferal assemblages do not allow precise zonal determination (Berggren in Laughton and others, 1972d) so that there is no possibility to determine whether a single unconformity or more occur and whether it is located within Zone NP13, at the NP13/NP12 zonal boundary, or at the NP13/NP14a zonal/subzonal boundary. Stratigraphic interpretation in Figure 24 is thus arbitrary, assuming continuity from Zone NP12 to NP13 and an unconformity at the NP13/NP14 zonal boundary.

Western North Atlantic.—

DSDP Site 613 (38° 46.26'N, 72° 30.43'W).—DSDP Site 613 is located in the offshore part of the Baltimore Canyon Trough on the New Jersey continental rise in 2309-m water depth (Poag and others, 1987a). A 300-m thick lower to middle Eocene section of chalk was recovered from this site, unconformable with upper Miocene siliceous mud. The stratigraphic interpretation of the lower to lower middle Eocene part of this section was discussed in Aubry (1991). This interpretation is extended to the upper part of the middle Eocene section (Fig. 26) and radiolarian (Palmer, 1987) and planktonic foraminiferal (Miller and Hart, 1987) stratigraphies are integrated with calcareous nannofossil stratigraphy (Poag and others, 1987a; Valentine, 1987; Bukry, 1987). While there are discrepancies between zonal assignments based on the radiolarians and calcareous microfossils (which suggest a need for revising the calibrations of middle Eocene radiolarian events to the GPTS), there is good agreement between calcareous nannofossil and planktonic foraminiferal zonal subdivisions. The two unconformities at ~482 m and ~437.9 m discussed in Aubry (1991) are well-delineated in Figure 26. The latter is associated with a slump zone between 437.92 and 438.70 m (Poag and others, 1987a). For the reasons given in Aubry (1991), the precise ages of these unconformities are difficult to determine, and the same arbitrary

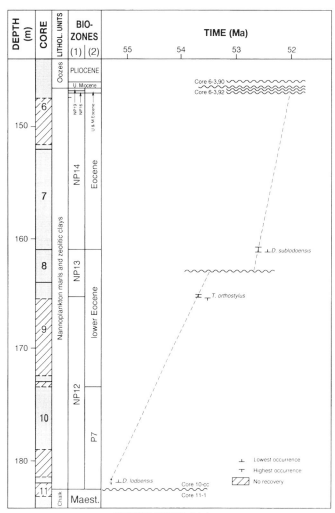

DSDP SITE 111

FIG. 25.—Stratigraphic interpretation of the Eocene section recovered from DSDP Hole 111A. Datums used to constrain the curve are given in Appendix D, Table 14. Lithologic units are from Laughton and others (1972). (1) from Laughton and others (1972d) and Perch-Nielsen (1972) and (2) from Berggren in Laughton and others (1972d).

choices made by Aubry (1991, Fig. 6) are followed in Figure 27.

DSDP Site 612 (38° 49.21'N, 72° 46.43'W).—DSDP Site 612 is located on the middle part of the New Jersey continental slope in 1404-m water depth (Poag and others, 1987b). Over 320 m of lower to middle Eocene porcellanitic chalk and nannofossil oozes overlying unconformably Maestrichtian chalks were recovered.

The stratigraphic interpretation of the lower part of this section, discussed in Aubry (1991), is extended to the entire lower to middle Eocene recovered from the site (Fig. 28). There is a very good agreement between the zonal subdivisions based on calcareous nannofossils (Poag and others, 1987b; Valentine, 1987), planktonic foraminifera (Miller and Hart, 1987), and radiolarians (Palmer, 1987). The unconformity in Core 37–3, 81 cm (Poag and Low, 1987) is well-delineated by juxtaposed biochronologic datums, while stratigraphic continuity of the up-

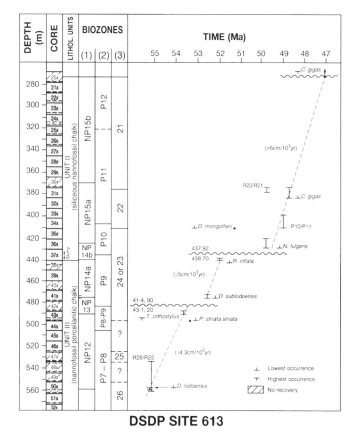

DSDP SITE 613

FIG. 26.—Stratigraphic interpretation of the lower and middle Eocene interval recovered from DSDP Site 613. Note that a slump occurs at the level of the inferred unconformity at ~435 m. Lithologic units are from Poag and others (1987a). Datums used to constrain the curve are given in Appendix D, Table 15. (1) from Poag and others (1987a) and Valentine (1987), (2) from Miller and Hart (1987) and (3) from Palmer (1987).

per lower Eocene is obvious. The offset of the LO of *D. lodoensis* (in Core 59–6) with respect to the HO of *T. orthostylus* (in Core 53X-cc) and younger datums was interpreted by Aubry (1991) as suggestive of increased sedimentation rate during Biochron NP13. Considering the constancy in sedimentation rate for a duration of over 3 my and the discontinuous record in Cores 60X and 61X, this offset more likely reflects a stratigraphic gap encompassing the upper part of Zone NP11 and the lower part of Zone NP12. It is probable that an unconformity also occurs in the upper part of the section, as indicated by the absence of Subzone NP15c (Poag and others, 1987b). Based on radiolarian stratigraphy (Poag and others, 1987b), it occurs between sections 5 and 6 in Core 27X. Bukry (1987) interpreted the stratigraphy of Cores 26X to 28X differently from Valentine (in Poag and others, 1987b), assigned it to Zone CP13c (~NP15c), and placed the CP13c/CP14a between Cores 25–5 and 26–1. Because Bukry's (1987) age determinations are based on a small number of samples and because radiolarian stratigraphy (Palmer, 1987) supports Valentine's (in Poag and others, 1987b) interpretation, the latter is preferred in this work (Figs. 27, 28). It should also be noted that unconformities of similar age occur at neighbouring DSDP Sites 605 and 613.

The age of the unconformity in Core 37–3, 81 cm and the duration of the associated hiatus were discussed in Aubry

(1991). The ages of the other two unconformities at this site are very difficult to determine precisely, and their position in Figure 27 is arbitrary.

DSDP Site 605 (38° 44.53′N; 72° 36.55′W).—DSDP Site 605 was drilled on the upper continental rise off New Jersey in 2194-m water depth (van Hinte and others, 1987). A 450-m thick section of lower and middle Eocene nannofossil limestone and chalk was recovered, unconformable with Pleistocene interbedded clay and silt. Its stratigraphic interpretation was discussed in Aubry (1991) and is summarized in Figure 29. It relies almost entirely on calcareous nannofossil stratigraphy (Applegate and Wise, 1987), the planktonic foraminifera being mostly poorly preserved (van Hinte and others, 1987), and the total NMR of the limestone and chalk being too low for reliable magnetostratigraphy (Bruins and Zijderveld, 1987). The unconformities at ~443 m (in Core 32) and at ~349.80 m (in Core 22) are well delineated by the offset of the biostratigraphic datums. A simpler interpretation for the stratigraphy around the NP14/NP15 zonal boundary than in Aubry (1991) is shown in Figure 27. There is no possibility with the biostratigraphic control presently available to evaluate the stratigraphic continuity between Cores 32–2 and 22–3. Both the NP14a/NP14b and NP14b/NP15a subzonal boundaries may be unconformable, as was proposed in Aubry (1991). However, since there is no lithologic evidence for a gap at the NP14a/NP14b subzonal boundary, Subzones NP14a and NP14b are considered to be in stratigraphic continuity in this work.

DSDP Site 387 (32° 19.2′N; 67° 40.0′W).—DSDP Site 387 lies on the western Bermuda Rise in 5117-m water depth (Tucholke and others, 1979a). A 190-m thick section of lower and middle Eocene claystone, mudstone, mud and ooze was recovered from this site. Poor recovery, large coring gaps and poor preservation of the microfossils hamper stratigraphic resolution. Except for Core 22 assigned to planktonic foraminiferal Zone P7 and Cores 12 to 9 assigned to the radiolarian *Thyrsocyrtis triacantha* Zone (Tucholke and others, 1979a), calcareous nannofossils provide the only means for stratigraphic interpretation (Fig. 30). Intensive reworking is obvious at this site, which probably accounts for the differences in the biozonal subdivision by Okada and Thierstein (1979) and by Bukry (1978b). The occurrence of *D. kuepperi, D. lodoensis, D sublodoensis, R. truncata, R. inflata* at levels assigned to Subzones NP15a, NP15b, NP15c, that of *Rhomboaster calcitrapa* and *R. cuspis* at intervals in Zone NP13 (Okada and Thierstein, 1979), and the presence of Cretaceous species in Cores 18 and 19 (Bukry, 1978b) are clear evidence of intensive reworking which confirms that lithologic subunit 3B represents fine grained turbidites (Tucholke and others, 1979a; McCave, 1979). Based exclusively on the biostratigraphic datums delineated by Okada and Thierstein (1979), it would appear that the lower-middle Eocene section recovered from DSDP Site 387 is essentially continuous (model (a) in Fig. 30), although the evidence is not compelling, in particular because of the unconformity in Core 23 at the contact between Zones NP10 and NP12, which prevents from using the lowest occurrence of *D. lodoensis* as a datum point. However, this is true only if the range of other species are ignored or invalidated. The report of *Sphenolithus furcatolithoides* in Cores 22 and 23 (Okada and Thierstein, 1979) is thought to result from a misprint. The occurrence of *R. inflata* in sample 20-CC as reported in Okada and Thierstein (1979, Table 6A) suggests that Sub-

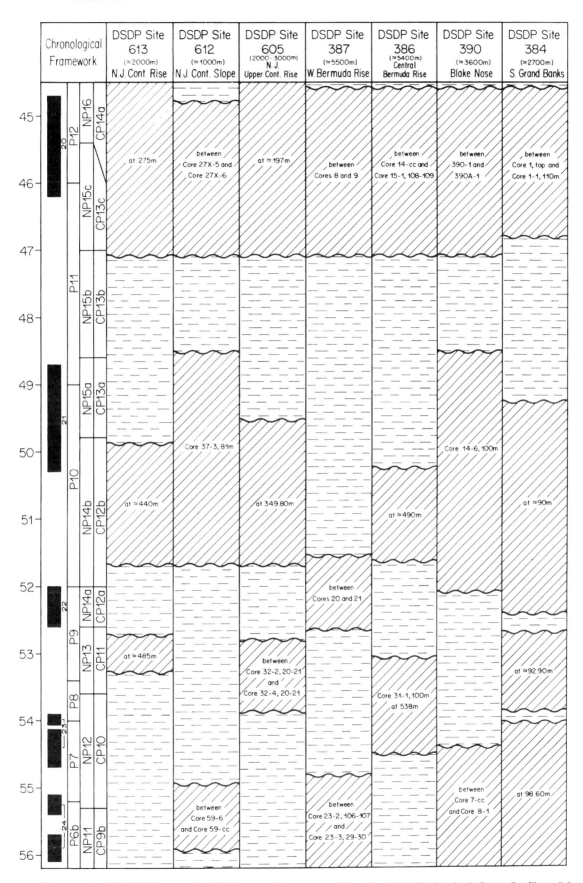

FIG. 27.—Temporal interpretation of 7 DSDP lower and middle Eocene sections recovered from the western North Atlantic Ocean. See Figure 9 for description.

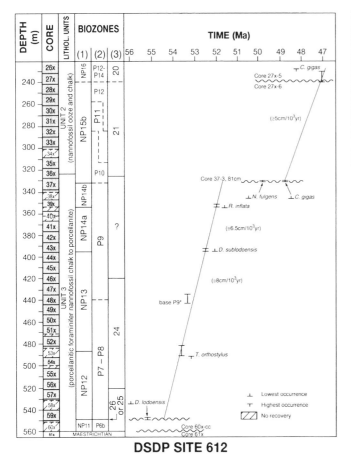

DSDP SITE 612

FIG. 28.—Stratigraphic interpretation of the lower and middle Eocene interval recovered from DSDP Site 612. Lithologic units are from Poag and others (1987b). Datums used to constrain the curve are given in Appendix D, Table 16. (1) from Poag and others (1987b) and Valentine (1987), (2) from Miller and Hart (1987) and (3) from Palmer (1987). * approximate datum based on the LO of *Morozovella caucasica* (Miller and Hart, 1987).

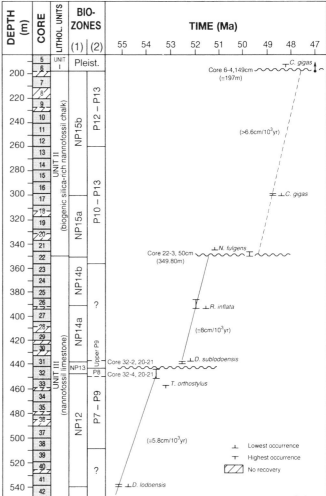

DSDP SITE 605

FIG. 29.—Stratigraphic interpretation of the lower and middle Eocene interval recovered from DSDP Site 605. Lithologic units are from van Hinte and others (1987). Datums used to constrain the curve are given in Appendix D, Table 17. (1) from Applegate and Wise (1987) and (2) from Olsson and Wise (1987).

zone NP14a may not be represented at this site (see also Bukry, 1978b, Fig. 2). The co-occurrence of *R. gladius*, a species thought to have evolved in early Biochron NP15 (see Aubry, 1983), with *R. inflata* in Core 18 and 19 suggests mixing. Bukry (1978b) also assigned Cores 19 and 18 to Subzone NP14b but noted the rarity of *R. inflata* and the absence of other diagnostic taxa in moderately well preserved assemblages. It is thus considered that the report of *R. gladius* at this level (Okada and Thierstein, 1979) may also result from misprinting. Finally, Bukry (1978b) assigned Core 9 to Subzone CP13b (= NP15b) which suggests that an unconformity occurs between Cores 8 and 9, the highest occurrence of *N. fulgens* being also in Core 9 (Okada and Thierstein, 1979). Radiolarian stratigraphy supports an NP15b subzonal assignement for Core 9 (the NP15b/NP15c subzonal boundary correlates with the radiolarian Zones R21/R22 boundary (cf. Berggren and others, 1985). It also supports the suggestion of an unconformity in Core 8. In addition to the fact that the boundary between radiolarian Zones R21 and R20 also occurs in this core, the *Podocyrtis ampla* Zone (Zone 20) constitutes only a thin interval in Cores 8 and 7. Based on Bukry's zonal assignments, it would appear that sed-

imentation was continuous throughout Biochrons NP14b to NP15b (model (b) in Fig. 30) and that an unconformity occurs at the NP13/NP14 boundary. This latter interpretation is followed in Figure 27.

DSDP Site 386 (31° 11.21′N, 64° 14.94′W).—DSDP Site 386 is located on the central Bermuda Rise, 140 km south-southeast of Bermuda in 4782-m water depth (Tucholke and others, 1979b). A 250-m thick section of lower and middle Eocene mudstones, claystones and turbidites was recovered. Stratigraphic interpretation relies upon calcareous nannofossil zonation since no planktonic foraminifera occur and radiolarian allow confident zonal assignment only for cores 18 to 14 (Tucholke and others, 1979b). Extensive coring gaps preclude a satisfactory stratigraphic interpretation of this section, further compromised by the discrepant zonal assignments by Okada and Thierstein (1979) and Bukry (1978b) in particular for Cores 23 to 30 (Fig. 31). The interpretation given in Figures 27 and

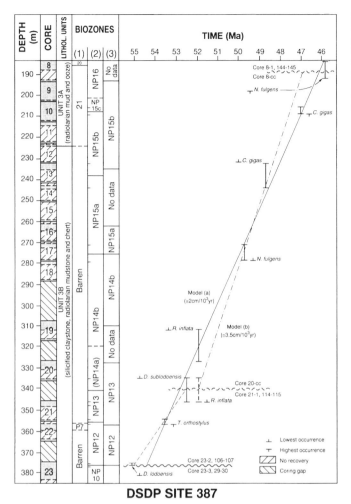

DSDP SITE 387

FIG. 30.—Stratigraphic interpretation of the lower and middle Eocene interval recovered from DSDP Site 387 and sedimentation rate curve(s). Lithologic units are from Tucholke and others (1979a). Datums used to constrain the curve are given in Appendix D, Table 18. (1) from Tucholke and others (1979a; planktonic foraminifera occur only in Core 22 and radiolarians above Core 12 (included)), (2) from Okada and Thierstein (1979), converted to Martini's (1971) zonal scheme and modified, and (3) from Bukry (1978b), converted to Martini's (1971) zonal scheme and modified. Model (a) based on Okada and Thierstein's (1979) biostratigraphic data, suggesting stratigraphic continuity from Core 9 to 23. Model (b) based on Bukry' (1978) zonal assignments, suggesting unconformities at the NP13/NP14 and NP15/NP16 zonal boundaries.

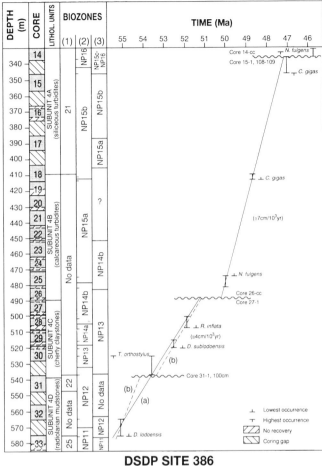

DSDP SITE 386

FIG. 31.—Stratigraphic interpretation of the lower and middle Eocene interval recovered from DSDP Site 386 and sedimentation rate curve(s). Lithologic units are from Tucholke and others (1979b). Datums used to constrain the curve are given in Appendix D, Table 19. (1) from Tucholke and others (1979b), (2) from Okada and Thierstein (1979), converted to Martini's (1971) zonal scheme and modified, and (3) from Bukry (1978b), converted to Martini's (1971) zonal scheme and modified. Sedimentation rate curve(s) based on the datum levels in Okada and Thierstein (1979).

31 is based on the datums delineated by Okada and Thierstein (1979), because the sedimentation rate curve derived from them reflects the lithologic changes that occur in the section. High sedimentation rates of ~7 cm/10³ yr characterize the middle Eocene calcareous and siliceous turbidites. The underlying upper lower and lower middle Eocene mudstones were deposited at slower rates of ~4 cm/10³ yr. The lower Eocene claystones, interpreted as turbidites (McCave, 1979), may have been deposited at similar rates (Fig. 31, model (a)), or faster (Fig. 31, model (b); see discussion below).

It is difficult to account for the lithologic change at ~490 m from cherty claystones (Subunit 4c) below to calcareous turbidites (Subunit 4b) above solely by an increase in sedimenta-

tion rates. It is likely that an unconformity occurs at this level (Fig. 31) as suggested by the biostratigraphic datums.

There may be stratigraphic continuity from Core 33 to 27, but the evidence from the sedimentation rate curve (Fig. 31, model (a)) is not conclusive. A thin stratigraphic gap may occur below the lowest occurrence of *D. sublodoensis*. There are both lithologic and biostratigraphic evidence that an unconformity occurs in Core 31–1, 100 cm. According to Tucholke and others (1979b, p. 210) the upper boundary of Subunit 4D is "marked by a possible erosional contact that dips at 20° between subunit 4D radiolarian mudstones and the overlying claystones." The presence of an unconformity at this level is further supported by the co-occurrence of *Reticulofenestra dictyoda* and *T. orthostylus* in Core 31–1, 4–5 cm. The former taxon first occurs in Zone NP13 (see discussion in Aubry, 1983), indicating that the base of Subunit 4C lies in Zone NP13 while the top of Subunit 4D lies in Zone NP12, further indicating that *T. orthostylus* is reworked in Subunit 4C.

The ages of the genetic surfaces of the unconformity at 490 m and that of the upper surface of the unconformity at ~538 m are determined by extrapolation using sedimentation rates of respectively 7 cm/10^3 yr and 4 cm/10^3 yr. There are no further constraints to determine the age of the older surface of the latter unconformity, thus arbitrarily located in Fig. 27.

DSDP Site 384 (40° 21.65'N, 51° 39.80'W).—DSDP site 384 was drilled on the J-Anomaly Ridge south of the Grand Banks in 3920-m water depth (Tucholke and others, 1979c). A 45-m thick section of lower and middle Eocene nannofossil oozes unconformable with upper Paleocene chalk was recovered. Planktonic foraminifera and radiolarians are well-preserved and abundant in these oozes, but although biozonal subdivisions based on these microfossils are outlined (Tucholke and others, 1979c), their detailed distribution is not given. As a result, the discrepancies between radiolarian and planktonic foraminiferal zonal assignments and between the latter and those based on calcareous nannofossils (Okada and Thierstein, 1979; Bukry, 1978b) cannot be resolved (Fig. 32). The stratigraphic interpretation of this lower-middle Eocene section thus relies primarily on calcareous nannofossil zonation, a reliance warranted by the good agreement between the zonal subdivisions by Bukry (1978b) and Okada and Thierstein (1979).

It is obvious that an unconformity occurs in Core 5–1, with Subzone NP14b being missing as well as parts of Subzones NP14a and NP15a. This unconformity probably corresponds to the sharp rise in $CaCO_3$ that occurs in Core 5 (Tucholke and others, 1979c). Following Okada and Thierstein (1979), part of Subzone NP14b is present, the co-occurrence of *N. fulgens* and *R. inflata* in Cores 4–6, 143–144 cm and 5–1, 49–50 cm being interpreted as downward contamination. Alternatively, reworking accounts for the occurrence of *R. inflata* in younger Subzone NP15a. While there is no other evidence of downhole contamination (e.g., *Sphenolithus furcatolithoides*) in Cores 5–1 and 4–6, intensive reworking in Cores 4 through 2 is indicated by the anomalous occurrences of *Discoaster kuepperi, D. lodoensis,* and *D. sublodoensis* at levels assigned to Subzones NP15a and NP15b.

The age of the younger genetic surface of the unconformity at ~90 m is extrapolated using a sedimentation rate of 1.5 cm/10^3 yr. It is clear from the juxtaposition of the HO of *Tribrachiatus orthostylus* and the LO of *Discoaster sublodoensis* that an unconformity also occurs at ~9.20 m. Because of the close stratigraphic succession of unconformities between 90 and 96.80 m, it is not possible to date their respective surfaces. They are located arbitrarily in Figure 27.

Finally, the presence of an unconformity in Core 1–1 is inferred from the thinness of the interval corresponding to Subzone NP15c (compared to Subzone NP15b) and assigned to planktonic foraminiferal Zone P12 (the P11/P12 zonal boundary lies in the upper part of Subzone NP15c).

DSDP Site 390 (30° 08.54'N, 76° 06.74'W).—DSDP Site 390 was drilled on the Blake Nose, a northeast-protruding spur of the Blake Plateau (Benson and others, 1978). A 65-m thick lower-middle Eocene section consisting of siliceous and calcareous oozes was recovered from Hole 390A, unconformably sandwiched between Pleistocene nannofossil oozes and foraminiferal sands and upper Paleocene interbedded ooze and chert. The integration of the biozonal subdivisions derived from the distribution of radiolarians (Weaver and Dinkelman, 1978), planktonic foraminifera (Gradstein and others, 1978) and calcareous nannofossils (Schmidt, 1978; Bukry, 1978c) provides a comprehensive interpretation of this lower-middle Eocene section (Fig. 33). The biozonal subdivisions based on calcareous nannofossils by Bukry (1978c) and Schmidt (1978) differ considerably (Fig. 33). The latter, which correlates well with the planktonic foraminifera and radiolarian zonal assignments, is followed here.

The sedimentation rate curve for the lower part of the section, relatively well-constrained by calcareous nannofossils and radiolarian datums, suggests stratigraphic continuity from Core 7 to the base of Core 4 (Fig. 33). At this level, a stratigraphic gap occurs as indicated by the direct superposition of calcareous nannofossil Subzones NP14a and NP15b and, except for a thin barren interval, that of radiolarian Zones 24 and 21. Core 4–5 is assigned to Subzone NP15b which agrees perfectly with assignment to radiolarian Zone R21 but disagrees with attribution to Zone P10. In fact, the assemblage listed for samples 4–2 to 4–5 (Benson and others, 1978) are not restrictive of Zone P10 and could represent Zone P11 (W. A. Berggren, pers. commun., 1992). The unconformity, located on the basis of biostratigraphic datums between Core 4–6, 129 cm, and Core 4–5, 119 cm, likely corresponds with the top of the interbedded interval extending from Core 4–6, 100 cm to Core 5–3. Contrary to the report by Benson and others (1978) stating that this interval with interbedded lithologies characteristic of Units 2 and 3 rep-

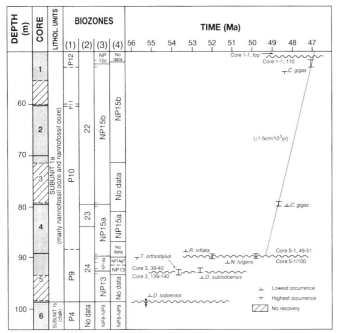

DSDP SITE 384

FIG. 32.—Stratigraphic interpretation of the lower and middle Eocene interval recovered from DSDP Site 384 and sedimentation rate curve(s). Lithologic units are from Tucholke and others (1979c). Datums used to constrain the curve are given in Appendix D, Table 20. (1) from Tucholke and others (1979c), (2) from Tucholke and others (1979c), (3) from Okada and Thierstein (1979), converted to Martini's (1971) zonal scheme and modified; and (4) from Bukry (1978b), converted to Martini's (1971) zonal scheme and modified. Sedimentation rate curve(s) based on the datum levels in Okada and Thierstein (1979) partly modified.

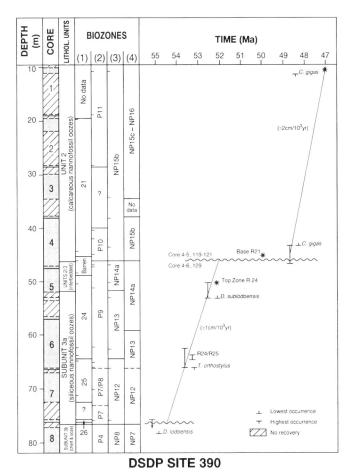

DSDP SITE 390

FIG. 33.—Stratigraphic interpretation of the lower and middle Eocene interval recovered from DSDP Hole 390A and sedimentation rate curve(s). Lithologic units are from Benson and others (1978). Datums used to constrain the curve are given in Appendix D, Table 21. (1) from Weaver and Dinkelman (1978), (2) from Gradstein and others (1978), (3) from Schmidt (1978), partly reinterpreted, and (4) from Bukry (1978c) converted to Martini's (1971) zonal scheme. Tick marks in columns (1), (2) and (3) indicate position of critical samples. The top of radiolarian Zone 24 and the base of radiolarian Zone 21 are used to constrain the stratigraphic interpretation of Core 4, sections 4 and 5, although neither was truly delineated because of the barren interval in Cores 4 and 5.

resents a gradational boundary between the two units, I suggest that it reflects sedimentary disturbance associated with the unconformity. There is evidence that mixing occurs in this interval. The occurrence of a single specimen of *Globigerinatheka* sp. in Core 4–6, 127–129 cm, which otherwise yields a planktonic foraminiferal assemblage characteristic of Zone P9, was interpreted as resulting from mechanical contamination (Benson and others, 1978). *Sphenolithus furcatolithoides*, a species which appeared during Biochron NP15a, occurs in Core 4–6, 130 cm within an assemblage otherwise diagnostic of Subzone NP14a (Schmidt, 1978). The co-occurrence of the two taxa suggests that Core 14–6 may belong to Zone P11 and Subzone NP15a yields mixed assemblages with the reworked forms dominant. The absence of *Nannotetrina fulgens*, the marker of the base of Zone NP15, in Core 4–6, 130 cm is not significant. This species is rare at this site and was not recorded in the lower

part of the interval assigned to Subzone NP15b. However, the sampling density in the interbedded interval is too low to resolve the possibility of mixing and mechanical contamination cannot be ruled out.

It is difficult to determine the hiatus associated with the unconformity. *Nannotetrina fulgens* is rare and is not associated with *Chiasmolithus gigas* throughout the range of this latter. In addition, *C. gigas* occurs discontinuously above Core 3–1, which may suggest reworking. Based on correlation with radiolarians and planktonic foraminifera, it is assumed that Subzone NP15b extends up to Core 1. Location of the surfaces of the unconformity in Figure 27 is based on this assumption.

ODP Site 628 (27° 31.85'N, 78° 18.95'W).—Site 628 is located on the northern slope of the Little Bahama Bank in 959-m water depth (Austin and others, 1986). Middle Eocene and possibly lower Eocene nannofossil ooze and chalk were recovered from Hole 628A, unconformable with lower Oligocene oozes and with upper Paleocene limestone. Drilling breccia (Core 31x) and drilling disturbance (Core 32x) in addition to poor recovery prevent a comprehensive stratigraphic interpretation of this ~20-m thick Eocene section. However, we note its stratigraphic complexity, with the mixing of assemblages of Zones CP12 (=NP14) and CP14 (=upper part Zone NP15 to lower part Zone NP16; Watkins and Verbeek, 1988).

Gulf of Mexico-Caribbean area.—

DSDP Site 94 (24° 31.64'N, 88° 28.16'W).—DSDP Site 94 (Worzel and others, 1973) is situated on the continental slope of the Yucatan Platform (Campeche Scarp) in 1793-m water depth.

One hundred and sixty meters of upper lower to middle Eocene foram-nanno chalks unconformable with upper Paleocene foram-rad nanno chalks were penetrated at this site. Stratigraphic interpretation of the lower and middle Eocene section (Fig. 34) is mostly derived from calcareous nannofossils and radiolarians. Despite poor recovery and extensive coring gaps, the good agreement between the datums provided by the calcareous nannofossils (Bukry, 1973a; Worzel and others, 1973) and the radiolarians (Sanfilippo and Riedel, 1973) suggests continuous sedimentation during the late early and early middle Eocene Epochs (biochrons NP13 to NP 14b) and during the later part of the middle Eocene Epoch. An unconformity between 460 and 470 m is inferred from the juxtaposition of the lowest occurrences of *Nannotetrina fulgens* and *Chiasmolithus gigas* at the level of the boundary between the Radiolarian Zones R21 (*Theocyrtis tuberosa*) and R22 (*Thyrsocyrtis bromia*). The upper surface of the unconformity lies in Subzone NP15b. I locate its lower surface in Subzone NP14b based on extrapolation using an estimated sedimentation rate of 3.5 cm/10^3 yr.

Planktonic foraminiferal stratigraphy (Worzel and others, 1973) suggests the absence of Zone P9 (and consequently an unconformity between 540 and 550 m). The slight offset between the R24/R25 zonal boundary and the lowest occurrence of *D. sublodoensis* may reflect a stratigraphic gap at this level, but because of the large coring gaps between 516 and 552 m, the evidence is not compelling. In addition, the report of *Globigerapsis* sp. cf. *G. kugleri* at a level lower than the lowest occurrence of *D. sublodoensis* and much lower than that of *R. inflata* is suspicious (the LO of *G. kugleri* defines the base of

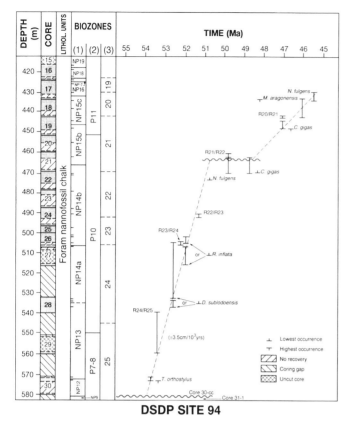

DSDP SITE 94

FIG. 34.—Stratigraphic interpretation of the lower and middle Eocene interval recovered from DSDP Site 94 and sedimentation rate curve(s). The unconformity at ~465 m is inferred. Lithologic units are from Wortzel and others (1973). Datums used to constrain the curve are given in Appendix D, Table 22. (1) mainly from Bukry (1973a) with complementary data from Wortzel and others (1973), (2) from Wortzel and others (1973) and (3) from Sanfilippo and Riedel (1973).

Zone P11, see Berggren and Miller, 1988). As the poorly documented planktonic foraminiferal data are difficult to reconcile with the data provided by the two other microfossil groups, a conservative interpretation for the stratigraphy of the lower to middle Eocene section recovered from DSDP Site 94 is given in Figure 35.

DSDP Site 536 (23° 29.39′N, 85° 12.58′W).—An 80-m thick Cenozoic section was recovered from DSDP Site 536 located at the base of the Campeche escarpment in 2790-m water depth (Buffler and others, 1984a). A 3-cm thick layer of upper lower Eocene hard nannofossil chalk was found at the top of Core 9, sandwiched between upper Oligocene nannofossil oozes and upper Paleocene nannofossil chalk (Buffler and others, 1984a). The contact between the lower Eocene and upper Paleocene sediments is marked by a hardground with a manganese crust. Planktonic foraminiferal Zone P9 was identified (Premoli Silva and McNulty, 1984) and the co-occurrence of *D. lodoensis* and *D. sublodoensis* (Buffler and others, 1984a) indicates calcareous nannofossil Subzone NP14a. There is no constraint as to the exact position of this Eocene chalk within this zonal interval, and its location in Figure 35 is arbitrary.

DSDP Site 538 (23° 50.95′N, 85° 09.93′W).—DSDP Site 538 lies on the top of the Catoche Knoll about 25 km northeast of

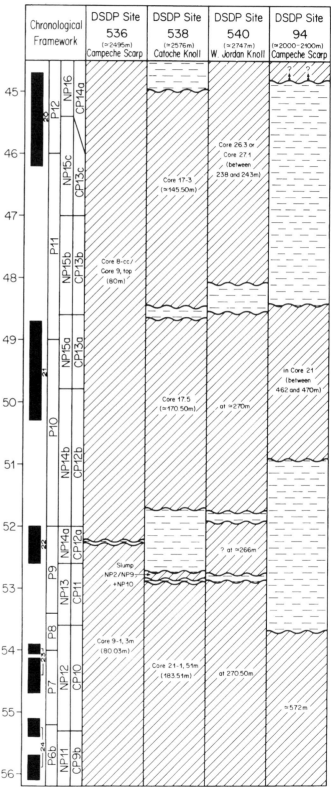

FIG. 35.—Temporal interpretation of 4 DSDP lower and middle Eocene sections recovered from the Gulf of Mexico. See Figure 9 for description.

the Campeche escarpment (Buffler and others, 1984b) in 2820 m of water.

The 40-m thick lower to middle Eocene section recovered from Hole 538A is remarkable by its lithologic heterogeneity with conspicuous bioturbation, stratigraphic gaps, dipping layers, slumps and hardgrounds represented by irregular, bored surfaces (as at level 538A-20–2, 60–65 cm, at the upper Paleocene/lower Eocene contact) or by black manganese-rich crusts (as at level 538A-21–1, 56–59 cm, at the Upper Cretaceous/lower Paleocene contact; Buffler and others, 1984b). The complex stratigraphic succession in Cores 20 and 21 which involves numerous unconformities and a Paleocene slump within a lower Eocene interval is summarized in Figure 36. This figure also shows the presence of middle Eocene unconformities inferred from calcareous nannofossil stratigraphy (planktonic foraminifera are rare and strongly dissolved in the middle Eocene interval, and zonal assignment was mainly inferred from calcareous nannofossil stratigraphy (Premoli Silva and McNulty, 1984). These inferred unconformities may be reflected by the few sedimentary features observed in Core 17 (e.g., the band with sharp contact, grading and a burrowed transitional upper contact in Section 3 where an abnormal zonal contact NP15b/NP16 is observed, and the 25° dipping clayey laminae in section 4 at the abnormal contact Subzone NP15a/NP15b or NP14b/NP15a). While there is obviously a stratigraphic gap around 149.50 m, it is not possible to determine whether only one or several unconformities occur between this depth and 153 m based on the available data. In Figure 36, Subzone NP15b is shown to overlie unconformably Subzone NP15a. However, for the sake of simplicity, both zones are drawn in stratigraphic continuity in Figure 35.

The estimated high rate of sedimentation (10 cm/10^3 yr) for the NP13/NP14 zonal interval reflects substantial reworking as indicated by the calcareous nannofossils (e.g., the abnormal co-occurrence of *D. lodoensis* and *D. kueppeli* with *R. inflata* in Subzone NP14b). The age of the lower surface of the intra middle Eocene unconformity at ~152.50 m and that of the upper surface of the unconformity at ~174.50 m can be estimated from sedimentation rates, but the ages of the lower surface of this latter and of both surfaces of the inferred unconformity at ~149.50 m are arbitrary (within the limits imposed by calcareous nannofossil zonal assignment).

DSDP Site 540 (23° 49.73'N, 84° 22.25'W).—DSDP Site 540 (Buffler and others, 1984c) is located on the eastern flank of a broad North-South erosional channel West of Jordan Knoll. Lower and middle Eocene sediments at this site are part of a lithologic section (Subunit IIc) which consists primarily of nannofossil ooze, chalk and limestone, characterized by numerous slump structures. The 27-m thick lower and middle Eocene interval (Fig. 37) is unconformably sandwiched between upper Eocene and upper Paleocene sediments (Buffler and others, 1984c). Because of strong dissolution of the planktonic foraminifera (Premoli Silva and McNulty, 1984), its biostratigraphic interpretation relies mostly on calcareous nannofossils (Buffler and others, 1984c; Lang and Watkins, 1984). The presence of *D. lodoensis* in Core 29-cc is tentatively regarded as indicative of Zone NP13, although the absence of *D. sublodoensis* may result from poor preservation at this level. There is some uncertainty regarding the exact location of the contact between upper and middle Eocene. According to Buffler and

others (1984c) and Premoli Silva and McNulty (1984), it lies in Core 27–1. Following Lang and Watkins (1984), it lies between Cores 26–3 and 27–1.

A comprehensive interpretation of this lower and middle Eocene section is hampered by both poor recovery and insufficient sampling coverage which result in large uncertainties on most datum levels (Fig. 37). Subzone NP14b may be thicker than shown in Figure 37 (if Core 28-cc is from a younger level in the core than shown). Subzones NP14a and NP15a may not have been recovered from sections 3 to 6 respectively in Cores 29 and 28. Thus the interpretation given in Figure 37 (and in Fig. 35) is speculative, although it appears reasonable if we consider (1) the lithologic heterogeneity of this strongly bioturbated section, (2) its extreme thinness (stratigraphic continuity would imply deposition at an average sedimentation rate of 0.3 cm/10^3 yr, an unlikely situation since reworking in Subzone NP14b is evident), (3) the common slump structures, (4) the highly discontinuous Cenozoic record at this site (see Fig. 5 in Lang and Watkins (1984), which shows a succession of 9 stratigraphic gaps in a 180-m thick lower Paleocene to Pleistocene section), and (5) the remarkably similar succession recovered from DSDP Site 538 (see above). Because of the lack of constraints on the ages of the inferred unconformities, the stratigraphic interpretation in Figure 35 is arbitrary, and based on an assumed rate of sedimentation of 10 cm/10^3 yr.

Stratigraphic Interpretation of DSDP and ODP Lower to Middle Eocene Sections Recovered from the South Atlantic Ocean

Eastern South Atlantic.—

DSDP Site 361 (35° 0.3.97'S, 15° 26.91'E).—DSDP Site 361 (Bolli and others, 1978a), located on the Cape Basin lower continental rise, west of Cape Aghulas, South Africa in 4547.5-m water depth, recovered a ~130-m thick lower to middle Eocene section. Largely due to extensive coring gaps, but also to poor preservation, mixing and poor recovery, the stratigraphic interpretation (Fig. 38) of this section is difficult and ambiguous. Poor preservation (dissolution) restricts the usefulness of the planktonic foraminifera (Toumarkine, 1978). However, Core 6-cc, (opus cited, in Fig. 5; but Core 7-cc in the text) was confidently assigned to lower Zone P10 equivalent based on the presence of *Morozovella aragonensis caucasica*, and Cores 6, section 3 to Core 5-cc, to upper P10 to P11 equivalent, based on the presence of *Morozovella aragonensis* (see Toumarkine and Luterbacher, 1985). Calcareous nannofossils are abundant, but heavy mixing (see Proto Decima and others, 1978, p. 582, Table 6) seems to have been a misleading factor in determining the zonal assignment of isolated samples, which likely explains the different calcareous nannofossil stratigraphic interpretations of the middle Eocene part of the section (in particular Cores 5 and 6) by Bukry (1978d) and Proto Decima and others (1978) (compare Bukry, 1978d, Fig. 2 with Proto Decima and others, 1978, Table 6). The latter interpretation is followed here.

There is no doubt that the upper lower to middle middle Eocene section recovered from DSDP Site 361 is riddled with unconformities. The older is located in the upper part of Core 8, at ~250.05 m, at the NP11/NP13 zonal contact (i.e., at the stratigraphic juxtaposition of the LO of *D. lodoensis* and the HO of *T. orthostylus*). The lower surface of the unconformity

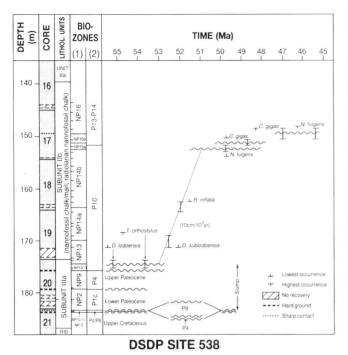

DSDP SITE 538

Fig. 36.—Stratigraphic interpretation of the lower and middle Eocene interval recovered from DSDP Hole 538A and sedimentation rate curve(s). Some unconformities are reflected by hardgrounds or sharp lithologic contacts; others are inferred from calcareous nannofossil stratigraphy. Lithologic units are from Buffler and others (1984b). Datums used to constrain the curve are given in Appendix D, Table 23. (1) Lang and Watkins (1984) and (2) from Buffler and others (1984b) and Premoli Silva and McNulty (1984).

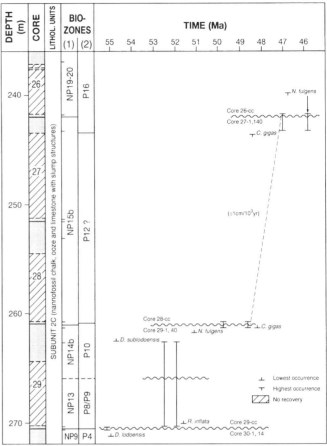

DSDP SITE 540

Fig. 37.—Stratigraphic interpretation of the lower and middle Eocene interval recovered from DSDP Site 540. Because of the uncertain position of the Core catchers in cores with poor recovery, there is a large uncertainty on the location of most datum levels. Lithologic units are from Buffler and others (1984c). Datums used to constrain the curve are given in Appendix D, Table 24. (1) Lang and Watkins (1984) and (2) from Buffler and others (1984c) and Premoli Silva and McNulty (1984).

lies in Zone NP11, its upper surface in Zone NP13. There is no data to further constrain the position of these surfaces in the zones; their location in Figure 39 is arbitrary.

The second unconformity occurs in the 9.5-m gap between Cores 7 and 8 and is indicated by the juxtaposed lowest occurrences of *D. sublodoensis* and *R. inflata*. Because of this coring gap, the thickness of Zone NP13 cannot be determined. The presence of a thin interval representing Subzone NP14a cannot be ruled out, but even if a normal zonal/subzonal succession NP13-NP14a-NP14b occurs at this site, the thinness of Zone NP13 and Subzone NP14a compared to the thickness of Subzone NP14b would be suggestive of an unconformity (unless sharp changes in sedimentation rates could be demonstrated). As for the older unconformity, there is no data to constrain the location of the lower and upper surfaces of the unconformity. As there is no evidence for the presence of Subzone NP14a, the lower surface is located within Zone NP13. It should be noted that estimation of the duration of this Eocene hiatus is dependant upon the duration (arbitrarily) assigned to the early Eocene hiatus.

The younger unconformity lies within the 19-m gap between Cores 5 and 6 as suggested by the juxtaposed lowest occurrences of *N. fulgens* and *C. gigas*, leading to the apparent contact between Subzones NP14b and NP15b, and their proximity to the highest occurrence of this latter species. Two propositions can be made. One is that Subzone NP15a is absent: the stratigraphic gap encompasses the upper part of Subzone NP14b, Subzone NP15a, and the lower part of Subzone NP15b. The

second is that Subzone NP15a is present: the stratigraphic gap encompasses the upper part of Subzone NP15a and the lower part of Subzone NP15b (in this case, the thicker Subzone NP15a, the thinner Subzone NP15b). While supporting the calcareous nannofossil zonal assignment, planktonic foraminiferal stratigraphy does not provide information which would help resolve the alternative. In the absence of positive evidence for the presence of Subzone NP15a, the first alternative is followed in Figure 39.

It is unsure whether the occurrence of *N. fulgens* in Core 4-cc corresponds to the highest occurrence of the species at DSDP Site 361. Mixing may have blurred its true range at this site, and no data are available for the overlying 40 m of sediments. A relatively high, minimal sedimentation rate of 3.8 cm/10³ yr would well reflect mixing of lower Eocene particles into the middle Eocene sediment.

DSDP Site 360 (35° 50.75'S, 18° 05.79'E).—Companion of DSDP Site 361, DSDP Site 360 (Bolli and others, 1978a) was also drilled on the Cape Basin continental rise but on its upper part in 2967-m water depth. As the site was terminated at a

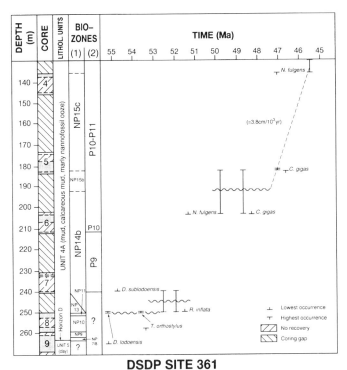

DSDP SITE 361

FIG. 38.—Stratigraphic interpretation of the lower and middle Eocene section recovered from DSDP Site 361. Three (inferred) unconformities occur, at the NP11/NP13 zonal contact, close to the NP13/NP14b and the NP14b/NP15b zonal contacts. Lithologic units are from Bolli and others (1978a). Datums used to constrain the curve are given in Appendix D, Table 25. (1) from Proto Decima and others (1978) and (2) from Toumarkine (1978). Correlation to "P" Zones based on Berggren and Miller (1988, Text Fig. 1).

level in the middle Eocene Series, it yields only few elements for comparison with DSDP Site 361. It is noteworthy, however, that at shallower DSDP Site 360, where Subzone NP15a was recorded, Subzone NP15b is at least 30 m thick (see Proto Decima and others, 1978, Table 5B), whereas at DSDP Site 361 it is less than 10 m thick.

DSDP Site 530 (19° 11.26′S, 9° 23.15′E).—DSDP Site 530 (Hay and others, 1984) lies on the abyssal floor in the southern corner of the Angola Basin beneath the gently northward flowing Benguela Current (<25 cm/s) in a water depth of 4639 m. An approximately 30-m thick upper lower to middle Eocene section was recovered from DSDP Hole 530A. Despite poor recovery, precise stratigraphic interpretation (Fig. 40) of this section can be made primarily based on detailed calcareous nannofossil stratigraphy (Steinmetz and Stradner, 1984). No planktonic foraminiferal stratigraphy is available for the lower to middle Eocene section from this hole. A magnetostratigraphic record is available, but no attempt was made at interpreting the magnetozones recorded above Core 40 (Keating and Herrera-Bervera, 1984; Steinmetz and others, 1984). Because precise depth data on the magnetic reversal stratigraphy is not provided by these authors, magnetic polarity stratigraphy as presented in Figure 40 was approximately reconstructed from Figure 4 in Keating and Herrera-Bervera (1984). The normal polarity interval in Core 40, section 1, and Core 39-cc, lies in the lower part of Zone NP12, and is identified as Chron C23n (in agree-

ment with Steinmetz and others, 1984, unlike Keating and Herrera-Bervera, 1984). The polarity reversal in the upper part of Core 40 is thus identified as the Chrons C23n/C23r boundary. Assuming that the magnetostratigraphic record in Cores 37 and 38 is reliable, correlation with calcareous nannofossil stratigraphy suggests the following. The normal polarity interval in Core 38, section 1, and Core 37-cc, results from the concatenation of the upper and lower parts of Chrons C22n and C21n, respectively, as indicated through correlation with Subzones NP14a and NP15a (see Berggren and others, 1985). Chron C21n extends up to ~470 m (i.e., Core 37, section 3, ~30 cm). The part of the interval with normal polarity in Core 37, section 2, which correlates with Subzone NP15c may be identified as Chron C20n. In turn, this would suggest that only the upper part of Subzone NP15c is present. If these identifications are correct, there should be a reversed polarity interval in Core 37, section 2 representing Chron C20r and correlating with Subzone NP15b (this cannot be verified with the available magnetic record).

The upper lower Eocene section, from Core 40, section 3, to Core 38, section 1, is continuous and was deposited at a slow rate of ~0.8 cm/10³ yr. There is a perfect agreement between calcareous nannofossil stratigraphy and identification of the magnetic reversal in Core 40, section 1, as the Chron C23r/C23n boundary.

In contrast to the upper lower Eocene sequence, the middle Eocene record, unconformable with the upper Eocene, is discontinuous. The record may be broken in a series of unconformity-bounded "sequences" as suggested by the "clastic limestone, chalk, marlstone and mudstone cycles typical of lithologic Unit 4" (Hay and others, 1984, Fig. 20). In addition, it includes at least 2 and perhaps 3, important stratigraphic gaps (Fig. 41). The older unconformity occurs at ~477 m and results in a NP14a/NP15a subzonal contact. The hiatus encompasses at least all of Chron C21r and the earlier part of Chron C21n. There is also an unconformity higher in the section as indicated by the very thin Subzone NP15b. If Subzone NP15b is in normal succession with Subzone NP15a, there is only one stratigraphic gap between the top of Subzone NP15b and the base of Subzone NP15c, and the hiatus encompasses most of Chron C20r. On another hand, Subzone NP15b may be unconformable with Subzone NP15a, resulting in 2 unconformities, one at ~473 m and the other at ~470 m. The older hiatus would correspond to the earlier part of Chron C20r, the younger one to the later part of Chron C20r. There is no evidence to determine which of the two interpretations is correct, and both are shown in Figure 39. It should be noted that whichever interpretation is followed, the amount of time not represented in the section remains the same. What differs is the distribution of the hiatuses and the position of the upper and lower surfaces of the unconformities.

DSDP Site 363 (19° 38.75′S, 09° 02.80′E).—DSDP Site 363 (Bolli and others, 1978b) was drilled near the crest of an isolated basement high on the North-facing escarpment of the Frio Ridge segment of the Walvis Ridge in 2248-m water depth.

The stratigraphic interpretation of the ~70-m thick section of upper lower and middle Eocene chalk recovered from this site is given in Figure 42. It is based mainly on calcareous nannofossil stratigraphy (Proto Decima and others, 1978; Bukry, 1978d) complemented by planktonic foraminiferal stra-

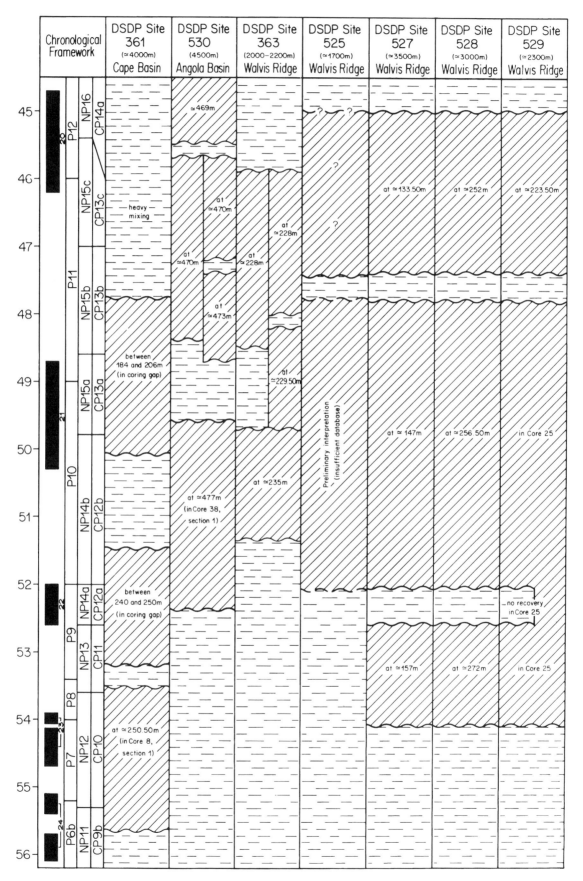

FIG. 39.—Temporal interpretation of 7 DSDP lower and middle Eocene sections recovered from the eastern South Atlantic Ocean. See Figure 9 for description.

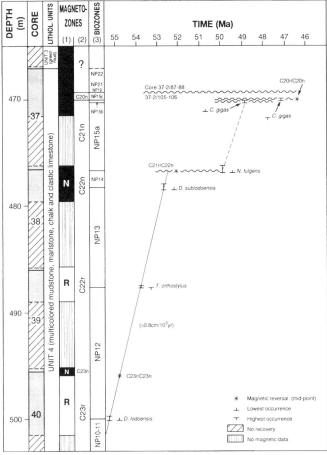

DSDP SITE 530

FIG. 40.—Stratigraphic interpretation of the lower and middle Eocene section recovered from DSDP Site 530A and sedimentation rate curve(s). In addition to the unconformity at ~144.65 m, three unconformities are inferred. Lithologic units are from Hay and others (1984). Biostratigraphic datums and magnetic reversals are given in Appendix D, Table 26. (1) from Keating and Herrero-Bervera (1984, as determined from their Fig. 4, p. 1215) and (2) from Steinmetz and Stradner (1984).

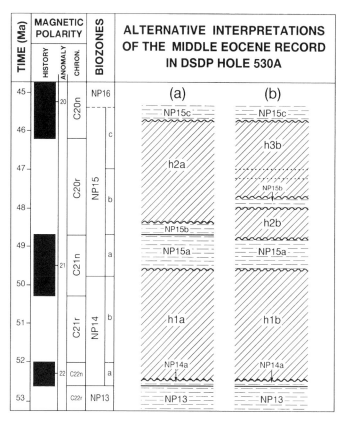

FIG. 41.—Alternative stratigraphic interpretations of the middle Eocene section recovered from DSDP Hole 530A. In interpretation (a), Zones NP15a and b are in stratigraphic continuity. A single stratigraphic gap occurs. In interpretation (b), Zones NP15a and b are unconformable. Two stratigraphic gaps occur. The importance of each gap depends upon the (upper, middle lower) part of Zone NP15b being present in Hole 530A. Dotted line in interpretation (b) indicates the youngest position of the interval assigned to Zone NP15b. The sum of the hiatuses h2b + h3b in interpretation (b) equals the single hiatus h2a in interpretation (a). Magnetobiostratigraphic correlations and estimated ages are from Berggren and others (1985).

tigraphy (Toumarkine, 1978). Calcareous nannofossil zonal interpretations by Bukry (1978d) and Proto Decima and others (1978) of the middle Eocene part of the section are in agreement, but there is some discrepancy for the upper lower Eocene interval. Because the interpretation by Proto Decima and others (1978) is based on a comprehensive study and is supported by planktonic foraminiferal stratigraphy (Toumarkine, 1978), it is followed here.

The sedimentation rate curve (Fig. 42) suggests continuous sedimentation throughout the late early Eocene (from Biochron NP12 to NP14a), and perhaps in the early middle Eocene Epoch. The stratigraphic propinquity of the lowest occurrence of *R. inflata* and *N. fulgens* is suggestive of an unconformity, but because of the 9.5 m gap between Cores 11 and 12, the stratigraphic interpretation around the NP14b/NP15a subzonal boundary at this site is unclear. If the lowest occurrence of *N. fulgens* is close to the top of Core 12, the unconformity is in position (a) in Fig. 42; if it close to the base of Core 11, it is in position (c) (Fig. 42). For any intermediate level, the uncon-

formity occurs within the coring gap (position (b), Fig. 42). Planktonic foraminiferal stratigraphy brings no clue because of the uncertainty in the zonal assignment of Cores 12 and 11. The unconformity is arbitrarily located in the stratigraphic gap at ~235 m (it may be as low as 240 m or as high as 230.5 m). Its lower surface lies within Subzone NP14b, and assuming constancy in sedimentation rate, it has an age estimate of ~51.3 Ma. Its upper surface lies in Subzone NP15a. It should be noted that while location of the lowest occurrence of *N. fulgens* in the lower part of the coring gap implies an unconformity, its location very close to the base of Core 11 could, alternatively, reflect a change in sedimentation rate, although there is no lithologic evidence for this (Bolli and others, 1978b).

The stratigraphic proximity of the lowest and highest occurrences of *C. gigas* and of the highest occurrence of *Morozovella aragonensis* is indicative of the presence of one or more unconformities. As for the middle Eocene interval recovered from DSDP Site 530 (see Fig. 40), depending whether Subzone NP15b is in stratigraphic continuity with Subzone NP15a or not, one or two unconformities occur in Core 11, section 5 (see Fig. 41 for comparison). Both alternatives are given in Figure 39.

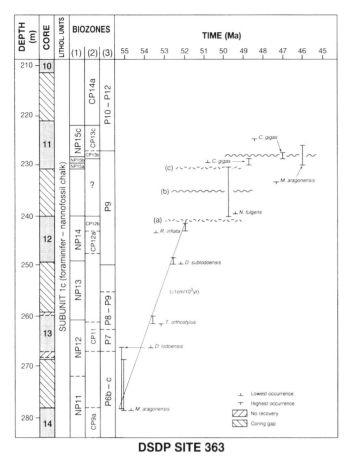

DSDP SITE 363

Fig. 42.—Stratigraphic interpretation of the lower and middle Eocene section recovered from DSDP Site 363 and sedimentation rate curve(s). Inferred unconformities occur at ~235 m. Lithologic units are from Bolli and others (1978b). Biostratigraphic datums are given in Appendix D, Table 27. (1) from Proto Decima and others (1978), (2) from Bukry (1978d) and (3) from Toumarkine (1978). Correlation to "P" Zones based on Berggren and Miller (1988, Text Fig. 1).

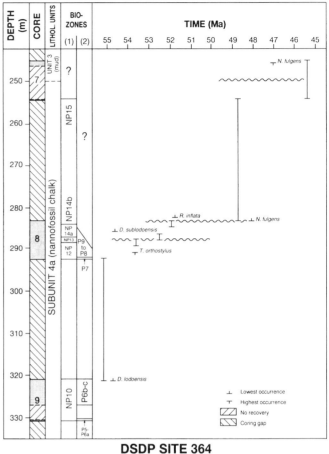

DSDP SITE 364

Fig. 43.—Stratigraphic interpretation of the lower and middle Eocene section recovered from DSDP Site 364. Lithologic units are from Bolli and others (1978c). Ages of biostratigraphic datums are given in Appendix D, Table 28. (1) from Proto Decima and others (1978) and (2) from Toumarkine, 1978. Correlation to "P" Zones based on Berggren and Miller (1988, Text Fig. 1).

DSDP Site 362 (19° 45.45'S, 10° 31.95'E).—DSDP Site 362 (Bolli and others, 1978b) is also located on the Walvis Ridge on the Abutment Plateau of the Frio Ridge segment, but it lies in 1336-m water depth. DSDP Hole 362A penetrated a 55-m thick upper lower to middle Eocene section. The hole was terminated as upper lower Eocene sediments were reached, and extensive coring gaps and poor recovery prevent a satisfactory interpretation of the lower to middle Eocene stratigraphic record at this site. However, it should be noted that in contrast to the less than 20-m thick Zone NP15 recovered from DSDP Site 363, Zone NP15 is 45 m thick at DSDP Site 362. Assuming that no unconformity occurs, the sedimentation rate during Biochron NP15 was ~1 cm/10³ yr.

DSDP Site 364 (11° 34.32'S, 11° 58.30'E).—DSDP Site 364 (Bolli and others, 1978c) lies on the seaward edge of the salt plateau on the continental slope of the Angola margin basin in 2439-m water depth. A ~70-m thick lower to middle Eocene section, unconformable with upper Oligocene and upper Paleocene sediments, was penetrated. Extensive coring gaps hamper a comprehensive interpretation of the lower to middle Eocene stratigraphic record at this site (Fig. 43), which, for this reason,

is not included in Fig. 39. It should be noted that the reduced thickness of Zones NP13 and Subzone NP14a suggests the presence of one or more unconformities (around the NP12/NP13, NP13/NP14a and/or NP14a/NP14b zonal/subzonal boundaries).

DSDP Sites 525, 527 to 529.—DSDP Sites 525 (Moore and others, 1984a), 527 (Moore and others, 1984b), 528 (Moore and others, 1984c) and 529 (Moore and others, 1984d) are part of a ~230-km long transect drilled from the crest of the Walvis Ridge (near 1000-m water depth) down its northern flank into the Angola Basin to a depth of 4400 m. These four sites are located ~800 km west of the Coast of South Africa and lie outside of the main flow of the Benguela Current, beneath the northward flowing surface current in the eastern part of the central tropical gyre (Moore and others, 1984a).

Discontinuous lower to middle Eocene sections were recovered from Holes 525A, 527, 528 and 529. While inferred unconformities are relatively easy to locate in these sections (Figs. 44 to 47), the ages of their upper and lower surfaces are difficult to estimate on the basis of the data presently available. Because of poor preservation, planktonic foraminifera provide limited,

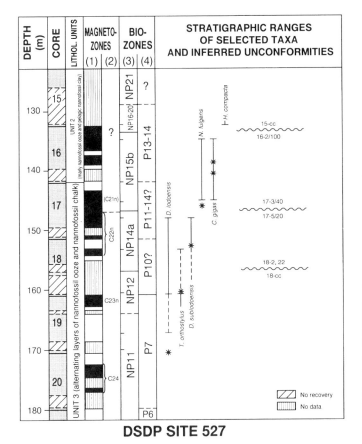

DSDP SITE 527

FIG. 44.—Stratigraphic interpretation of the lower and middle Eocene section recovered from DSDP Site 527. Lithologic units are from Moore and others (1984b). Stratigraphic ranges are given in Appendix D, Table 29. (1) from Moore and others (1984), (2) Proposed (re)interpretation, (3) from Manivit (1984), partly re-interpreted, and (4) from Boersma (1984).

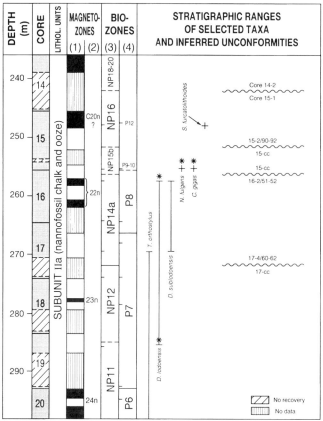

DSDP SITE 528

FIG. 45.—Stratigraphic interpretation of the lower and middle Eocene section recovered from DSDP Site 528. Lithologic units are from Moore and others (1984c). Stratigraphic ranges are given in Appendix D, Table 30. (1) from Shackleton and others (1984), (2) Proposed (re)interpretation, (3) from Manivit (1984) and (4) from Boersma (1984).

often tentative, biozonal information (Boersma, 1984). Calcareous nannofossils (Manivit, 1984) provide the main stratigraphic control, but their stratigraphic significance cannot be fully exploited due to scarce sampling and lack of range charts. Magnetostratigraphy is available for Holes 527 and 528. However, in this latter hole, poor recovery and drilling disturbance obscure the magnetic signal. Despite these limitations, the stratigraphy of the lower to middle Eocene record from these sites can be interpreted relatively confidently. The sites where better stratigraphic resolution is possible are discussed first.

DSDP Hole 527 (28° 02.49′S 01° 45.80′E).—A ~45-m thick lower to middle Eocene section was recovered from Hole 527, the deepest hole in the transect (4428-m water depth).

At least six normal polarity intervals were delineated in Cores 20 to 16 (Fig. 44). Correlation of the lower two, in mid Core 20 and in the upper part of Core 19, respectively with Zone NP11 and NP12 warrants their identification as Chrons C24n (probably C24Bn) and C23n, in agreement with Shackleton and others (1984). The normal polarity interval in Core 18, section 2, correlates with Subzone NP14a characterized by the co-occurrence of *D. sublodoensis* and *D. lodoensis*, without *R. inflata* (see Manivit, 1984). The normal interval in Core 17, section 2

to 5 (or to cc) is interpreted as Chron C21n by Shackleton and others (1984). If that were correct, this interval would correlate with Subzones NP14b and NP15a. Although it is unclear whether Subzone NP15a occurs or not in this section, the fact remains that Subzone NP14b is not represented in this hole. An inferred unconformity is therefore located in Core 17 between Subzone NP14a and Zone NP15. There are conflicting data regarding which part of Zone NP15 is present. Manivit (1984) reports the presence of *C. gigas* as low as 145.40 m (Core 17–3, 40 cm) while Backman (pers. commun. in Shackleton and others, 1984) places the LO of this species at 140.77 m. Depending on which interpretation is correct, the normal polarity interval in Core 17 may result from the concatenation of Chrons C22n (which extends from Core 18, section 2 to Core 17, section 4) and Chron C21n (if Subzone NP15a is present). If Subzone NP15a is absent, the upper part of this normal polarity interval (Core 17, sections 2 and 3) is not interpretable. Identification of the two normal polarity intervals in Core 16 as Chrons C20n and C18n (Shackleton and others, 1984) is not supported by calcareous nannofossil stratigraphy (Fig. 44). It must be pointed out that, should the magnetostratigraphic interpretation be more reliable than the calcareous nannofossil stratigraphy, it would remain that the magnetostratigraphic rec-

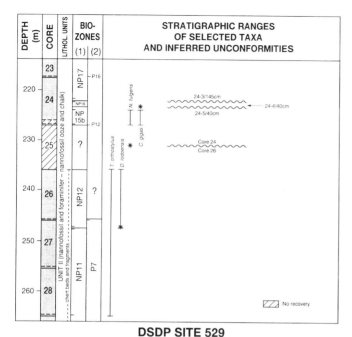

DSDP SITE 529

FIG. 46.—Stratigraphic interpretation of the lower to middle Eocene section recovered from DSDP Site 529. Lithologic units are from Moore and others (1984d). Stratigraphic ranges are given in Appendix D, Table 31. (1) from Manivit (1984), partly re-interpreted, and (2) from Boersma (1984).

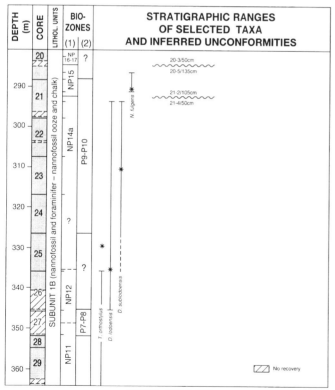

DSDP SITE 525

FIG. 47.—Stratigraphic interpretation of the lower to middle Eocene section recovered from DSDP Hole 525A. Lithologic units are from Moore and others (1984a). Stratigraphic ranges are given in Appendix D, Table 32. (1) from Manivit (1984) and (2) from Boersma (1984).

ord itself suggests that important stratigraphic gaps occur. Following Shackleton and others (1984), Chron C19n is missing and the stratigraphic intervals corresponding to Chrons C20n and C20r are anomalously thin.

Based on the available calcareous nannofossil stratigraphy (and without conflicting data from magnetostratigraphy), the lower to middle Eocene interval in Hole 527 is interpreted as follows:

— a relatively continuous NP11-NP12 zonal interval (~20 m-thick) underlies unconformably Subzone NP14a. It is not clear whether Zone NP13 is present or not, the HO and the LO of respectively *T. orthostylus* and *D. sublodoensis* being placed at different levels in Manivit (1984) and in Shackleton and others (1984). If it does, it is reduced to a thin veneer in Core 18;

— Subzone NP14a underlies unconformably Zone NP15, either Subzone NP15a (following Shackleton and others, 1984) or Subzone NP15b (following Manivit, 1984). We note that if the LO of *C. gigas* is at 140.77 m, it is coincident with the boundary between lithologic Units 2 and 3 (Fig. 44), which suggests a possible unconformity at this level (see discussion below); and

— Subzone NP15b underlies unconformably Zone NP16.

There are no conflicting data from the magnetostratigraphy, but there is little agreement with the planktonic foraminiferal stratigraphy which suggests much younger zonal assignments than the calcareous nannofossils (Fig. 44). Since zonal assignments based on planktonic foraminifera were not based on the occurrence of zonal marker taxa but were mostly tentative (Boersma, 1984), greater confidence is placed on the calcareous nannofossil stratigraphy. Alternatively, this may suggest intensive reworking of the calcareous nannofossils.

While the unconformities can be relatively precisely located, it is not possible to determine confidently the age of their upper and lower surfaces because of the uncertainties both in the magnetostratigraphic and the biostratigraphic records (the locations in the cores of most magnetic reversal boundaries and zonal boundaries are unknown). Thus the interpretation in Figure 39 is tentative with regard to the extent of the hiatuses. It is also based on the simplest situation in which Subzone NP15a and Zone NP13 are missing from the record.

DSDP Site 528 (28° 31.49'S, 02° 19.44'E).—A 40-m thick lower to middle Eocene section (Fig. 45) was recovered from Hole 528, located at intermediate depth between Sites 529 and 527.

Six normal polarity intervals were recorded between 245 and 300 m (Cores 15 to 20, see Shackleton and others, 1984, Fig. 1). The upper four were identified by Shackleton and others (1984) as Chrons C20n (between 245 and 248.59 m), C21n (between 257.05 and 258.24 m), C22n (centered on 261.40 m) and C24n (top at 277.81 m). The last three assignments are incompatible with calcareous nannofossil stratigraphy (see also Backman, 1986). The two normal polarity intervals in Core 16 correlate with Subzone NP14a and both represent Chron C22n rather than Chrons C21n and C22n (note that Chron C22n also appears to include a reversed interval in Hole 527). The next lower interval with normal polarity is problematic. If its top is located at 277.81 m as indicated in the text (Shackleton and others, 1984, Table 1), it correlates with Zone NP12. The magnetic reversal boundary at 277.81 m then represents the Chron C22r/C23n boundary rather than the Chron C23r/C24n boundary, located in Core 20 at ~294 m (see Fig. 45). On the other hand, if this normal polarity interval is located in Core 17 (Shackleton and others, 1984, Fig. 1), it correlates with Zone NP14a and may be part of Chron C22n. The younger normal polarity interval (in Core 15) may represent the upper part of Chron C20n (since it correlates with Zone NP16), but could correspond to Chron C19n or C18n as well.

Based on calcareous nannofossil stratigraphy (and easily supported by magnetostratigraphy although the pitfall of circular reasoning cannot be avoided), the lower to middle Eocene interval recovered from DSDP Hole 528 is interpreted as follows:

— a ~20-m thick, apparently continuous NP11—NP12 zonal interval unconformable with Subzone NP14a (~15 m-thick). There is no evidence that Zone NP13 is present. As at Hole 527, if it occurs, it corresponds to a thin veneer in Core 17;
— Subzone NP14a is unconformable with Subzone NP15b as indicated by the absence of Subzone NP14b and the simultaneous lowest occurrences of *N. fulgens* and *C. gigas*. and
— in turn, Subzone NP15b is unconformable with Zone NP16.

The thicknesses of the magnetozones are poorly established as a result of poor recovery, drilling disturbance and low inclination. Biostratigraphic determination is based on coarse sampling. Hence, a precise location of the unconformities in the section and a valid determination of the thickness of each biozone are not possible. As for Hole 527, it is extremely difficult to determine the age of both surfaces of the unconformities and the extent of the hiatuses. The interpretation given in Figure 39 is consequently tentative.

DSDP Hole 529 (28° 55.83' S, 02° 46.08' E).—A 40-m thick lower to middle Eocene section (Fig. 46) was recovered from DSDP Hole 529 drilled at an intermediate depth between Sites 525 and 528 in 3035-m water depth.

No magnetostratigraphy is available, but a detailed calcareous nannofossil stratigraphic study (Manivit, 1984) allows the following interpretation:

— Cores 28 to 26 correspond to a 30-m thick, apparently continuous, NP11—NP12 zonal interval; and
— Core 24 yields a thin, discontinuous middle Eocene section. Zone NP17 overlies unconformably Zone NP16 (~1 m thick), itself unconformable with Subzone NP15b. Because of no recovery in Core 25, the thickness of Subzone NP15b is unknown. Subzone NP15b and Zone NP12 may be directly unconformable, or Subzone NP14a may be intercaled. Both interpretations are given in Figure 39.

DSDP Hole 525A (2'° 04.24' S; 02° 59.12' E).—This is the shoalest site of the transect, located near the crest of a north-northwest south-southeast trending block in 2467-m water depth.

It is the site of the transect where the thickest (80 m) lower to middle Eocene section was recovered. No magnetostratigraphy is available, and the biostratigraphic control is limited (Fig. 47). A normal zonal succession NP11—NP12 is observed. It is unknown whether Zone NP13 is present. There is an unconformity in Core 21 as indicated by the absence of

Subzone NP14b. Which part of Zone NP15 is present remains uncertain. The suggestion that an unconformity may occur at ~284 m following Manivit's (1984) remark of a marked contrast between the calcareous nannofossil assemblages in samples 20–3, 50–51 cm (282.6 m) and 20–5, 135 cm (286.45 m) is contradicted by the report of the LO and FO of *Chiasmolithus gigas* at 283.89 m and 279.10 m, respectively, by Shackleton and others (1984). While the stratigraphic record in this hole is generally similar to that of the other sites in the transect, less information is available to delineate reasonably well the hiatuses. Therefore, this site is not included in Figure 39.

Western South Atlantic Ocean.—

DSDP Site 356 (28° 17.22' S, 41° 05.28' W).—DSDP Site 356 (Supko and others, 1977a) was drilled on the southern edge of the São Paulo Plateau at a water depth of 3175 m. It is located in the zone between the escarpment of the São Paulo Plateau and the area of diapirs that underlies most of the Plateau (Supko and others, 1977a, Fig. 2), and it is not known whether this contains also a very thin layer of evaporitic sediments.

A ~170-m thick lower to middle Eocene section, unconformable with upper Paleocene and lower Miocene sediments, was penetrated at this site. Its stratigraphic interpretation based on calcareous nannofossils and planktonic foraminifera is given in Figure 48. Interpretation from the latter (Boersma, 1977) are equivocal due to poor preservation and lack of index species. Alternative calcareous nannofossil stratigraphic interpretations are available (Perch-Nielsen, 1977; Bukry, 1977; Fig. 48). Both support the occurrence of an unconformity near the lower/middle Eocene boundary as indicated by the stratigraphic proximity of chronologically well-separated datums (HO of *Discoaster kuepperi, D. lodoensis, Rhabdosphaera inflata,* LO's of *Nannotetrina fulgens, R. inflata*). I suggest that this inferred unconformity occurs at 223.60 m (Core 10, Section 4, 60 cm; see Supko and others, 1977a, p. 175), and corresponds to the boundary between lithologic units 2 and 3. Its lower and upper surfaces are located respectively in the lowermost and uppermost parts of Biozone NP14b. The hiatus is shorter than Biochron NP14b (=2.2 my). It is arbitrarily estimated to be 1.5 my long. The formation of this unconformity probably involved reworking as indicated by the occurrence in Core 10 of "many contaminants" among the planktonic foraminifera (Boersma, 1977, p. 573).

The record may yield a second, older unconformity. Biochron NP14a is 0.6 my long. The LO's of *Discoaster sublodoensis* and *R. inflata* in the core cannot be established confidently because of the coring gap between Cores 10 and 11 and no recovery in sections 5 and 6 of Core 10. However, if it is assumed that (1) the levels where these species are first reported by Perch-Nielsen (1977) represent their lowest occurrences in the core, and (2) Subzone NP14a was sedimented at the same rate as Zone NP13 (>6 cm/10^3 yr following Perch-Nielsen's (1977) interpretation), Subzone NP14a is slightly over 3-m thick (and corresponds to Core 10, sections 5, 6, and cc), rather than 6-m thick as would be expected. The unprecise location of the LO of *R. inflata* prevents determination of both the thickness of the lower part of Subzone NP14b below the unconformity at 223.60 m and the thickness of Subzone NP14a. The 9.5-m coring gap impairs the delineation of the NP13/NP14a zonal boundary.

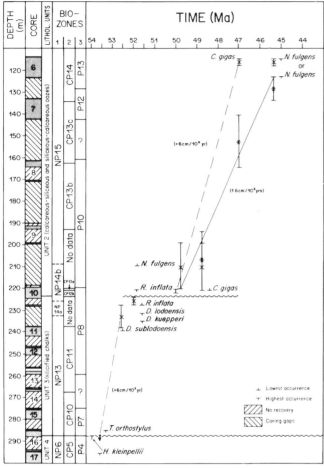

DSDP SITE 356

Fig. 48.—Stratigraphic interpretation of the lower and middle Eocene section recovered from DSDP Site 356 and sedimentation rate curve. An (inferred) unconformity occurs at 223.80 m at the boundary between lithologic Units 2 and 3. Lithologic units are from Supko and others (1977a). Datums used to constrain the curve(s) are given in Appendix D, Table 33. (1) from Perch-Nielsen (1977ˢ), (2) from Bukry (1977*) and (3) from Boersma (1977).

Thus the stratigraphic record around the lower/middle Eocene boundary at DSDP Site 356 may yield two unconformities, one firmly established in Subzone NP14b, the other uncertain in Subzone NP14a or at the boundary between Zone NP13 and Subzone NP14a. Both interpretations are given in Figure 49. It should be noted that there is no constraints as to the extent of the hiatus associated with the older unconformity if it exists, and its interpretation in Figure 49 is highly speculative. Interpretation of the middle Eocene record in Figure 49 follows Bukry's (1977) interpretation which appears to be more consistent for this interval than that of Perch-Nielsen (1977; see Fig. 48).

DSDP Hole 515B (26° 14.32'S, 36° 30.19'W).—DSDP Hole 515B (Barker and Johnson, 1983a) was drilled in the southern Brazil Basin about 200 km north of the northern exit of the Vema Channel at a water depth of 4252 m. It is located near the crest of a sediment drift and lies within a broad field of sediment waves.

A 20-m thick interval of lower Eocene zeolite-rich calcareous mudstone was recovered, unconformable with Oligocene terrigenous mudstones. The stratigraphic interpretation of this interval is given in Figure 50 based on calcareous nannofossil and planktonic foraminiferal zonations and magnetostratigraphy (Barker and Johnson, 1983a). Because of extremely poor preservation, the magnetobiostratigraphic control is limited, and no biozonal and magnetic information is available for the sediments immediately below the unconformity at 617 m (in sections 1, 2 and the upper 70 cm of section 3 in Core 56). There is a general agreement between biostratigraphic datums and magnetic reversals as shown in Figure 50, and the estimate of 0.46 cm/10³ yr for the early Eocene sedimentation rate (Fig. 50, model (a)) compares well with the estimate of 0.4 cm/10³ yr by Barker and Johnson (1983a). Following model (a) in Figure 50, the lower surface of the unconformity at 617 m is located in Figure 49 in the upper part of Zone NP13. Assuming (1) a constant sedimentation rate of 0.46 cm/10³ yr throughout the early Eocene Epoch, and (2) that the sediment in the upper part of Core 56 belongs to Zone NP13 (the simplest case is arbitrarily accepted here although more complex cases cannot be ruled out), the interval assigned to Zone NP13 represents a duration of 0.86 my (Biochron NP13 is estimated to be 1.1 my long, Berggren and others, 1985). The unconformity at 617 m (base Core 55) is well-defined. It is overlain by a lag deposit which consists of subrounded to angular grains of fine sand with quartz, fish teeth, glauconite, biotite and heavy minerals. This unconformity corresponds to a sharp and strongly discordant acoustic reflector which extends throughout the Brazil Basin (Barker and Johnson, 1983a).

There is a possibility that an older unconformity occurs in this section. If primarily constrained by the magnetic reversals, the sedimentation rate curve (model (b) in Fig. 50) suggests that an unconformity occurs at the NP13/NP13 zonal boundary. The proximity of the P8/P9 to the NP12/NP13 zonal boundary may support this interpretation, indicating that the lower part of Zone NP13 is absent. In the absence of definitive evidence, both interpretations of this stratigraphic interval are given in Figure 49. *DSDP Hole 516F (30° 16.59'S, 35° 17.10'W).*—DSDP Hole 516F (Barker and Johnson, 1983b) was drilled on the Rio Grande Rise in-1313 m water depth. A 370-m thick, apparently continuous, Eocene section was recovered. The upper lower to middle middle Eocene section represents a 140-m thick interval which comprises two lithologic units separated by an allochthonous 15-m thick lower Maestrichtian limestone thought to have been "emplaced as a semi-consolidated ooze or chalk by downslope slumping" (Barker and Johnson, 1983b, p. 162).

The stratigraphic interpretation of this lower to middle Eocene interval is given in Figure 51. There is little agreement between calcareous nannofossil (Barker and others, 1983b; Wei and Wise, 1989) and planktonic foraminiferal (Pujol, 1983) biozonal subdivisions. The latter appears unreliable, however, because of poor preservation and scarcity/absence of index species (Pujol, 1983). Biozonal assignment derived from the detailed analysis of the distribution of the calcareous nannofossil species by Wei and Wise (1989) is re-interpreted (see Fig. 51). In particular, Zone NP 14b is believed to be absent, based on the absence of *R. inflata* (a species which should occur in these shallow (~800 m, see below) marine deposits and is reported from nearby DSDP Site 357 (see below)) and the continuous

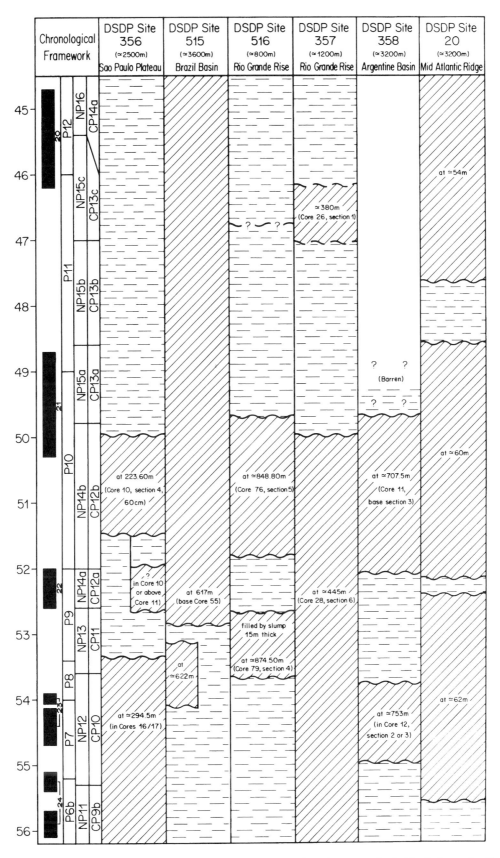

FIG. 49.—Temporal interpretation of 6 DSDP lower and middle Eocene sections recovered from the western South Atlantic Ocean. See Figure 9 for description.

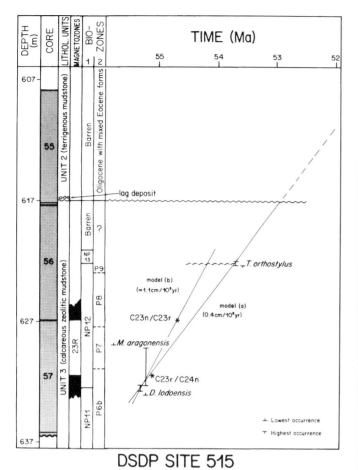

DSDP SITE 515

FIG. 50.—Stratigraphic interpretation of the lower Eocene section recovered from DSDP Site 515B and sedimentation rate curve. An unconformity occurs at 617 m at the boundary between lithologic Units 2 and 3. Lithologic units are from Barker and Johnson (1983a). Datums used to constrain the curve are given in Appendix D, Table 34. Magnetozones, (1) and (2) from Barker and Johnson (1983a).

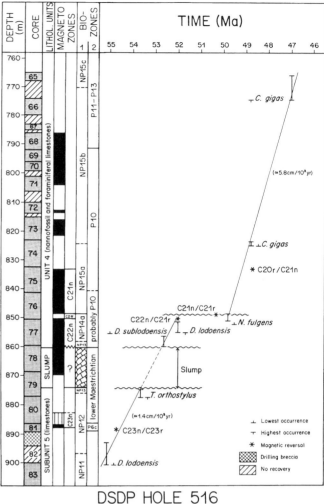

DSDP HOLE 516

FIG. 51.—Stratigraphic interpretation of the lower to middle Eocene section recovered from DSDP Site 516F and sedimentation rate curve(s). An inferred unconformity occurs at ~850 m. Another unconformity occurs within Zone NP13, as indicated by the slump of Lower Maestrichtian limestone (see text for further discussion and Fig. 49). Lithologic units are from Barker and Johnson (1983b). Datums used to constrain the curve are given in Appendix D, Table 35. (1) from Wei and Wise (1983) and (2) from Pujol (1983). Magnetozones from Berggren and others (1983), reinterpreted herein. Position of magnetic reversals taken from Berggren and others (1984, Fig. 2).

occurrence of *D. lodoensis* up to a level close to the lowest occurrence of *N. fulgens* (without evidence of reworking). Magnetostratigraphy is available (Barker and others, 1983b; Berggren and others, 1983), but only part of it is interpretable. Based on the revised calcareous nannofossil stratigraphy given herein, identification of the magnetozones by Berggren and others (1983) in the interval between 780 and 890 m is revised as follows. The normal polarity interval in the upper part of Core 81 which correlates with mid-Zone NP12 likely represents Chron C23n (rather than C24n in Berggren and others, 1983). A thick normal polarity interval extends from Core 77 to Core 79, section 1. Its upper part, in Core 77 and Core 78, section 1, correlates with Subzone NP14a and the uppermost part of Zone NP13. It is assignable to Chron C22n (it was identified as Chron C21n in Berggren and others, 1983). The lower part of this normal polarity interval reflects re-magnetization (or some Maestrichtian Chron). In agreement with Berggren and others (1983), the normal polarity interval in Cores 75 and 76, which correlates with Subzone NP15a, represents Chron C21n. The normal polarity intervals in Cores 73, 72 and 71 through 68 probably result from overprinting.

It results from this revision that the thin reversed interval in Core 77, section 5 and cc, which correlates with Subzone NP14a, represents the lower part of Chron C21r or a thin reversed interval in Chron C22n (the NP14b/NP15 zonal boundary occurs well within Chron C21n; if the interval represents the lower part of Chron C21r, there is a minor discrepancy with the correlations in Berggren and others (1985) where the Chrons C22n/C21r magnetic reversal does not lie in Subzone NP14a but is correlated with the NP13/NP14a zonal boundary; on another hand Chron C22n includes reversed intervals at DSDP Sites 527 an 528 (see Figures 44 and 45)). This implies that Chron C21r is truncated and that an unconformity occurs at the C21r/C21n (or the C22n/C21n) magnetic reversal boundary in Core 77 (i.e., at ~848.80 m in Hole 516F). The presence of this inferred unconformity is supported by the sedimentation rate curves which show an excellent agreement between calcareous nannofossil datums and magnetic polarity reversals. The middle Eocene curve indicates that Chron C21n is also truncated and that the upper surface of the unconformity lies in Subzone NP15a, very close to its base. The lower surface of the unconformity lies in the lower part of Chron C21r. The high value (5.8 cm /10³ yr) of the middle Eocene sedimentation rate is in agreement with the fact that Unit 2 is of turbiditic nature (Barker and others, 1983b).

Two abnormal contacts occur in Cores 78 and 79, as the allochtonous Maestrichtian limestone is intercalated within the upper lower Eocene limestones. As shown by the sedimentation rate curve (Fig. 51), the 15-m thick slump does not add to the thickness of the lower Eocene section. Instead, embedded in an interval assigned to Zone NP13, it merges with it. This implies that over 1-my worth of early Eocene sedimentation is unaccounted for. There is no reason to suspect a sudden change in sedimentation rate. Rather, this suggests an unconformity and a 1-my long hiatus. The slump may have come to rest on this unconformity, or may have contributed to its formation by pushing ahead 15 m of upper lower Eocene oozes.

The lower to middle Eocene record at DSDP Hole 516F thus comprises two unconformities, one in the upper lower Eocene, the other in the lower middle Eocene. The older hiatus is estimated to be slightly over 1 my long, the younger one ~2 my long (Fig. 49).

DSDP Site 357 (30° 00.25'S, 35° 33.59'W).—A 165-m thick middle Eocene section (NP14b to NP16), unconformable with lower upper Paleocene (NP5) sediments, was recovered from DSDP Site 357 (Supko and others, 1977b) drilled on the northern flank of the Rio Grande Rise in a water depth of 2086 m.

Because of intermittent coring and conflicting information obtained from planktonic foraminifera (Boersma, 1977) and calcareous nannofossils (Bukry, 1977; Perch-Nielsen, 1977), the stratigraphic interpretation (Fig. 52) of this middle Eocene section is extremely difficult. A provisional interpretation is proposed here, based on the integration of the data from Bukry (1977) and Perch-Nielsen (1977). Both sets of data agree reasonably well except for the age of the oldest middle Eocene sediments recovered, assigned to Zone NP15 by Perch-Nielsen, to Zone CP12b (=NP14b) by Bukry (1977). It seems reasonable to assign a thin interval above the unconformity at ~445 m to Subzone NP14b.

It is not the purpose of this paper to discuss unconformities younger than ~46 Ma, and no attempt has been made to trace a sedimentation rate curve for the chalk above Core 26. An unconformity probably occurs near the top of Core 26 or the base of Core 25, or may correspond to the base of the volcanic breccia. Alternatively (and likely) two (or more) unconformities occur in this interval. The current stratigraphic resolution available for this site does not allow resolution of the stratigraphic continuity of this part of the section. The interpretation of the upper part of the middle interval recovered from DSDP Site 357 in Figure 49 is thus provisional.

DSDP Site 358 (37° 39.31S, 35° 57.82'W).—DSDP Site 358 (Supko and others, 1977c) lies in the eastern part of the Argentine Basin, an area presently under the weak eastern arm (south flowing) of the Argentine bottom gyre, in a water depth of 4490 m.

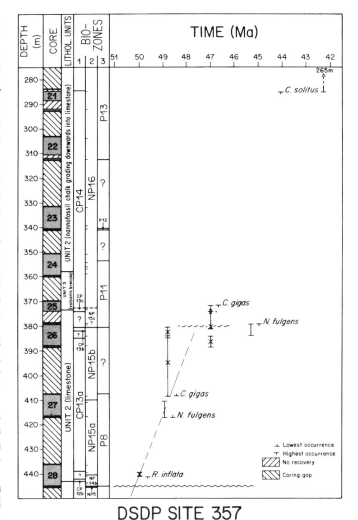

DSDP SITE 357

FIG. 52.—Stratigraphic interpretation of the middle Eocene section recovered from DSDP Site 357 and sedimentation rate curve. An (inferred) unconformity occurs within Subzone NP15b. Lithologic units are from Supko and others (1977b). Datums used to constrain the sedimentation rate curve are given in Appendix D, Table 36. (1) from Bukry (1977), (2) revised interpretation based on the data provided by Bukry (1977) and Perch-Nielsen (1977), and (3) from Boersma (1977).

There is little doubt that the 60-m thick lower to lower middle Eocene record recovered from this site comprises at least three unconformities. However, because of a 40-m coring gap which encompasses most of the upper lower Eocene Series, and of poor preservation of the microfossils, its stratigraphic interpretation (Fig. 53) is extremely difficult. The only stratigraphic control is provided by the calcareous nannofossils, the planktonic foraminifera being strongly dissolved (Boersma, 1977). Calcareous nannofossil assemblages are also impoverished because of solution effects, and biostratigraphic subdivision is ambiguous (see Perch-Nielsen, 1977, Table 20). In addition, although there is a general agreement in the biostratigraphic interpretations by Perch-Nielsen (1977) and Bukry (1977), there are discrepancies in the details. Thus while the presence of unconformities is clear, the extent of their associated hiatus is speculative.

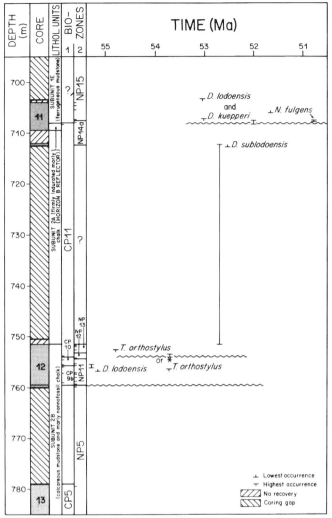

DSDP SITE 358

FIG. 53.—Stratigraphic interpretation of the lower to middle Eocene section recovered from DSDP Site 358. Two inferred unconformities occur, the younger at the NP14a/NP15 zonal boundary, the older around Zone NP12. In addition, a lower upper Paleocene (NP5)/lower Eocene (NP11) unconformity occurs at the base of Core 12 at ~759 m. Lithologic units are from Supko and others (1977c). Datums are given in Appendix D, Table 37. (1) from Bukry (1977) and (2) from Perch-Nielsen (1977).

The oldest unconformity occurs within subunit 2B at ~759.50 m between section 6 and cc in Core 12. The stratigraphic gap encompasses the lower part of the lower Eocene and most of the upper Paleocene Series.

The youngest unconformity is at ~707.50 m at the NP14a/NP15 zonal boundary and at the lithologic boundary between Subunits 1E and 2A, this latter corresponding to the Horizon B reflector (Supko and others, 1977c). Bukry (1977) did not record Zone CP12 (= NP14) at DSDP Site 358 (which suggests a NP13/NP15 abnormal contact, and hence, a longer hiatus). However, because of the greater number of samples studied by Perch-Nielsen (1977), the interpretation by this latter author is followed here. The lower surface of the unconformity lies in Subzone NP14a; there is no data to constrain the location of its

upper surface in Zone NP15 (Biochron NP15 is over 4 my long, see Berggren and others, 1985), since sediments immediately above the NP14/NP15 zonal boundary are barren. Arbitrarily and conservatively, the upper surface of the unconformity is located in Subzone NP15a.

The third unconformity is in Core 13 at ~753 m and is indicated by the unusual thinness of Zone NP12 relative to Zone NP13. The unconformity may be within Zone NP12 or may straddle the NP12/NP13 (or NP11/NP12) zonal boundary. The uncertain location of the lowest occurrence of *D. sublodoensis* hampers any realistic determination of the hiatus.

The interpretation of the lower to middle Eocene interval at DSDP Site 358 in Figure 49 should be regarded as very tentative with regard to the duration of the hiatuses.

DSDP Site 20 (28° 31.47'S, 26° 50.73'W).—DSDP Site 20 is located on the mid-Atlantic Ridge in 4506-m water depth (Maxwell and others, 1970). Lower and middle Eocene chalks and oozes were recovered from Hole 20C. The complexity of this section is well-illustrated by the stratigraphic succession in Core 5 which, in less than 7 m, comprises the Paleocene/Eocene (Thanetian/Ypresian) and the lower/middle Eocene (Ypresian/Lutetian) boundaries (Maxwell and others, 1970, Fig. 16B). The stratigraphic interpretation in Figure 54 is primarily based

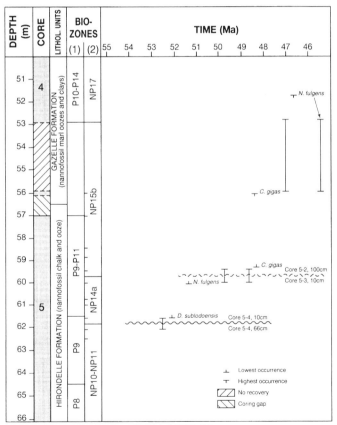

DSDP SITE 20

FIG. 54.—Stratigraphic interpretation of the lower middle Eocene section recovered from DSDP Hole 20C. Lithologic units are from Maxwell and others (1970). Datums are given in Appendix D, Table 38. (1) from Blow (1970) and (2) from Maxwell and others (1970), Gartner (1970), and Bukry and Bramlette (1970), combined.

on calcareous nannofossil zonation (as deduced from Maxwell and others,1970; Bukry and Bramlette, 1970; Gartner, 1970). Discrepancies in age determinations based on planktonic foraminifera (Blow, 1970) and calcareous nannofossils (Maxwell and others, 1970) may partly result from reworking (as suggested by the occurrence of *D. lodoensis* in Core 5–1, 148 cm at a level assignable to Zone NP15, that of *T. orthostylus* in Core 5–4, 100 cm at a level assignable to Zone NP14). They may also reflect a complex sedimentary history which cannot be retraced from the data presently available. The stratigraphic interpretation of this site (Figs. 49, 54) is therefore somewhat preliminary. However, the absence of Zones NP12 and NP13 indicates the presence of an unconformity in the upper part of Core 5–4 with a hiatus greater than 2.7 my. Although the description of the calcareous nannofossil assemblages in Core 5–2 is succinct, it seems that only the lower part of Zone NP14 (Subzone NP14a) is present. This suggests that an unconformity occurs between sections 2 and 3 of Core 5, with a hiatus of at least 2.2 my. Finally, the data available suggest that only Subzone NP15b occurs in this interval. It is not possible to estimate the ages of the genetic surfaces of the inferred unconformities based on calcareous nannofossil stratigraphy. The interpretation given in Figure 49, although arbitrary, accounts for the thinness of the lower and middle Eocene biozones represented in Hole 20C.

Subantarctic Atlantic Ocean.—

Four, thick lower to middle Eocene sections were recovered in the Subantarctic South Atlantic. Insufficiently detailed biostratigraphy due to lack of the index taxa and/or to too sparce sampling, and poor recovery at Sites 698, 699 and 700 hamper comprehensive stratigraphic/temporal interpretations of these sections which have been regarded as continuous (e.g., Cieselski and others, 1988a, b, c; Nocci and others, 1991). Yet, there is sufficient biostratigraphic evidence to demonstrate that these sections are discontinous. It is clear that a stratigraphic gap at the 4 sites encompasses Zone NP12 (and probably parts of Zones NP11 and NP13).

*ODP Site 698 (51° 27.51′S, 33° 05.96′W).—*Located near the eastern edge of the shallowest part of the northeast Georgia Rise in 2128-m water depth, ODP Hole 698A recovered 70 m of lower middle and lower Eocene foraminiferal bearing nannofossil ooze (Cieselski and others, 1988a). The lack of detailed calcareous nannofossil stratigraphy and of magnetostratigraphy and poor recovery renders the interpretation of this site speculative. Following Crux 's (1991) description of the range of the marker species, the upper part of the section belongs to Zone NP13 and Subzone NP14a (4.50 to ~20 mbsf). The absence of overlap in the ranges of *Discoaster lodoensis* and *Tribrachiatus orthostylus* (Cieselski and others, 1988a) suggests that Zone NP11 underlies unconformably Zone NP13, with a hiatus at ~20 mbsf. This intepretation is supported by the lowest occurrence of *Morozovella aragonensis* at ~35mbsf (Nocci and others, 1991). The FAD of this species is associated with Chron C24n (Berggren and others, this volume).

*ODP Site 699 (51° 32.537′S, 30° 40.619′W).—*Site 699 is located on the northeastern slope of northeast Georgia Rise in a water depth of 3705.5 m (Cieselski and others, 1988b). An over 100-m thick lower to middle Eocene section of nannofossil chalk was recovered from Hole A. Despite a clean magnetic

signal (Hailwood and Clement, 1991) allowing integrated magnetobiostratigraphy, poor recovery between 470 and 496 m and between 430 and 410 m renders the constrution of a sedimentation rate curve meaningless. Also the planktonic foraminiferal stratigraphy (Nocci and others, 1991) in this hole cannot be used to complement the calcareous nannofossil stratigraphy (Crux, 1991), because the delineation of the zonal boundaries is improperly based on the expected correlation either to magnetostratigraphy or to calcareous nannofossil stratigraphy rather than based on direct evidence from the planktonic foraminifera (see Nocci and others, 1991, p. 254, 255).

Zone NP12 appears to be absent (or reduced to a thin veneer), and I suggest the presence of an unconformity between Core 50X-4 and 51X-cc. The magnetic reversal at ~464.60 m suggested by Hailwood and Clement (1991) to correspond to the Chrons C22n/C22r boundary thus likely reflects the concatenation of Chron C24r and possibly (?)Chron C22n. Chron C22n may extend from ~457 to 431 mbsf based on the correlation with Subzone NP14a (co-occurrence of *Discoaster sublodoensis*, *D. lodoensis* and *D. kuepperi*, see Crux, 1991, Table 2; in Holes 527, 528 Chron C22n is seen to include a reversed polarity interval, see Figs. 44, 45). The delineation of the P9/P10 zonal boundary at ~440 mbsf (Nocci and others, 1991) would suggest that the Chron C21r/C22n boundary occurs close to this level and that the interval with normal polarity between ~434.68 mbsf and 431 mbsf (Hailwood and Clement, 1991) represents Chron C21n. However, delineation of the base of P10 at ~440 mbsf is improper because it is not based on direct evidence from the planktonic foraminifera but on the basis of the magnetostratigraphy.

Lack of magnetostratigraphic control above 430 mbsf, poor recovery, and insufficiently detailed calcareous nannofossil stratigraphy results in uncertain stratigraphic interpretation of the remainder of the middle Eocene section. However, the thinness of Subzone NP14b compared to Subzone NP14a may indicate an additional stratigraphic gap in Core 45 or 46. For obvious reasons, no temporal interpretation of this site is given.

*ODP Site 700 (51° 31.977′S, 30° 16.688′S).—*An over 160-m thick lower and middle Eocene section was recovered from Hole 700B drilled in the western region of the East Georgia Basin in 3601-m water depth (Cieselski and others, 1988c).

Because of poor recovery and insufficient biostratigraphic control, a stratigraphic interpretation of this section would be highly speculative. For the reason given above, the planktonic foraminiferal zonal subdivisions do not complement the calcareous nannofossil stratigraphy. Only the interval with normal polarity between ~152 and 159.50 mbsf and that with reversed polarity between 169 and 175.50 mbsf can be confidently identified. Their correlation with Zone NP14a (see Cieselski and others, 1988c) and Zone NP13, respectively, indicates Chron C22n and C22r, in agreement with Hailwood and Clement (1991). The magnetic reversal which occurs in Core 20R, questionably identified as the Chron C23n/C23r by Hailwood and Clement (1991), may well represent part of Chron C24. The interval with normal polarity in Core 9 may represent Chron C21n. If this interpretation is correct, it likely implies a stratigraphic gap in Core 14, with Chron C21r represented by a very thin interval. The available data do not allow a determination whether the lower Eocene section is continuous or contains unconformities.

ODP Site 702 (50° 56.786'S, 26° 22.127'S).—Hole 702B was drilled in the central part of the Islas Orcadas Rise in 3083.4-m water depth (Cieselski and others, 1988d). A 140-m thick lower and middle Eocene section was recovered from this hole. The lack of overlap of the ranges of *Tribrachiatus orthostylus* and *Discoaster lodoensis* (Crux, 1991) indicates that Zone NP12 is absent. The inferred unconformity between Core 22x-2, 20–21 cm and Core 22x-cc likely corresponds to the lithologic contact at 202.45 mbsf (Core 22x-5, 15 cm; Cieselski and others, 1988c). The gap encompasses Zone NP12 and parts of Zone NP11 and NP13.

The biozonal subdivisions (Crux, 1991) and the magnetic reversal stratigraphy (Clement and Hailwood, 1991) are difficult to reconcile in this section. The interval with normal polarity between ~98.70 and 113.40 mbsf is identified as Chron C20n by Clement and Hailwood (1991), unsupported by correlation with Subzone NP15b (see Crux, 1991). The interval with normal polarity between 157.30 and ~169.50 mbsf may correspond to Chron C21n, in agreement with Clement and Hailwood (1991). If this is correct, the thiness of Subzone NP14a may indicate an unconformity in Core 21x.

THE UPPER LOWER AND LOWER MIDDLE EOCENE STRATIGRAPHIC
RECORD DURING THE INTERVAL BETWEEN ~CHRON C24N.1–2N AND
CHRON C20N IN THE ATLANTIC OCEAN: DISCUSSION

This study reveals a number of unexpected features, some of which strongly contradict the conventional understanding of the deep-sea stratigraphic record. In this discussion, I will examine only the features that are relevant to the relationship between stratigraphy and geological time. These concern the continuity of the deep-sea stratigraphic record, the fluctuations in accumulation rates in oceanic settings, the relationship between sedimentary sequences on the shelves and in the deep sea, and the significance of stratigraphic versus chronologic resolution. The latter two points, further discussed elsewhere, are only briefly examined here. Prior to discussing these points, preliminary remarks are necessary to elucidate the precepts which have guided this analysis and to clarify the limitations on the stratigraphic and temporal interpretations.

Preliminary Remarks

I have attempted to interpret the lower and middle Eocene sections in a rigourous manner by taking advantage of all magneto- and biostratigraphic information available. While there may be some danger in compiling biostratigraphic information from different sources, my analysis relies most heavily upon calcareous nannofossil stratigraphy, and I believe that the zonal markers in this group for the interval considered are sufficiently distinctive to minimize discrepancies due to differences in taxonomic (species) concepts between authors. I have thus delineated certain subzones, as NP14a, b and NP15a, b, c, which may not have been differentiated in the original biostratigraphic analysis of the sections. These subzones are mostly equivalent to those described by Bukry (1975) and later codified as CP12a, b and CP13a, b, c, by Okada and Bukry (1980). They have been recognized in a sufficient number of sections to warrant their usefulness. In turn, I have used this revised biostratigraphic information to evaluate, and reinterpret if necessary, the magnetostratigraphic pattern observed in these sections. There is some

circular reasoning involved in interpreting a section from magnetobiostratigraphic correlations when the magnetic reversal pattern is interpreted based on biozonal determinations. I see no escape from this dilemma. However, I emphasize the point made above that the early and early middle Eocene magnetochrons are sufficiently long, and sufficiently distinctive in their association with biochronal subdivisions, to reduce considerably if not annul any possible ambiguity. Yet there are regional differences in the stratigraphic expression of magnetochrons, such as the occurrence of a reversed polarity interval within the magnetozone best identified as Chron C22n at several high latitude sections (see above).

The delineation of unconformities in this work is largely dependant upon the determination of the relative spacings between sea floor anomalies given in Berggren and others (1985). As pointed out above, the revision to the composite magnetic reference section by Cande and Kent (1992) and numerically calibrated by Cande and Kent (1995) produces differences in the chron durations (Table 2) that are negligible compared to the duration of the hiatuses identified in this study. Whereas the FADs and LADs of most of the calcareous nannofossil species used in this work are magnetochronologically well calibrated, a few are as yet poorly dated. The worst case concerns the FAD of *Rhabdosphaera inflata* whose timing in Chron C21r is poorly established mostly because of lack of continuous sections with definite magnetobiostratigraphic correlations through the lower/middle Eocene boundary (this work). To a lesser degree of uncertainty, the other unsettled cases are the FAD and LAD of *Chiasmolithus gigas*, clearly associated with Chron C20r as shown in DSDP Hole 523 (Poore and others, 1983). These relationships have not been confirmed at other sites although there are indications that they might be correct (see DSDP 548 and 549, Figs. 17, 18, respectively). It is important to recognize that the vagueness on the timing of these three events (FAD of *R. inflata*, FAD and LAD of *C. gigas*) have no bearing on the delineation of unconformities in this study because the hiatuses are of a duration much greater than the limits of uncertainty in these datums. However, in some cases, it results in an increased uncertainty in the age of the lower or upper surface of an unconformity (and hence in the estimated duration of the hiatus).

The accuracy of the stratigraphic interpretation of the sections given above varies considerably between sections. The greatest confidence is placed in the interpretation of the sections (1) for which magnetobiostratigraphic correlations are available, (2) on which detailed biostratigraphy has been conducted, and (3) which are sufficiently continuous to allow the construction of a reliable sedimentation rate curve. Among these, such sections as DSDP Sites 550, 549 and 530 (Figs. 16, 17, 40, respectively), none of which served as reference in the construction of the magnetobiochronologic framework of Berggren and others (1985), indicate that accumulation rates are constant over given intervals of time. This, in turn, allows confident reliance on stratigraphic interpretations based solely on sufficiently detailed biostratigraphy from such sections as DSDP Sites 404, 400, 548, 398 and 612 (Figs. 12, 13, 18, 22, 28, respectively). In such sections, it is possible to rather precisely locate the inferred unconformities.

The quality of the stratigraphic (and chronologic) interpretation deteriorates when the magneto- and/or biostratigraphic resolution decreases (mostly as a result of poor recovery in the cores), resulting in large uncertainties (up to ~10 m) on the location of the inferred unconformities, as for instance in DSDP Sites 552 and 370 (Figs. 10, 21, respectively). However, it should be noted that whereas uncertainty in the position of a zonal boundary leads to uncertain location of an unconformity, it does not result in alternative solutions in terms of stratigraphic continuity. The quality also diminishes when sections are insufficiently extended between unconformities as in the lower Eocene at DSDP Sites 111 and 384 (Figs. 25, 32, respectively) and the middle middle Eocene at many other sites, thus providing insufficient constraints for the construction of a sedimentation rate curve. Also, whereas the delineation of unconformities in a section which yields an incomplete biozonal succession requires little demonstration, the quality of the chronologic interpretation may be considerably limited by the impossibility to construct a sedimentation rate curve, for instance at DSDP Sites 527 and 528 (Figs. 44, 45, respectively). In a few cases, contradictory biostratigraphic reports have forced me to chose between alternative stratigraphic interpretations as in sections at DSDP Sites 387 and 386 (Figs. 30, 31, repectively). Finally, I have proposed alternative stratigraphic interpretations when necessary, as in ODP Hole 647 and DSDP Hole 363 (Figs. 23, 42, respectively). Overall, most of the sections studied above lend themselves to reliable, if not precise, stratigraphic interpretation, and although additional magneto-biostratigraphic studies would add precision, they probably would not fundamentally modify the interpretations that I propose.

Loss of sediment at core breaks may contribute to artificial stratigraphic gaps or to gaps larger than reality (de Menocal and others, 1991). Moore (1972) estimated that 35% of continuously cored sections may be lost at core breaks. Recent investigations on Plio-Pleistocene sections following the thorough examination by Ruddiman and others (1987) of the hydraulic piston corer (HPC) related problem, report average losses of 0.20 to 1 m at core breaks (de Menocal and others, 1991), but as large as 1.90 m (Murray and Prell, 1991) and 2.7 m (Farrell and Janecek, 1991). Farrell and Janecek (1991) suggested 12% loss in sequences cored with the advanced piston corer (APC), and Alexandrovich and Hays (1989) estimated that 15% of the DSDP Hole 504 Pliocene section was not sampled by the HPC. Multiple coring, from which composite depth sections are reconstructed, is now common practice to insure complete sampling of upper Neogene sections (e.g., Raymo and others, 1989; Farrell and Janecek, 1991).

The sections which are discussed here were mostly rotary cored, and there is little possibility to evaluate the loss at core gaps. As an example, if we assume a loss of 30% between 2 cores and a sedimentation rate of 1 cm/10³ yr (a reasonable value for lower Eocene sections), an artificial hiatus of 0.630 my is created. With greater sedimentation rates, the artificial hiatus shortens. A number of unconformities in this work are placed at core gaps. This is mainly when there was insufficient stratigraphic control due to poor recovery. Assuming that a maximum of 12 m of section could be unsampled and using the nominal sedimentation rate of 1 cm/10³ yr this may result in an artificial hiatus of 1.2 my. I cannot, therefore, exclude the possibility that some of the hiatuses described in this work are artificial as a result of incomplete sampling and poor recovery as for instance at DSDP Site 552 (Fig. 10). The evidence, however, supports the widespread development of unconformities. First, in sections with good recovery and stratigraphic control, unconformities can be firmly identified, for example in DSDP Sites 366, 612, 605 and 516 (Figs. 28, 29, 51, respectively). Second, many of the unconformities which I infer occur at boundaries between lithologic units/subunits, as in DSDP Sites 401, 549 and 386 (Figs. 15, 17, 31, respectively). Third, most of the hiatuses I describe are of greater duration than 2 my, which is well above the greatest probable duration of an artificially produced hiatus (see above). An example of the problem is given by DSDP 404 (Fig. 12). In this section, I suggest the presence of an unconformity between Core 9 and Core 10, both of which have little recovery. There is an uncertainty of almost 9 m in the position of the HO of *Tribrachiatus orthostylus* due to poor recovery (I have located core catchers at the base of the cores following the past DSDP policy). If we add a loss of 6 m of sediment at the core break, the uncertainty is 15 m, which represents a 0.2 my interval while the hiatus associated with the inferred unconformity is ~1 my. I suspect that the main consequence of neglecting sediment loss at core breaks (which loss I cannot estimate) would be an artificial increase in the duration of the hiatuses rather than a creation of artificial hiatuses (with possible rare exceptions as mentioned above). If stratigraphic gaps were artificially created at core breaks, it would be serendipitious that they create the particular pattern that is discussed below.

The temporal interpretation of sections represents an important step in the study of the stratigraphic record because it is the only means by which we can demonstrate the geological time that unconformities represent (compare Figs. 4, 5 with Fig. 6). In some ways, temporal interpretations are a prerequisite for sound stratigraphic correlations. The quality of the temporal interpretation of a section is narrowly dependent on the quality of its stratigraphic interpretation, and it decreases exponentially with decreasing precision in the stratigraphic interpretation. Good temporal resolution can be achieved based on close sampling and integrated stratigraphy (e.g., Aubry, 1993a). In this work, because of the very nature of this study, there is much greater uncertainty in the temporal interpretations of many sections than there is on the stratigraphic interpretations on which they are based. This results from the fact that it was rarely possible to determine precisely the age of both surfaces of the inferred unconformities, the location of the unconformities being arbitrarily determined within given contraints. The ages of the surfaces of the upper lower Eocene unconformities are more confidently estimated than those of the middle Eocene unconformities because of difficulties in positioning the latter (e.g., middle Eocene surfaces in DSDP Hole 530A, Figs. 40, 41). Unless otherwise indicated, when alternative temporal interpretations were possible, I have favoured the simplest—not necessarily the correct one—for the sake of simplicity. The uncertainty on the age estimate of each surface varies between a few tens of thousands of years to less than two hundred thousand years in most cases, an uncertainty that is negligible compared to the duration of the hiatuses (1 to several million years).

In summary, the temporal interpretations of the lower and middle Eocene stratigraphic records in various regions of the Atlantic Ocean given here (Figs. 9, 14, 20, 24, 27, 35, 39, 49) are the best that I can propose from the data presently available. Whereas it is unlikely (with a few possible exceptions) that complementary studies would establish stratigraphic continuity of sections which I interpret as discontinuous, such analyses would lead to a refinement in the stratigraphic interpretation, in particular the more accurate determination of the location of the unconformities, hence of the ages of the bounding surfaces and the extent of the hiatuses. Yet, they may reveal more complex temporal patterns than shown here.

Stratigraphic Completeness of the Deep-Sea Record

One of the most striking features that emerges from the temporal maps presented here for different regions of the Atlantic Ocean (Figs. 9, 14, 20, 24, 27, 35, 39 and 49) concerns the importance of the hiatuses. Studied at a fine scale and interpreted in a rigorous manner, the mid-Paleogene deep-sea record as we know it from the DSDP and OPD programs is not the mostly uninterrupted sedimentary record that has been generally assumed. In the 20-my long interval considered, there is almost as much geological time represented by unconformities as there is time represented by sediments. Of great significance is the fact that sediments and unconformities are not distributed at random, but organized in such a manner that relatively long intervals of continuous deposition alternate with relatively long hiatuses (I do not imply that all hiatuses result from non-deposition, and I do not intend to discuss the processes involved in their formation in this paper). With a few exceptions, the hiatuses are longer than 2 my; their duration is greater (often by several times) than the length of the entire Pleistocene Epoch (1.81 my long, Hilgen 1991; Berggren and others, 1995). This implies that there is a large-scale mechanism(s) which fashions the stratigraphic record, superimposed on the small-scale mechanisms which fashion the Pleistocene sedimentary record.

Comparison between maps established for different regions of the Atlantic Ocean also shows that there is a pattern in the distribution of the sedimentary sequences and the hiatuses. Continuous sequences occur mostly in the early Eocene whereas hiatuses occur mostly in the late early and early middle Eocene Epochs. I recognize that the database that supports such an important statement is rather limited, the geographic coverage being restricted (Fig. 7). There are areas such as the western central Atlantic (see Appendix A) that are not represented in this study. Also, it seems probable that closely spaced sections (as is mostly the case in this study, see Fig. 7) might yield similar stratigraphic successions, which would introduce a strong bias. This is well indicated for example by the succession observed along the ~230 km transect from the crest of the Walvis Ridge to the Angola Basin (Fig. 39). Yet, it is remarkable that the broad pattern of sedimentary sequences and hiatuses is observed in distant regions, as between the eastern and western North Atlantic (Figs. 9, 14, 27), and between the eastern and western South Atlantic (Figs. 39, 49). This generalized observation is obviously not pertinent to the Gulf of Mexico-Caribbean area (Fig. 35), and it may not be relevant to the subantarctic South Atlantic (see above). Whereas the data may not be conclusive for the eastern central Atlantic (Fig. 20) and are insufficient for the Labrador Sea (Fig. 24), there are some hints that it may apply to the western central Atlantic as well. At DSDP Site 355 in the western central Brazil Basin (Supko and others, 1977d), a ~100 m-thick section of lower Eocene pelagic mud equivalent to Zones NP12 and NP13, and repre-

senting less than 3 my, underlies a ~76-m middle Eocene section, representing ~5 my, which apparently lacks radiolarian Zones R23 and R20 (Supko and others, 1977d; see Appendix C).

Although they conducted a study at a coarser temporal scale than I do here and with an even more restricted coverage for the Atlantic Ocean (but including all oceans), Moore and others (1978) arrived at conclusions similar to those presented here. In addition to ". . . noting that the distribution patterns of the hiatuses . . . are surprisingly coherent. . ." (Moore and others, 1978, p. 117), they observed ". . . a distinct minimum in hiatus abundance in the early Eocene . . ." (p. 119) and a ". . . secondary hiatus maximum . . . near the base of the middle Eocene . . ." (p. 120). The temporal maps which I present here can be seen as a fine scale illustration of the conclusions of Moore and others (1978).

Moore and others (1978) pointed to differences in the hiatus pattern in different oceans. The temporal maps in the present work also show regional differences, even though the data base is not large enough to document them clearly. The lower and middle Eocene stratigraphic record of the Gulf of Mexico-Caribbean area (Fig. 35) shows no similarity with the same record in other regions of the North Atlantic and with the South Pacific. In fact this is an area where the early Eocene Epoch is poorly represented. This may be also the case, to a lesser extent, in the Subantarctic South Atlantic, but verification is needed. Finally, it should be noted that no two sections in these maps are alike, and that whereas there is overlap of the hiatuses, the age of most bounding surfaces of "correlative" unconformities, varies between sections.

Unsteadiness/Steadiness of Accumulation Rates

Much of the data in this work support the view that in any one locality/section in the deep-sea, sedimentation rates are relatively steady rather than highly variable through time. Relative consistency of accumulation rates is particularly striking for the lower Eocene sections at DSDP Sites 550, 549, 366, 612, 390 and 530 (Figs. 16, 17, 19, 28, 33, 40, respectively). This is not to say that accumulation rates are absolutely constant over small-scale intervals of 10,000 yr or less (see Sadler, 1981), but that their average is constant over large-scale intervals of hundred of thousands of years. The sedimentary history which emerges from this study is that represented in Figure 5, where continuous sedimentary units deposited at steady rates are separated by hiatuses. This is in agreement with the sedimentary history of the nine sites drilled during DSDP leg 14 (Hays and others, 1972) as expressed by accumulation rate curves derived by Berger and Von Rad (1972, Fig. 7), although these latter were established with the much lower chronologic resolution provided by the earlier biochronologic framework of Berggren (1972b). This is also in agreement with accumulation rate curves which have been established in many other DSDP and ODP reports, although in these the curves are drawn on a broader scale. Working at the same fine scale as in this study, I have illustrated similar sedimentary history in the Neogene of Jamaica and of the northeastern Gulf of Mexico (Aubry, 1993a, b).

Whereas it has been possible to determine early Eocene accumulation rates in many sections in this study, it has been rarely possible to estimate those for the middle Eocene sequences, which are insufficiently represented. However, none of the sections discussed in this work resembles the model in Figure 4, in which demonstrable stratigraphic continuity is associated with sharp changes in accumulation rates. This sharply contrasts with the conclusions of Shackleton and others (1984) who studied the accumulation rates at the five DSDP sites, 525 to 529, which form a short depth transect along the Walvis Ridge. Four of these sites (525 and 527 to 529) have been included in the present study and their lower and middle Eocene records have been interpreted as discontinuous (Figs. 39, 44 to 47). In contrast, the magnetostratigraphic records they portrayed for Holes 527 and 528, indicate that Shackleton and others (1984) interpreted the same records as continuous but with considerable variations in accumulation rates (Shackleton and others, 1984, Fig 3). These authors cautioned on the potential misidentification of the reversals in their study, particularly around Chrons C23n and 24n, but in fact the greatest source of error was around Chrons C22n and 21n. Shackleton and others (1984, p. 623) indicated that ". . . around Anomalies 23 and 24, we encountered considerable difficulty in reaching a definite solution, and the problem was exacerbated by the obvious changes in sedimentation rates around this time. . . ." However, if the interpretation by Shackleton and others (1984) was correct, all zones (and subzones) from NP11 to NP16 should occur in Holes 527 and 528, and correlation between biozones and magnetic reversals for the middle Eocene interval at these sites should be as straightforward as they are for the lower Eocene interval. I recognize that recovery was not complete at these sites, but the similarity in the stratigraphic succession at Sites 527, 528 and 529 is striking. It is most unlikely that the absence of the same biozones in all 4 holes could result from poor recovery, and Manivit (1984) also remarked on the absence of rhabdoliths in the middle Eocene sediments at all sites from Leg 74. It is difficult to invoke ecologic factors to explain the absence of *Rhabdosphaera inflata* (the LO of this species characterizes the base of Subzone NP14b) when this species occurs at the same latitude and in similar ecologic settings, for instance at nearby DSDP Sites 363 and 364 (Proto Decima and others, 1978; see above). It should be noted that the stratigraphic succession at Site 529, where the stratigraphic gaps are obvious, supports my interpretation of the stratigraphic succession at the other 3 sites.

In Figure 55, I compare Shackleton and others's (1984) interpretation of the lower and middle Eocene section in Hole 527 with the one I propose. I have replotted the chronology data (Fig. 55, right side) that Shackleton and others (1984) used to examine changes in accumulation rates during the early and middle Eocene Epochs (Tables 4, 5). From these, I have calculated sedimentation rates (cm/10^3 yr) for temporal increments between tie points (Table 5). This procedure differs from that in Shackleton and others (1984) who calculated accumulation rates (g/cm^2/10^3 yr) using shipboard density data. The two curves produced (compare Fig. 55, right, herein with Shackleton and others, 1984, Fig. 3) are quite similar although, for various minor reasons, they do not perfectly match. Both records show, in particular, a low centered on 46–47 Ma and a peak centered on 55 Ma. Also, the differences between the low and the peak values are in the same proportion (50 fold) in the two records.

The temporal interpretation given in Figure 55 (left) follows the stratigraphic interpretation given above (Fig. 44). For the lower Eocene (NP11-NP12 zonal) interval, I have followed Shackleton and others (1984) in identifying the Chron C24Ar/C24Bn reversal (55.66 Ma) at 172.93 m and the Chron C23n/C23r reversal (54.70 Ma) at 163.88 m. This indicates a sedimentation rate of 0.94 cm/10^3 yr. I have shown that sediments of Subzone NP14a are unconformably superposed on those of Zone NP12 (although there is the possibility that a thin veneer belonging to Zone NP13 was unsampled). As the FAD of *Discoaster sublodoensis* is in the early part of Chron C22n, I place the unconformity at the reversal boundary at 154.73 m (the upper surface of the unconformity is thus younger than 52.62 Ma). The age of the lower surface of the unconformity is determined by calculating the time needed for deposition of the interval 163.88 m-154.73 m using the local sedimentation rate of 0.94 cm/10^3 yr. This places the surface in Biozone NP12 slightly older than the NP12/NP13 chron boundary which is consistent with the observed data. I also consider that the sediments of Subzones NP14a and NP15b are probably directly unconformable. The unconformity is arbitrarily located at 147 m (equidistant between the lower sample with *Chiasmolithus gigas* and the next lower sample witout it). As Subzone NP14a correlates essentially with Chron C22n, this means that these strata were deposited at a minimum rate of 1.1 cm/10^3 yr. In addition, the interval assigned to Subzone NP15b is unconformable with that of Subzone NP14a and that of Zone NP16. The two unconformities above are approximated at 147 and 133.60 m, respectivley. As Biochron NP15b is 1.8 my long, the beds in Subzone NP15b in this section were deposited at rates greater than 0.78 cm/10^3 yr, and a rate of 1 cm/10^3 yr is assumed here. It can be seen (Fig. 55) that the sharp changes in accumulation rates deduced from the method applied by Shackleton and others (1984) to derive a chronology for the lower and middle Eocene record in Hole 527 are not reproduced using the integrated method that I propose here. Instead, I suggest that the section was deposited at quite steady sedimentation (and therefore accumulation) rates. In Figure 55, I have also plotted the number of ichthyoliths/g of sediment as given in Shackleton and others (1984, Table 7) which reveals no obvious relationship between the abundance of ichthyoliths and sedimentation rates in Hole 527.

Changes in accumulation rates have been used to estimate changes in productivity (Shackleton and others, 1984; Brummer and van Eijden, 1992). It is clear from the discussion above that the relationship is not simple. Brummer and van Eijden (1992) depicted sharp changes in accumulation rates during the Cenozoic Era in the Indian Ocean which were based, as for the Walvis Ridge transect, on the assumption of linearity between stratigraphic record and geological time (see their Table 2). Unfortunately, interpretations of the geological record, paleoceanographic reconstructions in particular, rarely take into account the fundamental principle so well expressed by Wheeler (1958): ". . . the temporal values of such significant events as non-deposition and erosion are reduced to zero in a section whose vertical dimension is adjusted to the thickness of the stratal record . . ." (Wheeler, 1958, p. 1047). Not only the significance of the erosional events is lost, but the stratigraphic record is distorted, stretched and compressed, in order to conform to a numerical chronology, and thereby it is falsified.

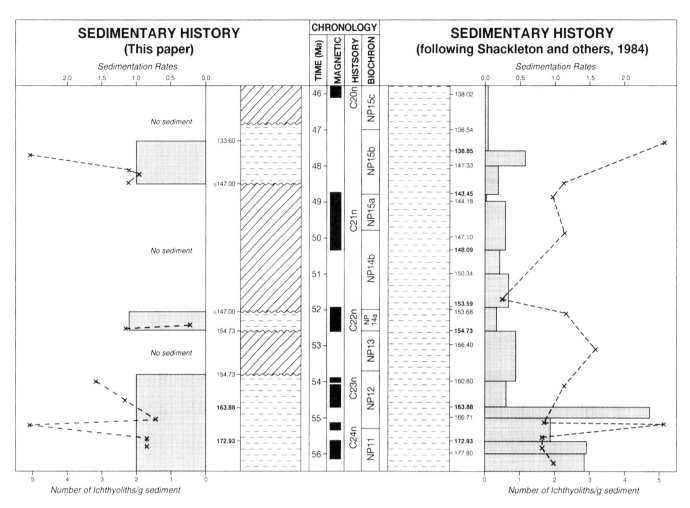

Fig. 55.—Compared interpretations of the variations of the early and middle Eocene sedimentation rates at Site 527. The chronology is that of Berggren and others (1985). Right: Variations of the sedimentation rates following the temporal interpretation (see Figs. 4, 5) by Shackleton and others (1984) of the lower and middle Eocene section recovered at the site. Left: Variations of the sedimentation rates following the temporal interpretation proposed herein. Dashed line shows variation in abundance (number/gram of sediment) of ichthyoliths in the section. Accumulation rates (here approximated by sedimentation rates) and ichthyolith abundance have been regarded as proxy indicators of past variations in productivity (right).

Relationship between Sedimentary Sequences on the Shelves and in the Deep Sea

The observation that in major areas of the Atlantic Ocean, the lower Eocene Series is well displayed whereas the middle Eocene Series is unconformity-riddled confirms my earlier observation (Aubry, 1991) of a contrast between a well-represented early Eocene Epoch and a poorly represented middle Eocene Epoch concurrently on the shelves and in the deep sea. This likely have implications for the mechanisms involved in eustatic changes, and the significance of the global cycle chart (Haq and others, 1988), but this is not the place to discuss them. However, I point to the contradistinction between the present conclusion and the report by Keller and others (1987) that, for the late Paleogene Period, unconformities in the deep sea correlate with depositional sequences on the shelves. The data which I present here indicate that the Atlantic oceanic record yields numerous sections that are largely correlative with the shallow water deposits of northwestern Europe which typify the Ypresian Stage. They also indicate that numerous stratigraphic

gaps which straddle the lower/middle Eocene boundary in the Atlantic Ocean overlap with the well known stratigraphic gap at the Ypresian/Lutetian boundary in the Paris Basin, and in other regions of northwestern Europe (see Aubry, 1991).

Chronologic Resolution Versus Stratigraphic Resolution

Chronologic/stratigraphic resolution is a multifaceted subject which is beyond the scope of this paper. I wish only to emphasize that *chronologic resolution* and *stratigraphic resolution* are two separate concepts although they are assumed by many to be the same.

I have shown in this study the impact that the IMBS has in stratigraphic analysis. Chronologic interpretation of a stratigraphic section is not merely the numerical dating of given stratigraphic levels through determining the position of magnetic reversals and of paleontologic and stable isotope (and any other) events. The interpretation of a stratigraphic section rests upon the comprehensive integration of various stratigraphic means in order to determine which magnetic reversals, which

TABLE 4.—TIE POINTS USED IN SHACKLETON AND OTHERS (1984) TO CONSTRAIN THE STRATIGRAPHIC/TEMPORAL INTERPRETATION OF THE LOWER AND MIDDLE EOCENE SECTION IN DSDP HOLE 527.

Shackleton and others (1984)			This paper	
Depth (mbsf)	Event	Age (Ma)	Level	Age (Ma)
138.85	HO *C. gigas* (hiatus?)	47.60	truncated upper range	>47.00
140.77	LO *C. gigas*	47.80	truncated lower range	<48.80
143.45	top Anomaly 21	48.75	questionable	—
148.09	base Anomaly 21	50.34	level within Chron C22n	>51.95
153.59	top Anomaly 22	51.95	level within Chron C22n	>51.95
154.73	base Anomaly 22	52.62	Chron C22n/C22r reversal boundary	52.62
163.88	base Anomaly 23	54.70	Chron C23n/C23r reversal boundary	54.70
172.93	top Anomaly 24b	55.66	acceptable, but may be also the Chron C23r/C24An reversal boundary	55.66

Note that the age of the FAD and LAD of *Chiasmolithus gigas* are 48.80 and 47.00 Ma, respectively, in Berggren and others (1985). Whereas the former datum is not used, the age of the latter is adjusted in Figure 55.

†TABLE 5.—PREDICTED TIE-POINTS USED IN SHACKLETON AND OTHERS (1984) TO ESTABLISH CHANGES IN ACCUMULATION RATES FOR INCREMENTS OF 1 MY (SEE FIG. 55). RESULTING ACCUMULATION RATES FOR ALL TEMPORAL INCREMENTS ARE GIVEN IN RIGHT COLUMN.

Age (Ma)	Depth (mbsf)	Predicted depth (mbsf)	Time increment (Ma)	Accumulation rate (cm/10³ yrs)
46.00		138.02		
47.00		138.54	47.00–46.00	0.05
47.60	138.85		47.60–47.00	0.05
48.00		141.33	48.00–47.60	0.58
48.75	143.45		48.75–48.00	0.21
49.00		144.18	49.00–48.75	0.03
50		147.10	50.00–49.00	0.29
50.34	148.09		50.34–50.00	0.29
51		150.34	51.00–50.34	0.22
51.95	153.59		51.95–51.00	0.34
52.00		153.68	52.00–51.95	0.02
52.62	154.73		52.62–52.00	0.16
53.00		156.40	53.00–52.60	0.44
54.00		160.80	54.00–53.00	0.44
54.70	163.88		54.70–54.00	0.32
55.00		166.71	55.00–54.70	2.35
55.66	172.93		55.66–55.00	0.94
56.00		177.80	56.00–55.66	1.45
57.00		192.11	57.00–56.00	1.43

the model section discussed above (Figs. 5, 6) and for many of the sites examined in this paper. Whereas in a continuous section, the numerical age of any level between two calibrated events can be estimated (= stratigraphic resolution), the numerical (estimated) age of any level between two closely spaced unconformities suffers from considerable uncertainty. The difficulty is particularly exacerbated by the fact that in such sections the sedimentation rates cannot be determined. The lowest bounding surface of an unconformity which truncates an extended complete section can be dated through extrapolation of sedimentation rates but other surfaces cannot. I have referred to this problem on several occasions in the course of the description of the oceanic sections above. Although the chronologic resolution as provided by the IMBS is available, it may not be applicable through the method I have described, and the stratigraphic resolution is then considerably decreased. This has considerable importance in the calibration of surfaces (seismic or unconformities) in the context of sequence stratigraphy. Future developments in Paleogene chronology will hopefully include a calibrated model for stable isotope excursions or shifts, which should improve stratigraphic resolution. Such effort is currently in progress for the late Paleocene and early Eocene.

GENERAL DISCUSSION AND CONCLUSIONS

Different views have been expressed regarding the completeness of the deep-sea stratigraphic record. Whereas some authors have argued that it is very incomplete (Moore and others, 1978; Sadler, 1981; Van Andel, 1981), others have asserted the opposite (e.g., Anders and others, 1987). A current view, Kidd and Hailwood (1993) for instance, holds that it is mostly complete, and certainly more complete than on shelves. The presence of unconformities in the deep-sea record has long been known, although most of the recognized gaps were initially those associated with the longest hiatuses, encompassing several biochrons and up to an entire period (or more; see Appendix B). Shorter hiatuses were later documented in various ocean basins (Edwards, 1973; Keller and Barron, 1983). The present study which takes advantage of the finest chronologic scale available today for the early Paleogene Period, documents an even greater level of incompleteness of the deep-sea record, at least in many areas. I have omitted consideration of paleodepth from this discussion, but it is striking that deeper (oceanic) and shallower (shelves) sections are about equally incomplete (cf. Sites 612, 386, 390, Fig. 27; Sites 525, 527, 529, Fig. 39). It is

paleontologic and which stable isotopic events have a temporal significance, to determine which intervals reflect continuous deposition, and where stratigraphic gaps are located. It is obvious that the quality of stratigraphic interpretation is entirely dependent upon the power of chronologic resolution. The closer the chronologic events in the IMBS, the more accurate the stratigraphic interpretation. In this respect, the strength of stable isotopic chronology (e.g., Shackleton and other, 1984; Miller and others, 1991) particularly when coupled with astrochronology (e.g., Hilgen, 1991) does not require further demonstration. I would argue, however, that magnetobiochronology already constitutes a powerful tool with practically infinite resolving power in continuous sections (see Aubry, 1991). In sections such as the lower Eocene interval at DSDP Sites 530 (between ~500 m to 478 m) and 550 (between 360 and ~320 m), stratigraphic resolution approximates temporal resolution. Such sections, particularly when accumulation rates are high, are those to which high-resolution chronology (such as the so called "late Neogene high-resolution stratigraphy") can be applied or from which it can be developed. On the other hand, stratigraphic resolution is extremely difficult to achieve through magnetobiochronology in truncated sections, as illustrated in

perhaps more remarkable, and somewhat of a paradox, that long hiatuses at deep sites overlap those in shallow-water sections (compare Figs. 14, 27 herein with Fig. 6 in Aubry, 1991; compare also Fig. 39 with Figs. 155, 159, 161 in Dingle and others, 1983).

The recognition of widespread unconformity-bounded units is the basis for allostratigraphy (North American Commission on Stratigraphic Nomenclature, 1983). Based on broad scale discontinuities, McGowran (1978, 1979) has subdivided the Maestrichtian and lower Paleogene record of the Indo-Pacific region into three allostratigraphic units and the Australian Cenozoic succession into 4 units or "sequences." Similarly, Poag (1993) has described 12 allostratigraphic units in the upper Cretaceous and Cenozoic record of the U.S. middle Atlantic continental margin. Martinsen and others (1993) have shown that the Mesaverde Group in southeastern Wyoming is comprised of 2 allostratigraphic units and observed that the stratigraphic record in tectonically influenced areas may be better described in term of allostratigraphy than in term of sequences (sensu van Wagoner and others, 1988). In the Jurassic Period, the stratigraphic sequences that Martire (1992) describes in the Rosso Ammonitico Veronese of northern Italy would probably be best referred to as allostratigraphic units. In my own work (1993a, b) I have described the Neogene stratigraphic record of Jamaica and the northeastern Gulf of Mexico in terms of allostratigraphic units. In these two regions the Neogene stratigraphic record consists of superposed units deposited at constant (as far as can be determined) rates in given sections separated by unconformities. It would be premature to describe allostratigraphic units in the Atlantic Ocean based on the present work, because allostratigraphic units seem to have a regional extent and our site coverage is insufficient. However, the data that I present here suggest that the oceanic record can be described in terms of fine-scale regional allostratigraphic units, in a manner similar to the Neogene stratigraphic record of Jamaica and the Gulf of Mexico. I support Poag (1993) in his belief that allostratigraphic units are invaluable in stratigraphic subdivision and consider that these (Aubry, 1994) are the fundamental element of stratigraphic architecture which can be traced from the (paleo)coast to the deep sea. What is now required is the precise dating of the surfaces which bound allostratigraphic units with a rigorous approach to temporal interpretation.

That allostratigraphy represents the fundamental nature of the stratigraphic record means that this record is inherently incomplete. It also means that geological time is unevenly represented by the marine stratigraphic record (on shelves *as well as* in the deep sea). As the examples show, the early Eocene Epoch is particularly well represented whereas numerous hiatuses encompasse the latest early and the earliest middle Eocene Epochs. A review of the literature shows similarly that in marine settings the late Paleocene Epoch is well represented whereas the latest Paleocene and earliest Eocene Epochs are not. After 4 years of diligent efforts IGCP Project 308 has not yet identified one section continuous from late Paleocene to early Eocene (Aubry and others, 1995; Berggren and Aubry, 1995). Because there is not a one to one relationship between geological time and stratigraphic record, I would argue against Harland and others (1990) and Harland (1992) that it remains essential to use a dual terminology to distinguish between temporal and stratigraphic terms. Geochronology is best defined as an ab-

stract reference temporal framework largely derived from the laborious construction of a virtual (ideal) stratigraphic composite reference section which constitutes chronostratigraphy. As such, chronostratigraphy itself is an abstract concept. As Wheeler (1958) so well expressed: ". . . complete time-stratigraphic integration involves the distinction between the objective preserved record and the partly subjective restored succession . . ." (Wheeler, 1958, p. 1051). Chronostratigraphy is the indispensable medium to establish the relative succession of events through time, and it yields key elements for isotopic dating (calibration points). However, chronostratigraphy is no temporal vernier. Except for specific intervals for which isotopic dating on closely spaced intervals can be performed (e.g., Montanari and others, 1988), chronostratigraphy is not constrained by a sufficiently detailed record of isotopic measurements ("absolute chronology" or "chronometry") to provide a satisfactory measure of duration. As shown by the vissicitudes of the Cretaceous time scale (Obradovich, 1993), chronostratigraphy alone does not constitute an efficient provider for a numerical chronology. As I have explained above, chronostratigraphy is not the only support of geochronology in the IMBS because duration in this context is measured as a function of an internal geodynamic process, sea floor spreading. Magnetostratigraphy yields information on duration exclusively in relation to the assumed constancy of sea floor spreading at least in the scale of events to which we refer. In astrochronology, duration is measured by reference to an external astronomical process in which sedimentary cyclicity yields age information exclusively in relation to Milankovich periodicities. In the light of the recent progress in understanding the nature of the stratigraphic record and in establishing a geochronologic framework, Ager's (1981) approach to stratigraphy/chronology is most tenable. He wrote: ". . . We have managed to confuse ourselves for the years with the jargon of lithostratigraphy, biostratigraphy, chronostratigraphy and the rest. In fact, it can well be argued that basically there are only two concepts—rocks and time—with the rest just an obfuscation of the nomenclature . . ." (Ager, 1981, p. 68). Without getting into an argument regarding the pros and cons of chronostratigraphic usage, I believe that, for the Cenozoic Era, we have come to a point in history where stratigraphy would benefit greatly from abandoning the concept of ideal sections in favor of establishing as well as possible the relationships between real sections and time. The stratigraphical/temporal analysis which can now be conducted on the Cenozoic stratigraphic record allows direct extraction of geological time, without the intermediary of a chronostratigraphic framework. This is what is required for best applying the fancy technologies available today and for understanding the fascinating discoveries on earth's past evolution.

As increasingly sophisticated methodologies and technologies are being applied to earth science research, the discipline of stratigraphy appears to be as fundamental as ever. Yet, stratigraphy is too often regarded as subjective and imprecise, more relevant to art than to science (see Shackleton, 1994). Based on the rigorous integration of as many stratigraphic means as possible, sound temporal interpretations of stratigraphic sections are essential in our search to elucidate the complexity of the earth's history. It is the key to correctly establishing fundamental links such as those between ocean circulation and climate, ocean productivity and climate or sea-level changes and productivity, if we are to establish cogent predictions regarding the future of the planet.

SUMMARY

Because the stratigraphic record is the only basis for reconstructing earth history, its interpretation must be rigorous and accurate. It is generally held that the relationship between the oceanic stratigraphic record and geological time is a straightforward one; the deep-sea stratigraphic record being considered by many to be essentially continuous. Abnormal patterns in the ranges of taxa are mostly attributed to ecologically controlled diachrony, and changes in thickness of specific intervals between sections are interpreted as reflecting variations in sedimentation/accumulation rates. An examination of these relationships shows that both, diachrony and differences in thickness, may result from unconformities as well. There is, however, a method to determine whether stratigraphic sections are continuous or not. It involves the recognition of condensed sections, delineation of (inferred) unconformities and dating of their surfaces. This is made possible through the integrated magnetobiochronologic scale (IMBS) derived from the geomagnetic polarity time scale (GPTS). For GPTS-based time scale(s), and unlike the classical time scales, the chronology provided by the sea floor lineation pattern is established independently from the chronostratigraphic framework. This allows a rigorous temporal analysis of the stratigraphic record. In the lower and middle Eocene stratigraphic record of the North and South Atlantic Ocean, temporal maps for nine geographic sectors, based on 51 DSDP and ODP sites, reveal a striking pattern of deposition and hiatuses such that the early Eocene Epoch is well represented in most of the Atlantic Ocean whereas the latest early and earliest middle Eocene Epochs correspond mostly to hiatuses. This indicates that the deep-sea record is far from being continuous, at least in this area, and supports previous findings by Moore and others (1978). This also suggests that the relationship between changes in sedimentation/accumulation rates as proxy indicator of changes in productivity is not a simple one as currently assumed and demonstrates the importance of rigorous temporal interpretations of stratigraphic sections. Whereas the stratigraphic record is currently interpreted in terms of superposed sequences, it may be interpreted more appropriately as consisting of imbricated allostratigraphic units.

ACKNOWLEDGMENTS

I am grateful to W. A. Berggren, D. V. Kent and J. Hardenbol for inviting me to present this work at the SEPM symposium on geochronology and for extended discussions. This work was initiated as part of the activities of the ODP Sea-level Working Group (1990–1992), and I am indebbeted to its members for stimulating discussions on the nature of the stratigraphic record. I have discussed this work with many colleagues who I would like to thank for their input. I am very thankful to J. A. Van Couvering, K. G. Miller and R. Norris for reviewing the manuscript. This is contribution no. 95045 of the ISEM.

To receive the appendices for this paper, contact SEPM (Society for Sedimentary Geology), 1731 East 71st Street, Tulsa, OK 74136-5108 or call 1-800-865-9765.

REFERENCES

AGER, D. V., 1981, The nature of the stratigraphical record: New York, Halsted Press, John Wiley and Sons, 122 p.

ALEXANDROVICH, J. M. AND HAYS, J. D., 1989, High-resolution stratigraphic correlation of ODP Leg 111 Holes 677A and 677B and DSDP Leg 69 Hole 504: College Station, Proceedings of the Ocean Drilling Program, Scientific Results, v. 121, p. 297–355.

ANDERS, M. H., KRUEGER, S. W., AND SADLER, P., 1987, A new look at sedimentation rates and the completeness of the stratigraphic record: Journal of Geology, v. 95, p. 1–14.

APPLEGATE, J. L. AND WISE, S. W., JR., 1987, Eocene calcareous nannofossils, Deep sea Drilling Project Site 605, upper continental rise off New Jersey, U.S.A.: Washington, D.C., Initial Reports of the Deep Sea Drilling Project, v. 93, p. 685–698.

ARMENTROUT, J. M., 1991, Paleontologic constraints on depositional modeling: Examples of integration of biostratigraphy and seismic stratigraphy, Pliocene-Pleistocene, Gulf of Mexico, *in* Weimer, P. and Link, M. H., eds., Seismic Facies and Sedimentary Processes of Submarine Fans and Turbidite Systems: New York, Springer-Verlag, p. 137–169.

AUBRY, M.-P., 1983, Biostratigraphie du Paléogène épicontinental de l'Europe du Nord-Ouest. Etude fondée sur les nannofossiles calcaires: Document des laboratoires de Géologie de Lyon, v. 89, 317 p.

AUBRY, M.-P., 1991, Sequence stratigraphy: Eustasy or tectonic imprint?: Journal of Geophysical Research, v. 96, p. 6641–6679.

AUBRY, M.-P., 1992, Late Paleogene calcareous nannoplankton evolution: A tale of climatic deterioration, *in* Prothero, D. R. and Berggren, W. A., Eocene-Oligocene Climatic and Biotic Evolution: Princeton, Princeton University Press, p. 272–309.

AUBRY, M.-P., 1993a, Neogene allostratigraphy and depositional history of the De Soto Canyon area, northern Gulf of Mexico: Micropaleontology, v. 39, p. 327–366.

AUBRY, M.-P., 1993b, Calcareous nannofossil stratigraphy of the Neogene formations of eastern Jamaica, *in* Robinson, E. R. and Wright, R. M., eds., Biostratigraphy of Jamaica: Boulder, Geological Society of America Memoir 182, p. 131–178.

AUBRY, M.-P., 1994, Interpreting the stratigraphic record (abs.), *in* Hesselbo, S., ed., Recent Advances in Sequence Stratigraphy: London, Geological Society of London, p. 15.

AUBRY, M.-P., BERGGREN, W. A., KENT, D. V., FLYNN, J. J., KLITGORD, K. D., OBRADOVICH, J. D., AND PROTHERO, R. R., 1988, Paleogene geochronology: An integrated approach: Paleoceanography, v. 3, p. 707–742.

AUBRY, M.-P., BERGGREN, W. A., STOTT, L. D., AND SINHA, A., 1995, The upper Paleocene-lower Eocene stratigraphic record and the Paleocene/Eocene boundary carbon isotope excursion, *in* Knox, R. W. O' B., Corfield, R., and Dunnay, R. E., eds., Correlation of the Early Paleogene in Northwest Europe: London, Geological Society of London Special Publication, in press.

AUSTIN, J. A., JR., SCHLAGER, W., PALMER, A. A., AND OTHERS, 1986, Site 628: College Station, Proceedings of the Ocean Drilling Program, Initial Results, v. 101, p. 213–270.

BACKMAN, J., 1984, Cenozoic calcareous nannofossil biostratigraphy from the northeastern Atlantic Ocean—Deep Sea Drilling Project Leg 81: Washington, D.C., Initial Reports of the Deep Sea Drilling Project, v. 81, p. 403–428.

BACKMAN, J., 1986, Late Paleocene to middle Eocene calcareous nannofossil biochronology from the Shatsky Rise, Walvis Ridge and Italy: Palaeogeography, Palaeoclimatology, Palaeoecology, v. 57, p. 43–59.

BACKMAN, J., MORTON, A. C., ROBERTS, D. G., BROWN, S., KRUMSIEK, K., AND McINTYRE, R. M., 1984a, Geochronology of the lower Eocene and upper Paleocene sequences of Leg 81: Washington, D.C., Initial Reports of the Deep Sea Drilling Project, v. 81, p. 877–882.

BACKMAN, J., WESTBERG-SMITH, M. J., BALDAUF, J. G., BROWN, S., BUKRY, D., EDWARDS, L., HARLAND, R., AND HUDDLESTUN, P., 1984b, Biostratigraphy of Leg 81 sediments—A high latitude record: Washington, D.C., Initial Reports of the Deep Sea Drilling Project, v. 81, p. 855–860.

BALDAUF, J. G., 1984, Cenozoic diatom biostratigraphy and paleoceanography of the Rockall Plateau region, North Atlantic, Deep Sea Drilling Project 81:

Washington, D.C., Initial Reports of the Deep Sea Drilling Project, v. 81, p. 439–478.

BARKER, P. F., 1983, Tectonic evolution and subsidence history of the Rio Grande Rise: Washington, D.C., Initial Reports of the Deep Sea Drilling Project, v. 72, p. 953–976.

BARKER, P. F., CARLSON, R. L., AND JOHNSON, D. A., AND OTHERS, 1983a, Site 515: Brazil Basin: Washington, D.C., Initial Reports of the Deep Sea Drilling Project, v. 72, p. 53–154.

BARKER, P. F., CARLSON, R. L., AND JOHNSON, D. A., 1983b, Site 516: Rio Grande Rise: Washington, D.C., Initial Reports of the Deep Sea Drilling Project, v. 72, p. 155–338.

BENSON, W. E., SHERIDAN, R. E., AND OTHERS, 1978, Sites 389 and 390: North rim of Blake Nose: Washington, D.C., Initial Reports of the Deep Sea Drilling Project, v. 44, p. 69–151.

BERGER, W. H., AND VON RAD, U., 1972, Cretaceous and Cenozoic sediments from the Atlantic Ocean: Washington, D.C., Initial Reports of the Deep Sea Drilling Project, v. 14, p. 787–954.

BERGGREN, W. A., 1969, Paleogene biostratigraphy and planktonic foraminifera of northern Europe, Proceedings of the First International Conference on Planktonic Microfossils (1967): Leiden, E. J. Brill, v. 1, p. 121–160.

BERGGREN, W. A., 1972a, Cenozoic biostratigraphy and paleobiogeography of the North Atlantic: Washington, D.C., Initial Reports of the Deep Sea Drilling Project, v. 12, p. 965–1001.

BERGGREN, W. A., 1972b, A Cenozoic time-scale—some implications for regional geology and paleobiogeography: Lethaia, v. 5, p. 195–215.

BERGGREN, W. A. AND AUBERT, J., 1976, Paleocene benthic foraminiferal biostratigraphy and paleoecology of Tunisia: Bulletin du Centre de Recherches de Pau-SNPA, v. 10, p. 379–469.

BERGGREN, W. A. AND AUBRY, M.-P., 1995, A late Paleocene-early Eocene NW European and North Sea magnetobiostratigraphic correlation network: A sequence stratigraphic framework, in Knox, R. W. O' B., Corfield, R., and Dunnay, R. E., eds., Correlation of the Early Paleogene in Northwest Europe: London, Geological Society of London Special Publication, in press.

BERGGREN, W. A., HAMILTON, N., JOHNSON, J. A., PUJOL, C., WEISS, W., CEPEK, P., AND GOMBOS, JR., A. M., 1983, Magnetobiostratigraphy of Deep Sea Drilling Project Leg 72, Sites 515–518, Rio Grande Rise (South Atlantic): Washington, D.C., Initial Reports of the Deep Sea Drilling Project, v. 72, p. 939–948.

BERGGREN, W. A., HILGEN, F. J., LANGEREIS, C. C., KENT, D. V., OBRADOVICH, J. D., RAFFI, I., RAYMO, M., AND SHACKLETON, N. J., 1995, Late Neogene (Pliocene-Pleistocene) chronology: New perspectives in high resolution stratigraphy: Geological Society of America Bulletin, in press.

BERGGREN, W. A., KENT, D. V., AND FLYNN, J. J., 1985, Paleogene geochronology and chronostratigraphy, in Snelling, N. J., ed., The Chronology of the Geological Record: London, Blackwell, Geological Society of London Memoir 10, p. 141–195.

BERGGREN, A. W., KENT, D. V., OBRADOVICH, J. D., AND SWISHER, C. C. III, 1992, Toward a revised Paleogene geochronology, in Prothero, D. R. and Berggren, W. A., eds., Eocene-Oligocene Climatic and Biotic Evolution: Princeton, Princeton University Press, p. 29–45.

BERGGREN, W. A. AND MILLER, K. G., 1988, Paleogene tropical planktonic foraminiferal biostratigraphy and magnetobiochronology: Micropaleontology, v. 34, p. 362–380.

BLACKWELDER, E., 1909, The valuation of unconformities: Journal of Geology, v. 17, p. 289–299.

BLECHSCHMIDT, G., 1979, Biostratigraphy of calcareous nannofossils: Leg 47B, Deep Sea Drilling Project: Washington, D.C., Initial Reports of the Deep Sea Drilling Project, v. 47, p. 327–360.

BLOW, H., 1970, Deep Sea Drilling Project, Leg 3, Foraminifera from selected samples: Washington, D.C., Initial Reports of the Deep Sea Drilling Project, v. 3, p. 629–661.

BLOW, H., 1979, The Cainozoic Globigerinada: A study of the morphology, taxonomy, evolutionary relationships and the stratigraphical distribution of some Globigerinada (mainly Globigerinacea)(3 vols.): Leiden, E. J., Brill, 1431 p.

BOERSMA, A., 1977, Cenozoic planktonic foraminifera—DSDP Leg 39 (South Atlantic): Washington, D.C., Initial Reports of the Deep Sea Drilling Project, v. 39, p. 567–590.

BOERSMA, A., 1984, Cretaceous-Tertiary planktonic foraminifers from the southeastern Atlantic, Walvis Ridge area, Deep Sea Drilling Project Leg 74: Washington, D.C., Inital Reports of the Deep Sea Drilling Project, v. 74, p. 501–523.

BOLLI, H. M., RYAN, W. B. F., AND OTHERS, 1978a, Cape Basin continental rise—Sites 360 and 361: Washington, D.C., Initial Reports of the Deep Sea Drilling Project, v. 40, p. 29–182.

BOLLI, H. M., RYAN, W. B. F., AND OTHERS, 1978b, Walvis Ridge—Sites 362 and 363: Washington, D.C., Initial Reports of the Deep Sea Drilling Project, v. 40, p. 183–356.

BOLLI, H. M., RYAN, W. B. F., AND OTHERS, 1978c, Angola continental margin—Sites 364 and 365: Washington, D.C., Initial Reports of the Deep Sea Drilling Project, v. 40, p. 357–455.

BRUINS, J., VAN HINTE, J. E., AND ZIJDERVELD, J. D. A., 1987, Upper Cretaceous to Paleocene magnetostratigraphy of Deep Sea Drilling Project Site 605, northwest Atlantic: Washington, D.C., Initial Reports of the Deep Sea Drilling Project, v. 93, p. 881–890.

BRUMMER, G. J. A. AND VAN EIJDEN, A. J. M., 1992, "Blue ocean" paleoproductivity estimates from pelagic carbonate mass accumulation rates: Marine Micropaleontology, v. 19, p. 99–117.

BUFFLER, R. T., SCHLAGER, W., AND OTHERS, 1984a, Site 536: Washington, D.C., Initial Reports of the Deep Sea Drilling Project, v. 77, p. 219–254.

BUFFLER, R. T., SCHLAGER, W., AND OTHERS, 1984b, Site 538: Washington, D.C., Initial Reports of the Deep Sea Drilling Project, v. 77, p. 279–336.

BUFFLER, R. T., SCHLAGER, W., AND OTHERS, 1984c, Sites 535, 539, and 540: Washington, D.C., Initial Reports of the Deep Sea Drilling Project, v. 77, p. 25–217.

BUKRY, D., 1972, Further comments on coccolith stratigraphy, Leg 12, Deep Sea Drilling Project: Washington, D.C., Initial Reports of the Deep Sea Drilling Project, v. 12, p. 1071–1083.

BUKRY, D., 1973a, Coccolith stratigraphy, Leg 10—Deep Sea Drilling Project: Washington, D.C., Initial Reports of the Deep Sea Drilling Project, v. 10, p. 385–406.

BUKRY, D., 1973b, Low-latitude coccolith biostratigraphic zonation: Washington, D.C., Initial Reports of the Deep Sea Drilling Project, v. 15, p. 127–149.

BUKRY, D., 1975, Coccolith and silicoflagellate stratigraphy, northwestern Pacific Ocean, Deep Sea Drilling Project Leg 32: Washington, D.C., Initial Reports of the Deep Sea Drilling Project, v. 32, p. 677–701.

BUKRY, D., 1977, Coccolith and silicoflagellate stratigraphy, South Atlantic Ocean, Deep Sea Drilling Project Leg 39: Washington, D.C., Initial Reports of the Deep Sea Drilling Project, v. 39, p. 825–839.

BUKRY, D., 1978a, Cenozoic coccolith and silicoflagellate stratigraphy, offshore northwest Africa, Deep Sea Drilling Project, Leg 41: Washington, D.C., Initial Reports of the Deep Sea Drilling Project, v. 41, p. 763–789.

BUKRY, D., 1978b, Cenozoic silicoflagellate and coccolith stratigraphy, northwestern Atlantic Ocean, Deep Sea Drilling Project Leg 43: Washington, D.C., Initial Reports of the Deep Sea Drilling Project, v. 44, p. 775–805.

BUKRY, D., 1978c, Cenozoic coccolith, silicoflagellate, and diatom stratigraphy, Deep Sea Drilling Project Leg 44: Washington, D.C., Initial Reports of the Deep Sea Drilling Project, v. 44, p. 807–863.

BUKRY, D., 1978d, Cenozoic silicoflagellate and coccolith stratigraphy, southeastern Atlantic Ocean, Deep Sea Drilling Project Leg 40: Washington, D.C., Initial Reports of the Deep Sea Drilling Project, v. 40, p. 635–649.

BUKRY, D., 1987, Eocene siliceous and calcareous phytoplankton, Deep Sea Drilling Project Leg 95: Washington, D.C., Initial Reports of the Deep Sea Drilling Project, v. 95, p. 395–415.

BUKRY, D. AND BRAMLETTE, M. N., 1970, Coccolith age determinations Leg 3, Deep Sea Drilling Project: Washington, D.C., Initial Reports of the Deep Sea Drilling Project, v. 3, p. 589–611.

CANDE, S. C. AND KENT, D. V., 1992, A new geomagnetic polarity time scale for the late Cretaceous and Cenozoic: Journal of Geophysical Research, v. 97, p. 13,917–13,951.

CANDE, S. C. AND KENT, D. V., 1995, Revised calibration of the geomagnetic polarity time scale for the Late Cretaceous and Cenozoic: Journal of Geophysical Research, v. 100, p. 6093–6095.

CATHLES, L. M. AND HALLAM, A., 1992, Stress-induced changes in plate density, Vail sequences, epiorogeny, and short-lived global sea-level fluctuations: Tectonics, v. 10, p. 659–671.

CIESELSKI, P. F., Kristoffersen, Y., AND OTHERS, 1988a, Site 698: College Station, Proceedings of the Ocean Drilling Program, Initial Reports, v. 114, p. 87–150.

CIESELSKI, P. F., KRISTOFFERSEN, Y., AND OTHERS, 1988b, Site 699: College Station, Proceedings of the Ocean Drilling Program, Initial Reports, v. 114, p. 151–254.

CIESELSKI, P. F., KRISTOFFERSEN, Y., AND OTHERS, 1988c, Site 700: College Station, Proceedings of the Ocean Drilling Program, Initial Reports, v. 114, p. 255–361.

CIESELSKI, P. F., KRISTOFFERSEN, Y., AND OTHERS, 1988d, Site 702: College Station, Proceedings of the Ocean Drilling Program, Initial Reports, v. 114, p. 483–548.

CLEMENT, B. M. AND HAILWOOD, E. A., 1991, Magnetostratigraphy of sediments from Sites 701 and 702: College Station, Proceedings of the Ocean Drilling Program, Scientific Results, v. 114, p. 359–366.

CRUX, J. A., 1991, Calcareous nannofossils recovered by Leg 114 in the subantarctic South Atlantic Ocean: College Station, Proceedings of the Ocean Drilling Program, Scientific Results, v. 114, p. 155–177.

DAVIES, T. A., WESER, O. E., LUYENDYCK, B. P., AND KIDD, R. B., 1975, Unconformities in the sediments of the Indian Ocean: Nature, v. 253, p. 15–19.

DEAN, W., HAY, W., AND SIBUET, J.-C., 1984, Geologic evolution, sedimentation, and paleoenvironments of the Angola Basin and adjacent Walvis Ridge: Synthesis of results of Deep Sea Drilling Project Leg 75: Washington, D.C., Initial Reports of the Deep Sea Drilling Project, v. 75, p. 509–544.

deMENOCAL, P., BLOEMENDAL, J., AND KING, J., 1991, A rock-magnetic record of monsoonal dust deposition over the Arabian Sea: Evidence for a shift in the mode of deposition at 2.4 Ma: College Station, Proceedings of the Ocean Drilling Program, Scientific Results, v. 117, p. 389–407.

DINGLE, R. V., SIESSER, W. G., AND NEWTON, A. R., 1983, Mesozoic and Cenozoic geology of southern Africa: Rotterdam, Balkema, A. A., 355 p.

DOWSETT, H. J., 1988, Diachrony of late Neogene microfossils in the southwest Pacific Ocean: Application of the graphic correlation method: Paleoceanography, v. 3, p. 209–222.

DOWSETT, H. J., 1989, Application of the graphic correlation method to Pliocene marine sequences: Marine Micropaleontology, v. 14, p. 3–32.

EDWARDS, A. R., 1973, Southwest Pacific regional unconformities encountered during Leg 21: Washington, D.C., Initial Reports of the Deep Sea Drilling Project, v. 21, p. 701–720.

EDWARDS, L. E., 1982, Quantitative biostratigraphy: The methods should suit the data, in Cubbitt, J. M. and Reyment, R. A., eds., Quantitative Stratigraphic Correlation: New York, John Wiley and Sons Ltd., p. 45–60.

EDWARDS, L. E., 1984, Insights on why graphic correlation (Shaw's method) works: Journal of Geology, v. 92, p. 583–597.

FARRELL, J. W., AND JANECEK, T. R., 1991, Late Neogene paleoceanography and paleoclimatology of the northeast Indian Ocean (Site 758) : College Station, Proceedings of the Ocean Drilling Program, Scientific Results, v. 121, p. 297–355.

FIRTH, J., 1989, Eocene and Oligocene calcareous nannofossils from the Labrador Sea, ODP Leg 105: College Station, Proceedings of the Ocean Drilling Program, Scientific Results, v. 105, p. 263–286.

GARTNER, S., 1970, Coccolith age determinations Leg 3, deep Sea Drilling Project: Washington, D.C., Initial Reports of the Deep Sea Drilling Project, v. 3, p. 613–627.

GRACIANSKY, P. C., DE, POAG, W. C., AND FOSS, G., 1985, Drilling on the Goban Spur: Objectives, regional geological setting, and operational summary: Washington, D.C., Initial Reports of the Deep Sea Drilling Program, v. 80, p. 5–13.

GRACIANSKY, P. C. DE, POAG, C. W., AND OTHERS, 1985a, Site 550: Washington, D.C., Initial Reports of the Deep Sea Drilling Program, v. 80, p. 251–355.

GRACIANSKY, P. C. DE, POAG, C. W., AND OTHERS, 1985b, Site 549: Washington, D.C., Initial Reports of the Deep Sea Drilling Program, v. 80, p. 33–122.

GRADSTEIN, F. M., AGTERBERG, F. P., BROWER, J. C., AND SCHWARZACHER, W. S., 1985, Quantitative stratigraphy: Dordrecht, D. Reidel Publishing, 589 p.

GRADSTEIN, F. M., BUKRY, D., HABIB, D., RENZ, O., ROTH, P. H., SCHMIDT, R. R., WEAVER, F. M., AND WIND, F., 1978, Biostratigraphic summary of DSDP Leg 44: western North Atlantic Ocean: Washington, D.C., Initial Reports of the Deep Sea Drilling Project, v. 44, p. 657–662.

HAILWOOD, E., 1978, A preliminary paleomagnetic stratigraphy for lower Eocene sediments at Site 366 (Sierra Leone Rise) and Miocene and Oligocene sediments at Site 368 (Cape Verde Rise), northwest African continental margin: Washington, D.C., Initial Reports of the Deep Sea Drilling Project, v. 41, p. 987–993.

HAILWOOD, E. A., 1979, Paleomagnetism of late Mesozoic to Holocene sediments from the Bay of Biscay and Rockall Plateau, drilled on IPOD Leg 48: Washington, D.C., Initial Reports of the Deep Sea Drilling Project, v. 48, p. 305–339.

HAILWOOD, E. A., 1989, Magnetotratigraphy: London, Geological Society of London Special Report 19, 84 p.

HAILWOOD, E. A., BOCK, W., COSTA, L., DUPEUBLE, P. A., MÜLLER, C., AND SCHNITKER, D., 1979, Chronology and biostratigraphy of northeast Atlantic sediments, DSDP Leg 48: Washington, D.C., Initial Reports of the Deep Sea Drilling Project, v. 48, p. 1119–1141.

HAILWOOD, E. A., AND CLEMENT, B. M., 1991, Magnetotratigraphy of Sites 699 and 700, east Georgia Basin: College Station, Proceedings of the Ocean Drilling Program, Scientific Results, v. 114, p. 337–357.

HAILWOOD, E. A., AND KIDD, R. B., 1993, High resolution stratigraphy: London, Geological Society of London Special Publication 70, 357 p.

HAQ, B. U., HARDENBOL, J., AND VAIL, P. R., 1988, Mesozoic and Cenozoic chronostratigraphy and sea-level changes, in Wilgus, C. K., Hastings, B. S., Kendall, C. G., Posamentier, H. W., Ross, C. A., and Van Wagoner, J. C., eds., Sea-level Changes: An Integrated Approach: Tulsa, Society of Economic Paleontologists and Mineralogists Special Publication 42, p. 71–108.

HARLAND, W. B., 1992, Stratigraphic regulation and guidance: A critique of current tendancies in stratigraphic codes and guides: Geological Society of America Bulletin, v. 104, p. 1231–1235.

HARLAND, W. B., ARMSTRONG, R. L., COX, A. V., CRAIG, L. E., SMITH, A. G., AND SMITH, D. G., 1990, A Geological Time Scale 1989: Cambridge, Cambridge University Press, 263 p.

HAY, W. W., 1972, Probabilistic stratigraphy: Eclogae Geologicae Helvetiae, v. 65, p. 255–266.

HAY, W. W., SIBUET, J.-C., AND OTHERS, 1984, Site 530: southeastern corner of the Angola Basin: Washington, D.C., Initial Reports of the Deep Sea Drilling Project, v. 75, p. 29–285.

HAYS, D. E., PIMM, A. C., AND OTHERS, 1972, Leg 14: Washington, D.C., Initial Reports of the Deep Sea Drilling Project, v. 14, 975 p.

HAZEL, J. E., EDWARDS, L. E., AND BYBELL, L. M., 1984, Significant unconformities and the hiatuses represented by them in the Paleogene of the Atlantic and Gulf coastal province, in Schlee, J. S., ed., Interregional Unconformities and Hydrocarbon Accumulation: Tulsa, American Association of Petroleum Geologists Memoir 36, p. 59–66.

HEAD, M. J. AND NORRIS, G., 1989, Palynology and dinocyst stratigraphy of the Eocene and Oligocene in ODP Leg 105, Hole 647A, Labrador sea: College Station, Proceedings of the Ocean Drilling Program, Scientific Results, v. 105, p. 515–550.

HEDBERG, H., ed., 1976, International Stratigraphic Guide. A Guide to Stratigraphic Classification, Terminology, and Procedure: New York, J. Wiley and Sons, 200 p.

HILGEN, F. G., 1991, Astronomical calibration of Gausss to Matuyama sapropels in the Mediterranean and implication for the Geomagnetic Polariy Time Scale: Earth and Planetary Science Letters, v. 104, p. 226–244.

HODDELL, D. A. AND WOODRUFF, F., 1994, Variations in the strontium isotopic ratio of sea water during the Miocene: Stratigraphic and geochemical implications: Paleoceanography, v. 9, p. 405–426.

HUDDLESTON, P. F., 1984, Planktonic foraminiferal biostratigraphy, Deep Sea Drilling Project Leg 81: Washington, D.C., Initial Reports of the Deep Sea Drilling Project, v. 81, p. 429–438.

HULSBOS, R. E., 1987, Eocene benthic foraminifers from the upper continental rise off New Jersey, Deep Sea Drilling Project 605: Washington, D.C., Initial Reports of the Deep Sea Drilling Project, v. 93, p. 525–538.

IACCARINO, S. AND PREMOLI SILVA, I., 1979, Paleogene planktonic foraminiferal biostratigraphy of DSDP Hole 398D, Leg 47B, Vigo Seamount, Spain: Washington, D.C. Initial Reports of the Deep Sea Drilling Project, v. 47, p. 237–253.

JOHNSON, D. A. AND NIGRINI, C. A., 1985, Time-transgressive late Cenozoic radiolarian events of the equatorial Indo-Pacific: Science, v. 230, p. 538–540.

JOHNSON, D. L., 1978, Cenozoic radiolaria from the eastern tropical Atlantic, DSDP, Leg 41: Washington, D.C., Initial Reports of the Deep Sea Drilling Project, v. 41, p. 763–789.

KAMINSKI, M., GRADSTEIN, F. M., AND BERGGREN, W. A., 1989, Paleogene benthic foraminifer biostratigraphy and paleoecology at Site 647, southern Labrador Sea: College Station, Proceedings of the Ocean Drilling Program, Scientific Results, v. 105, p. 705–730.

KATZ, M. AND MILLER, K. G., 1991, Early Paleogene benthic foraminiferal assemblages and stable isotopes in the southern Ocean: College Station, Proceedings of the Ocean Drilling Program, Scientific Results, v. 144, p. 481–512.

KEATING, B. H. AND HERRERA-BERVERA, E. , 1984, Magnetostratigraphy of Cretaceous and early Cenozoic sediments of Deep Sea Drilling Project Site 530, Angola Basin: Washington, D.C., Initial Reports of the Deep Sea Drilling Project, v. 75, p. 1211–1218.

KELLER, G. AND BARRON, J. A., 1983, Paleoceanographic implications of Miocene deep-sea hiatuses: Geological Society of America Bulletin, v. 94, p. 590–613.

KELLER, G., HERBERT, T., DORSEY, R., D'HONDT, S., JOHNSSON, M., AND CHI, W. R., 1987, Global distribution of late Paleogene hiatuses: Geology, v. 15, p. 199–203.

KIDD, R. B., AND HAILWOOD, E. A., 1993, High resolution stratigraphy in modern and ancient marine sequences: ocean sediment cores to Palaeozoic outcrops, in Hailwwod, E. A. and Kidd, R. B., eds., High Resolution Stratigraphy: London, Geological Society of London, Special Publication 70, p. 1–8.

KOVACH, W. L., 1993, Multivariate techniques for biostratigraphical correlation: Journal of the Geological Society of London, v. 150, p. 697–705.

KRASHENINNIKOV, V., 1979, Stratigaphy and planktonic foraminifers of Cenozoic deposits of the Bay of Biscay and Rockall Plateau, DSDP Leg 48: Washington, D.C., Initial Reports of the Deep Sea Drilling Project, v. 48, p. 431–450.

KRASHENINNIKOV, V. A. AND PFLAUMANN, U., 1978, Zonal stratigraphy and planktonic foraminifers of Paleogene deposits of the Atlantic Ocean to the West of Africa: Washington, D.C., Initial Reports of the Deep Sea Drilling Project, v. 41, p. 581–611.

KRUMSIECK, K. AND ROBERTS, D. G., 1984, Paleomagnetics of Tertiary sediments from the Southwest Rockall Plateau, Deep Sea Drilling Project Leg 81: Washington, D.C., Initial Reports of the Deep Sea Drilling Project, v. 81, p. 837–851.

LANCELOT, Y., SEIBOLD, E., AND OTHERS, 1978a, Site 366: Sierra Leone Rise: Washington, D.C., Initial Reports of the Deep Sea Drilling Project, v. 41, p. 21–161.

LANCELOT, Y., SEIBOLD, E., AND OTHERS, 1978b, Site 370: Deep basin off Morocco: Washington, D.C., Initial Reports of the Deep Sea Drilling Project, v. 41, p. 421–491.

LANG, T. H. AND WATKINS, D., 1984, Cenozoic calcareous nannofossils from Deep Sea Drilling Project Leg 77: Biostratigraphy and delineation of hiatuses: Washington, D.C., Initial Reports of the Deep Sea Drilling Project, v. 77, p. 629–648.

LAUGHTON, A. S., BERGGREN, W. A., AND OTHERS, 1972a, Site 118: Washington, D.C., Initial Reports of the Deep Sea Drilling Project, v. 12, p. 673–751.

LAUGHTON, A. S., BERGGREN, W. A., AND OTHERS, 1972b, Site 119: Washington, D.C., Initial Reports of the Deep Sea Drilling Project, v. 12, p. 753–901.

LAUGHTON, A. S., BERGGREN, W. A., AND OTHERS, 1972c, Site 112: Washington, D.C., Initial Reports of the Deep Sea Drilling Project, v. 12, p. 161–253.

LAUGHTON, A. S., BERGGREN, W. A., AND OTHERS, 1972d, Site 111: Washington, D.C., Initial Reports of the Deep Sea Drilling Project, v. 12, p. 33–159.

LOUTIT, T. S., HARDENBOL, J., VAIL, P. R., AND BAUM, G. R., 1988, Condensed sections: The key to age determination and correlation of continental margin sequences, in Wilgus, C. K., Hastings, B. S., Kendall, C. G., Posamentier, H. W., Ross, C. A., and Van Wagoner, J. C., eds., Sea-level Changes: An Integrated Approach: Tulsa, Society of Economic Paleontologists and Mineralogists Special Publication 42, p. 183–213.

MANIVIT, H., 1984, Paleogene and Upper Cretaceous calcareous nannofossils from Deep Sea Drilling Project Leg 74: Washington, D.C., Initial Reports of the Deep Sea Drilling Project, v. 74, p. 475–499.

MARTINI, E., 1971, Standard Tertiary and Quaternary calcareous nannoplankton zonation, in Farinacci, A., ed., Proceedings of the Second Planktonic Conference: Roma, Tecnoscienza, p. 739–785.

MARTINSEN, O. J., MARTINSEN, R. S., AND STEIDTMANN, J. R., 1993, Mesaverde Group (Upper Cretaceous), southeastern Wyoming: Allostratigraphy versus sequence stratigraphy in a tectonically active area: American Association of Petroleum Geologists Bulletin, v. 77, p. 1351–1373.

MARTIRE, L., 1992, Sequence stratigraphy and condensed pelagic sediments. An example from the Rosso Ammonitico Veronese, northeastern Italy: Palaeogeography, Palaeoclimatology, Palaeoecology, v. 94, p. 169–191.

MASSON, D. G., MONTADERT, L., AND SCRUTTON, R. A., 1985, Regional geology of the Goban Spur continental margin: Washington, D.C., Initial Reports of the Deep Sea Drilling Project, v. 80, p. 1115–1139.

MAXWELL, A. E., AND OTHERS, 1970, Site 20: Washington, D.C., Initial Reports of the Deep Sea Drilling Project, v. 3, p. 319–366.

McCAVE, N., 1979, Diagnosis of turbidites at Sites 386 and 387 by particle counter size analysis of the silt (2–40 µm) fraction: Washington, D.C., Initial Reports of the Deep Sea Drilling Project, v. 43, p. 395–405.

McGOWRAN, B., 1978, Maestrichtian to Eocene foraminiferal assemblages in the northern and eastern Indian Ocean region: Correlations and historical patterns, in Heitzler, J. R., Bolli, H. M., and Sclater, J. G., eds, Indian Ocean Geology and Biostratigraphy: Washington, D.C., American Geophysical Union, p. 417–458.

McGOWRAN, B., 1979, The Tertiary of Australia: Foraminiferal overview: Marine Micropaleontology, v. 4, p. 235–264.

MILLER, F. X., 1977, The graphic correlation method in biostratigraphy, in Kauffman, E. G. and Hazel, J. E., eds., Concepts and Methods of Biostratigraphy: Strousbourg, Dowen, Hutchinson and Ross, p. 165–186.

MILLER, K. G., CURRY, W. B., AND OSTERMAN, D. R., 1985, Late Paleogene (Eocene to Oligocene) benthic foraminiferal oceanography of the Goban Spur region, Deep Sea Drilling Project Leg 80: Washington, D.C., Initial Reports of the Deep Sea Drilling Program, v. 80, p. 505–538.

MILLER, K. G., FEIGENSON, M. D., WRIGHT, J. D., AND CLEMENT, B. M., 1991, Miocene isotope reference section, Deep Sea Drilling Project Site 608: An evaluation of isotopic and biostratigraphic resolution: Paleoceanography, v. 6, p. 33–52.

MILLER, K. G. AND HART, M. B., 1987, Cenozoic planktonic foraminifers and hiatuses on the new Jersey slope and rise: Deep Sea Drilling Project Leg 95, northwest Atlantic: Washington, D.C., Initial Reports of the Deep Sea Drilling Project, v. 95, p. 253–265.

MILLER, K. G. AND KATZ, M. E., 1987, Eocene benthic foraminiferal biofacies of the New Jersey transect: Washington, D.C., Initial Reports of the Deep Sea Drilling Project, v. 95, p. 267–298.

MONTADERT, L., ROBERTS, D. G., AND OTHERS, 1979a, Sites 405 and 406: Washington, D.C., Initial Reports of the Deep Sea Drilling Project, v. 48, p. 211–273.

MONTADERT, L., ROBERTS, D. G., AND OTHERS, 1979b, Sites 403 and 404: Washington, D.C., Initial Reports of the Deep Sea Drilling Project, v. 48, p. 165–209.

MONTADERT, L., ROBERTS, D. G., AND OTHERS, 1979c, Sites 399, 400 and Hole 400A: Washington, D.C., Initial Reports of the Deep Sea Drilling Project, v. 4, p. 35–71.

MONTADERT, L., ROBERTS, D. G., AND OTHERS, 1979d, Site 401: Washington, D.C., Initial Reports of the Deep Sea Drilling Project, v. 48, p. 73–123.

MONTANARI, A., DEINO, A. L., DRAKE, R. E., TURRIN, B. D., DE PAOLO, D. J., ODIN, G. S., CURTIS, G. H., ALVAREZ, W., AND BICE, D. M., 1988, Radioisotopic dating of the Eocene-Oligocene boundary in the pelagic sequence of the northeastern Apennines, in Premoli Silva, I., Coccioni, R., and Montanari, A., eds., International Union of Geological Sciences Commission on Stratigraphy: Ancona, International Subcommission on Paleogene Stratigraphy Report, p. 195–208.

MOORE, T. C., JR., 1972, DSDP: successes, failures, proposals: Geotimes, v. 17, p. 27–31.

MOORE, T. C., JR., RABINOWITZ, P. D., JR., BORELLA, P., BOERSMA, A., AND SHACKLETON, N., 1984a, Introduction and explanatory notes: Washington, D.C., Initial Reports of the Deep Sea Drilling Project, v. 74, p. 3–39.

MOORE, T. C., JR., RABINOWITZ, P. D., BORELLA, P. E., SHACKLETON, N. J., AND BOERSMA, A., 1984b, History of the Walvis Ridge: Washington, D.C., Initial Reports of the Deep Sea Drilling Project, v. 74, p. 873–894.

MOORE, T. C., JR., RABINOWITZ, P. D., AND OTHERS, 1984c, Site 525: Washington, D.C., Initial Reports of the Deep Sea Drilling Project, v. 74, p. 41–160.

MOORE, T. C., JR., RABINOWITZ, P. D., AND OTHERS, 1984d, Site 527: Washington, D.C., Initial Reports of the Deep Sea Drilling Project, v. 74, p. 237–306.

MOORE, T. C., JR., RABINOWITZ, P. D., AND OTHERS, 1984e, Site 528: Washington, D.C., Initial Reports of the Deep Sea Drilling Project, v. 74, p. 307–405.

MOORE, T. C., JR., RABINOWITZ, P. D., AND OTHERS, 1984f, Site 529: Washington, D.C.,, Initial Reports of the Deep Sea Drilling Project, v. 74, p. 407–465.

MOORE, T. C., JR., VAN ANDEL, TJ. H., SANCETTA, C., AND PISIAS, N., 1978, Cenozoic hiatuses in pelagic sediments: Micropaleontology, v. 24, p. 113–138.

MORGAN, G. E., 1979, Paleomagnetic results from DSDP Site 398: Washington, D.C., Initial Reports of the Deep Sea Drilling Project, v. 47, p. 599–611.

MÜLLER, C., 1979, Calcareous nannofossils from the North Atlantic (Leg 48): Washington, D.C., Initial Reports of the Deep Sea Drilling Project, v. 48, p. 589–639.

MÜLLER, C., 1985, Biostratigraphic and paleoenvironmental interpretation of the Goban Spur region based on a study of calcareous nannoplankton: Washington, D.C., Initial Reports of the Deep Sea Drilling Program, v. 80, p. 573-599.

MURRAY, D. W. AND PRELL, W. L., 1991, Pliocene to Pleistocene variations in calcium carbonate, organic carbon, and opal on the Owen Ridge, northern Arabian Sea: College Station, Proceedings of the Ocean Drilling Program, Scientific Results, v. 117, p. 343–363.

MURRAY, J. W., 1979, Cenozoic biostratigraphy and paleoecology of Sites 403 to 406 based on the foraminifers: Washington, D.C., Initial Reports of the Deep Sea Drilling Project, v. 48, p. 415–430.

MURRAY, J. W., 1984, Paleogene and Neogene benthic foraminifers from Rockall Plateau: Washington, D.C., Initial Reports of the Deep Sea Drilling Project, v. 81, p. 503–534,

NOCCI, M., AMICI, E., AND PREMOLI SILVA, I., 1991, Planktonic foraminiferal stratigraphy and paleoenvironmental interpretation of Paleogene faunas from the subantarctic transect, Leg 114: College Station, Proceedings of the Ocean Drilling Program, Scientific Results, v. 114, p. 233–279.

NORTH AMERICAN COMMISSION ON STRATIGRAPHIC NOMENCLATURE, 1983, North American stratigraphic code: American Association of Petroleum Geologists Bulletin, v. 67, p. 841–875.

OBRADOVICH, J., 1993, A Cretaceous time scale in Caldell, W. G. E. and Kauffman, E. G., eds., Evolution of the Western Interior Basin: St. Johns, Geological Association of Canada Special Paper 39, p. 379–396.

ODIN, G. S., 1982, Numerical Dating in Stratigraphy: New York, John Wiley and Sons, 1040 p.

OKADA, H., AND BUKRY, D. 1980, Supplementary modification and introduction of code numbers to the low-latitude coccolith biostratigraphic zonation: Marine Micropaleontology, v. 5, p. 321–325.

OKADA, H. AND THIERSTEIN, H. P., 1979, Calcareous nannoplankton — Leg 43, Deep Sea Drilling Project: Washington, D.C., Initial Reports of the Deep Sea Drilling Project, v. 43, p. 507–573.

OLSSON, R. K. AND WISE, S. W., JR., 1987, Upper Paleocene to middle Eocene depositional sequences and hiatuses in the New Jersey Atlantic margin: Washington, D.C., Cushman Foundation for Foraminiferal Research Special Publication, v. 24, p. 99–112.

PALMER, A. A., 1987, Cenozoic radiolarians from Deep Sea Drilling Project Sites 612 and 613 (Leg 95, New Jersey Transect) and Atlantic slope project Site ASP 15: Washington, D.C., Initial Reports of the Deep Sea Drilling Project, v. 95, p. 339–357.

PERCH-NIELSEN, K., 1972, Remarks on Late Cretaceous to Pleistocene coccoliths from the North Atlantic: Washington, D.C., Initial Reports of the Deep Sea Drilling Project, v. 12, p. 1003–1069.

PERCH-NIELSEN, K., 1977, Albian to Pleistocene calcareous nannofossils from the western South Atlantic, DSDP Leg 39: Washington, D.C., Initial Reports of the Deep Sea Drilling Project, v. 39, p. 699–823.

PIMM, A. C. AND SCLATER, J. C., 1975, Early Tertiary hiatuses in the northeastern Indian Ocean: Nature, v. 252, p. 362–365.

POAG, W. C., 1993, Allostratigraphy of the U. S. middle Atlantic continental margin — Characteristics, distribution, and depositional history of principal unconformity-bounded upper Cretaceous and Cenozoic sedimentary units: Washington, D.C., United States Geological Survey Professional Paper 1542, 81 p.

POAG, W. AND LOW, D., 1987, Unconformable sequence boundaries at Deep Sea Drilling Project Site 612, New Jersey transect: Their characteristics and stratigraphic significance: Washington, D.C., Initial Reports of the Deep sea Drilling Project, v. 95, p. 453–498.

POAG, W. C., WATTS, A., AND OTHERS, 1987a, Site 613: Washington, D.C., Initial Reports of the Deep Sea Drilling Project, v. 95, p. 155–241.

POAG, W. C., WATTS, A., AND OTHERS, 1987b, Site 612: Washington, D.C., Initial Reports of the Deep Sea Drilling Project, v. 95, p. 31–153.

POORE, R. Z., TAUXE, L., PERCIVAL, S. F., Jr., LABRECQUE, J. L., WRIGHT, R., PETERSON, N. P., SMITH, C. C., TUCKER, P., AND HSU, K. J., 1983, Late Cretaceous-Cenozoic magnetostratigraphic and biostratigraphic correlations of the South Atlantic Ocean: DSDP Leg 73: Palaeogeography, Palaeoclimatology, Palaeoecology, v. 42, p. 127–149.

PRELL, W. L., IMBRIE, J., MARTINSON, D. G., MORLEY, J. J., PISIAS, N. G., SHACKLETON, N. J., AND STREETER, H. F., 1986, Graphic correlation of oxygen isotope stratigraphy application to the late Quaternary: Paleoceanography, v. 1, p. 137–162.

PREMOLI SILVA, I. AND McNULTY, C. L., 1984, Planktonic foraminifers and Calpionellids from Gulf of Mexico sites, Deep Sea Drilling Project Leg 77: Washington, D.C., Initial Reports of the Deep Sea Drilling Project, v. 77, p. 547–587.

PROTO DECIMA, F., MEDIZZA, F., AND TODESCO, L., 1978, Southeastern Atlantic Leg 40 calcareous nannofossils: Washington, D.C., Initial Reports of the Deep Sea Drilling Project, v. 40, p. 571–634.

PUJOL, C., 1983, Cenozoic planktonic foraminiferal biostratigraphy of the southwestern Atlantic (Rio Grande Rise): Deep Sea Drilling Project Leg 72: Washington, D.C., Initial Reports of the Deep Sea Drilling Project, v. 72, p. 623–673.

RAYMO, M. E., RUDDIMAN, W. F., BACKMAN, J., CLEMENT, B. M., AND MARTINSON, D. G., 1989, Late Pliocene variation in northern hemisphere ice sheets and North Atlantic deep water circulation: Paleoceanography, v. 4, p. 413–446.

ROBERTS, D., SCHNITKER, D., AND OTHERS, 1984, Sites 552–553: Washington, D.C., Initial Reports of the Deep Sea Drilling Project, v. 81, p. 31–233.

RONA, P. A., 1973, Worldwide unconformities in marine sediments related to eustatic changes of sea-level: Nature Physical Science, v. 244, p. 25–26.

RUDDIMAN, W. F., CAMERON, D., AND CLEMENT, B. M., 1987, Sediment disturbance and correlation of offset holes drilled with the hydraulic piston corer: Leg 94: Washington, D.C., Initial Reports of the Deep Sea Drilling Project, v. 94, p. 615–634.

SADLER, P., 1981, Sediment acumulation rates and the completeness of stratigraphic sections: Journal of Geology, v. 89, p. 569–584.

SANFILIPPO, A. AND RIEDEL, W. R., 1973, Cenozoic radiolaria (exclusive of Theoperids, Artostrobiids and Amphipyndacids) from the Gulf of Mexico, Deep Sea Drilling Project — Leg 10: Washington, D.C., Initial Reports of the Deep Sea Drilling Project, v. 10, p. 475–611.

SANFILIPPO, A. AND RIEDEL, W. R., 1979, Radiolaria from the northeastern Atlantic Ocean, DSDP Leg 48: Washington, D.C., Initial Reports of the Deep Sea Drilling Project, v. 48, p. 493–511.

SCHMIDT, R. R., 1978, Calcareous nannoplankton from the western North Atlantic: Washington, D.C., Initial Reports of the Deep Sea Drilling Project, v. 44, p. 703–729.

SCHNITKER, D., 1979, Cenozoic deep water benthic foraminifers, Bay of Biscay: Washington, D.C., Initial Reports of the Deep Sea Drilling Project, v. 48, p. 377–413.

SHACKLETON, N. J., 1994, The evolution of high-resolution stratigraphy: EOS, Transactions, American Geophysical Union, v. 75, p. 52.

SHACKLETON, N., AND MEMBERS OF THE SHIPBOARD SCIENTIFIC PARTY, 1984, Accumulation rates in Leg 74 sediments: Washington, D.C., Initial Reports of the Deep Sea Drilling Project, v. 74, p. 621–644.

SHAW, A. B., 1964, Time in Stratigraphy: New York, McGraw Hill, 365 p.

SIBUET, J.-C. AND RYAN, W. B. F., 1979, Site 398: Evolution of the west Iberian passive continental margin in the framework of the early evolution of the North Atlantic Ocean: Washington, D.C., Initial Reports of the Deep Sea Drilling Project, v. 47, p. 761–775.

SIBUET, J.-C., RYAN, W. B. F., AND OTHERS, 1979, Site 398: Washington, D.C., Initial Reports of the Deep Sea Drilling Project, v. 47, p. 25–233.

SNYDER, S. W. AND WATERS, V. J., 1985, Cenozoic planktonic foraminiferal biostratigraphy of the Goban Spur region, Deep Sea Drilling Project Leg 80: Washington, D.C., Initial Reports of the Deep Sea Drilling Program, v. 80, p. 439–471.

SOUTHAM, J. R., HAY, W. W., AND WORSLEY, T. R., 1975, Quantitative formulation of reliability in stratigraphic correlation: Science, v. 188, p. 357–359.

SRIVASTAVA, S. P., ARTHUR, M., CLEMENT, B., AND OTHERS, 1987, Site 647: College Station, Proceedings of the Ocean Drilling Program, Initial Reports, p. 675–905.

STEINMETZ, J., BARRON, E. J., BOERSMA, A., KEATING, B., McNULTY, C., SANCETTA, C., AND STRADNER, H., 1984, Summary of biostratigraphy and magnetostratigraphy of Deep Sea Drilling Project Leg 75: Washington, D.C., Initial Reports of the Deep Sea Drilling Project, v. 84, p. 449–458.

STEINMETZ, J. AND STRADNER, H., 1984, Cenozoic calcareous nannofossils from Deep Sea Drilling Project Leg 75, southeast Atlantic Ocean: Washington, D.C., Initial Reports of the Deep Sea Drilling Project, v. 84, p. 671–753.

SUPKO, P. R., PERCH-NIELSEN, K., AND OTHERS, 1977a, Site 356: Sao Paulo Plateau: Washington, D.C., Initial Reports of the Deep Sea Drilling Project, v. 39, p. 141–230.

SUPKO, P. R., PERCH-NIELSEN, K., AND OTHERS, 1977b, Site 357: Rio Grande Rise: Washington, D.C., Initial Reports of the Deep Sea Drilling Project, v. 39, p. 231–327.

SUPKO, P. R., Perch-NIELSEN, K., AND OTHERS, 1977c, Site 358: Argentine Basin: Washington, D.C., Initial Reports of the Deep Sea Drilling Project, v. 39, p. 329–371.

SUPKO, P. R., PERCH-NIELSEN, K., AND OTHERS, 1977d, Site 355: Brazil Basin: Washington, D.C., Initial Reports of the Deep Sea Drilling Project, v. 39, p. 101–140.

THIEDE, J. AND VAN ANDEL, T. H., 1977, The paleoenvironment of anaerobic sediments in the late Mesozoic South Atlantic Ocean: Earth and Planetary Science Letters, v. 33, p. 301–309.

TJALSMA, R. C., 1983, Eocene to Miocene benthic foraminifers from DSDP Site 516, Rio Grande Rise, South Atlantic: Washington, D.C., Initial Reports of the Deep Sea Drilling Project, v. 72, p. 731–755.

TJALSMA, R. C. AND LOHMAN, G. P., 1983, Paleocene-Eocene Bathyal and Abyssal Benthic Foraminifera From the Atlantic Ocean: New York, Micropaleontology Special Publication 4, 90 p.

TOUMARKINE, M., 1978, Planktonic foraminiferal biostratigraphy of the Paleogene of Sites 360 to 364 and the Neogene of Sites 362A, 363 and 364 Leg 40: Washington, D.C., Initial Reports of the Deep Sea Drilling Project, v. 40, p. 679–721.

TOUMARKINE, M. AND LUTERBACHER, L., H., 1985, Paleocene and Eocene planktic foraminifera, in Bolli, H. M., Saunders, J. B., and Perch-Nielsen, K., eds., Plankton Biostratigraphy: Cambridge, Cambridge University Press, p. 87–154.

TOWNSEND, H., 1985, The paleomagnetism of sediments acquired from the Goban Spur on Deep Sea Drilling Project 80: Washington, D.C., Initial Reports of the Deep Sea Drilling Program, v. 80, p. 389–414.

TUCHOLKE, B. E. AND MOUNTAIN, G. S., 1979, Seismic stratigraphy, lithostratigraphy and paleosedimentation patterns in the North American basin, in Talwani, M., Hay, W. W., and Ryan, W. W. F., eds., Deep Drilling Results in the Atlantic Ocean: Continental Margins and Paleoenvironnement: Washington, D.C., American Geophysical Union, Maurice Ewing Series, v. 3, p. 58–86.

TUCHOLKE, B. E. AND MOUNTAIN, G. S., 1986, Tertiary paleoceanography of the western North Atlantic Ocean, in Vogt, P. R. and Tucholke, B. E., eds., The Geology of North America, The Western North Atlantic Region: Boulder, Geological Society of America, v. M, p. 631–650.

TUCHOLKE, B. E., VOGT, P. R., AND OTHERS, 1979a, Site 387: Cretaceous to Recent sedimentary evolution of the western Bermuda Rise: Washington, D.C., Initial Reports of the Deep Sea Drilling Project, v. 43, p. 323–391.

TUCHOLKE, B. E., VOGT, P. R., AND OTHERS, 1979b, Site 386: Fracture Valley sedimentation on the central Bermuda Rise: Wash ington, D.C., Initial Reports of the Deep Sea Drilling Project, v. 43, p. 195–321.

TUCHOLKE, B. E., VOGT, P. R., AND OTHERS, 1979c, Site 384: The Cretaceous/ Tertiary boundary, Aptian reefs, and the J-Anomaly Ridge: Washington, D.C., Initial Reports of the Deep Sea Drilling Project, v. 43, p. 107–154.

TUCHOLKE, B. E. AND VOGT, P. R., 1979, Western North Atlantic: Sedimentary evolution and aspects of tectonic history: Washington, D.C., Initial Reports of the Deep Sea Drilling Project, v. 43, p. 791–825.

VALENTINE, P., 1987, Lower Eocene calcareous nannofossil biostratigraphy beneath the Atlantic slope and upper rise off New Jersey — New zonation based on Deep Sea Drilling Project Sites 612 and 613: Washington, D.C., Initial Reports of the Deep Sea Drilling Project, v. 95, p. 359–394.

VAIL, P. R., MITCHUM, R. M., JR., TODD, R. G., WIDMIER, J. M., THOMPSON, S. III, SANGREE, J. B., BUBB, J. N., AND HATLELID, W. G., 1977, Seismic stratigraphy and global sea-level changes, in Payton, C. E., ed., Seismic Stratigraphy — Applications to Hydrocarbon Exploration: Tulsa, American Association of Petroleum Geologists Memoir 26, p. 49–212.

VAN ANDEL, T. H., 1981, Consider the incompleteness of the geological record: Nature, v. 294, p. 397–398.

van HINTE, J. E., WISE, S. W., JR., AND OTHERS, 1987, Sites 604 and 605: Washington, D.C., Initial Reports of the Deep Sea Drilling Project, v. 93, p. 277–413.

VAN WAGONER, J. C., POSAMENTIER, H. W., MITCHUM, R. M., JR., VAIL, P. R., SARG, J. F., LOUTIT, T. S., AND HARDENBOL, J., 1988, An overview of the fundamentals of sequence stratigraphy and key definitions, in Wilgus, C. K., Hastings, B. S., Kendall, C. G. St. C., Posamentier, H. W., Ross, C. A., and Van Wagoner, J. C., eds., Sea-level Changes: An Integrated Approach: Tulsa, Society of Economic Paleontolgists and Mineralogists Special Publication 42, p. 39–45.

WATKINS, D. K. AND VERBEEK, J. W., 1988, Calcareous nannofossil biostratigraphy from Leg 101, northern Bahamas: College Station, Proceedings of the Ocean Drilling Program, Scientific Results, v. 101, p. 63–85.

WEAVER, F. M. AND DINKELMAN, M. G., 1978, Cenozoic radiolarians from the Blake Plateau and the Blake-Bahama Basin, DSDP Leg 44: Washington, D.C., Initial Reports of the Deep Sea Drilling Project, v. 44, p. 865–885.

WEI, W. AND WISE, S. W., JR., 1989, Paleogene calcareous nannofossil magnetobiochronology: Results from South Atlantic DSDP Site 516: Marine Micropaleontology, v. 14, p. 119–152.

WEI, W. AND WISE, S. W., 1990, Middle Eocene to Pleistocene calcareous nannofossils recovered by Ocean Drilling Program Leg 113 in the Weddell Sea: College Station, Proceedings of the Ocean Drilling Program, Scientific Results, v. 113, p. 639–666.

WEI, W. AND WISE, S. W., 1992, Eocene-Oligocene calcareous nannofossil magnetobiochronology of the southern Ocean: Newsletter in Stratigrahy, v. 26, p. 119–132.

Westberg-SMITH, M. J. AND RIEDEL, W. R., 1984, Radiolarians from the western margin of the Rockall Plateau: Deep Sea Drilling Project Leg 81: Washington, D.C., Initial Reports of the Deep Sea Drilling Project, v. 81, p. 479–501.

WHEELER, H. E., 1958, Time-Stratigraphy: Bulletin of the American Association of Petroleum Geologists, v. 42, p. 1047–1063.

WORZEL, J. L., BRYANT, W., AND OTHERS, 1973, Site 94, Initial Reports of the Deep Sea Drilling Project: Washington, D.C., p. 195–258.

MAGNETOSTRATIGRAPHY OF UPPER PALEOCENE THROUGH LOWER MIDDLE EOCENE STRATA OF NORTHWEST EUROPE

JASON R. ALI AND ERNIE A. HAILWOOD

Department of Oceanography, University of Southampton, Southampton, SO17 1BJ, United Kingdom

ABSTRACT: Ormesby Clay Formation mudstones in eastern Norfolk are the oldest upper Paleocene deposits (Chron C26r) of southern England. The base of the type Thanet Sand Formation (= start of the Thanetian Stage) is approximately 0.65 my younger (C26n). C25n is missing from the stratigraphic record across much of southern England indicating an hiatus >0.5 my between the Ormesby Clay-Thanet Sand Formations and the Lambeth Group. However, in Central London the basal unit of the Lambeth Group, the Upnor Formation, contains a record of C25n. The bulk of the Lambeth Group, the Harwich Formation and lower London Clay Formation, were all deposited during Chron C24r. The NP9/10 boundary, a commonly used Paleocene/Eocene boundary marker, is correlated with the unconformity (approximately 0.4 my) that separates the Lambeth Group from the Harwich Formation. The start of C24n.3n is positioned at the base of Division B of the London Clay and Ieper Clay Formations. In the Hampshire Basin, an unconformity of at least 0.66 my is marked by the absence of a record of Chron C22n.

INTRODUCTION

The lower Paleogene deposits of northwest Europe have long been used as global chronostratigraphic reference sections. Unfortunately, the biostratigraphic record of these sequences is far from ideal with many intervals lacking key microfossils. Correlating the Thanetian and Ypresian stage stratotypes with the global marine record has proved difficult. Problems in defining the Paleocene/Eocene boundary in the region have provided an impetus for many biostratigraphic workers to reexamine critical intervals from sections around the southern margin of the North Sea Basin.

In addition to biostratigraphy, magnetostratigraphy is a powerful tool for subdividing the geological record. In this paper, we synthesize a large body of data that has been collected by the Paleomagnetism Group at Southampton University, from the lower Tertiary sections of southern England and northwest mainland Europe (Townsend and Hailwood, 1985; Cox and others, 1985; Aubry and others, 1986; Knox and others, 1990; Ali and others, 1993; Knox and others, 1994; Ali and Jolley, in press; Ali and others, in press; Ellison and others, in press). The studies have provided twelve chronostratigraphic markers for a 15-my early Tertiary interval. Particularly useful markers help define: (1) the age of the type Thanetian (Chron C26n to C25r); (2) the lower and upper bounds of Chron C24r, which brackets the Paleocene/Eocene boundary; and (3) the 'Chron C22n hiatus' which coincides with the Ypresian/Lutetian stage boundary in the Hampshire Basin.

DEPOSITIONAL SETTING

During the early Tertiary the ancestral North Sea extended across much of southeast England and northwest mainland Europe (see Ziegler, 1982). As the sea level of the North Sea fluctuated, the shoreline periodically advanced and retreated across the region. The basin was probably isolated from the main ocean during sea-level lowstands. The upper Paleocene through lower middle Eocene sediments preserved in southern England and mainland northwest Europe comprise outer neritic, inner neritic, marginal-marine and fluviatile deposits.

MAGNETOCHRON CORRELATIONS

The locations of the sections investigated are shown in Figure 1. The magnetostratigraphy of the upper Paleocene and lowermost Eocene of southern England, northern France and west-

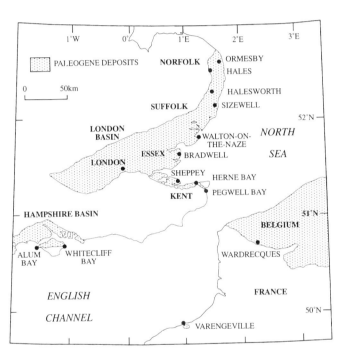

FIG. 1.—Map showing the outcrop and subcrop of the lower Paleogene deposits of southeast England, northern France and western Belgium, and the location of boreholes and outcrops referred to in this paper.

ern Belgium is summarised in Figure 2. Data from eleven localities are presented, some representing composite sections (Kent, Essex and Belgium), others individual outcrops or boreholes. The geomagnetic polarity time-scale of Berggren and others (this volume), based on Cande and Kent (1992), has been used to age-calibrate the magnetochron boundaries. Martini's (1971) nannofossil scheme has been positioned against this scale using the fossil zone-magnetochron correlations of Berggren and others (1985) and Berggren and others (this volume).

Chron C26r

The oldest upper Paleocene sediments in southern England occur at subcrop in the Ormesby and Hales boreholes, east Norfolk. There, the reverse polarity lower part of the Ormesby Clay Formation is correlated with Chron C26r. The base of the Ormesby Clay Formation at both Ormesby and Hales has been

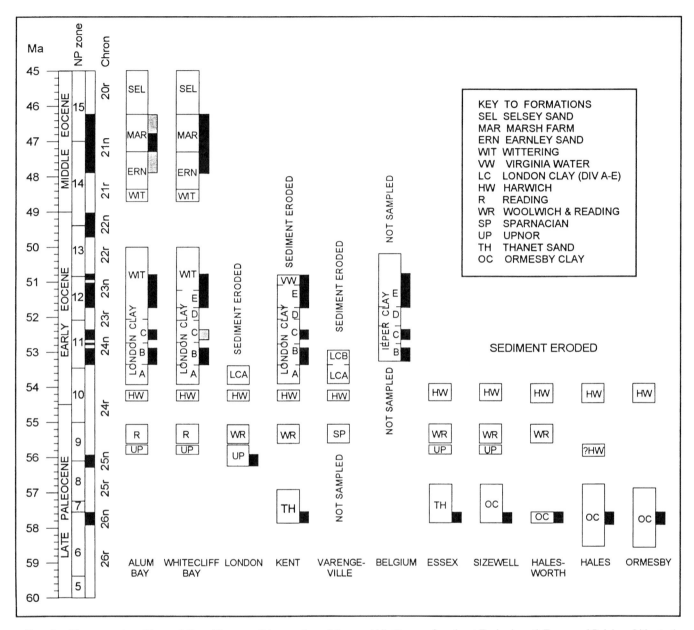

Fig. 2.—Magnetostratigraphic correlation of the upper Paleocene through lower middle Eocene of southeast England, north France and Belgium. Lithostratigraphic nomenclature from King (1981); Edwards and Freshney (1986); King (1990); Ellison and others (1994); and Jolley (in press). The sections are age-calibrated using the geomagnetic polarity time scale of Berggren and others (this volume). The nannoplankton zonation scheme is from Martini (1971). Magnetochron-nannofossil biozone correlations are based on Berggren and others (1985 and this volume). In the magnetostratigraphy columns, black represents normal polarity and white reverse polarity. Where normal polarity zones are anticipated, but have not yet been established, they are shown in gray. Magnetostratigraphic zonations: Alum Bay and Whitecliff Bay, Isle of Wight, (Aubry and others, 1986); London (Ellison and others, in press); Kent (Aubry and others, 1986; Ali and others, 1993); Varengeville, France, and Belgium (Ali and others, 1993); Essex (Knox and others, 1994; Ali and others, in press); Sizewell and Halesworth (Ali and Jolley, in press); Hales (Knox and others, 1990; Ali and Jolley, in press); and Ormesby (Cox and others, 1985; Ali and Jolley, in press).

assigned broadly similar ages, represented by palynomoph association sequences T1 and T2 respectively (Ali and Jolley, in press). The difference in thickness of C26r sediments between the two sections indicates relative condensation at Hales. Assuming a fixed accumulation rate throughout deposition of the Ormesby Clay Formation at Hales, Ali and Jolley (in press) calculate the base of the Ormesby Clay in Norfolk to be approximately 650 ky older than the oldest sections of the Thanet Sand Formation (Chron C26n).

Chron C26n

At Ormesby and Hales, Norfolk, the start of Chron C26n is within palynomorph association sequence T6 (Jolley, 1992). In all sections between Halesworth (Suffolk) and Kent, the basal Tertiary sediments carry a normal polarity magnetisation. The majority of the sections that have been studied for palynomorphs indicate that the base of the Thanet Sand Formation (and the Ormesby Clay in Sizewell) is at a higher level within

T6 (Jolley, 1992). Nannoplankton studies of the basal Thanet Sand Formation at Bradwell (Knox and others, 1994) indicate an NP6 age for these levels, which is similar to that reported from the type section in Kent (Aubry and others, 1986; Siesser and others, 1987). The normal polarity magnetozone identified at the base of the Thanet Sand and in the lower half of the Ormesby Clay Formation is thus correlated with magnetic Chron C26n.

The end of Chron C26n is recorded at a number of sections through the Ormesby Clay and Thanet Sand Formation. In the Thanet Sand Formation at Bradwell (Essex), the end of this chron corresponds to the upper part of nannofossil zone NP6 (Knox and others, 1994). Palynological studies (Jolley, 1992) indicate a mid T7 (T7c) age for the end of Chron C26n in all of the sections between Norfolk and Kent. The tops of the Thanet Sand and Ormesby Clay Formations are thought to have been eroded prior to deposition of younger sediments (Curry, 1981). In the sections through these formations north of London, the youngest preserved sediments are of T9 age (Jolley, 1992). In London and north Kent, the top of the Thanet Sand Formation is assigned to palynomorph association sequence T8, indicating a slightly deeper level of erosion.

Chron C25n

Chron C25n is associated with the upper part of nannofossil zone NP8 and the lower part of NP9 (Berggren and others, 1985). Hamilton and Hojjatzadeh (1982), Aubry and others (1986), and Siesser and others (1987) indicate that the record of nannofossil zones NP8 and NP9 in southern England is rather poor. Aubry and others (1986) deduced that Chron C25n was not preserved in southern England; the lack of such a record resulted from either a non-sequence or complete erosion of material deposited during that interval. The non-sequence/erosional event was located at the unconformity which separates the Thanet Sand Formation from the Lambeth Group Upnor Formation.

Biostratigraphic and paleomagnetic studies of the upper Paleocene at Ormesby and Hales revealed a similar gap in the sedimentary record (Cox and others, 1985; Knox and others, 1990; although the paper of Cox and others, 1985, predates that of Aubry and others, 1986, their interpretation relied heavily on the unpublished conclusions of the latter work). Knox and others (1990) recognised that the Hales Clay (lower Harwich Formation) was younger than the Lambeth Group. This implies that the unconformity separating the Ormesby Clay and Harwich Formations in east Norfolk represents a greater hiatus there than in northern Suffolk, less than 30 km to the south.

Recent studies by Ellison and others (in press) and Ali and Jolley (in press) report a record of Chron C25n in the Upnor Formation in central London. There, the Upnor Formation comprises two distinct sediment packages separated by a pebble bed. The lowermost unit, which records the Chron C25n magnetisation, extends across only a few square kilometers. It forms an isolated remnant preserved beneath sediments deposited during the main Upnor Formation transgression (early Chron C24r).

Chron C24r

The main body of the Upnor Formation is reversely magnetised (early Chron C24r). The Woolwich, Reading and Har-wich Formations and the lower part of the London Clay Formation were all deposited during the 2.56-my Chron C24r. According to Berggren and others (1985) and Berggren and others (this volume), these deposits straddle the Paleocene/Eocene boundary. Ali and Jolley (in press) position the Paleocene/Eocene boundary in southeast England just below the base of the Harwich Formation, at the unconformity which separates the formation from the Lambeth Group.

Chron C24n

In earlier geomagnetic polarity time scales (e.g., Berggren and others, 1985), the normal polarity parts of Chron C24 were allocated to two sub-chrons (C24BN and C24AN). In the time scales of Cande and Kent (1992) and Berggren and others (this volume), the normal polarity portions of Chron C24 are allocated to 3 sub-chrons. Chrons C24n.3n and C24n.1n are 0.46 and 0.31 my in length respectively. Chron C24n.2n is of much shorter duration (0.04 my). Aubry and others (1986) and Ali and others (1993) correlated two normal polarity magnetozones in the lower and middle parts of both the London Clay and Ieper Formations with Chrons C24BN and C24AN of Berggren and others (1985). In this paper, we take Chrons C24BN and C24AN to be the equivalents of Chrons C24n.3n and C24n.1n respectively.

Chron C24n.3n

The start of Chron C24n.3n coincides with the base of the London Clay Formation Division B at Alum Bay, Whitecliff Bay (Aubry and others, 1986) and Sheppey, Kent (Ali and others, 1993). At these sections, the normal polarity magnetozone correlated with Chron C24n.3n terminates within Division B2. At Varengeville, north France, a reverse polarity magnetisation is associated with the glauconitic clay bed marking the base of Division B. However, Ali and others (1993) considered this magnetization to be a much delayed post-depositional remanence acquired during a later reverse polarity interval.

In Belgium, a record of the middle and later part of Chron C24n.3n is preserved in the type section of the Ieper Clay Formation Wardrecques Member (Ali and others, 1993). This normal polarity interval is from a level equivalent to Division B of the London Clay Formation (King, 1990). As far as we are aware, intervals corresponding to the basal part of Division B are not exposed in Belgium, and it is not possible to locate the start of Chron C24n.3n there.

Chron C24n.1n

At Sheppey (Kent), the base of Chron C24n.1n is positioned at 8 m above the base of London Clay Formation Division C1 and its top is placed 8 m below the Division C/D junction. In Belgium, Chron C24n.1n extends from the lower part of Division C1 (which is highly condensed here) to the upper part of Division C2. This magnetozone terminates about 2.5 m above the Division C1/C2 boundary. At Whitecliff Bay the record of Chron C24n.3n is considered unreliable.

Chron C23n

Aubry and others (1986) correlated the Wittering magnetozone of Townsend and Hailwood (1985) in the Hampshire Basin with Chron C23n. This magnetochron was identified at Bracklesham Bay, Alum Bay and Whitecliff Bay. Ali and others (1993) repositioned the base of the magnetozone representing Chron C23n at Whitecliff Bay at 3.5 m below the base of the Wittering Formation, close to the Division D/E junction. At Whitecliff Bay the end of this magnetochron is at about 30 m above the base of the Wittering Formation.

At Sheppey (Kent), the base of Chron C23n is positioned 3.35 m below the London Clay Formation Division D/E boundary. The top of this magnetochron was not identified in the section at Sheppey; Chron C23n must terminate at a level above the lower part of the Virginia Water Formation. In Belgium, Chron C23n commences 1.4 m above the base of the Aalbeke Clay Member (the base of the Aalbeke Member is tentatively equated with the base of Division E (Ali and others 1993) and terminates in the lower part of the Egem Sand Member.

Chron C22n

Middle Eocene magnetostratigraphic data are presently available only from sections in the Hampshire Basin. Equivalent levels in Belgium and north France have not yet been investigated. In the London Basin, all Tertiary sediments younger than about 50 Ma (Chron C22r) have been eroded.

Aubry and others (1986) suggested that Chron C22n was missing from the lower Paleogene of the Hampshire Basin. Aubry and others (1986) proposed to locate the hiatus (>0.66 my) at the level of the Whitecliff Bay (lignite) Bed (= Plint's (1983) WB8) in the upper part of the Wittering Formation. This interpretation implies that Chron C22R is recorded in the sediments below the Whitecliff Bay Bed, and Chron C21R is recorded in the sediments above. However, these levels are barren of nannoplankton, and Ali and others (1993) suggested that the unconformity could be positioned at two other alternative levels. The first is in the topmost part of Plint's Unit WB6, which is an intensely glauconitic, highly bioturbated sand, indicating major condensation or a break in deposition. It is overlain very sharply by the base of Unit WB7, represented by a blocky silty clay with weak silt laminae. The second is between the Wittering Formation and the Earnley Sand Formation. This contact is a discontinuity (parasequence) boundary, marked by an interburrowed junction on which rests a thin pebbly glauconitic sand.

Chron C21n

At Whitecliff Bay, Chron C21n starts in the upper part of the Earnley Sand and continues to the top of the Marsh Farm Formation (a laminated glauconitic silty clay with thin sands previously included in the upper part of the Earnley Sand Formation — see lithostratigraphic revision of Edwards and Freshney, 1986).

CONCLUSIONS

The late Paleocene through early middle Eocene (60–45 Ma) deposits of southeast England are now placed within a magnetobiostratigraphic framework, that provides important constraints on our understanding of (1) how the originally defined Thanetian and Ypresian stage stratotypes relate to the global marine record and (2) the nature and timing of the Thanetian, Ypresian and Lutetian depositional sequences of the southern North Sea Basin. During this interval, deposition of the shallow marine and continental sediments along the southwest margin of the North Sea was punctuated by three hiatuses. The hiatuses at around 56.5 and 54.5 Ma both represent time intervals of more than 0.4 my. In the Hampshire Basin an hiatus of between 0.66 and 2.5 my corresponds with the early/middle Eocene boundary (at around 49 Ma).

ACKNOWLEDGMENTS

The Natural Environment Research Council is acknowledged for financial support to JRA during his Ph.D studies (1985–88). We are grateful to various colleagues for helpful and stimulating discussions during the development of the research, in particular Marie-Pierre Aubry, Bill Berggren, Richard Ellison, Norman Hamilton, Nick Johnston, David Jolley, Chris King, Robert Knox and Tony Morigi. Kate Davies helped with the drafting of figures. Reviews by Robert Knox and Dennis Kent and the comments of Dana Ulmer-Scholle have greatly improved the manuscript.

REFERENCES

ALI, J. R. AND JOLLEY, D. W., in press, Chronostratigraphic framework for the Thanetian and lower Ypresian deposits of southern England, *in* Knox, R. W. O'B., Corfield, R,. and Dunay, R. E., eds., Correlation of the Early Paleogene Strata in Northwest Europe: London, Geological Society of London Special Publication.

ALI, J. R., KING, C., AND AND HAILWOOD, E. A., 1993, Magnetostratigraphic calibration of early Eocene depositional sequences in the southern North Sea Basin, *in* Hailwood, E. A. and Kidd, R. B., eds., High Resolution Stratigraphy: London, Geological Society of London Special Publication 70, p. 99–125.

ALI, J. R., HAILWOOD, E. A. AND KING, C., in press, 'Oldhaven magnetozone' of East Anglia: a revised interpretation, *in* Knox, R. W. O'B., Corfield, R., and Dunnay, R. E., eds., Correlation of the Early Paleogene Strata in Northwest Europe: London, Geological Society of London Special Publication.

AUBRY, M.-P., HAILWOOD, E. A., AND TOWNSEND, H. A., 1986, Magnetic and calcareous-nannofossil stratigraphy of the lower Palaeogene formations of the Hampshire and London Basins: Journal of the Geological Society London, v. 143, p. 729–35.

BERGGREN, W. A., KENT, D. V., AND FLYNN, J. J., 1985, Palaeogene geochronology and chronostratigraphy, *in* Snelling, N. J., ed., Geochronology of the Geological Record: London, Geological Society London Memoir 10, p. 141–195.

CANDE, S. AND KENT, D. V., 1992, A new geomagnetic time scale for the Late Cretaceous and Tertiary: Journal of Geophysical Research, v. 97, p. 13,917–13,951.

COX, F. C., HAILWOOD, E. A., HARLAND, R., HUGHES, M. J., JOHNSTON, N., AND KNOX, R. W. O'B., 1985, Palaeocene sedimentation and stratigraphy in Norfolk, England: Newsletters on Stratigraphy, v. 14, p. 169–185.

CURRY, D., 1981, Thanetian, *in* Pomerol, C., ed., Stratotypes of Palaeogene Stages: Paris, Memoire Hors-Serie du Bulletin d'Information des Geologues du Bassin de Paris 2, p. 255–265.

EDWARDS, R. A. AND FRESHNEY, E. C., 1987, Lithostratigraphical classification of the Hampshire Basin Palaeogene deposits (Reading Formation to Headon Formation): Tertiary Research, v. 8, p. 43–73.

ELLISON, R. A., JOLLEY, D. W., KING, C., AND KNOX, R. W. O'B., 1994, A revision of the lithostratigraphical classification of the early Palaeogene strata in the London Basin and East Anglia: Proceedings of the Geologist's Association, v. 105, p. 187–197.

ELLISON, R. A., KNOX, R. W. O'B., ALI, J. R., JOLLEY, D. W., AND HINE, N. M., in press, Towards a higher stratigraphic resolution of the upper Paleocene strata of the London Basin, *in* Knox, R. W. O'B., Corfield, R., and Dunay, R. E., eds., Correlation of the Early Paleogene Strata in Northwest Europe: London, Geological Society of London Special Publication.

HAMILTON, G. B. AND HOJJATZADEH, M., 1982, Cenozoic calcareous nanno-fossils—a reconnaissance, *in* Lord, A. R., ed., A Stratigraphical Index of Calcareous Nannofossils: Chichester, Ellis Horwood Ltd, p. 136–167.

JOLLEY, D. W., 1992, Palynofloral association sequence stratigraphy of the Palaeocene Thanet beds and equivalent sediments in eastern England: Review of Palaeobotany and Palynology, v. 74, p. 207–237.

JOLLEY, D.W., in press, The earliest Eocene sediments of eastern England: an ultra high resolution palynological correlation, *in* Knox, R. W. O'B., Corfield, R., and Dunay, R. E., eds., Correlation of the Early Paleogene Strata in Northwest Europe: London, Geological Society of London Special Publication.

KING, C., 1981, The Stratigraphy of the London Clay and Associated Deposits: London, Tertiary Research Special Paper 6, 158 p.

KING, C., 1990, Eocene stratigraphy of the Knokke borehole (Belgium): Toelicht, Verhand, Geologische en Mijnkaarten van Belgie, v. 29, p. 67–102.

KNOX, R. W. O'B., HINE, N. M., AND ALI, J. R., 1994, New information on the age and sequence stratigraphy of the type Thanetian of Southeast England: Newsletters on Stratigraphy, v. 30, p. 45–60.

KNOX, R. W. O'B., MORIGI, A. N., ALI, J. R., HAILWOOD, E. A., AND HALLAM, J. R., 1990, Early Palaeogene stratigraphy of a cored borehole at Hales, Norfolk: Proceedings of the Geologist's Association, v. 101, p. 135–151.

MARTINI, E., 1971, Standard Tertiary and Quaternary calacareous nannoplankton zonation, *in* Farinacci, A., ed., Proceedings of the II Planktonic Conference, Roma 1970: Rome, Edizioni Tecnoscienza, p. 739–785.

PLINT, A. G., 1983, Facies, environments and sedimentary cycles in the Middle Eocene Bracklesham Formation of the Hampshire Basin: evidence for global sea level changes?: Sedimentology, v. 30, p. 625–653.

SIESSER W. G., WARD, D. J., AND LORD A. R., 1987, Calcareous nannoplankton biozonation of the Thanetian Stage (Palaeocene) in the type area: Journal of Micropalaeontology, v. 6, p. 85–102.

TOWNSEND, H. A. AND HAILWOOD, E. A., 1985, Magnetostratigraphic correlation of Palaeogene sediments in the Hampshire and London Basins, southern UK: Journal of the Geological Society of London, v. 142, p. 1–27.

ZIEGLER, P. A. 1982, Geological atlas of western and central Europe: Amsterdam, Shell International Petroleum.

GEOCHRONOLOGY, BIOSTRATIGRAPHY AND SEQUENCE STRATIGRAPHY OF A MARGINAL MARINE TO MARINE SHELF STRATIGRAPHIC SUCCESSION: UPPER PALEOCENE AND LOWER EOCENE, WILCOX GROUP, EASTERN GULF COASTAL PLAIN, U.S.A.

ERNEST A. MANCINI AND BERRY H. TEW

Department of Geology, University of Alabama, Tuscaloosa, AL 35487

ABSTRACT: In the eastern Gulf Coastal Plain, four, third-order unconformity-bounded depositional sequences are recognized for upper Paleocene and lower Eocene marginal marine to marine shelf strata of the Wilcox Group. In this area, the Wilcox Group includes the Nanafalia Formation (TAGC-2.1 depositional sequence), the Tuscahoma Sand (in part, the TAGC-2.1 and the TAGC-2.2 and TAGC-2.3 sequences), and the Hatchetigbee Formation (TAGC-2.4 depositional sequence). These cycles are interpreted to result from changes in sea level and coastal onlap during the late Paleocene and early Eocene epochs. The Nanafalia Formation and lower part of the Tuscahoma Sand are within the upper Paleocene (Selandian Stage) *Planorotalites pseudomenardii* Range Zone, whereas the middle and upper parts of the Tuscahoma Sand have been assigned to the upper Paleocene (Selandian Stage) *Morozovella velascoensis* Interval Zone. The Hatchetigbee Formation contains planktonic foraminifera diagnostic of the lower Eocene (Ypresian Stage) *Morozovella subbotinae* Interval Zone.

The Paleocene-Eocene Epoch boundary (ca. 54–55 Ma) occurs in the Wilcox Group and coincides with the lithostratigraphic contact of the upper Paleocene Tuscahoma Sand with the lower Eocene Hatchetigbee Formation. Palynological studies of the marginal marine strata of the uppermost clay and silt beds of the Tuscahoma Sand and the lowermost sand beds of the Hatchetigbee Formation indicate late Paleocene and early Eocene ages, respectively, for these formations.

Planktonic foraminiferal zone boundaries generally occur at depositional sequence boundaries dividing the upper Paleocene and lower Eocene marginal marine to marine shelf strata in the eastern Gulf Coastal Plain. Factors that affect the stratigraphic position of biozone boundaries in these strata include the presence of unconformities and associated biostratigraphic discontinuities resulting from nondeposition or erosion, the presence or absence of lowstand systems tract strata, the differential amounts and rates of sedimentation associated with paleobathymetry and/or distance from the shoreline at various depositional sites, differential subsidence within the depositional basin, and paleoenvironmental conditions. The development of a major fluvial-dominated delta complex in this area during Paleocene and early Eocene time had a significant influence on depositional conditions.

An integrated geochronologic, biostratigraphic, and sequence stratigraphic approach provides a useful mechanism for correlation of marginal marine and marine shelf strata of the Wilcox Group in the eastern Gulf Coastal Plain. The use of physical surfaces associated with depositional sequences, such as sequence bounding unconformities, transgressive surfaces, and surfaces of maximum transgression/sediment starvation, in conjunction with the first and last occurrences of age diagnostic microfossils and radiometric data, provides a framework for local, regional, and worldwide correlation.

INTRODUCTION

Planktonic foraminiferal zone boundaries in the Paleogene strata of the eastern Gulf Coastal Plain generally correspond to erosional surfaces resulting from the early phase of marine transgressions or to erosional or nondepositional sequence boundaries (Mancini and Tew, 1991). Further, major turnovers in benthic foraminiferal assemblages (Gaskell, 1991) and benthic molluscan faunas (Dockery, 1986) were reported to occur at erosional sequence boundaries in Gulf Coast marginal marine and shelf deposits. This relationship differs from the observations of Baum and Vail (1988) and Olsson (1991), which indicated that most major planktonic biozone boundaries occur within condensed sections associated with maximum transgression (maximum sea-level highstand) within depositional sequences. In addition, a review of the Paleogene planktonic foraminiferal zone boundaries depicted on the chart for Mesozoic and Cenozoic chronostratigraphy and cycles of sea-level change published by Haq and others (1988) suggests that these boundaries in the Paleogene occur more frequently within depositional sequences than at sequence boundaries. It should be noted, however, that the Paleogene strata studied by Dockery (1986), Gaskell (1991) and Mancini and Tew (1991) represent marginal marine to marine shelf deposition, whereas the studies of Baum and Vail (1988) and Haq and others (1988) address basinal marine to marine shelf environments. The Paleogene strata described by Olsson (1991) are recognized as shelf deposits.

The purpose of this paper is to explore possible explanations for the observation that planktonic foraminiferal zone boundaries occur more frequently at sequence boundaries in marginal marine to marine shelf deposits of the eastern Gulf Coastal Plain than in basinal marine to marine shelf environments as described by other authors. The interval selected for study includes strata of the upper Paleocene (Selandian) to lower Eocene (Ypresian) Wilcox Group of southern Alabama. The Wilcox Group includes the upper Paleocene Nanafalia Formation and Tuscahoma Sand and the lower Eocene Hatchetigbee Formation (Fig. 1). The Midway Group underlies the Wilcox Group, and the Midway/Sabine Gulf Coast provincial stage boundary is at the base of the Wilcox interval. The Wilcox Group is overlain by the Claiborne Group, and the Sabine/Claiborne provincial stage boundary coincides with the contact of these units. In this paper, we discuss the geochronology, biostratigraphy, and sequence stratigraphy of the Wilcox stratigraphic succession and the utility of these factors as an integrated approach to local, regional, and worldwide correlation.

GEOLOGIC SETTING

The depositional history of the western part of the eastern Gulf Coastal Plain during the late Paleocene and early Eocene epochs was dominated by deltaic and marginal marine sedimentation (Fig. 2). A major, fluvial-dominated delta complex developed during Paleocene and early Eocene time in Mississippi and southwest Alabama, and this system influenced deposition in the study area throughout the Paleogene (Galloway, 1968; Cleaves, 1980; Mancini, 1983). Subsidence and sediment accommodation were accentuated as a result of withdrawal and movement of the Jurassic Louann Salt in the Mississippi Interior Salt Basin, which is the dominant structural feature in this area. The source area for deltaic sediment was to the northwest,

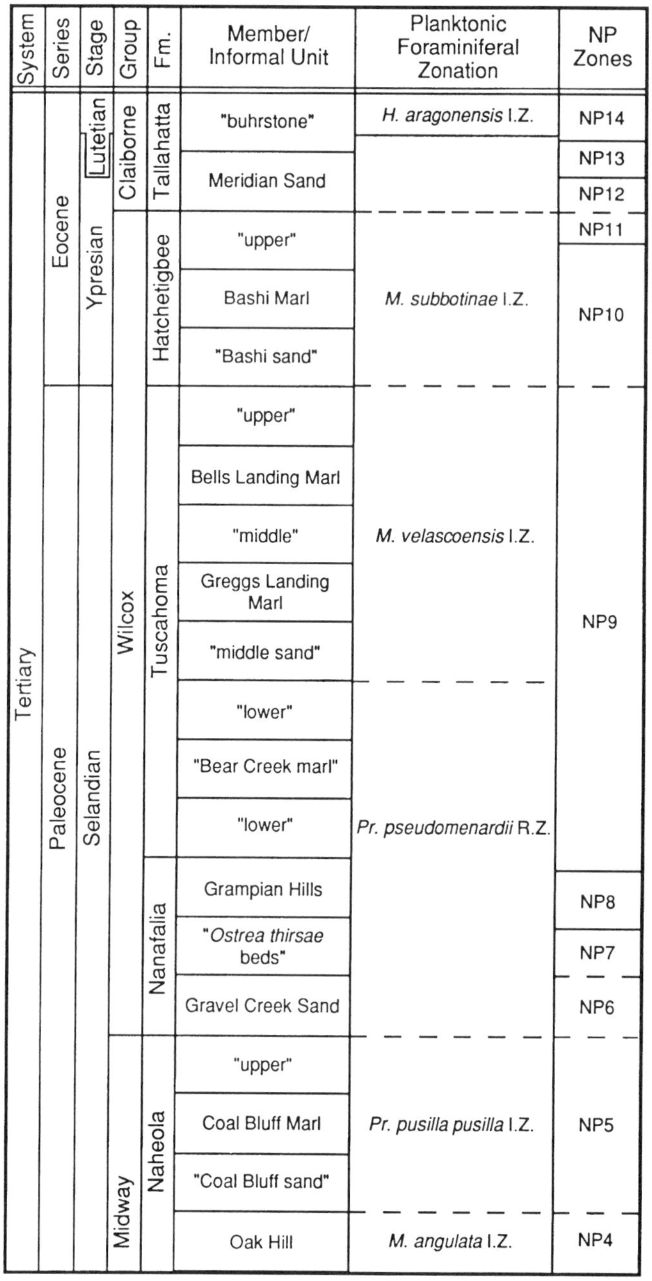

System	Series	Stage	Group	Fm.	Member/ Informal Unit	Planktonic Foraminiferal Zonation	NP Zones
Tertiary	Eocene	Lutetian	Claiborne	Tallahatta	"buhrstone"	H. aragonensis I.Z.	NP14
					Meridian Sand		NP13
							NP12
		Ypresian	Wilcox	Hatchetigbee	"upper"		NP11
					Bashi Marl	M. subbotinae I.Z.	NP10
					"Bashi sand"		
				Tuscahoma	"upper"		
					Bells Landing Marl		
					"middle"	M. velascoensis I.Z.	
					Greggs Landing Marl		
					"middle sand"		NP9
					"lower"		
					"Bear Creek marl"		
					"lower"	Pr. pseudomenardii R.Z.	
	Paleocene	Selandian		Nanafalia	Grampian Hills		NP8
					"Ostrea thirsae beds"		NP7
					Gravel Creek Sand		NP6
			Midway	Naheola	"upper"		
					Coal Bluff Marl	Pr. pusilla pusilla I.Z.	NP5
					"Coal Bluff sand"		
					Oak Hill	M. angulata I.Z.	NP4

FIG. 1.—Paleocene and lower Eocene lithostratigraphy and biostratigraphy for the eastern Gulf Coastal Plain.

and sediment influx was related to Laramide orogenic pulses in the Rocky Mountain region (Galloway, 1989, 1990). Sedimentation in south-central Alabama and the Florida panhandle was dominated by the presence of a persistent Paleogene stable platform that developed over Paleozoic basement rocks. This regional paleogeographic setting resulted in a thick, siliciclastic-dominated, upper Paleocene and lower Eocene stratigraphic succession in Mississippi and southwest Alabama and a somewhat thinner, carbonate-dominated interval to the east (Mancini and Tew, 1993).

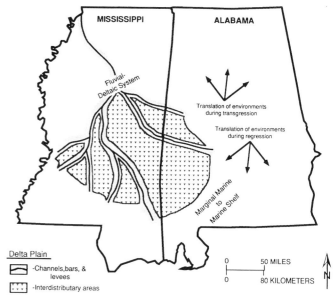

FIG. 2.—Paleocene and early Eocene paleogeography for the eastern Gulf Coastal Plain.

STRATIGRAPHY

In southern Alabama, the Wilcox Group consists of the Nanafalia, Tuscahoma, and Hatchetigbee Formations (Table 1). The Nanafalia Formation disconformably overlies the Naheola Formation of the Midway Group. The Naheola-Nanafalia (Midway/Wilcox) contact is an unconformity of regional extent, and cross-bedded sand of the Gravel Creek Sand Member of the Nanafalia Formation disconformably overlies lignitic clay of the Coal Bluff Marl Member of the Naheola Formation along this surface.

TABLE 1.—PALEOCENE AND EOCENE LITHOSTRATIGRAPHY FOR THE EASTERN GULF COASTAL PLAIN.

Formation	Member/Informal Unit	Lithologic Description
Tallahatta Formation	Meridian Sand Member	Micaceous, cross-bedded sand
Hatchetigbee Formation 76 m	"upper"	Interlaminated clay and silt, sand and lignite
	Bashi Marl Member	Glauconitic, fossiliferous sand and marl
	"Bashi sand"	Micaceous, cross-bedded sand
Tuscahoma Sand 107 m	"upper"	Interlaminated clay and silt, sand and lignite
	Bells Landing Marl Member	Glauconitic, fossiliferous sand and marl
	"middle"	Interlaminated clay and silt
	Greggs Landing Marl Member	Glauconitic, fossiliferous sand and marl
	"middle sand"	Micaceous, cross-bedded sand
	"lower"	Clay and silt interbedded with sand
	"Bear Creek Marl"	Glauconitic, fossiliferous sand and marl
	"lower"	Clay and silt interbedded with sand
Nanafalia Formation 56 m	Grampian Hills Member	Clay interbedded with glauconitic sand and marl
	"Ostrea thirsae beds"	Glauconitic, fossiliferous sand and marl
	Gravel Creek Sand Member	Micaceous, cross-bedded sand
Naheola Formation 34 m	"upper"	Interlaminated clay and silt, sand and lignite
	Coal Bluff Marl Member	Glauconitic, fossiliferous sand and marl
	"Coal Bluff sand"	Micaceous, cross-bedded sand
	Oak Hill Member	Interlaminated clay and silt, sand and lignite

The Nanafalia Formation is divided into three members (Fig. 1). The lower Gravel Creek Sand Member generally consists of white to yellow, micaceous, cross-bedded, medium- to coarse-grained sand. This unit is discontinuous in distribution and varies in thickness, reaching a maximum of approximately 15 m. The middle member, the "*Ostrea thirsae* beds," includes 11 to 14 m of calcareous, glauconitic, fossiliferous sand and silty marl. The upper Grampian Hills Member is 24 to 34 m thick and consists of green to gray, indurated clay interbedded with glauconitic sand and marl.

The Tuscahoma Sand conformably overlies the Nanafalia Formation. The Tuscahoma consists of 107 m of interlaminated silty clay, silt, fine-grained sand, and lignite. Glauconitic, fossiliferous, fine-grained sand and marl are present in the lower part of the formation below cross-bedded, fine- to medium-grained sand that contains angular blocks of clay and is discontinuous in distribution. The Greggs Landing Marl Member, the lower of two formally defined fossiliferous marine units, is present near the middle of the formation. The Greggs Landing Marl Member is comprised of fossiliferous, calcareous, glauconitic, quartzose, fine-grained sand and marl that is about 1.8 m thick. The Greggs Landing Marl is conformably overlain by gray, sandy clay and gray, very fine-grained sand, which disconformably underlies the Bells Landing Marl Member. This member consists of 2.7 m of highly fossiliferous, calcareous, glauconitic, fine-grained sand and marl. The upper Tuscahoma beds, which conformably overlie the Bells Landing Marl Member, include cross-bedded sand, laminated clay and silt, and several thin lignite beds.

The Hatchetigbee Formation overlies the Tuscahoma Sand along a regional unconformity. The Hatchetigbee is about 76 m thick and is divided into the lower Bashi Marl Member and an upper unnamed member. Above the unconformity, discontinuous cross-bedded sand at the base of the Bashi Marl Member overlies the Tuscahoma Sand. In absence of this sand unit, which is informally referred to as the "Bashi sand," the Bashi Marl Member directly overlies the Tuscahoma along the unconformity. The Bashi Marl Member consists of 1.8 to 10.7 m of greenish-gray, fossiliferous, glauconitic, calcareous sand and marl. The upper, unnamed member includes 61 to 76 m of gray, carbonaceous, micaceous, laminated clay and silt, and cross-bedded, fine-grained sand.

The Tallahatta Formation of the Claiborne Group disconformably overlies the Hatchetigbee Formation. The Hatchetigbee-Tallahatta contact is a regional unconformity, and discontinuous cross-bedded sand of the Meridian Sand Member of the Tallahatta Formation overlies the upper, unnamed member of the Hatchetigbee along this surface.

BIOSTRATIGRAPHY

The Paleogene planktonic foraminiferal zonation used herein includes the zonation of Bolli (1957, 1966, 1972), which was later modified by Stainforth and others (1975), and the zonation of Blow (1979). The calcareous nannoplankton zonation used is that of Martini (1971).

The "*Ostrea thirsae* beds" and Grampian Hills Members of the Nanafalia Formation contain *Planorotalites pseudomenardii* (Bolli) (Berggren, 1965; Mancini, 1984) and therefore have been assigned to the Selandian *Planorotalites pseudomenardii*

Range Zone of Stainforth and others (1975) and the P4 Zone of Blow (1979) (Gibson and others, 1982; Mancini, 1984). The "*Ostrea thirsae* beds" have been reported to rest within the NP7 and NP8 Zones of Martini (1971) on the basis of the occurrences of *Heliolithus kleinpelli* Sullivan and *Discoaster mohleri* Bukry and Percival in the lower part of this member, the presence of *Heliolithus riedeli* Bramlette and Sullivan, and the absence of *Discoaster multiradiatus* Bramlette and Riedel in the upper part of these beds (Gibson and others, 1982; Siesser, 1983). The occurrence of *H. riedeli* in the lower part of the Grampian Hills and the appearance of *D. multiradiatus* in the upper beds of this member indicate that the Grampian Hills is within the NP8 and NP9 Zones of Martini (1971) according to Gibson and others (1982). No calcareous microfossils have been reported from the Gravel Creek Sand Member.

The glauconitic sand and marl of the lower Tuscahoma Sand have been reported to rest within the Selandian *Planorotalites pseudomenardii* Range Zone (Mancini and Oliver, 1981; Mancini, 1984) and the P4 Zone (Gibson and others, 1982) based on the occurrence of *P. pseudomenardii*. The Greggs Landing Marl and the Bells Landing Marl Members have been assigned to the Selandian *Morozovella velascoensis* Interval Zone (Berggren, 1965; Mancini and Oliver, 1981). The Greggs Landing Marl Member has been reported to rest within the P5 Zone and the Bells Landing Marl Member within the P6 Zone (Mancini, 1984). *Morozovella velascoensis* (Cushman) occurs in both of these members, while *P. pseudomenardii* is absent. *Morozovella occlusa* Loeblich and Tappan and *Morozovella subbotinae* (Morozova) are present in the Bells Landing Marl Member, and *Acarinina mckannai* (White) is restricted to the Greggs Landing Marl Member. The "lower" Tuscahoma glauconitic sand and marl, the Greggs Landing Marl Member, and the Bells Landing Marl Member have been assigned to the NP9 Zone based on the presence of *D. mohleri* and *D. multiradiatus* (Gibson and others, 1982; Siesser, 1983). Palynomorphs recovered from the upper part of the Tuscahoma Sand in eastern Mississippi indicate that these strata are upper Paleocene (Ingram, 1991).

The Bashi Marl Member and marl beds in the upper, unnamed member of the Hatchetigbee Formation have been assigned to the Ypresian *Morozovella subbotinae* Interval Zone (Berggren, 1965; Mancini and Oliver, 1981; Mancini, 1984) and the P6 Zone based on the presence of *Morozovella subbotinae*, *Morozovella acuta* (Toulmin), *Morozovella aequa* (Cushman and Ponton), and *Pseudohastigerina wilcoxensis* (Cushman and Ponton) and the absence of *Morozovella occlusa*, *Morozovella velascoensis,* and *Morozovella aragonensis* Nuttall. The Bashi has been assigned to the NP9 Zone of Martini (1971) based on the presence of *Discoaster mohleri* and *D. multiradiatus* (Siesser, 1983) and the NP10 Zone of Martini (1971) on the basis of the occurrences of *Transversopontis pulcher* (Deflandre), *Discoaster binodosus* Martini, *D. mediosus* Bramlette and Sullivan, *Campylosphaera dela* (Bramlette and Sullivan), *Tribrachiatus contortus* (Stradner), and *T. bramlettei* (Bronnimann and Stradner) (Gibson and Bybell, 1981; Gibson and others, 1982; Siesser, 1983). Palynomorphs recovered from the "Bashi sand" in eastern Mississippi indicate an early Eocene age for these strata (Ingram, 1991).

GEOCHRONOLOGY

Six samples of glauconitic sand (Coal Bluff, *Ostrea thirsae* beds, Grampian Hills, Bells Landing, upper Tuscahoma, and

Bashi) were analyzed for conventional potassium-argon (K-Ar) age determination. Results from these analyses are shown on Table 2, which also compares the K-Ar age dates with the numerical limits (from Haq and others, 1988) that have been placed on the planktonic foraminiferal zone to which each of the units has been assigned. As illustrated, when analytical error is considered, glauconite dates from all units, with the exception of the Bells Landing Marl Member and perhaps the Grampian Hills Member, fall within or overlap the expected range and therefore are likely to closely approximate the time of deposition. The K-Ar date from the Bells Landing is much older than anticipated, is misplaced in terms of the expected stratigraphic progression of age dates, and, therefore, is considered anomalous, possibly due to the presence of extraneous Ar in the glauconite sample as discussed by Harris and Fullagar (1991).

SEQUENCE STRATIGRAPHY

Cyclic changes in global sea level and associated relative changes in coastal onlap during the late Paleocene and early Eocene epochs have been recognized by Baum and Vail (1988) and Haq and others (1988). Baum and Vail (1988) described five regional unconformities in the Gulf and Atlantic Coastal Plain areas that they used to divide the Wilcox Group into five, third-order depositional sequences: the TP2.1, which consisted of the Nanafalia Formation; the TP2.2, which included the lower and middle Tuscahoma; the TP2.3, which included the upper Tuscahoma; the TE1.1, which included the lower Hatchetigbee; and the TE1.2, which included the upper Hatchetigbee. Haq and others (1988) recognized five global unconformities that they used to divide upper Selandian and lower Ypresian strata into five, third-order depositional sequences. The sequences of the Haq and others (1988) include the TA2.1 (58.5 to 55 Ma), TA2.2 (55 to 54.5 Ma), TA2.3 (54.5 to 54.2 Ma), TA2.4 (54.2 to 53 Ma), and TA2.5 (53 to 52 Ma). A comparison of the late Paleocene and early Eocene depositional sequences recognized in this study to those described by Baum and Vail (1988) and Haq and others (1988) is provided in Figure 3.

Four, third-order unconformity-bounded depositional sequences, which apparently resulted from late Paleocene and early Eocene changes in sea level and coastal onlap, are recognized in this study. These sequences are here designated the TAGC-2.1 (Tejas A; Gulf Coast), TAGC-2.2, TAGC-2.3, and TAGC-2.4 depositional sequences. These sequences correspond to the TP2.1, TP2.2 , TP2.3, and TE1.1 depositional sequences in an earlier scheme of Mancini and Tew (1991).

The TAGC-2.1 type 1 sequence, with a lower type 1 sequence boundary, consists of the Gravel Creek Sand Member of the Nanafalia Formation (lowstand systems tract), "*Ostrea thirsae* beds" of the Nanafalia Formation (transgressive systems tract),

TABLE 2.—K-AR RADIOMETIC AGE DETERMINATIONS FOR PALEOCENE AND EOCENE GLAUCONITIC SANDS.

Unit	Location	Foraminiferal Zone	Stage	K-Ar Age Date[1] (ma)	Age of Zone[2] (ma)
Bashi Marl Mbr. Hatchetigbee Fm.	Highway 69 south of Bashi Creek, Clarke Co., Ala. sec. 9, T. 11 N., R. 1 E.,	*Morozovella subbotinae* Interval Zone	Ypresian	53.4 ± 1.4	54.0 to 52.7
"upper" Tuscahoma Tuscahoma Sand	Wal-Mart Shopping Center, Meridian, Miss., sec. 16, T. 6 N., R. 16 E.	*Morozovella velascoensis* Interval Zone	Selandian	54.5 ± 1.4	55.0 to 54.0
Bells Landing Marl Mbr. Tuscahoma Sand	Bells Landing, Monroe Co., Ala. sec. 36, T. 10 N., R. 6 E.	*Morozovella velascoensis* Interval Zone	Selandian	60.4 ± 1.5	55.0 to 54.0
Grampian Hills Mbr. Nanafalia Fm.	Gravel Creek, Wilcox Co., Ala., secs. 21 & 22, T. 11 N., R. 7 E.	*Planorotalites pseudomenardii* Range Zone	Selandian	58.3 ± 1.5	58.8 to 55.0
"*Ostrea thirsae* beds" Nanafalia Fm.	Gravel Creek, Wilcox Co., Ala., secs. 21 & 22, T. 11 N., R. 7 E.	*Planorotalites pseudomenardii* Range Zone	Selandian	56.3 ± 1.5	58.8 to 55.0
Coal Bluff Marl Mbr. Naheola Fm.	Gravel Creek, Wilcox Co., Ala., secs. 21 & 22, T. 11 N., R. 7 E.	*Planorotalites pusilla pusilla* Interval Zone	Selandian	58.2 ± 1.5	60.2 to 58.8

[1] Analyses performed by GeoChron Laboratories, Inc., Cambridge, Massachusetts. Decay constants used were $\lambda_\beta = 4.962 \times 10^{-10}$ yr^{-1} and $(\lambda_\Sigma + \lambda'_\Sigma) = 0.581 \times 10^{-10}$ yr^{-1}.

[2] Approximate age ranges from Haq and others (1988).

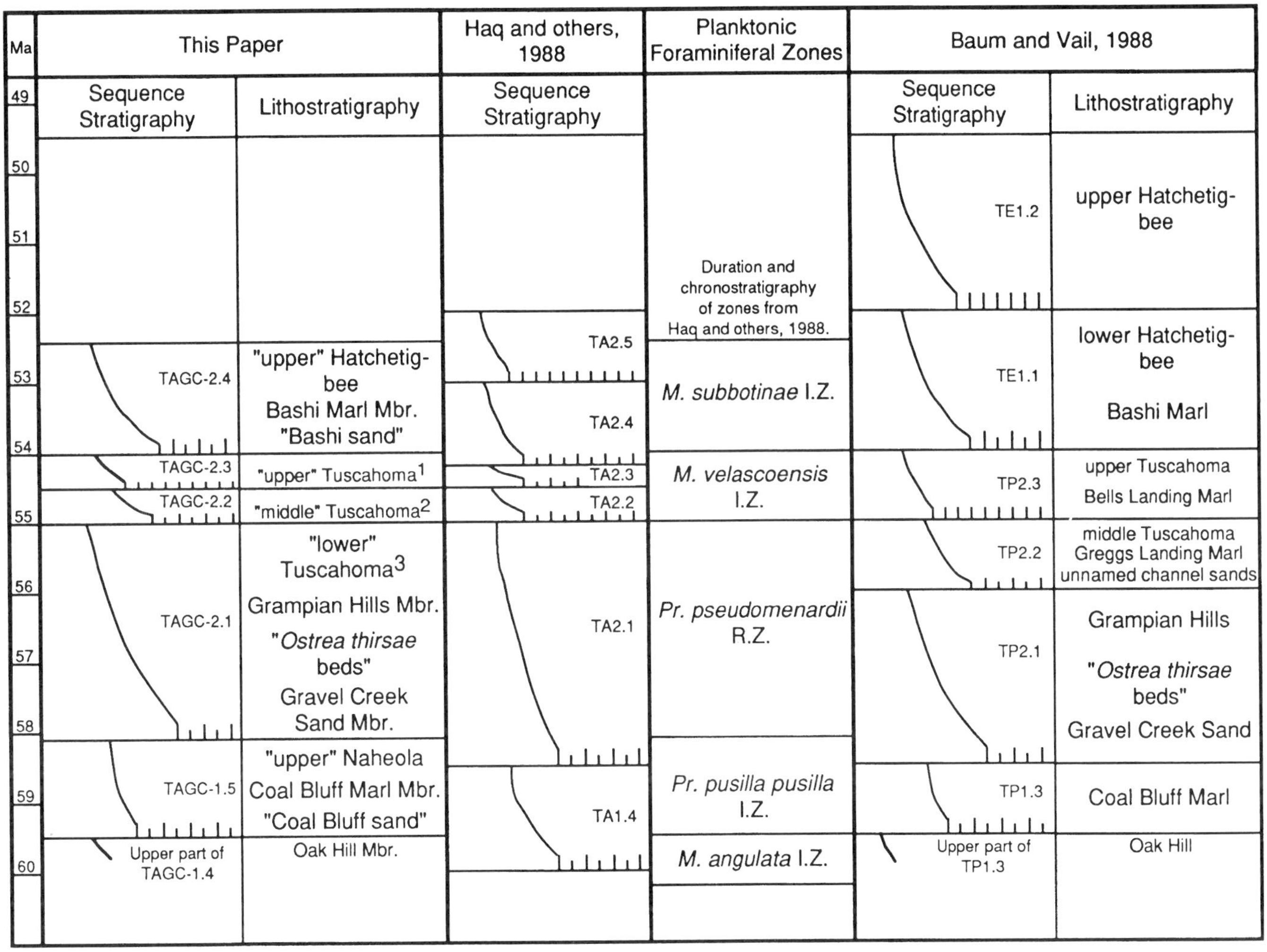

Ma	This Paper		Haq and others, 1988	Planktonic Foraminiferal Zones	Baum and Vail, 1988	
	Sequence Stratigraphy	Lithostratigraphy	Sequence Stratigraphy		Sequence Stratigraphy	Lithostratigraphy
49						
50					TE1.2	upper Hatchetigbee
51				Duration and chronostratigraphy of zones from Haq and others, 1988.		
52			TA2.5		TE1.1	lower Hatchetigbee
53	TAGC-2.4	"upper" Hatchetigbee Bashi Marl Mbr. "Bashi sand"	TA2.4	M. subbotinae I.Z.		Bashi Marl
54	TAGC-2.3	"upper" Tuscahoma[1]	TA2.3	M. velascoensis I.Z.	TP2.3	upper Tuscahoma Bells Landing Marl
55	TAGC-2.2	"middle" Tuscahoma[2]	TA2.2		TP2.2	middle Tuscahoma Greggs Landing Marl unnamed channel sands
56	TAGC-2.1	"lower" Tuscahoma[3] Grampian Hills Mbr. "Ostrea thirsae beds" Gravel Creek Sand Mbr.	TA2.1	Pr. pseudomenardii R.Z.	TP2.1	Grampian Hills "Ostrea thirsae beds" Gravel Creek Sand
57						
58						
59	TAGC-1.5	"upper" Naheola Coal Bluff Marl Mbr. "Coal Bluff sand"	TA1.4	Pr. pusilla pusilla I.Z.	TP1.3	Coal Bluff Marl
60	Upper part of TAGC-1.4	Oak Hill Mbr.		M. angulata I.Z.	Upper part of TP1.3	Oak Hill

[1] Includes the Bells Landing Marl Member and informal "upper" unit of the Tuscahoma Sand.

[2] Includes the informal "middle sand" unit, Greggs Landing Marl Member, and informal "middle" unit of the Tuscahoma Sand.

[3] The informal "Bear Creek marl" is included in the informal "lower" Tuscahoma Sand.

FIG. 3.—Comparison of Paleocene and lower Eocene, third-order, coastal onlap cycles recognized for the eastern Gulf Coastal Plain to regional cycles of Baum and Vail (1988) and global cycles of Haq and others (1988) for time-equivalent strata.

and the Grampian Hills Member of the Nanafalia Formation and the lower beds of the Tuscahoma Sand (highstand systems tract) (Fig. 4). For our purposes, a sequence boundary is considered a type 1 unconformity if field mapping reveals regionally extensive valley incision along the boundary with subsequent lowstand fill of the incised topography. The TAGC-2.1 is interpreted as a type 1, third-order sequence based on the regional and erosional nature of the lower bounding surface, the presence of lowstand systems tract fill of the incised topography, and the widespread occurrences of strata of the component systems tracts. Fourth- and fifth-order sequences (parasequences) are recognized in these depositional sequences, but parasequences do not have the regional characteristics typical of third-order sequences. For example, unnamed glauconitic sand beds in the Grampian Hills Member that are exposed at Gravel Creek and south of Camden, Wilcox County, Alabama, and Tuscahoma glauconitic sand beds that are exposed at Fatama

and Bear Creek, Wilcox County, Alabama, are interpreted to be associated with fourth-order cycles of relative sea-level change. This interpretation is based on the observation that the bounding surfaces of these strata are indistinct and are of a local nature. Further, the component systems tracts cannot be resolved. The alternating glauconitic sand and marl beds in the "Ostrea thirsae beds" are interpreted as fifth-order parasequences. These transgressive parasequences lack distinct bounding surfaces and component systems tracts.

The TAGC-2.2 type 1 sequence is comprised of the lower Tuscahoma sand beds (lowstand systems tract), Greggs Landing Marl Member of the Tuscahoma Sand (transgressive systems tract), and the middle Tuscahoma clay and silt beds (highstand systems tract).

The TAGC-2.3 sequence consists of the Bells Landing Marl Member of the Tuscahoma Sand (transgressive systems tract) and the upper Tuscahoma beds (highstand systems tract). Be-

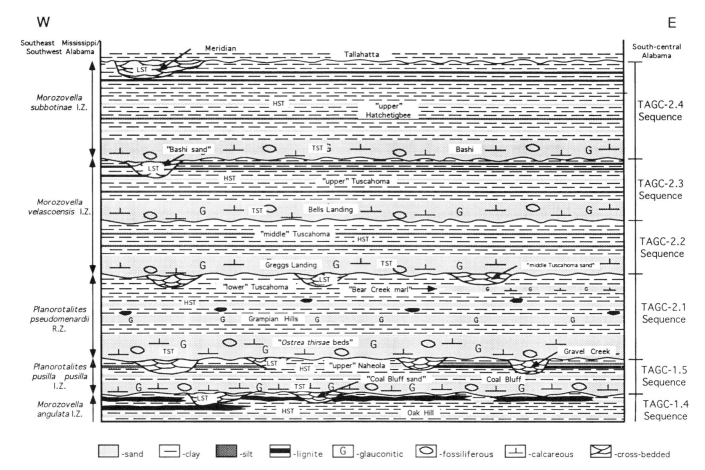

FIG. 4.—Schematic cross section illustrating lithofacies relationships, planktonic foraminiferal zones, and unconformity-bounded depositional sequences for the Wilcox Group and related strata (LST = lowstand systems tract deposits; TST = transgressive systems tract deposits; HST = highstand systems tract deposits).

cause of the absence of a lowstand systems tract, the TAGC-2.3 is interpreted as a type 2 sequence. Numerous lignite beds are present in the upper Tuscahoma, and these beds cap mud-dominated parasequences. These parasequences are interpreted as fourth- or fifth-order sequences due to the lack of distinct bounding surfaces and component systems tracts.

The TAGC-2.4 type 1 sequence is comprised of the "Bashi sand" beds (lowstand systems tract), Bashi Marl Member (transgressive systems tract), and upper unnamed member (highstand systems tract) of the Hatchetigbee Formation. Numerous shell beds (*Venericardia* spp.) occur in the upper Hatchetigbee, and these beds occur at the base of mud-dominated, fourth-order parasequences in the highstand systems tract. The unnamed shell beds at Hatchetigbee Bluff have indistinct bounding surfaces which are of a local nature, and the component systems tracts associated with these shell beds are not well developed.

LIGNITE OCCURRENCE

Lignite occurs in the highstand systems tract strata of the TAGC-2.1, TAGC-2.3, and TAGC-2.4 depositional sequences and of the two underlying sequences which include the Oak Hill and Coal Bluff Members of the Naheola Formation (TAGC-1.4 and TAGC-1.5 sequences, Fig. 4). Interestingly, the

five sequences containing lignite or lignitized wood are overlain by lowstand systems tract deposits, which in shelfal areas are associated with renewed sea-level rise and infilling of the incised topography after major falls. These deposits are not coeval with the lowstand systems tract strata of more basinal areas, which are deposited during the sea-level fall which produced the incised topography.

The Oak Hill lignite (TAGC-1.4 sequence) is characterized by a freshwater swamp flora, including arborescent angiosperm pollen of the hazelnut, birch, black gum, and pecan families. Associated with these trees is a variety of ferns and mosses, including *Sphagnum*. This floral assemblage indicates that the Oak Hill lignite accumulated in interdistributary swamps in a delta-plain setting. The Coal Bluff lignite (TAGC-1.5 sequence) contains a brackish water, coastal marsh flora, including herbaceous angiosperm pollen of the lily, amaryllis, and sedge families. Bisaccate gymnosperm pollen of the pine family and fungal elements are an integral part of this assemblage. Marsh environments in which the Coal Bluff lignite accumulated are interpreted to have occupied interdistributary coastal areas associated with a prograding deltaic system. The lignitized wood found in the Grampian Hills Member of the Nanafalia Formation (TAGC-2.1) sequence is believed to have formed in coastal marshes. The "upper" Tuscahoma lignite (TAGC-2.3 sequence) and the Hatchetigbee lignite (TAGC-2.4 sequence) are characterized by brackish to freshwater, mixed marsh/swamp to swamp flora, including arborescent angiosperm pollen of the hazelnut, birch, black gum, and pecan families, a variety of ferns and mosses, herbaceous angiosperm pollen of the lily, amaryllis, and sedge families, and fungal elements. Swamp and marsh/swamp environments were established in interdistributary areas in a delta-plain setting, and these environments were the sites of the accumulation and preservation of organic matter for the formation of lignite.

The observed repetitive pattern of occurrence of Paleocene and lower Eocene lignite beds within the regressive highstand systems tract components indicates that these strata were deposited in connection with cyclical falls in relative sea level during this interval of geologic time. These lignite beds occur within parasequences (fourth- or fifth-order sequences) in the mud-dominated, regressive intervals of third-order depositional sequences. The occurrence of these lignite beds in the highstand systems tract beds of depositional sequences is consistent with the work of Pasley and Hazel (1990), who studied the relationship of changes in organic matter relative to systems tracts in upper Eocene and lower Oligocene strata of the Gulf Coast. Pasley and Hazel (1990) found that the organic material in highstand systems tract strata is well preserved and consists primarily of structured terrestrial matter, while the organic matter in transgressive systems tract strata is amorphous (algal), nonstructured, and degraded.

Plant material recovered from the lignite beds of this study, when considered in combination with lithofacies, sequence stratigraphy, and paleogeographic relationships, indicates that these beds accumulated in swamp, marsh, and mixed marsh/swamp environments. These environments were established on interdistributary areas in delta-plain depositional settings during times of significant and extensive progradation and outbuilding of fluvial-dominated delta systems entering the Gulf Coast basin from the northwest (Fig. 5). Common features in a delta-plain environmental setting include active distributary bars and

channels and a wide variety of freshwater to brackish interdistributary and coastal environments, including swamps, marshes, tidal flats, and interdistributary bays (Bhattacharya and Walker, 1992). In fluvial-dominated delta plains, interdistributary areas are generally enclosed and exhibit shallow-water environments that are either quiet or stagnant. Further, interdistributary areas are generally the sites of accumulation of fine-grained material associated with suspension sedimentation during and immediately after flood events (Elliott, 1986); however, the generally fine-grained aspect of the interdistributaries may exhibit intermittent occurrences of coarser grained material associated with crevasse splays or channels (Bhattacharya and Walker, 1992). Commonly, as interdistributaries fill with sediment and become vegetated, marshes, swamps, and mixtures of these environments are established (Bhattacharya and Walker, 1992). Given the proper climatic conditions, these environments are conducive to the accumulation and preservation of organic material to form lignite (Elliott, 1986). The presence of peat, lignite, or coal in association with both modern and ancient, fluvial-dominated, delta-plain sequences is well established (Coleman and Gagliano, 1964; Coleman and Wright, 1975; Ferm and Horne, 1979).

In the study area, the highstand systems tracts that contain lignite beds are all overlain by well-developed sequence boundaries (unconformities) that are interpreted to have been formed during major sea-level falls and at sea-level lowstands (type 1 sequence boundaries in the sense of Van Wagoner and others, 1988). In the Paleogene succession of the eastern Gulf Coastal Plain, type 1 sequence boundaries can be recognized by the regional characteristics of the unconformity surface and by the presence of lowstand systems tract strata with their unique sedimentary properties overlying the boundary (Mancini and Tew, 1993). These sequence boundaries generally exhibit relief on the unconformity surface associated with shelf erosion or fluvial incision along the surface during sea-level lowstand. Subsequent to erosion and incision and during the initial part of sea-

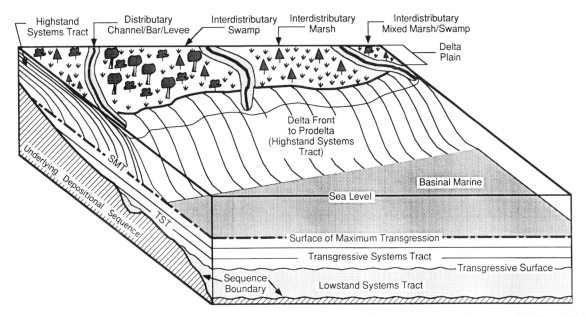

FIG. 5.—Schematic diagram illustrating mode of lignite formation in the Paleocene and lower Eocene strata of the eastern Gulf Coastal Plain.

level rise associated with the next cycle, shallow marine shelf, barrier, and/or marginal marine sand bodies were deposited as lowstand systems tract sediment that fills topographically low areas on the eroded and incised surface. These sand bodies, which include the "Coal Bluff sand" of the Coal Bluff Member of the Naheola Formation, the Gravel Creek Sand Member of the Nanafalia Formation, the "middle Tuscahoma sand" of the Tuscahoma Sand, the "Bashi sand" of the Hatchetigbee Formation, and the Meridian Sand Member of the Tallahatta Formation, are discontinuous in areal distribution due to being confined to topographic depressions. Lowstand systems tract sand bodies are overlain by the transgressive surface of the depositional sequence, a surface that is typically marked by sediment clasts, quartz pebbles, phosphate grains, shell hash, and shark teeth (Mancini and Tew, 1993). In areas where these sand bodies are absent, the transgressive surface overlies the highstand systems tract of the preceding depositional sequence, and in this situation, the transgressive surface and the sequence boundary are coincident. Where this is the case, it is possible for lignite beds to be directly and disconformably overlain by fossiliferous, glauconitic marine shelf sand and marl of the transgressive systems tract. For example, the lignite in the Oak Hill Member of the Naheola Formation is directly and disconformably overlain by glauconitic sand and marl of the Coal Bluff Marl Member of the Naheola Formation at Gravel Creek, Wilcox County, Alabama.

The widespread occurrence of fluvial-dominated, interdistributary swamp and marsh depositional settings in which organic matter accumulated to form lignite, in association with overlying type 1 sequence boundaries related to major falls in sea level and sea-level lowstand, suggests that the establishment and stabilization of Paleocene and lower Eocene lignite-forming environments occurred during intervals of extensive delta progradation and regression of the shoreline. During accumulation of the highstand systems tracts of the depositional sequences that contain lignite, sediment was delivered into the area by fluvial systems draining areas to the northwest. This sediment was deposited as the sediment-carrying capacity of the rivers decreased upon entering the low-lying, basin-marginal delta systems. As the deltas were augmented by progressive sediment accumulation, the effects of relative sea-level rise during an individual sea-level cycle were overwhelmed, resulting in a net loss of sediment accommodation and concomitant overall basinward progradation of the shoreline (regression). As the delta systems migrated basinward, delta-plain depositional environments became extensive in distribution, and interdistributary marsh and swamp settings in which organic matter could be concentrated became widespread. Cyclical flooding of the established coastal marshes and swamps, related to small scale fluctuations in water depth that resulted from the interaction of eustasy, subsidence, and sediment distribution and influx, was followed by periods of progradation and regression, as recorded by the superimposition of upward-shallowing, lignite-capped, mud-dominated parasequences on the overall regressive stratigraphic succession. Highstand systems tract deposition within a particular sequence culminated with a major relative sea-level fall and subsequent lowstand that resulted in a lowering of base level, an abrupt basinward shift in coastal onlap, and the carving of a type 1 sequence boundary as discussed above.

Following sea-level fall and the subsequent accumulation of the lowstand deposits, significant relative sea-level rise ensued, resulting in marine inundation of the area previously occupied by the delta system and associated environments. The sea-level rise generated the transgressive surface of the depositional sequence. Overlying the transgressive surface, marine transgression and stratal retrogradation resulted from continued sea-level rise. These transgressive systems tract marine shelf, glauconitic sand and marl beds have regional distribution (e.g., Coal Bluff Marl Member, "Ostrea thirsae beds," Greggs Landing Marl Member, Bells Landing Marl Member, Bashi Marl Member, etc.). The transition from transgression (increasing water depths) to regression (decreasing water depths) within a depositional sequence is commonly within a condensed section, which reflects maximum water depths and associated sediment starvation on the shelf. Well-developed, condensed sections, however, can be recognized only in depositional sequences in which sea-level rise resulted in relatively deeper water settings, such as middle to outer shelf environments, at maximum transgression. In inner-shelf, shallower water settings, the surface of maximum flooding and sediment starvation may be represented only by the transition from transgression to regression. Above this flooding surface, progressive progradation and regression and upward shallowing characterize the highstand systems tract as depositional facies translate basinward. In areas where fluvial-deltaic systems dominate the deposition in a basin and associated delta plains are established, the potential for lignite formation in interdistributary environments is renewed as a result of organic matter accumulation and preservation.

DISCUSSION

In the eastern Gulf Coastal Plain, numerous factors affect the stratigraphic position of biozone boundaries in upper Paleocene and lower Eocene strata. Important among these are: (1) the presence of unconformities and associated biostratigraphic discontinuities resulting from nondeposition or erosion; (2) the presence or absence of lowstand systems tract deposits associated with type 1 depositional sequences; (3) the differential amounts and rates of sedimentation associated with paleobathymetry and/or distance from the shoreline at various depositional sites; (4) differential subsidence within the depositional basin; and (5) paleoenvironmental conditions.

In this area, in cases where lowstand, incised-valley fill deposits occur below the transgressive surface as the basal components of the depositional sequences, biozone boundaries generally coincide with transgressive surfaces within the respective sequences. Lowstand deposits consist of marginal marine sand bodies that are generally devoid of age diagnostic, calcareous fossils. Therefore, within a particular depositional sequence, the first appearance of age diagnostic microfossils necessary for biozonal assignment is generally in the transgressive systems tract deposits immediately above the transgressive surface or the first flooding surface overlying maximum regression. Lowstand systems tract deposits generally have discontinuous distributions, and in the absence of these sequence components, the initial transgressive surface and the biozone boundary coincide with the basal sequence-bounding unconformity. Therefore, the lowstand systems tract strata are stratigraphically constrained by the transgressive surface with the coincident biozonal boundary where these deposits are present and by the sequence-bounding unconformity with coincident transgressive surface and biozonal boundary where these strata are absent. This observation implies that these typically barren rocks are generally correlative to the chronozone that encompasses the sediments immediately above the transgressive surface within the depositional sequence even though they cannot be assigned to a specific biozone because of the absence of age diagnostic species.

The TAGC-2.1 depositional sequence illustrates this point (Fig. 4). The TAGC-2.1 is a type 1 sequence that includes discontinuously distributed, lowstand systems tract, incised-valley fill deposits as the lowermost component. The type 1 unconformity at the base of the sequence is coincident with the Midway/Wilcox Group contact, and erosional scour associated with the unconformity resulted in relief on the upper surface of the underlying TAGC-1.5 sequence. The incised-valley fill deposits (Gravel Creek Sand Member of the Nanafalia Formation) are restricted to topographically low areas on the upper surface of the highstand systems tract of the TAGC-1.5 sequence. In topographically high areas where the incised-valley fill strata are absent, the transgressive systems tract deposits ("Ostrea thirsae beds" of the Nanafalia Formation) of the TAGC-2.1 sequence directly overlie the sequence boundary. Where present, the incised-valley fill sand bodies are overlain by a distinct transgressive surface, which, in turn, is overlain by the transgressive systems tract of the sequence. The transgressive strata immediately overlying the sequence boundary and/or transgressive surface can be assigned to the planktonic foraminiferal *Planorotalites pseudomenardii* Range Zone (P4 Zone, in part), and the lower biozonal boundary for this zone coincides with the sequence boundary and/or the transgressive surface. In this sequence, the stratigraphic placement of the planktonic foraminiferal zone boundary relative to the sequence boundary, the transgressive surface, and the lithostratigraphic group contact is significantly influenced by the presence or absence of intervening lowstand systems tract, incised-valley fill deposits.

The TAGC-2.2 and TAGC-2.4 depositional sequences also reflect the relationships outlined above. The incised-valley fill deposits of the TAGC-2.2 sequence (lower Tuscahoma sand) and of the TAGC-2.4 sequence (lower Hatchetigbee "Bashi sand") are discontinuous in distribution and are restricted to depressions on the upper surface of the highstand systems tracts of the TAGC-2.1 sequence and TAGC-2.3 sequence, respectively. In topographically high areas, the transgressive deposits (Greggs Landing Marl Member of the Tuscahoma Sand) of the TAGC-2.2 sequence or transgressive deposits (Bashi Marl Member of the Hatchetigbee Formation) of the TAGC 2.4 sequence directly overlie the respective sequence boundaries. The transgressive strata immediately overlying the sequence boundary and/or transgressive surface of the TAGC 2.2 sequence and of the TAGC 2.4 sequence can be assigned to the planktonic foraminiferal *Morozovella velascoensis* Interval Zone (P5 Zone, in part) and *Morozovella subbotinae* Interval Zone (P6 Zone, in part), respectively.

The TAGC-2.3 depositional sequence (middle to upper Tuscahoma) is a type 2 sequence which lacks a lowermost shelf margin systems tract component. Therefore, the transgressive systems tract (Bells Landing Marl Member of the Tuscahoma Sand) directly overlies the sequence boundary and/or trans-

gressive surface of this sequence. These strata can be assigned to the *Morozovella velascoensis* Interval Zone (P6 Zone, in part).

As illustrated by these examples, planktonic foraminiferal zone boundaries occur at sequence boundaries dividing depositional sequences in upper Paleocene and lower Eocene, marginal marine to marine shelf strata of the eastern Gulf Coastal Plain. On the other hand, in the depositional sequences discussed by Olsson (1991) for the Paleogene stratal succession of the Atlantic Coastal Plain of New Jersey, planktonic foraminiferal zone boundaries occur within the condensed section deposits, generally near the point of maximum transgression. However, the sequences described by Olsson (1991) generally lack recognizable, initial transgressive systems tract strata; therefore, late transgressive or condensed section deposits directly overlie highstand systems tract strata of an underlying depositional sequence. In this case, the maximum flooding surface, transgressive surface, and sequence boundary generally coincide. This relationship has been observed also by Loutit and others (1988) and by Mancini and Tew (1991) for Paleocene strata of the eastern Gulf Coastal Plain in situations where the terrigenous sedimentation rate is very low at the site of deposition and where relative sea-level rise overwhelms the preexisting highstand systems tract strata due to a combination of subsidence and eustasy. Apparently, these conditions prevailed throughout most of the Paleogene in the Atlantic Coastal Plain area and resulted in age diagnostic, planktonic foraminiferal species first occurring in what appear to be condensed section deposits near the point of maximum transgression in depositional sequences rather than in initial transgressive deposits as in the eastern Gulf Coastal Plain (Fig. 6). Interestingly, even though the zone boundary is present within condensed section or late transgressive deposits, due to the lack of initial transgressive deposits, the biozone boundary still generally coincides with the underlying sequence boundary. It is possible that the transgressive systems tracts and condensed sections of the depositional sequences of the Atlantic Coastal Plain have been amalgamated into thin stratigraphic intervals in which it is impossible to resolve the difference between early and late transgressive systems tract strata. The present-day continental margin of the Atlantic Coastal Plain is characterized by a steeper profile and narrower width (inner margin of coastal plain to 200-m isobath) relative to the Gulf Coastal Plain. This situation, in combination with low sedimentation rates, probably persisted throughout the Paleogene. Thus, at a particular depositional locus along the Atlantic margin, a cyclical fluctuation in sea level (up or down) would have had a much more profound effect on paleobathymetry than the same change along the Gulf margin. Transgressions would appear to be more "rapid" and a full suite of transgressive systems tract strata would probably not be preserved and, even if present, would be hard to distinguish from the overlying, deeper water, sediment-starved interval of the condensed section because of the "rapidity" of the upward paleobathymetric increase. This scenario would explain the observations of Olsson (1991).

As mentioned previously, the paleoenvironmental conditions resulting in deposition of marginal marine, lowstand systems tract deposits in the upper Paleocene and lower Eocene strata of the eastern Gulf Coastal Plain limit the occurrence of age diagnostic marine organisms. Therefore, the presence or ab-

Shelf

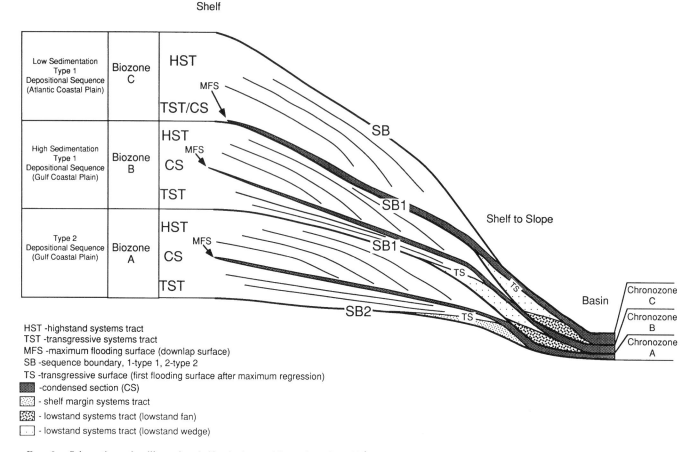

HST -highstand systems tract
TST -transgressive systems tract
MFS -maximum flooding surface (downlap surface)
SB -sequence boundary, 1-type 1, 2-type 2
TS -transgressive surface (first flooding surface after maximum regression)
■ -condensed section (CS)
▨ - shelf margin systems tract
▨ - lowstand systems tract (lowstand fan)
☐ - lowstand systems tract (lowstand wedge)

FIG. 6.—Schematic section illustrating shelf to basin transition and stratigraphic relationships of biozones, chronozones, and physical surfaces to sequence-bounding unconformities and their respective component systems tracts.

sence of a species may be due to environmental conditions rather than to evolutionary considerations. This should not be the case in deep sea environments that are characterized by stable marine conditions. It follows that marine sediments deposited in deep sea environments should contain a more continuous progression of age diagnostic planktonic species if the critical parameters controlling the presence or absence of these species in other depositional settings are environmental conditions. In the deeper marine settings, sedimentation would not be as strongly affected by sea-level fluctuations that have much more pronounced effects further landward. Indeed, basinal environments should be represented by a more generally continuous stratigraphic record of slow, constant sedimentation (i.e., a record of stacked condensed sections, Fig. 7) and/or the most basinward parts of systems tracts (Loutit and others, 1988). Thus, the observations of Baum and Vail (1988) and Haq and others (1988) that most major planktonic zonal boundaries occur in condensed sections or within depositional sequences in more distal settings follow logically the findings of this study of the more updip sections.

APPLICATION FOR CORRELATION AND PALEOGEOGRAPHIC
RECONSTRUCTION

An integrated geochronologic, biostratigraphic, and sequence stratigraphic approach is useful for correlation of marginal ma-

rine and marine shelf strata of the Wilcox Group and for reconstruction of the paleogeographic setting of the eastern Gulf Coastal Plain during deposition of these strata. In general, occurrences of age diagnostic calcareous nannoplankton and planktonic foraminifera necessary for biozonal assignments are confined to glauconitic sand and marl beds of the transgressive systems tracts and condensed sections of the third-order, unconformity-bounded, depositional sequences recognized in this study. Thus, based on biozonal assignment within each sequence, these strata can be placed within a particular chronozone and used for local, regional, and worldwide chronostratigraphic correlation. Further, in the Gulf Coastal Plain depositional basin, radiometric age assignments derived from these marine strata, when used in combination with physical correlation surfaces with inherent chronostratigraphic significance identified from sequence stratigraphic analysis, can be used as a framework to provide an understanding of regional lithofacies and lithostratigraphic relationships. This understanding can then be used to make interpretations of the depositional environments that were operative at a discrete time horizon during the accumulation of an individual, unconformity-bounded depositional sequence, thus providing a mechanism to make time-slice paleogeographic reconstructions. Physical surfaces associated with a depositional sequence that have chronostratigraphic significance include the upper and lower sequence

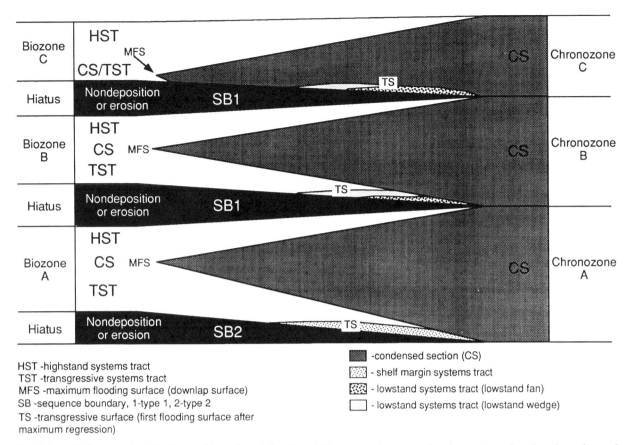

HST -highstand systems tract
TST -transgressive systems tract
MFS -maximum flooding surface (downlap surface)
SB -sequence boundary, 1-type 1, 2-type 2
TS -transgressive surface (first flooding surface after
maximum regression)

■ -condensed section (CS)
▦ - shelf margin systems tract
▨ - lowstand systems tract (lowstand fan)
□ - lowstand systems tract (lowstand wedge)

Fig. 7.—Schematic chronostratigraphic diagram illustrating relationships of biozones to chronozones in reference to condensed sections of unconformity-bounded depositional sequences in a shelf to basin transition.

bounding unconformities, the transgressive surface, and the surface of maximum transgression/sediment starvation. These bracketing physical surfaces can be used as constraints to interpret relative time lines at various points within the sequence. Interpreted time correlative points can then be used in conjunction with the lithofacies, lithostratigraphic, and biostratigraphic relationships for production of paleogeographic maps illustrating the regional distribution of paleoenvironments and rock types at various time horizons within relative sea-level cycles.

The surface of maximum transgression within a depositional sequence is an important datum for correlation because this surface represents a synchronous horizon throughout the eastern Gulf Coastal Plain. The utility of the point of maximum transgression in a particular cycle of sea-level fluctuation as a tool for correlation of Gulf Coastal Plain strata was noted by Israelsky (1949), who used the maximum depth in bathymetric cycles as determined by analysis of foraminifera, for correlation. This practice is currently employed by geologists mapping subsurface strata in the Gulf of Mexico area; however, these workers now refer to Israelsky's (1949) point of maximum depth as the maximum flooding surface in a depositional sequence (Armentrout and others, 1990; Vail and Wornhardt, 1990). Krumbein and Sloss (1963) also indicated that the point of change from transgression to regression can be used for "correlation by position in the bathymetric cycle" and that the turnaround point is regionally correlative and essentially synchronous.

The surface of maximum transgression or maximum flooding surface within a depositional sequence is also the surface of maximum sediment starvation (Loutit and others, 1988). The surface of maximum sediment starvation and the condensed section within which it is contained indicate very low sedimentation rates, and, frequently, this surface delineates a submarine hiatus that represents a period of extremely slow deposition or erosion that is described as a marine disconformity or discontinuity. Within a sea-level cycle, the maximum flooding surface represents the time of greatest accommodation of the shelf, maximum landward encroachment of the shoreline, and, generally, is the turnaround point in sedimentation from retrogradation or aggradation to progradation. Thus, this physical surface delineates the transition from the transgressive systems tract to the highstand systems tract within an individual depositional sequence. Because of the low rates of sedimentation associated with the late phase deposition of the transgressive systems tract, these strata, including the condensed section, are often rich in authigenic minerals, including authigenic glauconite (Loutit and others, 1988). Therefore, in addition to the biostratigraphic data which can be recovered from the marine strata of the transgressive systems tract for chronostratigraphic correlation, it is possible to obtain radiometric data for geochronometry from glauconite grains. The inherent limitations of these data, however, must be addressed and fully integrated into all chronostratigraphic interpretations.

The lower and upper bounding surfaces of a depositional sequence, as well as the first transgressive surface, are useful for both physical correlation and as chronostratigraphic constraints. Although these physical surfaces are diachronous along their traces, they are event markers, and in all cases separate older rocks below from younger rocks above (Loutit and others, 1988). Thus, the physical framework provided by these surfaces, when integrated with the geochronologic and chronostratigraphic data derived from biostratigraphy and radiometry, yield a very useful framework which can fundamentally advance the understanding of regional lithofacies and paleogeography. Further, local and regional relationships can be better understood in terms of global events and provide a mechanism by which to make worldwide correlations.

CONCLUSIONS

1. The late Paleocene and early Eocene depositional history of the western part of the eastern Gulf Coastal Plain was dominated by fluvial-deltaic, marginal marine, and marine shelf sedimentation.
2. Upper Paleocene strata in the eastern Gulf Coastal Plain consist of the Nanafalia and Tuscahoma Formations of the Wilcox Group, and lower Eocene strata include the Hatchetigbee Formation of the Wilcox Group.
3. The Nanafalia Formation and lower part of the Tuscahoma Sand rest within the late Paleocene (Selandian Stage), *Planorotalites pseudomenardii* Range Zone, while the middle and upper parts of the Tuscahoma Sand have been assigned to the late Paleocene (Selandian Stage) *Morozovella velascoensis* Interval Zone. The Hatchetigbee Formation is contained in the early Eocene (Ypresian Stage), *Morozovella subbotinae* Interval Zone.
4. With the exception of the Bells Landing Marl Member of the Tuscahoma Sand and perhaps the Grampian Hills Member of the Nanafalia Formation, K-Ar glauconite age dates determined from selected units (Coal Bluff, "*Ostrea thirsae* beds," Bells Landing, upper Tuscahoma, and Bashi) fall within or overlap the expected range determined by comparison to planktonic foraminiferal zonal ranges on the time scale of Haq and others (1988) and, therefore, are likely to closely approximate the time of deposition. The K-Ar date from the Bells Landing is much older than anticipated, is misplaced in terms of the expected stratigraphic progression of age dates, and, therefore, is considered anomalous.
5. Four, third-order unconformity-bounded depositional sequences, which resulted from late Paleocene and early Eocene changes in sea level and coastal onlap are recognized for the Wilcox strata in the eastern Gulf Coastal Plain.
6. Lignite beds generally occur within parasequences in mud-dominated, regressive intervals in the highstand systems tracts of the depositional sequences, and the lignites are interpreted to form in interdistributary coastal marshes and swamps in a delta-plain setting.
7. In the eastern Gulf Coastal Plain, planktonic foraminiferal zone boundaries occur at depositional sequence boundaries dividing the sequences recognized in the upper Paleocene and lower Eocene strata.
8. Only the synchronous maximum flooding surfaces associated with condensed section deposits have chronostratigraphic application for worldwide correlation for late Paleocene and early Eocene strata. Although sequence boundaries and transgressive surfaces have chronostratigraphic significance, these physical surfaces are diachronous, and therefore are only useful for local and regional correlation. The use of these diachronous surfaces for global correlation will produce conflicting results and therefore depositional cycles should be dated by synchronous events (maximum flooding surfaces).

9. An integrated geochronologic, biostratigraphic, and sequence stratigraphic approach provides a useful mechanism for correlation of marginal marine to marine shelf strata of the Wilcox Group in the eastern Gulf Coastal Plain.

REFERENCES CITED

ARMENTROUT, J. M., ECHOLS, R. J., AND LEE, T. D., 1990, Patterns of foraminiferal abundance and diversity: Implications for sequence stratigraphic analysis, *in* Sequence Stratigraphy as an Exploration Tool-Concepts and Practices in the Gulf Coast: Houston, Program and Extended and Illustrated Abstracts, 11th Annual Research Conference, Gulf Coast Section Society of Economic Paleontologists and Mineralogists Foundation, p. 53–58.

BAUM, G. R. AND VAIL, P. R., 1988, Sequence stratigraphic concepts applied to Paleogene outcrops, Gulf and Atlantic basins, *in* Wilgus, C. K., Hastings, B. S., Kendall, C. G. St. C., Posamentier, H. W., Ross, C. A., and Van Wagoner, J. C., eds., Sea-level Changes: An Integrated Approach: Tulsa, Society of Economic Paleontologists and Mineralogists Special Publication 42, p. 309–327.

BERGGREN, W. A., 1965, Some problems of Paleocene-Lower Eocene planktonic foraminiferal correlations: Micropaleontology, v. 11, p. 278–300.

BATTACHARYA, J. P. AND WALKER, R. G., 1992, Deltas, *in* Walker, R. G. and James, N. P., eds., Facies Models: Response to Sea Level Change: Toronto, Geological Association of Canada, p. 157–177.

BLOW, W. H., 1979, The Cainozoic Globigerinida: Leiden, E. J. Brill, 1413 p.

BOLLI, H. M., 1957, The genera *Globigerina* in the Paleocene-lower Eocene Lizard Springs Formation of Trinidad, B.W.I.: United States National Museum Bulletin, v. 215, p. 61–81.

BOLLI, H. M., 1966, Zonation of Cretaceous to Pliocene marine sediments based on planktonic foraminifera: Associacion de Venezolana Geologia, Mineria y Petroleo Boletin de Informacion, v. 9, p. 3–32.

BOLLI, H. M., 1972, The genus *Globigerinatheka* Bronnimann: Journal of Foraminiferal Research, v. 2, p. 109–136.

CLEAVES, A. W., 1980, Depositional systems and lignite prospecting models: Wilcox Group and Meridian Sandstone of northern Mississippi: Gulf Coast Association of Geological Societies Transactions, v. 30, p. 283–307.

COLEMAN, J. M. AND GAGLIANO, S. M., 1964, Cyclic sedimentation in the Mississippi River deltaic plain: Gulf Coast Association of Geological Societies Transactions, v. 14, p. 67–80.

COLEMAN, J. M. AND WRIGHT, L. D., 1975, Modern river deltas: Variability of processes and sand bodies, *in* Broussard, M. L., ed., Deltas, Models for Exploration: Houston, Houston Geological Society, p. 99–149.

DOCKERY, D. T., III, 1986, Punctuated succession of Paleogene mollusks in the northern Gulf Coastal Plain: Palaios, v. 1, p. 582–589.

ELLIOTT, T., 1986, Deltas, *in* Reading, H. G., ed., Sedimentary Environments and Facies, second edition: Oxford, Blackwell Scientific Publications, p. 113–154.

FERM, J. C. AND HORNE, J. C., 1979, Carboniferous Depositional Environments of the Appalachian Region: Columbia, Carolina Coal Group, Department of Geology, University of South Carolina, 760 p.

GALLOWAY, W. E., 1968, Depositional systems of the lower Wilcox Group, north-central Gulf Coast Basin: Gulf Coast Association of Geological Societies Transactions, v. 18, p. 275–289.

GALLOWAY, W. E., 1989, Genetic sequences in basin analysis II: Application to northwest Gulf of Mexico Cenozoic basin: American Association of Petroleum Geologists Bulletin, v. 73, p. 143–154.

GALLOWAY, W. E., 1990, Paleogene depositional episodes, genetic stratigraphic sequences, and sediment accumulation rates NW Gulf of Mexico basin, *in* Sequence Stratigraphy as an Exploration Tool: Concepts and Practices in the Gulf Coast: Houston, Gulf Coast Section, Society of Economic Paleontologists and Mineralogists Foundation, Eleventh Annual Research Conference, p. 165–176.

GASKELL, B. A., 1991, Extinction patterns in Paleogene benthic foraminiferal faunas: Relationship to climate and sea level: Palaios, v. 6, p. 2–16.

GIBSON, T. G. AND BYBELL, L. M., 1981, Facies changes in the Hatchetigbee Formation in Alabama-Georgia and the Wilcox-Claiborne Group unconformity: Gulf Coast Association of Geological Societies Transactions, v. 31, p. 301–306.

GIBSON, T. G., MANCINI, E. A., AND BYBELL, L. M., 1982, Paleocene to middle Eocene stratigraphy of Alabama: Gulf Coast Association of Geological Societies Transactions, v. 32, p. 449–458.

HARRIS, W. B. AND FULLAGAR, P. D., 1991, Middle Eocene and late Oligocene isotopic dates of glauconitic mica from the Santee River area, South Carolina: Southeastern Geology, v. 32, p. 1–19.

HAQ, B. L., HARDENBOL, J., AND VAIL, P. R., 1988, Mesozoic and Cenozoic stratigraphy and eustatic cycles, *in* Wilgus, C. K., Hastings, B. S., Kendall, C. G. St. C., Posamentier, H. W., Ross, C. A., Van Wagoner, J. C., eds., Sea-level Changes: An Integrated Approach: Tulsa, Society of Economic Paleontologists and Mineralogists Special Publication 42, p. 71–108.

INGRAM, S. L., 1991, The Tuscahoma-Bashi section at Meridian, Mississippi: First notice of lowstand deposits above the Paleocene-Eocene TP2/TE1 sequence boundary: Mississippi Geology, v. 11, p. 9–14.

ISRAELSKY, M. C., 1949, Oscillation chart: American Association of Petroleum Geologists Bulletin, v. 33, p. 92–98.

KRUMBEIN, W. C. AND SLOSS, L. L., 1963, Stratigraphy and Sedimentation, 2nd ed.: San Francisco, W. H. Freeman, 660 p.

LOUTIT, T. S., HARDENBOL, J., VAIL, P. R., AND BAUM, G. R., 1988, Condensed sections: The key to age dating and correlation of continental margin sequences, *in* Wilgus, C. K., Hastings, B. S., Kendall, C. G. St. C., Posamentier, H. W., Ross, C. A., and Van Wagoner, J. C., eds., Sea-level Changes: An Integrated Approach: Tulsa, Society of Economic Paleontologists and Mineralogists Special Publication 42, p. 183–213.

MANCINI, E. A., 1983, Depositional setting and characterization of the deep-basin Oak Hill lignite deposit (middle Paleocene) of southwest Alabama: Gulf Coast Association of Geological Societies Transactions, v. 33, p. 329–337.

MANCINI, E. A., 1984, Biostratigraphy of Paleocene strata in southwestern Alabama: Micropaleontology, v. 30, p. 268–291.

MANCINI, E. A. AND OLIVER, G. E., 1981, Late Paleocene planktic foraminifers from the Tuscahoma marls of southwest Alabama: Micropaleontology, v. 27, p. 204–225.

MANCINI, E. A. AND TEW, B. H., 1991, Relationships of Paleogene stage and planktonic foraminiferal zone boundaries to lithostratigraphic and allostratigraphic contacts in the eastern Gulf Coastal Plain: Journal of Foraminiferal Research, v. 21, p. 48–66.

MANCINI, E. A. AND TEW, B. H., 1993, Eustasy versus subsidence: Lower Paleocene depositional sequences from southern Alabama, eastern Gulf Coastal Plain: Geological Society of America Bulletin, v. 105, p. 13–17.

MARTINI, E., 1971, Standard Tertiary and Quaternary calcareous nannoplankton zonation, *in* Farinacci, A., ed., Second International Planktonic Conference Proceedings, 1970: Rome, Tecnoscienza, p. 739–785.

OLSSON, R. K., 1991, Cretaceous to Eocene sea level fluctuations on the New Jersey margin: Sedimentary Geology, v. 70, p. 195–208.

PASLEY, M. A. AND HAZEL, J. E., 1990, Use of organic petrology and graphic correlation of biostratigraphic data in sequence stratigraphic interpretations: Example from the Eocene-Oligocene boundary section, St. Stephens Quarry, Alabama: Gulf Coast Association of Geological Societies Transactions, v. 40, p. 661–683.

SIESSER, W. G., 1983, Paleogene Calcareous Nannoplankton Biostratigraphy: Mississippi, Alabama and Tennessee: Jackson, Mississippi Department of Natural Resources, Bureau of Geology Bulletin 125, 61 p.

STAINFORTH, R. M., LAMB, J. L., LUTERBACHER, H., BEARD, J. H., AND JEFFORDS, R. M., 1975, Cenozoic Planktonic Foraminiferal Zonation and Characteristics of Index Forms: Lawrence, Kansas University Paleontological Contributions, Article 62, 425 p.

VAIL, P. R., AND WORNHARDT, W. W., 1990, Well log-seismic sequence stratigraphy: An integrated tool for the 90's: Sequence Stratigraphy as an Exploration Tool-Concepts and Practices in the Gulf Coast: Houston, Program and Extended and Illustrated Abstracts, 11th Annual Research Conference, Gulf Coast Section Society of Economic Paleontologists and Mineralogists Foundation, p. 379–388.

VAN WAGONER, J. C., POSAMENTIER, H. W., MITCHUM, R. M., VAIL, P. R., SARG, J. F., LOUTIT, T. S., AND HARDENBOL, J., 1988, An overview of the fundamentals of sequence stratigraphy and key definitions, *in* Wilgus, C.K., Hastings, B. S., Kendall, C. G. St. C., Posamentier, H. W., Ross, C. A., and Van Wagoner, J. C., eds., Sea-level Changes: An Integrated Approach: Tulsa, Society of Economic Paleontologists and Mineralogists Special Publication 42, p. 125–154.

THE UPPER BOUNDARY OF THE EOCENE SERIES: A REAPPRAISAL BASED ON DINOFLAGELLATE CYST BIOSTRATIGRAPHY AND SEQUENCE STRATIGRAPHY

HENK BRINKHUIS AND HENK VISSCHER

Laboratory of Palaeobotany and Palynology, University of Utrecht, Heidelberglaan 2, 3584 CS The Netherlands

ABSTRACT: The IUGS Commission on Stratigraphy recently acccepted and ratified a proposal to establish an Eocene/Oligocene (E/O) boundary stratotype section and point (GSSP) in the pelagic Massignano section (central Italy), characterized by the last occurrence of the Hantkeninidae (planktonic foraminifera). However, detailed dinoflagellate cyst zonation of the E/O transition interval in the Mediterranean area indicates that this newly selected 'golden spike' horizon correlates to the middle part of the classic (marginal marine) Priabonian type section in northeast Italy. The coeval level occurs well below two earlier proposed E/O boundary stratotypes at Priabona.

In this paper, it is outlined that selection of either one of the proposed boundary-stratotypes at Priabona is more suitable than a hantkeninid-based E/O boundary. These boundary levels can be traced in different lithofacies by recognition of either the TA4.3/4.4 sequence boundary, or the last occurrence of the dinoflagellate cyst *Areosphaeridium diktyoplokus*. Particularly, a sequence stratigraphic concept of the E/O boundary enables a reasonable correlation of the upper boundary of the Priabonian with the base of the Rupelian (i.e., base Oligocene) as recognized in its type-area in Belgium. The acceptance of the position of the 'golden spike' at Massignano as the top of the Priabonian Stage thus creates a new Priabonian/Rupelian boundary problem at the very time old uncertainties and controversies on the mutual delimitation of these stages are becoming resolved. Since only stage boundaries should serve to define chronostratigraphic units of higher rank, it is concluded that the IUGS Commission on Stratigraphy should re-evaluate the status of the Massignano section as the E/O boundary GSSP.

INTRODUCTION

In global chronostratigraphic classification, the Priabonian Stage is generally accepted to represent the youngest stage of the Eocene Series. Consequently, the upper Priabonian boundary should mark the Eocene/Oligocene (E/O) boundary. Biostratigraphic deliniation of the upper boundary of the Priabonian in its type-area around Priabona (Vicentinian Alps, northeast Italy, Fig. 1) is mainly based on last occurrences (LOs) of the Discocyclinidae and *Nummulites fabianii* s.l. (larger foraminifera; see, e.g., Hardenbol, 1968; Setiawan, 1983; Barbin, 1986, 1988; Barbin and Bignot, 1986; Fig. 2). The criteria by which these proposed E/O boundaries may be recognized in coeval successions elsewhere are highly debated in the literature (see e.g., Hardenbol, 1968; Cita, 1969; Setiawan, 1983; Berggren and others, 1985; Barbin, 1986, 1988; Barbin and Bignot, 1986; Pomerol and Premoli-Silva, 1986a, b). Accurate correlation has been hampered by (1) the sparcity of chronostratigraphically important calcareous plankton in the marginal-marine type and reference sections of the stage, and (2) the apparent unsuitability of the type section of the Priabonian for obtaining reliable paleomagnetic signals.

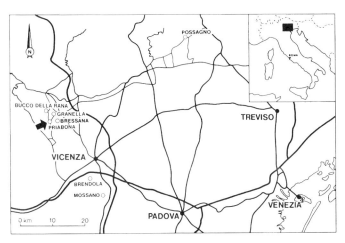

FIG. 1.—The region north of Vicenza (NE Italy), with the locations of the Priabonian type and reference sections.

Recently, a decision to select the section near Massignano (central Italy) as the global E/O boundary-stratotype section and point (GSSP) became ratified by the IUGS Commission of Stratigraphy during the International Geological Congress in Kyoto, Japan, 1992 (see also Cowie, 1986; Odin and Montanari, 1988, 1989). Without any reference to boundaries at a stage level, the E/O Series boundary in the pelagic succession of Massignano is considered to be characterized by the LO of the Hantkeninidae (planktonic foraminifera; e.g., Nocchi and others, 1988a, b). The extinction of the hantkeninids corresponds to the planktonic foraminiferal P17/P18 zonal boundary (Blow, 1969; redefined by Nocchi and others, 1988a, b; see also Premoli Silva and others, 1988a, b; Fig. 3). The extinction does not coincide with any major event in the calcareous nannoplankton record; it occurs within Martini's (1971) NP21 nannoplankton zone and below Okada and Bukry's (1980) nannoplankton CP16A/16B subzonal boundary (e.g., Coccioni, 1988; Odin and Montanari, 1988). In terms of magnetostratigraphy, the LO of the hantkeninids occurs at a level just above the youngest normal polarity excursion within Chron C13R and is believed to be reasonably synchronous on a global scale (e.g., Poore and others, 1982; Nocchi and others, 1988a, b). Radiometric measurements from Massignano suggest an age of 33.7 ± 0.5 Ma for the hantkeninid based E/O junction (Montanari and others, 1988; Odin and Montanari, 1989; compare Fig. 3).

The planktonic foraminiferal record from Massignano and other pelagic sections in central Italy has provided a valuable biostratigraphic background for long-range correlation of latest Eocene and earliest Oligocene deeper-marine deposits. Unfortunately, this background remains unsuitable for correlations with the relatively nearby marginal-marine Priabonian type and reference sections (e.g., Nocchi and others, 1988a, b; Barbin, 1988); it has not provided substantial evidence that the newly accepted E/O boundary at Massignano would correspond to formerly proposed positions of the upper boundary of the Priabonian Stage in its type-area. Rather, Odin and Montanari (1988, p. 261), the authors of the formal proposal to accept the Massignano section as E/O GSSP, state that "... in order to directly correlate the Massignano section to the shallow water sequences, such as the type-Priabonian, more detailed geochemical and magnetostratigraphic studies, and radioisotopic

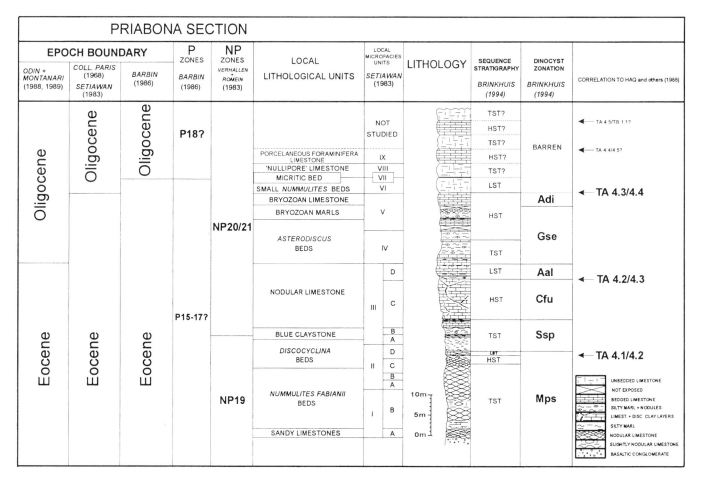

FIG. 2.—Schematic composite representation of the Priabona section (type section of the Priabonian Stage), with biostratigraphic data from Verhallen and Romein (1983) and Barbin (1986) and microfacies units after Setiawan (1983). Dinoflagellate cyst zones after Brinkhuis and Biffi (1993) and Brinkhuis (1994). Sequence stratigraphic interpretation and proposed correlation with the cycles of Haq and others (1988) from Brinkhuis (1994). Note different positions of the E/O boundary.

age-dating should be undertaken in a near future in the southern Alpine (Monte Lessini) region, where shallower pelagic and neritic facies are well represented, even through drilling an ad-hoc bore-hole near Priabona . . ."

The only marine plankton that have the potential for establishing correlations between successions from different marine depositional environments are dinoflagellate cysts. Reconnaissance studies by Gruas-Cavagnetto (1974, 1986), and Gruas-Cavagnetto and Barbin (1988) of the sections around Priabona indicated diverse dinoflagellate cyst assemblages. Until recently, no detailed and well-calibrated dinoflagellate zonal scheme was available for the E/O transition sequence in the Mediterranean realm. The stratigraphic potential of dinoflagellate cysts, therefore, could not be used in previous attempts to perform biostratigraphic correlations between the Eocene and Oligocene Series of central and northeast Italy.

To establish a detailed E/O dinoflagellate zonation for the Mediterranean area, Brinkhuis and Biffi (1993) have investigated a number of otherwise well-documented central Italian sections, including Massignano. The resulting high-resolution zonation is calibrated against the planktonic foraminifera and calcareous nannoplankton zonations as well as magnetostratig-raphy (Fig. 3). Subsequently, the established dinoflagellate zones could be recognized in the Priabonian type-section at Priabona and a reference-section at Bressana (Brinkhuis, 1994; Fig. 4). The biochronologic value of the resulting correlations between central and northeast Italy is corroborated by sequence stratigraphic and paleoclimatologic analysis (Brinkhuis, 1994). A discussion of the chronostratigraphic implications with respect to considerations on the definition of the E/O boundary, however, remained outside the scope of that study.

The implications are considerable. Dinoflagellate zonation, combined with sequence stratigraphy now provides the possibility of selecting an upper Priabonian boundary that can be inter-regionally traced in different sedimentary settings. In this paper, therefore, the status of the Priabonian Stage, and especially its upper boundary, is reassessed in the light of the new biostratigraphic information derived from the dinoflagellate record, supplemented with information from sequence stratigraphic analysis.

THE TYPE-PRIABONIAN

Fossiliferous deposits around Priabona, first described by Suess (1868), together with several other sections in northeast

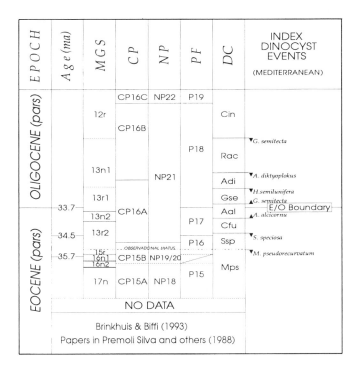

FIG. 3.—Dinoflagellate cyst zonation of sections in central Italy with schematic composite diagram of the correlatable bio- and magnetostratigraphies, from Brinkhuis and Biffi (1993). PF = standard planktonic foraminifer zones (after Blow, 1969, and Nocchi and others, 1988a, b); NP = standard nannoplankton zones (after Martini, 1971); CP = standard nannoplankton zones (after Okada and Bukry, 1980); MGS = magnetostratigraphy obtained from the Massignano section; all PF, NP, CP and MGS-data from papers in Premoli-Silva and others (1988a); DC = dinoflagellate cyst zonation (Brinkhuis and Biffi, 1993); E/O boundary positioned according to the Massignano GSSP.

Italy (Fig. 1), formed the basis for the recognition of the Priabonian Stage by Munier-Chalmas and De Lapparent (1893). No type section was designated. The stage was placed in the Upper Eocene Series.

Decisions at the 'Colloque sur l'Eocène', Paris, 1968

For a long time, the value of the Priabonian as a realistic unit for global chronostratigraphic classification was disputed (see reviews by Setiawan, 1983; Barbin, 1986; Brinkhuis, 1994). Eventually, at the 'Colloque sur l'Eocène', held in Paris (1968), following a proposal by Hardenbol (1968), the Priabonian Stage was accepted as the youngest stage of the Eocene Series. It was also decided to select the composite section at Priabona as the type section of the stage. Mainly on the basis of the extinction of the Discocyclinidae but without substantial support from other (micro)fossil categories, the top of the 'Bryozoan Beds' (now subdivided into Bryozoan Marls and Bryozoan Limestone after Barbin, 1986; Fig. 2) was taken to represent the top of the type-section and, implicitly, the upper boundary of the Priabonian Stage (Hardenbol, 1968; Cita, 1969). A basaltic conglomerate forms the base of the succession at Priabona. It was recognized that the lower part of the Priabonian Stage s.l. is not represented in the type section (Cita, 1969).

The sections at Granella, Bressana (formerly Ghenderle; see Hardenbol, 1968; Setiawan, 1983), Brendola, Mossano, and Possagno in northeast Italy (Fig. 1) were designated as refer-

ence sections during the Eocene Colloquium (Cita, 1969). The stage was defined as being equivalent to the entire "post-Lutetian Upper Eocene Series", corresponding to (1) the *Nummulites fabianii* s.l. Zone (larger foraminifera), (2) the *Globigerapsis semiinvoluta, Globorotalia cerrozoazulensis,* and *Globigerina gortanii* Zones (planktonic foraminifera; P15 to partial P18 zones of the standard Paleogene scheme of Blow, 1969), and (3) the upper part of the *Discoaster tani nodifera* and the *Isthmolithus recurvus* Zones (calcareous nannoplankton; now corresponding to Paleogene standard zones NP16–20 of Martini, 1971).

In the years following the Eocene Colloquium, the concepts of other Eocene and Oligocene stages were further refined. The inclusion of the Bartonian Stage into the standard chronostratigraphic scale, above the Lutetian and below the Priabonian Stage, became accepted despite uncertainties and controversies with respect to its delimitation (e.g., Curry and others, 1978; Hardenbol and Berggren, 1978; Pomerol, 1981; Berggren and others, 1985). The Rupelian became accepted as the oldest Oligocene standard stage, immediately succeeding the Priabonian Stage. The current concept of the Rupelian Stage is based on a number of lithological units in northern Belgium (e.g., Vandenberghe, 1981; Vandenberghe and Vanechelpoel, 1987). No composite stratotype has so far been proposed (Janssen, 1993). The Rupelian Stage in its type-area is principally characterized by the Boom clay. Planktonic foraminifera and nannoplankton from the basal part of this unit are indicative of zones P18 and NP22 (e.g., Berggren and others, 1985; Vandaniels and others, 1993). The lowermost Rupelian units recognized in Belgium are the Berg and Ruisbroek sands. These units are badly exposed and the classic reference outcrops no longer exist. Results of a recent study of dinoflagellates from the Ruisbroek sand (Stover and Hardenbol, 1993) indicates that this unit is correlative with the top of the *Reticulatosphaera actinocoronata* (Rac) Zone of Brinkhuis and Biffi (1993) or overlying zones on the basis of the occurrence of *Wetzeliella gochtii* in the unit. The calibrations of the (Mediterranean) zonal scheme of Brinkhuis and Biffi (1993) suggest the first occurrence (FO) of this species to be correlative to the lower part of the P18 zone and the NP21/22 zonal boundary. Hence, the sand units are likely to represent the lowermost Oligocene. Unfortunately, there is no further biostratigraphic information available to estimate more precisely the position of the sand units relative to Paleogene standard zonations.

The Priabonian Type and Reference Sections

After the acceptance of the Priabonian Stage at the Eocene Colloquium, more information on its type- and reference sections became available through various studies of for example Herb and Heckel (1973) on nummulites, Cruas-Cavagnetto (1974) on palynomorphs, Sirotti (1978) on Discocyclinids, Jossen (1982) on calcareous nannoplankton, and Picoli and Savazzi (1983) on molluscs. Unfortunately, these studies did not result in a better understanding of the potential of the Priabonian Stage in global chronostratigraphy. In addition, it became apparent that in most of the reference sections (notably Granella, Brendola, and Mossano) further biostratigraphic analysis is seriously hampered by unfavourable lithologies, tectonic complications, and weathering. In the well-exposed and well-studied

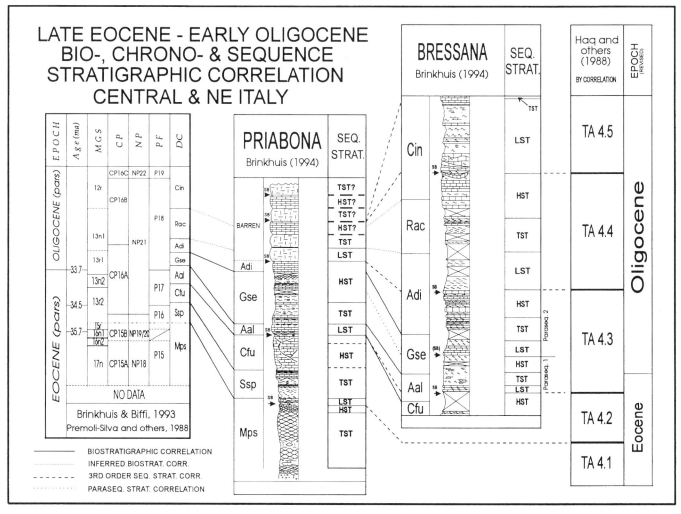

FIG. 4.—Chrono- and sequence stratigraphic interpretation of the Priabona and Bressana sections (NE Italy) and correlation to central Italy (after Brinkhuis, 1994; see also Figs. 2, 3).

outcrop at Possagno (e.g., Bolli, 1975; Grünig and Herb, 1980; Oberhänsli and others, 1984), the upper part of the section could not be biostratigraphically characterized; the E/O transition is poorly known and may be represented by a hiatus (Cita, 1975).

The monograph by Setiawan (1983) on the foraminifera and microfacies of the sections at Priabona, Bressana, and a new locality, Bucco della Rana, provided the first detailed account of sedimentary facies and paleoenvironments of the type area. Verhallen and Romein (1983) presented information on the calcareous nannoplankton distribution in the Priabona and Bressana sections. Setiawan (1983) succeeded in correlating the type section with two reference sections using foraminifera and microfacies composition. He could not confirm the correlation with the standard microplankton zones, however, as suggested in the diagnosis of the Priabonian Stage formulated at the Eocene Colloquium. Verhallen and Romein (1983) identified the NP19/20 boundary in the type section and the Bressana reference section, but were unable to recognize the E/O boundary.

According to Setiawan (1983, p. 102), only the larger foraminifera ". . . seem to show a clear-cut boundary . . ." (i.e., the LO of the discocyclinids). In accordance with the conclusions

of the Eocene colloquium, he suggested the E/O boundary to occur at the top of the Bryozoan Beds. Furthermore, he stated that the extinction of the discocyclinids was not environmentally defined at Priabona, because they ". . . disappeared at the same time . . ." in the more offshore deposited beds (local microfacies unit IV) of the Bressana section (Setiawan, 1983, p. 102). The succeeding fauna, mainly composed of small representatives of *Nummulites*, was suggested to have an Oligocene aspect.

More recently, Barbin (1986) studied the Priabonian type section and some of the reference sections. Barbin and co-workers concentrated on virtually every (micro)fossil group (including foraminifera, calcareous nannoplankton, ostracods, dinoflagellate cysts, and bryozoans), and also tried to incorporate paleomagnetic and geochemical data. Parts of Barbin's thesis were separately published (e.g., Barbin and Bignot, 1986; Barbin, 1988; Barbin and Guernet, 1988; Gruas-Cavagnetto and Barbin, 1988; Braga and Barbin, 1988), and were followed by additional data on pollen and spores from several reference sections (Gruas-Cavagnetto and Barbin, 1989) and nannoplankton from the Brendola section (Barbin, 1989).

Like Setiawan (1983), Barbin and co-workers observed only a few, poorly preserved calcareous plankton of low diversity. They also found that the paleomagnetic signal in the type-Priabonian section was too weak and the stable isotope data too influenced by diagenesis to provide any relevant stratigraphic information. In spite of the many uncertainties, Barbin provided a new definition of the Priabonian Stage s.l. His synthesis was based on (1) all available data from the Priabonian type area, (2) general comparisons with various other Upper Eocene-Lower Oligocene sections, such as the Contessa Quarry section from central Italy (Lowrie and others, 1982) and various DSDP-sites (e.g., Poore and others, 1982), and (3) an application of the sea level curve of Vail and others (1977).

Barbin (1986) considered the Priabonian Stage to correspond to calcareous nannoplankton zones NP17 (*pars*) to NP21 (*pars*), to planktonic foraminifera zones P15 to P18 (*pars*), and to dinoflagellate cyst zone W13 of Châteauneuf and Gruas-Cavagnetto (1978), all regarded equivalent to the larger foraminifera *Nummulites fabianii* s.l. Zone. In contrast to Hardenbol (1968) and Setiawan (1983), Barbin and co-workers suggested that the E/O boundary occurred at Priabona at the thin Micritic Bed (Fig. 2), about three meters above the top of the Bryozoan Limestone (i.e., local microfacies unit VII of Setiawan, 1983). It is not clear whether the boundary should be at the bottom or the top of the bed. Biostratigraphically, the position of the new boundary was estimated on the basis of the LO of *Nummulites fabianii* s.l. in the underlying small *Nummulites* beds, and the presence of Early Oligocene foraminifera (e.g., *Nummulites fichteli*) in beds overlying the Nullipore limestone at Priabona (Barbin, 1986, 1988; Barbin and Bignot, 1986). The LO of *Nummulites fabianii* s.l. was found to succeed that of the Discocyclinidae and was without evidence nor discussion equated to the LO of the Hantkeninidae in pelagic successions.

Barbin's interpretations resulted in a proposal to redefine the position of the top of the Priabonian type-section. The Micritic Bed was proposed as the E/O boundary stratotype (Barbin and Bignot, 1986). Also at Bressana, Barbin (1986) positioned the E/O boundary considerably higher than previously suggested by Setiawan (1983). On the basis of unclear lithololologic considerations, he tentatively selected the incoming channel system (see Fig. 4) to mark the series junction.

APPLICATION OF THE MEDITERRANEAN E/O DINOFLAGELLATE ZONATION

In contrast to the search for stratigraphically significant calcareous plankton, and despite statements to the contrary (Pomerol and Premoli Silva, 1986b), palynological studies by Gruas-Cavagnetto indicated that dinoflagellate cysts could be useful in long-range correlation (Gruas-Cavagnetto, 1974, 1986; Gruas-Cavagnetto and Barbin, 1988). Although only a few samples were studied from the Priabonian type section and reference sections, some of the reported assemblages could be confidently assigned to the *Charlesdowniea clathrata angulosa* Zone (W13 Biozone) of Châteauneuf and Gruas-Cavagnetto (1978), originally defined in the Upper Eocene sediments of the Paris Basin.

Brinkhuis (1994) has identified the dinoflagellate zones of Brinkhuis and Biffi (1993) from the pelagic E/O transition from central Italy in the Priabonian type area (compare Figs. 2–4). In this zonation there are eight successive interval zones, de-

fined at the Massignano and Monte Cagnero sections of central Italy (Fig. 3). The stratigraphic resolution of the dinoflagellate zones proved to be considerably higher than that of the zones based on planktonic foraminifera and calcareous nannoplankton. The resolution can be further increased by taking characteristic interzonal events into consideration. Both at Priabona and Bressana, the resulting dinoflagellate-based correlation between central and northeast Italy enables the recognition of the position of the E/O boundary at Massignano, characterized by the LO of the Hantkeninidae (P17/18 boundary, *sensu* Nocchi and others, 1988a, b). For discussion purposes, the latter boundary concept was provisionally followed by Brinkhuis and Biffi (1993) and Brinkhuis (1994).

At Massignano and other sections from central Italy, it was demonstrated that the LO of the Hantkeninidae corresponds to a level just below the boundary between the *Achomosphaera alcicornu* (Aal) and *Glaphyrocysta semitecta* (Gse) dinoflagellate Zones, defined by the first occurrence of *Glaphyrocysta semitecta* in the Mediterranean area (Brinkhuis and Biffi, 1993). At Priabona the Aal/Gse zonal boundary occurs in the middle part of the section (Figs. 2, 4). This implies that at Priabona an E/O boundary based on the LO of hantkeninids would approximate the boundary between the Nodular Limestone and the *Asterodiscus* Beds and the boundary between the local microfacies units III and IV of Setiawan (1983; Fig. 2). This level is considerably lower than the earlier proposed positions of the upper boundary of the Priabonian. Of these positions, only the top of the Bryozoan limestone is palynologically diagnosed. It can be assigned to the *Areosphaeridium diktyoplokus* (Adi) Zone (Figs. 2, 4). The succession overlying the Bryozoan Limestone is palynologically barren so that the Micritic Bed cannot be directly assigned to a dinoflagellate zone.

At Bressana the Aal/Gse zonal boundary is found in the presently exposed basal part of the section (Fig. 4). This would imply an essentially Oligocene age for this classic Priabonian reference section. The previously suggested E/O boundary based on the extinction of discocyclinids at the top of the Bressana microfacies unit VI (Setiawan, 1983) occurs just below the boundary between the *Areosphaeridium diktyoplokus* (Adi) and *Reticulatosphaera actinocoronata* (Rac) dinoflagellate Zones, defined by the LO of *Areosphaeridium diktyoplokus*. An E/O boundary determined by the incoming of a channel system (Barbin, 1986) would occur within the succeeding *Corrudinium incompositum* (Cin) Zone (Fig. 4).

The lower part of the Priabona section belongs to the *Melitasphaeridium pseudorecurvatum* (Mps) Zone of Brinkhuis and Biffi (1993; Fig. 2). The base of this zone, however, has not yet been defined. Brinkhuis (1994) suggested that the basal part of the Priabona section could correlate with the early part of (planktonic foraminifera) Zone P16 and (nannoplankton) zones NP19/20 and CP15B (Figs. 2, 4). Since the Bartonian-Priabonian boundary is frequently taken to be correlative with the P14/15 (*sensu* Blow, 1969) and the NP17/18 boundaries (despite discrepancies; see discussion in Berggren and others, 1985), the dinoflagellate record is not in conflict with the generally accepted view that the lowermost part of the Priabonian Stage s.l. is not represented in its type section. In the Bressana section, as presently exposed, the basal part belongs to the *Cordosphaeridium funiculatum* (Cfu) Zone and correlates with the lower part of P17.

Sequence Stratigraphic Analysis

It has been shown by Brinkhuis and Biffi (1993) and Brinkhuis (1994) that temporal and spatial distribution patterns of dinoflagellate cysts, with supplementary information from prasinophyte algae and pollen, can provide proxy variables indicative of fluctuations of sea level and sea-surface temperature (SST). Sequence stratigraphic analysis of the Italian E/O sections and their palynomorph content indicated that successive systems tracts, calibrated against the dinoflagellate zonation, can be inter-regionally traced (Brinkhuis, 1994). Parts of sections that do not yield dinoflagellates, such as the succession overlying the Bryozoan Limestone at Priabona, can be sequence stratigraphically interpreted using lithological characteristics. The inferred changes in sea level and SST may be reconciled with the third-order cycles TA4.1 through TA4.5 of Haq and others (1988; Figs. 2, 4).

Haq and others (1988) placed the Priabona type section in their latest Eocene cycles TA4.2 and TA4.3, in the upper part of their Priabonian s.l. They suggested that the type section does not include the TA4.3/4.4 boundary, which represents the E/O boundary in their scheme. On the basis of the new information available, however, there is every indication that at Priabona three recognized sequence boundaries mark the onset of the cycles TA4.2, 4.3, and 4.4 (Brinkhuis, 1994; Figs. 2, 4). Both in central and northeast Italy, the intervals of sea level lowstand that characterize the third-order cycle boundaries seem to correspond to periods with a relatively cool SST.

As at Massignano and other localities in central Italy, at Priabona and Bressana the TA4.2/4.3 transition approximates the boundary between the *Cordosphaeridium funiculatum* (Cfu) and *Achomosphaera alcicornu* (Aal) zones, below the level of the recently accepted E/O boundary based on the LO of the hantkeninids. At Bressana, the TA4.3/4.4 transition occurs just below the boundary between the *Areosphaeridium diktyoplokus* (Adi) and *Reticulatosphaera actinocoronata* (Rac) zones. Because of the absence of dinoflagellate cysts in the Small *Nummulites* Beds in the Priabonian type section, the actual LO of *Areosphaeridium diktyoplokus* (indicative of the Adi/Rac zonal boundary) could not be detected. On the basis of sequence stratigraphic considerations, however, the sequence boundary marking the onset of TA4.4 is considered to occur at the top of the Bryozoan Limestone. Thus, at Priabona as well as Bressana, the TA4.3/4.4 cycle boundary appears to be correlatable to the E/O boundary characterized by the extinction of the Discocyclinidae. The Micritic Bed, proposed by Barbin and Bignot (1988) to represent the E/O boundary, is thought to mark the onset of the earliest Oligocene transgression and to correlate with the extinction of *Areosphaeridium diktyoplokus* (and the Adi/Rac zonal boundary; Brinkhuis, 1994; Figs. 2, 4).

THE UPPER BOUNDARY OF THE PRIABONIAN STAGE

The foregoing considerations may lead to the following basic assumptions, relevant to a re-evaluation of the Priabonian type section in relation to the discussion on the definition of the E/O boundary:

1. At the Colloque sur l'Eocène, outcrops at Priabona were designated to constitute the stratotype of the Priabonian Stage.

2. The outcrops at Priabona include the E/O transition interval.
3. Following the decisions at the Colloque sur l'Eocène, the top of the Bryozoan Limestone can be regarded by implication as the earliest selected boundary stratotype of the upper boundary of both the Priabonian Stage and the Eocene Series.
4. An alternative position of the upper Priabonian (and the Eocene Series) boundary is the Micritic Bed (Barbin and Bignot, 1986).
5. Biostratigraphic correlation by dinoflagellate cysts enables the recognition of the approximate position of the E/O boundary at Priabona as recently accepted at Massignano in central Italy (i.e., between the Nodular Limestone and *Asterodiscus* Beds at Priabona).
6. The position of the hantkeninid based E/O boundary (from Massignano) is well below the Micritic Bed and the top of the Bryozoan Limestone unit.
7. The top of the Bryozoan Limestone is considered to represent the TA4.3/4.4 sequence boundary, while the overlying Micritic Bed is considered to represent the subsequent transgressive surface.

According to chronostratigraphic convention, a series should be defined in terms of the lower boundary of it oldest stage and the upper boundary of its youngest stage (Hedberg, 1976). Thus, the E/O boundary should conform to the Priabonian/Rupelian boundary. Furthermore, it is desirable that the upper boundary stratotype of one stage be the lower boundary stratotype of the immediately overlying stage. With respect to the Priabonian/Rupelian boundary, however, it seems unrealistic to follow the latter recommendation. Because of unfavourable facies and poor outcrop conditions, the Berg and Ruisbroek sands in Belgium are unsuitable for selection as the Priabonian/Rupelian boundary stratotype(s). Consequently, the E/O boundary must agree with the definition of the boundary stratotype of the upper boundary of the Priabonian Stage. Such a procedure is also recommended by the IUGS Commission on Stratigraphy (Cowie, 1986).

The Top of the Bryozoan Limestone

At the time of the Colloque sur l'Eocène, chronostratigraphic convention did not emphasize the definition of stage boundaries, and the concept of boundary stratotypes had not yet been introduced. In retrospect, however, the top of the Bryozoan Limestone (*sensu* Barbin, 1986, i.e., the upper part of the Bryozoan Beds of Hardenbol, 1968, and Setiawan, 1983) may be regarded as the Colloquium's definition of the upper boundary of the Priabonian Stage. This level is lithologically characterized by the boundary between local microfacies units V and VI of Setiawan (1983). In terms of the larger foraminifera record, the level is located above the actual LO of the Discocyclinidae but below the LO of *Nummulites fabianii* s.l. in the Small *Nummulites* Beds (Setiawan, 1983; Barbin, 1986). Dinoflagellate biostratigraphy indicates that the Bryozoan Limestone belongs to the *Areosphaeridium diktyoplokus* (Adi) Zone. Furthermore, the top of the Bryozoan Limestone can be considered to represent the sequence boundary marking the onset of the eustatic third-order cycle TA4.4 of Haq and others (1988; see Brinkhuis, 1994; Figs. 2, 4).

Our current state of knowledge indicates that an upper boundary of the Priabonian Stage defined by the top of the Bryozoan Limestone, may be traced in many parts of the world by means of biostratigraphic and sequence stratigraphic studies. The dinoflagellate record enables a correlation with the pelagic E/O succession in central Italy, where the Adi Zone has been described from the early part of Chron 13N, well above the P17/18 boundary, but approximate to the CP16A/16B subzonal boundary (Brinkhuis and Biffi, 1993).

In Belgium, the Rupelian clays yield rich and diversified dinoflagellate cyst assemblages (Stover and Hardenbol, 1993). Asssemblages from the underlying Berg sand (Brinkhuis, unpubl. data) and Ruisbroek sand (Stover and Hardenbol, 1993) show compositional characters that suggest correlation with the upper part of the Rac Zone in Italy.

As a result of sequence stratigraphic interpretation, the onset of third-order cycle TA4.4 is now being recognized in many Paleogene basins (Vail, pers. commun., 1991). The beginning of this cycle represents one of the earliest periods of major Paleogene glaciation, an event that can be recognized on a world-wide scale as a distinct positive shift in $\delta^{18}O$ (e.g., Pomerol and Premoli Silva, 1986b; Miller and others, 1987; Kennett and Barker, 1990). Regional sequence stratigraphic analysis of the Rupelian Stage in Belgium suggests that the corresponding sequence boundary may be detected at the bottom of the 'type-Rupelian' (the Berg and/or Ruisbroek sand; Vandenberghe and Vanechelpoel, 1987; Vanechelpoel and others, 1992; Stover and Hardenbol, 1993). These basal Rupelian sand units are considered to represent the onlapping or 'back-stepping' transgressive systems tracts of cycle TA4.4. The preceding lowstand phase is considered to be missing in the Rupelian Basin as a result of its shallow marine nature (Stover and Hardenbol, 1993). However, further analysis of the data presented by Stover and Hardenbol (1993), and comparison to the Italian information, indicates that the TA4.3/4.4 sequence boundary may possibly occur at the base of the Watervliet Member of the Zelzate Formation in Belgium, underlying the classic Rupelian units. The Watervliet Member, part of the disused Tongrian Stage, may tentatively be regarded to represent the lowstand phase of cycle TA4.4 on the basis of the data of Stover and Hardenbol (1993), with the LO of *Areosphaeridium diktyoplokus* (i.e., top Adi Zone) at its base. Moreover, Stover and Hardenbol (1993, p. 10) themselves assign part of the Watervliet Member to the *Reticulatosphaera actinocoronata* (Rac) Zone, indicating that they consider at least part of this unit to be of early Oligocene age.

The Micritic Bed

The Micritic Bed, proposed by Barbin and Bignot (1986) as the alternative E/O boundary stratotype at Priabona, overlies the LO of *Nummulites fabianii* s.l. and would thus comply with the biostratigraphic diagnosis of the Priabonian Stage as discussed at the Colloque sur l'Eocène. The presence of *Nummulites fichteli* in overlying strata might be indicative of the *Nummulites fichteli* Zone, tentatively regarded to correspond to the Rupelian Stage by Drooger and Laagland (1986). Sequence stratigraphically, it is considered to reflect the initiation of the transgressive phase ('transgressive surface'), following the onset of cycle TA4.4 (Brinkhuis, 1994). The bed does not yield

dinoflagellate assemblages or other diagnostic microfossils. However, by combining sequence stratigraphic interpretation and dinoflagellate biostratigraphy, it was proposed that this horizon correlates with the extinction of *Areosphaeridium diktyoplokus* and the Adi/Rac zonal boundary (Brinkhuis, 1994).

In many other parts of the world, the LO of *Areosphaeridium diktyoplokus* plays an important role in palynostratigraphic estimates of the position of the E/O transition interval in Paleogene basins (Maier, 1959; Weyns, 1970; Benedek and Müller, 1976; Williams, 1977; Soper and Costa, 1978; Ioakim, 1979; Goodman and Ford, 1983; Bujak, 1984; Williams and Bujak, 1985; Benedek, 1986; Costa and Manum, 1988; Matsuoka and Bujak, 1988; Stover and others, 1988; Head and Norris, 1989; Zaporcek, 1989; Köthe, 1990; Mohr, 1990; Stover and Hardenbol, 1993). In the cycle chart of Haq and others (1988), as well as in some regional dinoflagellate zonations (e.g., Matsuoka and Bujak, 1988; Köthe, 1990), the LO of *Areosphaeridium diktyoplokus* is explicitly used to mark the E/O boundary. The relevance of the event for correlations between the Priabonian and Rupelian type sections, following the work of Stover and Hardenbol (1993), was already discussed above. *Areosphaeridium diktyoplokus* has been found as far North as the Bering Sea (Bujak, 1984; Matsuoka and Bujak, 1988) and the Barentsz Sea (Brinkhuis, unpubl. data) and as far South as the Weddell Sea (Mohr, 1990), suggesting a cosmopolitan distribution of the species. Although further studies are needed to confirm SST-independency of the occurrences, the LO of *Areosphaeridium diktyoplokus* seems to be a valuable event in constraining the upper boundary of the Priabonian Stage in varied depositional environments.

The Top of the Nodular Limestone

With the aid of dinoflagellate biostratigraphy, it would be possible to propose a new boundary stratotype at Priabona at a position that closely correlates with the E/O GSSP at Massignano. From a lithological point of view, the most obvious position would be the boundary between the Nodular Limestone and the *Asterodiscus* Beds. This corresponds to the boundary between local microfacies units III and IV of Setiawan (1983), the transgressive surface following the sequence boundary that marks the TA4.2/4.3 transition (Brinkhuis, 1994).

A boundary stratotype at the top of the Nodular Limestone would have the advantage that the P17/18 boundary (*sensu* Nocchi and others, 1988a, b), as characterized by the LO of the Hantkeninidae at Massignano, could, of course, continue to serve as the diagnostic E/O boundary criterion in deeper marine settings. However, in marked contrast to the higher levels discussed above, it has to be realized that this level does not correspond to any major event in global environmental change. The acceptance of this horizon implies that the current sequence stratigraphic concept of Haq and others (1988), in which the E/O boundary is considered to mark the onset of cycle TA4.4 has to be abandoned. Moreover, the level cannot be recognized in the 'type-Rupelian', since it represents an older horizon (see also discussion in Stover and Hardenbol, 1993).

CONCLUSIONS AND RECOMMENDATION

The present understanding of the Priabonian Stage confirms the chronostratigraphic value of the either the top of the Byozoan Limestone, proposed at the Colloque sur l'Eocène (1968) or the Micritic Bed, proposed by Barbin and Bignot (1986) as the boundary stratotype of the upper boundary of the latest Eocene stage. Through biostratigraphic and sequence stratigraphic studies, selection of either one of these boundaries can now be approximated throughout the world in a variety of shallow- and deeper-marine depositional settings and represents the onset of cycle TA 4.4. More importantly, it appears that current studies must lead to the conclusion that the top of the Bryozoan Limestone can be reasonably correlated with the base of the Rupelian Stage in Belgium.

As a result of these findings, the disagreement on the mutual delimitation of the Priabonian and Rupelian stages as well as the definition of the E/O boundary may be settled satisfactorily. Moreover, an Eocene/Oligocene series boundary defined by the top the Bryozoan Limestone of the 1968-accepted type-Priabonian section correlates with the many physical, chemical, and biological reflections of a major episode of global change at the E/O junction.

Unlike the proposers of the E/O GSSP at Massignano, who envisaged correlations with the type-Priabonian section to be attempted after acceptance of the new boundary, we want to emphasize the importance of settling the E/O boundary problem at a stage level, since stage boundaries should ". . . serve to define not only stages but also chronostratigraphic units of higher rank . . ." (Hedberg, 1976, p. 71). Proposing an E/O series boundary without reference to the corresponding stage boundary is an impractical approach; one neglects the principle that stages should be the basic working units in standard chronostratigraphy.

The formal acceptance of the E/O GSSP places the E/O boundary below the top of the Bryozoan Limestone or the Micritic Bed. This procedure thus created a new Priabonian/Rupelian boundary problem at the very time the old uncertainties and controversies on the mutual delimitation of these stages are becoming resolved. We therefore believe that the recent acceptance and ratification of the proposal does not lead to a better understanding of Paleogene stratigraphy. This drawback has to be taken into consideration by the geological community when judging the advantages and disadvantages of the accepted E/O boundary stratotype at Massignano.

ACKNOWLEDGMENTS

Comments by G. L. Williams, R. Z. Poore, and D. S. Ulmer-Scholle significantly improved the manuscript. The first author acknowledges support from The Netherlands Foundation for Earth Science Research (AWON) and financial aid from The Netherlands Organization for the Advancement of Scientific Research (NWO) and the LPP Foundation. This is a publication of The Netherlands Research School of Sedimentary Geology, no. 950713.

REFERENCES

BARBIN, V., 1986, Le Priabonien dans sa région-type (Vicentin, Italie du Nord). Stratigraphie; Micropaléontologie; Essai d'intégration dans l'échelle chronostratigraphique: Unpublished Ph.D. Dissertation, University Pierre and Marie Curie, Paris, 227 p.
BARBIN, V., 1988, The Eocene-Oligocene transition in shallow-water environment: the Priabonian stage type area (Vicentin, northern Italy), *in* Premoli Silva, I., Coccioni, R., and Montanari, A., eds., The Eocene-Oligocene Boundary in the Marche-Umbria Basin (Italy): Ancona, International Union of Geological Sciences Commision on Stratigraphy; International Subcommission on Paleogene Stratigraphy Report, p. 163–171.
BARBIN, V., 1989, Calcareous nannofossil assemblages from the Brendola section (Priabonian stage stratotype area, northern Italy): Marine Micropaleontology, v. 14, p. 327–338.
BARBIN, V. AND BIGNOT, G. 1986, New proposal for an Eocene-Oligocene boundary according to microfacies from the Priabonian-type section, *in* Pomerol, C. and Premoli Silva, I., eds., Terminal Eocene Events: Amsterdam, Elsevier, Developments in Paleontology and Stratigraphy, v. 9, p. 49–52.
BARBIN, V. AND GUERNET, C., 1988, Contribution à l'étude du Priabonien dans sa région-type: Les Ostracodes: Revue de Micropaléontologie, v. 30, p 209–231
BENEDEK, P. N., 1986, Ergebnisse der Phytoplankton Untersuchungen aus dem Nordwestdeutschen Tertiär, *in* Tobien, H., ed., Nordwestdeutschland im Tertiär: Berlin-Stuttgart, Beiträge zur regionalen geologie der Erde, Gebr. Bornträger, v. 18, p. 157–185.
BENEDEK, P. N. AND MÜLLER, C., 1976, Die Grenze Unter-/Mittel-Oligozän am Doberg bei Bünde/Westfalen, I. Phyto- und Nannoplankton: Neues Jahrbuch für Geologie und Paläontologie, Monatshefte, v. 9, p. 129–144.
BERGGREN, W. A., KENT, D. V., AND FLYNN, J. J., 1985, Jurassic to Paleogene: Part 2. Paleogene geochronology and chronostratigraphy, *in* Snelling, N. J., ed., The Chronology of the Geological Record: Oxford, Blackwell Scientific Publishing, Geological Society of London Memoir 10, p. 141–195.
BLOW, W. H., 1969, Late Middle Eocene to Recent planktonic foraminiferal biostratigraphy, *in* Brönniman, R. and Renz, N. H., eds., Proceedings of the First International Conference on Planktonic Microfossils, Geneva, 1967, v. 1: Leiden, E. J. Brill, p. 199–422.
BOLLI, H. M., ed., 1975, Monografia Micropaleontologica sul Paleocene e l'Eocene di Possagno, Provincia di Treviso, Italia: Schweizerische Paläontologische Abhandlungen, v. 97, 221 p.
BRAGA, G. AND BARBIN, V., 1988, Les bryozoaires du Priabonien stratotypique (province Vicenza, Italie): Revue de Paléobiologie, v. 7, p. 495–556.
BRINKHUIS, H., 1994, Late Eocene to Early Oligocene dinoflagellate cysts from the Priabonian type-area (northwest Italy); Biostratigraphy and Palaeoenvironmental interpretation: Palaeogeography, Palaeoclimatology, Palaeoecology, v. 107, p. 121–163.
BRINKHUIS, H. AND BIFFI, U., 1993, Dinoflagellate cyst stratigraphy of the Eocene/Oligocene transition in central Italy: Marine Micropaleontology, v. 22, p. 131–183.
BUJAK, J. P., 1984, Cenozoic dinoflagellate cysts and acritarchs from the Bering Sea and northern Pacific, DSDP Leg 19: Micropaleontology, v. 30, p. 180–212.
CHÂTEAUNEUF, J. -J. AND GRUAS-CAVAGNETTO, C., 1978, Les zones de Wetzeliellaceae (Dinophyceae) du Bassin de Paris: Bulletin de Bureau de Recherches Géologique et Minières, v. 2, IV, p. 59–93.
CITA, M. B., 1969, Le Paléocène et l'Eocène de l'Italy du Nord: Mémoires de Bureau de Recherches Géologique et Minières, v. 69, p. 417–429.
CITA, M. B., 1975, Stratigrafia della Sezione di Possagno, *in* H.M. Bolli, ed., Monografia Micropaleontologica sul Paleocene e l'Eocene di Possagno, Provincia di Treviso, Italia: Schweizerische Paläontologische Abhandlungen, v. 97, p. 9–33.
COCCIONI, R., 1988, The genera *Hantkenina* and *Cribrohantkenina* (foraminifera) in the Massignano section (Ancona, Italy), *in* Premoli Silva, I., Coccioni, R., and Montanari, A., eds., The Eocene-Oligocene Boundary in the Marche-Umbria Basin (Italy): Ancona, International Union of Geological Sciences Commision on Stratigraphy; International Subcommission on Paleogene Stratigraphy Report, p. 81–96.
COSTA, L. I. AND MANUM, S. B., 1988, The description of the interregional zonation of the Paleogene (D 1—D 15) and the Miocene (D 16—D 20), *in* Vinken, K. L. and Renier, P. K., eds., The Northwest European Tertiary Basin, Results of the International Geological Correlation Programme, Project No 124: Geologisches Jahrbuch Reihe A, v. 100, p. 321–342.
COWIE, J. W., 1986, Guidelines for Boundary Stratotypes: Episodes, v. 9, p. 78–82.
CURRY, D., ADAMS, C. G., BOULTER, M. C., DILLEY, F. C., EAMES, F. E., FUNNELL, B. M., AND WELLS, M. K., 1978, A correlation of Tertiary rocks in the British Isles: London, Geological Society of London Special Report 12, p. 1–72.
DROOGER, C. W. AND LAAGLAND, H., 1986, Larger foraminiferal zonation of the European-Mediterranean Oligocene: Proceedings of the Koninklijke Nederlandse Akademie van Wetenschappen, Series B, v. 89, p. 135–148.
GOODMAN, D. K. AND FORD, L. N., JR., 1983, Preliminary dinoflagellate biostratigraphy for the Middle Eocene to Lower Oligocene from the southwest

Atlantic Ocean: Washington D.C., Initial Reports of the Deep Sea Drilling Project, v. 71, p. 859–877.

GRUAS-CAVAGNETTO, C., 1974, La palynoflore et le microplancton du Priabonien de sa localité type (prov. Vicenza, Italie): Bulletin de Societé de Géologie de France, v. 7, p. 86–90.

GRUAS-CAVAGNETTO, C., 1986, Étude Paléontologique; Les Palynomorphes, *in* Barbin, V., Le Priabonien dans sa région-type (Vicentin, Italie du Nord). Stratigraphie; Micropaléontologie; Essai d'intégration dans l'échelle chronostratigraphique: Unpublished Ph.D. Dissertation, University Pierre and Marie Curie, Paris, p. 119–123.

GRUAS-CAVAGNETTO, C. AND BARBIN, V., 1988, Les dinoflagellés du Priabonien stratotypique (Vicentin, Italie); mise en evidence du passage Eocène/Oligocène: Revue de Paléobiologie, v. 71, p. 163–198.

GRUAS-CAVAGNETTO, C. AND BARBIN, V., 1989, La palynoflore (spores et pollens) du Priabonien stratotypique (Vicentin, Italie du Nord): Revue de Paléobiologie, v. 8, p. 95–120.

GRÜNIG, A. AND HERB, R., 1980, Paleoecology of Late Eocene benthonic Foraminifera from Possagno (Treviso-Northern Italy): Washington, D.C., Cushman Foundation Special Publications, v. 18, p. 68–85.

HAQ, B. U., HARDENBOL J., AND VAIL, P. R., 1988, Mesozoic and Cenozoic chronostratigraphy and cycles of sea level change, *in* Litz, B. H., ed., Sea-level Changes: An Integrated Approach: Tulsa, Society of Economic Paleontologists and Mineralologists, Special Publication 42, p. 71–108.

HARDENBOL, J., 1968, The Priabonian type section (a preliminary note): Mémoires de Bureau de Recherches Géologique et Minières, v. 58, p. 629–635.

HARDENBOL, J. AND BERGGREN, W. A., 1978, A new Paleogene numerical time scale, *in* Cohee, G. V., Giaessner, M. F., and Hedberg, H. D., eds., The Geologic Time Scale: Tulsa, American Association of Petroleum Geology Studies in Geology 6, p. 213–234.

HEAD, M. J. AND NORRIS, G., 1989, Palynology and dinocyst stratigraphy of the Eocene and Oligocene in ODP-Leg 105, Hole 647A, Labrador Sea: College Station, Proceedings of the Ocean Drilling Program, Scientific Results, v. 105, p. 515–550.

HEDBERG, H. D., ed., 1976, International Stratigraphic Guide; A guide to stratigraphic classification, terminology and procedure: New York, John Wiley and Sons, 200 p.

HERB, R. AND HEKEL, H., 1973, Biostratigraphy, Variability and Facies Relations of some Upper Eocene Nummulites from Northern Italy: Eclogae Geologieae Helvetiae, v. 66, p. 419–445.

IOAKIM, C., 1979, Étude comparative des dinoflagellés du Tertiaire inférieur de la Mer du Labrador et de la Mer du Nord: Unpublished Ph.D. Dissertation, University Pierre and Marie Curie, Paris, 300 p.

JANSSEN, A. W., 1993, 'Working Group Rupelian', intermediate progress report: Bulletin van de Belgische Vereniging voor Geologie, v. 102, p. 243–247 (published in 1994).

JOSSEN, J. A., 1982, Les nannofossiles calcaires de Priabona: Revue de Paléobiologie, v. 1, p. 39–51.

KENNETT, J. P. AND BARKER, P. F., 1990, Latest Cretaceous to Cenozoic climate and oceanographic developments in the Weddell Sea, Antarctica: An Ocean-Drilling Perspective: College Station, Proceedings of the Ocean Drilling Program, Scientific Results, v. 113, p. 937–960.

KÖTHE, A., 1990, Paleogene dinoflagellates from northwest Germany: Geologisches Jahrbuch, Reihe A, v. 118, 111 p.

LOWRIE, W., ALVAREZ, W., NAPOLEONE, G., PERCH-NIELSEN, K., PREMOLI SILVA, I., AND TOUMARKINE, M., 1982, Paleogene magnetic stratigraphy in Umbrian pelagic carbonate rocks: The Contessa sections, Gubbio: American Geological Society Bulletin, v. 93, p. 414–432.

MAIER, D., 1959, Planktonuntersuchungen in tertiären und quartären marinen Sedimenten. Ein Beitrag zur Systematik, Stratigraphie und Ökologie der Coccolithophorideen, Dinoflagellaten und Hystrichosphaerideen vom Oligozän bis zum Pleistozän: Neues Jahrbuch für Geologie und Paläontologie, Abhandlungen, v. 107, p. 278–340.

MARTINI, E., 1971, Standard Tertiary and Quaternary calcareous nannoplankton zonation, *in* Farinacci, A., ed., Proceedings of the 2nd Planktonic Conference: Roma, Tecnoscienza, p. 739–785.

MATSUOKA, K. AND BUJAK, J. P., 1988, Cenozoic dinoflagellate cysts from the Navarin Basin, Norton Sound and St. George Basin, Bering Sea: Bulletin of the Faculty of Liberal Arts, Nagasaki University, Natural Science, v. 29, p. 1–147.

MILLER, K. G., FAIRBANKS, R. G., AND MOUNTAIN, G. S., 1987, Tertiary oxygen isotope synthesis, sea level history and continental margin erosion: Paleoceanography, v. 2, p. 1–19.

MOHR, B. A. R., 1990, Eocene and Oligocene sporomorphs and dinoflagellate cysts from Leg 113 drill sites, Weddell Sea, Antarctica: College Station,

Proceedngs of the Ocean Drilling Program, Scientific Results, v. 113, p. 595–612.

MONTANARI, A., DEINO, A. L., DRAKE, R. E., TURRIN, B. D., DEPAOLO, D. J., ODIN, G. S., CURTIS, G. H., ALVAREZ, W., AND BICE, D. M., 1988, Radioisotopic dating of the Eocene-Oligocene boundary in the pelagic sequence of the northeastern Apennines, *in* Premoli Silva, I., Coccioni, R., and Montanari, A., eds., The Eocene-Oligocene Boundary in the Marche-Umbria Basin (Italy): Ancona, International Union of Geological Sciences Commision on Stratigraphy; International Subcommission on Paleogene Stratigraphy Report, p. 195–208.

MUNIER-CHALMAS, E. AND DE LAPPARENT, A., 1893, Note sur la nomenclature des terraines sédimentaires: Bulletin de Societé Géologique de France, v. 3, p. 438–488 (published in 1894).

NOCCHI, M., MONECHI, S., COCCIONI, R., MADILE, M., MONACO, P., ORLANDO, M., PARISI, G., AND PREMOLI SILVA, I., 1988a, The extinction of the Hantkeninidae as a marker for recognizing the Eocene-Oligocene boundary: a proposal, *in* Premoli Silva, I., Coccioni, R., and Montanari, A., eds., The Eocene-Oligocene Boundary in the Marche-Umbria Basin (Italy): Ancona, International Union of Geological Sciences Commision on Stratigraphy; International Subcommission on Paleogene Stratigraphy Report, p. 249–252.

NOCCHI, M., PARISI, G., MONACO, P., MONECHI, S., AND MADILE, M., 1988b, Eocene and Early Oligocene micropaleontology and paleoenvironments in SE Umbria, Italy: Palaeogeography, Palaeoclimatology, Palaeoecology, v. 67, p. 181–244.

OBERHÄNSLI, H., GRÜNIG, A., AND HERB, R., 1984, Oxygen and carbon isotope study in the Late Eocene sediments of Possagno (northern Italy): Rivista Italiana de Paleontologia e Stratigrafia, v. 89, 377–399.

ODIN, G. S. AND MONTANARI, A., 1988, The Eocene-Oligocene boundary at Massignano (Ancona, Italy): a potential boundary stratotype, *in* Premoli Silva, I., Coccioni, R., and Montanari, A., eds., The Eocene-Oligocene Boundary in the Marche-Umbria Basin (Italy): Ancona, International Union of Geological Sciences Commision on Stratigraphy; International Subcommission on Paleogene Stratigraphy Report, p. 253–263.

ODIN, G. S. AND MONTANARI, A., 1989, Age radiométrique et stratotype de la limite Éocène-Oligocène: Comptes Rendues de Academie Scientifique de Paris, 309, series II, p. 1939–1945.

OKADA, H. AND BUKRY, D., 1980, Supplementary modification and introduction of code numbers to the low latitude coccolith biostratigraphy zonation: Marine Micropaleontology, v. 5, p. 321–324.

PICCOLI, G. AND SAVAZZI, E., 1983, Five shallow benthic molusc faunas from the upper Eocene (Baron, Priabona, Garoowa, Nanggulan, Takashima): Bolletino della Società Paleontologica Italiana, v. 22, p. 31–47.

POMEROL, C., ed., 1981, Stratotypes of Paleogene stages: Paris, Bulletin d'Information des Géologues de Bassin de Paris Mémoir 2, 305 p.

POMEROL, C. AND PREMOLI SILVA, I., eds., 1986a, Terminal Eocene Events: Amsterdam, Elsevier, Developments in Stratigraphy 9, 420 p.

POMEROL, C. AND PREMOLI SILVA, I., 1986b, The Eocene-Oligocene transition: events and boundary, *in* Pomerol, C. and Premoli Silva, I., eds, Terminal Eocene Events: Amsterdam, Elsevier, Developments in Stratigraphy 9, p. 1–24.

POORE, R. Z., TAUXE, L., PERCIVAL, S. F., AND LABRECQUE, J. L., 1982, Late Eocene-Oligocene magnetostratigraphy and biostratigraphy at South Atlantic DSDP Site 522: Geology, v. 10, p. 508–511.

PREMOLI SILVA, I., COCCIONI, R., AND MONTANARI, A., eds., 1988a, The Eocene-Oligocene Boundary in the Marche-Umbria Basin (Italy): Ancona, International Union of Geological Sciences Commision on Stratigraphy; International Subcommission on Paleogene Stratigraphy Report, 268 p.

PREMOLI SILVA, I., ORLANDO, M., MONECHI, S., MADILE, M., NAPOLEONE, G., AND RIPEPE, M., 1988b, Calcareous plankton biostratigraphy and magnetostratigraphy at the Eocene-Oligocene transition in the Gubbio area, *in* Premoli Silva, I., Coccioni, R., and Montanari, A., eds., The Eocene-Oligocene Boundary in the Marche-Umbria Basin (Italy): Ancona, International Union of Geological Sciences Commision on Stratigraphy; International Subcommission on Paleogene Stratigraphy Report, p. 137–161.

SETIAWAN, J. R., 1983, Foraminifera and microfacies of the type Priabonian: Utrecht Micropaleontological Bulletins, v. 29, 161 p.

SIROTTI, A., 1978, Discocyclinidae from the Priabonian Type-section (Lessini Mountains, Vicenza, Northern Italy): Bolletino della Società Paleontologica Italiana, v. 17, p. 49–67.

SOPER, N. J. AND COSTA, L. I., 1978, Palynological evidence for the age of Tertiary basalts and post-basaltic sediments at Kap Dalton, central East Greenland: Rapporter Grønlands Geologiske Undersøgelse, v. 80, p. 123–127.

STOVER, L. E. AND HARDENBOL, J., 1993, Dinoflagellates and depositional sequences in the Lower Oligocene (Rupelian) Boom clay Formation, Belgium: Bulletin van de Belgische Vereniging voor Geologie, v. 102, p. 5–78 (published in 1994).

STOVER, L. E., WILLIAMS, G. L., AND EATON, G. L., 1988, Morphology of the Palaeogene dinoflagellate cyst genus *Areosphaeridium* Eaton 1971 (abs.): Brisbane, VII International Palynological Conference, p. 157.

SUESS, E., 1868, Über die Gliederung des Vicentinischen Tertiär Gebirges: Sitzungsberichte der Königlichen Akademie der Wissenschaften, v. 58, p. 265–280.

VAIL, P. R., MITCHUM, R. M., JR., TODD, R. G., WIDMIER, J. M., THOMPSON, S., III, SANGREE, J. B., BUBB, J. N., AND HATLELID, W. G., 1977, Seismic stratigraphy and global changes of sea level, *in* Payton, C. E., ed., Seismic Stratigraphy — Applications to Hydrocarbon Exploration: Tulsa, American Association of Petroleum Geologists Memoir 26, p. 49–212.

VANDANIELS, C. H., GRAMANN, F., AND KÖTHE, A., 1993, The Oligocene "Septarienton" of Lower Saxony. Biostratigraphy of an equivalent of the Boom Formation of Belgium, with special considerations to its upper and lower boundaries: Bulletin van de Belgische Vereniging voor Geologie, v. 102, p. 79–89 (published in 1994).

VANDENBERGHE, N., 1981, Rupelian, *in* Pomerol, C., ed., Stratotypes of Paleogene stages: Paris, Bulletin d'Information des Géologues de Bassin de Paris Mémoir 2, p. 203–217.

VANDENBERGHE, N. AND VANECHELPOEL, E., 1987, Field guide to the Rupelian Stratotype: Bulletin van de Belgische Vereniging voor Geologie, v. 96, p. 325–337.

VANECHELPOEL, E., VANDENBERGHE, N., AND LAENEN, B., 1992, Cyclostratigraphy of the Boom Clay Formation (Rupelian, Belgium) (abs.): Dyon, Meeting on Sequence Stratigraphy of European Basins, p. 428–429.

VERHALLEN, P. J. J. M. AND ROMEIN, A. J. T., 1983, Calcareous nannofossils from the Priabonian stratotype and correlations with the Parastratotypes, *in* Setiawan, J. R., ed., Foraminifera and Microfacies of the Type-Priabonian: Utrecht Micropaleontological Bulletins, v. 29, p. 163–173.

WEYNS, W., 1970, Dinophycées et acritarches des "Sables de Grimmertingen" dans leur localité-type, et les problèmes stratigraphiques de Tongrien: Bulletin de Societé Belge de Géologie, Paléontologie et Hydrologie, v. 79, p. 247–268.

WILLIAMS, G. L., 1977, Dinocysts: their classification, biostratigraphy and palaeoecology, *in* Ramsay, A. T. S., ed., Oceanic Micropalaeontology: London, Acadamic Press, p. 1231–1325.

WILLIAMS, G. L. AND BUJAK, J. P., 1985, Mesozoic and Cenozoic dinoflagellates, *in* Bolli, H. M., Saunders, J. B., and Perch-Nielsen, K., eds., Plankton Stratigraphy: Cambridge, Cambridge University Press, p. 847–964.

ZAPORCEK, N. J., 1989, Upper Eocene and Lower Oligocene Palynocomplexes and phyto-plankton from Borehole no. 1., Landzar (Armenia, U.S.S.R.), *in* Phanerozoic Paleoflora and stratigraphy: Moscow, Editions of the Soviet Academy of Sciences, p. 85–103 (in Russian).

GEOCHRONOLOGY AND MAGNETOSTRATIGRAPHY OF PALEOGENE NORTH AMERICAN LAND MAMMAL "AGES": AN UPDATE

DONALD R. PROTHERO

Department of Geology, Occidental College, Los Angeles, CA 90041

ABSTRACT: Laser-fusion ^{40}Ar/^{39}Ar dating and magnetostratigraphy have significantly changed our conception of the temporal duration and correlation of Paleogene North American land mammal "ages." The Wood Committee (1941) originally divided the Paleocene Epoch into five land mammal "ages." Current age estimates of their time spans are: Puercan, 65–63.8 Ma; Torrejonian (including the "Dragonian"), 63.8–61 Ma; Tiffanian, 61–56 Ma; and Clarkforkian, 56–55.2 Ma. The Paleocene/Eocene boundary, long placed in the Clarkforkian, occurs in the earliest Wasatchian, based on correlations using mammals, pollen, and terrestrial carbon isotopes.

The Wood Committee (1941) divided the North American Eocene Epoch into four land mammal "ages": Wasatchian (originally thought to be early Eocene), Bridgerian (thought to be middle Eocene), and Uintan and Duchesnean (both once thought to be late Eocene). The earliest Wasatchian is now considered Paleocene age, and the Wasatchian/Bridgerian boundary is about 50.4 Ma in age. The Bridgerian, Uintan and Dudnesnean land mammal "ages" are all middle Eocene age. The Bridgerian/Uintan boundary occurs in magnetic Chron C21n, about 47 Ma. The Uintan/Duchesnean boundary occurs within Chron C18n, and lies above an ash dated at about 40 Ma. The Duchesnean/Chadronian boundary lies within Chron C16n, about 37 Ma.

Finally, the Wood Committee (1941) divided their concept of North American Oligocene sequence into three land mammal "ages": the Chadronian, Orellan and Whitneyan (supposedly early, middle, and late Oligocene). The Chadronian/Orellan transition occurs just above a date of 33.9 Ma, late in Chron C13r; it is slightly younger than the Eocene/Oligocene boundary, and this makes the Chadronian mostly late Eocene, not early Oligocene age. The Orellan/Whitneyan boundary occurs in the middle of Chron C12r, just below a date of 31.8 Ma. The Whitneyan/Arikareean boundary occurs within Chron C11n, above a date of 30.0 Ma. Consequently, the Orellan and Whitneyan are both early Oligocene, and most of the Arikareean (long considered early Miocene) is late Oligocene age. These new age estimates and correlations differ greatly from the time scales published as recently as 1987.

INTRODUCTION

In North America, the only practical method of correlation and dating of most Cenozoic terrestrial deposits has been with land mammals. At the turn of the century, William Diller Matthew and Henry Fairfield Osborn (Matthew, 1899; Osborn and Matthew, 1909; Osborn, 1907, 1910, 1929) attempted to create biostratigraphic zonations of the North American terrestrial sequence, based on principles followed by European stratigraphers at the time (see review by Tedford, 1970). But the good beginning established by Osborn and Matthew was lost, since the next generation of vertebrate paleontologists virtually ignored their pioneering work.

Unlike most marine invertebrates, fossil mammals typically occur in localized fossiliferous horizons, or in isolated pockets or quarries without stratigraphic superposition. As a result, traditional biostratigraphic methods developed by European invertebrate paleontologists (based on detailed stratigraphic ranges of fossils in measured sections) were not widely followed by North American vertebrate paleontologists (nor are they followed by European vertebrate paleontologists, then or now). In 1937 the Vertebrate Paleontology Section of the Paleontological Society appointed a committee to clarify the confusion over correlation, and adopt a terminology that could be widely used. Known as the Wood Committee (Wood and others, 1941), it was chaired by Horace E. Wood II, and included four other vertebrate paleontologists (Edwin H. Colbert, John Clark, Glenn L. Jepsen, and Chester Stock), plus paleobotanist Ralph Chaney, and invertebrate paleontologist J. B. Reeside.

The Wood Committee's "Provincial Ages" were a complex hybrid of local rock units and time units delineated by index taxa, characteristic taxa, and first and last occurrences of mammalian genera. As Tedford (1970) and Woodburne (1977, 1987) have pointed out, these units were not true geochronological ages, which must be based on biostratigraphic zones and stages (according to western stratigraphic codes, such as the 1983 North American Code of Stratigraphic Nomenclature). Since

they were not standard stratigraphic ages, the North American land mammal "ages" should properly be put in quotes in most publications. Instead, the Wood Committee's methods resemble a system called "biochronology," first proposed by H. S. Williams (1901), which attempts to reconstruct the sequence of occurrences of taxa without documenting every event in a local stratigraphic section. Demonstration of stratigraphic superposition was noted when available, but detailed biostratigraphic work in the tradition of European invertebrate paleontologists was not considered essential to their "provisional" system.

In spite of their loose characterization, the land mammal "ages" of the Wood Committee worked quite well for over forty years, mostly because mammals do evolve and disperse very rapidly (Savage, 1977). However, problems eventually arose. For example, the Chadronian land mammal "age" was originally defined both on the co-occurrence of the horse *Mesohippus* and brontotheres, and also on the limits of the Chadron Formation. At the time, the last occurrence of brontotheres was thought to coincide with the top of the Chadron Formation, so there was no conflict. When Morris Skinner discovered brontothere specimens in rocks correlative with the overlying Orella Member of the Brule Formation (which typified the Orellan), the difficulty with defining the Chadronian both biochronologically and lithostratigraphically became apparent (Emry and others, 1987; Prothero, 1982; Evanoff and others, 1992). Yet many of Skinner's contemporaries could not accept this evidence, since the misconception that rocks units could be treated as time units was widespread among paleontologists of that generation.

Since the 1950s, a newer generation of paleontologists (see Savage, 1955, 1962, 1977; Tedford, 1970; Woodburne, 1977, 1987) has tried to bring vertebrate paleontology back to more rigorous classical biostratigraphic methods. Trained in modern stratigraphic thinking, vertebrate biostratigraphers now appreciate the possibility that rock units can be time-transgressive over distance (Shaw, 1964; Prothero, 1990), and so they rarely confuse rock units with time units. In addition, we have come to realize that detailed stratigraphic zonations of mammal fossils provides much higher resolution of time than collections whose only stratigraphic information is the formation they came from. As discussed by Woodburne (1977), such detailed zonation could potentially subdivide the Cenozoic Era into increments of time of 300,000 years or less. In recent years, portions of the North American continental Tertiary have been formally subdivided by proper biostratigraphic methods. However, the zonation of the entire Cenozoic Era is still in progress, and biochronological methods are still widely used.

Although mammalian paleontologists are increasingly moving toward classical biostratigraphic procedures, there are still problems. For example, biostratigraphic "zonations" proposed by Gingerich (1980, 1983), Sloan (1987), and Gunnell (1989) do not meet all the criteria established by the North American Code of Stratigraphic Nomenclature. Typically, these "zones" do not have formally proposed type sections (required in Article 54e), and in some cases the actual local stratigraphic ranges of key taxa are not clearly indicated. Consequently, they are still "biochrons" based on the abstract first and last occurrences of taxa, not true biostratigraphic zones and stages, which must be based on local ranges of fossils in specific sections.

Even the review of the Paleocene land mammal chronology by Archibald and others (1987), which attempted to rigorously define a number of different biostratigraphic zones, did not tie these to specific levels in a local biostratigraphic zonation. Instead, these authors relied on the Wood Committee's practice of referring a number of different localities to a specific "zone" and listing taxa of biochronological importance. Rather than follow the stricter criteria of the North American Stratigraphic Code, they adopted lineage-zones and interval-zones in the looser sense of the International Stratigraphic Guide (Hedberg, 1976), which were not tied to local biostratigraphic zonations. Archibald and others (1987, p. 25) acknowledge that "these ages (and zones) are based on faunal content that in many instances cannot be defined with precision in type sections . . . for the most part these units cannot be regarded as stages. This is, of course, one of the goals for the future."

In addition to higher-resolution biostratigraphy, other techniques have come along to improve terrestrial correlation. The original application of K-Ar dating by Evernden and others (1964) to the North American terrestrial record provided an independent test of the Wood Committee sequence and showed it to be substantially correct. For over 25 years, K-Ar methods continued to refine the chronology, and provided numerical estimates of the age for most of the sequence (Savage, 1977; Savage and Russell, 1983; Woodburne, 1987). However, in the last five years, the development of high-precision single-crystal $^{40}Ar/^{39}Ar$ dating (McDougall and Harrison, 1988; Swisher, this volume) showed that many of the classical K-Ar dates must be recalibrated. In some cases, they have significantly changed the chronology that was accepted for decades (Swisher and Prothero, 1990; Prothero and Swisher, 1992; Berggren and others, this volume).

Another breakthrough came from the application of magnetic stratigraphy to terrestrial sections. Unlike any other method of correlation, magnetic stratigraphy can supply many globally synchronous, numerically dated time horizons to terrestrial sections (Lindsay and others, 1987; Prothero, 1988, 1990; Opdyke, 1990). The combination of both magnetic stratigraphy and $^{40}Ar/$ ^{39}Ar dating has provided much higher temporal resolution and precision than was thought possible just 20 years ago. More importantly, magnetic stratigraphy is the only technique that allows direct correlation with the global polarity record and thus with the marine time scale. This in turn allows us to make direct comparison between global climatic changes, diversity fluctuations, and mass extinctions and the North American terrestrial record for the first time.

THE NORTH AMERICAN TERRESTRIAL PALEOGENE "AGES"

Paleocene Epoch

In 1941, the Wood Committee created five land mammal "ages" for an interval they considered approximately equivalent to the Paleocene Epoch (Fig. 1). The first two, the Puercan and Torrejonian, were based on faunas from the San Juan Basin in

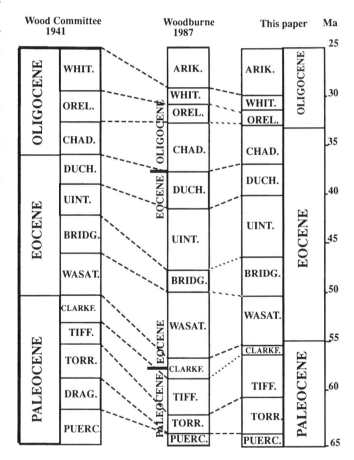

FIG. 1.—Comparison of the correlation of Paleogene North American land mammal "ages" from their original formulation by the Wood Committee (1941) to the present. K-Ar dating was unavailable in 1941, so the Wood Committee had no means of estimating numerical ages. Note that the Arikareean had become mostly late Oligocene age by the time of Woodburne (1987), and the Dragonian had been abandoned. The most important recent changes (besides the new numerical age estimates) is the shifting of the Chadronian from the early Oligocene to the late Eocene age, and the placement of the Paleocene/Eocene boundary in the early Wasatchian, rather than the Clarkforkian. Abbreviations are as follows: ARIK. = Arikareean; BRIDG. = Bridgerian; CHAD. = Chadronian; CLARKF. = Clarkforkian; DRAG. = Dragonian; DUCH. = Duchesnean; OREL. = Orellan; PUERC. = Puercan; TIFF. = Tiffanian; TORR. = Torrejonian; UINT. = Uintan; WASAT. = Wasatchian; WHIT. = Whitneyan.

New Mexico. The controversial Dragonian "age" was based on the limited Dragon Canyon local fauna from the North Horn Formation of central Utah. Although correlative faunas were later found in New Mexico (Tomida, 1981), most paleontologists now consider the Dragonian to be the beginning of the Torrejonian (Tomida and Butler, 1980; Tomida, 1981; Sloan, 1987; Archibald and others, 1987). The Tiffanian was originally based on faunas from the San Juan Basin in Colorado, but since then the Tiffanian has become much better known from faunas in Montana and Wyoming. The Clarkforkian land mammal "age" was originally based on faunas from the Clark's Fork Basin on the Montana-Wyoming border. Originally considered latest Paleocene age by the Wood Committee (1941), in recent years it was thought to span the Paleocene/Eocene boundary based on correlations of plesiadapids from Europe and North America (Gingerich, 1976; Gingerich and Rose, 1977; Rose, 1981). As we shall see in the next section, the Paleocene/Eocene boundary is now thought to occur in the earliest Wasatchian, so not only the Clarkforkian but also the earliest Wasatchian are now considered Paleocene age.

North American Paleocene chronology was most recently summarized by Archibald and others (1987). Most of the Paleocene Epoch has now been subdivided into a series of biostratigraphic "zones," abbreviated "Pu1, Pu2, Pu3, To1" ("Puercan 1, 2, 3, Torrejonian 1") and so on (Fig. 2). This scheme was originally introduced by Gingerich (1976, 1980, 1983) based on a local zonation of adapid primates and plesiadapids from Tiffanian and Clarkforkian of the northern Bighorn Basin of Wyoming, and expanded to the Puercan and Torrejonian strata by Archibald and others (1987) and by Sloan (1987).

Archibald and Lofgren (1990) added an additional "interval-zone," Pu0, for the earliest Paleocene beds above the K/T boundary in eastern Montana (called the "Bugcreekian" by Sloan and others, 1986). As noted above, these "zones" do not meet the criteria of the North American Code of Stratigraphic Nomenclature since they lack type sections. Although there are some difficulties with this zonation (see Schankler, 1980, 1981), the scheme has been modified with additional mammalian groups and has been widely adopted (Archibald and others, 1987). There are four Puercan "zones," three Torrejonian "zones," six Tiffanian "zones," and three Clarkforkian "zones" (plus Wasatchian 0, which is also latest Paleocene).

Relatively few Paleocene radiometric dates are available. The most recent dates on the Cretaceous/Tertiary boundary place its age at 65 Ma (Berggren and others, this volume). Revisions to the dating of the magnetic polarity time scale and new dates on the Eocene (Berggren and others, this volume) place the Paleocene/Eocene boundary at about 55 Ma. Magnetic polarity stratigraphy (Fig. 2) has been studied in the key sections of the Bighorn Basin of Wyoming and Montana (Butler and others, 1980, 1984, 1987; Clyde and others, 1994; Tauxe and others, 1995), the Crazy Mountain Basin in Montana (Butler and others, 1987), the San Juan Basin in New Mexico and Colorado (Butler and others, 1977; Lindsay and others, 1978, 1981; Taylor and Butler, 1980; Butler and Lindsay, 1985; Butler and others, 1987), Dragon Canyon in Utah (Tomida and Butler, 1980), and the Big Bend region in Texas (Rapp and others, 1983).

Based on magnetic stratigraphy from the San Juan and Crazy Mountain Basins, Butler and Lindsay (1985) and Butler and others (1987) squeezed the Puercan into Chron C29n and latest C29r (64–65 Ma), so Puercan "zones" Pu0-Pu3 are each about 250,000 years in duration. In the San Juan and Crazy Mountain Basins, Torrejonian "zone" To1 occurs in Chron C28n (62.5–63.5 Ma), To2 in C27r (61.3–62.5 Ma), and To3 in Chron C27n (61.0–61.3 Ma), so the three Torrejonian "zones" To1-To3 range from 0.3–1.2 million years in duration. The Torrejonian/Tiffanian boundary occurs early in Chron C26r, about 60.5 Ma. Based on magnetic stratigraphy from the northern Bighorn Basin, Butler and others (1980, 1984) showed that Ti2 occurs early in Chron C26r (about 59–60 Ma), and Ti3 in late Chron C26r and C26n (57.5–58.5 Ma). Ti4 occurs in earliest Chron C25r (57.0–57.3 Ma), and Ti5 in early Chron C25n (56.1–56.3 Ma). The Tiffanian/Clarkforkian boundary occurs in Chron C25n, about 56.2 Ma, so the five Tiffanian "zones" range from 0.2–1.0 million years in duration. The Clarkforkian/Wasatchian boundary occurs in the middle of Chron C24r, about 55.5 Ma, so the three Clarkforkian "zones" are each about 200,000 years in duration.

Eocene Epoch

The major difficulty in establishing the correlation of the North American terrestrial chronology with the marine-based global time scale is the lack of direct interfingering of mammal-bearing terrestrial deposits with marine sequences. Fortunately, the type areas of many of the European marine stages interfinger with mammal-bearing beds in both the Paris and London Basins (Savage and Russell, 1983). This allows direct correlation of the European Eocene mammalian chronology with the global time scale. During intervals of faunal interchange with North

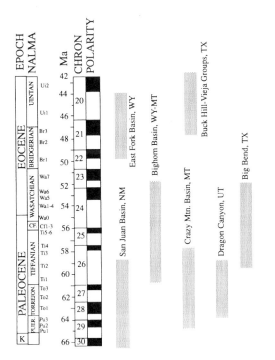

FIG. 2.—Calibration of early Paleogene North American land mammal "ages" with the revised magnetic polarity time scale of Berggren and others (this volume). Subdivisions of the "ages" (i.e., Pu1, Pu2, and so on) are discussed in the text. On the right, the temporal span (based on magnetostratigraphy) of some important Paleocene and early Eocene sections is shown.

America (such as in the early Eocene, but not the middle or late Eocene), we can correlate the North American sequence with the epoch stratotype sequences based in Europe. We also have magnetic stratigraphy and a large number of radiometric dates for many areas in the North American Eocene sequence. New radiometric dates, however, have greatly changed our notions of the temporal correlation of the North American sequences of middle and late Eocene age (Swisher and Prothero, 1990; Prothero and Swisher, 1992).

The Wood Committee recognized four land mammal "ages" in North America (Figs. 1–3), which they thought were approximately correlative with the European Eocene sequence. The Wasatchian land mammal "age" was named for the Wasatch Formation in basins of southern Wyoming. The Bridgerian land mammal "age" got its name from the faunas of the Bridger Formation in southwest Wyoming, and the Uintan and Duchesnean land mammal "ages" were named for the Uinta and Duchesne River Formations in the Uinta Basin of northeast Utah. The Wood Committee thought that the Wasatchian was early Eocene, the Bridgerian middle Eocene, and the Uintan and Duchesnean were late Eocene age. The next "age," the Chadronian (named after the Chadron Formation in the High Plains, especially Nebraska and South Dakota), was considered early Oligocene age.

As mentioned in the previous section, the Wood Committee and most subsequent authors placed the Paleocene/Eocene boundary within or at the end of the Clarkforkian in North America (Gingerich, 1976; Gingerich and Rose, 1977; Rose, 1981; Gunnell, 1989). However, a number of recent lines of

evidence suggest that the boundary actually falls within the earliest Wasatchian. The original correlations of Gingerich and Rose were based primarily on the interpretation of adapid primates and plesiadapids. Other taxa, such as pantodonts (Lucas, 1984, 1989, 1993; Rea and others, 1990) and omomyid primates (Beard and Tabrum, 1991) suggested that the boundary occurs in the earliest Wasatchian. Another datum is the first appearance of *Platycarya* pollen in the earliest Wasatchian of the northern Bighorn Basin (Wing, 1984; Wing and others, 1991). This palynological datum occurs at the NP9/NP10 nannofossil zone boundary in the Gulf Coast (Frederiksen, 1980), which is 300,000 years older than the Paleocene/Eocene boundary (Berggren, 1993). These correlations were borne out by the recent detection of the striking carbon isotopic event near the Paleocene/Eocene boundary (Rea and others, 1990; Kennett and Stott, 1991) in earliest Wasatchian terrestrial carbon isotopes extracted from paleosols and mammalian teeth and bones (Koch and others, 1992).

Yet some mammalian paleontologists (Gingerich, 1989; Gunnell and others, 1993) continue to place the Paleocene/Eocene boundary at the Clarkforkian/Wasatchian boundary. The key to the entire controversy lies in the interpretation of the Sparnacian mammalian faunas of Europe. In the past, European mammalian paleontologists had considered the Sparnacian the beginning of the Eocene Epoch, because its mammals were more similar to later Eocene faunas, and differed radically from the underlying upper Paleocene Thanetian faunas (Savage and Russell, 1983). The discovery of a new earliest Wasatchian fauna (Wa0) in North America with strong similarities to the Sparnacian in Europe supported that correlation (Gingerich, 1989, p. 83–87). Gunnell and others (1993) argued that the beginning of the Clarkforkian/Wasatchian transition was the time of greatest faunal turnover, and therefore it seemed to be the best place to mark the boundary.

But the Paleocene/Eocene boundary is defined on the basis of marine taxa in European type sections, not on the basis of North American or European mammalian faunas. Where the Clarkforkian/Wasatchian boundary in North America (or the Thanetian/Sparnacian boundary in Europe) falls in relation to marine stratotypes must be demonstrated, not asserted. One cannot assume the coincidence of mammalian faunal turnover and European epoch boundaries. For instance, the "Grande Coupure" in Europe was long thought to represent the Eocene/Oligocene boundary because of the great faunal turnover, but recent work (Hooker, 1992) has demonstrated that it actually occurred in the early Oligocene, about 1 my *after* the Eocene/Oligocene boundary.

Marine stratigraphers have long ago shown that the European Sparnacian correlates with late Paleocene nannoplankton zone NP9 (Costa and Downie, 1976; Costa and others, 1978; Berggren and others, 1985, this volume; Aubry and others, 1988). It is at least one or two sequences lower (and about 1 million years older) than the Paleocene/Eocene boundary as denoted by the base of the London Clay or the Argile d'Ypres (Berggren, 1993), long recognized as the base of the Ypresian and therefore the base of the Eocene Epoch (Berggren and others, 1985; Aubry and others, 1988).

Tauxe and others (1995) described the magnetic stratigraphy of the Willwood Formation in the southern Bighorn Basin of Wyoming, and also argued that the Paleocene/Eocene boundary

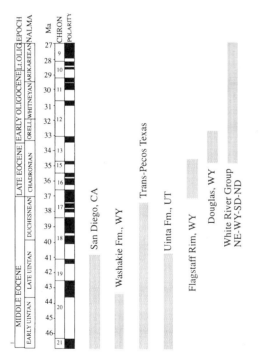

FIG. 3.—Calibration of late Paleogene North American land mammal "ages" with the revised magnetic polarity time scale of Berggren and others (this volume). On the right, the temporal span (based on magnetostratigraphy) of some important middle-late Eocene and Oligocene sections is shown. Abbreviations as in Figure 1.

might correlate with the Clarkforkian/Wasatchian boundary. The only direct evidence to support their argument is that the *Platycarya* datum apparently occurs 35 meters *below* the first occurrence of a Wa0 fauna in the southern Bighorn Basin, although it occurs 160 meters *above* the Wa0 mammals in the northern Bighorn Basin. Thus, Tauxe and others (1995) suggest that the range of *Platycarya* (and thus the Paleocene/Eocene boundary) has been extended downward. However, as we saw above, the *Platycarya* datum and the NP9/NP 10 boundary are not at the Paleocene/Eocene boundary, but about 300,000 years older (Berggren, 1993).

The most recent summary of most of the Eocene North American land mammal "ages" (Krishtalka and others, 1987) did not attempt to formally divide the entire interval into biostratigraphic stages. However, biostratigraphic zonations for the Wasatchian and Bridgerian have been proposed by Savage (1977), Gingerich (1976, 1980, 1983), Gingerich and Simons (1977), Schankler (1980), Stucky (1984), and Gunnell (1989). Some of these biostratigraphic schemes (especially those of Savage, 1977, Schankler, 1980, and Stucky, 1984) follow the 1983 North American Stratigraphic Code in specifying type sections, but others do not. Currently, the Wasatchian is divided into eight "zones," labeled Wa0 to Wa7 (Gingerich, 1983, 1989). Clyde and others (1994) found that Wa0 to lower Wa5 occurred in Chron C24r (53.5–55 Ma) in the northern Bighorn Basin, and that upper Wa5 and Wa6 correlated with C24n (52.3–53.4 Ma). Wa7 occurred in C23r (51.6–52.3 Ma). These ages are in good agreement with a new ^{40}Ar/^{39}Ar date on Wa6–7 of 52.8 ± 0.3 Ma (Woodburne and Swisher, this volume).

No magnetic stratigraphy for the Wasatchian-Bridgerian transition has ever been published. Good sections are available in the Wind River, Green River, and Huerfano Basins (see Krishtalka and others, 1987, p. 86–87), so eventually it should be possible to tie these areas to the magnetic polarity time scale. Based on K-Ar dates of 50.5 Ma on the latest Wasatchian and 50.3 Ma on the earliest Bridgerian in the Wind River Formation (Evernden and others, 1964; Krishtalka and others, 1987, p. 93, dates T and U), the boundary probably occurs around 50.4 Ma, or within Chron C22r (Woodburne, this volume). Recently, Clyde and others (1995) reported on the magnetic stratigraphy of the Wasatchian-Bridgerian transition in the Green River Basin, Wyoming. They found this transition in Chron C22r and suggested that it occurs between 49.7 Ma and 50.7 Ma.

After the Wasatchian, North America was separated from direct faunal interchange with Europe, and the correlations become more indirect. As a result, the most effective technique has been radiometrically-dated magnetic stratigraphy. The magnetic stratigraphy of the type Bridgerian in the Bridger Basin of Wyoming has been studied, but not published (Jerskey, pers. commun., 1981), and is now being restudied (Flynn, pers. commun., 1994). The magnetic stratigraphy of the classic Uintan and Duchesnean sections in the Uinta Basin of Utah was described by Prothero and Swisher (1992) and Prothero (1995a). Bridgerian and Uintan beds have been studied in California and Wyoming by Flynn (1986) and by McCarroll and others (1993), and in Texas by Walton (1992) and Prothero (1995b). Uintan-Duchesnean beds in the Sand Wash Basin of Colorado (Stucky and others, 1995), the Sespe Formation of Ventura County, California (Prothero and others, 1995), the Poway Group in San Diego County, California (Walsh and others, 1995), the Galisteo Formation in central New Mexico (Prothero and Lucas, 1995) and several units in western Montana (Tabrum and others, 1995) have also been sampled.

According to Flynn (1986), the Bridgerian-Uintan transition occurs early in Chron C20r, about 46 Ma. However, recent studies of the critical sections in San Diego suggest that the transition might lie within Chron C21n, about 47 Ma (Walsh, 1995; Walsh et al., 1995). The Uintan-Duchesnean transition occurs within Chron C18n, about 40 Ma (Prothero and Swisher, 1992; Prothero, 1995a, 1995b; Prothero and others, 1995). The latest stratigraphic correlations of the middle Eocene Bartonian and late Eocene Priabonian stages in Europe (Berggren and others, 1985; Aubry and others, 1988) place the Priabonian/Bartonian boundary in Chron C17n1 (Berggren and others, this volume), so that the middle Eocene includes not only the Bridgerian, but also the Uintan and Duchesnean (Figs. 1, 3).

Gunnell (1989) named two "zones" (Ui1, the *Epihippus* assemblage "zone," and Ui2, the camelid-canid appearance "zone") for the early and late Uintan. However, unlike earlier Eocene zones, these are not based on recent detailed biostratigraphic work, but simply formalize the distinction between the faunas of Uinta Formation Member "B" (upper part of the Wagonhound Member of Wood, 1934) and Member "C" (Myton Member of Wood, 1934). More detailed biostratigraphic work to subdivide the seven million years of the Uintan would be valuable, but the stratigraphic data on the existing collections from the Uinta Basin are too imprecise, and the area is too poorly fossiliferous to make significant new collections with good stratigraphic data (Prothero and Swisher, 1992; Prothero, 1995a). The biostratigraphy of existing Uinta Formation collections was summarized by Prothero (1995a). Ultimately, it would be better to subdivide the Uintan by precisely dating stratigraphically superposed faunas in Texas (from the Vieja and Buck Hill Groups in the Big Bend region), Colorado (the Sand Wash Basin), Wyoming (the Washakie Basin), and California (the Sespe and San Diego sections). However, this effort is also hampered by the high degree of endemism of Uintan faunas, making true first and last occurrences difficult to distinguish from local geographic effects (Lillegraven, 1979; Flynn, 1986; Krishtalka and others, 1987).

The Duchesnean has always been the most controversial land mammal "age" of the Eocene Epoch (Fig. 1). Originally considered the last of the Eocene ages by Wood and others (1941), opinions have fluctuated from placing it in the early Oligocene (Scott, 1945) or back in the late Eocene (Simpson, 1946), and back and forth several times since then. Gazin (1955, 1956, 1959) assigned the faunas from the lower Duchesne River Formation (Randlett and Halfway faunas) to the late Uintan. The poor quality of the overlying Lapoint fauna (the "classic" Duchesnean) and its similarity to some Chadronian faunas led several workers to reduce the Duchesnean to a subage of the Chadronian, or drop it altogether (Wilson, 1978, 1984, 1986; Emry 1981). In recent years, opinion has swung back toward recognizing a distinct Duchesnean, with important correlatives in Saskatchewan, South Dakota, Wyoming, Montana, Texas, New Mexico, Oregon, and California (Krishtalka and others, 1987; Kelly, 1990; Lucas, 1992).

Even though the Duchesnean spans almost three million years, efforts to subdivide it into biostratigraphic zones have been controversial since the faunas are so sparse and often endemic to the many scattered localities (Lucas, 1992). Several authors have attempted to recognize an early and late Duchesnean in Texas (Wilson, 1984, 1986) and California (Kelly,

1990). Further radiometric dating and magnetic stratigraphy will probably provide the best test of the age of these faunas, and may eventually help establish a biostratigraphic zonation for the Duchesnean.

The most surprising conclusion of recent dating, however, concerns the Chadronian land mammal "age." Considered early Oligocene age by the Wood Committee, it was K-Ar dated by Evernden and others (1964) at between 32 and 36 Ma. Since the Eocene/Oligocene boundary was generally placed around 36.5 Ma (Berggren and others, 1985), the correlation of the Chadronian with the early Oligocene seemed secure. However, the European Eocene/Oligocene boundary has been recently redated at 33.5 Ma (Berggren and others, this volume). At the same time, redating of K-Ar-dated ashes by $^{40}Ar/^{39}Ar$ methods have shown that the Chadronian spans an interval from 33–37 Ma (Swisher and Prothero, 1990; Prothero and Swisher, 1992). This places the Chadronian in the late Eocene, and the Eocene/Oligocene boundary appears to fall just below the Chadronian/Orellan boundary, not at the Duchesnean/Chadronian boundary, as long thought.

At the time of the Wood Committee report, the biostratigraphy of the Chadronian was very poorly understood. Many of the taxa (especially brontotheres and oreodonts) were badly oversplit taxonomically, and the stratigraphic data on the fossils from the classic collections in the Chadron Formation were inadequate (Emry and others, 1987). Attempts to divide the Chadronian based on lithostratigraphy of the Chadron Formation in Nebraska (Schultz and Stout, 1955) or South Dakota (Clark, 1937, 1954; Clark and others, 1967) were unsuccessful (Emry, 1973; Emry and others, 1987). Since that time, however, Emry (1973, 1992) has carefully documented the mammalian biostratigraphy in the thickest and most fossiliferous Chadronian sequence at Flagstaff Rim, Wyoming. Based on this work, Emry (1992; Emry and others, 1987) suggested criteria for recognizing early, middle, and late Chadronian, but no formal biostratigraphic zonation was proposed. Such a zonation is now in progress (Prothero and Emry, 1995).

Oligocene Epoch

The geochronology of the Eocene/Oligocene boundary has long been controversial. For decades, the dates produced by K-Ar methods on volcanics and by various isotopes in marine glauconites produced highly contradictory estimates ranging from 32 to 38 Ma (Hardenbol and Berggren, 1978; Berggren and others, 1985; Berggren, 1986; Aubry and others, 1988; Obradovich, 1988; Odin, 1978, 1982; Curry and Odin, 1982; Odin and Curry, 1985; Glass and Crosbie, 1982; Glass and others, 1986; summarized in Prothero, 1990, 1994b, and Berggren and others, 1992). However, recent $^{40}Ar/^{39}Ar$ dating of volcanic ashes in the deep marine sections in Gubbio and Massignano, Italy, directly tied to marine microfossils and magnetic stratigraphy (Montanari, 1988, 1990; Montanari and others, 1985, 1988; Odin and others, 1988, 1991) and of terrestrial volcanic ashes in North America (Swisher and Prothero, 1990; Prothero and Swisher, 1992) have resolved the controversy by eliminating many erroneous age estimates. The emerging consensus places the Eocene/Oligocene boundary at about 33.5 Ma (Berggren and others, this volume).

In addition to recalibrating the Eocene/Oligocene boundary, new $^{40}Ar/^{39}Ar$ methods provided radiometric dates on much of the rest of the North American Oligocene land mammal sequence for the first time. The Wood Committee (Wood and others, 1941) named three land mammal "ages" which they thought were approximately Oligocene in age (Fig. 1): the Chadronian, the Orellan (based on the Orella Member of the Brule Formation in Nebraska), and the Whitneyan (based on the Whitney Member of the Brule Formation in Nebraska). They informally considered the Chadronian land mammal "age" to be early Oligocene, the Orellan to be middle Oligocene, and the Whitneyan to be late Oligocene age. The Arikareean (based on the Arikaree Group in Nebraska) was thought to be early Miocene. As we have seen above, the Chadronian is now considered late Eocene age. The Orellan/Whitneyan boundary occurs at about 32 Ma, within Chron C12r, and the Whitneyan/Arikareean boundary within Chron C11n, about 30 Ma (Swisher and Prothero, 1990; Prothero and Swisher, 1992). This makes both the Orellan and Whitneyan early Oligocene, and most of the Arikareean becomes late Oligocene age (since the European type Oligocene has only two stages, the Rupelian and Chattian, there is no "middle" Oligocene). Magnetic stratigraphy has now been completed on virtually all the important Orellan, Whitneyan and early Arikareean outcrops from the White River and Arikaree Groups (Prothero, 1982; Prothero and others, 1983, Prothero, 1985a, 1985b; Prothero and Swisher, 1992; Evanoff and others, 1992; Prothero, 1995c; Tedford and others, 1995), as well as important localities in Montana (Prothero, 1984; Tabrum and others, 1995) and California (Prothero, 1991; Prothero and others, 1995).

The Oligocene/Miocene boundary is less well constrained than the boundaries discussed above. When the Wood Committee considered the Arikareean to be early Miocene, their concept of the Arikareean was based on the Agate Springs fauna, which is latest Arikareean; it does have taxa in common with the early Miocene in Europe. However, all of the underlying units in the Arikaree Group have a very different fauna, and the entire Arikareean appears to span an interval from 30 Ma to 21 Ma, almost nine million years—by far the longest land mammal "age" (Tedford and others, 1987, 1995). In addition, the definition of the Oligocene/Miocene boundary in Europe has fluctuated over the years but now seems stabilized around 23–24 Ma (Berggren and others, 1985, this volume). Recent $^{40}Ar/^{39}Ar$ dating now places the early Arikareean Gering Fauna between 28 and 30 Ma (Tedford and others, 1995), and most of the Monroe Creek Fauna also appears to be late Oligocene age (Tedford and others, 1987, 1995). The late Arikareean Harrison Fauna and the overlying latest Arikareean Agate Springs Quarry (in the Upper Harrison Formation of Peterson, 1909) may be earliest Miocene.

Unlike the detailed biostratigraphic "zonation" now in place for most of the Paleocene and parts of the Eocene (Gingerich, 1983; Gunnell, 1989; Archibald and others, 1987; Krishtalka and others, 1987), a detailed Oligocene biostratigraphy is just now being published. For decades, it was impossible to subdivide the Orellan or Whitneyan land mammal "ages," since most of the early collections had very poor stratigraphic data. The stratigraphic scheme of Schultz and Stout (1955, 1961) was actually based on lithostratigraphic units, and did not have biostratigraphic resolution within the units. Outlines of the biostratigraphic potential of the North American Oligocene sequences were presented by Emry and others (1987), but no detailed zonation was proposed. Thanks to decades of work by Morris Skinner, Bob Emry, and collectors of the Frick Laboratory, however, there are now large collections of White River mammals with stratigraphic data zoned to the nearest foot from volcanic ashes. These collections allowed Prothero (1982) to propose a preliminary biostratigraphy which divided the Orellan into four zones, and recognize one zone for the early Whitneyan. Korth (1989) also proposed a biostratigraphic zonation for the Orellan, but it is based on University of Nebraska collections which do not have the resolution to subdivide the lithostratigraphic units. Hence, Korth's (1989) "faunal zones" are really based on lithostratigraphic boundaries.

A finely resolved biostratigraphy for the Orellan and Whitneyan has long been in preparation, but its publication has been delayed by the lack of up-to-date systematic revisions of many of the key taxa, especially oreodonts, leptomerycids, and ischyromyid rodents. Such revisions are now in press (various papers in Prothero and Emry, 1995), and that volume suggests a formal biostratigraphic zonation for the Chadronian through early Arikareean (Prothero and Emry, 1995).

Rensberger (1971, 1973, 1983; Fisher and Rensberger, 1972) proposed a formal biostratigraphic zonation of rodents from the late Oligocene-early Miocene John Day Formation of central Oregon. The magnetostratigraphy of these beds has also been published (Prothero and Rensberger, 1985), although it may need further revision when new ^{40}Ar/^{39}Ar dates are analyzed by Carl Swisher. Unfortunately, some of the key rodent taxa used in Rensberger's zonation do not occur in the classic Arikareean faunas in the High Plains, so his biostratigraphic zones have proven useful only in Oregon, Montana, and South Dakota (Tedford and others, 1987). When the magnetic stratigraphy of the sequences containing the "type" Arikareean fauna in Nebraska is published (Hunt and MacFadden, pers. commun.), it will be possible to overcome these difficulties, and establish biostratigraphic zones for the Arikareean throughout the western United States.

CONCLUSION

Despite its limitations, correlation by fossil mammals is still the only practical method of dating most Cenozoic terrestrial deposits. In retrospect, the pioneering work of the Wood Committee (1941) was remarkably accurate in most of its correlations. However, new technologies not available until the last decade have significantly changed some of the important correlations. In particular, the Paleocene/Eocene boundary has moved up from the middle of the Clarkforkian into the early part of the Wasatchian, and the Eocene/Oligocene boundary shifted from the base to the top of the Chadronian. The Oligocene/Miocene boundary has moved up to include most of the Arikareean. Although these changes may seem minor to the non-specialist, they require major adjustments in the thinking of several generations of paleontologists who were trained to equate Chadronian with early Oligocene, Arikareean with early Miocene, and so on. Those who have long talked about huge "Oligocene" brontotheres must now get used to the fact that there are probably no Oligocene brontotheres at all (not even in Asia, see Berggren and Prothero, 1992). As uncomfortable as that may be, it is required by the data that have now emerged. More importantly, such major shifts in the time scale strongly affect all studies of evolutionary patterns, rates of sedimentation, climatic changes, and other geologic processes that depend upon a particular correlation scheme or version of the time scale (e.g., Berggren and Prothero, 1992; Prothero, 1994a).

In the past, some vertebrate paleontologists have tried to salvage their outmoded concepts by referring to the Duchesnean as the "North American late Eocene," or the Arikareean as "North American early Miocene," or similar evasions. This is indefensible, since the Lyellian epochs are strictly a European marine concept, and the global time scale is based on European, not North American, chronostratigraphy. The duration of the epochs in North America is only known by correlation to an independent North American chronology, and not by redefining the European epochs in North American terms to rescue obsolete notions. Indeed, the strength of the original North American land mammal chronology lies in its independence. The relative sequence of land mammal "ages" remains the same, regardless of where the European epoch boundaries fall, as the Wood Committee (1941) realized.

Although North American vertebrate paleontologists are making encouraging attempts to construct formal biostratigraphies consistent with standard biostratigraphic methods, some problems remain. Considering all the detailed work done by some paleontologists to construct their "zonations," it is rather surprising that they did not finish the job and propose their "zones" in compliance with the North American Stratigraphic Code. These same paleontologists would be appalled if someone did not follow the International Code of Zoological Nomenclature and proposed taxa that had no type specimen and became *nomina nuda*, yet they have done something analogous with their "biostratigraphic zonations." In some places where the fossils occur in isolated quarries without much vertical biostratigraphic ranges, the reason for this non-compliance with the North American Stratigraphic Code is potentially excusable. However, Gingerich (1976; Gingerich and Rose, 1977), Sloan (1987), and Archibald and others (1987) abundantly document much of the stratigraphic detail necessary for formal type sections, yet do not take this final step.

With the detailed documentation now available, we should soon be able to replace the informal biochronological schemes used by Archibald and others (1987), Krishtalka and others (1987), and Emry and others (1987) with formal range-zone biostratigraphy. When a formal biostratigraphic basis for all the North American land mammal "ages" is established, they will become true stratigraphic stages. At that point, we will no longer need to apologize for our system with the ubiquitous quotes around the word "age." More importantly, tying the informal biochronology to specific sections will make our task of locating magnetic polarity zones and radiometric dates much easier. Half a century of ignoring standard biostratigraphic practice is enough!

ACKNOWLEDGMENTS

I thank Bill Berggren for inviting me to the 1993 SEPM symposium for which this chapter was written. I thank Lisa Tauxe for faxing me a preprint of her paper. Dave Archibald, John Flynn, Dave Krause, Jay Lillegraven, Richard Stucky, and Mike Woodburne provided careful and thoughtful reviews of this manuscript, although they do not necessarily endorse all the conclusions.

REFERENCES

ARCHIBALD, J. D., CLEMENS, W. A., GINGERICH, P. D., KRAUSE, D. W., LIND-SAY, E. H., AND ROSE, K. D., 1987, First North American land mammal ages of the Cenozoic Era, *in* Woodburne, M. O., ed., Cenozoic Mammals of North America: Geochronology and Biostratigraphy: Berkeley, University of California Press, p. 24–76.

ARCHIBALD, J. D. AND LOFGREN, D. L., 1990, Mammalian zonation near the Cretaceous-Tertiary boundary: Boulder, Geological Society of America Special Paper 243, p. 31–50.

AUBRY, M.-P., BERGGREN, W. A., KENT, D. V., FLYNN, J. J., KLITGORD, K. D., OBRADOVICH, J. D., AND PROTHERO, D. R, 1988, Paleogene geochronology: an integrated approach: Paleoceanography, v. 3, p. 707–742.

BEARD, K. C. AND TABRUM, A. R., 1991, The first early Eocene mammal from eastern North America: An omomyid primate from the Bashi Formation, Lauderdale County, Mississippi: Mississippi Geology, v. 11, p. 1–6.

BERGGREN, W. A., 1986, Geochronology of the Eocene/Oligocene boundary, *in* Pomerol, C. and Premoli-Silva, I., eds., Terminal Eocene Events: Amsterdam, Elsevier, p. 349–356.

BERGGREN, W. A., 1993, NW European and NE Atlantic Paleocene-Eocene boundary interval: bio- and sequence stratigraphy and geochronology: Journal of Vertebrate Paleontology, v. 13, supplement to no. 3, p. 36A.

BERGGREN, W. A., KENT, D. V., AND FLYNN, J. J., 1985, Paleogene geochronology and chronostratigraphy, *in* Snelling, N. J., ed., The Chronology of the Geological Record: London, Memoir of the Geological Society of London, v. 10, p. 141–195.

BERGGREN, W. A., KENT, D. V., OBRADOVICH, J. D., AND SWISHER, C. C., III, 1992, Toward a revised Paleogene geochronology, *in* Prothero, D. R. and Berggren, W. A., eds., Eocene-Oligocene Climatic and Biotic Evolution: Princeton, Princeton University Press, p. 29–45.

BERGGREN, W. A. AND PROTHERO, D. R., 1992, Eocene-Oligocene climatic and biotic evolution: an overview, *in* Prothero, D. R. and Berggren, W. A., eds., Eocene-Oligocene Climatic and Biotic Evolution: Princeton, Princeton University Press, p. 1–28.

BUTLER, R. F., GINGERICH, P. D., AND LINDSAY, E. H., 1980, Magnetic polarity stratigraphy and Paleocene-Eocene biostratigraphy of Polecat Bench, Wyoming: University of Michigan Papers in Paleontology, v. 24, p. 95–98.

BUTLER, R. F., GINGERICH, P. D., AND LINDSAY, E. H., 1984, Magnetic polarity stratigraphy and biostratigraphy of Paleocene and lower Eocene continental deposits, Clark's Fork Basin, Wyoming: Journal of Geology, v. 89, p. 299–316.

BUTLER, R. F., KRAUSE, D. W., AND GINGERICH, P. D., 1987, Magnetic polarity stratigraphy and biostratigraphy of middle-late Paleocene continental deposits of south-central Montana: Journal of Geology, v. 95, p. 647–657.

BUTLER, R. F. AND LINDSAY, E. H., 1985, Mineralogy of magnetic minerals and revised magnetic polarity stratigraphy of continental sediments, San Juan Basin, New Mexico: Journal of Geology, v. 93, p. 535–554.

BUTLER, R. F., LINDSAY, E. H., JACOBS, L. L., AND JOHNSON, N. M., 1977, Magnetostratigraphy of the Cretaceous-Tertiary boundary in the San Juan Basin, New Mexico: Nature, v. 267, p. 318–323.

CLARK, J., 1937, The stratigraphy and paleontology of the Chadron Formation in the Big Badlands of South Dakota: Annals of the Carnegie Museum of Natural History, v. 25, p. 261–350.

CLARK, J., 1954, Geographic designation of the members of the Chadron Formation in South Dakota: Annals of the Carnegie Museum of Natural History, v. 33, p. 197–198.

CLARK, J., BEERBOWER, J. R., AND KIETZKE, K. K., 1967, Oligocene sedimentation, stratigraphy, paleoecology, and paleoclimatology in the Big Badlands of South Dakota: Fieldiana Geology Memoir, v. 5, p. 1–158.

CLYDE, W. C., STAMATAKOS, J., AND GINGERICH, P. D., 1994, Chronology of the Wasatchian land-mammal age (early Eocene): magnetostratigraphic results from the McCullough Peaks section, northern Bighorn Basin, Wyoming: Journal of Geology, v. 102, p. 367–377.

CLYDE, W. C., STAMATAKOS, J., ZONNEVELO, J.-P., GUNNELL, G. P., AND BARTELS, W. S., 1995, Timing of the Wasatchian-Bridgerian (Early-Middle Eocene) faunal transition in the Green River Basin, Wyoming: Journal of Vertebrate Paleontology, v. 15(3), p. 24A.

COSTA, L. I., DENISON, C., AND DOWNIE, C., 1978, The Palaeocene/Eocene boundary in the Anglo-Paris Basin: Journal of the Geological Society of London, v. 135, p. 261–264.

COSTA, L. I. AND DOWNIE, C., 1976, The distribution of the dinoflagellate *Wetzeliella* in the Palaeogene of north-western Europe: Palaeontology, v. 19, p. 591–614.

CURRY, D. AND ODIN, G. S., 1982, Dating of the Paleogene, *in* Odin, G. S., ed., Numerical Dating in Stratigraphy: New York, John Wiley, p. 607–630.

EMRY, R. J., 1973, Stratigraphy and preliminary biostratigraphy of the Flagstaff Rim area, Natrona County, Wyoming: Smithsonian Contributions to Paleobiology, v. 18.

EMRY, R. J., 1981, Additions to the mammalian fauna of the type Duchesnean, with comments on the status of the Duchesnean: Journal of Paleontology, v. 55, p. 563–570.

EMRY, R. J., 1992, Mammalian range zones in the Chadronian White River Formation at Flagstaff Rim, Wyoming, *in* Prothero, D. R. and Berggren, W. A., eds., Eocene-Oligocene Climatic and Biotic Evolution: Princeton, Princeton University Press, p. 106–115.

EMRY, R. J., BJORK, P. R., AND RUSSELL, L. S., 1987, The Chadronian, Orellan, and Whitneyan land mammal ages, *in* Woodburne, M. O., ed., Cenozoic Mammals of North America: Geochronology and Biostratigraphy: Berkeley, University of California Press, p. 118–152.

EVANOFF, E., PROTHERO, D. R. AND LANDER, R. H., 1992, Eocene-Oligocene climatic change in North America: The White River Formation near Douglas, east-central Wyoming, *in* Prothero, D. R. and Berggren, W. A., eds., Eocene-Oligocene Climatic and Biotic Evolution: Princeton, Princeton University Press, p. 116–130.

EVERNDEN, J. F., SAVAGE, D. E., CURTIS, G. H., AND JAMES, G. T., 1964, Potassium-argon dates and the Cenozoic mammalian chronology of North America: American Journal of Science, v. 262, p. 145–198.

FISHER, R. V. AND RENSBERGER, J. M., 1972, Physical stratigraphy of the John Day Formation, central Oregon: University of California Publications in Geological Sciences, v. 101, p. 1–33.

FLYNN, J. J., 1986, Correlation and geochronology of middle Eocene strata from the western United States: Palaeogeography, Palaeoclimatology, Palaeoecology, v. 55, p. 335–406.

FREDERIKSEN, N. O., 1980, Paleogene sporomorphs from South Carolina and quantitative correlation with the Gulf Coast: Palynology, v. 4, p. 125–179.

GAZIN, C. L., 1955, A review of the upper Eocene Artiodactyla of North America: Smithsonian Miscellaneous Collections, v. 128, p. 1–96.

GAZIN, C. L., 1956, The geology and vertebrate paleontology of upper Eocene strata in the northeastern part of the Wind River Basin, Wyoming, Part 2. The mammalian fauna of the Badwater area: Smithsonian Miscellaneous Collections, v. 131, no. 8, p. 1–35.

GAZIN, C. L., 1959, Paleontological exploration and dating of the early Tertiary deposits adjacent to the Uinta Mountains: Intermountain Association of Petroleum Geologists Guidebook, v. 10, p. 131–138.

GINGERICH, P. D., 1976, Cranial anatomy and evolution of early Tertiary Plesiadapidae (Mammalia, Primates): University of Michigan Papers on Paleontology, v. 15, p. 1–141.

GINGERICH, P. D., 1980, Evolutionary patterns in early Cenozoic mammals: Annual Review of Earth and Planetary Sciences, v. 8, p. 407–424.

GINGERICH, P. D., 1983, Paleocene-Eocene faunal zones and a preliminary analysis of Laramide structural deformation in the Clarks Fork Basin, Wyoming: Wyoming Geological Asssociation Guidebook, v. 34, p. 185–195.

GINGERICH, P. D., 1989, New earliest Wasatchian mammalian fauna from the Eocene of northwestern Wyoming: composition and diversity in a rarely sampled high-floodplain assemblage: University of Michigan Papers on Paleontology, v. 28, p. 1–97.

GINGERICH, P. D. AND ROSE, K. D., 1977, Preliminary report on the Clark Fork mammal fauna, and its correlation with similar faunas in Europe and Asia: Géobios Mémoir Spécial, v. 1, p. 39–45.

GINGERICH, P. D. AND SIMONS, E. L., 1977, Systematics, phylogeny, and evolution of early Eocene Adapidae (Mammalia, Primates) in North America: Contributions from the Museum of Paleontology, University of Michigan, v. 24, p. 245–279.

GLASS, B. P. AND CROSBIE, J. R., 1982, Age of the Eocene/Oligocene boundary based on extrapolation from North American microtektite layer: Bulletin of the American Association of Petroleum Geologists, v. 66, p. 471–476.

GLASS, B. P., HALL, C. D., AND YORK, D., 1986, $^{40}Ar/^{39}Ar$ laser-probe dating of North American tektite fragments from Barbados and the age of the Eocene–Oligocene boundary: Chemical Geology (Isotope Geoscience Section), v. 59, p. 181–186.

GUNNELL, G. F., 1989, Evolutionary history of Microsyopoidea (Mammalia, ?Primates) and the relationship between Plesiadapiformes and Primates: University of Michigan Papers in Paleontology, v. 27, p. 1–155.

GUNNELL, G. F., BARTELS, W. S., AND GINGERICH, P. D., 1993, Paleocene-Eocene boundary in continental North America: biostratigraphy and geochronology, northern Bighorn Basin, Wyoming: New Mexico Museum of Natural History Bulletin, v. 2, p. 137–144.

HARDENBOL, J. AND BERGGREN, W. A., 1978, A new Paleogene numerical time scale, *in* Cohee, G. V., Glaessner, M. F., and Hedberg, H. D., eds., Contributions to the Geologic Time Scale: Tulsa, American Association of Petroleum Geologists Studies in Geology 6, p. 213–234.

HEDBERG, H. D., 1976, International Stratigraphic Guide: New York, Wiley, 200 p.

HOOKER, J. J., 1992, British mammalian paleocommunities across the Eocene-Oligocene transition and their environmental implications, *in* Prothero, D. R. and Berggren, W. A., eds., Eocene-Oligocene Climatic and Biotic Evolution: Princeton, Princeton University Press, p. 494–515.

KELLY, T. S., 1990, Biostratigraphy of Uintan and Duchesnean land mammal assemblages from the middle member of the Sespe Formation, Simi Valley, California: Contributions to Science of the Natural History Museum of Los Angeles County, v. 419, p. 1–42.

KENNETT, J. P. AND STOTT, L. D., 1991, Abrupt deep-sea warming, paleoceanographic changes and benthic extinctions at the end of the Palaeocene: Nature, v. 353, p. 225–229.

KOCH, P. L. AND ZACHOS, J. C., 1993, Chemostratigraphic correlation of marine and continental strata across the Paleocene/Eocene boundary: Journal of Vertebrate Paleontology, v. 13, p. 45A.

KOCH, P. L., ZACHOS, J. C., AND GINGERICH, P. D., 1992, Correlation between isotope records in marine and continental carbon reservoirs near the Palaeocene/Eocene boundary: Nature, v. 358, p. 319–322.

KORTH, W. W., 1989, Stratigraphic occurrence of rodents and lagomorphs in the Orella Member, Brule Formation (Oligocene), northwestern Nebraska: Contributions to Geology, University of Wyoming, v. 27, p. 15–20.

KRISHTALKA, L., STUCKY, R. K., WEST, R. M., MCKENNA, M. C., BLACK, C. C., BOWN, T. M., DAWSON, M. R., GOLZ, D. J., FLYNN, J. J., LILLE-GRAVEN, J. A., AND TURNBULL, W. A., 1987, Eocene (Wasatchian through Duchesnean) biochronology of North America, *in* Woodburne, M. O., ed., Cenozoic Mammals of North America: Geochronology and Biostratigraphy: Berkeley, University of California Press, p. 77–117.

LILLEGRAVEN, J. A., 1979, A biogeographical problem involving comparisons of late Eocene terrestrial vertebrate faunas of western North America, *in* Gray, J. and Boucot, A. J., eds., Historical Biogeography, Plate Tectonics, and the Changing Environment: Corvallis, Oregon State University Press, p. 333–347.

LINDSAY, E. H., BUTLER, R. F., AND JOHNSON, N. M., 1981, Magnetic polarity zonation and biostratigraphy of Late Cretaceous and Paleocene continental deposits, San Juan Basin, New Mexico: American Journal of Science, v. 281, p. 390–435.

LINDSAY, E. H., JACOBS, L. L., AND BUTLER, R. F., 1978, Biostratigraphy and magnetic polarity stratigraphy of upper Cretaceous and Paleocene terrestrial deposits, San Juan Basin, New Mexico: Geology, v. 6, p. 425–429.

LINDSAY, E. H., OPDYKE, N. D., JOHNSON, N. M., AND BUTLER, R. F., 1987, Mammalian chronology and the magnetic polarity time scale, *in* Woodburne, M. O., ed., Cenozoic Mammals of North America: Geochronology and Biostratigraphy: Berkeley, University of California Press, p. 269–284.

LUCAS, S. G., 1984, Systematics, biostratigraphy, and evolution of early Cenozoic Coryphodon (Mammalia, Pantodonta): Unpublished Ph.D. Dissertation, Yale University, New Haven, 648 p.

LUCAS, S. G., 1989, Fossil mammals and the Paleocene-Eocene boundary in Europe, North America, and Asia: Abstracts, 28th International Geological Congress, v. 2, p. 355.

LUCAS, S. G., 1992, Redefinition of the Duchesnean land mammal "age," late Eocene of western North America, *in* Prothero, D. R. and Berggren, W. A., eds., Eocene-Oligocene Climatic and Biotic Evolution: Princeton, Princeton University Press, p. 88–105.

LUCAS, S. G., 1993, Mammalian biochronology of the Paleocene-Eocene boundary in North America, Europe, and Asia: Journal of Vertebrate Paleontology, v. 13, supplement to no. 3, p. 47A.

MATTHEW, W. D., 1899, A provisional classification of the fresh-water Tertiary of the West: Bulletin of the American Museum of Natural History, v. 12, p. 19–75.

MCCARROLL, S. M., FLYNN, J. J., AND TURNBULL, W. D., 1993, Biostratigraphic and magnetic polarity correlations of the Washakie Formation, Washakie Basin, Wyoming: Journal of Vertebrate Paleontology, v. 13, supplement to no. 3, p. 49A.

MCDOUGALL, I. AND HARRISON, T. M., 1988, Geochronology and Thermochronology by the ^{40}Ar/^{39}Ar Method: New York, Oxford University Press, 212 p.

MONTANARI, A., 1988, Geochemical characterization of volcanic biotites from the upper Eocene–upper Miocene pelagic sequence of the northeastern Ap-

ennines, *in* Premoli-Silva, I., Coccioni, R., and Montanari, A., eds., The Eocene–Oligocene Boundary in the Marche-Umbria Basin (Italy): Ancona, International Sub-commission on Paleogene Stratigraphy, Eocene/Oligocene Boundary Meeting, Special Publication, p. 209–227.

MONTANARI, A., 1990, Geochronology of the terminal Eocene impacts; an update: Boulder, Geological Society of America Special Paper, v. 247, p. 607–616.

MONTANARI, A., DEINO, A. L. DRAKE, R. E., TURRIN, B. D., DEPAOLO, D. J., ODIN, G. S., CURTIS, G. H., ALVAREZ, W. AND BICE, D., 1988, Radioisotopic dating of the Eocene-Oligocene boundary in the pelagic sequences of the northeastern Apennines, *in* Premoli-Silva, I., Coccioni, R., and Montanari, A., eds., The Eocene–Oligocene Boundary in the Marche-Umbria Basin (Italy): Ancona, International Sub-commission on Paleogene Stratigraphy, Eocene/Oligocene Boundary Meeting, Special Publication, p. 195–208.

MONTANARI, A., DRAKE, R., BICE, D. M., ALVAREZ, W., CURTIS, G. H., TURRIN, B., AND DEPAOLO, D. J., 1985, Radiometric time scale for the upper Eocene and Oligocene based on K/Ar and Rb/Sr dating of volcanic biotites: Geology, v. 13, p. 596–599.

NORTH AMERICAN COMMISSION ON STRATIGRAPHIC NOMENCLATURE, 1983, North American stratigraphic code: American Association of Petroleum Geologists Bulletin, v. 67, p. 841–875.

OBRADOVICH, J. D., 1988, A different perspective on glauconite as a chronometer for geologic time scale studies: Paleoceanography, v. 3, p. 757–770.

ODIN, G. S., 1978, Isotopic dates for the Paleogene time scale, *in* Cohee, G. V., Glaessner, M. F., and Hedberg, H. D., eds., Contributions to the Geologic Time Scale: Tulsa, American Association of Petroleum Geologists Studies in Geology, v. 6, p. 247–257.

ODIN, G. S., ed., 1982, Numerical Dating in Stratigraphy: New York, John Wiley and Sons, 1040 p.

ODIN, G. S. AND CURRY, D., 1985, The Paleogene time-scale: radiometric dating versus magnetostratigraphic approach: Journal of the Geological Society of London, v. 142, p. 1179–1188.

ODIN, G., GUISE, P., REX, D. C., AND KREUZER, H., 1988, K-Ar and ^{39}Ar/^{40}Ar geochronology of late Eocene biotites from the northeastern Apennines, *in* Premoli-Silva, I., Coccioni, R., and Montanari, A., eds., The Eocene–Oligocene Boundary in the Marche-Umbria Basin (Italy): Ancona, International Sub-commission on Paleogene Stratigraphy, Eocene/Oligocene Boundary Meeting, Special Publication, p. 239–245.

ODIN, G. S., MONTANARI, A., DEINO, A., DRAKE, R., GUISE, P. G., KREUZER, H., AND REX, D. C., 1991, Reliability of volcano-sedimentary biotite ages across the Eocene-Oligocene boundary (Apennines, Italy): Chemical Geology (Isotope Geoscience Section), v. 86, p. 203–224.

OPDYKE, N. D., 1990, Magnetic stratigraphy of Cenozoic terrestrial sediments and mammalian dispersal: Journal of Geology, v. 98, p. 621–637.

OSBORN, H. F., 1907, Tertiary mammal horizons of North America: Bulletin of the American Museum of Natural History, v. 23, p. 237–254.

OSBORN, H. F., 1910, The Age of Mammals in Europe, Asia, and North America: New York, MacMillan and Company, 635 p.

OSBORN, H. F., 1929, The titanotheres of ancient Wyoming, Dakota, and Nebraska: United States Geological Survey Monograph, v. 55, p. 1–953.

OSBORN, H. F. AND MATTHEW, W. D., 1909, Cenozoic mammal horizons of western North America: United States Geological Survey Bulletin, v. 361, p. 1–138.

PETERSON, O. A., 1909, A revision of the Entelodontidae: Memoirs of the Carnegie Museum, v. 9, p. 41–158.

PROTHERO, D. R., 1982, Medial Oligocene magnetostratigraphy and mammalian biostratigraphy: testing the isochroneity of mammalian biostratigraphic events: Unpublished Ph.D. Dissertation, Columbia University, 284 p.

PROTHERO, D. R., 1984, Magnetostratigraphy of the early Oligocene Pipestone Springs locality, Jefferson County, Montana: Contributions to Geology, University of Wyoming, v. 23, p. 33–36.

PROTHERO, D. R., 1985a, Chadronian (early Oligocene) magnetostratigraphy of eastern Wyoming: implications for the Eocene-Oligocene boundary: Journal of Geology, v. 93, p. 555–565.

PROTHERO, D. R., 1985b, Correlation of the White River Group by magnetostratigraphy, *in* Martin, J. E., ed., Fossiliferous Cenozoic Deposits of Western South Dakota and Northwestern Nebraska: Rapid City, Dakoterra, Museum of Geology, South Dakota School of Mines, v. 2, p. 265–276.

PROTHERO, D. R., 1988, Mammals and magnetostratigraphy: Journal of Geological Education, v. 34, p. 227–236.

PROTHERO, D. R., 1990, Interpreting the Stratigraphic Record: New York, W.H. Freeman, 410 p.

PROTHERO, D. R., 1991, Magnetic stratigraphy of Eocene-Oligocene mammal localities in southern San Diego County, *in* Abbott, P. L. and May, J. A., eds., Eocene Geologic History, San Diego Region: Pacific Section, Society of Economic Paleontologists and Mineralogists Guidebook, v. 68, p. 125–130.

PROTHERO, D. R., 1994a, The late Eocene-Oligocene extinctions: Annual Reviews of Earth and Planetary Sciences, v. 22, p. 145–165.

PROTHERO, D. R., 1994b, The Eocene-Oligocene Transition: Paradise Lost: New York, Columbia University Press, 291 p.

PROTHERO, D. R., 1995a, Magnetostratigraphy and biostratigraphy of the middle Eocene Uinta Formation, Uinta Basin, Utah, *in* Prothero, D. R. and Emry, R. J., eds., The Terrestrial Eocene-Oligocene Transition in North America: New York, Cambridge University Press, p. 3–24.

PROTHERO, D. R., 1995b, Magnetostratigraphy of Eocene-Oligocene transition in Trans-Pecos Texas, *in* Prothero, D. R. and Emry, R. J., eds., The Terrestrial Eocene-Oligocene Transition in North America: New York, Cambridge University Press, p. 174–183.

PROTHERO, D. R., 1995c, Magnetostratigraphy of the White River Group, High Plains of North America, *in* Prothero, D. R. and Emry, R. J., eds., The Terrestrial Eocene-Oligocene Transition in North America: New York, Cambridge University Press, p. 247–262.

PROTHERO, D. R., DENHAM, C. R., AND FARMER, H. G., 1983, Magnetostratigraphy of the White River Group and its implications for Oligocene geochronology: Palaeogeography, Palaeoclimatology, Palaeoecology, v. 42, p. 151–166.

PROTHERO, D. R. AND EMRY, R. J., 1995, The Terrestrial Eocene-Oligocene Transition in North America: New York, Cambridge University Press, 670 p.

PROTHERO, D. R., HOWARD, J., AND DOZIER, T. H. H., 1995, Stratigraphy and paleomagnetism of the upper middle Eocene to lower Miocene (Uintan-Arikareean) Sespe Formation, Ventura County, California, *in* Prothero, D. R., and Emry, R. J., eds., The Terrestrial Eocene-Oligocene Transition in North America: New York, Cambridge University Press, p. 125–142.

PROTHERO, D. R. AND LUCAS, S. G., 1995, Magnetostratigraphy of the Duchesnean portion of the Galisteo Formation, New Mexico, *in* Prothero, D. R., and Emry, R. J., eds., The Terrestrial Eocene-Oligocene Transition in North America: New York, Cambridge University Press, p. 184–190.

PROTHERO, D. R. AND RENSBERGER, J. M., 1985, Magnetostratigraphy of the John Day Formation, Oregon, and the North American Oligocene-Miocene boundary: Newsletters in Stratigraphy, v. 15, p. 59–70.

PROTHERO, D. R. AND SWISHER, C. C., III, 1992, Magnetostratigraphy and geochronology of the terrestrial Eocene-Oligocene transition in North America, *in* Prothero, D. R. and Berggren, W. A., eds., Eocene-Oligocene Climatic and Biotic Evolution: Princeton, Princeton University Press, p. 46–73.

RAPP, S. D., MACFADDEN, B. J., AND SCHIEBOUT, J. A., 1983, Magnetic polarity stratigraphy of the early Tertiary Blacks Peak Formation, Big Bend National Park, Texas: Journal of Geology, v. 91, p. 555–572.

REA, D. K., ZACHOS, J. C., OWEN, R. M., AND GINGERICH, P. D., 1990, Global change at the Paleocene-Eocene boundary: climatic and evolutionary consequences of tectonic events: Palaeogeography, Palaeoclimatology, Palaeoecology, v. 79, p. 117–128.

RENSBERGER, J. M., 1971, Entoptychine pocket gophers (Mammalia, Geomyoidea) of the early Miocene John Day Formation, Oregon: University of California Publications in Geological Sciences, v. 90, p. 1–163.

RENSBERGER, J. M., 1973, Pleurolicine rodents (Geomyoidea) of the John Day Formation, Oregon, and their relationships to taxa from the early and middle Miocene of South Dakota: University of California Publications in Geological Sciences, v. 102, p. 1–95.

RENSBERGER, J. M., 1983, Successions of meniscomyine and allomyine rodents (Aplodontidae) in the Oligo-Miocene John Day Formation, Oregon: University of California Publications in Geological Sciences, v. 124, p 1–157.

ROSE, K. D., 1981, The Clarkforkian land-mammal age and mammalian faunal composition across the Paleocene-Eocene boundary: University of Michigan Papers on Paleontology, v. 26, p. 1–197.

SAVAGE, D. E., 1955, Nonmarine lower Pliocene sediments in California, geochronologic-stratigraphic classification: University of California Publications in Geological Sciences, v. 31, no. 1, p. 1–26.

SAVAGE, D. E., 1962, Cenozoic geochronology of the fossil mammals of the Western Hemisphere: Revista Museo Argentino Ciencias Naturales, v. 8, p. 53–67.

SAVAGE, D. E., 1977, Aspects of vertebrate paleontological stratigraphy and geochronology, *in* Kaufman, E. G. and Hazel, J. E., eds., Concepts and Methods in Biostratigraphy: Stroudsburg, Dowden, Hutchinson, and Ross, p. 427–442.

SAVAGE, D. E. AND RUSSELL, D. E., 1983, Mammalian Paleofaunas of the World: Reading, Addison Wesley Publishing Company, 432 p.

SCHANKLER, D. M., 1980, Faunal zonation of the Willwood Formation in the central Bighorn Basin, Wyoming: University of Michigan Papers on Paleontology, v. 24, p .99–114.

SCHANKLER, D. M., 1981, Local extinction and ecological re-entry of early Eocene mammals: Nature, v. 293, p. 135–138.

SCHULTZ, C. B. AND STOUT, T. M., 1955, Classification of the Oligocene sediments in Nebraska: Bulletin of the University of Nebraska State Museum, v. 4, p. 17–52.

SCHULTZ, C. B. AND STOUT, T. M., 1961, Field conference on the Tertiary and Pleistocene of western Nebraska: Lincoln, Special Publication of the University of Nebraska State Museum, v. 2, p. 1–54.

SCOTT, W. B., 1945, The Mammalia of the Duchesne River Oligocene: Transactions of the American Philosophical Society, v. 34, p. 209–253.

SHAW, A. B., 1964, Time in Stratigraphy: New York, McGraw-Hill, 365 p.

SIMPSON, G. G., 1946, The Duchesnean fauna and the Eocene-Oligocene boundary: American Journal of Science, v. 244, p. 52–57.

SLOAN, R. E., 1987, Paleocene and latest Cretaceous mammal ages, biozones, magnetozones, rates of sedimentation, and evolution: Boulder, Geological Society of America Special Paper 209, p. 165–200.

SLOAN, R. E., RIGBY, J. K., JR., VAN VALEN, L. M., AND GABRIEL, D., 1986, Gradual dinosaur extinction and simultaneous ungulate radiation in the Hell Creek Formation, Montana: Science, v. 232, p. 629–633.

STUCKY, R. K., 1984, The Wasatchian-Bridgerian land mammal age boundary (early to middle Eocene) in western North America: Annals of the Carnegie Museum, v. 53, p. 347–382.

STUCKY, R. K., PROTHERO, D. R., LOHR, W. G., AND SNYDER, J., 1995, Magnetostratigraphy and biostratigraphy of the earliest Uintan Sand Wash Basin, northwest Colorado, *in* Prothero, D. R. and Emry, R. J., eds., The Terrestrial Eocene-Oligocene Transition in North America: New York, Cambridge University Press, p. 40–51.

SWISHER, C. C., III AND PROTHERO, D. R., 1990, Single-crystal ^{40}Ar/^{39}Ar dating of the Eocene-Oligocene transition in North America: Science, v. 249, p. 760–762.

TABRUM, A. R., PROTHERO, D. R., AND GARCIA, D., 1995, Magnetostratigraphy and biostratigraphy of the Eocene-Oligocene transition in western Montana, *in* Prothero, D., R. and Emry, R. J., eds., The Terrestrial Eocene-Oligocene Transition in North America: New York, Cambridge University Press, p. 263–294.

TAUXE, L., GEE, J., GALLET, Y., PICK, T., AND BOWN, T., 1994, Magnetostratigraphy of the Willwood Formation, southern Bighorn Basin, Wyoming: Earth and Planetary Science Letters, v. 125, p. 159–179.

TAYLOR, L. H. AND BUTLER, R. F., 1980, Magnetic polarity stratigraphy of Torrejonian sediments, Nacimiento Formation, San Juan Basin, New Mexico: American Journal of Science, v. 280, p. 97–115.

TEDFORD, R. H., 1970, Principles and practices of mammalian geochronology in North America: Proceedings of the North American Paleontological Convention, v. 2F, p. 666–703.

TEDFORD, R. H., GALUSHA, T., SKINNER, M. F., TAYLOR, B. E., FIELDS, R. W., MACDONALD, J. R., RENSBERGER, J. M., WEBB, S. D., AND WHISTLER, D. P., 1987, Faunal succession and biochronology of the Arikareean through Hemphillian interval (late Oligocene through earliest Pliocene Epochs) in North America, *in* Woodburne, M. O., ed., Cenozoic Mammals of North America: Geochronology and Biostratigraphy: Berkeley, University of California Press, p. 152–210.

TEDFORD, R. H., SWINEHART, J., SWISHER, C. C., III, PROTHERO, D. R., KING, S. A., AND TIERNEY, T. E., 1995, The Whitneyan-Arikareean transition in the High Plains, *in* Prothero, D., R. and Emry, R. J., eds., The Terrestrial Eocene-Oligocene Transition in North America: New York, Cambridge University Press, p. 295–317.

TOMIDA, Y., 1981, "Dragonian" fossils from the San Juan Basin and the status of the "Dragonian" land mammal "age," *in* Lucas, S.G., Rigby, J.K., Jr. and Kues, B.S., eds., Advances in San Juan Basin Paleontology: Albuquerque, University of New Mexico Press, p. 222–241.

TOMIDA, Y., AND BUTLER, R. F., 1980, Dragonian mammals and Paleocene magnetic polarity stratigraphy, North Horn Formation, central Utah: American Journal of Science, v. 280, p. 787–811.

WALSH, S. A., 1995, The Bridgerian/Uintan boundary and the status of the "Shoshonian" (earliest Uintan) land mammal "subage," *in* Prothero, D. R. and Emry, R. J., eds., The Terrestrial Eocene-Oligocene Transition in North America: New York, Cambridge University Press, p. 52–59.

WALSH, S. A., PROTHERO, D. R., AND LUNDQUIST, D., 1995, Uintan magnetostratigraphy and biostratigraphy, southern San Diego County, California, *in* Prothero, D. R., and Emry, R. J., eds., The Terrestrial Eocene-Oligocene

Transition in North America: New York, Cambridge University Press, p. 105–139.

WALTON, A. H., 1992, Magnetostratigraphy of the lower and middle members of the Devil's Graveyard Formation (middle Eocene), Trans-Pecos Texas, *in* Prothero, D. R. and Berggren, W. A., eds., Eocene-Oligocene Climatic and Biotic Evolution: Princeton, Princeton University Press, p. 74–83.

WILLIAMS, H. S., 1901, The discrimination of time-values in geology: Journal of Geology, v. 9, p. 570–585.

WILSON, J. A., 1978, Stratigraphic occurrence and correlation of early Tertiary vertebrate faunas, Trans-Pecos Texas, Part 1: Vieja area: Texas Memorial Museum Bulletin, v. 25, p. 1–42.

WILSON, J. A., 1984, Vertebrate fossil faunas 49 to 36 million years ago and additions to the species of *Leptoreodon* found in Texas: Journal of Vertebrate Paleontology, v. 4, p. 199–207.

WILSON, J. A., 1986, Stratigraphic occurrence and correlation of early Tertiary vertebrate faunas, Trans-Pecos Texas: Agua Fria-Green Valley areas: Journal of Vertebrate Paleontology, v. 6, p. 350–373.

WING, S. L., 1984, A new basis for recognizing the Paleocene/Eocene boundary in Western Interior North America: Science, v. 226, p. 439–441.

WING, S. L., BOWN, T. M., AND OBRADOVICH, J. D., 1991, Early Eocene biotic and climatic change in interior western North America: Geology, v. 19, p. 1189–1192.

WOOD, H. E., II, 1934, Revision of the Hyrachyidae: Bulletin of the American Museum of Natural History, v. 67, p. 182–295.

WOOD, H. E., II, CHANEY, R. W., JR., CLARK, J., COLBERT, E. H., JEPSEN, G. L., REESIDE, J. B., AND STOCK, C., 1941, Nomenclature and correlation of the North American continental Tertiary: Geological Society of America Bulletin, v. 52, p. 1–48.

WOODBURNE, M. O., 1977, Definition and characterization in mammalian chronostratigraphy: Journal of Paleontology, v. 51, p. 220–234.

WOODBURNE, M. O., ed., 1987, Cenozoic Mammals of North America: Geochronology and Biostratigraphy: Berkeley, University of California Press, 336 p.

CENOZOIC SOUTH AMERICAN LAND MAMMAL AGES: CORRELATION TO GLOBAL GEOCHRONOLOGIES

JOHN J. FLYNN

Department of Geology, Field Museum, Chicago, IL 60605

AND

CARL C. SWISHER, III

Berkeley Geochronology Center, Berkeley, CA 94709

ABSTRACT: The radiation of a diverse array of endemic marsupials, edentates, primates, rodents and "ungulates" has been exceedingly useful for developing a detailed biochronologic sequence of about 20 South American Land Mammal Ages (SALMAs), covering much of Cenozoic time. Independent chronologic controls on this terrestrial South American Cenozoic record have increased dramatically in the past 25 years. There now are numerous radioisotopic dates (including many new laser fusion ^{40}Ar/^{39}Ar analyses) and magnetochronologic studies, especially for the Neogene Period, leading to better resolution of the ages of the Riochican SALMA, the new Chilean Tinguiririca faunal interval, Deseadan SALMA, Santacrucian SALMA, "Friasian" SALMA, Colombian LaVenta sequence, Huayquerian SALMA, Ensenadan SALMA, and Uquian SALMA. ^{40}Ar/^{39}Ar dating and magnetostratigraphic studies are continuing actively, but current data indicate that there are significant temporal gaps in the SALMA sequence, most notably representing the early Paleocene, most of the Eocene, part of the early Oligocene, much of the early Miocene, and part of the late Miocene Epochs. In particular, the Tiupampan, Casamayoran, Mustersan, Divisaderan, and Colhuehuapian Cenozoic SALMAs (as well as the late Cretaceous [?Campanian] "Alamitian" SALMA), lack either magnetic polarity stratigraphies or radioisotopic dating, are temporally constrained only by the ages of superposed intervals or weakly justified "stage-of-evolution" arguments and thus remain relatively poorly constrained geochronologically. Best estimates for the approximate durations (question marks indicate very poor geochronologic control, and therefore highly provisional age estimates) of the SALMAs are: Lujanian, 10,000–800,000 years ago; Ensenadan, 0.8–1.2 Ma; Uquian, 1.5–3.0 Ma; Chapadmalalan, 3.4–4.0 Ma; Montehermosan, 4.0–6.8 Ma; Huayquerian, 6.8–9.0 Ma; Chasicoan, 9.0–10.0(?) Ma; Mayoan, (?)10.0–11.8 Ma; Laventan, 11.8–13.8 Ma; Colloncuran, 14.0–15.5 Ma; Friasian, uncertain; Santacrucian, 16.3–17.5 Ma; Colhuehuapian, (?)19–21(?) Ma; Deseadan, 24.5–29 Ma; New SALMA ("Tinguirirican"), 31.5–36 Ma; Divisaderan, (?)40–42(?) Ma; Mustersan, (?)45–48(?) Ma; Casamayoran, (?)51–54(?) Ma; Riochican 55.5–57 Ma; Itaboraian, 57.5–59 Ma; Peligran, 61–62.5 Ma; and Tiupampan, 63–64.5 Ma.

The South American faunal and floral record can be combined with available geochronologic information to evaluate timing and pattern of major biotic and environmental changes and events. It is clear that Cenozoic terrestrial biotas responded to both global and regional, physical and biotic, changes and events, including major plate tectonic reorganizations and associated biogeographic events (e.g., Mesozoic-early Cenozoic Gondwanan [and subsidiary North American] continental connections and biogeographic relationships, final separation of Antarctica during the ?Eocene Epoch, formation of the Isthmus of Panama in the Pliocene Epoch, etc.), global Eocene-Oligocene boundary events (including climate change, initiation of oceanic deep water flow through the Drake Passage, onset of major Antarctic ice cap formation, expansion of open habitats [wooded grasslands and grasslands], increase in hypsodont mammalian taxa, and major clade extinction, origination, and diversification), phases of Andean uplift, Pliocene-Pleistocene Epoch glaciation/climatic changes, etc.

INTRODUCTION

South America was an island continent for much of the past 65 million years, yielding a highly endemic terrestrial fauna and flora. The radiation of a diverse array of endemic marsupials, edentates, primates, rodents and "ungulates" has made biostratigraphic correlation to other continents extremely difficult. That endemism, so problematic for correlation outside South America, has been exceedingly useful for developing a detailed sequence of about 20 South American Land Mammal Ages (SALMAs), covering much of Cenozoic time, but with some marked gaps. These SALMAs have proven very useful in temporal correlation even though all but one (the Laventan, see below) are informal biochronologic units (see discussions in Tedford, 1970; Savage, 1977; Woodburne, 1977, 1987; Lillegraven and McKenna, 1986) rather than formal geochronologic units based on formal chronostratigraphic units. Remarkable strides have been made in integrating Land Mammal Ages into geochronologic studies since Evernden and others (1964) and Patterson and Pascual (1968) published their landmark papers. Nevertheless, understanding of the temporal sequence and range of applicability (intracontinental, or only regional) of SALMAs has lagged well behind that available for Mammal Ages of North America and Europe. By 1964, Evernden and others had produced numerous K-Ar dates for the Cenozoic terrestrial sequences of Europe and North America, but for the South American "Tertiary" Period there was only a single such date (from Argentine Patagonian sequences representing the Santacrucian Land Mammal Age). As on other continents, there were no paleomagnetic stratigraphies, and there were very few

sequences with intertonguing marine and terrestrial strata available for independent marine biochronologic control. Since then, extensive work has greatly enhanced age control on the major South American terrestrial biochronology and correlation of the SALMAs to standard global geochronologies. This improved chronologic control provides the temporal framework for calibrating South American biotic, environmental, and geologic events. It also clarifies understanding of the relative influences of regional and global physical and biotic changes on South American terrestrial ecosystems.

Chronologic controls on the terrestrial South American Cenozoic record have increased dramatically in the past 25 years. There now are numerous radioisotopic dates (including many new laser fusion ^{40}Ar/^{39}Ar analyses) and magnetochronologic studies for the Cenozoic SALMA sequence, especially for the Neogene Period. The Tiupampan, Casamayoran, Mustersan, Divisaderan, and Colhuehuapian Cenozoic SALMAs (as well as the late Cretaceous [?Campanian] "Alamitian" SALMA, characterized by the only diverse Mesozoic mammal fauna known from South America, the Los Alamitos Fauna, Bonaparte, 1990; Pascual and Ortiz Jaureguizar, 1992), lack either magnetic polarity stratigraphies or radioisotopic dating, are temporally constrained only by the ages of superposed intervals or weakly justified "stage-of-evolution" arguments and thus remain relatively poorly constrained geochronologically. ^{40}Ar/^{39}Ar dating and magnetostratigraphic studies of South American terrestrial sequences are continuing actively. These promise to provide more precise temporal correlation for many already dated sequences, new temporal controls on currently undated sequences, and much better calibration of the entire SALMA sequence.

We use the anglicized spelling Tiupampan SALMA, rather than Tiupamp*ian*, for the "Tiupampense", as the original stem (Tiupampa) already ends in an "-a", requiring only simple addition of an "-n" suffix, rather than insertion of an "-i-" preceding the "-an" ending.

RADIOISOTOPIC DATING

Recent radioisotopic dating of Cenozoic terrestrial deposits in South America generally has consisted of K-Ar and fission-track dating, either alone, or together in the same sequences to provide some test of concordancy. Unfortunately, in some cases the analyses have not been sufficiently abundant or reliable to permit stabilization of boundary ages and temporal durations for most SALMAs (see Table 1). Only recently have these techniques (each with well-known limitations and deficiencies) been supplemented or supplanted by $^{40}Ar/^{39}Ar$ dating, and the chronologic calibration of many of the SALMAs is stabilizing.

In $^{40}Ar/^{39}Ar$ dating, radiogenic ^{40}Ar and ^{39}Ar are measured on the same sample split by mass spectrometry. Therefore, one avoids problems (minimize sample heterogeneity, measure smaller samples, avoid measurement errors [e.g., weighing errors] on separate splits, etc.) associated with related K-Ar techniques using separate sample splits and two different analytical chemical measurement techniques (gas mass spectrometry for argon and wet-chemistry techniques for potassium). The $^{40}Ar/^{39}Ar$ dating of the South American sequence has focused on single-crystal laser fusion techniques, which increase precision of analytical dates by up to an order of magnitude (<0.1% error versus 1–5% error using other techniques; e.g., see Odin, 1982; Flynn and others, 1984; Faure, 1986). The single-crystal laser fusion technique is based on dating of individual mineral crystals within a single sample rather than dating whole-rocks or bulk mineral separates as in alternative K-Ar or bulk step-heating $^{40}Ar/^{39}Ar$ techniques. Although sample preparation time is significant, because samples are small less time is required to "clean-up" the mass spectrometer, and an individual date can be obtained in less than 20 minutes. Therefore, multiple individual grains can be analyzed from a single sample, or rare phenocrysts can be dated from volcanic rocks not datable using other techniques. Replicate grain analyses permit direct calculation, rather than estimation, of analytical error (including error due both to analytical techniques/apparatus and within sample heterogeneity) based on "reproducibility of multiple analyses"; identification of multiple grain populations within a sample; and rejection of older detrital contaminants or altered grains.

There now are numerous radioisotopic dates for the SALMA sequence. Many new laser fusion $^{40}Ar/^{39}Ar$ analyses have been generated for the new Eocene-Oligocene Tinguiririca faunal interval, late Oligocene Deseadan, early-middle Miocene Santacrucian, Miocene "Friasian" (in quotation marks to emphasize that great confusion surrounds the validity and temporal scope of sequences [Mayoan, Friasian, Colloncuran] often lumped within a Friasian SALMA), and middle Miocene Colombian LaVenta fauna. Older, less precise, and more variable (among horizons) K-Ar dates are available from many sequences including units below the Paleocene Peligran-Riochican, the mid-

dle Miocene Bolivian Quebrada Honda sequence, and Bolivian strata containing a late Miocene Montehermosan correlative fauna (Figs. 1, 2).

It is clear from Figure 1 that there are significant temporal gaps in the SALMA sequence, most notably representing the early Paleocene, most of the Eocene, part of the early Oligocene, much of the early Miocene and part of the late Miocene Epochs. In contrast to the temporal continuity of much of the North American Land Mammal Age sequence, most SALMAs are separated from one another by a temporal gap. In fact in contrast to the portrayal in some earlier studies (e.g., Marshall, 1985), the SALMAs may only very incompletely represent the temporal span of the Cenozoic Era— there currently may be more unrepresented time or "gaps" than temporal intervals represented by SALMA sequences. Current efforts by many research groups are focusing on filling these gaps through discovery of new faunal sequences, better collecting of intervals previously considered poorly fossiliferous but which lie in superposition with well-known SALMA assemblages, and gathering new geochronologic information to better constrain the ages of the known SALMA assemblages. A few SALMA sequences, such as the Huayquerian-Montehermosan, appear to be conformably superposed (Butler and others, 1984). Some intervals that had been portrayed as separated by temporal gaps, such as the Santacrucian-"Friasian", now appear to be proximate with little or no time between them. New data indicate that other SALMAs may be separated by even larger temporal gaps than had been suspected previously, such as between the Deseadan and Santacrucian (the poorly understood Colhuehuapian fills an unknown portion of this time).

There is wide variance in quality, precision, and number of dated samples for the age calibration of the SALMAs. For example, the Paleocene Riochican (and the underlying new assemblages/SALMAs noted above) is dated only by 3 low-precision K-Ar determinations from underlying horizons (within the marine Salamanca Formation/Group)—whole rock K-Ar basalt dates from the base of (64.0 ± 0.8 Ma) and probably below (62.8 ± 0.8 Ma) the marine Salamanca Formation/Group, and a vitric tuff (62.5 ± 5.0 Ma) within the Salamanca Formation/Group (Marshall and others, 1981). In contrast, the late Oligocene Deseadan has yielded more than 40 K-Ar age determinations (although precision is variable, and both the oldest and youngest previously reported determinations are invalid; Swisher, unpubl. data), and 72 high-precision analyses have been produced recently for "Friasian" aged sequences using single-crystal laser fusion $^{40}Ar/^{39}Ar$ dating techniques (most of these come from the Colloncuran and sequences containing the somewhat younger "Rio Negro faunas"; Madden and others, 1991). Other reasonably well-dated SALMAs include an assemblage representing a new unnamed latest Eocene-early Oligocene SALMA (based on the Tinguiririca Fauna from central Chile, Wyss and others, 1990, 1993, 1994; Charrier and others, in press), the early-middle Miocene Santacrucian, the late Miocene Huayquerian-Montehermosan, and late Cenozoic strata from Bolivia (Marshall and others, 1992; MacFadden and Anaya, 1993).

The Tiupampan, Peligran, Itaboraian, Riochican, Casamayoran, Mustersan, Divisaderan, Colhuehuapian, Chasicoan, Montehermosan (except possibly in Bolivia), Chapadmalalan,

TABLE 1.—PREVIOUS AGE ESTIMATES (in Ma) FOR SOUTH AMERICAN LAND MAMMAL AGES

SALMA	Metc77	M&P78	MHP83	Metc86	PP68	PP68*
	0.0*	0.1		0.1*		
LUJANIAN						
	0.7*	0.3	0.3*			
ENSENADEN						
	1.8*	1.0	1.0*	0.7*		
UQUIAN						
	3.5*	2.3 X(1)X	2.5*	1.6* (1)X		
CHAPADMALALAN						
	4.0*	2.3	3.0*	1.6* x		
MONTEHERMOSAN						
	6.0*	4.5	5.0*	2.8*		
HUAYQUERIAN						
	8.0*	9.5	9.0*	4.0*		
CHASICOAN						
	10.0*	10.5*	10.5*	5.2*		
hiatus[1]						
	12.0*	12.0*	12.0*	xxxx		
FRIASIAN						
	15.8*	16.0*	16.0*	15.0	10.4*	12.0*
SANTACRUCIAN						
	22.5*	23.0* xxxxx	22.5* xxxxx	18.0	16.2*	18.0*
marine Mt. León						
	23.0*	24.0*	22.5*	18.0	23.7*	23.7*
COLHUEHUAPIAN						
	25.0*	25.0	24.0*	19.0*	28.0*	28.0*
hiatus						
	34.0*	34.0	27.0*	21.0*	32.3*	32.0*
DESEADAN						
	37.5*	37.5* ?????	37.5* xxxxx	34.0* xxxxx	36.0*	36.0*
hiatus						
	40.0*	41.0*	41.0*	34.0?	?????	36.0*
DIVISADERAN						
	43.0*	43.0*	43.0*	36.0*	?????	
hiatus						
	45.0*	46.0*	46.0*		41.0*	
MUSTERSAN						
	48.0*	48.5*	48.0*		52.0* xxxxx	
hiatus						
	50.0*	50.0*	50.5*		52.0*	
CASAMAYORAN						
	53.5*	53.5*	55.0*		57.8*	
RIOCHICAN						
	57.0*	57.0*	62.0* xxxxx		60.7*	
hiatus						
	61.0*	62.0*	62.0*			
marine Salamancan						
	65.0*	65.0*	64.0*			

Metc77 = Marshall and others, 1977
M&P78 = Marshall and Pascual, 1978
MHP83 = Marshall and others, 1983
Metc86 = Marshall and others, 1986a, 1986b
PP68 = Patterson and Pascual, 1968
PP68* = Patterson and Pascual, 1968 as presented in Marshall and others, 1986a, 1986b (referred to as Patterson and Pascual, 1972 in those papers)
NOTES:
1)— M&P78, include Chapadmalalan within Montehermosan.
2) hiatus[1]- represents earlier concepts; in this paper we consider the Mayoan and Laventan to occupy part or all of this previously presumed hiatus.
3) xxxxx and ????? refer to absence (xxxxx) or questionable presence (?????) of a hiatus or marine incursion in the time scales cited.
4) *indicates age approximation we inferred from time scale figures in the papers cited, as no numerical tables of boundary ages or SALMA durations given therein.

and Lujanian SALMAs have not been directly dated radioisotopically.

MAGNETIC POLARITY STRATIGRAPHY AND INTEGRATIVE GEOCHRONOLOGY

Overview

Magnetic polarity stratigraphies, generally associated with radioisotopic dates, now exist for all or parts of the Riochican SALMA (and other parts of the early Paleocene Epoch represented in Patagonia), the new Eocene-Oligocene Chilean Tinguiririca faunal interval, late Oligocene Deseadan SALMA, early-middle Miocene Santacrucian SALMA, reportedly "Late Miocene" (Micaña) and "Friasian"-aged (Quebrada Honda) faunas from Bolivia, Miocene-aged Friasian SALMA type, middle Miocene Colombian La Venta sequence, late Miocene Huayquerian SALMA, late Miocene-early Pliocene(?) Montehermosan SALMA, Pliocene Chapadmalalan SALMA, late Pliocene-early Pleistocene Uquian SALMA, and Pleistocene Ensenadan and Lujanian SALMAs. The Tiupampan, Casamayoran, Mustersan, Divisaderan, and Colhuehuapian SALMAs, and classical Argentine Patagonian "Friasian" faunas

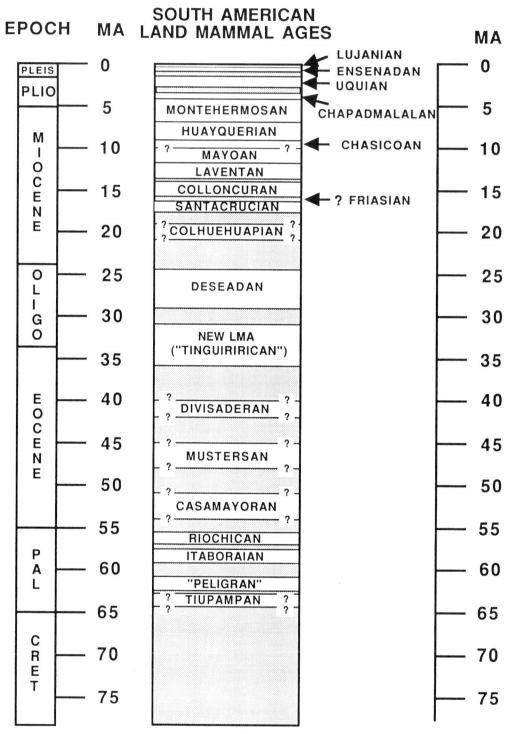

FIG. 1.—Geochronology of South American Land Mammal Ages. Ages for Epoch boundaries follow Berggren and others (this volume). Question marks at SALMA boundaries indicate very poor geochronologic control, and thus the ages for these SALMAs or boundaries must be considered highly provisional. "Tinguiririan" is used informally to represent a new, currently undescribed land mammal age characterized by the Tinguiririca Fauna of Chile (see Wyss and others, 1993, 1994). The question mark preceding the Friasian refers to the uncertain temporal placement and validity of the SALMA (see text). Best estimates for the approximate durations of the SALMAs are: Lujanian, 10,000–800,000 years ago; Ensenadan, 0.8–1.2 Ma; Uquian, 1.5–3.0 Ma; Chapadmalalan, 3.4–4.0 Ma; Montehermosan, 4.0–6.8 Ma; Huayquerian, 6.8–9.0 Ma; Chasicoan, 9.0–10.0(?) Ma; Mayoan, (?)10.0–11.8 Ma; Laventan, 11.8–13.8 Ma; Colloncuran, 14.0–15.5 Ma; Friasian, uncertain; Santacrucian, 16.3–17.5 Ma; Colhuehuapian, (?)19–21(?) Ma; Deseadan, 24.5–29 Ma; New SALMA ("Tinguiririan"), 31.5–36 Ma; Divisaderan, (?)40–42(?) Ma; Mustersan, (?)45–48(?) Ma; Casamayoran, (?)51–54(?) Ma; Riochican 55.5–57 Ma; Itaboraian, 57.5–59 Ma; Peligran, 61–62.5 Ma; Tiupampan, 63–64.5 Ma.

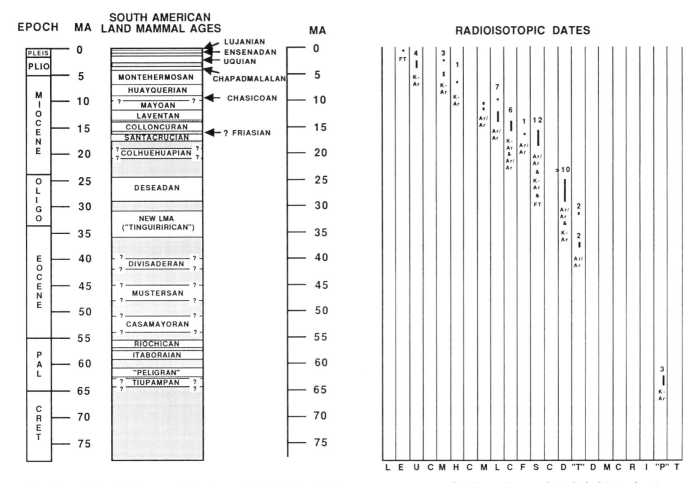

FIG. 2.—Radioisotopic dates constraining the age of the SALMAs. The dates represent a summary of the dates, given as points (single dates) or bars (range of several dates), we currently accept as the most accurate available constraints on the age of the SALMAs. Most represent dates from within sequences containing the particular SALMA (columns [oldest SALMA to right] in right side of figure, with first letter abbreviation of SALMA), although some dates are derived from horizons above or below faunal assemblages characterizing the SALMA, thereby providing minimum or maximum age constraints. Numbers above the points or bars represent number of dates. FT = fission-track date, K-Ar = [40]Potassium-[40]Argon date, Ar/Ar = single-crystal laser fusion [40]Ar/[39]Ar date. The SALMA chart on the left is that of Figure 1. Published sources for dates include: Ensenadan, MacFadden and others (1983); Uquian, Marshall and others (1982a); Montehermosan and Huayquerian, Butler and others (1984; see also, Marshall and others, 1979); Mayoan, Madden and others (1991); Laventan, Flynn and others (in press); Colloncuran, Madden and others (1991); Friasian, Flynn and Swisher (unpubl. data); Santacrucian, Marshall and others (1977, 1983, 1986a), Swisher and others (unpubl. data); Deseadan, Marshall and others (1983, 1986b), Swisher (unpubl. data); "Tinguiririan", Wyss and others (1993); below Riochican-"Peligran", Marshall and others (1981).

currently lack magnetic polarity stratigraphies. Throughout this paper we have standardized all magnetic polarity chronology terminology to correspond to that in Cande and Kent (1992, 1995) and Berggren and others (this volume). Figure 3 summarizes available magnetic polarity stratigraphy coverage for the SALMA sequence.

Although integrative geochronology studies (incorporating radioisotopic dating [especially [40]Ar/[39]Ar dating], magnetic polarity stratigraphy, and biostratigraphy) of the South American terrestrial record are becoming more common, previous studies have been scattered geographically and temporally.

Paleogene Record

Overview.—

The South American early Paleogene record is woefully constrained geochronologically. It is much more poorly sampled

than the Neogene record by geochronologic studies integrating multiple correlation techniques, largely because of the paucity of radioisotopically datable horizons in known fossiliferous sequences. Integrative Paleogene studies include: 1) the Riochican SALMA (and superposed assemblages from older parts of the sequences, see below) from localities near the type sequence in Argentine Patagonia, including paleomagnetics (within Riochican and older strata), K-Ar dates (from within and below the marine Salamanca Formation underlying Peligran through Riochican strata), and Danian forams (from the Salamanca Formation) (late Danian, Bertels, 1979; Marshall and others, 1981; early Danian, Bonaparte and others, 1993 based on data in Toumarkine and Lutherbacher, 1985); 2) a new, but currently unnamed, land mammal age typified by the Tinguiririca Fauna from the central Chilean Andes, including paleomagnetics, K-Ar dates, and [40]Ar/[39]Ar dates (Wyss and others, 1990, 1993, 1994; Charrier and others, 1990, in press); and 3) the Deseadan

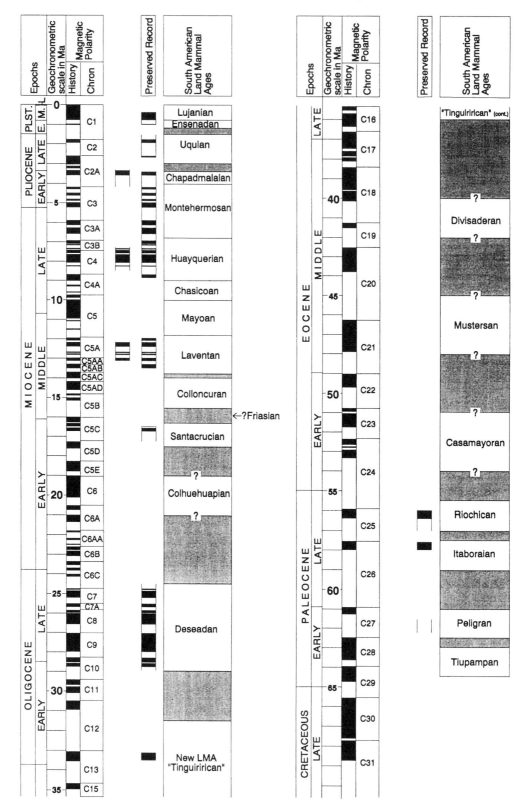

Fig. 3.—Correlation of Cenozoic South American Land Mammal Age (SALMA) biochronology with Magnetic Polarity Time Scale (MPTS, "Epoch"/ "Geochronometric Scale in Ma"/"Magnetic Polarity" columns), from Berggren and others (this volume, based on Cande and Kent, 1995). Portions of the Cenozoic Era for which magnetostratigraphy permits correlation of SALMAs to the MPTS indicated by black (normal polarity) and white (reversed polarity) shading in "Preserved Record" column; intervals for which such correlation is not presently available indicated by blank areas in the same column. The left "Preserved Record" column in the Miocene-Pliocene interval portrays magnetostratigraphic correlations of Bolivian sequences of uncertain biochronologic correlation to the standard SALMA sequence (Quebrada Honda, Micaña, Inchasi; see text). The question mark next to the "Preserved Record" for the Peligran indicates the uncertain correlation of this magnetozone to the standard MPTS.

SALMA from Salla, Bolivia, presenting paleomagnetics, fission-track dates, and K-Ar dates (MacFadden and others, 1985; Naeser and others, 1987; McRae, 1990), and studies of the type and related Deseadan SALMA sequences from Argentina, including many older K-Ar dates (Marshall and others, 1983, 1986b) and new ^{40}Ar/^{39}Ar dates (Swisher, unpubl. data), but no paleomagnetism.

Paleocene (Riochican SALMA and older units) Epoch.—

For many years, the Paleocene record was represented by a single SALMA, the Riochican. Recent work suggests that there may be four, or more, distinct Paleocene SALMAs.

The oldest Cenozoic SALMA appears to be the early Paleocene Tiupampan SALMA. The new Chulpas fauna from Peru (Crochet and Sigé, 1993) has only been described preliminarily; it may be earliest Paleocene in age and slightly older than the Tiupampan SALMA. Neither the Tiupampan nor the Chulpas fauna have associated, direct geochronological constraints.

Short magnetostratigraphic sections of the Salamanca and Rio Chico Formations at Punta Peligro and Cerro Redondo in Argentine Patagonia, representing the Paleocene Riochican SALMA, have been correlated to Chrons C26r to C25n (Marshall and others, 1981). Bonaparte and others (1993) disputed part of this correlation, especially for the basal stratigraphic interval sampling the "Banco Negro Inferior" of the Salamanca Formation (which contains the fauna they used to propose a new SALMA, the Peligran; fauna considered either Tiupampan or possibly a distinct Peligran SALMA by Bond and others, 1994; tentatively considered Tiupampan by Pascual and others, 1992a, b and Bond and others, pers. commun. 1994), suggesting that while the majority of the longer sequence at Cerro Redondo (which includes the classical Riochican assemblages) might in fact span Chrons C26r to C25n, there may be problems with interpreting the basal part of that sequence and the very short section at Punta Peligro (both of which include the "Banco Negro Inferior").

The newly proposed Peligran SALMA (typified by the faunal assemblage from the "Banco Negro Inferior", Bonaparte and others, 1993) is considered broadly early to middle Paleocene age. The typifying fauna occurs in strata underlying (probably unconformably) Riochican faunas and overlying parts of the Salamanca Formation with Danian Foraminifera, overlies strata yielding low precision K-Ar dates of about 62.5–64 Ma, and occurs within a reversed polarity interval of uncertain correlation (correlated to Chron C26r by Marshall and others, 1981, part of which possibly may be Chron C27r instead, according to Bonaparte and others, 1993). Although Marshall and others (1981) included the "Banco Negro Inferior" in the Rio Chico Formation, Bonaparte and others (1993) pointed out that it is better assigned to the Salamanca Formation. (Bond and others, pers. commun. 1994, similarly call it the Hansen Member of the Salamanca Group, but without specifying a formation.) Bonaparte and others (1993) also suggested that there may be a missing normal polarity zone in the lower part of the Rio Chico Formation as the stratigraphic thickness between the reversed horizons in the "Banco Negro Inferior" and the first reliable reversed horizon in the Rio Chico Formation is greater than the thickness of the two normal polarity intervals identified from the upper part of the section, and that there is an unconformity between the Rio Chico and Salamanca Formations.

Therefore, they conclude that there is no reason that the "Banco Negro Inferior" (and thus the Peligran) *must* be younger than Chron C27n (as proposed by Marshall and others, 1981) because: (1) there is an unconformity of unknown duration, (2) there may be a missing normal polarity zone, (3) the foraminiferal faunas from the underlying parts of the Salamanca Formation may be older than previously thought (e.g., as old as early Danian [Bonaparte and others, 1993 based on data in Toumarkine and Lutherbacher, 1985] rather than late Danian [Bertels, 1975]), and (4) the radioisotopic dates from the underlying Salamanca Formation suggest possible correlation of the magnetozones to older chrons if one correlates to newer time scales (Cande and Kent, 1992). We would further point out that the old dates cited in point 4 above are poorly constrained and may not be useful for any precise correlation. Bonaparte and others (1993) suggested instead that the Peligran SALMA falls within Chron C27r. Unfortunately, there are only preliminary studies of the fauna and poor geochronological constraint for the "Banco Negro Inferior," and a lack of independent geochronologic evidence from the Tiupampan SALMA in Bolivia. This makes it difficult to determine the temporal relationships between the Tiupampan and Peligran SALMAs and to decide if indeed the Peligran SALMA is younger than the Tiupampan SALMA (as suggested by Bonaparte and others, 1993) or the "Peligran" SALMA is temporally equivalent to the Tiupampan SALMA (as suggested tentatively by Pascual and others, 1992a, b; Bond and others, pers. commun. 1994) and therefore invalid.

The same weak temporal constraints exist for the unnamed ("indeterminada"), post-Peligran and pre-Riochican mammal age of Bonaparte and others (1993). This unnamed interval apparently was considered by them to be equivalent to what has been called the "Itaboraian" SALMA in the past (e.g., Marshall, 1985), a usage they rejected because the Itaborai fauna of Brazil may be temporally mixed. According to Bonaparte and others (1993), typifying faunas for the new unnamed interval would include those from levels near the base of the Rio Chico Formation (or Group, Pascual and others, 1994, and Bond and others, pers. commun. 1994) containing *Carodnia* and *Kibenikhoria* (called the *Kibenikhoria* Faunal Zone and *Carodnia* Faunal Zone in Pascual and others, 1992b, and Bond and others, 1994, pers. commun. 1994), and probably the Las Flores fauna from south of Lago Colhue-Huapi. Pascual and others (1994) and Bond and others (1994, pers. commun. 1994) presented a more finely subdivided early Paleocene sequence in Patagonia (in superposition, from older to younger): (1) Hansen Member of the Salamanca Group (no formation specified; = "Banco Negro Inferior") containing faunas considered Peligran by Bonaparte and others (1993); (2) Peñas Coloradas Formation of the Rio Chico Group, equivalent to the *Carodnia* Faunal Zone of Pascual and others (1992b), and containing faunas that (together with the fauna of the *Kibenikhoria* Faunal Zone, see below) characterize the unnamed SALMA of Bonaparte and others (1993); (3) Las Flores Formation of the Rio Chico Group, containing the *Kibenikhoria* Faunal Zone of Pascual and others (1992b) and producing a rich new undescribed fauna almost certainly assignable to a valid Itaboraian SALMA (Goin, pers. commun. 1994) and would not require erecting the unnamed new mammal age of Bonaparte and others (1993); and (4) Koluel Kaike Formation of the Rio Chico Group, characterized by *Ernestokokenia* and containing classical Riochican SALMA assemblages.

Eocene (Casamayoran-Divisaderan? SALMAs) Epoch.—

Both the Casamayoran and Mustersan SALMAs generally have been considered early and late Eocene in age, respectively, as they overlie Paleocene units or underlie Oligocene or younger units, although there is no direct independent geochronological constraint available for either.

The Divisaderan is a problematic SALMA, previously considered to be of Late Eocene (e.g., Pascual and others, 1965), Early Oligocene (e.g., Pascual and Ortiz-Jaureguizar, 1990), Early Oligocene to questionably as young as Middle Oligocene (Marshall and others, 1983), or only of uncertain "more or less middle Cenozoic" (Simpson, 1970) age. It is typified by a depauperate fauna from Divisadero Largo, Argentina, characterized by an unusual mixture of "archaic" (Casamayoran-Mustersan SALMA) and "advanced" (Deseadan and younger SALMA) taxa (Simpson and others, 1962). Unfortunately, there are no older or younger faunas in direct superposition, and there is no independent geochronological information from the Divisaderan SALMA. Many workers have emphasized similarities to younger faunas, and considered the Divisaderan SALMA to be latest pre-Deseadan, or even the earliest part of the Deseadan, in age (Simpson and others, 1962; MacFadden and others, 1985; Marshall and others, 1986b). However, Bond (1991) suggested that the Divisaderan SALMA is much older than the Deseadan SALMA, and that it might only be of regional utility. Recent discovery and dating of the pre-Deseadan Tinguiririca Fauna make it clear that the Divisaderan SALMA is not part of the Deseadan SALMA, is older than the Deseadan SALMA, and probably also predates the Tinguiririca Fauna (see below; Wyss and others, 1990, 1993, 1994).

Eocene-Oligocene (Tinguiririca Fauna) Interval.—

A stratigraphic sequence from the central Chilean Andes has produced the Tinguiririca Fauna, which represents a new but yet unnamed SALMA, and the only South American Eocene radioisotopic dates (Wyss and others, 1990, 1993, 1994; Charrier and others, 1990, in press). Magnetostratigraphic sampling of this part of the sequence has produced a single normal polarity interval, tentatively correlated to Chron C12n (or conceivably C13n, Charrier and others, in press). The Chilean Tinguiririca Fauna is: 1) clearly biochronologically older than the Deseadan SALMA, and 2) at least as young as 31.5–32 Ma (and possibly extends as old as 36–37 Ma) and the faunal-bearing part of the sequence thus is latest Eocene-early Oligocene in age.

Marshall and Sempere (1993) and Sempere and others (1994) have incorrectly assigned the Tinguiririca Fauna to the problematic Divisaderan SALMA, based on an unsubstantiated notion of a "Divisideran aspect" of the fauna and supposed agreement between the radioisotopic dates for the Tinguiririca Fauna (Wyss and others, 1993, 1994) and a proposed 30–36 million year age for the Divisaderan SALMA (citing Marshall and others, 1986b). However, this correlation is certainly invalid; there are almost no taxa (at familial or lower taxonomic levels) in common between the Tinguiririca and Divisaderan faunas. Sempere and others (1994) provide no explicit justification for considering the Tinguiririca Fauna to be of Divisaderan "as-

pect" (nor do they define this ambiguous term), and there are no radioisotopic dates or any other direct geochronological constraints from the Divisaderan SALMA (or the South American SALMA sequence for the bracketing 30-my interval between about 65 Ma and 35 Ma). For further discussion, see Wyss and others (1993, 1994) and Charrier and others (in press).

Oligocene (Deseadan SALMA) Epoch.—

The age of the Deseadan SALMA has been very poorly understood, even though it has been considered an extremely important temporal interval because it followed a long gap in the fossil record, contained the first more "modern" mammalian assemblages, and had the oldest known occurrence of South American primates (previously of South American rodents also, although an older occurrence has now been documented, Wyss and others, 1993). As pointed out by Wyss and others (1994), the recent (Marshall and Sempere, 1993) suggestion to subdivide the Deseadan SALMA into parts predating and postdating the arrival of caviomorph rodents in South America is invalid. The Deseadan SALMA conventionally was considered of early Oligocene (e.g., Simpson, 1940; Patterson and Pascual, 1968) age. However, consideration of radioisotopic dates has yielded widely varying and even completely non-overlapping estimates of temporal duration for the Deseadan SALMA, spanning approximately as old as 34–37.5 Ma (Marshall and others, 1977), to 27–37.5 Ma (Marshall and others, 1983), to 21.5–34 Ma (Marshall, 1985; Marshall and others, 1986a, b), to as young as 21–29 Ma (MacFadden, 1985; MacFadden and others, 1985; Sempere and others, 1990, 1994).

In addition, correlation of the Salla, Bolivia Deseadan SALMA magnetochronology has been more problematic, and no magnetostratigraphies are available from classical Deseadan sequences in Argentina. Radioisotopic dates clearly indicate that at least the base of the Salla sequence is late Oligocene age. However, K-Ar dates conflict with fission-track dates within this section, causing MacFadden and others (1985) and McRae (1990) to suggest radically different correlations for the magnetostratigraphy. We consider the fission-track dates inaccurate; the available K-Ar dates (from the lower third of the section) span about 25–28 Ma and are concordant with superposition (within analytical errors). The K-Ar dates support MacFadden and others' (1985) correlation of the Salla Deseadan sequence to Chrons C10r to C6Cr rather than McRae's (1990) or Sempere and others (1990) correlation to the latest Oligocene-early Miocene Chrons C6Cr to C6r.

The new geochronologic results from the pre-Deseadan Tinguiririca Fauna (see above) dictate that the base of the Deseadan SALMA *must* be younger than 31–32 Ma, which is consistent with the K-Ar dates and magnetostratigraphic correlation of the Deseadan strata from Salla, Bolivia (see above and MacFadden and others, 1985). In addition, one of us (Swisher, unpubl. data) currently is redating Deseadan samples from Pico Truncado and Scarritt Pocket, Argentina (using $^{40}Ar/^{39}Ar$ analyses on splits from the same samples used by Marshall and others, 1977, for K-Ar dating). Those $^{40}Ar/^{39}Ar$ results indicate that the Deseadan SALMA sequence in Argentina spans only about 27–29 Ma. Both the older dates from Pico Truncado and the younger dates from Scarritt Pocket were inaccurate, although the oldest Deseadan date from Pico Truncado has been rejected without reanalysis or explicit justification by Sempere and others (1994).

The Deseadan SALMA currently is known to span from about 24.5–29 Ma based on data from Bolivia and Argentina and with certainty only 26–29 Ma in Argentine sequences, although there are Deseadan faunas that have not yet been dated. Thus, current geochronological constraints indicate that the Deseadan SALMA is restricted almost entirely to the Late Oligocene Epoch, rather than its usual assignment to the Early Oligocene Epoch (Marshall and others, 1986b, and references therein) or more recent placement as spanning the Late Oligocene-Early Miocene Epochs (Sempere and others, 1990, 1994).

Neogene Record

Overview.—

The South American Neogene record is much better sampled and chronologically constrained than the Paleogene record, except in the early Miocene Epoch. For the Neogene Period, there have been integrative studies of: (1) the type and other Santacrucian sequences from Argentine Patagonia, including paleomagnetics, K-Ar dates, fission-track dates, [40]Ar/[39]Ar dates, and marine biochronology (from the underlying Monte León Formation, Bertels, 1979; Marshall and others, 1977, 1983, 1986a; Bown and others, 1988; work in progress by Bown, Swisher, Buckley and Fleagle); (2) the type Friasian SALMA from Rio Frias/Rio Cisnes, Chile, paleomagnetics and at least one [40]Ar/[39]Ar date (Flynn and others, 1989a, b, unpubl. data); (3) studies in progress (Argentina-U.S. collaboration including Mazzoni, Vucetich, Franchi, Swisher, Kay, Madden) from classical "Friasian" sequences (particularly the Colloncuran and the "Rio Negro-type" faunas) from Argentine Patagonia, including extensive [40]Ar/[39]Ar dating, but no paleomagnetics yet (Madden and others, 1989, 1991); (4) the Miocene La Venta faunal sequence from the Magdalena River valley, Colombia (previously considered assignable to the "Friasian" or Santacrucian SALMAs, but clearly representing a new mammalian biochronologic interval, the Laventan Stage/Age; Madden and others, in press), detailing extensive paleomagnetics, K-Ar dates, and [40]Ar/[39]Ar dates (Flynn and others, in press); (5) a sequence from Quebrada Honda, Bolivia (fauna equivocally considered assignable to the "Friasian" or Santacrucian SALMAs, but geochronological data suggest it is younger than either), presenting paleomagnetics and several K-Ar dates (MacFadden and others, 1990); (6) two Miocene Bolivian faunas (Hoffstetter, 1986) dated with [40]Ar/[39]Ar techniques (analytical data as yet unpublished, see MacFadden and Anaya, 1993) at about 15 Ma (Cerdas-Atocha) and between 12.8 and 6.8 Ma (Quehua); (7) a long, continuous sequence with superposed Huayquerian to Montehermosan SALMA assemblages from Catamarca Province, Argentina, providing paleomagnetics and a series of K-Ar dates (most dates rejected by Marshall and others, 1979, 1983, and Butler and others, 1984, without detailing of explicit rejection criteria, as contaminated or suffering from excess [40]Ar); (8) a late Miocene sequence from Micaña in the Eastern Cordillera of Bolivia, temporally correlative with the Huayquerian SALMA (but mammalian fauna not assigned with certainty to that SALMA), summarizing two fission-track dates on zircons (one from the fossiliferous sequence and one from unconformably overlying strata) and paleomagnetics (MacFadden and others, 1990); (9) the type sequence for the Uquian SALMA from Jujuy Province, Argentina detailing paleomagnetics and several

K-Ar dates (Marshall and others, 1983); and (10) Ensenadan strata from Tarija, Bolivia incorporating paleomagnetics and one zircon fission-track date (MacFadden and others, 1983).

Santacrucian SALMA (early-middle Miocene Epoch).—

The Santacrucian SALMA has been studied by Marshall and others (1986a), who suggested it spanned from about 15–18 Ma. The 21.7 Ma sample from Felton's Estancia (Evernden and others, 1964; Marshall and others, 1986a) is being redated using [40]Ar/[39]Ar techniques, and this earlier K-Ar date clearly is much too old to be consistent with revised chronologies and other data. Work in progress by T. Bown, J. Fleagle, G. Buckley, C. Swisher and others, is refining the Santacrucian geochronology. Preliminary results from sequences at Felton's Estancia, Monte León and Monte Observación suggest that the Santacrucian SALMA is known to range, with certainty, only from about 16–17.5 Ma. The available Santacrucian magnetostratigraphies span a relatively short temporal interval with part of a reversed interval near the base followed by a normal interval and a long reversed interval. This short, non-diagnostic magnetostratigraphic pattern is difficult to correlate uniquely to the time scale and depends heavily on the accuracy of the available time scales and on precision of radioisotopic dating control within the sections. Available constraints suggest correlation either to Chrons C5Cr to C5Cn.2r (or C5Cr to C5Br, if the lower normal actually represents a compressed C5Cn.1–3 with no recovery of intervening reversals) or Chrons C5Dr to C5Cr. The former correlation is favored if one uses the Cande and Kent (1992) time scale and available [40]Ar/[39]Ar constraints of about 16.3–16.8 Ma for the entire section (Swisher, unpubl. data), although either of the two variants beginning with Chron C5Cr at the base imply dramatic sedimentation rate shifts within several discrete sections, and/or possibly lack of preservation of several reversed polarity zones in a high sedimentation rate section. If the Cande and Kent (1992) time scale is miscalibrated by a few hundred thousand years in this interval (and is closer to the Berggren and others, 1985, time scale), or if extreme fluctuations in sedimentation rates seem unlikely, the latter correlation may be favored.

"Friasian" SALMA (?early-middle Miocene Epoch).—

The "Friasian" SALMA currently is one of the most perplexing problems in the SALMA sequence. Faunas considered "Friasian" in age come from a variety of localities previously assigned to several different biochronologic units ("Mayoan", "Friasian", "Colloncuran"), which may be temporally sequential and continuous, sequential but with hiatuses between them, overlapping, or even equivalent. The faunas have not been reviewed recently, but work in progress by several international collaborations promises to refine our understanding soon. The type sequence of the Friasian SALMA, at Rio Frias/Cisnes, Chile has a magnetostratigraphy spanning several polarity zones (Flynn and Swisher, unpubl. data). Although this section has not been correlated yet to the GMPTS, a single [40]Ar/[39]Ar date of about 16.5 Ma (not "ca. 17 Ma" as reported in Marshall and Salinas, 1990) from low in the section (conformably below most of the fossiliferous horizons) suggests that: (1) the type Friasian assemblage may overlap with, or closely follow, in time the Santacrucian SALMA, and (2) the type Friasian spans several magnetozones and correlates temporally, at least in part, with "Colloncurense" faunas from Argentina (Flynn and Swisher, unpubl. data). We do not believe the type Friasian sequence to be temporally correlative with early Santacrucian faunas, as suggested by Marshall and Salinas (1990) based on their: (1) study of marsupials (although no marsupials were known, for comparison, from possible "Friasian" faunas in Argentina), (2) brief survey and no description of the type Friasian rodents, (3) stratigraphy, and (4) incorrect citation of an age for a sample then undergoing analysis by C. Swisher.

Study of "Friasian" Argentine faunas currently is underway by an Argentine-US collaboration (Mazzoni, Vucetich, Franchi, Swisher, Madden, Kay). Their work has produced at least twenty-five $^{40}Ar/^{39}Ar$ analyses on an ignimbrite from 4 different localities, all bracketed by "Colloncuran" faunas. Dates from this ignimbrite cluster tightly at around 15.7 Ma (Madden and others, 1989, 1991; Swisher, Mazzoni, and others, unpubl. data), rather than 14.5 Ma suggested by previous K-Ar dates (Marshall and others, 1977, 1983, 1986a), indicating that "Colloncuran" faunas are close in age to, but possibly distinguishably younger than, the Santacrucian SALMA. With an apparent temporal gap, "Rio Negro-type" faunas may be 2–3 my younger than "Colloncuran" faunas. No "Mayoan" faunas have yet been dated directly.

The La Venta Fauna (middle Miocene Epoch).—

The La Venta Fauna from Colombia has been variably considered to be either "Friasian" (e.g., Stirton, 1953; Hirschfeld and Marshall, 1976) or Santacrucian (e.g., Kay and others, 1987) age. An extensive study on the geochronology (including magnetostratigraphy and $^{40}Ar/^{39}Ar$ dating, Flynn and others, in press) and mammalian fauna (see Kay and others, in press) of a Honda Group section spanning more than 1000 m makes La Venta one of the best constrained mammal-bearing sequences in the world. The geochronologic data clearly indicate that the stratigraphic sequence containing the La Venta fauna spans about 2 my, from about 13.8 to 11.8 Ma (Flynn and others, in press), almost certainly representing Chrons C5An.1n-C5ABn. The La Venta faunal interval does not correlate temporally with the Santacrucian, type Friasian, or other "Friasian" SALMA assemblages (Flynn and others, in press). This temporal span falls partly or entirely within what appears to be a hiatus within the known sequence of Patagonian "Friasian" faunas (between the older Colloncuran and the younger Mayoan) (Madden and others, in press). However, $^{40}Ar/^{39}Ar$ dating of those sequences is on-going, and the boundaries of the relevant SALMA assemblages are not yet precisely determined, so it is not yet possible to determine if the La Venta Fauna is temporally discrete or partly overlaps in time with Colloncuran or Mayoan-aged faunal assemblages. Madden and others (in press) erect a new formal chronostratigraphic unit, the Laventan Stage, based on a stratotype (containing the La Venta Fauna) within the Honda Group, Huila Province, Colombia. The Laventan Stage is the first formal chronostratigraphic unit within South America to be based on mammal-bearing strata. It is of documented regional scope and may be more broadly applicable across all of South America, although the endemism of the tropical La Venta Fauna may make it difficult to correlate to high-latitude faunas (which typify most SALMAs). The Laventan Age, the geochronologic unit corresponding to the time interval represented by the Laventan Stage, may be considered of continental (or

even global) scope, by definition. As with any formal chronostratigraphic or geochronologic unit, the ability to use biostratigraphic data to *recognize* the temporal unit is dependent on the geographic and environmental range of the defining and characterizing taxa; however, this does not preempt correlation, recognition, and use of the units using other chronologic criteria. Practical recognition of the Laventan in high-latitude areas of South America may only be possible using non-biochronologic information (e.g., radioisotopic dating, magnetic polarity chronology, etc.), because of the paucity of temporally-restricted taxa present in both tropical and high-latitude faunas.

The Quebrada Honda Fauna (middle Miocene Epoch).—

MacFadden and others (1990) correlated a magnetostratigraphy from Quebrada Honda in Bolivia to Chrons C5AAn to C5An.1r. The faunal assemblage from this sequence is difficult to correlate to the standard SALMA sequence, perhaps because of faunal provinciality or the fact that temporally equivalent faunas are not well-represented in the classical Patagonian sequences, and the Quebrada Honda Fauna has been suggested to be probably "Friasian" in age. The temporal correlation indicated by the magnetostratigraphic correlation and available radioisotopic dates suggest that the Quebrada Honda Fauna is younger than standard "Friasian" SALMAs and may be temporally correlative with the Colombian La Venta Fauna (which represents a new SALMA, perhaps applicable only at low latitudes; see below).

Huayquerian-Montehermosan-Chapadmalalan SALMAs (late Miocene-Pliocene Epochs).—

Correlation problems arise even in intensively studied temporal intervals. One example includes the Huayquerian to Montehermosan interval, and the location of the boundary between those SALMAs. The Huayquerian to Montehermosan SALMA interval is relatively well-sampled paleomagnetically in Bolivia and NW Argentina but not in the type sequences in Argentina. The classical sequences at Chiquimil and Corral Quemado, Catamarca Province, NW Argentina containing Huayquerian and Montehermosan faunas in superposition have produced both radioisotopic dates (but only using K-Ar dating techniques) and lengthy magnetostratigraphies (Marshall and others, 1979, 1983; Butler and others, 1984) spanning continuously approximately from Chron C4An to Chron C2Ar (of Cande and Kent, 1992, see Fig. 3; mid-Chron 8 to late Gilbert Chron, in old terminology; Butler and others, 1984; see also Marshall and others, 1979, 1983, for earlier interpretations of less complete data).

Sequences from the Bolivian Altiplano containing faunas considered to be "early" Huayquerian (although it is not clear whether this assignment is based only on faunal composition, or inferences based on both radioisotopic dates and fauna), Montehermosan, possibly early Ensenadan, and possibly early Lujanian in age, have recently produced numerous $^{40}Ar/^{39}Ar$ dates (complementing and refining older K-Ar dates) and some associated magnetostratigraphic information (Servant-Vildary and Blanco, 1984; Thouveny and Servant, 1989; Marshall and others, 1992 and references cited therein).

However, results from the NW Argentine and Bolivian studies yield somewhat different estimates for the age of the Huayquerian/Montehermosan boundary. Marshall and others

(1979) considered the boundary to be about 5.0 Ma, based on preliminary magnetostratigraphy and K-Ar dates in Argentina and K-Ar dates from below the Bolivian Montehermosan fauna. Butler and others (1984) tentatively considered the boundary to be about 6.0 Ma, based on more extensive magnetostratigraphic data and K-Ar dates. Marshall and others (1992, Fig. 4, p. 1) also placed the boundary at 6.0 Ma, apparently because ". . . faunas of Montehermosan age are bracketed below by the Toba 76 and Cota Cota tuffs (*ca.* 5.4 Ma) . . . ," and early Huayquerian faunas are underlain by a 9.0 Ma tuff. However, dates are only available from horizons above Huayquerian faunas or below Montehermosan faunas; the only dates available from directly within strata containing Montehermosan faunas are a series of highly variable dates from the Bolivian Remedios Formation (or correlatives, Marshall and others, 1992). Our evaluation of all the available radioisotopic dates and location of the boundary within the early part of Chron C3Ar (Butler and others, 1984, using chron terminology and the time scale of Cande and Kent, 1992, 1995, and Berggren and others, this volume) indicates the best estimate for the Huayquerian/Montehermosan boundary is about 6.8 Ma, and certainly older than 6.5 Ma.

MacFadden and others (1990) sampled a late Miocene (associated fauna of uncertain correlation to the SALMA sequence) section at Micaña, Bolivia, which they correlated to Chrons C4r.1n to C4n.1n ("Chron 7") and possibly including Chron C4r.2r ("youngest part of Chron 8") at the base and the oldest part of Chron C3Br ("oldest part of Chron 6") at the top. Although the mammalian fauna is of uncertain correlation to the SALMA, temporal correlation of this sequence would indicate that it is Huayquerian in age.

New $^{40}Ar/^{39}Ar$ dates from late Cenozoic strata in Bolivia (Marshall and others, 1992) provide the first direct (albeit from overlying strata) radioisotopic control on the age (minimally 2.65–2.9 Ma) of the younger boundary of the Montehermosan SALMA.

Orgeira and Valencio (1984) and Orgeira (1990) reported the magnetostratigraphy of late Cenozoic strata of coastal Argentina, including the Chapadmalal Formation (considered Montehermosan *sensu lato* in age by them, but probably better considered either undifferentiated Montehermosan-Chapadmalalan or Chapadmalalan alone). There, the Chapadmalalan (or late Montehermosan *sensu lato*) sequence was of early Chron C3n.1r to C2An.3n ("Gilbert to early Gauss") age. This age span indicates that there is a temporal overlap between the age of the lower Chapadmalal Formation in that area and Montehermosan sequences elsewhere (Butler and others, 1984). This suggests either that the Chapadmalal Formation in that area is really Montehermosan (*sensu stricto*) in age or, if the formation is Chapadmalalan, that there is a temporal overlap between the Chapadmalalan and Montehermosan SALMAs. Either interpretation reinforces previous observations that the Montehermosan and Chapadmalalan are faunally similar and close in age.

MacFadden and others (1993) recently reported a magnetostratigraphy associated with the early to middle Pliocene Inchasi local fauna from Bolivia. The Inchasi mammalian fauna is of undifferentiated Montehermosan-Chapadmalalan age, but lack of North American immigrant taxa suggests Inchasi is pre-Uquian. Based on the known age ranges of the Uquian and Montehermosan-Chapadmalalan faunas elsewhere, as well as

corroborating radioisotopic constraints on sequences from presumed correlative sequences at Ayo Ayo several hundred kilometers northwest (Thouveny and Servant, 1989; Marshall and others, 1992), MacFadden and others (1993) correlated the Inchasi magnetostratigraphy to Chrons C3n.1n to C2An.2r ("late Gilbert [Cochiti event] through early Gauss [Mammoth event] magnetochrons"), spanning from about 4–3.3 Ma (using the time scale of Cande and Kent, 1992).

Uquian SALMA (late Pliocene to Pleistocene Epochs).—

Study of the type Uquian SALMA (Marshall and others, 1982a) generated two purportedly reliable K-Ar dates (about 2.8 Ma) and a magnetostratigraphy correlated to latest Chron C2An.1n to the older part of Chron C1r.2r "from just below the Gauss/Matuyama boundary [latest Chron 3n1] into the older part of the long reversed interval in the later half of the Matuyama Epoch [Chron 2r2]" (Marshall and others, 1982a; brackets added). Studies of Uquian sequences, in superposition above Montehermosan (*sensu lato*, including Chapadmalalan) strata, from coastal Buenos Aires Province, Argentina by Orgeira (1987, 1990) indicated an older age, spanning much of Chron C2An (Chron C2An.3n- C2An.1n; "Gauss"). The almost complete lack of temporal overlap between the type and coastal Argentine Uquian sequences suggests either a long time span (and discontinuous stratigraphic representation from NW to coastal Argentina) or possible correlation problems with the magnetostratigraphies of one or both sections. Further study will be required to resolve this issue.

Ensenadan SALMA (Pleistocene Epoch).—

The Ensenadan SALMA has been sampled at Tarija in Bolivia (MacFadden and others, 1983; see also Servant and Blanco, 1984, on the Charaña Formation of Bolivia), where it spans Chron C1r.1n to early Chron C1n, "the Jaramillo normal subchron [Chron 2n1] to the early Brunhes [Chron 1]" (MacFadden and others, 1983; brackets added). Orgeira (1990) considered the Ensenadan SALMA to span Chrons C1r to C2r ("Matuyama") based on studies of the Miramar Formation in coastal Buenos Aires Province, Argentina. Questionably Ensenadan strata were sampled in several very short sections in Buenos Aires, Argentina (Nabel and Valencio, 1981; Valencio and Orgeira, 1983); they were largely of reversed polarity and of "Matuyama" age, with a maximum span of Chron C1r to Chron 2r. However, these short Ensenadan(?) sections have no radioisotopic dates, weak biostratigraphic control, and are difficult to correlate on the basis of pattern alone (e.g., the sections might represent only Chron C1r including the short normal interval of Chron C1r.1n, rather than spanning Chron C1r through C2r lacking preservation of the short normals of Chrons C1r.1n and C2r.1n). There is either a small temporal gap between the Uquian and Ensenadan strata so far sampled paleomagnetically, representing the later part of Chron C1r.2r (or, in older terminology, the late part of Chron 2r2, preceding the Jaramillo normal subchron), or less likely, significant temporal overlap between the earlier part of the Ensenadan SALMA (if it extends into Chron C2) and the Uquian SALMA.

Lujanian SALMA (Pleistocene Epoch).—

Orgeira (1990) reported a Chron C1r to C1n (late "Matuyama" to "Brunhes") age for Lujanian sequences from coastal

Buenos Aires Province, Argentina. A very short section of the "Bonaerense" age (Valencio and Orgeira, 1983; latest Cenozoic and probably equivalent to the Lujanian SALMA), also from Buenos Aires Province, is of normal polarity and assignable to Chron C1n ("Brunhes").

MARINE INTERTONGUES IN TERRESTRIAL SEQUENCES

Unfortunately, there are very few time intervals or geographic areas from the Cenozoic sequence of South America in which fossiliferous marine strata intertongue with fossiliferous terrestrial sequences. The few marine sequences that do intertongue, generally represent very shallow-water environments that have not yielded good microplankton faunas or floras.

Three Cenozoic South American transgressive phases are well-represented in epicontinental marine sequences from several high-latitude Patagonian basins (Zinsmeister, 1981). One of the most famous epicontinental marine sequences is the "Patagoniano" or "Patagonian Stage" (Simpson, 1940, including "Formation Patagonienne" and "Etage superpatagoneen" of the "Formation Santacruzienne" of Ameghino, 1900–1902; see Marshall and others, 1983), considered of latest Oligocene or early Miocene age, but now considered to represent at least three distinct units "ranging in age from Eocene to early Miocene" (Marshall and others, 1983, p. 27, citing pers. commun. from Zinsmeister; see also Zinsmeister, 1981). These strata include sequences now called the Monte León Formation (Camacho, 1974; middle to late Oligocene, upper Rupelian and Chattian; forams, Bertels, 1970, 1975; molluscs, Camacho, 1974; Zinsmeister, 1981; Marshall and others, 1983); an unnamed unit probably correlative to the Monte León Formation (overlying Colhuehuapian SALMA sequences; late middle to early late Miocene mollusc *Nodipecten* sp.; Smith and Zinsmeister, 1982; Marshall and others, 1986b); "Estratos con *Monophorastery* y *Venericor*" (late Eocene molluscs, Camacho, 1974; see also Zinsmeister, 1981; Marshall and others, 1983); and San Julian Formation (early [and middle?] Eocene molluscs, Camacho, 1974; but considered late Eocene to early Oligocene in age, upper Priabonian and lower Rupelian by Bertels, 1970, 1975, based on forams; see discussion in Zinsmeister, 1981). The marine Monte León Formation (middle to late Oligocene or possibly Early Miocene forams, Bertels, 1979) conformably underlies classical Santacrucian SALMA strata in Argentine Patagonia. Colhuehuapian SALMA strata in Chubut Province, Argentina are conformably overlain by an unnamed marine unit that is older than late middle to early late Miocene in age (based on occurrence of the mollusc *Nodipecten* sp.; see above—Smith and Zinsmeister, 1982, p. 235, as cited in Marshall and others, 1986b, p. 941). Although the Colhuehuapian SALMA generally is considered older than the Santacrucian SALMA and younger than the Deseadan SALMA, based on composition of the mammalian faunas, there is no locality in which assemblages assigned to all three ages occur in direct superposition. With respect to marine biochronology evidence, all that is certain is that Colhuehuapian SALMA assemblages stratigraphically overlie Deseadan SALMA assemblages in Chubut, the Colhuehuapian SALMA is older than late middle-early late Miocene (based on the age of the overlying marine unit), and the Santacrucian SALMA is younger than middle-late Oligocene or possibly early Miocene (based on the

age of the underlying marine Monte León Formation). By themselves, those data do not preclude the Colhuehuapian and Santacrucian SALMAs from overlapping in time. Zinsmeister (1981) reported on a small marine macroinvertebrate fauna collected by Elmer Riggs of the Field Museum from the San Julian Formation at Punta Casamayor. At that locality, San Julian Formation marine strata unconformably overlie mammal-bearing strata of the Casamayoran SALMA (*Notostylops* beds of Ameghino). Given the conflict between ages assigned to the San Julian Formation based on molluscs (early-middle Eocene, Camacho, 1974) and planktonic forams (late Eocene-early Oligocene, Bertels, 1979), all that can be said with confidence is that the age of the Casamayoran SALMA is likely Eocene, older than late Eocene to early Oligocene, and possibly older than early to middle Eocene.

The Riochican SALMA (and the older Itaboraian or unnamed interval, and Peligran from the same sequence, discussed above) is the other mammalian biochronologic interval constrained in age by associated marine units. In its type area but not at the type section, the mammal-bearing Rio Chico Formation/Group is underlain conformably (Marshall and others, 1981, 1983) or unconformably (Bonaparte and others, 1993) by the marine Salamanca Formation/Group. The Salamanca Formation/Group contains planktic foraminifera indicative of late Danian age (Bertels, 1970, 1975, 1979), and the oyster *Odontogryphaea* is said to indicate a poorly resolved Thanetian or Ypresian age (Stenzel, 1945). Data in Toumarkine and Lutherbacher (1985) suggested that the available biostratigraphic evidence from the parts of the Salamanca Formation/Group underlying the Peligran to Riochican assemblages would permit a correlation as old as early Danian Age (Bonaparte and others, 1993).

SALMA CHRONOLOGY, 1994

Our summary of the best estimate of temporal calibration and correlation for the SALMA sequence is shown in Figure 1. This chronology is based on our evaluation of all available geochronologic data from superpositional stratigraphy, mammalian biochronology, marine biochronologies (both macroinvertebrate and microplanktonic), magnetic polarity stratigraphies, and radioisotopic dating (emphasizing new ^{40}Ar/^{39}Ar results). Given the great pulse of activity in geochronologic studies currently in progress in the South American Cenozoic terrestrial record, we expect significant future advances in refining and modifying this summary chronology.

REGIONAL/GLOBAL EVENTS REFLECTED IN THE SOUTH AMERICAN TERRESTRIAL RECORD

The South American faunal and floral record can be combined with available geochronologic information to evaluate timing and pattern of major biotic and environmental changes and events. It is clear that Cenozoic terrestrial biotas responded to both global and regional biogeographic, climatic, tectonic, plate movement, sea-level, ecologic, and environmental changes and events (see Pascual and Ortiz Jaureguizar, 1990). These include major plate tectonic reorganizations and associated biogeographic events (e.g., Mesozoic-early Cenozoic Gondwanan [and subsidiary North American] continental connections and biogeographic relationships, final separation of

Antarctica during the ?Eocene Epoch, formation of the Isthmus of Panama in the Pliocene Epoch, etc.), global Eocene/Oligocene boundary events (including climate change, initiation of oceanic deep water flow through the Drake Passage, onset of major Antarctic ice cap formation, expansion of open habitats [wooded grasslands and grasslands], increase in hypsodont mammalian taxa, and major clade extinction, origination, and diversification), phases of Andean uplift, Pliocene-Pleistocene Epoch glaciation/climatic changes, etc. General summaries of major patterns in the evolution of South American mammalian diversity are presented in Patterson and Pascual (1968), Stehli and Webb (1985), Marshall and Cifelli (1990), and Pascual and Ortiz Jaureguizar (1990, 1992). Many of the Late Cenozoic events have received extensive attention in the past (e.g., Webb, 1978; Marshall and others, 1982b; papers in Stehli and Webb, 1985; Marshall, 1988), so we focus below on brief summaries of some of the older biologic and geologic patterns and events.

Modern and fossil floras clearly indicate affinities between floras from South America and other Gondwanan continents, reflecting their Mesozoic connections. Crane and Lidgard (1989) and Lidgard and Crane (1990) also noted that angiosperms appear to originate in the tropics during the Cretaceous Period, then diversify and spread to more temperate latitudes globally. More detailed analysis of the South American record will provide intriguing insights into biotic endemism, the timing and biogeography of angiosperm diversification, and the plate tectonic and biogeographic relationships among Mesozoic continents.

There is but a single well known South American Mesozoic mammal fauna, the late Cretaceous Los Alamitos Fauna of Argentina. Additionally, there are three other poorly known mammal faunas that have been considered of Cretaceous age. As one might expect, the scarce, but tantalizing evidence from these Cretaceous and early Cenozoic mammal faunas reflects geographic connections to other "Gondwanan" continents (particularly Antarctica and Australia; little evidence is available from Africa). However, some taxa from those South American faunas also have relationships with taxa and faunas from northern Pangean continents, suggesting either a broad Pangean distribution for many late Cretaceous-early Cenozoic taxa, or more likely a more complex temporal pattern to the biogeographic history of South America.

There is a single mammal taxon known from a single pocket in the type section of the Early Cretaceous (Late Neocomian, based on charophytes and ostracods from intergradationally underlying marine units) La Amarga Formation, Neuquen Province, Argentina (Bonaparte, 1990; Hopson and Rougier, 1993, and references therein). However, this taxon, *Vincelestes* is represented by exquisite material from at least 9 individuals (6 complete skulls, 17 lower jaws, postcrania), and it is of great significance for understanding therian mammal anatomy, phylogeny, and evolutionary diversification because it is the sister-taxon to Theria (marsupials plus placentals) (Bonaparte and Rougier, 1987; Rougier and others, 1992; Hopson and Rougier, 1993).

The Los Alamitos Fauna (Los Alamitos Formation) from Rio Negro Province, Argentina is the only diverse mammalian fauna from the Mesozoic Era of South America. It has been considered to characterize an "Alamitian" SALMA (e.g., Bonaparte, 1990; Pascual and Ortiz Jaureguizar, 1992). The Los Alamitos

Fauna is considered of Late Cretaceous (Campanian) age because it unconformably overlies volcanic rocks of the Jurassic Marfil Formation to terrestrial deposits of the Middle Cretaceous Chubut Group and is overlain by the Maastrichtian-Danian marine beds of the Roca Formation (Bonaparte, 1990). The mammalian fauna includes about 14 species of non-tribosphenic mammals, including one species of triconodont, one or two multituberculates, 5 species of symmetrodont therians, 5 species of eupantotheres, and the strange endemic multituberculate *Gondwanatherium* (Bonaparte, 1990; Krause and Bonaparte, 1990; Krause and others, 1992). This fauna has a high degree of endemism with respect to all early through late Cretaceous faunas from Laurasian continents (Bonaparte, 1990; Pascual and Ortiz Jaureguizar, 1992). Bonaparte (1990) pointed out that in Laurasia tribosphenic mammal groups strongly dominated non-tribosphenic mammals in diversity by the Cretaceous Period, whereas there are no tribosphenic mammals at all in this Late Cretaceous South American fauna. Further, there clearly was an endemic radiation within all the groups of non-tribosphenic mammals, indicated particularly by specialized taxa and high-level taxonomic differentiation in the multituberculates, symmetrodonts, and dryolestoids, and that these patterns suggest "long-term isolation of Gondwana and Laurasia, perhaps from the Late Jurassic until the Late Cretaceous" (Bonaparte, 1990). Because there are so few mammal faunas of comparable age known from other Gondwanan continents (Africa, Australia, Antarctica, India) and none from elsewhere in South America, it is unclear at what geographic level the endemism indicates isolation. It may be that there was strong biogeographic distinction within South America, or of South America relative to Laurasia, or a broader Gondwanan/Laurasian isolation, or endemism of South American faunas with respect to both those of Laurasia and those of other Gondwanan continents.

The Laguna Umayo local fauna of Peru contains just 3 confidently identified mammalian species and several other taxa, all of which are known only from tooth fragments, including (Sigé, 1972; see discussion in Marshall and others, 1983; Muizon, 1991) a periptychid condylarth (Van Valen, pers. commun. 1994, suspects this specimen probably is not a periptychid and that all South American records, including his own reports, Van Valen, 1988, of the family Periptychidae are dubious, which would make the family endemic to North America) or notoungulate, a peradectid marsupial (a geographically widespread group), a pediomyid marsupial (a group common in the Late Cretaceous of North America), and an indeterminate didelphid marsupial and an indeterminate condylarth. Van Valen (pers. commun. 1994) suspects that the Laguna Umayo Fauna is Cenozoic rather than Cretaceous age.

The new Chulpas Fauna (Crochet and Sigé, 1993) occurs in superposition about 200 m above the Laguna Umayo Fauna, in the same local section of the Umayo Formation of Peru. Preliminary study (Crochet and Sigé, 1993) of the fauna lists six species of marsupials (including a caroloameghiniine peradectid, the only taxon formally identified to genus or species; the oldest recorded caenolestid; and several didelphids) and four eutherians (a "proteutherian" and several indeterminate notoungulates). Crochet and Sigé (1993) considered the Chulpas fauna to be of uncertain age, close to the Cretaceous/Paleocene boundary, but likely of early Paleocene age, and probably older than the Tiupampa Fauna and Tiupampan SALMA (based on indirect sedimentologic evidence and stratigraphic correlation).

The Tiupampan SALMA (early [?earliest] Paleocene), Tiupampa Fauna (Muizon, 1991) of Bolivia contains at least 3 mammalian taxa ("Proteutheria", Pantodonta, up to 3 "condylarth" ungulate families; although Van Valen, pers. commun. 1994, and in Bonaparte and others, 1993, considers all Tiupampa condylarths to belong to a single endemic condylarth taxon, within Mioclaenidae) that document a biogeographic connection between North America (and possibly Asia) and at least part of South America during the late Mesozoic-early Cenozoic, as well as several South American endemic groups (but interestingly, no edentates, an endemic and diverse South American group), but no taxa that would strongly suggest a broader Gondwanan distribution (although coeval early Paleocene deposits are not well known on other Gondwanan continents).

The Punta Peligro local fauna from the "Banco Negro Inferior" of the Salamanca Formation is the oldest Cenozoic fauna from the classical mammal-bearing sequences of Patagonia. This fauna characterizes the Peligran SALMA (of Bonaparte and others, 1993), although the fauna was tentatively considered by Pascual and others (1992a, b) to be equivalent to the Tiupampan, at least in age. Interestingly, the mammalian assemblage: (1) contains a monotreme, suggesting biogeographic relationships to Australia and presumably Antarctica also (Pascual and others, 1992a, b), (2) yields a multituberculate belonging to an endemic late Mesozoic-early Cenozoic South American or Gondwanan radiation, and distinct from coeval clades from northern hemisphere continents (Krause and Bonaparte, 1990; Krause and others, 1992; Bonaparte and others, 1993, Pascual and others, 1993), and (3) has several condylarths suggesting biogeographic and phylogenetic relationships to both the Bolivian Tiupampan fauna, and more distantly to the Paleocene of North America (Bonaparte and others, 1993). Like the Tiupampan assemblages, this Patagonian Paleocene fauna lacks any evidence of the presence of edentates. The Punta Peligro local fauna also preserves non-mammalian vertebrates, including isolated fish vertebrae, 3 or more species of chelid turtles, a possible pipid frog and 2 species of leptodactylid frogs, an alligatorid and a questionable crocodylid. Most of these taxa have relationships with Cretaceous to Cenozoic South American groups, and support general South American or Gondwanan biogeographic affinities of the Punta Peligro fauna, although the crocodiles have Laurasian affinities and probably represent immigrants from Central or North America as crocodiles are not yet known from the Mesozoic of South America (Bonaparte and others, 1993).

Recent (Godthelp and others, 1992) description of an early Eocene assemblage from Australia (predating its plate tectonic separation from Antarctica), including a "condylarth-like" placental mammal and a diverse marsupial fauna (most are plesiomorphic, but one "resembles a family of marsupials from the Casamayoran (Early Eocene) of South America", Godthelp and others, 1992), also suggests biogeographic affinities between Australia and other Gondwanan continents during the late Mesozoic through early Cenozoic Eras.

New discoveries from the ?late Eocene of Antarctica (e.g., astrapothere, ?litoptern and toxodont "ungulates", polydolopid marsupials, tardigrade edentate; Woodburne and Zinsmeister, 1982, 1984; Case and others, 1988; Bond and others, 1989;

Carlini and others, 1990; Pascual and Ortiz Jaureguizar, 1991; Hooker, 1992; Marenssi and others, 1994) indicate a biogeographic association between South America and Antarctica at least until well into the Cenozoic Era. However, the generic-level endemism of the Antarctic taxa, relative to the coeval nearby faunas from Argentine Patagonia, suggests that significant land-bridge connections between the two continents were disrupted prior to the late Eocene Epoch, consistent with available plate tectonic and oceanographic evidence suggesting continuous surface water flow through the Drake Passage by the late Eocene and deep water flow by the early Oligocene Epoch.

Previously, Primates (platyrrhine New World monkeys) and Rodentia (caviomorph hystricognaths) were observed to first appear in the South American fossil record during the Deseadan SALMA (late Oligocene). Recent (Wyss and others, 1993) discovery of the earliest South American caviomorph in beds (the Tinguiririca Fauna, Chile) older than those producing the oldest South American primates complicates the picture slightly. The origin of both groups has been widely debated, but it appears likely, based on a variety of evidence, to have been Africa (Wyss and others, 1993, Flynn and others, 1995, and references cited therein). It is conceivable that some type of "sweepstakes" dispersal of these taxa from Africa to South America is related to the dramatic climatic and oceanographic changes occurring around the Eocene/Oligocene boundary, including increased latitudinal temperature gradients and increased intensity of south Atlantic oceanic surface water gyre circulation. It is possible that both primates and rodents arrived simultaneously through the same "sweepstakes" dispersal event (and therefore the apparent lag between the first appearance of rodents and primates in the fossil record may be due to the low preservation potential of Neotropical primates), or alternatively that their immigration occurred at different times. Additionally, evidence from the new central Chilean Andes Tinguiririca Fauna, of latest Eocene-early Oligocene age, suggests that changes from forest-dominated biomes to more open habitats (woodlands and savanna grasslands?) occurred earlier in South America than had been believed previously and earlier than on other continents, possibly in response to regional climatic changes due to an early phase of Andean uplift (Wyss and others, 1993, 1994; MacFadden and others, 1994; see also Patterson and Pascual, 1968).

Beginning in the Miocene but accelerating during completion of the Isthmus of Panama and initiation of Arctic ice caps at about 3–3.5 Ma, dramatic faunal changes associated with the Great American Biotic Interchange set the stage for creation of the modern South American terrestrial ecosystems. The large-scale faunal and floral provinciality (areas of endemism) seen in modern faunas and floras (due to climatic gradients, geographic and ecologic barriers, historical biogeography, and other causes) existed by the middle to late Miocene Epoch and may have been present even earlier (e.g., there are faunal differences between classical Patagonian ?early Eocene Casamayoran SALMA assemblages and correlatives from the Lumbrera Formation and Salta Group of NW Argentina, Pascual and others, 1981, Pascual and Ortiz Jaureguizar, 1990). The strongest evidence for major biotic provinciality by the mid-late Miocene Epoch comes from the biota of the La Venta sequence in the Magdalena River Valley of Colombia (Kay and others, in press, Flynn and others, in press). The La Venta Fauna clearly is very different from any described elsewhere in South America, with no species and very few genera shared between it and well-known assemblages from classical sequences in Patagonia and elsewhere. This suggests that, like today, Miocene tropical and temperate South America represented different biogeographic regions ("provinces") with only a few taxa of broad geographic distribution in common between them. The current paucity of fossil assemblages precludes determining precise geographic boundaries between these areas of endemism (as well as how Andean assemblages would relate to them) or temporal patterns of change in response to major biotic and environmental transformations. However, South America presents promising opportunities for future investigation of such patterns, for a number of reasons: (1) it covers a broader latitudinal range than any other continent, (2) it already has the best fossil record for any continent spanning the tropic to temperate zones, (3) it has great past and present environmental and topographic heterogeneity, (4) there is a high likelihood of significant fossil discoveries throughout the tropics, temperate regions, and the Andes, and (5) there is great potential for increased geochronologic calibration of fossiliferous sequences.

ACKNOWLEDGMENTS

As with any review paper, this study builds on the earlier efforts and the active on-going work of many scientists from throughout the world. We gratefully acknowledge reviews and constructive commentary on the manuscript from Bruce MacFadden, Leigh Van Valen, and André Wyss. In addition, we have benefitted greatly from access to unpublished information and discussions (in the field, conferences, and labs) with Mariano Bond, Gregory Buckley, Alfredo Carlini, Reynaldo Charrier, Javier Guerrero, Dennis Kent, Richard Madden, Christian de Muizon, Rosendo Pascual, Marcello Reguero, and Guillomar Vucetich. The research was supported by NSF grants (to JF) BSR-8896178/8614133 and DEB-9317943.

REFERENCES

AMEGHINO, F., 1900–1902, L'age des formations sedimentaires de Patagonie: Anales de la Sociedad Científica, Argentina, v. 50, p. 109–130, 145–165, 209–229 (1900); v. 51, p. 20–39, 65–91 (1901); v. 52, p. 189–197, 244–250 (1901), v. 54, p. 161–180, 220–249, 283–342 (1902).

BERGGREN, W. A., KENT, D. V., FLYNN, J. J., AND VANCOUVERING, J. A., 1985, Cenozoic geochronology: Geological Society of America Bulletin, v. 96, p. 1407–1418.

BERTELS, A., 1970, Sobre el "Piso Patagoniano" y la representacion de la epoca del Oligoceno en Patagonie austral, Republica Argentina: Revista Asociación Geológica Argentina, v. 25, p. 495–501.

BERTELS, A., 1975, Bioestratigrafía del Paleogeno en la Republica Argentina: Revista Espanola de Micropaleontologia, v. 7, p. 429–450.

BERTELS, A., 1979, Paleobiogeografía de los foraminíferos del Cretácico superior y Cenozoico de América del Sur: Ameghiniana, v. 16, p. 273–356.

BONAPARTE, J. F., 1990, New Late Cretaceous mammals from the Los Alamitos Formation, northern Patagonia: National Geographic Research, v. 6, p. 63–93.

BONAPARTE, J. F. AND ROUGIER, G. W., 1987, Mamíferos del Cretácico inferior de Patagonia: IV Congreso Latinoamericano Paleontología, Bolivia, v. 1, p. 343–359.

BONAPARTE, J. F., VAN VALEN, L. M., AND KRAMARTZ, A., 1993, La fauna local de Punta Peligro, Paleoceno Inferior, de la Provincia del Chubut, Patagonia, Argentina: Chicago, Evolutionary Monographs, no. 14, p. 1–61.

BOND, M., 1991, Sobre las capas de supuesta edad Divisaderense en los "Estratos de Salla", Bolivia, in Suarez-Soruco, R., ed., Fósiles y Facies de Bolivia, 1, Vertebrados: Revista Técnica de Yacimientos Petroliferos Fiscales Bolivianos, v. 12, p. 701–705.

BOND, M., CARLINI, A. A., GOIN, F. J., LEGARRETA, L., ORTIZ JAUREGUIZAR, E., PASCUAL, R., PRADO, J. L., AND ULIANA, M. A., 1994, Episodes in South American land mammal evolution and sedimentation: Testing their apparent concomitance [sic] in a Paleocene succession from Central Patagonia: VI Congreso Argentino de Paleontología y Bioestratigrafía, Resumenes, p. 20.

BOND, M., PASCUAL, R., REGUERO, M. A., SANTILLANA, S., AND MARENSSI, S., 1989, Los primeros "ungulados" extinguidos sudamericanos de la Antártida: Ameghiniana, v. 26, p. 240.

BOWN, T. M., LARRIESTRA, C. N., POWERS, D. W., NAESER, C. W., AND TABBUTT, K., 1988, New information on age, correlation, and paleoenvironments of fossil platyrrhine sites in Argentina: Abstracts, Journal of Vertebrate Paleontology, v. 8, supplement to no. 3, p. 9A.

BUTLER, R. F., MARSHALL, L. G., DRAKE, R. E., AND CURTIS, G. H., 1984, Magnetic polarity stratigraphy and 40K-40Ar dating of Late Miocene and Early Pliocene continental deposits, Catamarca Province, Northwest Argentina: Journal of Geology, v. 92, p. 623–636.

CAMACHO, H. H., 1974, Bioestratigrafía de las formaciones marinas del Eoceno y Oligoceno de la Patagonia: Anales de la Academia [Nacional] de Ciencias Exactas Fisicas y Naturales de Buenos Aires, v. 26, p. 39–57.

CANDE, S. C. AND KENT, D. V., 1992, A new Geomagnetic Polarity Time Scale for the late Cretaceous and Cenozoic: Journal of Geophysical Research, v. 97, p. 13,917–13,951.

CANDE, S. C. AND KENT, D. V., 1995, Revised calibration of the Geomagnetic Polarity Time Scale for the Late Cretaceous and Cenozoic: Journal of Geophysical Research, v. 100, p. 6093–6095.

CARLINI, A. A., PASCUAL, R., REGUERO, M. A., SCILLATO-YANÉ, G. J., TONNI, E. P., AND VIZCAÍNO, S. F., 1990, The first Paleogene land placental mammal from Antarctica: Its paleoclimatic and paleobiogeographical bearing: Fourth International Congress on Systematics and Evolutionary Biology, Abstracts, p. 325.

CASE, J. A., WOODBURNE, M. O., AND CHANEY, D. S., 1988, A new genus of polydolopid marsupial from Antarctica, in Feldmann, R. M. and Woodburne, M. O., eds., Geology and Paleontology of Seymour Island, Antarctic Peninsula: Boulder, Geological Society of America Memoirs, v. 169, p. 505–521.

CHARRIER, R., WYSS, A. R., FLYNN, J. J., NORELL, M. A., NOVACEK, M. J., SWISHER, C. C., ZAPATTA, F., AND MCKENNA, M. C., in press, New evidence for late Mesozoic-early Cenozoic evolution of the Chilean Andes in the upper Tinguiririca Valley (35°S), central Chile: Journal of South American Earth Sciences.

CHARRIER, R., WYSS, A. R., NORELL, M. A., FLYNN, J. J., NOVACEK, M. J., MCKENNA, M. C., SWISHER, C. C. III, FRASINETTI, D., AND SALINAS, P., 1990, Hallazgo de mamiferos fosiles del Terciario Inferior en el sector de Termas del Flaco, Cordillera Principal, Chile Central: Implicaciones paleontologicas, estratigraficas y tectonicas: Concepcion, Chile, II Simposio sobre el Terciario de Chile Central, Universidad de Concepción, p. 73–84.

CRANE, P. R. AND LIDGARD, S., 1989, Angiosperm diversification and paleolatitudinal gradients in Cretaceous floristic diversity: Science, v. 246, p. 675–678.

CROCHET, J.-Y. AND SIGÉ, B., 1993, Les Mammifères de Chulpas (Formation Umayo, transition Crétacé-Tertiare, Pérou): Données préliminaires: Documents des Laboratoires de Géologie Lyon, no. 125, p. 97–107.

EVERNDEN, J. F., SAVAGE, D. E., CURTIS, G. H., AND JAMES, G. T., 1964, Potassium-Argon dates and the Cenozoic mammalian chronology of North America: American Journal of Science, v. 262, p. 145–198.

FAURE, G., 1986, Principles of Isotope Geology (2nd ed.): New York, Wiley, 589 p.

FLYNN, J. J., GUERRERO, J., AND SWISHER, C. C. III, in press, Geochronology, in Kay, R. F., Madden, R. H., Flynn, J. J., and Cifelli, R. L., eds., Vertebrate Paleontology in the Neotropics: The Miocene Fauna of La Venta, Colombia: Washington, D.C., Smithsonian Institution Press.

FLYNN, J. J., MACFADDEN, B. J., AND MCKENNA, M. C., 1984, Land-Mammal Ages, faunal heterochrony, and temporal resolution in Cenozoic terrestrial sequences: Journal of Geology, v. 92, p. 687–705.

FLYNN, J. J., MARSHALL, L. G., AND GUERRERO, J., 1989a, Constraints on the age of "Friasian" (Miocene) faunas: Abstracts, Journal of Vertebrate Paleontology, v. 8, p. 20A.

FLYNN, J. J., MARSHALL, L. G., GUERRERO, J., AND SALINAS, P., 1989b, Geochronology of middle Miocene ("Friasian" Land Mammal Age) faunas from Chile and Colombia: Geological Society of America Abstracts with Programs, v. 21, p. A133.

FLYNN, J.J., WYSS, A.R., CHARRIER, R., AND SWISHER, C.C., 1995, An early Miocene anthropoid skull from the Chilean Andes: Nature, v. 373, p. 603–607.

GODTHELP, H., ARCHER, M., CIFELLI, R., HAND, S. J., AND GILKESON, C. F., 1992, Earliest known Australian Tertiary mammal fauna: Nature, v. 356, p. 514–516.

HIRSCHFELD, S. E. AND MARSHALL, L. G., 1976, Revised faunal list of the La Venta fauna (Friasian- Miocene) of Colombia, South America: Journal of Paleontology, v. 50, p. 433–436.

HOFFSTETTER, R. A., 1986, High Andean mammalian faunas during the Plio-Pleistocene, in Vuilleumier, F. and Munesterio, M., eds., High Altitude Sub-Tropical Biogeography: Oxford, Oxford University Press, p. 218–245.

HOOKER, J. J., 1992, An additional record of placental mammal (Order Astrapotheria) from the Eocene of West Antarctica: Antarctic Science, v. 4, p. 107–108.

HOPSON, J. A. AND ROUGIER, G. W., 1993, Braincase structure in the oldest known skull of a therian mammal: Implications for mammalian systematics and cranial evolution: American Journal of Science, v. 293-A-A, p. 268–299.

KAY, R. F., MADDEN, R. H., FLYNN, J. J., AND CIFELLI, R. L., eds., in press, A History of Neotropical Fauna: Vertebrate Paleobiology of the Miocene of Tropical South America: Washington, D.C., Smithsonian Institution Press.

KAY, R. F., MADDEN, R. H., PLAVCAN, J. M., CIFELLI, R. L., AND GUERRERO DIAZ, J., 1987, Stirtonia victoriae, a new species of Miocene Colombian primate: Journal of Human Evolution, v. 16, p. 173–196.

KRAUSE, D. W. AND BONAPARTE, J. F., 1990, The Gondwanatheria, a new suborder of Multituberculata from South America: Journal of Vertebrate Paleontology, v. 10, suppl. no. 3, p. 31A.

KRAUSE, D. W., KIELAN-JAWOROWSKA, Z., AND BONAPARTE, J. F., 1992, Ferugliotherium Bonaparte, the first known multituberculate from Gondwanaland: Journal of Vertebrate Paleontology, v. 12, p. 351–376.

LIDGARD, S. AND CRANE, P. R., 1990, Angiosperm diversification and Cretaceous floristic trends: A comparison of palynofloras and leaf macrofloras: Paleobiology, v. 16, p. 77–93.

LILLEGRAVEN, J. A. AND McKENNA, M. C., 1986, Fossil mammals from the "Mesa Verde" Formation (Late Cretaceous, Judithian) of the Bighorn and Wind River Basins, Wyoming, with definitions of Late Cretaceous North American Land-Mammal "Ages": American Museum Novitates, no. 2840, p. 1–68.

MACFADDEN, B. J., 1985, Drifting continents, mammals, and time scales: Current developments in South America: Journal of Vertebrate Paleontology, v. 5, p. 169–174.

MACFADDEN, B. J. AND ANAYA, F., 1993, Geochronology and paleoecology of Miocene mammals from the Bolivian Andes: Ameghiniana, v. 30, p. 350.

MACFADDEN, B. J., ANAYA, F., AND ARGOLLO, J., 1993, Magnetic polarity stratigraphy of Inchasi: A Pliocene mammal-bearing locality from the Bolivian Andes deposited just before the Great American Interchange: Earth and Planetary Science Letters, v. 114, p. 229–241.

MACFADDEN, B. J., ANAYA, F., PEREZ, H., NAESER, C. W., ZEITLER, P. K., AND CAMPBELL, K. E. JR., 1990, Late Cenozoic paleomagnetism and chronology of Andean basins of Bolivia: Evidence for possible oroclinal bending: Journal of Geology, v. 98, p. 541–555.

MACFADDEN, B. J., CAMPBELL, K. E. JR., CIFELLI, R. E., SILES, O., JOHNSON, N. M., NAESER, C. W., AND ZEITLER, P. K., 1985, Magnetic polarity stratigraphy and mammalian fauna of the Deseadan (late Oligocene-early Miocene) Salla beds of northern Bolivia: Journal of Geology, v. 93, p. 223–250.

MACFADDEN, B. J., SILES, O., ZEITLER, P., JOHNSON, N. M., AND CAMPBELL, K. E. JR., 1983, Magnetic polarity stratigraphy of the Middle Pleistocene (Ensenadan) Tarija Formation of southern Bolivia: Quaternary Research, v. 19, p. 172–187.

MACFADDEN, B. J., WANG, Y., CERLING, T. E., AND ANAYA, F., 1994, South American fossil mammals and carbon isotopes: A 25 million-year sequence from the Bolivian Andes: Palaeogeography, Palaeoclimatology, Palaeoecology, v. 107, p. 257–268.

MADDEN, R. H., GUERRERO, J., KAY, R. F., FLYNN, J. J., AND SWISHER, C. C., in press, The Laventan Stage: A new chronostratigraphic unit for the Miocene of South America, in Kay, R. F., Madden, R. H., Flynn, J. J., and Cifelli, R. L., eds., A History of Neotropical Fauna: Vertebrate Paleobiology of the Miocene of Tropical South America: Washington, D.C., Smithsonian Institution Press.

MADDEN, R. H., KAY, R. F., LUNDBERG, J. G., AND SCILLATO-YANE, G., 1989, Vertebrate paleontology, stratigraphy, and biochronology of the Miocene of southern Ecuador: Abstracts, Journal of Vertebrate Paleontology, v. 9, suppl. to no. 3, p. 31A.

MADDEN, R. H., KAY, R. F., VUCETICH, G., SWISHER, C. C., FRANCHI, M., AND MAZZONI, M., 1991, The Friasian of Patagonia: Abstracts, Journal of Vertebrate Paleontology, v. 11, suppl. to no. 3, p. 44A-45A.

MARENNSI, S. A., REGUERO, M. A., SANTILLANA, S. N., AND VIZCAINO, S. F., 1994, Review: Eocene land mammals from Seymour Island Antarctica: palaeobiogeographical implications: Antarctic Science, v. 6, p. 3–15.

MARSHALL, L. G., 1985, Geochronology and land-mammal biochronology of the Transamerican faunal interchange, in Stehli, F.G. and Webb, D., eds., The Great American Biotic Interchange: New York, Plenum Press, p. 49–85.

MARSHALL, L. G., 1988, Land mammals and the Great American Interchange: American Scientist, v. 76, p. 380–388.

MARSHALL, L. G. AND CIFELLI, R. L., 1990, Analysis of changing diversity patterns in Cenozoic Land Mammal Age faunas, South America: Palaeovertebrata, v. 19, p. 169–210.

MARSHALL, L. G. AND PASCUAL, R., 1978, Una escala temporal radiometrica preliminar de las edades-Mamifero del Cenozoico medio y tardio sudamericano: Obra del Centenario del Museo de la Plata, Tomo V, p. 11–28.

MARSHALL, L. G. AND SALINAS, P., 1990, Stratigraphy of the Rio Frias Formation (Miocene), along the Alto Rio Cisnes, Aisen, Chile: Revista de Geologia, Chile, v. 17, p. 57–78.

MARSHALL, L. G. AND SEMPERE, T., 1993, Evolution of the neotropical Cenozoic land mammal fauna in its geochronologic, stratigraphic and tectonic context, in Goldblatt, P., ed., Biological Relationships Between Africa and South America: New Haven, Yale University Press, p. 329–392.

MARSHALL, L. G., BUTLER, R. F., DRAKE, R. E., AND CURTIS, G. H., 1981, Calibration of the beginning of the age of mammals in Patagonia: Science, v. 212, p. 43–45.

MARSHALL, L. G., BUTLER, R. F., DRAKE, R. E., AND CURTIS, G. H., 1982a, Geochronology of Type Uquian (Late Cenozoic) Land Mammal Age, Argentina: Science, v. 216, p. 986–989.

MARSHALL, L. G., BUTLER, R. F., DRAKE, R. E., CURTIS, G. H., AND TEDFORD, R. H., 1979, Calibration of the Great American Interchange: Science, v. 204, p. 272–279.

MARSHALL, L. G., CIFELLI, R. L., DRAKE, R. E., AND CURTIS, G. H., 1986b, Vertebrate paleontology, geology, and geochronology of the Tapera de Lopez and Scarritt Pocket, Chubut Province, Argentina: Journal of Paleontology, v. 60, no. 4, p. 920–951, 12 figs.

MARSHALL, L. G., DRAKE, R. E., CURTIS, G. H., BUTLER, R. F., FLANAGAN, K. M., AND NAESER, C. W., 1986a, Geochronology of type Santacrucian (middle Tertiary) Land Mammal Age, Patagonia, Argentina: Journal of Geology, v. 94, no. 4, p. 449–457.

MARSHALL, L. G., HOFFSTETTER, R., AND PASCUAL, R., 1983, Mammals and stratigraphy: Geochronology of the continental mammal-bearing Tertiary of South America: Montpellier, Palaeovertebrata, Mémoire Extraordinaire 1983, p. 1–93.

MARSHALL, L. G., PASCUAL, R., CURTIS, G. H., AND DRAKE, R. E., 1977, South American geochronology—radiometric time scale for middle to late Tertiary mammal-bearing horizons in Patagonia: Science, v. 195, p. 1325–1328.

MARSHALL, L. G., SWISHER, C. C. III, LAVENU, A., HOFFSTETTER, R., AND CURTIS, G. H., 1992, Geochronology of the mammal-bearing late Cenozoic on the northern Altiplano, Bolivia: Journal of South American Earth Sciences, v. 5, p. 1–19.

MARSHALL, L. G., WEBB, S. D., SEPKOSKI, J. J., AND RAUP, D. M., 1982b, Mammalian evolution and the Great American Interchange: Science, v. 215, p. 1351–1357.

MCRAE, L. E., 1990, Paleomagnetic isochrons, unsteadiness, and uniformity of sedimentation in Miocene intermontane basin sediments at Salla, Eastern Andean Cordillera, Bolivia: Journal of Geology, v. 98, p. 479–500.

MUIZON, C. DE, 1991, La fauna de mamíferos de Tiupampa (Paleoceno Inferior, Formación Santa Lucía), Bolivia, in Suárez Soruco, R., ed., Fósiles y Facies de Bolivia, 1, Vertebrados: Santa Cruz, Revista Técnica de Yacimientos Petroliferos Fiscales Bolivianos, v. 12, p. 575–624.

NABEL, P. E. AND VALENCIO, D. A., 1981, La magnetoestratigrafía del Ensenadense de la ciudad de Buenos Aires: Su significado geologico: Revista Asociación Geológica Argentina, v. 36, p. 7–18.

NAESER, C. W., McKEE, E. H., JOHNSON, N. M., AND MACFADDEN, B. J., 1987, Confirmation of a late Oligocene-early Miocene age of the Deseadan Salla Beds of Bolivia: Journal of Geology, v. 95, p. 825–828.

ODIN, G. S, 1982, Numerical Dating in Stratigraphy: Chichester, Wiley-Interscience, 1040 p.

ORGEIRA, M. J., 1987, Estudio paleomagnético de sedimentos del Cenozóico tardío en la Costa Atlántica Bonaerense: Revista Asociación Geológica Argentina, v. 42, p. 362–376.

ORGEIRA, M. J., 1990, Paleomagnetism of late Cenozoic fossiliferous sediments from Barranca de los Lobos (Buenos Aires Province, Argentina). The magnetic age of the South American Land Mammal Ages: Physics of the Earth and Planetary Interiors, v. 64, p. 121–132.

ORGEIRA, M. J. AND VALENCIO, D. A., 1984, Estudio paleomagnético de los sedimentos asignados al Cenozoico tardío aflorantes en la Barranca de los Lobos, Provincia de Buenos Aires: Noveno Congreso Geologico Argentino, Actas, v. 4, p. 162–173.

PASCUAL, R. AND ORTIZ JAUREGUIZAR, E., 1990, Evolving climates and mammal faunas in Cenozoic South America: Journal of Human Evolution, v. 19, p. 23–60.

PASCUAL, R. AND ORTIZ JAUREGUIZAR, E., 1991, El Ciclo Faunístico Cochabambiano (Paleoceno temprano): Su incidencia en la historia biogeográfica de los mamíferos sudamericanos, in Suárez-Soruco, R., ed., Fósiles y Facies de Bolivia: Santa Cruz, Revista Técnica de Yacimientos Petroliferos Bolivianos, v. 12, p. 559–574.

PASCUAL, R. AND ORTIZ JAUREGUIZAR, E., 1992, Evolutionary pattern of land mammal faunas during the Late Cretaceous and Paleocene in South America: A comparison with North American pattern: Annales Zoologici Fennici, v. 28, p. 245–252.

PASCUAL, R., ARCHER, M., ORTIZ J., E., PRADO, J., GODTHELP, H., AND HAND, S., 1992a, First discovery of monotremes in South America: Nature, v. 356, p. 704–706.

PASCUAL, R., ARCHER, M., ORTIZ J., E., PRADO, J., GODTHELP, H., AND HAND, S. J., 1992b, The first non-Australian monotreme: An early Paleocene South American platypus (Monotremata, Ornithorhynchidae), in Augee, M. L., ed., Platypus and Echidnas: Sydney, The Royal Zoological Society of New South Wales, p. 2–15.

PASCUAL, R., BOND, M., AND VUCETICH, M. G., 1981, El Subgrupo Santa Bárbara (Grupo Salta) y sus vertebrados. Cronología, paleoambientes y paleobiogeografía: Actas VIII Congreso Geológico Argentino, v. 3, p. 743–758.

PASCUAL, R., CARLINI, A. A., AND GOIN, F. J., 1994, Paleogene land mammal-bearing localities in Central Patagonia, Argentina: Trelew, Field Trip Guide, IV Congreso de Paleontología y Bioestratigrafía, p. 1–50.

PASCUAL, R., GOIN, F. J., ORTIZ J., E., CARLINI, A. A., AND PRADO, J. L., 1993, Ferugliotherium and Sudamerica, Multituberculata and Gondwanatheria. One more evolutionary process occurred in isolation: X Jornadas Argentinas de Paleontologia Vertebrados, p. 334.

PASCUAL, R., ORTEGA HINOJOSA, E. J., GONDAR, D., AND TONNI, E. P., 1965, Las Edades del Cenozoico mamalífero de la Argentina, con especial atención a aquellas del territorio Bonaerense: Anales del Comité de Investigaciones Científicas, Buenos Aires, v. 6, p. 165–193.

PATTERSON, B. AND PASCUAL, R., 1968, The fossil mammal fauna of South America, in Keast, A., Erk, F. C., and Glass, B., eds., Evolution, Mammals, and Southern Continents: Albany, State University of New York Press, p. 247–309.

ROUGIER, G. W., WIBLE, J. R., AND HOPSON, J. A., 1992, Reconstruction of the cranial vessels in the Early Cretaceous mammal Vincelestes neuquenianus: Implications for the evolution of the mammalian cranial vascular system: Journal of Vertebrate Paleontology, v. 12, p. 188–216.

SAVAGE, D. E., 1977, Aspects of vertebrate paleontological stratigraphy and geochronology, in Kauffman, E. ed., Concepts and Methods in Biostratigraphy: Stroudsburg, Dowden, Hutchinson and Ross, p. 427–442.

SEMPERE, T., HÉRAIL, G., OLLER, J., AND BONHOMME, M. G., 1990, Late Oligocene-early Miocene major tectonic crisis and related basins in Bolivia: Geology, v. 18, p. 946–949.

SEMPERE, T., MARSHALL, L. G., RIVANO, S., AND GODOY, E., 1994, Late Oligocene-Early Miocene compressional tectosedimentary episode and associated land-mammal faunas in the Andes of central Chile and adjacent Argentina (32–37°S): Tectonophysics, v. 229, p. 251–264.

SERVANT-V., S. AND BLANCO, M., 1984, Les diatomées fluvio-lacustres plio-pléistocènes de la Formation Charaña (Cordillère Occidentale des Andes,

Bolivie): Cahiers Office de la Recherche Scientifique et Technique d'Outre-Mer Série Géologique, v. 14, p. 55–102.

SIGÉ, B., 1972, La faunule de mammifères du Crétacé supérieur de Laguna Umayo (Andes péruviennes): Paris, Bulletin du Muséum national d'Histoire naturelle, 3rd ser., no. 99, Sciences de la Terre, v. 19, p. 375–409.

SIMPSON, G. G., 1940, Review of the mammal-bearing Tertiary of South America: Proceedings of the American Philosophy Society, v. 83, p. 649–709.

SIMPSON, G. G., 1970, Addition to knowledge of Groeberia (Mammalia, Marsupialia) from the Mid-Cenozoic of Argentina: Breviora, no. 362, p. 1–17.

SIMPSON, G. G., MINOPRIO, J. L., AND PATTERSON, B., 1962, The mammalian fauna of the Divisadero Largo Formation, Mendoza, Argentina: Bulletin of the Museum of Comparative Zoology, v. 127, p. 239–293.

SMITH, J. T. AND ZINSMEISTER, W. J., 1982, Paleogeographic implications of a tropical eastern Pacific Nodipecten from the Tertiary of Patagonia: Geological Society of America, Abstracts, v. 14, p. 235.

STEHLI, F. G. AND WEBB, S. D., eds., 1985, The Great American Biotic Interchange: New York, Plenum Press, 532 p.

STENZEL, H. B., 1945, Stratigraphic significance of the Patagonian Odontogryphaeas: Geological Society of America Bulletin, v. 56, p. 1202.

STIRTON, R. A., 1953, Vertebrate paleontology and continental stratigraphy in Colombia: Bulletin of the Geological Society of America, v. 64, p. 603–622.

TEDFORD, R. H., 1970, Principles and practices of mammalian geochronology in North America: Proceedings of the North American Paleontological Convention, v. F, p. 666–703.

THOUVENY, N. AND SERVANT, M., 1989, Paleomagnetic stratigraphy of Pliocene continental deposits of the Bolivian altiplano: Palaeogeography, Palaeoclimatology, Palaeoecology, v. 70, p. 331–334.

TOUMARKINE, M. AND LUTHERBACHER, H., 1985, Paleocene and Eocene planktic Foraminifera, in Bolli, H. M., Saunders, J. B., and Perch-Nielsen, K., eds., Plankton Stratigraphy: Cambridge, Cambridge University Press, p. 87–154.

VALENCIO, D. A. AND ORGEIRA, M. J., 1983, La magnetostratigrafía del Ensenadense y Bonaerense de la ciudad de Buenos Aires, Parte II: Revista Asociación Geológica Argentina, v. 38, p. 24–33.

VAN VALEN, L., 1988, Paleocene dinosaurs or Cretaceous ungulates in South America: Evolutionary Monographs, v. 10, p. 1–79.

WEBB, S. D., 1978, A history of savanna vertebrates in the new world, Part II, South America and the Great Interchange: Annual Review of Ecology and Systematics, v. 9, p. 393–426.

WOODBURNE, M. O., 1977, Definition and characterization in mammalian chronostratigraphy: Journal of Paleontology, v. 51, p. 229–234.

WOODBURNE, M. O., 1987a, Introduction, in Woodburne, M. O., ed., Cenozoic Mammals of North America: Geochronology and Biostratigraphy: Berkeley, University of California Press, p. 1–8.

WOODBURNE, M. O., 1987b, Principles, classification, and recommendations, in Woodburne, M. O., ed., Cenozoic Mammals of North America: Geochronology and Biostratigraphy: Berkeley, University of California Press, p. 9–17.

WOODBURNE, M. O. AND ZINSMEISTER, W. J., 1982, Fossil land mammal from Antarctica: Science, v. 218, p. 284–286.

WOODBURNE, M. O. AND ZINSMEISTER, W. J., 1984, The first land mammal from Antarctica and its biogeographic implications: Journal of Paleontology, v. 58, p. 913–948.

WYSS, A. R., FLYNN, J. J., NORELL, M. A., SWISHER, C. C. III, NOVACEK, M. J., MCKENNA, M. C., AND CHARRIER, R., 1994, Paleogene mammals from the Andes of central Chile: A preliminary taxonomic, biostratigraphic, and geochronologic assessment: American Museum Novitates, no. 3098, p. 1–31.

WYSS, A. R., FLYNN, J. J., SWISHER, C. C. III, NORELL, M. A., CHARRIER, R., NOVACEK, M. J., AND MCKENNA, M. C., 1993, South America's oldest rodent and recognition of a new interval of mammalian evolution: Nature, v. 365, p. 434–437.

WYSS, A. R., NORELL, M. A., FLYNN, J. J., NOVACEK, M. J., CHARRIER, R., MCKENNA, M. C., FRASINETTI, D., SALINAS, P., AND MENG, J., 1990, A new early Tertiary mammal fauna from central Chile: Implications for stratigraphy and tectonics: Journal of Vertebrate Paleontology, v. 10, p. 518–522.

ZINSMEISTER, W. J., 1981, Middle to Late Eocene invertebrate fauna from the San Julian Formation at Punta Casamayor, Santa Cruz Province, Southern Argentina: Journal of Paleontology, v. 55, p. 1083–1102.

LAND MAMMAL HIGH-RESOLUTION GEOCHRONOLOGY, INTERCONTINENTAL OVERLAND DISPERSALS, SEA LEVEL, CLIMATE, AND VICARIANCE

MICHAEL O. WOODBURNE
Department of Earth Sciences, University of California, Riverside, CA 92521
AND
CARL C. SWISHER, III
Berkeley Geochronology Center, Berkeley, CA 94709

ABSTRACT: The generally well developed and understood stratigraphic record associated with fossil mammals in North America is combined with independent chronological data sets that foster the development of high-resolution geochronology in nonmarine sequences. An updated chronology for all North American mammal ages (or subdivisions) is utilized to examine the tempo and mode of overland mammal immigration/emigration episodes during the Cenozoic Era. In addition to the thirty or more "background" dispersals involving only a few taxa, ten major immigration/emigration episodes are recorded during the Cenozoic Era in North America. All are important for evaluating the dispersal pattern, as well as for mammal age boundary definition. For the Paleogene interval, major immigration/emigration episodes define the following mammal ages (or intervals): Clarkforkian, Wasatchian, late Uintan, and Chadronian, with the Wasatchian and late Uintan being especially noteworthy. The interval that embraces the late Arikareean mammal age is the first immigration episode of the Neogene interval, but the events recognized in the early and late Hemingfordian mammal ages, respectively, are the most impressive. An important, "medial" Clarendonian emigration is reflected in the North American basis for the Old World *"Hipparion"* Datum. The events that define the beginning of the Hemphillian and late Blancan mammal ages also are founded on important immigrant first occurrences, but for the first time in the Neogene interval involve taxa from South America as well as from Asia. At other times, either there is no effective immigration or, if present, it involves only a few (four or less) taxa (= "background"). In certain intervals, apparently lowered sea level had no effect on dispersal, but in an even larger number of cases immigration took place in spite of what appears to have been times of relative sea-level highstand. Thus tectonic, climatic, and other factors must be considered to account for North America's dispersal history during the Cenozoic Era.

INTRODUCTION

In a reassessment of the bases for the North American Land Mammal Ages (Woodburne, 1987), contributors commented that certain times during the Cenozoic Era reflected a greater amount of change in mammal faunas than were recognized at others. In a number of instances, the revolution in taxa (that form the bases upon which mammal ages may be recognized) was derived at least as much, if not more so, from changes in endemic groups than upon changes driven by immigrations. Nevertheless, boundaries between mammal ages (or subdivisions) are mostly based on immigrations of new taxa. The text therefore is biased in that direction, and the impact on endemic faunas derived from immigrants certainly is an aspect important to the study of faunal dynamics at all levels. However, so are responses of land mammal faunas to changes in climate, topography, and the like, and taxonomic reorganizations that stem from all factors are, and should be, utilized in defining or characterizing mammal ages or their subdivisions.

The present work updates the chronologic framework for fossil mammal ages in North America. The effort then focuses on times of immigration or emigration in order to assess the role played in these processes by changes in sea-level or other factors. A major question is whether or not overland dispersals were causally related to times of lowered sea level. Opdyke (1990) suggests that there is a gross correspondence between mammal age boundaries in North America and the Old World (mainly Europe) and lowstands of sea level during the Cenozoic Era. This was based on an hypothesis that both American and European mammal ages should have concurrent boundaries in that each were based on immigrations driven by sea-level changes (as represented by changes in oxygen isotope ratios of seawater). The basic hypothesis requires some reconsideration in that, whereas most North American mammal ages (and subdivisions) are now defined on one or more immigrant taxa (more so for the Neogene than Paleogene intervals), those of Europe mostly are not and, especially for the Neogene, inter-

continental-scale dispersals are discounted (DeBruijn and others, 1992; see also Fahlbusch, 1991). Further, as seen herein, the role of sea level is relatively minor in this regard, except for later Miocene into the Pleistocene epochs. Other factors thus must contribute to faunal interchange. Among these, the influence of climatic events is difficult to ascertain closely, except perhaps during the Pliocene and Pleistocene epochs. Therein, a number of arvicoline rodent dispersals are propelled from an unknown source or sources in eastern Asia to sites in Europe and North America, where new taxa arrived at about the same times (Repenning, 1987; Lundelius and others, 1987, p. 222, Biharian; Fejfar and Repenning, 1992; Repenning and others, 1990; Repenning and Brouwers, 1992; Repenning, 1992). Whether related to sea-level changes or not, it appears that a significant lag in dispersal time from place of origin to place of introduction transpired in a number of cases. Contemporaneity in dispersal may be inhibited by sea level but also by factors such as climate, tectonism, and the biology of the organisms, themselves.

Organization

This report utilizes all aspects of the mammal age correlation network affected by recent modifications in the Magnetic Polarity Time Scale (MPTS; Cande and Kent, 1992; Berggren and others, this volume), and advances in faunal as well as stratigraphic, radioisotopic and paleomagnetic data achieved since the summary of North American mammal ages provided in Woodburne (1987). The purpose of this report is two-fold: (A) to update the calibration and correlation of mammal ages in North America and (B) to utilize the new framework to evaluate the dispersal history of land mammals on that continent during the Cenozoic Era. In that some type of faunal interchange appears to have taken place in North America every few million years or less, all data are relevant. Thus, comments in the text address all parts of the mammal age chronology and expand on those episodes reflecting major immigrations/emigrations. In light of the general improvements mentioned above, explanations for the figures relate the mammal age chronology to both the MPTS and to the Cenozoic sea-level patterns described by Haq and others (1988) and Bartow (1992). The discussion evaluates patterns of dispersal, sea-level fluctuations, climatic factors, dispersal lag, and vicariance. For convenience, the treatment is sequential within the epoch level.

Terminology, Abbreviations, and Conventions

Age calibration.—

Calibration of the Magnetic Polarity Time Scale (MPTS) follows Berggren and others (this volume). Radioisotopic calibrations rely, in part, on those recently developed by the ^{40}Ar/^{39}Ar total-fusion methodology (Swisher and others, 1993; Ma (megannum) in the radioisotopic time scale).

Faunal terminology.—

The distinction between Faunas (F.) and Local Faunas (L.F.) follows Tedford and others (1987, p. 155). Corridor, Filter and Sweepstakes, terms to characterize broad, more limited, and only likely, aspects of overland dispersal of fossil mammals, follow Simpson (1953). Medial is used in place of middle for age terms (early, medial, and late), in contrast to lower, middle, and upper for lithostratigraphic terms. Disperals terms used in this paper are: LSD; lowest stratigraphic datum (occurrence) of a taxon in a local section (Lindsay and others, 1984), FAD (first appearance datum) composed of a regional array of LSDs (Berggren and Van Couvering, 1974), HSD (highest stratigraphic datum or occurrence) of a taxon in a local section (Lindsay and others, 1984), and LAD (last appearance datum) composed of a regional array of HSDs (Berggren and Van Couvering, 1974). It is understood here that use of terms such as Wasatchian, Blancan, etc., refer to the equivalent North American Land Mammal Age (for example, Woodburne, 1987).

Sea-level changes.—

Figures 1–7 summarize the relative sea-level history during the Cenozoic Era, modified from Haq and others (1988) and Bartow (1992), but differs primarily in that we have readjusted the sea-level episodes to the new MPTS calibrations supplied by Berggren and others (this volume). The methodology of Haq and others (1988) utilizes sequence- and seismic-stratigraphic precepts in which sequences of types 1, 2, and 3 (for example, Hallam, 1992) may be recognized. In Figures 1–7, supercycles (second-order intervals) generally correspond to Type 1 sequences; cycles (third-order intervals) to Type 2 sequences. Type 1 boundaries are those that reflect withdrawal of the sea to beyond the depositional shoreline break, so that much or all of the continental shelf may be exposed, and fluvial incision of the shelf may occur. These boundaries mostly (but not exclusively) occur at supercycle intervals (Figs. 1–7), which typically are composed of sequences of second-order rank that embrace temporal intervals of from 10 to 80 my. Type 2 boundaries result from less extensive sea-level falls during which neither the shoreline break nor the shelf is exposed. The sequences upon which these are based are commonly expressed as cycles (Figs. 1–7) and are third-order boundaries which record even shorter intervals, at a scale of from 1–3 my (Christie-Blick and others, 1990, p. 131).

We have not re-evaluated any of the primary data which formed the bases of the correlations developed by Haq and others (1988). We recognize the considerations regarding the utility of global sea-level curves evaluated in Christie-Blick and others (1990), but follow their recommendation that first- and second-order sea-level lowstands may be treated as eustatic, but that because of the short time-scales involved, third- and higher-order fluctuations cannot be evaluated due to correlation difficulties. We note in passing, however, that each supercycle begins with a third-order cycle, if not a smaller division, and that somehow the correlation of these is not in doubt. Williams (1988) discusses differences in proposed magnitude of sea-level changes as based on seismic stratigraphic (larger) versus magnitudes derived from the δ^{18}O record (smaller), and Cloetingh (1988) considers a tectonic (intraplate stress) origin of third-order cycles. Miall (1992) asserts that the global reality of third-order cycles remains to be demonstrated; Dickinson (1993) suggests that the record for these, as for other events, degrades systematically with geologic age. For these reasons, our discussion focuses mostly on the Type 1 sea-level changes that are utilized by Haq and others (1988) as defining supercycle boundaries; we place less emphasis on fluctuations identified as cycles (mostly third-order; Type 2). We also follow Christie-Blick and others (1990) in noting that the coastal onlap curves provided by Haq and others (1988) likely relate more to timing than to magnitude of sea-level change and that the eustatic curves therein appear to be smoothed versions of the onlap curves.

Dispersals.—

These are allocated as either "background" (typically represented by four or fewer taxa) or as Interchange Events. The latter are considered to have been especially significant for reasons given in each case. Unless otherwise detailed, dispersal during "background" episodes is assumed to have been via a filter regime, across the North Atlantic prior to the Wasatchian Age; across Beringia thereafter. The direction and magnitude of dispersal is treated as pertinent for Interchange Events.

DISPERSAL RECORD FOR NORTH AMERICA IN THE PAST 70 MA

Cretaceous Period

Cretaceous-Paleogene Boundary.—

For this paper the boundary criteria of Olsson and Liu (1993) and Berggren and others (this volume) are followed, as is the thesis that the marine record (especially associated with the Chicxulub "event") is dated at 65 Ma (Swisher and others, 1992), and that the marine realm date compares well with the approximately 65 Ma age for the boundary in nonmarine strata of Montana (Swisher and others, 1993).

Late Cretaceous Period.—

The oldest unequivocal marsupials (*Pariadens kirtklandi*; Cifelli and Eaton, 1987) have been recovered from Cenomanian deposits of Utah, and as data improve, there is a potential for refined correlation to or from North America in the later Cretaceous Period (Cenomanian to Maestrichtian). The data summarized in Simpson (1953), Lillegraven and others (1979), McKenna (1981), Case and Woodburne (1986), Van Valen (1988), Marshall and Cifelli (1990), Krause and Maas (1990), Muizon and Marshall, (1991), Muizon (1991), Jaillard and others (1993), suggest that some of the potential is being realized, but still the situation yet provides no closely correlated dispersal scenarios relative to North America.

Paleocene Epoch

The upper and lower limits of the Paleocene Epoch vary historically depending upon the group of fossils (plants, invertebrates, vertebrates) under consideration (Savage and Russell, 1983; Aubry and others, 1988; Gingerich, 1989). This report follows the recommendations of Berggren and others (this volume) and is shown in Figure 1. For North American mammals,

the Paleocene Epoch begins with the Puercan mammal age, and continues through Clarkforkian time. The Clarkforkian is recognized as Interchange Event #1 (Fig. 1).

North America still (Archibald and others, 1987, p. 66; Russell and others, 1982; Russell and Zhai, 1987) contains the unequivocally oldest Paleocene mammal record. Prior to the Clarkforkian, few, if any, North American taxa are suggested as being of overseas origin, although the diversity of early Paleocene (Puercan, Pu1, Fig. 1) mammal faunas in North America is not easily explained by endemic origin from those of the Late Cretaceous Epoch, and Muizon (1991) suggests that some marsupials (didelphids) colonized North America from the south in late Paleocene time. Stucky (1992, Table 24.1) indi-

PALEOCENE TIME SCALE

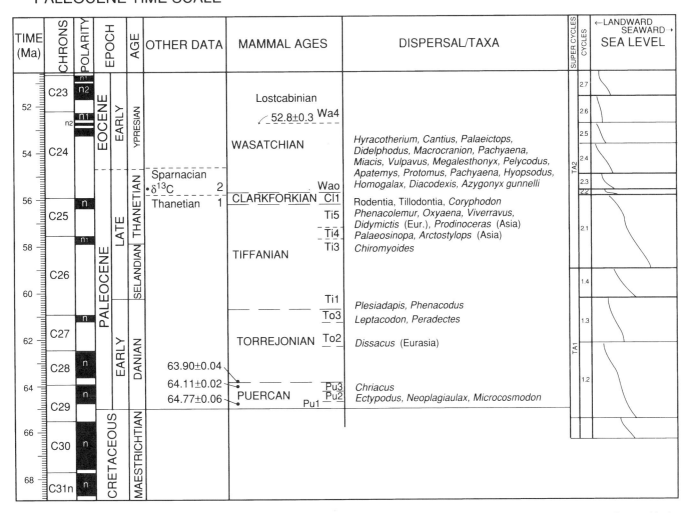

Fig. 1.—Paleocene time scale, including Late Cretaceous and early Eocene epochs, after Berggren and others (this volume). Mammal ages discussed in the text. Sea-level pattern is after Haq and others (1988) and Bartow (1992). Heavy lines indicate Type 1 unconformities, mostly at second-order boundaries (e.g., between Supercycles TA1 and TA2). Sometimes, Type 1 unconformities are at third-order boundaries (e.g., between TA1.3 and TA1.4). Heavy numbers (1, 2, etc. refer to Interchange Events; see text). Unless otherwise specified, dispersal is usually between North America and Europe in the Paleocene through Wasatchian epochs; between Asia (or Eurasia) and North America thereafter.

Note limited interchange in the Puercan through Tiffanian mammal ages, the paucity of that being important in view of the strong sea-level lowstand at about medial Tiffanian (Type 1; *Terminology*; beginning of supercycle TA2 = TA2.1). Note that all five taxa listed as appearing in Ti5 are considered to co-occur at or near its base. Graphic limitations may suggest otherwise.

The first major mammalian overland dispersal is recorded at the beginning of the Wasatchian Stage/Age (Interchange Event 2), with an important precursor in the Clarkforkian Stage/Age (Interchange Event 1). Graphic limitations result in the list of Wa0 taxa having an apparent vertical range in time. All Wa0 taxa, from *Hyracotherium* through *Azygonyx gunnelli*, first appear at Wa0.

cates that the multituberculates *Ectypodus, Neoplagiaulax* and *Microcosmodon* are shared with Europe in the medial Puercan (Pu2), and the arctocyonian, *Chriacus* in the late Puercan (Pu3). Archibald and others (1987, p. 66–68) suggest some degree of immigration to North America during the Torrejonian; Stucky (1992) indicates that *Dissacus* is shared with Europe and Asia in the medial Torrejonian (To2); *Peradectes* and *Leptacodon* are shared with Europe in the late Torrejonian (To3). Archibald and others (1987, p. 55) indicate that virtually all taxa first appearing in the basal Tiffanian are plausible derivatives from the Torrejonian, as do Kraus and Maas (1990, p. 87). According to Stucky (1992), the primate, *Plesiadapis,* and condylarth, *Phenacodus,* are shared with Europe in the early Tiffanian (Ti1), the primate, *Chiromyoides* (medial, Ti3), the cimolestan, *Palaeosinopa* (late, Ti4), the primate, *Phenacolemur,* the carnivores, *Viverravus* and *Didymictis,* and the creodont, *Oxyaena,* comprise the remaining taxa shared with Europe (Ti5). Whereas an allochthonous origin for some of these mammals is potentially possible, overland interchange involving North America is minor, at best, during most of the Paleocene Epoch, even though a major (Type 1) sea-level lowstand (TA 2.1, Fig. 1) appears to have occurred in about medial Tiffanian. TA 2.1 also is too early for the Asian introduction of *Arctostylops* (Arctostylopida, Ti4) and *Prodinoceras* (Dinocerata,Ti5) to late Paleocene North America.

Calibration and correlation.—

In North America, the base of the Puercan mammal age is correlated to the MPTS as within Chron 29r (Archibald and others, 1987, p. 42, Figs. 3.2), now considered to be 65.0 my old (also, Swisher and others, 1993). Figure 1 indicates that the Puercan through Wasatchian mammal ages can be correlated to MPTS as shown, but at the moment supporting radioisotopic ages are rare. Recent additions from Swisher and others (1993) indicate that the Pu1-Pu2 transition is calibrated at about 64.77 ± 0.06 Ma; that the Pu3-Torrejonian transition is calibrated as between about 63.90 ± 0.04 Ma and 64.11 ± 0.02 Ma within Chron 28R.

Interchange Event #1.—

Clarkforkian mammal age.—The first well documented correlation to parts of the outside world comes with the Clarkforkian Stage/Age (1, Fig. 1) as proposed by Rose (1980, 1981). Archibald and others (1987) segregate this interval into three units and recognize its base (Rodentia/*P. cookei* Interval-Subzone [Cf1] of the Ti6 [Tiffanian 6]-Cf1 Lineage-Zone) upon the immigrant first occurrence of the Rodentia. The group presumably entered from Asia, but a chronologically closely controlled Asian source has not been identified, other than generally from the Eurymyloidea of the Paleocene (Krause and Maas, 1990, p. 87). Tillodontia, *Haplomylus,* and *Coryphodon* also apparently occur synchronously with the Rodentia at the base of Cf1. Whereas the Tillodontia have been proposed as having had an Asian source (Shanghuanian; Krause and Maas, 1990, p. 88), the place of origin of *Haplomylus* and *Coryphodon* is less certain. Krause and Maas (1990, p. 87, Table 6) suggest that *Haplomylus* (along with some other first-appearing taxa, *Planetetherium* and *Apheliscus*) are endemic, having originated in the Tiffanian. *Coryphodon* may have had an Asian or a European

origin. Other taxa (*Leipsanolestes, Protentomodon* and *Prosthecion*) are too poorly known to be of use in this analysis.

Late Paleocene or early Eocene age?.—There is some discussion (Archibald and others, 1987, p. 66, 67) as to whether the Clarkforkian Stage/Age contains the Paleocene/Eocene boundary as recognized in mammalian faunas, or whether this boundary is within, or at the base of, the Wasatchian Stage/Age. Based on data cited here, Clarkforkian is taken as late Paleocene age in the mammalian time scale (Fig. 1; see also Gingerich, 1989, p. 86) but the base of the Wasatchian yet may be even later Paleocene (see below).

Route and characterization.—Filter route, plausibly via Beringia or the North Atlantic for rodents and tillodonts and *Coryphodon.* McKenna (1983a, b) notes that the Euramerica corridor (across the northern Atlantic Ocean) was present in late Paleocene as well as early Eocene time. Also at this time (and for much of the Cenozoic), an overland connection was present across Beringia from northeastern Asia to northwestern North America, but its northern paleolatitude likely resulted in a limiting dispersal pattern between these areas, at least for the Paleogene interval (McKenna, 1983b). Note also that until the end of the Eocene Epoch, the Turgai Straits separated central Asia from western Europe south of the Baltic, hindering Eurasian dispersals in earlier Paleogene time. Figure 1 shows that the Clarkforkian correlates about with sea-level lowstands TA2.2 and 2.3, both Type 1 unconformities (*Terminology*), but during an interval of relatively high sea level (Haq and others, 1988).

Calibration and correlation.—The Clarkforkian is not directly calibrated. Correlation to the MPTS suggests that the base of the Clarkforkian is embraced within Chron C25n (Butler and Lindsay, 1985; Archibald and others, 1987, Fig. 3.2, p. 63, or about 56.05 Ma in the time scale of Berggren and others (this volume). The prime Clarkforkian correlatives in the European succession are those of Cernaysian (late Thanetian) age (Archibald and others, 1987, p. 67; Lucas, 1989; Gingerich, 1989). Berggren and Aubry (1995) indicate that the base of the Sparnacian Stage (and thus likely the base of the correlative Wasatchian) correlates to near the base of Chron C24r, or to an age of about 55.8 Ma. If the Clarkforkian begins at about 56.35 Ma and ends within Chron C24r (Butler and Lindsay, 1985), the Clarkforkian has a duration of only about 0.55 my (Berggren and Aubry, 1995). These correlations suggest that the 55.7 Ma age interpolated for the NP9/NP10 calcareous nannoplankton Zone boundary (Gunnell and others, 1993, p. 141) conflicts with that (ca 55.0 Ma) proposed by Berggren and Aubry (1995). Evidently some adjustment is required.

The Paleocene/Eocene boundary remains in a state of flux, however. Berggren and Aubry (1995) suggest that the transition corresponded to (or embraced):

1. Major climatic change (warming of deep ocean waters to nearly 10°C and to nearly 20°C as shown by increase in $\delta^{18}O$ values and decrease in $\delta^{13}C$ values);
2. Weakening of atmospheric circulation as reflected by the reduction in wind-blown aerosol grain size;
3. Faunal extinction (30–40% of deep-sea foraminifera);
4. 1–3 correspond to about medial foraminiferan Zone P5 and calcareous nannofossil Zone NP9/10 boundary which, according to the correlation chart (Berggren and Aubry, 1995, Fig. 12) correlates to about 55 Ma.

5. The Sparnacian Stage is equivalent to the uppermost Pale-ocene Series in the London Basin (pre-Oldhaven beds/Wool-wich/Reading beds). This is correlative with the basal Eo-cene Series in the Vertebrate Paleontology literature; correlative to within Zone NP9, about 0.8 my older than NP9/10 boundary and about 1.2 my older than the base of the Ypres Clay. The base of the Ypresian Stage is considered to be about 54.6 Ma, and thus about 0.2 my younger than the base of the London Clay. Berggren and Aubry (1995) provisionally use the base of the London Clay (about 54.6 Ma) as equivalent to the Paleocene/Eocene boundary.

6. A decrease of ~4‰ in $\delta^{13}C$ of marine carbonates at several sites and in mammalian teeth at base of Wasatchian and in organic matter above basal Sparnacian. The age of the $\delta^{13}C$ decrease is estimated at ~55.5 Ma.

On the other hand, Aubry and others (1995) show that there are two $\delta^{13}C$ spikes in Chron C24r; one at about mid NP9 and P5 and another in lower NP10 and P6; thus the age of any boundary based on $\delta^{13}C$ spikes is uncertain to at least 0.5 my. These authors conclude that there are numerous hiatuses in deep-sea cores relevant to the location therein of the Paleocene/Eocene boundary, but accept an age of ~55.5 Ma for the "clas-sic" $\delta^{13}C$ excursion.

Berggren and Aubry (1995) indicate that the base of the Ypresian is estimated at 54.6 Ma, and the P5/6 boundary at 54.7 Ma and that the Paleocene/Eocene boundary is contained within a span that ranges in age from 54.8–55.0 Ma, 54.8 Ma for the base of the London Clay (basal Eocene by definition) and 55.0 Ma for NP9/10 boundary. The $\delta^{13}C$ spike is estimated at 55.5 Ma. For the purposes of this paper, we take the Paleo-cene/Eocene boundary as 54.8 Ma (base of London Clay). Basal Wasatchian (with $\delta^{13}C$ spike) at 55.5 Ma is late Paleocene in age. This is effectively consistent with Berggren and others (this volume); contra Clyde and others, 1994, p. 375).

Eocene Epoch

Mammal faunas of Wasatchian through Chadronian age are embraced within the Eocene Epoch (Figs. 1, 2). Of these, the basal Wasatchian, late Uintan, and basal Chadronian record sig-nificant immigrations to or from North America.

Interchange Event #2.—

Wasatchian Stage/Age.—The Wasatchian Stage/Age (Savage, 1977) reflects a major overland dispersal to or from North America and western Europe (recognized in the base of the Sparnacian Stage in Europe; also McKenna, 1975). In addition, the degree of generic-level similarity between these areas (Ho-larctica in modern terms) was greater then (Wa0; Figs. 1, 2) than at any time in the Cenozoic Era before or since (for ex-ample, Savage and Russell, 1983, p. 67). McKenna (1983a, b) indicates that the Wasatchian-equivalent of the Eureka Sound Formation, northern Ellesmere Island, both shares and lacks a number of taxa found both in western North America and the London and Paris basins. Nevertheless, the general area that included northeastern Canada, Baffin Island, Greenland, Elles-mere Island, Spitsbergen and Fennoscandia is the most likely overland corridor in the Wasatchian across what later became the present Atlantic Ocean (see also Savage and Russell, 1983). Novacek and others (1991) indicate that data that are equivocal

as to the direction of dispersal, and that the place of origin of many of the taxa is uncertain.

Taxa that *first appear* at the base of the Wasatchian include Perissodactyla (*Hyracotherium*), Artiodactyla, Adapidae [pri-mates], Omomyidae [primates], Hyaenodontidae [creodonts], Sciuravidae [rodents], *Palaeictops, Didelphodus, Macrocran-ion, Pachyaena, Miacis, Vulpavus,* and *Megalesthonyx* (Krish-talka and others, 1987, p. 85), plus *Coryphodon (sic), Haplo-mylus (sic), Pelycodus, Apatemys, Palaeosinopa (sic), Prototomus, Viverravus (sic), Pachyaena, Hyopsodus, Homo-galax,* and *Diacodexis* (Savage, 1977). Gingerich (1989) nom-inates the Sandcouleean as a new subage (Wo) at the base of the Wasatchian, and provides additional first occurrences: *Can-tius* (adapid primate), *Azygonyx gunnelli* (tillodont), four genera of hyaeanodontid creodont, and removes *Coryphodon* from the list (occurs in Clarkforkian faunas; Rose, 1980, 1981).

Route and characterization.—Dispersal at this time is via a Cor-ridor, with a wide range of taxa, via Greenland, Spitzbergen, and adjacent land masses (McKenna, 1975, 1983a, b). After this time, the North Atlantic Ocean connection is severed. Henceforward all dispersals from or to North America, Europe or Asia appear to be via Beringia. Note, however, that this dis-persal corridor is located in high paleolatitudes (McKenna, 1983b) and may have had limited throughput at times. Figure 1 indicates that the base of the Wasatchian correlates approxi-mately with Type 1 (see *Terminology*) sea-level lowstands, TA 2.2–2.3 (Haq and others, 1988) in an interval of relatively high sea-level.

Calibration and correlation.—Interregional correlations sug-gest that the base of the Eocene Epoch is contained within Chron C24n.3r, or about 54.8 Ma as discussed previously (also Berggren and others, this volume; Berggren and Aubry, 1995). Importantly, the magnetic data are consistent with the polarity signature obtained from early Wasatchian-aged strata in North America (Butler and Lindsay, 1985). Wing and others (1991) indicate that rocks of early Lostcabinian (= late Wasatchian) age in Wyoming yield a $^{40}Ar/^{39}Ar$ age of 52.8 ± 0.3 Ma (Fig. 2). Otherwise, Wasatchian-aged strata are not directly cali-brated. Wing (1984) proposes that the Wasatchian begins near the top of nannofossil Zone NP9 and post-dates the beginning of the Sparnacian Stage, correlated to about the beginning of NP9. In this scenario, the Wasatchian is mainly a Ypresian cor-relate, and the Sparnacian is wholly late Paleocene (not Eo-cene). This is approximately similar to Berggren and Aubry (1995), who indicate that the Clarkforkian/Wasatchian bound-ary is about 0.5-≤1.0 my older than the Paleocene/Eocene boundary. Swisher and Knox (1991) estimate an age of 55 my for the NP 9/10 boundary, or basal Eocene according to nan-nofossil data; Berggren and Aubry, 1995). Gingerich (1989) correlates earliest Wasatchian mammal faunas (Wo) with those found in Sparnacian strata of the London-Paris Basin. In that Wo faunas occur in Wasatchian-aged strata that contain the $\delta^{13}C$ spike, the rocks may be as old as 55.7 Ma (see above and Figs. 1, 2). If early Sparnacian faunas (Meudon) in the Paris Basin are correlative with those of Zone Wo (Gingerich, 1989, p. 84–86), then the basal Wasatchian appears to pre-date the base of the Eocene by about 0.5-≤1.0 my. This correlation preserves that of Swisher and Knox (1991), and much of Wing (1984), Beard and Tabrum (1991), and Stucky (1992).

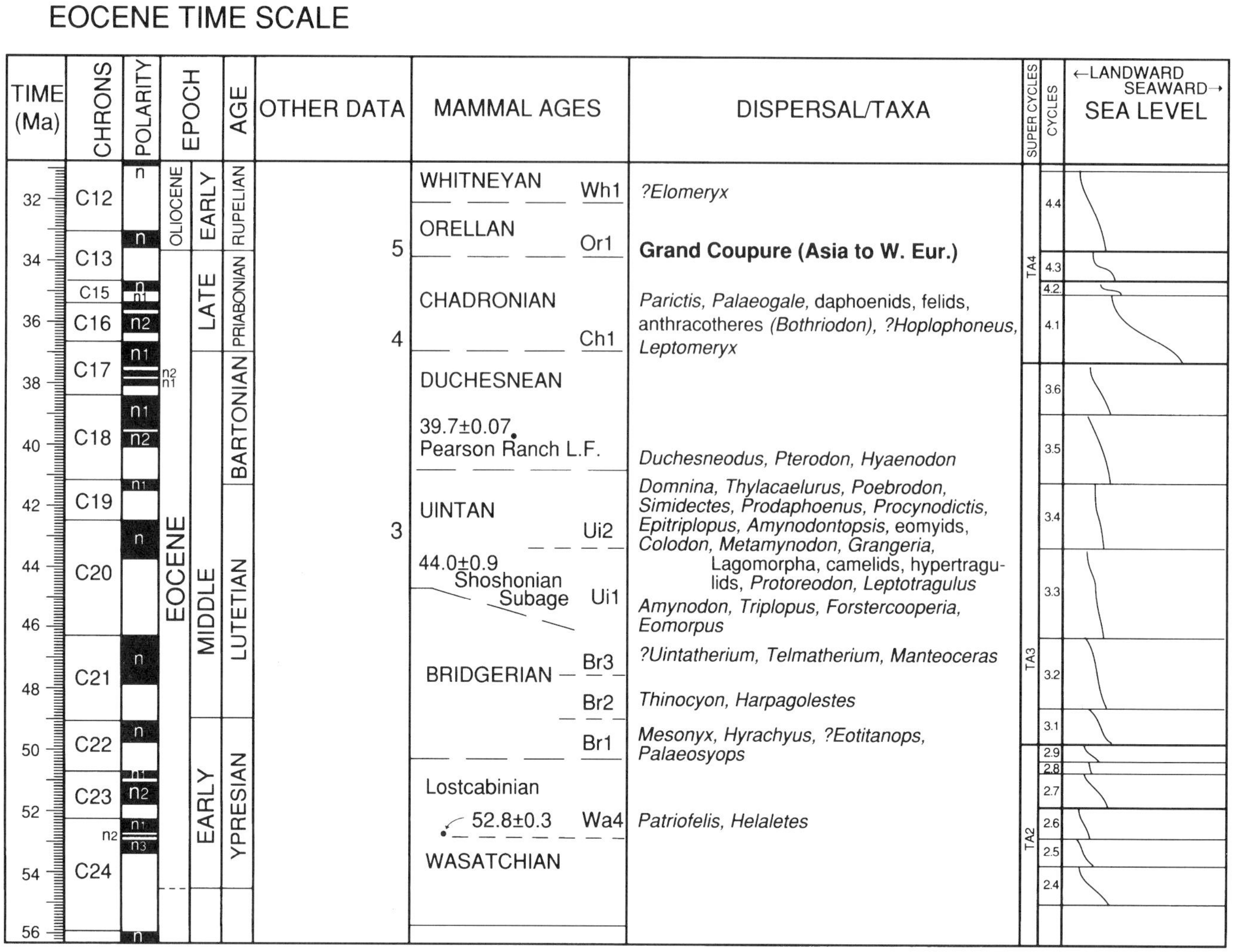

EOCENE TIME SCALE

TIME (Ma)	CHRONS	POLARITY	EPOCH	AGE	OTHER DATA	MAMMAL AGES	DISPERSAL/TAXA	SUPER CYCLES	CYCLES	←LANDWARD SEAWARD→ SEA LEVEL
32	C12	n	OLIGOCENE EARLY	RUPELIAN		WHITNEYAN — Wh1	?Elomeryx		4.4	
34	C13	n			5	ORELLAN — Or1	**Grand Coupure (Asia to W. Eur.)**	TA4	4.3	
	C15	n n1	LATE	PRIABONIAN		CHADRONIAN	*Parictis, Palaeogale,* daphoenids, felids,		4.2	
36	C16	n2			4		anthracotheres *(Bothriodon),* ?*Hoplophoneus,*		4.1	
		n1				— Ch1	*Leptomeryx*			
38	C17	n2 n1				DUCHESNEAN			3.6	
40	C18	n1 n2	BARTONIAN			39.7±0.07 Pearson Ranch L.F.	*Duchesneodus, Pterodon, Hyaenodon*		3.5	
42	C19	n1	MIDDLE			UINTAN	*Domnina, Thylacaelurus, Poebrodon, Simidectes, Prodaphoenus, Procynodictis, Epitriplopus, Amynodontopsis,* eomyids, *Colodon, Metamynodon, Grangeria,*		3.4	
44	C20	n	EOCENE	LUTETIAN	3	— Ui2	Lagomorpha, camelids, hypertragulids, *Protoreodon, Leptotragulus*		3.3	
46						44.0±0.9 Shoshonian Subage — Ui1	*Amynodon, Triplopus, Forstercooperia, Eomorpus*	TA3		
48	C21	n				BRIDGERIAN — Br3 — Br2	?*Uintatherium, Telmatherium, Manteoceras* *Thinocyon, Harpagolestes*		3.2	
50	C22	n	EARLY	YPRESIAN		— Br1	*Mesonyx, Hyrachyus,* ?*Eotitanops, Palaeosyops*		3.1 2.9 2.8 2.7	
52	C23	n2 n2				Lostcabinian 52.8±0.3 — Wa4	*Patriofelis, Helaletes*	TA2	2.6 2.5	
54	C24	n1 n3 n				WASATCHIAN			2.4	
56		n								

FIG. 2.—Eocene time scale, including late Paleocene and early Oligocene epochs, after Berggren and others (this volume). Mammal ages discussed in the text. Sea-level pattern is after Haq and others (1988) and Bartow (1992). 3, 4 and 5 refer to Interchange Events. Interchange Event 3 is strongest for North America; Event 5 affects only Eurasia. Unless otherwise specified, dispersal is from Asia to North America. All taxa that are listed for a given interval (e.g., Ui2) are considered to begin at its base; indented taxa also are part of the basal list and are so shown due to graphic limitations.

Late Wasatchian through Bridgerian faunas.—

From the late Wasatchian through Bridgerian time (Fig. 2), Stucky (1992) indicates that thirteen genera among ten families and six orders are found in North America that have Palaearctic affinities. In the late Wasatchian (Wa4, Fig. 2), there are two non-chiropteran genera, found in two families: *Patriofelis* (creodont [As]) and *Helaletes* (helaletid perissodactyl [As]). In early Bridgerian time (Br1, Fig. 2), there are four genera in three families: *Mesonyx* (creodont [As]), *Hyrachyus* [Eur, ?As], ?*Eotitanops* [As], and *Palaeosyops* [As]. Medial Bridgerian taxa (Br2, Fig. 2) include two genera in two families: *Thinocyon* (creodont [As]) and *Harpagolestes* (mesonychid [As]). In late Bridgerian time (Br3, Fig. 2), there are three genera in two families: *Uintatherium* (dinoceratan [As]), *Telmatherium,* and *Manteoceras* (perissodactyl [As]). In the foregoing, [Eur] = Europe; [As] = Asia. The episodes of immigration during an approximately seven million year interval (from 46 to 53 Ma) may be correlative with sea-level changes as discussed in the

explanation to Figure 2. None represent major pulses, however, and sea level was generally high during the Eocene Epoch (Fig. 2).

Uintan mammal age.—

The early Uintan (Fig. 2) records very few allochthonous land mammals, but a major immigration is reflected in the later part of the age. The marine-nonmarine interdigitations of strata with early Uintan land mammals and improved correlations of magnetic polarity signatures make the Uintan an important cornerstone to calibration of the MPTS. Salient features in this regard are summarized below. Many of the radioisotopic data summarized for Uintan faunas in Krishtalka and others (1987) are considered too old or otherwise suspect (Prothero and Swisher, 1992; see also Prothero, 1991).

Shoshonian (Ui1, Fig. 2) is the earliest subage of the Uintan mammal age, and defines the Bridgerian/Uintan boundary (see Flynn, 1986). The base of the Shoshonian is defined on the LSD of the Asian immigrant amynodontid rhinoceros, *Amynodon*. Stucky (1992) adds *Triplopus*, *Forstercooperia*, and *Eomoropus*. See Flynn (1986) for characteristic taxa. Although many are new to the Uintan, none of them appear to have been allochthonous immigrants.

Calibration and correlation.—Flynn (1986) describes the magnetic polarity data from Uintan-age strata in the San Diego area, as well as from sections in the type areas of the Bridgerian and Uintan mammal ages in Wyoming. The results suggest a correlation of part of the Uintan mammal age with the MPTS. Whereas the California succession provides a valuable opportunity for correlating mammal-bearing rocks to marine strata (and thus the marine chronologic systems), the geologic setting demonstrably represents a dynamic, regionally transgressive and regressive situation with few, if any, distinctive isochronous marker beds to assist in correlation.

Prothero and Swisher (1992) develop a consensus of data to support the biochronological correlations offered by Flynn (1986), and concur that the basal Uintan (and Shoshonian) correlates best to Chron 20r in the MPTS, as proposed by Flynn (1986). In Figure 2, the Shoshonian correlates to about Type 2 sea-level lowstand TA3.3.

Interchange Event #3.—

Late Uintan faunas.—This interval (Ui2, Fig. 2) records the greatest influx of allochthonous land mammals since the Wasatchian. Krishtalka and others (1987, p. 89) show that the late Uintan is based on the fauna from Uinta C, from the Myton Member of the Uinta Formation, and from the Brennan Basin Member of the Duchesne River Formation. Stucky (1992) reports that late Uintan immigrants, likely from Asia, include *Amynodontopsis*, *Colodon*, *Metamynodon*, and *Grangeria*. Other immigrants (most likely from Asia; Krishtalka and others, 1987) include Camelidae (artiodactyls) and perhaps Canidae (carnivores). Later in the Uintan, but still within the interval (and included as late Uintan by Stucky, 1992), new forms include *Domnina*, *Thylacaelurus*, *Poebrotherium* (*Poebrodon*), *Prodaphoenus*, *Simidectes*, *Procynodictis*, and *Epitriplopus* (Krishtalka and others, 1987). Stucky (1992) adds Lagomorpha. Thus the familial or subfamilial range of immigrants includes leporid rabbits, miacid carnivores, colodontine, eomoropine, and metamynodontine rhinos, camelids, agriochoerids (*Protoreodon*), protoceratids (*Leptotragulus*), and hypertragulids (Webb, 1977; Webb and Taylor, 1980).

Route and characterization.—Dispersal probably was via a filter to corridor route due to the number and taxonomic diversity of the colonizing fauna. Note that European lagomorphs are first recorded in the Frohnstetten faunal level, considered (Russell and others, 1982, p. 42, Fig. 15) as just pre-Oligocene Epoch, and pre-traditional *Grand Coupure*. In China, Wang (1992) reports lagomorphs from the Urtin Obo fauna considered by her as early Oligocene, but as late Eocene Epoch by Berggren and Prothero (1992:21). In any case, Russell and Zhai (1987, p. 404) indicate that Lagomorpha first occur in Asian faunas of medial Eocene age, correlated with the Bridgerian mammal age, consistent with the late Uintan FAD in North America.

Calibration and correlation.—The late Uintan clearly records an important immigration event, correlative with the Type 2 sea-level lowstand TA3.4. Prothero and Swisher (1992, p. 55) correlate the lower part of Uinta C to a normal polarity interval,

considered likely to be Chron 20n, or about 43 Ma as correlated by Berggren and others (this volume). Prothero (1991, p. 128) reports upon a ^{40}Ar/^{39}Ar date of 42.18 Ma for a late Uintan site in California, correlated to the top of Chron C20n, which Berggren and others (this volume) show as being about 42.5 Ma.

Duchesnean mammal age.—

Whereas the Duchesnean (Du1, Fig. 2) is considered to reflect the important modernization of land mammal faunas in North America (Krishtalka and others, 1987), it records only a few immigrants. Krishtalka and others (1987, p. 84, 90) define the base of the Duchesnean mammal age on the first occurrence of *Duchesneodus*, *Brachyhyops*, *Hyaenodon*, *Simimeryx*, *Poabromylus*, *Hyracodon*, and *Agriochoerus*. Krishtalka and others (1987, p. 90) suggest that at least some of those taxa entered by immigration from Asia. The base of the Duchesnean is estimated to be about 42 Ma old (Krishtalka and others, 1987), but this is revised herein to about 40.5 Ma.

Lucas (1992) redefines the base of the Duchesnean on the sole immigrant occurrence of *Duchesneodus*; Stucky (1992) shows that Eurasian taxa, *Pterodon* and *Hyaenodon*, also are likely Duchesnean immigrants. Other taxa new to the Duchesnean (mostly a few insectivores and carnivores, as well as some rhinocerotoid perissodactyls and camelid and entelodont artiodactyls) are endemically derived. The Asian rodent, *Pseudocylindrodon*, is considered as first appearing in late Uintan faunas according to Stucky (1992), who also shows that *Ardynomys* is a late Duchesnean immigrant, but see Emry and others (1987, p. 133) who suggest that *Ardynomys* is a Chadronian taxon that originated in North America. In any case, the role of immigrants in revising the tone of land mammal faunas relative to those of the late Uintan seems relatively small.

Calibration and correlation.—Lucas (1992) reiterates that the age of the Duchesnean is based on the Halfway and Lapoint faunas of the Dry Gulch Creek and Lapoint members of the Duchesne River Formation, Utah. Lucas (1992, p. 98) suggests that the fauna from the Halfway Member of the Duchesne River Formation, Utah and those of Slim Buttes, South Dakota; Pearson Ranch, California; and the Skyline/Cotter channels, Texas, are of early Duchesnean age, and older than the Lapoint Tuff and associated stratigraphically higher faunas and correlatives (Lapoint, Utah; Tonque, New Mexico; Porvenir, Texas). Walton (1992) shows that the Skyline/Cotter channels (Bandera Mesa Member) unconformably overlie the middle member of the Devil's Graveyard (= Pruett) Formation in the Trans-Pecos region of Texas, the uppermost part of which has yielded a K-Ar age of 42.7 ± 1.1 Ma. The error range in this age allows the possibility that the associated, uppermost, normal magnetochron in the middle member could correlate with Chron C19n, and also that the Skyline channel fauna (Duchesnean) could be even younger.

Prothero and Swisher (1992, p. 49) indicate that the Lapoint Tuff yields a ^{40}Ar/^{39}Ar age of 39.74 ± 0.07 Ma. Fossils above the tuff are considered to be of Duchesnean age. Prothero and Swisher (citing Krishtalka and others, 1987, p. 90 but not Fig. 4.3) indicate that faunas older than the Lapoint Tuff are considered Uintan. Lucas (1992), however, includes the limited Halfway fauna in the Duchesnean, so in his terms, the base of the Duchesnean pre-dates somewhat the age of the Lapoint Tuff. Prothero and others (1992) indicate that the Pearson

Ranch L.F. (California) correlates to Chron C18n, permissible to be older than the date on the Lapoint Tuff, but younger than about 40.2 Ma in the MPTS (Berggren and others, this volume). The age for the base of the Duchesnean chosen here, about 40.5 Ma, takes into account all of the uncertainties expressed. According to this, the beginning of the Duchesnean correlates to near the middle of Chron C18r and to a point somewhat younger than the Type 2 sea-level lowstand of TA3.5.

Interchange Event #4.—

*Chadronian mammal age.—*Emry and others (1987, p. 132–133) indicate that the base of the Chadronian (Ch1, Fig. 2) may be defined on the joint first occurrence of [*Agriochoerus*], [*Archaeotherium*], *Daphoenictis, Dinictis, Hoplophoneus,* "*Ictops,*" [[*Metamynodon*]], *Merycoidodon, Palaeolagus, Poebrotherium, Hesperocyon, Parictis, Palaeogale,* and *Leptomeryx* ([   ] = taxa considered to appear in the late Duchesnean or [[   ]] late Uintan; Stucky, 1992). Emry and others (1987, p. 132–133) also state that the place of origin of many of these taxa is uncertain, but some apparently are Asian (or Palaearctic) in source: *Parictis, Palaeogale,* ?daphoenids, felids, anthracotheres (*Bothriodon* = ?*Aepinacodon*), ?*Hoplophoneus* and *Leptomeryx.* In addition to the above immigrants, Lucas (1992, p. 98) bases the beginning of the Chadronian on the first occurrence of other endemic taxa: Sciuridae (*Protosciurus*), *Daphoenocyon, Penetrigonias, Stibarus,* Tayassuidae (*Perchoerus*), Anthracotheriidae ([*Heptacodon*]), *Bathygenys, Montanatylopus, Pseudoprotoceras,* and *Hypisodus.* Taxa identified by Stucky (1992) are included within the above lists, except that *Agriochoerus, Archaeotherium,* and *Heptacodon* are considered as late Duchesnean forms. Emry (1992) does not modify this list, but points out that the beginnings of the Duchesnean and Arikareean mammal ages, respectively, are more important in terms of faunal innovation, than is the beginning of the Chadronian (also Emry and others, 1987). Data reviewed herein suggest that the late Uintan saw a more important immigration of taxa than did the base of the Duchesnean or Chadronian.

*Route and characterization.—*The trans-Beringian dispersal route probably was a filter because of the "small and peculiarly assorted part of the fauna" (Emry and others, 1987, p. 132).

*Calibration and correlation.—*Swisher and Prothero (1990) and Prothero and Swisher (1992) recalibrate the Oligocene strata in the Big Badlands of North America, relying heavily on tuffs associated with faunal sites in the Chadron Formation of the Flagstaff Rim, Wyoming. These authors suggest that the base of the Chadronian is most likely dated at about 37 Ma, but Emry (1992) indicates that it could be older based on the amount of evolutionary change shown by some taxa (*Leptomeryx*) in the lower (and currently not radioisotopically dated) part of the section at Flagstaff Rim. Until the boundary is modified as per Emry (above), the base of the Chadronian appears to be slightly younger than the beginning of the Type 1 sea-level lowstand TB4.1.

Interchange Event #5: the Eocene-Oligocene Boundary

Oligocene Epoch.—

*Grande Coupure.—*For mammalian paleontologists, the beginning of the Oligocene Epoch in the Old World has long been associated with the *Grande Coupure,* a concept coined by Stehlin (1909) to recognize an episode in which the land mammal fauna of Europe was sharply reorganized. The *Grande Coupure* saw endemic and archaic European taxa replaced suddenly in the rock record by a host of new forms: amynodont (*Cadurcotherium*) and rhinocerotoid (*Hyracodon* and other rhinocerotoid) rhinoceroses, titanotheres (?), chalicotheriids, entelodonts (*Entelodon*), anthracotheres (*Bothriodon, Anthracotherium, Brachyodus*), and palaeochoerids (considered to be tayassuid peccaries; for example, Savage and Russell, 1983). Savage and Russell (1983, p. 154) point out that the faunal change was not instantaneous. Nevertheless, this episode represents an important immigration, most components of which came from Asia, and fewer from North America. Even though about 60% of the indigenous land mammals of Europe became extinct during the time of the *Grande Coupure* (Berggren and Prothero, 1992, p. 19), this was not solely due to immigration of new forms, but coincided with major climatic deterioration, as well. The dispersal/sea-level scenario is also complex in that not only is Beringia involved, but also retreat of the seas in the Turgai straits and tectonic activity in the mid Alpine area (Heissig, 1979). Groups that became extinct or suffered diminution included Primates, pseudosciurid rodents, Artiodactyla, Perissodactyla, creodonts, condylarths, and proteutherians (Hooker, 1992; Legendre and others, 1991).

The traditional view has been that the *Grande Coupure* coincided with comparable changes in the land mammal faunas of North America and Asia (Savage and Russell, 1983, p. 190–192: "The resurgence of similarity between land mammals of Eurasia (especially striking for Europe) and North America at the beginning of the Oligocene or at the end of the Eocene is noteworthy"). The proposition that these changes were globally correlative is discussed further below (see also Discussion: *Vicariance*). The first question is the age of the *Grande Coupure* and the beginning of the Oligocene Epoch.

Berggren and others (1985b, p. 164) state that the *Grande Coupure* consists of a "rapid but demonstrably time-transgressive appearance of some 10–13 new mammalian families . . . between the late Eocene (Gypse de Montmartre in the Paris Basin) and the early Oligocene (Ronzon in the Haute-Loire, Soumailles in Lot-et-Garonne . . .)." In terms of the European chronology, the first immigrants of the *Grande Coupure* are recorded in relatively widespread faunas in England and western Europe at the level of Hoogbutsel and correlative faunas (Soumaille; MP [Mammal Paleogene] interval 21) or at about the time of the boundary between nannofossil Zones NP21/22 (Hooker, 1992, p. 508). In addition to those cited above, taxa entering the region at about this time include modern insectivores (Erinaceidae, Heterosoricidae), rodents (Cricetidae, Aplodontidae, Eomyidae, new sciurids, dipodids, Castoridae), mustelid and viverrid carnivores, lagomorphs, and new families of perissodactyls (mainly rhinocerotids, amynodontids, chalicotheriids) (Legendre and others, 1991). R. H. Tedford (pers. commun., 1994) suggests that the following could be added to this list: Ursidae (*Amphicynodon*), Amphicynodontidae (*Cynodictis*), Nimravidae (*Nimravus*), Felidae (*Proaelurus*) and the feliform, *Palaeogale.*

Whereas some taxa appeared in faunas of late Eocene age (the anthracothere, *Elomeryx cripsus*; La Debruge; Lagomorpha, Frohnstetten; Russell and others, 1982), the general consensus appears to be that the main elements of the *Grande Cou-pure* and the base of the Oligocene coincide at the beginning of the Rupelian (= basal Stampian and locally, Sannoisian) Stage. Berggren and others (this volume) correlate this with the upper part of Chron C13r, near the base of foraminiferal Zone P18, and at the top of the Priabonian Stage, and near the middle of nannofossil Zone NP 21. This would result in the beginning of the *Grande Coupure* being somewhat (perhaps 0.5 my) younger than the beginning of the Oligocene Epoch (Berggren and Prothero, 1992, p. 19). If the Lattorfian localities such as Frohnstetten are taken as basal Oligocene (and thus faunas of late Ludian age, as well, are earliest Oligocene) and correlative to some part of NP 21, then earliest Oligocene could equate with earliest *Grande Coupure*.

Swisher and Prothero (1990) indicate that the base of the Oligocene Epoch is about 34 Ma old (33.9 Ma), coincident with the Chadronian/Orellan boundary in North America, and cor-relate this to a position near the top of Chron 13r. This is com-parable to the age selected by Berggren and others (this volume) of about 33.7 Ma for the beginning of the Oligocene. It may be important to the *Grande Coupure* (Asia to western Europe; but not to North America) that this is coeval with the Type 1 sea-level lowstand TA4.4.

Orellan mammal age.—The base of the Orellan (Or1, Figs. 2, 3) corresponds to the base of the Oligocene Epoch and about to the time of the *Grande Coupure, but shows no major faunal exchange* with Palaearctica at this time. Emry and others (1987, p. 142–143) show that the Orellan taxa differ slightly, and mostly at the species level, from Chadronian forms. Swisher and Prothero (1990) and Prothero and Swisher (1992) show that the Orellan begins at about 34 Ma.

Whitneyan mammal age.—Likewise, the Whitneyan shows *no important faunal* exchange with Palaearctica. The beginning of the Whitneyan (Wh1, Figs. 2, 3) can be recognized by the first occurrence of endemic taxa, such as *Leptauchenia* and *Scotti-mus lophatus* (Emry and others, 1987, p. 144), to which might be added *Miospermophilus, Palaeocastor, Protospermophilus, Leidymys, Paciculus, Geringia, Kirkomys, Diceratherium, Elomeryx* (a Palaearctic immigrant?), *Protoceras, Nannotra-gulus, Chaenohyus, Pithexistes* and *Promesoreodon* (Stucky, 1992, Table 24.2). Of these taxa, only *Promesoreodon* is re-stricted to the Whitneyan, although Emry and others (1987, p. 145) suggest a number of potentially diagnostic and restricted species. Prothero and Swisher (1992) indicate that the Orellan/Whitneyan boundary is about 32.2 Ma old.

Arikareean mammal age.—The Arikareean begins at about 30 Ma (subsequent to the beginning of the Type 2 sea-level low-stand TA4.5; Ar1, Fig. 3) and encompasses a time of major sea-level lowstand at about 28–30 Ma (supercycle TB1), during which a limited number of land mammals were introduced into North America. Rather than being manifested by a single im-migration event, the Arikareean records a number of spaced episodes that may generally correspond to a time of relative lowstand. Thus all segments of the Arikareean are discussed here.

Tedford and others (1987, p. 184, Fig. 6.3) define the begin-ning of the Arikareean by the immigrant occurrence of the aplo-dontid rodent, *Allomys*; the zapodid rodent, *Plesiosminthus*; er-inaceid hedgehogs; the galericine *Ocajila*; and the talpine moles. This rather limited roster of immigrants serves to un-

derscore the general faunal continuity (White River chrono-fauna; Emry and others, 1987, p. 145, 146) with preceding mammal ages (see above). Work in progress revises the LSD of these taxa and indicates that they first occur in local sections in Nebraska and South Dakota at various levels within the 28–30 my span at the beginning of the Arikareean (R. H. Tedford, pers. commun., 1994). In any case, the main immigration event occurs in *the late Arikareean* (early Miocene), and the FAD of *Allomys* (*Alwoodia*; Rensberger, 1983) is shown (below) to slightly post-date the base of the Arikareean in Oregon and in South Dakota (where it still may be earlier than in Oregon (Fisher and Rensberger, 1972, Fig. 7).

Calibration and correlation.—Swisher and Prothero (1990) and Prothero and Swisher (1992) recalibrate the strata in North America conventionally considered to be of Oligocene age. These authors suggest that the base of the Arikareean is most likely dated at about 29.0 Ma, and this is followed in Prothero (1991, Fig. 3, revised), with faunas above the Rockyford Ash (Prothero, 1991, Fig. 3, revised) suggested as showing that ear-liest Arikareean faunas are correlative with Chron C10n. The Rockyford Ash in South Dakota may be equivalent to the Non-pareil Ash Zone of Nebraska dated at 30.0 ± 0.2 Ma (Prothero and others, 1991, p. 51A; compare differences in correlation to the MTPS with that of Prothero, 1991; R. H. Tedford, pers. commun., 1994, indicates that the Rockyford Ash is best cor-related with the middle ash of the Nonpareil Ash Zone, which lies within chron C11n.2n or at about 30 Ma). Other radioiso-topic dates reported in Prothero (1991) and Prothero and others (1991) are consistent with the Gering Formation having been deposited mostly during the time of Chron C9r. Mason and Swisher (1989) report a K/Ar age of 28.2 ± 0.02 Ma for a tuff that underlies the early Arikareean South Mountain L.F. of California.

New radioisotopic dating of the John Day Formation (Swisher, unpubl. data) indicates, however, that the base of the Arikareean mammal age is older than previously thought. In the John Day Formation, Oregon, the Picture Gorge Ignimbrite (PGI) has been dated at about 28.7 Ma by means of the ⁴⁰Ar/³⁹Ar method. Rensberger (1983; Fig. 55), but not Prothero and Rensberger (1985), show that the PGI falls within the *Menis-comys* Concurrent Range-Zone, and above the "unnamed aplo-dontid zones" (Prothero and Rensberger, 1985, Fig. 1) of the Turtle Cove Member of the John Day Formation that yields early Arikareean fossils (John Day "Fauna," Tedford and oth-ers, 1987, Fig. 6.2). If the Rockyford Ash (and its proposed equivalent, Nonpareil Ash Zone) marks the approximate age of the geochron of the Whitneyan/Arikareean boundary, then fos-sils in the "unnamed aplodontid zones" of the John Day For-mation likely are of earliest Arikareean age. Swisher (unpubl. data) has obtained a radioisotopic age (⁴⁰Ar/³⁹Ar method) of about 29.7 Ma on an ash in the upper part of the normal mag-netozone correlated by Prothero and Rensberger (1985, Fig. 5) as Chron C8n. We suggest here that it is much more likely that this normal magnetozone corresponds, in part, to Chron C11n.2n (Berggren and others, this volume). These correlations are comparable to those proposed in Prothero (1991, Fig. 3, revised) in which the faunas of the Gering Formation, Nebraska range in age from somewhat younger than 28.3 ± 0.5 Ma (Twin Sisters Pumice; revised from 27.7 ± 0.6 Ma in Tedford and others, 1987, #39, Appendix A) to somewhat younger than the

OLIGOCENE TIME SCALE

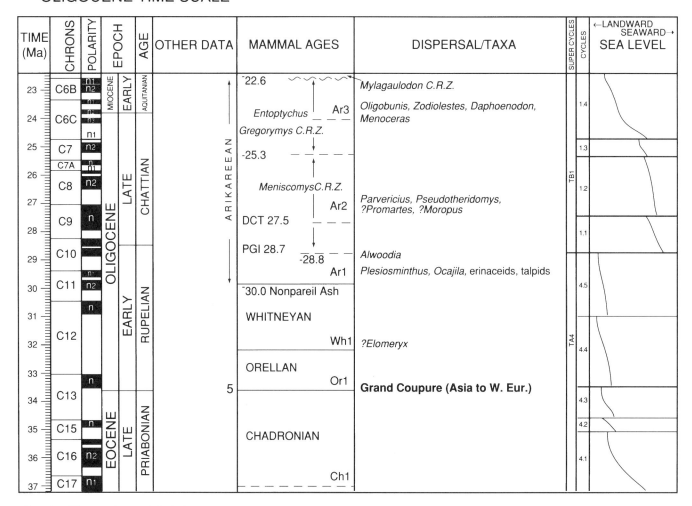

Fig. 3.—Oligocene time scale, including late Eocene and early Miocene epochs, after Berggren and others (this volume). Mammal ages discussed in the text. Sea-level pattern is after Haq and others (1988) and Bartow (1992). The Orellan through most of the Arikareean mammal ages correlate with the Oligocene Epoch as used here. 5 refers to an Interchange Event. Unless otherwise specified, dispersal is from Asia to North America. Lists of multiple taxa separated by commas are considered to begin at the base of the indicated interval. See text for radioisotopic ages. PGI = Picture Gorge Ignimbrite; DCT = Deep Creek Tuff; *Meniscomys* C.R.Z. (Concurrent Range Zone), and *Mylagaulodon* C.R.Z. are from the John Day Formation, Oregon. *Entoptychyus-Gregorymys* C.R.Z. is from the John Day, and rock units in the Great Plains, as explained in the text. Ar3 is early late Arikareean (see Fig. 4).

28.6 ± 0.3 Ma date on the Roundhouse Rock Ash in the underlying Brule Formation. This also suggests that the older of the dates (28.8 ± 0.6 Ma) reported by Tedford and others (1987, Fig. 6.2, #38, Appendix A) for the Helvas Canyon Member of the Gering Formation is the more reasonable. For this report, we assign an age of about 30 Ma to the base of the Arikareean. In likely correlating to Chron C11n.2n, the base of the Arikareean corresponds to within the generally regressive phase of the Type 2 lowstand TA4.5, which nevertheless occurs in a generally high sea-level stand, possibly accounting for the limited early Arikareean immigration episode described above (see also Ar1, Fig. 3).

Late early Arikareean faunas.—Tedford and others (1987, p. 184, Fig. 6.2) suggest that the later part of the early Arikareean (Ar2; Fig. 3) can be defined on the immigrant occurrence of the hedgehog, *Parvericius*; the eomyid rodent, *Pseudotheridomys*; possibly the mustelid, *Promartes*; and possibly the chali-

cothere, *Moropus* (two of four taxa are equivocal). This event is based on faunas of the Monroe Creek Formation, Nebraska and correlatives in South Dakota in part coeval with the *Meniscomys* Concurrent Range-Zone and partly with the *Entoptychus-Gregorymys* Concurrent Range-Zone of Fisher and Rensberger (1972).

Calibration and correlation.—Prothero and Rensberger (1985) have correlated the *Meniscomys* Concurrent Range-Zone of the John Day Formation to magnetic Chron 7An (near its middle) to the lower part of Chron 6Cr, calibrated (Berggren and others, this volume) as about 25.5–24.4 Ma old. Rensberger (1983, Fig. 55) shows that the *Meniscomys* Concurrent Range-Zone begins a short interval stratigraphically below the Picture Gorge Ignimbrite, dated (*in* Fisher and Rensberger, 1972) at about 26 Ma, but we suggest (above) that the PGI is more likely 28.7 Ma old.

In that the late early Arikareean boundary correlates to about the middle of the *Meniscomys* Concurrent Range-Zone (Tedford and others, 1987, Fig. 6.2), it likely would fall near the age of the Deep Creek Tuff. This tuff has been dated (^{40}Ar/^{39}Ar method; Swisher, unpubl. data) at about 27.5 Ma. If accurate, this age for the Deep Creek Tuff forces a major downward revision of the age of the *Meniscomys* Concurrent Range-Zone, consistent with the revised age of the beginning of the Arikareean discussed above. The new age for the Deep Creek Tuff indicates that the mixed-polarity magnetozone correlated by Prothero and Rensberger (1985, Figs. 6, 7) to Chron C7Ar is more likely attributed to Chron C9n and perhaps to Chron C9r. If the age of the Deep Creek Tuff can be applied to the base of the late early Arikareean, that corresponds to about the middle of Chron C9n, and to near the base of the Type 1 sea-level lowstand TB1.2.

The *Entoptychus-Gregorymys* Concurrent Range-Zone is correlated on paleomagnetic evidence by Prothero and Rensberger (1985) from within Chron 6Cr to Chron C6Cn.2n or from about 24.4 to 23.7 Ma (Berggren and others, this volume). The strong unconformity that separates the Kimberly Member of the John Day Formation from the overlying Haystack Valley Member results (in part) in the upper limit of the *Entoptychus-Gregorymys* Concurrent Range-Zone still being approximately known. The Haystack Valley Member, which also has not yet been sampled for paleomagnetic data, contains the boundary between the *Entoptychus-Gregorymys* Concurrent Range-Zone and the next-successive *Mylagaulodon* Concurrent Range-Zone of latest Arikareean to early Hemingfordian age (Fisher and Rensberger, 1972, Fig. 5; Prothero and Rensberger, 1985, Fig. 1, but not Fig. 5 or 6; Tedford and others, 1987, Fig. 6.2).

C. C. Swisher (unpubl. data) has obtained a radioisotopic age (^{40}Ar/^{39}Ar method) of about 25.3 Ma on a tuff at the base of the *Entoptychus-Gregorymys* Concurrent Range-Zone as portrayed by Prothero and Rensberger (1985, Fig. 5; see the tuff indicated between PL [*Entoptychus planifrons* level] and EB [*Entoptychus basilaris* level]). Swisher (unpubl. data) also has obtained an age of about 22.6 Ma for a sample above the normal magnetozone near the top of *the Entoptychus-Gregorymys* Concurrent Range-Zone. Prothero and Rensberger (1985) suggest that the *Entoptychus-Gregorymys* Concurrent Range-Zone correlates to Chrons C6Cr to C6C (in part). We suggest that a correlation is more likely from Chron C7n.2n to Chron C6Bn.1n-2n, or from about 25.3 to 23 Ma in the time scale of Berggren and others, this volume). The *Mylagaulodon* Concurrent Range-Zone is thus no older than about 22.6 Ma.

The Oligocene-Miocene Boundary

Cande and Kent (1992, p. 13,933), and Berggren and others (this volume) indicate that the Chattian/Aquitanian boundary corresponds to a level within "the middle part of chron C6Cn," and estimate an age of 23.8 Ma for it. In terms of its radioisotopic correlation, the late early Arikareean mammal age effectively represents the terminal record of Oligocene land mammal faunas in North America, the age of which apparently falls within the span of the *Entoptychus-Gregorymys* Concurrent Range-Zone.

Miocene Epoch.—

Early late Arikareean faunas.—Tedford and others (1987, Fig. 6.3, p. 185) define this interval (Ar3, Figs. 3, 4) by the first

immigrant occurrence of the mustelid carnivores, *Oligobunis* and *Zodiolestes* (the latter now interpreted as a mustelid; R. H. Tedford, pers. commun., 1994), the amphicyonid carnivore, *Daphoenodon*, and the rhinoceratid, *Menoceras*. These taxa are drawn largely from the faunas of the Harrison Formation of Nebraska and equivalents. The late Arikareean apparently begins within the span of the *Gregorymys-Entoptychyus* Concurrent Range-Zone of Fisher and Rensberger (1972).

Calibration and correlation.—Tedford and others (1987, Fig. 6.2) show that elements of the fauna from the Harrison Formation are included within the span of the *Entoptychus-Gregorymys* Concurrent Range-Zone in Oregon and occur stratigraphically above and below an ash dated at about 22 Ma in Nebraska (Tedford and others, 1987, Fig. 6.2, and #41, Appendix A). Above, we suggest that the *Entoptychus-Gregorymys* Concurrent Range-Zone is calibrated at between 25.3 and 22.6 Ma. We suggest that the late Arikareean (Ar3, Fig. 3) began about 24 Ma, correlative with about the middle of Chron C6Cn. On this basis, the late Arikareean may have coincided to within the waning phase of the Type 1 sea-level lowstand TB1.4 (Ar3, Fig. 3).

Late late Arikareean faunas.—Tedford and others (1987, Fig. 6.3, p. 185) define this interval by the first immigrant occurrence of the hemicyonine ursid, *Cephalogale*, the amphicyonids *Cynelos* and *Ysengrinia*, the blastomerycine, *Blastomeryx*, and the dromomerycid, *Barbouromeryx*, with the fauna from the Agate Springs quarries of Nebraska being typical of the age (Ar4, Fig. 4). The later part of the late Arikareean apparently begins within the later part of the *Gregorymys-Entoptychyus* Concurrent Range-Zone, and ends within the *Mylagaulodon* Concurrent Range-Zone of Fisher and Rensberger (1972).

Calibration and correlation.—Fisher and Rensberger (1972), Prothero and Rensberger (1985, Fig. 1, but not Figs. 5 and 6) and Tedford and others (1987, Fig. 6.2) show that the boundary between the *Entoptychus-Gregorymys* and *Mylagaulodon* concurrent range-zones occurs within the Haystack Valley Member of the John Day Formation, Oregon and that this correlates about with the Agate Springs Local Fauna of Nebraska which pre-dates a fission-track age of about 19 Ma (Tedford and others, 1987, #51, Appendix A). Above, we suggest that the base of the *Mylagaulodon* Concurrent Range-Zone is no older than about 22.6 Ma. We follow Tedford and others (1987, Figs. 6.2, 6.3) in correlating the base of the late late Arikareean as 21 Ma. This corresponds to within the Type 2 sea-level lowstand interval TB1.5 (Ar4, Fig. 4).

Interchange Event #6

Hemingfordian mammal age.—Tedford and others (1987, Fig. 6., p. 186) indicate that the beginning of the Hemingfordian (He1; Fig. 4) can be defined by the immigrant first occurrence of the soricine shrew, *Antesorex*, and the limnoecine shrew, *Angustidens*, the plesiosoricid, *Plesiosorex*, the ochotonid lagomorph, *Oreolagus*, the carnivores, *Amphicyon*, *Hemicyon* (*Phoberocyon*), *Ursavus* (a bear), the leptarctine mustelids, *Craterogale*, and *Leptarctus*, the seal-like mustelid, *Potamotherium*, the procyonids, *Edaphocyon* and *Amphictis*, the rhino, *Brachypotherium*, and the dromomerycid, *Aletomeryx*. This represents a major incursion of Palaearctic taxa, recorded in faunas of the Runningwater Formation, Nebraska and correla-

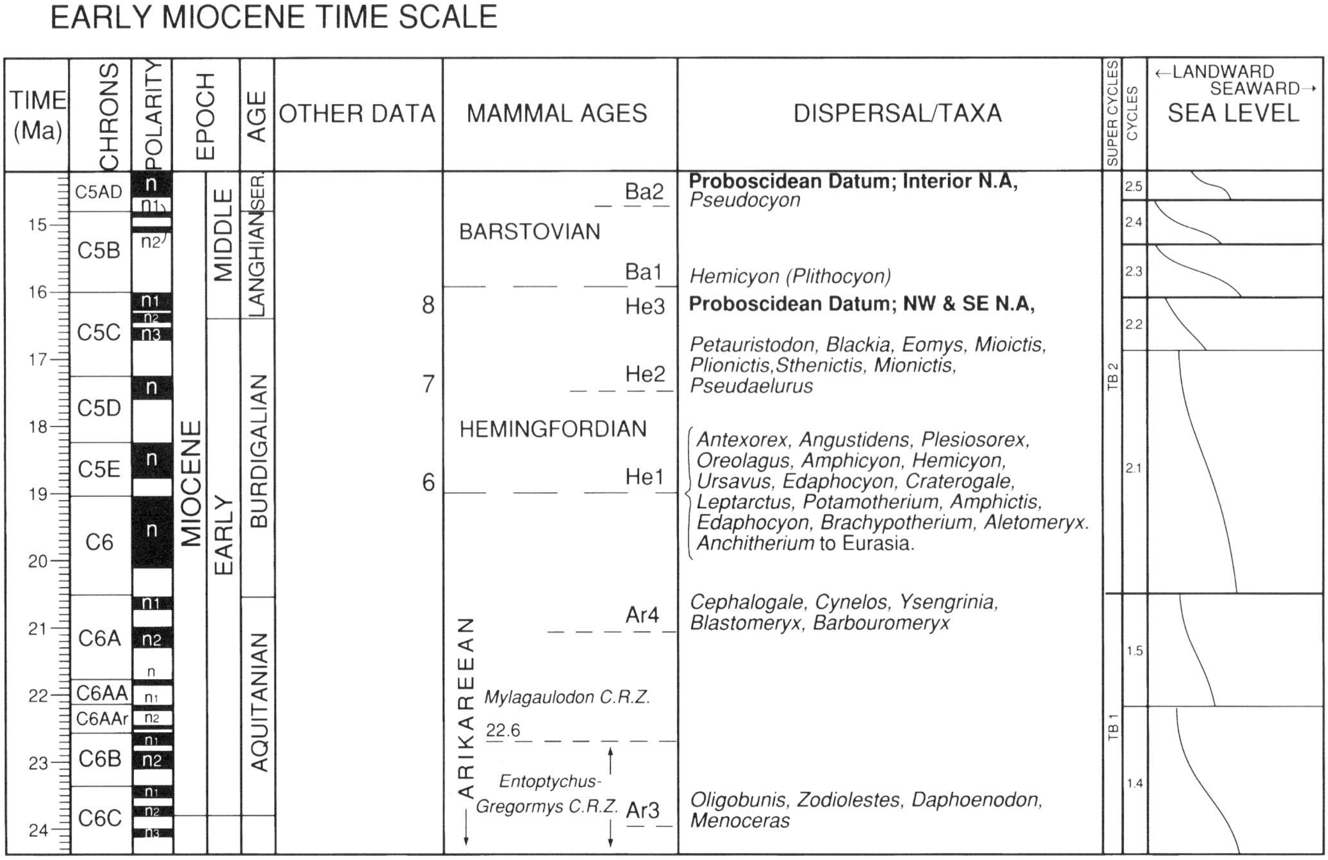

EARLY MIOCENE TIME SCALE

TIME (Ma)	CHRONS	POLARITY	EPOCH	AGE	OTHER DATA	MAMMAL AGES	DISPERSAL/TAXA	SUPER CYCLES	CYCLES	←LANDWARD SEAWARD→ SEA LEVEL
15	C5AD	n, n1, n2	MIDDLE / LANGHIAN / SER.			Ba2 / BARSTOVIAN	**Proboscidean Datum; Interior N.A,** *Pseudocyon*		2.5 / 2.4	
16	C5B					Ba1	*Hemicyon (Plithocyon)*		2.3	
17	C5C	n1, n2, n3			8	He3	**Proboscidean Datum; NW & SE N.A,** *Petauristodon, Blackia, Eomys, Mioictis, Plionictis, Sthenictis, Mionictis, Pseudaelurus*	TB 2	2.2	
18	C5D	n	EARLY MIOCENE / BURDIGALIAN		7	He2 / HEMINGFORDIAN	*Antexorex, Angustidens, Plesiosorex, Oreolagus, Amphicyon, Hemicyon, Ursavus, Edaphocyon, Craterogale, Leptarctus, Potamotherium, Amphictis, Edaphocyon, Brachypotherium, Aletomeryx. Anchitherium* to Eurasia.		2.1	
19	C5E	n			6	He1				
20	C6	n								
21	C6A	n1, n2, n	AQUITANIAN	ARIKAREEAN		Ar4	*Cephalogale, Cynelos, Ysengrinia, Blastomeryx, Barbouromeryx*		1.5	
22	C6AA / C6AAr	n1, n2				*Mylagaulodon C.R.Z.* 22.6		TB 1		
23	C6B	n1, n2				*Entoptychus- Gregormys C.R.Z.*			1.4	
24	C6C	n1, n2, n3				Ar3	*Oligobunis, Zodiolestes, Daphoenodon, Menoceras*			

FIG. 4.—Early Miocene time scale, including part of the middle Miocene Epoch, after Berggren and others (this volume). Mammal ages discussed in the text. Sea-level pattern is after Haq and others (1988) and Bartow (1992). 6, 7 and 8 refer to Interchange Events. Events 6 and 7 record major dispersals to North America. Lists of multiple taxa separated by commas are considered to begin at the base of the indicated interval.

tives, but does not appear to correlate with a major sea-level lowstand.

A Type 1 sea-level lowstand was present at about 20 Ma (TB2.1; Fig. 4). The equid, *Anchitherium*, apparently dispersed to the Old World about this time, a rare instance of a North American emigration in the early Neogene interval. In the Old World, *Anchitherium* marks the beginning of Mammal Neogene (MN) 3 (Mein, 1989), suggested as MN 3b (Ginsburg, 1989); Mammal Unit II in China (Qiu, 1989), the Orleanian mammal age (Savage and Russell, 1983, p. 226), and within the Orleanian (Mein, 1989), correlated at about 18.0 ± Ma, but less than 22 Ma (Steininger and others, 1989, p. 28). If the *Anchitherium* FAD occurs in later Orleanian faunas (Ginsburg, 1989) and at about the beginning of Chron C6N (Steininger and others, 1989, Fig. 1), the alignment shown in Figure 4 is plausible.

Route and characterization.—The early Hemingfordian dispersal route apparently was a trans-Beringian corridor as reflected in the taxonomic breadth of the immigrant elements.

Calibration and correlation.—This follows Tedford and others (1987, # 51, p. 198, and Fig. 6.2) who show that an ash that yields a fission-track age of 19.2 ± 0.5 Ma occurs within the Marsland Formation, above which are faunas of early Hemingfordian age in the overlying Runningwater Formation. The 21 my age of the base of the Hemingfordian (Tedford and others,

1987; Fig. 6.2, 6.3) is inconsistent with the text. The text is followed here.

Interchange Event #7

Late Hemingfordian mammal age.—Tedford and others (1987, Fig. 6., p. 187) define this interval (He2, Fig. 4) as beginning with the immigrant first occurrence of petauristine squirrels *Petauristodon* and *Blackia*, the eomyid rodent, *Eomys*, the musteline mustelid carnivores, *Mioictis, Plionictis* and *Sthenictis*, the lutrine mustelid, *Mionictis*, the earliest Felidae (cats), *Pseudaelurus*, and by the endemic rhinoceroses, *Teleoceras* and *Aphelops*. These taxa are typical of the classical Sheep Creek Fauna, Nebraska. This is a relatively strong immigration episode, but it apparently transpired during the waning phase of interval TA2.1 and about 1 my before a Type 2 lowstand at TA2.2.

Route and characterization.—The route was via Beringia, and likely was a corridor due to the taxonomic breadth of the immigrants. This is one of the major Palaearctic immigrations in the Miocene Epoch.

Calibration and correlation.—Tedford and others (1987, Fig. 6.2) suggest that the late Hemingfordian began about 18 Ma and pre-dated somewhat the fauna from the Box Butte Formation in Nebraska, which contains taxa that are more primitive than those from the Sheep Creek Formation. The Box Butte Formation is not directly calibrated, but the Sheep Creek Formation contains an ash that occurs stratigraphically above the Sheep Creek Fauna, dated by fission-track means at 17.5 ± 0.5 and 16.5 ± 0.6 Ma (Tedford and others, 1987, p. 198, average of 17.0 ± 0.6). It is well known that fission-track ages commonly are younger than K/Ar or ^{40}Ar/^{39}Ar ages from the same sample. Equid species from the Box Butte Fauna (Galusha, 1975) include *"Merychippus" primus*, *"M." tertius* (Hulbert and MacFadden, 1991), and a form possibly close to *Merychippus insignis* (MOW, pers. observ.). MacFadden and others (1991) indicate that the LSD of *"M." primus* in Florida is about 16.2 ± 1.4 Ma old, based on ^{87}Sr/^{86}Sr ratios correlated to the marine chronostratigraphy (Veizer, 1989). This is permissive of the circa 17.0 my minimum age for that species in Nebraska. MacFadden and others (1991) also indicate that a sample of *Parahippus leonensis*, the nominal out-group taxon for merychippines from the Seaboard site in Florida, is about 17.7 ± 1.4 Ma old. *P. leonensis* also is known from the upper part of the Runningwater Formation of Nebraska (Hulbert and Mac-Fadden, 1991), considered by Tedford and others (1987) as containing the youngest fauna of early Hemingfordian age there. Miller (1980) records the presence of *Aletomeryx* in the Upper Cady Mountains L.F. of California, and MacFadden, Woodburne and Opdyke (1990; Fig. 7) indicate that this local fauna likely correlates with Chron C5Cr (circa 16.7–17.3 Ma; Fig. 4). Tedford and others (1987: 168) indicate that the Box Butte Formation contains a LSD of *Aletomeryx*. Based on all of the above, the Box Butte fauna, and thus the beginning of the late Hemingfordian, is taken as about 17.5 my old.

MacFee and Iturralde-Vinent (1994) describe remains of a megalonychid sloth from deposits of late early Miocene age (ca 17–18 Ma) in central Cuba, and suggest that the fundamental relationships of the group are to be sought in sloths of Santacrucian age in South America (18–15 Ma; Marshall, 1985). MacFee and Iturralde-Vinent (1994) suggest that marine barriers may have been present to isolate Cuba at this time (Straits of Florida; Yucatan Channel; see also Erickson and Pindell, 1993, for a discussion of major subsidence in northeastern Venezuela in the Oligocene and Miocene epochs). Nevertheless, the Cuban sloth record is a tantalizing glimpse of what may have been a growing nucleus of West Indian edentate evolution and potential outward dispersal in the Neogene interval (see early Hemphillian, below).

Latest Hemingfordian-Clarendonian mammal ages.—This interval (Figs. 4 and 5) records few dispersal events. The Proboscidean Datum and the Hipparion Datum are the main important episodes.

Interchange Event #8

The Proboscidean Datum.—

Latest Hemingfordian mammal age.—The Proboscidea apparently dispersed to North America at this time (He3, Fig. 4), and are associated with a late Hemingfordian fauna in Nevada (Morea, 1981), calibrated at about 16.2 Ma (Swisher, unpubl. data). This immigration episode possibly pertains to the somewhat earlier late Hemingfordian event, but could have occurred

slightly later (about coeval with the Type 1 sea-level low at TB2.3). Bryant (1991) records an early Barstovian occurrence of the group in Florida, likely about 15.8 Ma old. As seen below, there apparently was a major lag in proboscidean dispersal to the more interior locations of the United States.

Barstovian mammal age.—As indicated in Tedford and others (1987, Fig. 6.3), the Barstovian begins with the immigration of the cricetid rodent, *Copemys* and the hemicyonine bear, *Plithocyon* (Ba1, Fig. 4). Work by E. H. Lindsay (1991) and R. E. Reynolds (pers. commun., 1990, *in* Woodburne, 1991, p. 72) suggest that *Copemys* is now recorded in strata of the Mud Hills (stratotype Barstovian) and Alvord Mountain, southern California, in strata associated with taxa of late Hemingfordian age (likely equivalent to the Red Division Fauna; Woodburne and others, 1990). *Plithocyon* occurs in early Barstovian faunas in Oregon and New Mexico as defined by Tedford and others (1987, p. 168).

Calibration and correlation.—Woodburne and others (1990) correlate the base of the Barstovian at 15.9 my, based on considerations of the rate of accumulation of the Barstow Formation, Mud Hills, southern California. MacFadden and others (1990) correlate the base of the Barstovian to within Chron C5B of Berggren and others (1985a, this volume), and Cande and Kent (1992). On this basis the Barstovian begins at about the Type 1 sea-level lowstand TB2.3. Note that a late Hemingfordian entry of *Copemys* may have coincided with the immigration of the taxa first found in the Sheep Creek Fauna of Nebraska (Tedford and others, 1987, p. 168).

Late Barstovian mammal age.—Tedford and others (1987) define this (Ba2, Fig. 5) upon the first occurrence of Proboscidea and the amphicyonid carnivore, *Pseudocyon*. See the above discussion of the Proboscidean Datum as to the zoogeographic situation of this group; it suggests that this event is subject to vagaries of biofacies and should be used with extreme caution.

Calibration and correlation of the late Barstovian Proboscidean Datum.—A LSD for Proboscidea is recorded in the Barstow Formation (Ba2, Figs. 4, 5), at or stratigraphically slightly below the Dated Tuff (14.8 ± 0.06 Ma). A chronologically similar LSD of Proboscidea occurs in the upper Pawnee Creek Formation of northeast Colorado (Tedford and others, 1987, p. 173), a short interval stratigraphically above an ash dated at circa 14.4 Ma (Tedford, pers. commun., 1990). Tedford and others (1987, Fig. 6.3) correlate the base of the late Barstovian at about 14.5 Ma, but this is revised here to be about 14.8 Ma. In the Barstow Formation, the Proboscidean LSD falls near the base of local magnetozones R10, or near the base of C5ACr (MacFadden and others, 1990) correlated as being about 14.2 Ma old (Cande and Kent, 1992), rather than circa 14.8 Ma. A perhaps more satisfactory correlation of this part of the magnetic polarity stratigraphy of the Barstow Formation is that magnetozone R10 is equivalent to Chron 5ADr of Cande and Kent (1992), the base of which is calibrated therein at 14.8 Ma. Under this interpretation, the age of the Proboscidean Datum in the Barstow Formation is about 14.8 Ma old, and correlates to Chron 5ADr in the MPTS of Cande and Kent (1992) and Berggren and others (this volume). Thus, it appears that proboscideans occur later in the interior locations as compared to more marginal sites in North America. In that the group apparently was already in the continent as of the latest Hemingfordian, the younger occurrences are irrelevant to a discussion of

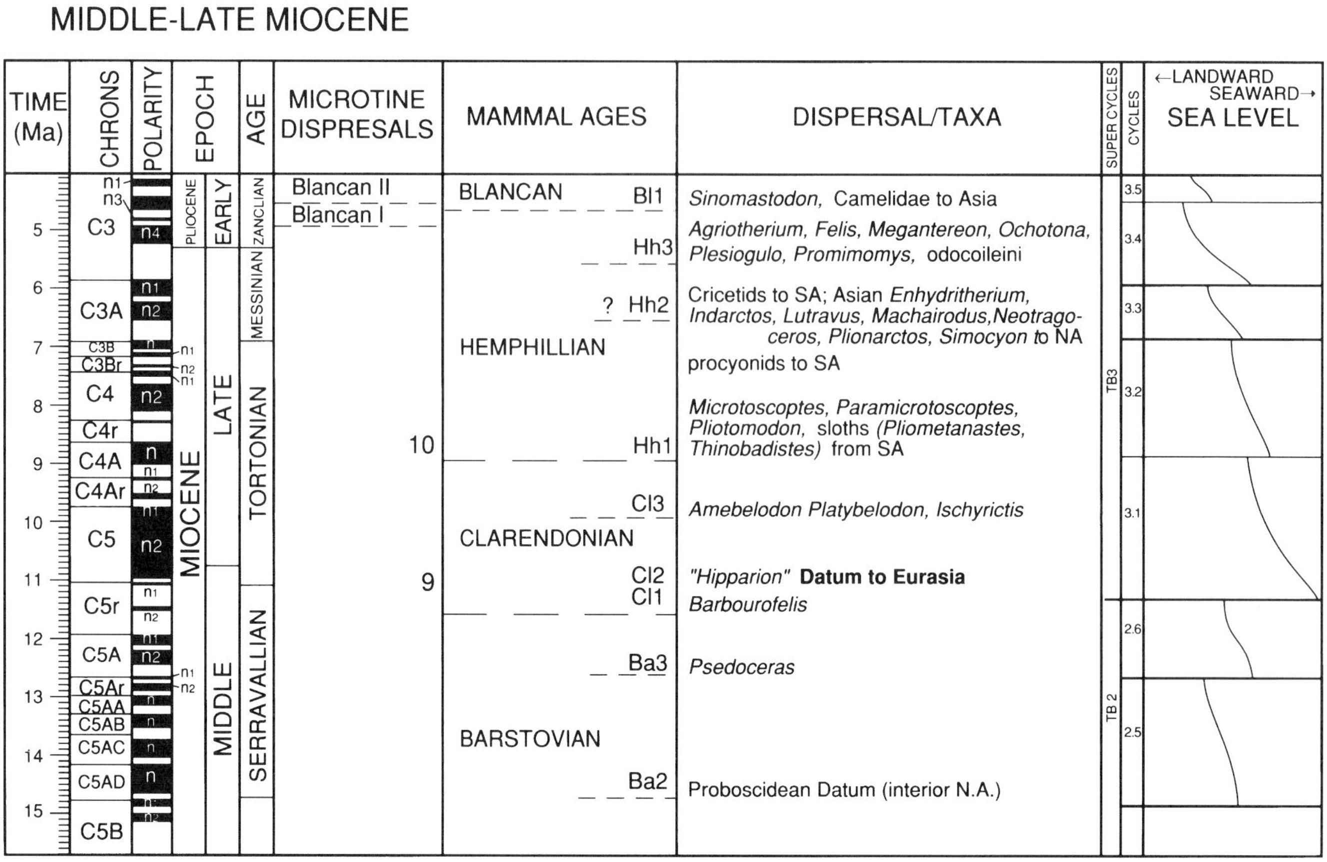

MIDDLE-LATE MIOCENE

TIME (Ma)	CHRONS	POLARITY	EPOCH	AGE	MICROTINE DISPRESALS	MAMMAL AGES		DISPERSAL/TAXA	SUPER CYCLES	CYCLES	←LANDWARD SEAWARD→ SEA LEVEL
5	C3		PLIOCENE EARLY	ZANCLIAN	Blancan II / Blancan I	BLANCAN	Bl1	Sinomastodon, Camelidae to Asia		3.5	
		n1 n3 n4		MESSINIAN			Hh3	Agriotherium, Felis, Megantereon, Ochotona, Plesiogulo, Promimomys, odocoileini		3.4	
6											
	C3A	n1 n2					? Hh2	Cricetids to SA; Asian Enhydritherium, Indarctos, Lutravus, Machairodus, Neotrago- ceros, Plionarctos, Simocyon to NA		3.3	
7	C3B C3Br	n1 n2 n1	MIOCENE LATE	TORTONIAN		HEMPHILLIAN		procyonids to SA	TB3	3.2	
8	C4 C4r	n2						Microtoscoptes, Paramicrotoscoptes, Pliotomodon, sloths (Pliometanastes, Thinobadistes) from SA			
9	C4A C4Ar	n n1 n2					Hh1				
10	C5	n2			10	CLARENDONIAN	Cl3	Amebelodon Platybelodon, Ischyrictis		3.1	
11	C5r	n1 n2		SERRAVALLIAN	9		Cl2 Cl1	"Hipparion" Datum to Eurasia Barbourofelis		2.6	
12	C5A	n1 n2									
13	C5Ar C5AA	n1 n2					Ba3	Psedoceras	TB2	2.5	
14	C5AB C5AC	n				BARSTOVIAN					
15	C5AD C5B	n					Ba2	Proboscidean Datum (interior N.A.)			

FIG. 5.—Middle and late Miocene time scale, including part of the Pliocene Epoch, after Berggren and others (this volume). Mammal ages discussed in the text. Sea-level pattern is after Haq and others (1988) and Bartow (1992). Microtine Dispersals follows Repenning and others (1990). 9 and 10 refer to Interchange Events. Event 9 is an unusual dispersal from North America. From Event 10 and younger, taxa are interchanged between North and South America, as well. Lists of multiple taxa separated by commas are considered to begin at the base of the indicated interval, including indented names.

intercontinental dispersal, but may be important in considering the speed of overland dispersal of land mammals, at least in some instances.

Late late Barstovian mammal age.—Tedford and others (1987, Fig. 6.3) indicate that this unit (Ba3, Fig. 5) begins at about 12.5 Ma, based on LSD of *Pseudoceras* in the Burge L.F. (Nebraska; Tedford and others, 1987, p. 169), but in Figure 6.2 (Tedford and others, 1987) the Burge Member (and fauna) of the Valentine Formation is correlated to somewhat less than 12.0 Ma.

Calibration and correlation.—The Burge Member of the Valentine Formation is unconformably overlain by the Cap Rock Member of the Ash Hollow Formation. Fission-track dates on zircons and glass from a vitric tuff at the base of the Cap Rock Member yield ages of 10.3 ± 0.3 Ma and 10.6 ± 0.6 Ma, respectively (Naeser and others, 1980), and an ash at the base of the Valentine Formation (well below the Burge Member) yielded a fission-track age (glass) of 13.6 ± 1.3 Ma (Boellstorf and Skinner, 1977). If this unit proves regionally useful, it would appear to correlate with the Type 1 sea-level lowstand TB2.6.

Whistler and Burbank (1992) provide an important synthesis of the geochronology, biochronology, and magnetostratigraphy of the Dove Spring Formation (Ricardo Group), adjacent to the

northwestern Mojave Desert, southern California. They conclude that the Burge Fauna closely resembles that of the *Cupidinimus avawatzensis/Paracosoryx furlongi* Assemblage-Zone of the Dove Spring Formation, the base of which is correlated to Chron C5Ar.2r or about 12.7 Ma in the Cande and Kent (1992) and Berggren and others (this volume) time scale. Whistler and Burbank (1992) also take the *C. avawatzensis/P. furlongi* Assemblage-Zone as marking the beginning of the Clarendonian in the Dove Spring Formation and correlate the Burge Fauna as early Clarendonian, as well. Important for discussions below, both faunas record LSDs of the equid, *Cormohipparion occidentale*, but the taxonomy of this species is under review by MOW, so both of these occurrences are referred to here as *Cormohipparion occidentale s.l.* For the purposes of this report, the definition of Clarendonian follows Tedford and others (1987), and the Burge Fauna and correlative part of the *C. avawatzensis/P. furlongi* Assemblage-Zone is retained in the Barstovian. Following Whistler and Burbank (1992), the base of the Burge Fauna is correlated at about 12.7 Ma, and thus also is the base of the late late Barstovian (= Chron C5Ar.2r).

Clarendonian mammal age.—Tedford and others (1987) begin the Clarendonian (Cl1, Fig. 5) with the immigrant first occurrence of the nimravid carnivore *Barbourofelis*. The genus is represented in Great Plains faunas equivalent to that (Minnechaduza Fauna) from the base of the Ash Hollow Formation, that unconformably overlies the Burge Member of the Valentine Formation.

Calibration and correlation.—Tedford and others (1987, Figs. 6.2, 6.3) indicate that the Clarendonian begins at about 11.5 Ma, consistent with the 10.3–10.6 Ma fission-track calibration of the Swallow Ash that occurs at the base of the Cap Rock Member of the Ash Hollow Formation, and with an early K-Ar age of 11.7 Ma for a tuff that stratigraphically underlies the Fish Lake Valley Fauna (Esmeralda Formation, Nevada; Evernden and others, 1964), considered to be an early Clarendonian correlate (Tedford and others, 1987, Fig. 6.2). An age of 11.5 Ma correlates to about the mid part of Chron 5r in the Cande and Kent (1992) and Berggren and others (this volume) time scale, which is included within about the upper third of the *C. avawatzensis/P. furlongi* Assemblage-Zone in the Dove Spring Formation (Whistler and Burbank, 1992; note that Whistler and Burbank follow the MPTS calibration in Berggren and others, 1985a, which differs in some respects from that of Cande and Kent, 1992 and Berggren and others, this volume). In any case, the boundary proposed by Tedford and others (1987) compares well with that which results from this correlation to the MPTS in the Dove Spring Formation; an age prior to that of the Minnechaduza Fauna in Nebraska that is contained within the base of the Ash Hollow Formation (Whistler and Burbank, 1992, Fig. 8). For purposes of this report, the Clarendonian begins at about 11.5 Ma (= mid part Chron C5r; = C5r.2n), somewhat older than the Type 1 sea-level lowstand TB3.1.

Interchange Event #9

The "Hipparion" Datum.—

In addition to the early Hemingfordian dispersal of *Anchitherium* (see above), one of the few known pre-Pliocene Neogene dispersals from North America to the Old World, rather than the more common opposite direction, involves hipparion horses (Cl2; Fig. 5). These horses began to colonize Old World land habitats about 11 Ma ago and radiated into a greatly diverse group (Forstén, 1968; Woodburne and Bernor, 1980; Woodburne and others, 1981; Bernor and others, 1990). The new estimate for the Old World FAD is based on Swisher (1996) who shows that this datum occurs near the base of Chron C5n.2n, correlated by Cande and Kent (1992) at 10.8 Ma. At present, 10.8 Ma appears to be the maximum permissable age for the "*Hipparion*" FAD in the Old World, although it actually might be slightly younger. Whistler and Burbank (1992, Fig. 4) indicate that the LSD of the *Cormohipparion occidentale* group in the Dove Spring Formation is about 12.7 Ma old, and that the LSD of *Hipparion tehonensis* in the Dove Spring Formation occurs at or just below the base of Chron C5n.2n, or at about 10.8 Ma. The "*Hipparion*" Datum of the Old World involves a different taxon than *Hipparion s.s.* Studies that deal with the problem (such as Woodburne and others, 1981; Bernor and others, 1990) indicate that the source of the Old World "*Hipparion*" datum is to be sought within the genus, *Cormohipparion* and that the most likely candidate is the *C. occidentale* group (s.l.) as used here. This morphological complex ranges in age from the late Barstovian (Burge Fauna and correlates in the Dove Spring Formation) to within the late Clarendonian (Xmas-Kat quarries in Nebraska and the Gidley Horse Quarry, Texas).

This interval ranges in age from about 12.7 to somewhat younger than 10 my. Based on correlations proposed in Whistler and Burbank (1992, Figs. 4, 8), the base of Chron C5n.2n is contained within the upper-most part of the *C. avawatzensis/P. furlongi* Assemblage-Zone in the Dove Spring Formation, the fauna of which correlates with, or is just older than, elements of the Minnechaduza Fauna, Nebraska and the MacAdams Quarry fauna of the Texas Panhandle. MOW is currently assessing the morphologies contained within samples here grouped as *Cormohipparion occidentale s.l.* Preliminary work suggests that at least three distinct species are represented by the sample contained within the geochron of that taxon. It remains to be determined which, if any, of these species is the best candidate for the origin of the Old World "*Hipparion*" Datum. It is interesting that Tedford and others (1987, Fig. 6.3) do not show any Old World taxa immigrating to North America at about 10.8 Ma, or within the early Clarendonian, but rather that immigrations are recorded at about 11.5 and 10.0 Ma. Based on a correlation of the "*Hipparion*" Datum to the mid to base of Chron C5n.2n, the ancestral species for that datum may have left North America during the span of the Type 1 sea-level lowstand of TB3.1 (Fig. 5).

Late Clarendonian mammal age.—Tedford and others (1987, Fig. 6.3, p. 190) indicate that the late Clarendonian (Cl3, Fig. 5) begins with the immigrant first occurrence of the gomphotheriid mastodonts, *Amebelodon* and *Platybelodon*, and the mustelid *Ischyrictis* (*Hoplictis*) (= *Beckia*). These taxa occur in deposits of this age from the Columbia Plateau to Florida that are equivalent to the upper part of the Ash Hollow Formation in Nebraska. Faunas from this interval include the Xmas and Kat quarries in Nebraska and the Gidley Horse Quarry in the Texas Panhandle.

Calibration and correlation.—Tedford and others (1987) indicate that this interval began about 10.0 Ma and report on fission-track ages (glass) from the Davis Ash, at the base of the Ash Hollow Formation, that range in age from 9.7 ± 1.2 Ma to 10.2 ± 0.3 Ma. Whistler and Burbank (1992, Fig. 8) indicate that the Xmas-Kat local faunas and the Gidley Horse Quarry L.F. correlate to within the upper half of the *Epicyon aphobus/Hipparion forcei* Assemblage-Zone in the Dove Spring Formation, and that the base of this interval correlates to about the base of the upper third of Chron 5n in the MPTS. Cande and Kent (1992) show that this is equivalent to about 10.0 Ma, consistent with the fission-track ages cited above (but see Prothero and others, 1993). As correlated, the late Clarendonian corresponds with the waning phase of the Type 1 sea-level lowstand TB3.1.

Interchange Event #10

Hemphillian mammal age.—

Tedford and others (1987, Figs. 6.2, 6.3, p. 175, 180) indicate that the Hemphillian begins (Hh1, Fig. 5) with the immigrant occurrence of taxa from two different zoogeographic regions. The arvicoline rodents, *Microtoscoptes*, *Paramicrotoscoptes*, and the cricetodontine, *Pliotomodon*, apparently originated in Eurasia (or at least outside of fossil-bearing areas in northern North America; R. H. Tedford, pers. commun., 1994), whereas the megalonychid sloths, *Pliometanastes* and *Thinobadistes* (and, possibly later, *Megalonyx*), entered from South America, even though the Panamanian isthmus still was incompletely formed. Hirschfeld (1981) reports that *Pliometanastes* occurs stratigraphically 4 m below a tuff dated at (K-Ar) as 8.19 ± 0.16 Ma, in the Mehrten Formation (Sierran foothills, California). Webb (1985b, p. 358) indicates that the entry of *Pliometanastes* and *Thinobadistes* is essentially coeval in North America (also Marshall and others, 1979).

Faunas equivalent in age to Hemphillian in South America, as well as those of later times, saw an increase in the number of northern taxa, as well, and these are discussed here. Thus, Interchange Event #10 begins about 9 Ma and lasts until about Interchange Event #11, the time of the Great American Interchange (e.g, Marshall and others, 1979, 1983, 1984, 1992; Marshall, 1985; Webb, 1985a, b; Marshall and Cifelli, 1990; MacFadden and others, 1993; Webb and Barnosky, 1989). At about 9.0 Ma or a little younger, the base of the Hemphillian correlates to about the base of Chron C4An and near the base of the Type 2 lowstand TB3.2. The younger dispersals southward (Hayquerian and Montehermosan; about 7.5 to 5.8 Ma for different genera of procyonids, and Montehermosan, or about 6.0 Ma or younger for cricetids; Marshall, 1985; Webb, 1985b; Marshall and Cifelli, 1990) may have coincided about with lowstands TB3.2 (Type 2), TB3.3 (Type 1) or TB3.4 (Type 1). Webb (1985b, p. 360) for discussion of the southward dispersal of procyonids at about 8–9 Ma, and the synchronous northward move of *Thinobadistes* and *Pliometanastes*. This is shown at Hh1 on Figure 5.

Route and characterization.—For the rodents, dispersal likely was via a trans-Beringian filter. For the sloths dispersal was via an interestingly fortuitous sweepstakes from and, for the procyonids and rodents, to South America. We suggest that this tectonically active area could have been an island-hopping pathway for the entry of edentates into the United States at about 9 Ma (perhaps from an evolutionary nucleus that already was present at about 17–18 Ma; MacFee and Iturralde-Vinent, 1994). It is interesting in this regard, that one of the oldest known occurrences of the mylodontid sloth, *Thinobadistes*, likely is found in the early Hemphillian Mixon Local Fauna of Florida (Webb, 1985b; Tedford and others, 1987, p. 191). Primitive megalonychids (*Pliometanastes*) are found in apparently coeval deposits in California (Hirschfeld, 1981) and elsewhere in North America (Tedford and others, 1987).

Calibration and correlation.—Tedford and others (1987, Fig. 6.3, p. 160, 162) indicate that the base of the Hemphillian is 9 Ma old, younger than the 9.6 Ma date below the Smith's Valley Fauna, Nevada (Evernden and others, 1964), and younger than a 9.4 ± 0.6 Ma K-Ar date (sanidine; Greene, 1973) from a welded tuff in the lowest part of the Drewsey Formation, Oregon, that is stratigraphically below the Otis Basin, Drinkwater and Bartlett Mountain local faunas of early Hemphillian age. Repenning (1987, p. 242–243, 252–253) reports other early Hemphillian sites in Wyoming, Idaho, and Oregon that bear either *Microtoscoptes* or *Goniodontomys* in stratigraphic settings consistent with the approximate 9 Ma age for the basal Hemphillian. Whistler and Burbank (1992, Figs. 4, 5, 8) indicate that the *Paronychomys/Osteoborus diabloensis* Assemblage-Zone in the Dove Spring Formation is a reasonable correlate of early Hemphillian faunas elsewhere, although defining taxa are not present. The base of this assemblage-zone correlates to the top of Chron C4Ar.1r in the Cande and Kent (1992)

time scale or about 9.1 Ma. All of the data are consistent with an approximate 9.0 Ma age for the base of the Hemphillian.

Late early Hemphillian faunas.—Tedford and others (1987, Figs. 6.2, 6.3) indicate that the late early Hemphillian is based on the immigrants, *Enhydritherium, Eomellivora, Indarctos, Lutravus, Machairodus, Neotragoceros, Plionarctos,* and *Simocyon* (Hh2, Fig. 5). These taxa nowhere occur all together in North America (but most are contemporaneous elements in late Miocene faunas of eastern Asia, R. H. Tedford, pers. commun., 1994). For example, *Simocyon* and *Plionarctos* occur in the Rattlesnake Fauna, Oregon; *Lutravus* is found at Thousand Creek, Nevada; and *Enhydritherium* occurs in the Withlacoochee 4A Local Fauna, Florida. *Neotragoceras*, a rare bovid, is found in the Aphelops Draw Fauna, Nebraska. The best local stratotype, at the base of the Hemphill beds, Texas records the joint occurrence of *Indarctos, Eomellivora,* and *Machairodus.* Cricetid rodents, that first occur at the base of the Montehermosan mammal age in South America (Marshall and Cifelli, 1990) at about 6.0 Ma (Marshall and others, 1992), may have left North America about in the late early Hemphillian. An age of about 6.0 Ma compares with the sea-level lowstand TB3.4, a Type 1 unconformity (Haq and others, 1988).

Route and characterization.—For Asian immigrants dispersal apparently was via a trans-Beringian filter. On the other hand, cricetid rodents entered South America via a trans-Panamanian sweepstakes route.

Calibration and correlation.—Tedford and others (1987, p. 197) indicate that the base of this part of the Hemphillian is about 6–7 my old. Faunas of this age are not recorded in the Dove Spring Formation (Whistler and Burbank, 1992) and calibration sites within North America are sparse (Tedford and others, 1987, Fig. 6.2). The age span correlates about with Chron C3A and possibly with a Type 1 lowstand, TB3.3, near the middle of Chron C3Ar, or about 6.7–6.8 Ma.

Late Hemphillian faunas.—Tedford and others (1987, Figs. 6.2, 6.3) begin this interval with the immigrant occurrence of *Agriotherium, Felis, Megantereon, Ochotona, Plesiogulo, Promimomys* and Odocoileini cervids (Hh3, Fig. 5). In the Texas panhandle stratotype, the Coffee Ranch Local Fauna is found in the upper part of the Hemphill beds that contain faunas typifying the Hemphillian.

Calibration and correlation.—Correlative early late Hemphillian faunas are represented by the Optima (Oklahoma), Uptegrove (Colorado), Santee, Bear Tooth and ZX Bar (Nebraska), and Warren and Pinole (California). The Pinole Tuff, which occurs stratigraphically a few feet above beds that produce the Pinole Local Fauna, is dated (K-Ar) at 5.3 ± 0.1 Ma (Evernden and others, 1964), and May (1981) considered likely a correlation with what is now designated as Chron C3n.4 in the MPTS of Cande and Kent (1992). This suggests that the Warren Local Fauna is about 5.2 Ma old. Tedford and others (1987, Figs. 6.2, 6.3) calibrate the beginning of the late Hemphillian at about 6 Ma, but a slightly younger age (5.5 Ma) is suggested here, still within the regressive phase of the Type 1 lowstand TB3.4.

Blancan mammal age.—Tedford and others (1987, Fig. 6.3, p. 92) indicate that the Blancan (Bl1, Figs. 5, 6) is defined by the immigrant first occurrence of Old World taxa, such as *Nebraskomys, Mimomys, Trigonictis, Lutra, Chasmoporthetes, Lynx, Ursus, Parailurus,* and the true cerivds, *Bretzia* and *Odocoileus.* Repenning (1987, Blancan I, Fig. 8.1) adds immigrant arvico-

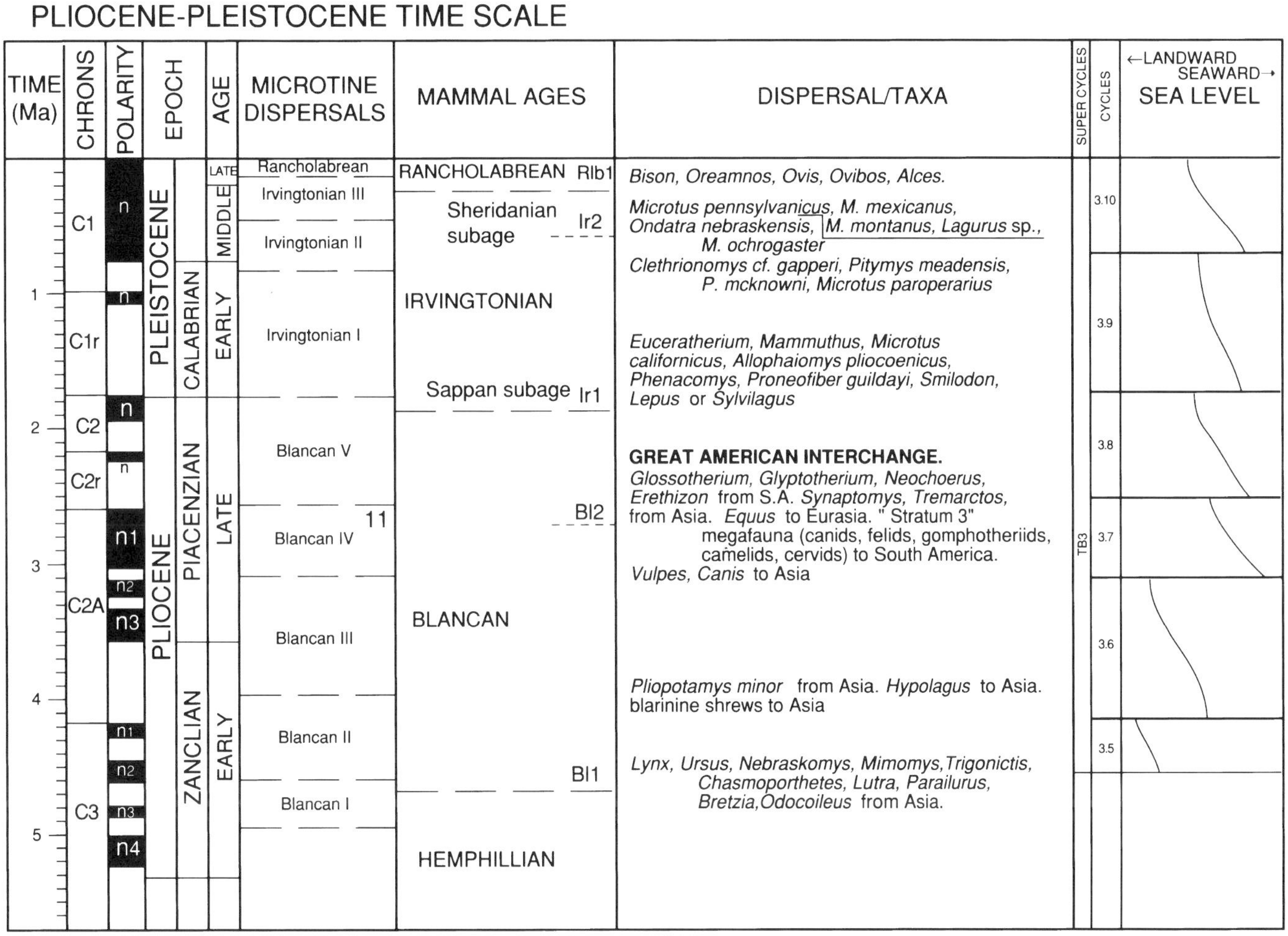

PLIOCENE-PLEISTOCENE TIME SCALE

FIG. 6.—Pliocene and Pleistocene time scale, after Berggren and others (this volume). Mammal ages discussed in the text. Sea-level pattern is after Haq and others (1988) and Bartow (1992). Microtine Dispersals follows Repenning and others (1990). 11 refers to The Great American Interchange. Throughout note North American record of dispersals to as well as from Asia and South America. Lists of multiple taxa separated by commas are considered to begin at the base of the indicated interval, including indented names.

line rodents, *Mimomys (Cosomys) sawrockensis*, *M. (Ophiomys) mcknighti*, and *Nebraskomys* cf. *rexroadensis*.

Route and characterization.—The record indicates a bidirectional trans-Beringian filter for Asian and North American taxa. The base of the Blancan apparently is nearly coeval with the Type 2 sea-level lowstand TB3.5. Flynn and others (1991) indicate that dispersals from North America to Asia include Camelidae and *Sinomastodon* (at about 4.5 Ma), Canidae and blarinine shrews (circa 4.2 Ma), and *Hypolagus* (circa 4.0 Ma). *Vulpes* and *Canis* first appear in the Yushe Basin, China, at about 3 Ma. Overall, the faunal exchange across Beringia appears to have been balanced, even though Asian evidence still is being developed.

Calibration and correlation.—Stratigraphic successions that span the Hemphillian/Blancan boundary are rare in North America. One of the best displayed is in the Palm Springs Formation of the Anza-Borrego Desert, southern California but, unfortunately, late Hemphillian taxa are rare. The early Blancan Layer Cake Fauna in the Palm Springs Formation begins at the base of the Cochiti subchron in the Gilbert Chron (Lundelius

and others, 1987, p. 216–217) correlated (Cande and Kent, 1992) at about 4.2 Ma. Lundelius and others (1987, p. 217) indicate that the Fox Canyon Local Fauna (Rexroad Formation, Kansas) and White Bluffs Local Fauna (Ringold Formation, Washington) are early Blancan correlatives, both on faunal as well as on magnetic polarity grounds. Repenning (1987, Fig. 8.1) correlates the White Bluffs L.F., with Blancan I of his nomenclature and the Fox Canyon L.F. with Blancan II.

If the correlations of Repenning (1987) are utilized, Blancan I begins near the base of Chron C3n.3r or at about 4.9 Ma. For the purposes of this report, the Blancan is considered to have begun about this time, based upon the principle that immigrant taxa are useful in boundary definition, with the two rodent genera consistent with the proposal of Tedford and others (1987). This boundary age correlates about with the final regressive phase of the Type 1 lowstand TB3.4, or near the base of TB3.5 (Type 2 unconformity; Haq and others, 1988).

Blancan II apparently begins about 4.6 Ma (Repenning and others, 1990), and reflects the endemic evolution of *Mimomys (Ogmodontomys) poaphagus*, *M. (Cosomys) primus*, *M.*

(*Ophiomys*) *mcknighti-taylori*, *Pliophenacomys finneyi*, *Pliolemmus antiquus*, and *Nebraskomys rexroadensis*. Flynn and others (1991) indicate that *Sinomastodon* has a LSD of about 4.5 Ma in the Yushe Basin, China along with Camelidae, both taxa of North American origin. Other North American taxa entering the Yushe Basin during Blancan II time include Canidae and blarinine shrews (about 4.2 Ma) and *Hypolagus* (circa 4.0 Ma; Flynn and others, 1991, p. 260; may correlate with base of Blancan III as used herein). In any case, the dispersals that transpired between about 4.0 and 4.5 my were bi-directional between North America and Asia and likely were more balanced than the record now shows (Flynn and others, 1991, p. 262–263).

Blancan III of Repenning (1987), and Repenning and others (1990) is defined on the FAD of *Pliopotamys minor* and is characterized on the endemic persistence of *Pliopotamys meadensis*, *Mimomys (Ogmodontomys) poaphagus*, *M. (Cosomys) primus*, *M. (Ophiomys) taylori*, *M. (Ophiomys) magilli*, *Pliopotamys antiquus*, *Pliophenacomys primaevus*, and *Nebraskomys mcgrewi*. As revised here, correlation to the MPTS indicates that Blancan III spans the interval from 4.0–3.1 my.

Blancan IV begins late within the early Blancan of Tedford and others (1987) and is defined on the endemic persistence [or evolution] of the microtines, *Mimomys (Cosomys) primus, M. (Ogmodontomys) poaphagus, [M. (Ophiomys) taylori-parvus], M. (Ophiomys) meadensis, Pliopotamys meadensis, Pliopotamys minor, [Pliophenacomys primaevus-osborni]*, and *Pliolemmus antiquus* (taxa in [ ] represent *in situ* evolution, rather than dispersal). Blancan IV begins at 3.1 Ma, and may correspond to the sea-level lowstand TB3.7, a Type 1 unconformity (Haq and others, 1988). Flynn and others (1991, p. 260) indicate that *Vulpes* and *Canis* are North American immigrants in the Yushe Basin at about 3 Ma.

The Rexroadian mammal age (= early Blancan) of Lundelius and others (1987, Fig. 7.3, p. 219) is based on faunas that are apparently no older than about 3.6 Ma (Hagerman, Idaho). The apparent FAD of *Pliopotamys* in this interval resembles Blancan III of Repenning (1987). On this basis, Rexroadian likely is not earliest Blancan.

Interchange Event #11

Late Blancan faunas.—According to Lundelius and others (1987) the late Blancan (Bl2, Fig. 6). begins with the immigrant first occurrences of neotropical taxa: the armadillio, *Glossotherium*, the glyptodont, *Glyptotherium*; and the rodents, *Neochoerus* and *Erethizon*. In addition, Eurasian immigrant rodents, *Synaptomys (Plioctomys)*, and the bear, *Tremarctos*, also appear in late Blancan faunas. Lindsay and others (1980) indicate that *Equus* entered Eurasia from North America subsequent to about 2.5 Ma.

This is the beginning of the Great American Interchange (for example, Webb, 1985a, b), generally considered to be recognized in South American faunas of Uquian and younger age. Webb (1985b) and Marshall and Cifelli (1990) provide excellent summaries of the nature of the interchange and the taxa involved. According to Webb (1985b) at least eight neotropical genera (in at least eight families) moved northward and 29 northern genera (in fifteen families) moved south at this time, from which latter a procyonid (Hayaquerian), two cricetids

(Montehermosan), a peccary and a mustelid (Chapadmalalan) may be subtracted, as having dispersed previously.

Blancan V microtine division (Repenning, 1987; Repenning and others, 1990) is defined on the FAD of *Synaptomys (Mictomys) vetus, S. (Synaptomys) rinkeri*, and characterized on the endemic co-occurrence of *S. (Mictomys) landesi, Ondatra idahoensis, Mimomys (Ogmodontomys) monohani, M. (Ophiomys) parvus, M. (Ophiomys) meadensis, Nebraskomys mcgrewi, Pliopotamys meadensis, Pliolemmus antiquus*, and *Pliophenacomys osborni*. Correlations herein suggest that Blancan V ranges in age from 2.7–1.8 my, and its base apparently is coeval with the sea-level lowstand TB3.8, a Type 1 unconformity (Haq and others, 1988).

Route and characterization.—The Panamanian land bridge was available now for neotropical forms to enter more northern sites and vice versa, and the trans-Beringia filter apparently was operative (bi-directionally), as well.

Calibration and correlation.—Taxa found in late Blancan faunas apparently made their way into these sites at about 2.5–2.7 Ma, which compares on faunal (Cita Canyon and Blanco, Texas; Tusker, Arizona; Grand View, Idaho) and geochronologic grounds to Blancan V of Repenning (1987; immigration of *Synaptomys [Mictomys] vetus* and *S. [Synaptomys] rinkeri*). Correlation to the upper part of the Gauss normal Chron in the MPTS (Lundelius and others, 1987, Fig. 7.3), and Repenning (1987, Fig. 8.1) corresponds to an age of about 2.7 Ma in the Cande and Kent (1992) time scale, to somewhat after the Type 1 sea-level lowstand TB3.8 (see also Keigwin, 1982). In this report, Bl2 and the Great American Interchange are taken as coeval at about 2.7 Ma.

Marshall and others (1992) suggest that the Uquian began at about 2.5 Ma. MacFadden and others (1993) refine this slightly to 2.6 Ma. See Cione and Tonni (1995) for a more detailed appraisal, however, and replacement of Uquian with Marplatan State.

Irvingtonian and Rancholabrean Mammal age.—As discussed further under *Climatic Factors*, these mammal ages record a number of fundamentally Asian immigrants, although Webb (1985b) shows that southern taxa continue to arrive, as well. At the same time, these episodes appear to be of less significance than the beginning of the Great American Interchange. Irvingtonian I and II dispersals appear to correlate with Type 1 unconformities (Haq and others, 1988).

DISCUSSION

This report summarizes data designed to update the timing and regional interrelationships of overland fossil mammal dispersals to or from North America. The results are compared with estimates of times of major sea-level lowering during the Cenozoic Era. The following section discusses these comparisons, and briefly explores the roles played by climate, vicariance, and dispersal lag.

The overall sea-level history pattern for the Cenozoic Era seems to reveal that many of the major episodes of overland dispersal of North American land mammals were not closely tied to eustatic sea-level changes. It may be worth noting in passing, however, that most post-early Eocene dispersals involved Beringia; lands adjacent to the Caribbean Sea became important (facilitating dispersals) only in the past 10–16 my.

ow interval from about 10.8 Ma to about 5 Ma (Fig. a number of interchanges that are important to land faunas in North and South America but, in terms of hese still are relatively minor when compared to the een in the Great American Interchange at about 2.7 h also is associated with a strong drop in sea level; Figs. 6, 7). As discussed below, however, the Great Interchange likely was as much effected by tectonic at raised the land) as by a global fall in sea level.

ne dispersal events recorded by Repenning (1987) nning and others (1990) generally correlate with low- sea level, on the basis of present evidence. Whereas not be of the scope accorded major recognition in the fact of this correspondence is noted here (Fig. an III (ca 4.0 Ma) is slightly younger than the Type 1 TB 3.6; Blancan V (ca 2.6 Ma) is coeval with the wstand TB 3.8; Irvingtonian I (ca 1.8 Ma) is coeval Type 1 lowstand TB 3.9; and Irvingtonian II (ca 0.8 but coeval with the Type 1 low stand 3.10. Blancan I Ma) does not appear to correspond with the beginning stand; Blancan II and IV do not reflect immigrations. ptine immigration episodes are taken up further in *Cli- tors*.

nary, of the major Cenozoic sea-level low stands only ning of supercycles TA4 and TB2 can be directly as- with overland dispersals having sufficient numbers dth) of taxa to be significant. The Great American ge, at cycle TB3.7, reflects the other strong interdis- olving both North and South America in a major way st time since the Late Cretaceous Period. Tectonism y played a helping hand here, at least.

of dispersal but high sea level.—

e Paleocene and early Eocene dispersals represented mammals in the Clarkforkian and Wasatchian are e most spectacular instances of this pattern (Fig. 1). short-lived (but Type 1) drops in sea level are recon- for these times (TA2.2, 2.4), the overall pattern is for el maximum (Haq and others, 1988, Fig. 14), in the which occurs the greatest Palaearctic homogeneity in nmal faunas ever seen (Fig. 7). In that the dispersal s) appear to have been located in the high polar lati- d in that the northern Atlantic Ocean was effectively ctual status of sea level may have been irrelevant to rsal potential at this time.

se, the late Uintan (ca 43 Ma; TA3.4) saw nearly as interchange, but still, sea-level was relatively high The pattern for the Clarkforkian-Wasatchian and late in strong contrast to that of the *Grand Coupure*. No nt land mammal interchange between Europe and nerica transpired at the time of the *Grand Coupure*, gh a Type 1 low stand is reflected at that time in cycle his is discussed further below.

Vicariance

ong dissimilarity of post-Wasatchian-Sparnacian land faunas in North America and the Old World represents icariant event, reflected by the opening of the northern Ocean. Prior to that time, juxtaposition of Holarctic land areas allowed the interchange of mammalian taxa but, pric to the Clarkforkian, at least, extrinsic factors such as climate i northern latitudes and the presence of the Turgai Strait ma have hindered dispersal (McKenna, 1983a, b). Notably, th Clarkforkian and Wasatchian interchanges transpired durin times of relative sea-level highstand (Figs. 1, 7), but this facto apparently was irrelevant. If the post-Wasatchian disruption c a formerly widely dispersed pattern can be said to illustrate relatively typical vicariant situation, the examples cited belov portray the "reverse," in which dispersals that would have bee expected to have taken place at a given time were actually de layed significantly.

The *Grande Coupure* at or near the beginning of the Oligc cene Epoch in western Europe (ca 33.5 Ma; Figs. 2, 7) repre sents a major reorganization of its land mammal faunas (Tabl 1). Importantly, new varieties of taxa had appeared in Nort America beginning in the late Uintan, and were present in Asi as well, from within medial to late Eocene time. Thus, th *Grande Coupure* recognizes the breakdown of a dispersal bai rier that not only resulted in withdrawal of the Turgai Strait but was associated with tectonic events in the mid-Alpine dis trict (Heissig, 1979), as well as in Beringia. In that the mod ernization of the Asian theater was manifested well within me dial Eocene time, data reflect a somewhat tardy time of dispers from Asia to North America in the late Uintan (ca 43 Ma), wit a second pulse at the beginning of the Chadronian (ca 37 Ma Note, however, that certain groups (aplodontid and eomyid ro dents [Duchesnean] and sciurid rodents [Chadronian]) may rep resent westward dispersals from North America to Europe (se *Dispersal Lag*). Nevertheless, Asia apparently retained its faun for about 14 my (48–34 Ma) before boiling over into wester Europe at the *Grande Coupure* at about 33.5 Ma. By that time the modernization of the North American mammal fauna wa already virtually completed.

In another theater, a Central American "holding pen," re vealing only hints at present, could be represented by the Wes Indian region. Presently meagre, but exciting, evidenc (MacFee and Iturralide-Vinent, 1994) suggest that this regio may have been a center of edentate evolution (perhaps popu lated about 17–18 Ma and first releasing some of its fauna a about 9 Ma).

The Great American Interchange at circa 2.7 Ma reflects major lowering of sea level (TB3.7–8), but also increased ele vation of the Panamanian region (Repenning and Ray, 1977 Keigwin, 1982). In fact, the record beginning approximatel with the early Hemphillian (ca 9 Ma) gives a glimpse of th importance of growing tectonic events in promoting sweep stakes to filter to, finally, corridor interchange across the neo tropical region that persists to the present time (Figs. 5, 6, 7).

Climatic Factors

The global climate in the early Cenozoic Era appears to hav been relatively equable and warm, perhaps rebounding some what from a Late Cretaceous cooling episode. As the late Eo cene Epoch is approached, however, climatic deterioration ac celerates and likely drives the major modernization of lan mammal faunas witnessed during that interval. It is less clea however, that dispersal events were directly affected by climat except in the Pliocene and Pleistocene epochs. Prentice an

Authors have commented that the sea-level curves of Haq and others (1988) derive much of their data from districts that border the Atlantic Ocean, and it is well recognized that the lands that bound the northern Pacific Ocean are now, as well as in the past, tectonically active. Beringia likely was emergent throughout much of the Cenozoic Era (see also Marincovich and others, 1990). Some of the land mammal data summarized here may reflect the effects of tectonism or other factors rather more than eustasy. These suggestions are derived from the following considerations.

Sea-level was Generally Higher during the Paleogene than during the Neogene Interval

This is born out by Haq and others (1988, Fig. 14) and numerous other sources (such as Hallam, 1992, Chap. 6).

Paleogene Period.—

For the Paleogene interval, notable apparent lowstands include supercycle TA2 (cycle 2.1, ca 59 Ma, "medial" Tiffanian, Figs. 1, 7) and supercycle TA3 (cycle 3.1, ca 49.5 Ma, early Bridgerian). The next Type 1 regression is recorded at TA4.1 (ca 37 Ma, early Chadronian, Figs. 2, 7). The last major Paleogene lowstand begins with supercycle TB1 (ca 28.5 Ma, correlates to post-early Arikareean, Figs. 3, 7), which likely reflects the opening of the Drake Passage between South America and east Antarctica as shown by sea-floor spreading evidence and data heralding the advent of major continental glaciation in Antarctica (Barker and Burrell, 1982; Kennett and Barker, 1990). As discussed below, only the lowstand at TA4.1 appears to be associated with an important overland interchange with North America. The *Grande Coupure*, a conspicuous immigration event in Europe, seems to have no counterpart in North America at the time in question (ca 33.5 Ma, coeval with the base of cycle TA4.4 which Haq and others (1988) recognize as a Type 1 boundary).

Neogene Period.—

After recovering from the late Oligocene sea-level minimum (supercycle TB1) at about 28.5 Ma, sea levels rise through the early Miocene Epoch (Figs. 3, 4). The Hemingfordian collectively records the most impressive land mammal immigrations in the early Neogene interval. Supercycle TB2 begins with the next sea-level drop, with cycle TB2.1 embracing the two Hemingfordian immigration episodes, at circa 19 and 17.5 Ma (Figs. 4, 7). Supercycle TB2 is unique in the Neogene record in having every (third-order) cycle boundary being of Type 1, according to Haq and others (1988), meaning that sea level was low enough to expose at least some of the continental shelf to erosion. The range in potential calibration of land mammal faunas at least allows the possibility that the late Hemingfordian immigration coincided with cycle boundary TB2.2, but Type 1 boundaries notwithstanding, the remainder of the time-scale embraced within supercycle TB2 records only limited faunal interchange with North America. Some of these are operationally significant, however. The Proboscidean Datum may equate with cycle TB2.3, as may the immigration of *Hemicyon (Plithocyon)* which appears at the beginning of the Barstovian at about 16 Ma.

The sea-level lowstand recognized as the beginning of supercycle TB3 apparently corresponds with the exit of a species of *Cormohipparion* at about 10.8 Ma, precursor to the prochoresis event designated as the Old World "*Hipparion*" Datum. Sea level remains relatively low at the beginning of the Hemphillian (correlates about with the beginning of cycle TB3.2) which, in addition to the usual immigration of taxa (rodents) from Asia, also sees the first interchange with the neotropics. A punctuated, but generally rising sea level from about 7 Ma (Fig. 5) culminates in a sea-level highstand at about 3 Ma (Fig. 6). This is just prior to the drop that corresponds with the beginning of the Great American Interchange (TB3.7), as well as the opening of the Bering Strait to the Arctic Ocean, and heralds generally low (but punctuated) sea levels during the late Pliocene and into the Pleistocene epochs.

Overland Dispersal of Land Mammals at Times Coincided with, and at others was Insensitive to, Sea-level Lowstands

Immigrations coincident with low sea levels.—

*Paleogene Period.—*According to Haq and others (1988), major (second-order) withdrawals of the sea are recognized as the beginning of supercycles TA1–4 and TB1. Supercycle TA1 is in the Late Cretaceous Period, and will not be discussed here. The greatest Paleogene sea-level drops are correlated as within the Tiffanian (TA2.1; Figs. 1, 7), and as slightly after the beginning of the Arikareean (Figs. 3, 7), at about 28.5 Ma (TB1.1) after which sea level (with punctuations) remained generally low until about 22–23 Ma. As indicated by Russell and Zhai (1987, p. 399) only endemic evolution of Asian land mammal faunas transpired during the time of TA2 (cycle TA2.1), and Russell and others (1982) likewise report only possible, rather than a strong episode of actual, interchange between North America and western Europe. Supercycle TA3 (cycle TA 3.1), of about early Bridgerian age (Figs. 2, 7), witnessed only a minor exchange. The greatest Paleogene immigration coeval with a major sea-level drop is associated with the beginning of the Chadronian, correlated as about equivalent to TA4.1 at 37 Ma (Figs. 2, 7). The Chadronian witnessed one of the least extensive of the major Paleogene faunal interchanges, however.

Thus only the initial phase of supercycle TA4 saw a significant episode of overland mammal interchange in North America, but this still was a relatively minor event as compared with others in the Paleogene (Wasatchian; late Uintan; see below). The absence of a major overland dispersal correlative with the beginning of supercycle TB1 is noteworthy, as this strong sea-level fall associated with the opening of the Drake Passage reflects a major tectonic event of the southern oceans.

*Neogene Period.—*The major Neogene sea-level lowstands are reflected at the beginning of supercycles TB2 and TB3, with that of TB2 apparently being the least extensive, although recording a high number of Type 1 unconformities. The interval embraced within cycle TB2.1 (Hemingfordian) saw the major Neogene dispersal, except for the Great American Interchange. According to the evidence (see text and Figs. 4, 7) the Hemingfordian immigration took place in two waves separated by about 1.5 my, but may not be compellingly related to sea-level change. The other major sea-level lowstand of the late Neogene, represented by TB3.1 at about 10.8 Ma (Figs. 5, 7), sees little more than a single-taxon emigration from North America ("*Hipparion*" Datum; above). The persistent (with fluctuations)

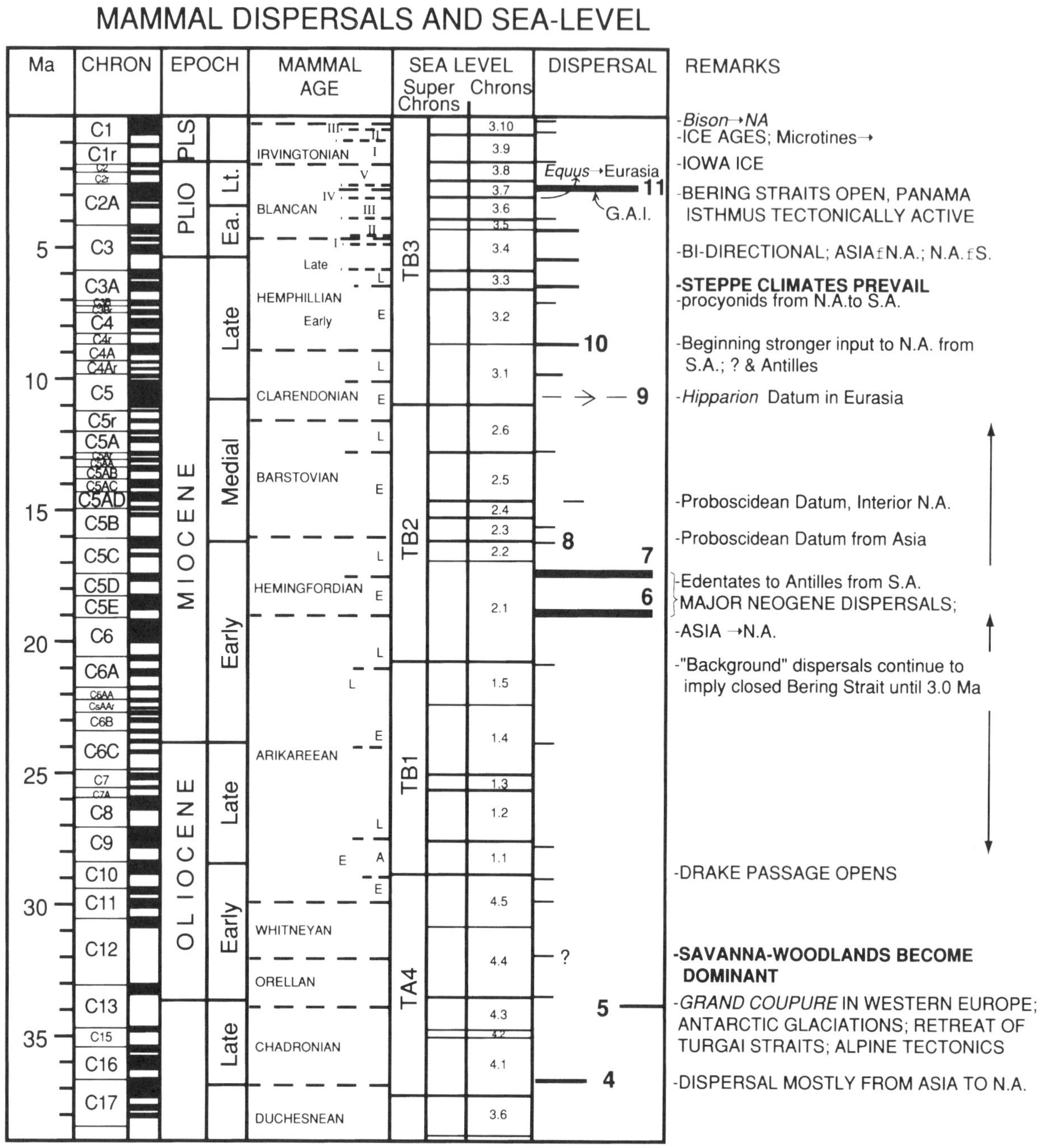

MAMMAL DISPERSALS AND SEA-LEVEL

Ma	CHRON	EPOCH	MAMMAL AGE	SEA LEVEL Super Chrons	Chrons	DISPERSAL	REMARKS
	C1 / C1r / C2 / C2A	PLIO (PLS / Ea. Lt.)	IRVINGTONIAN (I–V) / BLANCAN (I–IV)	TB3	3.10 / 3.9 / 3.8 / 3.7 / 3.6 / 3.5	11	-Bison→NA -ICE AGES; Microtines→ -IOWA ICE Equus →Eurasia -BERING STRAITS OPEN, PANAMA ISTHMUS TECTONICALLY ACTIVE — G.A.I.
5	C3 / C3A / C4 / C4A	MIOCENE Late	Late HEMPHILLIAN (Early)	TB3	3.4 / 3.3 / 3.2		-BI-DIRECTIONAL; ASIA±N.A.; N.A.±S. -STEPPE CLIMATES PREVAIL -procyonids from N.A.to S.A.
10	C4A / C5 / C5r	MIOCENE Late	CLARENDONIAN		3.1	10 / 9	-Beginning stronger input to N.A. from S.A.; ? & Antilles -Hipparion Datum in Eurasia
15	C5A / C5AB / C5AC / C5AD / C5B	MIOCENE Medial	BARSTOVIAN	TB2	2.6 / 2.5 / 2.4 / 2.3 / 2.2	8 / 7	-Proboscidean Datum, Interior N.A. -Proboscidean Datum from Asia
20	C5C / C5D / C5E / C6 / C6A	MIOCENE Early	HEMINGFORDIAN		2.1	6	-Edentates to Antilles from S.A. MAJOR NEOGENE DISPERSALS; -ASIA →N.A. -"Background" dispersals continue to imply closed Bering Strait until 3.0 Ma
25	C6AA / C6B / C6C / C7 / C8 / C9	OLIGOCENE Late	ARIKAREEAN (A = FAD)	TB1	1.5 / 1.4 / 1.3 / 1.2 / 1.1		-DRAKE PASSAGE OPENS
30	C10 / C11 / C12	OLIGOCENE Early	WHITNEYAN / ORELLAN	TA4	4.5 / 4.4	?	-SAVANNA-WOODLANDS BECOME DOMINANT
35	C13 / C15 / C16 / C17	OLIGOCENE Late	CHADRONIAN / DUCHESNEAN		4.3 / 4.2 / 4.1 / 3.6	5 / 4	-GRAND COUPURE IN WESTERN EUROPE; ANTARCTIC GLACIATIONS; RETREAT OF TURGAI STRAITS; ALPINE TECTONICS -DISPERSAL MOSTLY FROM ASIA TO N.A.

FIG. 7.—Summary of North American mammal ages, dispersal history, sea-level events, and other factors during the Cenozoic Era. Isotopic, paleomagnetic, and epoch chronologies after Berggren and others (this volume). Sea-level pattern is after Haq and others (1988). Microtine Dispersals follow Repenning and others (1990; Early, Late in the Hemphillian and I–IV in the Blancan, I–III in the Irvingtonian). For other mammal ages, E = early; L = late; A = FAD of *Alwoodia* in the early Arikareean. Shoshonian is an early Uintan subage. G.A.I. = Great American Interchange. Bold face numbers 1–11 are major Interchange Events, as discussed in the text. Supercycle boundaries are separated by Type 1 unconformities; base of cycles shown by comparably heavy lines also are Type 1 boundaries; others are Type 2 (see text).

Floras and climate.—These are generally subtropical until the Oligocene Epoch. Thereafter savanna-woodlands dominate, at least in the continental interior. Steppe floras prevail after about 5–7 Ma. Climatic cooling toward the end of the Eocene Epoch leads to modernization of land mammal taxa as seen in Asia and North America and finally reflected in the *Grande Coupure* in the Oligocene Epoch of western Europe. Climatic fluctuations of the past 3 my are reflected in dispersals of microtine rodents, and some other taxa.

Dispersals.—Taxonomic breadth at the time in question is comparable to horizontal length and vertical thickness of the heavy lines; shortest lines indicate "background" level (four or fewer taxa). The relatively small number of "background" dispersals in the Paleocene Epoch may be due in part to the paucity of the record in Europe, but few of the North American taxa apparently require an exotic source at this time. Perhaps the boreal climatic barrier, as well as that of the Turgai Straits, was effective in separating Asian, European, and American theaters in the Paleocene. The Type 1 boundary at the base of Supercycle TA2 has no associated land mammal interchange in the "medial Tiffanian."

The Clarkforkian and Wasatchian Interchange Events (1, 2) record minor, then major, overland dispersals between Europe and North America, with Tillodontia of the Clarkforkian possibly being of Asian origin. Sea-level was generally high at this time, but cycle TA2.4 begins with a Type 1 unconformity.

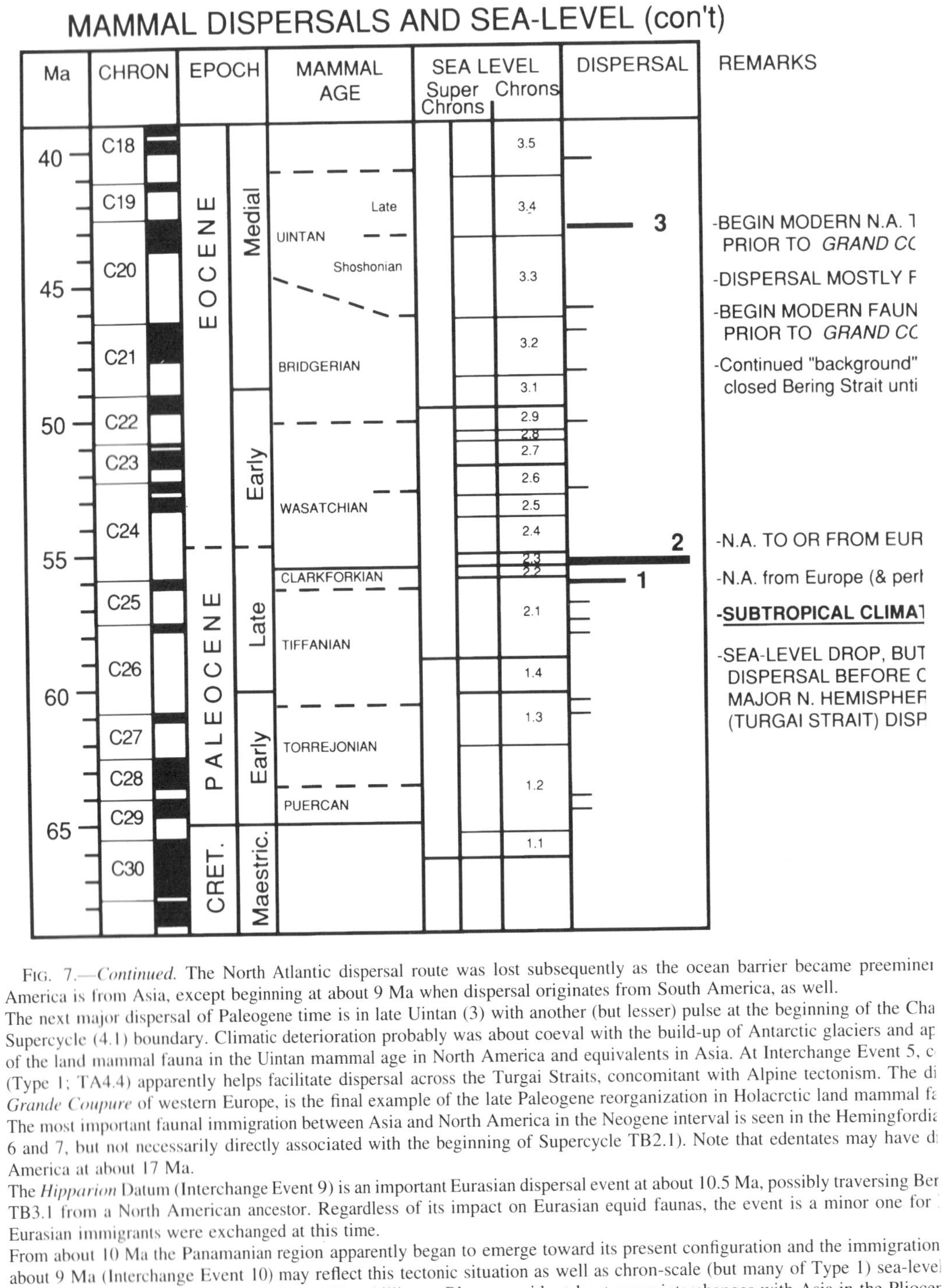

MAMMAL DISPERSALS AND SEA-LEVEL (con't)

Ma	CHRON	EPOCH	MAMMAL AGE	SEA LEVEL Super Chrons	Chrons	DISPERSAL	REMARKS
40	C18 / C19	EOCENE Medial	UINTAN (Late / Shoshonian)		3.5 / 3.4	3	-BEGIN MODERN N.A. T... PRIOR TO *GRAND C(* -DISPERSAL MOSTLY F...
45	C20 / C21	EOCENE Medial	BRIDGERIAN		3.3 / 3.2 / 3.1		-BEGIN MODERN FAUN... PRIOR TO *GRAND C(* -Continued "background" closed Bering Strait unti...
50	C22 / C23	EOCENE Early	WASATCHIAN		2.9 / 2.8 / 2.7 / 2.6 / 2.5 / 2.4	2	-N.A. TO OR FROM EUR...
55	C24 / C25	PALEOCENE Late	CLARKFORKIAN / TIFFANIAN		2.3 / 2.2 / 2.1	1	-N.A. from Europe (& per| -SUBTROPICAL CLIMA1
60	C26 / C27 / C28	PALEOCENE Early	TORREJONIAN		1.4 / 1.3 / 1.2		-SEA-LEVEL DROP, BUT... DISPERSAL BEFORE C... MAJOR N. HEMISPHEF... (TURGAI STRAIT) DISP
65	C29 / C30	CRET. Maestric.	PUERCAN		1.1		

FIG. 7.—*Continued.* The North Atlantic dispersal route was lost subsequently as the ocean barrier became preemine[n...] America is from Asia, except beginning at about 9 Ma when dispersal originates from South America, as well.

The next major dispersal of Paleogene time is in late Uintan (3) with another (but lesser) pulse at the beginning of the Cha[...] Supercycle (4.1) boundary. Climatic deterioration probably was about coeval with the build-up of Antarctic glaciers and ap[...] of the land mammal fauna in the Uintan mammal age in North America and equivalents in Asia. At Interchange Event 5, c[...] (Type 1; TA4.4) apparently helps facilitate dispersal across the Turgai Straits, concomitant with Alpine tectonism. The di[...] *Grande Coupure* of western Europe, is the final example of the late Paleogene reorganization in Holarctic land mammal fa[...] The most important faunal immigration between Asia and North America in the Neogene interval is seen in the Hemingfordia[...] 6 and 7, but not necessarily directly associated with the beginning of Supercycle TB2.1). Note that edentates may have d[...] America at about 17 Ma.

The *Hipparion* Datum (Interchange Event 9) is an important Eurasian dispersal event at about 10.5 Ma, possibly traversing Ber[...] TB3.1 from a North American ancestor. Regardless of its impact on Eurasian equid faunas, the event is a minor one for [...] Eurasian immigrants were exchanged at this time.

From about 10 Ma the Panamanian region apparently began to emerge toward its present configuration and the immigration [...] about 9 Ma (Interchange Event 10) may reflect this tectonic situation as well as chron-scale (but many of Type 1) sea-level[...] North America at the pulses shown in the Hemphillian to Blancan, with at least some interchanges with Asia in the Plioce[...] In the Great American Interchange (G.A.I.; Interchange Event 11at about 2.7 Ma), more taxa went south than came no[...] nevertheless. Note the advent of steppe conditions from about 5–7 Ma.

At about 3 Ma, the Bering Strait opens to set the stage for Northern Hemisphere glaciations, with the first major (in Iowa) cor[...] recorded at about 2.1 Ma. See text for mega- and microfauna that disperse to North America during these times. At least for[...] my seems to show the best relation to climatic control.

Tectonism vs. Sea-level.—Tectonism probably is grossly constant along western border of Americas during the Cenozoic [...] ("background dispersals"). A major dispersal gap persists between North and South America from Cretaceous to about mec[...] disperse to the Antilles) or later. The Panamanian Isthmus begins evolution toward its modern expression by at least 9–7 Ma (sl[...] southward). The Panamanian Isthmus is established by at least 2.7 Ma (Great American Interchange). Tectonic activity in t[...] establishment of the *Grande Coupure* about 34 Ma. Dispersals in the late Uintan, Chadronian and Hemingfordian mamma[...] instances in which the role of sea level can be assessed effectively without other complicating factors and of these, only the Cl[...] episodes apparently are associated with important sea-level drops.

	North America				Asian			Europe		
	Eocene			Olig.	Eocene		Olig.	Eocene		Olig.
Taxon	Uin.	Duch.	Chad.	Orel.	M. Eoc.	L. Eoc.	E. Olig.	M. Eoc. 14–16[mp]	L. eoc. 17–20	E. Olig. 21–22
Insectivora[a]										
Talpidae			X[c]							X[a]
Erinaceidae	X[s]									X[a]
Heterosoricidae			X[c]							X[a]
Soricidae		X[s]								X[a]
Primates[a]									X[a]	
Lagomorpha	X[s]	X[s]	X[s]	X[s]	X[z]	X[z]	X[z]		X[a]	
Rodents[a]										
Ischyromyidae[a]									X	
Aplodontidae[a]	X[s]									X[i]
Cedromus[s]			X							
Pipestoneomys[s]			X[l]							
Oligopetes[c]										X
Eomyidae	X[s,c]		X[c]							X[a]
Sciuridae[a]										X[i]
Castoridae			X[s]				X[z]			later[a]
Agnotocastor			X[s]				X[c]			
Propalaeocastor							X[c]			
Stenofiber										X[c]
Cricetidae			X[c]			X[z]	X[z]			X
Artiodactyls										
Cebochoeridae[a]								X[c]	X	X
Choeropotamidae[a]								X[c]	X	
Haplobunodontidae[a]								X[c]	X	
Xiphodontidae[a]									X	X
Anthracotheriidae[a]			X	X		X[c]		X[c]	X	X[l]
Entelodontidae[a]		X[s]	X[s]	X[s]		X[c]	X[l,c]			X[a]
Suidae[a]										X
Perissodactyls[a]										
Brontotheriidae	X[c]	X[c]	X[c,s]		X[c,z]	X[c,z]	X[c,z]			X[c]
Eomoropidae		X[s]	X[s]			X[c,z]	X[c,z]			
Chalicotheriidae				Mio.		X[z]	X[z]	X[c]	X[c]	X[l]
Hyrachyidae	X[c]	X[c]	X[c]	X[c]	X[z]	X[z]	X[z]	X[c]	X[c]	
Hyracodontidae	X[s]	X[s]	X[s]		X[z]	X[z]	X[z]	X[c]	X[c]	X[r]
Amynodontidae		X[s]	X[s]	X[s]	X[z]	X[z]	X[z]			X[c]
Rhinocerotidae			X[s]	X[s]						X[r]

Authors have commented that the sea-level curves of Haq and others (1988) derive much of their data from districts that border the Atlantic Ocean, and it is well recognized that the lands that bound the northern Pacific Ocean are now, as well as in the past, tectonically active. Beringia likely was emergent throughout much of the Cenozoic Era (see also Marincovich and others, 1990). Some of the land mammal data summarized here may reflect the effects of tectonism or other factors rather more than eustasy. These suggestions are derived from the following considerations.

Sea-level was Generally Higher during the Paleogene than during the Neogene Interval

This is born out by Haq and others (1988, Fig. 14) and numerous other sources (such as Hallam, 1992, Chap. 6).

Paleogene Period.—

For the Paleogene interval, notable apparent lowstands include supercycle TA2 (cycle 2.1, ca 59 Ma, "medial" Tiffanian, Figs. 1, 7) and supercycle TA3 (cycle 3.1, ca 49.5 Ma, early Bridgerian). The next Type 1 regression is recorded at TA4.1 (ca 37 Ma, early Chadronian, Figs. 2, 7). The last major Paleogene lowstand begins with supercycle TB1 (ca 28.5 Ma, correlates to post-early Arikareean, Figs. 3, 7), which likely reflects the opening of the Drake Passage between South America and east Antarctica as shown by sea-floor spreading evidence and data heralding the advent of major continental glaciation in Antarctica (Barker and Burrell, 1982; Kennett and Barker, 1990). As discussed below, only the lowstand at TA4.1 appears to be associated with an important overland interchange with North America. The *Grande Coupure*, a conspicuous immigration event in Europe, seems to have no counterpart in North America at the time in question (ca 33.5 Ma, coeval with the base of cycle TA4.4 which Haq and others (1988) recognize as a Type 1 boundary).

Neogene Period.—

After recovering from the late Oligocene sea-level minimum (supercycle TB1) at about 28.5 Ma, sea levels rise through the early Miocene Epoch (Figs. 3, 4). The Hemingfordian collectively records the most impressive land mammal immigrations in the early Neogene interval. Supercycle TB2 begins with the next sea-level drop, with cycle TB2.1 embracing the two Hemingfordian immigration episodes, at circa 19 and 17.5 Ma (Figs. 4, 7). Supercycle TB2 is unique in the Neogene record in having every (third-order) cycle boundary being of Type 1, according to Haq and others (1988), meaning that sea level was low enough to expose at least some of the continental shelf to erosion. The range in potential calibration of land mammal faunas at least allows the possibility that the late Hemingfordian immigration coincided with cycle boundary TB2.2, but Type 1 boundaries notwithstanding, the remainder of the time-scale embraced within supercycle TB2 records only limited faunal interchange with North America. Some of these are operationally significant, however. The Proboscidean Datum may equate with cycle TB2.3, as may the immigration of *Hemicyon (Plithocyon)* which appears at the beginning of the Barstovian at about 16 Ma.

The sea-level lowstand recognized as the beginning of supercycle TB3 apparently corresponds with the exit of a species of *Cormohipparion* at about 10.8 Ma, precursor to the prochoresis event designated as the Old World "*Hipparion*" Datum. Sea level remains relatively low at the beginning of the Hemphillian (correlates about with the beginning of cycle TB3.2) which, in addition to the usual immigration of taxa (rodents) from Asia, also sees the first interchange with the neotropics. A punctuated, but generally rising sea level from about 7 Ma (Fig. 5) culminates in a sea-level highstand at about 3 Ma (Fig. 6). This is just prior to the drop that corresponds with the beginning of the Great American Interchange (TB3.7), as well as the opening of the Bering Strait to the Arctic Ocean, and heralds generally low (but punctuated) sea levels during the late Pliocene and into the Pleistocene epochs.

Overland Dispersal of Land Mammals at Times Coincided with, and at others was Insensitive to, Sea-level Lowstands

Immigrations coincident with low sea levels.—

*Paleogene Period.—*According to Haq and others (1988), major (second-order) withdrawals of the sea are recognized as the beginning of supercycles TA1–4 and TB1. Supercycle TA1 is in the Late Cretaceous Period, and will not be discussed here. The greatest Paleogene sea-level drops are correlated as within the Tiffanian (TA2.1; Figs. 1, 7), and as slightly after the beginning of the Arikareean (Figs. 3, 7), at about 28.5 Ma (TB1.1) after which sea level (with punctuations) remained generally low until about 22–23 Ma. As indicated by Russell and Zhai (1987, p. 399) only endemic evolution of Asian land mammal faunas transpired during the time of TA2 (cycle TA2.1), and Russell and others (1982) likewise report only possible, rather than a strong episode of actual, interchange between North America and western Europe. Supercycle TA3 (cycle TA 3.1), of about early Bridgerian age (Figs. 2, 7), witnessed only a minor exchange. The greatest Paleogene immigration coeval with a major sea-level drop is associated with the beginning of the Chadronian, correlated as about equivalent to TA4.1 at 37 Ma (Figs. 2, 7). The Chadronian witnessed one of the least extensive of the major Paleogene faunal interchanges, however.

Thus only the initial phase of supercycle TA4 saw a significant episode of overland mammal interchange in North America, but this still was a relatively minor event as compared with others in the Paleogene (Wasatchian; late Uintan; see below). The absence of a major overland dispersal correlative with the beginning of supercycle TB1 is noteworthy, as this strong sea-level fall associated with the opening of the Drake Passage reflects a major tectonic event of the southern oceans.

*Neogene Period.—*The major Neogene sea-level lowstands are reflected at the beginning of supercycles TB2 and TB3, with that of TB2 apparently being the least extensive, although recording a high number of Type 1 unconformities. The interval embraced within cycle TB2.1 (Hemingfordian) saw the major Neogene dispersal, except for the Great American Interchange. According to the evidence (see text and Figs. 4, 7) the Hemingfordian immigration took place in two waves separated by about 1.5 my, but may not be compellingly related to sea-level change. The other major sea-level lowstand of the late Neogene, represented by TB3.1 at about 10.8 Ma (Figs. 5, 7), sees little more than a single-taxon emigration from North America ("*Hipparion*" Datum; above). The persistent (with fluctuations)

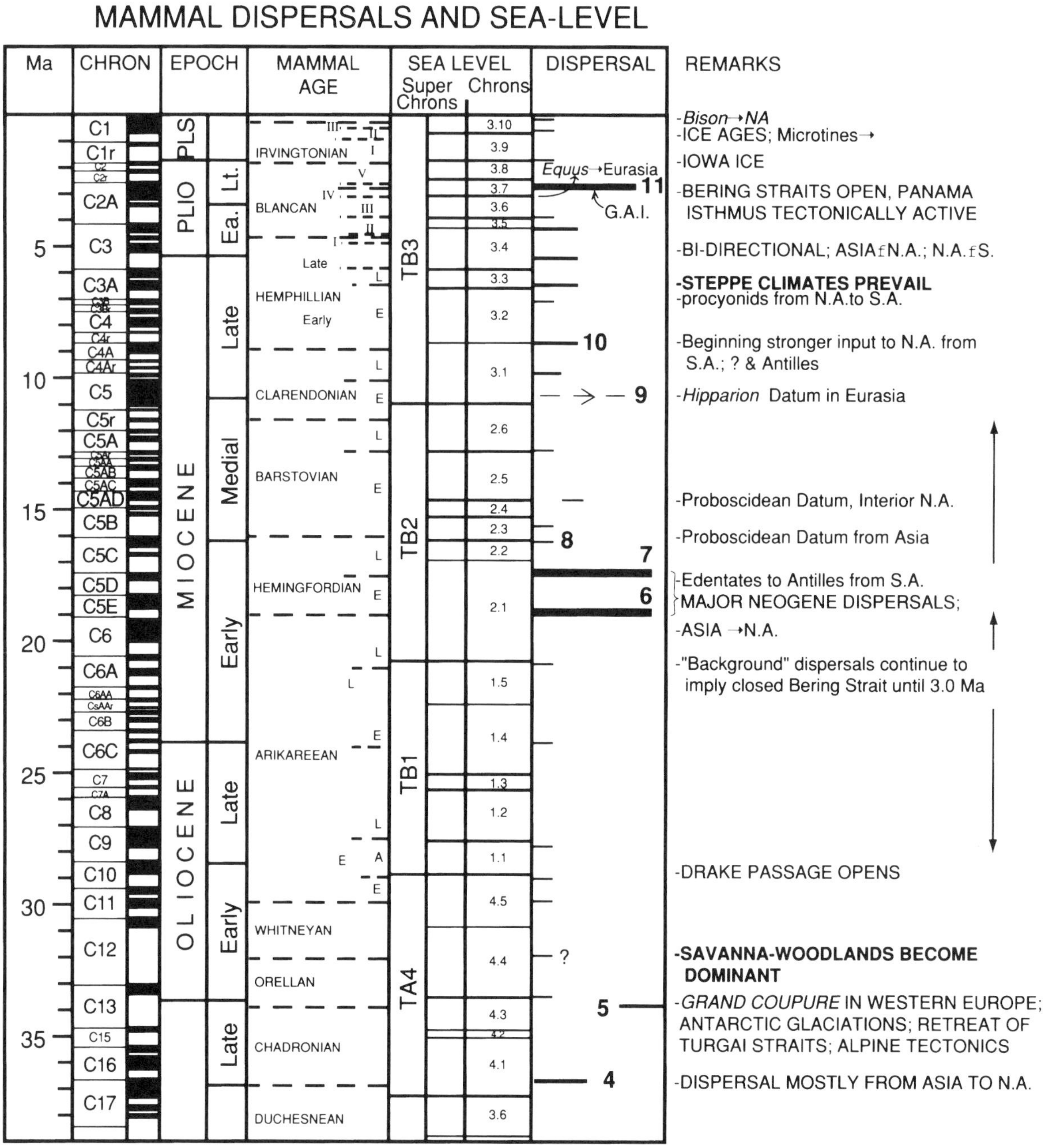

MAMMAL DISPERSALS AND SEA-LEVEL

Ma	CHRON	EPOCH	MAMMAL AGE	SEA LEVEL Super Chrons / Chrons	DISPERSAL	REMARKS
	C1	PLS	IRVINGTONIAN III / II / I	TB3 — 3.10 / 3.9		-*Bison*→NA -ICE AGES; Microtines→
	C1r	PLIO		3.8	*Equus*→Eurasia — 11	-IOWA ICE
	C2 / C2r			3.7		
	C2A	Ea. Lt.	BLANCAN V / IV / III / II / I	3.6 / 3.5	G.A.I.	-BERING STRAITS OPEN, PANAMA ISTHMUS TECTONICALLY ACTIVE
5	C3		Late	3.4		-BI-DIRECTIONAL; ASIA£N.A.; N.A.£S.
	C3A / C3B		HEMPHILLIAN	3.3		**-STEPPE CLIMATES PREVAIL** -procyonids from N.A. to S.A.
	C4 / C4r	Late	Early E	3.2		
	C4A / C4Ar			3.1	— 10	-Beginning stronger input to N.A. from S.A.; ? & Antilles
10	C5		CLARENDONIAN L / E	2.6	— → — 9	-*Hipparion* Datum in Eurasia
	C5r / C5A	MIOCENE Medial		2.5		
	C5AB / C5AC		BARSTOVIAN E	2.4 / 2.3		
15	C5AD / C5B			2.2	— 8 / 7	-Proboscidean Datum, Interior N.A. -Proboscidean Datum from Asia
	C5C		HEMINGFORDIAN L / E	2.1	6	-Edentates to Antilles from S.A. MAJOR NEOGENE DISPERSALS;
	C5D / C5E					-ASIA →N.A.
20	C6	Early		1.5		-"Background" dispersals continue to imply closed Bering Strait until 3.0 Ma
	C6A		L	1.4		
	C6AA / C6Ar / C6B					
	C6C		ARIKAREEAN E	1.3		
25	C7 / C7A	OLIGOCENE Late		1.2		
	C8		L	1.1		
	C9		E / A	4.5		-DRAKE PASSAGE OPENS
	C10		E	4.4		
30	C11	Early	WHITNEYAN		?	**-SAVANNA-WOODLANDS BECOME DOMINANT**
	C12		ORELLAN	4.3 / 4.2	5 —	-*GRAND COUPURE* IN WESTERN EUROPE; ANTARCTIC GLACIATIONS; RETREAT OF TURGAI STRAITS; ALPINE TECTONICS
35	C13 / C15	Late	CHADRONIAN	4.1	4	-DISPERSAL MOSTLY FROM ASIA TO N.A.
	C16 / C17		DUCHESNEAN	3.6		

FIG. 7.—Summary of North American mammal ages, dispersal history, sea-level events, and other factors during the Cenozoic Era. Isotopic, paleomagnetic, and epoch chronologies after Berggren and others (this volume). Sea-level pattern is after Haq and others (1988). Microtine Dispersals follow Repenning and others (1990; Early, Late in the Hemphillian and I–IV in the Blancan, I–III in the Irvingtonian). For other mammal ages, E = early; L = late; A = FAD of *Alwoodia* in the early Arikareean. Shoshonian is an early Uintan subage. G.A.I. = Great American Interchange. Bold face numbers 1–11 are major Interchange Events, as discussed in the text. Supercycle boundaries are separated by Type 1 unconformities; base of cycles shown by comparably heavy lines also are Type 1 boundaries; others are Type 2 (see text).

Floras and climate.—These are generally subtropical until the Oligocene Epoch. Thereafter savanna-woodlands dominate, at least in the continental interior. Steppe floras prevail after about 5–7 Ma. Climatic cooling toward the end of the Eocene Epoch leads to modernization of land mammal taxa as seen in Asia and North America and finally reflected in the *Grande Coupure* in the Oligocene Epoch of western Europe. Climatic fluctuations of the past 3 my are reflected in dispersals of microtine rodents, and some other taxa.

Dispersals.—Taxonomic breadth at the time in question is comparable to horizontal length and vertical thickness of the heavy lines; shortest lines indicate "background" level (four or fewer taxa). The relatively small number of "background" dispersals in the Paleocene Epoch may be due in part to the paucity of the record in Europe, but few of the North American taxa apparently require an exotic source at this time. Perhaps the boreal climatic barrier, as well as that of the Turgai Straits, was effective in separating Asian, European, and American theaters in the Paleocene. The Type 1 boundary at the base of Supercycle TA2 has no associated land mammal interchange in the "medial Tiffanian."

The Clarkforkian and Wasatchian Interchange Events (1, 2) record minor, then major, overland dispersals between Europe and North America, with Tillodontia of the Clarkforkian possibly being of Asian origin. Sea-level was generally high at this time, but cycle TA2.4 begins with a Type 1 unconformity.

MAMMAL DISPERSALS AND SEA-LEVEL (con't)

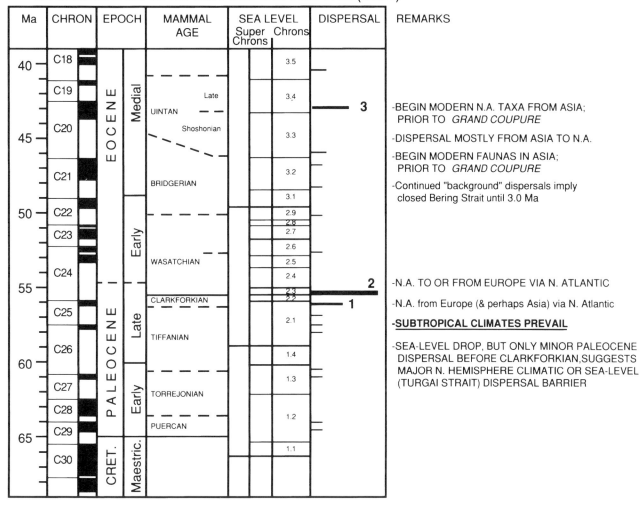

FIG. 7.—*Continued.* The North Atlantic dispersal route was lost subsequently as the ocean barrier became preeminent. Subsequently dispersal to North America is from Asia, except beginning at about 9 Ma when dispersal originates from South America, as well.

The next major dispersal of Paleogene time is in late Uintan (3) with another (but lesser) pulse at the beginning of the Chadronian (4), there associated with a Supercycle (4.1) boundary. Climatic deterioration probably was about coeval with the build-up of Antarctic glaciers and apparently sparked the modernization of the land mammal fauna in the Uintan mammal age in North America and equivalents in Asia. At Interchange Event 5, contemporaneous falling of sea level (Type 1; TA4.4) apparently helps facilitate dispersal across the Turgai Straits, concomitant with Alpine tectonism. The dispersal event, as recognized as the *Grande Coupure* of western Europe, is the final example of the late Paleogene reorganization in Holarctic land mammal faunas.

The most important faunal immigration between Asia and North America in the Neogene interval is seen in the Hemingfordian (two pulses; Immigrations Events 6 and 7, but not necessarily directly associated with the beginning of Supercycle TB2.1). Note that edentates may have dispersed to the Antilles from South America at about 17 Ma.

The *Hipparion* Datum (Interchange Event 9) is an important Eurasian dispersal event at about 10.5 Ma, possibly traversing Beringia at the beginning of Supercycle TB3.1 from a North American ancestor. Regardless of its impact on Eurasian equid faunas, the event is a minor one for North America and, apparently, no Eurasian immigrants were exchanged at this time.

From about 10 Ma the Panamanian region apparently began to emerge toward its present configuration and the immigrations from South America beginning at about 9 Ma (Interchange Event 10) may reflect this tectonic situation as well as chron-scale (but many of Type 1) sea-level fluctuations. Asian taxa also enter North America at the pulses shown in the Hemphillian to Blancan, with at least some interchanges with Asia in the Pliocene Epoch being relatively balanced. In the Great American Interchange (G.A.I.; Interchange Event 11 at about 2.7 Ma), more taxa went south than came north, but the exchange is dramatic, nevertheless. Note the advent of steppe conditions from about 5–7 Ma.

At about 3 Ma, the Bering Strait opens to set the stage for Northern Hemisphere glaciations, with the first major (in Iowa) continental ice sheet in North America recorded at about 2.1 Ma. See text for mega- and microfauna that disperse to North America during these times. At least for microtines, the record of the last 3 my seems to show the best relation to climatic control.

Tectonism vs. Sea-level.—Tectonism probably is grossly constant along western border of Americas during the Cenozoic Era, at least since Wasatchian time ("background dispersals"). A major dispersal gap persists between North and South America from Cretaceous to about medial Miocene time (when edentates disperse to the Antilles) or later. The Panamanian Isthmus begins evolution toward its modern expression by at least 9–7 Ma (sloths disperse northward; procyonids southward). The Panamanian Isthmus is established by at least 2.7 Ma (Great American Interchange). Tectonic activity in the Alps also is associated with the establishment of the *Grande Coupure* about 34 Ma. Dispersals in the late Uintan, Chadronian and Hemingfordian mammal ages may be the remaining few instances in which the role of sea level can be assessed effectively without other complicating factors and of these, only the Chadronian and early Hemingfordian episodes apparently are associated with important sea-level drops.

sea-level low interval from about 10.8 Ma to about 5 Ma (Fig. 5) records a number of interchanges that are important to land mammal faunas in North and South America but, in terms of breadth, these still are relatively minor when compared to the diversity seen in the Great American Interchange at about 2.7 Ma (which also is associated with a strong drop in sea level; TB3.7–8; Figs. 6, 7). As discussed below, however, the Great American Interchange likely was as much effected by tectonic events (that raised the land) as by a global fall in sea level.

Microtine dispersal events recorded by Repenning (1987) and Repenning and others (1990) generally correlate with low-stands of sea level, on the basis of present evidence. Whereas these may not be of the scope accorded major recognition in this study, the fact of this correspondence is noted here (Fig. 6): Blancan III (ca 4.0 Ma) is slightly younger than the Type 1 lowstand TB 3.6; Blancan V (ca 2.6 Ma) is coeval with the Type 1 lowstand TB 3.8; Irvingtonian I (ca 1.8 Ma) is coeval with the Type 1 lowstand TB 3.9; and Irvingtonian II (ca 0.8 Ma) is about coeval with the Type 1 low stand 3.10. Blancan I (circa 5.0 Ma) does not appear to correspond with the beginning of a low stand; Blancan II and IV do not reflect immigrations. The microtine immigration episodes are taken up further in *Climatic Factors*.

In summary, of the major Cenozoic sea-level low stands only the beginning of supercycles TA4 and TB2 can be directly associated with overland dispersals having sufficient numbers (and breadth) of taxa to be significant. The Great American Interchange, at cycle TB3.7, reflects the other strong interdispersal involving both North and South America in a major way for the first time since the Late Cretaceous Period. Tectonism apparently played a helping hand here, at least.

Times of dispersal but high sea level.—

The late Paleocene and early Eocene dispersals represented by land mammals in the Clarkforkian and Wasatchian are among the most spectacular instances of this pattern (Fig. 1). Although short-lived (but Type 1) drops in sea level are reconstructed for these times (TA2.2, 2.4), the overall pattern is for a sea-level maximum (Haq and others, 1988, Fig. 14), in the face of which occurs the greatest Palaearctic homogeneity in land mammal faunas ever seen (Fig. 7). In that the dispersal corridor(s) appear to have been located in the high polar latitudes, and in that the northern Atlantic Ocean was effectively closed, actual status of sea level may have been irrelevant to the dispersal potential at this time.

Likewise, the late Uintan (ca 43 Ma; TA3.4) saw nearly as strong an interchange, but still, sea-level was relatively high (Fig. 2). The pattern for the Clarkforkian-Wasatchian and late Uintan is in strong contrast to that of the *Grand Coupure*. No significant land mammal interchange between Europe and North America transpired at the time of the *Grand Coupure*, even though a Type 1 low stand is reflected at that time in cycle TA4.4. This is discussed further below.

Vicariance

The strong dissimilarity of post-Wasatchian-Sparnacian land mammal faunas in North America and the Old World represents a major vicariant event, reflected by the opening of the northern Atlantic Ocean. Prior to that time, juxtaposition of Holarctic land areas allowed the interchange of mammalian taxa but, prior to the Clarkforkian, at least, extrinsic factors such as climate in northern latitudes and the presence of the Turgai Strait may have hindered dispersal (McKenna, 1983a, b). Notably, the Clarkforkian and Wasatchian interchanges transpired during times of relative sea-level highstand (Figs. 1, 7), but this factor apparently was irrelevant. If the post-Wasatchian disruption of a formerly widely dispersed pattern can be said to illustrate a relatively typical vicariant situation, the examples cited below portray the "reverse," in which dispersals that would have been expected to have taken place at a given time were actually delayed significantly.

The *Grande Coupure* at or near the beginning of the Oligocene Epoch in western Europe (ca 33.5 Ma; Figs. 2, 7) represents a major reorganization of its land mammal faunas (Table 1). Importantly, new varieties of taxa had appeared in North America beginning in the late Uintan, and were present in Asia, as well, from within medial to late Eocene time. Thus, the *Grande Coupure* recognizes the breakdown of a dispersal barrier that not only resulted in withdrawal of the Turgai Straits, but was associated with tectonic events in the mid-Alpine district (Heissig, 1979), as well as in Beringia. In that the modernization of the Asian theater was manifested well within medial Eocene time, data reflect a somewhat tardy time of dispersal from Asia to North America in the late Uintan (ca 43 Ma), with a second pulse at the beginning of the Chadronian (ca 37 Ma). Note, however, that certain groups (aplodontid and eomyid rodents [Duchesnean] and sciurid rodents [Chadronian]) may represent westward dispersals from North America to Europe (see *Dispersal Lag*). Nevertheless, Asia apparently retained its fauna for about 14 my (48–34 Ma) before boiling over into western Europe at the *Grande Coupure* at about 33.5 Ma. By that time, the modernization of the North American mammal fauna was already virtually completed.

In another theater, a Central American "holding pen," revealing only hints at present, could be represented by the West Indian region. Presently meagre, but exciting, evidence (MacFee and Iturralide-Vinent, 1994) suggest that this region may have been a center of edentate evolution (perhaps populated about 17–18 Ma and first releasing some of its fauna at about 9 Ma).

The Great American Interchange at circa 2.7 Ma reflects a major lowering of sea level (TB3.7–8), but also increased elevation of the Panamanian region (Repenning and Ray, 1977; Keigwin, 1982). In fact, the record beginning approximately with the early Hemphillian (ca 9 Ma) gives a glimpse of the importance of growing tectonic events in promoting sweepstakes to filter to, finally, corridor interchange across the neotropical region that persists to the present time (Figs. 5, 6, 7).

Climatic Factors

The global climate in the early Cenozoic Era appears to have been relatively equable and warm, perhaps rebounding somewhat from a Late Cretaceous cooling episode. As the late Eocene Epoch is approached, however, climatic deterioration accelerates and likely drives the major modernization of land mammal faunas witnessed during that interval. It is less clear, however, that dispersal events were directly affected by climate, except in the Pliocene and Pleistocene epochs. Prentice and

TABLE 1.—REGIONAL AND TEMPORAL DISTRIBUTION OF TAXA OF THE *GRANDE COUPURE*.

Taxon	North America Eocene Uin.	North America Eocene Duch.	North America Eocene Chad.	North America Olig. Orel.	Asian Eocene M. Eoc.	Asian Eocene L. Eoc.	Asian Olig. E. Olig.	Europe Eocene M. Eoc. 14–16[mp]	Europe Eocene L. eoc. 17–20	Europe Olig. E. Olig. 21–22
Insectivora[a]										
Talpidae			X[c]							X[a]
Erinaceidae	X[s]									X[a]
Heterosoricidae			X[c]							X[a]
Soricidae		X[s]								X[a]
Primates[a]									X[a]	
Lagomorpha	X[s]	X[s]	X[s]	X[s]	X[z]	X[z]	X[z]		X[a]	
Rodents[a]										
Ischyromyidae[a]									X	
Aplodontidae[a]	X[s]									X[i]
Cedromus[s]			X							
Pipestoneomys[s]			X[l]							
Oligopetes[c]										X
Eomyidae[c]	X[s,c]		X[c]							X[a]
Sciuridae[a]										X[i]
Castoridae			X[s]				X[z]			later[a]
Agnotocastor			X[s]				X[c]			
Propalaeocastor							X[c]			
Stenofiber										X[c]
Cricetidae			X[c]			X[z]	X[z]			X
Artiodactyls										
Cebochoeridae[a]								X[e]	X	X
Choeropotamidae[a]								X[e]	X	
Haplobunodontidae[a]								X[e]	X	
Xiphodontidae[a]									X	X
Anthracotheriidae[a]			X	X		X[c]		X[e]	X	X[l]
Entelodontidae[a]		X[s]	X[s]	X[s]		X[c]	X[l,c]			X[a]
Suidae[a]										X
Perissodactyls[a]										
Brontotheriidae	X[c]	X[c]	X[c,s]		X[c,z]	X[c,z]	X[c,z]			X[c]
Eomoropidae		X[s]	X[s]			X[c,z]	X[c,z]			
Chalicotheriidae				Mio.		X[z]	X[z]	X[c]	X[c]	X[r]
Hyrachyidae	X[c]	X[c]	X[c]	X[c]	X[z]	X[z]	X[z]	X[c]	X[c]	
Hyracodontidae	X[s]	X[s]	X[s]		X[z]	X[z]	X[z]	X[c]	X[c]	X[r]
Amynodontidae		X[s]	X[s]	X[s]	X[z]	X[z]	X[z]			X[c]
Rhinocerotidae			X[s]	X[s]						X[r]

Uin. = Uintan; Duch. = Duchesnean; Chad. = Chadronian; Orel. = Orellan (N.A. mammal ages).

In Carroll (1988), L. Olig. NA = Chadronian; L.Olig. Eur. = *Grande Coupure*. Where Carroll (1988) differs from Stucky (1992), Stucky is followed.

[a]After Legengre and others (1991). Primates last appearance in Europe.

[mp]Mammal Paleogene reference levels[a]

[s]After Stucky (1992). Erinaceidae = *Dartonius, Creotarsus, Talpavoides, Talpavus*, from Wasatchian (at least). Aplodontidae in Uintan = *Eohaplomys*[s], *Spurimus*[c]

[c] After Carroll (1988; *Trimylus* = Heterosoricidae; also[s]).

[i]Includes *Sciurodon, Trigonmys*[c] (Aplodontidae), *Heteroxerus, Palaeosciurus*[s] (Sciuridae).

[l]later in age

[r]Russell and others (1982)

[z]Russell and Zhai (1987)

Eomyidae = *Protadjidaumo* (Uintan) and several N.A. Chadronian and younger North American genera[s]. *Cupressimus*[s] from Europe[a].

Cebochoerids, choeropotamids, haplobunodontids, xiphodontids, anthracotheriids are archaic hold-overs.

Entelodontidae includes *Brachyhyops* in Duch.[s], *Eoentelodon* in Lt. Eoc., As[c].

Suidae are Tayassuidae[c] (*Palaeochoerus*).

Brontotheriidae; major Asian-N.A. radiation; only one genus, *Menodus*, shared with Europe.

Eomoropidae mainly Asian-N.A., no *Grande Coupure* representative, even in Oligocene.

Hyrachyids well represented prior to *Grande Coupure* in Europe, but not subsequently.

Hyracodontids have major tie between Asia and N.A. in late Eocene, but not with *Grand Coupure*.

Metamynodontids have major tie between Asia and N.A. in late Eocene; *Cadurcodon* in medial Oligocene Asia, shared with *Grande Coupure*.

Rhinocerotids represented by *Ronzotherium* in *Grande Coupure*, but not shared elsewhere.

Matthews (1988, p. 965–966) favor a world which supported an ice volume at least as great as that of today beginning about 40 Ma; prior to then (50–65 Ma), the $\delta^{18}O$ record for planktonic tropical foraminifera suggests ice-free conditions (at least intermittently, if not completely). Webb (1977) reviews the overall situation for North America and proposes that the earlier subtropical forests gave way to the savannah-woodland environments, with local aridity at about the beginning of Oligocene time, but steppe-adapted faunas not appearing until about 5 Ma as aridity and climatic deterioration made major advances preliminary to Northern Hemisphere Pliocene and Pleistocene cooling. The main pre-Pliocene effects of climate on overland dispersals relative to North America are difficult to document, especially during early (pre-Clarkforkian-Wasatchian-Sparnacian) Paleogene time. After the late Oligocene Epoch, climatic changes apparently produce sea-level low stands with mixed results, as summarized above.

The influence of climate on Pliocene and Pleistocene land mammal faunal dynamics has an extensive literature, and the overland dispersals that take place in late Pliocene through

Pleistocene times commonly are attributed to sea-level low stands associated with continental glaciation in the Northern Hemisphere.

Repenning (1987) and Repenning and others (1990) record 10 episodes of microtine dispersal and evolution that are effectively synchronous across Holarctica during the past circa 6 my (see also Figs. 5, 6). Of these, two (Blancan II and IV) are based on endemic evolution, rather than upon immigration, and many (see below) can be attributed to climatic causes, at least subsequent to the past 3 Ma. On the other hand, the cause for the dispersals prior to that time is more problematic, and Flynn and others (1991, p. 261) point this out. Nevertheless, a bi-directional and possibly relatively balanced dispersal scenario between Asia and North America was present at least by the Pliocene Epoch (Flynn and others, 1991), a phenomenon not limited then or later to microtines.

Once begun, however, there is little doubt as to the role played by climate in these overland dispersals, and one important aspect of this is the concept of "disharmonious" associations of Pleistocene taxa, articulated by Lundelius and others (1987, p. 227). Faunas are "disharmonious" when the species of which they are composed are not compatible in their ecologic preferences at the present time. In addition to interpreting a fauna as having "cool" (glacial) or "warm" (interglacial) affinities, some are mixed when temperature tolerance is judged from living counterparts. Thus, Pleistocene climates were not always neatly comparable to those of the present day, at least in terms of their affect on then-living taxa, and this could play a part in affecting past dispersals or subsequent regional habitation of species.

Repenning and Brouwers (1992, p. 4L) propose that the ultimate cause of the glaciations of the past 3 my is rooted in the tectonic elevation of Central America since about the last 15 my, with resultant disruption of equatorial oceanic circulation and warming of the North Atlantic, increased precipitation over the Arctic region and the development of continental ice sheets. In that scenario, the development of cooler climates has many factors (see also Keigwin, 1982, and literature cited therein for primary ideas in this regard, but also Willard and others, 1993, and Emslie and Morgan, 1994).

Whatever the ultimate cause, Repenning and Brouwers (1992) characterize the interval, that began about 3.0 Ma and lasted until the present, as comprising three major cooling cycles, of which two transpired prior to the time of major continental glaciation that began about 0.85 Ma (the Ice Ages). In all intervals, the overall pattern consists of times of increased glacial activity centered around the Arctic Ocean, with glacial maxima and other advances corresponding generally with sea-level low stands at various scales. The dispersal of microtine rodents during this interval generally was driven by climatic factors, and usually transpired from the Arctic Ocean Borderland in the north to more southerly locations in Europe or in North America.

The chronology of these events is based on a combination of paleomagnetic, radioisotopic, climatic and faunal data. Within certain limits, the immigrations define episodes that began about 3.0, 2.6, 2.0, 0.85, 0.45, and 0.15 Ma. Of these, the first four appear to correspond approximately to sea-level low cycles: 3.0 Ma (cycle 3.7), 2.8 Ma (cycle 3.8), 1.8 Ma (cycle 3.9), and 0.8 Ma (cycle 3.10). At the same time, the dispersal sce-

narios summarized by Repenning and Brouwers (1992) are more complex than this, and greater variations in world sea-level history are shown than are reflected directly in dispersals. Hence, Antarctic, as well as Arctic, climatic events apparently are reflected in the sea-level record of the last 3 my. A brief summary of climatic events and aspects of microtine rodent faunal dynamics is summarized below, chiefly from Repenning and Brouwers (1992).

Depending upon their climatic adaptations, microtine rodents either dispersed in synchrony with the advancing ice sheets, or were prevented by them from doing so. Apparently driven by cooler climates, the bog lemmings, *Plioctomys* and *Mictomys*, immigrated to North America at about 2.6 Ma. According to Repenning and Brouwers (1992, p. 15, 29), this was coincident with the first major extension of the Scandinavian Ice Sheet (as evidenced by ice-rafted debris at ODP Site 644 in the Norwegian Sea, beginning of density stratification of the Arctic Ocean and thinning of boreal forests, leading to tundra and permafrost). These microtines (beginning of Blancan V, Fig. 6) heralded the cool climates in the United States (Texas, Arizona, Idaho) that culminated in the first continental glaciation at about 2.1 Ma (glacial till in Iowa). All of this developed during the "First Cooling Cycle," which began about 3.0 Ma and lasted until about 2.0 Ma, and saw initially ice-free conditions in the Arctic Ocean. During the same time, *Allophaiomys*, evolved in the more temperate districts of eastern Asia, and in North America the archaic rabbit, *Hypolagus*, was being replaced by leporines (*Lepus*; Repenning and Brouwers, 1992, p. 19R).

By 1.9 Ma, northern hemisphere climates had become much warmer, and the "Second Cooling Cycle" had begun. A few sites record the last co-occurence of *Hypolagus* and *Lepus* at this time. At about 1.9 Ma (Irvingtonian I; Fig. 6) the temperate-adapted microtine, *Allophaiomys*, dispersed from Beringia to North America (and northward from the Black Sea region to Holland; Repenning and Brouwers, 1992, p. 28), apparently synchronous with the Tiglian warm forest in the Netherlands. This is consistent with warm climates having prevailed in Beringia at about 2.0 Ma (or slightly earlier), by which time *Allophaiomys* had evolved and was pushed southward as climate deteriorated. At 1.7 Ma, cool-climates again prevailed, as identified by an increase in ice-rafted debris in the Norwegian Sea, the apparent beginning of the Eburonian (cold) flora in Holland, and the southward dispersal from Beringia to North America of the heather vole, *Phenacomys*. According to Repenning and Brouwers (1992, p. 31), *Phenacomys* is first known in Beringia at about 2.4 Ma, but did not disperse from there until about 0.7 my later, to co-exist with *Allophaiomys* in North America (but not in Beringia). In fact, *Allophaiomys* persists during most of the Matuyama Chron (ca 2.0–0.8 Ma) from southern Saskatchewan to southern Texas and from the Rocky Mountains east to the Atlantic Coast. This suggests that the interval experienced generally warmer climates than those of the glacial maximum at 2.1 Ma, but still cool enough to shelter *Phenacomys*. Similarly, the microtine, *Lasiopodomys*, is a cool-adapted form first known, in this case, in Beringian deposits about 1.7 Ma old, but not seen in more southern localities in North America until the beginning of the Ice Ages at about 0.85 Ma.

At about 1.1 Ma, the cool-adapted microtine, *Microtus* (evolved from *Allophaiomys*), is recorded in Beringian sites of about that age (Krestovka) but is not found in the United States east of the Rocky Mountains until about 0.45 Ma. This suggests that even though cool climates were re-established in the northern hemisphere at about 0.85 Ma, regional climates in the United States were either subtropical or overly moist (and preferred by *Allophaiomys*).

Northern hemisphere Ice Age glaciations began about 0.85 Ma, at which time (Irvingtonian II) *Lasiopodomys* and *Terricola* disperse to North America from Beringia and correlate with faunas of "Nebraskan" age, but with no contemporary influx of mammal megafauna. *Lasiopodomys* becomes extinct in North America at about 0.83 Ma, but still lives in China. The last microtine migration to North America from Beringia is recorded at about 0.45 Ma, with *Microtus* sp., cf. *M. montanus* (Irvingtonian III). Bison enters North America at about 0.15 Ma at the beginning of the Rancholabrean (Fig. 6), but had been present in China as early as about 2 Ma, having originated in an unknown northern Asian district (Flynn and others, 1991).

Dispersal Lag

The *Grande Coupure* represents one of the most important examples of dispersal lag in Holarctica, especially when the question of modernization of Holarctic land mammal faunas is considered. Table 1 shows the regional and temporal distribution of the main taxa or groups that occur in western Europe at the time of the *Grande Coupure*. Regarding the Insectivora, the Talpidae, Erinaceidae and Soricidae, at least, have occurrences in North America that pre-date the *Grande Coupure*; lagomorphs are similar, but also have a major (founder) presence in Asia in Paleogene time (Russell and Zhai, 1987). Among the rodents, aplodontids and eomyids are known from at least the Duchesnean in North America (but interestingly not in Asia), sciurids have a Chadronian occurrence in North America (but are not represented in Asia), as do castorids (but these, in contrast, are known in the early Oligocene of Asia). Among artiodactyls, a number of archaic families (from within the Eocene, at least) still are present in the *Grande Coupure* (cebochoerids, choeropotamids, haplobunodontids, xiphodontids, and anthracotheriids), but two are new (entelodontids and suids). Entelodontids are well known in Eocene faunas of Asia and North America. There still is the question of the relationship between the Chadronian tayassuids in North America and the stem suid (or palaeochoerid) of Europe (none of these families is known in the early Oligocene or earlier faunas of Asia). Perissodactyls show a major pattern of relationship between Asia and North America in the Eocene and into the Oligocene epochs, with only minor representation (and virtually only at the time of the *Grande Coupure*) in Europe. Examples include the Brontotheriidae (major radiation in both Asia and North America, many shared genera [7 of 18–20 in each region], with only one genus, *Menodus* shared with Europe). The Eomoropidae is largely a late Eocene to early Oligocene group in Asia and North America, but is never seen in Europe, even after the *Grande Coupure*. Chalicotheriids are fundamentally European and Asian in the Eocene, but have no North American record until the late Oligocene. Among the tapiroids, the isectolophids are only known from the early and medial Eocene; hyrachyids (helaletids) are represented in all three regions in the early to late Eocene, but the early Oligocene *Colodon* of North America and Asia is not seen in Europe. Lophialetids, deperetellids, and lophiodontids

(effectively Eocene groups) have no particular relevance here, but the rhinocerotoids show a familiar pattern. Asia is the major place of (?founding and) diversity of the hyracodontids, with definite ties to North America in the late Eocene (*Prohyracodon, Forstercooperia, Triplopus*). There is a minor showing in Europe (less than five genera), but only *Indricotherium* is shared (with Asia, early Miocene). Among the amynodontids, the usual Asian theater shares *Metamynodon* and *Amynodon* with North America in the late Eocene (both genera persist until the early Oligocene, at least). Although the family is known (two genera) in Europe, only *Cadurcodon* is shared (medial Oligocene) with Asia. Finally, the rhinocerotid, *Ronzotherium*, of the *Grande Coupure* represents a family known from the Chadronian of North America on the basis of four genera. Thus, apparently more so than North America, Asia was a potential source for European taxa prior to the *Grande Coupure*. Even with the relaxation of barriers that had kept the mammals in those areas from reaching westward until then, it is perhaps as, if not more, interesting to ask why immigration to western Europe was not more dramatic! Further, a number of the families cited above have taxa of Oligocene age that are not represented in Europe after the *Grande Coupure*. Regarding dispersal lag, Asian taxa began to populate North America at least by the late Uintan, circa 43 Ma, with a second pulse at the beginning of the Chadronian (37 Ma), whereas the taxa that eventually arrived in Europe at the time of the *Grande Coupure* (mostly from Asia, but also with input from North America [possibly some or all of: talpids, erinaceids, soricids, aplodontids, eomyids]) finally arrived at circa 33.5 Ma.

At a latitudinal scale, of course, is the long-term, and mostly effective, isolation of the land mammal faunas of South America. In addition to the well-known scenarios of Cenozoic sweepstakes input, and the late Pliocene climactic trans-Panamanian Great American Interchange, the West Indian edentate fauna may record another dispersal lag of significant magnitude (in this case, immigration to North America from the south apparently is delayed for nearly 8 my, from circa 17 Ma to about 9 Ma).

In North America, a less dramatic, but still locally important, dispersal lag appears to be represented by the Proboscidean Datum. The group is first recorded in faunas of latest Hemingfordian or early Barstovian age in peripheral (northwest or southeastern) North America, but is seen in late Barstovian faunas in the continental interior about 1 my later. Finally, instances of dispersal lag at a relatively fine scale are provided by the circa 1.7 Ma immigration of *Phenacomys* to North America after a delay of about 0.7 my from its first occurrence in Beringia and a comparable but differently timed delay of *Lasiopodomys* from its first occurrence in Beringia (ca 1.7 Ma) to its immigrant presence in the United States at about 0.85 Ma. *Bison* apparently dispersed southward from northern Asia into China about 2.0 Ma, but did not reach North America until much later (ca 0.15 Ma).

Dispersal and boundary definition.—Many workers have suggested that boundaries between biochronological units (mammal ages or subdivisions) are usefully defined on the basis of first appearances; hence the numerous citations of FADs, LSDs or evolutionary first occurrences seen in these pages and, for example, in Woodburne (1977). One of the noteworthy aspects of the present discussion, however, is the illustration of dispersal lag at various scales for a variety of taxa. Whereas we re-spect the principle that boundaries must be defined in order to be useful and that various kinds of first occurrences are poten-tially valuable criteria, the spectre of temporally significant pro-choresis still must be discounted at whatever scale is pertinent for consideration. Clearly, it is important to determine the time of origin of a taxon, so as to be able to set a lower limit on the potential time of its first occurrence in an immigrant setting, but it also becomes progressively more important, than is at least generally stated, to document the time(s) of its dispersal along the route. The purpose of this discussion is to reinforce the need not only (1) to make a (hopefully single-taxon) defi-nition of a boundary in its type section, but also (2) to carefully evaluate (rather than assume) its age in progressively more pe-ripheral sites in the eventual total geographic range of the taxon. *Endnotes.*—Webb and Opdyke (1995) utilize positive oxygen isotope excursions of marine waters based on planktonic fora-minifera as a proxy for both changes in sea level and changes in climate. In the present paper, there does not seem to be a simple correlation between sea-level low stands and overland mammalian dispersal to or from North America during much of the Cenozoic Era. This leaves us with a consideration of climate as the major cause, but difficult to pin-point except in general terms (see Webb and Opdyke, 1995). The case linking climate with dispersal appears to be best for the Pliocene and Pleistocene epochs where the record may be more detailed than at other times (including deep sea core data of ice rafting; pollen data for climatic setting of land floras, Repenning and Brou-wers, 1992). It may be most important, however, that evidence clearly supports the finding that the Bering Strait actually was *open* beginning at about 3 Ma. Marincovich and others (1990) suggest on invertebrate evidence that the Bering Strait was closed during most of the Cenozoic Era. If that was the case, we might expect to see generally continuous land mammal ex-change across Beringia throughout that time. Perhaps we do. Figure 7 shows that at least from about 65 Ma, small-scale immigrations are seen in the Cenozoic record at frequent inter-vals. Previously in the text, these have been relegated to "back-ground" status, as not showing major immigrations. Perhaps one should ask why such "background" dispersals are present, at all. Is it important that the frequency of the "background" dispersals is higher (and thus more common) than the recur-rence of supercycles that are purported to reflect major eustatic changes in sea level? If the "background" dispersals show gen-eral continuity across Beringia for much of the Cenozoic Era (with the North Atlantic route being important up through the Wasatchian), what does the general pre-Pleistocene asynchrony between major sea-level low stands and major overland disper-sals show us? If dispersal was generally possible, then why were there so relatively few times of major exchange as portrayed on Figure 7? Flynn and others (1991) comment on the role played by competition in limiting dispersals that otherwise would ap-pear to have been possible in the last 2–5 my. In that context it may be significant that most (or many) colonizers at both the "background" and major episodes are carnivores which gener-ally have larger home ranges, than herbivores (at least for the Neogene which means that some herbivores were strongly mo-bile, and likely adventurous ecologically, as well). Tectonism played a major role in the elevation of the Panamanian isthmus by about 2.7 Ma, and this may have begun as early as circa 9

Ma. What has been the role of tectonism along the margin of the northern circum-Pacific for the past 65 my? One of the most interesting aspects of the above discussions is the magnitude and range in timing of dispersal lag in taxon distribution. It is not trivial that in past times North American late Eocene faunas with modern taxa were considered to be of Oligocene age due to their general similarity to modern forms of the *Grande Coupure.*

SUMMARY

Data summarized herein allow a refined correlation of mam-mal ages in North America and permit a comparison between immigrations of land mammals recorded therein and the global sea-level history for the Cenozoic Era. Whereas dispersals are promoted by low stands of sea level at certain times in this era, the pattern apparently is mostly one of opportunism, regardless of sea level. Tectonic factors contribute to important vicariant modifications in land mammal ranges, and some major effects are seen in significant time-lags of otherwise expected dispersal events. Climate in North America apparently played only an indirect role in influencing Cenozoic land mammal dispersal. Major exceptions appear to be the preventing of northern hem-isphere (North Atlantic) dispersals in the Paleogene prior to the late Paleocene and early Eocene epochs, and at least some of the changes in sea level stimulated by episodes of global cool-ing subsequent to the onset of Antarctic glaciations from about 40 Ma. The most detailed record of climatic influence on the overland dispersal of mammals is seen during the past 3.0 my in the Arctic Borderland and adjacent areas.

ACKNOWLEDGMENTS

We thank W. A. Berggren, D. V. Kent, and J. A. Hardenbol for organizing this symposium and for inviting our participa-tion. The manuscript benefited from comments offered by L. B. Albright, J. D. Archibald, J. A. Bartow, W. A. Berggren, D. A. Osleger, R. K. Stucky, R. H. Tedford, and S. D. Webb. R. MacFee kindly made available his manuscript on Antillean ground sloths; S. D. Webb made similar courtesies on his manu-script with N. D. Opdyke. We accept full responsibility for errors.

REFERENCES

ARCHIBALD, J. D., GINGERICH, P. D., LINDSAY, E. H., CLEMENS, W. A., KRAUSE, D. W., AND ROSE, K. D., 1987, First North American land mammal ages of the Cenozoic Era, *in* Woodburne, M. O., ed., Cenozoic Mammals of North America; Geochronology and Biostratigraphy: Berkeley, University of California Press, p. 24–76.

AUBRY, M.-P., BERGGREN, W. A., KENT, D. V., FLYNN, J. J. , KLITGORD, K. D., OBRADOVICH, J.D., AND PROTHERO, D. R., 1988, Paleogene geochronology: an integrated approach: Paleoceanography v. 3, p. 707–742.

AUBRY, M.-P., BERGGREN, W. A., STOTT, L., AND SINHA, A., 1995, The upper Paleocene-lower Eocene stratigraphic record and the Paleocene/Eocene boundary carbon isotope excursion; Implications for geochronology, *in* Knox, R. O'B., Corfield, R., and Dunnay, R. E., eds., Correlations of the Early Paleogene in Northwestern Europe: London, Geological Society of London, Special Publication, in press.

BARKER, P. F. AND BURRELL, J., 1982, The influence upon Southern Ocean circulation, sedimentation, and climate of the opening of Drake Passage, *in* Craddock, C., ed., Antarctic Geoscience: Madison, University of Wisconisin Press, p. 377–385.

BARTOW, J. A, 1992, Neogene time scale for southern California, and Paleo-gene time scale for southern California: Washington, D.C., United States Geological Survey Open File Report, 92–212.

BEARD, K. C. AND TABRUM, A. R., 1991, The first early Eocene mammal from eastern North America: an omomyid primate from the Bashi Formation, Lauderdale County, Mississippi: Mississippi Geology, v. 11, p. 1–6.

BERGGREN, W. A. AND AUBRY., M.-P, 1995, A late Paleocene-early Eocene NW European and North Sea magnetobiostratigrapic correlation network: A sequence stratigraphic approach, *in* Knox, R. O'B., Corfield, R., and Dunnay, R. E., eds., Correlations of the Early Paleogene in Northwestern Europe: London, Geological Society of London, Special Publication, in press.

BERGGREN, W. A., KENT, D. V., FLYNN, J. J., AND VAN COUVERING, J. A., 1985a, Cenozoic geochronology: Geological Society of America Bulletin, v. 96, p. 1407–1418.

BERGGREN, W. A., KENT, D. V., FLYNN, J. J., AND VAN COUVERING, J. A., 1985b, Jurassic to Paleogene: Part 2. Paleogene geochronology and chronostratigraphy, *in* Snelling, N. J., ed., The Chronology of the Geological Record: London, Geological Society Memoir 10, Blackwell Scientific Publications, p. 141–195.

BERGGREN, W. A. AND PROTHERO, D. R., 1992, Eocene-Oligocene climatic and biotic evolution: an overview, *in* Prothero, D. R. and Berggren, W. A., eds., Eocene-Oligocene Climatic and Biotic Evolution: Princeton, Princeton University Press, p. 1–28.

BERGGREN, W. A. AND VAN COUVERING, J. A., 1974, The Late Neogene; biostratigraphy, geochronology, and paleoclimatology of the last 15 million years in marine and continental sequences: Palaeogeography, Palaeoclimatology and Palaeoecology, v. 16: p. 1–216.

BERNOR, R. L., TOBIEN, H., AND WOODBURNE, M. O., 1990. Patterns of Old World hipparionine evolutionary diversification and biogeographic extension, *in* Lindsay, E. H., Fahlbusch, V., and Mein, P., eds., European Neogene Mammal Chronology: New York, Plenum Press, p. 263–319.

BOELLSTORF, J. AND SKINNER, M. F., 1977, A fission-track date from post-Rosebud early Valentine Rocks: Lincoln, Proceedings Nebraska Academy of Science, 87th Annual Meeting, p. 39–40.

BRYANT, J. D, 1991, New early Barstovian (middle Miocene) vertebrates from the upper Torreya Formation, eastern Florida Panhandle: Journal of Vertebrate Paleontology, v. 11, p. 472–489.

BUTLER, R. F. AND LINDSAY, E. H., 1985, Mineralogy of magnetic minerals and revised magnetic polarity stratigraphy of continental sediments, San Juan Basin, New Mexico: Journal of Geology, v. 94, p. 535–554.

CANDE, S. C. AND KENT, D. V., 1992, A new geomagnetic polarity time-scale for the Late Cretaceous and Cenozoic: Journal of Geophysical Research, v. 97, p. 13,917–13,951.

CARROLL, R. L., 1988, Vertebrate Paleontology and Evolution: New York, W. H. Freeman, 698 p.

CASE, J. A. AND WOODBURNE, M. O., 1986, South American marsupials: a successful crossing of the Cretaceous-Tertiary boundary: Palaios, v. 1, p. 413–416.

CHRISTIE-BLICK, N., MOUNTAIN, G. S., AND MILLER, K. G., 1990, Seismic stratigraphic record of sea-level changes, *in* Revelle, R. R., panel leader, Sea Level Change, Studies in Geophysics: Washington, D. C., National Academy Press, p. 116–140.

CIFELLI, R. L. AND EATON, J. G., 1987, Marsupial mammal from earliest Late Cretaceous of western U.S: Nature, v. 325, p. 520–522.

CIONE, A. AND TONNI, E., 1995, Chronostratigraphy and "Land Mammal Ages" in the Cenozoic of southern South America: principles, practices, and the "Uquian" problem: Journal of Paleontology, v. 69, p. 135–159.

CLOETINGH, S., 1988, Intraplate stresses: a tectonic cause for third-order cycles in apparent sea level, *in* Wilgus, C. K., Hastings, B. S., Ross, C. A., Posamentier, H., Van Wagoner, J., and Kendall, C. G. St. C., eds., Sea-Level Changes: an Integrated Approach: Tulsa, Society of Economic Paleontologists and Mineralogists Special Publication 42, p. 19–30.

CLYDE, W. C., STAMATAKOS, J., AND GINGERICH, P. D., 1994, Chronology of the Wasatchian Land-Mammal Age (early Eocene): Magnetostratigraphic results from the McCullough Peaks Section, northern Bighorn Basin, Wyoming: Journal of Geology, v. 102, p. 367–377.

DEBRUIJN, H., DAAMS, R., DAXNER-HOCK, G., FAHLBUSCH, V., GINSBURG, L., MEIN, P., AND MORALES, J., 1992, Report of the RCMNS working group on fossil mammals, Reisensburg 1990: Newsletters in Stratigraphy, v. 26, p. 65–118.

DICKINSON, W. R., 1993, Comment. Exxon global cycle chart: an event for every occasion? Comments and Reply: Geology, v. 21, p. 282–283.

EMSLIE, S. D. AND MORGAN, G. S., 1994, A catastrophic death assemblage and paleoclimatic implications of Pliocene seabirds of Florida: Science, v. 264, p. 684–685.

EMRY, R. J., 1992, Mammalian range zones in the Chadronian White River Formation at Flagstaff Rim, Wyoming, *in* Prothero, D. R. and Berggren,

W. A., eds., Eocene- Oligocene Climatic and Biotic Evolution: Princeton, Princeton University Press, p. 106–115.

EMRY, R. J., RUSSELL, L. S., AND BJORK, P. R., 1987, The Chadronian, Orellan, and Whitneyan North American land mammal ages, *in* Woodburne, M. O., ed., Cenozoic Mammals of North America: Geochronology and Biostratigraphy: Berkeley, University of California Press, p. 118–152.

ERICKSON, J. P. AND PINDELL, J. L., 1993, Analysis of subsidence in northeastern Venezuela as a discriminator of tectonic models for northern South America: Geology, v. 21, p. 945–948.

EVERNDEN, J. F., SAVAGE, D. E., CURTIS, G. H., AND JAMES, G. T., 1964, Potassium-argon dates and the Cenozoic mammalian chronology of North America: American Journal of Science, v. 62, p. 145–198.

FAHLBUSCH, V., 1991, The meaning of MN-zonation: considerations for a subdivision of the European continental Tertiary using mammals: Newsletters in Stratigraphy, v. 23, p. 159–173.

FEJFAR, O. AND REPENNING, C. A., 1992, Holarctic dispersal of arvicolids (Rodentia, Cricetidae): Courier Forscher Institute Senckenberg, v. 153, p. 205–212.

FISHER, R .V. AND RENSBERGER, J. M., 1972, Physical stratigraphy of the John Day Formation, central Oregon: University of California Publications in Geological Sciences, v. 101, p. 1–95.

FLYNN, J. J., 1986, Correlation and geochronology of middle Eocene strata from the western United States: Palaeogeography, Palaeoclimatology and Palaeoecology, v. 55, p. 335–406.

FLYNN, L. J., TEDFORD, R. H., AND ZHANXIANG, Q., 1991, Enrichment and stability in the Pliocene mammalian fauna of north China: Paleobiology, v. 17, p. 246–265.

FORSTÉN, A. M., 1968, Revision of Palaearctic Hipparion: Acta Zoologica Fennica, v. 119, p. 1–134.

GALUSHA, T., 1975, Stratigraphy of the Box Butte Formation, Nebraska: Bulletin American Museum of Natural History, v. 156, p. 1–68.

GINGERICH, P. D., 1989, New earliest Wasatchian mammalian fauna from the Eocene of northwestern Wyoming: composition and diversity in a rarely sampled high-floodplian assemblage: University of Michigan Papers on Paleontology, v. 28, p. 1–97.

GINSBURG, L., 1989, The faunas and stratigraphical subdivisions of the Orleanean in the Loire Basin (France), *in* Lindsay, E. H., Fahlbusch, V., and Mein, P., eds., European Neogene Mammal Chronology: New York, Plenum Press, NATO ASI Series, p. 157–176.

GREENE, R. C., 1973, Petrology of the Welded Tuff of Devine Canyon, southeastern Oregon: Washington D.C., United States Geological Survey Professional Paper 797, 26 p.

GUNNELL, G. F., BARTELS, W. S., AND GINGERICH, P. D., 1993, Paleocene-Eocene boundary in continental North America: Biostratigraphy and geochronology, northern Bighorn Basin, Wyoming, *in* Lucas, S. G., and Zidek, J., eds., Vertebrate Paleontology in New Mexico: New Mexico Museum of Natural History and Science, Bulletin 2, p. 137–144.

HALLAM, A., 1992, Phanerozoic Sea-Level Changes: New York, Columbia University Press, 266 p.

HAQ, B. U., HARDENBOL, J., AND VAIL, P. R., 1988, Mesozoic and Cenozoic chronostratigraphy and cycles of sea-level change, *in* Wilgus, C. K., Hastings, B. S., Ross, C. A., Posamentier, H., Van Wagoner, J., and Kendall, C. G. St. C., eds., Sea-level Changes: An Integrated Approach: Tulsa, Society of Economic Paleontologists and Mineralogists Special Publication 42, p. 71–108.

HEISSIG, K., 1979, Die hypothetische Rolle Sudosteuropas bei den Saugetierwanderungen im Eozän und Oligozän: Neues Jahrbuch Geologie u. Palaeontologie Monatshefte 1979, p. 83–96.

HIRSCHFELD, S. E., 1981, Pliometanastes protistus (Edentata, Megalonychidae) from Knight's Ferry, California, with discussion of early Hemphillian megalonychids: PaleoBios, v. 36, 16 p.

HOOKER, J. J., 1992, British mammalian paleocommunities across the Eocene-Oligocene transition and their environmental implications, *in* Prothero, D. R. and Berggren,W. A., eds., Eocene-Oligocene Climatic and Biotic Evolution: Princeton, Princeton University Press, p. 494–515.

HULBERT, R. C., JR. AND MACFADDEN, B. J., 1991, Morphological transformation and cladogenesis at the base of the adaptive radiation of Miocene hypsodont horses: American Museum Natural History Novitates, v. 3000, 61 p.

JAILLARD, E., CAPPETTA, H., ELLENBERGER, P., FEIST, M., GRAMBAST-FESSARD, N., LEFRANC, J.-P., AND SIGÉ, B., 1993, Sedimentology, biostratigraphy, and correlation of the Late Cretaceous Vilquechico Group of southern Peru: Cretaceous Research, v. 14, p. 623–661.

KEIGWIN, L., 1982, Isotopic paleoceanography of the Caribbean and east Pacific: role of Panama uplift in late Neogene time: Science v. 217, p. 350–353.

KENNET, J. P. AND BARKER, P. F., 1990, Latest Cretaceous to Cenozoic climate and oceanographic developments in the Weddell Sea, Antarctica: an ocean drilling perspective: Proceedings of the Oceanic Drilling Project, v. 113, p. 121–151.

KRAUSE, D. W. AND MAAS, M .C., 1990, The biogeographic origins of late Paleocene-early Eocene mammalian immigrants to the Western Interior of North America, in Bown, T. M. and Rose, K. D., eds., Dawn of the Age of Mammals in the Northern Part of the Rocky Mountain Interior, North America: Boulder, Geological Society of America Special Paper 243, p. 71–105.

KRISHTALKA, L., WEST, R. M., BLACK, C. C., DAWSON, M. R., FLYNN, J. J., TURNBULL, W. D., STUCKY, R. K., MCKENNA, M. C., BOWN, T. M., GOLZ, D. J., AND LILLEGRAVEN, J. A., 1987, Eocene (Wasatchian through Duchesnean) biochronology of North America, in Woodburne, M. O., ed., Cenozoic Mammals of North America: Geochronology and Biostratigraphy: Berkeley, University of California Press, p. 77–117.

LEGENDRE, S., CROCHET, J.-Y., GODINOT, M., HARTENBARGER, J.-L., MARANDAT, B., REMY, J. A., SIGÉ, B., SUDRE, J., AND VAINEY-LIAUD, M., 1991, Évolution de la diversité des faunes de mammifères d'Europe occidentale au Paléogène (MP11 à MP 30): Bulletin Societe géologique d'France, v. 162, p. 867–874.

LILLEGRAVEN, J. A., KIELAN-JAWOROWSKA, Z., AND CLEMENS, W. A., 1979, Mesozoic Mammals. The first two-thirds of mammalian history: Berkeley, University of California Press, 311 p.

LINDSAY, E. H., 1991, Identification of land mammal age boundaries: Journal of Vertebrate Paleontology, v. 11, Supplement to no. 3, p. 43A.

LINDSAY, E. H., OPDYKE, N. D., AND JOHNSON, N. M., 1980, Pliocene dispersal of the horse Equus and late Cenozoic mammalian dispersal events: Nature, v. 287, p. 135–138.

LINDSAY, E. H., OPDYKE, N. D., AND JOHNSON, N. M., 1984, Blancan-Hemphillian land mammal ages and late Cenozoic mammal dispersal events: Annual Review Earth and Planetary Science, v. 12, p. 445–488.

LUCAS, S. G., 1989, Fossil mammals and Paleocene-Eocene boundary in Europe, North America and Asia: 28th International Geological Congress, Abstracts, vol. 2, p. 335.

LUCAS, S. G., 1992, Redefinition of the Duchesnean Land Mammal "Age," Late Eocene of Western North America, in Prothero, D. R. and Berggren, W. A., eds., Eocene-Oligocene Climatic and Biotic Evolution: Princeton, Princeton University Press, p. 88–105.

LUNDELIUS, E. L., JR., DOWNS, T., LINDSAY, E. H., SEMKEN, H. A., ZAKREZEWSKI, R. J., CHURCHER, C. S., HARRINGTON, C. R., SCHULTZ, G. E., AND WEBB, S. D., 1987, The North American Quaternary Sequence, in Woodburne, M. O., ed., Cenozoic Mammals of North America: Geochronology and Biostratigraphy: Berkeley, University of California Press, p. 211–235.

MACFADDEN, B., BRYANT, J. D., AND MUELLER, P. A., 1991, Sr-isotopic, paleomagnetic, and biostratigraphic calibration of horse evolution: evidence from the Miocene of Florida: Geology, v. 19, p. 242–245.

MACFADDEN, B., ANAYA, F., AND ARGOLLO, J., 1993, Magnetic polarity stratigraphy at Inchasi: a Pliocene mammal-bearing locality from the Bolivian Andes deposited just before the Great American Interchange: Earth and Planetary Science Letters, v. 114, p. 229–241.

MACFADDEN, B. J., SWISHER, C. C., III, OPDYKE, N. D., AND WOODBURNE, M. O., 1990, Paleomagnetism, geochronology, and possible tectonic rotation of the middle Miocene Barstow Formation, Mojave Desert, California: Geological Society of America Bulletin, v. 10, p. 478–493.

MACFADDEN, B.J., WOODBURNE, M.O., AND OPDYKE, N.D., 1990, Paleomagnetism and Neogene clockwise rotation of the Northern Cady Mountains, Mojave Desert of Southern California: Journal of Geophysical Research, v. 95, p. 4597–4608.

MACFEE, R. D. E., AND ITURRALIDE-VINENT, M., 1994, First Tertiary land mammal from Greater Antilles: an early Miocene sloth (Xenarthra, Megalonychidae) from Cuba: American Museum of Natural History Novitates, v. 3094, p. 1–13.

MARINCOVICH, L., JR., BROUWERS, E. M., HOPKINS, D. M., AND MCKENNA, M. C., 1990, Late Mesozoic and Cenozoic paleogeographic history of the Arctic Ocean Basin, based on shallow-water marine faunas and terrestrial vertebrates, in Grantz, A., Johnson, L., and Sweeney, J. F., eds., The Arctic Ocean region: Boulder, Geological Society of America. The Geology of North America, v. L, p. 403–426.

MARSHALL, L. G., 1985, Geochronology and land-mammal biochronology of the Transamerican faunal interchange, in Stehli, F. G. and Webb, S. D., eds.,

The Great American Biotic Interchange: New York, Plenum Press, p. 49–85.

MARSHALL, L. G., BERTA, A., HOFFSTETTER, R., PASCUAL, R., ODREMAN-RIVAS, O.A., BOMBIN, M., AND MONES, A., 1984, Mammals and stratigraphy: geochronology of the continental mammal-bearing Quaternary of South America: Palaeovertebrata, Memoir Extraordinaire, p. 1–76.

MARSHALL, L. G., BUTLER, R. F., DRAKE, R. E., CURTIS, G. H., AND TEDFORD, R. H., 1979., Calibration of the Great American interchange: Science, v. 204, p. 272–279.

MARSHALL, L. G., AND CIFELLI, R. L., 1990, Analysis of changing diversity patterns in Cenozic land mammal age faunas, South America: Palaeovertebrata, v. 19, p. 169–210.

MARSHALL, L . G., HOFFSTETTER, R., AND PASCUAL, R., 1983, Mammals and stratigraphy: geochronology of the continental mammal-bearing Tertiary of South America: Palaeovertebrata, Memoir Extraordinaire, 93 p.

MARSHALL, L G., SWISHER, C.C., III, LAVENU, A., HOFFSTETTER, R., AND CURTIS, G. H., 1992, Geochronology of the mammal-bearing late Cenozoic on the northern altiplano, Bolivia: Journal of South American Earth Sciences, v. 5, p. 1–19.

MASON, M. A. AND SWISHER, C. C., III, 1989, New evidence for the age of the South Mountain Local Fauna, Ventura County, California: Los Angeles Museum Contributions in Science, v. 410, p. 1–9.

MAY, S. R., 1981, Geology and mammalian paleontology of the Horned Toad Hills, Mojave Desert, California: Unpublished M.S. Thesis, University of California, Riverside, 290 p.

MCKENNA, M. C. 1975. Fossil mammals and early Eocene North Atlantic land continuity: Annals of the Missouri Botanical Garden, v. 62, p. 335–353.

MCKENNA, M. C, 1981, Early history and biogeography of South America's extinct land mammals, in Ciochon, R. L. and Chiarelli, A. B., eds., Evolutionary Biology of the New World Monkeys and Continental Drift: New York, Plenum Press, p. 43–77.

MCKENNA, M. C., 1983a, Cenozoic Palaeogeography of North Atlantic Land Bridges, in Bott, M. H. P., Saxov, A., Talwani, M., and Thiede, J., eds., Structure and Development of the Greenland-Scotland Ridge: New York, Plenum Press, p. 351–399

MCKENNA, M. C., 1983b, Holarctic landmass rearrangement, cosmic events, and Cenozoic terrestrial organisms: Annals of the Missouri Botanical Garden, v. 70, p. 459–489.

MEIN, P., 1989, Updating of MN Zones, in Lindsay, E. H., Fahlbusch, V., and Mein, P., eds., European Neogene Mammal Chronology: New York, Plenum Press, NATO ASI Series, p. 73–90,

MIALL, A. D., 1992, Exxon global cycle chart: an event for every occasion?: Geology, v. 20, p. 787–790.

MILLER, S. T., 1980, Geology and mammalian biostratigraphy of a part of the northern Cady Mountans, California: Washington, D.C., United States Geological Survey Open-File Report 80–978, p. 1–121.

MOREA, M. F., 1981, The Massacre Lake local fauna (Mammalia, Hemingfordian) from north-western Washoe County, Nevada. Unpublished Ph.D. Dissertation, University of California, Riverside, 247 p.

MUIZON, C., DE, 1991, Le faune de mamiferos de Tiupampa (Paleoceno inferior, Formacion Santa Lucia), Bolivia, in Suarez-Coruco, R., ed., Fosiles y Facies de Bolivia, Vol. 1 Vertebrados: Revista Technica de YPFB, v. 12, p. 525–624.

MUIZON, C. AND MARSHALL, L. G., 1991, Nouveaux condylarthres du Paléocène inferieur de Tiupampa (Bolivie): Paris, Bulletin Muséum national d'Histoire naturelle, 4th ser., v. 13, p. 201–227.

NASER, C. W., IZETT, G. A., AND OBRADOVICH, J. D., 1980, Fission-track and K-Ar ages of natural glasses: United States Geological Survey Bulletin 1489, 31 p.

NOVACEK, M., FERRISQUIA-VILLAFRANCA, I., FLYNN, J. J, WYSS, A. R., AND NORELL, M., 1991, Wasatchian (early Eocene) mammals and other vertebrates from Baja California, Mexico: the Lomas Las Tetas de Cabra Fauna: American Museum of Natural History Bulletin 208, 88 p.

OLSSON, R. K. AND LIU, C., 1993, Controversies on the placement of Cretaceous-Paleogene boundary and the K/P mass extinction of planktonic foraminifera: Palaios, v. 8, p. 127–139.

OPDYKE, N. D., 1990, Magnetic stratigraphy of Cenozoic terrestrial sediments and mammalian dispersal: Journal of Geology, v. 98, p. 621–637.

PRENTICE, M. L. AND MATTHEWS, R. K., 1988, Cenozoic ice-volume history: development of a composite oxygen isotope record: Geology, v. 16, p. 963–966.

PROTHERO, D. R., 1991, Magnetic stratigraphy of Eocene-Oligocene mammal localities in southern San Diego County, in Abbott, P. L. and May, J. A., eds., Eocene Geologic History San Diego Region: Pacific Section, Society of Economic Petrologists and Mineralogists, v. 68, p. 125–130.

PROTHERO, D. R., DOZIER, D. R., AND HOWARD, J., 1992, Magnetostratigraphy of the Uintan-Arikareean (middle Eocene-late Oligocene) Sespe Formation, Ventura County, California: Journal of Vertebrate Paleontology, v. 12, Supplement to no. 3, p. 48A.

PROTHERO, D. R. AND RENSBERGER, J. M., 1985, Preliminary magnetostratigraphy of the John Day Formation, Oregon, and the North American Oligocene-Miocene boundary: Newsletters in Stratigraphy, v. 15, p. 59–70.

PROTHERO, D. R. AND SWISHER, C. C., III, 1992, Magnetostratigraphy and geochronology of the terrestrial Eocene-Oligocene transition in North America, *in* Prothero, D. R. and Berggren, W. A., eds., Eocene-Oligocene Climatic and Biotic Evolution: Princeton, Princeton University Press, p. 46–73.

PROTHERO, D. R., TIERNEY, I., SWISHER, C. C., III, AND SWINEHART, J. B., 1991, Magnetostratigraphy and geochronology of the late Oligocene Gering Formation, Nebraska: implications for the Arikareean: Journal of Vertebrate Paleontology, v. 11, supplement to no. 3, p. 72A.

PROTHERO, D. R. AND WILSON, E. L., 1993, Magnetostratigraphy of the Tejon Hills, southern San Joaquin valley, California, and implications for Clarendonian correlations: Journal of Vertebrate Paleontology, v.13, supplement to No. 3, p. 53A.

QIU, Z., 1989, The Chinese Neogene mammalian biochronology— its correlation with the European Neogene mammalian zonation, *in* Lindsay, E. H., Fahlbusch, V., and Mein, P., eds., European Neogene Mammal Chronology: New York, Plenum Press, NATO ASI Series, p. 527–556.

RENSBERGER, J. M., 1983, Successions of Meniscomyine and Allomyine rodents (Aplodontidae) in the Oligo-Miocene John Day Formation, Oregon: University of California Publications in Geological Sciences, v. 124, p. 1–157.

REPENNING, C. A., 1987, Biochronology of the microtine rodents of the United States, *in* Woodburne, M. O., ed., Cenozoic Mammals of North America: Geochronology and Biostratigraphy: Berkeley, University of California Press, p. 236–238.

REPENNING, C. A., 1992, Allophaiomys and the age of the Olyor Suite, Krestovka Sections, Yakutia: United States Geological Survey Bulletin 2037, 98 p.

REPENNING, C. A. AND BROUWERS, E. M., 1992, Late Pliocene-early Pleistocene ecologic changes in the Arctic Ocean Borderland: United States Geological Survey Bulletin 2036, 37 p.

REPENNING, C. A., FEJFAR, O., AND HEINRICH, W.-D., 1990, Arvicolid rodent biochronology of the Northern Hemisphere, *in* Fejfar, O. and Heinrich, W.-D., eds., International Symposium on Evolution, Phylogeny and Biostratigraphy of Arvicolids (Rodentia, Mammalia): Prague, Geological Survey, p. 385–418.

REPENNING, C. A. AND RAY, C. E., 1977, The origin of the Hawaiian monk seal: Biological Society of Washington Proceedings, v. 89, p. 667–688.

ROSE, K. D., 1980, Clarkforkian land-mammal age: revised definition, zonation and tentative intercontinental correlation: Science, v. 208, p. 744–746.

ROSE, K. D., 1981, The Clarkforkian land-mammal age and mammalian fauna composition across the Paleocene-Eocene boundary: University of Michigan Papers on Paleontology, v. 26, p. 1–197.

RUSSELL, D. E., HARTENBARGER, J.-L., POMEROL, C., SEN, S., SCHMIDT-KITTLER, N., AND VAINEY-LIAUD, M., 1982, Mammals and stratigraphy: the Paleogene of Europe: Palaeovertebrata, Mémoire Extraordinaire, 77 p.

RUSSELL, D. E. AND ZHAI, REN-JIE, 1987, The Paleogene of Asia: mammals and stratigraphy: Mémoirs du Muséum national d'Histoire naturelle, C, v. 12, 488 p.

SAVAGE, D. E., 1977, Aspects of vertebrate paleontological stratigraphy and geochronology, *in* Kauffman, E. G. and Hazel, J. E., eds., Concepts and Methods of Biostratigraphy: Stroudsburg, Dowden, Hutchison, and Ross, p. 427–442.

SAVAGE, D. E. AND RUSSELL, D. E., 1983, Mammalian Paleofaunas of the World: New York, Addison-Wesley Publishing Company, Inc., 432 p.

SCHULTZ, C. B., MARTIN, L. D., TANNER, L. G., AND CORNER, R. G., 1978, Provincial land mammal ages for the North American Quaternary: Transactions Nebraska Academy of Sciences, v. 5., p. 59–64.

SIMPSON, G. G., 1953, Evolution and Geography: Condon Lectures, Oregon State System of Higher Education, 35 p.

STEHLIN, H. G., 1909., Remarques sur le faunules de mammifères des couches éocènes et oligocènes de Bassin de Paris: Bulletin Societe géologique de France, v. 9, p. 488–520.

STEININGER, F., BERNOR, R. L., AND FAHLBUSCH, V., 1989, European Neogene marine/continental chronologic correlations, *in* Lindsay, E. H., Fahlbusch, V., and Mein, P., eds., European Neogene Mammal Chronology: New York, Plenum Press, NATO ASI Series, p. 15–46

STUCKY, R. K., 1992., Mammalian faunas in North America of Bridgerian to early Arikareean "ages" (Eocene and Oligocene), *in* Prothero, D. R. and Berggren, W. A., eds., Eocene-Oligocene Climatic and Biotic Evolution: Princeton, Princeton University Press, p. 464–493

SWISHER, C. C., III, 1996, 40Ar/39Ar dating of circum-Mediterranean hipparionine sites and a revised calibration of Vallesian and Turolian MN Zones, *in* Bernor, R. L., Fahlbusch, V., and Mittmann, H.-W. S., eds., The Evolution of Western Eurasian Later Neogene Mammal Faunas: New York, Columbia University Press, in press.

SWISHER, C. C., III, DINGUS, L., AND BUTLER, R. F., 1993, 40Ar/39Ar dating and magnetostratigraphic correlation of the terrestrial Cretaceous-Paleogene boundary and Puercan mammal age, Hell Creek-Tullock formations, eastern Montana: Canadian Journal of Earth Sciences, v. 30, p. 1981–1996.

SWISHER, C. C., III, and KNOX, R. W. O'B., 1991, The age of the Paleocene/ Eocene boundary: ^{40}Ar/^{39}Ar dating of the lower part of NP10, North Sea Basin and Denmark: Brussels, International Geological Correlation Project 308; Paleocene/Eocene Boundary, p. 16.

SWISHER, C. C., III, AND PROTHERO, D. R., 1990, Single-crystal ^{40}Ar/^{39}Ar dating of the Eocene-Oligocene transition in North America: Science, v. 249, p. 760–762.

SWISHER, C .C., III, GRAJALES-NISHIMURA, J. M., MONTANARI, A., MARGOLIS, S. V., CLAEYS, P., ALVAREZ, W., RENNE, P., CEDILLO-PARDO, E., MAURRASSE, F. J.-M., CURTIS, G. H., SMIT, J., AND McWILLIAMS, M. O., 1992, Coeval ^{40}Ar/^{39}Ar ages of 65.0 million years ago from Chicxulub crater melt rock and Cretaceous-Tertiary boundary tectites: Science, v. 257, p. 954–958.

TEDFORD, R. H., SKINNER, M. F., FIELDS, R. W., RENSBERGER, J. M., WHISTLER, D. P., GALUSHA, T., TAYLOR, B. E., MacDONALD, J. R., AND WEBB, S. D., 1987, Faunal succession and biochronology of the Arikareean through Hemphillian interval (late Oligocene through earliest Pliocene epochs) in North America, *in* Woodburne, M. O., ed., Cenozoic Mammals of North America: Geochronology and Biostratigraphy: Berkeley, University of California Press, p. 153–210.

VAN VALEN, L., 1988, Paleocene dinosaurs or Cretaceous ungulates in South America: Evolutionary Monographs, v. 10, 79 p.

VEZIER, J., 1989, Strontium isotopes in seawater through time: Annual Review Earth and Planetary Sciences, v. 7, p. 141–167.

WALTON, A. H., 1992., Magnetostratigraphy of the lower and middle members of the Devil's Graveyard Formation (middle Eocene), Trans-Pecos, Texas, *in* Prothero, D. R. and Berggren, W. A., eds., Eocene-Oligocene Climatic and Biotic Evolution: Princeton, Princeton University Press, p. 74–87.

WANG, B., 1992, The Chinese Oligocene: a preliminary review of mammalian localities and local faunas, *in* Prothero, D. R. and Berggren, W. A., eds., Eocene-Oligocene Climatic and Biotic Evolution: Princeton, Princeton University Press, p. 529–547.

WEBB, S. D., 1977., A history of savanna vertebrates in the New World. Pt. I: North America: Annual Reviews of Ecological Systematics, v. 8, p. 355–380.

WEBB, S. D., 1985a, Main pathways of mammalian diversification in North America, *in* Stehli, F. G. and Webb, S. D., eds., The Great American Biotic Interchange: New York, Plenum Press, p. 201–218.

WEBB, S. D., 1985b, Late Cenozoic mammal dispersals between the Americas, *in* Stehli, F. G. and Webb, S. D., eds., The Great American Biotic Interchange: New York, Plenum Press, p. 357–38.

WEBB, S. D. AND BARNOSKY, A., 1989, Faunal dynamics of Pleistocene mammals: Annual Review of Earth and Planetary Science, v. 17, p. 413–438.

WEBB, S. D. AND OPDYKE, N. D., 1995, Global Climatic Influence on Cenozoic land mammal faunas, *in* Kennett, J. and Stanley, S., eds., Effects of Past Global Change on Life: National Academy of Sciences, Studies in Geophysics, p. 184–208.

WEBB, S. D. AND TAYLOR, B. E., 1980, The phylogeny of hornless ruminants and a description of the cranium of Archaeomeryx: American Museum of Natural History Bulletin 167, p. 117–158.

WHISTLER, D. P. AND BURBANK, D. W., 1992, Miocene biostratigraphy and biochronology of the Dove Spring Formation, Mojave Desert, California, and characterization of the Clarendonian mammal age (late Miocene) in California: Geological Society of America Bulletin, v. 104, p. 644–658.

WILLARD, D. A., CRONIN, T. M., ISHMAN, S. E., AND LITWIN, R. J., 1993, Terrestrial and marine records of climatic and environmental changes during the Pliocene in subtropical Florida: Geology, v. 21, p. 679–682.

WILLIAMS, D. F., 1988, Evidence for and against sea-level changes from the stable isotope record of the Cenozoic, *in* Wilgus, C.K., Hastings, B.S., Ross, C.A., Posamentier, H., Van Wagoner, J., and Kendall, C.G. St. C., eds., Sea-Level Changes: an Integrated Approach: Tulsa, Society of Economic Paleontologists and Mineralogists Special Publication 42, p. 32–36.

WING, S. L., 1984, A new basis for recognizing the Paleocene/Eocene boundary in the western Interior of North America: Science, v. 226, p. 439–441.

WING, S. L., BOWN, T. M., AND OBRADOVICH, J. D., 1991, Early Eocene biotic and climatic change in western North America: Geology, v. 19, p. 1189–1192.

WOODBURNE, M. O., 1977, Definition and characterization in mammalian chronostratigraphy: Journal of Paleontology, v. 51, p. 220–234.

WOODBURNE, M. O., 1987, Cenozoic Mammals of North America: Geochronology and Biostratigraphy: Berkeley, University of California Press, 336 p.

WOODBURNE, M. O., 1991, The Mojave Desert Province, *in* Woodburne, M. O., Reynolds, R. E., and Whistler, D. P., eds., Inland southern California; the last 70 million years: San Bernardino County Museum Association Quarterly, v. 38, p. 60–77.

WOODBURNE, M. O. AND BERNOR, R. L., 1980, On superspecific groups of some Old World hipparionine horses: Journal of Paleontology, v. 54, p. 1319–1348.

WOODBURNE, M. O., MACFADDEN, B. J., AND SKINNER, M. F., 1981, The North American "hipparion" datum and implications for the Neogene of the Old World: Geobios, v. 14, p. 493–524.

WOODBURNE, M. O., TEDFORD, R. H., AND SWISHER, C. C., III, 1990, Lithostratigraphy, biostratigraphy, and geochronology of the Barstow Formation, Mojave Desert, southern California: Geological Society of America Bulletin, v. 102, p. 457–477.

AGES OF KEY FOSSIL ASSEMBLAGES IN THE LATE NEOGENE TERRESTRIAL RECORD OF NORTHERN CHINA

LAWRENCE J. FLYNN

Peabody Museum, Harvard University, Cambridge, MA 02138

ZHANXIANG QIU

Institute of Vertebrate Paleontology and Paleoanthropology, Academia Sinica, P.O. Box 643, Beijing, People's Republic of CHINA

NEIL D. OPDYKE

Department of Geology, University of Florida, Gainesville, FL 32611

AND

RICHARD H. TEDFORD

Department of Vertebrate Paleontology, The American Museum of Natural History, Central Park West at 79th Street, New York, NY 10024

ABSTRACT: We discuss how the chronology of terrestrial sediments can be refined when sound systematics of the contained fossils is combined with paleomagnetic data. In the absence of other means of dating Neogene deposits of China, strata can be ordered in time on the basis of their contained fossils. Biochronology alone attains good but unquantified accuracy in a biogeographically restricted system, such as China; dating the biochronology requires relationship in time to other systems of known age. This can be achieved by building a long composite of successive assemblages, faunal correlation at times of low endemism, identification of tie points in the chronology by using index fossils, and age estimation through paleomagnetic correlation. Late Neogene examples from China illustrate how that sequence of ordered faunas can be placed in a time scale through faunal comparison with the Yushe Basin (Shanxi Province) magnetostratigraphic sequence, and how the increased information content can be applied to new problems in geochronology. New data from Yushe Basin identify an Ertemte-like assemblage at Jiayucun in reversely magnetized rock. This is correlated with early Gilbert Chron. The Jingle and Youhe reference faunas are correlated with Gauss Chron Yushe assemblages. Nihewan-like assemblages are correlated with the Matuyama Chron, some being pre-Olduvai Subchron, and therefore Pliocene in age.

INTRODUCTION

Paleontology offers a classical approach to geochronology; a persistent problem in paleontology is precise dating of terrestrial sequences. It can be a difficult problem when local assemblages share few (sometimes no) fossil elements with reference faunas from other faunal regions. This is typically the case when the only comparison for the sequence in question is a set of assemblages from another biogeographic province. China offers an excellent example. Terrestrial deposits are widespread, assemblages span long intervals yielding a fairly dense fossil record but with high endemism, and few instances of independent dating are now available.

Dating independent of the terrestrial vertebrates, be it radioisotopic, paleomagnetic, or by superpositional constraints from marine deposits, is an essential check on the local chronology, and given special conditions, can offer a scale of precision that the vertebrates cannot. In China, marine deposits are few and their stratigraphic relationships are generally irrelevant to dating terrestrial deposits, especially Neogene basins. Radioisotopic dates are rare, exceptions being the early Miocene localities Shanwang and Sihong (Xiacaowan). Paleomagnetic characteristics of fossiliferous rocks are useful only when sufficiently well dated through biochronology. Magnetic data, then, offer one line of independent dating, but without fossils they are ambiguous. The most fruitful approach to dating terrestrial deposits in northern Asia is by a marriage of biostratigraphic and paleomagnetic data.

CHRONOLOGY AND THE FOSSIL RECORD

Using the fossil record to derive a chronology requires a set of assumptions that should be understood. First, evolution is seen as a process operating in time, and thus evolution interpreted through the fossil record has some relevance to inferring passage of time; it applies to relative dating of strata. This is the concept of stage-in-evolution as a chronological tool. Of course, the phylogeny of any lineage used in dating should be sufficiently understood to provide a useful level of inference to be drawn from the fossils. Generally, single lineages offer some insight into relative dating, but not absolute chronology; strata containing a derived taxon are likely (not guaranteed) to be younger than strata containing its primitive relative. Such conclusions should be framed as probability statements; they assume that differences reflect time (not ecology), that migration was ("geologically") rapid, and that primitive relatives did not survive as contemporaries with derived taxa. Depending on the taxon in question and the inference made, there is always some chance that relative dating based on a single taxon is wrong.

Relative dating based on fossils has increased probability of accuracy when multiple taxa are employed. Norell and Novacek (1992) have noted the generally good correlation of fossil occurrences and cladistic estimates of first appearances, given good superpositional and morphologic data. Dating based on whole assemblages has been practiced in Europe and North America for a century. In North America a system of biochrons (Land Mammal Ages) was established and used with success in the internal system of the North American bioprovince without accurate knowledge of their limits in time. For China, paleontologists have seriated Neogene faunas and begun to recognize biochrons based on them (Chiu and others, 1979; Li and others, 1984; Qiu, 1990; Qiu and Qiu, 1995). Problems persist in dating the sequence.

Despite an absence of independent dates, the sequence can be placed approximately in a chronological time frame by means of faunal data. This approach has limited success in the case of China since faunal endemism was rather high in eastern Asia during much of the Cenozoic Era. Dating Miocene Asian assemblages is usually done by reference to Europe; late Neogene faunas can be related to those of North America in some cases. Correlations based on poorly defined taxa are inaccurate; shared multiple taxa refined at the species level offer stronger correlations. Fortunately for continents such as Asia, endemism has not remained constant through time. Intervals of low endemism, as in early middle Miocene time, can be closely compared to the European record and these constrain correlations for intervals of high endemism. A long interval of high endemism can be subdivided for correlation within the province, in this case North China, based on changes in faunas and relative age from "stage-in-evolution" of lineages.

Appearance of a widespread exotic taxon, such as proboscideans or horses, offers a distinctive datum for correlation. Such index fossils are easily recognized and common, and appear over a wide region, even more than one continent, over a relatively short interval. Although exotics constrain correlations, demands for improved resolution in dating render as less dependable arguments based on "stage-in-evolution" and the assumption of simultaneous appearance of immigrants. The Late Miocene "*Hipparion* datum" is useful, but remains imprecise at the scale of 10^5 years, and the hypothesis of rapid dispersal throughout the Old World remains untested at that time scale. The Early Miocene "proboscidean datum" is even less well constrained.

China has a wealth of long stratigraphic sections, many of which are under study and a few of which are being sampled for magnetic stratigraphy. In combination with biochronology, magnetic reversal stratigraphy has the potential to resolve dating within a biogeographic province on the scale of 10^5 yr (see Flynn and others, 1990, for discussion of magnetochronologic resolution of the Siwalik Group of the Indian subcontinent).

Generally, vertebrate faunas are cohesive over large land masses, and through time. Similarity of assemblages across some continents is high over short time spans. Taxa or lineages and, retrospectively, communities can be characterized for large areas, up to continent size, that are called biogeographic provinces. In provinces that were isolated for a long period of time, the land mammals within them evolved independently. Such settings were "theaters of evolution" (Simpson, 1953). Rather detailed biostratigraphies and, by inference, biochronologies can be constructed for theaters of evolution, but the geochronology of these remains imprecise without independent dating. For the island continent of South America, its detailed biochronology floated in time until paleomagnetic and radioisotopic data became available to constrain correlation tie points (see MacFadden, 1990). Despite a persistent dearth of radioisotopic data, research in China has progressed rapidly; now long magnetic sequences are needed to constrain chronologies. The following review of work on late Neogene deposits explores: (1) the dating of the long sequence of Chinese biochrons in terms of the global magnetic time scale and (2) the contribution of refined correlations, based on paleomagnetic data, to the internal biochronological framework of the northern Chinese biogeographic province.

CHINESE LATE NEOGENE FAUNAS

Qiu (1990) summarizes the biostratigraphic work undertaken in China this century and presents the rationale behind the development of local Neogene biochrons. At least for later Neogene history, it is clear that North China and South China represent different biotic provinces (Chiu and others, 1979) and that the province including North China extends through Mongolia and southern Siberia to eastern Kazakhstan (Tedford and

others, 1991). North China contains a rich Miocene record of vertebrate evolution (see Fig. 1), but the chronology of Miocene faunas remains imprecise. Qiu (1990) chronicles how the Chinese late Miocene "*Hipparion* faunas" were unfortunately linked to assemblages correlated with the Ponto-Caspian Pontian Stage and how they came to be considered Pliocene age. By the 1980's, refinement of European (and American) correlations and linkage of the "*Hipparion* red clays" with the nonmarine Pikermian Stage of the Mediterranean relegated the Chinese units to the late Miocene. There are many localities in China recording high faunal diversity and long term faunal stability in the late Miocene. These are typified by a set of local faunas around Baode in northwestern Shanxi Province, which characterize the Baodean age and represent most of late Miocene time.

Knowledge of the history of terrestrial faunas in northeastern Asia was incomplete well into the second half of this century. Other than the *Hipparion* faunas and Pleistocene assemblages, the roots of the modern fauna seemed enigmatic. Work by Teilhard and Piveteau (1930) in northern Hebei Province had suggested that the Nihewan fauna likely was intermediate in age, but this could not be resolved for lack of faunas from adjacent rock units. To the French Jesuits Licent and Trassaert (1935), Yushe Basin in southeastern Shanxi Province was a sort of "Rosetta Stone," because they recognized demonstrably superposed faunas of "*Hipparion* red clay" type, intervening assemblages, and overlying Nihewan type. These were called Zones 1, 2, and 3, respectively, and provided the key to other correlations in North China. These zones were conceived of as having chronostratigraphic value (explicitly used as such by Teilhard and Trassaert, 1937, under the rubric Zones I, II, III). The general temporal relationships of the Yushe zones remain accurate, although their contained faunas were incompletely known and poorly constrained in time.

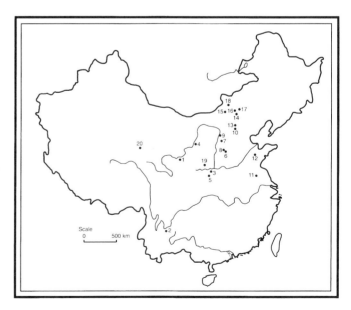

FIG. 1.—Location map for Neogene localities discussed in the text. 1. Lanzhou, 2. Lufeng, 3. Youhe, 4. Wuzhong, 5. Bahe, 6. Gaozhuang (Yushe Basin), 7. Jingle, 8. Mazegou (Yushe Basin), 9. Baode, 10. Daodi, 11. Sihong, 12. Shanwang, 13. Xiashagou, 14. Ertemte, 15. Harr Obo, 16. Tunggur, 17. Bilike, 18. Amuwusu, 19. Lingtai, and 20. Qaidam.

In building a mammal biochronology for China, Li and others (1984; our Fig. 2) recognized two Pliocene biochrons, an older Jinglean and younger Youhean. The earliest Pleistocene was thought to be represented by the fauna collected near Nihewan in Hebei Province. Qiu (1990) agreed in fundamentals but recognized in his Mammal Unit IV a Miocene fauna that was stable over a very long time. He considered the faunal succession around Gaozhuang (Yushe Basin) to represent early Pliocene time (Mammal Unit V) better than the isolated locality of Jingle, and associated Youhe with Nihewan (Xiashagou, sensu stricto) in his Mammal Unit VI. Qiu and Qiu (1995) utilize reference faunas, rather than the more abstract Mammal Units. They designate Baodean, Yushean, and Nihewanian biochrons that correspond in most essentials to Mammal Units IV-VI. One important detail is that their Nihewanian straddles the Pliocene/

Pleistocene boundary and includes part of what previously was conceived of as a Youhean biochron.

This general ordering of faunas is robust because faunal differences are strong and correlated mainly with the passage of time. The reference faunas are widely scattered over the last 7 my. As other assemblages are interpolated into this seriation, the ordering becomes less robust, because the assemblages are closer in age with fewer faunal differences. The Ertemte local fauna of Inner Mongolia illustrates this point. It is generally considered late Miocene age, but apparently postdates the local faunas around Baode. How much younger Ertemte may be has been unclear (Fahlbusch and others, 1983; Repenning and others, 1990; Qiu, 1990). Recent workers (e.g., Zheng and Li, 1990) have proceeded to insert assemblages and presumed time units as for Ertemte, into the coarse-scale biochronologic schemes of Li and others (1984) and Qiu (1990). We do not deal with these. Our goal here is to resolve time limits for the major biochrons and date reference faunas using that time frame.

The Institute of Vertebrate Paleontology and Paleoanthropology (IVPP) recognized Yushe Basin, Shanxi Province, as the best area to develop a precise, datable, late Neogene biostratigraphy for North China. Over a series of field seasons, and later with American colleagues, the biostratigraphy was refined and tied to a paleomagnetic section. The focus of the following discussion is the effect of the resultant data on the previously conceived late Neogene biochronology of North China.

Yushe Basin Biostratigraphy

Qiu and others (1987) and Qiu (1987) presented the fruits of the initial phases of research by IVPP in Yuncu subbasin, one of the components of the complex Yushe Basin, including its rich fauna and local stratigraphy. The adopted lithostratigraphic nomenclature differs from that of Cao and others (1985) for the neighboring Zhangcun subbasin to the south, because the different sedimentary sequence of that area originated under independent depositional conditions. This work also laid much of the paleobiological systematic foundation for general biochronologic correlations. The Yushe large mammal fauna was dominated by deer, antelopes and gazelles, by hipparionine horses and elephant-like proboscideans, and by hyaenid and felid carnivores. Small mammals included shrews and moles, the pika *Ochotona* and at least one hare, flying and ground squirrels, gerbils, hamsters, mice, and the ubiquitous gopher-like zokors (myospalacines such as *Prosiphneus*).

The Yuncu subbasin includes over 800 m of mainly fluviatile deposits filling a large river valley cut in non-marine Triassic sediments. Fossil vertebrate horizons are numerous but vary in richness. The lowermost 200 m (Fig. 3) record the commencement of basin sedimentation. Basal coarse, colluvial debris is succeeded by fluviatile deposits consisting of loosely consolidated sands and silts with gravel units. The clay component increases upward and the sequence is capped by a resistant lacustrine marl. This basal cycle in Yuncu subbasin constitutes the Mahui Formation. It is only moderately fossiliferous, with the coarse basal 50 m being virtually barren. Its large mammal assemblages include the last records in North China of the browsing horse *Sinohippus*, the hyaena *Adcrocuta*, and the bear *Indarctos*, the first North China elephantine mastodont, *Stego-*

A

EPOCH	MAMMAL AGE	
	Li, Wu, Qiu, 1984	Qiu and Qiu, 1995
PLEIST	NIHEWANIAN	NIHEWANIAN
PLIOCENE	YOUHEAN	YUSHEAN
	JINGLEAN	
MIOCENE	BAODEAN	BAODEAN

B

EPOCH	TEILHARD - TRASSAERT ZONE	FORMATION
PLIOCENE	ZONE III	HAIYAN
	ZONE II	MAZEGOU
		GAOZHUANG
MIOCENE	ZONE I	MAHUI

FIG. 2.—Relationships of (A) mammal age biochrons and (B) Yushe Basin stratigraphic units to epoch boundaries. The biostratigraphic zonation of Teilhard and Trassaert (1937) was based on Yushe stratigraphy. These provide the basis for revision of the late Neogene biochronology of Li and others (1984). In B, dashes indicate imprecision in zone boundaries; wavy lines indicate disconformities.

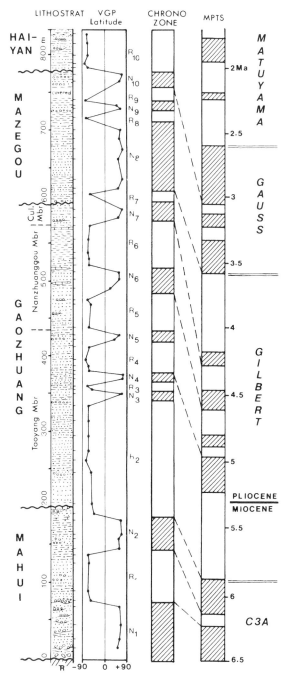

FIG. 3.—Yuncu subbasin magnetostratigraphy (Tedford and others, 1991) adapted to the polarity time scale of this volume.

tive throughout. Large mammal fossils include advanced hipparionine horses, various new deer and antelope, the appearance of *Pliohyaena*, and first records of camelids and canids in Eurasia. Small mammals are characterized by different species of genera present in the Mahui Formation and first records of advanced genera like the murine *Chardinomys* (see Wu and Flynn, 1992) and, higher in the sequence, two voles.

The succeeding 200-m thick Mazegou Formation is lithologically distinctive in its thinner fining-upward sequences and greater quantity of clays. The Mazegou Formation appears to rest conformably on the Gaozhuang Formation, but a hiatus is implied at the basal contact by magnetic data. The fauna of this unit resembles that of the upper part of the Gaozhuang Formation, but contrasts at the species level. Notable first appearances are *Ochotonoides*, *Felis*, *Lynx*, *Homotherium*, *Vulpes*, *Canis*, *Dama*, *Rusa*, several bovids and the mammoth *Archidiskodon*.

An angular unconformity separates the northwesterly dipping Mazegou Formation from the overlying horizontal Haiyan Formation. The fauna of this 80-m thick, partly lacustrine unit differs greatly from that of the Mazegou Formation (see Flynn and others, 1991, for a discussion of relative faunal change in the Yushe sequence). Zokors with ever-growing cheek teeth, *Marmota*, *Borsodia*, and advanced species of Mazegou genera distinguish the small mammals. The horse *Equus* and carnivores *Cuon* and *Megantereon* are notable. Giant deer, *Eucladoceros* and *Elaphurus*, as well as large bovids, *Bison* and *Budorcas* (the takin), appear in the Haiyan Formation. Reddish loess blanketing the Yushe Group yields mid-Pleistocene small mammals.

Chronology and Correlations

The magnetic data for the Mahui Formation establish a dominantly normal polarity, previously identified as Chron 5, presently the younger part of C3A. The most productive part of the formation for fossils is the upper part, which places most of the fossils, including *Stegodon* and *Prosiphneus murinus*, at about 6 Ma. The overlying Taoyang Member of the Gaozhuang Formation is mainly reversed, which corresponds to the older portion of the Gilbert Chron. Fossils are not abundant, but important latest Miocene first occurrences are the camel *Paracamelus* and two lineages of Canidae from North America, and the bamboo rat *Rhizomys* (*Brachyrhizomys*) from southern Asia. The Miocene/Pliocene boundary corresponds to moderate faunal turnover (Flynn and others, 1991), including replacement of *Gazella gaudryi* by *G. blacki*.

The early Pliocene Nanzhuanggou and Culiugou Members of the Gaozhuang Formation are much more fossiliferous than underlying strata and include first occurrences of the hare *Hypolagus* from North America, the modern bear *Ursus*, and rodents *Germanomys* and *Mimomys*, from western or northern Asia. The later reversed portion of the Gilbert Chron is not recorded in our interpretation, which is part of the rationale for recognizing a significant hiatus at the disconformable contact with the overlying deposits.

The Mazegou Formation is dominantly normal, and correlated with the Gauss, although much of the long n1 of that chron, the lower part of the Matuyama chron, and the unconformable contact with the Haiyan Formation are not exposed. The pika *Ochotonoides* appears at circa 3.5 Ma and the mam-

don, and diverse pigs, giraffes, and deer. The common Mahui gazelle is *Gazella gaudryi*. Small mammals include an early North China murine record, the gerbil *Pseudomeriones*, the hamster *Neocricetodon*, and *Prosiphneus murinus*.

In Yuncu subbasin the Yushe sequence continues with the superposed Gaozhuang Formation, 400 m thick and with a disconformable (locally unconformable) basal contact. This formation includes three fining upward units that we designate Taoyang, Nanzhuanggou, and Culiugou Members. The thick Taoyang Member is poorly fossiliferous; the others are produc-

moth *Archidiskodon* at circa 3 Ma. The succeeding Haiyan Formation is mostly reversely magnetized. Its correlation to the magnetic time scale is early Matuyama, late Pliocene, which is consistent with records of the local zokor *Youngia tingi* and of *Equus* and *Bison*, widely recognized as members of an immigrant pre-Pleistocene fauna.

Magnetochronology.—

Lindsay and others (1987) discuss the theoretical aspects of correlating local magnetic data to the polarity time scale, the usefulness of distinctive magnetic "fingerprints" and the synergistic value of complementary data. The composite Yushe Basin polarity sequence (Fig. 3), contains enough character in relative lengths of magnetozones to suggest the correlation; the basal normal couplet is Chron C3A, the dominantly reversed strata with short normals equals Gilbert, the dominantly normal rock is Gauss, and the reversed Haiyan is Matuyama. The superposed magnetozones constrain each other in possible correlations. It is reasonable to query why the ends of the sequence are correlated as they are: why not shift the whole correlation down or up? In the special case of late Neogene magnetochronology, shifting upward to a great degree is not a reasonable option, as the long normal Brunhes is a limiting factor in itself. However, why not correlate the Haiyan with the later Matuyama? These options can be ruled out on faunal grounds.

The fauna of the Mahui Formation includes advanced hipparionine horses and other elements that demonstrate faunal similarity with Baode and general correlation to the late Miocene Turolian faunas of western Asia and Europe. For example, Late Miocene Palaearctic index fossils found at Yushe include the hyaenas *Adcrocuta* and *Ictitherium*. Furthermore, Mahui sediments yield *Stegodon* and murine rodents, which are not recorded at Baode. While we are inclined to see these as additions consistent with an age younger than that of typical Baode assemblages, we cannot rule out that Yushe, in a coastal mountainous setting, experienced a different paleoecology than in the interior at Baode, which permitted these elements to thrive in Yushe Basin. The correlation to Chron C3A is corroborated strongly by the Taoyang Member records of Canidae and Camelidae. These families were indigenous elements of the North American fauna for the last 40 Ma, and arrived in the Palaearctic only at the end of Miocene time. Their first records outside of North America, other than in Yushe, are in the late Miocene Venta del Moro fauna of Spain. Opdyke and others (1990) postulated a Chron 5 correlation, but subsequent analysis (Opdyke, in prep.) demonstrates that the Venta del Moro fauna occurs in reversely magnetized sediment correlated as basal Gilbert. It is unreasonable to assign an age older than basal Gilbert Chron to the Taoyang Member, and unreasonable to assign an age in excess of Chron C3A for the Mahui Formation, given the number of species shared in common between the two units (see Flynn and others, 1991).

Correlations of short magnetozones may be suspect in individual cases, but these are not likely to cause gross misalignment of the composite, due to the relatively great thickness of the Yushe sequence. We see lithological evidence for breaks in deposition, which we interpret to truncate some magnetozones. Toward the top of the sequence, in the reversely magnetized Haiyan Formation, the index fossil *Equus* appears, which nowhere is known in excess of circa 2.5 Ma (see Lindsay and others, 1980), meaning that the Haiyan Formation could not be pre-Matuyama. It is reasonable to postulate and test a correlation of the Haiyan Formation to post-Olduvai Matuyama. We rule this out based on absence in the Haiyan Formation of *Allophaiomys*, a small mammal indicator of Pleistocene (Olduvai and younger) age, and on faunal correlation to Nihewan, where similar assemblages are correlated with the earlier Matuyama chron.

Local biostratigraphy and correlation.—

In 1991, we continued work on the Yushe Basin in the region south of the town of Yushe itself, in what we term the Yushe subbasin. Our goal was to complement the record of Yuncu subbasin through data from formerly contiguous deposits that are now separated tectonically. In this we were successful, because our new magnetic data (Fig. 4) place the richest faunas from Yushe subbasin in the early Gilbert Chron, at just the time when our Yuncu record suffered from poor sampling.

In the Yushe subbasin, we prospected 34 fossil localities spanning 120 m of sediment above the Triassic basement, and found mainly Mahui Formation large mammals. A few elements collected earlier this century are of imprecise provenance, but represent taxa known otherwise from the Gaozhuang Formation. The greatest concentration of fossils in the area we studied was high in the section, particularly around the town of Jiayucun, a locality known to Young (1927). Magnetic polarity data (Fig. 4) show the lower half of the section to be normally magnetized, the upper half reversed. Coupling this observation with the fauna, we consider the lower strata to have been deposited during Chron C3A and coeval with the Mahui Formation. At Jiayucun, where the rocks are reversed, there is no apparent depositional hiatus. We thus interpret these strata as earliest Gilbert Chron (latest Miocene), equivalent to basal Gaozhuang Formation. We believe the hiatus in the Yuncu subbasin to be slight, but it is conceivable that some of the upper part of the Yushe subbasin depositional sequence corresponds to a stratigraphic gap in the Yuncu subbasin.

By combining faunal similarity with magnetic information, this intrabasin correlation is made accurate. The Jiayucun fauna represents a part of the fossil record that is poorly sampled in our previous work. The terminal Miocene small mammal faunal list for Yushe Basin is expanded, demonstrating a number of species shared with the classic (Schlosser, 1924) localities of Ertemte and Harr Obo, Inner Mongolia, including *Yanshuella primaeva*, *Ochotona lagrelli*, *Pliopetaurista rugosa*, *Karnimata hipparionum*, *Apodemus orientalis*, and *Micromys chalceus*. Interestingly, key Ertemte elements such as *Microtoscoptes*, *Microtodon*, and *"Prosiphneus" eriksoni* are missing from Yushe. We take this to indicate different environmental preferences for these rodents. In any case, Ertemte, which previously seemed to defy correlation to the Yushe sequence, can now be determined as terminal Miocene age and equal to basal Gaozhuang strata, above Chron C3A and about the age of the Spanish site Venta del Moro. In our correlation, Ertemte is pre-Thvera Subchron, probably in excess of 5.5 Ma. Information from Yushe and other magnetostratigraphic sections (see below) suggest that Ertemte-like faunas occur in rock that can be correlated to an interval of earliest Gilbert Chron up to the Sidufjal Subchron, circa 5.9 to 4.8 Ma at broadest limits. This is in general agreement with the conclusion of Repenning and others (1990) who

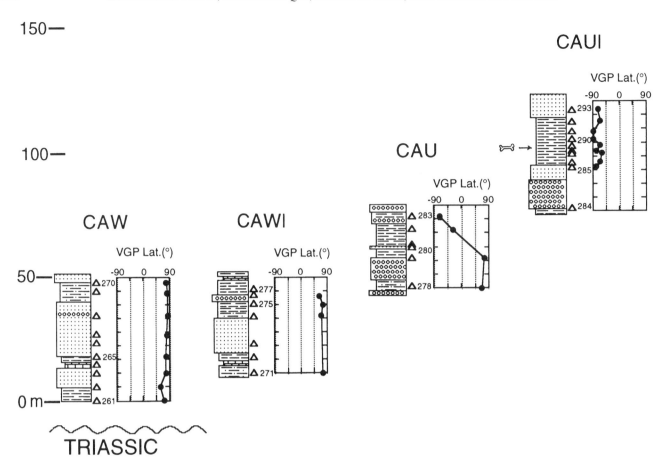

Fig. 4.—Yushe subbasin magnetostratigraphy indicating normal polarity below and reversed polarity above, which is correlated with the C3A-Gilbert boundary. The bone symbol indicates fossil localities of Jiayucun, which are early Gilbert in age and yield Ertemte faunal elements.

recognized an Ertemtean age, except that those authors, based on biochronological comparisons, projected a somewhat younger interval.

Recent work by a Chinese-Japanese team (Huang and others, 1993; Zheng, 1994) in Lingtai County, eastern Gansu Province, 700 km southwest of Yushe, has begun to delineate another Pliocene biostratigraphic sequence correlated with the magnetic polarity time scale. This provides the opportunity to test the local range zones of Yushe small mammal taxa in an interior setting. The fossiliferous fluviatile and lacustrine Red Clay in Lingtai County rests on Cretaceous sediments and is overlain by the early Matuyama Wucheng Loess.

Lingtai and Yushe share over 25% of the same species. Most of the shared taxa show extensions at Lingtai over their Yushe ranges, but only a few have range extensions of 50% or more. The correspondence is sufficiently robust that conclusions on general aspects of faunal turnover (Flynn and others, 1991; Tedford and others, 1991) are not negated. Differences in the faunal lists very likely reflect the different ecological setting (most clearly depicted for the large mammals), and the greater number of specimens collected at Lingtai.

Overlapping ranges of Ertemte small mammals found at Lingtai sites occur in the early Gilbert beneath the Sidufjal Subchron of the early Pliocene. This probably represents the top of their concurrent range zone. As at Yushe, the Lingtai micro-

fauna lacks the high crowned *Microtoscoptes* that characterizes the coeval Ertemte assemblage in Mongolia.

Similar composites in Mongolia and Siberia.—

Mongolian and Russian colleagues have developed the biostratigraphy and magnetostratigraphy for sections in western Mongolia and adjacent Siberia. Repenning and others (1990) and Tedford and others (1991) use fauna to correlate these successions with the Yushe sequence. The magnetochronology, as interpreted by Pevzner and others (1982), is largely corroborative, although classes of reliability in their magnetic data are not distinguished. For western Mongolia Pevzner and others (1982) correlate Ertemte-like faunas that include *Microtoscoptes* to both late Chron C3A and to early Gilbert. Although open to criticism of details, these correlations are robust overall, partly because they are long. Short sections in long composites tend to constrain each other in the overall sequence; the longer the sequence the fewer the plausible correlations.

Late Pliocene assemblages.—

The reference fauna of Jingle, once taken as representative for the early Pliocene fauna of North China (Li and others, 1984) contains many micromammals that we encounter in the Yuncu subbasin Mazegou Formation and consider correlative with Gauss Chron. Chen (1994) presents the paleomagnetic data from the Hefeng section at Jingle, which shows normal polarity at the base where the Jingle fauna is found. The disconformably overlying Wucheng Loess is mainly reversely magnetized, although the Olduvai Subchron is recognized within it. *Equus* occurs in the early Matuyama beneath the Olduvai Subchron of the Wucheng Loess. Above the Olduvai, the advanced vole *Allophaiomys* is recorded.

The late Cenozoic fluviatile and lacustrine deposits exposed in the vicinity of Nihewan in northern Hebei Province provided the first substantial evidence of the nature of the mammal fauna contemporaneous with the early phase of loess deposition (Barbour and others, 1926). Study of these complex deposits over a large area has shown that they were laid down in adjacent basins that did not have fully comparable histories. Outcrops in the eastern Yangyuan Basin around Nihewan village comprise the type area for the Nihewan Formation. A rich fauna collected from the vicinity of nearby Xiashagou Village (Teilhard and Piveteau, 1930) came from unspecified levels near the middle of the unit. Recollecting and lithostratigraphic study in 1974 by the Nihewan Cenozoic Group, and subsequent magnetostratigraphic work (Dong and others, 1986; Wang, 1988) has indicated that the fossiliferous beds occur in the late Gauss and early Matuyama (including the Olduvai Subchron). The small collection recovered in 1974 (e.g., *Equus*, wolves, giant deer) contained only 22% of the original faunal list.

Recently the focus of study has shifted to the outcrops in the Yuxian Basin along the lower Huliuhe River southwest of Nihewan. The local sequence begins with dipping redbeds that contain Pliocene assemblages of late Gilbert into early Gauss Chron. These rocks are unconformably overlain, but with little evidence of hiatus, by flat-lying fluviatile and lacustrine deposits formerly considered Nihewan Formation, but now subdivided (Zheng and Cai, 1991). The lowest unit, the Daodi Formation, lies totally within the late part of the Gauss Chron and contains a microfauna resembling the Mazegou and Jingle assemblages (see Cai, 1987, 1989). The overlying Dongyaozitou Formation (restricted sense of Zheng and Cai, 1991) is a widely traceable set of cross-bedded sands and gravels that appears to represent the early Matuyama and contains at the base a sparse large mammal fauna (with the woolly rhinoceros *Coelodonta*, typical of the Nihewan fauna, but unknown from older strata in China). A vole, cf. *Mimomys youhenicus* from the lower Dongyaozitou, supports Li and others (1984) placement of the Youhe Local Fauna from southern Shanxi in the late Pliocene.

The Olduvai Subchron apparently lies within the lower part of the restricted Nihewan Formation, overlying the Dongyaozitou Formation in the Yuxian Basin. The large mammal fauna shares 80% of its species with that from the typical Nihewan site at Xiashagou, including the horses *Equus sanmeniensis* and *Proboscidipparion sinense*. The rich micromammal fauna includes a vole near *Allophaiomys pliocaenicus*, a widespread taxon whose earliest Eurasian occurrence is at or just below the Olduvai Subchron (Repenning, 1992). The referred Nihewan Formation of the Yuxian Basin, like that in the Yangyuan Basin, spans the late Matuyama Chron, and perhaps crosses into the Brunhes. Fossil mammals from that part of the section (Zheng and Cai, 1991) include *Microtus* and *Lasiopodomys,* which support these correlations.

The large mammal fauna from the Haiyan Formation of Yushe Basin closely resembles at the species level that from Nihewan. The small mammals also agree, including the moderately advanced zokor *Youngia tingi,* but the Haiyan microfauna lacks the derived vole *Allophaiomys,* which is consistent with assignment to early Matuyama Chron.

Faunal Correlations

For North China, an increase in the time resolution of Neogene sediments has proceeded through coordinated use of magnetic and faunal data. The likely temporal relationships of key fossil assemblages in North China can be summarized by reference to the Yushe composite sequence (Fig. 5). The Yuncu Mahui Formation spans a relatively short interval of time in the late Miocene. It is close in age to many of the classic Baodean localities, with faunal differences (see Tedford and others, 1991) explained by hypothesizing a younger age for Mahui sites based on records of *Stegodon* and murids that presently are taken as regional first appearances. Alternatively, paleoecological differences could account for the minor faunal differences between Baode and Mahui assemblages if they were contemporaneous.

Now we can argue that Ertemte correlates with Yushe deposits of the early part of the Gilbert Chron; it is terminal Miocene age as proposed by Fahlbusch and others (1983), but perhaps younger in absolute years (less than 6 Ma) than many workers have suspected. Ertemte is older than Repenning and others (1990) hypothesized. If Ertemte is considered late Baodean biochron (presently the faunal evidence is weak for this), then Baodean time must extend up to the Miocene/Pliocene boundary, which lies well up in the Gaozhuang Formation. If not, Ertemte would be early Yushean age, and that biochron would commence in the latest Miocene.

Jingle, Daodi and similar localities turn out to be contemporaneous with the Gauss Chron and correlative with sites in

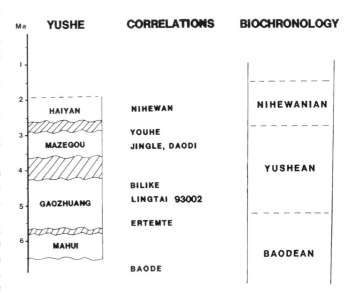

FIG. 5.—Yushe Basin stratigraphy applied as a tool in ordering other North China localities, and dating the corresponding biochronology. The top of the Haiyan Formation could be as young as 1.9 Ma, which would imply a larger hiatus at its base. Biochron limits are dashed, because they depend on refined definitions, which will clarify such issues as whether Nihewanian should include Youhe-like faunas, and whether Ertemte should be considered Yushean.

the Mazegou Formation. The Youhe fauna resembles these, but in fact is not rich. It is distinguished in producing two rare species of *Mimomys* and the elephantid *Archidiskodon*. *Archidiskodon* is reported from the upper part of the Mazegou Formation, with which we correlate the Youhe fauna. Thus the Jingle fauna is late Pliocene epoch and the Youhe fauna is later in age, but not latest Pliocene. It turns out that very few studied sites in China date to early Pliocene epoch. Likely exceptions other than localities from upper members of the Gaozhuang Formation, are in the Lingtai area, Gansu, and Bilike, Inner Mongolia (Qiu, 1988), which produces a fauna comparable to Yushe sites of circa 4.5 Ma, such as YS 50.

The Pliocene faunas of Yushe are homogeneous, despite some species substitution, from greater than 5 Ma to circa 3 Ma (Flynn and others, 1991). For this reason, we argue that a single biochron, the Yushean of Qiu and Qiu (1995), should be recognized for this interval.

Earlier we presented arguments as to why we consider the Haiyan Formation correlative with the early Matuyama Chron and why we consider the Nihewan fauna equivalent to it or slightly younger in age. This would place the Yushean-Nihewanian boundary at least as low as the Mazegou-Haiyan Formation hiatus, or older than 2.5 Ma. The precise position of the boundary can be argued, depending on whether the Youhe fauna is considered similar to that of Nihewan and therefore placed in the Nihewanian. This would extend the Nihewanian back to 3 Ma. In either case the Nihewanian biochron is late Pliocene into early Pleistocene, and a Nihewan-like assemblage should not be used as an indicator of Pleistocene time.

FUTURE PROSPECTS

We have a good start on documenting the geochronology and biochronology of the late Neogene terrestrial rocks and faunas of North China. Refining the entire Neogene record is largely a matter of building downward. Extending the magnetostratigraphic and faunal record back into earlier late Miocene deposits is important in building the framework for defining the beginning of the Baodean biochron. What that corresponding age is and whether a distinct preceding Bahean biochron can be resolved is not known, but is knowable. There are good prospects for starting this in the Baode-Fugu region, northwestern Shanxi and adjacent Shaanxi Provinces, in deposits on both sides of the Yellow River (Fig. 1). Strata there cover considerable time and have the potential to constrain the dating and faunal change for late Miocene time. The classic Lantian area, although tectonically active, should also be compared to the stratigraphic record in the Baode area. The beginning of the late Miocene may be spanned in the Lantian area along the Bahe River, in Ningxia at Wuzhong, or in the Qaidam Basin (Fig. 1). The potential to date the first introduction to Asia of hipparionine horses is compelling. The middle Miocene record can best be assembled in Inner Mongolia, as witnessed by localities such as Tunggur and Amuwusu (see Qiu, 1988).

For the early Miocene, we have a start in the Lanzhou Basin, which spans the Oligocene/Miocene boundary and extends well into younger Miocene strata. Around Lanzhou, IVPP researchers have conducted several seasons of field work, and we have added some investigation funded by the National Geographic Society. We have collected bulk samples of microvertebrates

from a dozen productive localities around the Oligocene/Miocene boundary, and located superposed fossil horizons, including two that are likely middle Miocene in age.

To complete an initial picture of Chinese Neogene vertebrate history, the faunal record in South China must be integrated with that to the North. Up to now, localities around Lufeng (Fig. 1) and Yuanmou have been dated only by faunal comparisons. It will be necessary for Chinese scientists to pull these more precisely into the emerging biochronology of North China.

Examining patterns of faunal change may well complement study of tectonic changes in the Tibetan Plateau. Clearly, these highlands had tremendous impact on paleoclimate and faunal change in central China and may be a proxy for dating uplift episodes. Furthermore, detailed comparisons with the Siwalik record of the Indian Subcontinent, south of the Tibetan Plateau, will be possible when the Chinese record becomes well dated. Finally, studies of basin tectonics and the timing of rotation or deformation will be enhanced by precise knowledge of the age of basin fill throughout central China. The vertebrate record has great potential to enhance all such studies as it becomes better documented.

ACKNOWLEDGMENTS

Our collaboration has succeeded through the efforts of many IVPP researchers as well as Arizonans Will Downs and Everett Lindsay. We are grateful for the support of Academia Sinica and the National Science Foundation (grant BSR 9020065), both on scholarly and financial levels. Charles Repenning generously helped the first author to comprehend the Yushe microtines in early phases of the research. He aided greatly in assessing faunal correlations and provided a thorough review of the manuscript. Ev Lindsay also invested patient effort in offering many improvements to the text.

REFERENCES

BARBOUR, G. B., LICENT, E., AND TEILHARD DE CHARDIN, P., 1926, Geological study of the deposits of the Sangkanho Basin: Geological Society of China Bulletin, v. 5, p. 263–278.

CAI, B., 1987, A preliminary report on the late Pliocene micromammalian fauna from Yangyuan and Yuxian, Hebei: Vertebrata Palasiatica, v. 25, p. 124–136.

CAI, B., 1989, Fossil lagomorphs from the late Pliocene of Yangyuan and Yuxian, Hebei: Vertebrata Palasiatica, v. 27, p. 170–181.

CAO, Z., XING, L., AND YU, Q., 1985, The magnetostratigraphic age and boundaries of the Yushe Formation: Bulletin of the Institute of Geomechanics, v. 6, p. 144–154.

CHEN, X., 1994, Stratigraphy and large mmmals of the "Jinglean" stage, Shanxi, China: Quaternary Sciences, no. 4, p. 339–352, 1 plate.

CHIU, C., LI, C., AND CHIU, C., 1979, The Chinese Neogene—A preliminary review of the mammalian localities and faunas: Annals Géologiques Pays Hélléniques, Tome hors séries 1, p. 263–272.

DONG, M., WANG, Y., WANG, S., LUO, B., WANG, Q, YUE, J., JIANG, M., AND Ge, S., 1986, A study of the Nihewan beds of the Yangyuan-Yuxian Basin, Hebei Province, China: Bulletin of the Chinese Academy of Geological Science, n. 15, p. 149–160.

FAHLBUSCH, V., QIU, Z., AND STORCH, G., 1983, Neogene mammalian faunas of Ertemteand Harr Obo in Nei Monggol, China-1. Report on field work in 1980 and preliminary results: Scientia Sinica, v. B26, p. 205–224.

FLYNN, L. J., PILBEAM, D., JACOBS, L. L., BARRY, J. C., BEHRENSMEYER, A. K., AND KAPPELMAN, J. W., 1990, The Siwaliks of Pakistan: Time and faunas in a Miocene terrestrial setting: Journal of Geology, v. 98, p. 589–604.

FLYNN, L. J., TEDFORD, R. H., AND QIU, Z., 1991, Enrichment and stability in the Pliocenemammalian fauna of North China: Paleobiology, v. 17, p. 246–265.

HUANG, W., ZHENG, S., ZONG, G., AND LIU, J., 1993, Pliocene mammals afrom the Leijiahe Formation of Lingtai, Gansu, China—Preliminary report on field work in 1972 and 1992: Northern Hemisphere Geo-Bio Traverse, n. 1, p. 29–37.

LI, C., WU, W., AND QIU, Z., 1984, Chinese Neogene: subdivision and correlation: Vertebrata Palasiatica, v. 22, p. 163–178.

LICENT, E. AND TRASSAERT, M., 1935, The Pliocene lacustrine series in central Shansi: Geological Society of China Bulletin, v. 14, p. 211–219.

LINDSAY, E. H., OPDYKE, N. D., AND JOHNSON, N. M., 1980, Pliocene dispersal of *Equus* and late Cenozoic mammalian dispersal events: Nature, v. 287, p. 135–138.

LINDSAY, E. H., OPDYKE, N. D., JOHNSON, N. M., AND BUTLER, R. F., 1987, Mammalian chronology and the magnetic polarity time scale, *in* Woodburne, M. O., ed., Cenozoic Mammals of North America: Geochronology and Biostratigraphy: Berkeley, University of California Press, p. 269–284.

MacFADDEN, B. J., 1990, Chronology of Cenozoic primate localities in South America: Journal of Human Evolution, v. 19, p. 7–21.

NORELL, M. A. and Novacek, M. J., 1992, Congruence between superpositional and phylogenetic patterns: comparing cladistic patterns with fossil records: Cladistics, v. 8, p. 319–337.

OPDYKE, N. D., MEIN, P., MOISSENET, E., PÉREZ-GONZALÉZ, A., Lindsay, E., and Petko, M., 1990, The magnetic polarity stratigraphy of the late Miocene sediments of the Cabriel Basin, Spain, *in* Lindsay, E. H., Fahlbusch, V., and Mein, P., eds., European Neogene Mammal Chronology: New York, Plenum Press, p. 507–514.

PEVZNER, M. A., VANGENEIM, E. A., ZHEGALLO, V. I., ZAZHIGIN, V. S., AND LISKUN, I. G., 1982, Correlation of the upper Neogene sediments of central Asia and Europe on the basis of paleomagnetic and biostratigraphic data: Akademii Nauk SSSR Izvestiya Seriya Geologicheskaya, n. 6, p. 5–16 (International Geology Review, v. 25, p. 1075–1084).

QIU, ZHANXIANG, 1987, Die Hyaeniden aus dem Ruscinium und Villafranchium Chinas: Münchener Geowissenschaftliche Abhandlungen, v. A9, p. 1–109.

QIU, ZHANXIANG, 1990, The Chinese Neogene mammalian biochronology — its correlation with the European Neogene mammalian zonation, *in* Lindsay, E. H., Fahlbusch, V., and Mein, P., eds., European Neogene Mammal Chronology: New York, Plenum Press, p. 527–556.

QIU, ZHANXIANG, HUANG W., AND GUO Z., 1987, The Chinese hipparionine fossils: Palaeontologia Sinica, new series, n. 25, p. 1–250.

QIU, ZHANXIANG AND QIU, ZHUDING, 1995, Chronological sequence and subdivision of Chinese Neogene mammalian faunas: Palaeogeography, Palaeoclimatology, Palaeoecology, v. 116, p. 41–70.

QIU, ZHUDING, 1988, Neogene micromammals of China, *in* The Palaeoenvironment of East Asia from the Mid-Tertiary, v. 2: Hong Kong, University of Hong Kong, p. 834–848.

REPENNING, C. A., 1992, *Allophaiomys* and the age of the Olyor Suite, Krestovka Sections, Yakutia: United States Geological Survey Bulletin 2037, p. 1–98.

REPENNING, C. A., FEJFAR, O., AND HEINRICH, W.-D., 1990, Arvicolid rodent biochronology of the Northern Hemisphere, *in* Fejfar, O. and Heinrich, W.-D., eds., International Symposium Evolution, Phylogeny, and Biostratigraphy of Arvicolids (Rodentia, Mammalia): Prague, Geological Survey, p. 385–417.

SCHLOSSER, M., 1924, Tertiary vertebrates from Mongolia: Palaeontologia Sinica, n. C1, p. 1–119.

SIMPSON, G. G., 1953, The Major Features of Evolution: New York, Columbia University Press, 434 p.

TEDFORD, R. H., FLYNN, L. J., QIU, Z., OPDYKE, N. D., AND DOWNS, W. R., 1991, Yushe Basin, China; paleomagnetically calibrated mammalian biostratigraphic standard from the late Neogene of eastern Asia: Journal of Vertebrate Paleontology, v. 11, p. 519–526.

TEILHARD DE CHARDIN, P. AND PIVETAU J., 1930, Les mammifères fossiles de Nihowan (Chine): Annales de Paléontologie, v. 19, p. 1–134.

Teilhard DE CHARDIN, P. AND TRASSAERT, M., 1937, The Proboscideans of southeastern Shansi: Palaeontologia Sinica, n. C13, p. 1–58.

WANG, S., 1988, Paleomagnetic stratigraphy of Nihewan beds, *in* Chen M., ed., Study on the Nihewan Beds: Beijing, Ocean Press, p. 117–123.

WU, W. AND FLYNN, L. J., 1992, New murid rodents from the Late Cenozoic of Yushe Basin, Shanxi: Vertebrata Palasiatica, v. 30, p. 17–38.

YOUNG, C. C., 1927, Fossile Nagetiere aus Nord-China: Palaeontologia Sinica, n. C5, p. 1–82.

ZHENG, S., 1994, Preliminary report on late Miocene-early Pleistocene micromammals collected from Lingtai of Gansu, China in 1992 and 1993: Northern Hemisphere Geo-Bio Traverse, n. 2, p. 44–56.

ZHENG, S. AND CAI, B., 1991, Fossil micromammals from Danangou of Yuxian, Hebei: Beijing, Contributions to the XIII INQUA, Beijing Scientific and Technological Publishing House, p. 100–131.

ZHENG, S. AND LI, C., 1990, Comments on fossil arvicolids of China, *in* Fejfar, O. and Heinrich, W.-D., eds., International Symposium Evolution, Phylogeny, and Biostratigraphy of Arvicolids (Rodentia, Mammalia): Prague, Geological Survey, p. 431–442.

SUBJECT INDEX

SUBJECT INDEX